Stable Radicals

Stable Radicals

Fundamentals and Applied Aspects of Odd-Electron Compounds

Editor

ROBIN G. HICKS
Department of Chemistry, University of Victoria, Canada

WILEY
A John Wiley and Sons, Ltd., Publication

Registered office
John Wiley & Sons Ltd, The Atrium, Southern Gate, Chichester, West Sussex, PO19 8SQ, United Kingdom.

For details of our global editorial offices, for customer services and for information about how to apply for permission to reuse the copyright material in this book please see our website at www.wiley.com.

Library of Congress Cataloging-in-Publication Data

Stable radicals : fundamentals and applied aspects of odd-electron compounds /
editor, Robin G. Hicks.
p. cm.
Includes bibliographical references and index.
ISBN 978-0-470-77083-2 (hardback)
1. Radicals (Chemistry) – Stability. 2. Electron mobility. I. Hicks, Robin G.
QD471.S73 2010
541′.224 – dc22

2010010754

A catalogue record for this book is available from the British Library.

ISBN 978-0-470-77083-2

Typeset in 10/12 Times by Laserwords Private Limited, Chennai, India
Printed and bound in the United Kingdom by Antony Rowe Ltd, Chippenham, Wiltshire

Contents

Preface

Valency is a core tenet of chemical structure and bonding. In essence, valency can be thought of as a "bond-forming capability" parameter that permits understanding of the chemical formula and properties of known compounds, the prediction of the existence of new ones, and the rationalization of the non-existence (or instability) of species with incompletely satisfied valencies. This last point provides a rationale for why radicals – with an unpaired electron and, as such, quintessential examples of subvalent compounds – are typically described as short-lived, highly reactive species. To be sure, the vast majority of radicals based on s- and p-block elements are indeed thermodynamically and kinetically unstable. Thus, it should not be surprising that, since Gomberg's paradigm-shifting discovery of the triphenylmethyl radical in 1900, the very existence of radicals stable enough to be observed or isolated has always been met with wonder (and at times skepticism) by the general chemistry community. The ongoing spate of publications highlighting the discovery of a new molecule that can be isolated but has an unpaired electron serves as evidence that stable radicals are generally regarded as exotic and esoteric compounds.

In fact, several kinds of stable molecules containing one (or more) unpaired electrons have been known for decades, and several more general classes of stable radicals have been developed in recent times. Moreover, applications which explicitly make use of stable radicals abound, spanning synthesis, materials science, and medicine. Many review articles and books have dealt with particular kinds or uses of stable radicals. However, the only attempt to provide unified coverage of the stable radical literature was a book by Forrester, Hay, and Thomson (*Organic Chemistry of Stable Free Radicals,* Academic Press) published in 1968. The present book is, in essence, an update on the huge developments in (and diversification of) the field of stable radicals over the four decades since the book by Forrester *et al*. My ambitions for this book were to bring together the diverse range of topics and interests whose common theme is stable *neutral* radicals (the omission of radical ions was a deliberate decision, made largely due to space limitations). Thus, Chapters 1–10 introduce readers to the many different kinds (organic, inorganic, old, new) of existing stable radicals, and Chapters 11–17 provide accounts of various applications of stable radicals, ranging from synthesis (oxidation catalysts, living radical polymerization) to materials science (magnetochemistry, battery components) to chemical biology/medicine (spin labeling, EPR imaging, redox biochemistry). I offer my sincere thanks to all of the contributors to each chapter who have contributed to what I hope will be viewed as an informative, interesting, and illuminating text.

Robin G. Hicks
Victoria, Canada

List of Contributors

D. Scott Bohle, Department of Chemistry, McGill University, 801 Sherbrooke St.W., Montreal, Canada H3A 2K6

Lawrence J. Berliner, Department of Chemistry and Biochemistry, University of Denver, 2190 E. Iliff Avenue, Denver, CO 80208, USA

Christian Brückner, Department of Chemistry, University of Connecticut, Storrs, CT 06269, USA

Tristram Chivers, Department of Chemistry, University of Calgary, Calgary, AB, Canada T2N 1N4

Michael K. Georges, Department of Chemical and Physical Sciences, University of Toronto at Mississauga, 3359 Mississauga Road, Mississauga, Ontario, Canada L5L 1C6

Sara Goldstein, Institute of Chemistry and the Accelerator Laboratory, The Hebrew University of Jerusalem, Jerusalem 91904, Israel

Robin G. Hicks, Department of Chemistry, University of Victoria, Victoria, British Columbia, Canada V8W 3V6

Keith U. Ingold, National Research Council, Ottawa, Ontario, Canada K1A 0R6

Hakim Karoui, Laboratoire Chimie Provence, UMR 6264, Aix-Marseille Université and CNRS, Centre de St Jérôme, 13397, Marseille Cédex 20, France

Valery V. Khramtsov, Dorothy M. Davis Heart and Lung Research Institute, 201 HLRI, 473 W 12th Ave, The Ohio State University, Columbus, OH 43210, USA

Jari Konu, Department of Chemistry, University of Calgary, Calgary, AB, Canada T2N 1N4

François Le Moigne, Laboratoire Chimie Provence, UMR 6264, Aix-Marseille Université and CNRS, Centre de St Jérôme, 13397, Marseille Cédex 20, France

Julie Lukkarila, Department of Chemical and Physical Sciences, University of Toronto at Mississauga, 3359 Mississauga Road, Mississauga, Ontario, Canada L5L 1C6

Yasushi Morita, Department of Chemistry, Graduate School of Science, Osaka University, Toyonaka, Osaka 560-0043, Japan

Shinsuke Nishida, Department of Chemistry, Graduate School of Science, Osaka City University, Osaka 558-8585, Japan

Hiroyuki Nishide, Department of Applied Chemistry, Waseda University, Tokyo 169–8555, Japan

Olivier Ouari, Laboratoire Chimie Provence, UMR 6264, Aix-Marseille Université and CNRS, Centre de St Jérôme, 13397, Marseille Cédex 20, France

Victor Ovcharenko, International Tomography Center, Novosibirsk, Russia

Imma Ratera, Institut de Ciència de Materials de Barcelona (ICMAB-CSIC), 08193 Bellaterra (Barcelona), Spain

Amram Samuni, Department of Molecular Biology, The Hebrew University of Jerusalem - Hadassah Medical School, Jerusalem 91120, Israel

Takeo Suga, Department of Applied Chemistry, Waseda University, Tokyo 169-8555, Japan

Andrea R. Szkurhan, Department of Chemical and Physical Sciences, University of Toronto at Mississauga, 3359 Mississauga Road, Mississauga, Ontario, Canada L5L 1C6

Fabrice Thomas, Inorganic Redox Chemistry, Department of Molecular Chemistry (UMR-5250), University of Grenoble, BP 53, 38041, Grenoble Cedex 9, France

Thomas T. Tidwell, Department of Chemistry, University of Toronto, Toronto, Ontario, Canada M5S 3H6

Paul Tordo, Laboratoire Chimie Provence, UMR 6264, Aix-Marseille Université and CNRS, Centre de St Jérôme, 13397, Marseille Cédex 20, France

Jaume Veciana, Institut de Ciència de Materials de Barcelona (ICMAB-CSIC), 08193 Bellaterra (Barcelona), Spain

Jay L. Zweier, Dorothy M. Davis Heart and Lung Research Institute, 611 HLRI, 473 W 12th Ave, The Ohio State University, Columbus, OH 43210, USA

1

Triarylmethyl and Related Radicals

Thomas T. Tidwell

Department of Chemistry, University of Toronto, Ontario, Canada

1.1 Introduction

1.1.1 Discovery of the triphenylmethyl radical

During the Nineteenth Century the understanding of the structure of organic compounds was beginning to evolve. The theory of free radicals had risen to prominence, and then fallen into disrepute. This changed abruptly with the bold announcement in 1900 by Moses Gomberg of the formation of the stable and persistent free radical triphenylmethyl **1**, with its radical character shown by its facile reaction with oxygen forming the peroxide **2** (Equation 1.1).[1] This had an immediate impact, and was a major landmark that set the stage for the rapid development of free radical chemistry in the Twentieth Century. Gomberg's work attracted the attention of the world chemical community, and led to careful scrutiny and the ultimate acceptance of this controversial discovery.[2]

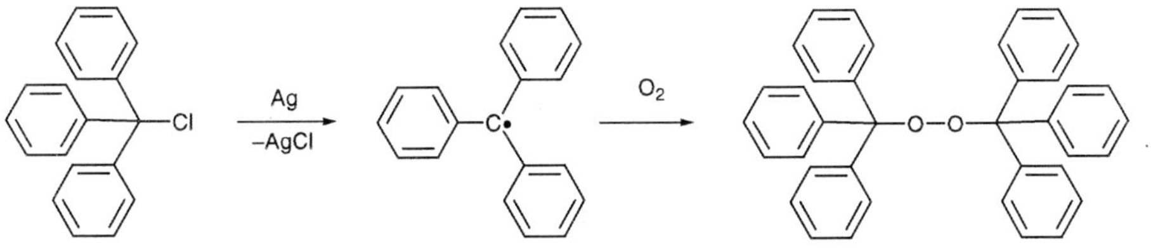

1 2 (1.1)

Gomberg treated triphenylmethyl chloride with silver or zinc metal and obtained a colored solution, which upon reaction with oxygen yielded peroxide **2**. The species in solution was confidently identified by Gomberg as the triphenylmethyl radical **1**, and he published his discovery in both German and English.[1] Over the next decade there was much dispute as to the identification of **1**, but Wilhelm Schlenk and coworkers in 1910 obtained tris(4-biphenylyl)methyl **3** as a deeply colored solid that was almost completely dissociated in solution, which confirmed the existence of **1**.[2i]

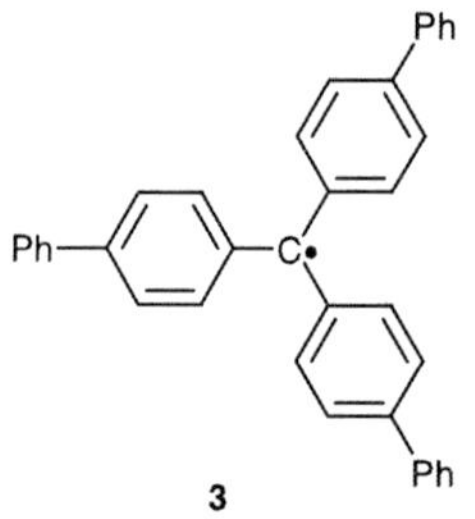

3

Upon removal of the solvent for the isolation of **1**, a solid dimer was obtained, for which the symmetrical head-to-head structure **4** as well as the unsymmetrical structure **5** (head-to-tail, Jacobson structure)[3a] and **6** (tail-to-tail, Heintschel structure)[3b,c] were given serious consideration. However, as recounted by McBride,[3d] the wrong structure for the dimer, namely the head-to-head structure **4**, became accepted for more than half a century, before this was corrected to the unsymmetrical structure **5** based on spectroscopic data.[3e] In retrospect, not only was the original evidence for the misidentified structure rather flimsy, but techniques, such as NMR, IR, and UV, were also widely available that would have permitted correction of this structure well before 1968. This provides a cautionary tale that skepticism and a critical look at the evidence available for supposed chemical truths is warranted even in the face of conventional wisdom.

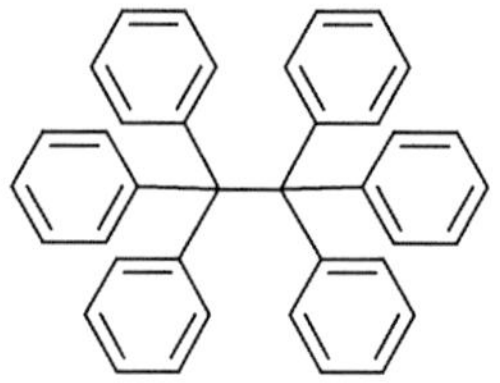

4

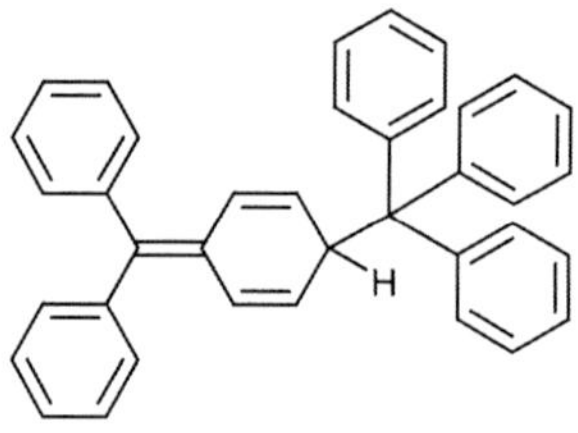

5

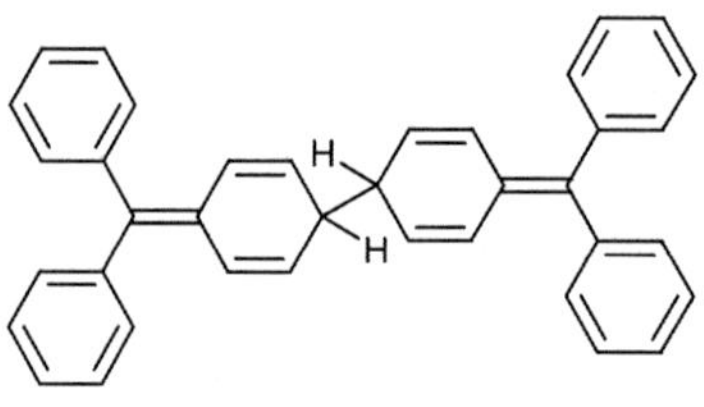

6

The preparation and identification of the stable triphenylmethyl radical was one of the great chemical discoveries of the Twentieth Century, but surprisingly this was not honored by the award of the Nobel Prize. As revealed by the investigation by Lennart Eberson in the Nobel archives, Gomberg was repeatedly nominated for the award but, due to a series of unfortunate circumstances, the nominations were not accepted.[3f] Despite Gomberg's confident assertion in his first paper that he had proof of the existence of the radical there was later some equivocation, and some contrary opinions, which were enough for the Nobel Committee not to approve the award. With the later preparation of the tris(4-biphenylyl)methyl radical **3** as a stable solid the uncertainty vanished,[2i] but then subsequent nominations were turned down either because both Gomberg and Schlenk were not nominated in the same year, or because too much time had elapsed since the initial discovery. Even in 1940 Gomberg was still being nominated, but without success. Gomberg's discovery was a clearly momentous discovery by a single individual, and although well recognized he did not receive the ultimate accolade he deserved. Paradoxically, Gerhard Herzberg was awarded the 1971 prize in chemistry "for his contributions to the knowledge of electronic structure and geometry of molecules, particularly free radicals", studies which had occupied him for 30 years.

By 1968 the study of stable free radicals had advanced to the stage that the book *Organic Chemistry of Stable Free Radicals* appeared,[4] but in Chapter 2 entitled "Triarylmethyls and Other Carbon Radicals" the introductory paragraph describing the discovery of triphenylmethyl in 1900 ended with the dispiriting words "The behaviour of such radicals was elucidated during the following twenty years, mainly by the work of Gomberg, Schlenk and Wieland, since then little new chemistry has come to light although numerous triarylmethyls have been prepared and more physical data are available." However six pages later, under "Dimerisation" there was the alarming statement "Much of what has been said in this section may require revision in light of the recent communication by Lankamp, Nauta and MacLean", which described the surprising but in retrospect completely predictable finding[3d,e] that the triphenylmethyl dimer had the head-to-tail structure **5**. The rapid development in triarylmethyl radical chemistry since 1968 also belies the tacit assumption of the authors noted above that such studies had become an intellectual backwater. Much of the rather extensive chemistry of triarylmethyl radicals described in this earlier review is not repeated here. The period from 1968 has been a new golden age for free radical chemistry, and this was given great impetus by the widespread use of electron paramagnetic resonance (EPR) spectroscopy, which led to a rapid development of free radical chemistry, and includes many advances in the study of triarylmethyl radicals.

1.1.2 Bis(triphenylmethyl) peroxide

The reaction of the triphenylmethyl radical with oxygen to form the peroxide discovered by Gomberg in 1900 (Equation 1.1) was a strong piece of evidence for the radical structure **1**.[5,6] The affinity of carbon-centered radicals for oxygen remains one of their defining characteristics, and was a striking chemical property that provided strong evidence for the proposed free radical character. The addition of oxygen to **1** had been shown to discharge the color,[5a] and this formation of an initial peroxy radical was later shown to be reversible.[5b]

Just as for the triphenylmethyl radical dimer, there were three conceivable structures for the peroxide corresponding to the dimer structural models, namely head-to-head (**2**), head-to-tail (**7**), and tail-to-tail (**8**). The head-to-head structure **2** proved to be correct, but substituted derivatives of **7** and **8** were later found.[6]

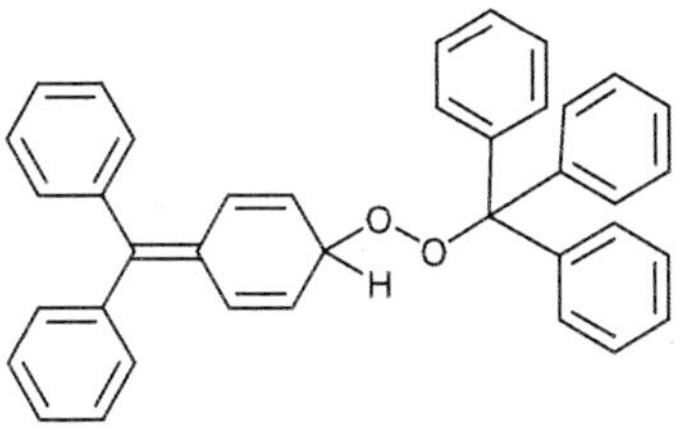

7

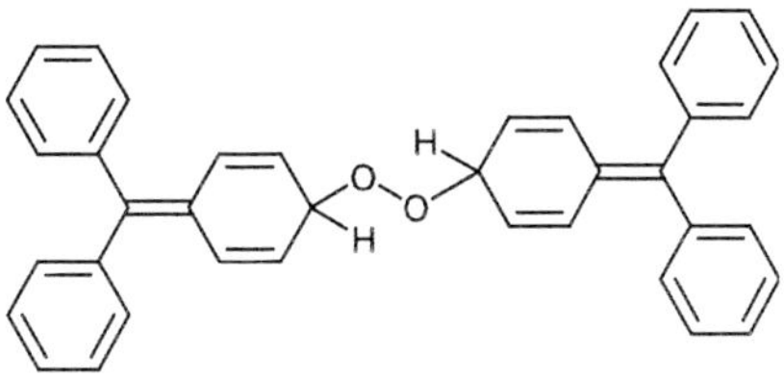

8

The tris(2,6-dimethoxyphenyl)methyl radical **9** was crystallized and found to have a structure with two of the aryl rings twisted out the plane of the central carbon by 61°, while the third was almost coplanar with an 11° twist angle.[6] Reaction of **9**, and of the bis(2,6-dimethoxyphenyl)phenylmethyl radical **10**, in solution in the presence of air resulted in crystallization of the substituted Heintschel and Jacobson peroxides **11** and **12**, respectively, with the structures proven by NMR, and by X-ray, respectively.[6]

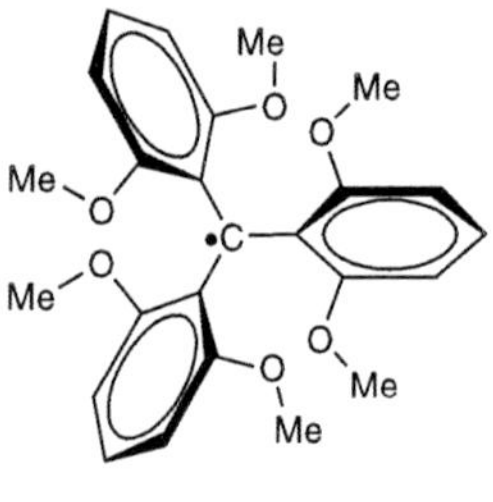

9

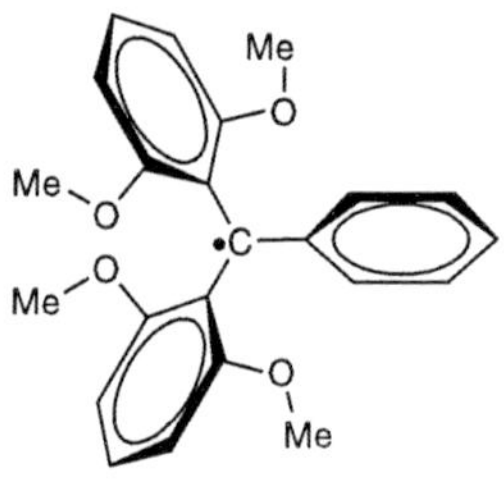

10

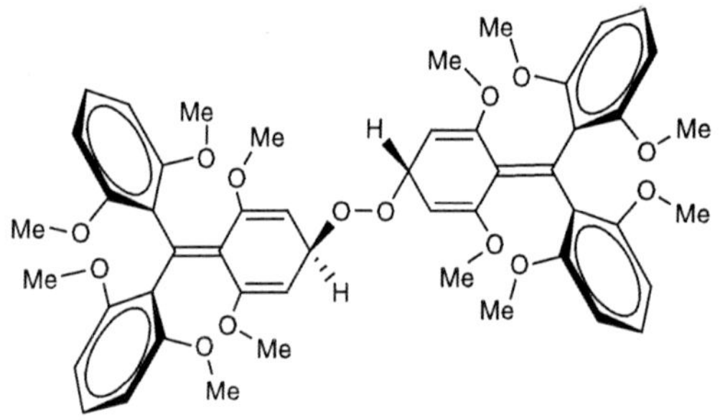

11

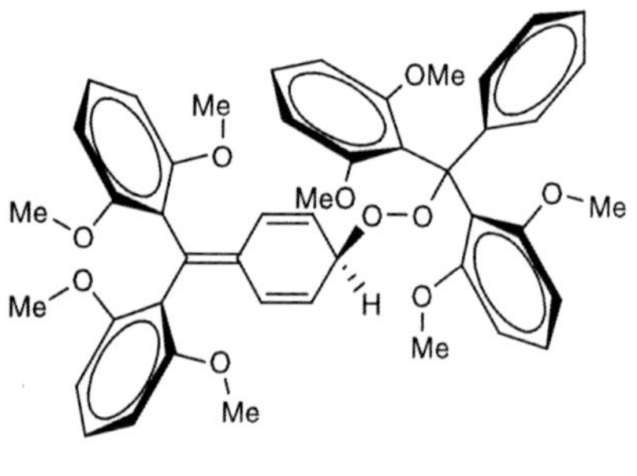

12

1.2 Free radical rearrangements

Substrates containing the triphenylmethyl group have been important in the discovery and elucidation of the first free radical rearrangements.[7] Wieland observed in 1911 that the Gomberg peroxide **2** rearranged upon heating to the pinacol ether **15** and attributed this to initial dissociation forming an intermediate oxyl radical **13**, and this rearranged forming the radical **14** which dimerized to give **15** (Equation 1.2).[7a] Recent experimental and computational studies indicate this reaction occurs via a phenyl-bridged intermediate.[7g]

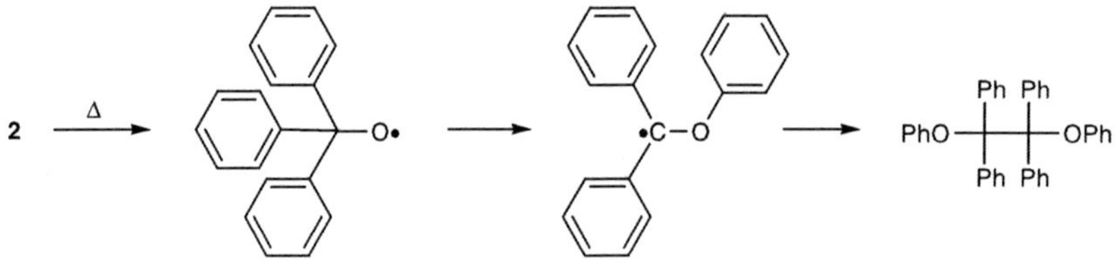

13 14 15

(1.2)

There was a long delay from this observation by Wieland of a radical rearrangement of Gomberg's peroxide until the generality of radical rearrangements was recognized. A key discovery by Urry and Kharasch in 1944 of the neophyl rearrangement (Equation 1.3)[7d] included the proposal for the neophyl radical intermediate **16** and radical **17** forming **18** and other products after work-up (Equation 1.3).[7c]

The McBay rearrangement of acyl radical **19** forming **20** leading to the benzyl radical by decarbonylation is a more recent example (Equation 1.4).[7e,f]

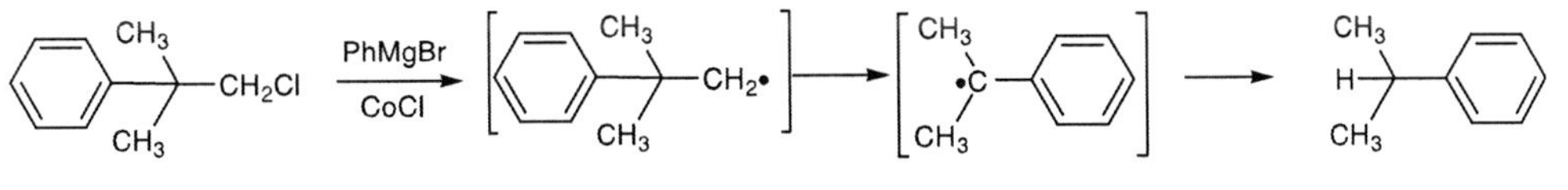

16 **17** **18**

(1.3)

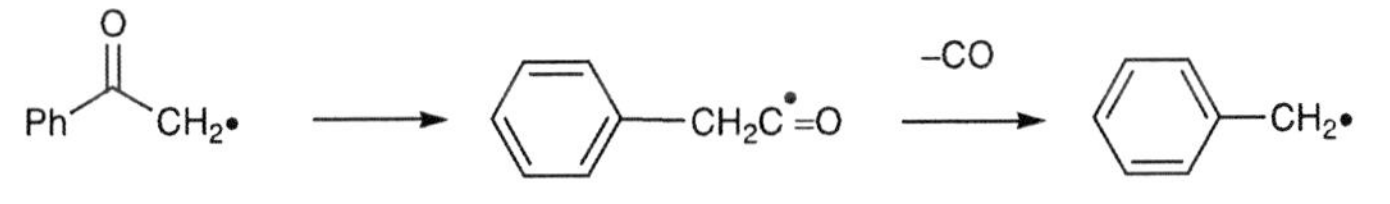

19 **20**

(1.4)

Thermolysis of bis(triphenylmethyl) peroxide **2** in the presence of triphenylmethyl radical resulted in the observation of luminescence from both benzophenone and triphenylmethyl radical.[8] The reaction was interpreted as involving chemiluminescent activation of the triphenylmethyl radical by triplet benzophenone, and was claimed to be the first observed chemical activation of an organic free radical. A mechanism was proposed involving disproportionation of the triphenylmethyloxyl radical **13** (Equation 1.5), but further substantiation of this mechanism appears warranted.

$$\underset{\mathbf{2}}{Ph_3COOCPh_3} \longrightarrow \underset{\mathbf{13}}{2Ph_3CO\bullet} \longrightarrow Ph_3COPh + Ph_2C{=}O^{*} \xrightarrow[-Ph_2C=O]{Ph_3C\bullet(S)} Ph_3C\bullet(T) \qquad (1.5)$$

1.3 Other routes to triphenylmethyl radicals

Gomberg had already used phenylazotriphenylmethane **21** in the classic preparation of tetraphenylmethane **22** in 1897 (Equation 1.6).[9a,b] Wieland, *et al*, in 1922,[9c] showed that $Ph_3C^{\bullet}$ **1** was formed in this reaction, and proposed that phenyl radicals were also formed (Equation 1.6). Triphenylmethyl radicals were generated thermally from the perester **23** even at 25 °C (Equation 1.7)[10a] and at much higher temperature from the significantly less reactive *N*-triphenylacetoxy 2,2,6,6-tetramethylpiperidine **25** (Equation 1.8).[10b] Reactivity studies demonstrated that both **23** and **25** reacted thermally via concerted two-bond cleavage (**24** and **26**, respectively) forming triphenylmethyl radicals **1** and carbon dioxide.

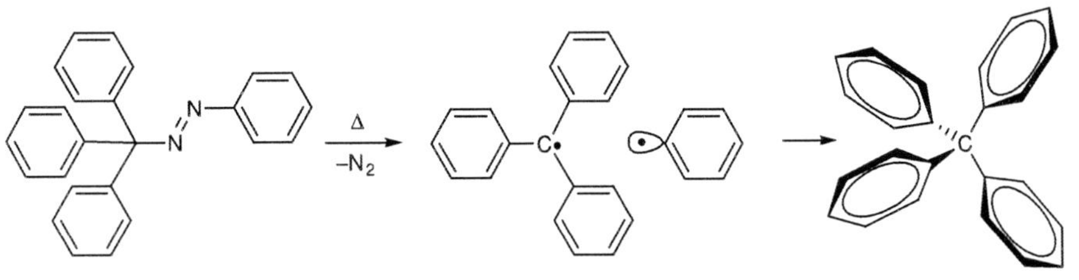

21 **1** **22**

(1.6)

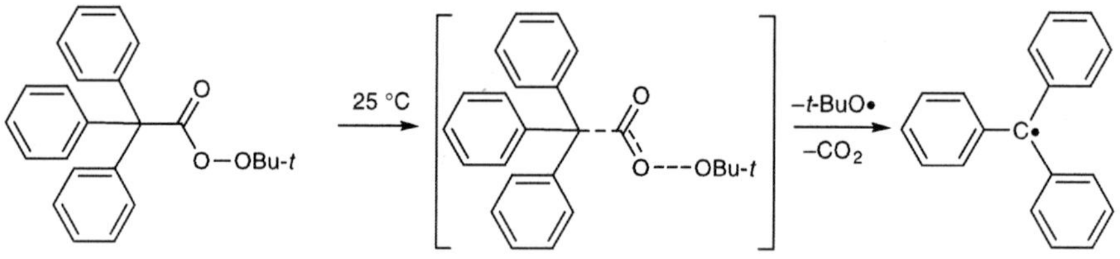

23 24 1 (1.7)

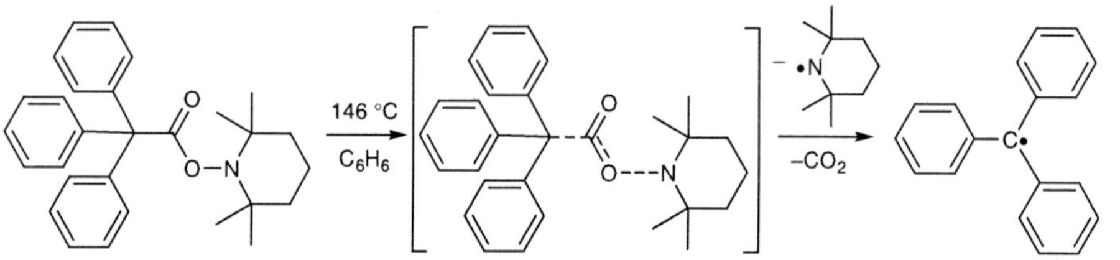

25 26 1 (1.8)

Many other dissociations of triarylmethyl substituted compounds can form triarylmethyl radicals,[10c] including, for example, formation of the radical $4\text{-}Me_2NC_6H_4(Ph)_2C^{\bullet}$ by photodissociation of $4\text{-}Me_2NC_6H_4(Ph)_2CSiMe_3$.[10d] Pulse radiolysis of trityl derivatives Ph_3CR (R = 2-naphthS, PhO, PhNH, PhNMe, and $PhCH_2$) in *n*-BuCl led to observation of $Ph_3C^{\bullet}$ by generation of the radical cation *n*-BuCl($^{+\bullet}$) by electron removal followed by free electron transfer (FET) forming $Ph_3CR(^{+\bullet})$, which dissociated to $Ph_3C^{\bullet}$ and to Ph_3C^+ by separate reaction channels.[10e] These were proposed to form from two different conformations of $Ph_3CR(^{+\bullet})$, both from a planar radical cation, which follows the thermodynamically favored pathway generating $R^{\bullet}$ and Ph_3C^+, and from a twisted radical cation, which produces the thermodynamically unfavorable product pair R^+ and $Ph_3C^{\bullet}$.[10e] Triphenylmethanol at high temperature and pressure in subcritical water gave $Ph_3C^{\bullet}$, as detected by EPR spectroscopy, and **1** was also formed in supercrtitical water, as shown by capture with hydrogen donors giving Ph_3CH.[10f] A number of triphenylmethyl derivatives have been found to have potent anticancer properties,[10g] although there is no suggestion that this is due to triphenylmethyl radicals.

A novel route for generation of triarylmethyl radicals was by reaction of 1-(haloalkenyl)-2-allenylbenzenes **27** in tetrahydrofuran (THF) at room temperature in the presence of mercury, resulting in cyclization to 2-naphthyldiphenylmethyl radical **28**, as detected by EPR (Equation 1.9).[11a] The radical formed a head-to-tail dimer involving the phenyl group, and was captured by oxygen to give the peroxide (Equation 1.9).[11a] Several related ene–yne cyclizations have been examined, including a route to a di(radical) **29**, which forms **30** (Equation 1.10).[11b–d]

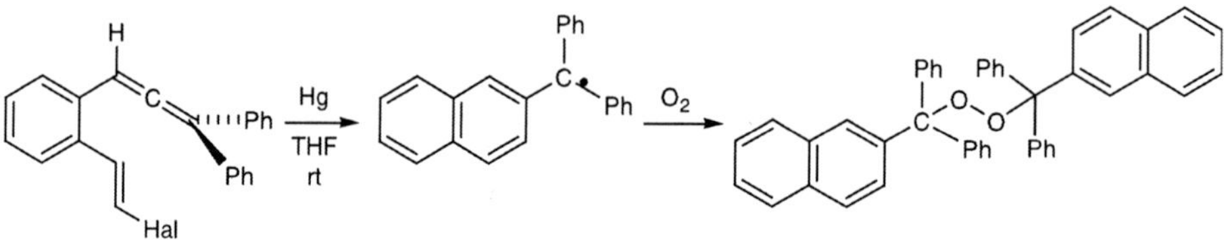

27 28 (1.9)

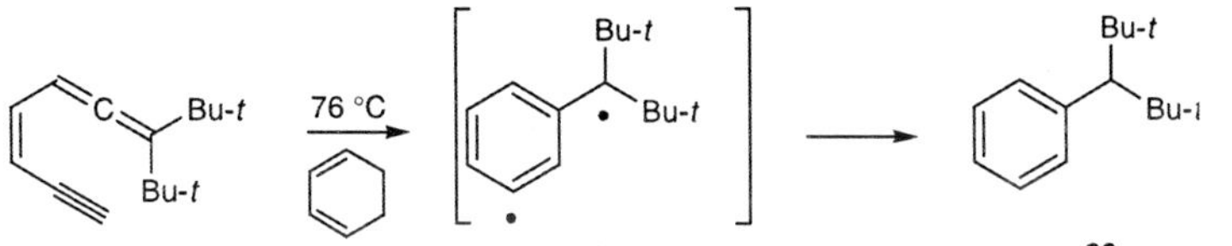

29 30

(1.10)

1.4 The persistent radical effect

The persistent radical effect[12] is an important phenomenon in radical chemistry that was first discovered through studies of triphenylmethyl radicals. One manifestation of this effect is seen in the reaction of carbon tetrachloride with phenyl radicals and triphenylmethyl radicals generated from phenylazotriphenylmethane **21**, which gives excellent yields of chlorobenzene and of 1,1,1-trichloro-2,2,2-triphenylethane **31** (Equation 1.11).[12]

$$\underset{\mathbf{21}}{Ph_3CN{=}NPh} \xrightarrow[-N_2]{\Delta} \underset{\mathbf{1}}{Ph_3C\bullet} + Ph\bullet \xrightarrow{CCl_4} PhCl + CCl_3\bullet \xrightarrow{\mathbf{1}} \underset{\mathbf{31}}{Ph_3CCCl_3} \quad (1.11)$$

The azo compound **21** decomposes into phenyl and trityl radicals, together with a molecule of nitrogen. The more reactive phenyl radical then abstracts chlorine from carbon tetrachloride and radical coupling of the trichloromethyl radicals with the persistent trityl radicals **1** produces **31**. It is striking that only *unsymmetrical* coupling takes place, and no appreciable amount of hexachloroethane is formed. The explanation is that formation of even trace amounts of hexachloroethane would also lead to formation of the dimer **5** of triphenylmethyl in equilibrium with the monomeric radical. Therefore, the concentration of triphenylmethyl will greatly exceed that of the transient trichloromethyl radical, and since the rates of reactions of trichloromethyl with itself and with triphenylmethyl are both close to the diffusion limit, unsymmetrical coupling will predominate. However, the concentration of triphenylmethyl will remain essentially constant, since for every trichloromethyl formed another triphenylmethyl is also produced. Experimental confirmation is provided by EPR examination of the reacting solution, which shows a high concentration of the triphenylmethyl radical.

This phenomenon was correctly analyzed in 1936 by Gomberg's student Bachmann, who studied the role of triphenylmethyl in the reaction of pentaphenylethane **32** solution at 100 °C (Equation 1.12).[12f] This ethane dissociates rapidly and reversibly into triphenylmethyl and diphenylmethyl, but dimerization of the diphenylmethyl radical **33** to form detectable quantities of tetraphenylethane (which is stable under the reaction conditions) seldom occurs.

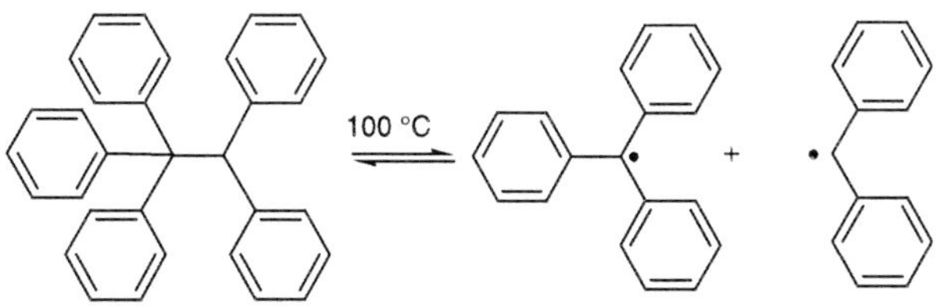

32 1 33

(1.12)

The decomposition of $Ph_3CN{=}NPh$ **21** in benzene has also been subjected to kinetic analysis[12g] and the effect of less persistent radicals, such as *t*-alkylperoxyls, which do decay irreversibly but only by relatively slow processes, was analyzed.[12a] A recent review on this subject is available.[12h]

1.5 Properties of triphenylmethyl radicals

The theoretical basis for the understanding of free radicals was first provided by G. N. Lewis in 1916.[13a,b] His clear recognition of the electron pair bond and the possibility of odd electron systems was heavily influenced by the work of pioneers such as Gomberg, Schlenk, and Wieland, who had shown remarkable prescience in formulating free radical structures without using the principles of electron pair bonding. The theoretical study of the triphenylmethyl radical was later provided by Erich Hückel at the 1933 Faraday conference on free radicals, which brought together many, but not all, of the pioneers in this field.[13c]

The electron-nuclear double resonance (ENDOR) and EPR spectra of triphenylmethyl **1**, (4-biphenylyl)diphenyl-methyl **34**), bis(4-biphenylyl)phenylmethyl **35**, and tris(4-biphenylyl)methyl **4** radicals were measured in 1968 by Maki *et al.*[13d] and the signals for specific protons were assigned based on the observed and calculated spectra. Computations of the spin density distributions were compared to the experimental values. Other spectroscopic measurements of triarylmethyl radicals include the laser induced fluorescence and emission spectra.[13e]

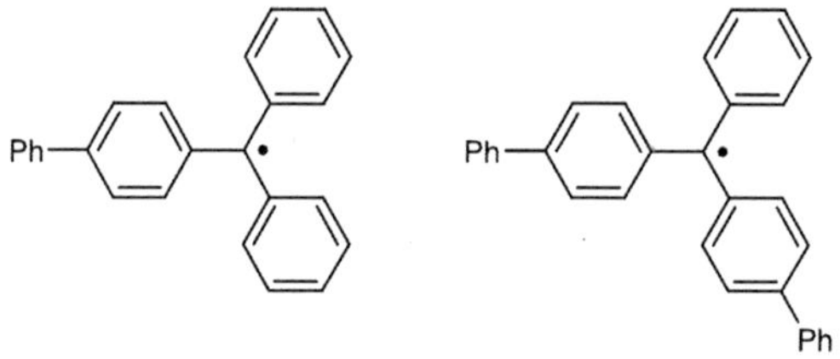

34 **35**

Studies of reactions of triarylmethyl radicals include kinetic measurements of hydrogen atom abstraction from thiols,[14a] electrochemical generation by oxidation of triaryl anions and the subsequent further oxidation to the carbocations,[14b] and reduction of carbocations to the radicals using cyclic voltammetry.[14c]

The oxidation/reduction potentials of triphenyl- or tributyl-substituted silicon-, germanium-, or tin-centered radicals **36** were measured in acetonitrile, tetrahydrofuran, or dimethylsulfoxide by photomodulated voltammetry and through oxidation of the corresponding anions using linear sweep voltammetry for comparison to those of the carbon analogues (Equation 1.13).[15] The order of reduction potentials follows Sn > Ge > C > Si, and is C > Si for the two oxidation potentials. Computational methods gave qualitative interpretation of the experimental results by considering a combination of effects, such as charge capacity, resonance stabilization, and solvation effects.

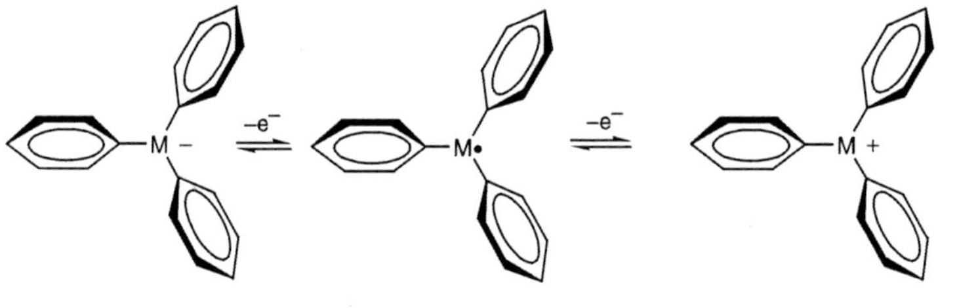

(1.13)

36 (M = C, Si, Sn, Ge)

1.6 Steric effects and persistent radicals

In his first paper Gomberg[1a] recognized that steric factors make a contribution to the stability of the triphenylmethyl radical, and Conant and Bigelow[16a] reported in 1928 the reversible dissociation at 50 °C of 1,2-(di-*tert*-butyl)tetraphenylethane **37** to the yellow radical **38** (Equation 1.14),[16a] which reacts rapidly with oxygen. The radical **38** generated in the presence of $Ph_3C^•$ forms the mixed dimer **39** (Equation 1.14).[16b] The di-*tert*-butylbenzyl radical **40** was stable for several days in solution at room temperature, and the EPR spectrum indicated the phenyl group was twisted into coplanarity with the singly occupied p orbital, preventing spin delocalization.[16c,d] Other work showed that purely aliphatic radicals without β hydrogens, such as di-*tert*-methyl **41**,[17a] and tri-*tert*-butylmethyl **42**,[17a] and even triisopropylmethyl **43**[17b] were sufficiently long lived in solution for direct observation by EPR. These radicals also did not dimerize but decayed by other undetermined routes. Such radicals with kinetic stability have been classified as persistent radicals.[17c]

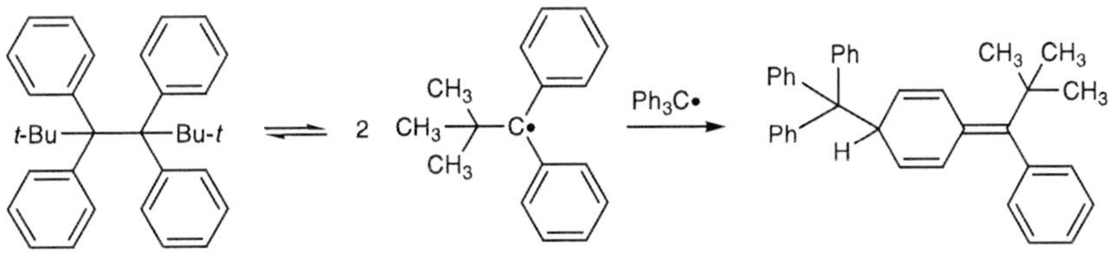

(1.14)

37 **38** **39**

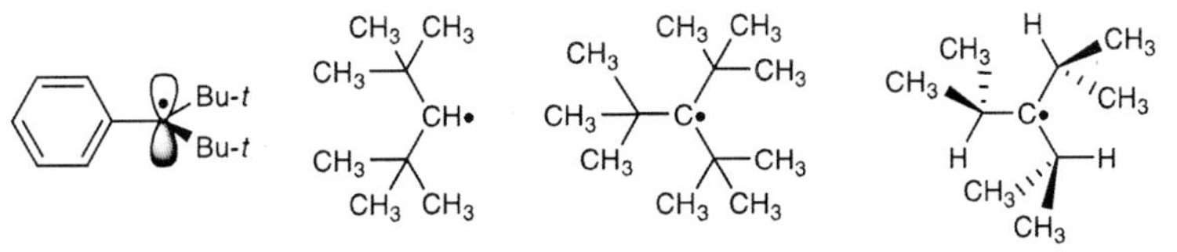

40 **41** **42** **43**

The importance of resonance stabilization of triarylmethyl radicals was emphasized by Pauling and Wheland in 1933.[17d] It had already been shown by Marvel and Munro that alkynyl groups with minimal steric barriers were also effective at promoting dissociation of the ethane **44** by stabilization of diarylmethyl radicals **45** (Equation 1.15).[17e]

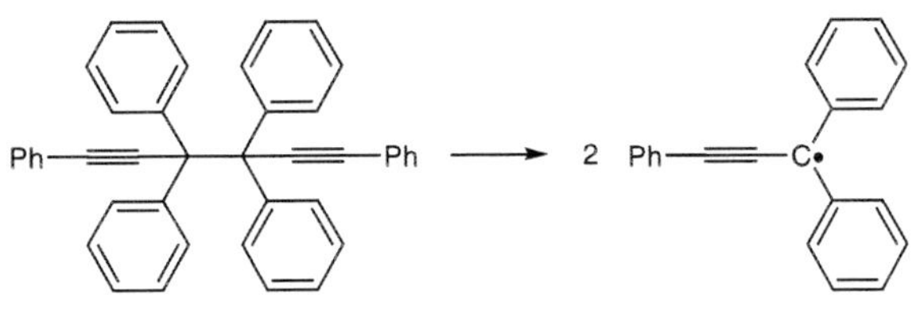

(1.15)

44 **45**

1.7 Substituted triphenylmethyl radicals and dimers

The effect of substituents in stabilizing triarylmethyl radicals has been measured by determining the equilibrium constants for dissociation of the dimers of a large number of such radicals by EPR spectroscopy

Table 1.1 *Equilibrium constants for the dimerization of triarylmethyl radical derivatives 46 and correlation with substituent constants*

R, R^1	σ	$\sigma^\bullet$	$\sigma^\bullet + 0.01\sigma$	log*K*	R, R^1	σ	$\sigma^\bullet$	$\sigma^\bullet + 0.01\sigma$	log*K*
H,H	0.00	0.0000	0.0000	−3.48	H,NO_2	0.78	0.0630	0.0710	−2.55
H,*t*-Bu	−0.20	0.0080	0.0055	−3.10	*t*-Bu,*t*-Bu	−0.20	0.0080	0.0055	−2.39
H,CF_3	0.54	−0.0086	−0.0019	−3.15	CF_3,CF_3	0.54	−0.0086	0.0019	−2.82
H,CN	0.70	0.040	0.0400	−2.66	CN,CN	0.70	0.0488	0.0488	−2.03
H,PhCO	0.42	0.0554	0.0554	−2.49	PhCO,PhCO	0.42	0.0554	0.0607	−2.05
H,Ac	0.50	0.0597	0.0597	−2.70	MeO,MeO	−0.28	0.0185	0.0150	−2.63
H,MeO	−0.28	0.0185	0.0185	−2.82	Ph,Ph	−0.01	0.0615	0.0614	−1.57
H,Ph	−0.01	0.0615	0.0614	−2.56	PhO,PhO	−0.32	0.0185	0.0145	−2.78
H,PhO	−0.32	0.0185	0.0145	−3.21	MeS,MeS	0.00	0.0630	0.0630	−1.72
H,MeS	0.00	0.0630	0.630	−2.78					

(Equation 1.16).[18a] A Hammett-like equation was found correlating log*K* with $\sigma^\bullet + 0.01\sigma$, which relates the monomer/dimer equilibrium constants with the radical stabilizing parameter $\sigma^\bullet$ with a minute contribution from the polar substituent constant σ (Table 1.1). Captodative effects were not found in this study,[18a] but an ENDOR study showed that radicals **46** (R = PhO; R^1 = CF_3, CN, phenyl) are stabilized with respect to dimerization to **47** by 0.6, 1.2, and <0.2 kcal/mol over the amounts expected on the basis of substituent additivity, suggesting some influence of captodative effects.[18b]

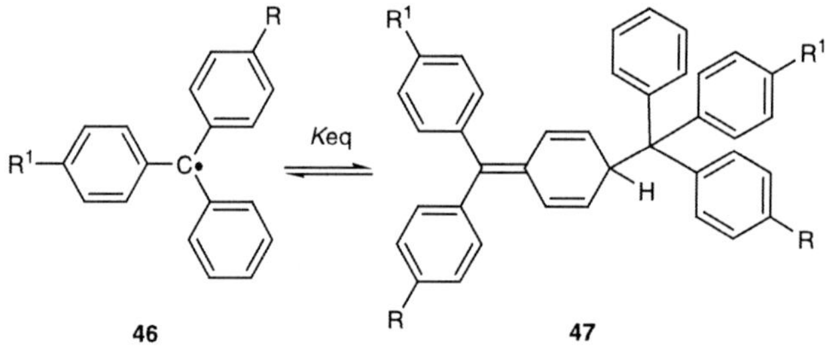

46 **47** (1.16)

Triarylmethyl radicals **9** and **48** fully substituted with 2,6-oxygen substituents on all three rings were prepared and characterized by EPR.[19a] The radical **9** had a twisted geometry as shown by EPR, and did not dimerize. As noted above, **9** was obtained in crystalline form and its X-ray structure determined; this was consistent with the interpretation of the EPR spectrum.[6a] The radical **49** with only two oxygen bridges was quite sensitive to oxygen and was extensively dissociated in the solid state, and the 2,6-hydrogens cause twisting out of the plane,[19b] while the dimer of **48** is undissociated in the solid, with a central C–C bond length of 1.63 Å measured by X-ray.[19a,c,d]

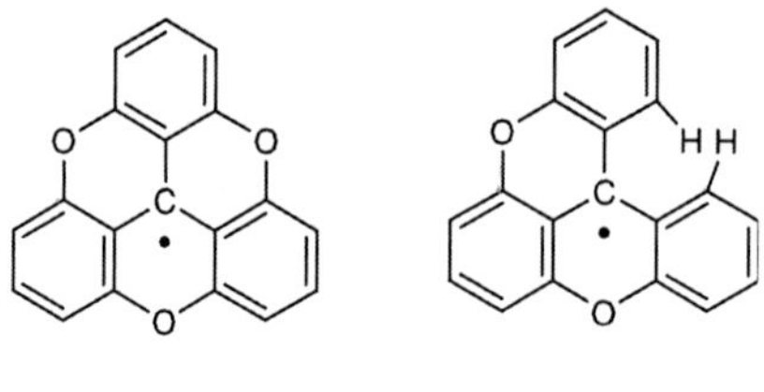

48 **49**

Other hexaphenylethane molecules with structures determined by X-ray include **50**, which is restricted from forming a head-to-tail dimer by the bulky substituents, with a central C–C bond length of 1.67(3) Å.[19e] A different type of hexaphenylethane derivative is typified by **51**, which had already been already prepared in 1933 for R = hydrogen,[19f] and has the central bond constrained by its intramolecular character. The central C–C bond of **51** (R = MeO) was measured as 1.670(3) Å,[19g] and related examples with similar intramolecular constriction have bond lengths that are even longer.[19h,i]

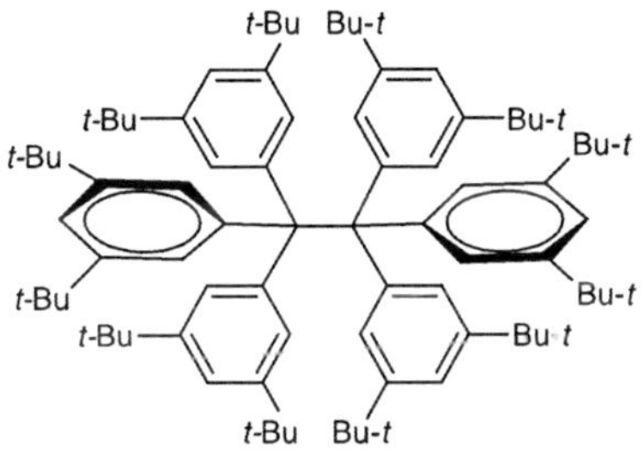

50

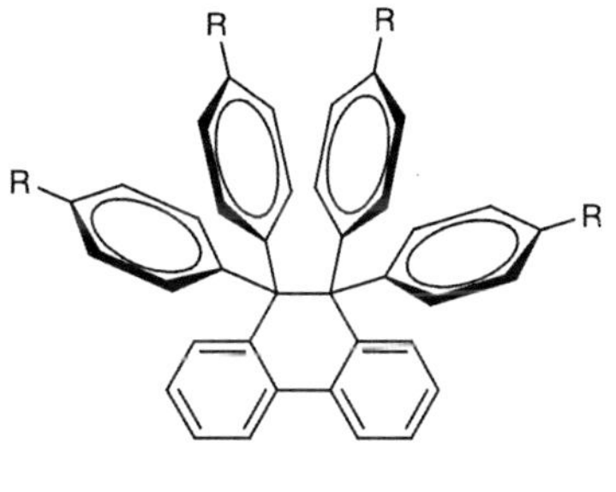

51

A variety of triarylmethyl radicals **52** with electron donating substituents were prepared by reduction of the corresponding carbocations using cyclic voltametry (Table 1.2), including the ferrocenyl substituted radical **53** (Equation 1.17).[20a] These showed strong downward shifts in the reduction potentials due to the strongly electron donating substituents, and reversible redox behavior was even observed in aqueous solution.

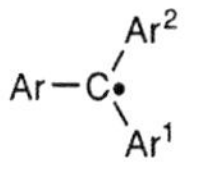

52

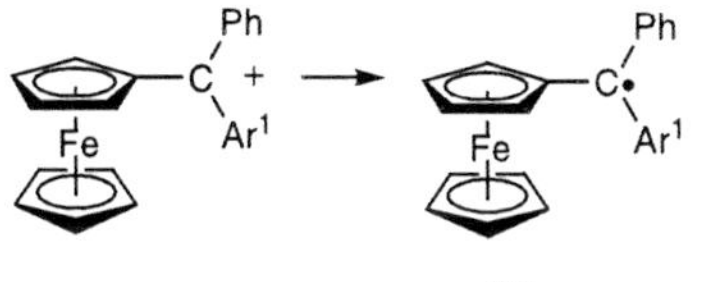

53

(1.17)

The bis(diphenylpentafulvene) iron complex **54** was also prepared, for which the structure was considered as either the bis(radical) **54a** or the bis(fulvene) **54b**.[20b] The fact that the X-ray and calculated structure (Figure 1.1) showed the exocyclic groups on the same side of the molecule was taken as favoring the

Table 1.2 *Triarylmethyl radicals **52** studied by cyclic voltammetry*

Ar	Ar$^{1'}$	Ar^2
2,4,6-$(MeO)_3C_6H_2$	2,4,6-$(MeO)_3C_6H_2$	2,4,6-$(MeO)_3C_6H_2$
2,4,6-$(MeO)_3C_6H_2$	2,6-$(MeO)_2C_6H_3$	2,6-$(MeO)_2C_6H_3$
2,6-$(MeO)_2C_6H_3$	2,6-$(MeO)_2C_6H_3$	2,6-$(MeO)_2C_6H_3$
4-$MeOC_6H_4$	2,6-$(MeO)_2C_6H_3$	2,6-$(MeO)_2C_6H_3$
2-$MeOC_6H_4$	2,6-$(MeO)_2C_6H_3$	2,6-$(MeO)_2C_6H_3$
Ph	2,6-$(MeO)_2C_6H_3$	2,6-$(MeO)_2C_6H_3$
Ferrocenyl	Ph	2,6-$(MeO)_2C_6H_3$
Ferrocenyl	2,6-$(MeO)_2C_6H_3$	H

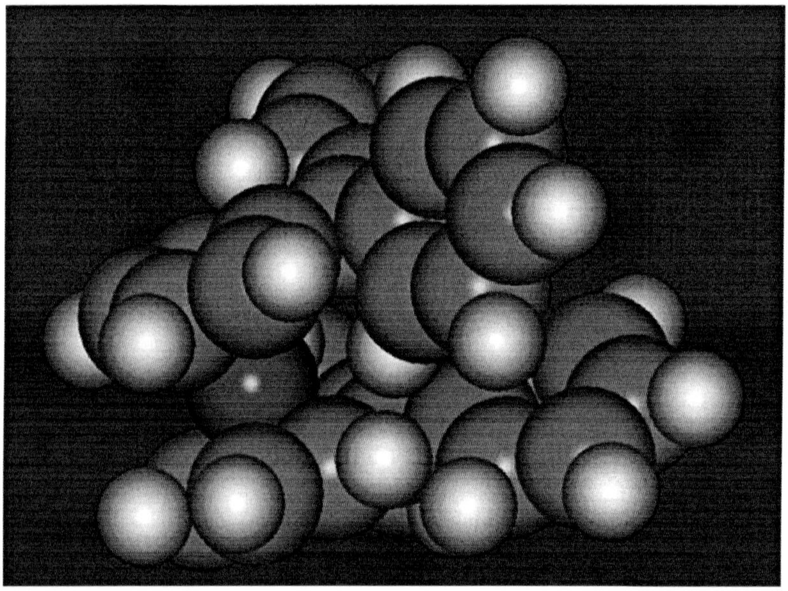

Figure 1.1 *Calculated structure of **54a**. (Reproduced by permission of Professor Mathias Tacke.)*

diradical structure, with some attraction between the two radical sites.[20b] Later computational studies agreed with this conclusion, and also found that the diradical was a ground state singlet.[20c]

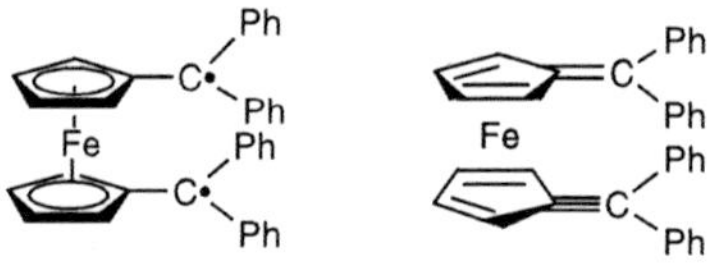

54a **54b**

1.8 Tris(heteroaryl)methyl and related triarylmethyl radicals

Tris(2-thienyl)methyl and tris(3-thienyl)methyl radicals generated by zinc reduction of the corresponding carbocations were directly observed by EPR and characterized by their distinctive spectra.[21a] Tris(2-thienyl)methyl radical **55** generated by reduction of the corresponding cation with zinc dimerizes to a mixture of the tail to tail dimer **56** and head to tail dimer **57** (Equation 1.18).[21b]

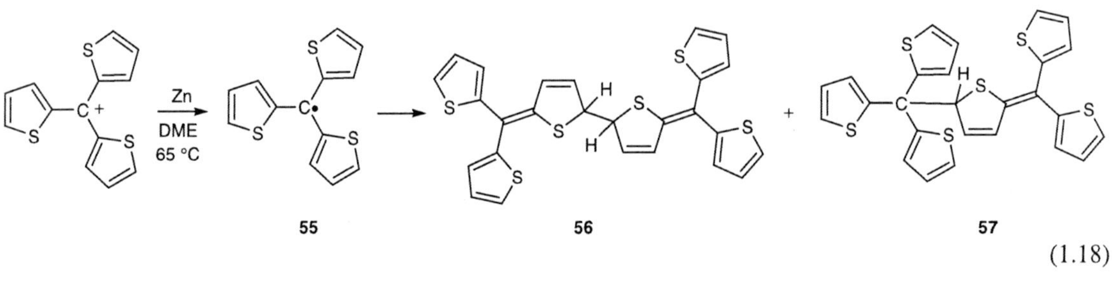

55 **56** **57**

(1.18)

The 3- and 4-pyridyldiphenylmethyl radicals **58** were generated by treatment of the corresponding triarylmethyl chlorides with silver in benzene at room temperature (Equation 1.19), while the corresponding conversion of the 2-pyridyl isomer required silver in tetrahydofuran with ultrasound activation.[22] All three radicals were observed by EPR and characterized by their distinctive spectra, and all formed dimers by head-to-tail coupling involving addition to a pyridyl ring. Dimers **59** and **60** from the 2- and 3-pyridyl radicals, respectively, resulted from hydrogen migration in the initial coupling products. The 4-pyridyl radical gave the *N*-trityl bonded dimer **61** that, upon heating at 70 °C, rearranged to **62**, proposed to form by dissociation to the radical followed by dimerization involving a phenyl ring and irreversible formation of the product by hydrogen migration. In the range −27 to −15 °C reversible formation of the initial dimers was observed. Bond dissociation energies were determined that were about 42 kJ/mol larger than for the corresponding dimer of triphenylmethyl radical, and the greater stability of the pyridyl dimers was attributed to reduced steric repulsions in the pyridyl dimers due to the absence of two hydrogen atoms. Other factors, such as lower stability of the pyridyl radicals or attractive interactions between the nitrogen atoms and the other groups in the dimers, may also be involved.

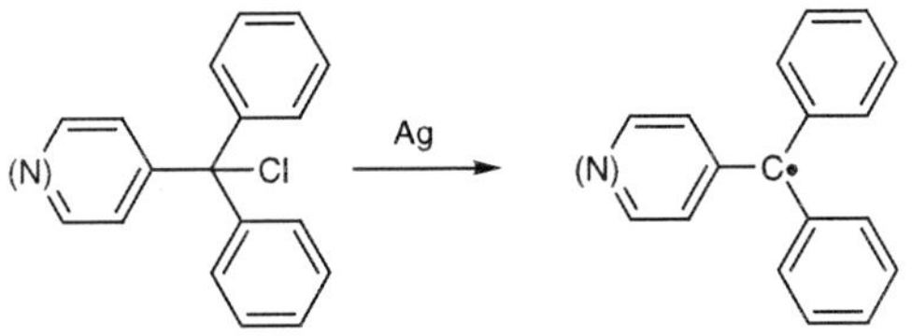

58

(1.19)

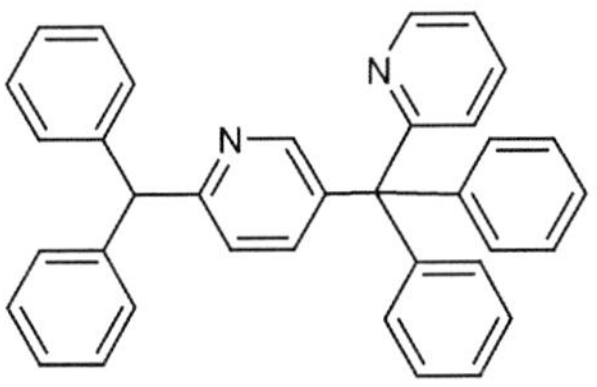

59

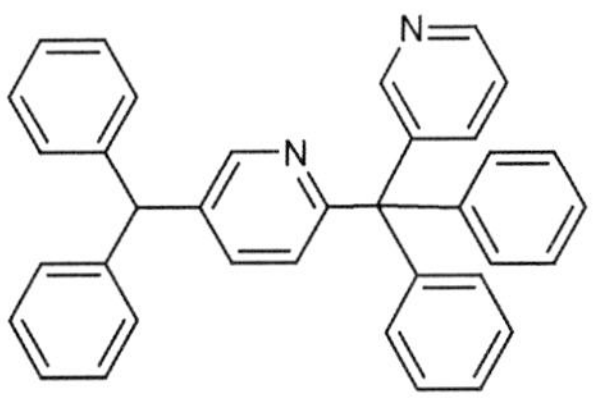

60

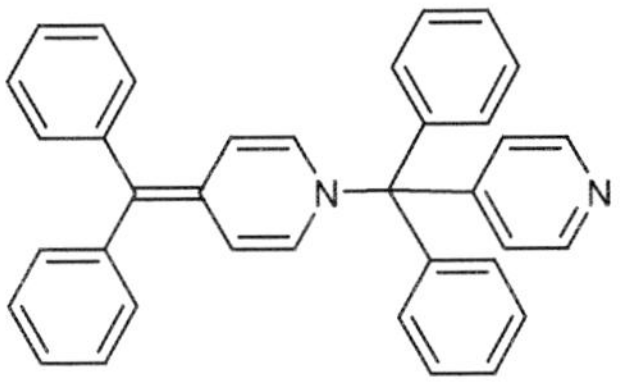

61

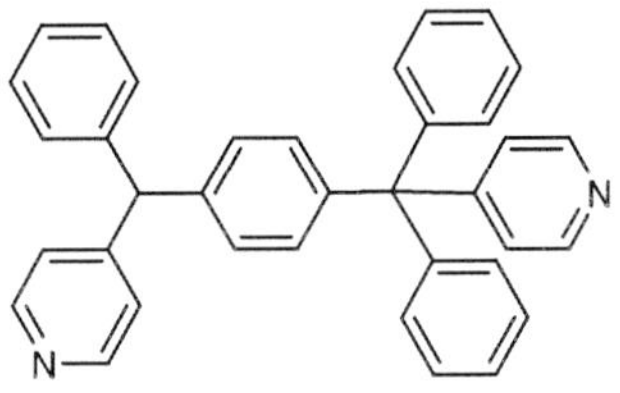

62

2-Triazolyldiarylmethanes **63** upon reaction with *n*-BuLi followed by iodine evidently gave unobserved free radicals **64**, based on the isolation of dimers **65** (Equation 1.20).[23] 1-Triazolyldiarylmethanes **66** upon reaction with *n*-BuLi followed by iodine gave the radicals **67**, which were the first triarylmethyl radicals reported with a directly attached heteroatom at the radical center. The radicals formed the dimers **68** as well as phenanthridines **69** (Equation 1.21).[24] The latter were proposed to form from loss of N_2 from **67** with cyclization (Equation 1.22). Radicals **71** generated from the corresponding triarylmethanes were detected by EPR, and gave dimers **72**, while **73** formed by loss of N_2 with cyclization (Equation 1.23).[24]

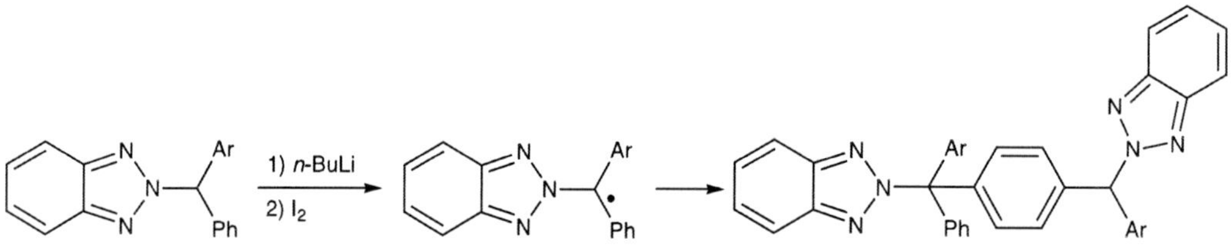

63 64 65 (1.20)

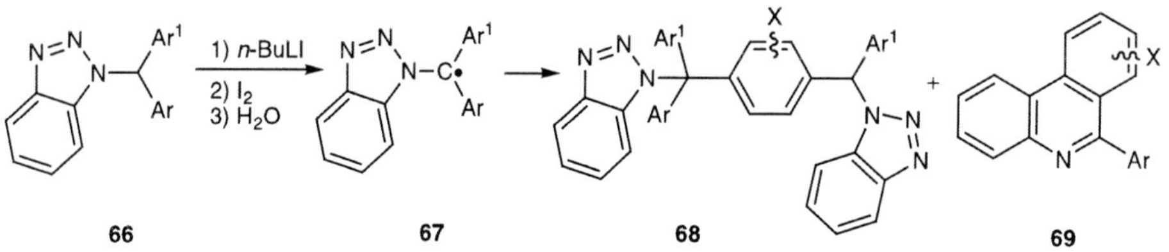

66 67 68 69 (1.21)

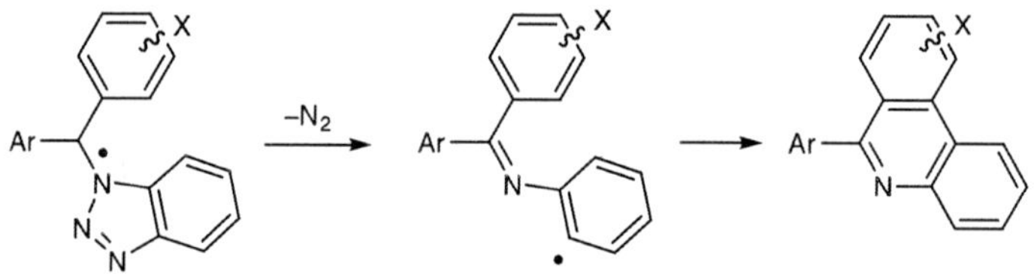

67 70 69 (1.22)

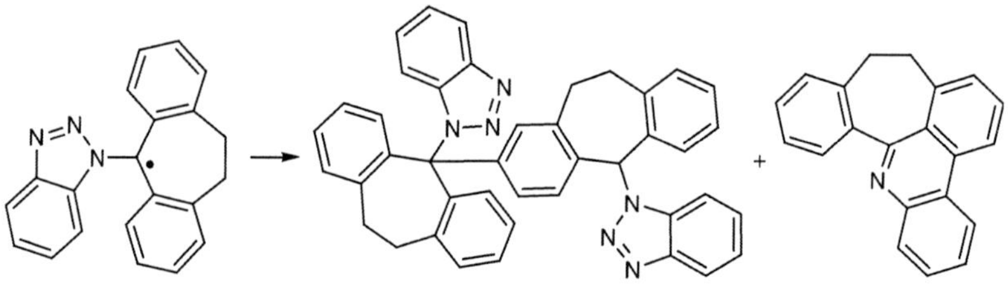

71 72 73 (1.23)

1.9 Delocalized persistent radicals: analogues of triarylmethyl radicals

The discovery by Gomberg of the triphenylmethyl radical opened the way for the discovery of a variety of related persistent radicals. Some of these are covered in greater detail elsewhere in this book. Schlenk and Mark reported the preparation of the equilibrating pentaphenylethyl radical **74**,[25a] which exists as a long lived species (Equation 1.24). Karl Ziegler and coworkers prepared the tetraphenylallyl radical **75**[25b] and the pentaphenylcyclopentadienyl radical **76**.[25c]

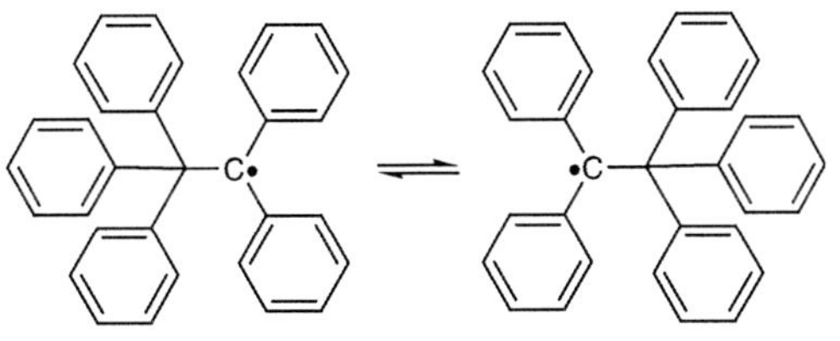

(1.24)

74

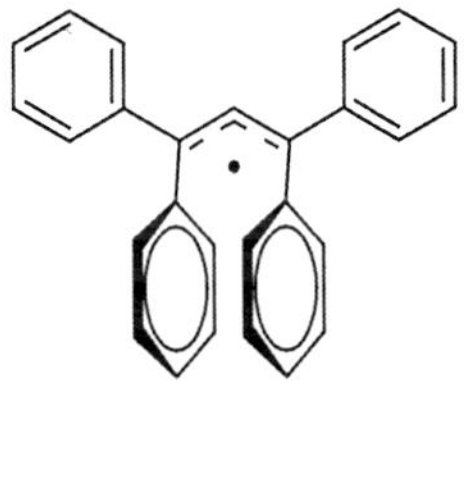

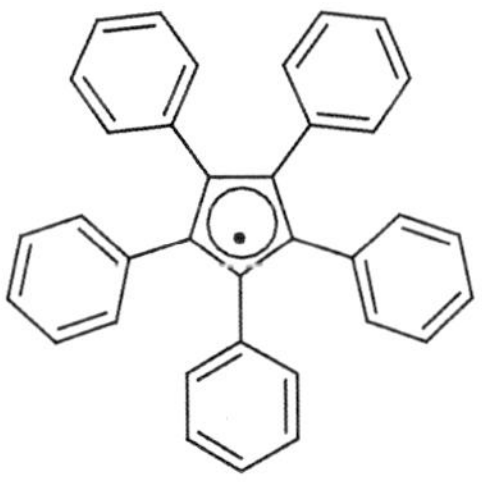

75 **76**

The long-lived radical **77** was first reported by Löwenbein and Folberth: it reversibly forms a dimer, and reacts only slowly with oxygen to give a peroxide.[25d,e] Radical **77** and its analogues have found extensive recent application as chain breaking antioxidants.[25f,g] These function by trapping peroxyl radicals and can completely suppress autoxidation. Diarylacetate radicals **78** have long been known and undergo reversible dimer formation[25h,i] and react with oxygen to form peroxides.[25j,k] Addition of the tetramethylpiperidinyloxyl radical (TO•) to diphenylketene gave the radical **79**, which reacted with oxygen forming the peroxide **80**, with the structure determined by X-ray crystallography (Equation 1.25).[25l]

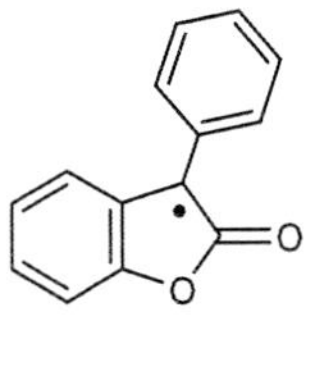

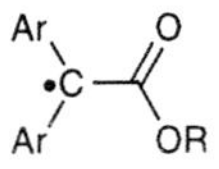

77 **78**

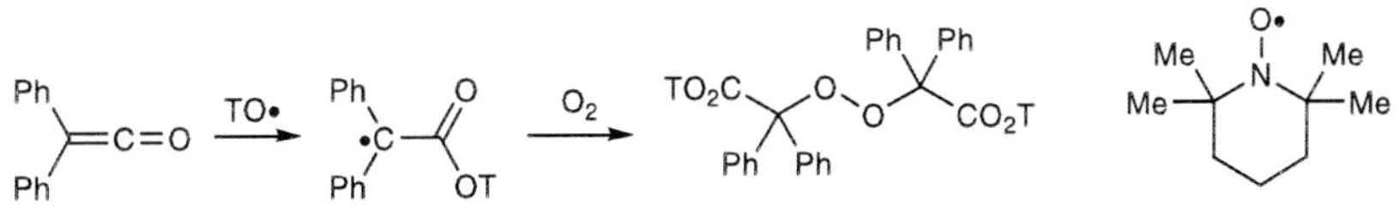

(1.25)

79 **80** TO•

The Koelsch radical **81** was prepared in 1931 but, somewhat surprisingly, even at this date, the idea of stable radicals met some resistance. Koelsch prepared and submitted for publication a report of the long lived radical, but the lack of reactivity of the radical towards oxygen seemed improbable, and the manuscript was rejected. In 1957, EPR spectroscopy of the same sample confirmed the identification, and the original manuscript was resubmitted and published.[26a] Nitrogen analogues **82** were reported in 2000.[26b] More recently, the Koelsch-type diradicals **83** and **84** have been reported and found to have singlet and triplet ground states, respectively.[26c,d]

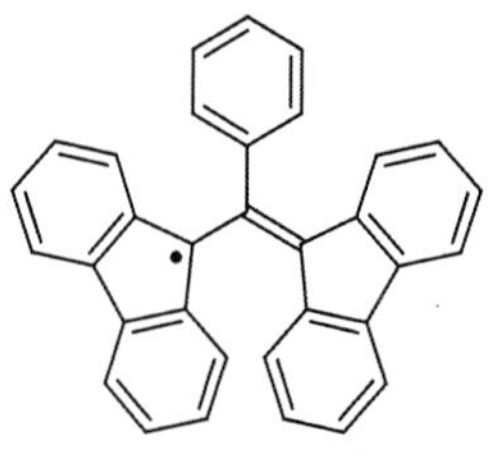

81

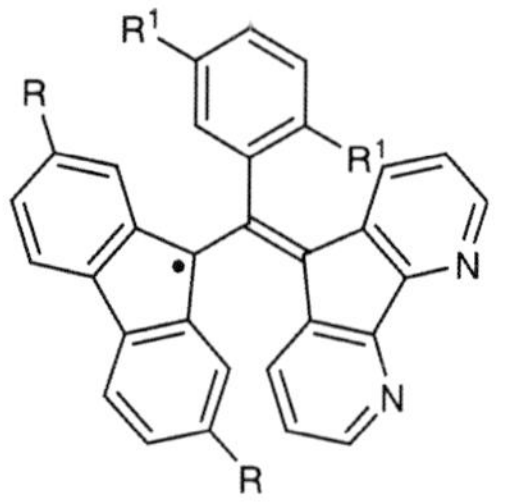

82

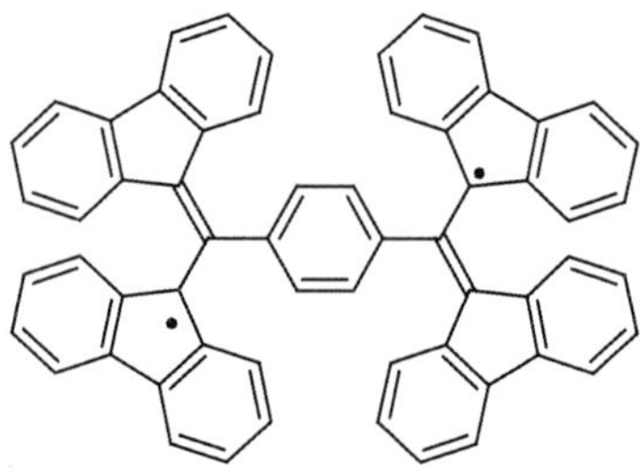

83

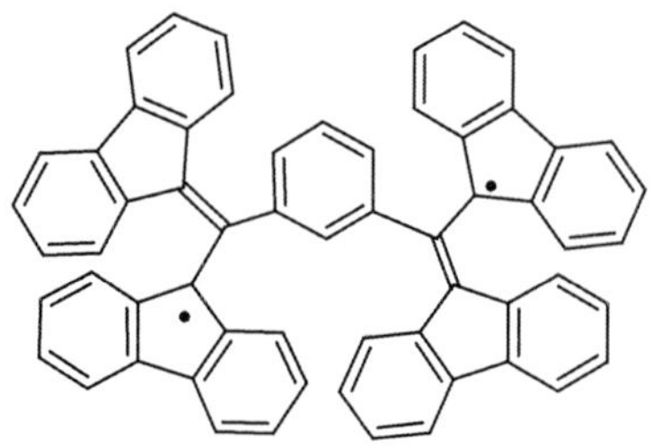

84

1.10 Tetrathiatriarylmethyl (TAM) and related triarylmethyl radicals

Specialized triarylmethyl radicals are finding extensive and increasing use in imaging applications using EPR spectroscopy because of their extraordinary stability in cells and tissues, narrow line widths resulting in high analytical resolution at concentrations in the μM range, and enhanced sensitivity to oxygen.[27] Derivatives which vary in their state of ionization as a function of pH show a corresponding dependence upon of their EPR spectra with pH and are being examined for a variety of biomedical applications.

Tetrathiatriarylmethyl (TAM) trityl radicals **86** and various structural variations were originally developed in industry for use as contrast agents in Overhauser magnetic resonance imaging (OMRI),[27–29] and were first reported in the patent literature.[27c–f] Their properties are discussed in detail in Chapter 16. Important characteristics of these materials are the narrow signals promoted by the absence of inhomogeneous hyperfine couplings. An earlier preparation of these compounds[29c] has been improved by a convenient large scale synthesis of the sodium salt of **86** (known as Finland trityl).[28] The oxygen substituted analogue **87** and the methyl deuterated derivative have also been studied, and together with **86** are soluble in water with single sharp lines in the EPR spectra.[27a] The radical **86** was formed in one step in quantitative yield by treatment of the tri(*tert*-butyl ester) **85** in neat trifluoroacetic acid at room temperature (Equation 1.26).[28] Other trityl alcohols formed radicals in a similar fashion.[28]

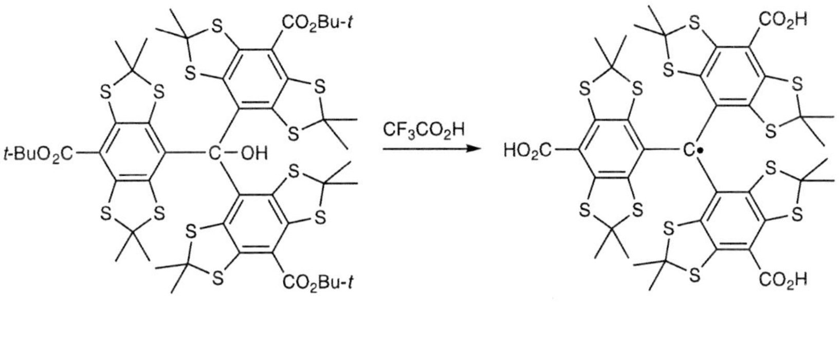

85 86 (1.26)

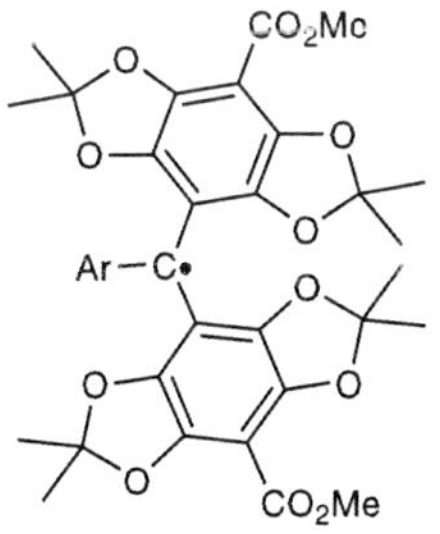

87

The presence of ionizable carboxyl groups on the TAM radicals leading to pH sensitivity of the EPR parameters has been examined, as has the presence of limited numbers of aryl ring protons leading to doublet or triplet EPR signals.[29a] These serve as dual function pH and oxygen probes. Dendritic TAM molecules were prepared in which encapsulation of the radical by the dendrimer enhances their stability and solubility in water over a wide pH range, and the copper(II) complex of **88** is a highly effective pH probe.[29f]

Amino substituted TAMs **90** and related derivatives were prepared by selective reduction of one or more carboxyl groups of **85** to hydroxylmethyl substituents, and these were converted to mesylates that were displaced by amino groups forming **89**. Conversion to radicals **90** was accomplished using trifluoroacetic acid (Equation 1.27), and provides derivatives useable in other pH ranges.[29b]

Collisions of these probe molecules reduce relaxation times of the probes and provide a measure of local oxygen concentration; to investigate the viscosity and frequency dependence on the EPR spectra, measurements of TAM radicals in glycerol/water ranging from 10 to 90 % concentration were carried out, including the use of deuterated solvents.[30a] The conformations of the TAM radicals in solution were investigated using ^{1}H, ^{2}H, and ^{13}C measurements,[30b] with hyperfine coupling tensors determined using X-band pulsed electron-nuclear double resonance (ENDOR) spectroscopy for two triarylmethyl (trityl) radicals used in EPR imaging and oximetry.[30b] Calculated geometries were used to predict the experimental hyperfine tensors, and these gave satisfactory agreement.[30b]

Rapid-scan EPR spectra of a 0.2 mM aqueous solution of deuterated **86** (trityl-CD_3) radical at 9.8 GHz using a rectangular resonator were obtained.[31a] Larger effects of magnetic field inhomogeneities were found for the extended structure of **86** than in smaller lithium phthalocyanine crystals.

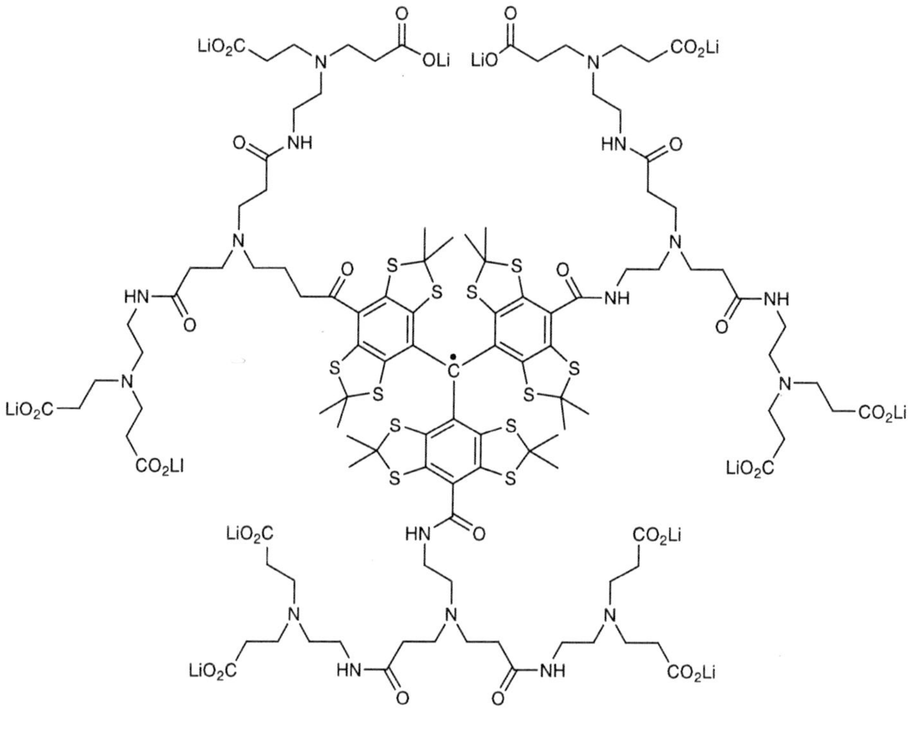

88

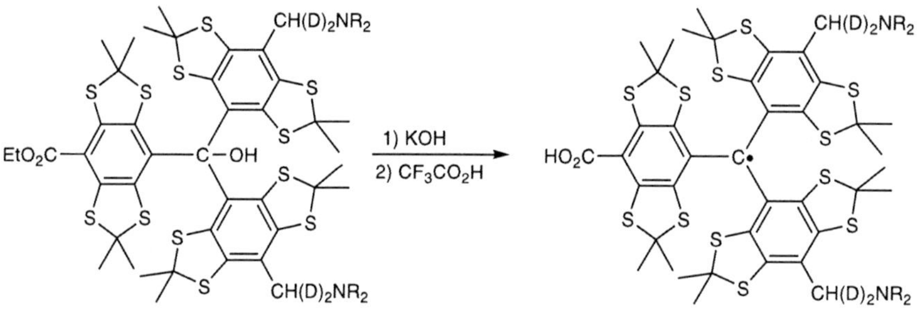

(1.27)

89 **90**

Continuous wave spectra at W-band of four triarylmethyl (trityl) radicals at 100 K in 1 : 1 water/glycerol exhibit rhombic EPR spectra.[31b] The results are consistent with assignment of the Raman process and a local mode as the dominant relaxation processes.

Low frequency (300 MHz) pulsed EPR with pulsed FID detection of triarylmethyl radicals injected in mice was used to obtain three-dimensional oxymetric images.[32a,b] Single-point imaging, a technique developed for solid state NMR, applied to pulsed EPR imaging yielded artifact-free spatial images.[32a] Continuous wave EPR imaging was used to obtain slice-selective images of free radicals without measuring three-dimensional (3D) projection data.[32c]

Notable among these widely studied radicals is **91**, which has been extensively derivatized for optimum properties.[27b,29a] This reacts with superoxide ($O_2^{\bullet -}$) radicals with an apparent second order rate constant of $3.1 \times 10^3\,M^{-1}\,s^{-1}$, and this was used for analysis of the superoxide.[33a,b] Assay of superoxide by monitoring the newly formed product absorption peak at 546 nm[33a] or by EPR[33b] gave good agreement with the widely used cytochrome *c* method of superoxide detection.[33a] The identity of the species giving rise to the observed absorption at 546 nm, which was stable for 12 hours, was not identified. Hydroperoxide was detected by ferrous oxidation-xylenol orange (FOX) assay,[33b] but mass spectrometric analysis did not detect a covalent adduct, and formation of an ion pair **92** of hydroperoxide anion and the trityl cation was proposed (Equation 1.28).[33a]

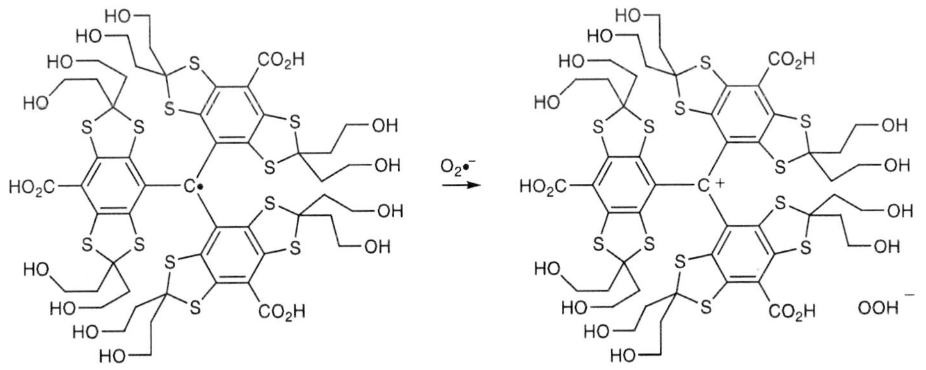

(1.28)

91 **92**

Dynamic nuclear polarization of the substrate by microwave irradiation of **91** permitted the measurement of the kinetics of trypsin catalyzed hydrolysis of N_α-benzoyl-L-arginine ethyl ester by ^{13}C NMR using stopped flow.[33c] Extension of this method to the study of other enzyme catalyzed reactions is anticipated. The dynamic nuclear polarization (DNP) of the triphenylmethyl radical in water has been compared to that of the 4-hydroxyTEMPO radical **93** and the latter showed considerably larger DNP enhancements up to 100 (9 GHz) and −20 (94 GHz) using continuous microwave irradiation.[33d]

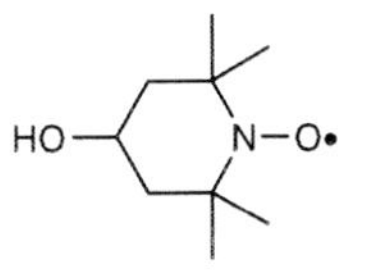

93

The degradation of tetrathiatriarylmethyl radicals **86** used as oximetry probes for EPR imaging applications due to reaction with oxygen was investigated experimentally and by using computational methods.[33e] Tricarboxylate salts bearing CO_2Na groups were the most stable, and it was proposed that degradation by oxygen attack on the TAMs preferentially occurred at unsubstitued 4-positions of the aryl rings leading to hydroperoxide intermediates. Radical pathways with addition of O_2 or $HO_2^{\bullet}$ to **86** forming **94** or **95** were considered (Equation 1.29), as well as ionic routes, such as the addition of the anion HO_2^- to the cation of **74** forming **80**.[33e]

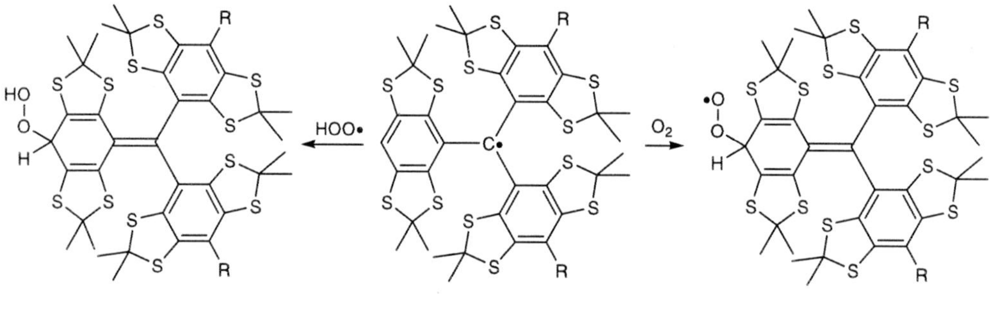

95 86 94

(1.29)

1.11 Perchlorinated triarylmethyl radicals

Perchlorination of aromatic compounds using a reagent composed of sulfur monochloride and aluminum chloride in sulfuryl chloride (SCM reagent) was introduced in 1960,[34a] and has proved very effective. In 1971 this method was applied to the formation of perchlorodiphenylmethyl **97** and perchlorotriphenylmethyl **99** radicals, both of which have lifetimes in air of decades and are characterized as *inert* carbon free radicals.[34b,c] Conversion of the chlorocarbon **96** to **97** could be effected by reduction with sodium iodide (Equation 1.30). More generally, radical **99** and its derivatives are prepared by deprotonation of the corresponding tris(pentachlorophenyl)methane **98** followed by oxidation of the anion with iodine (Equation 1.31).[34b,c] The *ortho*-chorines on the aryl rings effectively shield the central radical site, preventing reactions of the chlorinated radicals (Figure 1.2). This family is reviewed in detail in Chapter 2.

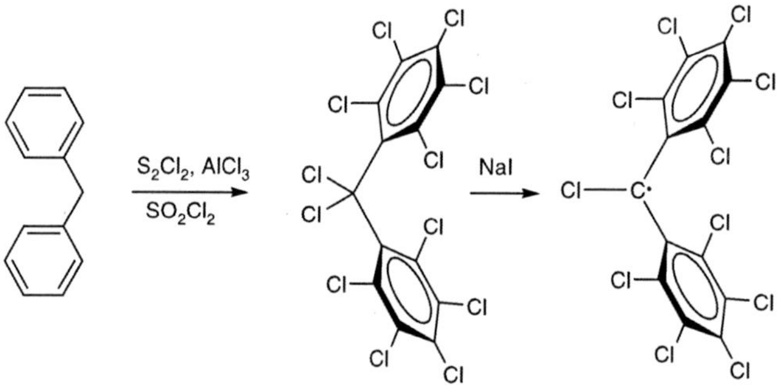

96 97

(1.30)

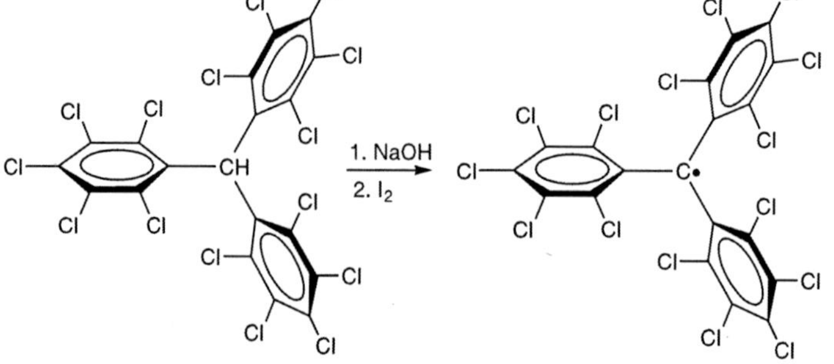

98 99

(1.31)

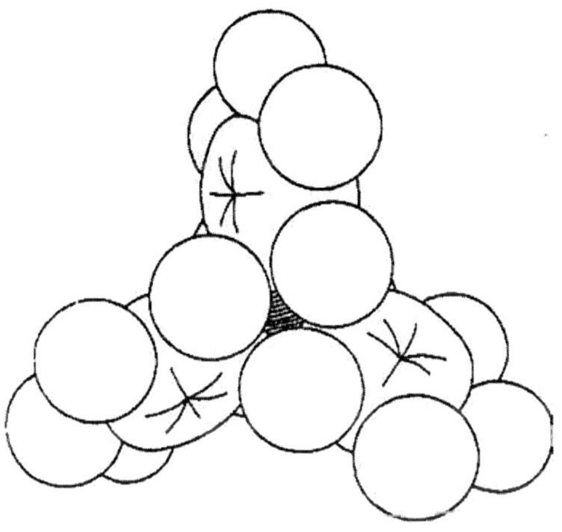

Figure 1.2 Model of the tris(pentachlorophenyl)methyl radical **99**. (Reprinted with permission from [34c]. Copyright 1985 American Chemical Society.)

Some of the properties of these chlorinated radicals include the formation of clathrate hosts for benzene[35a] and the use of partially chlorinated amino-substituted derivatives in magnetic materials.[35b] Their EPR properties have been determined,[36a] and monofunctionalized derivatives have been prepared and characterized.[36c] These studies have been extended to perchlorinated diradical **100**[34b] and to corresponding radical cations and radical anions, which have single electron transfer equilibria,[34e,f] and to other highly but not completely perchlorinated radicals, including the monoradical **101**[35b] and the diradical **102**.[36d]

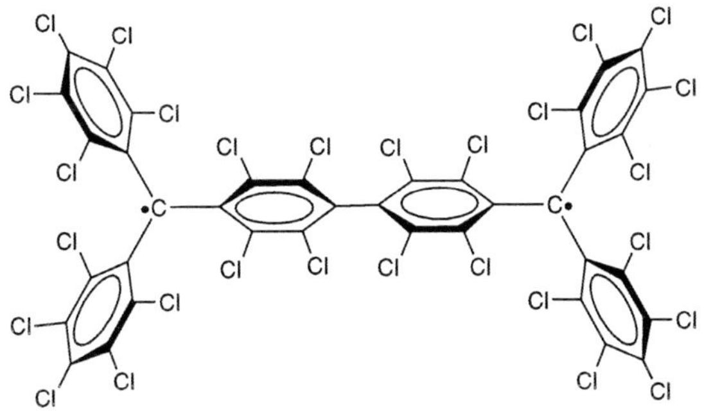

100

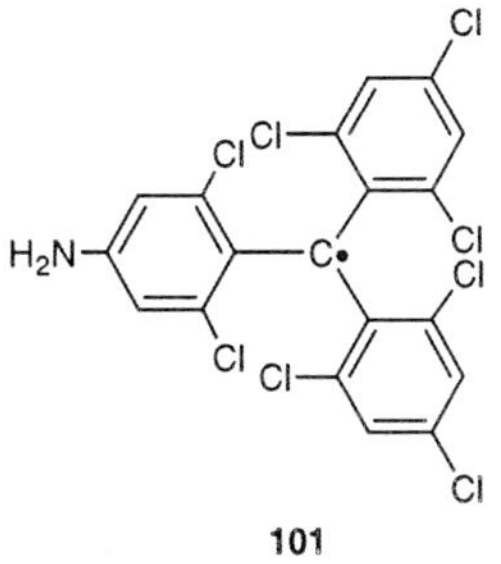

101

There is essentially no delocalization of the free electron into the twisted perchlorophenyl rings of these perchloroaryl radicals, but because of the steric shielding normal atom transfer reactions or radical combination reactions are strongly inhibited. While these radicals are remarkably inert to almost all reagents, they are susceptible to electron transfer, either reduction to carbanions or oxidation to carbocations.[34e,f]

Perchorinated triarylmethyl radical **103** combining perchorotriphenylmethyl and bis(4-methoxyphenyl) amine moieties was used to study electron transfer and charge transfer properties. The absorption spectrum of this highly soluble compound was studied in 13 solvents. The electronic coupling and the dipole moments of the ground and excited states were also determined.[37a]

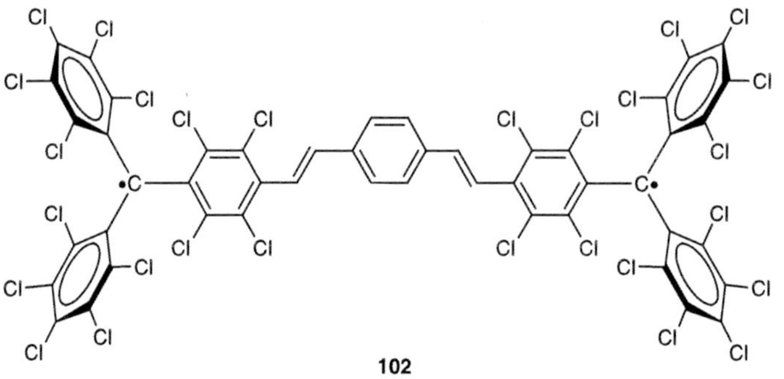

102

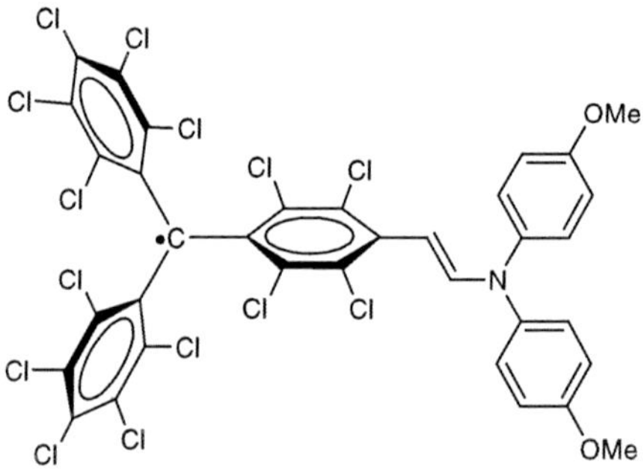

103

The dependence of the UV spectra of the transition from radical **104** to the zwitterion **105** was examined as the solvent composition was varied and the process changed from the normal to the inverted Marcus region (Equation 1.32).[37b] The redox properties of the two species were also examined.

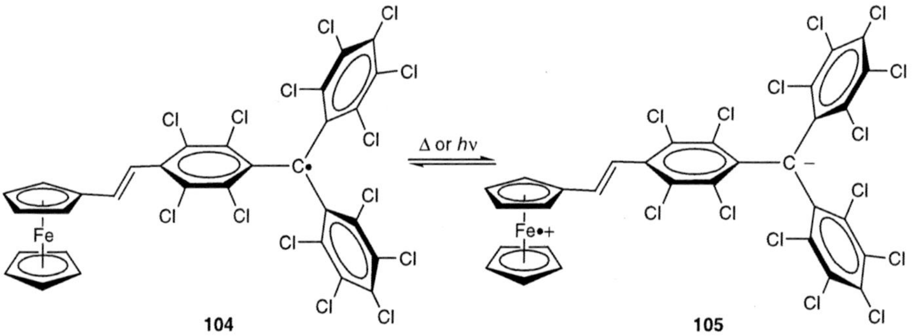

104 **105**

(1.32)

Perfluorinated triphenylmethyl radical **106** was generated by reduction of the corresponding carbocation and gave an EPR spectrum measured in benzene (Equation 1.33).[38a] The product was recrystallized from benzene giving material with a melting point of 158 °C containing an undetermined amount of free radicals as detected by EPR. Tris(4-nitro-2,3,5,6-tetrafluorophenyl)methyl radical **107** was formed by oxidation of the corresponding carbanion with potassium permanganate and characterized by its EPR spectrum (Equation 1.34).[38b]

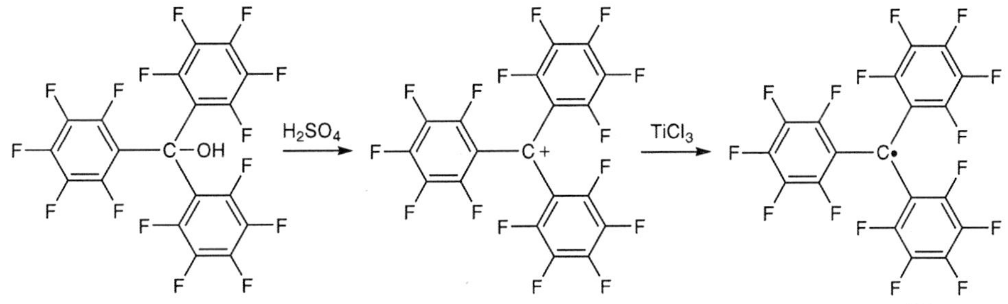

106

(1.33)

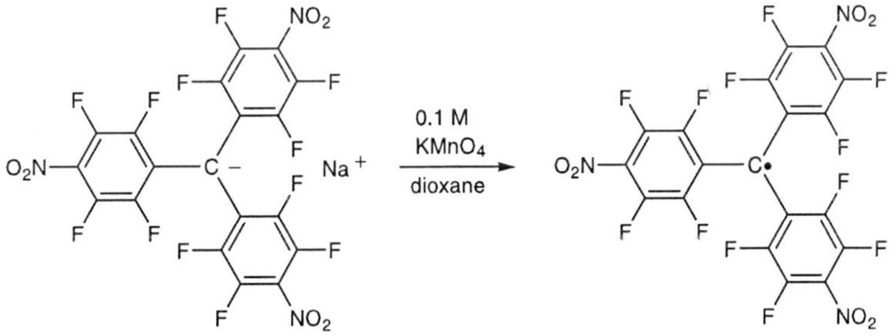

107

(1.34)

1.12 Other triarylmethyl radicals

Earlier studies of less substituted tris(4-halophenyl)methyl radicals indicated these underwent dehalogenation upon formation of unsymmetrical dimers. For example, reaction of 4-bromophenyl(diphenyl)methyl chloride with silver evidently proceeds with formation of radical **108**, and after dimerization to **109** there is loss of the allylic bromine forming **110**, which was captured as the peroxide (Equation 1.35).[39a] Previously it had been found that treatment of 4-bromophenyl(diphenyl)methyl chloride with silver resulted in loss of all the chlorine and half the bromine, in agreement with this process. Similarly 4-fluorophenyl(diphenyl)methyl chloride lost fluorine on treatment with silver.[39b,c]

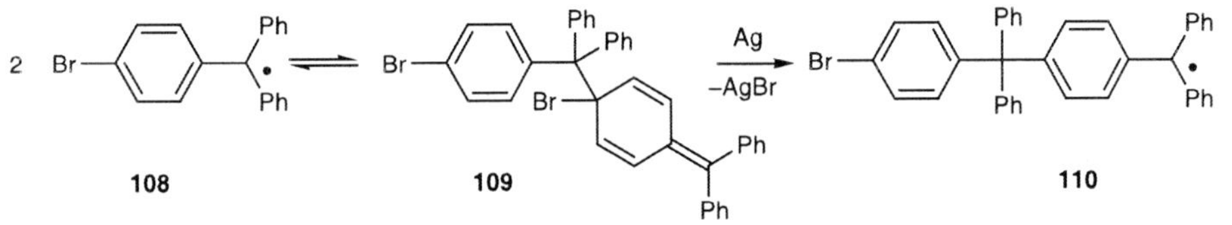

108 109 110

(1.35)

Galvinoxyl-based radical **111** is a stable free radical inert to oxygen that is used as a spin trap for reactive free radicals; it is obtained by oxidation of the galvinol **112** (Equation 1.36).[40a,b] Reduction of **112** yields the triarylmethyl radical **113**, which has quite different properties from **111**.[40c] A variety of analogues of **113** have been prepared.[40c]

Cyclophane derived triarylmethyl bromides **114** (n = 6,7,8,9) reacted with the reducing agent obtained from titanium tetrachloride and lithium aluminum hydride ($LiAlH_4$) to form dimers **116** evidently derived from triarylmethyl radicals **115** (Equations 1.37 and 1.38).[41] The dimers either were accompanied by or converted to the fully aromatized isomers.

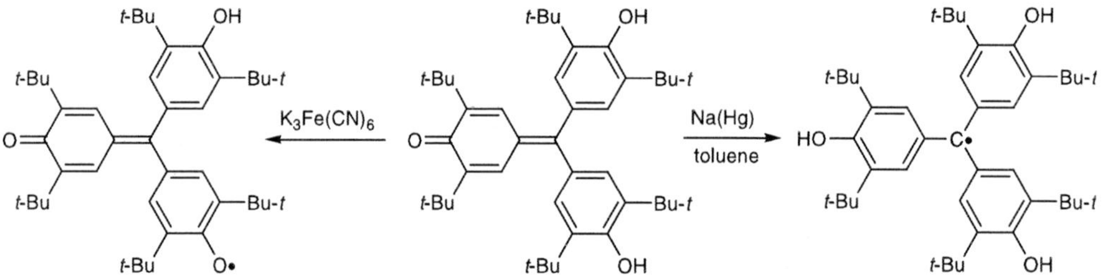

(1.36)

111 **112** **113**

111 **112**

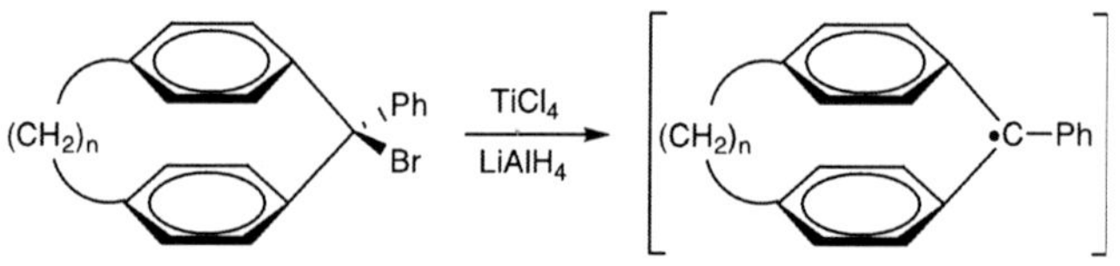

(1.37)

114 **115**

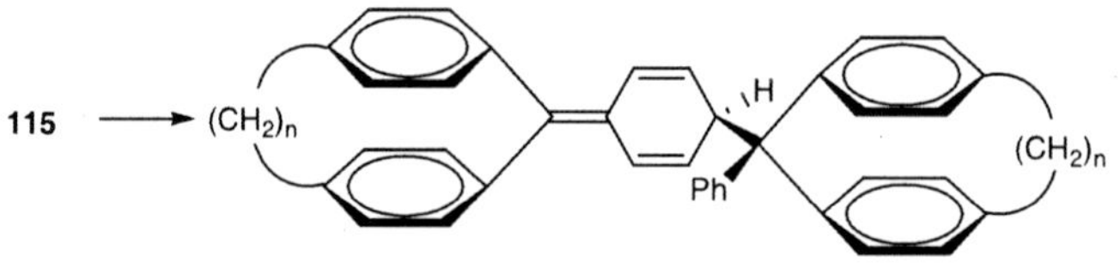

(1.38)

116

1.13 Diradicals and polyradicals related to triphenylmethyl

As noted in Section 1.7 a number of hexaarylethanes, such as **51**, which are potential diradicals but are constrained to close intramolecular geometries, have been prepared but do not have diradical character.[19e–i] However, **54** is proposed as a singlet diradical.[20b–d] Stable chlorinated diradicals, such as **102**, have also been prepared.[34b]

Tetraphenyl-*para*-xylylene **117** (Thiel radical) was prepared by Thiel and Balhorn in 1904 as an isolable species[42a] and has been characterized by X-ray.[42b] This species, however, has some diradical character (**117b**) and continues to attract attention after more than 100 years.

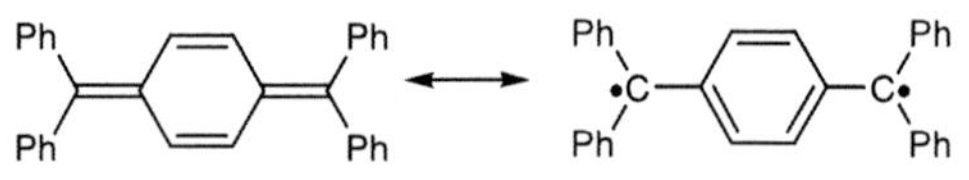

117a **117b**

The first stable diradical **118**, which is analogous to triphenylmethyl, was reported by Schlenk in 1915 (Schlenk diradical)[43a] and promoted the expansion into the field of molecules containing multiple radical sites. The analogous 1,3,5-trisradical **119** (Leo triradical) was reported in 1937.[43b] This marked the beginning of the study of "high spin" molecules, which are gaining increasing attention because of the potential application of these materials as molecular magnets.[43c] The chemistry of triarylmethyl di- and polyradicals and related species in the design of magnetic clusters and organic magnets has been frequently reviewed.[44]

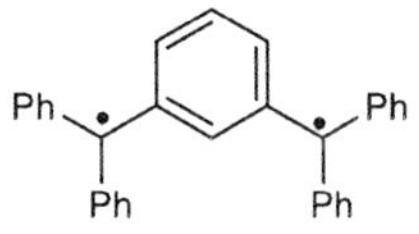

118

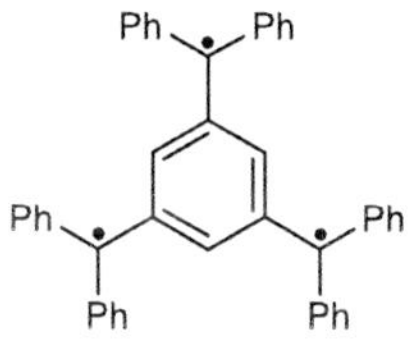

119

Many diradicals which share the topology of the Schlenk radical have been prepared, including **120** and **121**.[45a,b] Triradicals **122** have also been prepared and characterized by EPR.[45c]

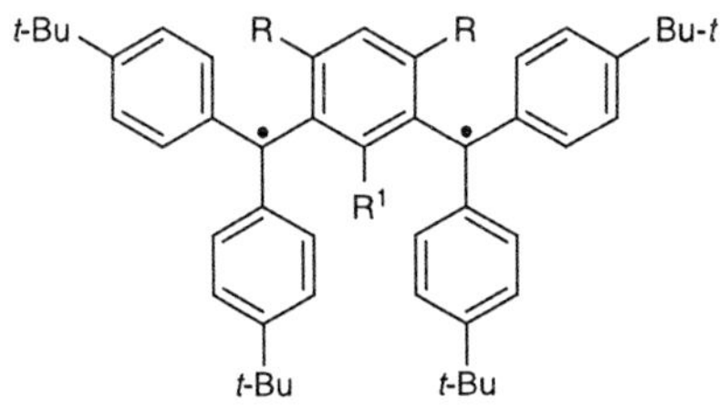

120

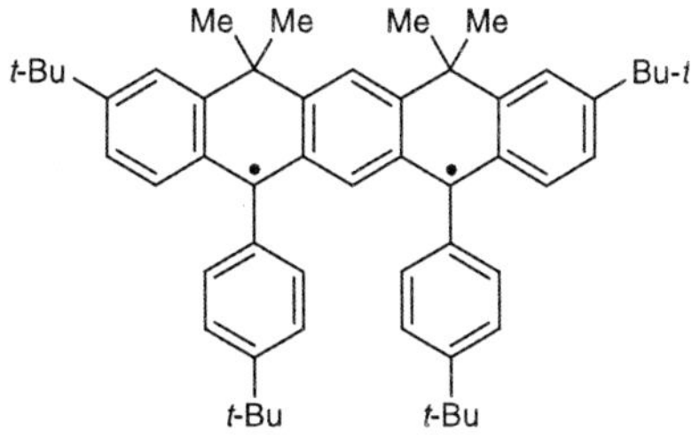

121

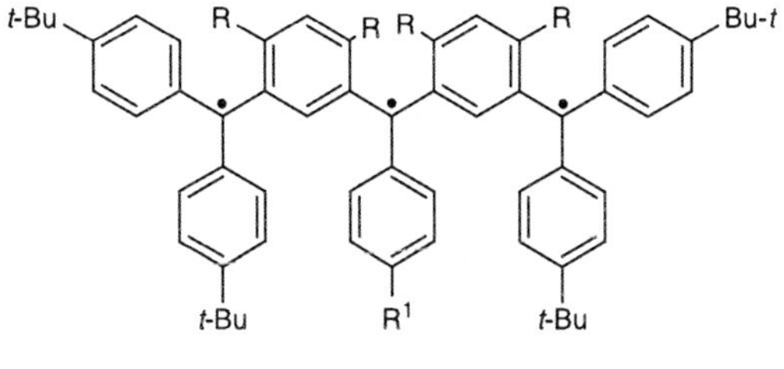

122

The tetraradical **123** was prepared by oxidation of the corresponding tetra-anion with molecular iodine, and identified by the characteristic quintet EPR spectrum. The zero field parameter $[D/hc]$ is less than that of the triplet radical **118** and the small value of $[E/hc]$ was interpreted as showing threefold symmetry for **123**.[45d] Analogous heptaradical **124** and decaradicals were also prepared.[45e] This work has been reviewed.[46]

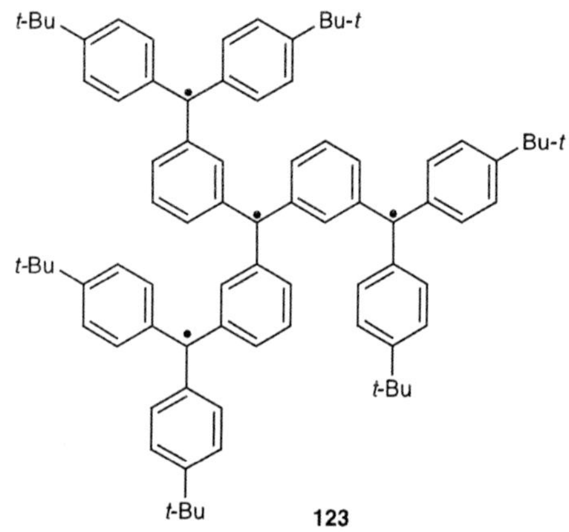

123

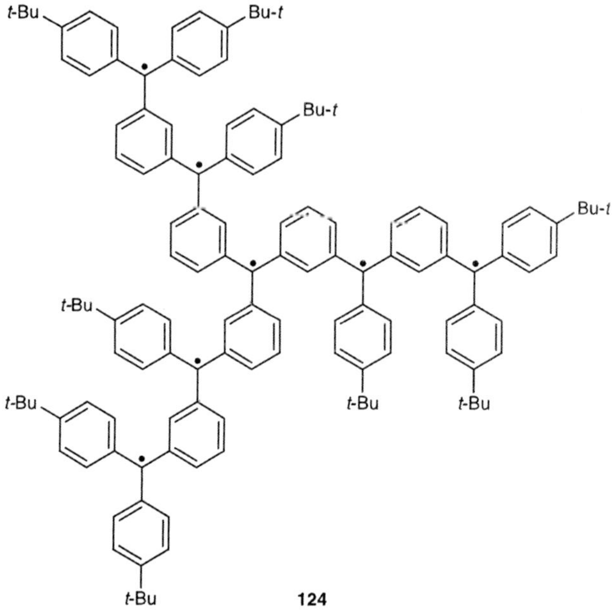

124

The Chichibabin hydrocarbon **125**[47a] is a further example of an extended linear polyaromatic with diradical character (**125a**) and is related to the Thiel radical **117**. It is unusual for a neutral, even-electron molecule with a closed shell electronic structure such as **125** to display magnetic activity and this species has remained of interest. In recent studies, extended viologens **126** have been compared to **125**.[47b,c] These yield ambiguous results for their electronic structure (**126b**) using EPR techniques, but **126** was recently found to have singlet character as determined using Raman spectroscopy.[47b,c] Earlier X-ray studies also favored a singlet structure for **125**.[42b] Other related diradicals include **127**,[47d] **128**,[47e] and the Muller hydrocarbon **129**.[47f]

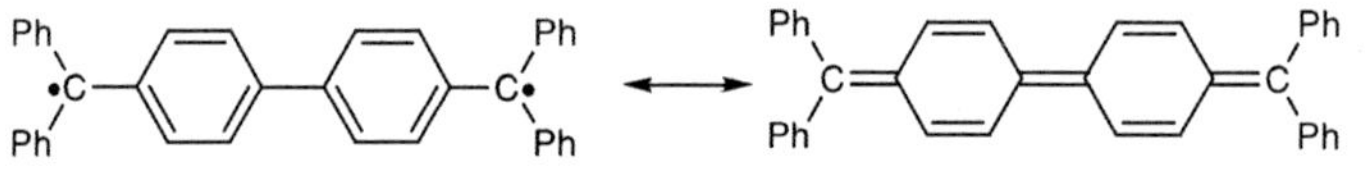

125a 125b

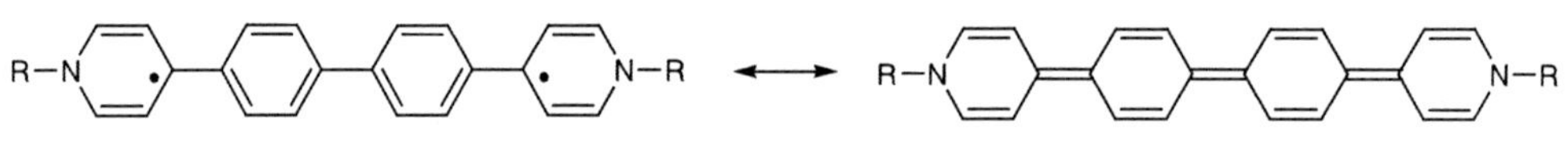

126a 126b

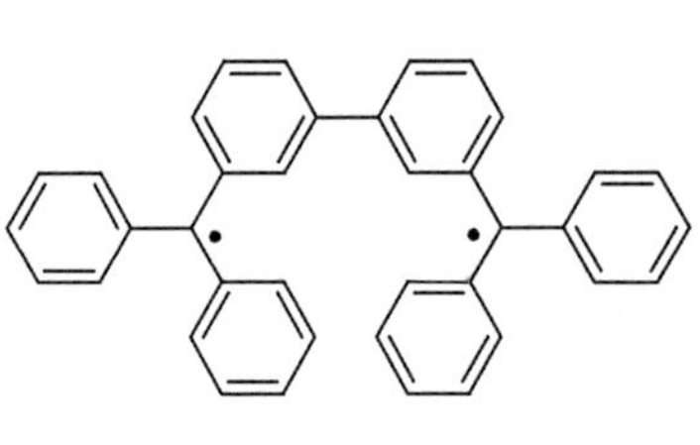

127

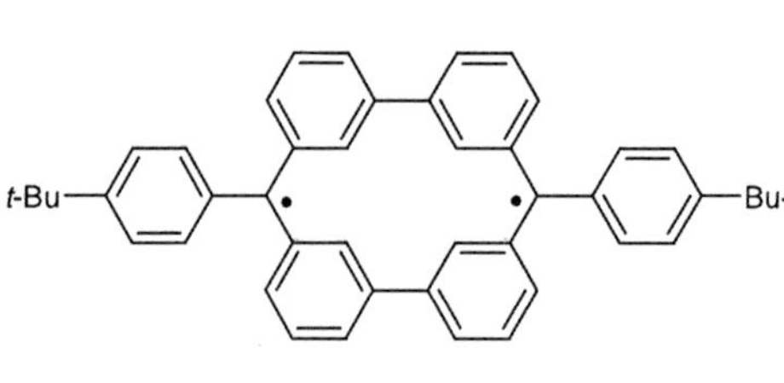

128

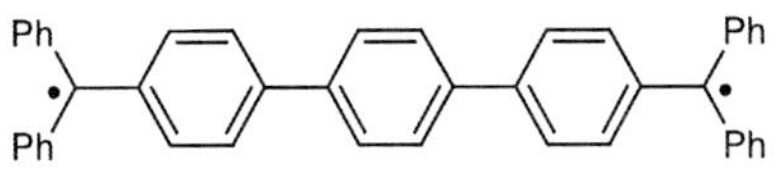

129

Poly(sulfonylphthalide) **130** was reduced with lithium to give poly(triarylmethyl) radical **131**, characterized by EPR and UV-Visible spectra as a diradical ground state singlet of the Chichibabin hydrocarbon type (Equation 1.39).[48] Poly(radical) **132** had similar character, while **133**, for which the formation of quinonoid structures is energetically unfavorable, led to polyradicals of the triarylmethyl type.[48]

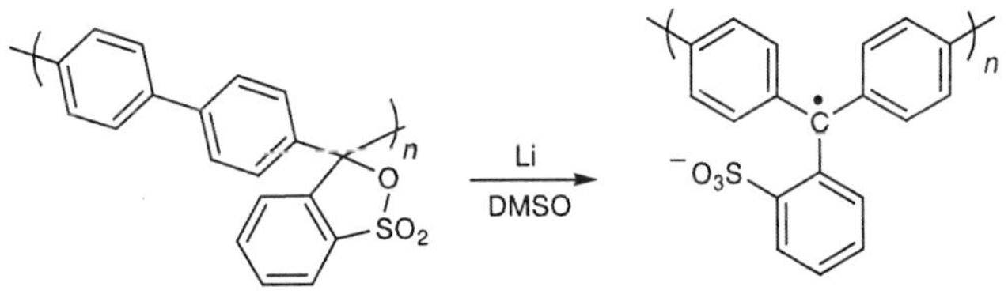

(1.39)

130 131

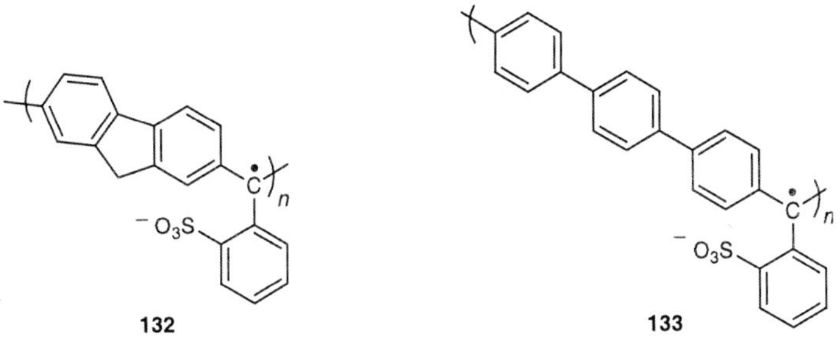

132 133

Polymer-based triarylmethyl radicals **135** in which the individual triarylmethyl groups are separated from one another by saturated hydrocarbon chains were generated by zinc or potassium treatment of the corresponding polymer with attached triphenylmethyl chlorides **134**, and were characterized by their UV and EPR spectra (Equation 1.40).[49a–e] Radical sites were found to be present on between 25 and 33 % of the triarylmethyl moieties.[49] These species behave as isolated radicals, with no direct spin interaction between the triarylmethyl groups.

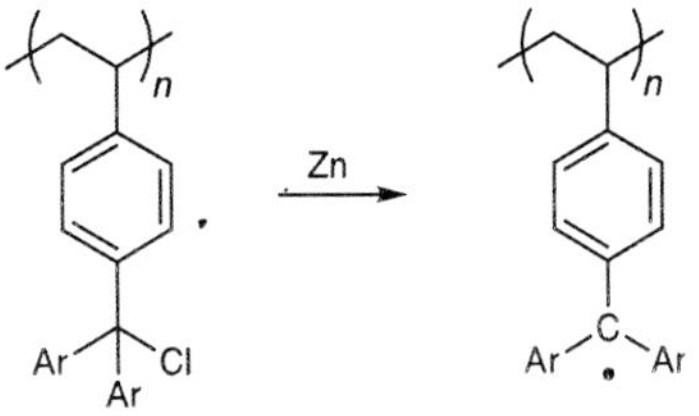

(1.40)

134 135

A number of further related diradicals of the type **136** and **137** with insulating groups between the radical sites were obtained from reactions of the triarylmethyl chlorides with silver.[49h]

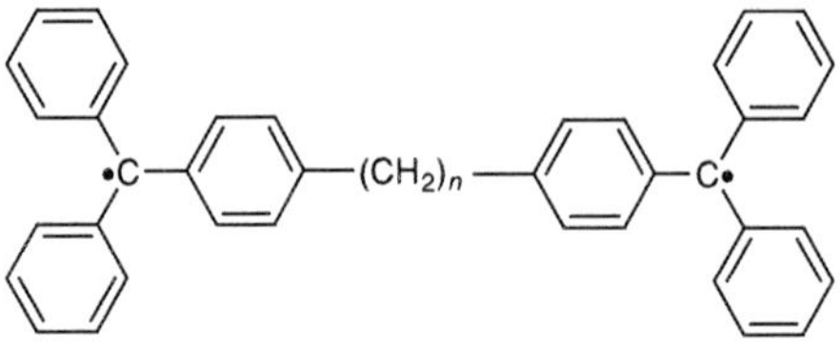

136

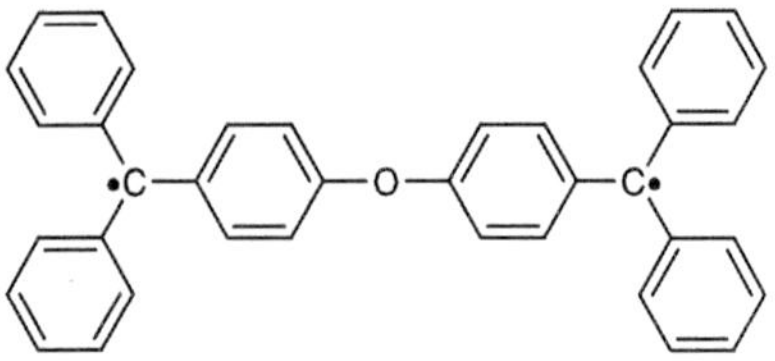

137

1.14 Outlook

Originally the discovery of stable triarylmethyl radicals was mainly of theoretical interest and helped lay the foundation for the electronic theory of organic chemistry. The realization of the existence of free radicals with variable levels of stability then led to many advances in chemical theory and practical application, as a variety of free radical reactions, such as polymerizations and substitutions, are of great industrial importance. Similarly, free radicals have been found to play a vital role in biological processes. Now, with the application of triaryl radicals in biology, these species are themselves of great utility, highlighting the inextricable connection between fundamental knowledge and useful application.

Acknowledgements

Financial support by the Natural Sciences and Engineering Research Council of Canada is gratefully appreciated.

References

1. M. Gomberg, *Ber. Dtsch. Chem. Ges.*, **33**, 3150–3163 (1900); *J. Am. Chem. Soc.*, **22**, 757–771 (1900).
2. (a) J. C. Bailar, Jr., *Biog. Mem. Nat. Acad. Sci. USA*, **41**, 141–173 (1970). (b) C. S. Schoepfle and W. E. Bachmann, *J. Am. Chem. Soc.*, **69**, 2921–2925 (1947). (c) T. T. Tidwell, *Chemical Intelligencer*, July 2000, 33–38 (2000). (d) T. T. Tidwell, *Adv. Phys. Org. Chem.*, **36**, 1–58 (2001). (e) T. T. Tidwell, *Angew. Chem. Int. Ed.*, **40**, 331–337 (2001). (g) G. B. Kaufman, *Chem. Educator*, **13**, 28–33 (2008). (h) N. M. Shishlov, *Russ. Chem. Rev*, **75**, 863–884 (2006). (i) W. Schlenk, T. Weickel, and A. Herzenstein, *Justus Liebigs Ann. Chem.*, **372**, 1–20 (1910).
3. (a) P. Jacobson, *Ber. Dtsch. Chem. Ges.*, **38**, 196–199 (1905). (b) E. Heintschel, *Ber. Dtsch. Chem. Ges.*, **36**, 320–322 (1903). (c) E. Heintschel, *Ber. Dtsch. Chem. Ges.*, **36**, 579 (1903). (d) J. M. McBride, *Tetrahedron*, **30**, 2009–2022 (1974). (e) H. Lankamp, W. Th. Nauta, and C. MacLean, *Tetrahedron Lett.*, **9**, 249–254 (1968). (f) L. Eberson, *Chem. Intelligencer*, **6** (3), 44–49 (2000).
4. A. R. Forester, J. M. Hay, and R. H. Thomas, *Organic Chemistry of Stable Free Radicals*, Academic Press, London, 1968.
5. (a) J. Schmidlin, *Ber. Dtsch. Chem. Ges.*, **41**, 2471–2479 (1908). (b) E. G. Janzen, F. J. Johnston, and C. L. Ayers, *J. Am. Chem. Soc.* **89**, 1176–1183 (1967).
6. (a) S.-H. Jang, P. Gopalan, J. E. Jackson, and B. Kahr, *Angew. Chem. Int. Ed.* **33**, 775–777 (1994). (b) B. Kahr, J. E. Jackson, D. L. Ward, *et al.*, *Acta Cryst. Sect. B* **48**, 324–329 (1992).
7. (a) H. Wieland, *Ber. Dtsch. Chem. Ges.*, **44**, 2550–2556 (1911). (b) D. E. Falvey, B. S. Khambatta, and G. B. Schuster, *J. Phys. Chem.* **1990**, 94, 1056–1059. (c) K. U. Ingold, M. Smeu, and G. A. DiLabio, *J. Org. Chem.*, **71**, 9906–9908 (2006). (d) W. H. Urry and M. S. Kharasch, *J. Am. Chem. Soc.*, **66**, 1438–1440 (1944). (e) P. H. Kasai, D. McLeod, Jr., and H. C. McBay, *J. Am. Chem. Soc.*, **96**, 6864–6868 (1974). (f) G. Brunton, H. C. McBay, and K. U. Ingold, *J. Am. Chem. Soc.*, **99**, 4447–4450 (1977). (g) G. A. DiLabio, K. U. Ingold, S. Lin, *et al.*, *Angew. Chem. Int. Ed.*, **49**, in press (2010).

8. R. G. Bulgakov, S. P. Kuleshov, L. I. Sharapova, *et al*., *Russ. Chem. Bull. Int. Ed*., **50**, 1194–1197 (2001).
9. (a) M. Gomberg, *Ber. Dtsch. Chem. Ges*., **30**, 2043–2047 (1897). (b) M. Gomberg, *J. Am. Chem. Soc*., **20**, 773–780 (1898). (c) H. Wieland, E. Popper, and H. Seefried, *Ber. Dtsch. Chem. Ges*., **55**, 1816–1834 (1922).
10. (a) J. P. Lorand and P. D. Bartlett, *J. Am. Chem. Soc*., **88**, 3294–3302 (1966). (b) H. Henry-Riyad and T. T. Tidwell, *J. Phys. Org. Chem*., **16**, 559–563 (2003). (c) P. Buckus, *Russ. Chem. Rev*., **39**, 112–129 (1970). (d) D. A. Tasis, M. G. Siskos, A. K. Zarkadis, *et al*., *J. Org. Chem*. **65**, 4274–4280 (2000). (e) A. Baidak, S. Naumov, and O. Brede, *J. Phys. Chem. A* **112**, 10200–10209 (2008). (f) K. Kobiro, M. Matsura, H. Kojima, and K. Nakahara, *Tetrahedron*, **65**, 807–810 (2009). (g) R. Palchaudhuri, V. Nesterenko, and P. J. Hergenrother, *J. Am. Chem. Soc*., **130**, 10274–10281 (2008).
11. (a) N. A. Porter, D. J. Hogenkamp, and F. F. Khouri, *J. Am. Chem. Soc*., **112**, 2402–2407 (1990). (b) K. K. Wang, Z. Wang, and P. D. Sattsangi, *J. Org. Chem*., **61**, 1516–1518 (1996). (c) B. Liu, K. K. Wang, and J. L. Petersen, *J. Org. Chem*., **61**, 8503–8507 (1996). (d) K. K. Wang, H.-R. Zhang, and J. L. Petersen, *J. Org. Chem*., **64**, 1650–1656 (1999).
12. (a) H. Fischer *J. Am. Chem. Soc*., **108**, 3925–3927 (1986). (b) R. F. Bridger and G. A. Russell, *J. Am. Chem. Soc*., **85**, 3754–3765 (1963). (c) G. Binsch and C. Rüchardt, *J. Am. Chem. Soc*., **88**, 173–174 (1966). (d) J. I. G. Cadogan, R. M. Paton, and C. Thomson, *Chem. Commun*., 614–615 (1969). (e) M. J. Perkins, *J. Chem. Soc*., 5932–5935 (1964). (f) W. E. Bachmann and F. Y. Wiselogle, *J. Org. Chem*., **1**, 354–382 (1936). (g) D. F. DeTar, *J. Am. Chem. Soc*., **89**, 4058–4068 (1967). (h) K.-S. Focsaneanu and J. C. Scaiano, *Helv. Chim. Acta*, **89**, 2473–2482. (2006).
13. (a) G. N. Lewis, *J. Am. Chem. Soc*., **38**, 762–785 (1916). (b) G. N. Lewis, *Proc. Natl. Acad. Sci. USA*, **2**, 586–592 (1916). (c) E. Hückel, *Trans. Faraday Soc*., **30**, 40–52 (1934). (d) A. H. Maki, R. D. Allendoerfer, J. C. Danner, and R. T. Keys, *J. Am. Chem. Soc*., **90**, 4225–4231 (1968). (e) T.-Y. D. Lin, C. P. Damo, J. R. Dunlop, and T. A. Miller, *Chem. Phys. Lett*., **168**, 349–354 (1990).
14. (a) T. H. Colle and E. S. Lewis, *J. Am. Chem. Soc*. **101**, 1810–1814 (1979). (b) S. Bank, C. L. Ehrlich, and J. A. Zubieta, *J. Org. Chem*. **44**, 1454–1458 (1979). (c) M. R. Feldman and W. C. Flythe, *J. Org. Chem*., **43**, 2596–2600 (1978).
15. A. H. Holm, T. Brinck, and K. Daasbjerg, *J. Am. Chem. Soc*., **127**, 2677–2685 (2005).
16. (a) J. B. Conant, and N. M. Bigelow, *J. Am. Chem. Soc*., **50**, 2041–2049 (1928). (b) W. P. Neumann and R. Stapel, *Chem. Ber*., **119**, 2006–2012 (1986). (c) K. Schreiner and A. Berndt, *Angew. Chem. Int. Ed*., **13**, 144–145 (1974). (d) K. Schreiner and A. Berndt, *Angew. Chem. Int. Ed Engl*., **14**, 366–367 (1975).
17. (a) G. D. Mendenhall, D. Griller, D. Lindsay, *et al*., *J. Am. Chem. Soc*., **96**, 2441–2447 (1974). (b) D. Griller, S. Icli, C. Thankachan, and T. T. Tidwell, *J. Chem. Soc. Chem. Commun*., 913–914 (1974). (c) D. Griller and K. U. Ingold, *Acc. Chem. Res*., **9**, 13–19 (1976). (d) L. Pauling and G. W. Wheland, *J. Chem. Phys*., **1**, 362–374 (1933). (e) H. E. Munro and C. S. Marvel, *J. Am. Chem. Soc*., **54**, 4445–4450 (1932).
18. (a) W. P. Neumann, A. Penenory, U. Stewen, and M. Lehnig, *J. Am. Chem. Soc*., **111**, 5489–5851 (1989). (b) U. Stewen and M. Lehnig, *Chem. Ber*., **122**, 2319–2322 (1989).
19. (a) M. J. Sabacky, C. S. Johnson, Jr., R. G. Smith, *et al*., *J. Am. Chem. Soc*., **89**, 2054–2058 (1967). (b) O. Neunhoffer and H. Haase, *Chem. Ber*., **91**, 1801–1805 (1958). (c) N. J. Peters, Ph. D. Dissertation, University of Illinois, Champaign-Urbana, 1980, quoted in ref. 19d. (d) K. M. Baldridge, T. R. Battersby, *et al*., *J. Am. Chem. Soc*., **119**, 7048–7054 (1997). (e) B. Kahr, D. Van Engen, and K. Mislow, *J. Am. Chem. Soc*., **108**, 8305–8307 (1986). (f) G. Wittig and H. Petri, *Justus Liebig's Ann. Chem*., **505**, 17–41 (1933). (g) H. Wang and F. P. Gabbaï, *Angew Chem. Int. Ed*., **43**, 184–187 (2004). (h) T. Suzuki, T. Takeda, H. Kawai, and K. Fujiwara, *Pure App. Chem*., **80**, 547–553 (2008). (i) H. Kawai, T. Takeda, K. Fujiwara, *et al*., *Chem. Eur. J*., **14**, 5780–5793 (2008).
20. (a) T. Erabi, T. Ohtsuki, E. Osaki, *et al*., *Bull. Chem. Soc. Jpn*., **73**, 2237–2242 (2000). (b) R. Teuber, R. Köppe, G. Linti, and M. Tacke, *J. Organomet. Chem*., **545–546**, 105–110 (1997). (c) S. Fox, J. P. Dunne, M. Tacke, and J. F. Gallagher, *Inorg. Chim. Acta*, **357**, 225–234 (2004). (d) S. Fox, J. P. Dunne, M. Tacke, and D. Schmitz, *Eur. J. Inorg. Chem*., 3039–3046 (2002).
21. (a) A. Mangini, G. F. Padulli, and M. Tiecco, *Tetrahedron Lett*., **9**, 4941–4944 (1968). (b) J. Nakayama, A. Ishii, M. Yamada, *et al*., *Tetrahedron Lett*., **31**, 2627–2630 (1990).
22. N. I. Tzerpos, A. K. Zarkadis, R. P. Kreher, *et al*., *J. Chem. Soc. Perkin Trans. 2*, 755–761 (1995).

23. A. R. Katritzky, B. Yang, and D. P. M. Pleynet, *J. Org. Chem.*, **63**, 9992–9994 (1998).
24. A. R. Katritzky, B. Yang, and N. S. Dalal, *J. Org. Chem.*, **63**, 1467–1472 (1998).
25. (a) W. Schlenk and H. Mark, *Ber. Dtsch. Chem. Ges.*, **55**, 2285–2299 (1922). (b) K. Ziegler, G. Bremer, F. Thiel, and F. Thielmann, *Justus Liebigs Ann. Chem.*, **434**, 34–78 (1923). (c) K. Ziegler and B. Schnell, *Justus Liebigs Ann. Chem.*, **445**, 266–282 (1925). (d) A. Löwenbein and W. Folberth, *Ber. Dtsch. Chem. Ges.*, **58**, 601–609 (1925). (e) H. H. Wasserman, T.-C. Liu, and E. R. Wasserman, *J. Am. Chem. Soc.*, **75**, 2056–2058 (1953). (f) M. Frenette, P. D. MacLean, L. R. C. Barclay, and J. C. Scaiano, *J. Am. Chem. Soc.*, **128**, 16432–16433 (2006). (g) C. Aliaga, D. R. Stuart, A. Aspee, and J. C. Scaiano, *Org. Lett.*, **7**, 3665–3668 (2005). (h) B. Kubiak, M. Lehnig, W. P. Neumann, *et al.*, *J. Chem. Soc., Perkin Trans. 2*, 1443–1447 (1992). (i) W. P. Neumann and R. Stapel, *Chem. Ber.*, **119**, 3422–3431 (1986). (j) B. Witten and F. Y. Wiselogle, *J. Org. Chem.*, **6**, 584–595 (1941). (k) W. Schlenk, H. Hillemann, and I. Rodloff, *Justus Liebigs Ann. Chem.*, **487**, 135–154 (1931). (l) W. Huang, H. Henry-Riyad, and T. T. Tidwell, *J. Am. Chem. Soc.*, **121**, 3939–3943 (1999).
26. (a) C. F. Koelsch, *J. Am. Chem. Soc.*, **79**, 4439–4441 (1957). (b) M. J. Plater, S. Kemp, and E. Lattmann, *J. Chem. Soc., Perkin Trans. 1*, 971–979 (2000). (c) H. Tukada, *J. Am. Chem. Soc.*, **113**, 8991–8992 (1991). (d) H. Tukada, and K. Mutai, *Tetrahedron Lett.*, **33**, 6665–6668 (1992).
27. (a) J. H. Ardenkjær-Larsen, I. Laursen, I. Leunbach, *et al.*, *J. Magn. Reson.*, **133**, 1–12 (1998). (b) M. C. Krishna, S. English, K. Yamada, *et al.*, *Proc. Natl. Acad. Sci. USA*, **99**, 2216–2221 (2002). (c) S. Andersson, F. Radner, A. Rydbeek, *et al.*, US Patent 5530140 (1996). (d) J. H. Ardenkjær-Larsen and I. Leunbach, PCT Int. Appl. Wo/9709633, (1997). (e) S. Andersson, F. Radner, A. Rydbeek, *et al.*, US Patent 5728370, (1998). (f) M. Thaning, PCT Int. Appl. Wo/9839277 (1998).
28. I. Dhimitruka, M. Velayutham, A. A. Bobko, *et al.*, *Bioorg. Med. Chem. Lett.*, **17**, 6801–6805 (2007).
29. (a) A. A. Bobko, I. Dhimitruka, J. L. Zweier, and V. V. Khramtsov, *J. Am. Chem. Soc.*, **129**, 7240–7241 (2007). (b) J. L. Dhimitruka, A. A. Bobko, C. M. Hadad, *et al.*, *J. Am. Chem. Soc.*, **130**, 10780–10787 (2008). (c) T. J. Reddy, T. Iwama, H. J. Halpern, and V. H. Rawal, *J. Org Chem.*, **67**, 4635–4639 (2002). (d) R. A. Wind and J.-H. Ardenkjaer-Larsen, *J. Magn. Reson.*, **141**, 347–354 (1999). (e) M. Joergensen, F. Rise, S. Andersson, *et al.*, PCT Int. Appl. (1991). (f) Liu, Y; Villamena, F. A., and Zweier, J. L., *Chem. Comm.*, **36**, 4336–4338 (2008).
30. (a) R. Owenius, G. R. Eaton, and S. S. Eaton, *J. Magn. Reson.*, **172**, 168–175 (2005). (b) M. K. Bowman, C. Mailer, and H. J Halpern, *J. Magn. Reson.*, **172**, 254–267 (2005).
31. (a) J. P. Joshi, G. R. Eaton, and S. S. Eaton, *Appl. Magn. Reson.*, **28**, 239–249 (2005). (b) A. J. Fielding, P. J. Carl, G. R. Eaton, and S. S. Eaton, *Appl. Magn. Reson.*, **28**, 231–238 (2005).
32. (a) S. Subramanian and M. C. Krishna, *Biol. Magn. Reson.*, **23** (Part A), 321–382 (2005). (b) S. Subramanian, N. Devasahayam, and M. C. Krishna, Proceedings of SPIE-Int. Soc. Opt. Eng., 644106/1-644106/17 (2007). (c) H. Sato-Akiba, H. Abe, H. Fujii, and H. Hirata, *Magn. Reson. Med.*, **59**, 885–890 (2008).
33. (a) V. K. Kutala, N. L. Parinandi, J. L. Zweier, and P. Kuppusamy, *Arch. Biochem. Biophys.*, **424**, 81–88 (2004). (b) C. Rizzi, A. Samouilov, V. K. Kutala, *et al.*, *Free Radicals Biol. Med.*, **35**, 1608–1618 (2003). (c) S. Bowen and C. Hilty, *Angew. Chem. Int. Ed.*, **47**, 5235–5237 (2008). (d) P. Höfer, G. Parigi, C. Luchinat, *et al.*, *J. Am. Chem. Soc.*, **130**, 3254–3255 (2008). (e) S. Xia, F. A. Villamena, C. M. Hadad, *et al.*, *J. Org Chem.*, **71**, 7268–7279 (2006).
34. (a) M. Ballester, C. Molinet, and J. Castañer, *J. Am. Chem. Soc.*, **82**, 4254–4258 (1960). (b) M. Ballester, J. Riera, J. Castañer, *et al.*, *J. Am. Chem. Soc.*, **93**, 2215–2225 (1971). (c) M. Ballester, *Acc. Chem. Res.*, **18**, 380–387 (1985). (d) M. Ballester, I. Pascual, C. Carreras, and J. Vidal-Gancedo, *J. Am. Chem. Soc.*, **116**, 4205–4210 (1994). (e) M. Ballester, I. Pascual, J. Riera, and J. Castañer, *J. Org. Chem.*, **56**, 217–225 (1991). (f) M. Ballester, J. Castañer, J. Riera, and I. Pascual, *J. Am. Chem. Soc.*, **106**, 3365–3566 (1984).
35. (a) J. Veciana, C. Carilla, C. Miravitlles, and E. Molins, *J. Chem. Soc. Chem. Commun.*, 812–814 (1987). (b) L. Teruel, Ll. Viadel, J. Carilla, *et al.*, *J. Org. Chem.* **61**, 6063–6066 (1996).
36. (a) H. R. Falle, G. R. Luckhurst, A. Horsfeld, and M. Ballester, *J. Chem. Phys.*, **50**, 258–264 (1969). (b) O. Armet, J. Veciana, J. Rovira, *et al.*, *J. Phys. Chem.*, **91**, 5608–5616 (1987). (c) M. Ballester, J. Castañer, A. Riera, *et al.*, *J. Org Chem.*, **47**, 259–264 (1982). (c) J. Bonvoisin, J.-P. Launay, C. Rovira, and J. Veciana, *Angew. Chem., Int. Ed. Engl.*, **33**, 2106–2109 (1994).

37. (a) A. Heckmann, C. Lambert, M. Goebel, and R. Wortmann, *Angew. Chem. Int. Ed.*, **43**, 5851–5856 (2004). (b) A. Heckmann and C. Lambert, *J. Am. Chem. Soc.*, **129**, 5515–5527 (2007). (c) I. Ratera, C. Sporer, D. Ruiz-Molina, *et al.*, *J. Am. Chem. Soc.*, **129**, 6117–6129 (2007).
38. (a) C. Trapp, C.-S. Wang, and R. Filler, *J. Chem. Phys.*, **45**, 3472–3474 (1966). (b) Filler, R., Fiebig, A. E., and Mandal, B. K., *J. Fluorine Chem.*, **102**, 185–188 (2000). (c) S. V. Kulkarni and C. Trapp, *J. Am. Chem. Soc.*, **92**, 4801–4808 (1970).
39. (a) K. I. Beynon and S. T. Bowden, *J. Chem. Soc.*, 4257–4262 (1957). (b) S. T. Bowden, and Watkins, T. F. *J. Chem. Soc.*, 1249–1257 (1940). (c) M. Gomberg, *J. Am. Chem. Soc.*, **36**, 1144–1170 (1914).
40. (a) G. M. Coppinger, *J. Am. Chem. Soc.*, **79**, 501–502 (1957). (b) P. D. Bartlett and T. Funahashi, *J. Am. Chem. Soc.*, **84**, 2596–2601 (1962). (c) B. Kirste, W. Harrer, and H. Kurreck, *J. Am. Chem. Soc.*, **107**, 20–28 (1985).
41. H. A. Staab, C. Kuo-chen, and A. Ruland, *Chem. Ber.* **115**, 1765–1774 (1982).
42. (a) J. Thiele and H. Balhorn, *Ber. Dtsch. Chem, Ges.*, **37**, 1463–1470 (1904). (b) L. K. Montgomery, J. C. Huffman, E. A. Jurczak, and M. P. Grendze, *J. Am. Chem. Soc.*, **108**, 6004–6011 (1986).
43. (a) W. Schlenk and M. Brauns, *Ber. Dtsch. Chem, Ges.*, **48**, 661–669 (1915). (b) M. Leo, *Ber. Dtsch. Chem, Ges.*, **70**, 1691–1695 (1937). (c) N. M. Shishlov, *Russ. Chem. Revs.*, **75**, 863–884 (2006).
44. (a) W. T. Borden (Ed) *Diradicals*, John Wiley & Sons, Inc., NY (1982). (b) H. Iwamura, *Adv. Phys. Org. Chem.*, **26**, 179–253 (1990). (c) A. Racja, *Adv. Phys. Org. Chem.*, **40**, 153–199 (2005).
45. (a) A. Rajca, S. Utamapanya, and J. Xu, *J. Am. Chem. Soc.*, **113**, 9235–9241 (1991). (b) A. Rajca and S. Utamapanya, *J. Org. Chem.*, **57**, 1760–1767 (1992). (c) A. Rajca and S. Utamapanya, *J. Am. Chem. Soc.*, **115**, 2396–2401 (1993). (d) A. Rajca, *J. Am. Chem. Soc.*, **112**, 5890–5892 (1990). (e) A. Rajca, S. Utamapanya, and S. Thayumanavan, *J. Am. Chem. Soc.*, **114**, 1884–1885 (1992).
46. A. Rajca, *Chem. Rev.* **94**, 871–893 (1994).
47. (a) A. E. Tschitschibabin, *Ber. Dtsch. Chem, Ges.*, **40**, 1810–1819 (1907). (b) J. Casado, S. Patchkovskii, M. Z. Zgierski, *et al.*, *Angew. Chem. Int. Ed.*, **47**, 1443–1446 (2008). (c) W. W. Porter, III, T. P. Vaid, and A. L. Rheingold, *J. Am. Chem. Soc.*, **127**, 16559–16566 (2005). (d) W. Schlenk, and M. Brauns, *Ber. Dtsch. Chem, Ges.*, **48**, 716–728 (1915). (e) A. Rajca, S. Rajca, and J. Wongsriratanakul, *Chem. Commun.*, 1021–1022 (2000). (f) R. Schmidt and H.-D. Brauer, *Angew. Chem. Int. Ed.*, **10**, 506–507 (1971).
48. (a) N. M. Shishlov, V. N. Khrustaleva, Sh. S. Akhmetzyanov, *et al.*, *Russ. Chem. Bull., Int. Ed.*, **52**, 385–390 (2003). (b) N. M. Shishlov, S. I. Maslennikov, Sh. S. Akhmetzyanov, *et al.*, *Dokl. Phys. Chem.*, **394**, 27–30 (2004). (c) N. M. Shishlov, K. Yu. Murinov, Sh. S. Akhmetzyanov, and V. N. Khrustaleva, *Russ. Chem. Bull.*, **48**, 1992–1994 (1999).
49. (a) D. Braun and R. J. Faust, *Angew. Chem. Int. Ed.*, **5**, 838 (1966). (b) D. Braun, *Pure Appl. Chem.*, **30**, 41–55 (1972). (c) D. Braun, *J. Polym. Sci., Polym. Symp.*, **24**, 7–13 (1968). (d) D. Braun and W. Euler, *Makromol. Chem.*, **177**, 1373–1386 (1976). (e) D. Braun, and P. Lehmann, *Makromol. Chem.*, **177**, 1387–1400 (1976). (f) D. Braun and P. Lehmann, *Makromol. Chem.*, **177**, 1673–1686 (1976). (g) D. Braun, U. Platzek, and H. J. Hefter, *Chem. Ber.*, **104**, 2581–2587 (1971). (h) G. J. Sloan and W. R. Vaughan, *J. Org. Chem.*, **22**, 750–761 (1957).

2

Polychlorotriphenylmethyl Radicals: Towards Multifunctional Molecular Materials

Jaume Veciana and Imma Ratera

Institut de Ciència de Materials de Barcelona (ICMAB-CSIC), Barcelona, Spain

2.1 Introduction

Polychlorotriphenylmethyl (PTM) radicals are open shell organic molecules with a net magnetic moment which are characterized by extremely high persistence and stability. They are composed of three totally or partially chlorinated phenyl rings connected to a central carbon atom with a sp^2 hybridization (Figure 2.1).[1] The steric strain that exist in a hypothetical planar conformation of this molecule, mainly originated by the bulky chlorine atoms, is released by the rotation of the aromatic rings around the bonds linking the central to the *ipso*-carbons, giving the molecule a propeller-like conformation. As a consequence, from the stereochemical point of view, such radicals exist in two different enantiomeric forms that can interconvert at high temperatures with an energy barrier of about 23 kcal/mol.[2] Following the Cahn–Ingold–Prelog nomenclature,[3] the two atropoisomeric forms are called *Plus* (*P*), when the rings present a clockwise torsion, and *Minus* (*M*) when the torsion is the opposite one.

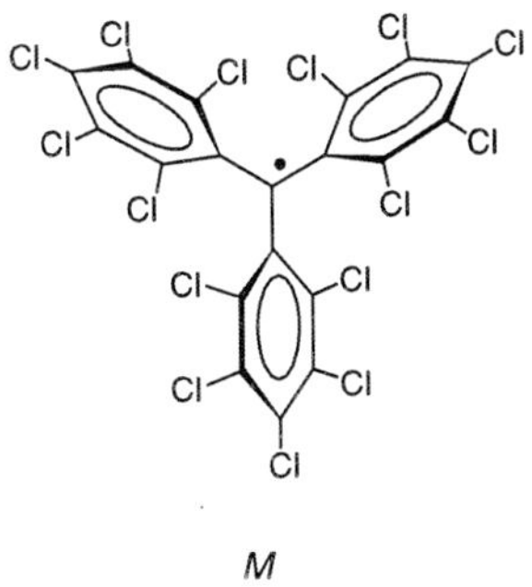

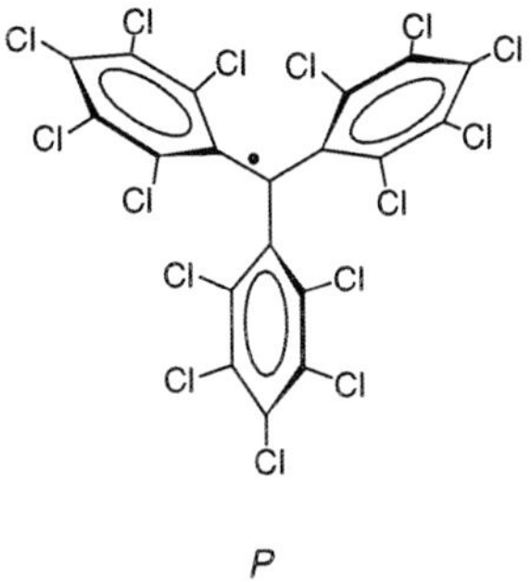

1

M *P*

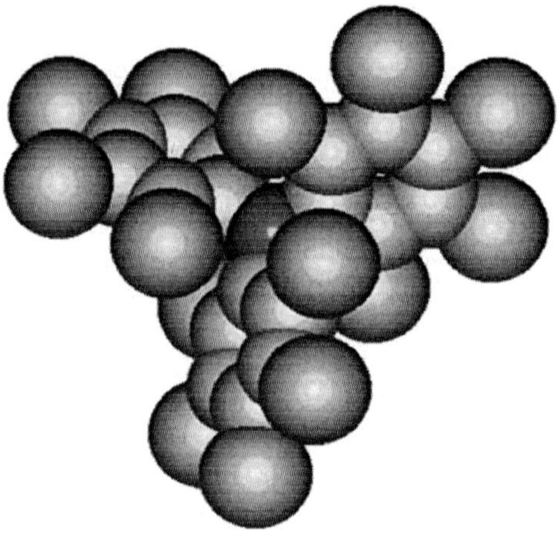

Figure 2.1 CPK representation of perchlorotriphenylmethyl radical (**1**•) showing the high steric shielding of the central carbon atom (in red) surrounded by the six bulky chlorine atoms at the ortho positions. A full-colour version of this figure appears in the Colour Plate section of this book.

Besides the generation of a molecular propeller-like conformation, the existence of the bulky chlorine atoms at the *ortho* positions also provides the main source of steric protection for the methyl carbon atom, which is the atom with the major spin density. A molecular model of the perchlorotriphenylmethyl radical (**1**•) is illustrated in Figure 2.1, showing that the central carbon atom is protected by three *ortho*-chlorine atoms at each side of the molecule. Indeed, the protection of the methyl carbon is translated into a large persistence and a very high chemical and thermal stability for this family of radicals.[4–8] Thus, unlike in the case of the non-chlorinated analogue, which is very reactive to oxygen and cannot be isolated as radical, in solution PTM radicals remain perfectly inert to oxygen and to many aggressive reagents, decomposing only in the presence of light. Consequently, such radicals behave as classical organic compounds and can be used as intermediates in many reactions and treated with usual separation or purification techniques, such as chromatography, crystallization, sublimation, and so on. These radicals, in solid state, decompose with high melting points,[9] usually near 300 °C, and are stable to light and air practically indefinitely.[10] Such astonishing thermal and chemical stabilities are the reason why the term of "inert free radicals" was coined for this family of radicals.

The synthesis of PTM radicals is usually achieved from the corresponding polychlorotriphenylmethanes through the formation of their carbanions with a base, like OH^-, followed by the oxidation of the latter with a weak oxidant, such as *p*-chloranil, I_2 or Ag^+, as exemplified in Scheme 2.1 for radical **1**•. The synthesis of the radical precursors generally consists in a sequence of Friedel–Crafts condensations at high temperature with an excess of the chlorinated benzene derivatives; such as is depicted in Scheme 2.1 for the hydrocarbon ***1H***.[6]

The high persistence of PTM radicals has allowed them to be functionalized at the *ortho* and *para* positions of the aromatic rings. Thanks to the efforts of Ballester and coworkers during the 1970s and 1980s, these radicals were functionalized using a great variety of reactions.[11] Although sometimes there is the need for extreme conditions, PTM functionalization has allowed the use of such radicals as building blocks of molecular materials with specific properties. One of the first applications of these radicals was the spin labeling of amino acids and peptides.[12] Since then, PTM radicals have been used as constitutive units of mixed valence systems that present intramolecular electron transfer phenomena,[13] bistable molecular switches,[14,15] and new valence tautomeric systems,[16] as well as of push–pull compounds with nonlinear

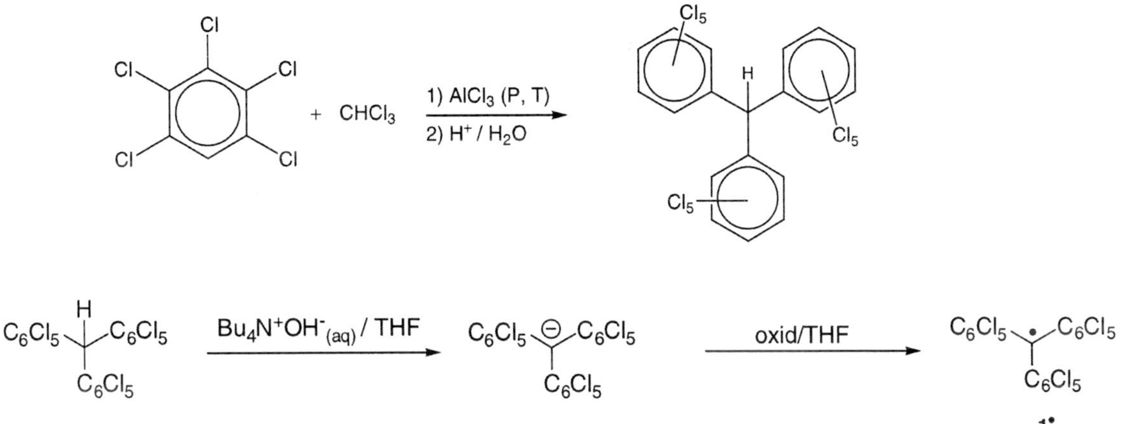

1*H* **1⁻** **1•**

Scheme 2.1 Synthesis of radical **1•** from its hydrocarbon precursor **1*H***, through the carbanionic derivative **1⁻**

optical (NLO) properties.[17] The particular structural and electronic characteristics of PTM radicals have permitted, not only organic open shell dendrimers with high-spin ground states and low-spin excited states, inaccessible even at room temperature, to be obtained[18,19] but also the ability to functionalize them with one or more carboxylic groups. These radicals have been used as molecular synthons to build up open shell supramolecular structures, either with a purely organic or a metal–organic nature, which show relevant porous and magnetic properties.[20] More recently, in the search for new molecular spintronic systems, PTM radicals have also been grafted onto surfaces, exhibiting interesting electronic transport properties.[21]

PTM radicals are also interesting because they are electroactive species from which it is possible to generate the corresponding carbanionic and carbocationic species. Indeed, as shown in Figure 2.2, the cyclic voltammetry of radical **1•** exhibits two reversible waves that correspond to the oxidation and reduction of this radical, giving rise to the corresponding carbanion $\mathbf{1}^-$ and carbocation $\mathbf{1}^+$, which are also quite stable species both in solution and in the solid state. The reduction of PTM radicals turns out to be more feasible than the corresponding oxidation, a fact that makes this type of radical an excellent electron acceptor units in push–pull systems with NLO properties and/or to achieve intramolecular electron transfer in molecular nanowires. Electroactivity of PTM radicals can be used for making switchable multifunctional molecular materials, since the magnetic, optic and electronic properties of these open shell molecules can be switched on and off with an external electrochemical stimuli. This chapter describes the chemical and physical properties of PTM radicals, mainly from the molecular material point of view. The main focus is on multifunctionality, giving different insights to the magnetic, electronic and optical properties of PTM based materials.

2.2 Functional molecular materials based on PTM radicals

In the last few years there has been much interest in the field of functional molecular materials.[22,23] This interest arises from their expected use in future emerging technologies as well as their use to improve already existing technologies. Functional molecular materials comprise purely organic or metal–organic building blocks that show interesting physical properties, for example conducting,[24] optical[25] or magnetic ones (Figure 2.3).[26]

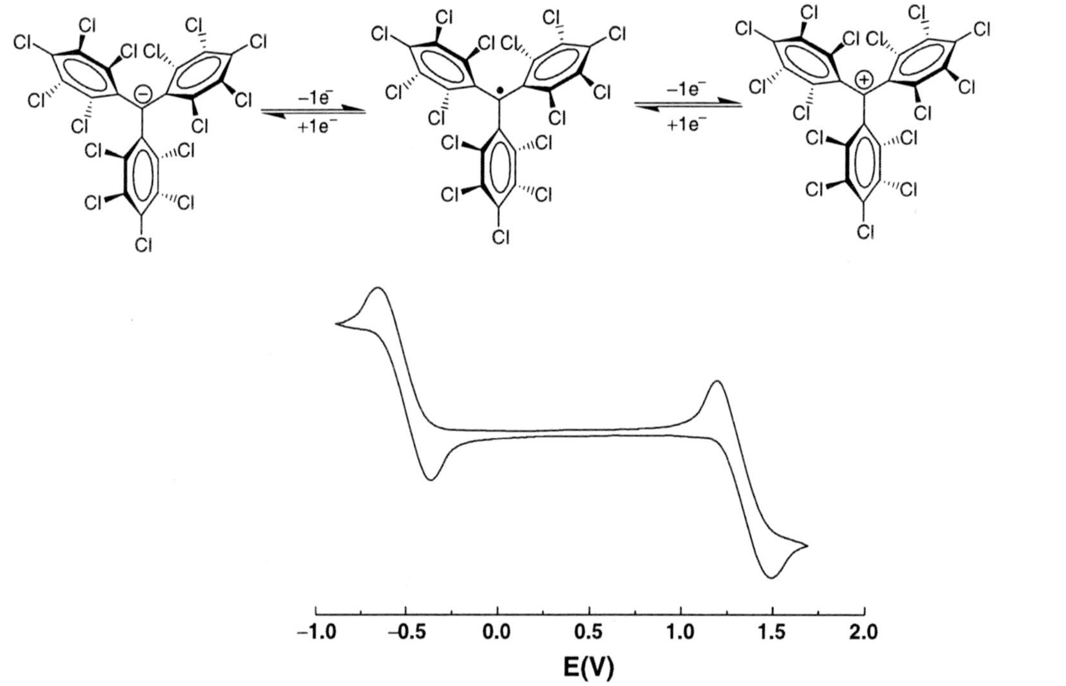

Figure 2.2 *Cyclic volammogram of radical* $\mathbf{1}^{\bullet}$ *in* CH_2Cl_2*, with 0.1 M n-*Bu_4NPF_6 *(vs Ag/AgCl) showing the reversible formation of two different ionic species that correspond to the oxidation and reduction of this radical.*

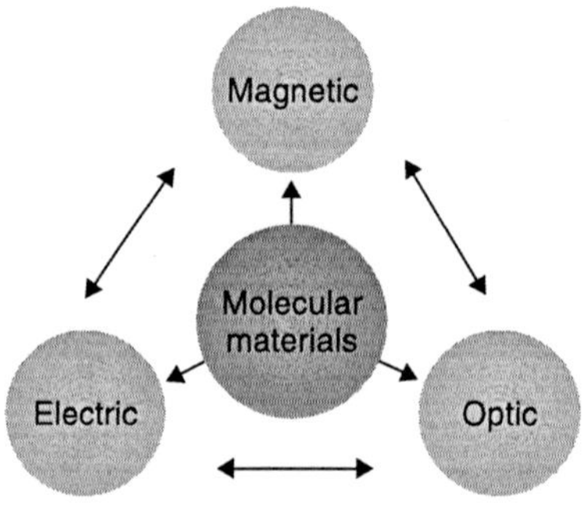

Figure 2.3 *Interplay between the different properties shown by functional molecular materials.*

Tunability and versatility of molecular chemistry offers a unique opportunity to design multifunctional molecular materials or devices – that is, molecule-based systems showing two or more functions simultaneously.[27] These new type of materials or devices may exhibit different combinations of properties, such as magnetism/conductivity, magnetism/optical or conductivity/NLO properties.[28] The presence of two properties in the same crystal lattice, or even better in the same molecule, might result, if there is a synergism between them, in new physical phenomena and novel applications that are difficult to achieve in conventional inorganic solids. The design of such multifunctional molecular systems that are able to change their functions by an external stimuli – like temperature, light and electric field – results in more appealing systems because of their potential applications as molecular-scale switches in nanotechnology.[29] A molecular-scale switch is a fundamental component of any true molecular magnetic/electronic/photonic device. To achieve a real switching effect, the molecule must be stable in two (or more) states exhibiting very different properties which may be interconverted reversibly in response to an external stimuli, such as redox, temperature, pressure or irradiation. To date, several molecular switches exhibiting changes in one property, such as color,[30] luminescence,[31] optical nonlinearity,[32] or magnetic properties,[33] have been reported. However, only very recently the number of useful properties being simultaneously modulated on a bistable molecule-based system has been extended to three properties (electrical, optical and magnetic), although in this case the changes in the properties have an intermolecular origin.[34]

PTM radicals are nice examples of molecular building blocks for obtaining such multifunctional molecular materials, since they permit to combine the intrinsic magnetic characteristics of these molecules with optical, conducting and electrochemical properties.

2.2.1 Materials with magnetic properties

The design of molecular materials, especially of those of a purely organic nature, with relevant magnetic properties has been one of the major challenges of the last decades. Due to the light nature of elements composing organic compounds, only the magnetic exchange interactions determine the magnetic behavior at temperatures well above 0.1 K. Other kinds of magnetic interactions (such as hyperfine or spin-orbit interactions) as well as sources of magnetic anisotropies can be considered as negligible above this temperature in most open shell organic materials. Consequently, magnetic organic systems may be described at zero applied magnetic fields by the effective spin Hamiltonian approach, which takes the form of Equation 2.1:

$$H = -2\Sigma J_{ij}\mathbf{S}_i \cdot \mathbf{S}_j \qquad (2.1)$$

where J_{ij} represents the effective exchange interaction parameter for the magnetic centres i and j – the spin-containing building blocks – which have total quantum spin numbers $\mathbf{S}_i$ and $\mathbf{S}_j$, respectively, and the summation runs over all the adjacent pairs of centers. According to the formalism of Equation 2.1, when J_{ij} is positive the two spins tend to be parallel to each other in the ground state and the magnetic interaction (or coupling) is ferromagnetic (FM). On the contrary, if J_{ij} is negative the two spins align in an antiparallel fashion in the ground state and the interaction is antiferromagnetic (AFM).

Most organic magnetic materials exhibit paramagnetic behaviors at high temperatures where the spins of the material behave independently of each other. However, as the temperature is decreased, the exchange interactions become comparable with the thermal energy of the system and the neighboring spins tend to align in accordance with the signs of their J_{ij} parameters, appearing either in a short or a long range ordering of the spins. Alignments of spins can be propagated along one, two or three dimensions of the solid, but only if the alignment occurs in three dimensions can a real long range magnetic ordering appear. In this case the solid exhibits a cooperative magnetic property below its critical temperature (T_C), behaving as a bulk magnet. Keeping in mind the characteristics previously described, the design of organic molecular materials showing bulk magnetic properties involves three main aspects or steps that must be taken in consideration carefully: (i) the stability of the spin-containing building blocks; (ii) the coupling routes or magnetic interaction mechanisms between the neighboring spin-containing building blocks; and (iii) the propagation of the magnetic interactions along the three dimensions of the material.

Substituted PTM radicals are stable species and, accordingly, they can be used as the constituent building blocks of molecular magnetic materials. Regarding the coupling routes, a ferromagnetic interaction may result only if an orthogonal arrangement (zero overall overlap integral) between the two singly-occupied molecular orbitals (SOMOs) of the two interacting subunits is produced. On the other hand, if such an integral is non-zero, the resulting magnetic interaction would be antiferromagnetic for non-degenerate orbitals. Magnetic interactions can take place either between neighboring molecules or, in the case that there are two (or more) spin-containing units in the same molecule, intramolecularly. Intermolecular interactions among organic radicals are mainly determined by the isotropic exchange interactions between the unpaired electrons located on the SOMOs of the nearest neighbouring molecules. The construction of magnetic materials requires that the structural subunits exhibit non-covalent interactions suitable to be controlled in a predictable manner which are able to promote right isotropic exchange interactions between the nearest neighboring open shell molecules. Up to now, the different types of non-covalent intermolecular interactions that have been used for the assembly of such molecular subunits are hydrogen bonding, transition metal ligation, stack-type alignment and bridging of ion radicals by their counterions. In the case that the assembly of spin-containing molecules produces the proper magnetic interactions and they are propagated along the three dimensions, the material will exhibit a bulk cooperative magnetic property.

Intramolecular magnetic interactions play a major role in high-spin organic molecules.[35] Thus, when a molecule has two high-lying orthogonal orbitals, close or equal in energy, with two less electrons than necessary for a closed shell structure, its ground state becomes a triplet, as dictated by Hund's rule. However, the actual nature of the interaction (ferromagnetic or antiferromagnetic) depends upon the symmetry and topology of the degenerate or nearly degenerate SOMOs' orbitals. One of the general approaches to obtain high-spin organic compounds is based on conjugated polyradicals with topologically polarized α spins. These compounds are designed using appropriate ferromagnetic coupling units able to align in parallel the spins of a pair (or more) of radical centers connected through such a unit. *m*-Phenylene has become the most widely used ferromagnetic organic coupler because it is highly dependable.

In this section attention is focused on the developments and studies of magnetic materials derived from PTM radicals. The magnetic properties of high-spin PTM radicals is described in detail as well as those of extended systems based on PTM radicals. The possibility to functionalize surfaces with PTM molecules as spin-containing units towards their use in molecular spintronics is also described.

2.2.1.1 High-spin PTM radicals

***m*-Phenylene Units as Intramolecular Ferromagnetic Couplers** With this aim, *m*-phenylene was used as a ferromagnetic organic coupler of PTM radical centers in line with the efforts to increase the number of persistent and stable high-spin radicals. Indeed, since the Schlenk hydrocarbon, first prepared in 1915, high-spin alignments in ground states have been successfully demonstrated for several 1,3-phenylene connected carbenes[36] and radicals based on triarylmethyl,[37] nitrogen-centred[38] and aminoxyl units.[39] The work done in this direction included the synthesis, characterization and study of physicochemical properties of molecules with two, three, or more triarylmethyl subunits, that are called bis-, tris-, or poly-triarylic systems or, more generically, extended triarylmethyl systems. The first PTM polyradicals reported were the diradical **2** and triradical **3**; both have a high-spin ground states with low-spin excited states, which are inaccessible even at room temperature.[18b] The synthesis of larger polyradicals like **4** and **5** with hyperbranched dendritic nature was also attempted. However, the numerous experimental problems found during the synthesis and purification of pentadecaradical **5**, due to the presence of abundant magnetic defects and numerous atropisomeric forms, suggested that the tetraradical **4** is the limit generation for this dendritic radical series.

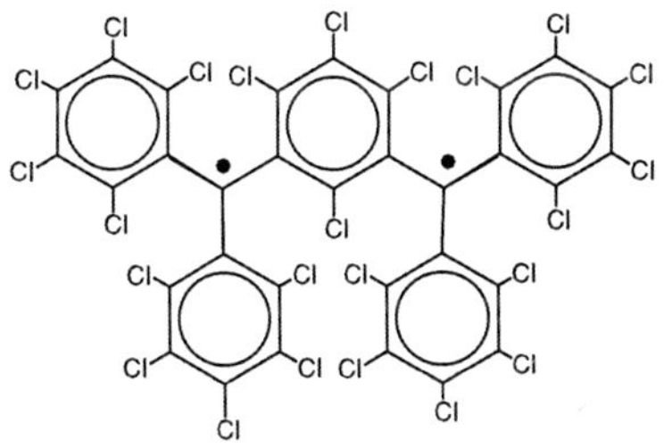

2

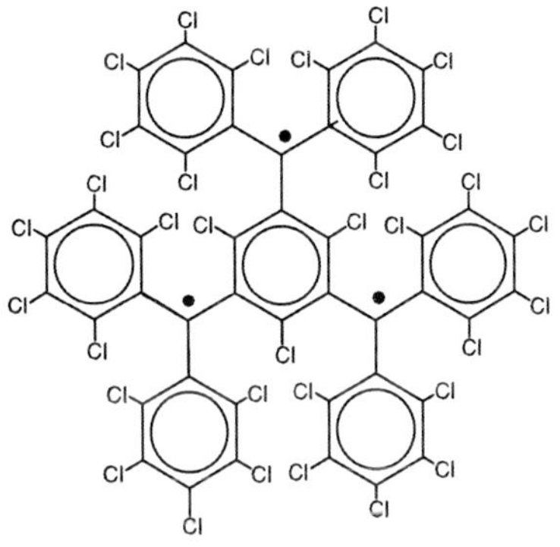

3

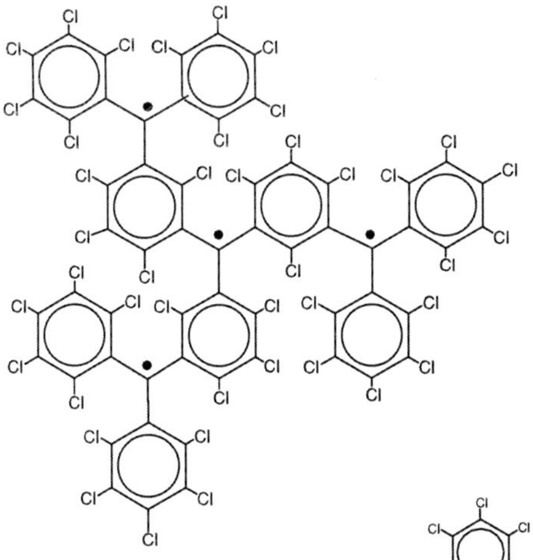

4

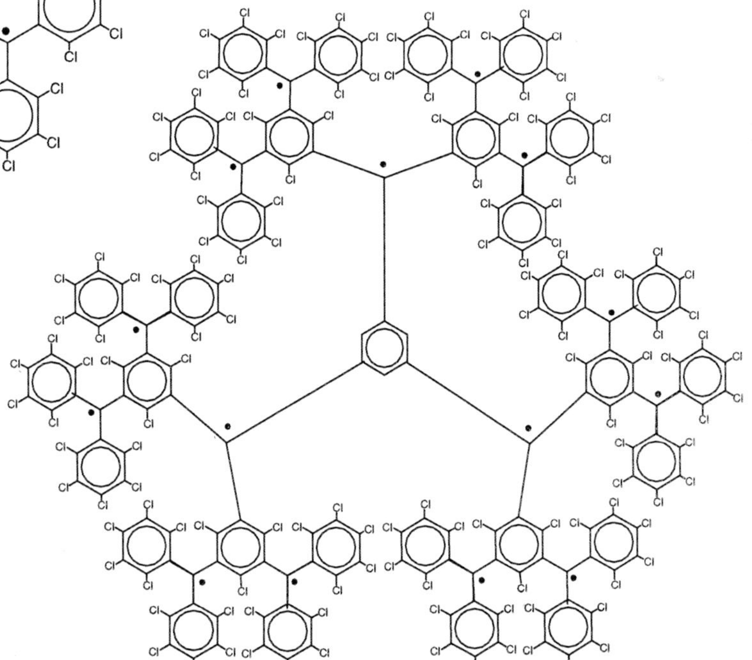

5

Biradical **2** and triradical **3** present the highest spin multiplicities – the triplet and quartet, respectively – in their ground states.[18a] Also worthy of remark is the fact that triradical **3** exists in two stereoisomeric forms with D_3 and C_2 symmetries. Such stereoisomeric forms exist in two enantiomeric isomers and are isolable as stable solids that show outstanding stabilities since they do not show any sign of decomposition up to 250 °C, even in air.[40]

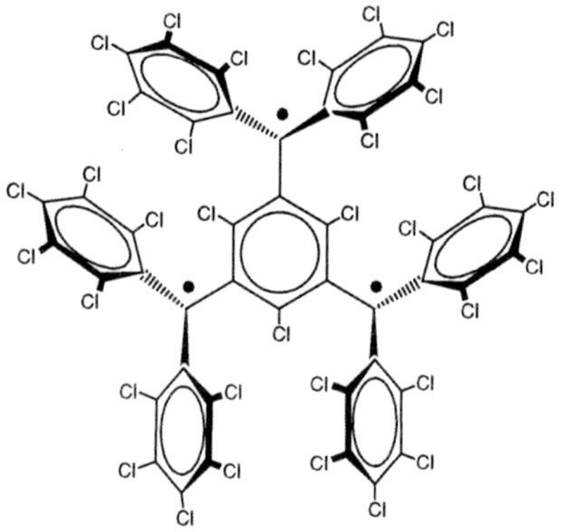

3a

***(M,M,P)*-1**

C_2 Symmetry

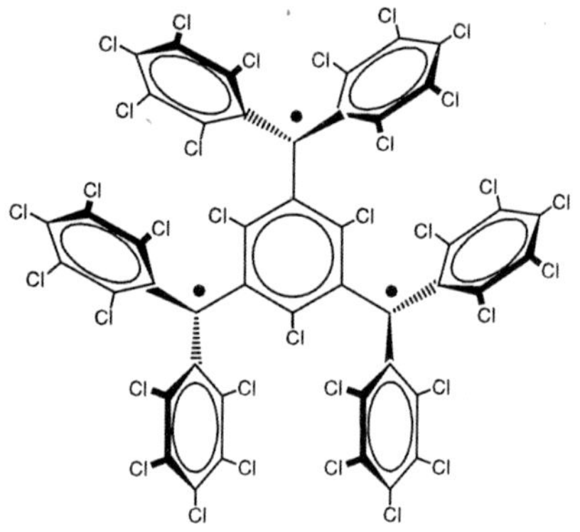

3b

***(M,M,M)*-1**

***D3* Symmetry**

A linear solvation free-energy study was performed with quartet **3** to study the influence of the molecular characteristics of the two diastereoisomers on their physicochemical properties. It was found that the different surface areas and fractal dimensionalities of the two diastereomeric forms of this rigid molecule are among the most important molecular parameters controlling some of their physicochemical properties. Their ability to interact with the surrounding media is a result mainly of cavitational effects.[37] Moreover, the two isolated solids have been characterized by single crystal X-ray diffraction. In this respect, the two solid state structures, each corresponding to the two different diastereomers, constituted the first reported examples of crystal structures of carbon-based high-spin polyradicals.[37]

Cyclic voltammetry of quartet **3** shows three reversible reduction waves at −0.19, −0.49 and −0.75 V versus SCE (Figure 2.4), corresponding to its reduction to the mono-, di- and trianionic species respectively, while diradical **2** revealed only two reversible waves at −0.22 and −0.54 V, corresponding to the reduction to the mono- and dianionic species respectively.

X-band Electron Paramagnetic Resonance (EPR) spectra of high-spin radicals **2** and **3** in dilute toluene solutions at room temperature show symmetrical signals centered at $g_{iso} = 2.0026$. In frozen media the EPR spectrum of triradical **3** showed a fine structure characteristic of two quartet species with distinct populations: a major quartet (**3a**, 83 %) with $|D|/hc = 0.0036\,\text{cm}^{-1}$, $|E|/hc = 0.0001\,\text{cm}^{-1}$, $g_{xx} = g_{yy} = 2.0029$ and $g_{zz} = 2.0012$, and a minor quartet (**3b**, 17 %) with $|D|/hc = 0.0050\,\text{cm}^{-1}$, $|E|/hc = 0\,\text{cm}^{-1}$, and $g_{xx} = g_{yy} = g_{zz} = 2.0024$. The observed rigid media EPR spectra of pure **3a** and **3b** quartets, as well as those spectra obtained by a simulation procedure, are shown in Figure 2.5. EPR data permit the assignment of the **3a** quartet to the enantiomeric pair with C_2 symmetry, that is to *(M,M,P)*-**3** + *(P,P,M)*-**3**, and the **3b** quartet to that with D_3 symmetry, that is to (M,M,M)-**3** + (P,P,P)-**3**. Rigid media EPR spectra of quartets ***C*₂-3** and ***D*₃-3** also showed weak signals corresponding to forbidden $\Delta m_s = \pm 2$ transitions (Figure 2.5, top). Variable temperature EPR experiments with pure $\boldsymbol{C_2}$**-3** and $\boldsymbol{D_3}$**-3** quartets were performed in the temperature range 130–200 K. Total intensities of their $\Delta m_s = \pm 1$ signals increase linearly with 1/T, indicating that quartets are the ground states of both stereoisomers and that the low-spin excited states of both isomers are thermally inaccessible in this temperature range.

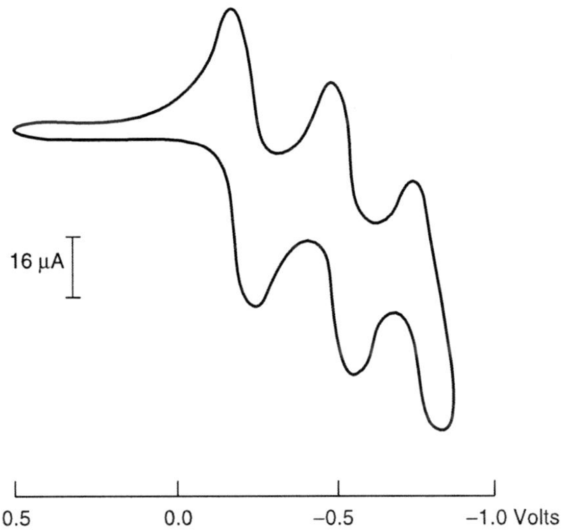

Figure 2.4 *Cyclic voltammogram of triradical **3** ($3 \cdot 10^{-4}$M) in CH_2Cl_2 using 0,1 M $Bu_4N^+PF_6^-$ as electrolyte (vs SCE). (Reprinted with permission from [18b]. Copyright 1993 American Chemical Society.)*

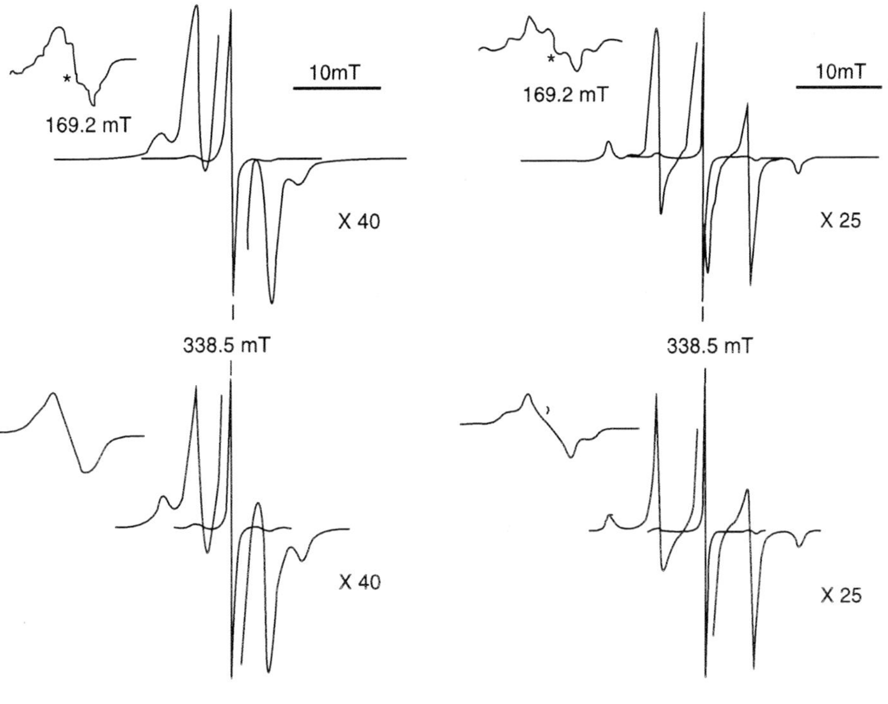

Figure 2.5 *Top: Experimental EPR first derivative spectra at 143 K in glassy toluene of pure samples of (left) **C_2-3** (**3a** quartet) and (right) **D_3-3** (**3b** quartet). Bottom: Simulated spectra for randomly oriented species. Insets show the observed and simulated signals corresponding to $\Delta M_s = \pm 2$ forbidden transitions of quartets. (Reprinted with permission from [18b]. Copyright 1993 American Chemical Society.)*

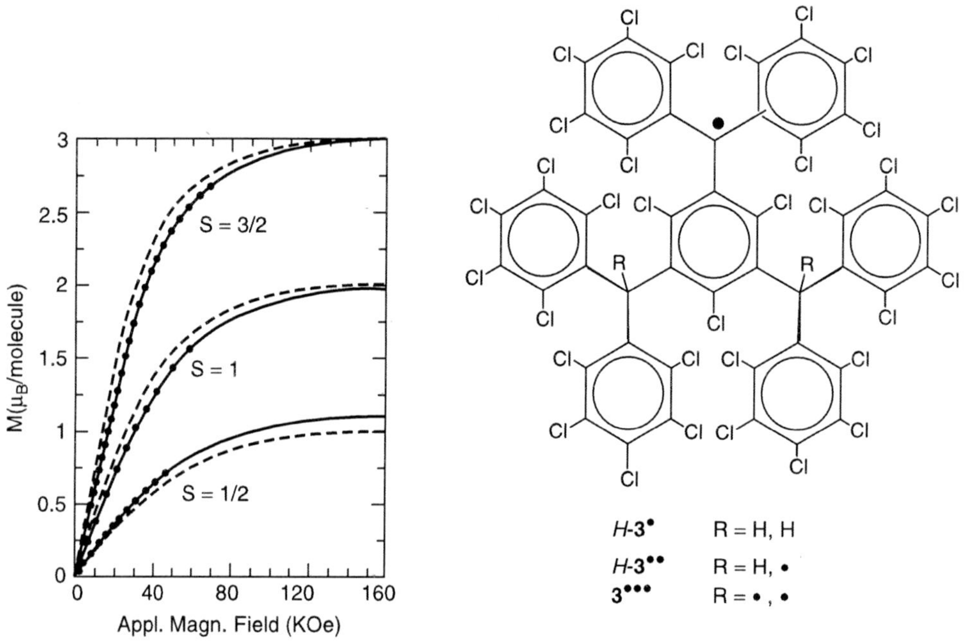

Figure 2.6 *Field strength dependencies of experimental molar magnetizations of radicals 3•••,* ***H-3••*** *and* ***2H-3•****. Solid lines are the calculated molar magnetization curves. Dashed lines are theoretical molar magnetization curves for $S = 3/2$, 1 and $1/2$ paramagnetic molecules. (Reprinted with permission from [18b]. Copyright 1993 American Chemical Society.)*

Magnetization measurements of radicals **3•••**, ***H*-3••** and ***H*-3•** at 4.2 K in the magnetic field range 0–5 T are depicted in Figure 2.6 as a function of the magnetic field H. The data can be compared with the theoretical magnetization curves of ideal paramagnets with $S = 3/2$, 1 and $1/2$, respectively. Calculated magnetization curves indicate that the ground states of radicals **3•••**, ***H*-3••** and ***H*-3•** are the quartet ($S = 3/2$), triplet ($S = 1$) and doublet ($S = 1/2$), respectively, in excellent agreement with EPR and susceptibility results.

Metallocene Units as Intramolecular Ferromagnetic Couplers One of the basic tools to realize high-spin molecules is the finding of bridges that may act as robust ferromagnetic couplers when they are connected to two (or more) open shell subunits. Metallocenes have been used very successfully as building blocks of molecular solids promoting intermolecular magnetic interactions.[41] Based on their electronic structures and rich chemistry, metallocenes appear to be promising candidates as ferromagnetic couplers. With this objective, diradical **6**, consisting of two PTM radical subunits connected by a 1,1′-metalloceneylene bridge, was constructed. The particular structure and topology of **6** were expected to lead to a non-negligible spin density on the metallocene moiety, making the magnetic coupling between the two organic radical subunits feasible. In addition, the location of both radical units far away from each other avoided any possibility of having intramolecular through-space contacts and, consequently, a significant direct through-space magnetic interaction.

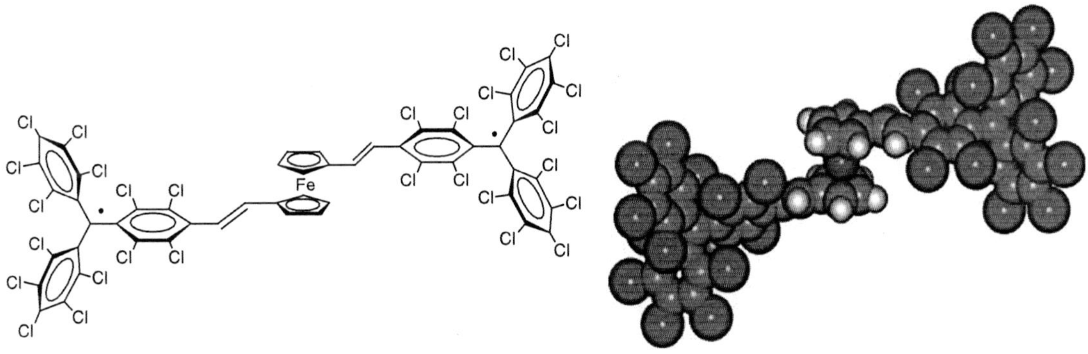

6

The EPR spectrum of diradical **6** in a frozen toluene/dichloromethane (1 : 1) mixture shows the characteristic fine structure of a triplet species in which the forbidden $\Delta m_s = \pm 2$ transition, characteristic of triplet species, is also observed at the half-field region of the spectrum. The intensity of such a signal (I_{pp}) was measured in the 4–100 K temperature region, and its temperature dependence indicated that the triplet is in the ground state with a thermally accessible excited singlet state. A singlet–triplet separation of +10 K (7 cm^{-1}) was obtained from the fitting of the $I_{pp}{\cdot}T$ versus T plot to the Bleaney–Bowers equation (Figure 2.7). In conclusion, the metallocene unit in **6** was shown to effectively transmit a ferromagnetic interaction between two PTM radicals.[42] This result was justified by ZINDO/1 semiempirical calculations[43] where it was observed that the SOMO orbitals are almost degenerate and coextensive, justifying the high-spin ground state experimentally observed.

Transmission of Magnetic Interactions Through Hydrogen Bonds Investigation of the transmission of magnetic interactions through hydrogen bonds has also been carried out with a PTM radical with

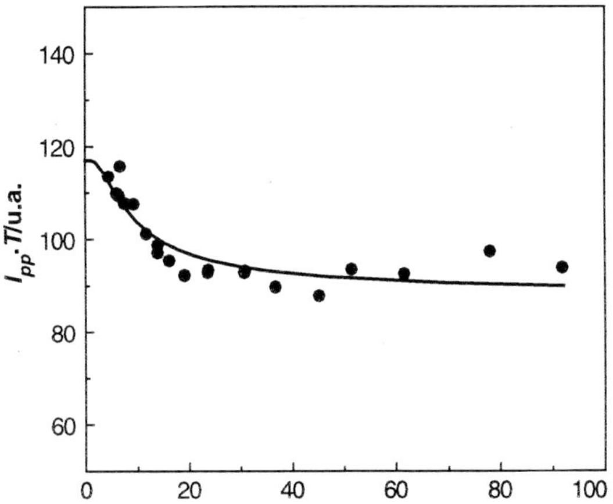

*Figure 2.7 Temperature dependence of $I_{pp}{\cdot}T$ of diradical **6**. The solid points represent the experimental data and the continuous lines the fit of experimental data to the Bleaney–Bowers equation. (Reprinted with permission from [43]. Copyright 2001, Elsevier.)*

a carboxylic acid group at the *para* position, radical **7**. The general tendency of carboxylic acids to form dimers, both in solution and in the solid state, suggested that **7** would be an ideal system to show whether the propagation of the magnetic exchange is efficient at a long distance as well as to determine its ferromagnetic or antiferromagnetic nature.

An extensive magnetic study, both in solution and in the solid state, with radical **7** showed that magnetic exchange interactions are transmitted efficiently through the hydrogen bonds.[44] Low temperature EPR experiments performed in solution with radical **7** did not lead to any conclusive results, since supramolecular aggregates are formed by weak non-covalent interactions, like Cl–Cl and π–π, distinct from the initially expected hydrogen bonds. In order to overcome this drawback, solid studies with radical **7** were performed. Radical **7** crystallized in two polymorphic phases, the α and β phases, depending on the solvent used for the crystallization. As shown in Figure 2.8, in an aprotic solvent, like dichloromethane, the α phase is favored, in which the radicals form nearly isolated $R^2{}_2(8)$ hydrogen bonded dimers, while in presence of ethanol the solvent molecules are intercalated by hydrogen bonds among the two radicals of dimeric entities pushing them far away. Moreover, the bulkiness of the two radicals prevents, in both phases, short contacts between the regions with high-spin densities inside each radical subunit of the dimers.

Solid state magnetic measurements (Figure 2.9) of the two phases of radical **7** provided evidence about the nature and strengths of magnetic interactions inside the dimeric entities. Magnetic data of the α phase clearly shows the presence of weak ferromagnetic interactions, while such interactions are opposite in the β phase (Figure 2.9). The suppression of the ferromagnetic interactions in the β phase, in which direct hydrogen bonded radical dimers are not present, gave further evidence that the hydrogen bonded dimers are the species uniquely responsible for this interaction. Therefore, it has been demonstrated that exchange through hydrogen bonded bridges can occur at a long distance, even in the cases of well delocalized radicals.

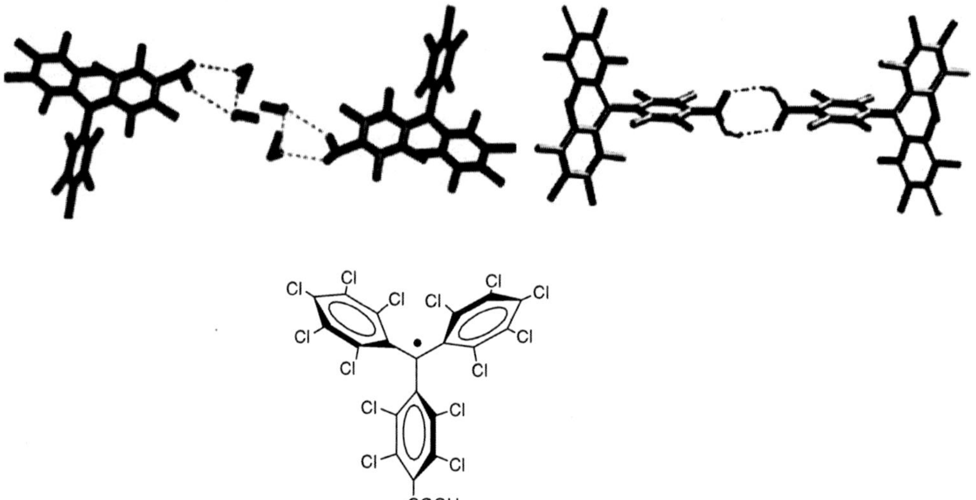

7

Figure 2.8 *View of hydrogen bonded dimers present in the* α *phase (left) and* β *phase (right) of radical 7. (Reprinted with permission from [44]. Copyright 2002 Wiley-VCH Verlag GmbH & Co. KGaA.)*

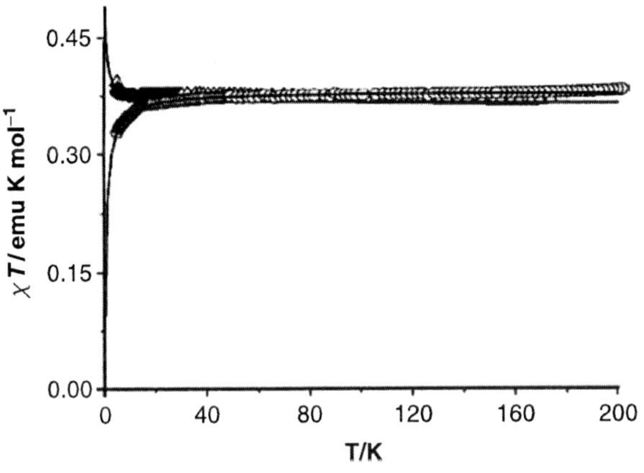

Figure 2.9 Temperature dependences of the magnetic susceptibility, χ, of polycrystalline samples of α phase (Δ) and β phase (○) of radical 7. (Reprinted with permission from [44]. Copyright 2002 Wiley-VCH Verlag GmbH & Co. KGaA.)

Transmission of Magnetic Interactions Through Coordinative Bonds A great deal of work in molecular magnetism has been focused on the so-called metal–radical approach that combines paramagnetic metal ions and organic radicals as ligating sites.[45] The electronic open shell character of PTM radicals was particularly appealing for this purpose, since they were expected to interact with transition metal ions enhancing the strength of magnetic interactions. Carboxylic acid substituted PTM radical **7** was an excellent coordinating ligand to obtain new metal complexes with different coordinative geometries following the metal–radical approach. Thus, the reaction of the PTM radical **7** with copper(II) metal ions gave a series of metal–radical complexes (**8, 9, 10** and **11**) which were structurally characterized (Scheme 2.2).[46]

In complexes **8–10** the copper(II) ions are coordinated to two radical units in slightly distorted square planar pyramidal geometries, while complex **11** shows a paddle-wheel copper(II) dimeric structure. Each copper ion in this complex is coordinated to one oxygen atom of a water molecule and with four oxygen atoms of different carboxylate groups, two of which belong to two radicals and another two to two acetate ions. The exchange couplings between copper(II)–radical subunits are antiferromagnetic for the three complexes **8, 9** and **10**, since a fitting to a linear three-spin model gave $J/k_B \approx -20\,\mathrm{K}$ for the three complexes. Magnetic properties of complex **11** are more interesting, since it is one of the scarce examples of a spin-frustrated system composed of organic radicals and metal ions. In this case, experimental data were fitted to a magnetic model based on a symmetrical butterfly arrangement to give a copper(II)–copper(II) exchange coupling of $J/k_B = -350.0\,\mathrm{K}$ and a copper(II)–radical exchange coupling of $J/k_B = -21.3\,\mathrm{K}$, similar to that observed for the copper(II)–radical interactions in complexes **8–10** (Figure 2.10).[47] Following the same strategy, a series of complexes with zinc(II), nickel(II) and cobalt(II) ions was also synthesized. The magnetic behaviors of the resulting complexes showed the presence of antiferromagnetic interactions of $J/k_B \approx -20\,\mathrm{K}$ between the PTM radical subunits and the first row transition ions.[47–49] Finally, lanthanide metal ion complexes, such as europium(III), and the carboxylate PTM radical derived from **7** were also reported.[50] In the same line, coordination capabilities of the PTM monosulfonate radical using different metals were also explored. In this case, the direct coordination of the organic radical through the sulfonate group provides weaker antiferromangetic couplings than with the carboxylated PTM derivatives.[51]

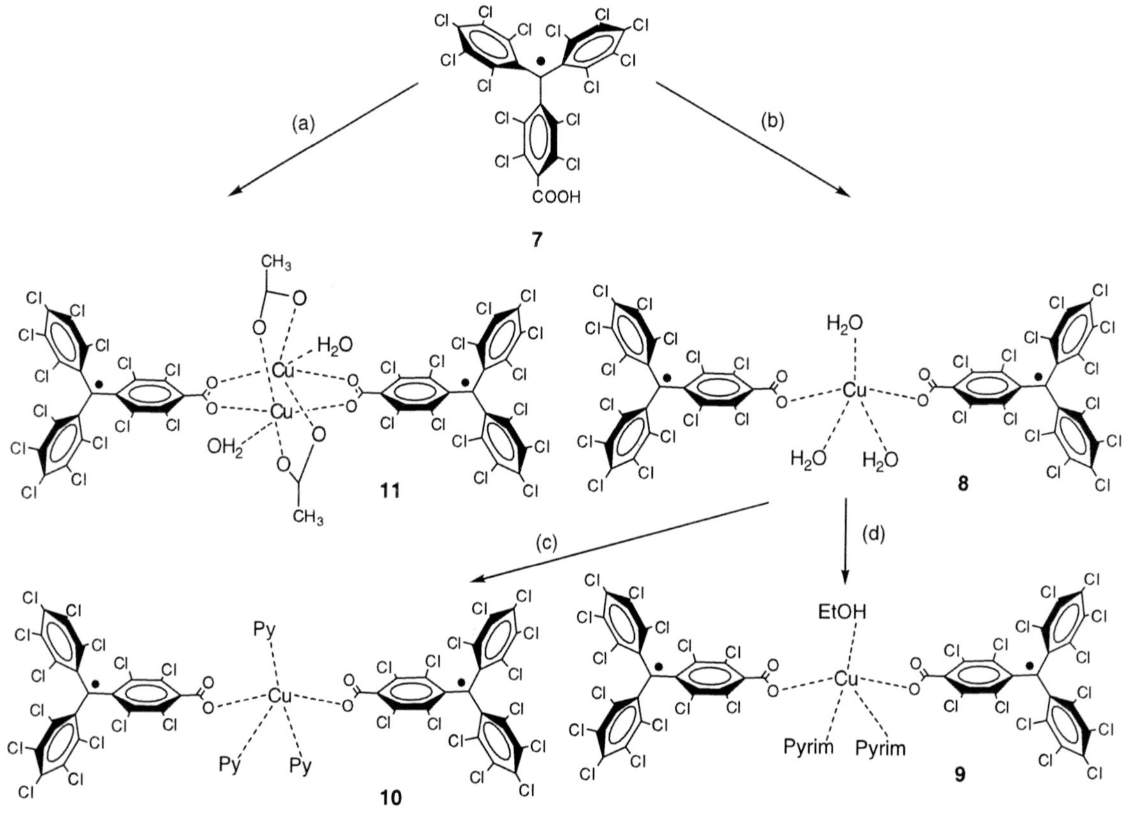

Scheme 2.2 Synthesis of copper complexes based on PTM radical 7. Reagents and conditions: (a) **7**/$Cu_2(O_2CCH_3)_4{\cdot}H_2O$ (4 : 1) in ethanol/water; (b) **7**/$Cu_2(O_2CCH_3)_4{\cdot}H_2O$ (2 : 1) in ethanol/water; (c) **7**/pyridine (excess) in ethanol/*n*-hexane/tetrahydrofuran; (d) **7**/pyrimidine (excess) in ethanol/*n*-hexane/tetrahydrofuran

2.2.1.2 Magnetic extended systems based on PTM radicals

The results presented in the preceding section demonstrate that carboxylic-substituted PTM radicals are excellent coordinating paramagnetic ligands able to transmit efficiently moderate metal–radical antiferromagnetic interactions.[52] This fact, together with the ability of hydrogen bonds to transmit magnetic interactions and link molecules with hydrogen bonding acceptor and donor groups, makes possible the obtaining of magnetic extended systems based on PTM radicals functionalized with several carboxylate/carboxylic acid groups. These magnetic extended systems resulted in two new families of open framework materials with either a metal–organic or purely organic nature; named as metal–organic radical open frameworks (MOROFs) or purely organic radical open frameworks (POROFs), respectively.

The use of organic molecules for achieving open framework materials has become an attractive prospect since the resulting materials are porous in many cases.[53] The use of coordination or crystal engineering techniques allows the systematic design of open framework structures, similar to those shown schematically in Figure 2.11, with a considerable range of pore sizes and functionalities using different organic ligands, such as cyano groups, N,N'-type ligands and polycarboxylic acids. Furthermore, the use of transition metal ions

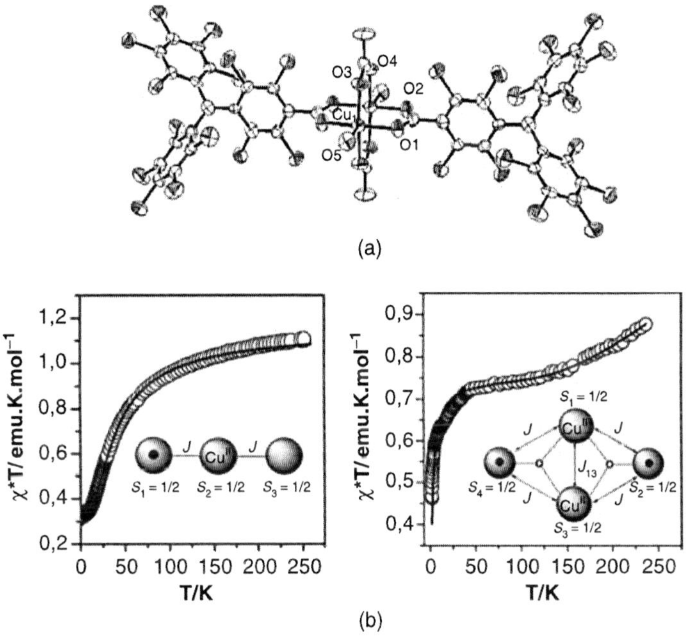

Figure 2.10 *(a) ORTEP view of $[Cu_2(\mathbf{7})_2(O_2CH_3)_2(H_2O)_2]$ (**11**); (b) Temperature dependence of the magnetic susceptibilities. The inset figures show the schematic arrangements of the metal ions and organic radicals. (Reprinted with permission from [47]. Copyright 2002 Royal Society of Chemisty.)*

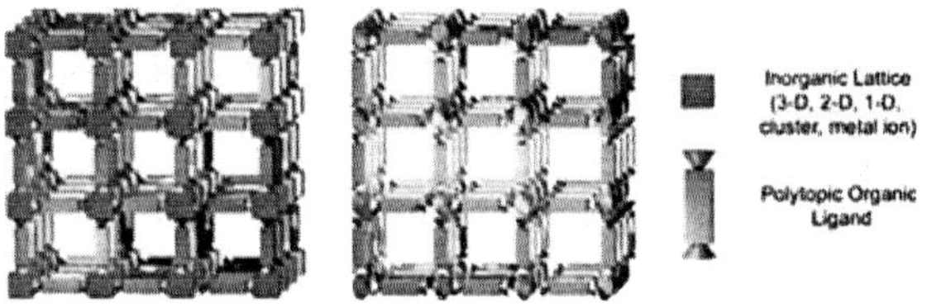

Figure 2.11 *Left: Coordination polymer created by an inorganic subunit (0-D clusters or isolated metal ions) connected through polytopic organic ligands. Right: Purely organic open framework formed by organic molecules interlinked through non-covalent bonds such as hydrogen bonds or π–π interactions. (Reprinted with permission from [53]. Copyright 2007 Royal Society of Chemistry.) A full-colour version of this figure appears in the Colour Plate section of this book.*

brings the possibility to obtain porous materials with additional electrical, optical or magnetic properties. Among them, the search for magnetic open framework structures has become a major objective due to their potential applications in the development of low density magnetic materials, magnetic sensors and intelligent or multifunctional materials. Indeed, a large number of pure inorganic materials, organic–inorganic hybrid materials and coordination polymers with magnetic properties have been reported thanks to the use of constitutive open shell transition metal ions within the framework of the structure. More limited, however, is the number of pure organic porous magnetic materials, mainly because of the limited range of free organic radicals with enough persistence and stability.

Metal–Organic Radical Open Frameworks The polytopic organic ligands, generally nitrogen and oxygen donor ligands, used in metal–organic open frameworks (MOFs), also named as coordination polymers, connect the 'inorganic' subunits along the space in the appropriate topology to originate connected void volumes in the resulting structure. As in many other aspects related to the molecular magnetism field, Kahn and coworkers pioneered the discovery of coordination polymers that combined both porosity characteristics and magnetic properties.[54] Such coordination polymers were made with closed shell organic multitopic ligands and isolated magnetically active metallic ions and it can be noted that when the distance between metal ions increased by enlarging the size of organic ligands, the magnetic couplings between them decreased exponentially, so limiting obtaining large pores.[55] One strategy to overcome this problem is the use of stable radicals with rigid and large sizes as polytopic ligands together with magnetically active transition metal ions.[56] The resulting structures must in principle exhibit larger magnetic couplings and dimensionalities in comparison with systems made up from diamagnetic polytopic coordinating ligands, since the organic radicals may become an *active* part of the magnetic structure as they have their own spin, rather than 'just' mediate magnetic interactions between adjacent metal ions.[57,58]

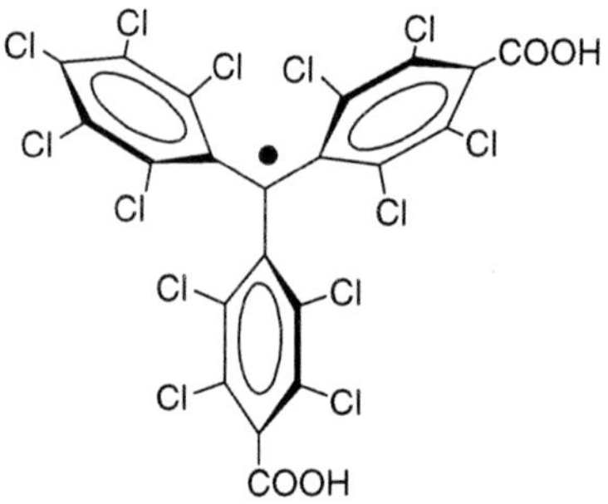

12

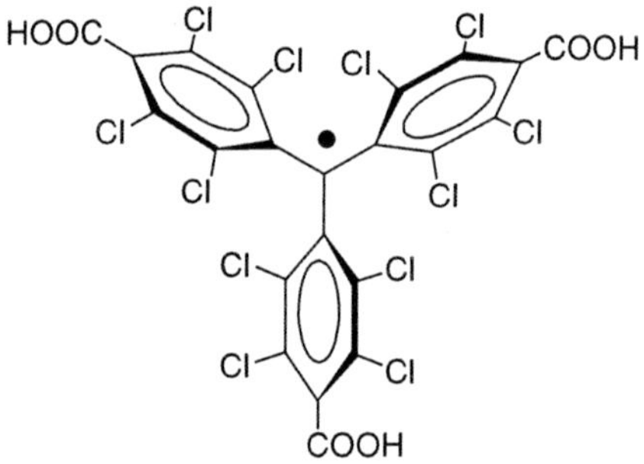

13

By using the di and tricarboxylic acid radicals **12** and **13**, a series of metal–organic radical open frameworks (**MOROF-n**) has been prepared. Interestingly, the crystal structure of Cu_3(**13**)$_2$(py)$_6$ $(C_2H_5OH)_2(H_2O)$ (**MOROF-1**) reveals a 2-D honeycomb (6,3) network composed of rings with six metal units and six **13** radicals connected between them (Figure 2.12). Such planar layers are piled up forming very large 1-D hexagonal pores, which measure 3.1 and 2.8 nm between opposite vertices.[59] To the best of our knowledge, these pores are one of the largest reported so far for a metal–organic open framework structure. Moreover, complex **MOROF-1** shows additional channels with square and rectangular shapes in the perpendicular directions with estimated sizes of 0.5×0.5 and 0.7×0.3 nm, respectively. The interconnected pores generate solvent-accessible voids in the crystal structure that amount to 65 % of the total unit cell volume. This extremely large void volume is responsible for the exotic properties (*vide infra*) of this material.

Accordingly, with the capability of carboxylic-substituted PTM radicals to transmit efficiently the magnetic interactions, each **13** ligand was able to magnetically bridge three copper(II) ions and, therefore, to extend the magnetic interactions across the infinite layers, giving rise to ferrimagnetic 2-D layers. Indeed, as shown in Figure 2.13, the smooth decrease of the χT values below 250 K is a clear signature of the presence of antiferromagnetic couplings between nearest neighboring copper(II) ions and **13** ligands within the 2-D layers. The minimum of χT corresponds to a short range order state where the spins of adjacent magnetic centers are antiparallel, provided there is no net compensation due to the 3 : 2 stoichiometry of the copper(II) ions and radical units. The huge increase of χT at lower temperatures indicates an increase of the correlation length of antiferromagnetically coupled units of copper(II) and **13** as randomizing thermal

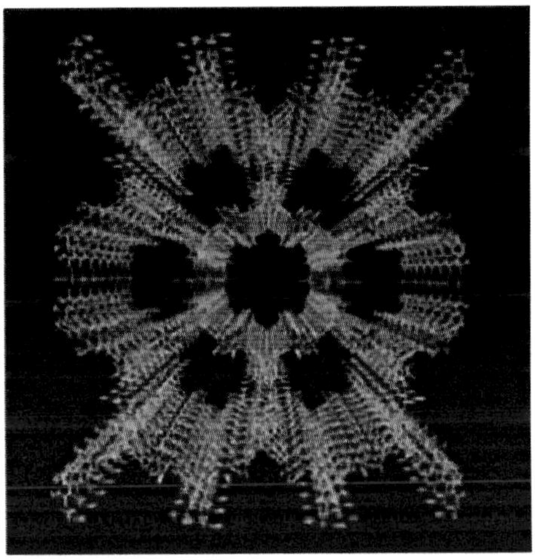

Figure 2.12 Illustration of the nanochannel-like structure of ***MOROF-1***. The superposition of the honeycomb (6,3) layers creates large pores of dimensions of 3.1 and 2.8 nm between opposite vertices. (Reprinted with permission from [56]. Copyright 2004 Royal Society of Chemistry.) A full-colour version of this figure appears in the Colour Plate section of this book.

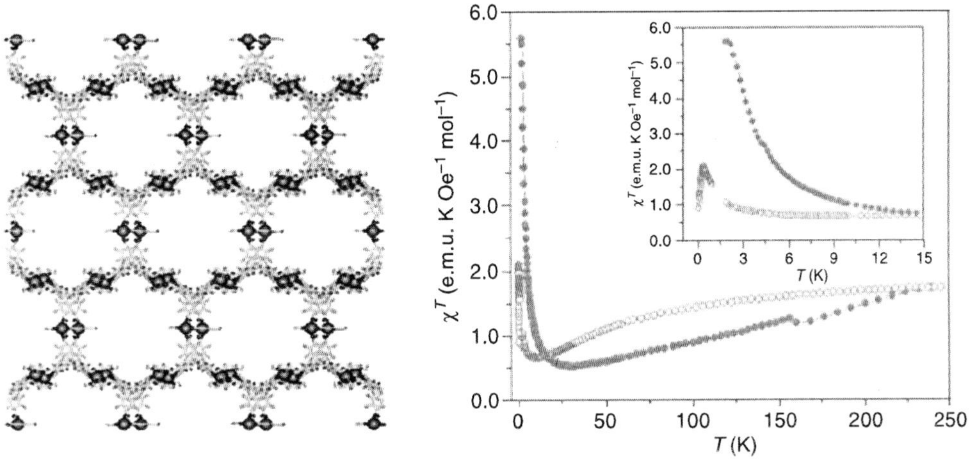

Figure 2.13 Left: A honeycomb (6,3) layer showing the organization of the copper ions (orange spheres) and the central methyl carbon (magenta spheres) of the **13** radicals. Right: χT as a function of the temperature for as-synthesized (•) and evacuated (○) ***MOROF-1***. Inset: Magnetic field dependence of the magnetization at 2 K. (Reprinted with permission from [56]. Copyright 2004 Royal Society of Chemistry.) A full-colour version of this figure appears in the Colour Plate section of this book.

effects are reduced yielding, finally, to a long range magnetic ordering below 4 K. This bulk magnetic ordering was confirmed by the magnetization curve at 2 K since it exhibits a very rapid increase, as expected for a bulk magnet, although no significant hysteretic behavior was observed because of the lack of anisotropic sources in the magnetic units. The magnetization value increases much more smoothly at higher fields up to a saturation value of 1.2 μ_B, which is very close to that expected for a saturated system with an S = ½ magnetic ground state. Thus, this molecular material can be considered as a soft magnet with an overall magnetic ordering below a T_C of 4 K.

A second remarkable feature of **MOROF-1** is the reversible 'shrinking–breathing' of its solid state structure upon solvent uptake and release. Indeed, when **MOROF-1** is removed from the mother liquor solution in which it was crystallized and exposed to air at room temperature, the crystalline material loses the ethanol and water guest molecules, very rapidly becoming an amorphous material with a volume decrease of around 30 % (Figure 2.14). Even more interesting is the fact that the evacuated sample of **MOROF-1** recovers its original crystallinity and up to 90 % of its original size after exposure to liquid or vapour ethanol solvent. The quasi-reversible process also occurs with methanol, but not with other organic solvents, therefore showing a large selectivity for small alcohols. Thus, apparently **MOROF-1** behaves as a sponge-like magnet material selective towards methanol and ethanol. The chemical, structural and morphological quasi-reversibility is also accompanied by changes in the magnetic properties that are macroscopically detected. Magnetic properties of an evacuated amorphous sample of **MOROF-1** show similar magnetic behavior to that shown by the as-synthesized crystals of **MOROF-1**, with the exception that its critical temperature is one order of magnitude lower, around 0.4 K. Thus, the most striking feature exhibited by **MOROF-1** is that the structural and chemical evolution of the material in the process of solvent inclusion can be completely monitored either by the magnetic properties or its X-ray diffraction pattern. When **MOROF-1** is again re-immersed in ethanol, a fast recovery of up to 60 % of the signal can be seen during the first minutes, whereupon the recovery of magnetic signal seems to be linear with the logarithm of time.

Two new supramolecular cobalt(II)-based coordination polymers, $[Co(\mathbf{13})(4,4'\text{-bpy})(H_2O)_3]\cdot 6C_2H_5OH\cdot 2H_2O$ (**MOROF-2**) and $Co_6(\mathbf{13})_4(py)_{17}(H_2O)_4\text{-}(C_2H_5OH)$ (**MOROF-3**) have also been reported.[60,49]

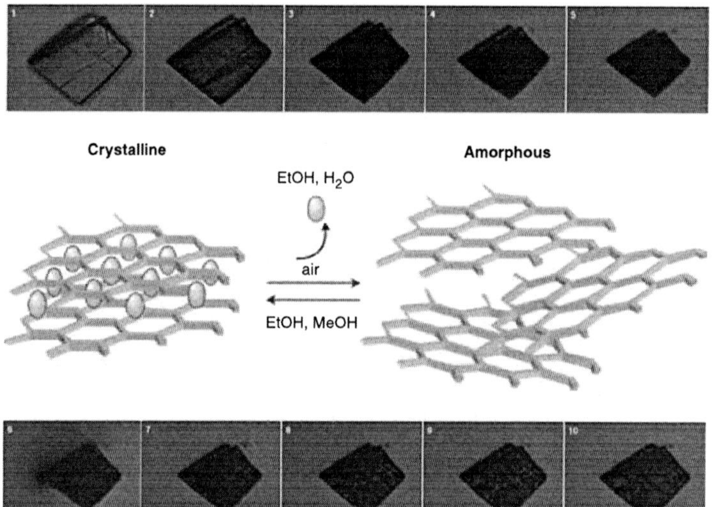

Figure 2.14 *Images of a single crystal of **MOROF-1** followed with an optical microscope. The top series show the 'shrinking' process, in which a crystal of **MOROF-1** exposed to the air experiences a volume decrease of around 30%. In the lower series, the same crystal exposed against to ethanol liquid begins to swell. The scheme represents the structural changes of **MOROF-1** in contact or not with ethanol or methanol solvent. (Reprinted with permission from [56]. Copyright 2004 Royal Society of Chemistry.) A full-colour version of this figure appears in the Colour Plate section of this book.*

*Figure 2.15 Illustration of the helical nanochannel-like structure of **MOROF-2**. (Reprinted with permission from [56]. Copyright 2004 Royal Society of Chemistry.) A full-colour version of this figure appears in the Colour Plate section of this book.*

The first coordination polymer shows a paramagnetic non-interpenetrated supramolecular network with an unprecedented $(6^3)\cdot(6^8\cdot8^1)$ topology. Additionally, **MOROF-2** exhibits helical nanochannels of dimensions 13.2 × 9.4 Å and a void volume of 54.5 % along with antiferromagnetic interactions (Figure 2.15). In contrast, **MOROF-3** exhibits a (6,3)-helical network with large 17.5 × 6.8 Å 1-D channels and a bulk magnetic ordering below 1.8 K. This was the first reported example of a nanochannel-like structure that in addition to exhibiting an unusual (6,3) helical coordination network showed mixed ferromagnetic and antiferromagnetic interaction between the cobalt(II) ions and the **13** radicals.

The hexacarboxylic PTM radical with six carboxylic acid groups at the *meta* positions of the three phenyl rings have been complexed with copper(II) ions to obtain a three-dimensional coordination polymer. The resulting 3-D structure can be described as two interpenetrating cubic nets. Magnetic properties of this metal radical organic open framework have been studied showing unexpected metal–radical ferromagnetic interactions.[61]

Pure Organic Radical Open Frameworks In addition to their demonstrated utility as ligands, carboxylic acid functionalized PTMs can also self-assemble into purely organic radical open frameworks (POROFs) with porous structures and interesting magnetic properties. Thus, radicals **12** and **13** were used to obtain extended hydrogen bonded networks, which in principle would show such a combination of properties. The advantages of these radicals were considerable: (i) their trigonal symmetry provided a typical template for getting channels held together by hydrogen bonds through the carboxylic groups; (ii) the molecular bulkiness and rigidity of PTM radicals was expected to prevent a close packing of molecular units; and (iii) besides their structural control, hydrogen bonds have been shown to favour magnetic coupling between bound radical molecules (Section 'Transmission of Magnetic Interactions Through Hydrogen Bonds').[62] These expectations were confirmed in the hydrogen bonded self-assemblies of **12** and **13** by the formation of robust porous extended networks. Indeed, the dicarboxylic **12** radical crystallizes in a robust porous 2-D extended network (**POROF-1**) with weak antiferromagnetic interactions.[63] The framework shows 1-D tunnels formed by narrowed polar windows (≈5 Å in diameter) and larger hydrophobic cavities, where a sphere 10 Å in diameter can fit inside them. The combination of supercages and windows gives way to solvent accessible voids in the crystal structure that amount up to 31 % of the total volume of the unit cell

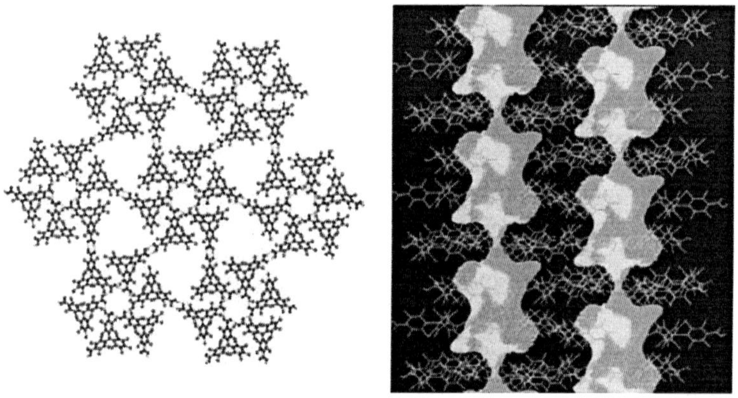

Figure 2.16 Left: Crystal structure of **POROF-1** obtained from the crystallization of radical **12** showing the 2-D hydrogen bonded layer. The repetitive $R_6^6(24)$ hydrogen bonded hexamers generate polar windows due to the presence of six carboxylic groups, whereas the linking of each hexamer with six more identical units in a hexagonal topology originates six trigonal-shaped hydrophobic voids. Right: Space-filling view along the b axis of the large nanocontainers formed along the one-dimensional channel. (Reprinted with permission from [63a]. Copyright 2004 American Chemical Society.) A full-colour version of this figure appears in the Colour Plate section of this book.

(Figure 2.16). Furthermore, it is remarkable that the structural rigidity of the framework permits the evacuation of the guest *n*-hexane solvent molecules at 373 K without collapsing the self-assembled molecules.

More recently, radical **13** provided another robust porous 2-D hydrogen bonded magnet (**POROF-2**). As shown in Figure 2.17, the stacking of these hydrogen bonded radical molecules generates a 3-D structure that exhibits tubular highly hydrophilic channels 5.2 Å in diameter with solvent-accessible voids that amount up to 15 % of the total unit cell volume. The lack of guest solvent molecules within the channels and its structural rigidity up to 573 K are excellent conditions to explode the porosity properties of this new porous material. No less remarkable are the magnetic properties of this material, which orders ferromagnetically at very low temperatures (Figure 2.17).[63]

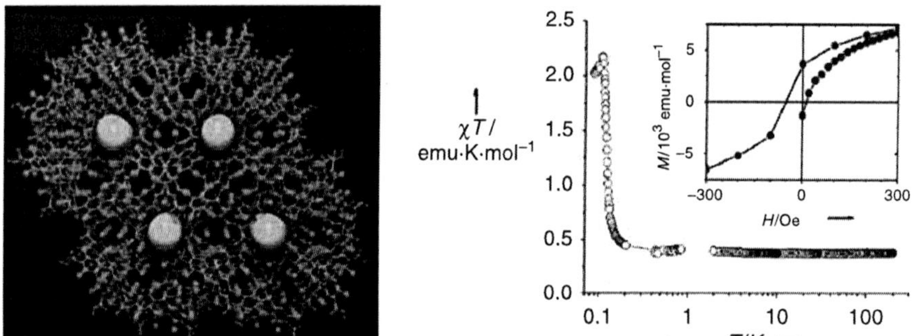

Figure 2.17 Left: Crystal packing of **POROF-2**, showing the tubular hydrogen bonded channels. The empty space is represented as yellow spheres. Right: Temperature dependence of χT up to 200 K, measured with an applied magnetic field of 200 Oe. Inset: magnetic hysteresis obtained at 0.08 K. (Reprinted with permission from [53]. Copyright 2007 Royal Society of Chemistry.) A full-colour version of this figure appears in the Colour Plate section of this book.

The magnetic properties of structures based on the hexacarboxylic acid-substituted PTM radical have also been reported. The study of its self-assembly in the solid state revealed that it is possible, through a tuning of the crystallization conditions, to control the structural dimensionality going from 0- to 2-D solid state structures in which the association of radicals through direct hydrogen bonds to form layers gives rise to the presence of weak ferromagnetic intermolecular interactions between radicals.[64]

2.2.2 Materials with electronic properties

The enormous current interest in molecule-based electronics arises from the potential use of such systems in molecular-scale devices; like nano-sized transistors, switches and wires.[65,66] The study of intramolecular electron transfer (IET) processes has been of prime interest in this area because the understanding and control of IET could lead to a better design and improvement of such molecule-sized electronic devices.[67] For such studies, the use of molecular wires with two electroactive units connected by a rigid bridge has definite advantages over systems with two freely diffusing units.[68] There are two kinds of such systems showing electronic properties (Scheme 2.3): one consisting of an electron donor (D) and an electron acceptor (A) group as electroactive units, called D-A dyads; the other composed of the same unit at both sides of the bridge but with different oxidation states, which are named as mixed-valence (M-V) compounds. According to the degree of electronic coupling between the two electroactive units through the bridge, the compounds showing IET can be classified in three classes. Class I contains those compounds in which the two units are electronically independent of any significant electronic coupling, while Class III comprises those systems in which there is a large electronic coupling with a complete delocalization of the electron along the whole molecular system. By contrast, Class II compounds show an intermediate situation, in which the electron is transferring between the two subunits.

M-V compounds and D-A dyads of Class II are excellent candidates for IET studies, since the rate and efficiency of such phenomena can be easily followed and studied from the position, intensity and width of the so-called intervalence transition (IVT) band, also named the charge transfer band, which usually appears at the near-infrared (NIR) region. IET phenomena may take place either through thermally or optically induced mechanisms, as shown schematically in Figure 2.18. To have a thermally-induced IET, the electron must move from one site to another along the nuclear coordinate overcoming the thermal energy barrier, E_{th}, which arises from the nuclear reorganization required for the passage of the electron from one site to another. This nuclear reorganization is produced both within the molecule as in the surrounding solvent molecules. On the contrary and according with the Frank–Condon principle, an optically-induced IET takes place, either directly from the ground state $[A{-}B^{+}]$ to a vibrationally excited level of the first

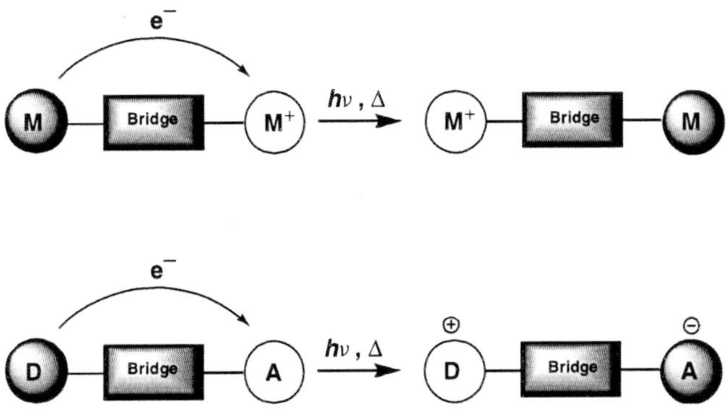

Scheme 2.3 Schematic representation of D-A dyads (bottom) and M-V compounds (top)

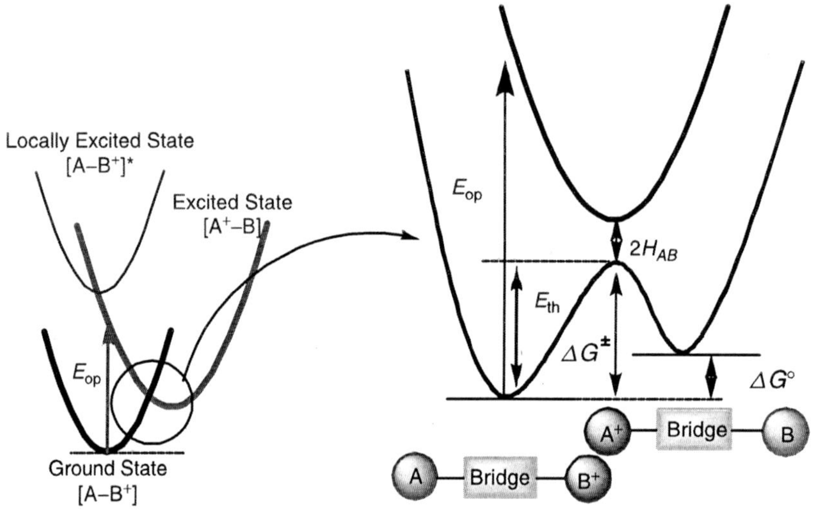

Figure 2.18 *Potential energy curve of a Class II system where the optically (E_{op}) and thermally (E_{th}) induced IET processes are shown. Different energetic terms involved in IET processes are also depicted: Energy difference between the ground and the first excited state ($\Delta G°$), the thermal energy barrier ($\Delta G^{\pm}$) and the coupling parameter H_{AB}.*

excited [A^+–B] state in a M-V compound or through a direct transition from the ground state [A–B^+] to a locally excited state [A–B^+]* in a D-A dyad. In both mechanisms neither the solvent nor the internal geometry of the molecular wire are allowed to relax during the IET. As a consequence, an IVT band corresponding to the vertical excitation is observed in the optical absorption spectrum of these compounds. From the position, intensity and width of such an IVT band, the IET can be easily characterized. In most cases, E_{op} is the energy required for the optically-induced electron transfer, $\Delta G^{\pm}$ is the activation barrier for the thermally-induced electron transfer, H_{AB} is the electronic coupling matrix between the diabatic potential energy surfaces of the two states and $\Delta G°$ is the free energy difference between both states; λ is the reorganization energy value, which is composed of two terms, one inner term, λ_i, corresponding to the reorganization inside the molecule and an outer term, λ_o, that depends on the surrounding media. According to Marcus–Hush theory, the parameters controlling the optical and thermal induced IET processes are closely interrelated by Equations 2.2 and 2.3:

$$\Delta G^{\neq} = \frac{(\lambda + \Delta G°)^2}{4\lambda} \tag{2.2}$$

$$E_{op} = \lambda + \Delta G° \tag{2.3}$$

The effective electronic coupling H_{AB} (expressed in cm^{-1}) between the two states can be determined based on Equation 2.4, where r is the effective separation of the two electroactive sites (in Å), ε_{max} is the maximum extinction coefficient (in $M^{-1} \cdot cm^{-1}$) of the IVT band and ν_{max} is the transition energy and $\Delta\nu_{1/2}$ is the full width at half height (both in cm^{-1}) of the IVT band.

$$H_{AB} = (2.05 \cdot 10^{-2}) \left[\frac{\varepsilon_{max}\Delta\nu_{1/2}}{\nu_{max}}\right]^{1/2} \frac{\nu_{max}}{r} \tag{2.4}$$

In the following, IET phenomena observed in D-A dyads derived from PTM radicals is described. In addition, M-V compounds derived from PTM radicals is discussed in which one of both PTM units is reduced to the anionic form enabling the generation of the mixed-valence species.

2.2.2.1 *Intramolecular electron transfer in mixed-valence species*

It is worth mentioning that there are very few examples of organic M-V molecules exhibiting IET phenomena together with high-spin ground states. Since the nature of the intramolecular magnetic couplings and IET phenomena are not at all independent of each other,[69] the study of molecular systems that simultaneously present both phenomena is a task of particular interest.[4–6,70] The study of organic M-V molecules with high-spin ground states was acomplished for the first time using the triradical **3**, which has the quartet ($S = 3/2$) as the ground state.[13]

Quartet **3** ranks among the most chemically and thermally stable organic high-spin molecules reported to date (Section 2.2.1) and the derived mixed-valence species, like the biradical anion $\mathbf{3}^{-\bullet\bullet}$ and the monoradical dianion $\mathbf{3}^{2-\bullet}$, obtained by its partial reduction, also showed a noticeable degree of chemical stability, allowing a reliable study of electron transfer properties (Scheme 2.4). Spectroelectrochemistry experiments performed with partially reduced solutions of quartet **3** showed two isosbestic points at 312.5 and 406.5 nm, together with a progressive decrease of the radical band at 368 nm and a concomitant increase of the band corresponding to the anion chromophore at 508 nm.[71] In addition to these changes, the appearance of an intense, and remarkably broad, band centered at 2900 nm (3450 cm^{-1}) was observed (Figure 2.19). Throughout the reduction process the intensity of the latter band increased at the first reduction stages and then started disappearing only after reaching an average reduction state of two electrons per molecule, that is after the formation of $\mathbf{3}^{2-\bullet}$ species. This result confirms that the broad band is an IVT band due to the optically induced IET processes occurring in the M-V species $\mathbf{3}^{-\bullet\bullet}$ and $\mathbf{3}^{2-\bullet}$.[72] EPR spectroscopy was used to study the magnetic characteristics of the M-V species $\mathbf{3}^{-\bullet\bullet}$ and $\mathbf{3}^{2-\bullet}$. For this purpose, aliquots of dichloromethane solutions of partially reduced **3** with average reduction states of $n = 1.17$ and 2.04 e^-/molecule were diluted with toluene and the spectra of the frozen glassy solutions were recorded. As expected, for the solution with $n = 2.04\,e^-$/molecule, containing the M-V radical $\mathbf{3}^{2-\bullet}$ with traces of silent EPR trianion $\mathbf{3}^{3-}$, only a single line corresponding to the $\mathbf{3}^{2-\bullet}$ species was observed. In contrast, the solution with $n = 1.17\,e^-$/molecule shows a signal with the characteristic fine structure of a non-axial triplet species superimposed onto a central narrow line. The central narrow line corresponds to the monoradical dianion $\mathbf{3}^{2-\bullet}$ while the symmetrical spectral features arise from the triplet species $\mathbf{3}^{-\bullet\bullet}$. The latter species is also responsible for the structureless line observed at the half-field region corresponding to the forbidden $\Delta m_s = \pm 2$ transition of a triplet species. In order to identify the ground state of $\mathbf{3}^{-\bullet\bullet}$, the temperature dependence of the $\Delta m_s = \pm 2$ signal was studied indicating a triplet ($S = 1$) ground state with a thermally inaccessible singlet ($S = 0$) state.

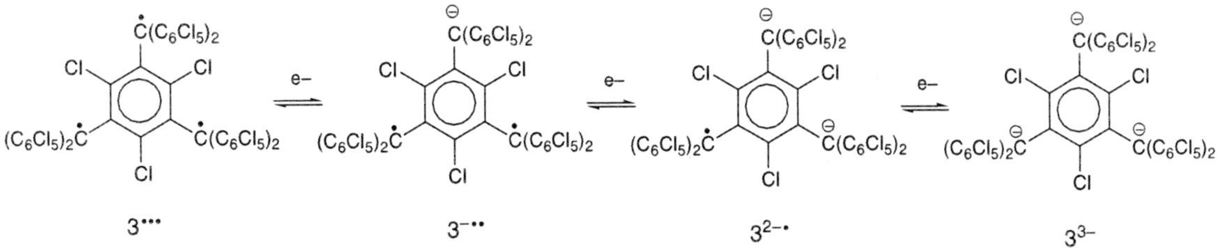

Scheme 2.4 Redox processes of triradical **3**

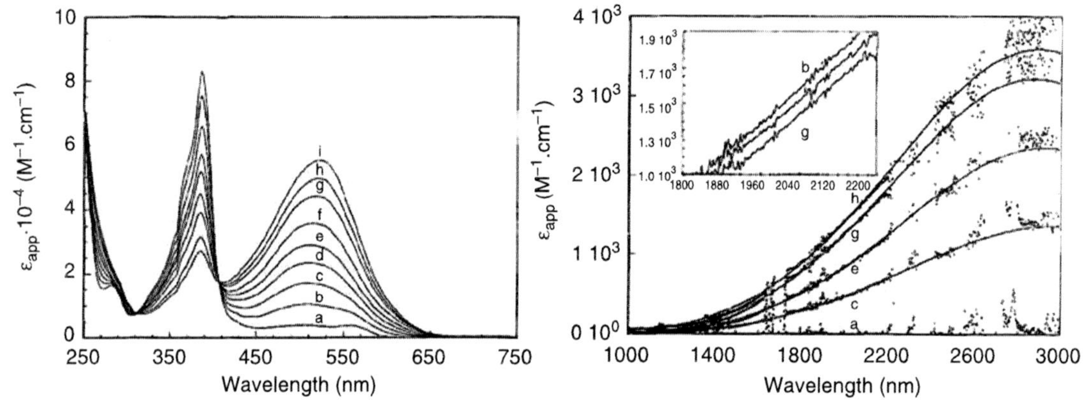

Figure 2.19 *Evolution of the electronic absorption spectra during the course of the electrochemical reduction of triradical 3. Left: Different traces in the visible range corresponding to distinct average reduction states of: (a) 0.00; (b) 0.29; (c) 0.56; (d) 0.73; (e) 0.90; (f) 1.17; (g) 1.49; (h) 1.71; and (i) 2.10 e^-/molecule. Right: Selected spectral traces observed in the NIR region. (Reprinted with permission from [13b]. Copyright 1996 Wiley-VCH Verlag GmbH & Co. KGaA.)*

To study longer range IET phenomena, diradical **14**, consisting of two PTM units connected by a 1,4-bis(1-ethynylphenyl) benzene bridge, was obtained together with its radical anion derivative $\mathbf{14}^{-\bullet}$.[73] The bridge of **14** not only ensures a long distance between the two PTM units and a high degree of molecular rigidity, but also a suitable electronic structure to promote an electronic coupling between the two PTM ends separated by a significantly large through-space radical-to-radical distance of 3.2 nm, as revealed by semiempirical AM1 calculations.[74]

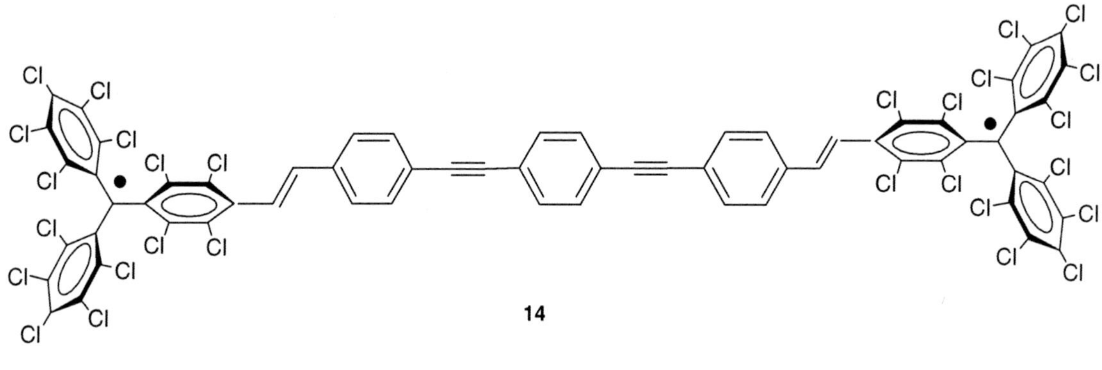

14

Magnetic susceptibility measurements on **14** show that the magnetic moment of this compound decreases with decreasing temperature, indicating that the ground state is a singlet with a thermally accessible triplet state. A separation of 1.5 K between both states was obtained from the fitting to the Bleaney–Bowers equation of the experimental χT versus T data.[75] The EPR spectrum of **14** at room temperature was in agreement with such an electronic structure. Thus, it shows three overlapped lines that correspond to the coupling of each of the two unpaired electrons with two equivalent ^{1}H nuclei, along with other satellite lines that correspond to the coupling of the two unpaired electrons with the naturally abundant ^{13}C nuclei of

PTM units. Computer spectral simulation gave an isotropic g-value of 2.0028 and the following hyperfine coupling constants: a(^{1}H) $\approx$ 0.88 G (2H), a($^{13}C_\alpha$) $\approx$ 13.0 G, a($^{13}C_{bridge}$) $\approx$ 6.3 G, and a($^{13}C_{ortho}$) $\approx$ 5.1G, which are approximately half the value of those found for related monoradical PTM species.[76] It is then possible to conclude that the two unpaired electrons of **14** are magnetically coupled with an exchange coupling constant, J, that fulfils the following condition: $J \gg a_i$. Furthermore, the radical **14** does not exhibit the forbidden $\Delta m_s = \pm 2$ transition characteristic of a triplet species, which is in accordance with the rapid fall down of the magnetic dipolar interaction with the effective distance of the two unpaired electrons, which for **14** is extremely large.[77] Despite the presence of two electronically active units, electrochemical studies (dichloromethane, nBu_4NPF_6 (0,1 M) vs SCE) show only one reversible two-electron reduction process at −0,13 V (vs SCE) due to the conjugation of the bridge. Spectroelectrochemical experiments performed during the reduction of diradical **14** to the radical anion **14**$^{-\bullet}$ evidenced the lack of any IVT band originated by an IET. Isotropic hyperfine coupling constants of **14**$^{-\bullet}$ were approximately double than those found for diradical **14**, demonstrating that the unpaired extra electron is strongly localized on one side of the molecule and the rate of transfer, if any, is very slow. An increase of temperature neither leads to a thermally activated IET process since any change in the spectrum is observed. This result, together with the lack of an IVT band, suggests that there is a strong localization of the extra electron in the radical anion **14**$^{-\bullet}$ species. In conclusion, in this rigid nano-sized M-V system there is a subtle interplay between the acceptor ability of the PTM unit and the size and nature of the bridge that makes feasible or not the electron transfer phenomena.

The influence of the bridge topology on the IET phenomena was also studied with the PTM diradicals **15** and **16** and their corresponding radical anions **15**$^{-\bullet}$ and **16**$^{-\bullet}$ using magnetic measurements, spectroelectrochemistry and variable temperature EPR.[78]

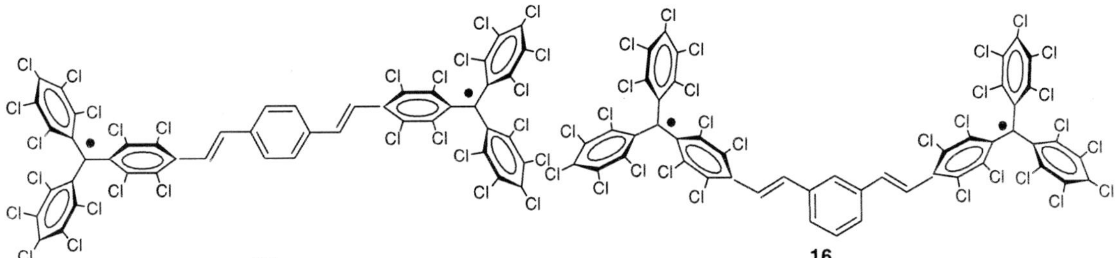

15 **16**

Magnetic susceptibility data of powder samples of diradicals **15** and **16** both follow the Curie–Weiss law with $\theta = -4$ K and −0.4 K for **15** and **16**, respectively, indicating the presence of very weak intra or intermolecular antiferromagnetic interactions. The magnetic moments at room temperature for both diradicals are $2.42 \pm 1\mu_B$, indicating the existence of non-interacting pair of doublets at this temperature. Experimental electronic coupling parameters, H_{AB}, were determined for both diradicals by means of spectroelectrochemistry experiments. Thus, during the reduction of both diradicals the evolution of the resulting spectra was similar, although only for the radical anion **15**$^{-\bullet}$ an IVT band, centred on 1400 nm, was observed. From the position and line width of this band it is possible to calculate the effective electronic coupling, H_{AB}, between the two PTM sites of **15**, which is found to be $H_{AB} = 121\,cm^{-1}$. The absence of any significant IVT band for radical anion **16**$^{-\bullet}$ is an indication that its effective electronic coupling is at least one order of magnitude smaller than in **15**. The thermally activated IET was also studied by variable temperature EPR for radical anions **15**$^{-\bullet}$ and **16**$^{-\bullet}$. Thus, the EPR spectrum of **15**$^{-\bullet}$ at 200 K displays two symmetrical lines arising from the hyperfine coupling of the unpaired electron with one hydrogen atom of the ethylene moiety and the resulting ^{1}H hyperfine coupling constant is very close to that of related monoradicals. This result clearly demonstrates that at this temperature the unpaired electron of radical anion **15**$^{-\bullet}$

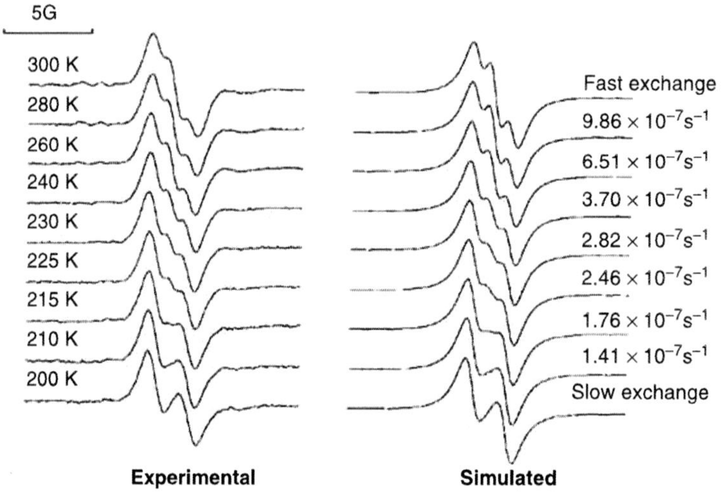

Figure 2.20 *Experimental (left) and simulated (right) EPR spectra of* **15**$^{-\bullet}$ *at different temperatures in dichloromethane. (Reprinted with permission from [78a]. Copyright 2001 Wiley-VCH Verlag GmbH & Co. KGaA.)*

is localized on the EPR timescale on only one-half of the molecule, that is on one *p*-phenylenevinylene-PTM moiety. When the temperature is increased, a new central line in the EPR spectra gradually emerges between the two initial ones (Figure 2.20). This evolution is consistent with the increase of the electron transfer on going from 200 K (the slow-exchange limit) to 300 K (the fast-exchange limit) due to a thermally activated IET between the two equivalent sites of this M-V species. Under these conditions, the unpaired electron of **15**$^{-\bullet}$ is coupled with two equivalent ^{1}H nuclei. The EPR spectra were simulated by using the jumping rates, k_{th}, given in Figure 2.20 and the resulting k_{th} values plotted using a linear Eyring plot [$\ln(k_{th}/T)$ vs $1/T$], from which the energy barrier of the thermal electron transfer, $\Delta G^{\pm}$, of 0.117 eV was obtained.

The EPR spectrum of radical anion **16**$^{-\bullet}$ is unchanged in the temperature range 150–300 K and consists of two symmetrical lines arising from the hyperfine coupling of the unpaired electron with one hydrogen atom of the ethylene moiety with a ^{1}H hyperfine coupling constant close to that of the corresponding monoradical anion at low temperature. This is a clear indication that the extra electron of radical anion **16**$^{-\bullet}$ is always localized on the EPR timescale on one-half of the molecule regardless the temperature. This result can be ascribed to the localization of frontier orbital in the latter radical anion **16**$^{-\bullet}$, because of the *meta* connectivity of this non-Kekulé molecule.

Finally, EPR spectra of diradicals **15** and **16** also provide information about the degree of electronic delocalization in such species. Thus, the absolute values of zero field splitting parameters for diradicals **15** and **16**, obtained from the simulated spectra in frozen trichloromonofluoromethane ($CFCl_3$), are $|D/hc| = 3.9 \times 10^{-4}$ and $2.3 \times 10^{-4}\,cm^{-1}$, respectively, with null $|E/hc|$ values for both diradicals. These parameters arise from the dipolar magnetic interactions between the two unpaired electrons and can be used to calculate the average interspin separation. Therefore, from the $|D/hc|$ parameter in cm^{-1} and Equation 2.5, average interspin separations of 19 and 22 Å were found for diradicals **15** and **16**, respectively. The average interspin separation found for diradical **15** is smaller than the nominal separation between the two alpha carbon atoms where most of the spin density of the two PTM units is localized. This result is in agreement with the existence of a certain degree of electron delocalization for the *para* connectivity of the divinylphenylene bridge that reduces the effective separation of the two spins in **15**. As expected, the average interspin separation found for **16** is closer to the nominal separation between the two alpha carbons due to the lower degree of electronic conjugation in *meta*.

$$r = \left[\frac{3g^2\beta^2}{2hc|D/hc|} \right]^{1/3} \tag{2.5}$$

2.2.2.2 *Intramolecular electron transfer in donor-acceptor species*

The excellent electron acceptor abilities of PTM radicals meant that they were used as electron acceptor units in D-A dyads. One remarkable example of such D-A dyads is the Fc-PTM compound **17**, in which a PTM unit is covalently bonded through a vinylene spacer to a ferrocene moiety. Indeed **17** was shown to exist in the solid state in two distinct electronic isomeric forms – one neutral form and another zwitterionic form with separated charges (Scheme 2.5) that may be reversibly interconverted.[79] Variable temperature Mössbauer spectra (Figure 2.21) show that the Fc-PTM dyad exists in a neutral form at low temperature, while the zwiterionic form is favored at higher temperatures, meaning this thermal conversion is completely reversible. Very recently, it has been found out that the coexistence in a crystalline sample of both the neutral and zwitterionic forms of Fc-PTM is a result of the bistability of crystals, induced by stabilizing electrostatic intermolecular interactions developed in the charge separated state.[80] From this finding it is concluded that novel bistable compounds can be generated in crystals of largely neutral (closed or open shell) D-A molecules with attractive electrostatic intermolecular interactions that operate in the charge separated state. This conclusion has opened a new avenue for the design of valence tautomeric molecules, since the variety of valence tautomeric species has up to now been mostly limited to transition metal complexes with quinone ligands.[81]

The solvent dependence of IET phenomena was studied using the D-A dyads **17** and the related species **18** containing a nonamethylferrocene group.[82] Both compounds exhibit broad absorptions in the NIR region with band maxima appearing around 1000 and 1500 nm for **17** and **18**, respectively. These bands correspond to the excitation of a neutral [DA] ground state to the charge separated [D^+A^-] state, indicative of an IET process. As shown in Figure 2.22, both bands show a large solvatochromic effect due to the different relative stabilization of neutral and charge separated states. Actually, symmetric M-V compounds exhibit negligible ΔG° values due to their symmetric structures and, therefore, the IET are governed by the so-called Marcus-normal region energy conditions, where ΔG° is smaller than the λ term (Equation 2.3).[83] On the other hand, asymmetric D-A dyads usually exhibit large ΔG° values, which may place them in the Marcus-inverted region that occurs when the ΔG° value is larger than λ.[84] In this particular region, the IET process occurs via a different kinetic pathway and unusual features associated with the environment of D-A dyads are quite common.[85] The solvatochromism of the asymmetric compounds **17** and **18** has

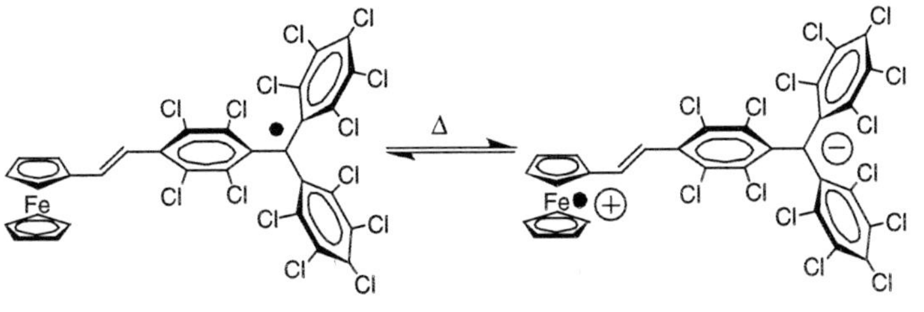

Scheme 2.5 Reversible interconversion between neutral and zwitterionic form of **17**

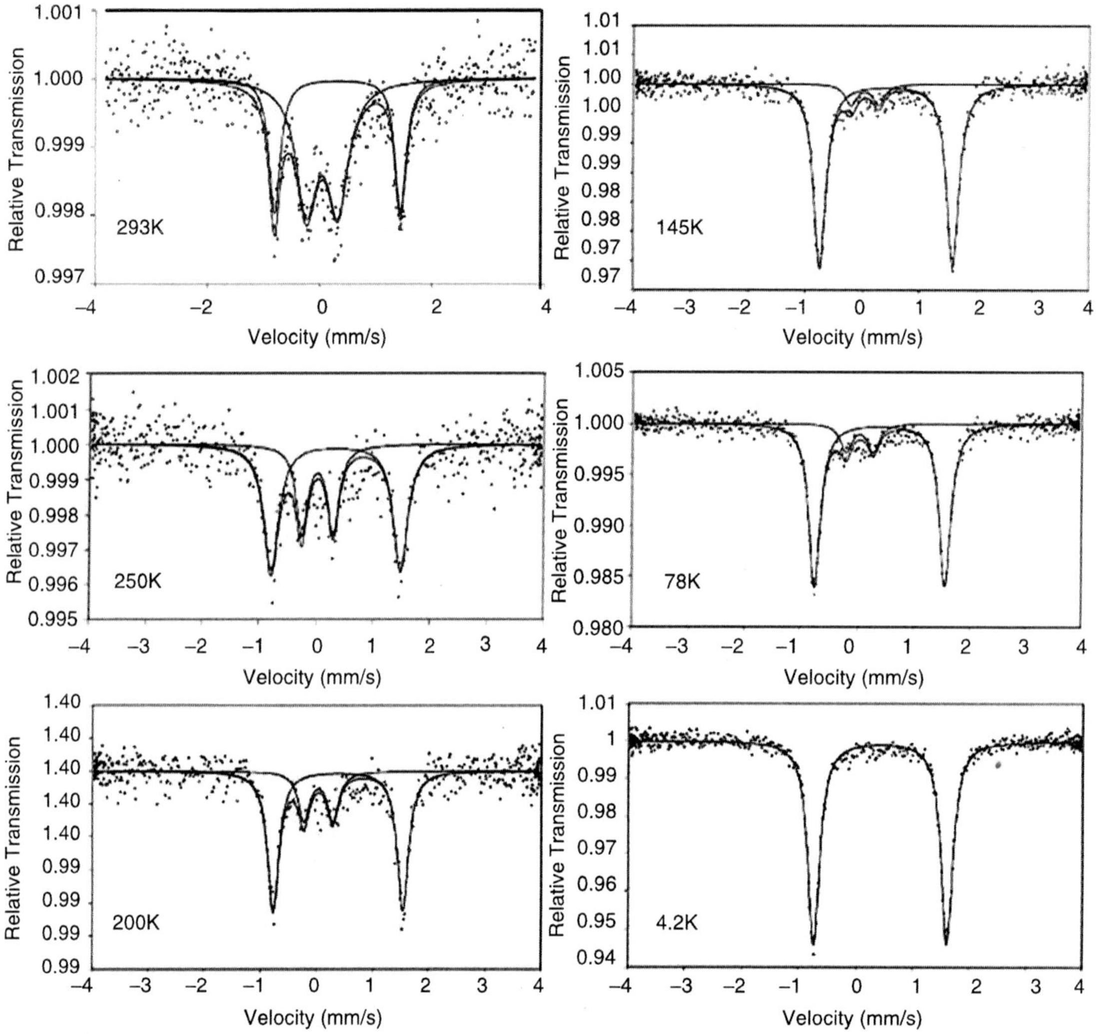

Figure 2.21 ^{57}*Fe Mössbauer spectra of D-A dyad* ***17*** *in the solid state as a function of temperature. (Reprinted with permission from [79]. Copyright 2003 American Chemical Society.)*

been studied and it was found out that for these D-A dyads is possible to shift from the Marcus-normal to the inverted region simply by changing the polarity of the solvent.

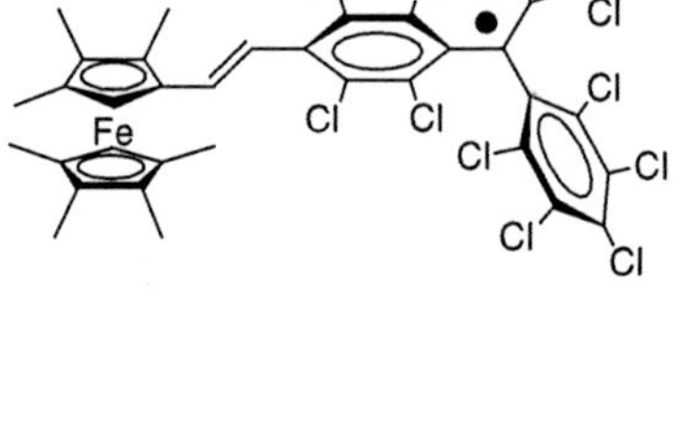

18

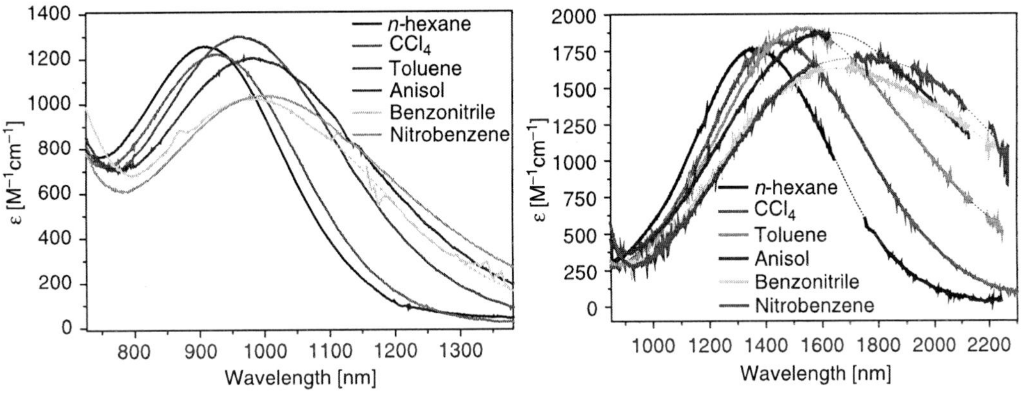

Figure 2.22 *IVT bands of D-A dyads **17** (left) and **18** (right) in some selected solvents. (Reprinted with permission from [82]. Copyright 2007 American Chemical Society.) A full-colour version of this figure appears in the Colour Plate section of this book.*

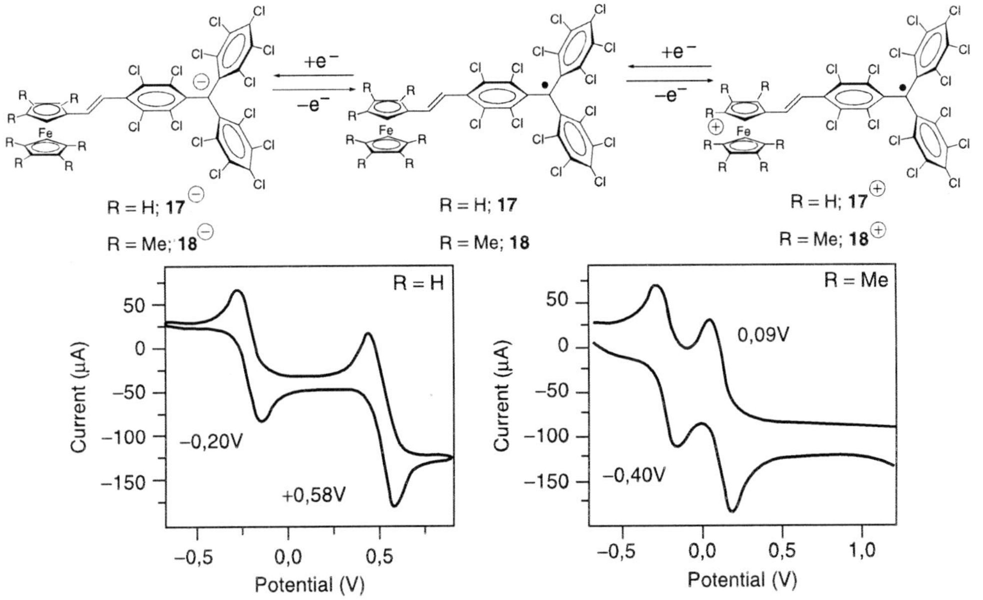

Figure 2.23 *Top: Reversible one-electron redox processes associated with the first oxidation and first reduction potentials of dyads **17** and **18**. Bottom: Cyclic voltammograms of dyads **17** and **18** in dichloromethane using (n-Bu_4N)PF_6 (0.1 M) as electrolyte (vs Ag/AgCl). (Reprinted with permission from [82]. Copyright 2007 American Chemical Society.)*

Dyads **17** and **18** show two reversible one-electron redox processes associated with the oxidation of the ferrocene and the reduction of the PTM subunits (Figure 2.23). The solvent dependence of these redox processes was also investigated, allowing the determination of the free energy difference between the neutral and charge separated states, ΔG°, also called redox asymmetry, in different solvents. The ΔG° values, along with the experimental E_{opt} spectroscopic data in different solvents (Figure 2.22), allow the estimate, using the total energy balance $E_{opt} = \lambda + \Delta G^\circ$ (Equation 2.3), of the total reorganization energy values, λ, and their solvent polarity dependence. Since ΔG° and λ are of the same order of magnitude but exhibit opposite trends in their solvent polarity dependence, a unique shift from the normal ($\Delta G^\circ < \lambda$) to the inverted Marcus region ($\Delta G^\circ > \lambda$) with the change in solvent polarity is found. This fact is of utmost importance since it indicates a crossing of the so-called inverted and normal Marcus regions, where striking differences in the electron transport mechanisms between the donor and acceptor moieties are present. Although it is known that solvent effects can influence strongly the relative values of ΔG° and λ terms in the context of IET in proteins and other biological environments,[86] there were no previous examples of molecules that can be shifted from the Marcus-normal to the inverted region simply by changing the polarity of the solvent.

Another interesting family of D-A dyands, studied by Lambert *et al.*,[87] consist of PTM radicals linked to different amine donor centers by various spacers. This family of D-A dyads are good model systems for the study of IET due to their uncharged nature in their ground states and their high solubility in solvents with low polarity. All of the dyads **19–25** show characteristic IVT bands at 11 000–13 000 cm^{-1} with weak but non-systematic solvatochromic behaviors. For this family of compounds a detailed band shape analysis of the IVT bands in the context of the Jortner's theory allowed the parameters controlling the IET to be extracted independently. Furthermore, the results show that the values of the averaged vibrational modes of radicals **20** and **23–25** with a spacer containing a carbon–carbon triple bond are distinctly higher than the values obtained for radicals **21** and **22**, where the triple bond is absent. Furthermore, a spectroelectrochemical investigation of these substituted PTM radicals shows that the IET can be optically induced in both the neutral D-A compounds and in their oxidized amine forms.

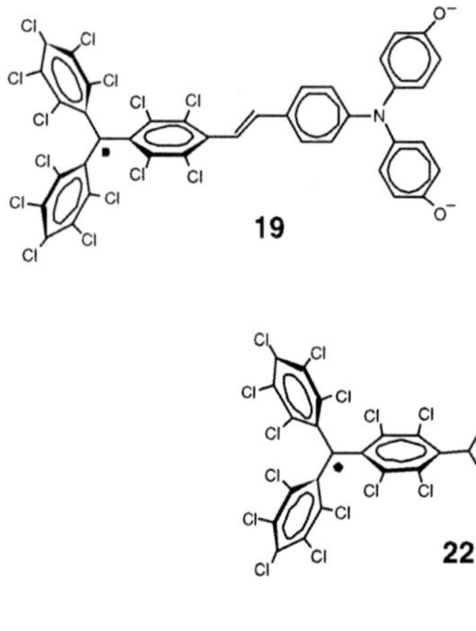
19

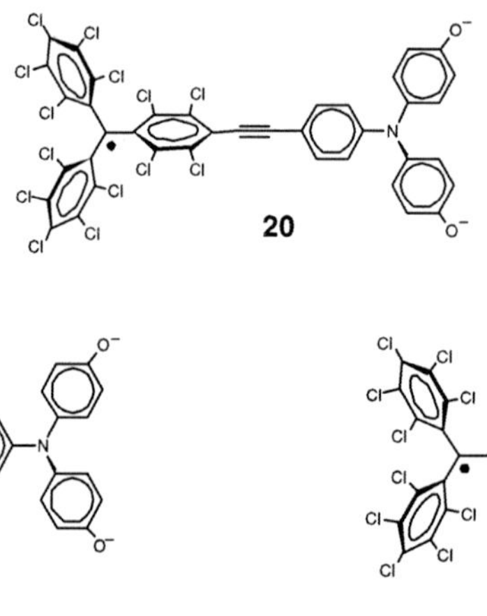
20

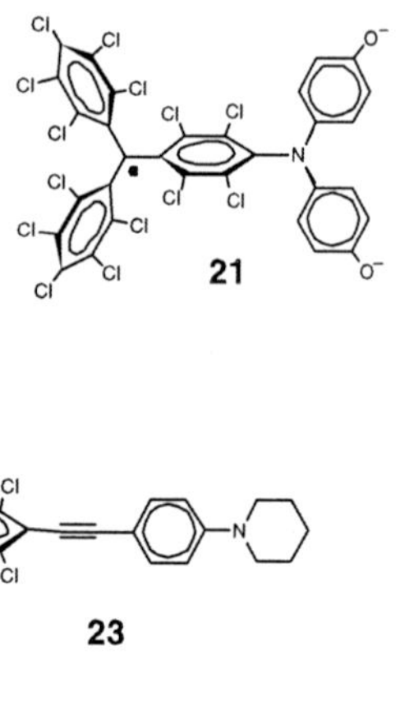
21

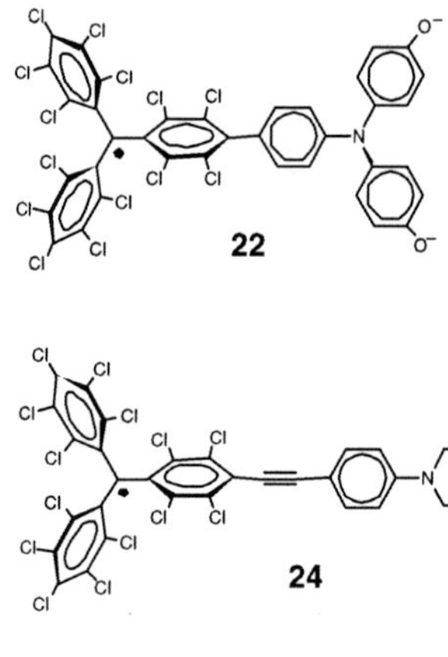
22

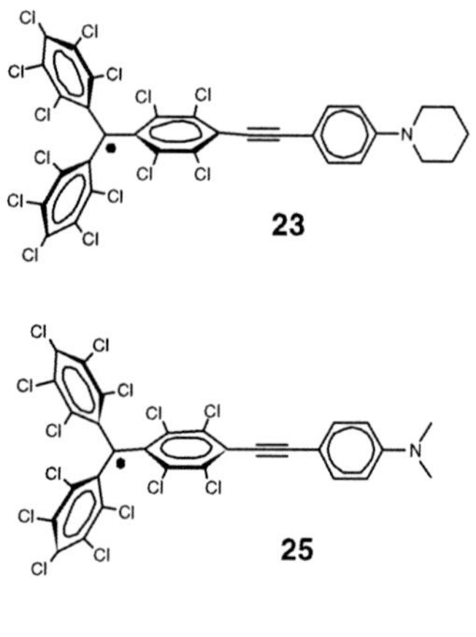
23

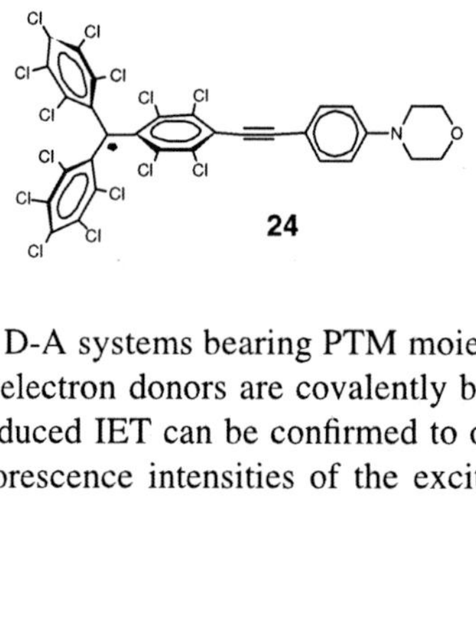
24

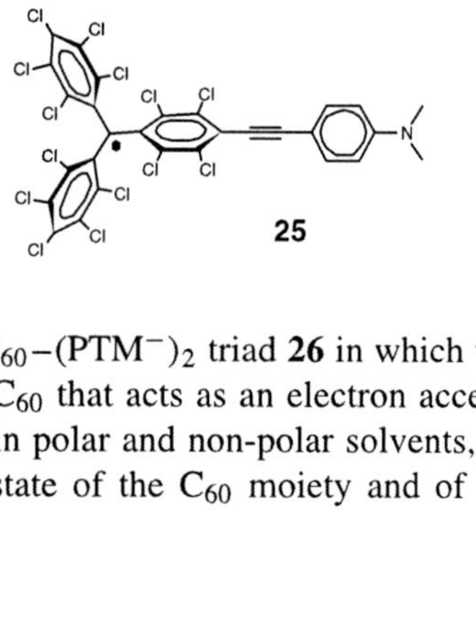
25

A final example of D-A systems bearing PTM moieties is the C_{60}–$(PTM^-)_2$ triad **26** in which two PTM anion units acting as electron donors are covalently bonded to a C_{60} that acts as an electron acceptor. For the triad **26**, photo-induced IET can be confirmed to occur, both in polar and non-polar solvents, from the quenching of the fluorescence intensities of the excited singlet state of the C_{60} moiety and of the PTM

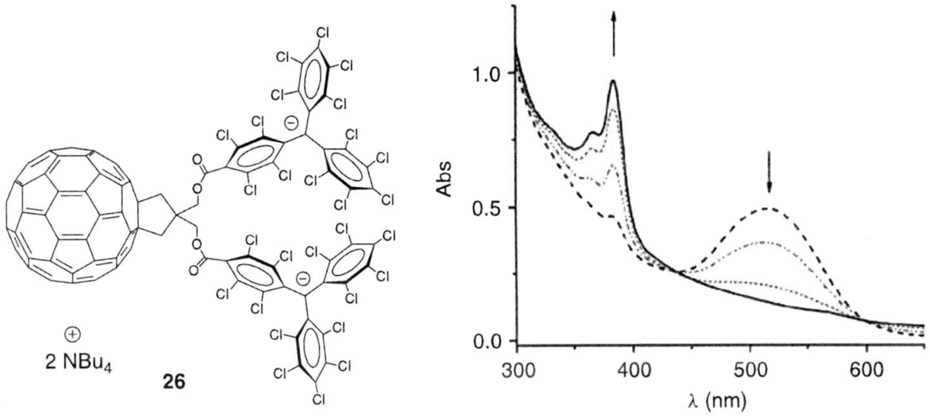

Figure 2.24 UV-Vis spectra obtained upon oxidation of bisanion **26** in dichloromethane. (Reprinted with permission from [88]. Copyright 2006 Royal Society of Chemistry.)

anion moiety occurring in the regions of 700–750 nm and 560–630 nm, respectively (Figure 2.24).[88] This result provides a new situation in D-A systems since the charge recombination takes place between the C_{60} radical anion and a neutral radical. The charge separation state was later on confirmed by the nanosecond transient absorption spectra in the visible and near-IR spectral regions. Thus, after the charge separation, a back electron transfer takes place with a relatively long lifetime of about 80 ns, which is comparable with those of other C_{60} based dyads and triads, for example C_{60}-fluorene-diphenylamine,[89] C_{60}-extended TTF[90] or C_{60}-long flexible bridge-TTF (TTF = tetrathiafulvalene).[91] Even though the properties of C_{60}-$(PTM^-)_2$ anion-based triads are not as good as the best ones found in some C_{60}-porphyrins and C_{60}-chlorine based dyads,[92] the present study provides a new opportunity for functionalized fullerenes with PTM anions. These compounds can be added to other known functional materials based on PTM radicals that work as acceptors and have permitted the development of multifunctional switchable molecular systems.[15]

2.2.2.3 *Spin transport in PTM radicals grafted on surfaces*

Molecular spintronics[93] is a new and exciting field that studies the spin transport phenomena in molecular systems and it has interesting potential applications, such as magnetic recording and memory devices.[94–96] The interest of organic molecules for the study of spin transport phenomena is based on their weak spin-orbit coupling and hyperfine interactions that allow a much longer spin coherence during the transport than in classical inorganic materials.[97] PTM radicals always show very weak spin-orbit couplings, as witnessed by their g-factors which are very close to that of a free electron[4–6,70] and, consequently, they have been proposed as possible molecules to favor the spin polarization conservation during the transport. Moreover, PTM derivatives are electroactive molecules susceptible to being reduced or oxidized to the anionic or cationic forms, allowing, in this way, the 'on–off' switching of their magnetic character.[78,15] To further investigate their electronic and spin transport properties at the molecular level, PTM radicals have been recently deposited on surfaces of different nature by means of different approaches and, in all cases, it was proved that the magnetic character of the molecule persists on the surface. Thus, PTM molecules have been grafted by physisorption on gold(111) and on graphite (highly ordered pyrolytic graphite, HOPG), and by chemisorption on gold(111) and silicon dioxide-based substrates (Figure 2.25), and the resulting surfaces have been characterized and studied by a variety of techniques.[98–102] The different natures of the substrates

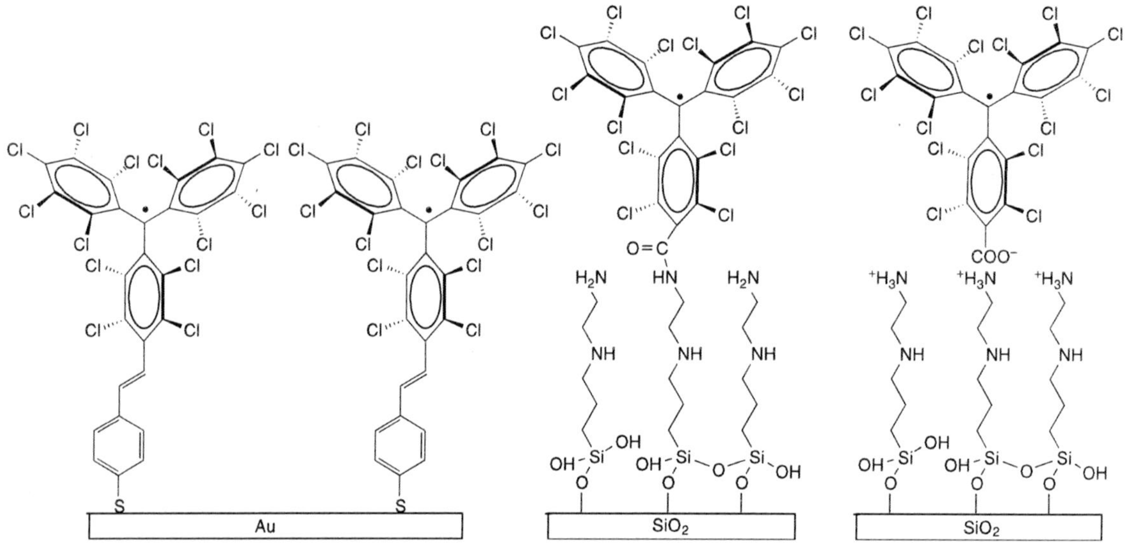

Figure 2.25 Approaches employed to anchor PTM radical derivatives on surfaces of different nature by chemisorptions, through covalent bonds or electrostatic interactions bonds. (Reprinted with permission from [93]. Copyright 2009 Royal Society of Chemistry.)

employed offer different advantages – while gold and graphite permit electroactivity and transport studies of the functionalized surfaces, glass substrates are compatible with the use of optical techniques.

The tris(trichlorophenyl)methyl radical (**27**) was chosen as a model compound to address the magnetic behaviour of a metal surface after physisorption. As confirmed by scanning tunneling microscopy (STM), molecules of **27** were adsorbed on the gold surface forming aggregates of a few molecules in ordered domains of periodic rows separated by 1.5 ± 0.1 nm, which is consistent with the molecular size of **27** (Figure 2.26). The paramagnetic character of the surface was confirmed by EPR spectroscopy, showing a sharp line with a weak magnetic anisotropic behavior, demonstrating unambiguously that the **27** radicals

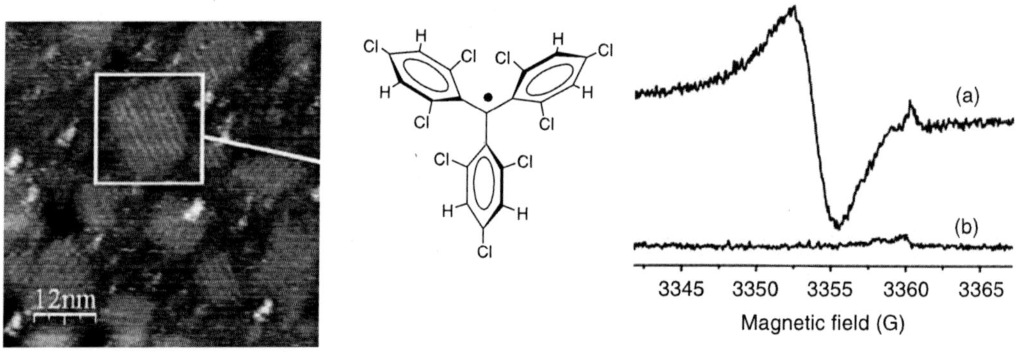

*Figure 2.26 Left: STM image (60 × 60 nm) of **27** physisorbed on gold(111). Tip sample bias 0.3 V and tunnelling current 15 pA. Right: (a) EPR spectrum of a gold surface decorated with physisorbed **27** radicals; (b) EPR spectrum of the gold substrate. (Reprinted with permission from [98]. Copyright 2009 Elsevier.)*

preserve their radical character on the surface and confirming, in addition, the small spin-orbit couplings under these conditions. These samples were also characterized by means of electron spin noise scanning tunneling microscopy (ESN-STM).[103] This technique combines the spatial resolution of the STM with the spectral resolution of EPR, and has recently been used to study small clusters of organic radicals on surfaces.[104] Analysis of the data collected using this technique leads to the conclusion that it is possible to detect the magnetic character of a decorated surface with **27** radicals.

Regarding the chemisorption of PTM radical derivatives on surfaces, different substrates have been investigated by preparing self-assembled monolayers (SAMs).[100–102] For this purpose, two different approaches were followed: (i) direct anchoring of the PTM radical on the surface, and (ii) growth of a prefunctionalized SAM which then reacts with the PTM radical derivative by the formation of either a covalent bond, a coordination bond or via electrostatic interactions (Figure 2.25). The magnetic properties were characterized by EPR in all cases, by cyclic voltammetry (CV) to study the electrochemical properties and by UV-Vis and fluorescent spectra for the optical properties. Spectroelectrochemical experiments have also been carried out proving that the PTM radical can be reversibly converted to the corresponding anion and, therefore, the functionalized surfaces can act as chemical switches in which the magnetic and optical properties can be used as read-out mechanisms (Section 2.3.2.2). Recently, it has been demonstrated that a gold surface grafted with a PTM radical derivative linked with a conjugated bond to the gold shows electron transport rates that are one order of magnitude larger than the surface grafted with the close shell counterpart of the PTM derivative. This important result offers promising perspectives for fabricating devices for molecular spintronics.[105]

2.2.3 Materials with optical properties

In the last few years, the interest to develop novel second order nonlinear optical (NLO) materials has considerably increased due to their potential application in emerging optoelectronic technologies. Traditionally, materials exhibiting second order NLO behaviors were inorganic crystals, such as lithium niobate (LiNbO3) and potassium dihydrogenphosphate (KDP). However, organic materials, such as organic crystals and polymers, have been shown to offer better nonlinear optical and physical properties, such as ultrafast response times, lower dielectric constants, better processability and a remarkable resistance to optical damage, as compared to inorganic materials.[106] Most of the efforts to discover new molecular chromophores having large NLO properties have been focused on closed shell electronic organic species. However, more recently, a large interest has been devoted to the investigation of materials having open shell electronic structures. As it has been recently pointed out by Marks *et al.*[107], species having open shell electronic states, such as organic radicals[108] or paramagnetic transition metal complexes,[109] can exhibit very large first order hyperpolarizabilities (β) in comparison with analogous closed shell systems, thanks to the presence of accessible low-lying charge transfer electronic states. In spite of this interest, only a few examples of organic open shell species showing second order hyperpolarizabilities have been described up to now, mostly due to the low stability of these species.[110]

Systems presenting NLO properties can be grouped into two categories: (i) 'push–pull' systems and (ii) octupolar systems. Donor-acceptor systems linked through a π backbone are one of the most developed 'push–pull' structures in the search for new compounds with efficient NLO responses.[111,112] The requirements for molecules exhibiting interesting NLO responses are best met by highly polarizable donor-acceptor (D-A) dyads, showing intramolecular electron transfer between the electron donating an electron withdrawing groups. Since the pioneering work of Green *et al.*,[113] who reported obtaining a ferrocene derivative with excellent NLO responses, there has also been considerable effort in using metallocenes as donor groups in NLO molecular materials.[114] In addition to 'push–pull' dipolar molecules, more recently there have appeared octupolar molecules exhibiting NLO responses.[115] These are nondipolar species whose second order NLO response is related to multidirectional charge transfer excitations, rather than to dipolar unidirectional excitations. The main advantage of octupolar molecules belonging to either planar T_d, D_{3h} or C_{3v} or non-planar D_3 or C_2 symmetry groups, when compared to dipolar 'push–pull' systems, lies on the looser structural requirements, like centrosymmetrical molecular arrangements, as well as in the improvement of the balance between transparency and second order NLO signals.[116]

In this section, the capability of PTM radicals to generate NLO optical responses, either as octupolar materials or as a component of a 'push–pull' system, is presented and the corresponding hyperpolarizability values is provided.[117]

2.2.3.1 *Nonlinear optical properties of octupolar systems*

Molecular nonlinear optical coefficients of PTM radicals **1** and **27–33** were measured, using the Hyper-Rayleigh Scattering (HRS) technique, in dichloromethane solution at room temperature irradiating with a laser light at 1064 nm. The molecular quadratic hyperpolarizability (β) values for all these radicals were calibrated against pure dichloromethane ($\beta = 0.43 \times 10^{-30}$ esu) used as solvent. Data reported in Figure 2.27 show that radicals **1** and **27–33** exhibit relatively high NLO responses with β values ranging from 118×10^{-30} to 755×10^{-30} esu. This fact is not surprising, since this family of radicals is structurally very similar to crystal violet, which also exhibits a large β value due to its octupolar symmetry.[118] In fact, all these compounds have the general formula $Ar_1Ar_2Ar_3Z$, where Z is an sp^2 hybridized carbon atom – a C^+ in crystal violet and $C^{\bullet}$ in PTM radicals – and Ar_i denotes an aromatic group with polarizable substitutents – a diethylamine (NEt_2) substituent in crystal violet and chlorine atoms in PTM

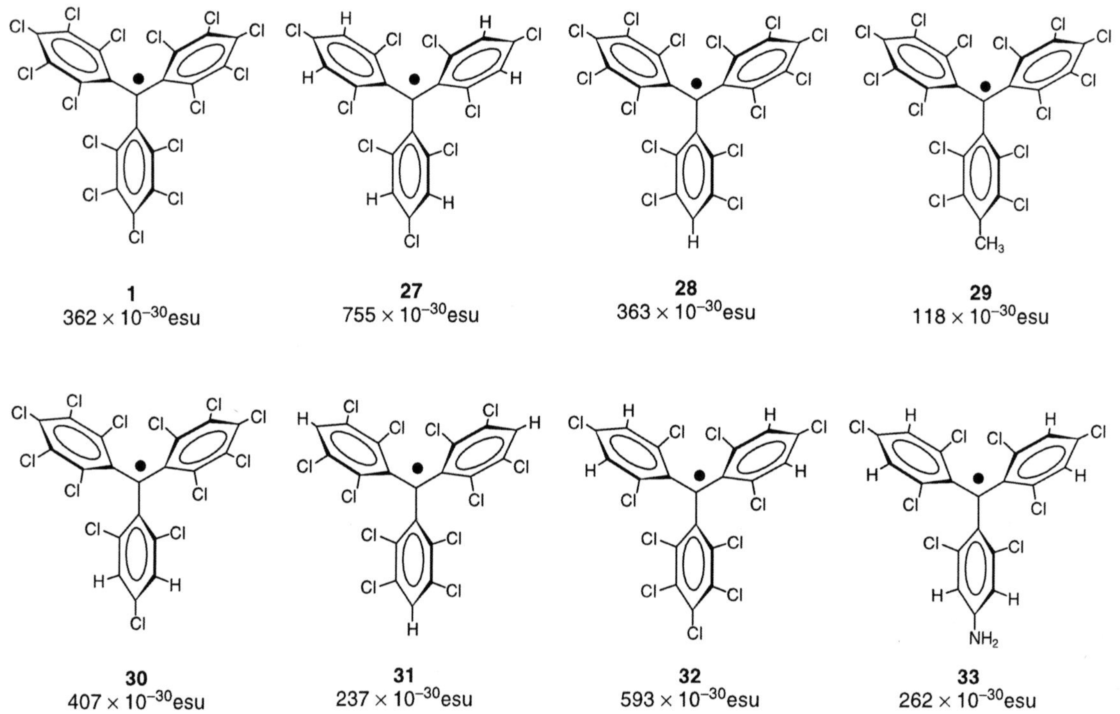

1 362×10^{-30}esu; **27** 755×10^{-30}esu; **28** 363×10^{-30}esu; **29** 118×10^{-30}esu

30 407×10^{-30}esu; **31** 237×10^{-30}esu; **32** 593×10^{-30}esu; **33** 262×10^{-30}esu

Figure 2.27 *Hyperpolarizability β values of PTM radicals **1, 27–33** measured with HRS technique.*

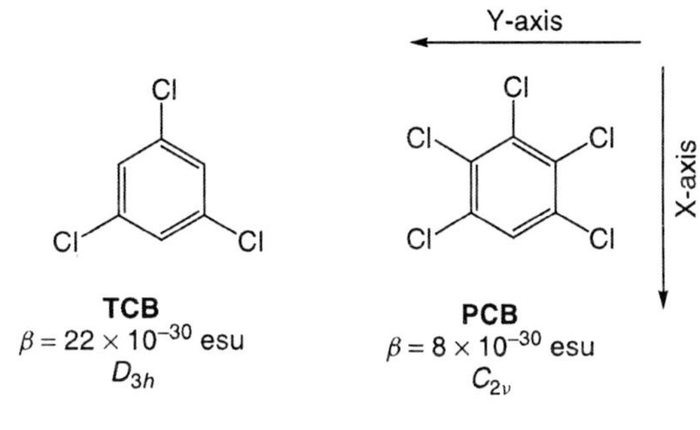

Figure 2.28 *Schematic representation of prototypical non-centrosymmetric planar D_{3h} and C_{2v} symmetry aromatic molecules: TCB, 1,3,5-trichlorobenzene and PCB, pentachlorobenzene.*

radicals. Moreover, the considerably large β values obtained point to an enhancement of the second order NLO activities according to their open shell electronic states.[119] However, it is not possible to extract an irrefutable conclusion about the influence of open shell character on NLO responses, since a closer look at radicals **1** and **27–33** reveals that they are composed of three individual subunits – thee aryl groups – each one having a NLO contribution. More precisely, 1,3,5-trichlorobenzene (TCB) and the pentachlorobenzene (PCB) compounds have also octupolar components and belong, respectively, to D_{3h} (pure octupolar system) and C_{2v} (with dipolar and octupolar contributions) group of symmetries (Figure 2.28). For this reason, PTM radicals give access to individual pure octupolar molecules, constituting individual octupolar subunits, for which the term of 'Super Octupolar' systems was coined. If an octupolar synergistic effect takes place, this would greatly contribute to enhance the NLO responses of this type of materials.

Along this line, the comparison between purely octupolar radical compounds **1** and **27** may be very interesting. Both systems have D_3 symmetry and are composed of three individual building blocks with NLO contributions based on PCB and TCB subunits, respectively. HRS measurements show that the β value of **1** is lower than that obtained for **27**. Interestingly, such values follow the trend found for the molecular hyperpolarizabilities derived from each individual octupolar building block (Figure 2.28), suggesting that a synergetic effect between individual building blocks may occur.

To give more insight to the relationship between the electronic configuration and the NLO responses of PTM radicals, cyclic volatmmetry studies of radicals **1, 27–33** in dichloromethane, using 0.1 M $n\mathrm{Bu_4NPF_6}$ as electrolyte, were performed and analysed. Interestingly, the representation of the hyperpolarizability β value in front of the reduction/oxidation potentials of radicals **1, 27–33** yields a linear dependence (Figure 2.29). Indeed, as the oxidation potential decreases, the hyperpolarizability β value increases whereas, on the contrary, the easier to reduce the radicals the lower results the β values. This confirms a direct correlation between the energy of the electronic transition and the 'Super Octupolar' NLO response.[120]

2.2.3.2 Nonlinear optical properties of 'push–pull' systems

The PTM radicals discussed so far are octupolar systems that exhibit relatively high hyperpolarizabilities. The incorporation of PTM radicals into 'push–pull' (dipolar) systems represents an alternative design strategy for NLO active systems.[121] With this aim, the hyperpolarizability β values of the series of radicals shown in Figure 2.30 have been measured using HRS techniques in dichloromethane solution at room

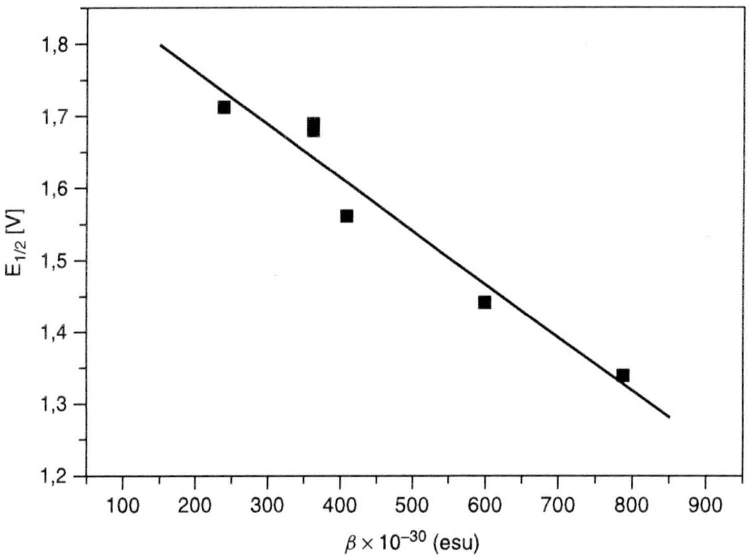

Figure 2.29 *Linear relationship between the redox potentials (oxidation potential) and the NLO response for the series of polychlorotryphenylmethyl radicals **1, 27–33**. (Reprinted with permission from [119]. Copyright 2003 Elsevier.)*

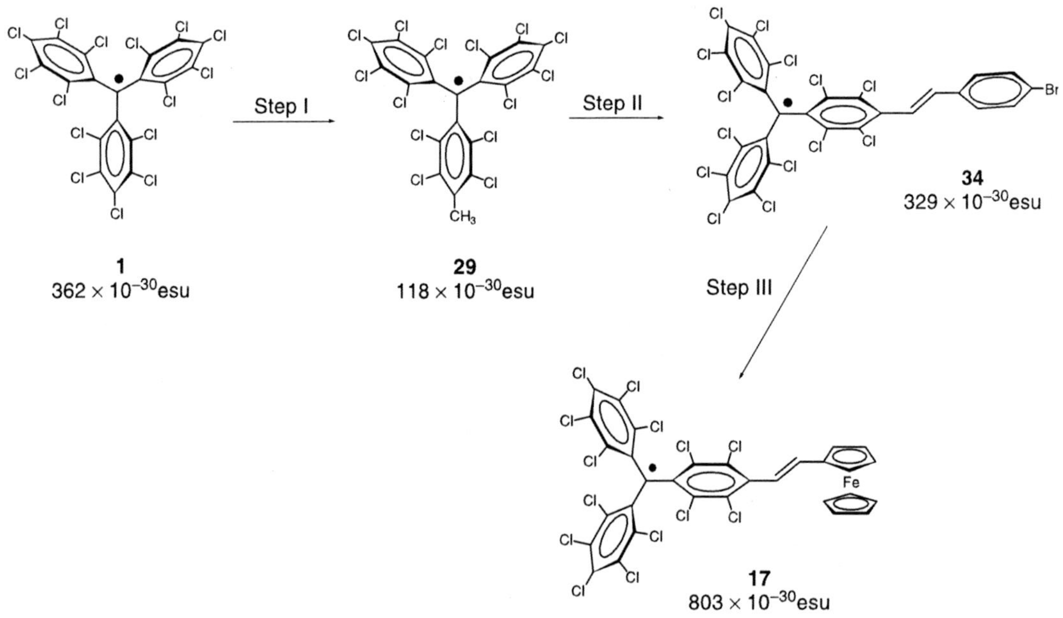

Figure 2.30 *Designing steps for increasing the NLO responses of PTM-based 'push–pull' systems.*

temperature. The starting point was a PTM radical with an ideal octupolar D_3 symmetry, whose NLO response has already been rationalized in the preceding section. Step I (breaking the symmetry) involves the introduction of a substituent directly connected to the octupolar system. This reduces the symmetry from D_3 to C_2, and, hence, a decrease of the hyperpolarizability value is expected. Thus, the introduction of a methyl group reduces the hyperpolarizability value β from $362 \cdot \times 10^{-30}$ esu for radical **1** to $118 \cdot \times 10^{-30}$ esu for radical **29**. Step II (generating a 'push–pull' system) involves the introduction of a vinylene bridge connecting the PTM to an electron donor substituent, like a *p*-bromophenyl group, and leads to the formation of a real dipolar 'push–pull' compound giving to an increase of NLO response up to $329 \cdot \times 10^{-30}$ esu. Finally, step III (tuning the NLO properties) involves the replacement of the electron donor group by a stronger one, such as a ferrocenyl, which increases the NLO response to $803 \cdot \times 10^{-30}$ esu.

2.3 Multifunctional switchable molecular materials based on PTM radicals

As already shown in the previous sections, PTM radicals are advantageous building blocks of functional materials exhibiting magnetic, electronic and optical properties. The intrinsic redox properties of PTM radicals together with the increasing interest for molecular-scale switches made interesting the exploration of switching systems based on PTM radicals. The possibility to use other external stimuli, such as light, has also been considered for switching of their functionalities.

2.3.1 Photo switchable molecular systems

Currently, there is a special interest in obtaining photomagnetic materials whose magnetic properties are controlled at will by means of external irradiation with light.[122–124] Such a control over the magnetic properties by an optical stimuli may have applications in magneto-optical devices. Different examples of photo-induced magnetization changes in purely organic materials have been described. For instance, Irie and Matsuda described a photochromic system that interconverts reversibly between a singlet and triplet states.[125] Latter on, Iwamura *et al.* reported a diradical bearing two stable nitroxide radicals connected through an isomerizable bridge in which the magnetic exchange coupling can be externally modified.[126] There are very few examples of supramolecular photomagnetic materials.[127] The construction of such systems requires that the structural subunits exhibit non-covalent interactions suitable to be controlled and changed in a predictable manner and at the same time transmit the magnetic interactions properly. Among them, hydrogen bonding has emerged as particularly useful and efficient molecular engineering tools, since they can control the structure of the interacting subunits. Switchable magnetic systems based on hydrogen bonded supramolecular entities whose properties may be systematically controlled by external stimuli are limited to date to only one example.[128] This compound is the radical **35**, which has a PTM subunit connected to a ferrocenyl subunit through a photoisomerizable conjugated imine bridge. Specifically, imine derivatives exhibit a *trans/cis* photoisomerization in which one of the isomers – the *cis* one – may be prone to intermolecular hydrogen bonds (Figure 2.31).

Indeed, the *trans* isomer of **35** exists in solution as a monomeric species while the *cis* isomer dimerizes forming a thermodynamically stabilized hydrogen bonded diradical species. In both species, the donor character of the ferrocene unit and the acceptor character of the radical unit ensure a significant spin delocalization over the imino bridge, so that once the supramolecular species is formed, a magnetic exchange coupling between the spin is established. Thus, the *trans* isomer (existing as two independent doublets) interconverts by irradiation into the dimeric species of the *cis* isomer, in which relatively strong antiferromagnetic interactions leading to a singlet ground state are developed.

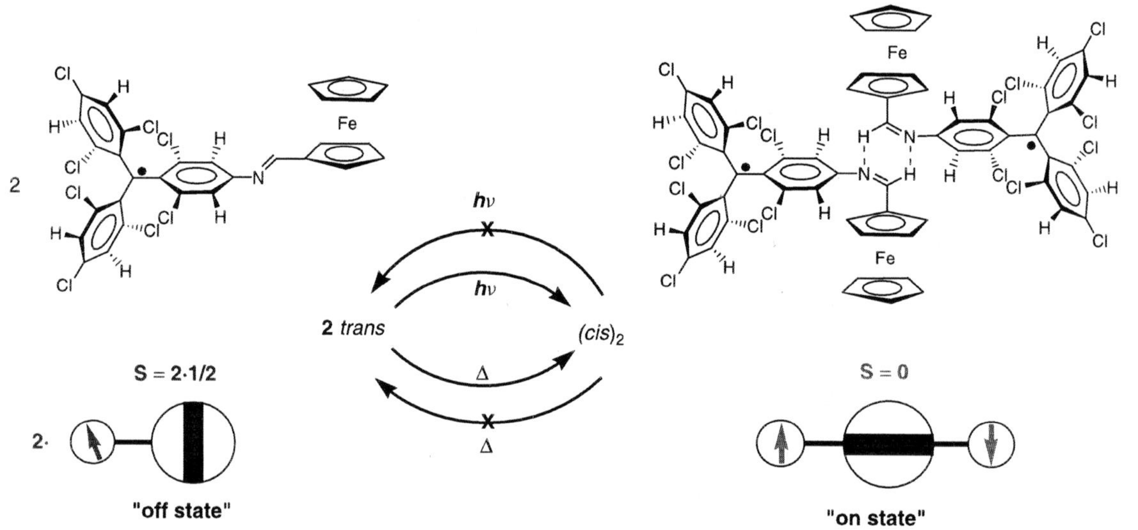

Figure 2.31 *Possible isomerizations of the PTM Schiff base radical **35**. (Reprinted with permission from [128a]. Copyright 2001 Wiley-VCH Verlag GmbH & Co. KGaA.)*

2.3.2 Redox switchable molecular systems

2.3.2.1 Switching multifunctional PTM radicals in solution

Presented here is a multifunctional redox switchable molecular system showing simultaneous changes of three different outputs – the linear optical, nonlinear optical and magnetic properties – which is based on the rich electrochemical behaviour of radical **18** (Scheme 2.6), previously described and for which the

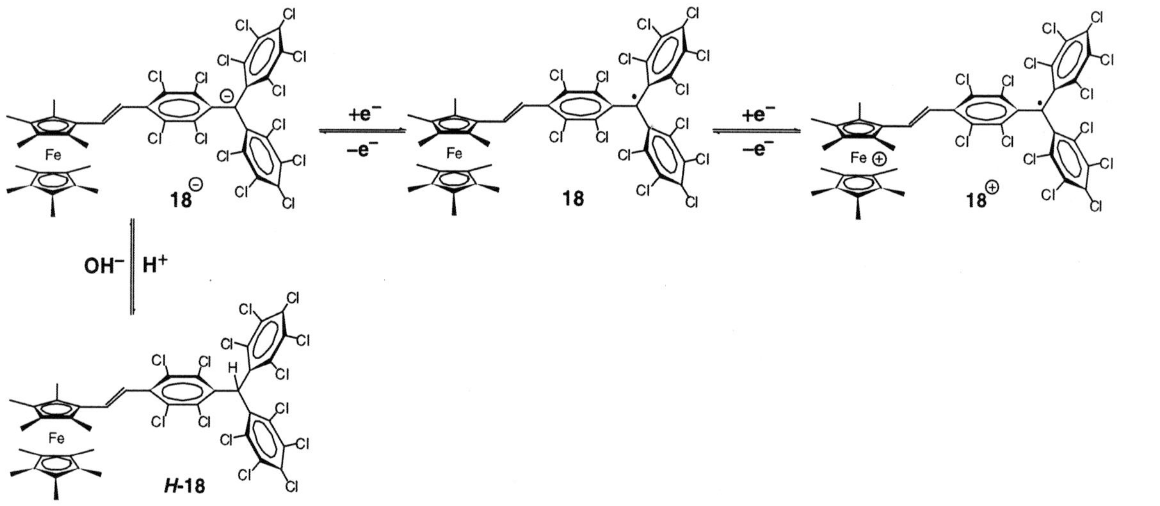

Scheme 2.6 The four states of the molecular switch: **18**$^-$, **18**, **18**$^+$ and *H*-**18**

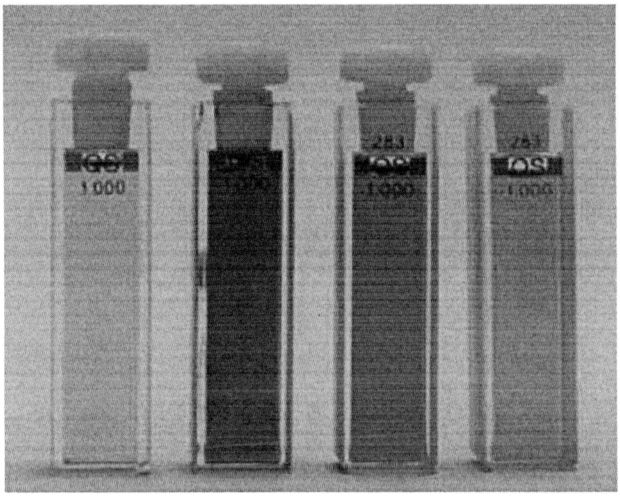

Figure 2.32 Colors shown by dichloromethane solutions of the different oxidation states of the radical **18**. From left to right: H-**18**, **[K(18crown-6)]$^+$ 18$^-$**, **18** and **18$^+$BF$_4$$^-$**. (Adapted with permission from [15]. Copyright Wiley-VCH Verlag GmbH & Co. KGaA.) A full-colour version of this figure appears in the Colour Plate section of this book.

PTM unit provides a source of magnetism and shows a NLO response (Section 2.2.3.2).[15] The reduction of **18** into its diamagnetic anionic form **18$^-$**, its oxidation into the ferrocenium radical derivative **18$^+$**, and the protonation of the carbanion species **18$^-$** to its conjugate acid form H-**18** provided four states to this switchable molecular system. Thus radical **18** can be interconverted between four forms with distinct physical properties, such as magnetism, color, NLO response, and so on.

In accordance with the different electronic structures, the four species **18, 18$^+$**, **18$^-$** and H-**18** show strikingly distinct optical properties (Figure 2.32). Thus, radical **18** shows the characteristic absorptions of conjugated PTM radicals and a broad IVT band, centered at 1520 nm, associated with an IET phenomenon (Section 2.2.2.2). The ferrocenium radical derivative **18$^+$** exhibits the typical band of a ferrocenium derivative with a significant enhancement of the intensity of the radical band at 385 nm, which accounts for the deep yellow color of this species (Figure 2.32). On other hand, the IVT band is absent in this ferrocenium radical derivative because of the lack of electron donor character of the organometallic unit when it is oxidized. Carbanion **18$^-$** shows an intense absorption at 534 nm, characteristic of PTM anions, which is responsible of its intense wine red colour. At the same time, the broad IVT band vanishes in **18$^-$** confirming the absence of the electron acceptor capability in the reduced PTM unit. Finally, the lack of the radical character in the H-**18** compound accounts for its pink colour, which is similar to those of non-amethylated ferrocene derivatives. Similarly to the striking different optical absorptions of **18, 18$^+$**, **18$^-$**, such species also show distinct NLO responses. The dynamic hyperpolarizabilities of these species, measured by hyper-Rayleigh scattering experiments with a laser of 800 nm, also gave remarkable differences. Thus, as suggested by its intense IVT band, the radical **18** gives a large NLO response with a β value of 545 ($\pm$30)$\cdot$ $\times$ 10^{-30} esu. This value is reduced almost ninefold to 66 ($\pm$7)$\cdot$ $\times$ 10^{-30} esu for **18$^+$**, and even more for the carbanion **18$^-$**, which has a β value of 30 ($\pm$3)$\cdot$ $\times$ 10^{-30} esu.

Electrochemical interconversions between the three **18, 18$^+$**, **18$^-$** species are completely reversible, as shown in Figure 2.33, by following the changes of their optical responses of their solutions. This result clearly demonstrates the switching capabilities of this multifunctional molecular device.

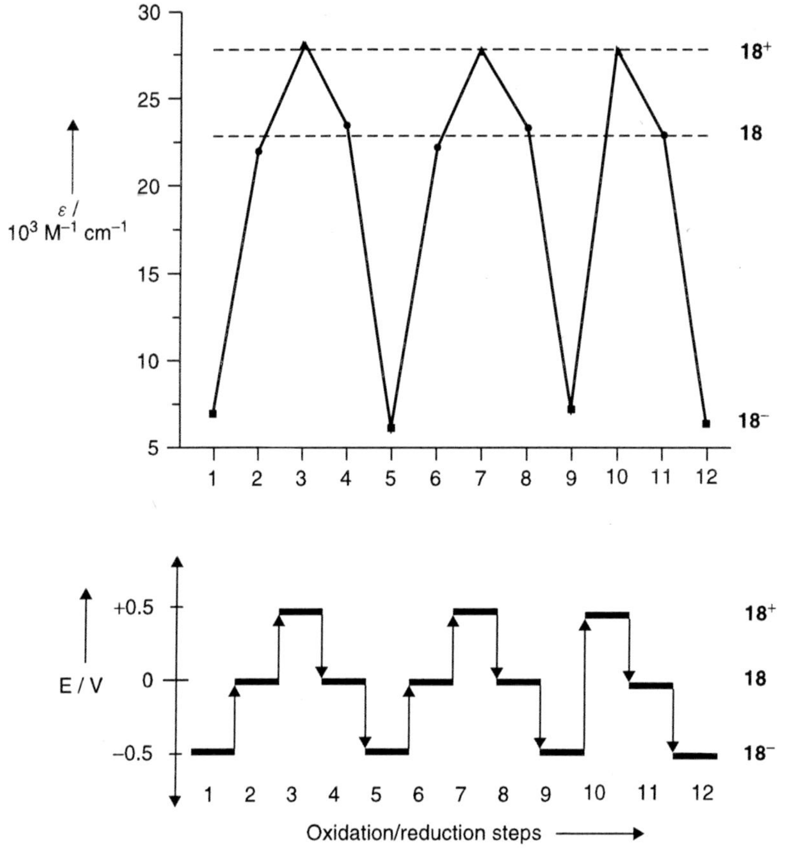

Figure 2.33 *Step-wise oxidations and reductions carried out with **18** in THF with a chronoamperometric technique monitoring the changes in the visible spectrum. Top: Changes observed at a wavelength of 385 nm where **18** (•) and **18**$^+$ (▲) exhibit the strongest absorption and **18**$^-$ (■) shows a very weak absorption. Bottom: Fixed potentials E used in the different steps of cyclic redox experiments. (Adapted with permission from [15]. Copyright Wiley-VCH Verlag GmbH & Co. KGaA.)*

The magnetic properties of the three studied species are also different. Thus, while the salt [K([18]crown-6)]$^+$**18**$^-$ is diamagnetic, the magnetic susceptibility between 4–300 K of compounds **18** and **18**$^+$**BF**$_4^-$ showed quasi-ideal paramagnetic behaviors with effective magnetic moments at 300 K of 1.72 and 2.50μ_B, respectively, as expected for systems with S = ½ and S = 2 × ½ units. In addition, the value of 2.50μ_B found for **18**$^+$**BF**$_4^-$ indicates that the magnetic interaction between the open shell ferrocenium moiety and the PTM radical unit is very weak with both spins being apparently uncoupled above 4 K. Analogous differences in the behavior of the three complexes were found in solution with EPR spectroscopy since **18**$^-$ is EPR silent while **18** and **18**$^+$ show different complex signals.

2.3.2.2 Switching multifunctional PTM radicals on surfaces

In order to move one step forward and use PTM-based multifunctional switchable systems for applications like molecular memory devices,[129] it is crucial to control their deposition on surfaces and demonstrate that they can be reversibly (magnetically, optically or electrochemically) switched between two (or more) stable states when grafted on the surfaces. In addition, it is required that such states exhibit different responses in order to be able to read the status of the switch. A strategy to scale down to molecular level memory devices is focused on the fabrication of charge storage devices by substrate immobilization and patterning of molecules that can be reversibly oxidized and reduced.[130] A simple and versatile technique to address all these molecular building blocks on surfaces is the preparation of self-assembled monolayers (SAMs),[131,132] which allows the functionalization of surfaces with a layer of molecules, two-dimensionally organized, that gives to the substrate new properties governed by the inherent characteristics of the molecules grafted on it.

There are extremely few examples of self-assembled monolayers based on organic radicals in the literature.[133] As described in Section 2.2.2.3, silicon oxide and gold surfaces have been grafted with PTM radicals making use of either covalent or non-covalent interactions.[85] The magnetic properties of the functionalized surfaces were characterized by EPR in all cases, by cyclic voltammetry (CV) to study the electrochemical properties and by UV-Vis and fluorescent spectra for the optical properties. The optical characterization corroborated the chemical nature of the monolayer, exhibiting an absorption band at 382 nm and fluorescent emission band at 690 nm, which are characteristic of the radical character of the grafted PTM molecules.[134] The EPR spectrum was also recorded to demonstrate the radical character of the PTM functionalized surfaces. The EPR showed a signal at $g = 2.0024$, with a line width of 5.2 Gauss, which is close to that observed for many other PTM radicals. Electrochemical experiments were also carried out proving that the grafted PTM radicals can be reversibly converted to the corresponding anion and, therefore, the functionalized surfaces act as chemical switches in which the magnetic and optical properties can be used as read-out mechanisms (Figure 2.34).

More interesting with respect to potential device applications is the possibility to locally address the PTM radical molecules on the surfaces in order to fabricate multifunctional switchable patterned surfaces. Thus, both the covalently and the electrostatic PTM bonded surfaces have been patterned by microcontact printing (Figure 2.35) and visualized by laser scanning confocal microscopy or fluorescence microscopy due to the fluorescent nature of the PTM molecules.

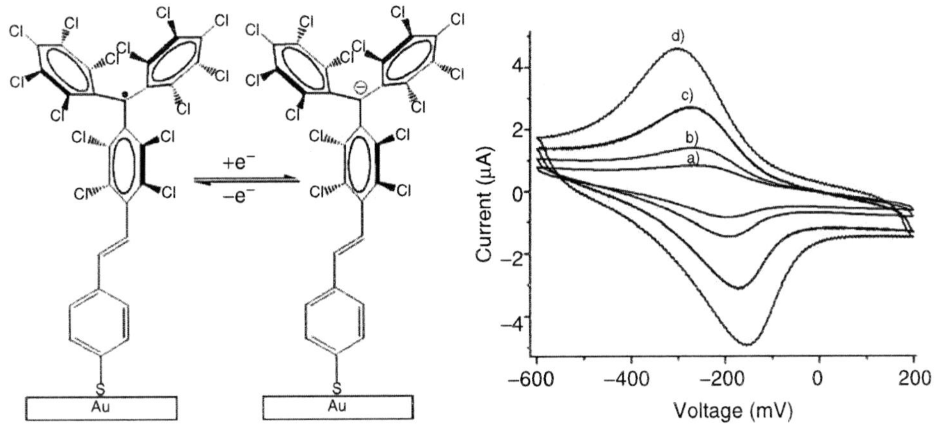

Figure 2.34 *Left: Scheme of the redox process on a gold surface with covalently bonded PTM radicals. Right: Cyclic voltammograms in dichloromethane, with 0.1 M n-Bu$_4$NPF$_6$ (vs Ag/AgCl) at different scan rates: a) 50, b) 100, c) 300 and d) 400 mV s^{-1}. (Reprinted with permission from [100]. Copyright 2008 American Chemical Society.)*

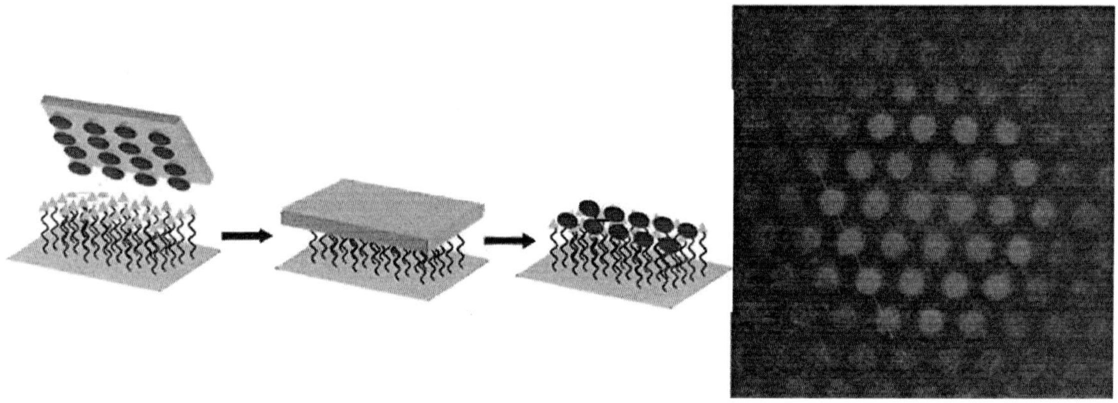

Figure 2.35 Left: Schematic representation of the microcontact printing of a functionalized PTM radical on an adhesive pre-functionalized SAM. Right: Confocal microscopy image ($\lambda_{exc} = 488$ nm) of a glass surface patterned with a PTM radical derivative. The diameter of the fluorescent dots is 5 μm. (Reprinted with permission from [101]. Copyright 2007 Wiley-VCH Verlag GmbH & Co. KGaA.) A full-colour version of this figure appears in the Colour Plate section of this book.

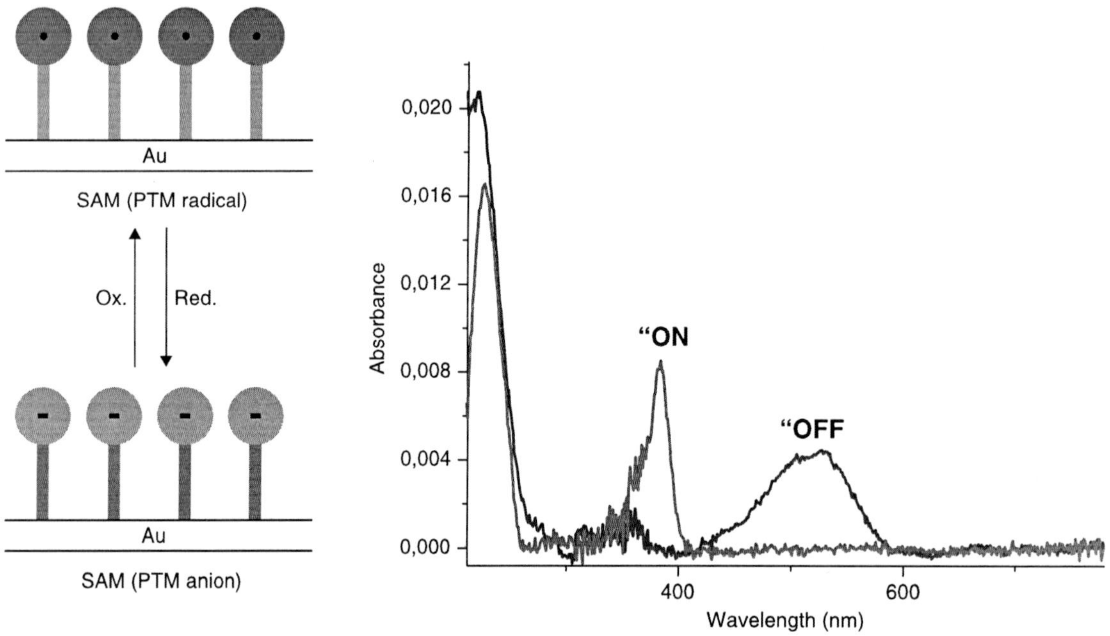

Figure 2.36 Left: Representation of the electroactive PTM SAM. Right: Absorbance spectra of the PTM anion SAM on silicon (OFF state, black line) before and after oxidation to the PTM radical SAM (ON state, red line). (Reprinted with permission from [101]. Copyright 2007 Wiley-VCH Verlag GmbH & Co. KGaA.) A full-colour version of this figure appears in the Colour Plate section of this book.

Spectroelectrochemical experiments have been carried out proving that the PTM radical can be reversibly reduced to the corresponding anion and, therefore, the functionalized substrates act as a surface chemical switch in which the magnetic and optical properties are used as read-out mechanisms. Figure 2.36 shows the UV-Vis spectrum corresponding to the reduction–oxidation process of electroactive PTM self-assembled monolayers. Upon reduction to the PTM anion, the characteristic PTM radical absorption band at 384 nm disappears, while a band centered at 525 nm is observed. This system behaves thus as a real bistable chemical switch, since it is possible to interconvert the self-assembled monolayer between two states which exhibit different properties: an OFF state associated with the PTM anion SAM (non-fluorescent and diamagnetic) and an ON state corresponding to the PTM radical SAM (fluorescent and paramagnetic).[92]

2.4 Conclusions

In this chapter the chemical and physical properties of PTM radicals have been described, mainly from the molecular materials point of view. The main focus is on multifunctionality, giving different insights to the magnetic, electronic and optical properties of PTM-based materials with representative examples for each of them.

Regarding the magnetic properties of PTMs, the magnetism of high-spin PTM radicals have been described in detail as have those of extended systems. The strengths of intramolecular ferromagnetic exchange couplings of PTM radicals through the *m*-phenylene unit are strong and, consequently, such polyradicals are robust high-spin molecules showing outstanding chemical and thermal stabilities. Metallocenes were also shown to act as effective magnetic couplers that effectively transmit ferromagnetic interactions between PTM radicals. Investigation of the transmission of magnetic interactions through hydrogen bonds has also been carried out with PTM radicals with carboxylic acid groups. This fact, together with the ability of hydrogen bonds to transmit magnetic interactions and link molecules with hydrogen bonding acceptor and donor groups, made possible the obtaining of magnetic extended systems based on PTM radicals functionalized with several carboxylate/carboxylic groups. These magnetic extended systems resulted in two new families of open framework materials with either a metal–organic or purely organic nature, named metal–organic radical open frameworks (MOROFs) or purely organic radical open frameworks (POROFs), respectively, with porous structures and interesting magnetic properties.

Regarding the electronic properties of PTMs, the most relevant results were obtained in the study of IET phenomena observed in D-A dyads, where the PTM radical unit acts as a strong electron acceptor unit due to its low reduction potential. In addition, M-V compounds derived from PTM radicals in which one of both PTM units is reduced to the anionic form enabling the generation of the mixed-valence species have also been discussed. Moreover PTM derivatives are electroactive molecules susceptible to reduction or oxidation to the anionic or cationic forms, allowing in this way the 'on–off' switching of their magnetic character.

Going one step further, the possibility to functionalize surfaces with PTM molecules as spin-containing units towards their use in molecular spintronics has also been described. PTM radicals show very weak spin-orbit couplings and, consequently, they have been proposed as possible molecules to favor the spin polarization conservation during the electronic or spin transport. In this line, PTM radicals have been recently deposited on surfaces of a different nature, proving for all cases that the magnetic character of the molecule persist on the surface.

Regarding the optical properties, the capability of PTM radicals to generate NLO optical responses, either as octupolar materials or as component of a 'push–pull' system, has been presented and the corresponding hyperpolarizability values have been provided.

Finally, the intrinsic redox properties of PTM radicals together with the increasing interest for molecular-scale switches made interesting the exploration of switching systems based on PTM radicals. Specifically, it has been demonstrated that PTM open shell D-A systems can exist in three stable oxidation states exhibiting different linear and nonlinear optical responses as well as distinct magnetic properties. The possibility to use other external stimuli, such as light, has also been considered for switching of their functionalities. The chemical flexibility of the PTM radicals demonstrates, too, the potential of preparing self-assembled multifunctional molecular switching devices on surfaces.

References

1. M. Ballester, *Acc. Chem. Res.*, 18, 380–387 (1985).
2. J. Veciana, and M. I. Crespo, *Angew. Chem. Int. Ed.*, 30, 74–76 (1991).
3. R. S. Cahn, C. Ingold, and V. Prelog, *Angew. Chem. Int. Ed. Engl.*, 5, 385–415 (1966).
4. N. Ventosa, D. Ruiz, C. Rovira, and J. Veciana, *Mol. Crys. Liq. Cryst.*, 232, 333–342 (1993).
5. M. Ballester, *Adv. Phy. Org.Chem.*, 25, 267–445 (1989).
6. O. Armet, J. Veciana, C. Rovira, *et al.*, *J. Phys. Chem.*, 91, 5608–5616 (1987).
7. M. Ballester, I. Pascual, J. Riera, and J. Castañer, *J. Org. Chem*, 56, 212–217 (1991).
8. V. M. Domingo, and J. Castañer, *Chem. Comun.*, 895–896 (1995).
9. J. Rius, C. Miravitlles, E. Molins, *et al.*, *Mol. Crys. Liq. Crys.*, 187, 155–163 (1990).
10. M. Ballester, J. Riera, J. Castañer, *et al.*, *J. Am. Chem. Soc.*, 93, 2215–2225 (1971).
11. (a) M. Ballester, J. Riera J. Castañer, *et al.*, *Synthesis*, 64–66 (1986). (b) M. Ballester, J. Veciana, J. Riera, *et al.*, *J. Org. Chem.*, 51, 2468–2472 (1986).
12. M. Ballester, J. Riera, J. Castañer, *et al.*, *J. Org. Chem.*, 48, 3716–3720 (1983).
13. J. Bonvoisin, J. P. Launay, C. Rovira, and J. Veciana, *Angew. Chem. Int. Ed. Engl.*, 33, 2106–2109 (1994). (b) J. Sedó, D. Ruiz, J. Vidal-Gancedo, *et al.*, *Adv. Mater.*, 8, 745–748 (1996).
14. I. Ratera, D. Ruiz-Molina, J. Vidal-Gancedo, *et al.*, *Angew. Chem. Int. Ed.*, 40, 919–922 (2001).
15. C. Sporer, I. Ratera, D. Ruiz-Molina, *et al.*, *Angew. Chem. Int. Ed.*, 43, 5266–5268 (2004).
16. (a) I. Ratera, D. Ruiz-Molina, F. Renz, *et al.*, *J. Am. Chem. Soc.*, 125, 1462–1463 (2003). (b) G. D'Avino, L. Grisanti, J. Guasch, *et al.*, *J. Am. Chem. Soc.*, 130, 12064–12072 (2008).
17. I. Ratera, S. Marcen, S. Montant, *et al.*, *Chem. Phys. Let.*, 363, 245–251 (2002).
18. (a) J. Veciana, C. Rovira, M. I. Crespo, *et al.*, *J. Am. Chem. Soc.*, 113, 2552–2561 (1991). (b) J. Veciana, C. Rovira, N. Ventosa, *et al.*, *J. Am. Chem. Soc.*, 115, 57–64 (1993).
19. D. Ruiz-Molina, J. Veciana, F. Palacio, and C. Rovira, *J. Org. Chem.*, 62, 9009–9017 (1997).
20. D. Maspoch, N. Domingo, D. Ruiz-Molina, *et al.*, *C. R. Chimie*, 8, 1213–1225 (2005).
21. N. Crivillers, M. Mas-Torrent, J. Vidal-Gancedo, *et al.*, *J. Am. Chem. Soc.*, 130, 5499–5506 (2008).
22. M. Kini, and H. H. Wang, Proc. NATO Advanced Research Workshop on Molecular Low Dimensional and Nanostructured Materials for Advanced Applications, Poznan, Poland, September 1-5, 2001, A. Graja, B. R. Bulka, and F. Kajzar (eds), Kluwer Academic, pp. 139–148 (2002).
23. (a) T. J. Marks, *Angew. Chem. Int. Ed. Engl.*, 29, 857–866 (1990). (b) J. S. Miller, *Adv. Mater.*, 2, 98–99 (1990). (c) J. S. Miller, *Adv. Mater.*, 5, 587–589 (1993). (d) J. S. Miller, *Adv. Mater.*, 5, 671–676 (1993). (e) A. Kraft, A. C. Grimsdale, and A. B. Holmes, *Angew. Chem. Int. Ed. Engl.*, 37, 402–428 (1998).
24. *Molecular Conductors*, special issue of *J. Mater. Chem.*, 5 (10) (1995). (b) P. Cassoux, and J. S. Miller, in *Chemistry of Advanced Materials: An Overview* (eds L. V. Interrante, and M. J. Hampden-Smith), Wiley-VCH Verlag GmbH, New York, p. 19–72 (1998).
25. *Optical Nonlinearities in Chemistry*, special issue of *Chem. Rev.*, 94 (1) 94–278 (1994).
26. (a) O. Kahn, *Molecular Magnetism*; Wiley-VCH Verlag GmbH, Weinheim (1993). (b) J. S. Miller, and A. J. Epstein, *Angew. Chem. Int. Ed. Engl.*, 33, 385–415 (1994).
27. (a) L. Capes, J. F. Létard, and O. Kahn, *Chem. Eur. J.*, 6 (12), 2246–2255 (2000). (b) S. Bénard, A. Léaustic, *et al.*, *Chem. Mater.*, 13, 3709–3716 (2001). (c) E. Coronado, J. R. Galan-mascarós, and C. Gomez-Garcia, *J. Chem. Soc., Chem. Commun.*, 205–210 (2000).
28. (a) P. Lacroix, *Chem. Mater.*, 13, 3495–3506 (2001). (b) P. G. Lacroix, and K. Nakanti, *Adv. Mater.*, 9, 1105–1108 (1997). (c) R. Andreu, I. Malfant, P. G. Lacroix, *et al.*, *Chem. Mater.*, 11, 840–848 (1999).

29. (a) B. L. Feringa, Molecular Switches, Wiley-VCH Verlag GmbH, Weinheim (2001). (b) J.-M. Lehn, Supramolecular Chemistry, Wiley-VCH Verlag GmbH, Weinheim (1995). (c) A. P. de Silva, and N. D. McClenaghan, *Chem. Eur. J.*, 10, 574–586 (2004). (d) R. L. Carroll, and C. B. Gorman, *Angew. Chem. Int. Ed.*, 41, 4378–4400 (2002).
30. (a) *Photochromism, Molecules and Systems* (Eds H. Dürr, and H. Bouas-Laurent), Revised Edition, Elsevier, Amsterdam (2003). (b) Special issue Photochromism: Memories and Switches (Ed. M. Irie), *Chem. Rev.*, 100, 1683–1684 (2000).
31. A. P. de Silva, H. Q. Gunaratne, T. Gunnlaugsson, *et al.*, *Chem. Rev.*, 97, 1515–1566 (1997).
32. (a) B. J. Coe, *Chem. Eur. J.*, 5, 2464–2471 (1999). (b) M. Malaun, Z. R. Reeves, R. L. Paul, *et al.*, *Chem. Commun.*, 49–50 (2001). (c) F. Paul, K. Costuas, I. Ledoux, *et al.*, *Organometallics*, 21, 5229–5235 (2002).
33. (a) F. Renz, H. Oshio, V. Ksenofontov, *et al.*, *Angew. Chem. Int. Ed.*, 39, 3699–3700 (2000). (b) P. Gütlich, Y. Garcpa, and T. Woike, *Coord. Chem. Rev.*, 839, 219–221 (2001). (c) O. Sato, *Acc. Chem. Res.*, 36, 692–700 (2003).
34. (a) M. E. Itkis, X. Chi, A. W. Cordes, and R. C. Haddon, *Science*, 296, 1443–1445 (2002). (b) J. S. Miller, *Angew. Chem. Int. Ed.*, 42, 27–29 (2003).
35. For a collection of reviews on high-spin organic molecules see: (a) J. Veciana, and H. Iwamura, *MRS Bull.*, 36, 41–51 (2000). (b) P. M. Lathi, Magnetic Properties of Organic Materials, Marcel Dekker, New York (1999).
36. (a) K. Matsuda, N. Nakamura, K. Takahashi, *et al.*, *J. Am. Chem. Soc.*, 117, 5550–5560 (1995).
37. (a) N. Ventosa, D. Ruiz, J. Sedo, *et al.*, *Chem. Eur. J.* 12, 3533–3546 (1999). (b) J. Sedo, N. Ventosa, D. Ruiz-Molina, *et al.*, *Angew. Chem. Int. Ed.*, 37, 330–333 (1998). (c) A. Rajca, S. Rajca, and J. Wongsriratanakul, *J. Am. Chem. Soc.*, 121, 6308-6309 (1999).
38. R. J. Bushby, D. R. McGill, K. M. Ng, and N. J. Taylor, *J. Chem. Soc. Perkin Trans.*, 2, 1405–1414 (1997).
39. F. Kanno, K. Inoue, N. Koga, and H. Iwamura, *J. Phys. Chem.*, 97, 13267–13272 (1993).
40. (a) J. Sedó, N. Ventosa, A. Molins, *et al.*, *J. Org. Chem.*, 66, 1567–1578 and 1579–1589 (2001).
41. J. S. Miller, and J. A. Epstein, *Angew. Chem. Int. Ed. Engl.*, 106, 385–415 (1994).
42. O. Elsner, D. Ruiz-Molina, J. Vidal-Gancedo, *et al.*, Chem.Commun., 579–580 (1999).
43. O. Elsner, D. Ruiz-Molina, I. Ratera, *et al.*, *J. Organomet. Chem.*, 637, 251–257 (2001).
44. D. Maspoch, L. Catala, P. Gerbier, *et al.*, *Chem. Eur. J.*, 8, 16, 3635–3645 (2002).
45. H. Iwamura, K. Inoue and T. Hayamizu, *Pure Appl. Chem.*, 68, 243–252 (1996).
46. D. Maspoch, D. Ruiz-Molina, K. Wurst, *et al.*, *Dalton Trans.*, 1073–1082 (2004).
47. D. Maspoch, D. Ruiz-Molina, K. Wurst, *et al.*, *Chem. Comm.*, 2958–2959 (2002).
48. D. Maspoch, D. Ruiz-Molina, K. Wurst, *et al.*, *Polyhedron*, 22, 1929–1934 (2003).
49. D. Maspoch, N. Domingo, D. Ruiz-Molina, *et al.*, *Chem. Commun.*, 5035–5037 (2005).
50. N. Roques, S. Perruchas, D. Maspoch, *et al.*, *Inorg. Chimica Acta*, 360, 3861–3869 (2007).
51. X. Ribas, D. Maspoch, K. Wurst, *et al.*, *Inorg. Chem.*, 45, 14, 5383–5392 (2006).
52. D. Maspoch, N. Domingo, D. Ruiz-Molina, *et al.*, *Inorg. Chem.*, 46, 5, 1627–1633 (2007).
53. D. Maspoch, D. Ruiz-Molina, and J. Veciana, *Chem. Soc. Rev.*, 36, 770–818 (2007).
54. (a) F. Lloret, M. Julve, R. Ruiz, *et al.*, *Inorg. Chem.*, 32, 27–31 (1993). (b) K. Turner, O. Kahn and L. Rabardel, *J. Am. Chem. Soc.*, 118, 6428–6432. (1996). (c) O. Kahn, J. Larionova and J. V. Yakhmi, *Chem. Eur. J.*, 5, 3443–3449 (1999).
55. (a) M. Eddaoudi, J. Kim, N. Rosi, *et al.*, *Science*, 295, 469–472 (2002). (b) H. Li, M. Eddaoudi, M. O'Keeffe and O. M. Yaghi, *Nature*, 402, 276–279 (1999).
56. D. Maspoch, D. Ruiz-Molina, and J. Veciana, *J. Mat. Chem.*, 14, 2713–2723 (2004).
57. A. Caneschi, D. Gatteschi, R. Sessoli and P. Rey, *Acc. Chem. Res.*, 22, 392–398 (1989).
58. (a) O. Kahn, *Molecular Magnetism*, Wiley-VCH Verlag GmbH, New York, (1993). (b) J. S. Miller and M. Drillon, *Magnetism: Molecules to Materials II*, Wiley-VCH Verlag GmbH, Weinheim (2001). (c) M. M. Turnbull, T. Sugimoto and L. K. Thompson, *Molecule-based Magnetic Materials: Theory, Techniques and Applications*, ACS Publications, Washington DC (1996). (d) K. Itoh and M. Kinoshita, *Molecular Magnetism: New Magnetic Materials*, Kodansha, Tokyo (2000). (e) J. Veciana, C. Rovira and D. B. Amabilino, *Supramolecular Engineering of Synthetic Metallic Materials: Conductors and Magnets*, Kluwer Academic Publishers, Dordrecht, The Netherlands (1999).

59. D. Maspoch, D. Ruiz-Molina, K. Wurst, *et al.*, *Nat. Mater.*, 2, 190–195 (2003).
60. D. Maspoch, D. Ruiz-Molina, K. Wurst, *et al.*, *Chem. Commun.*, 1164–1165 (2004).
61. N. Roques, D. Maspoch, F. Luix, A. *et al.*, *J. Mat. Chem.*, 18, 98–108 (2008).
62. D. Maspoch, L. Catala, P. Gerbier, *et al.*, *Chem. Eur. J.*, 8, 3635–3645 (2002).
63. (a) D. Maspoch, N. Domingo, D. Ruiz-Molina, *et al.*, *J. Am. Chem. Soc.*, 126, 730–731 (2004). (b) D. Maspoch, N. Domingo, D. Ruiz-Molina, *et al.*, *Angew. Chem. Int. Ed.*, 43, 1828–1832 (2004). (c) N. Roques, D. Maspoch, A. Datcu, *et al.*, *Polyhedron*, 26, 1934–1948 (2007). (d) D. Maspoch, N. Domingo, N. Roques, *et al.*, *Chem. Eur. J.*, 13, 8153–8163 (2007).
64. N. Roques, D. Maspoch, K. Wurst, *et al.*, *Chem. Eur. J.*, 12, 9238–9254 (2006).
65. (a) R. Metzger, *Acc. Chem. Res.*, 32, 950–957 (1999). (b) J. M. Tour, *Acc. Chem. Res.*, 33, 791–804 (2000). (c) R. Lloyd and C. B. Gorman, *Angew. Chem. Int. Ed.*, 41, 4378–4400 (2000). (d) J. M. Tour, M. Kozaki and J. M. Seminario, *J. Am. Chem. Soc.*, 120, 8486–8493 (1998).
66. J. M. Seminario and J. M. Tour, *Molecular Electronics – Science and Technology* (eds A. Aviram and M. A. Ratner), New York Academy of Science, New York, p. 69 (1998).
67. K. Prassides (ed.), *Mixed Valence Systems: Applications in Chemistry, Physics and Biology*,), NATO AS1 Series, Kluwer Academic Publishers, Dordrecht, The Netherlands 1991.
68. (a) J. M. Verhoeven, M. N. Paddon-Row, N. S. Hush, *et al.*, *Pure Appl. Chem.*, 58, 1285–1290 (1986). (b) P. Pasman, N. W. Koper, J. W. Verhoeven, and J. R. Neth, *Chem. Soc.*, 102, 55–56 (1983).
69. Electron transfer processes between inorganic magnetic centres are spin dependent and the special term of double exchange has been coined for it. Double exchange and trapping ideas implemented by electron exchange theory are currently being actively explored. See: (a) P. W. Anderson, and H. Hasegawa, *Phys. Rev.*, 100, 675–681 (1955). (b) L. Noodleman, *Inorg. Chem.* 27, 3677–3679 (1988). (c) G. Blondin, and J.-J. Girerd, *Chem. Rev.*, 90, 1359–1376 (1990).
70. M. Ballester, *Acc. Chem. Res.* 380–387 (1985).
71. J. Veciana, J. Riera, J. Castañer, and N. Ferrer, *J. Organomet. Chem.*, 297, 131–141 (1985).
72. J. E. Sutton, P. M. Sutton, and H. Taube, *Inorg. Chem.* 18, 1017–1021 (1979).
73. O. Elsner, D. Ruiz-Molina, J. Vidal-Gancedo, *et al.*, *NanoLetters*, 1, 3, 117–120 (2001).
74. (a) A. El-Ghayoury, A. Harriman, A. Khatyr, and R. Ziessel, *Angew. Chem. Int. Ed. Engl.*, 39, 185–189 (2000). (b) C. Lambert, G. Nöll, *J. Am. Chem. Soc.*, 121, 8434–8442 (1999).
75. R. L. Carlin, *Magnetochemistry*, Springer-Verlag, Berlin, 1986, p. 71.
76. (a) S. F. Nelsen, H. Q. Trean, and M. A. Nagy, *J. Am. Chem. Soc.*, 120, 298–304 (1998). (b) D. Ruiz-Molina, J. Sedó, C. Rovira, and J. Veciana, in *Handbook of Advanced Electronic and Photonic Materials* (ed. H. S. Nalwa), Academic Press, New York, 2000, Vol. 3, and references therein.
77. J. R. Bolton, J. A. Weil, and J. E. Wertz, *Electron Paramagnetic Resonance*, John Wiley & Sons, Inc., 1994.
78. (a) C. Rovira, D. Ruiz-Molina, O. Elsner, *et al.*, *Chem. Eur. J.*, 7, 1, 240–250 (2001). (b) V. Lloveras, J. Vidal-Gancedo, D. Ruiz-Molina, *et al.*, *Faraday Discussions*, 131, 291–305 (2006).
79. I. Ratera, D. Ruiz-Molina, F. Renz, *et al.*, *J. Am. Chem. Soc.*, 125, 1462–1463 (2003).
80. G. D'Avino, L. Grisanti, J. Guasch, *et al.*, *J. Am. Chem. Soc.*, 130, 12064–12072 (2008).
81. (a) R. M. Buchanan, and C. G. Pierpont, *J. Am. Chem. Soc.*, 102, 4951–4957 (1980). (b) C. Roux, D. M. Adams, J. P. Itie, *et al.*, *Inorg. Chem.*, 35, 2846–2852 (1996). (c) D. Ruiz-Molina, J. Veciana, K. Wurst, *et al.*, *Inorg. Chem.*, 39, 617–619 (2000). (d) D. Ruiz, J. Yoo, I. Guzei, *et al.*, *Chem. Commun.*, 2089–2090 (1998). (e) S. H. Bodnar, A. Caneschi, A. Dei, *et al.*, *Chem. Commun.*, 20, 2150–2151 (2001).
82. I. Ratera, C. Sporer, D. Ruiz-Molina, *et al.*, *J. Am. Chem. Soc.*, 129, 6117–6129 (2007).
83. (a) C. Creutz, and H. Taube, *J. Am. Chem. Soc.*, 91, 3988–3989 (1969). (b) C. Creutz, *Prog. Inorg. Chem.*, 30, 1–73 (1983). (c) K. D. Demandis, C. M. Hartshorn, and T. J. Meyer, *Chem. Rev.*, 101, 2655–2686 (2001). (d) S. F. Nelsen, M. A. Tran, and H. Q. Nagy, *J. Am. Chem. Soc.*, 120, 298–304 (1998). (e) J. Bonvoisin, J.-P. Launay, W. Verbouwe, *et al.*, *J. Phys. Chem.*, 100, 17079–17082 (1996). (f) C. Lambert, G. Nöll, and J. Schelter, *Nat. Mater.*, 1, 69–73 (2002). (g) N. Gautier, F. Dumur, V. Lloveras, *et al.*, *Angew. Chem. Int. Ed.*, 42, 2765–2768 (2003).
84. (a) S. Utamapanya, and A. Rajca, *J. Am. Chem. Soc.*, 113, 9242–9251 (1991). (b) L. Hviid, A. M. Brouwer, M. N. Paddon-Row, and J. W. Verhoeven, *Chem. Phys. Chem.*, 2, 232–235 (2001). (c) A. S. D. Sandanayaka,

H. Sasabe, Y. Araki, *et al.*, *Phys. Chem. A*, 108, 5145–5155 (2004). (d) G. L. Closs, and J. R. Miller, *Science*, 240, 440–447 (1988).
85. (a) R. A. Marcus, *J. Chem. Phys.*, 24, 966–978 (1956). (b) R. A. Marcus, *Annu. Rev. Phys. Chem.*, 15, 155–196 (1964). (c) R. A. Marcus, and N. Sutin, *Biochim. Biophys. Acta*, 811, 265–322 (1985).
86. (a) J. M. Kriegl, and G. Ulrich Nienhaus, *Proc. Natl. Acad. Sci. USA*, 101, 123–128 (2004). (b) M. Cascella, A. Magistrato, I. Tavernelli, *et al.*, Proc. *Natl. Acad. Sci. USA*, 103, 19641–19646 (2006). (c) A. Warshel, Z. T. Chu, and W. W. Parson, *Science*, 246, 112–116 (1989).
87. A. Heckmann, and C. Lambert, *J. Am. Chem. Soc.*, 129, 5515–5527 (2007).
88. S. Chopin, J. Cousseau, E. Levillain, *et al.*, *J. Mater. Chem.*, 16, 112–121 (2006).
89. H. Luo, M. Fujitsuka, Y. Araki, *et al.*, *J. Phys. Chem. B*, 107, 9312–9318 (2003).
90. M. A. Herranz, N. Martin, J. Ramey and M. D. Guldi, *Chem. Commun.*, 2968–2969 (2002).
91. S. Chopin, Z. Gan, J. Cousseau, *et al.*, *J. Mater. Chem.*, 15, 2288–2296 (2005).
92. (a) M. Yamazaki, Y. Araki, M. Fujitsuka and O. Ito, *J. Phys. Chem. A*, 105, 8615–8622 (2001). (b) F. D'Souza, G. R. Deviprasad, M. E. Zandler, *et al.*, *J. Phys. Chem. B*, 106, 4952–4962 (2002).
93. M. Mas-Torrent, N. Crivillers, V. Mugnaini, *et al.*, *J. Mat. Chem.*, 19, 1691–1695 (2009).
94. H. Ago, *Nature*, 401, 572–574 (1999). (b) L. E. Hueso, J. M. Pruneda, V. Ferrari, *et al.*, *Nature*, 445, 410–413 (2007).
95. (a) Z. H. Xiong, D. Wu, Z. Valy Vardeny and J. Shi, *Nature*, 427, 821–824 (2004). (b) T. X. Wang, H. X. Wei, Z. M. Zeng, *et al.*, *Appl. Phys. Lett.*, 88, 242505-1–242505-3 (2006). (c) F. L. Bloom, W. Wagemans, M. Kemerink and B. Koopmans, *Phys. Rev. Lett.*, 99, 257201-1–257201-4 (2007). (d) L. E. Hueso, I. Bergenti, A. Riminucci, *et al.*, *Adv. Mater.*, 19, 2639–2642 (2007).
96. (a) T. S. Santos, J. S. Lee, P. Migdal, *et al.*, *Phys. Rev. Lett.*, 98, 16601-1–16601-4 (2007). (b) J. R. Petta, S. K. Slater and D. C. Ralph, *Phys. Rev. Lett.*, 93, 136601-1–136601-4 (2004). (c) W. Wang and C. Richter, *Appl. Phys. Lett.*, 89, 153105-1–153105-4 (2006).
97. A. R. Rocha, V. M. García-Suárez, S. W. Bailey, *et al.*, *Nat. Mater.*, 4, 335–339 (2005).
98. V. Mugnaini, M. Fabrizioli, I. Ratera, *et al.*, *Solid State Sci.*, 11, 956–960 (2009).
99. N. Crivillers, S. Furukawa, A. Minoia, *et al.*, *J. Am. Chem. Soc.*, 131, 17, 6246–6252 (2009).
100. N. Crivillers, M. Mas-Torrent, J. Vidal- Gancedo, *et al.*, *J. Am. Chem. Soc.*, 130, 5499–5506 (2008).
101. N. Crivillers, M. Mas-Torrent, S. Perruchas, *et al.*, *Angew. Chem. Int. Ed.*, 46, 2215–2219 (2007).
102. O. Shekhah, N. Roques, V. Mugnaini, *et al.*, *Langmuir*, 24, 6640–6648 (2008).
103. Y. Manassen, R. J. Hamers, J. E. Demuth and A. J. Castellano, *Phys. Rev. Lett.*, 62, 2531–2534 (1989).
104. P. Messina, M. Mannini, A. Caneschi, *et al.*, *J. Appl. Phys.*, 101, 053916-1–053916-14 (2007).
105. N. Crivillers, C. Munuera, M. Mas-Torrent, *et al.*, *Adv. Mater.*, 21, 1177–1188 (2009).
106. Ashwell G. J. and Bloor, D. (eds) *Optical Materials for Non-linear Optics III*, Royal Society of Chemistry, London, (1993) and references therein.
107. A. Di Bella, I. Fragalà, T. J. Marks, and M. A. Ratner, *J. Am. Chem. Soc.*, 118, 12747–12751 (1996).
108. (a) C. Nicoud, G. Serbutoviez, G. Puccetti, *et al.*, *J. Chem. Phys. Lett.*, 175, 257–261 (1990). (b) L. Angeloni, A. Caneschi, L. David, *et al.*, *J. Mater. Chem.*, 4, 1047–1053 (1994).
109. S. Di Bella, I. Fragalà, I. Ledoux, and T. J. Marks, *J. Am. Chem. Soc.*, 117, 9481–9485 (1995).
110. (a) M. Takahashi, S. Yamada, H. Matsuda, *et al.*, *Chem. Commun.*, 1853–1854 (1997). (b) P. M. Lundquist, S. Yitzchaik, T. J. Marks, *et al.*, *Phys. Rev. B*, 55, 14055–14058 (1997). (c) P. G. Lacroix, R. Clément, K. Nakatani, *et al.*, *Science*, 263, 658–660 (1994). (d) L. Angeloni, A. Caneschi, L. David, *et al.*, *J. Mater. Chem.*, 4, 1047–1053 (1994).
111. (a) J. L. Oudar, *J. Chem. Phys*, 67, 446–457 (1977). (b) J. L. Oudar, and D. S. Chemala, *J. Chem. Phys.*, 67, 446–457 (1977).
112. S. J. Lalama, and A. F. Garito, *Phys. Rev. A*, 20, 1179–1194 (1979).
113. M. L. H. Green, S. R. Marder, M. E. Thompson, *et al.*, *Nature*, 330, 360–362 (1987).
114. (a) S. Barlow, H. E. Bunting, C. Ringham, *et al.*, *J. Am. Chem. Soc.*, 121, 3715–3723 (1999). (b) N. J. Long, *Angew. Int. Ed. Engl.*, 34, 21–38 (1995). (c) T. Verbiest, S. Houbrechts, M. Kauranen, *et al.*, *J. Mater. Chem*, 7, 2175–2189 (1997). (d) B. J. Coe, T. A. Hamor, C. J. Jones, *et al.*, *J. Chem. Soc. Dalton Trans.*, 673–684 (1995).

115. (a) J. Zyss, *Nonlinear Opt.*, 1, 3–18 (1991). (b) J. Zyss, and I. Ledoux, *Chem. Rev.*, 94, 77–105 (1994).
116. (a) J. Zyss, *J. Chem. Phys.*, 98, 6583–6599 (1993). (b) C. Dhenaut, I. Ledoux, I. D. W. Samuel, *et al.*, *Nature*, 374, 339–342 (1995).
117. K. Clays, and A. Persoons, *Phys. Rev. Lett.*, 66, 2980–2983 (1991).
118. (a) J. Zyss, T. Chauvan, C. Dhenaut, and I. Ledoux, *Chem. Phys.*, 177, 281–296 (1993). (b) T. Verbiest, K. Clays, C. Samyn, *et al.*, *J. Am. Chem. Soc.*, 116, 9320–9323 (1994). (c) C. Lambert, E. Schmalzlin, K. Meerholz, and C. Brauchle, *Chem. Eur. J.*, 4, 512–521 (1998). (d) S. Brasselet, F. Cherioux, P. Audebert, and J. Zyss, *Chem. Mater.*, 11, 1915–1920 (1999).
119. I. Ratera, D. Ruiz-Molina, C. Sporer, *et al.*, *Polyhedron*, 22, 1851–1856 (2003).
120. I. Ratera, S. Marcen, S. Montant, *et al.*, *Chem. Phys. Lett.*, 363, 245–251 (2002).
121. I. Ratera, D. Ruiz-Molina, C. Sánchez, *et al.*, *Synth. Met.*, 121, 1834–1835 (2001).
122. (a) M. Verdaguer, *Science*, 272, 698–699 (1996). (b) S. Schmitt, and K. Hafner, *Mol. Cryst. Liq. Cryst.*, 348, 1–6 (2000).
123. (a) P. Gütlich, A. Hauser, and H. Spiering, *Angew. Chem. Int. Ed. Engl.*, 33, 2024–2054 (1994) and references therein. (b) K. Nagai, T. Iyoda, A. Fujishima, and K. Hashimoto, *Solid State Commun.*, 102, 809–812 (1997).
124. S. Nakatsuji, M. Mizumoto, A. Takai, *et al.*, *Mol. Cryst. Liq. Cryst.*, 348, 1–2 (2000).
125. (a) K. Matsuda, and M. Irie, *Tetrahedron Lett.*, 41, 2577–2580 (2000). (b) K. Matsuda, and M. Irie, *Chem. Lett.*, 16–17 (2000).
126. K. Hamachi, K. Matsuda, T. Itoh, and H. Iwamura, *Bull. Chem. Soc. Jpn.*, 71, 2937–2943 (1998).
127. (a) P. M. Lahti (ed.) *Magnetic Properties of Organic Materials*, Marcel Dekker, NY, 1999. (b) J. Miller, and M. Drillon (eds) *Magneto Science: Molecules to Materials* Vols. 1 and 2, Wiley-VCH Verlag GmbH, 2001. (c) J. Veciana, C. Rovira, and D. B. Amabilino (eds), *Supramolecular Engineering of Synthetic Metallic Materials*, NATO ASI Series, Vol. 518, 1998.
128. (a) I. Ratera, D. Ruiz-Molina, J. Vidal-Gancedo, *et al.*, *Angew. Chem. Int. Ed.*, 40, 919–922 (2001). (b) I. Ratera, D. Ruiz-Molina, J. Vidal-Gancedo, *et al.*, *Chem. Eur. J.*, 10, 603–616 (2004).
129. (a) J. V. Barth, G. Constantini, and K. Kern, *Nature*, 437, 671–679 (2005). (b) B. D. Gates, Q. Xu, M. Stewart, *et al.*, *Chem. Rev.*, 105, 1171–1196 (2005).
130. A. D. Shukla, A. Das, and M. E. van der Boom, *Angew. Chem. Int. Ed.*, 44, 3237–3240 (2005).
131. J. C. Love, L. A. Estroff, J. K. Kriebel, *et al.*, *Chem. Rev.*, 105, 1103–1170 (2005).
132. A. Ulman, *Chem. Rev.*, 1996, 96, 1533–1554 (1996).
133. (a) M. M. Matsushita, N. Ozaki, T. Sugawara, *et al.*, *Chem. Lett.*, 6, 596–597 (2002). (b) Y. Kashiwagi, K. Uchyama, F. Kurashima, *et al.*, *Anal. Scien.*, 15, 907–909 (1999).
134. (a) M. A. Fox, E. Gaillard, and Ch. Chen, *J. Am. Chem. Soc.*, 109, 7088–7094 (1987). (b) T. L. Chu, and S. I. Weissmann, *J. Chem. Phys.*, 22, 21–25 (1954).

3

Phenalenyls, Cyclopentadienyls, and Other Carbon-Centered Radicals

Yasushi Morita[1] and Shinsuke Nishida[2]

[1]*Osaka University, Department of Chemistry, Graduate School of Science, Osaka, Japan*
[2]*Osaka City University, Department of Chemistry, Graduate School of Science, Osaka, Japan*

3.1 Introduction

Studies on stable neutral radicals have a history of about 110 years, since Gomberg's discovery of the triphenylmethyl radical in 1900.[1] Physical properties of neutral radicals give substantial far reaching effects and inspiration for recent researches of molecule-based functional materials.[2] One of the most characteristic properties of organic neutral radicals is the magnetism induced by interactions of unpaired electrons. For the realization of molecule-based magnetic materials, it is necessary to align spins due to the unpaired electrons with adequate interactions through π electrons or the lone pair of heteroatoms in the solid state.[2] 2,2,6,6-Tetramethylpiperidin-*N*-oxyl (TEMPO) and α-nitronylnitroxide derivatives, synthesized a half-century ago as neutral radicals stable enough to handle in air, have importantly contributed to the development of the molecule-based magnetism and spin sciences, and even now play important roles in other various research fields as stable spin sources.[2] For further progress and development of spin sciences opened up by the stable neutral radicals, the design and syntheses of stable neutral radicals with novel molecular and electronic structures are essential issues. To achieve these projects, the use structural diversity, which is one of the most important virtues of the organic molecules, should be used to the maximum.

The molecular skeleton of stable neutral radicals has two aspects: a "container" for stable handling of spin and a "joystick" for modulation and control of electronic structure. For design of novel neutral radicals, it is vital to utilize these two aspects effectively. For example, stabilization of neutral radicals by steric protections not only suppresses the dimerization reaction and the reaction with oxygen, but also greatly influences crystal structures and, thus, intermolecular spin–spin interactions. Introduction of heteroatoms is effective for the control of intermolecular electronic interactions through lone pair electrons (Chapters 5 and 6). Furthermore, heteroatomic modulation affects the distribution of coefficient of singly occupied molecular orbitals (SOMOs) and spin density on the molecular skeleton by the changes in π-topological symmetries,[3] and influences the molecular orbital energy levels (redox ability).

Phenalenyl is a "spin delocalized" hydrocarbon radical with an unpaired electron extensively delocalized on the planar π-conjugated network. The electronic structure can be significantly modulated and perturbed by the choice of substituents and their introduction positions. These features are intrinsic in nature for spin delocalized radical systems, that are difficult to realize in spin localized ones such as TEMPO and α-nitronylnitroxide derivatives. Taking advantage of the extensively delocalized spin on a large π-conjugated molecular skeleton, various novel and intriguing natures (or properties) that were previously unknown have been realized.

This chapter begins with a review of the design strategies, synthetic methods, spin delocalized natures, and related physical properties of the families of phenalenyl-based stable neutral radicals with both planar and curved molecular/electronic structures. Later, cyclopentadienyl radicals with a 5π electronic system and other kinds of carbon-centered neutral radicals are reviewed.

3.2 Open shell graphene

Graphene is a planar sheet-like gigantic π-conjugated electronic system comprising condensed six-membered carbon rings. It had been presumed not to exist in an isolated state because of low stability. Graphene was first prepared in 2004 by the mechanical exfoliation of highly orientated pyrolytic graphite, and ambipolar field-effect transistor (FET) properties were disclosed by Novoselov and Geim.[4a] Since this epoch making study, graphene has drawn much attention as a next generation electronic material to replace current silicon devices. The prepared graphenes, however, are mixtures with various different structures and have not been isolated as a single (monodisperse) material. For further development of new electronic materials based on graphene and elucidation of their electronic properties, fundamental research utilizing structurally well defined molecules with π-conjugated electronic systems is important. Organic chemists view graphene as a gigantic planar π-conjugated condensed-polycyclic aromatic hydrocarbon. From this viewpoint, well known planar fused-polycyclic aromatic hydrocarbons can be recognized as a structurally well defined small graphene, and its chemistry can also be viewed as "graphene chemistry". Therefore, organic chemistry has the potential to make great contributions to the elucidation of the nature of graphene and to the development of novel functional materials.

To view graphene more realistically from an organic chemistry perspective, cutting out a single structural molecule from a graphene sheet offers an intriguing top-down approach. Clipping modes can be categorized as either of two types, armchair and zigzag. Figure 3.1a shows the armchair-type clipping mode.[4b] The connection of the terminal edges of the clipping motif affords curved π-conjugated molecules, such as fullerene C_{60} and carbon nanotubes, and a stacking of the planar graphene sheets gives graphite. All of these fused-polycyclic π-conjugated molecules and molecular aggregates possess closed shell structures. In a sharp contrast, triangular motifs with the zigzag mode give π-conjugated molecules possessing unpaired electron(s) (open shell structures) as a logical consequence (Figure 3.1b).

Here the possibility of a bottom-up approach, namely conventional organic synthesis, to π-conjugated systems designed by the top-down approaches outlined above is considered. In the case of armchair-type (closed shell) molecules, Müllen has synthesized the structurally well defined largest hydrocarbon molecule containing 222 carbon atoms (Figure 3.2a)*.[5] On the other hand, π-conjugated molecules designed by the zigzag clipping mode with triangular motifs are generally known as non-Kekulé polynuclear benzenoid

*Unless otherwise stated, spin density distribution calculations throughout this chapter are made at the UBLYP/6-31G**//UBLYP/6-31G** level of theory. Black and gray regions denote positive and negative spin densities, respectively.

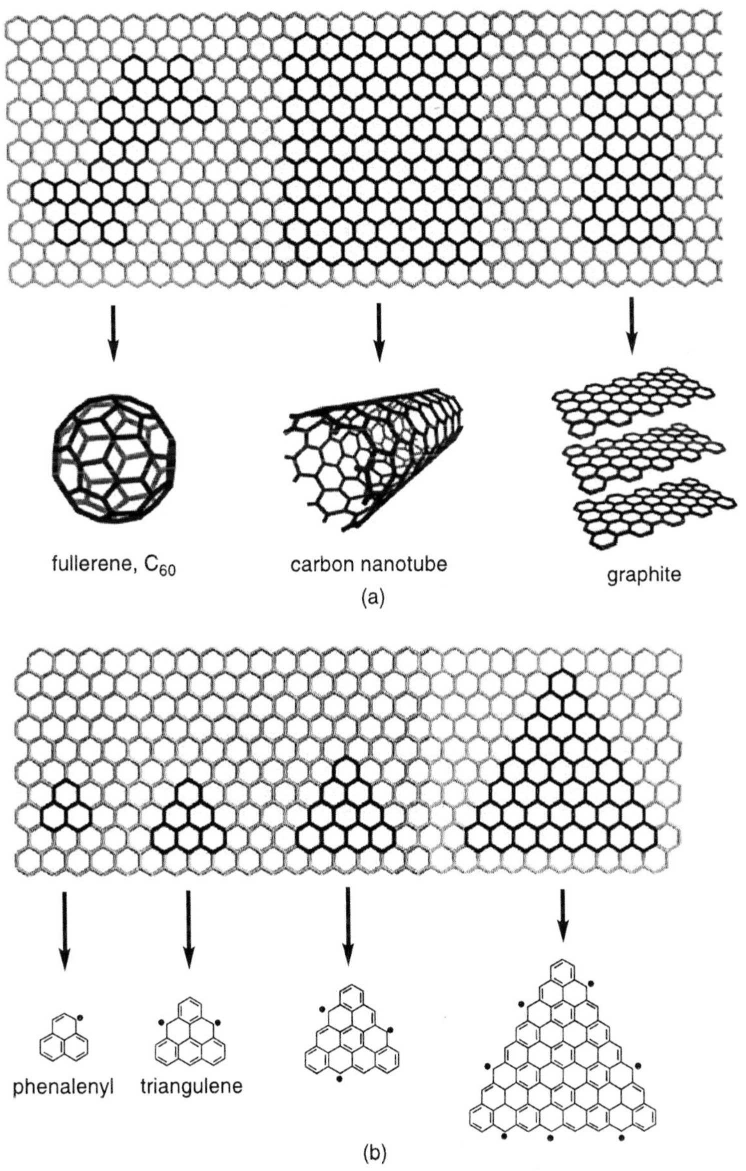

Figure 3.1 *(a) Closed and (b) open shell π-conjugated molecular systems designed by the clipping modes of armchair and zigzag from graphene.*

(PNB) molecules. These molecular systems can be termed "open shell graphene" and have been little investigated from both theoretical and experimental aspects.

Triangulene (or Clar's hydrocarbon, Figure 3.1b) is the most fundamental non-Kekulé PNB, and its chemical identification has been a long standing issue since Clar's proposal.[6] Introduction of three *tert*-butyl substituents onto the triangulene skeleton succeeds in partial stabilization of this hydrocarbon, enabling an experimental determination of its ground state triplet nature (*vide infra*).[7] In general, the synthesis

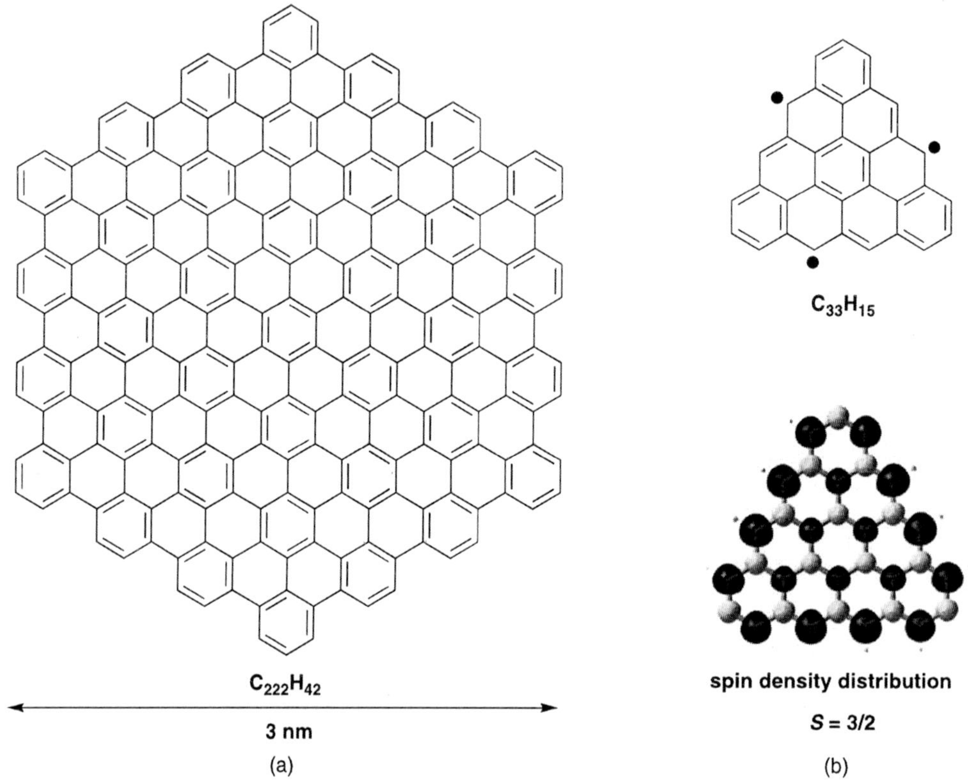

Figure 3.2 (a) Gigantic π-conjugated hydrocarbon with closed shell structure prepared by the bottom-up approach by Müllen[5] and (b) non-Kekulé polynuclear benzenoid (PNB) molecule with open shell structure (open shell graphene), and its calculated spin density distribution. Black and gray regions denote positive and negative spin densities, respectively.

and isolation of such non-Kekulé PNB molecules are extremely difficult for two main reasons; firstly, the high instability due to the high reactivity at the edge sides of the molecule bearing most of the spin densities (Figure 3.2b),[7] and, secondly, the necessity to construct large fused π-electronic structures in which two or more sides are shared by many neighboring ring systems. Despite these difficulties, the synthesis and elucidation of electronic structures of open shell graphene subunits are very attractive, with interest in their physical properties based on (1) the interplay between the unpaired electron(s) and the extensively delocalized π-electron systems and charge fluctuation, and (2) the possibility that these species may provide a qualitatively novel materials challenge in the field of molecule-based functional materials.

3.3 Phenalenyl

Phenalenyl (**PLY**) is the smallest odd-alternant π-conjugated hydrocarbon neutral radical with a fused-polycyclic planar structure[8] and is also the smallest open shell graphene. The spin density distribution and molecular structure of **PLY** are shown in Figure 3.3, compared with those of TEMPO and the α-nitronylnitroxide derivative as typical stable neutral radicals. The molecular and electronic structural

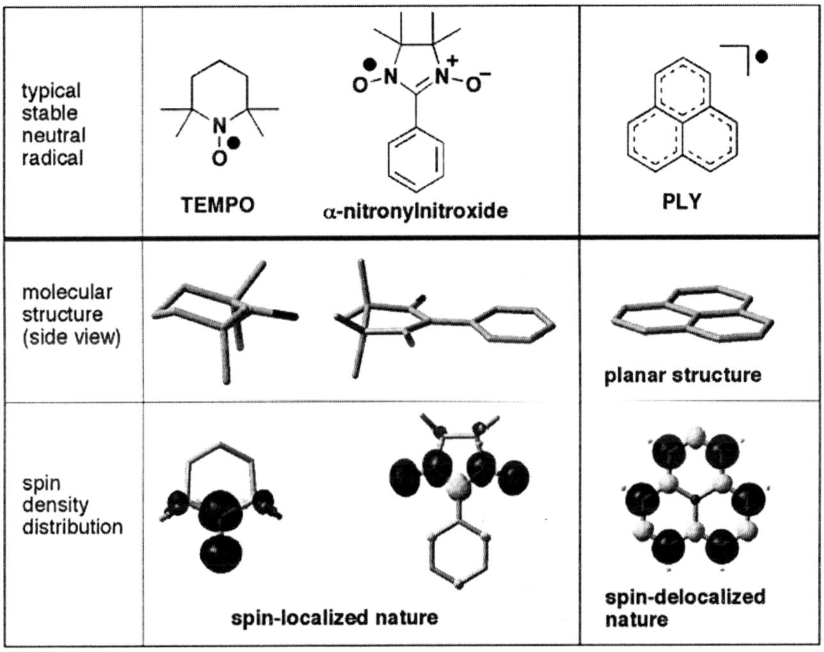

Figure 3.3 Molecular structures and spin density distributions of TEMPO, phenyl-substituted α-nitronylnitroxide, and phenalenyl (**PLY**) UBLYP/6-31G**//UBLYP/6-31G** level of theory. Black and gray regions denote positive and negative spin densities, respectively.

features of TEMPO and α-nitronylnitroxide are characterized by the heteroatomic spin center of N–O moieties, where most of the spin densities are localized ("spin localized nature"), with adjacent tertiary carbon substituents. In sharp contrast, **PLY** possesses a planar, rigid structure, where the spin spreads over the whole molecular skeleton ("spin delocalized nature"). As can be interpreted by resonance structures, most of the spin density of **PLY** exists on its α positions (Figure 3.4a). This feature is well corroborated by the distribution of the non-bonding molecular orbital (NBMO) obtained from the Hückel–MO method. Furthermore, **PLY** exhibits a high amphoteric redox ability deduced by the NBMO, and provides corresponding anion, neutral radical, and cation species with thermodynamically high stabilities (Figure 3.4c). The sum of the oxidation potential (E^{ox}) and reduction potential (E^{red}) (obtained from cyclic voltammetry (CV) experiments in solution), $E^{sum} = E^{ox} + (-E^{red})$, represents the amphotericity of a given molecule. This value is generally used as a convenient experimental measure for evaluation of a HOMO–LUMO (highest occupied molecular orbital–lowest unoccupied molecular orbital) gap for closed shell molecules, but provides an estimate of coulombic repulsion between two electrons on the same orbital for open shell molecules. **PLY** has an E^{sum} value of 1.6 V, which is relatively small among hydrocarbons.

All of these features of the phenalenyl system have been attracting special attention in the fields of physical chemistry. Notably, Haddon claimed, in 1975, the potential use of **PLY** as a component for molecule-based conductors.[9] However, since the kinetic instability of **PLY** causes an immediate dimerization by intermolecular σ-bond formation as well as reactivity towards oxygen, all experimental effort to isolate the pristine **PLY** as solid state has been in vain (Figure 3.4d).[10] Thus, phenalenyl chemistry for the past a half-century has been limited mainly to solution studies under degassed and sealed conditions.

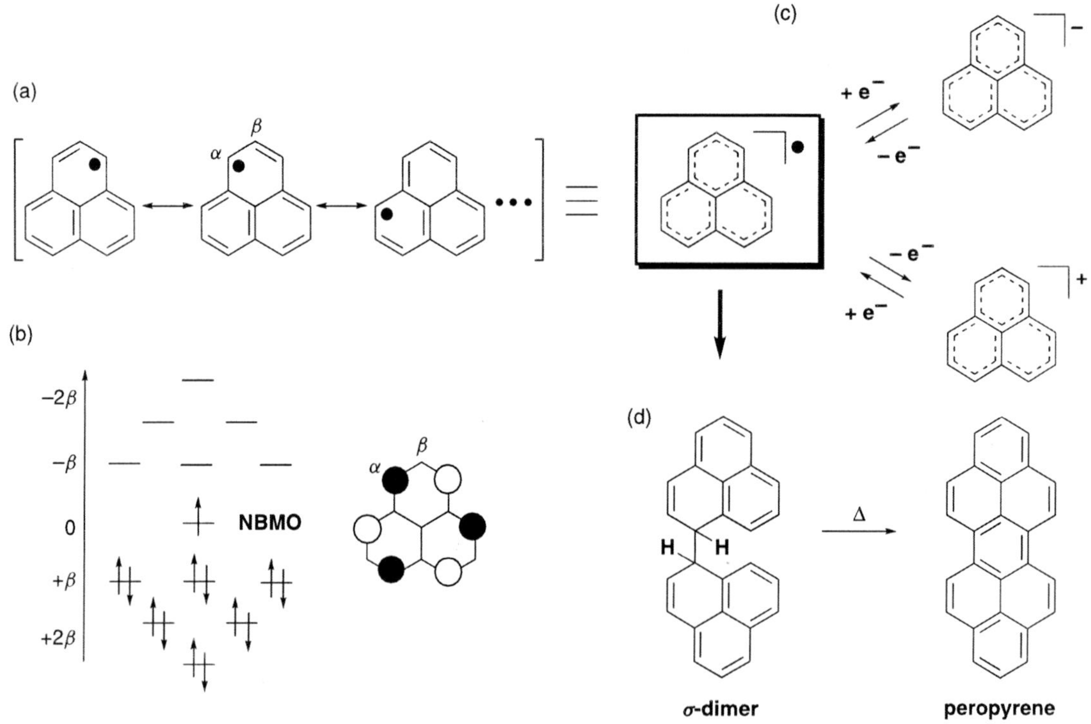

*Figure 3.4 (a) Resonance structures, (b) MO diagram and NBMO picture calculated by the Hückel–MO method, (c) amphoteric redox processes, and (d) σ-dimerization reaction of phenalenyl (**PLY**) and chemical structure of peropyrene.*

Besides these physical chemistry studies, continuous synthetic efforts for the stabilization of the phenalenyl by chemical modifications have enriched phenalenyl chemistry and aroused the phenalenyl-based material sciences recently. These studies have helped to realize, for example, not only isolation of the neutral radical as stable solid in air but also the developments of exotic magnetic and optical properties by the introduction of substituents into the phenalenyl skeleton and by the extension of the π-conjugated system: enhancement of spin multiplicities and the redox abilities, switching of the electronic structures depending on redox states, creation of single component organic (semi)conductors, and application for the electrode active material in secondary batteries (Figure 3.5). In the following sections, recent significant progress in neutral radicals based on the phenalenyl system are described, emphasizing synthetic challenges for novel open shell molecules, topological symmetries of the spin distributions, and dynamic spin behaviors inherent in the spin delocalized nature of this radical type.

3.4 2,5,8-Tri-*tert*-butylphenalenyl radical

As the first approach for the stabilization of the phenalenyl, introduction of primary alkyl groups, such as methyl and trimethylene groups, were carried out.[11] Murata and Nakasuji claim an electronic effect of electron donating and electron accepting groups attached to the phenalenyl skeleton.[12a] Although isolation in the solid state is not achieved, the capto-dative effect of the substituents enhances the stability of the

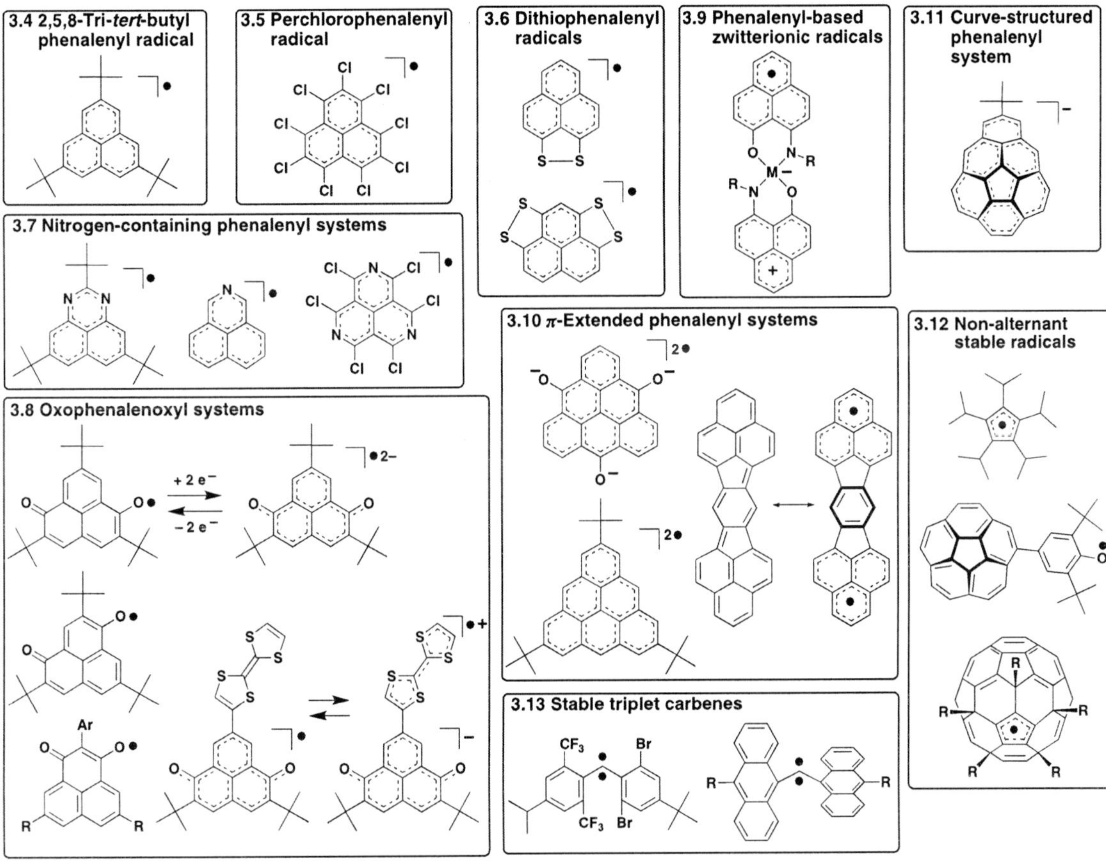

Figure 3.5 *Phenalenyl derivatives and other carbon-centered neutral radicals evolving from recent material challenges.*

phenalenyl in solution and provides a high reversibility of the redox processes with a significant decrease of E^{sum} values ($1.09 \sim 1.17$ V).[12a] Furthermore, 1,4,7-tris(propylamino) and 1,4,7-tris(dimethylamino) derivatives of the phenalenyl system also have low E^{sum} values of 0.58 and 0.69 V, respectively.[12b] In 1999, Nakasuji synthesized 2,5,8-tri-*tert*-butylphenalenyl (**TBPLY**) having three *tert*-butyl groups at the β positions of the phenalenyl skeleton, and succeeded for the first time in the isolation of the phenalenyl in the solid state in air (Scheme 3.1).[13a] This synthetic work contributed to the elucidation of the solid state properties of the phenalenyl, such as crystal and molecular structures and magnetic properties, unrevealed for a half-century. For the following reasons it is assumed that the perturbation by *tert*-butyl groups of the electronic structure of the phenalenyl is small: firstly, the *tert*-butyl group has a small electronic effect in comparison with other substituents containing heteroatoms and, secondly, the connecting positions of all *tert*-butyl groups are β positions, which have negligible spin densities. Thus, **TBPLY** can be viewed as the most electronically fundamental neutral radical system in the phenalenyl system (*vide infra*).

The synthetic method for obtaining **TBPLY** is depicted in Scheme 3.1.[13a] Bromination of 2,7-di-*tert*-butylnaphthalene followed by lithiation affords the formyl derivative. After the Reformatsky reaction and subsequent reductive elimination of hydroxyl group, demethylation with lithium iodide was carried out.

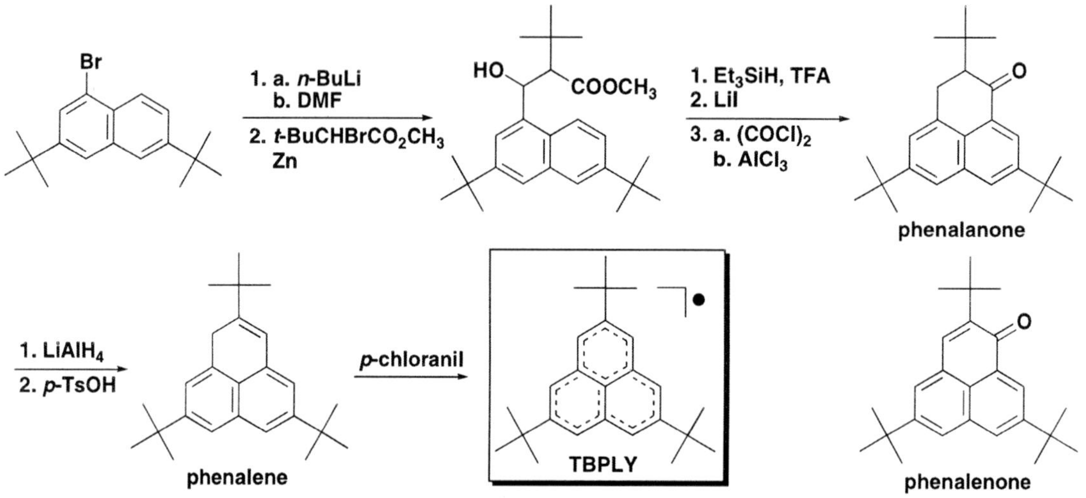

Scheme 3.1 Synthetic method for **TBPLY**

Friedel–Crafts acylation of the carboxyl derivative yields the phenalanone derivative, and then the radical precursor, phenalene, is obtained as pale yellow crystals by reduction with lithium aluminum hydride ($LiAlH_4$) and dehydration. Oxidation of the phenalene by *p*-chloranil in degassed toluene gives **TBPLY** as deep blue crystals. This crystal possesses high stability under an argon atmosphere, but changes in air in one week to a mixture of the phenalenone derivative and structurally unidentified compounds by aerial oxidation. The decomposition temperature of **TBPLY** under argon atmosphere is 232 °C.

The Electron Paramagnetic Resonance (EPR) spectrum of **TBPLY** in a toluene solution shows the signals attributable to six protons at the α positions in the phenalenyl skeleton, and the hyperfine coupling constant (hfcc) for these protons is $-0.620\,\mathrm{mT}$ (Figures 3.6a and 3.6b).[13a] The observed g-value is 2.0028,

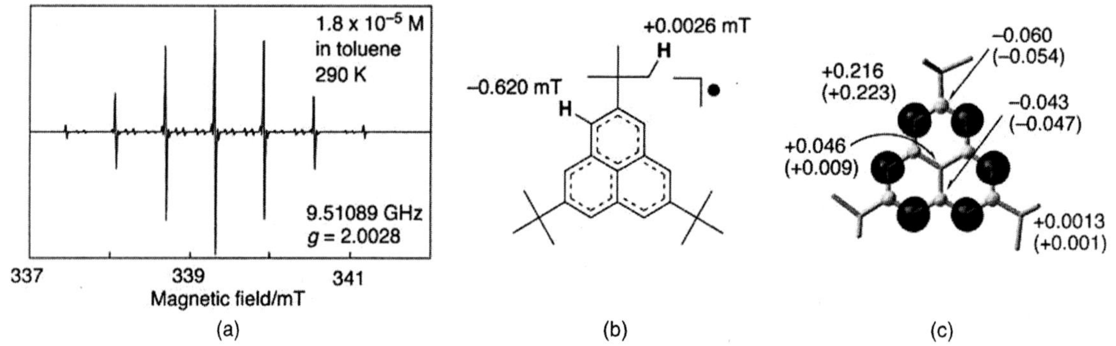

Figure 3.6 *(a) EPR spectrum of **TBPLY** in a toluene solution at 290 K; (b) observed ^{1}H hfccs. The relative sign of the ^{1}H hfccs of* tert-butyl *groups and α-positions were determined using ^{1}H-ENDOR and paramagnetic ^{1}H NMR spectroscopy (in $CDCl_3$ at 300 K).[13b] (c) Experimental and theoretical (in parentheses) spin density distributions of **TBPLY**. Experimental values were determined by the observed hfccs with the help of McConnell, Heller–McConnell, and Fraenkel–Karplus equations.[13b] Calculation carried out at the SVWN/6-31G**//SVWN/6-31G** level of theory.*

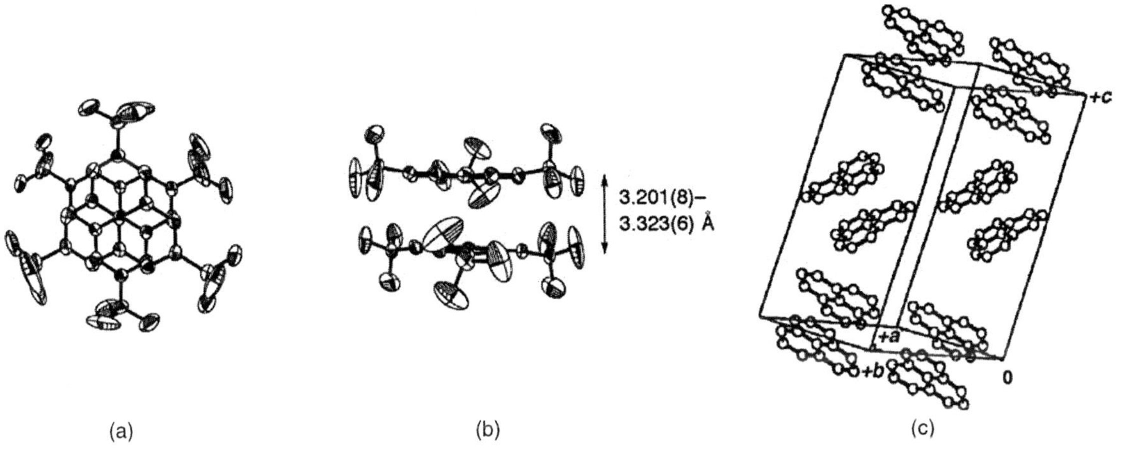

Figure 3.7 Crystal structure of **TBPLY**. (a) Top view, (b) side view, and (c) a herringbone motif of the crystal. (c Reprinted with permission from [13a]. Copyright 1999 American Chemical Society.)

being similar to that of the pristine **PLY** (2.0027).[8i,11b] The experimental and theoretical spin density distributions (Figure 3.6c) indicate that **TBPLY** possesses a characteristic spin delocalized nature and spin polarized structure, in which large positive and small negative spin densities locate at the six α carbons and β carbons with a C_3-symmetry, respectively.

The first X-ray crystal structure analysis of an odd-alternant hydrocarbon neutral radical with a fused polycyclic π-conjugated system was carried out on a single crystal of **TBPLY** obtained by recrystallization from hexane at −30 °C (Figure 3.7).[13a] The molecule has D_{3h} symmetry, and the phenalenyl skeleton shows a nearly planar geometry. The C–C bond lengths in the phenalenyl ring range from 1.374 to 1.421 Å. The molecule forms a π-dimeric pair in a staggered arrangement of the *tert*-butyl groups to avoid their steric repulsion. In such an arrangement, the maximum overlaps of the radical SOMOs are expected. The interplanar distances in the π-dimeric pair range from 3.201 to 3.323 Å, which is shorter than the sum of van der Waals radius of carbon atom (3.4 Å). The π-dimeric pair is in a herringbone motif in the crystal, and no effective interdimeric interaction is found. The magnetic susceptibility measurement of a polycrystalline sample of **TBPLY** (2–350 K) shows a large antiferromagnetic intermolecular exchange interaction ($2J/k_B = -2000$ K), indicating a ground state spin singlet in the π-dimeric pair.[13] In the solid state electronic spectrum, a broad and strong absorption band attributable to the strong intermolecular transition in the π-dimeric pair is observed in the low energy region (510–700 nm).

Interestingly, the strong interaction between **TBPLY**s observed in the solid state is also found in solution at low temperature, leading to a thermochromic phenomenon.[14,15] The red purple color of the hexane solution in **TBPLY** at room temperature gradually changes to blue at low temperature (~200 K), as seen in the increment of the absorbance of 530–670 nm with a lowering of temperature (Figure 3.8a), and this color change occurs reversibly. The chemical species exhibiting each color are assigned as the neutral radical monomer at room temperature and the π-dimeric pair at low temperature.[14,15] The temperature dependence of the electronic structure originating from the neutral radical was also observed by EPR measurements.[14,15] The molecular weight of the π dimer formed at low temperature was unequivocally detected by cold-spray ionization mass spectrometry (CSI-MS), by which substances can be ionized at much lower temperatures than conventional ESI-MS (Figure 3.8b).[15] Notably, this is the first experimental observation of the molecular weight of the open shell molecule aggregates formed in solution.

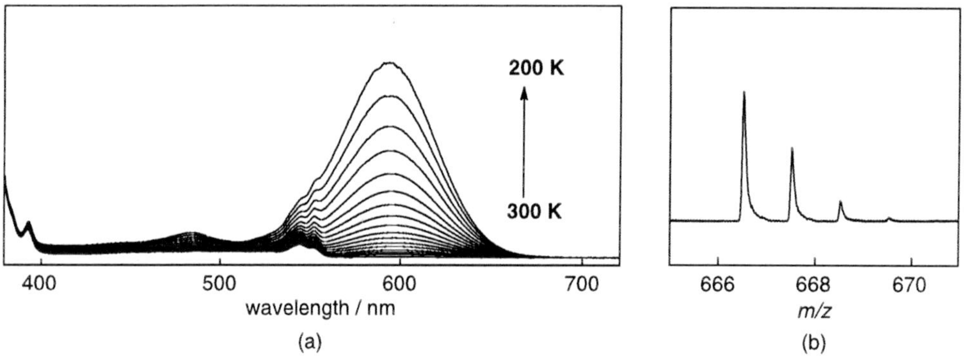

Figure 3.8 *(a) Temperature-dependent electronic spectra showing thermochromism of* ***TBPLY*** *in a hexane solution. (b) Cold-spray ionization mass spectrometry (CSI-MS) spectrum of the π dimer formed in a solution state at low temperature.*

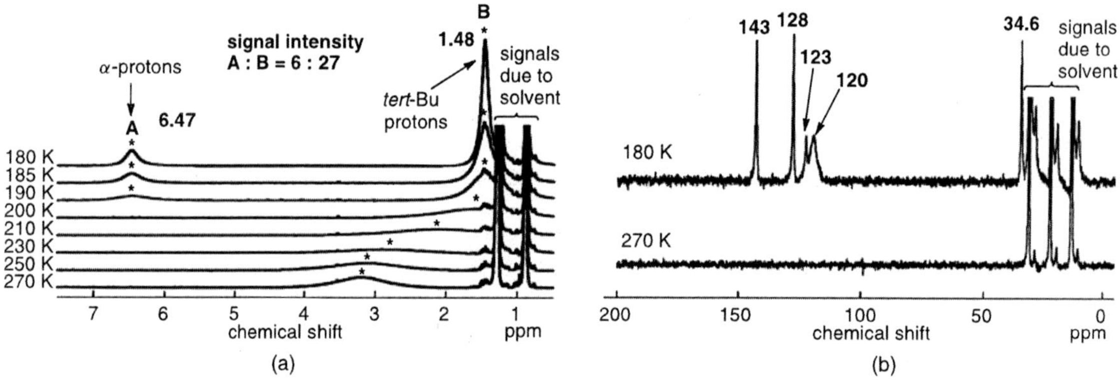

Figure 3.9 *Temperature dependences of (a) ^1H and (b) ^{13}C NMR spectra of* ***TBPLY*** *in hexane-d_{14}. These spectra clearly show a formation of a highly symmetrically structured dimer at low temperature. (Reprinted with permission from [15]. Copyright 2006 American Chemical Society.)*

The definite determination of the π-dimeric structure with a high symmetry was accomplished by low temperature ^{1}H and ^{13}C NMR measurements and chemical shift simulations by quantum chemical calculations (Figure 3.9).[15] In the ^{1}H NMR spectrum in hexane-d_{14} at a high temperature (270 K), a broad signal attributed to the protons of the *tert*-butyl groups is observed at around 3.2 ppm due to the downfield shift by an effect of radical spin. This broad signal shifts to the high field region with a decreasing temperature, indicating the formation of the π dimer (Figure 3.9a). At low temperature regions ($T < 190$ K), the π dimer becomes dominant and the *tert*-butyl protons shift to 1.48 ppm (**B** in Figure 3.9a), and a signal originating from the α protons on the phenalenyl ring emerges in the aromatic region (6.47 ppm, **A** in Figure 3.9a). The integral intensity ratio of α protons to the *tert*-butyl protons is 6:27. In the ^{13}C NMR spectra, almost no signal is found at room temperature due to the presence of paramagnetic species. At low temperature, four signals attributable to the phenalenyl skeleton are clearly observed, since the paramagnetic spins are quenched by the π dimerization (Figure 3.9b). A definite π-dimeric structure is confirmed by the good agreement between the observed chemical shifts at low temperature and those obtained by quantum chemical calculations. These experiments were the first solution NMR observations of a face-to-face radical π dimer, and prove that NMR spectroscopy is an effective and fascinating method to observe a dynamic behavior of the open shell molecular systems.[15]

The formation of the π-dimeric pair at low temperature raises the issue of the aromaticity of the radical aggregates from the viewpoint of fundamental chemistry.[15] NICS (nucleus-independent chemical shifts) values of the monomer and the π dimer of **TBPLY** have been examined by quantum chemical calculations.[16] As shown in Figure 3.10a, the ring center of the π dimer becomes more aromatic (−7.1 ppm) than that of the monomer (−3.8 ppm). The trend of the enhancement is more pronounced in the interior of the π dimer. These results indicate a great enhancement of the aromaticity due to the dimerization. Since the contribution of spin paramagnetism is already subtracted in the radical NICS values, this change reflects a susceptible probe for the π dimerization of **TBPLY**. This "aromaticity generation" is for the first time demonstrated in the **TBPLY** system, and is inherent in spin delocalized organic open shell molecules, unlike closed shell molecules and open shell molecules with spin localized nature. In the electrostatic potential surface calculated for the π dimer of **TBPLY**, a negative electron density is largely populated at the internal region, contrasting with the external region, despite the whole neutrality of the dimer (Figure 3.10b).[15] The imbalanced electron distribution is totally consistent with the anomalous NICS distribution, suggesting a share of unpaired electrons and an existence of a 12-center-2-electron long C–C bond[15] or a 12-centered covalent bond.[14a]

TBPLY possesses reversible two-stage amphoteric redox ability.[13a] Generation and elucidation of the electronic structures of the corresponding cation and anion species of **TBPLY** have been carried out. The BF_4^- salt of the cation can be prepared from the phenalene derivative (precursor of **TBPLY**) and is a

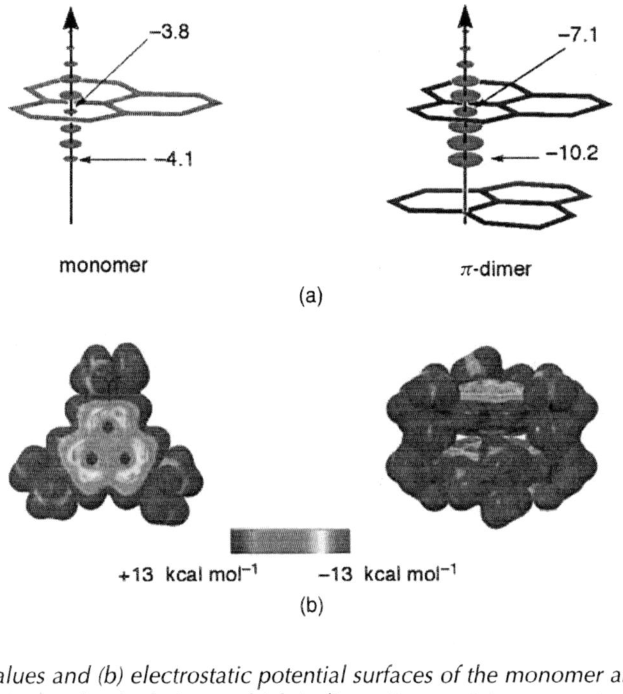

Figure 3.10 *(a) NICS values and (b) electrostatic potential surfaces of the monomer and the π dimer of **TBPLY** calculated by quantum chemical calculations, which indicate "aromaticity generation" by the π dimerization. (Reprinted with permission from [15]. Copyright 2006 American Chemical Society.) A full-colour version of part (b) of this figure appears in the Colour Plate section of this book.*

stable enough crystal in air to undergo X-ray crystal structure analysis.[13a] The $B(C_6F_5)_4^-$ salt of the cation is also prepared and structurally solved.[14a] The anion species was also generated in a degassed solution by reaction of the phenalene with alkaline metals or alkaline hydrides, and is stable in a degassed solution.[13a] Isolation of the anion in air as crystal, however, is in vain due to an immediate decomposition. Cyclic voltammetry measurement of the cation salt of **TBPLY** in acetonitrile shows two reversible redox waves at +0.27 V and −1.26 V (vs SCE), which correspond to the reduction processes from the cation to the neutral radical and from the neutral radical to the anion, respectively ($E^{sum} = 1.53$ V). In contrast to **TBPLY**, the redox processes of the pristine **PLY** are irreversible due to an occurrence of chemical reactions, such as σ dimerization. These results indicate that the *tert*-butyl groups in **TBPLY** effectively increase the kinetic stabilities of the three redox species.

3.5 Perchlorophenalenyl radical

With the aim of developing molecular conductors, Haddon has been actively conducting chemical modifications of the phenalenyl system. Since his pioneering work on a phenalenyl derivative possessing a *peri*-fused five-membered ring with a S–S bond in 1978 (*vide infra*),[10a,17] synthetic routes to phenalenyl radical derivatives having alkoxy,[18a] hydroxyl,[18b,c] amino,[18b,c] N–S–N groups,[19a] and the spiro-conjugated to phosphazene[19b] or metal elements (*vide infra*) have been described. In 1987, the perchlorophenalenyl radical (**PCPLY**) with nine chlorine atoms at all α and β carbons of the phenalenyl skeleton was designed and prepared (Scheme 3.2),[20] and Haddon's continuous efforts realized a crystal structure analysis of **PCPLY** in 2001.[21]

The synthetic route for **PCPLY** is depicted in Scheme 3.2. Perchloroacenaphthylene, obtained by the reaction of acenaphthylene with phosphorus pentachloride, is converted to the cyclopropane derivative by the addition of dichlorocarbene. The phenalenyl cation is obtained by allyl-rearrangement under

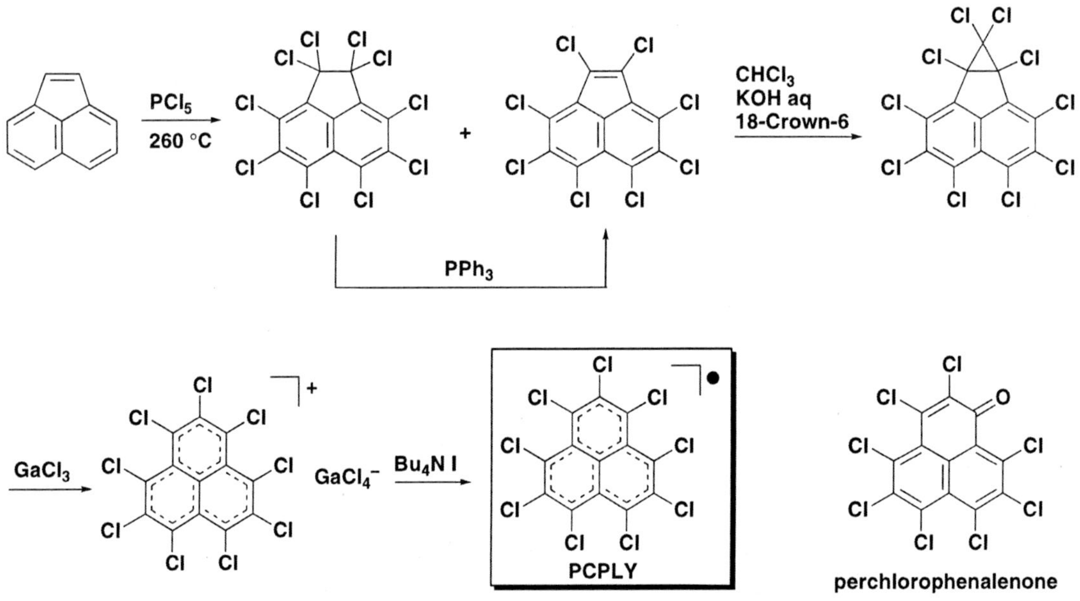

Scheme 3.2 Synthetic method for **PCPLY**

Lewis acid ($GaCl_3$) conditions. Treatment of the cation with tetrabutylammonium iodide (Bu_4NI) gives **PCPLY** as shiny black hexagonal crystals. This neutral radical possesses a high stability in air, while perchlophenalenone is formed by reaction with oxygen. The melting point of **PCPLY** is 160 °C in air with decomposition and is 248 °C under an argon atmosphere.

Characterization of the electronic structure of **PCPLY** was facilitated by EPR spectroscopy in solution. The spectrum in benzene shows fifteen peaks with uniform separation with the hfcc (0.06 mT) attributable to the ^{35}Cl and ^{37}Cl nuclei at the α positions.[22] The hfccs attributable to chlorine nuclei on the β positions are not observed because they are masked in the spectral line width. The spin density distribution obtained by the quantum chemical calculation based on the crystal structure is shown in Figure 3.11.

PCPLY is easily reduced to the corresponding anion due to the electron withdrawing effect of nine chlorine substituents. Cyclic voltammetry measurement of the neutral radical in dichloromethane shows two reversible redox waves at +1.14 V and −0.08 V (vs SCE), which respectively correspond to the oxidation and reduction processes forming the cation and anion. Compared to the corresponding redox potentials of the pristine **PLY** in acetonitrile observed at +0.7 V and −0.9 V (vs SCE),[17] redox potentials of **PCPLY** show a great cathodic shift.

Single crystals of **PCPLY** were obtained by repeated sublimation under high vacuum (5×10^{-5} Torr).[21,23] This molecule has threefold rotational symmetry and a ruffled structure with a propeller shape (Figure 3.12). The reason is that the intramolecular Cl–Cl distances at the *peri* positions

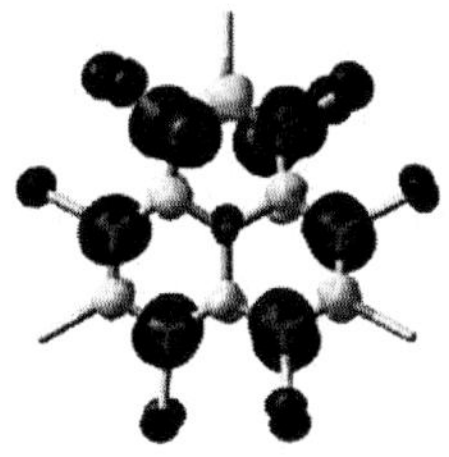

Figure 3.11 *Spin density distribution of **PCPLY**.*

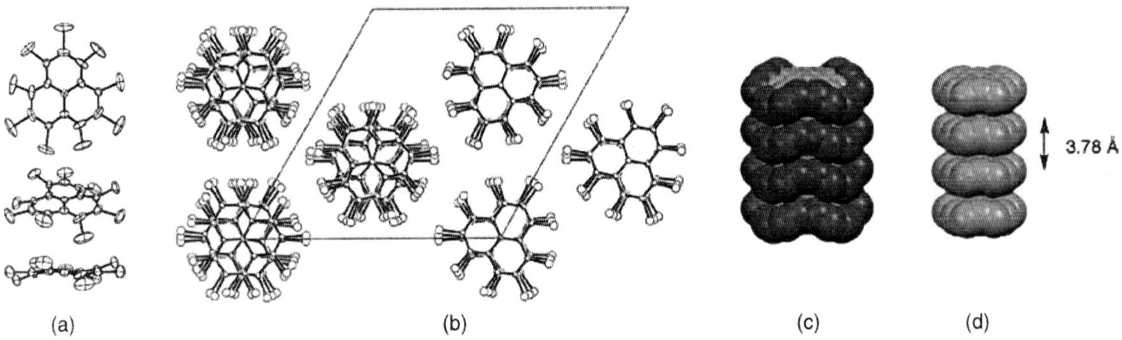

Figure 3.12 *(a) Molecular structure of **PCPLY** with a ruffled structure. (b) Projection along the c axis. Two sets of the three independent stacks are shown. The set on the left shows all six molecular orientations for each of the three stacks. The set on the right shows only three orientations for each stack; the other three are generated by the c glide operations. (c) Columnar structure of **PCPLY**. (d) Columnar structure without chlorine atoms. (a,b Reprinted with permission from [21]. Copyright 2001 American Chemical Society.)*

are only 3.02–3.03 Å even after distortion, despite the fact that the van der Waals radius of chlorine is only 1.80 Å. The molecule stacks with a quasi-superimposed motif to form a trimer, which is further stacked with the rotation by about 60° to construct a one-dimensional columnar structure with a hexagonally close-packed array. The intermolecular distance within a stack is 3.78 Å in the crystals, and is relatively long compared to the usual Cl–Cl distance (3.5–3.6 Å). The reason for this long intermolecular separation in **PCPLY** is due to the non-planarity of the phenalenyl skeleton inhibiting the formation of 60° rotated stacking motif as seen in **TBPLY** crystal. The magnetic susceptibility of **PCPLY** solid exhibits Curie paramagnetism in the temperature range 100–380 K until antiferromagnetic ordering at below 100 K. This solid is a Mott–Hubbard insulator with a conductivity at room temperature of 10^{-10} S/cm due to the isolation of the individual molecules.[21]

3.6 Dithiophenalenyl radicals

To obtain stable neutral radicals with less steric hindrance and further delocalization of an unpaired electron, Haddon and coworkers designed phenalenyl radicals having dithio substituent(s), such as dithiophenalenyl (**DTPLY**),[10a,17] tetrathiophenalenyl (**TTPLY**),[24] and hexathiophenalenyl (**HTPLY**).[24] **DTPLY** was prepared in 1978, and solution EPR and electrochemical measurements were reported.[17] Continuous studies over 30 years led to the X-ray single crystal structure analysis of **DTPLY** and elucidation of its solid state properties.[10a] **HTPLY**, having a threefold symmetrical molecular skeleton, has not been prepared so far, and is expected to have an extensively spin delocalized nature on the whole molecular skeleton involving the S–S moieties with a high molecular symmetry (*vide infra*). Furthermore, this molecule may form effective SOMO–SOMO and S–S interactions in the solid state and, therefore, is expected to exhibit intriguing magnetic and electronic functionalities based on the multidimensional network.

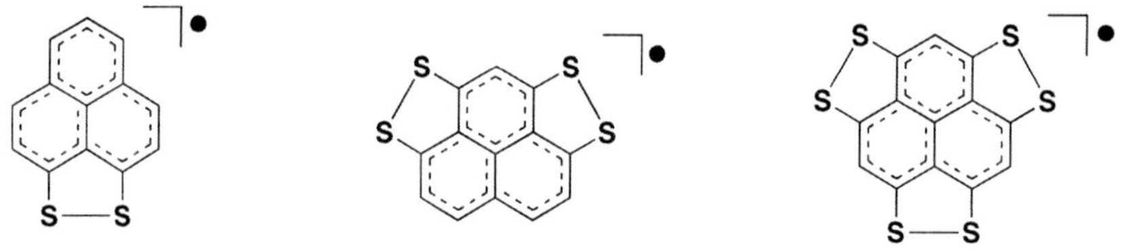

DTPLY was prepared according to Scheme 3.3. The cation salt is prepared by the reaction of 9-pentoxyphenalenone with an excess of phosphorus pentasulfide in toluene followed by reflux in methanol. Counter-anion exchange and reduction with decamethylmanganocene ($Cp^*{}_2Mn$) give **DTPLY** as a black

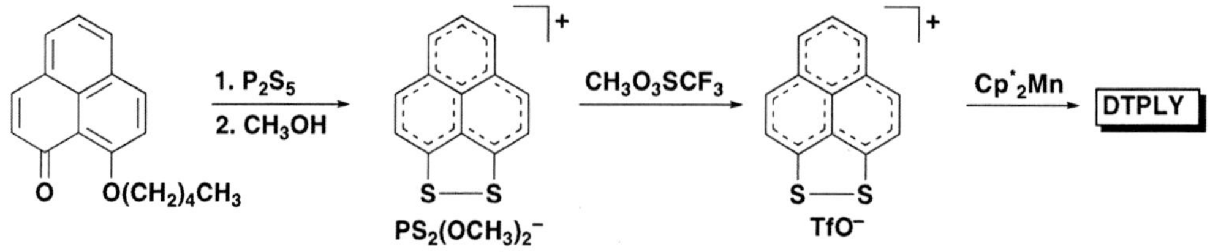

Scheme 3.3 Synthetic method for **DTPLY**

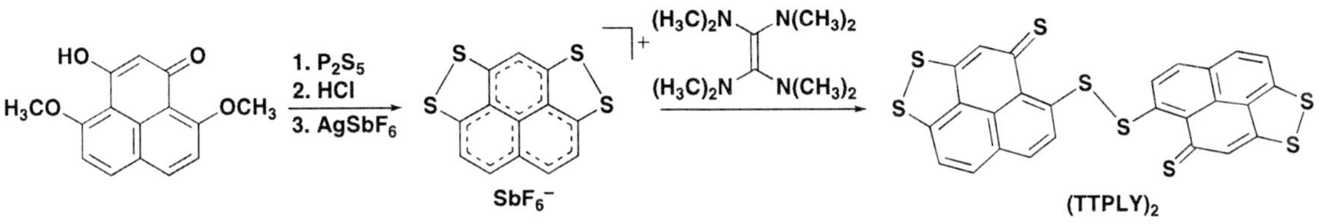

Scheme 3.4 Synthetic method for the **TTPLY** cation and a closed shell dimer (**TTPLY**)$_2$

microcrystalline solid which survives in air for up to 24 hours. In contrast, in solution the radical readily decomposes, even in the absence of oxygen.

The SbF_6 salt of the **TTPLY** cation is prepared by the reaction of 4,9-dimethoxy-3-hydroxyphenalenone[18a] with phosphorus pentasulfide followed by the treatment with hydrogen chloride and metathesis with $AgSbF_6$ (Scheme 3.4). The reduction with tetrakis(dimethylamino)ethylene gives a closed shell dimer of **TTPLY**, (**TTPLY**)$_2$, as black needle crystals. X-ray crystal structure analysis reveals an existence of a S–S σ bond formed by the reaction between two **TTPLY** (Figure 3.13). While the solid sample is silent in an EPR measurement, a weak EPR signal due to **TTPLY** is observed after dissolution in toluene. The resulting neutral radical shows a high stability in an oxygen-free solvent condition.

The spin density distributions of **DTPLY, TTPLY**, and **HTPLY** obtained by quantum chemical calculations are illustrated in Figure 3.14. The unpaired electron of each derivative delocalizes not only on the phenalenyl skeleton but also on the sulfur atoms. The calculation also shows that most of the spin on the phenalenyl skeleton is found on the α positions, similar to that found in pristine **PLY**. The redox

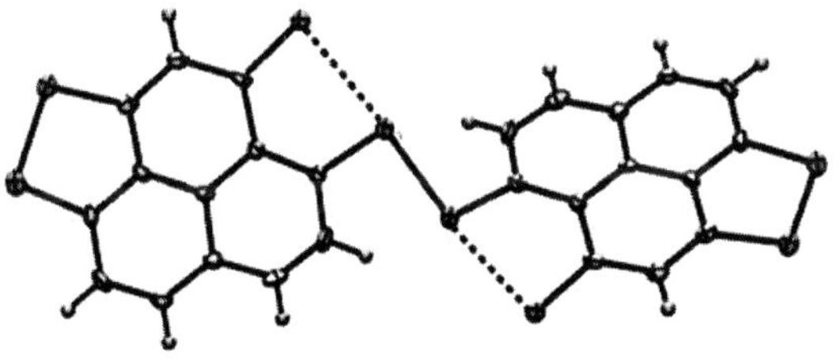

Figure 3.13 *Crystal structure of the closed shell dimer of **TTPLY**, (**TTPLY**)$_2$, with S–S σ bond. (Reprinted with permission from [24]. Copyright 2008 American Chemical Society.)*

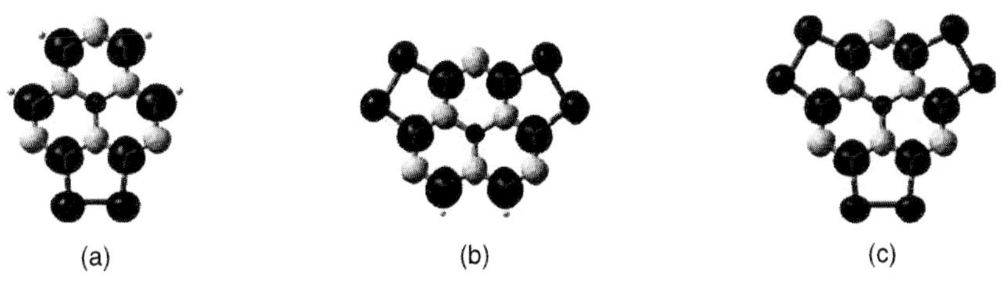

Figure 3.14 *Spin density distributions of (a) **DTPLY**, (b) **TTPLY**, and (c) **HTPLY**. (a,c Reprinted with permission from [10a]. Copyright 2007 American Chemical Society.)*

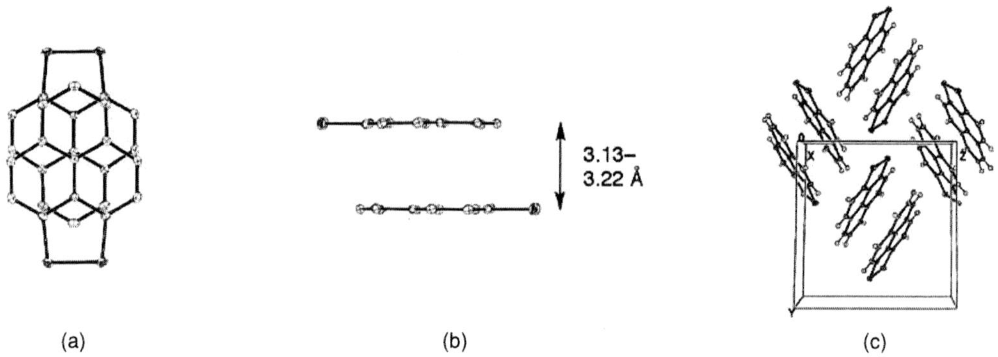

Figure 3.15 (a) π Dimeric structure from the top view. (b) Side view showing the 3.13–3.22 Å distance within the π dimer. (c) packing structure of **DTPLY** crystal with a herringbone motif. (a,c Reprinted with permission from [10a]. Copyright 2008 American Chemical Society.)

properties of **DTPLY** and **TTPLY** derivatives were studied through cyclic voltammetry studies on the corresponding cations in acetonitrile. For the PF_6^- salt of the **DTPLY** cation, three redox waves are observed, at −0.22, −0.77, and −1.55 V (vs SCE).[17] The TfO^- salt of the **DTPLY** cation also shows redox peaks at −0.29, −0.80, and −1.53 V (vs SCE).[10a] These redox events correspond to the reductions from the cation to the neutral radical, from the neutral radical to the anion, and from the anion to the dianion, respectively.[17] **TTPLY** shows two redox processes.[24] Cyclic voltammetry measurement on the SbF_6^- salt reveals a reversible reduction wave ($E_{1/2} = -0.34$ V vs SCE) from the cation to the neutral radical and the irreversible reduction wave ($E_p = -0.71$ V vs SCE) from the neutral radical to the anion. These molecules possess smaller gaps between the reduction potentials from the neutral radicals to the anions and the oxidation ones from the neutral radicals to the cations than that of the pristine **PLY**.

In the crystalline state, **DTPLY** possesses a planar structure with C_{2v} symmetry and forms a π dimer in a face-to-face fashion with a 180° rotation between the two radicals (Figure 3.15).[10a] This π-dimeric structure shows an almost complete superposition of six α-carbon atoms on the phenalenyl skeletons, similar to that of **TBPLY**, maximizing the intermolecular SOMO–SOMO overlap. The C–C distance within the π dimer (3.13–3.22 Å) is shorter than those of **TBPLY** (3.201–3.323 Å). The packing structure with a herringbone motif is also the same to that of **TBPLY**. It is noteworthy that **DTPLY** does not form a σ dimer structure in spite of the absence of bulky substituents. This is probably because the spin delocalization on the whole molecular skeleton involving the sulfur atoms greatly increases the stability of the neutral radical. Due to the strong π dimerization and absence of effective electronic contacts between π dimers, room temperature conductivity of compressed pellet sample is $<10^{-6}$ S/cm.

As demonstrated in studies on **DTPLY** and **TTPLY**, the phenalenyl derivatives having alkoxy or oxo groups are important synthetic intermediates for the introduction of dithio substituent(s) into the phenalenyl skeleton. In the light of this synthetic viewpoint, studies on cationic species, such as the 1,9-diethoxyphenalenyl cation and 1,3,4,9-tetramethoxyphenalenyl cation, conducted in the 1970s and 1980s had great foresight.[17–20] Very recently, the synthesis and X-ray crystal structure analysis of the phenalenyl cation with six methoxy groups at all α-carbon atoms were reported (Figure 3.16).[25] This cation has a high stability even in air and, therefore, is an important intermediate for the synthesis of **HTPLY**.

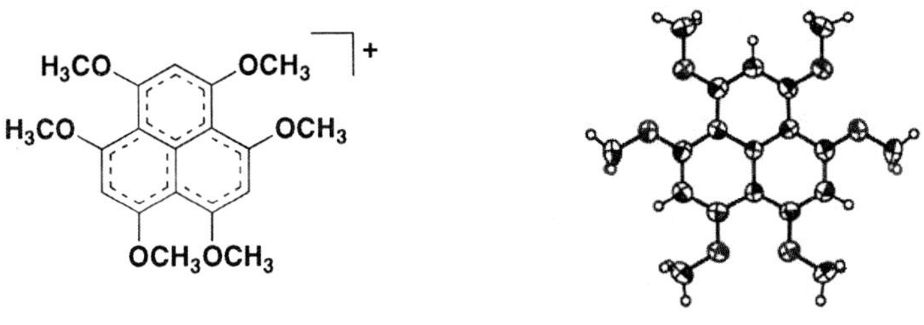

(a) (b)

Figure 3.16 (a) Chemical structure and (b) X-ray crystal structure of the 1,3,4,6,7,9-hexamethoxyphenalenyl cation.

3.7 Nitrogen-containing phenalenyl systems

3.7.1 Molecular design and topological isomers

The incorporation of nitrogen atoms into a persistent hydrocarbon radical skeleton is an intriguing concept for elucidating the effects on electronic structure and stability of the radical in air due to the higher electronegativity of nitrogen relative to carbon. The degree of modulation effect in the phenalenyl system is expected to depend on the incorporated positions of nitrogen atoms because the coefficient of the SOMO only exists on its α positions. Furthermore, incorporation of nitrogen atoms may be a powerful tool for the construction of multidimensional network structures by coordination to metal ions and also by hydrogen bonding interactions with proton donors.

Taking the molecular symmetry into account, nitrogen-incorporated phenalenyl (**APLY**) derivatives are classified as α and β type by the positions of the nitrogen atoms (Figure 3.17). The group illustrated in Figure 3.17a is major prototypes of the α-type **APLY**. The diazaphenalenyl derivatives are topological isomers of each other, in which only the incorporated positions of nitrogen atoms are different. Quantum chemical calculations suggest that their spin density distributions are basically similar to that of pristine **PLY**. Furthermore, the spin densities on nitrogen atoms of the diaza- and triazaphenalenyls decrease in comparison with those on carbon atoms in each structure (Figure 3.17a).

Three derivatives of the β-type **APLY**s are shown in Figure 3.17b. In these systems, perturbation to the electronic structures by nitrogen atoms is expected to be small because the nitrogen atoms locate on positions on the phenalenyl skeleton bearing small and negative spin densities. However, their lone pair electrons are insusceptible to the steric effects of the hydrogen atoms at neighboring *peri* positions because the difference of the directionality of their lone pair electrons from that of the α-type **APLY**s. Thus, from the viewpoint of materials science challenges, there are many interesting possibilities in terms of the construction of network structures based on the abilities of the nitrogen atoms to coordinate to metal ions and to engage in hydrogen bonding.

3.7.2 2,5,8-Tri-*tert*-butyl-1,3-diazaphenalenyl

The synthetic method of tri-*tert*-butylated 1,3-diazaphenalenyl (**TBDAP**) is depicted in Scheme 3.5.[26] Di-*tert*-butylated 1,8-diaminonaphthalene, obtained by reduction of the corresponding dinitro derivative, is condensed with pivalaldehyde and dehydrogenated in the presence of palladium carbon, giving a

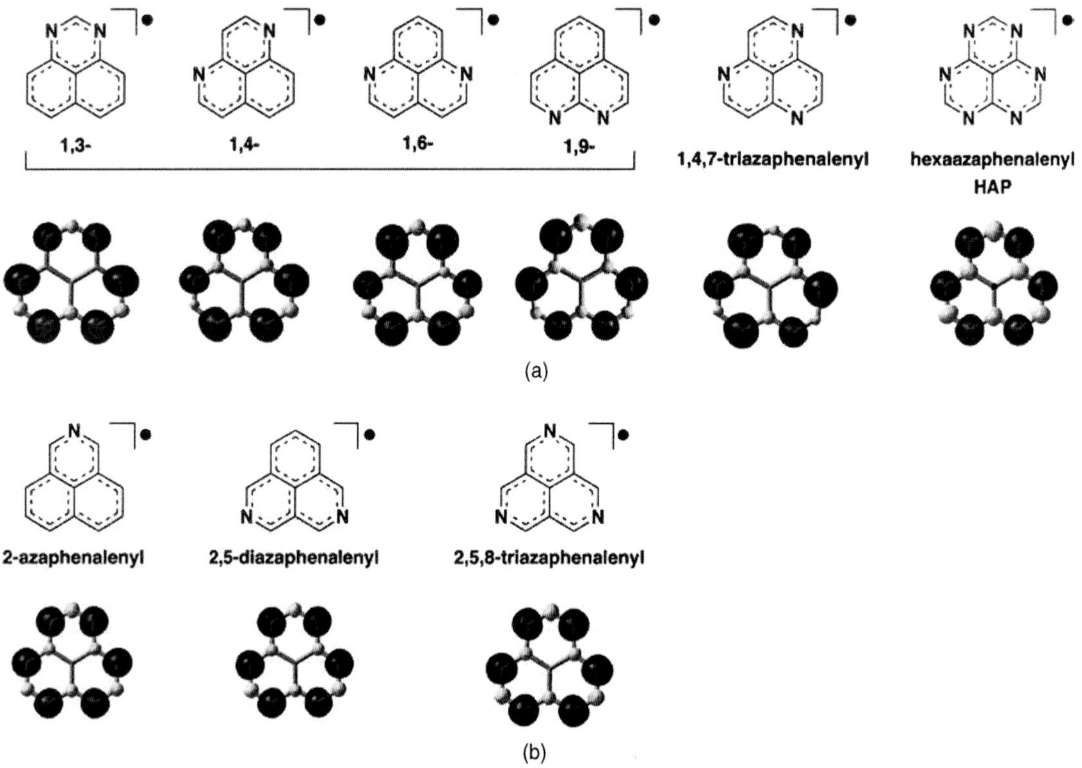

Figure 3.17 *Topological isomers of the azaphenalenyl (**APLY**) system. (a) α-incorporated **APLY**s, (b) β-incorporated **APLY**s, and their spin density distributions. Calculations carried out at the SVWM/6-31G**//SVWM/6-31G** level of theory.*

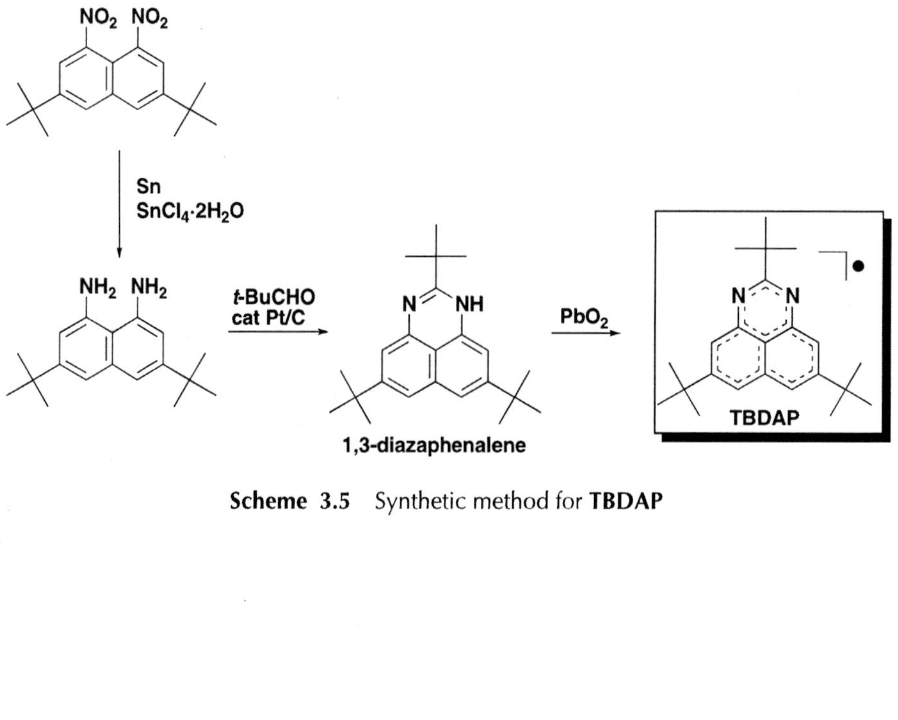

Scheme 3.5 Synthetic method for **TBDAP**

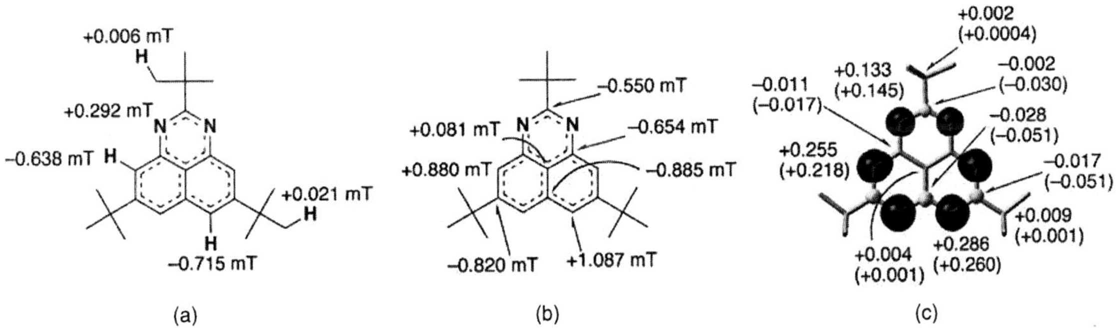

Figure 3.18 *Observed hfccs of (a)* ^1H, ^{14}N, *and (b)* ^{13}C; *(c) experimental and theoretical (in parentheses) spin density distributions of* ***TBDAP****, in which black and gray regions denote positive and negative spin densities, respectively. Calculations carried out at the SVWN/6-31G**//SVWN/6-31G** level of theory.*

1,3-diazaphenalene,[27] a radical precursor. Oxidation with lead(IV) oxide (PbO_2) in degassed benzene affords **TBDAP** as dark green crystals. Interestingly, the crystals are extremely stable in air with no significant decay of the color for years, which shows a high contrast with **TBPLY**. The decomposition point in air is 164–166 °C.

Solution EPR studies and quantum chemical calculations of **TBDAP** reveal that the spin densities on the nitrogen atoms slightly decrease and those of the carbon atoms on the α positions increase in comparison to **TBPLY**, being in a good agreement with the prediction (Figure 3.18). The deviation of the electronic structure from C_3-symmetry gives a clue to the nature of the σ dimer (Figure 3.20) with a C–C bond in solution at low temperature.[28] This contrasts the case of **TBPLY**, which forms a highly symmetrical π dimer in the solution at low temperature.[15]

Single crystals of **TBPLY** show temperature-dependent, continuous color changes (Figure 3.19).[29] The crystal at 100 K is colorless (or extremely pale green). On increasing the temperature, the crystal color gradually and monochromatically turns to green and then gradually deepens at higher temperature, becoming dark green at 300 K. This monochromatic change of the crystal color gives a color thermometer with a high degree of resolution. In a sharp contrast, the crystal of **TBPLY** does not show such a color-changing phenomenon. The thermochromic behavior of the purely organic open shell molecule in the crystalline state is very rare, and known examples occur only by structural phase transitions that show discontinuous color changes.[30] Furthermore, it is noteworthy that the temperature-dependent color change seen in the **TBDAP** single crystal is the entirely unknown dynamic behavior not only in organic compounds but also in metal complexes. This unconventional dynamic behavior in the monochromatic continuous crystal-color change is fully corroborated by the temperature-dependent polarized electronic spectra of the single crystal (Figures 3.19b and 3.19c).[29]

To elucidate the detailed molecular level mechanism from a structural viewpoint, X-ray crystal structure analyses at several temperatures were carefully carried out. The experiments reveal the coexistence of two dimers of different bonding natures, σ and π dimers, in the single crystal (Figure 3.20).[29] Furthermore, the ratio of σ and π dimers is temperature-dependent, indicating a thermal equilibrium between the two types. The temperature-dependent polarized electronic spectra of the single crystal reveal that the σ and the π dimer are colorless and dark green, respectively (Figures 3.19b and 3.19c). The continuous changes in the ratio of the two dimers give a rationale for the monochromatic continuous color change of the crystal in the absence of structural phase transitions. In addition, the color originates from the relative positional relationship between the radicals in the π dimer, and thus the **TBDAP** single crystal possesses polarization

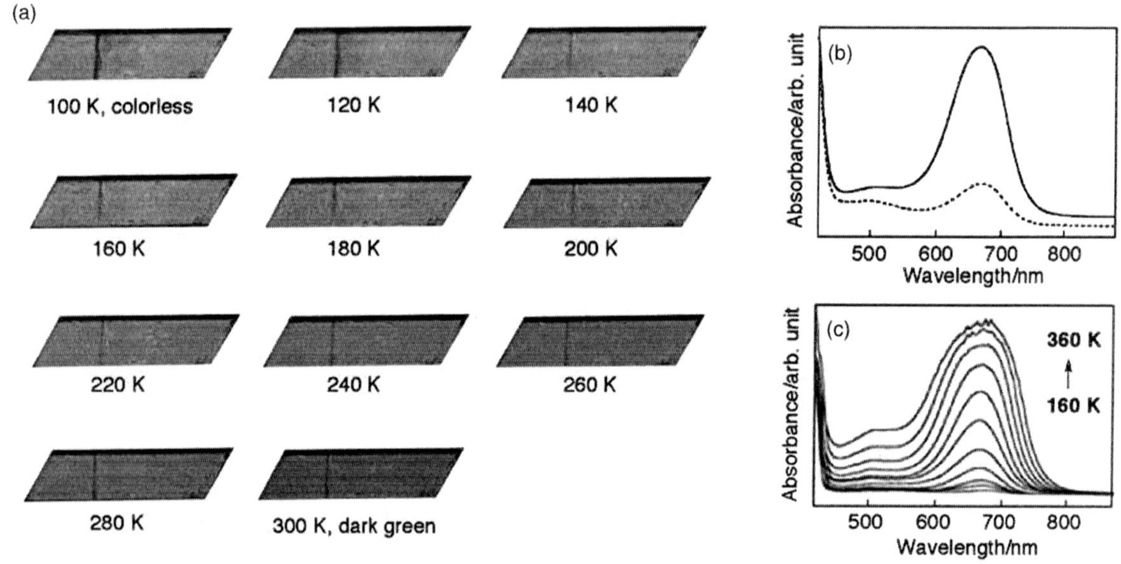

Figure 3.19 (a) Pictures of ***TBDAP*** single crystal showing a continuous color change from colorless to dark green depending on the temperature. (b) Polarized absorption spectra of a ***TBDAP*** single crystal with polarized light parallel to the c axis (solid line) and perpendicular to the c axis (dashed line) on the crystal plane (010) at 300 K. (c) Temperature-dependent polarized absorption spectra of a ***TBDAP*** single crystal with polarized light parallel to the c axis at 160–360 K. (Reprinted by permission from Macmillan Publishers Ltd,[29] copyright 2008.) A full-colour version of part (a) of this figure appears in the Colour Plate section of this book.

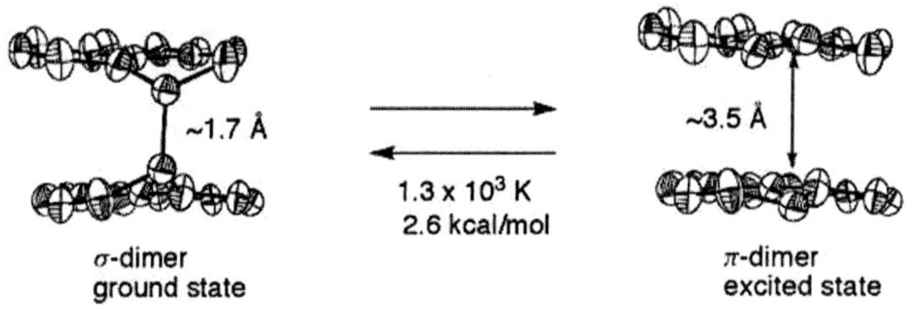

Figure 3.20 Crystal structures of the σ and π dimer of ***TBDAP*** and the experimentally obtained energy gap between two states. tert-*Butyl groups are omitted for clarity.*

properties exhibiting the color change only for light from a specific direction. The energy gap between the two dimer states is experimentally determined to be 1.3×10^3 K (2.6 kcal/mol) by the temperature-dependent changes of the oscillator strength in visible area, being in a good agreement with the difference in the total energy of 3.0 kcal/mol between the optimized σ and π dimers calculated by the B3LYP/6-31G* method. This value is as same as the rotation barrier energy in the C–C bond of ethane (2.9 kcal/mol). By the EPR observation of the thermally accessible triplet state, a complete energy diagram of **TBDAP** in the crystalline state is unequivocally determined to have three states: two singlet states (corresponding to the σ and π dimers) and an excited π dimer triplet state with the energy gap of 1.3×10^3 K and 4.2×10^3 K. Note that this is the first example where the energy levels of open shell organic molecules with three

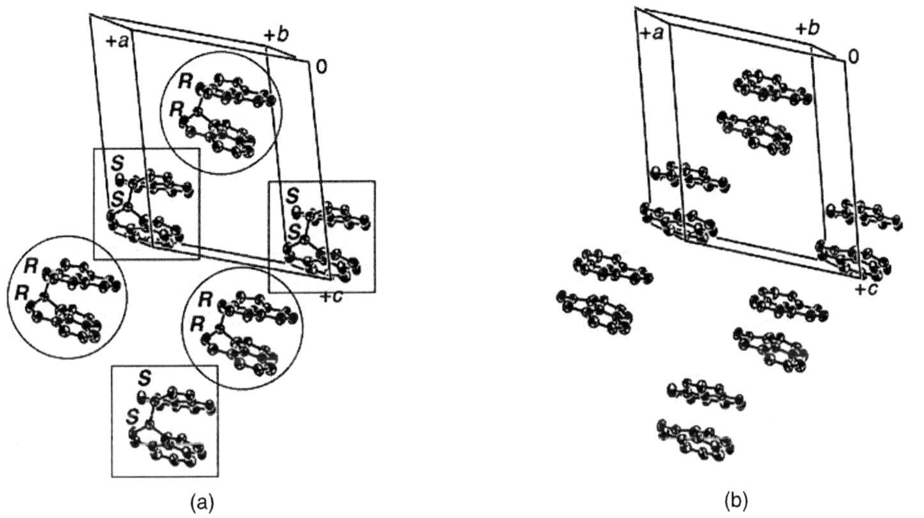

Figure 3.21 *Packing diagram of (a) σ dimer and (b) π dimer of **TBDAP** extracted from disordered arrangement. The circles and squares represent the same enantiomers in the crystal.* tert-*Butyl groups are omitted for clarity.*

energy states have been experimentally located. These unconventional features with high stability of the radical are applicable to the micro-optical temperature sensor and the adaptive polarizing optics that cover a wide range of temperature (100–430 K).

The **TBDAP** dimers form a 1-D columnar structure in the crystal (Figures 3.21 and 3.22),[26] in sharp contrast to the herringbone motif for **TBPLY**. In the σ-dimeric columnar structure extracted from the disordered arrangement, each σ dimer is chiral and has a dipole moment (Figure 3.21a). However, the ratio of the optical isomers is 1:1, as expected for an achiral space group ($P2_1/n$: No.14), and the dipole

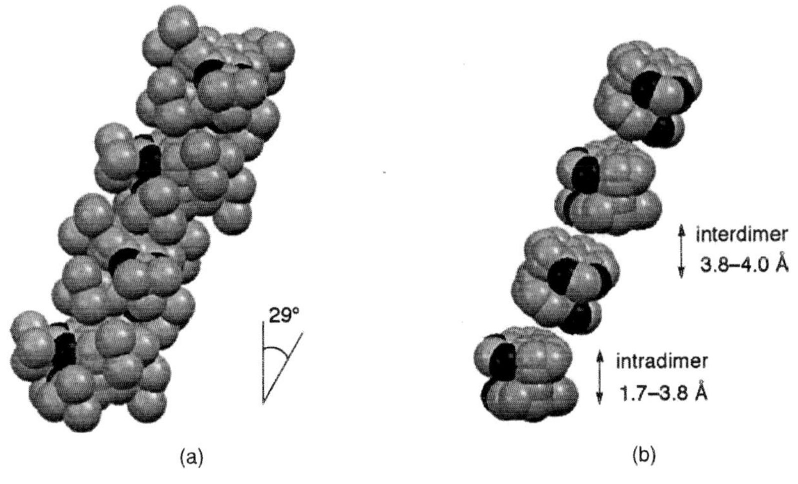

Figure 3.22 *1-D columnar structures of **TBDAP** crystal, (a) with and (b) without* tert-*butyl groups.*

moments are canceled by the symmetry of the dimers in the crystal. The finding of the columnar motif for **TBDAP** derived by the heteroatomic modification is of great interest in crystal engineering and organic material sciences.

3.7.3 Hexaazaphenalenyl derivatives

1,3,4,6,7,9-Hexaazaphenalenyl (**HAP**) is a highly symmetrical molecule with full nitrogen substitution in all α sites of the phenalenyl. The lone pair electrons on the six nitrogen atoms are directed radially and can form coordination and hydrogen bonds to the outside (Figure 3.23a). Quantum chemical calculations suggest that the HOMO energy level of the anion state decreases substantially in this molecular system in comparison with the pristine **PLY** anion (Figure 3.23b). Such a marked decrease in the HOMO energy is not seen in other azaphenalenyl anions, indicating the exceptionally higher stability of the **HAP** anion among the phenalenyl anions. In 1964, Tomlin first described a preparation of the **HAP** system in his Ph.D. thesis; however, the structural and physical properties were not examined.[31] In 2005, 1,3,4,6,7,9-hexaazaphenalene, one of the **HAP** systems, was prepared according to Tomlin's method with some modifications (Scheme 3.6), and the potassium salt of the anion is isolated as stable crystals in air at room temperature.[32] As mentioned above, the stabilities of **PLY** and **TBPLY** anions in air are very low in contrast to the corresponding cation species. Therefore, the **HAP** anion is the first phenalenyl one isolated as air-stable crystals. Cyclic voltammetry of the potassium salt gives an irreversible oxidation wave at around +1.2 V (vs Fc/Fc^+), which is assigned to the oxidation of the anion to the neutral radical. This potential is much higher than that of the pristine **PLY** (−0.9 V vs SCE[17]), consistent with the decrease of the HOMO energy of the **HAP** anion suggested by quantum chemical calculation.[32] In a copper (II)

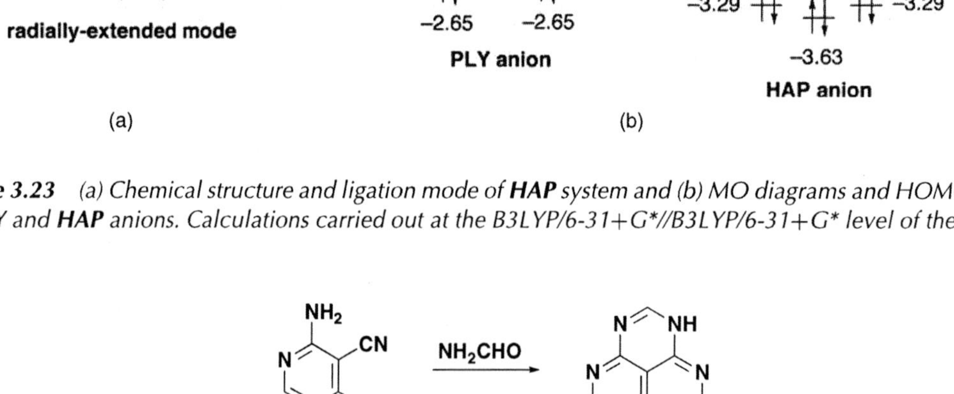

*Figure 3.23 (a) Chemical structure and ligation mode of **HAP** system and (b) MO diagrams and HOMO pictures of **PLY** and **HAP** anions. Calculations carried out at the B3LYP/6-31+G*//B3LYP/6-31+G* level of theory.*

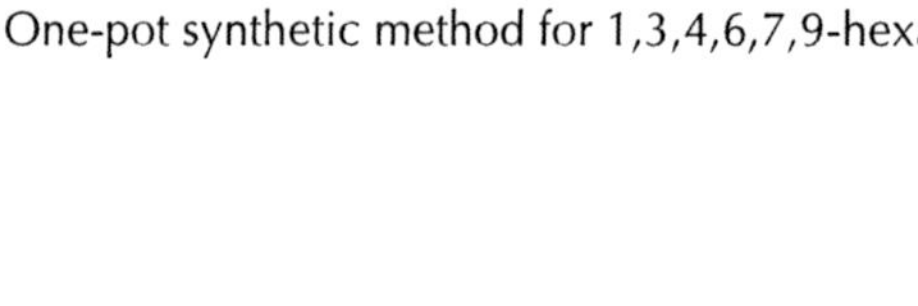

Scheme 3.6 One-pot synthetic method for 1,3,4,6,7,9-hexaazaphenalene

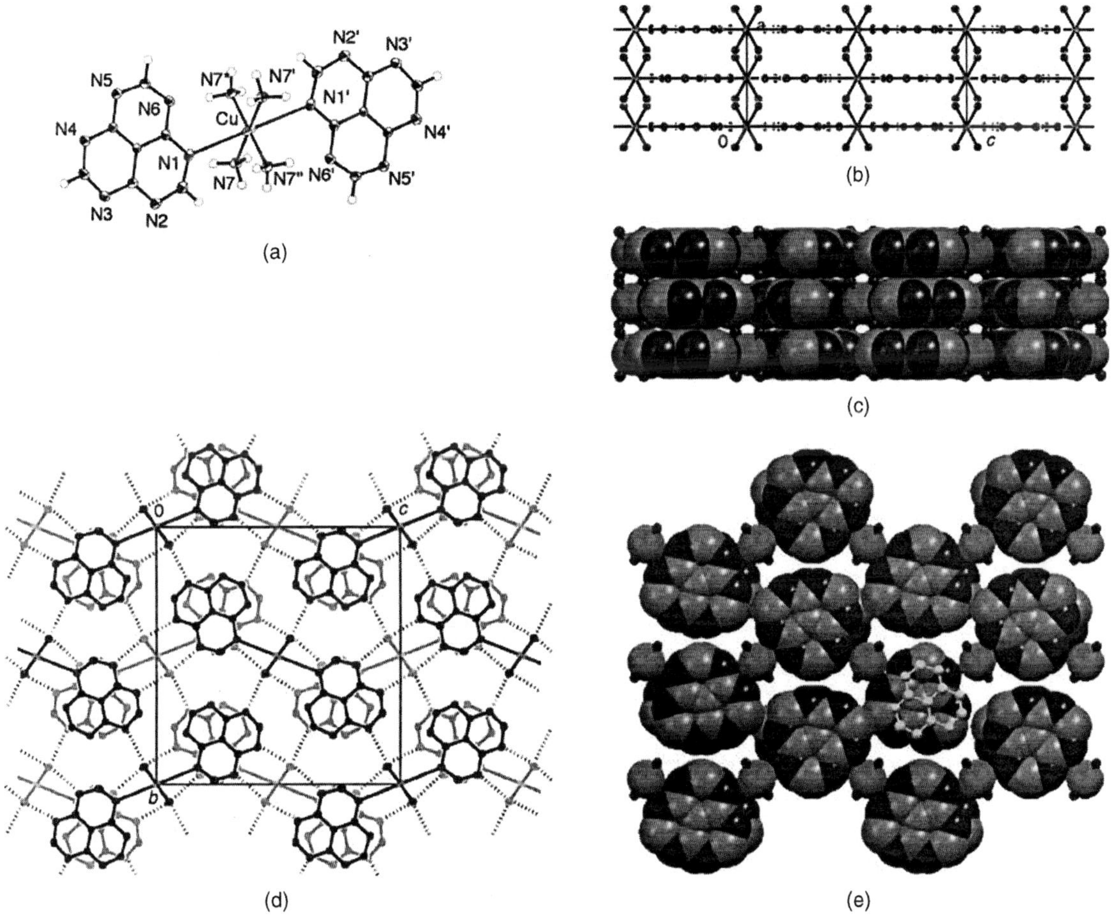

Figure 3.24 ORTEP representation of copper (II) complex of **HAP** anion. (a) molecular structure, (b) packing diagrams, (c) packing structure in ac-plane, and (d) (e) bc-plane. Dashed lines denote hydrogen bond. The light colored molecular framework belongs in the next sheet of the forefront network. In the packing structure of (e), HAP skeleton is represented by the white color. (a,b,d Reprinted with permission from [32]. Copyright 2005 American Chemical Society.)

complex, two **HAP** anions coordinate to a copper atom in a *trans*-fashion, and the anions are uniformly stacked (Figure 3.24). The resulting 1-D columns are connected with each other by multiple hydrogen bonds through ammonia molecules coordinated to the copper atom, resulting in a 3-D network. These results indicate that the **HAP** anion works as an adhesive ligand that is suitable for in-plane networking as well as out-of-plane π stacking.

3.7.4 β-Azaphenalenyl derivatives

For the construction of 2-D and 3-D interactions used by the coordination of nitrogen atoms to metal ions, Rubin designed and synthesized the 2-azaphenalenyl (**2APLY**)[33] and perchloro-2,5,8-triazaphenalenyl (**PTAZ**) radicals.[34] Because the nitrogen atoms are incorporated into the β positions with very small

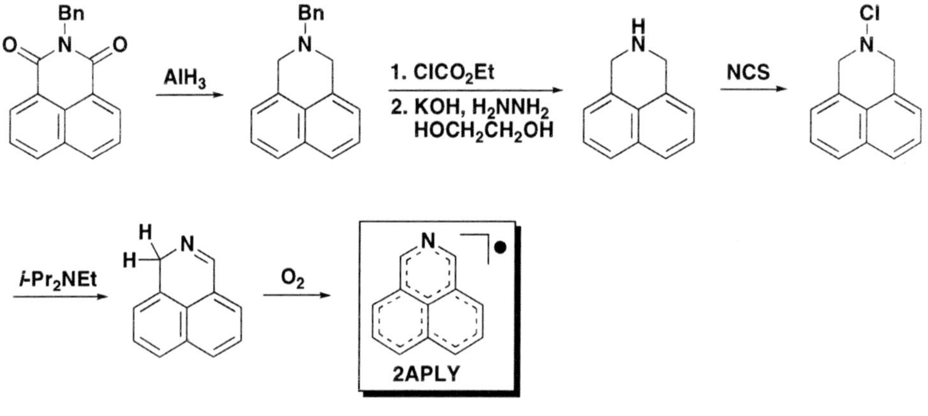

Scheme 3.7 Synthetic method for **2APLY**

coefficients of the SOMO, perturbations on their electronic structures are expected to be much smaller than those of α-type **APLY**s.

The synthetic method for **2APLY** is shown in Scheme 3.7.[33] Reduction of the carbonyl groups of the 1,8-naphthalimide derivative and the following deprotection of the benzyl group give the amine derivative. *N*-Chlorination, dehydrochlorination and oxidation with oxygen give **2APLY** as a yellow–green solution. While this neutral radical is stable in a degassed solution for months, the radical stability with air in the solution and in the solid state is low, hampering its isolation as a pure form.

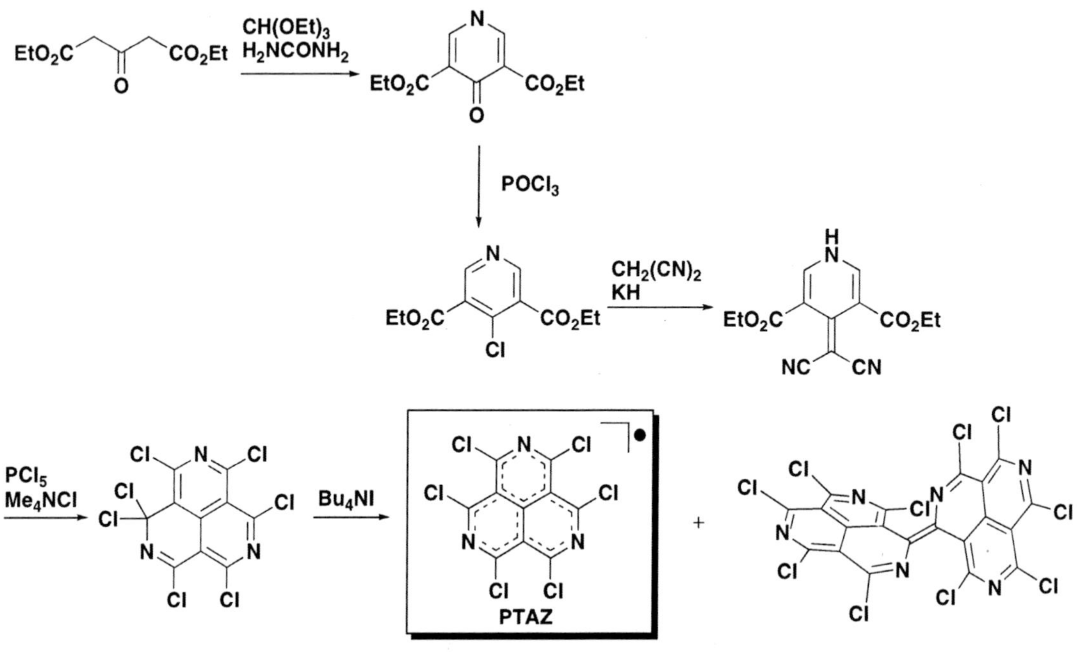

Scheme 3.8 Synthesis of **PTAZ** and the dechlorinated closed shell dimer

The synthesis of **PTAZ** is achieved by an intramolecular double cyclization reaction as a key step (Scheme 3.8).[34] The chloropyridine derivative, prepared in two steps from diethyl 1,3-acetonedicarboxylate, is converted to the dicyanomethylene derivative. The triazaphenalene skeleton is constructed by the solid state reaction with phophorus pentachloride (PCl_5) and tetramethylammonium chloride (Me_4NCl) through chlorination and cyclization in one pot, and is obtained as a moisture-sensitive compound. Reduction of the triazaphenalene with Bu_4NI gives a mixture of **PTAZ** and its dechlorinated closed shell dimer having a C=C double bond. The neutral radical is purified by silica gel column chromatography and isolated as a dark blue solid with a high stability in air.

The spin density distributions of these neutral radicals are determined by EPR measurements and density functional theory (DFT) calculations (Figure 3.25). The results show large spin densities on their α carbons similar to pristine **PLY**.

The crystal structures of these β-azaphenalenyl neutral radicals were not elucidated. However, the copper(II) diimine chelate complex of the closed shell σ dimer, with a C–C single bond formed by dimerization at the α positions of the two **2APLY**, was obtained by means of crystallization of **2APLY** with $Cu(hfac)_2$ (hfac = hexafluoroacetylacetonato) (Figure 3.26).

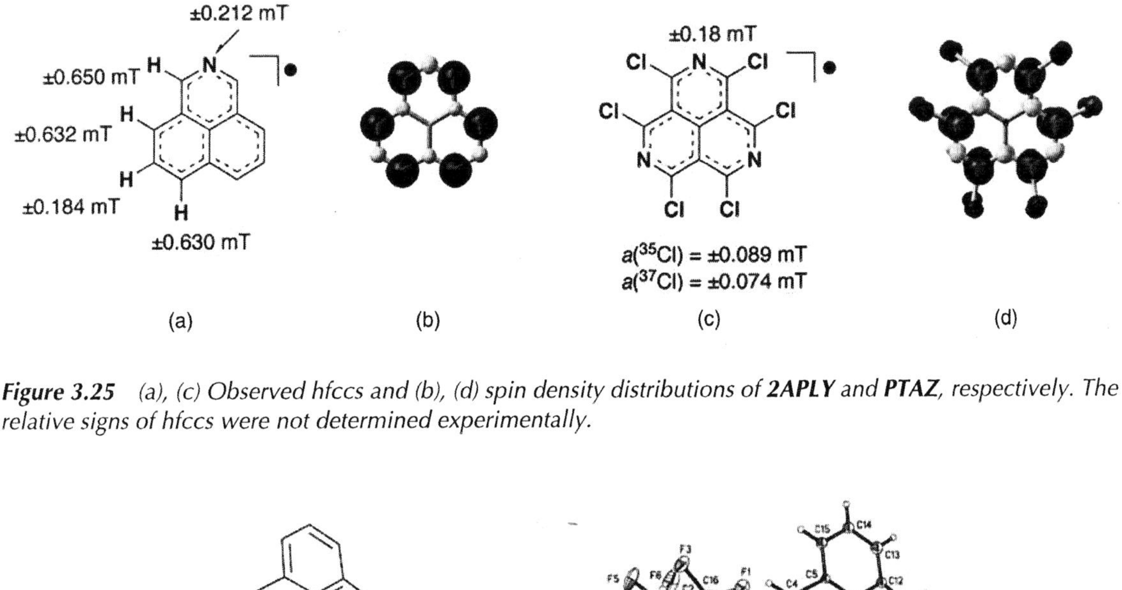

Figure 3.25 (a), (c) Observed hfccs and (b), (d) spin density distributions of **2APLY** and **PTAZ**, respectively. The relative signs of hfccs were not determined experimentally.

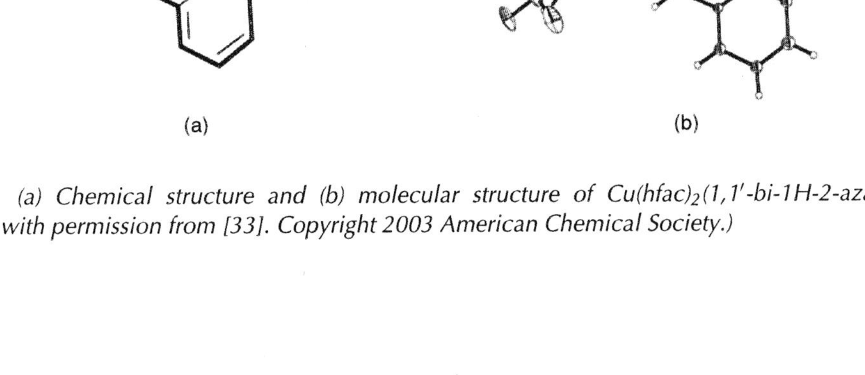

Figure 3.26 (a) Chemical structure and (b) molecular structure of $Cu(hfac)_2$(1,1′-bi-1H-2-azaphenalene). (b Reprinted with permission from [33]. Copyright 2003 American Chemical Society.)

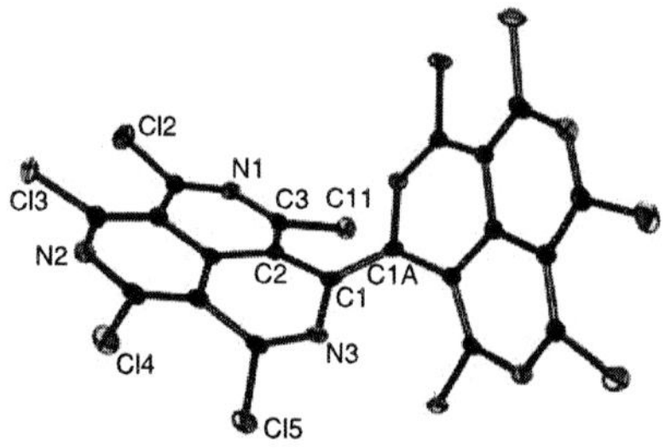

Figure 3.27 Crystal structure of the dechlorinated closed shell dimer of ***PTAZ****. (Reprinted with permission from [34]. Copyright 2005 American Chemical Society.)*

Analysis of the magnetic susceptibility of **PTAZ** indicates that the radical forms a π dimer similar to the case of **TBPLY** in the solid state. The color of the dechlorinated closed shell dimer of **PTAZ** is reddish in the crystal, but it turns to deep blue upon dissolution. X-ray crystal structure analysis shows the bent nature of the double bond between the two triazaphenalene units (Figure 3.27). Therefore, the color change may be due to an increase of bond distortion in solution or free rotation of the strained central C=C bond.

The dechlorinated σ bond formation nature of **PTAZ** may indicate that the β-incorporated **APLY** derivatives have similar spin distributions as the pristine **PLY**. However, as mentioned above, **PCPLY** does not form an σ bond dimer between the neutral radicals. Thus, it is likely that the modulation effect in the electronic structure by the introduction of nitrogen atoms appears in **PTAZ**.

3.8 Oxophenalenoxyl systems

3.8.1 Molecular design and topological isomers

All of the phenalenyl derivatives having *tert*-butyl, chloro-, and dithio- substituents, as well as the nitrogen-containing ones described above, have more or less phenalenyl-type electronic structures with large positive spin densities at the α positions. On the other hand, introduction of two oxygen atoms into the phenalenyl skeleton as oxo- and oxyl- functional groups produces the novel neutral radicals termed "oxophenalenoxyl" (**OPO**) (Figure 3.28). A variety of topological isomers exists depending on the positions of the oxygen substituents, and in general these neutral radical systems have electronic structures dramatically different from that of the **PLY** system. A β, β-substituted isomer is represented as a triradical structure, implying low stability of this system. In contrast, α, β- and α, α-substituted isomers give monoradical structures. Among them, α, α- isomers are expected to possess high stabilities by extensive delocalization of unpaired electrons (except for **3OPO**) with unique topological symmetries illustrated by resonance structures (Figure 3.29a). The quantum chemical calculations for **OPO** indicate that most of the spin densities exist on the β carbons and the two oxygen atoms in all **OPO** topological isomers, showing the remarkable difference from that of the **PLY** system (Figure 3.29b). Furthermore, the distribution natures change significantly depending on the positions of two oxygen atoms, which is supported by the resonance structures and local aromaticities by NICS calculations (Figure 3.29a,c).[16] The dependence of the spin distributions on the positions of oxygen substituents has been termed "spin topological symmetry control".[35,36]

The molecular orbital pictures and energy diagrams of **PLY, 4OPO** and **6OPO** are illustrated in Figure 3.30. In sharp contrast to the **PLY** system, the coefficients of the SOMO in the **OPO** systems are mainly centered on the β carbons and the oxygen atoms. Furthermore, significant lowerings of the

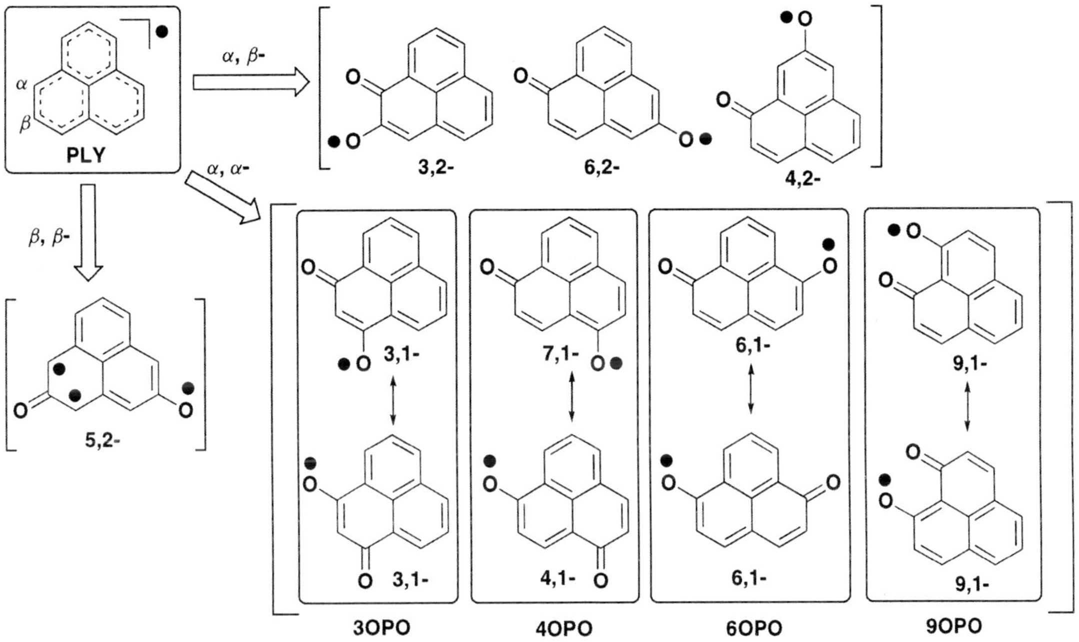

Figure 3.28 Topological isomers of the oxophenalenoxyl (**OPO**) system.

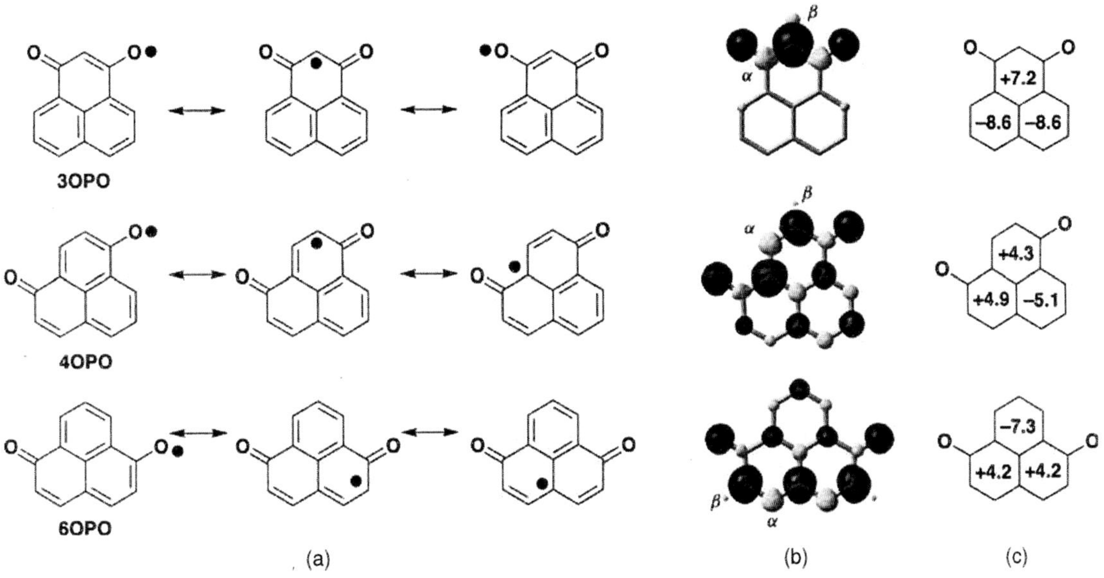

Figure 3.29 (a) Representative local spin structures and (b) spin density distributions. (c) NICS values of ***3OPO***, ***4OPO***, and ***6OPO***.

SOMO levels of the **OPO** systems compared to that of the **PLY** system imply high electron accepting abilities of the **OPO** systems. The energy gaps between the SOMO and LUMO in the **OPO** systems are significantly smaller than that of the **PLY** system, indicating multistage electron accepting abilities which may be key factors for applications as electrode active materials in secondary batteries (*vide infra*). As a matter of fact, cyclic voltammetry measurements of **3OPO, 4OPO**, and **6OPO** systems show two-stage one-electron reduction behaviors (*vide infra*). Importantly, the coefficients of the SOMOs of the resulting radical dianions (LUMO in the neutral radical systems) mainly exist on the α positions of the phenalenyl skeletons, implying the phenalenyl-type electronic structure (Figure 3.30). This electronic feature is also supported by quantum chemical calculations (Figure 3.31). Such a phenomenon – showing a remarkable change in the electronic structures depending on the redox states – is a rare event, and is termed "redox-based spin diversity".[36,37a,38] Considering all of these significant electronic features, **OPO**s are viewed as completely novel neutral radical systems, unlike the **PLY** system, and have a high potentiality for creating a qualitatively different molecular functionality from that obtained by conventional phenoxyl radical.

3.8.2 3-Oxophenalenoxyl (3OPO) system

The synthetic method for **3OPO** is depicted in Scheme 3.9. Condensations of 1,8-naphthalic anhydride with arylacetic acid derivatives gives 2-aryl-substituted 3-hydroxyphenalenones as **3OPO** precursors. Oxidation of these hydroxyl compounds affords the closed shell σ dimers of **3OPO** derivatives.[39] These experimental observations are well understood by the major contribution of the spin density distribution on the 2-position

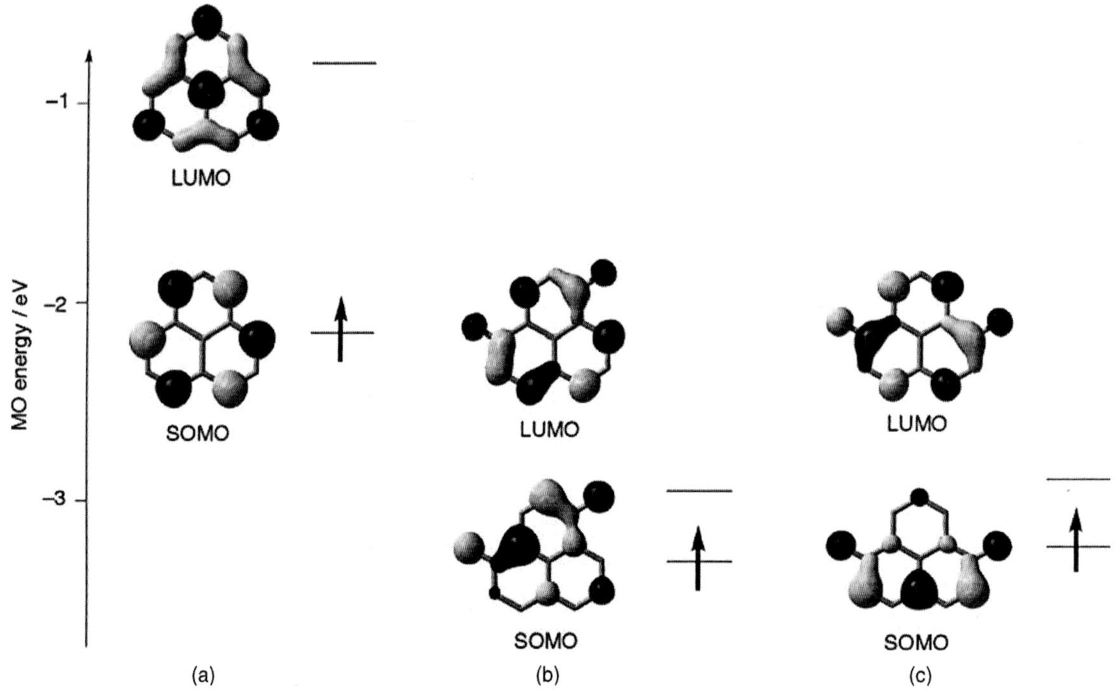

*Figure 3.30 Molecular orbital pictures and energy diagrams of (a) **PLY**, (b) **4OPO**, and (c) **6OPO**. Calculations carried out at the ROBLYP/6-31G**//ROBLYP/6-31G** level of theory.*

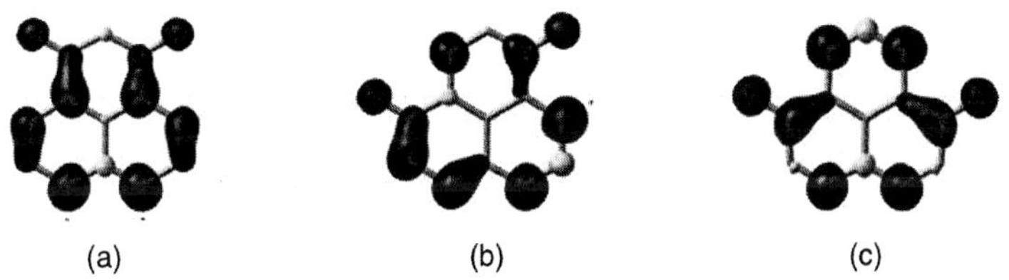

Figure 3.31 Spin density distributions of the radical dianions of (a) ***3OPO***, (b) ***4OPO***, and (c) ***6OPO*** systems.

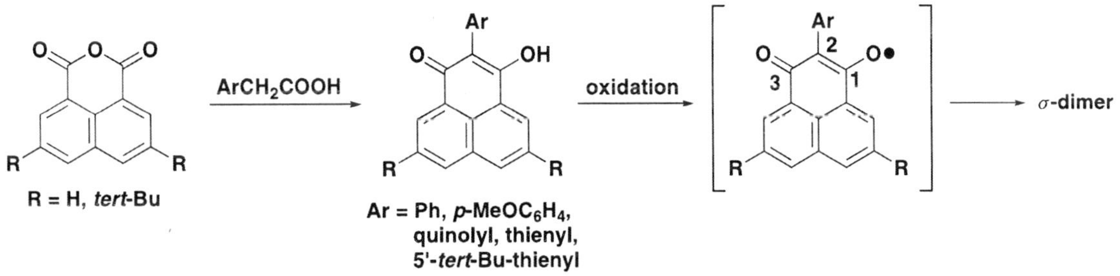

Scheme 3.9 Synthetic method for 3-hydoxyphenalenone derivatives and their σ-dimerization reactions upon oxidation

of **3OPO** (Figure 3.29). These chemical reactivities were extensively investigated with various substituents of the **3OPO** system, indicating a general trend in the formation of the closed shell σ dimers at the 2-positions. Pyrolysis of the σ dimer gives the corresponding neutral radical of **3OPO** in solution, although this radical could not be isolated.[40]

For the stabilization of **3OPO**, suppression of reactivity at the 2-position is the most effective approach. Thus, chemical modification for the decrease in spin density at the 2-position by spin delocalization into the π system substituted at 2-positions was investigated. Phenyl, *p*-methoxyphenyl, and quinolyl derivatives result in σ-dimeric structures between the carbon atoms at the 2-positions (Figure 3.32a,b).[39a] In sharp contrast, in the σ dimer of the thienyl derivatives, the carbon atoms on the thiophene rings take part in bond

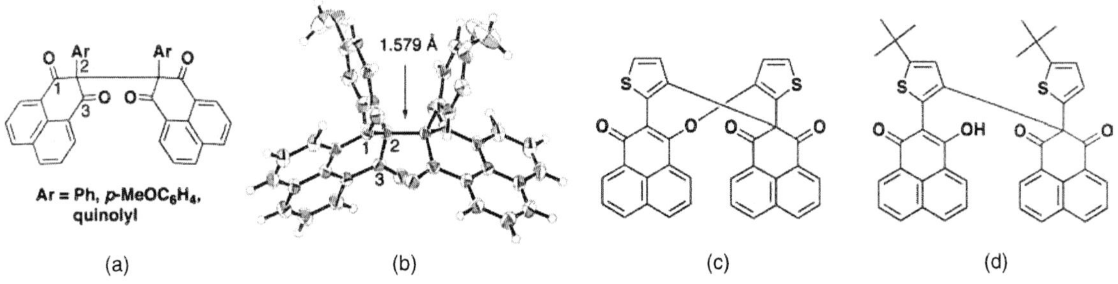

Figure 3.32 σ-Dimeric structures of (a) phenyl, p-methoxyphenyl, and quinolyl derivatives, (b) X-ray crystal structure of p-methoxyphenyl derivative. (c) and (d) σ dimeric structures of thienyl derivatives. (b Reprinted from [39a], Copyright 2005, with permission from Elsevier.)

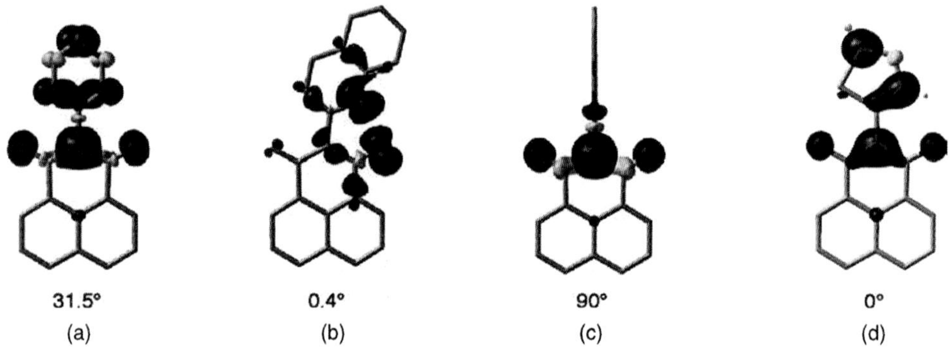

Figure 3.33 *Spin density distributions and torsion angles of (a) 2-phenyl, (b), (c) 2-quinolyl, and (d) 2-thienyl substituted **3OPOs**. (a,c Reprinted from [39a], Copyright 2005, with permission from Elsevier.)*

formation between the radicals (Figure 3.32c,d). This result indicates that the degree of spin delocalization into the thiophene ring seems to be relatively large because thiophene is more likely to be coplanar with the phenalenone skeleton than six-membered rings such as the phenyl and quinolyl derivatives.[39b]

Quantum chemical calculations also support this kind of stereoelectronic effects (Figure 3.33). While the phenyl derivative has a torsion angle of 31.5°, the thienyl derivative possesses a coplanar structure, which gives rise to a considerable amount of spin delocalization into the thienyl moiety and, correspondingly, a decrease of the spin density at the 2-position. Furthermore, the quinolyl derivative exhibits significant changes in its electronic structure depending on the torsion angle (Figure 3.33b,c). This study demonstrates an effective manipulation of the degree of spin delocalized nature of open shell π-electronic systems by invoking functional substitution, and provides important criteria to design novel spin delocalized π radicals for realization of novel spin-mediated molecular functional materials. From these points of view, utilization of intermolecular bond formation for molecular memory is an intriguing issue[41]: that is, the weak bonding nature seen in the π-bonded dimer of **TBDAP** can be applied to a dynamic memory in an operation, while the controllable strong bonding nature seen in the σ-bonded dimer of **3OPO** can be applied to a static memory to save data.

3.8.3 4- and 6-Oxophenalenoxyl (4OPO, 6OPO) systems

In a sharp contrast to **3OPO**, the **4OPO** and **6OPO** derivatives are extensively spin delocalized and are expected to have higher thermodynamic stabilities than the **3OPO** derivatives (Figure 3.34).[35] However,

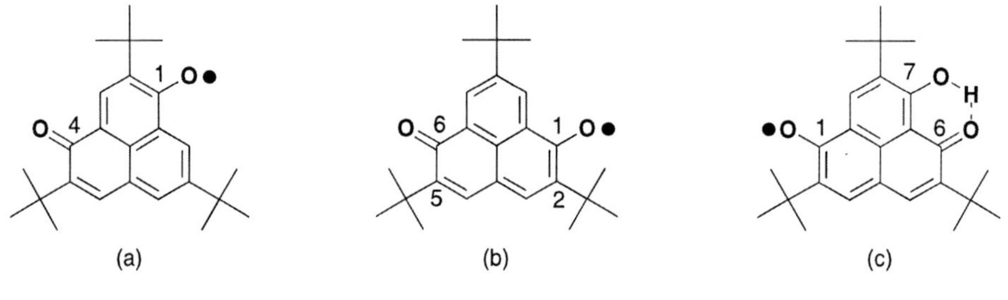

Figure 3.34 *Representative local spin structures of (a) **TB4OPO**, (b) **TB6OPO**, and (c) 7-hydroxy-**TB6OPO**.*

in order to isolate them as stable species in air, kinetic stabilization by steric protection at the appropriate positions is a key issue. From this viewpoint, **6OPO** system with methyl substituents at the 2, 5-positions was first designed. However, this neutral radical could not be isolated due to the generation of benzyl-type radical and subsequent σ dimerization.[42] The *tert*-butyl group, sterically more congested group than methyl group, greatly improves the stabilities of **4OPO** and **6OPO**, enabling them to be handled in air.[35,43] Taking advantage of this strategy, 7-hydroxy-**TB6OPO** was also designed and synthesized as the first neutral radical having an intramolecular hydrogen bond coupled with an extremely delocalized spin system.[44]

These neutral radicals were prepared by the methods depicted in Scheme 3.10. Among the synthetic steps, *tert*-butylation of 2,7-dimethoxynaphthlene, the first step of the method, is the most important key reaction. The Friedel–Crafts type introduction of *tert*-butyl groups into the electron rich aromatic compounds is not easy, because one-electron oxidation reaction of the substrate easily occurs and thus inhibits the Friedel–Crafts reaction. This difficulty is overcome by the use of *t*-butyl alcohol and trifluoroacetic acid (CF_3COOH) as solvents. Thus, tri-*tert*-butylated 4,9-dimethoxyphenalanone ("phenalanone" in Scheme 3.10), a common synthetic intermediate of three kinds of neutral radicals, is prepared in five steps. Reduction of the carbonyl group followed by demethylation and aerobic oxidation gives the 6-hydroxyphenalenone. Treatment of this compound with lead(IV) oxide affords **TB6OPO** as a deep green solid. **TB4OPO**, a topological isomer of **TB6OPO**, is also obtained as a green solid in four steps from the phenalanone derivative. The key reaction step is the regioselective reductive demethoxylation reaction of the dimethoxyphenalenone derivative. This reaction is furnished with lithium aluminum hydride in the presence of copper(I) iodide. 7-Hydroxy-**TB6OPO** is also obtained as dark grayish green powder from

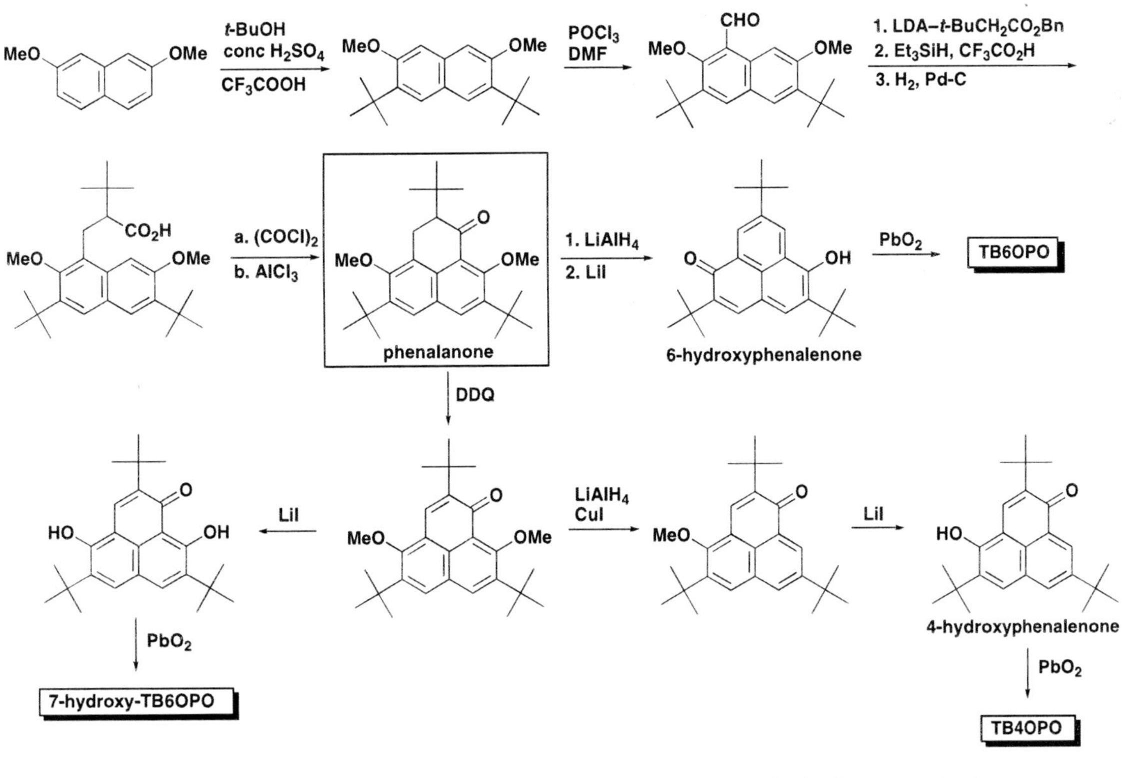

Scheme 3.10 Synthetic methods for **TB4OPO**, **TB6OPO**, and 7-hydroxy-**TB6OPO**

the same common synthetic intermediate by DDQ (dichlorodicyanobenzoquinone) oxidation, demethylation and oxidation. All of these **OPO**s are very stable in air in the solid state for a few months. In addition, 8-hydro[43] and 8-(p-halophenyl)[45] derivatives of **6OPO** were also designed and prepared. Both neutral radicals are stable enough to handle in air in the solid state.

The electronic structures of **TB4OPO** and **TB6OPO** were investigated using solution EPR/^{1}H-ENDOR/TRIPLE studies. Figure 3.35 shows the EPR spectra of **4OPO** and **6OPO**, and the hfccs determined in terms of these spectra and quantum chemical calculations. The spin delocalized natures and "spin topological symmetry control" phenomena predicted by theoretical calculations are experimentally demonstrated.

Figure 3.36b–d shows the possible tautomers of 7-hydroxy-**TB6OPO** depending on the delocalized nature of the unpaired electron based on the **6OPO, 4OPO**, and **9OPO** π-electronic structures, respectively.

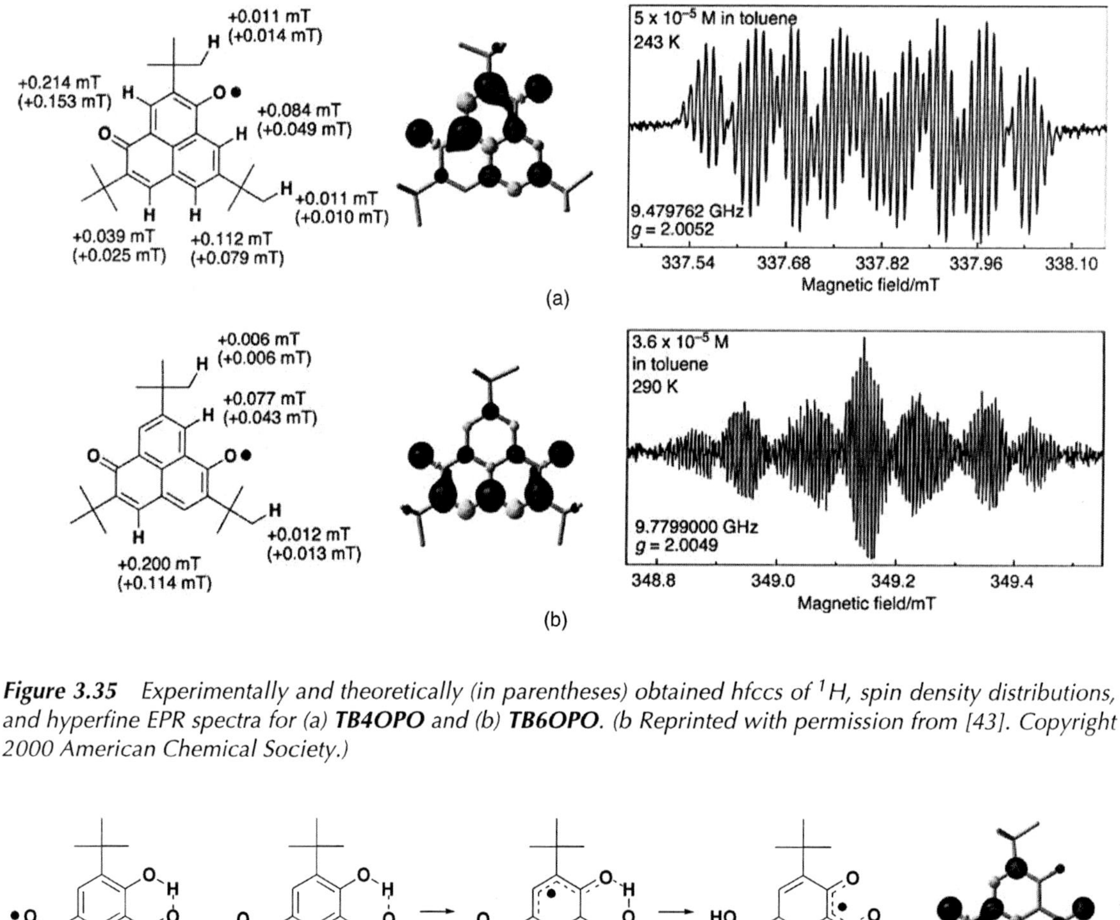

Figure 3.35 Experimentally and theoretically (in parentheses) obtained hfccs of ^{1}H, spin density distributions, and hyperfine EPR spectra for (a) ***TB4OPO*** *and (b)* ***TB6OPO****. (b Reprinted with permission from [43]. Copyright 2000 American Chemical Society.)*

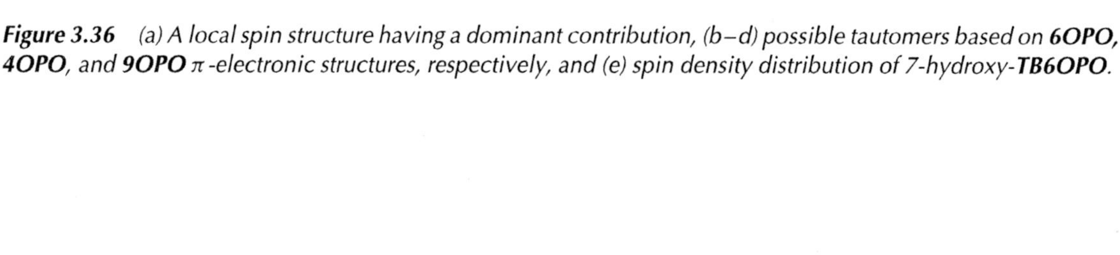

Figure 3.36 (a) A local spin structure having a dominant contribution, (b–d) possible tautomers based on ***6OPO, 4OPO****, and* ***9OPO*** *π-electronic structures, respectively, and (e) spin density distribution of 7-hydroxy-****TB6OPO****.*

The solution EPR/^{1}H-ENDOR/TRIPLE and diluted solution IR spectra, with the aid of quantum chemical calculations, reveal the dominant contribution of the **6OPO**-type electronic structure with an intramolecular hydrogen bond between the 1- and 9-positions (Figure 3.36a), although the radical might exist as a mixture of tautomers. An intramolecular proton transfer gives unique dynamic dielectric properties, as is well demonstrated by a 9-hydroxyphenalenone system (closed shell molecule).[46] Thus, the possible interconversion between structures **b** and **c** may contribute to the control of magnetic properties in the solid state, as well as to the realization of dynamic spin nature.

7-Hydroxy-**TB4OPO**, a topological isomer of 7-hydroxy-**TB6OPO**, was also designed and prepared (Figure 3.37a).[47] This neutral radical also has several possible tautomeric structures based on the **4OPO** electronic structure (Figure 37b–d). Solution EPR measurements and quantum chemical calculation confirm this electronic structure. Interestingly, the radical has a hydrogen bonding capability directed outwards from the molecular skeleton, raising the possibility of the construction of *inter*molecular hydrogen bonding interactions.

To investigate a heteroatomic chemical modification in **6OPO**, a 7,9-diaza derivative of **6OPO** system (**DAOPO**) was designed and generated by a treatment of a corresponding anion with DDQ oxidation in a toluene solution (Figure 3.38).[48] The position of the two oxygen and nitrogen atoms raises the possibility that this radical could be used as a bridging ligand for intermolecular networks. EPR/^{1}H-, ^{14}N-ENDOR/TRIPLE spectroscopies and quantum chemical calculations reveal that **DAOPO** possesses the **6OPO**-type electronic structure. Interestingly, exposure of the EPR tube of this radical degassed solution to sunlight at −78 °C induces a rapid decomposition of the radical within 10 minutes, although in the dark the neutral radical is stable at −30 °C for a few weeks. Thin layer chromatography indicates

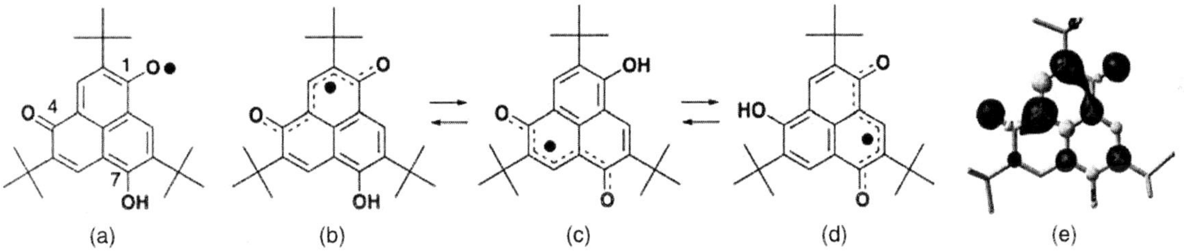

Figure 3.37 (a) Local spin structure, (b–d) possible tautomers, and (e) spin density distribution of 7-hydroxy-TB4OPO.

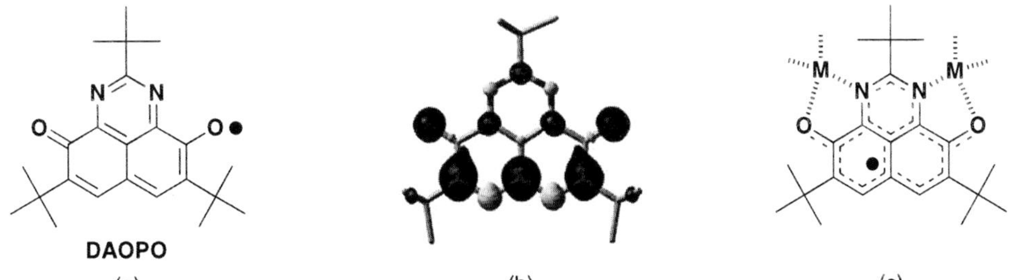

*Figure 3.38 (a) A local spin structure, (b) spin density distribution, and (c) ligation motif of **DAOPO**.*

a transformation of the radical to the corresponding hydroxyphenalenone derivative, a radical synthetic precursor. A considerable decrease of the calculated SOMO–LUMO energy gap (0.24 → 0.09 eV) as well as a lowering of SOMO energy level (−2.99 → −3.27 eV) in comparison with those of the **6OPO** may influence this instability to sunlight.

3.8.4 Redox-based spin diversity

The redox activities of the **OPO** systems were investigated by cyclic voltammetry measurements. Each of the anionic salts shows two-stage one-electron redox waves (Figure 3.39).[36,37a] All of the redox processes for the **TB4OPO** and **TB6OPO** systems are reversible, while the oxidation wave (E_p^{OX}) from the anion to the neutral radical of *p*-methoxyphenyl derivative of **3OPO** system, **MP3OPO**, is irreversible. This is probably due to the σ dimerization of **3OPO** system (*vide supra*). Furthermore, the reduction potentials from the neutral radical to the anion in the **OPO** systems are similar to those of tetracyanoquinodimethane (TCNQ) and *p*-chloranil, typical electron acceptor molecules, indicating high electron accepting abilities. In these experimental conditions, all the chemical species except for the **3OPO** neutral radical possess high stabilities.

Chemical generation of the radical dianions of **MP3OPO**, **TB4OPO**, and **TB6OPO** systems and investigations of their electronic structures have been conducted. Treatment of the neutral radicals with alkali metals such as potassium and sodium give the radical dianions as stable chemical species in degassed sealed conditions even at room temperature (Figure 3.40). EPR/^{1}H-ENDOR/TRIPLE measurements show totally different hfccs from those of the corresponding neutral radicals, experimentally revealing the "redox-based spin diversity" nature of these species (*vide supra*).[36,37a,38] Quantum chemical calculations and the "MO-based VB method" (molecular orbital-based valence bond method)[37b] also corroborate these views with contributing weights of any possible resonance structures. In order to observe these unique electronic properties in other systems, there are two major necessities: (1) high stabilities in each oxidation state, and (2) a significant difference in the orbital symmetry between two frontier orbitals. For these limitations, not only polycyclic aromatic hydrocarbons, such as naphthalene and anthracene, but also the multistage amphoteric redox systems based on the phenalenyl (*vide infra*) are excluded.[49] This novel concept attracts much attention in terms of the topological symmetry of spin density distribution versus charge fluctuations and stabilization effect of the **OPO** systems.[36,37a,38]

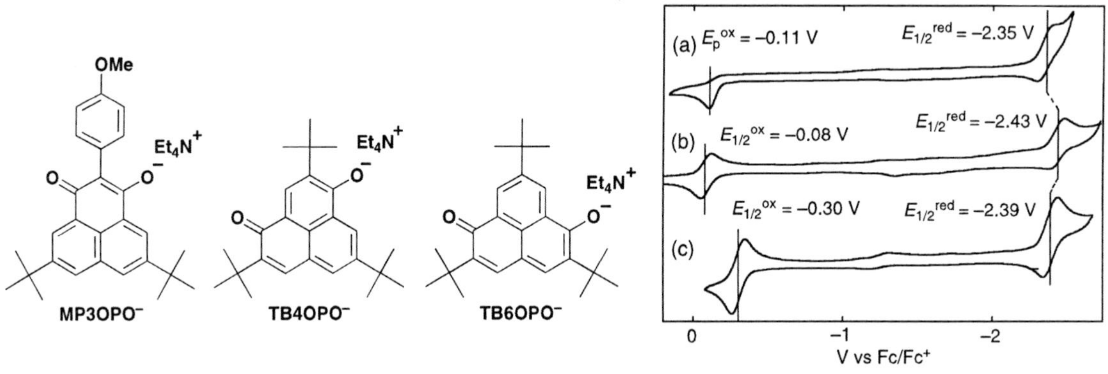

*Figure 3.39 Cyclic voltammograms for (a) $Et_4N^+ \cdot$ **MP3OPO**$^-$ (5 mM), (b) $Et_4N^+ \cdot$ **TB4OPO**$^-$ (3 mM), and (c) $Et_4N^+ \cdot$ **TB6OPO**$^-$ (10 mM) in acetonitrile solutions at 300 K.*

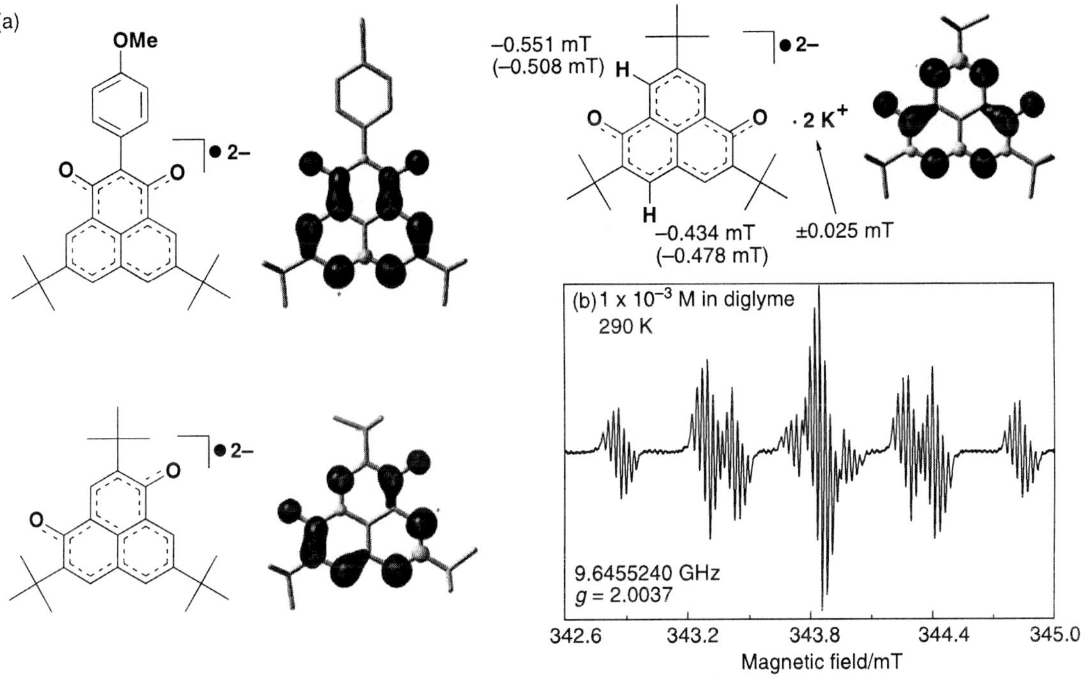

*Figure 3.40 (a) Chemical structures and spin density distributions of the radical dianions of **MP3OPO, TB4OPO,** and **TB6OPO** (potassium salt) with observed and calculated (in parentheses) hfccs of ^1H and $^{39}K^+$. (b) EPR spectrum of the potassium salt of **TB6OPO** radical dianion in a diglyme (bis(2-methoxyethyl) ether) solution. (b Reprinted with permission from [37a]. Copyright 2002 American Chemical Society.)*

The observed hfccs include couplings attributable to ^{39}K and ^{23}Na nuclear spins (counter cations), indicative of tight-ion-pairing structures. DFT calculations suggest extremely delocalized natures of the two anion charges over the **OPO** skeleton likely to the neutral radical systems, probably leading to their high stabilities (*vide infra*).[36,37a,38] As expected by their dramatic change of electronic structures, color changes also occurred depending on the redox states, meaning that the **OPO** system is electrochromic.

3.8.5 Molecular crystalline secondary battery

Reversible redox processes in a solution can be compared to charge/discharge processes of an electrode active material in a secondary battery. An organic molecule with multistage redox ability can, therefore, accommodate a large number of electrons in the molecular skeleton. Thus, there is a unique possibility to realize a large discharge capacity in the secondary battery system by utilizing such organic molecules as electrode active materials. In 2002, this idea of a secondary battery utilizing phenalenyl and **OPO** derivatives was independently claimed by different authors.[26,37a,38] Then, the novel secondary battery based on **TB6OPO** as a cathode active material was developed and termed "molecular crystalline secondary battery" (Figure 3.41a,b).[50] Importantly, this secondary battery shows a step-wise charge/discharge behavior and a comparable performance in discharge capacity (182 Ah/kg) to commercially available lithium ion secondary batteries (150 ~ 170 Ah/kg) (Figure 3.41c). This step-wise process and discharge

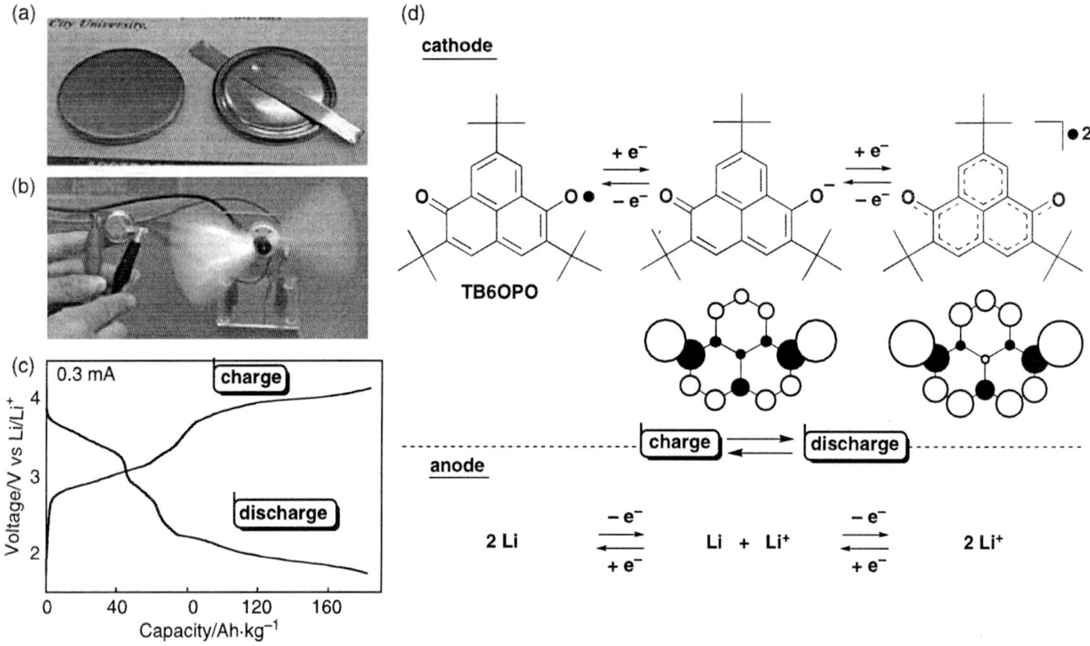

Figure 3.41 *(a) The coin cell of the "molecular crystalline secondary battery" containing* ***TB6OPO*** *as a cathode active material and metal lithium as an anode active material; (b) a propeller working using the force of the battery; (c) the charge/discharge curves at a current density of 0.27 mA/cm² in the voltage range of 2.0–4.0 V; and (d) the possible electrochemical reactions in the cathode and the anode under charge/discharge processes. Calculated charge distributions of the anion and the radical dianion of* ***6OPO*** *are shown below the corresponding chemical structures.*

capacity unequivocally demonstrate that *all* of the **TB6OPO** molecules in the cathode participate in the charge/discharge processes.

Delocalization greatly contributes to the high stability of anion and radical dianion species in the cathode (Figure 3.41d). The use of organic molecule, unlike polymer and metal complex, as a cathode active material has expanded to include closed shell electron donor and acceptor molecules, such as tetrathiafluvalene (TTF) and TCNQ derivatives, which show discharge capacities of 200 ~ 260 Ah/kg.[51] The relationship between the physical properties of the molecule (i.e. solubility and redox potential and on-site coulombic repulsion measured by cyclic voltammetry in a solution) and discharge voltage and cyclability have been carefully investigated.[51] These studies clearly demonstrated a conceptual advance in design criteria of electrode active materials with an increased discharge capacity. These studies also contribute significantly to practical considerations of avoiding transition metal elements and increasing energy density. A molecular level understanding of redox processes (charge/discharge processes) occurring inside the secondary battery in terms of a variety of experimental measurements is the next object for the design of more sophisticated "molecular crystalline secondary battery". Energy storage devices having multiple functionalities based on redox-driven changes in physical properties of the molecule are anticipated for next generation molecular devices.

3.8.6 Spin-center transfer and solvato-/thermochromism

As mentioned above, **6OPO** has an electron accepting ability similar to *p*-chloranil. Thus construction of intra/intermolecular interactions with an electron donating group/molecule affords intriguing possibilities as the functional (donor–acceptor) organic material. From this viewpoint, **6OPO** derivatives linked with TTF or ethylenedithio-TTF (EDT-TTF), TTF-**6OPO**, TTF-*p*-Ph-**6OPO** and EDT-TTF-*p*-Ph-**6OPO** (shown below), have been designed and prepared and their intramolecular interactions evaluated.[52,53] It was found that **TTF-6OPO** is unique in its ability of modulated intramolecular electron transfer (IET) by a moderate change of external environments, such as solvent and temperature in a solution, leading to the "spin center transfer" accompanied with solvato-/thermochromism.[52b]

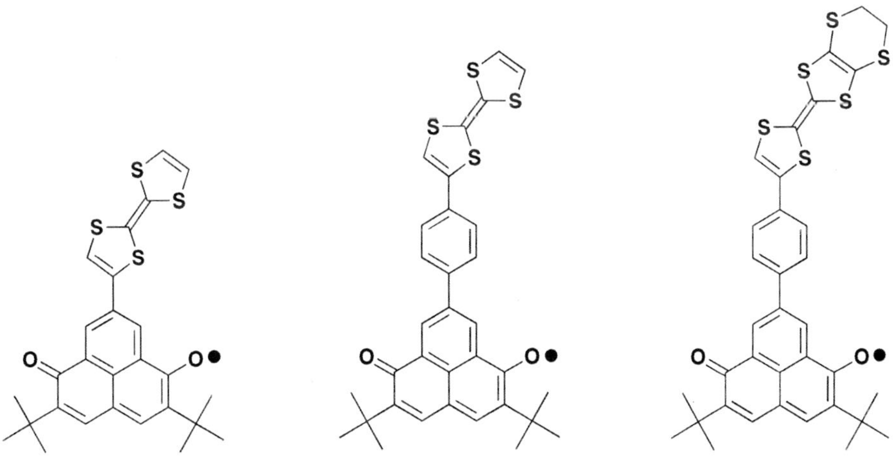

TTF-**6OPO** TTF-*p*-Ph-**6OPO** EDT-TTF-*p*-Ph-**6OPO**

The synthesis of TTF-**6OPO** was accomplished by the method depicted in Scheme 3.11. Iodination of 2,7-dimethoxyphenalanone, followed by cross-coupling with a tributyltin derivative of TTF, afforded a TTF-linked phenalenone derivative. A 1,4-reduction, demethylation, and aerobic oxidation of the phenalenone gave a hydroxyphenalenone, a radical precursor. Finally, the radical was prepared by treatment with lead(IV) oxide and obtained as a black powder with a reasonable stability in air at room temperature for weeks. The radical is very stable in degassed solution.

The spin transfer evident in TTF-**6OPO** is depicted in Figure 3.42a.[52b] This unconventional dynamic property involves two chemical species, neutral radical and zwitterionic radical, and is induced by an effective and reversible IET. In a dichloromethane solution, TTF-**6OPO** is in the neutral radical state, in which most of the spin localizes on the whole molecular skeleton of **6OPO** moiety. In sharp contrast, in trifluoroethanol (CF_3CH_2OH) solution TTF-**6OPO** is in the zwitterionic radical state, in which most of the spin localizes on the TTF moiety as radical cation species. This spin switching behavior is accompanied with a solvatochromism from orange to dark green color. The two kinds of chemical species are unequivocally characterized by extensive EPR measurements. More interestingly, by controlling the ratio of dichloromethane and trifluoroethanol, temperature-dependent spin center transfer was demonstrated with 100 % interconversion within a 50 K temperature differential (0.1 kcal/mol !). This phenomenon is accompanied by thermochromism (Figure 3.42b,c). This unprecedented phenomenon is the first spin switching phenomenon in a purely organic open shell molecular system,[54] while this type of phenomenon induced

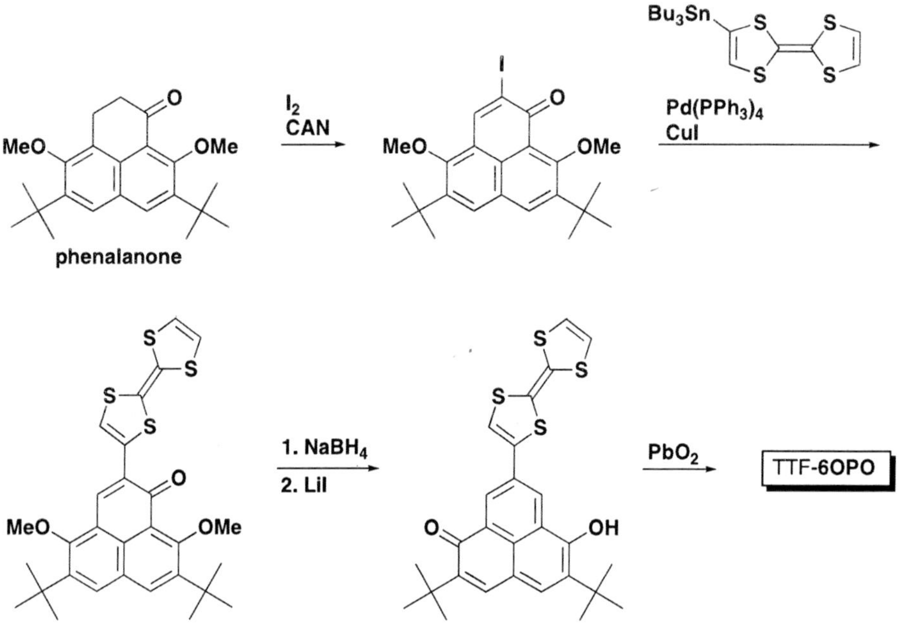

Scheme 3.11 Synthetic method for TTF-**6OPO** (CAN = cerium(IV) ammonium nitrate)

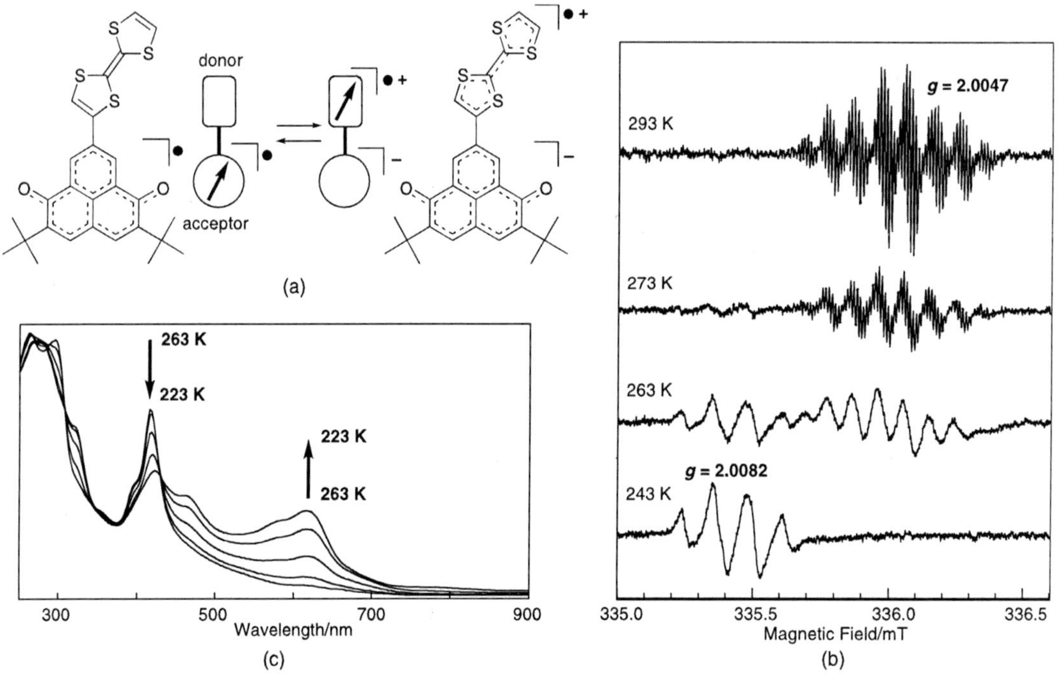

Figure 3.42 *(a) Spin center transfer in a purely organic system between neutral radical and zwitterionic radical; (b) temperature dependence of the liquid phase EPR spectra in a CH_2Cl_2/CF_3CH_2OH (199:1) solution (7.7 × 10^{-5} M); and (c) temperature-dependent UV-Vis spectra in a CH_2Cl_2/CF_3CH_2OH (99:1) solution (8.0 × 10^{-5} M) of TTF-**6OPO**. (b,c Reproduced with permission from [52b]. Copyright Wiley-VCH Verlag GmbH & Co. KGaA.)*

by IET is well known as "valence tautomerism" in a metal complex, especially in cobalt-semiquinone complexes.[55] This phenomenon is speculated to arise from intermolecular hydrogen bond between the carbonyl moieties in TTF-**6OPO** and the solvent trifluoroethanol, which is absent in dichloromethane solution. Characterization of the solid state properties and possible switching by applying external stimuli, such as an electrostatic field or light, is of particular interest in creating novel molecular functionality. In particular, the realization of a "photo-induced spin center transfer" is also intriguing from the viewpoint of the solar cell application.[55,56]

3.9 Phenalenyl-based zwitterionic radicals

Among the topological isomers of hydroxyphenalenones, synthetic precursors of the **OPO**, 9-hydoroxyphenalenone have a hydroxyl group and a carbonyl group at the adjacent *peri* positions, similar to acetylacetone. Thus, the corresponding anion derivative can act as a chelate ligand for metal ions. In the 1970s, 9-hydoroxyphenalenone-based metal complexes, **[PLY(O,O)]$_2$M** (**M** = transition metal ions such as Co^{2+}, Zn^{2+}, Mn^{2+}, Ni^{2+}, and Fe^{2+} or main group elements such as Na^+, Mg^{2+}, Be^{2+}, and Al^{3+}), were prepared.[57a–e] Each phenalenyl moiety in these metal complexes is in monoanionic (nonradical) state. In 1986, Haddon focused on **[PLY(O,O)]$_2$Be** and **[PLY(O,O)]$_2$B**, and for the first time succeeded in generating zwitterionic radicals by electrochemical reduction.[57f] Haddon's subsequent studies led to successful isolation of a series of spiro-conjugated bisphenalenyl boron complexes **[PLY(O,N)]$_2$$^+B^-$**,[58–64] **[PLY(N,N)]$_2$$^+B^-$**,[65] and **[PLY(O,O)]$_2$$^+B^-$**.[66] The negative charge on the boron atom is formally neutralized by the positive charge on one of the phenalenyl ligands, giving the zwitterionic radical structure. Furthermore, because of the perpendicular nature of the π-conjugated planes of the two phenalenyl moieties, a spiro-conjugated interaction between the molecular orbitals derived from the NBMO of the pristine **PLY** permits full delocalization of spin and charge. Among these complexes, the hexyl derivative **[PLY(O,N)]$_2$$^+B^-$** ($R^1$ = *n*-Hex) was the first phenalenyl-based single-component molecular conductor, showing a room temperature conductivity of $\sigma = 0.05$ S/cm (Figure 3.43), although there are no intermolecular contacts shorter than van der Waals radius (3.4 Å for C···C).[58,67,68] This complex shows a weak antiferromagnetic interaction with Weiss constant of $\theta = -10$ K.

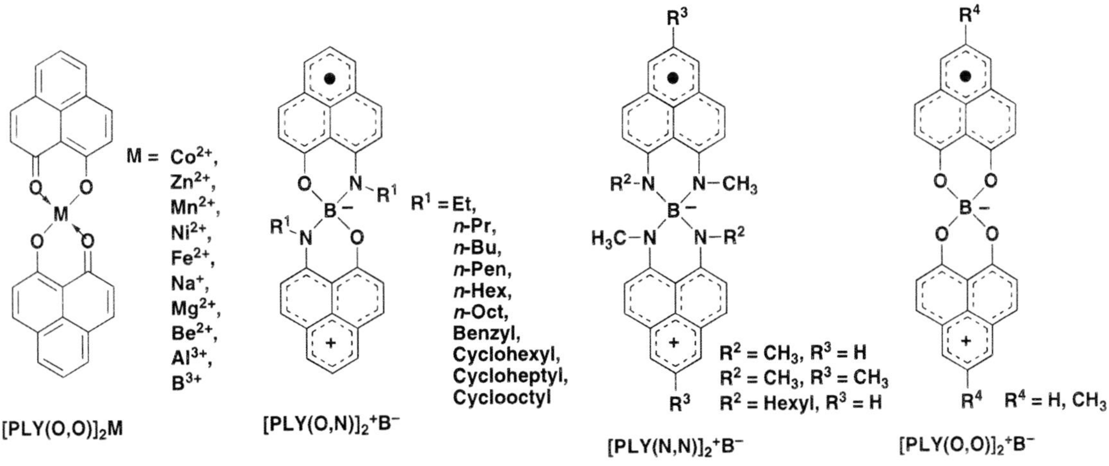

The ethyl derivative **[PLY(O,N)]$_2$$^+B^-$** ($R^1$ = ethyl) and the butyl derivative **[PLY(O,N)]$_2$$^+B^-$** ($R^1$ = *n*-butyl) show "magneto-optical-electronic bistability", in which remarkable changes in magnetic, optical, and

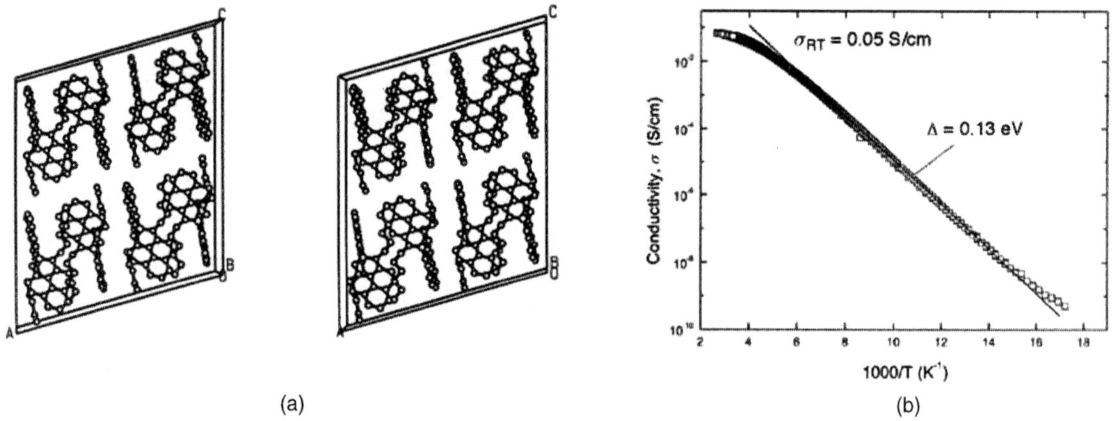

Figure 3.43 (a) Stereo view along b-axis *and (b) electric conductivity of* ***[PLY(O,N)]$_2$$^+B^-$*** *($R^1$ = n-Hex) in the crystal. (a,b Reprinted with permission from [58]. Copyright 1999 American Chemical Society.)*

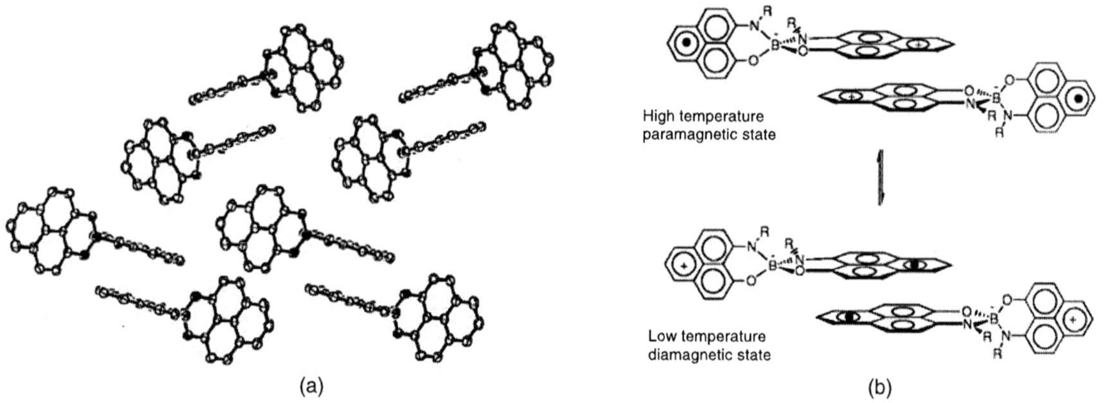

Figure 3.44 (a) Crystal structure of and (b) interconversion between the diamagnetic π dimer (low temperature form) and the paramagnetic π dimer (high temperature form) of ***[PLY(O,N)]$_2$$^+B^-$*** *($R^1$ = ethyl).*

electrical properties in the crystalline state depend on temperature.[59] The n-butyl derivative possesses a high electrical conductivity of 0.14 S/cm at around 347 K[59,60] as single-component molecular conductor,[67,68] and forms π dimers in the crystal without shorter contacts between π dimers than the sum of van der Waals radius of carbon atom (Figure 3.44a). At temperatures below 320 K, each complex possesses a spin in the π dimer regions and shows an antiferromagnetic interaction. In a sharp contrast, above 350 K each complex shows paramagnetic behavior because the spins reside mainly on the phenalenyl moieties not directly involved in the π dimer (Figure 3.44b). This dynamic spin behavior induces a simultaneous elongation of the intradimer distance from 3.2 to 3.3 Å and significant changes in IR transmittance spectra (Figures 3.45a–c). Figures 3.45d–f show the temperature dependence of IR transmittance, conductivity, and magnetic susceptibility of **[PLY(O,N)]$_2$$^+B^-$** ($R^1$ = n-butyl). These three physical property changes occur within the temperature range 320–350 K and are hysteretic in nature, indicating the bistability of the system. An occurrence of the phase transition above room temperature is a rare event, and in general

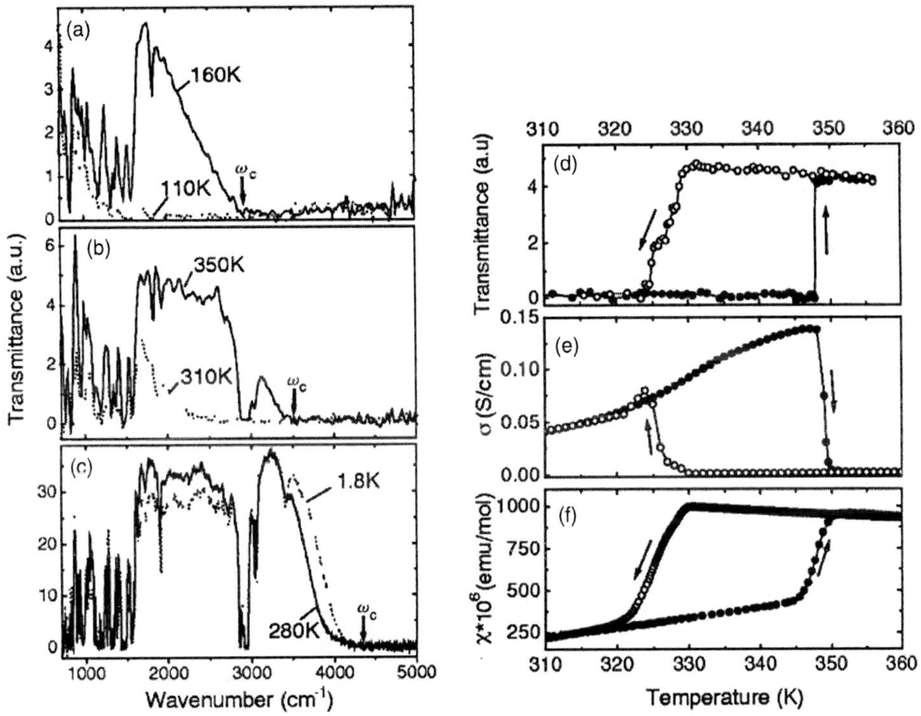

Figure 3.45 *IR transmittance of single crystals of (a)* ***[PLY(O,N)]$_2$$^+B^-$*** *($R^1$ = ethyl), (b)* ***[PLY(O,N)]$_2$$^+B^-$*** *($R^1$ = n-butyl), and (c)* ***[PLY(O,N)]$_2$$^+B^-$*** *($R^1$ = n-Hex) at the high temperature (solid lines) and low temperature (dotted lines) states. The arrows indicate cut-off frequencies of the transmittance (ω_C). Bistability of (d) IR transmittance at 3.85μm, (e) conductivity, and (f) magnetic susceptibility, arising from the hysteretic phase transition in* ***[PLY(O,N)]$_2$$^+B^-$*** *($R^1$ = n-butyl). The open circles represent the cooling part of the temperature cycle, and the closed circles represent the heating part. The arrows indicate the direction of the temperature sweep.*

the temperature region does not attract attention in standard measurements for molecule-based magnetic materials. Thus, these studies underscore the importance of magnetic measurements in the high temperature range above room temperature.[59c] These materials are applicable for information storage devices and molecular memory.[41]

Importantly, these boron complexes have high stabilities and do not form σ-dimer structures in the crystalline states in spite of the lack of bulky substituents on the phenalenyl skeletons.[61,62] The crystal structures greatly depend on the kinds of alkyl substituents at the nitrogen atoms coordinating the metal ions, and the phenalenyl moieties form the π dimer, π chains, and π steps structures. The magnetic susceptibility measurements of all metal complexes (except for **[PLY(O,N)]$_2$$^+B^-$** ($R^1$ = cyclooctyl)) show relatively weak antiferromagnetic intermolecular exchange interactions (1/10 ~ 1/100) in comparison with **TBPLY** (−2000 K). Extended Hückel theory (EHT) band structure calculations suggest that **[PLY(O,N)]$_2$$^+B^-$** ($R^1$ = cyclohexyl) shows 3-D metallic behavior with a bandwidth of ~0.5 eV. However, the complex has an activated conductivity. The conducting behavior of this class of zwitterionic radicals, including this cyclohexyl derivative, is considered in terms of a "resonating valence bond theory".[63,66]

Extension from the bisphenalenyl to trisphenalenyl metal complexes having silicon[69] or germanium[69b] as central metals (shown below) is intriguing from the viewpoint of multistage redox ability and lowering of the on-site coulombic repulsion. Recently, Haddon reported the first synthesis and isolation of the neutral radical of the trisphenalenyl silicon complex, which shows a weak partial π-dimer structure and weak antiferromagnetic interaction ($\theta = -0.75\,\text{K}$).[70] Further studies on the trisphenalenyl metal complex are expected.

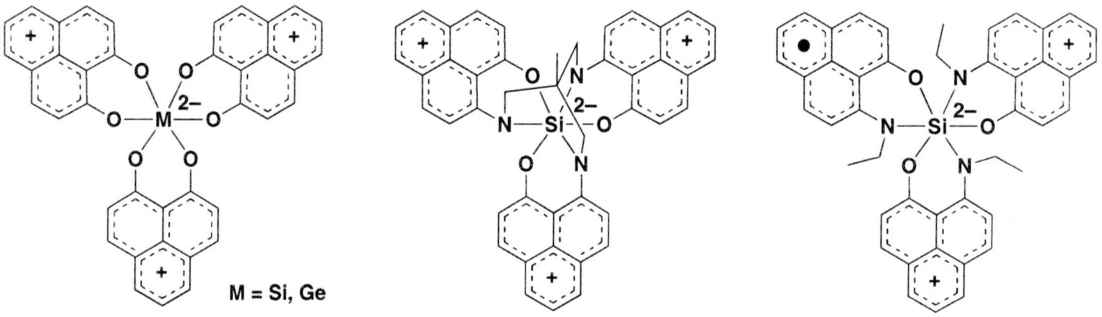

3.10 π-Extended phenalenyl systems

3.10.1 Triangulenes

Extension of the π-conjugated electronic network of the phenalenyl system leads to the enhancement of the spin delocalized nature and spin multiplicities. As mentioned previously (Section 3.2), zigzag "clipping" of a triangular motif from graphene affords non-Kekulé polynuclear benzenoid (PNB) hydrocarbons (π-extended phenalenyl system termed "open shell graphene") possessing multiple unpaired electrons (Figure 3.1b and 3.2b). Most importantly, these molecular systems are expected to be in high spin states in the ground states due to the topological degeneracy of NBMOs with non-disjoint nature according to theoretical calculations.[7,13b,71a]

Triangulene (or Clar's hydrocarbon) possesses three phenalenyl substructures with C_3 molecular symmetry and a pair of degenerate NBMOs, and has not been detected so far (Figure 3.1b, Scheme 3.12). In 1953, Clar reported that generation of the triangulene gives polymerized products, revealing the kinetic instability of the molecule.[6d] Thus, the generation and chemical identification of non-Kekulé PNB hydrocarbon not only are long standing issues in chemistry but also attract much attention from the viewpoint of recent material challenges in the field of molecule-based magnets.[2d–j] In 1993, Bushby attempted to make a series of non-Kekulé PNBs with heteroatomic modification by chemical reduction of the corresponding diketones, and generated trioxytriangulene, a triangulene with three oxide groups at the positions having large coefficients of NBMOs (Scheme 3.12).[71] This diradical trianion species has a high stability in a degassed solution even at room temperature, and its ground state spin multiplicity is determined to be a triplet based on its EPR spectroscopy in a frozen solution. The introduction of heteroatoms lifts the degeneracy of the NBMOs, causing a kinetic exchange interaction to dominate over dynamic spin polarization in the parent π systems.

In 2001, a hydrocarbon triangulene derivative, tri-*tert*-butyltriangulene, was designed in which three *tert*-butyl groups are introduced on the carbon sites with nodes in the NBMOs for protecting the reactive carbon sites as well as minimizing the electronic perturbation (Schemes 3.12 and 3.13, Figure 3.46a).[7,13b] The diradical molecule was generated by a treatment of dihydrotriangulene in a toluene solution with p-chloranil. The triplet state EPR spectrum is observed as a superposition of spectra due to monoradical

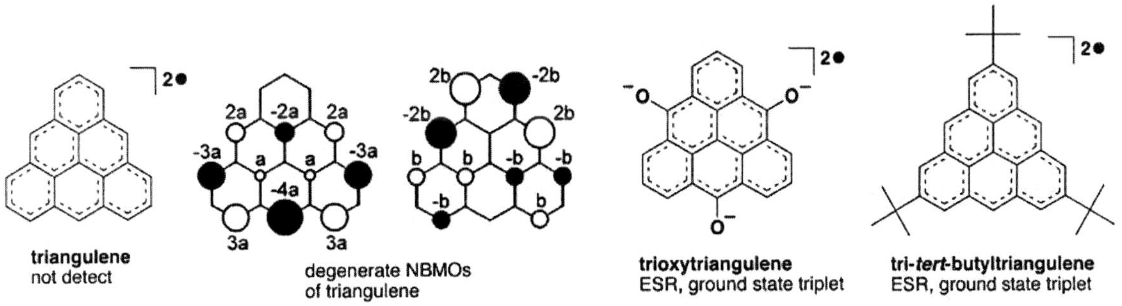

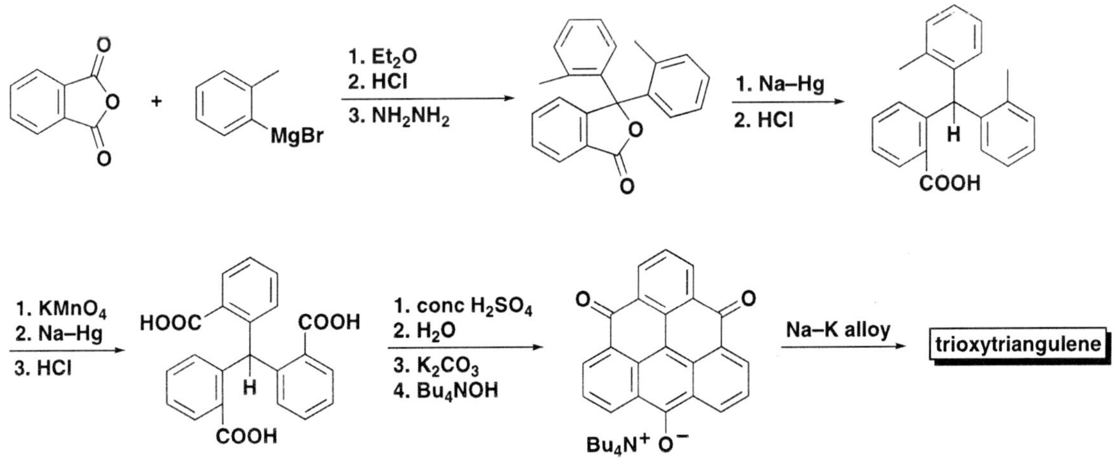

Scheme 3.12 Synthetic method for trioxytriangulene

species (Figure 3.46b). By the spectral simulation, the spin Hamiltonian parameters are determined to be $S = 1$, $g = 2.003$, $|D|/hc = 0.0073\,\text{cm}^{-1}$, and $|E|/hc \sim 0\,\text{cm}^{-1}$, which are attributed to the tri-*tert*-butyltriangulene, a genuine hydrocarbon structure with a threefold rotation axis. The linear dependence of the triplet signal intensity I on $1/T$ obviously shows that the triplet is the ground state (Figure 3.46c). The result is in a good agreement with the theoretical prediction for pristine triangulene. The resulting diradical does not have a stability high enough to survive in a solution state, and thus oligomerization at the carbon sites with large spin densities occurs at a temperature of more than 273 K. This study illustrates a partial protection of the triangulene diradical system, and further stabilization by additional substitutions is necessary for the isolation of the genuine non-Kekulé PNBs in the crystalline state.

Recent developments in high spin chemistry from experimental and theoretical sides are based on the conceptual proposal of π-topological symmetry rules, which govern the spin alignment of open shell genuine hydrocarbon π systems.[3] A high spin polymeric system was designed by 2-D π extension of triangulene (Figure 3.47a).[7b,13b] Band structure calculations based on crystal orbital approach on such an infinite structure indicate the generation of a flat π-non-bonding crystal orbital (π-NBCO) band, which delocalizes over the whole system (Figure 3.47b). The system is expected to have a super high spin ground state because dynamic π-spin polarization mechanism occurs and the N-fold π-NBCO is occupied

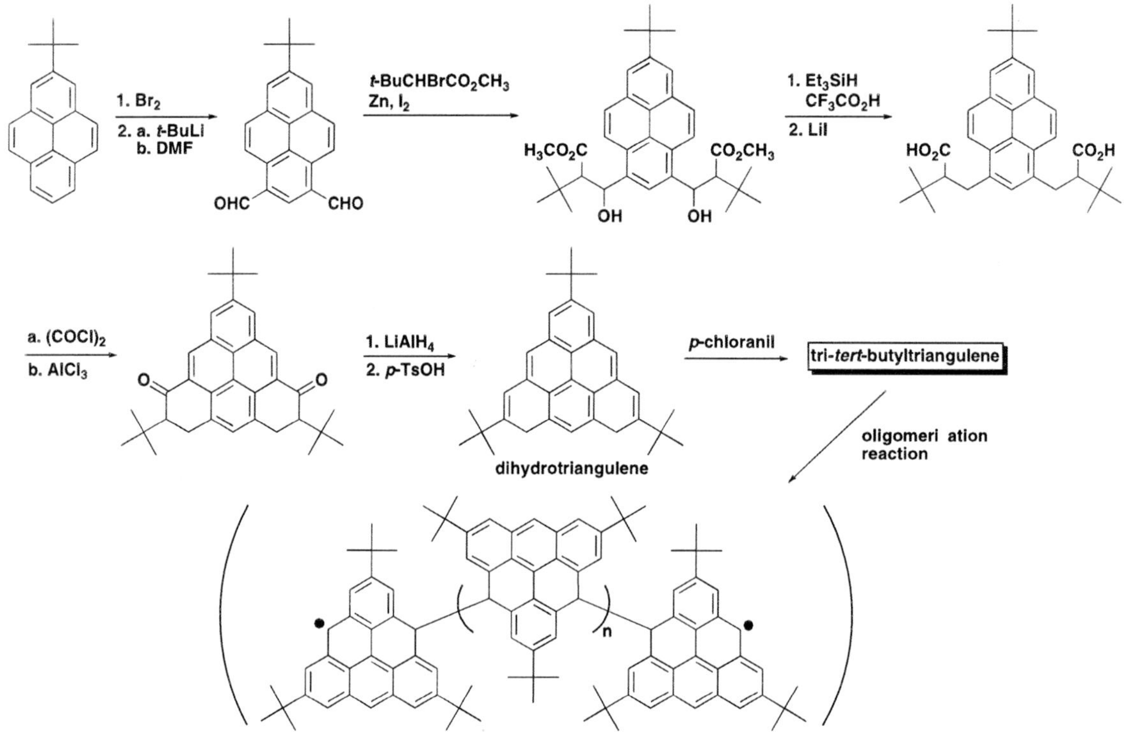

Scheme 3.13 Synthetic method for tri-*tert*-butyltriangulene and its oligomerization reaction

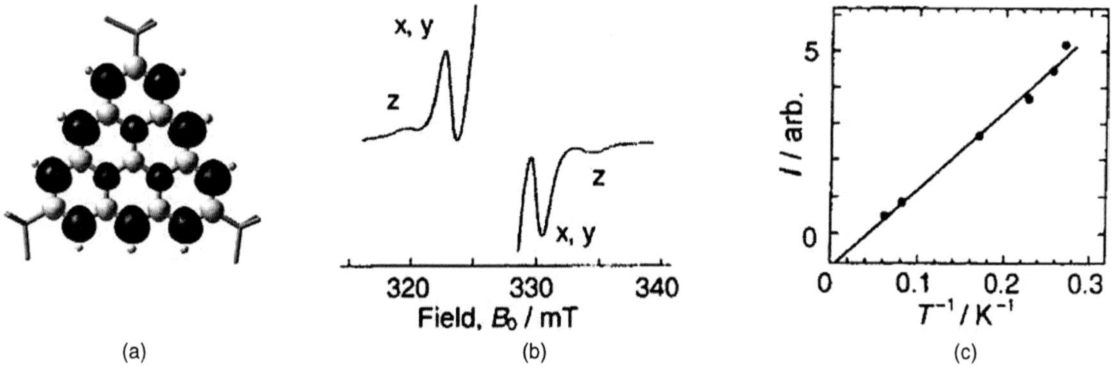

Figure 3.46 *(a) Spin density distribution of tri-*tert*-butyltriangulene. Calculations carried out at the UBLYP/6-31G//UBLYP/6-31G level of theory. (b) Observed triplet state EPR spectrum with monoradical species in a frozen toluene matrix at 123 K; x, y, and z denote the canonical absorption peaks. (c) Plot of the temperature dependence of the triplet signal intensity I versus 1/T from 3.7–16 K. (b,c Reprinted with permission from [7a]. Copyright 2001 American Chemical Society.)*

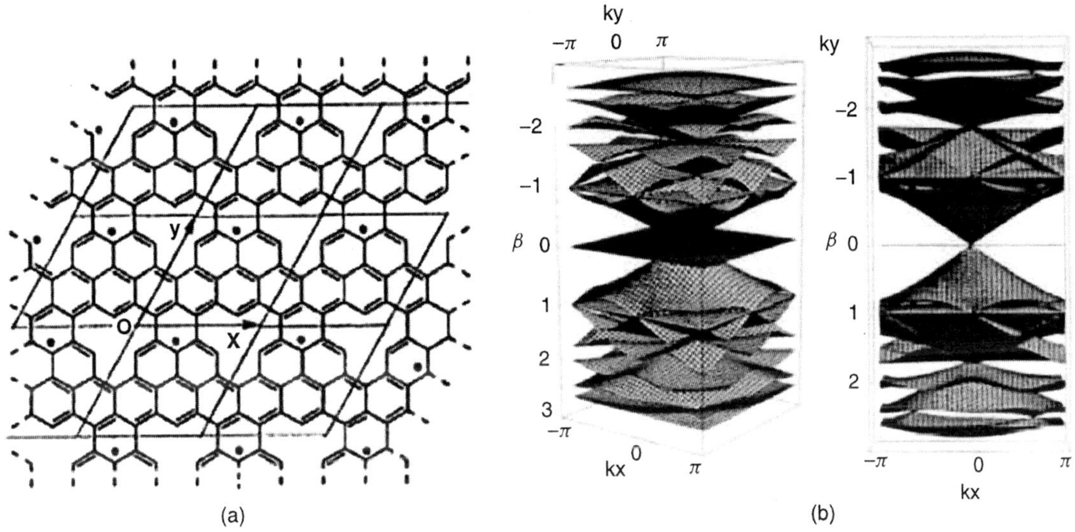

Figure 3.47 (a) The structure of a 2-D triangulene-based super high spin polymeric π system. (b) The band structure calculated for (a) in terms of a crystal orbital approach. (Reprinted from [7b], Copyright 2001, with permission from Elsevier.)

by N ferromagnetically coupled spins. The structural rigidity is advantageous for maintaining the planarity of the π-conjugated electron network,[7b,13b]

3.10.2 Trioxytriangulene with redox-based spin diversity nature

Trioxytriangulene, prepared by Bushby, possesses three **6OPO** substructures, and it can be viewed as a planar π-extended derivative of the **6OPO** radical dianion. Quantum chemical calculation indicates that the corresponding neutral radical possesses an electronic structure similar to that of the **6OPO** neutral radical, being significantly different from those of the radical dianion and the diradical trianion (trioxytriangulene) (Figure 3.48). Accordingly, this molecular system has a high potential to show the redox-based spin diversity nature similar to the **OPO** systems.[72] Furthermore, the expected high redox ability of this system arouses interest as an electrode active material in the secondary battery. To realize this possibility, appropriate chemical modifications to improve the stability of the molecular system are certainly necessary.[72]

3.10.3 Bis- and tris-phenalenyl system and singlet biradical characters

As mentioned above, the phenalenyl system possesses amphoteric redox ability ($E^{\text{sum}} = 1.6\,\text{V}$), readily generating the cation and anion species by one-electron oxidation/reduction (Figure 3.4c). To construct a phenalenyl-based hydrocarbon with lower E^{sum} values, chemical modifications to conquer the instability of the neutral radical system are essential. In this context, Murata and Nakasuji have designed bis- and tris-phenalenyl systems since the early 1970s, in which two or three phenalenyl ring systems are connected with π-conjugated systems (Figure 3.49).[73]

These molecular systems possess Kekulé structures in the neutral states (except for **TBzD**). The amphoteric four-stage redox behavior of **IDPL**, having a p-quinodimethane skeleton in the center

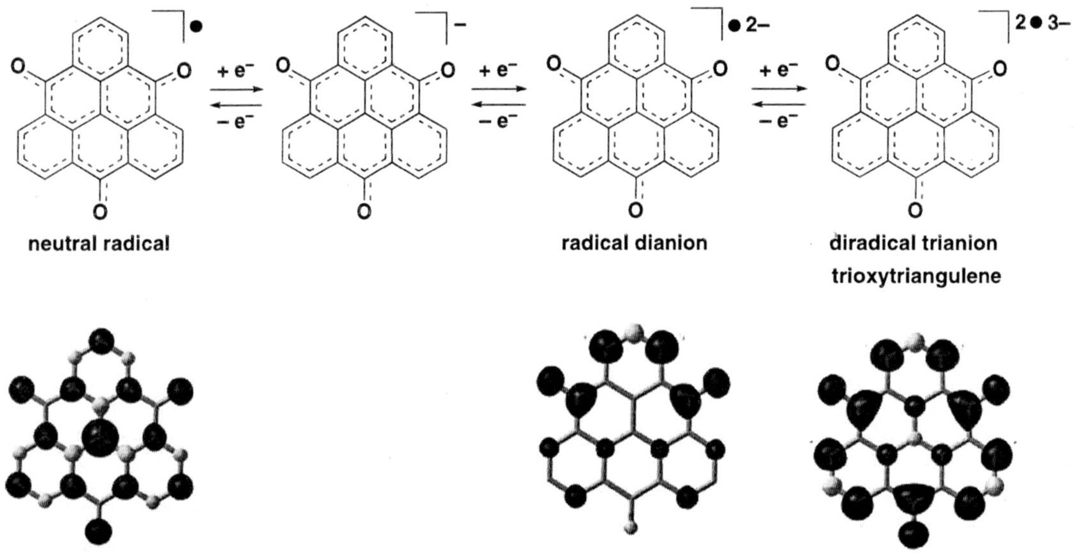

Figure 3.48 *Spin density distributions of neutral radical, radical dianion, and diradical trianion (trioxytriangulene). Unbalanced spin density distribution in the radical dianion may be caused by the distorted optimized structure from threefold symmetrical structure.*

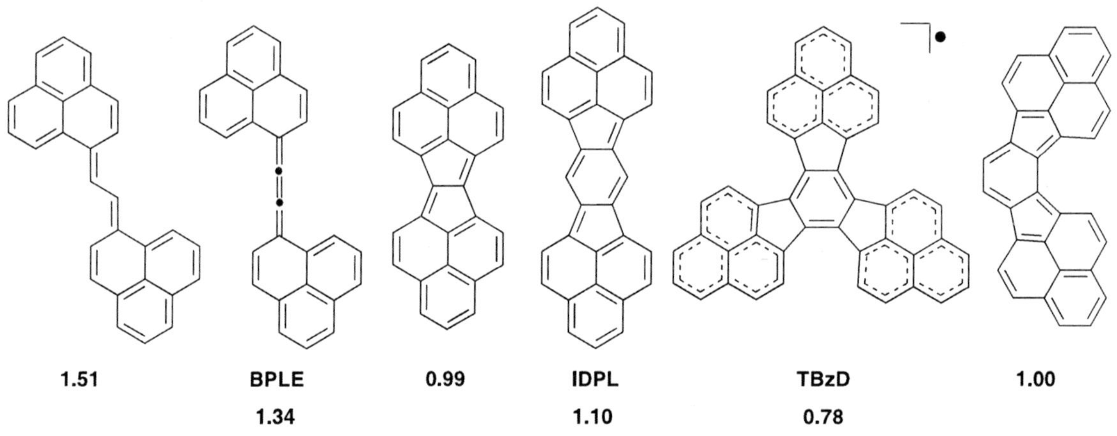

Figure 3.49 *Bis- and tris-phenalenyl systems showing amphoteric multistage redox natures.* E^{sum} *values are cited under the chemical structures.*

of molecule, is illustrated in Figure 3.50a.[73e] This multistage redox nature is understandable by the synergistic effect of destabilization of the neutral state due to the existence of *p*-quinodimethane structure and the stabilization of charged redox states due to the formation and aromatization of the benzene π system, as well as the amphoteric redox ability inherent in the phenalenyl system. Interestingly, **TBzD**, possessing three phenalenyl moieties with a central benzene ring, shows an amphoteric six-stage redox behavior in a solution state, although the molecule has low stability in air.[73f]

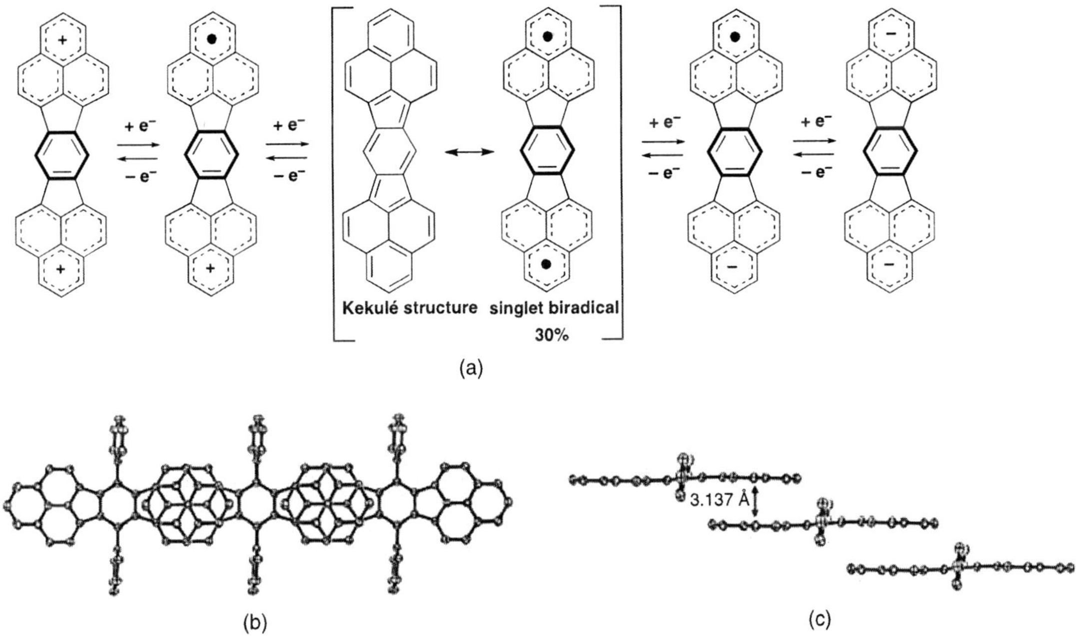

Figure 3.50 *(a) Amphoteric four-stage redox behavior and singlet biradical character in the neutral state of* ***IDPL*** *estimated by CASSCF(2,2)/6-31G//RB3LYP/6-31G** and broken symmetry UB3LYP/6-31G** calculations. 1-D chain with a slipped stack arrangement in the crystal of Ph$_2$-**IDPL**. (b) top view, (c) side view. (b,c Reproduced with permission from [74b]. Copyright Wiley-VCH Verlag GmbH & Co. KGaA.)*

Recently, in the bisphenalenyl neutral system, the singlet biradical state was found to be a significant resonance form (Figure 3.50a center).[74] The degree of the contribution in **IDPL** was estimated to be 30 % by CASSCF(2,2)//6-31G//RB3LYP/6-31G** and broken symmetry UB3LYP/6-31G** calculations.[74b] This large singlet biradical character highly influences the formation of intermolecular interactions and thus crystal packing structure. Ph$_2$-**IDPL**, having two phenyl groups in the central six-membered ring (Figure 3.51a), forms a 1-D chain in the crystal with a slipped stack arrangement and an average $\pi-\pi$ distance of 3.14 Å, which is substantially shorter than the van der Waals contact of carbon atoms (3.4 Å) (Figure 3.50c).[74b] The overlap mode of the phenalenyl rings is quite similar to that of **TBPLY** in the crystal or the solution state at low temperature, maximizing the SOMO–SOMO interaction between the molecules (Figures 3.7, 3.10, 3.50b). The band dispersions along the $\pi-\pi$ stacking direction are estimated to be 0.54 eV (valence band) and 0.51 eV (conduction band) by EHT. These values reflect the extremely close $\pi-\pi$ contacts, which are comparable to the band dispersions of the molecular conductors. The HOMO–LUMO gap of Ph$_2$-**IDPL** is 1.15 eV by electrochemical methods. The room temperature conductivity measured by the two-probe method using a compressed pellet is 10^{-5} S/cm with an activation energy of 0.3 eV. This conductivity is the highest value among structurally well defined, single component hydrocarbons in a neutral state.[74b,67,68]

A bisphenalenyl derivative connected with a naphthoquinoid structure, Ph$_4$-**NDPL**, possessing 50 % singlet biradical character, was also prepared and isolated as air-stable crystals (Figure 3.51a).[74c] The absorption band attributed to intramolecular HOMO–LUMO transition was observed at 865 nm, which is significantly red-shifted from the corresponding absorption band of Ph$_2$-**IDPL** (746 nm) (Figure 3.51b). This result indicates that the HOMO–LUMO gap of Ph$_4$-**NDPL**, with larger biradical character, is smaller

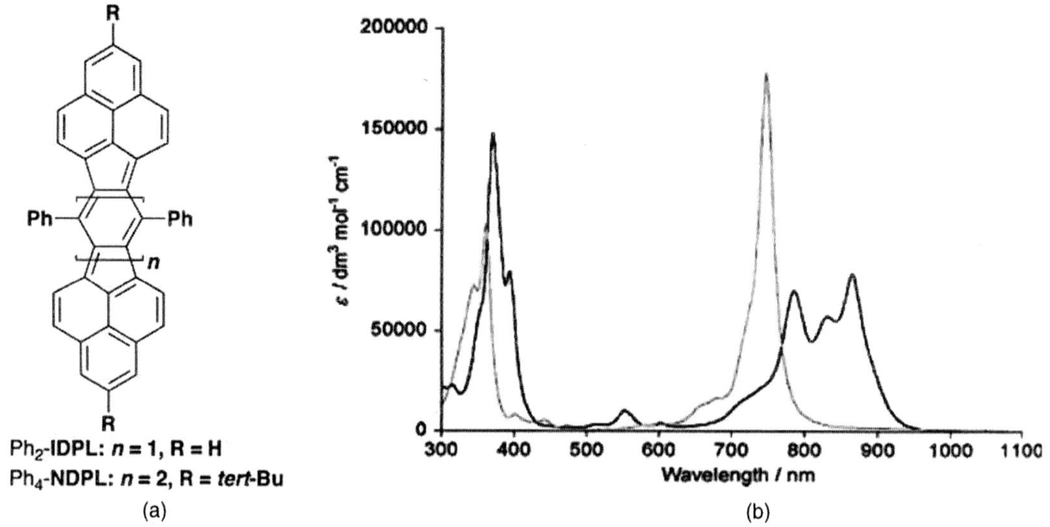

Figure 3.51 (a) Chemical structures and (b) solution absorption spectra of Ph_4-**NDPL** (black line) and Ph_2-**IDPL** (gray line). (Reprinted with permission from [74c]. Copyright 2007 American Chemical Society.)

than that of Ph_2-**IDPL**. The high stability of such hydrocarbons possessing contributions of singlet biradical characters makes a complete contrast to most singlet biradicaloids, which can only be detected in matrixes at low temperature due to their extremely high reactivity.

A phenyl-substituted derivative of **BPLE** (Figure 3.49), Ph_2-**BPLE**, possesses a contribution of a singlet biradical structure (14 %) and gives a charge transfer (CT) complex with F_4-TCNQ (Figure 3.52).[74d] Ph_2-**BPLE** exists in the charge transfer complex as a radical cationic state, Ph_2-**BPLE**$^{\bullet+}$, which is represented by two canonical structures (Figure 3.52b). Interestingly, $\pi-\pi$ overlap of the phenalenyl moieties of Ph_2-**BPLE**$^{\bullet+}$ is large, with an average distance of 3.34 Å in a staggered stacking motif, indicating a maximizing SOMO–SOMO interaction similar to **PLY** (Figures 3.52c,d). The band structure calculation with EHT gives a relatively large dispersion (0.42 eV) for a half-filled band along the $\pi-\pi$ stacking direction. The room temperature (290 K) resistivity ρ_{RT} is about 0.7 Ω cm with metallic-like behavior down to 280 K. Below 280 K, the charge transfer complex shows a transition into a semiconductive state as a result of bond formation between the phenalenyl and F_4-TCNQ carbon atoms.[74d]

The bisphenalenyl neutral hydrocarbon with singlet biradical character shows strong two-photon absorption (TPA) properties. The measurement of the TPA cross-section of Ph_2-**IDPL** by the open-aperture Z-scan method with a femtosecond optical parametric amplifier shows the large TPA cross-sectional value of 5490 GM (Figure 3.53) utilized by resonance enhancement and sequential TPA process.[75] The widely accepted guiding principles of the molecular design for large TPA is the introduction of electron donor and acceptor groups to the ends of a π-conjugated system as well as the extension of electronic communication through π conjugation. Thus, the TPA experiment of the singlet biradical hydrocarbons based on the bisphenalenyl system provides new insights into design criteria.

The electronic properties, including amphoteric multistage redox ability and strong intermolecular electronic communication, of Ph_2-**IDPL** lead to ambipolar carrier transport properties similar to graphene.[4a,76] Organic field-effect transistors (OFETs) showing ambipolar properties attract considerable interest from the viewpoint of fundamental science for organic semiconductors and also with respect to device applications for light-emitting transistors and logic circuits. The OFET characteristics of Ph_2-**IDPL** were measured in

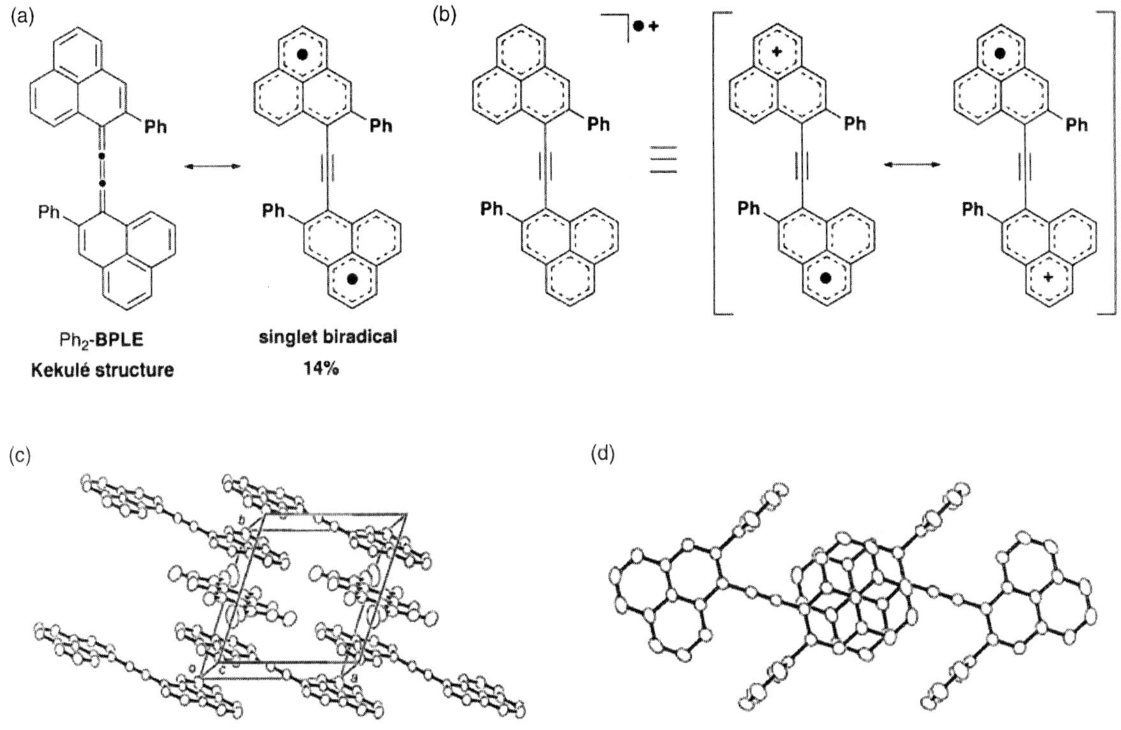

Figure 3.52 *(a) Resonance structures and calculated singlet biradical character of Ph_2-**BPLE**; (b) resonance structures of Ph_2-**BPLE**$^{\bullet+}$; (c) crystal structure of Ph_2-**BPLE**–F_4-TCNQ CT complex at room temperature; and (d) overlap pattern of Ph_2-**BPLE**$^{\bullet+}$ moieties. (c,d Reproduced with permission from [74d]. Copyright Wiley-VCH Verlag GmbH & Co. KGaA.)*

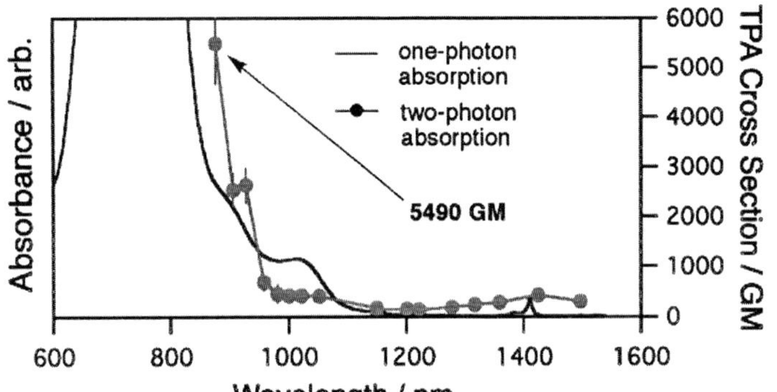

Figure 3.53 *One-photon absorption spectrum (black solid line, left) and two-photon absorption spectrum (circle, right) of Ph_2-**IDPL** in trichlormethane.*

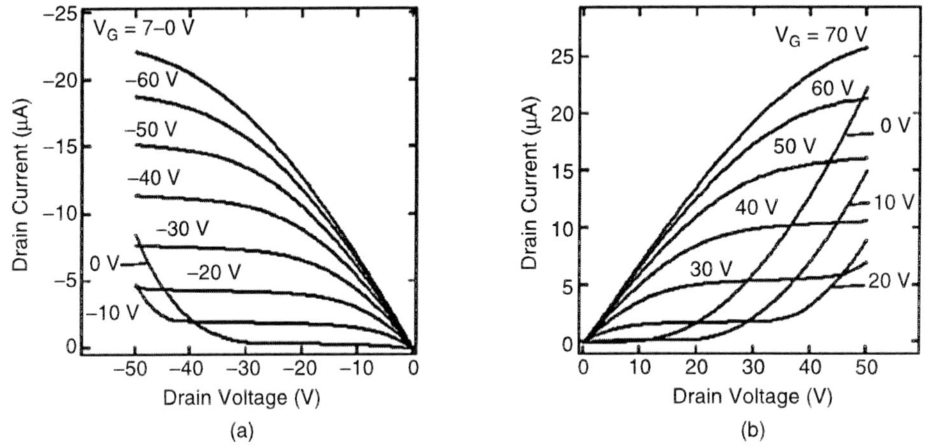

Figure 3.54 *Ambipolar carrier transporting properties of Ph_2-**IDPL**. (a) Negative and (b) positive gate biases. (Reprinted with permission from [76]. Copyright 2007, American Institute of Physics.)*

vacuum at room temperature by using Ph_2-**IDPL** film fabricated on the silica by vacuum evaporation. Figure 3.54 shows the $I_D - V_D$ characteristics of Ph_2-**IDPL** film, where I_D and V_D represent drain current and drain voltage, respectively, revealing good ambipolar transport properties with balanced hole and electron mobilities in the order of 10^{-3} cm^2/Vs. Because of the narrow band gap of Ph_2-**IDPL** film (0.8 eV), it is considered that not only the hole mobility but also the electron mobility show high values even when a gold electrode is used. The results fully exemplify the applicability of the phenalenyl system for ambipolar semiconductive materials.

3.11 Curve-structured phenalenyl system

Exotic π-conjugated networks of nonplanar polycyclic aromatic hydrocarbons, that is bowl-shaped, belt-shaped, or Möbius molecules, have drawn much attention not only in organic chemistry but also in materials science.[77] In order to approach the electronic structures and intra/intermolecular interactions of the nonplanar electronic system in terms of spin nature, bowl-shaped neutral radicals based on corannulene, a substructure of C_{60},[78] were designed and prepared (*vide infra*).[79] This novel bowl-shaped π radical possesses intrinsically 3-D molecular and electronic structures, and as such occupies an intermediate geometry between the planar π radical such as phenalenyl systems and tetrahedral σ radical (Figure 3.55).

Taking advantage of the corannulene system, a curved phenalenyl anion was designed and prepared as the first example of a nonplanar phenalenyl system (Figure 3.56a).[80] The anion possesses a six-membered *peri*-annelated corannulene featuring full π conjugation, a substructure of C_{70} with calculated average of bowl depth and POAV (π-orbital axis vector) of 0.90 Å and 8.3 Å, respectively.[80] The anion has a high stability in a degassed solution state. Notably, NMR studies and quantum chemical calculations demonstrate the salient features of the phenalenyl anion-type electronic structure are retained on the curved-surface π system (Figures 3.56b,c). Furthermore, the "curved aromaticity" of this unique π-conjugated system was evaluated using the NICS method and resonance structure studies.[80,81] This result enters curved-phenalenyl chemistry into the world of general phenalenyl chemistry, and contributes to the conceptual advance in electronic communication between an odd-alternant phenalenyl system and a non-alternant five-membered ring system (*vide infra*).

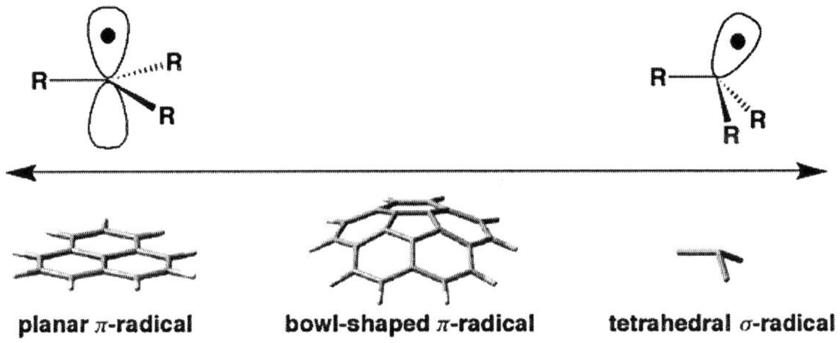

Figure 3.55 *Schematic drawings of planar π-radical, bowl-shaped π-radical, and tetrahedral σ-radical.*

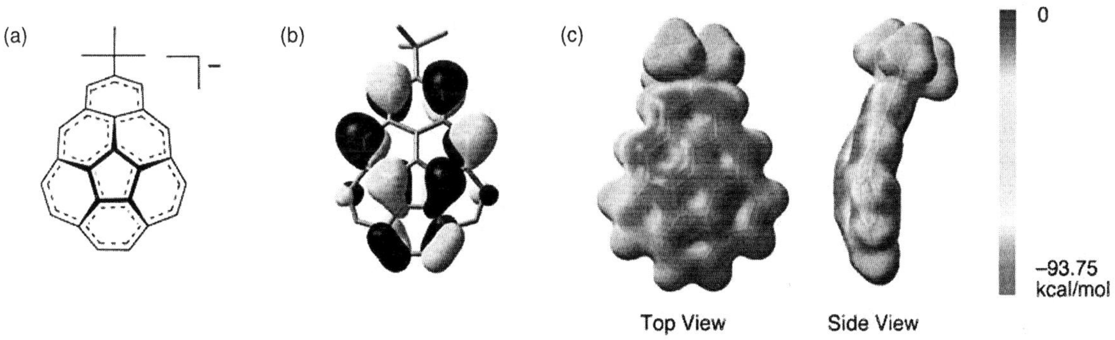

Figure 3.56 *(a) Chemical structure, (b) calculated HOMO picture, and (c) electrostatic potential surfaces of phenalenyl-fused corannulene anion. Calculations carried out at the RB3LYP/6-311G**//RB3LYP/6-311G** level of theory. (b,c Reprinted with permission from [80]. Copyright 2008 American Chemical Society.) A full-colour version of part (c) of this figure appears in the Colour Plate section of this book.*

Neutral radical derivatives having one and five phenalenyl systems incorporated into the corannulene system have been designed (Figure 3.57).[80] Interestingly in these molecular systems, three types of spin density distribution nature, α-, β-, and cyclopentadienyl-types, are seen in terms of resonance structures (Figures 3.57a,c). Quantum chemical calculations suggest a coexistence of the three types of spin density distribution (Figures 3.57b,d). These unique electronic features certainly rely on the existence of the five-membered π-electronic system. The synthesis and characterization of these molecules will lead to new synthetic methods for constructing gigantic curved π-conjugated skeletons consisting of odd-alternant phenalenyl and non-alternant five-membered ring systems, and will also be a good testing ground for elucidation of electronic structures and intrinsically 3-D intra/intermolecular interactions of the novel class of neutral radical systems (*vide infra*).

3.12 Non-alternant stable radicals

3.12.1 Cyclopentadienyl radicals

The cyclopentadienyl radical possesses 5π electrons in a five-membered ring and is categorized as a non-alternant hydrocarbon neutral radical. One-electron reduction or oxidation of this radical gives an

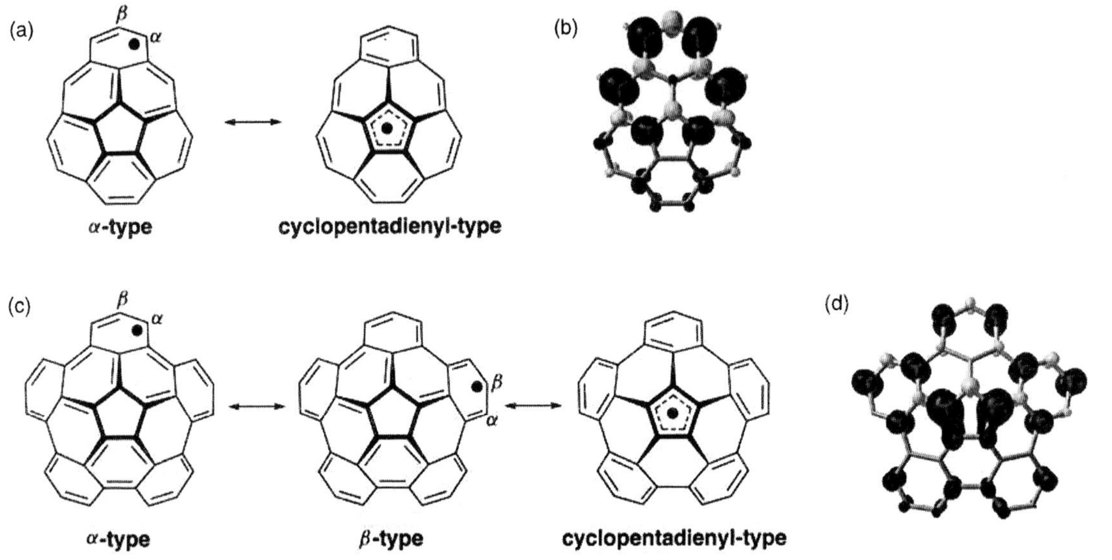

Figure 3.57 *(a), (c) Representative local spin structures and (b), (d) calculated spin density distributions of mono-phenalenyl-fused and penta-phenalenyl-fused corannulene neutral radicals. Calculations carried out at the UBLYP/6-31G*//UBLYP/6-31G level of theory.*

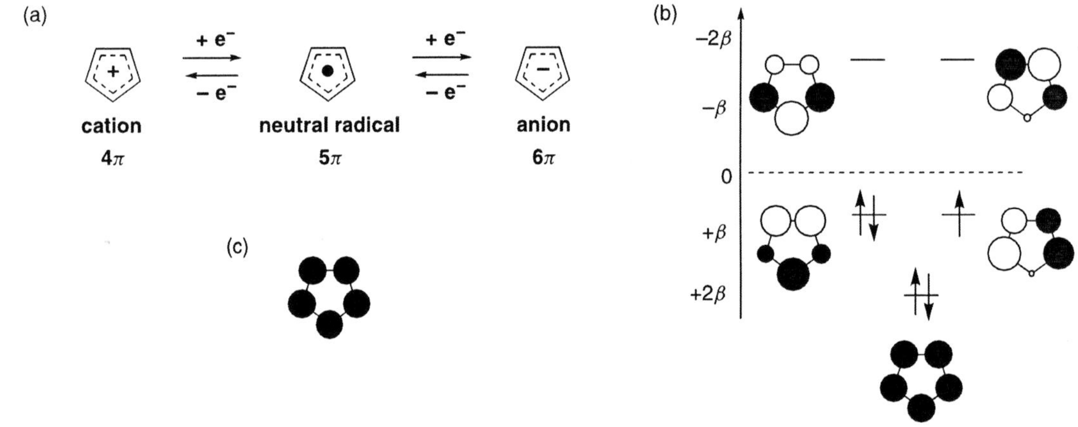

Figure 3.58 *(a) Redox behaviors, (b) molecular orbital diagram, and (c) spin density distribution of cyclopentadienyl radical calculated by the Hückel–molecular orbital method.*

anion (6π electrons) or a cation (4π electrons), showing aromaticity and antiaromaticity, respectively (Figure 3.58a). The frontier molecular orbitals of the cyclopentadienyl radical calculated by the Hückel-MO method are shown in Figure 3.58b. This radical has no NBMO in contrast to the phenalenyl system, and thus an unpaired electron resides in one of the two degenerate HOMOs. Quantum chemical calculation indicates an extensive spin delocalization on the whole molecular skeleton without spin polarization.

In general, the stabilities of cyclopentadienyl radicals are not high. The pristine radical[82,83d,f,g] (including monodeuterated[83c–g] and pentadeuterated[84]), monoalkyl,[83a–d,f,g] mono(alkylsilyl),[83b,d] pentahalogenated (fluorine, chlorine, bromine),[85] pentamethyl,[86] and pentakis(methoxycarbonyl)[87] derivatives are only studied spectroscopically in a degassed solution, in the gaseous phase, or in a matrix. Pentaaryl derivatives possess relatively high stabilities than those of the above derivatives.[88]

Sitzmann prepared and isolated a pentaisopropyl derivative in 1991, and reported the first X-ray crystal structure analysis of a cyclopentadienyl radical (Figure 3.59).[89] The yellow–green air-sensitive crystals were obtained by sublimation in high vacuum. Solution EPR measurements at 293 K yielded hyperfine splittings due to the methyl and the methyne protons and the methyl carbon nuclei (Figure 3.59b). Interestingly, temperature-dependent EPR spectra indicate that hfcc attributable to the methyne protons decreases substantially with decreasing temperature, while hfcc due to the methyl protons does not change (Figure 3.59b). This change in hfcc is ascribable to the temperature-dependent conformational change of the isopropyl groups. In the crystal, the cyclopentadienyl ring system possesses a planar structure with equivalent C–C bond lengths (1.40–1.41 Å) (Figure 3.59c). The radical forms a slipped stack 1-D columnar structure with a tilt angle of about 23° (Figure 3.59d). Probably due to the steric hindrance of the isopropyl groups, the distance between the π planes is 5.28 Å, indicating no electronic communication within the structure. Electrochemical measurements in acetonitrile show a reversible reduction process from the neutral radical to the anion (−1.91 V vs. Fc/Fc^+) and a quasi-reversible oxidation process from the neutral radical to the cation (+0.58 V vs. Fc/Fc^+).

In 1996, Rubin prepared and isolated a pentakis(trialkylsilylethynyl)cyclopentadienyl radical (shown below) in the process of the synthetic study of a metallocene derivative.[90] This neutral radical is stable enough to purify by silica gel column chromatography and survives in air for weeks in the solid state and for days even in a solution state, showing a sharp contrast to those of the pentaaryl and the

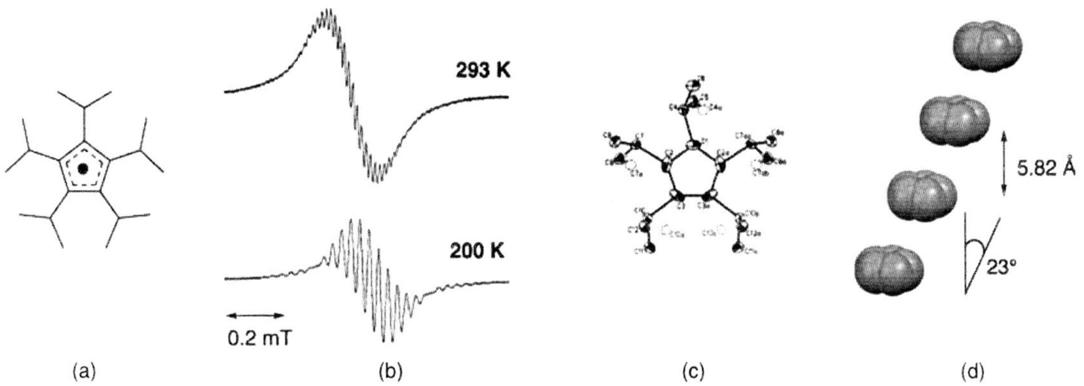

Figure 3.59 (a) Chemical structure, (b) EPR spectra in a toluene solution, (c) X-ray crystal structure, and (d) 1-D columnar structure of pentaisopropylcyclopentadienyl radical in which there is no electronic communication. (b,c Reprinted with permission from [89b]. Copyright 1993 American Chemical Society.)

pentaisopropyl derivatives. Electrochemical measurement of the radical in dichloromethane revealed a lower oxidation potential from the neutral radical to the cation (+0.15 V vs. Fc/Fc$^+$) than that of pentaisopropyl derivative, being attributable to the significant propargylic and allenic resonance stabilization by the silylethynyl moieties.

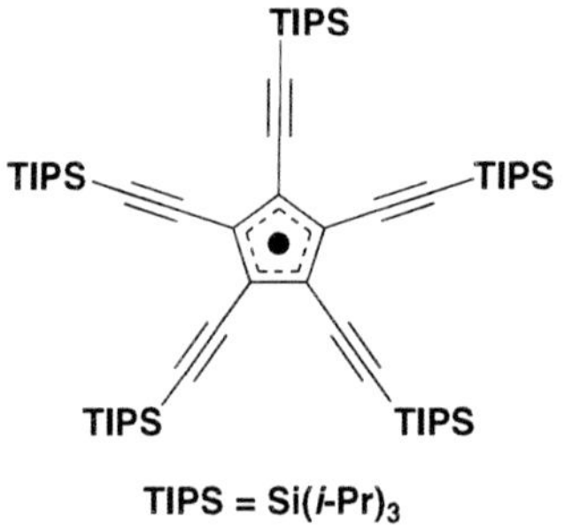

In 1997, Janiak reported the crystal structure of pentaphenylcyclopentadienyl radical (Ziegler's radical) (Figure 3.60).[88g] Although the analysis data are, unfortunately, of low quality because of the disorder of the phenyl moieties, it turns out that the cyclopentadienyl ring possesses a planar structure.

In 2004, Kitagawa and Komatsu prepared and isolated a cyclopentadienyl radical annelated with two homoadamantene skeletons (Figure 3.61a), and reported the X-ray crystal structure of this radical at 100 K (Figures 3.61b,c).[91] The radical survives only for a few hours in air in the solid state, but remains unchanged for months under vacuum. In a solution state, the radical is highly air-sensitive and slowly abstracts hydrogen atom from solvents. The substantial bond alternations are found in the C–C bonds of the cyclopentadienyl ring, and indicate a localization of the unpaired electron, being in agreement with the geometry calculated by the DFT method. EPR spectra in a solution state suggest a rapid equilibrium between the two allylic states. From these results, the radical can be viewed as an allyl radical analogue incorporated into the five-membered ring system. Cyclic voltammetry in dichloromethane shows a reversible oxidation wave at +0.30 V (vs. Fc/Fc$^+$) and an irreversible reduction wave at -1.78 V (vs. Fc/Fc$^+$), indicating that the radical is more easily oxidized and reduced than the pentaisopropyl derivative. The lowering of the oxidation potential represents a cation stabilizing effect of $\sigma-\pi$ conjugation induced by the annelation of the homoadamantene skeletons.

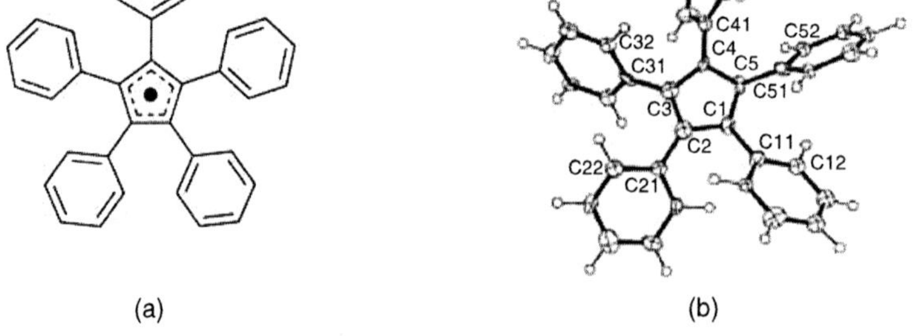

Figure 3.60 *(a) Chemical structure and (b) X-ray crystal structure of pentaphenylcyclopentadienyl radical. (b Reprinted with permission from [88g]. Copyright 1997 American Chemical Society.)*

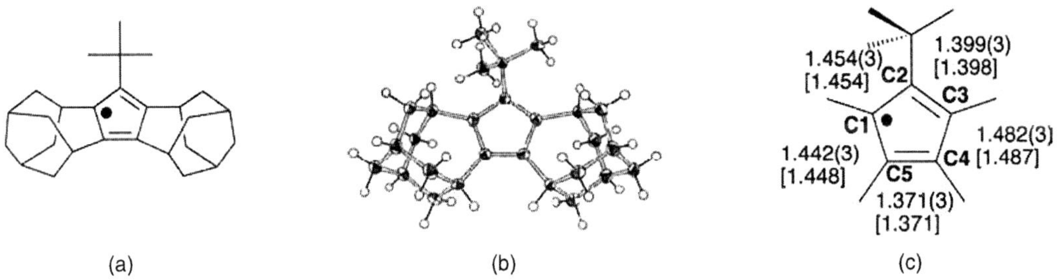

Figure 3.61 (a) Chemical structure, (b) X-ray crystal structure, and (c) selected bond lengths (Å) of the cyclopentadienyl ring moiety annelated with two homoadamantenes. Calculations carried out at the UB3LYP/6-31G* level of theory; calculation are given in square brackets. (b,c Reprinted with permission from [91]. Copyright 2004 American Chemical Society.)

3.12.2 Cyclopentadienyl radicals within a larger π-electronic framework

The cyclopentadienyl radical incorporated into larger π-electronic frameworks is an intriguing molecular system from the viewpoints of possible electronic modulation and spin delocalization. Fullerene derivatives have a high potentiality for realizing such molecular systems. The pioneering work based on the fullerenes were performed by Wasserman and Wudl in 1991 and 1997, disclosing the cyclopentadienyl radicals on the C_{60} and monoaza-C_{60} surfaces, $\mathbf{R_5C_{60}}^\bullet$ (R = $PhCH_2$, phenyl) and $\mathbf{C_{59}N}^\bullet$, respectively (shown below).[92] Although these neutral radicals unfortunately are not so stable enough to be isolated (and thus their studies are limited to the spectroscopic measurements in solution states), the synthetic access to these radicals provide challenging issues on the elucidation of their electronic structures.

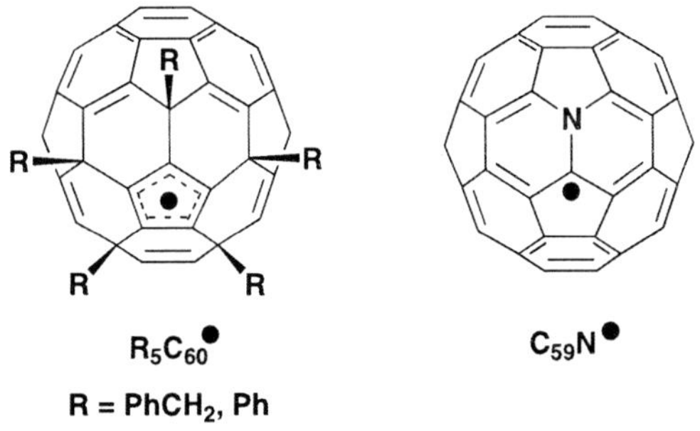

In the light of novel this class of open shell molecule, corannulene derivatives having a neutral radical substituent **Cor-R**$^\bullet$ were prepared and isolated as air-stable powder/crystals (Figure 3.62).[79] An unbalanced delocalization of spin and curved aromaticity[80,81] were revealed from EPR measurements and DFT calculations.[79c] Furthermore, magnetic susceptibility measurements and crystal structure analyses of the curve-structured neutral radicals reveal intermolecular magnetic interactions ($J/k_B = -22.5$ K) between the carbon atoms possessing large amounts of spin densities on the corannulene π surface. The high stability, highly spin delocalized and spin polarized natures as well as the unique geometric relationship between the planar π radical and tetrahedral σ radical (*vide supra*) provide indispensable opportunities for exploring new aspects of spin chemistry: that is, spin delocalized and spin polarized natures in the neutral diradical derivative based on the corannulene system,[93] and dynamic spin behavior as a function of bowl-to-bowl inversion.

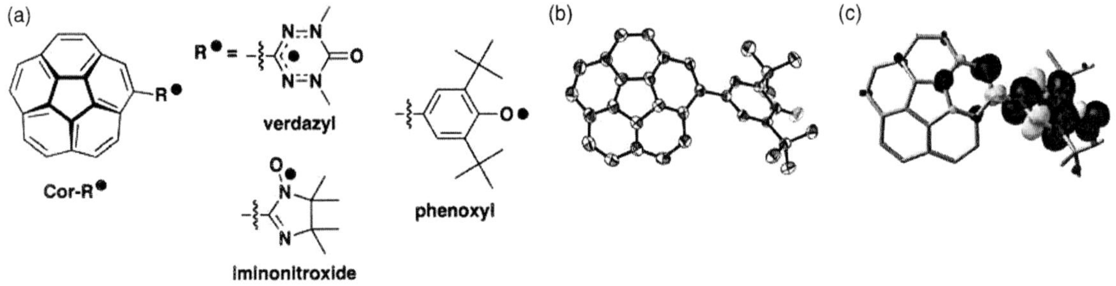

Figure 3.62 (a) Chemical structure of **Cor-R**•, (b) X-ray crystal structure, and (c) spin density distribution of **Cor-phenoxyl**. Calculations carried out at the UB3LYP/6-31G** level of theory, showing the unbalanced delocalization of spin. (b Reproduced with permission from [79c]. Copyright Wiley-VCH Verlag GmbH & Co. KGaA.)

3.13 Stable triplet carbenes

Carbenes are neutral divalent derivatives of carbon, and the carbenic center has two non-bonded electrons that can be spin-paired (singlet carbene) or unpaired (triplet carbene). Singlet carbenes are stabilized thermodynamically by heteroatom substituents directly connected to the carbenic center. The high stability of singlet carbene enables their molecular structures to be determined by X-ray crystallography at room temperature. In sharp contrast, triplet carbenes[94] are regarded as one of the most effective spin sources for molecular magnetics based on organic neutral radicals.[95] However, triplet carbenes are less readily stabilized thermodynamically than their singlet counterparts.

In 1999, Tomioka reported a diphenylcarbene derivative **DPCa** as the first triplet carbene which can survive more than one hour in a solution state at room temperature (half-life $t_{1/2} \sim 10$ minutes) and can be stored in a bottle at −40 °C (Figure 3.63a).[94a,96] The high stability was realized by steric protection of the carbenic center with bromine and trifluoromethyl (CF_3) groups at *ortho* positions of the phenyl groups. Especially, trifluoromethyl groups play a key role for stabilization of the triplet carbene in terms of their steric bulk. More importantly, the C–F bond is almost the only type of chemical bond unreactive toward a carbenic center.[97] Taking advantage of this feature, **DPCb** and **DPCc** possessing two trifluoromethyl

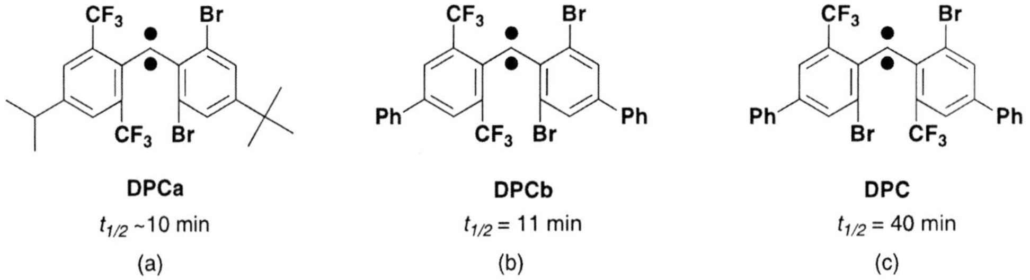

Figure 3.63 (a) The first stable diphenylcarbene derivative with a half-life of ~10 minutes. Bis(p-phenyl)-substituted diphenylcarbenes with two bromine and two CF_3 groups in (b) unsymmetrical and (c) symmetrical fashions.

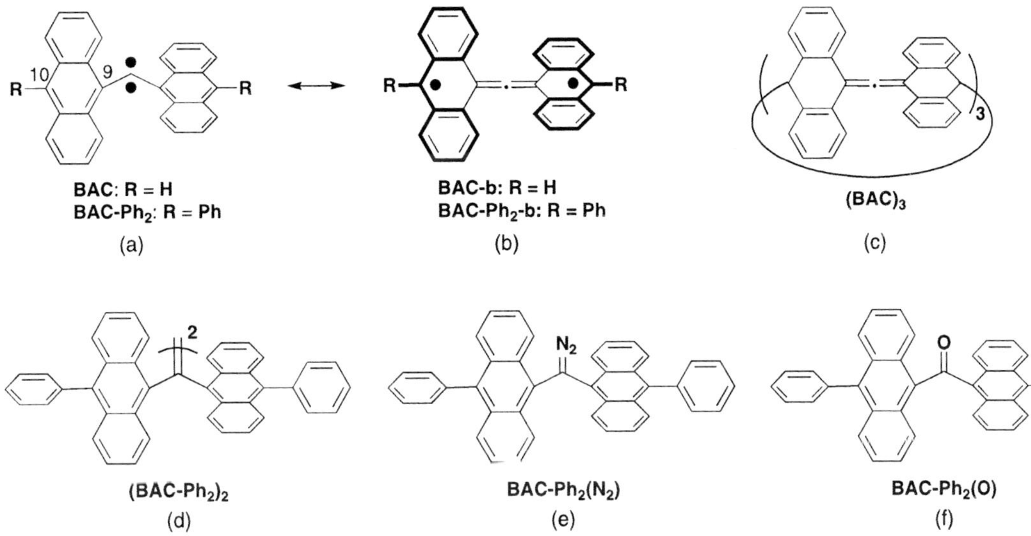

Figure 3.64 *(a) Chemical structures of bis(9-anthryl)carbenes, **BAC** (lifetime 0.5μs) and **BAC-Ph$_2$** (half-life 19 minutes). (b) Diradical type resonance structures, **BAC-b** and **BAC-Ph$_2$-b**, with allene-bridged bis(triphenylmethyl) diradical structures. (c) Chemical structures of carbene trimer, **(BAC)$_3$**, and (d) of carbenic dimer, **(BAC-Ph$_2$)$_2$**. (e) Chemical structures of **BAC-Ph$_2$(N$_2$)** as the precursor of **BAC-Ph$_2$** and (f) of ketone **BAC-Ph$_2$(O)** generated by irradiation of **BAC-Ph$_2$(N$_2$)** in the presence of oxygen. The phenyl groups effectively prevent the spins from "leaking out" and thus suppress the reaction at 10-positions.*

and two bromine groups in unsymmetrical and symmetrical fashions were designed and generated[94b] (Figures 3.63b,c).

Spin delocalization from the carbenic center in triplet carbenes was investigated in bis(9-anthryl)carbene **BAC** by Wasserman using EPR spectroscopy, elucidating a possible structure with a collinear divalent carbon (allenic carbon) and perpendicularly arranged anthryl groups (Figures 3.64a,b).[98] These electronic effects are expected to provide a thermodynamic stability in addition to the kinetic stability through shielding the carbenic center with four *peri*-protons of the anthryl groups. In spite of these favorable stereoelectronic effects, **BAC** possesses low stability[99] with a lifetime of 0.5 μs.[100] This is because **BAC** tends to form a trimer **(BAC)$_3$**[101] by a contribution of the diradical resonance structure **BAC-b** (Figures 3.64b,c). In 2001, Tomioka reported a bis(9-anthryl)carbene derivative with phenyl groups, **BAC-Ph$_2$**, showing a half-life of 19 minutes in a solution state at room temperature by suppression of the trimerization reaction.[100] A resonance structure **BAC-Ph$_2$-b** can be regarded as a bis(triphenylmethyl radical) connected by an allenic bond (Figure 3.64b). The contributions of the carbene and the diradical structures for **BAC-Ph$_2$** are estimated by chemical reactivity with help of EPR spectroscopy. **BAC-Ph$_2$** forms a carbene dimer **(BAC-Ph$_2$)$_2$** by double bond formation at the carbenic center (Figure 3.64d). Furthermore, irradiation of **BAC-Ph$_2$(N$_2$)**, a precursor of **BAC-Ph$_2$**, in the presence of oxygen gives a ketone **BAC-Ph$_2$(O)** (Figures 3.64e,f). Consequently, these experimental results suggest that the electronic contribution of the carbenic character is more important than that of the diradical one, although the spins are still extensively delocalized.

Tomioka's successful stabilization method for the triplet carbene enabled the design and synthesis of stable high spin polycarbenes in ambient conditions (Figure 3.65).[102,103] Tris**BAC**, tricarbene system, was designed based on bis(9-anthryl)carbene, and was generated in 2-methyltetrahydrofuran at 77 K with

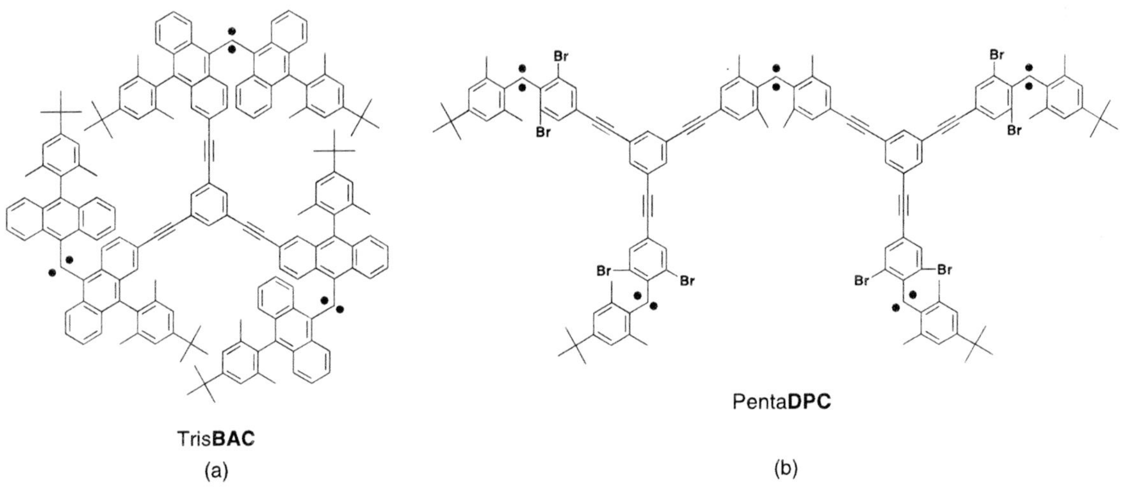

Figure 3.65 Chemical structures of (a) tricarbene, Tris**BAC**, and (b) pentacarbene, Penta**DPC**, based on bis(9-anthryl)carbene.

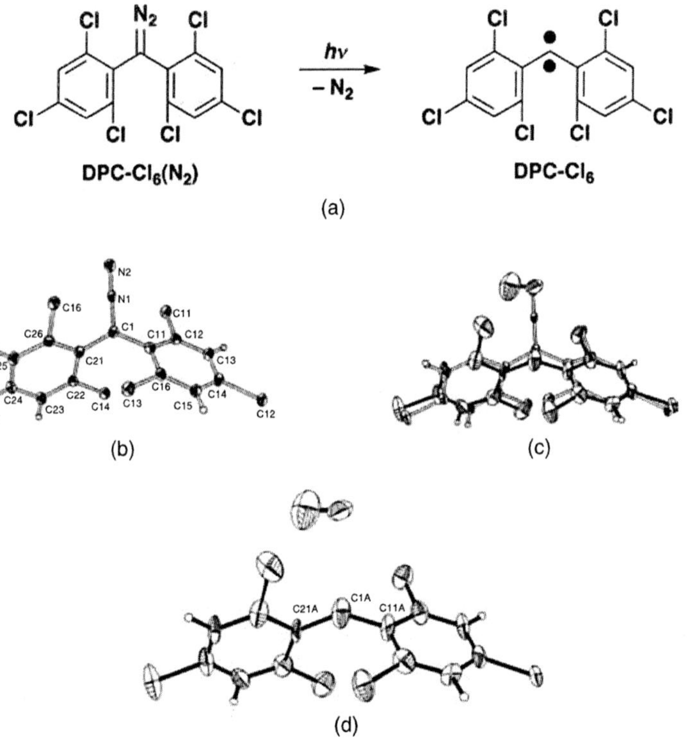

Figure 3.66 (a) Generation of ***DPC-Cl$_6$*** by photo-irradiation of the corresponding precursor ***DPC-Cl$_6$(N$_2$)*** in the single crystal. (b) Perspective view of ***DPC-Cl$_6$(N$_2$)***. (c) Perspective view of a disordered structure of ***DPC-Cl$_6$(N$_2$)*** (open line) and ***DPC-Cl$_6$*** (solid line) with occupancy factors of 0.803(5) and 0.197(5), respectively. (d) Perspective view of ***DPC-Cl$_6$*** and a dinitrogen molecule (with ***DPC-Cl$_6$(N$_2$)*** omitted) in part c. (b,c,d Reprinted with permission from [106]. Copyright 2001 American Chemical Society.)

a half-life of 13 h.[102a] EPR and SQUID measurements show Tris**BAC** possesses a weak ferromagnetic interaction between the carbenic units. Penta**DPC** was designed by a dendrimer approach based on diphenylcarbene derivatives.[102b,104] This polycarbene has five carbenic units connected in non-disjoint manner ($S = 2$, $J > 0$), and thus shows ferromagnetic interaction between five carbenic centers ($S = 4.4$ at 2.0 K).[105]

In 2001, Kawano, Tomioka, and Ohashi reported the first X-ray crystal structure of a transient triplet carbene trapped in a crystal.[106] The X-ray crystallographic analysis was carried out by photo-irradiation of a single crystal of the diazo derivative, **DPC-Cl$_6$(N$_2$)**, below 80 K (Figures 3.66a,b). The difference electron density map after irradiation shows new peaks around the molecule, being assignable to a dinitrogen molecule and the carbene **DPC-Cl$_6$** (Figures 3.66c,d). For the *in situ* X-ray crystallography, low temperature conditions (<80 K) and careful selection of the wavelength of the irradiating light are crucially important to suppress loss of crystallinity and thermal motion of dinitrogen molecules trapped in the crystal. This *in situ* X-ray crystallography reveals that the carbene is in the triplet ground state judged by the carbenic angle and the bond distances between the carbenic carbon and the phenyl carbons with help of IR spectroscopy and DFT calculation.[106,107]

3.14 Conclusions

Phenalenyl chemistry began a half-century ago. The comprehensive book written by Forrester, Hay, and Thomson published in 1968 contains only eight papers concerning the phenalenyl system.[2b] In the early period of the past five decades, the main emphasis was on the experimental elucidation of an extremely spin delocalized nature in a degassed solution state. The solid state properties of the phenalenyl system, however, remained unrevealed for a long time because of their high instability in air. More recently, enthusiastic efforts on chemical modification for stabilization by means of steric and electronic effects lead to isolation of the phenalenyl system as pure crystalline materials, greatly fertilizing phenalenyl chemistry from the viewpoints of both fundamental and applied aspects.

Haddon's claim[9] for molecular conductors based on the phenalenyl system was marvelously realized by his own fascinating idea.[58–66] The studies on azaphenalenyl (**APLY**), oxophenalenoxyl (**OPO**), and other derivatives surely demonstrate a high potentiality for chemical modifications to create a variety of exotic properties, which may be far beyond what was imagined a half-century ago: "aromaticity generation", "temperature-dependent continuous color change", "spin topological symmetry control", "redox-based spin diversity", "spin center transfer", "two-photon absorption", and "ambipolar FET properties" and so on. Notably, these properties are based on the highly delocalized spin nature inherent in the phenalenyl system. Recent curve-structured phenalenyl chemistry and applications for the electrode active material in the secondary battery[50,51] also give new conceptual advances in material sciences. In order to pursue a qualitatively new material challenge, it is still indispensable to design and synthesize novel π radicals having unique electronic/molecular structures. The spin delocalized π radical having chirality, photoactivity, liquid crystallinity, π interaction with metal ion are intrinsically novel molecular systems and fundamentally important target systems in near future. The fused-polycyclic π radicals with gigantic π-electronic networks (i.e. open shell graphene) and manupulation of their multiple spin states are intriguing for the application to the molecular spin quantum computing and quantum information processing.[108]

Acknowledgements

We thank Professor Takeji Takui (Osaka City University), Professor Kazunobu Sato (Osaka City University), Dr Tsuyoshi Murata (Osaka University), Dr Shuichi Suzuki (Osaka City University), and Mr Akira Ueda (Osaka University) for helpful and encouraging discussions.

References

1. a) M. Gomberg, *J. Am. Chem. Soc.*, **22**, 757–771 (1900); b) M. Gomberg, *J. Am. Chem. Soc.*, **23**, 496–502 (1901); c) M. Gomberg, *Chem. Rev.*, **1**, 91–141 (1924).
2. a) Stability and persistency are the most important issue for organic radicals, see: D. Griller, and K. U. Ingold, *Acc. Chem. Res.*, **9**, 13–19 (1976); Overviews of stable radicals, see: b) A. R. Forrester, J. M. Hay, and R. H. Thomson, *Organic Chemistry of Stable Free Radicals*, Academic Press, London and New York, 1968; c) R. G. Hicks, *Org. Biomol. Chem.*, **5**, 1321–1338 (2007); Overviews of molecule-based magnetic materials on the base of organic stable radicals, see: d) P. M. Lahti (Ed.), *Magnetic Properties of Organic Materials*, Marcel Dekker, New York, 1999; e) P. Day, and A. E. Underhill (Eds), *Metal–Organic and Organic Molecular Magnets*, The Royal Society of Chemistry, Cambridge, 1999; f) J. Veciana, C. Rovira, and D. B. Amabilino (Eds), *Supramolecular Engineering of Synthetic Metallic Materials*, Kluwer Academic Publishers, Dordrecht, 1999; g) K. Itoh, and M. Kinoshita (Eds), *Molecular Magnetism*, Kodansha, and Gordon and Breach Science Publishers, Tokyo, 2000; h) J. S. Miller, and M. Drillon (Eds), *Magnetism: Molecules to Materials, Vol. I-V*, Wiley-VCH Verlag GmbH, Weinheim, 2001–2005; i) W. Linert, and M. Verdaguer (Eds), *Molecular Magnets Recent Highlights*, Springer-Verlag, Vienna, 2003; j) T. L. Makarova, and F. Palacio (Eds), *Carbon-Based Magnetism*, Elsevier B.V., Amsterdam, 2006.
3. a) K. Itoh, *Chem. Phys. Lett.*, **1**, 235–238 (1967); b) N. Mataga, *Theor. Chim Acta*, **10**, 372–376 (1968); c) K. Itoh, *Pure Appl. Chem.*, **50**, 1251–1259 (1978); d) H. Iwamura, *Pure Appl. Chem.*, **58**, 187–196 (1986); e) T. Takui, Y. Teki, K. Sato, and K. Itoh, *Mol. Cryst. Liq. Cryst.*, **272**, 1–22 (1995).
4. a) K. S. Novoselov, A. K. Geim, S. V. Morozov, *et al.*, *Science*, **306**, 666–669 (2004); b) A. K. Geim, and K. S. Novoselov, *Nature Mater.*, **6**, 183–191 (2007).
5. C. D. Simpson, J. D. Brand, A. J. Berresheim, *et al.*, *Chem. Eur. J.*, **8**, 1424–1429 (2002).
6. a) E. Clar, *Aromatische Kohlenwasserstoffe*, p. 311, Springer-Verlag, Berlin, 1941; b) E. Clar, *Aromatische Kohlenwasserstoffe 2nd Ed.*, pp 93–97 and pp 461–465, Springer-Verlag, Berlin, 1952; c) E. Clar, *Polycyclic Hydrocarbon*, Vol. 1, pp 24–31, Academic Press, New York, 1964; d) E. Clar, and D. G. Stewart, *J. Am. Chem. Soc.*, **75**, 2667–2672 (1953); e) E. Clar, and D. G. Stewart, *J. Am. Chem. Soc.*, **76**, 3504–3507 (1954).
7. a) J. Inoue, K. Fukui, T. Kubo, *et al.*, *J. Am. Chem. Soc.*, **123**, 12702–12703 (2001); b) K. Fukui, J. Inoue, T. Kubo, *et al.*, *Synth. Met.*, **121**, 1824–1825 (2001). See also: ref 13b.
8. Pioneering studies on the phenalenyl chemistry, see: a) V. Boekelheide, and C. E. Larrabee, *J. Am. Chem. Soc.*, **72**, 1245–1249 (1950); b) D. H. Reid, *Chem. Ind. (London)*, 1504–1505 (1956); c) P. B. Sogo, M. Nakazaki, and M. Calvin, *J. Chem. Phys.*, **26**, 1343–1345 (1957); d) J. E. Bennett, *Nature*, **188**, 485–486 (1960); e) F. C. Stehling, and K. W. Bartz, *J. Chem. Phys.*, **34**, 1076–1077 (1961); f) J. E. Bennett, *Proc. Chem. Soc.*, 144–145 (1961); g) D. H. Reid, M. Fraser, B. B. Molloy, *et al.*, *Tetrahedron Lett.*, 530–535 (1961); h) D. H. Reid, *Quart. Rev., Chem. Soc.*, **19**, 274–302 (1965); i) F. Gerson, *Helv. Chim. Acta*, **49**, 1463–1467 (1966). see also: j) I. Murta, in *Topics in Nonbenzenoid Aromatic Chemistry*, (eds T. Nozoe, R. Breslow, K. Hafner, *et al.*), Vol 1, pp 159–190, Hirokawa, Tokyo (1973); k) R. Biehl, M. Plato, and K. Möbius, *J. Chem. Phys.*, **63**, 3515–3522 (1975); l) R. Biehl, C. Hass, H. Kurreck, *et al.*, *Tetrahedron*, **34**, 419–424 (1978); m) W. Broser, H. Kurreck, S. Oestreich-Janzen, *et al.*, *Tetrahedron*, **35**, 1159–1166 (1979); n) C. Hass, B. Kirste, H. Kurreck, and G. Schlömp, *J. Am. Chem. Soc.*, **105**, 7375–7383 (1983). See also: Ref. 2b (pp. 31 and pp 77–78) and 6b (pp. 426–433).
9. R. C. Haddon, *Nature*, **256**, 394–396 (1975).
10. It is generally known that σ dimer of the phenalenyl undergoes dehydrogenation by heat to produce peropyrene. Crystal structure of peropyrene derivative is recently reported, see: a) L. Beer, S. K. Mandal, R. W. Reed, *et al.*, *Cryst. Growth Des.*, **7**, 802–809 (2007). See also: b) S. Pogodin, and I. Agranat, *J. Am. Chem. Soc.*, **125**, 12829–12835 (2003).
11. a) D. H. Reid, *Tetrahedron*, **3**, 339–352 (1958); b) F. Gerson, E. Heilbronner, H. A. Reddoch, *et al.*, *Helv. Chim. Acta*, **50**, 813–821 (1967); c) I. C. Lewis, and L. S. Singer, *J. Phys. Chem.*, **73**, 215–218 (1969); d) B. Kirste, W. Broser, K. Grein, *et al.*, *Chem. Ber.*, **118**, 3464–3480 (1985).
12. a) 1-Cyano-4,5,6-tri(methoxy)phenalenyl and 1-cyano-6-(methylthio)phenalenyl possess small E^{sum} values of 1.09 and 1.17V, respectively, see: K. Nakasuji, M. Yamaguchi, I, Murata, *et al.*, *J. Am. Chem. Soc.*, **111**,

9265–9267 (1989); b) 1,4,7-Tris(alkylamino) derivatives give hydrogen-bonded charge-transfer complexes, see: Y. Morita, E. Miyazaki, T. Yokoyama, *et al.*, *Synth. Met.*, **135–136**, 617–618 (2003).
13. a) K. Goto, T. Kubo, K. Yamamoto, *et al.*, *J. Am. Chem. Soc.*, **121**, 1619–1620 (1999); b) K. Fukui, *An Electron-Nuclear Magnetic Resonance Study of Highly Symmetric Organic Open-Shell Systems with Homo- and Hetero-Atomic π-Conjugation and Their High-Spin Assemblages*, Ph.D. Thesis, Osaka City University, Japan, January 2001; c) Theoretical study on magnetic interactions of π-dimer of **PLY**, see: Y. Takano, T. Taniguchi, H. Isobe, *et al.*, *J. Am. Chem. Soc.*, **124**, 11122–11130 (2002).
14. a) D. Small, V. Zaitsev, Y. Jung, *et al.*, *J. Am. Chem. Soc.*, **126**, 13850–13858 (2004); b) D. Small, S. V. Rosokha, J. K. Kochi, and M. Head-Gordon, *J. Phys. Chem. A*, **109**, 11261–11267 (2005); c) V. Zaitsev, S. V. Rosokha, M. Head-Gordon, and J. K. Kochi, *J. Org. Chem.*, **71**, 520–526 (2006).
15. S. Suzuki, Y. Morita, K. Fukui, *et al.*, *J. Am. Chem. Soc.*, **128**, 2530–2531 (2006).
16. NICS calculation for open shell systems: a) V. Gogonea, P. v. R. Schleyer, and P. R. Schreiner, *Angew. Chem., Int. Ed.*, **37**, 1945–1948 (1998); b) P. R. Serwinski, and P. M. Lahti, *Org. Lett.*, **5**, 2099–2102 (2003). See also: Ref. 35.
17. R. C. Haddon, F. Wudl, M. L. Kaplan, *et al.*, *J. Am. Chem. Soc.*, **100**, 7629–7633 (1978).
18. a) R. C. Haddon, A. M. Hirani, N. J. Kroloff, and J. H. Marshall, *J. Org. Chem.*, **48**, 2115–2117 (1983); b) W. Peebles, R. M. Pagni, and R. C. Haddon, *Tetrahedron Lett.*, **30**, 2727–2730 (1989); c) R. M. Pagni, W. Peebles, R. C. Haddon, and S. V. Chichester, *J. Org. Chem.*, **55**, 5595–5601 (1990).
19. a) M. L. Kaplan, R. C. Haddon, A. M. Hirani, *et al.*, *J. Org. Chem.*, **46**, 675–678 (1981); b) R. C. Haddon, S. L. Mayo, S. V. Chichester, and J. H. Marshall, *J. Am. Chem. Soc.*, **107**, 7585–7591 (1985).
20. R. C. Haddon, S. V. Chichester, S. M. Stein, *et al.*, *J. Org. Chem.*, **52**, 711–712 (1987).
21. P. A. Koutentis, Y. Chen, Y. Cao, *et al.*, *J. Am. Chem. Soc.*, **123**, 3864–3871 (2001).
22. EPR sample of the neutral radical is generated by reduction of the corresponding cation with mercury in a degassed benzene solution, see: Ref. 20.
23. P. A. Koutentis, R. C. Haddon, R. T. Oakley, *et al.*, *Acta Cryst.*, **B57**, 680–691 (2001).
24. L. Beer, R. W. Reed, C. M. Robertson, *et al.*, *Org. Lett.*, **10**, 3121–3123 (2008).
25. Y. Morita, and S. Suzuki, unpublished result.
26. Y. Morita, T. Aoki, K. Fukui, *et al.*, *Angew. Chem. Int. Ed.*, **41**, 1793–1796 (2002).
27. a) 2-Pyrimidine-substituted 1,3-diazaphenalene derivative forms chelete complex with cobalt (II) ion, see: Y. Morita, S. Suzuki, K. Fukui, *et al.*, *Polyhedron*, **22**, 2215–2218 (2003); b) 1,3- and 1,6-diazaphenalene derivatives are useful building blocks for highly conductive charge-transfer complexes possessing hydrogen-bonding networks, see: T. Murata, Y. Morita, K. Fukui, *et al.*, *Bull. Chem. Soc. Jpn.*, **79**, 894–913 (2006).
28. Y. Morita, S. Suzuki, unpublished result.
29. Y. Morita, S. Suzuki, K. Fukui, *et al.*, *Nature Mater.*, **7**, 48–51 (2008).
30. a) W. Fujita, and K. Awaga, *Science*, **286**, 261–262 (1999); b) W. Fujita, K. Awaga, H. Matsuzaki, and H. Okamoto, *Phys. Rev. B*, **65**, 064434 (2002). See also: Ref. 59a.
31. C. D. S. Tomlin, *The Synthesis and Properties of s-Hexaazaphenalene*. Ph.D. Thesis, Princeton University, New Jersey, October 1964. See also: *Chem. Abstr.*, **63**, 16354g (1965).
32. S. Suzuki, Y. Morita, K. Fukui, *et al.*, *Inorg. Chem.*, **44**, 8197–8199 (2005).
33. S. Zheng, J. Lan, S. I. Khan, and Y. Rubin, *J. Am. Chem. Soc.*, **125**, 5786–5791 (2003).
34. S. Zheng, J. D. Thompson, A. Tontcheva, *et al.*, *Org. Lett.*, **7**, 1861–1863 (2005).
35. Y. Morita, J. Kawai, K. Fukui, *et al.*, *Org. Lett.*, **5**, 3289–3291 (2003).
36. Y. Morita, S. Nishida, J. Kawai, *et al.*, *Pure Appl. Chem.*, **80**, 507–517 (2008).
37. a) Y. Morita, S. Nishida, J. Kawai, *et al.*, *Org. Lett.*, **4**, 1985–1988 (2002); b) "*MO-based VB method*" gives a quantitative evaluation for each resonance structure (local spin structure), see: Refs 13b, 38, 43.
38. Y. Morita, S. Nishida, J. Kawai, *et al.*, *Polyhedron*, **22**, 2209–2213 (2003).
39. a) Y. Morita, S. Nishida, K. Fukui, *et al.*, *Polyhedron*, **24**, 2194–2199 (2005); b) S. Nishida, Y. Morita, T. Ohba, *et al.*, *Tetrahedron*, **63**, 7690–7695 (2007).
40. K. Hatanaka, Y. Morita, T. Ohba, *et al.*, *Tetrahedron Lett.*, **37**, 873–876 (1996).
41. Molecular systems with reversible switching ability between two different states by external stimuli can be used as rewritable memories. In contrast, the systems with irreversible change can be used as permanent memories,

see: a) S. Kobatake, S. Takami, H. Muto, *et al.*, *Nature*, **446**, 778–781 (2007); b) V. Balzani, A. Credi, and M. Venturi, Memories, Logic Gates, and Related Systems, in *Molecular Devices and Machines* 2nd Edn, Wiley-VCH Verlag GmbH, Weinheim, 2008; c) B. L. Feringa (Ed.), *Molecular Switches* 2nd Edn, Vol 1–2 Wiley-VCH Verlag GmbH, Weinheim, 2008.

42. K. Hatanaka, Y. Morita, T. Ohba, *et al.*, *Tetrahedron Lett.*, **37**, 877–880 (1996).
43. Y. Morita, T. Ohba, N. Haneda, *et al.*, *J. Am. Chem. Soc.*, **122**, 4825–4826 (2000).
44. Y. Morita, S. Maki, K. Fukui, *et al.*, *Org. Lett.*, **3**, 3099–3102 (2001).
45. Y. Morita, J. Kawai, S. Nishida, *et al.*, *Synth. Met.*, **137**, 1217–1218 (2003).
46. T. Mochida, A. Izuoka, T. Sugawara, *et al.*, *J. Chem. Phys.*, **101**, 7971–7974 (1994).
47. Y. Morita, S. Suzuki, unpublished result.
48. Y. Morita, S. Suzuki, J. Kawai, *et al.*, *Synth. Met.*, **137**, 1209–1210 (2003).
49. Galvinoxyl is an interesting neutral radical having a possibility for showing the "redox-based spin diversity" nature. However, low stability of the radical dianion species may hamper its realization, see: a) A. I. Prokof'ev, S. P. Solodovnikov, G. A. Nikiforov, and V. V. Ershov, *Izv. Akad. Nauk SSSR, Ser. Khim.*, 324–328 (1971); b) S. G. Kukes, N. N. Bubnov, A. I. Prokof'ev, *et al.*, *Izv. Akad. Nauk SSSR, Ser. Khim.*, 2375–2378 (1973).
50. a) Y. Morita, T. Okafuji, and M. Satoh, Jpn. Kokai Tokkyo Koho JP 2007227186 A 20070906, 2007; b) S. Nishida, Y. Morita, M. Moriguchi, *et al.*, Molecular crystalline secondary batteries: properties of phenalenyl-based multi-stage redox systems as cathode active organic materials. Presented at the 1st Russian-Japanese Workshop "Open Shell Compounds and Molecular Spin Devices", Novosibirsk, Russia, June 30–July 2, 2007.
51. Preliminary results on molecular crystalline secondary batteries based on closed-shell electron-donor and -acceptor molecules as cathode-active materials are reported. Y. Morita, S. Nishida, N. Akutagawa, *et al.*, Molecular crystalline secondary batteries: novel secondary batteries based on organic molecules as cathode-active materials. Presented at the 19th Symposium on Fundamental Organic Chemistry, Osaka, Japan, October 3–5, C28, 2008.
52. a) Y. Morita, J. Kawai, N. Haneda, *et al.*, *Tetrahedron Lett.*, **42**, 7991–7995 (2001); b) S. Nishida, Y. Morita, K. Fukui, *et al.*, *Angew. Chem. Int. Ed.*, **44**, 7277–7280 (2005).
53. If one oxidation state from the TTF moiety is realized, relatively strong intramolecular spin interactions between the **6OPO** neutral radical and the donor radical cation due to a spin polarization effect through the π-electron network can be expected. See: Ref. 52a.
54. An intra- and inter-molecular spin-site exchange natures depending on temperature are realized in metal complexes having ferrocene and metallofullerene moieties. In these systems, high redox abilities of metal complex moieties play vital roles. See: a) I. Ratera, D. Ruiz-Molina, F. Renz, *et al.*, *J. Am. Chem. Soc.*, **125**, 1462–1463 (2003); b) C. Sporer, I. Ratera, D. Ruiz-Molina, *et al.*, *Solid State Sci.*, **11**, 786–792, (2009); c) T. Tsuchiya, K. Sato, H. Kurihara, *et al.*, *J. Am. Chem. Soc.*, **128**, 14418–14419 (2006).
55. Photo-induced valence tautomerism based on transition metal complexes, see: a) D. M. Adams, B. Li, J. D. Simon, and D. N. Hendrickson, *Angew. Chem. Int. Ed.*, **34**, 1481–1483 (1995); b) O, Sato, J. Tao, and Y.-Z. Zhang, *Angew. Chem. Int. Ed.*, **46**, 2152–2187 (2007); c) O, Sato, A. Cui, R. Matsuda, *et al.*, *Acc. Chem. Res.*, **40**, 361–369 (2007), and references therein.
56. "Photo-induced spin-center transfer" in purely organic open-shell system is recently achieved in the **6OPO** derivative. Y. Morita, S. Nishida, unpublished result.
57. a) I. C. Paul, and G. A. Sim, *Proc. Chem. Soc.*, 352–353 (1962); b) R. Aderjan, H. Breer, H. J. Keller, and H. H. Rupp, *Z. Naturforsch.*, **28B**, 164–167 (1973); c) Y. Demura, T. Kawato, H. Kanatomi, and I. Murase, *Bull. Chem. Soc. Jpn.*, **48**, 2820–2824 (1975); d) R. Neidlein, and Z. Behzadi, *Chem.-Ztg.*, **100**, 388–389 (1976); e) R. Neidlein, and Z. Behzadi, *Chem.-Ztg.*, **102**, 150–152 (1978); f) R. C. Haddon, S. V. Chichester, and J. H. Marshall, *Tetrahedron*, **42**, 6293–6300 (1986).
58. X. Chi, M. E. Itkis, B. O. Patrick, *et al.*, *J. Am. Chem. Soc.*, **121**, 10395–10402 (1999).
59. a) M. E. Itkis, X. Chi, A. W. Cordes, and R. C. Haddon, *Science*, **296**, 1443–1445 (2002); b) X. Chi, M. E. Itkis, K. Kirschbaum, *et al.*, *J. Am. Chem. Soc.*, **123**, 4041–4048 (2001); c) J. S. Miller, *Angew. Chem. Int. Ed.*, **42**, 27–29 (2003).
60. X. Chi, M. E. Itkis, R. W. Reed, *et al.*, *J. Phys. Chem. B*, **106**, 8278–8287 (2002).

61. **[PLY(O, N)]$_2$$^+$ B$^-$** (R^1 = *n*-Oct) forms a σ dimer between two molecules when the crystallization is conducted under visible light irradiation condition, see: P. Liao, M. E. Itkis, R. T. Oakley, *et al*., *J. Am. Chem. Soc*., **126**, 14297–14302 (2004).
62. In **[PLY(O, N)]$_2$$^+$ B$^-$** (R^1 = Cyclooctyl), the reduction of the corresponding cation with cobaltocene gives the π dimer in the crystal, while the reduction with decamethylnickelocene gives the σ dimer, see: S. K. Pal, M. E. Itkis, F. S. Tham, *et al*., *J. Am. Chem. Soc*., **129**, 7163–7174 (2007).
63. S. K. Pal, M. E. Itkis, F. S. Tham, *et al*., *Science*, **309**, 281–284 (2005).
64. Other bisphenalenyl boron derivatives, see: **[PLY(O, N)]$_2$$^+$ B$^-$** (R^1 = *n*-Pen), a) X. Chi, M. E. Itkis, F. S. Tham, *et al*., *Int. J. Quant. Chem*., **95**, 853–865 (2003); **[PLY(O, N)]$_2$$^+$ B$^-$** (R^1 = Benzyl), b) S. K. Pal, M. E. Itkis, R. W. Reed, *et al*., *J. Am. Chem. Soc*., **126**, 1478–1484 (2004)
65. S. K. Mandal, M. E. Itkis, X. Chi, *et al*., *J. Am. Chem. Soc*., **127**, 8185–8196 (2005).
66. S. K. Mandal, S. Samanta, M. E. Itkis, *et al*., *J. Am. Chem. Soc*., **128**, 1982–1994 (2006).
67. Other examples of single-component zwitterionic radical conductors without metal ions, see: a) Y. Tsubata, T. Suzuki, T. Miyashi, and Y. Yamashita, *J. Org. Chem*., **57**, 6749–6755 (1992); b) O. Neilands, *Mol. Cryst. Liq. Cryst*., **355**, 331–349 (2001); c) K. Balodis, S. Khasanov, C. Chong, *et al*., *Synth. Met*., **133–134**, 353–355 (2003); d) T. Suzuki, S. Miyanari, H. Kawai, *et al*., *Tetrahedron*, **60**, 1997–2003 (2004); e) T. Murata, Y. Morita, Y. Yakiyama, *et al*., *Chem. Commun*., 4009–4011 (2007).
68. Single-component neutral radical conductors, see: a) A. W. Cordes, R. C. Haddon, R. T. Oakley, *et al*., *J. Am. Chem. Soc*., **113**, 582–588 (1991); b) M. P. Andrews, A. W. Cordes, D. C. Douglass, *et al*., *J. Am. Chem. Soc*., **113**, 3559–3568 (1991); c) T. M. Barclay, A. W. Cordes, R. C. Haddon, *et al*., *J. Am. Chem. Soc*., **121**, 969–976 (1999); d) L. Beer, J. F. Britten, O. P. Clements, *et al*., *Chem. Mater*., **16**, 1564–1572 (2004); e) J. L. Brusso, K. Cvrkalj, A. A. Leitch, *et al*., *J. Am. Chem. Soc*., **128**, 15080–15081 (2006); f) A. A. Leitch, R. W. Reed, C. M. Robertson, *et al*., *J. Am. Chem. Soc*., **129**, 7903–7914 (2007).
69. a) S. Samanta, M. E. Itkis, R. W. Reed, *et al*., *Synth. Met*., **154**, 285–288 (2005); b) S. K. Pal, F. S. Tham, R. W. Reed, and R. T. Oakley, *Polyhedron*, **24**, 2076–2083 (2005).
70. S. K. Pal, M. E. Itkis, F. S. Tham, *et al*., *J. Am. Chem. Soc*., **130**, 3942–3951 (2008).
71. a) G. Allinson, R. J. Bushby, and J.-L. Paillaud, *J. Am. Chem. Soc*., **115**, 2062–2064 (1993); b) G. Allinson, R. J. Bushby, and J.-L. Paillaud, *J. Mater. Sci., Mater. Elect*., **5**, 67–74 (1994); c) G. Allinson, R. J. Bushby, J.-L. Paillaud, and M. Thornton-Pett, *J. Chem. Soc., Perkin Trans. 1*, 385–390 (1995); d) R. J. Bushby, D. R. McGill, and K. M. Ng, *NATO ASI Ser., Ser. C*, 181–204 (1996); e) G. Allinson, R. J. Bushby, M. V. Jesudason, *et al*., *J. Chem. Soc., Perkin Trans. 2*, 147–156 (1997).
72. Introduction of *tert*-butyl groups into the trioxytriangulene system enables isolation of the neutral radical species as a highly air-stable crystal. The neutral radical is termed "trioxotriangulene" by Y. Morita. The stabilization provides four-stage reversible redox ability. See: a) Ref. 50b, and b) Y. Morita, A new trend in purely organic open-shell molecules: π-extensions and multi-step redox abilities of phenalenyl systems. Presented at the 12th International Symposium on Novel Aromatic Compounds (ISNA-12), Awaji Island, Japan, July 22–27, 2007. The change in spin topological symmetry in the trioxotriangulene system depending on redox process is also observed, see: c) Y. Morita, K. Fukui, K. Nakasuji, *et al*., Trioxotriangulene: 25 pi-conjugated neutral radical with high stability and multi-redox ability in a solution and solid states. Presented at the International Symposium on Reactive Intermediates and Unusual Molecules (ISRIUM), Ascona, Switzerland, August 19–24, 2007.
73. a) I. Murata, K. Nakasuji, and H. Kume, *Tetrahedron Lett*., 3405–3408 (1973); b) K. Nakasuji, K. Yoshida, and I. Murata, *J. Am. Chem. Soc*., **104**, 1432–1433 (1982); c) K. Nakasuji, K. Yoshida, and I. Murata, *Chem. Lett*., 969–970 (1982); d) K. Nakasuji, K. Yoshida, and I. Murata, *J. Am. Chem. Soc*., **105**, 5136–5137 (1983); e) I. Murata, S. Sasaki, K.-U. Klabunde, *et al*., *Angew. Chem. Int. Ed. Engl*., **30**, 172–173 (1991); f) T. Kubo, K. Yamamoto, K. Nakasuji, *et al*., *Angew. Chem. Int. Ed. Engl*., **35**, 439–441 (1996); g) T. Kubo, K. Yamamoto, K. Nakasuji, and T. Takui, *Tetrahedron Lett*., **42**, 7997–8001 (2001); h) K. Nakasuji, and T. Kubo, *Bull. Chem. Soc. Jpn*., **77**, 1791–1801 (2004).
74. a) T. Kubo, M. Sakamoto, M. Akabane, *et al*., *Angew. Chem. Int. Ed*., **43**, 6474–6479 (2004); b) T. Kubo, A. Shimizu, M. Sakamoto, *et al*., *Angew. Chem. Int. Ed*., **44**, 6564–6568 (2005); c) T. Kubo, A. Shimizu, M. Uruichi, *et al*., *Org. Lett*., **9**, 81–84 (2007); d) T. Kubo, Y. Goto, M. Uruichi, *et al*., *Chem. Asian J*., **2**, 1370–1379 (2007).

75. K. Kamada, K. Ohta, T. Kubo, *et al.*, *Angew. Chem. Int. Ed.*, **46**, 3544–3546 (2007).
76. M. Chikamatsu, T. Mikami, J. Chisaka, *et al.*, *Appl. Phys. Lett.*, **91**, 043506 (2007).
77. a) F. Vögtle, *Top. Curr. Chem.*, **115**, 157–159 (1983); b) D. Ajami, O. Oeckler, A. Simon, and R. Herges, *Nature*, **426**, 819–821 (2003); c) A. Hirsch, and M. Brettreich, *Fullerenes: Chemistry and Reactions*, Wiley-VCH Verlag GmbH, New York, 2005.
78. The first synthesis of corannulene, see: a) W. E. Barth, and R. G. Lawton, *J. Am. Chem. Soc.*, **88**, 380–381 (1966). Recent overviews of bowl-shaped hydrocarbons including corannulene derivative, see: b) Y.-T. Wu, and J. S. Siegel, *Chem. Rev.*, **106**, 4843–4867 (2006); c) V. M. Tsefrikas, and L. T. Scott, *Chem. Rev.*, **106**, 4868–4884 (2006).
79. a) Y. Morita, S. Nishida, T. Kobayashi, *et al.*, *Org. Lett.*, **6**, 1397–1400 (2004); b) S. Nishida, Y. Morita, T. Kobayashi, *et al.*, *Polyhedron*, **24**, 2200–2204 (2005); c) Y. Morita, A. Ueda, S. Nishida, *et al.*, *Angew. Chem. Int. Ed.*, **47**, 2035–2038 (2008).
80. S. Nishida, Y. Morita, A. Ueda, *et al.*, *J. Am. Chem. Soc.*, **130**, 14954–14955 (2008). As curve-structured phenalenyl system, C_{74}, a family of fullerene, is studied theoretically, see: V. I. Kovalenko, and A. R. Khamatgalimov, *Chem. Phys. Lett.*, **377**, 263–268 (2003).
81. Chemical term of "spherical aromaticity" is also discussed on the fullerene derivative, see: a) M. Bühl, and A. Hirsch, *Chem. Rev.*, **101**, 1153–1183 (2001); b) Z. Chen, and R. B. King, *Chem. Rev.*, **105**, 3613–3642 (2005).
82. a) A. G. Harrison, L. R. Honnen, H. J. Dauben, Jr., and F. P. Lossing, *J. Am. Chem. Soc.*, **82**, 5593–5598 (1960); b) S. Ohnishi, and I. Nitta, *J. Chem. Phys.*, **39**, 2848–2849 (1963); c) P. J. Zandstra, *J. Chem. Phys.*, **40**, 612 (1964); d) R. W. Fessenden, and S. Ogawa, *J. Am. Chem. Soc.*, **86**, 3591–3592 (1964); e) G. R. Liebling, and H. M. McConnell, *J. Chem. Phys.*, **42**, 3931–3934 (1965).
83. a) R. F. Pottie, and F. P. Lossing, *J. Am. Chem. Soc.*, **85**, 269–271 (1963); b) M. Kira, M. Watanabe, and H. Sakurai, *J. Am. Chem. Soc.*, **99**, 7780–7785 (1977); c) M. Kira, M. Watanabe, and H. Sakurai, *Chem. Lett.*, 973–976 (1979); d) P. J. Barker, A. G. Davies, and J. D. Fisher, *J. Chem. Soc., Chem. Commun.*, 587–588 (1979); e) P. J. Barker, and A. G. Davies, *J. Chem. Soc., Chem. Commun.*, 815–816 (1979); f) M. Kira, M. Watanabe, and H. Sakurai, *J. Am. Chem. Soc.*, **102**, 5202–5207 (1980); g) P. J. Barker, A. G. Davies, and M-W. Tse, *J. Chem. Soc., Perkin Trans. 2*, 941–948 (1980).
84. L. Yu, J. M. Williamson, and T. A. Miller, *Chem. Phys. Lett.*, **162**, 431–436 (1989).
85. a) F. Graf, and Hs. H. Günthard, *Chem. Phys. Lett.*, **8**, 395–398 (1971); b) P. Bachmann, F. Graf, and Hs. H. Günthard, *Chem. Phys.*, **9**, 41–56 (1975); c) T, Chen, F. Graf, and Hs. H. Günthard, *Chem. Phys.*, **75**, 165–173 (1983); d) T. Chen, and Hs. H. Günthard, *Chem. Phys.*, **97**, 187–203 (1985).
86. a) A. G. Davies, and J. Lusztyk, *J. Chem. Soc., Chem. Commun.*, 554–555 (1980); b) A. G. Davies, and J. Lusztyk, *J. Chem. Soc., Perkin Trans. 2*, 692–696 (1981).
87. a) A. G. Davies, J. P. Goddard, M. B. Hursthouse, and N. P. C. Walker, *J. Chem. Soc., Dalton Trans.*, 1873–1877 (1986); b) P. N. Culshaw, J. C. Walton, L. Hughes, and K. U. Ingold, *J. Chem. Soc., Perkin Trans. 2*, 879–886 (1993).
88. a) K. Ziegler, and B. Schnell, *Justus Liebigs Ann. Chem.*, **445**, 266–282 (1925); b) W. Broser, P. Siegle, and H. Kurreck, *Chem. Ber.*, **101**, 69–83 (1968); c) K. Möbius, H. van Willigen, and A. H. Maki, *Mol. Phys.*, **20**, 289–304 (1971); d) A. A. Kuznetsov, S. N. Novikov, and A. N. Pravednikov, *Izv. Akad. Nauk SSSR Ser. Khim.*, **2**, 297–303 (1979); e) W. Kieslich, and H. Kurreck, *J. Am. Chem. Soc.*, **106**, 4328–4335 (1984); f) R. H. Lowack, and K. P. C. Vollhardt, *J. Organomet. Chem.*, **476**, 25–32 (1994); g) C. Janiak, R. Weimann, and F. Görlitz, *Organometallics*, **16**, 4933–4936 (1997); h) J.-Y. Thépot, and C. Lapinte, *J. Organomet. Chem.*, **656**, 146–155 (2002).
89. a) H. Sitzmann, and R. Boese, *Angew. Chem. Int. Ed.*, **30**, 971–973 (1991); b) H. Sitzmann, H. Bock, R. Boese, *et al.*, *J. Am. Chem. Soc.*, **115**, 12003–12009 (1993).
90. N. Jux, K. Holczer, and Y. Rubin, *Angew. Chem. Int. Ed.*, **35**, 1986–1990 (1996).
91. T. Kitagawa, K. Ogawa, and K. Komatsu, *J. Am. Chem. Soc.*, **126**, 9930–9931 (2004).
92. a) P. J. Krusic, E. Wasserman, P. N. Keizer, *et al.*, *Science*, **254**, 1183–1185 (1991); b) J. C. Hummelen, B. Knight, J. Pavlovich, *et al.*, *Science*, **269**, 1554–1556 (1995); c) M. Keshavarz-K., R. González, R. G. Hicks, *et al.*, *Nature*, **383**, 147–150 (1996); d) K. Hasharoni, C. Bellavia-Lund, M. Keshavarz-K., *et al.*, *J. Am. Chem.*

Soc., **119**, 11128–11129 (1997); e) H. Iikura, S. Mori, M. Sawamura, and E. Nakamura, *J. Org. Chem.*, **62**, 7912–7913 (1997); f) O. Vostrowsky, and A. Hirsch, *Chem. Rev.*, **106**, 5191–5207 (2006); g) B. Tumanskii, and O. Kalina (Eds), *Radical Reactions of Fullerenes and their Derivatives*, Kluwer Academic Publishers, Dordrecht, 2001.

93. Synthesis and spin nature of the neutral diradical with *curved* π-surface, see: A. Ueda, Y. Morita, S. Nishida, *et al.*, Intramolecular spin-spin interaction through a curved and non-alternant hydrocarbon: corannulene with phenoxyl radical moieties. Presented at the 12th International Symposium on Novel Aromatic Compounds (ISNA-12), Awaji Island, Japan, July 22–27, 2007.
94. a) K, Hirai, and H. Tomioka, *J. Am. Chem. Soc.*, **121**, 10213–10214 (1999); b) T. Itoh, Y. Nakata, K. Hirai, and H. Tomioka, *J. Am. Chem. Soc.*, **128**, 957–967 (2006).
95. a) N. Koga, and H. Iwamura, in *Carbene Chemistry* (ed. G. Bertrand), Fontis Media, Lausanne, 2002, pp 271–296; b) K. Matsuda, N. Nakamura, K. Takahashi, *et al.*, *Molecule-Based Magnetic Materials*, ACS Symposium Series 644, American Chemical Society, Washington, 1996, p 142.
96. Tomioka's works on triplet diphenylcarbene derivatives possessing lifetime over minutes, see: a) H. Tomioka, M. Hattori, K. Hirai, and S. Murata, *J. Am. Chem. Soc.*, **118**, 8723–8724 (1996); b) H. Tomioka, *Acc. Chem. Res.*, **30**, 315–321 (1997); c) H. Tomioka, in *Advances in Carbene Chemistry* (ed. U. H. Brinker), JAI Press, Greenwich, CT, 1998, Vol. 2, pp 175–214. d) H. Tomioka, J. Nakajima, H. Mizuno, *et al.*, *J. Am. Chem. Soc.*, **117**, 11355–11356 (1995); e) H. Tomioka, M. Hattori, K. Hirai, *et al.*, *J. Am. Chem. Soc.*, **120**, 1106–1107 (1998); f) H. Tomioka, H. Mizuno, H. Itakura, and K. Hirai, *Chem. Commun.*, 2261–2262 (1997).
97. In general triplet carbenes are reactive enough to attack even C–H bonds, see: H. Tomioka, and K. Taketsuji, *Chem. Commun.*, 1745–1746 (1997).
98. The electronic and molecular structures are estimated by the *D* and *E* values obtained by EPR spectroscopy, see: E. Wasserman, V. J. Kuck, W. A. Yager, *et al.*, *J. Am. Chem. Soc.*, **93**, 6335–6337 (1971).
99. Kinetic studies and product analyses of the reaction of **BAC**, see: D. J. Astles, M. Girard, D. Griller, *et al.*, *J. Org. Chem.*, **53**, 6053–6057 (1988).
100. H. Tomioka, E. Iwamoto, H. Itakura, and K. Hirai, *Nature*, **412**, 626–628 (2001).
101. Chemical structure of **(BAC)**$_3$ is unequivocally determined in terms of ^{1}H, ^{13}C, and 2D NMR spectroscopies, FAB-MS, and X-ray single crystallographic analysis, see: Y. Takahashi, M. Tomura, K. Yoshida, *et al.*, *Angew. Chem. Int. Ed.*, **39**, 3478–3480 (2000).
102. a) Y. Tsuchiya, M. Matsuno, T. Itoh, *et al.*, *Bull. Chem. Soc. Jpn.*, **78**, 2037–2050 (2005); b) K. Hirai, E. Kamiya, T. Itoh, and H. Tomioka, *Org. Lett.*, **8**, 1847–1850 (2006); c) T. Itoh, K. Hirai, and H. Tomioka, *Bull. Chem. Soc. Jpn.*, **80**, 138–157 (2007).
103. Other examples of polycarbenes: a) N. Nakamura, K. Inoue, H. Iwamura, *et al.*, *J. Am. Chem. Soc.*, **114**, 1484–1485 (1992); b) N. Nakamura, K. Inoue, and H. Iwamura, *Angew. Chem., Int. Ed. Engl.*, **32**, 872–874 (1993); c) K. Matsuda, N. Nakamura, K. Takahashi, *et al.*, *J. Am Chem. Soc.*, **117**, 5550–5560 (1995); d) K. Matsuda, N. Nakamura, K. Inoue, *et al.*, *Chem. Eur. J.*, **2**, 259–264 (1996); e) K. Matsuda, N. Nakamura, K. Inoue, *et al.*, *Bull. Chem. Soc. Jpn.*, **69**, 1483–1494 (1996).
104. Half-life and/or lifetime of the carbene are not mentioned in the literature, see: Ref. 102b.
105. Generation of dendric hexakis- and dodekakis-(diphenylcarbene) derivatives are also reported. They show low spin-states due to the connectivities with both non-disjoint ($S = 2$, $J > 0$) and disjoint ($J \sim 0$) manners, see: T. Itoh, T. Maemura, Y. Ohtsuka, *et al.*, *Eur. J. Org. Chem.*, 2991–3003 (2004).
106. M. Kawano, K. Hirai, H. Tomioka, and Y. Ohashi, *J. Am. Chem. Soc.*, **123**, 6904–6908 (2001).
107. X-ray single crystal data of various diphenylcarbene derivatives trapped in the crystals are recently reported, see: M. Kawano, K. Hirai, H. Tomioka, and Y. Ohashi, *J. Am. Chem. Soc.*, **129**, 2383–2391 (2007).
108. Molecular spin quantum computing studied by pulsed ENDOR technique, see: a) R. Rahimi, K. Sato, K. Furukawa, *et al.*, *Int. J. Quantum Info.*, **3** (Supp), 197–204 (2005); b) K. Sato, R. Rahimi, N. Mori, *et al.*, *Physica E*, **40**, 363–366 (2007); c) R. Rahimi, K. Sato, D. Shiomi, and T. Takui, in *Modern Magnetic Resonance* (ed. G. A. Webb), pp 643–650, Springer-Verlag, 2007; d) K. Sato, S. Nakazawa, R. Rahimi, *et al.*, *J. Mater. Chem.*, 19, 3739–3754, 2009; e) K. Sato, S. Nakazawa, R. D. Rahimi, *et al.*, in *Molecular Realizations of Quantum Computing 2007* (eds M. Nakahara, Y. Ota, and R. Rahimi), Kinki University Series on Quantum Computing, World Scientific, 2009, pp 58–162.

4

The Nitrogen Oxides: Persistent Radicals and van der Waals Complex Dimers

D. Scott Bohle

Department of Chemistry, McGill University, Montreal, Canada

4.1 Introduction

Although many persistent radicals contain nitrogen–oxygen moieties, the radicals which result from the direct combination of nitrogen and oxygen form a unique class of simple gaseous radicals. In particular, nitric oxide (NO) and nitrogen dioxide (NO_2) are monomeric radicals under standard conditions with a long-standing history and remain of considerable biological, industrial, and chemical importance. The chemistry of these ostensibly simple species has been recently reinvigorated by two orthogonal efforts to understand nitric oxide's important role in biology, and in efforts to control nitrogen oxide emission (described as NO_x) from coal burning power plants and from internal combustion engines. An important consequence of this renaissance in nitrogen oxide chemistry is a reappraisal of the importance of the transient diamagnetic dimers and cross adducts which form. The title and subject of this chapter stems from this growing appreciation.

The origins of the chemistry of nitrogen oxides are associated with many of the founders of this field and includes Davy, Lavoisier, Dalton, von Bayer, Eyring, Sidgwick, and Taube among many others. Given this history there are now many authoritative reviews of the diverse aspects of this chemistry (Table 4.1). The chemistry of the nitrogen oxides figured prominently in the last comprehensive single monograph in this area, Sidgwick's "*The Organic Chemistry of Nitrogen*".[1] With the discovery of nitric oxide's biology there is now an extensive literature related to its biological chemistry, but there now exists an unfortunate gap between the initial and current literature. While some of this disjunction relates to the remarkably sensitive analytical methods for nitric oxide analysis and notions of working "with biologically relevant conditions," it is perhaps the perceived age of this prior literature which limits contemporary consideration of prior results. The stage is set for frequent efforts that fall into the category of "rediscovery of the wheel."

An important factor in the chemistry of nitric oxide and nitrogen dioxide is the strength and prevalence of van der Waals' complex formation. Their self and cross dimers are examples of particularly strong

Table 4.1 *Seminal literature on NO chemistry and biochemistry*

Author	Subject	Year	Ref.
Green	Electronic Structure	1966	
Yost	Physical Chemistry	1946	[210]
Ragsdale	NO Reaction Chemistry	1973	[145]
Legzdins	Nitrosyl Complexes	1992	[158]
Eisenberg	Nitrosyl Complexes	1975	[211]
Jolly	Inorganic Chemistry of Nitrogen	1964	[54]
McCleverty	Nitrosyl Complexes	1979	[212]
Keefer	Diazeniumdiolates	2003	[213]
Huie	Chemistry of NO_2	1994	[214]
Tsukahara	NO Oxidation Kinetics	1999	[123]
Heicklen	NO Photochemistry	1968	[111]
Western	NO Dimer in Molecular Beams	1981	[36]
Addison	NO_2 as a Solvent	1980	[215]
Various	NO Chemistry	2002	[138]
Rees and Williams	NO_x Organic Radical Reactions	1968	[216]
Sloss	NO_x Emission Control	1992	[217]
Bonner and Hughes	Aqueous Chemistry of N(I) and N(II)	1988	[156]
Menzel	Biological Chemistry of NO_2	1999	[218]
Butler	Non-heme iron nitrosyls in biology	2002	[193]

van der Waals complexes (Equations 4.1–4.3), and the role of these complexes in their chemistry is an important theme of this chapter. Many other small molecules complexes with nitric oxide have been characterized in the gas phase and these include carbon dioxide,[2] argon,[3] hydrogen,[4] hydrogen fluoride,[5] neon,[6] ethylene,[7] and ethane.[8] Much of the evidence from these studies comes from gas phase spectroscopy,[4] matrix isolation, or supersonic gas beam[9] experiments and methods, but the results and conclusions from these studies are likely to bear on all of nitric oxide's chemistry. These types of complexes which result from "sticky collisions" produce new intermediates and adducts which need to be considered if the kinetics and mechanisms of nitric oxide's chemistry are to be understood.[10]

(4.1)

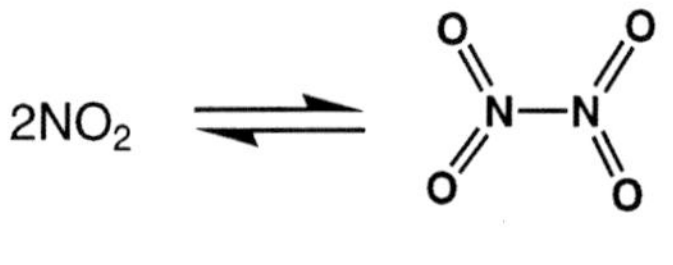

(4.2)

(4.3)

$$3NO \longrightarrow NO_2 + N_2O \quad (4.4)$$

4.2 Synthetic access

The numerous methods for the synthesis of nitric oxide and nitrogen dioxide will not be exhaustively surveyed here, but both are important commercially available industrial gases. Because of their reactivity, however, in particular the nitric oxide disproportionation reaction (Equation 4.4),[11] cylinders of nitric oxide are invariably contaminated with nitrous oxide and nitrogen dioxide. Considerable thought in the use of these cylinders is required, especially for any exacting analytical, physical, or biological applications. At a minimum, passing the gas through a strong base is required to remove the bulk of the nitrogen dioxide, but it is difficult to remove the final traces of nitrogen dioxide. Extra fractionation steps are required to remove the nitrous oxide. Failure to heed these precautions has led to some monumental misunderstandings in nitric oxide's biology.[12] Indeed, several chemistry texts mistakenly describe solid nitric oxide as blue in color when in fact it is colorless[13]; the blue color is due to contamination by dinitrogen trioxide.[14] As a consequence, the prudent approach is to prepare these gases directly as needed. The direct stoichiometric reduction of nitrite by acidic aqueous solution of ferrous ions (Equation 4.5),[15] or the acid-promoted disproportionation of nitrite,[16] are useful for atmospheric pressure experiments where water is not an issue,[17] and if these conditions are problematic then the dry chromium trioxide reduction of nitrite/nitrate (Equation 4.6) is a useful alternative.[18] For nitrogen dioxide the thermal decomposition of plumbous nitrate (Equation 4.7)[19] or dinitrogen pentoxide[20] are facile.

$$Fe^{+2} + NO_2^- + 2H^+ \longrightarrow Fe^{+3} + NO + H_2O \tag{4.5}$$

$$3KNO_2 + KNO_3 + Cr_2O_3 \xrightarrow{\sim 600\ K} 2K_2CrO_4 + 4NO \tag{4.6}$$

$$2Pb(NO_3)_2 \xrightarrow{\sim 700\ K} 2PbO + 4NO_2 + O_2 \tag{4.7}$$

For mechanistic studies, isotopic labelling with ^{18}O or ^{15}N are powerful tools,[21,22] which are readily adapted to a host of applications using any of these methods. As an example of the power in the use of isotopes, note that it is possible to specifically label either of the nitrogen atoms with ^{15}N in nitrous oxide.[23]

4.3 Physical properties

Being open shell stable room temperature gases, both nitric oxide and nitrogen dioxide have received sustained considerable attention and are physically well characterized species. Table 4.2 shows some of this data for the two monomeric gaseous species. In the course of the pioneering physical chemistry in this area it was suggested that nitric oxide, as was known for nitrogen dioxide, dimerizes at low temperatures and in the solid state (Equation 4.1). For example, although paramagnetic as a gas, nitric oxide is diamagnetic as a solid or liquid.[24,25] Also, the observed and calculated values for S° differ by 3.1 J/deg, which is within experimental error of $\frac{1}{2}$ R ln2, and this is explained in terms of a disorder of $[NO]_2$ dimers in the crystal.[26] In contrast, the nitrogen dioxide dimer, dinitrogen tetroxide (Equation 4.2), is well characterized. The van der Waals adduct of nitric oxide and nitrogen dioxide, dinitrogen trioxide (N_2O_3, Equation 4.3), is also a well characterized adduct.

Self and cross dimerization reactions of nitric oxide and nitrogen dioxide are both of considerable importance. The resulting species, dinitrogen dioxide (N_2O_2) from nitric oxide, dinitrogen tetroxide (N_2O_4) from nitrogen dioxide, and their cross dimerization product, dinitrogen trioxide are all relatively weakly bound

Table 4.2 *Physical properties of NO and NO_2*

Property[a]	Nitric Oxide	Nitrogen Dioxide
Melting point (K)	109.64	261.90[b]
Boiling point (K)	121.41	294.25[b]
Heat of fusion ΔH (kJ mol^{-1})	2.300	14.652[b]
Heat of vaporization ΔH (kJ mol^{-1})	13.776	38.116[b]
Heat of formation ΔH°_{273} (kJ mol^{-1})	90.0	33.32
Free energy of formation ΔG°_{273} (kJ mol^{-1})	86.40	17.506
Standard entropy S°_{273} (J deg^{-1})	207.5	240.5
Magnetic moment μ°_{273} (B.M.)	1.82	1.73
Ionization potential (eV)	9.25	11
Electron affinity (eV)	0.9	4.0
Dipole moment (D)	0.16	0.39
Color	Colorless	Red

[a]Data: [210, 219, 220].
[b]Values for nitrogen dioxide correspond to an equilibrium mixture of monomer and dimer at these temperatures.

Table 4.3 *The van der waals complexes of NO and NO_2*

Complex	N–N (Å)	D_e (kJ mol^{-1})	UV-Vis (nm)	Ref.
N_2O_2	2.18	10.25	$\lambda_{max} \sim$ 232–236 nm	[221]
N_2O_3	1.85	39.75	Blue, 684, 225 nm	[222, 223]
N_2O_4	1.75	57.32	Colorless, 268, 240 nm	[224]

diamagnetic complexes with weak to very weak central N–N bonds (Table 4.3). An interesting characteristic of both dinitrogen dioxide and dinitrogen trioxide is that their direduction gives dianions with shorter formally N=N double bonds, corresponding to hyponitrite and Angeli's salts, respectively.[27,28] A similar reduction of dinitrogen tetroxide results in N–N cleavage and the formation of two equivalents of nitrite.

4.4 Structural chemistry of the monomers and dimers

Microwave spectroscopy, vibrational spectroscopy, and diffraction (neutron, electron, and X-ray) have all been used to characterize the gaseous and solid state structures of nitric oxide and nitrogen dioxide, and their dimers (Figure 4.1 and Table 4.4). Although many of these studies used cutting edge technology for their time, few of these structural studies have been reinvestigated with more modern methods. As a result, the errors in some of the determinations are larger than are typically found in more modern determinations.

4.4.1 Nitric oxide and dinitrogen dioxide

Diffraction of the nitric oxide dimer[29–32] at 114 K in the solid state was originally interpreted as being due to a head-to-tail rectangular dimer[30] and this was then re-refined[29] with least square techniques to the

Table 4.4 Structural data for NO, NO_2 and their dimers

Compound/Method	Observations	Ref.
Nitric oxide		
Infrared spectroscopy	Gas phase, *d*(N–O) 1.15 Å	[229]
Infrared spectroscopy	Gas phase, *d*(N–O) 1.1508 Å	[230]
Near Infrared spectroscopy	Gas phase, *d*(N–O) 1.1503 Å	[231]
Microwave spectroscopy	Gas phase, *d*(N–O) 1.1540 Å	[232]
	Gas phase, *d*(N–O) 1.1509 Å	[233]
Dinitrogen Dioxide		
Infrared spectroscopy	Band shape analysis gas phase, 77–150 K, *d*(N–N) 1.75 Å, ∠NNO 90°	[234]
X-ray diffraction	Solid state, 114 K, *d*(N–N, O–O) 2.40(1) Å, *d*(N–O) 1.12(2) Å, ∠NNO 90°	[30]
	Solid state,[b] 114 K, *d*(N–N) 2.18(3) Å, d(O–O) 2.62(3) Å	[29]
Microwave specotroscopy	Gas phase, *d*(N–N) 2.237(2) Å, *d*(N–O) 1.161(6) Å, *d*(O–O) 2.506(3) Å, ∠NNO 99.6(4)°	[35]
Microwave/Radiofrequency Spectroscopy	Gas phase, *d*(N–N) 2.33(12) Å, *d*(N–O) 1.15(1) Å, *d*(O–O) 2.506(3) Å, ∠NNO 95(5)°	[36]
Nitrogen dioxide		
Electron diffraction	Gas phase, 348–363 K, *d*(N–O) 1.21(2) Å, ∠ONO 130(2)°	[235]
Electron diffraction	Gas phase, 418 K, *d*(N–O) 1.20(2) Å, ∠ONO 132(3)°	[236]
	Gas phase, 294, 480, and 691 K. For 480 and 691 K dimensions are: *d*(N–O) 1.1995(3) and 1.2006(3) Å, and ∠ONO 133.7(1) and 133.7(2)°.[c]	[50]
Infrared spectroscopy	Gas phase, 294 K, *d*(N–O) 1.188(2) Å, ∠ONO 134.1(3)°	[237]
Dinitrogen tetroxide		
X-ray diffraction	Cubic lattice, 238 K, *d*(N–O) 1.17(3) Å, *d*(N–N) 1.64(3) Å, ∠ONO 126(1)°	[44]
	Monoclinic lattice,[a] *d*(N–O) 1.21 Å, *d*(N–N) 1.75 Å, ∠ONO 135°	[46, 238]
	Cubic lattice, 25, 145, and 260 K. At 145 K: *d*(N–O) 1.209(7) Å, *d*(N–N) 1.726(9) Å, ∠ONO 133.1(7)°	[45]
Neutron diffraction	Cubic lattice, 20 K, *d*(N–O) 1.1893(5) Å, *d*(N–N) 1.7561(9) Å, ∠ONO 134.40(7)°	[47]
Electron diffraction	Gas phase, 298 ± 5 K, *d*(N–O) 1.180 Å, *d*(N–N) 1.750 Å, ∠ONO 133.7°	[48]
	Gas phase, 252 ± 1 K, *d*(N–O) 1.191(2) Å, *d*(N–N) 1.785(8) Å, ∠ONO 135.4(6)°	[49]
	Gas phase, 294 ± 1 K, *d*(N–O) 1.192(3) Å, *d*(N–N) 1.777(6) Å, ∠ONO 134.6(4)°	[50]
Infrared Spectroscopy	Gas phase, 50 ± 5 K, *d*(N–N) 1.756(10) Å, *d*(N–O) 1.196(5), ∠ONO Set to values for NO_2	[239]

[a] Temperature not specified.
[b] A rerefinement of the original data.[30]
[c] ESDs from 1000 refinements on observed data and do not include systematic total errors.

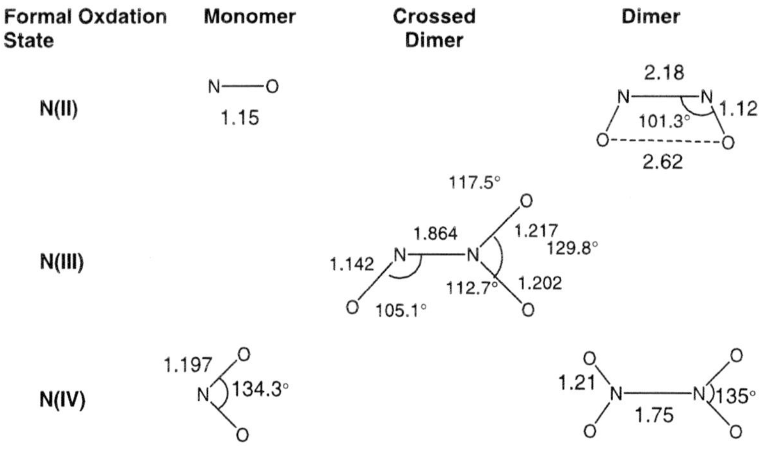

ᵃ All metric dimensions given in Angstroms and in degree

Figure 4.1 *Key structural data for NO, NO_2 and their dimers*[a].

trapezoidal dimer shown in Figure 4.1. The bond lengths estimated for the dimer are 2.18(3) Å for the N–N bond, 2.62(3) Å for the O–O interaction, and 1.12(2) Å for N–O bonds. Unfortunately, this solid state structure does not solve the issue of the dimer's structure as it should. More recent determinations from microwave[33–36] and infrared spectroscopy[33,37–39] are consistent with a C_{2v} geometry for the dimer but the poor agreement of the gas condensed phase vibrational data suggest that the solid state environment perturbs the gas phase structure.[40] Moreover, the relatively flat potential energy surface for the dimer suggests that no single structure may describe the dimer in different phases.[41]

Lipscomb also applied some of the same X-ray diffraction techniques to the elucidation of the structure of dinitrogen trioxide at 163 K, which gave an improbable tetragonal unit cell with such large dimensions of $a = b = 16.4$ Å and $c = 8.86$ Å, that volume considerations alone suggest Z = 32. A space group was not unambiguously determined and electron density projections along the c axis were badly disordered.[42] A microwave spectrum of dinitrogen trioxide in the gas phase has been determined at 200 K, and the structure, shown in Figure 4.1, indicates a long 1.85(3) Å N–N bond length.[43]

4.4.2 Nitrogen dioxide and dinitrogen tetroxide

The structural characterization of the dinitrogen tetroxide dimer is much more well established by diffraction methods. Two phases have been characterized by X-ray diffraction[44–46] and the more stable cubic structure has also been characterized by neutron diffraction.[47] In the gas phase, mixtures of nitrogen dioxide monomer and its dimer have also been characterized by electron diffraction[48–50] Surprisingly, dimer formation does not appreciably alter the N–O bond lengths or the O–N–O bond angle in the monomer.

The sustained interest in all of these van der Waals complexes is their weak long N–N bonds. In spite of the weakness of this bond, all diffraction methods at all temperatures result in models of dinitrogen tetroxide which are consistently planar structures with a high barriers for rotation, 40 kJ mol^{-1}, around the N–N bond.[50] On the much shorter time scales of vibrational spectroscopy, the point group symmetry of dinitrogen tetroxide is clearly D_{2h}. The temperature dependence of the N–N bond length has also been investigated, and the collected results for all spectroscopies are shown in Figure 4.2. Cartwright was the

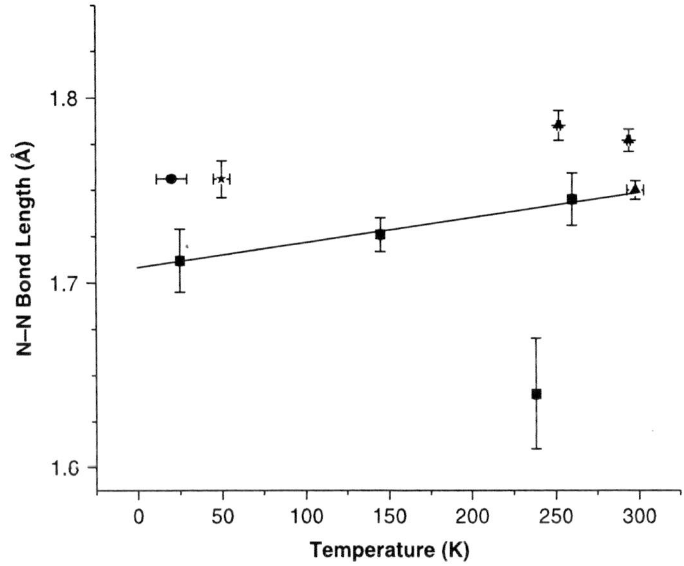

Figure 4.2 N–N bond lengths (Å) in N_2O_4 as a function of temperature for (■) X-ray diffraction, (•) neutron diffraction, (▲) electron diffraction and (∗) for IR data. The best fit line for the X-ray diffraction data described by Cartwright[45] is also shown.

first to investigate this and employed variable temperature X-ray diffraction to determine these dimensions in the solid state. Although other weak intermolecular interactions such element hydrogen[51] and halogen[52] bond lengths are known to vary with temperature, for covalent species or van der Waals complexes such as dinitrogen tetraoxide this temperature dependence has not been well studied. If Broadley's 1949 X-ray diffraction result with a very short 1.64(3) Å N–N bond length is removed, there is a small apparent increase in N–N bond length with temperature in Figure 4.2. Clearly more data at intermediary temperatures are required before this correlation is well established, but the overall affect has the magnitude found in other systems. A conclusion is that close lying non-bonding or antibonding excited states do not appear to be thermally populated, and this result finds support from recent femtosecond spectroscopy of the dimer.[53]

(4.8)

4.5 Electronic structure of nitrogen oxides

In the molecular orbital description for nitric oxide the valence ground state configuration is $\sigma^2\sigma_{\mathrm{lp}}^2\sigma_{\mathrm{lp}}^2$ $\pi^4\pi^{*1}$ $X\ ^2\Pi$, which suggests a bond order of 2.5 with the unpaired electron in a π^* orbital with substantial localization on the nitrogen. Although this is a good start to understanding nitric oxide's electronic structure, and it is certainly better than valence bond descriptions[54] (Equation 4.8) (as a double bond and a three electron bond between the atoms),[55] simple MO theory does not explain the magnetic susceptibility or the EPR data in a satisfactory way. A plot of the variable temperature magnetic susceptibility of gaseous nitric oxide is shown in Figure 4.3, along with a theoretical fit from a model developed by

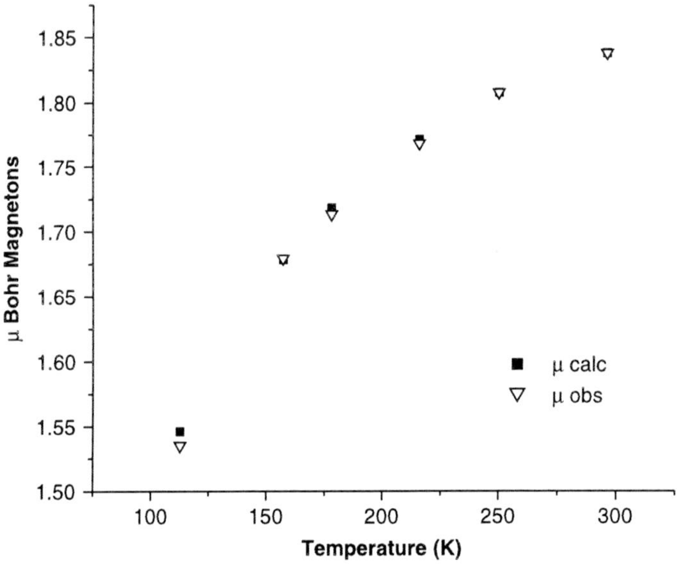

Figure 4.3 *Observed (open triangles) and calculated (filled squares) magnetic moments from the Van Vleck distribution law model.*

Van Vleck, Equation 4.9.[56–58]

$$\mu_{eff} = 2\sqrt{\frac{1-e^{-x}+xe^{-x}}{x+xe^{-x}}} \qquad x = \frac{h\Delta\nu}{kT} = \frac{173}{T} \tag{4.9}$$

Occupation of the π^* orbital results in a net combination of angular momentum of the electron and the orbital where the spin either opposes the orbital momentum, with a result of $1 - \frac{1}{2}$ or $\frac{1}{2}$ or combines with it to give $1 + \frac{1}{2}$ or $\frac{3}{2}$. These two doublet states are designated $^2\Pi_{1/2}$ and $^2\Pi_{3/2}$ respectively and differ in energy by 1.47 kJ mol^{-1} or 124.2 cm^{-1}. At room temperature slightly more than a third of the molecules are in the lower energy $^2\Pi_{1/2}$ state. The resultant magnetic susceptibility is then the combination of these two states and the expectation values are $\mu = 0$ B.M. for $^2\Pi_{1/2}$ and $\mu = 2$ B.M for $^2\Pi_{3/2}$. In addition to affecting the magnetic susceptibility and magnetic moment for nitric oxide, the combined orbital momenta broaden and shift the EPR resonance of the unpaired electron from g_e, so that with conventional cavity instruments an EPR resonance is not observable for nitric oxide and cooling to low temperatures gives the diamagnetic dimer. Adduct formation with either surfaces[59] or metal ions[60] breaks the π^* degeneracy and leads to observable EPR resonances.[61] Turner's early molecular photoelectron spectra[62] for nitric oxide have since been followed by many studies,[63–73] including those of the dimer under a variety of conditions.[40,74–76]

Low lying excited states for nitric oxide include a Rydberg $\sigma^2\sigma_{lp}^2\sigma_{lp}^2\pi^4\pi^{*0}\sigma_{ryd}^1 A\ ^2\Sigma$ state from the excitation of an electron to an s-type Rydberg orbital, and two, $^4\Pi$ and B $^2\Pi$ states, which arise from a $\pi \rightarrow \pi^*$ excitation and correspond to $\pi^3\pi^{*2}$ configurations. Collectively these three excited states lie between 4.7 and 5.6 eV in higher energy,[77,78] and need to be considered for understanding the electronic structure of the nitric oxide, its dimer, and its van der Waals complexes. In all, 16 valence electronic states need to be considered for the dimer.[41]

On the other hand, nitrogen dioxide's bent structure is readily accounted for by a Walsh bending model from a linear $D_{\infty h}$ point group to C_{2v}.[79] The resulting unpaired electron is largely localized on the nitrogen but there are many low lying excited states and configuration interaction with these is required for a complete description of the electronic structure. Nevertheless, $S = 1/2$ remains a good quantum number and the magnetic susceptibility for gaseous nitrogen dioxide at 293 K is 1.72 B.M. and its temperature dependence is well accounted for by its equilibrium with the diamagnetic monomer.[80,81] The X-band EPR spectrum of gaseous nitrogen dioxide at atmospheric pressure is a broad unresolved band which becomes a triplet with $A_N = 50.9$ gauss at pressures between 14 and 1 mm Hg P_{NO2}.[82] At lower pressures, 0.02 mm Hg, the spectrum is described as having more than 200 lines.[83] At 4.2 K, and doped into solid argon, nitrogen dioxide gives an axial EPR spectrum with $g_{\|} = 1.9920(5)$ and $g_{\perp} = 2.0033(5)$ with an observed coupling of 53.4 gauss.[84] At these temperatures the photolytic generation of nitrogen dioxide in a sodium nitrite crystal gives a rhombic signal with $g_1 = 2.0057, g_2 = 2.0015$, and $g_3 = 1.9910$.[85] More recent studies have demonstrated that nitrogen dioxide forms adducts/reacts with many solvents.[86,87] There is a rich literature for the photoelectron spectroscopy of nitrogen dioxide[88–90] and dinitrogen tetroxide.[91–97] A remarkable high resolution He-I study ($h\nu = 21.22$ eV) of gas phase nitrogen dioxide has exceptional vibronic coupling in all of the observed bands.[98]

Early theoretical attempts to understand the electronic structure of the nitric oxide dimer were unsuccessful in accounting for the geometry, the vibrational data or its low D_e.[99–102] The subsequent efforts to determine the electronic structure of the dimer have been well described in East and Bondybey's analyses,[40,41] but surprisingly many problems remain to be resolved for all of the dimers, neutral and charged ($[N_2O_2]^+$, $[N_2O_2]^-$) alike. Even one of the most exhaustive post-Hartree-Fock studies by Roos *et al*. considered only a *cis* $(NO)_2$ geometry.[103] As noted by Schaefer *et al*. theory[104] has had a "muddled relationship to experiment." in this domain. Certainly part of the problem is due to the differing geometric stabilities of the neutral versus the anionic and cationic dimers. Thus, although the neutral *cis* dimer is circa 45 kJ mol^{-1} more stable than the *trans* in an inert gas matrix, the *trans* is more stable by circa 6.0 kJ mol^{-1} for the cation and the *cis* and *trans* are almost equi-energetic for the anion. As a consequence, interpreting the results of ionizing experiments such as photoelectron spectroscopy are complicated by geometry considerations for the product states.

At first glance the nitrogen oxides are small molecules and should be ideally suited for modern *ab initio* calculations. But for a wide array of theoretical methods this class of compounds remains a challenge to accurately model and predict the structure, spectroscopy, and reactivity of their dimers. New codes and algorithms are often checked against the nitrogen oxides to assess improvements or advances in methodology. Clearly the combination of weak bonds, numerous lone pairs, and low lying spin states make this area as challenging for theoreticians as it is for experimentalists.

4.6 Reactivity of nitric oxide and nitrogen dioxide and their van der Waals complexes

Central to any discussion of the reactivity of nitric oxide and nitrogen dioxide is consideration of the dimerization equilibria in Equations 4.1 and 4.2. Both reactions are readily reversible and rapidly established; their kinetic and thermodynamic data are contrasted in Table 4.5. The rate of nitric oxide dimerization and its reverse must be rapid reactions, since there is no spectroscopic evidence for the presence of dimers in reactions such as nitric oxide's oxygenation. Earlier reports suggested[54] that, based on determinations of the second virial coefficients and Berthelot's equations, there was no evidence for associative dimerization

Table 4.5 Kinetic and thermodynamic data for NO and NO_2 dimerizations

Property	NO	NO_2	Ref.
Equilibrium constant, K			
at 273 K	6.6×10^{-5} atm	8.8 atm	[54, 123]
at 87 K[225]	0.015 M		
Forward rate constant, k_f	$\sim 10^9$ L mol^{-1} s^{-1}	5×10^8 L mol^{-1} s^{-1}	[226]
Reverse rate constant, k_r	$\sim 10^8$ s^{-1}	1.0×10^6 s^{-1}	[227]
		1.7×10^5 s^{-1}	[228]

of nitric oxide between 273 K and the boiling point.[105] However Guggenheim's lucid reanalysis of this earlier data clearly demonstrates that nitric oxide does indeed dimerize if McGlashan and Potter's principle of corresponding states[106] is taken into account.[107,108] Many gas phase kinetic studies for nitric oxide clearly indicate the presence of rate laws which are bimolecular in [NO], and thus frequently termolecular overall.[109] Eyring ascribed this to "sticky collisions" between nitric oxide molecules,[110] and it has been estimated that the nitric oxide dimer lifetime is between 10^5 and 10^6 times that of the duration of a collision.[107] Clearly these dimers have sufficient lifetimes to be kinetically and mechanistically significant.

$$2NO + O_2 \longrightarrow 2NO_2 \quad (4.10)$$

$$2NO + Cl_2 \longrightarrow 2ClNO \quad (4.11)$$

$$2NO + Br_2 \longrightarrow 2BrNO \quad (4.12)$$

$$2NO + H_2 \longrightarrow N_2O + H_2O \quad (4.13)$$

$$\text{rate} = -\frac{d[NO]}{dt} = k[NO]^2[X_2] \quad (4.14)$$

$$X_2 = O_2, Cl_2, Br_2, H_2$$

4.7 The kinetics of nitric oxide's termolecular reactions

Many of the best examples of termolecular rate laws and reactions come from nitric oxide's reactions with other gases.[109] Oxygen, chlorine, bromine, and (in part) hydrogen (Equations 4.10–4.13) all follow the rate law shown in Equation 4.14. There is sustained effort to understand the origin of this rate law with the mechanism of nitric oxide's reaction with oxygen being the most extensively investigated. In addition to being unusual third order gas phase reactions, they also have a negative temperature coefficient, so the reaction slows at higher temperatures.[111] The problem is that since three body collisions have insufficient frequency to bring the two molecules of nitric oxide and one oxygen molecule together to the transition state, a multistep mechanism needs to be proposed with both a transient ONOO van der Waal's adduct as in Equation 4.15 and the nitric oxide dimer in Equation 4.17 having been invoked. In distinguishing

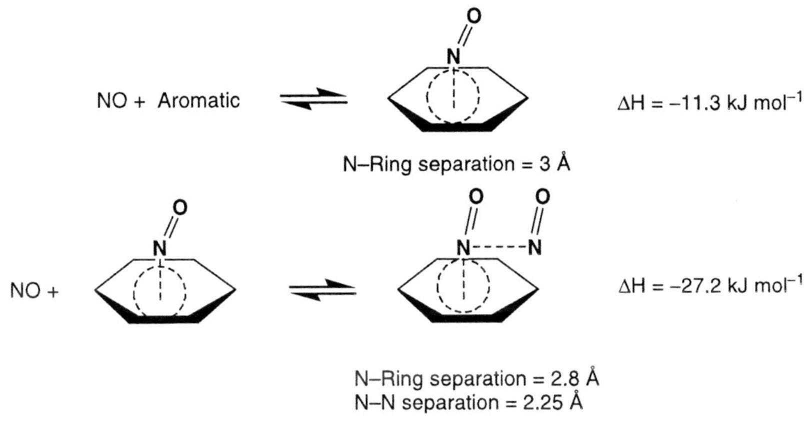

Figure 4.4 Houk's theoretically proposed aromatic van der Waals stabilization.[115]

these mechanisms it is noted that, although there is considerable evidence for the dimer in the gas and condensed phases, direct evidence for the ONOO radical remains elusive.[112–114] Heicklen found evidence for the operation of both mechanisms,[111] with perhaps the best argument for the common operation of the dimer mechanism being Occam's razor, as only one common mechanism is consistent with all of the data. It has been theoretically suggested that aromatic rings can stabilize the nitric oxide dimer as in Figure 4.4 and that this would then promote the reaction of the dimer at room temperature and under biological conditions.[115] While the modest stabilization offered by these rings might very well promote dimerization, there is no necessary connection to promoting the reactivity of the dimer with other reagents, if it is already in a van der Waals complex with a ring. Recent kinetic evidence shows that nitric oxide's reaction with secondary amines is not accelerated by aromatic solvents and the complex and proposal in Figure 4.4 remains largely hypothetical.[116]

$$\text{Monomer} \quad NO + O_2 \rightleftharpoons [OONO] \tag{4.15}$$

$$NO + [OONO] \xrightarrow[\text{r.d.s.}]{k_1} 2NO_2 \tag{4.16}$$

$$\text{Dimer} \quad 2\,NO \rightleftharpoons N_2O_2 \tag{4.17}$$

$$N_2O_2 + O_2 \xrightarrow[\text{r.d.s.}]{k_2} 2NO_2 \tag{4.18}$$

Nitric oxide's reactions with a number of other substrates also have rate laws, which are bimolecular in nitric oxide. Ford found that the oxidation of triphenylphosphine to triphenylphosphine oxide is bimolecular in nitric oxide and unimolecular in triphenylphosphine.[117] Based on solvent effects he favored a rate-determining step between the dimer and triphenylphosphine. A similar reaction with triethylphosphite[118] also followed these kinetics, as did the oxidation of a $Cp_2Zr(PMe_3)_2$ with nitric oxide.[119] The reaction of nitric oxide with sulfite and bisulfite in water has multiterm rate laws which were rationalized by terms for hydrated nitric oxide.[120,121] Theoretically and experimentally there is little support for proposed hydrated nitric oxide steps. *Ab initio* methods predict less than 2 kJ mol^{-1} stabilization for any direct van der Waals complexes,[115] and there is no spectroscopic or reaction evidence for any such species. For example, oxygen scrambling in the NO/$^{18}OH_2$ system is not detectable under the conditions for "hydrated nitric oxide" in these mechanisms. A more likely explanation for these kinetics is the operation of a competitive dimer/monomer rate law, as has been observed for the addition of secondary amines to nitric oxide.[116]

The question of mechanism is of considerable importance in all of nitric oxide's biological, environmental, and atmospheric chemistry, where low concentrations make dimerization pathways very unlikely. If transient radical intermediates such as ONOO are present, then their chemistry is of considerable importance. These mechanistic issues remain an active area of research for scientists interested in free radicals in biology.[122,123]

4.8 Biochemical and organic reactions of nitric oxide

Unlike any other development in its chemistry, the recognition of nitric oxide's important roles in biology has triggered an avalanche of new research.[124–126] The initial discoveries[124,127,128] in the early 1980s were greeted with considerable initial skepticism by the biomedical community.[129] Commonplace opinion was that, since nitric oxide was a radical, how could it have any lifetime or specificity in a cell? Three independent lines of research rapidly led to the acceptance and recognition of its biological role. Nitric oxide's role in understanding nitroglycerine vasodilation[130] was the initial discovery, but equally important were Hibb's pioneering study into the macrophage requirements for phagolysosome activity,[131–133] and Tannenbaum and Marletta's studies into metabolic levels of nitrite and nitrate.[134,135] All of these studies laid the foundation for understanding nitric oxide's constitutive and inducible physiological roles. Subsequent studies helped recognize the biosynthetic source of nitric oxide from the oxygenation of arginine,[136,137] and the normal nondietary basal levels of nitrite and nitrate.[135] It is now recognized that the main nitric oxide pathway (Figure 4.5 and Equation 4.19), have heme enzymes directly involved in the generation and sensing of nitric oxide.

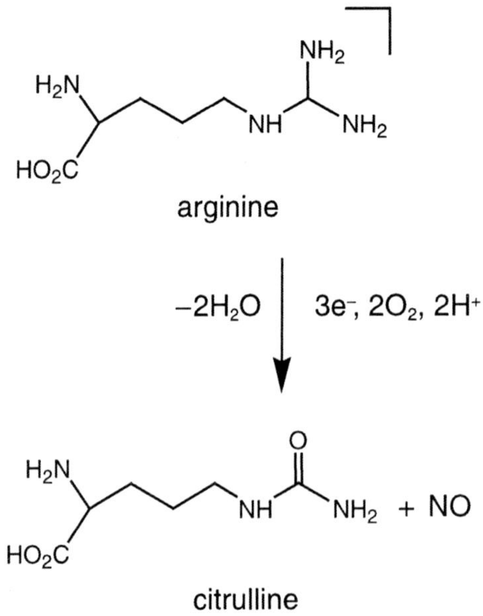

(4.19)

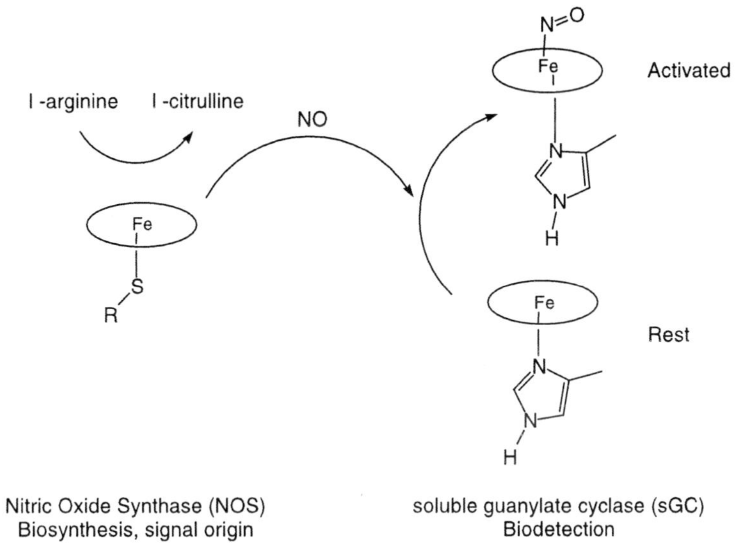

Figure 4.5 *Nitric oxide messenging pathway involving the heme enzymes nitric oxide synthase, a p-450, and soluble guanyl cyclase, sGC. Activation of sGC leads to upregulation of cGMP production from GTP.*

Nitric oxide's role and biological fate continues to be an important theme of bioinorganic chemistry. Metalloproteins are among the best characterized targets for nitric oxide messenging systems in biology, and oxygenated hemoglobin and myoglobin are important sinks for nitric oxide in blood. As a consequence, our understanding of nitric oxide's coordination and bioinorganic chemistry is well developed,[138] and overshadows the more difficult and general question of what other non-metal-based substrates will nitric oxide interact with in its biochemistry. Although many metalloproteins have been demonstrated to bind nitric oxide *in vitro* (Table 4.6), demonstrating this *in vivo* is a significant challenge. Much of this activity and lingering uncertainty relates to a fundamental aspect of nitric oxide's biochemistry: as a simple diatomic

Table 4.6 *Representative metalloprotein nitric oxide interactions*[193]

Protein	Metal center	Proposed Function	Ref.
aconitase	Iron sulfur	NO inhibits citric acid cycle	[240]
Rieske center	Iron sulfur	Respiratory chain inhibition	[241]
Fe-SOD	Non-heme iron	Free radical formation and control	[242]
Methane monooxygenase	Oxo-bridged diiron	NO inhibits methane oxidation	[243]
Iron regulatory protein-1	Iron sulfur	Inhibition of iron sensitive gene expression	[244]
Glutathione reductase	Active thiolate binding	Indirect inhibition of oxidative stress repair	[245]
Bacterioferritin	Heme iron and oxo-bridged diiron	Iron trafficking and storage	[246]
Soluble Guanylate cyclase	Heme	Upregulates cGMP production	[128]
Nitrite reductase	Copper	Regulates nitrogen oxyanion metabolism	[247]

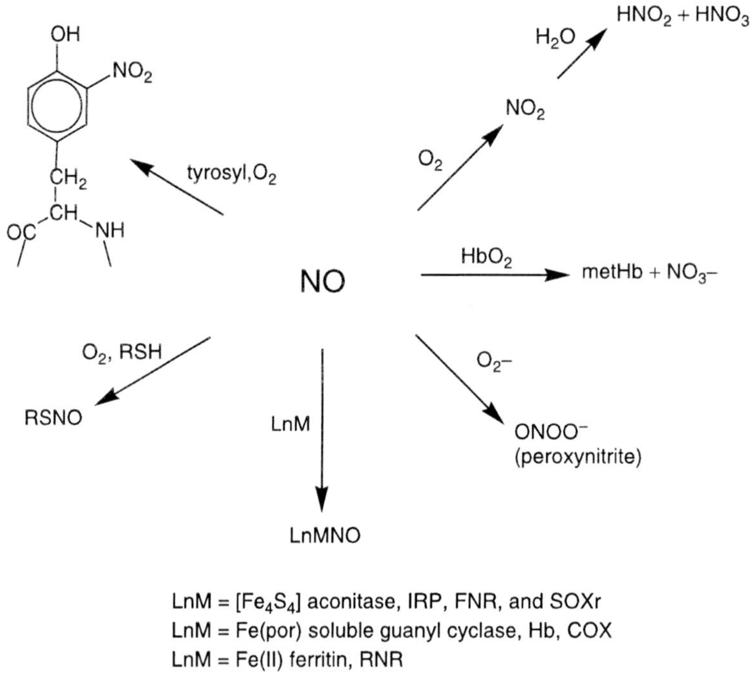

Figure 4.6 *A chemical hierarchy of nitric oxide's biological targets.*

gas it does not need active transport to cross cell and organelle membranes, and since it readily permeates most macromolecules it is impossible to use conventional genomic or molecular biology techniques to determine its target cells or macromolecules. The biochemical universe of possible targets is wide open to nitric oxide and there is no "consensus sequence" or structure to search or screen for its targets with genomic or proteomic techniques. So to answer the question "what are nitric oxide's biological targets?", its chemistry and those species with which it reacts most rapidly have to be understood. Based on these latter considerations, a general set of possible targets is shown in Figure 4.6.[139] Of these reactions, its reaction with superoxide is the fastest,[140] and that with oxyhemoglobin the most pervasive.[141] The reaction of nitric oxide with thiol to give RSNO requires an electron acceptor.[142]

4.9 General reactivity patterns

A summary of the general reactions for monomeric nitric oxide is shown in Figure 4.7.

4.9.1 Oxidation

Nitric oxide is readily oxidized and a variety of stable nitrosonium salts including tetrafluoroborate, perchlorate, and hydrosulfate are available.[143] Indeed, $[NO]HSO_4$ is an intermediate in the lead chamber process for the manufacture of sulfuric acid. Although the nitrosonium salts are hydrolytically sensitive, they can be handled briefly in solvents such as methanol.[144] Base-promoted addition of water, alcohols, and amines leads to either nitrite, alkylnitrites, or nitrosamines. Nitrosonium salts are also good outer sphere one-electron oxidants and are useful in cases where the gaseous nitric oxide product can easily be

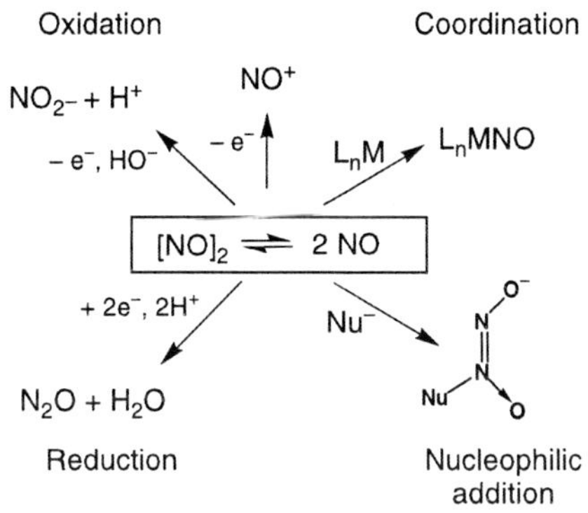

Figure 4.7 *General reactions of nitric oxide and its dimer.*

removed from the other products. In addition to one-electron oxidation, nitric oxide is readily oxidized by halogens and pseudohalogens to give X–N=O type molecular compounds.[145]

4.9.2 Reduction

The reduction of nitric oxide to give the nitroxyl anion (NO^-), has been studied by a number of high energy methods, including pulse radiolysis and laser flash photolysis.[146–150] Nitroxyl is often proposed to result from the hydrolysis of Angeli's salt ($Na_2N_2O_3$) or Piloty's acid ($PhSO_2NHOH$) and these putative sources of nitroxyl have been intensively investigated. As with almost all reactions which are thought to generate nitroxyl, these two reagents generate primarily nitrous oxide (Equation 4.20) as the ultimate product. Care needs to be taken when using these reagents and interpreting results from their use: in acid conditions nitric oxide is a known product of Angeli's salt hydrolysis,[151–154] and Bonner has found that the putative NO^- generating step in Piloty's acid decomposition is reversible.[155] Perhaps the single exception to the general observation that nitroxyl dimerization[156] leads to nitrous oxide is the reduction of nitrite by a sodium/mercury amalgam, which gives some *trans*-hyponitrite[157] as well as nitrous oxide (Equation 4.21). However in this case there may very well be other key steps which lead to the N–N bond formation aside from nitroxyl dimerization.

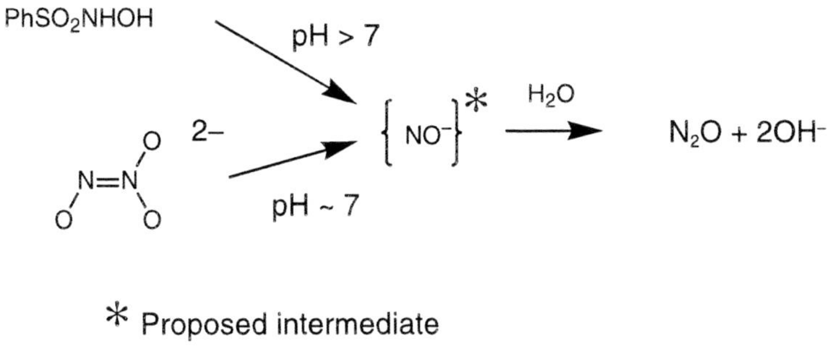

(4.20)

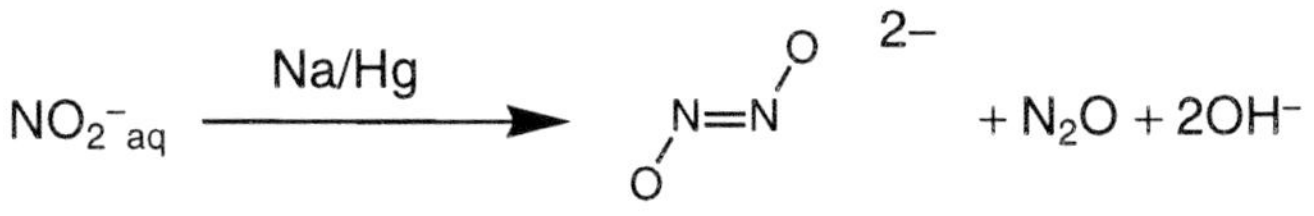

(4.21)

4.9.3 Coordination

Most transition metals form stable nitrosyl complexes[158] and, as noted above, these are of considerable importance in the bioinorganic chemistry of nitric oxide. Although the analogy of the NO^+ and CO ligands is often made, and on a frontier orbital level has considerable merit, there are two phenomenological characteristics that distinguish metallonitrosyls from metallocarbonyls: (1) metallonitrosyls readily adopt either linear or bent configurations, and (2) metallonitrosyls often form only mononitrosyl complexes. Di-,[159] tri-,[160] and tetra-nitrosyl[161] complexes are rare. Both traits are in contrast with carbonyl complexes, which invariably adopt linear geometries and which frequently form homoleptic and polycarbonyl complexes with many transition metals. Very few ligands bind in two such distinct geometries as does nitric oxide. The bent versus linear binding of nitric oxide is a textbook example of ligand amphoteric character: the ability of a ligand to both π-accept and donate depending upon the electronic environment of the metal.[162] In a handful of complexes both geometries are known to be present.[144,163–165] These geometries can be naively described as involving the binding of NO^+ (linear) and NO^- (bent), but this inadequate description of the metal–nitrosyl interaction is largely superseded by the Feltham–Enemark analysis[60] which considers the metal–nitrosyl geometry as resulting from the number of π electrons delocalized in the M–NO framework, $\{MNO\}^n$. If $n > 6$ then the more stable geometry is bent and this is certainly the case where $n = 8$. It is significant that most examples of stable metalloprotein nitrosyls in biology bind nitric oxide in complexes with a formal $\{FeNO\}^7$, which is just at the margin of this bent versus linear duality.

The binding of nitric oxide to many complexes frequently involves electron transfer in either a formal or a step-wise mechanism. There are few kinetic studies in this area, and so it is not possible to predict if either the nitric oxide or dinitrogen dioxide react faster with a complex. A recent study by Karlin suggests that in the case of a bimetallic nitric oxide reductase model reaction occurs for the dimer and on the same coordination site on a metal.[166] This is an area where numerous questions remain unanswered. The coordination sphere reactions of nitric complexes have however been carefully reviewed.[167,168]

4.9.4 Addition of nucleophiles

Nitric oxide and its dimer are modest electrophiles and react pairwise with a range of nucleophiles to give diazeniumdiolates (Figure 4.8). This functional group has been variously termed methoxazonyl[169] or *N*-nitrosohydroxylamines and has been recently termed diazeniumdiolates.[170,171] Although only recently named, the diazeniumdiolates have an important and early history in chemistry. Davy used nitric oxide's reaction with bisulfite (Equation 4.22) to distinguish nitric oxide from the other nitrogen oxides in 1802.[172] The chemistry of these species has been well reviewed; both Ragsdale's[145] 1973 and Keefer's[171] 2002 reviews have excellent discussions of this chemistry. Remarkably there are more than twelve diazeniumdiolates found in natural products (Figure 4.9 and Table 4.7), and phenyldiazeniumdiolate (PhN(O)=NO^-) has found use as an analytical reagent termed cupferron.[173–174] The biosynthesis of these natural products remains largely unexplored, but most likely do not involve nitric oxide directly. This remains an area

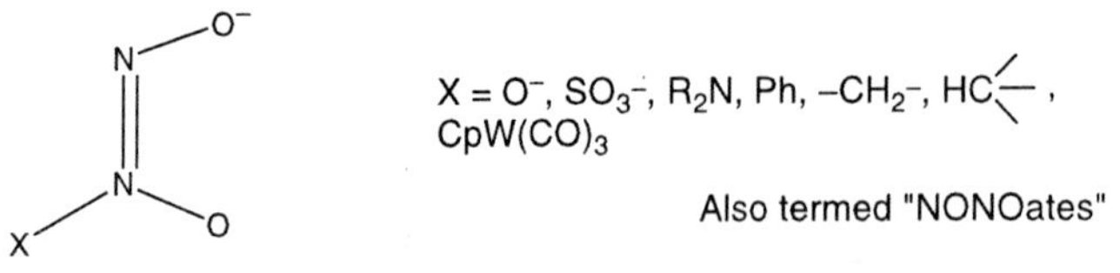

Figure 4.8 Structure and range of known diazeniumdiolates.[194,195]

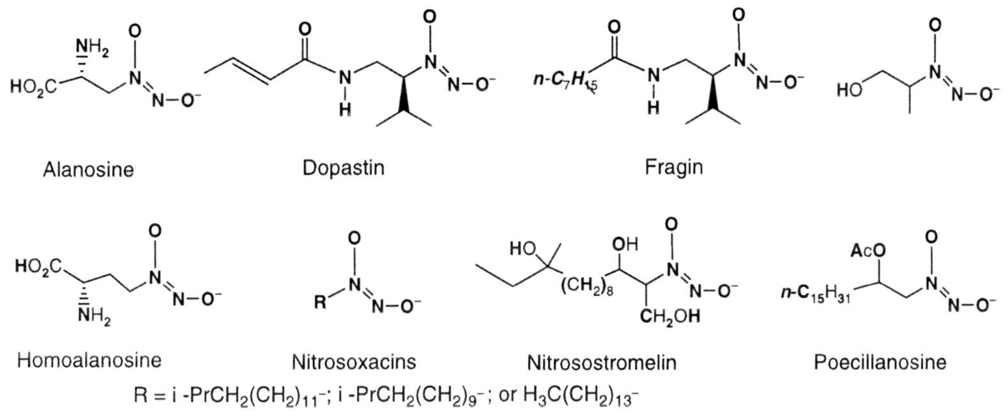

Figure 4.9 *Naturally occurring diazeniumdiolates and their biochemistry.*[196–209]

Table 4.7 *Known biochemistry of natural diazeniumdiolates*

Diazeniumdiolate	Source	Recognized Pharmacology	Known inhibitor of	Ref.
Alanosine	*Streptomyces A*	Cancer Chemotherapy, (Phase II) Antibiotic, Antiviral	Adenylsuccinate synthase	[197, 199–201]
Dopastin	*Streptomyces?*		Dopamine β–hydroxylase	[196, 198–200]
Fragin	*Pseudomonas*	Herbicide		[202–206]
Homoalanosine		Herbicide		[199]
Nitrosoxacins	*Streptomyces?*	Antibiotic	5-lipoxygenase	[207]
Nitrosostromelin	*Streptomyces*		Stromelysin	[207]
Poecillanosine	*Marine sponges*			[208, 209]

of considerable theoretical and practical activity, with much of this interest stemming from the use of dialkylaminediazeniumdiolates as nitric oxide donors in biomedical applications.

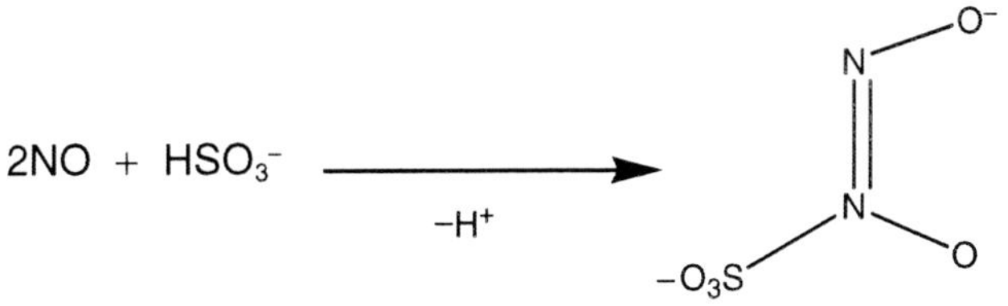 (4.22)

Under basic conditions secondary amines react rapidly with nitric oxide to give diazeniumdiolates, which are readily isolated as their dialkylammonium salts (Equation 4.23).[175] The retro reaction in neutral conditions returns nitric oxide,[176] with their rate of nitric oxide release being dependent upon the nitrogen substituents.[177] With the exception of three cyclic cases (Figure 4.10),[178–180] the diazeniumdiolates which

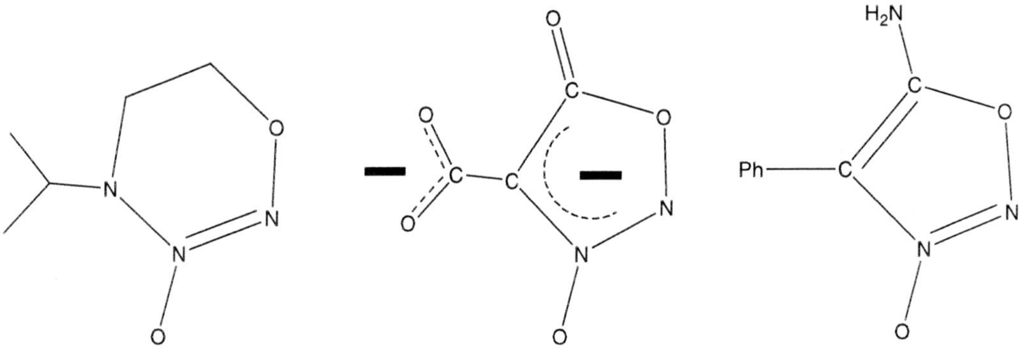

Figure 4.10 *The known E-diazeniumdiolates.*

result from the addition of nucleophiles (Equation 4.23) adopt the Z geometry.[170] Recently, a series of tetradentate amine derivatives with a terminal diazeniumdiolate were proposed to have a coordinated E-diazeniumdiolate (Figure 4.10) but this was not structurally or spectroscopically confirmed.[181] *Ab initio* calculations indicate that the Z isomers are more stable than the E isomers by between 4–10 kJ mol^{-1}, but the barriers for interconversion[179] by nitroso group rotation are often high, 120–160 kJ mol^{-1}. The origin of the stereoselectivity in these reactions remains an important question. A priori there is little reason for the apparent thermodynamic control in these reactions unless it corresponds to the same product from the kinetic control. As with nitric oxide oxidation, two mechanisms have been proposed involving (i) rate limiting nucleophile addition to the dimer (Equations 4.24–4.26) or (ii) reaction of the second NO to the pre-equilibrated $HR_2NNO^{\bullet}$ radical (Equations 4.27–4.29). Initial mechanistic studies[175] were confused by oxygen contamination[145] and subsequent studies have demonstrated that both mechanisms operate, with a strong solvent dependence.[116] The current interpretation is that at higher pressures and with methanol as a solvent, a bimolecular mechanism (Equations 4.24–4.26) predominates.

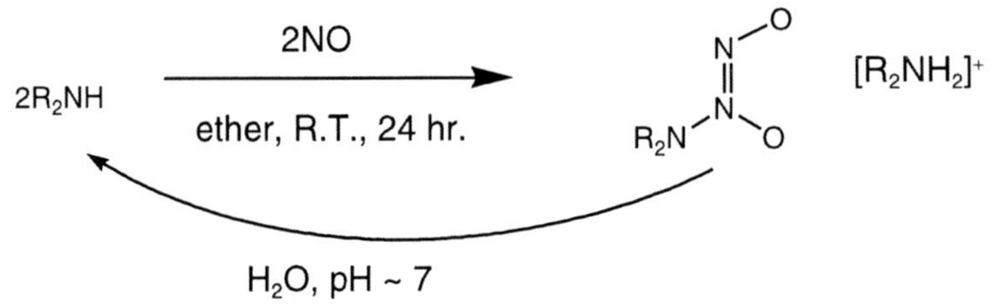

(4.23)

$$\text{Dimer} \quad 2\,NO \rightleftharpoons N_2O_2 \tag{4.24}$$

$$N_2O_2 + R_2NH \xrightarrow[\text{r.d.s.}]{k_2} HR_2NN_2O_2 \tag{4.25}$$

$$HR_2NN_2O_2 + R_2NH \longrightarrow [R_2NN_2O_2]^-[R_2NH]^+ \tag{4.26}$$

$$\text{Monomer} \quad NO + R_2NH \rightleftharpoons [HR_2NNO] \tag{4.27}$$

$$NO + [HR_2NNO] \xrightarrow[\text{r.d.s.}]{k_1} HR_2NN_2O_2 \tag{4.28}$$

$$HR_2NN_2O_2 + R_2NH \longrightarrow [R_2NN_2O_2]^-[R_2NH]^+ \tag{4.29}$$

4.9.5 General organic reactions

An important reason why the organic chemistry of nitric oxide is so sparse is that many of the products are complex mixtures of salts. These frequently require careful recrystallization steps and exacting control of conditions to give clean, reproducible products. There is a wealth of new chemistry to be discovered from the addition reactions of nitric oxide, with a particularly telling example being the range of salts which result from the diazeniumdiolation of ketones.[182] Pressure, temperature, alkali metal, and stoichiometry are all critical parameters which need to be controlled in this chemistry. A particularly complex example of this chemistry is the diazeniumdiolation of benzylcyanide, which gives either bis diazeniumdiolates, a 1,2,3-oxadiazole, or an oxime (Equation 4.30).[178,183,184] Depending upon conditions of temperature, stoichiometry, pressure, and alkali metal, it is possible to isolate the oxime or the dianion in high yields.[178] Recently, the reactions of nitric oxide with organic radicals has been reviewed.[185]

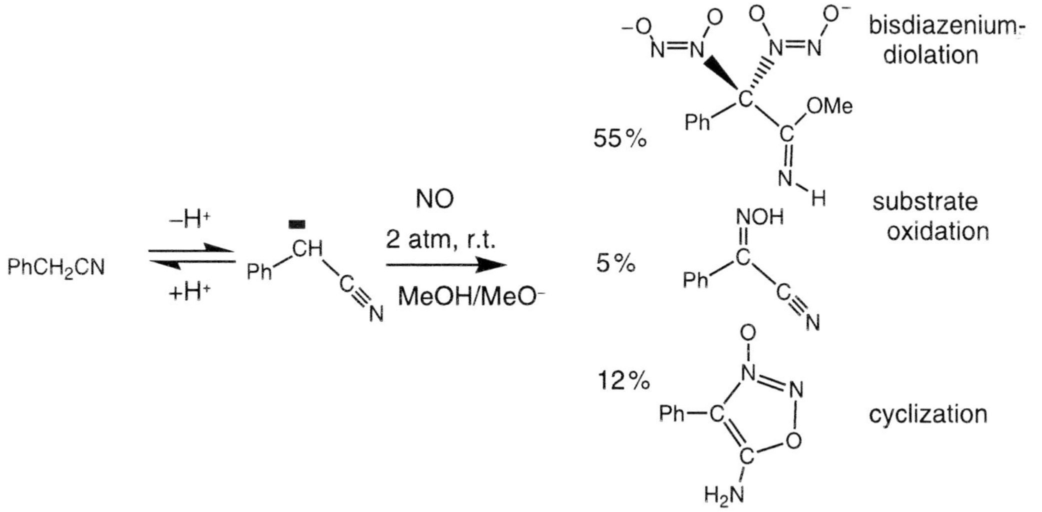

(4.30)

4.9.6 Reactions with other nucleophiles

The range of nucleophiles which give stable diazeniumdiolates from the addition of nitric oxide (Figure 4.8 and Equation 4.22) is surprisingly small. Hydrides, cyanide, most tertiary phosphines, tertiary amines, azide, and ammonia all fail to return diazeniumdiolates from the addition of nitric oxide. The phosphorus adducts, if formed, rapidly eliminate nitrous oxide and form the phosphine oxide. Apart from Pelouze's adduct of nitric oxide and bisulfite, sulfur-based diazeniumdiolates are not isolated from the reactions of mercaptans and nitric oxide.

4.10 The colored species problem in nitric oxide chemistry

Sporadic and diverse reports over the last century have described colored species forming during the manipulation and/or reaction of nitric oxide with various substrates. It is clear that the deep blue solutions and solids often condensed at low temperature are due to dinitrogen trioxide and result from the reaction of nitric oxide and nitrogen dioxide. The later is often a contaminant of commercial samples of nitric oxide, but of course it can also arise from the adventitious reaction of dioxygen with nitric oxide. However, there are related, but harder to rationalize, reports of red and red-blue adducts at low temperature. These often appear in the presence of Lewis acids. Characteristic features of these adducts are their ease of dissociation and reactivity. There is thus only limited characteristic data relating to these adducts with Seel's determination of a λ_{max} of 580 nm and an $\varepsilon = 40\,M^{-1}cm^{-1}$,[186] and Laane's vibrational spectroscopy characterization being particularly good examples.[187] In the later case, the vibrational data from the four nitric oxide isotopomers was interpreted in terms of the formation of an asymmetric nitric oxide dimer (ONON). While it is not clear how the Lewis acid would promote the formation of this isomer over the more frequently observed symmetric dimer, the Lewis acids may stabilize the asymmetric isomer and allow for its observation.

4.11 Conclusions

Although the spectroscopic, physical, and theoretical basis for nitric oxide dimerization is now well understood and characterized, the practical consequences of this propensity in terms of nitric oxide's chemistry and biology remain important problems. Ultimately, these are mechanistic problems, and it should be no surprise that this chemistry remains so poised. The inorganic chemistry of nitrogen is filled with spectacular examples of kinetic control and it is in its oxides that mechanistic issues are critical. In the late 1880s the mechanism of nitric oxide's oxidation was a source of considerable controversy between two rival camps[188,189] using classical product analysis to advocate different mechanisms. A century later, in the 1990s, similar difficulties were encountered in attempts to understand the chemistry of peroxynitrite.[190–192] While the community waits for a resolution to this vexing discussion, which may very well require new and orthogonal methods to untangle, large areas of this ostensibly simple chemistry remain to be explored. And there are many lessons in this chemistry for all chemists interested in transient and persistent radicals. Given these difficulties in understanding simple diatomic and triatomic radicals, the chemistry of larger homologues may well pose even more nuance and problems.

References

1. N. V. Sidgwick, *The Organic Chemistry of Nitrogen*, Clarendon Press, Oxford, 1966.
2. M. Zhou, L. Zhang and Q. Qin, *J. Am. Chem. Soc.*, **122**, 4483–4488 (2000).
3. N. Shafizadeh, P. Brechignac, M. Dyndgaard, *et al*., *J. Chem. Phys.*, **108**, 9313–9326 (1998).
4. R. E. Miller, *Science*, **240**, 447–53 (1988).
5. C. R. Dennis, C. J. Whitham, R. J. Low and B. J. Howard, *Chem. Phys. Lett.*, **282**, 421–428 (1998).
6. Y. Kim and H. Meyer, *Int. Rev. Phys. Chem.*, **20**, 219–282 (2001).
7. D. S. King and J. C. Stephenson, *J. Chem. Phys.*, **82**, 5286–8 (1985).

8. S. E. Daire, J. Lozeille, S. D. Gamblin and T. G. Wright, *J. Phys. Chem.*, **104**, 9180–9183 (2000).
9. B. J. Howard, *Faraday Disc. Chem. Soc.*, **71**, 23–9 (1981).
10. D. A. Micha, *Acc. Chem. Res.*, **6**, 138–44 (1973).
11. S. F. Agnew, B. I. Swanson, L. H. Jones and R. L. Mills, *J. Phys. Chem.*, **89**, 1678–1682 (1985).
12. L. Jia, C. Bonaventura, J. Bonaventura and J. S. Stamler, *Nature*, **380**, 221–226 (1996).
13. F. A. Cotton and G. Wilkinson, *Advanced Inorganic Chemistry*, 3rd edn., John Wiley and Sons, Inc., New York, 1972.
14. J. Mason, *J. Chem. Educ.*, **52**, 445–447 (1975).
15. A. A. Blanchard, *Inorganic Syntheses* (Ed. W. C. Fernelius), McGraw Hill, New York, 1946.
16. O. Bostrup, *Inorganic Syntheses* (Ed. H. E. J. Holtzclaw), McGraw Hill, New York, 1966.
17. D. S. Bohle, E. S. Sagan, W. H. Koppenol and R. Kissner, *Inorganic Syntheses*, **34**, 36–42 (2004).
18. J. D. Ray and R. A. Ogg, Jr., *J. Am. Chem. Soc.*, **78**, 5993 (1956).
19. G. Starck, *Elementa*, **50**, 213–21 (1967).
20. A. Pedler and F. H. Pollared, *Inorganic Syntheses* (Ed. T. Moeller), McGraw Hill, New York, 1957.
21. M. Anbar and H. Taube, *J. Am. Chem. Soc.*, **77**, 2993–4 (1955).
22. M. Anbar and H. Taube, *J. Am. Chem. Soc.*, **76**, 6243–7 (1954).
23. K. Clusius and H. Schumacher, *Helv. Chim. Acta.*, **129**, 1137–1144 (1957).
24. J. H. Van Vleck, *Physical Review*, **31**, 587–613 (1928).
25. E. Bauer and A. Piccard, *J. Phys. et Radium*, **1**, 97–122 (1920).
26. H. L. Johnston and W. F. Giauque, *J. Am. Chem. Soc.*, **51**, 3194 (1929).
27. N. Arulsamy, D. S. Bohle, J. A. Imonigie and E. S. Sagan, *J. Am. Chem. Soc.*, **122**, 5539–5549 (2000).
28. A. Snis and I. Panas, *Chem. Phys.*, **221**, 1–10 (1997).
29. W. N. Lipscomb, R. E. Wang, W. R. May and E. L. Lippert, *Acta. Cryst.*, **14**, 1100–1101 (1961).
30. W. J. Dulmage, E. A. Meyers and W. N. Lipscomb, *Acta. Cryst.*, **6**, 760–764 (1953).
31. W. J. Dulmage, E. A. Meyers and W. N. Lipscomb, *J. Chem. Phys.*, **19**, 1432–1433 (1951).
32. W. N. Lipscomb, *J. Chem. Phys.*, **54**, 3659–3660 (1971).
33. M. D. Brookes, A. R. W. McKellar and T. Amano, *J. Mol. Spec.*, **185**, 153–157 (1997).
34. S. G. Kukolich, *J. Mol. Spec.*, **98**, 80–6 (1983).
35. S. G. Kukolich, *J. Am. Chem. Soc.*, **104**, 4715–16 (1982).
36. C. M. Western, P. R. R. Langridge-Smith, B. J. Howard and S. E. Novick, *Mol. Phys.*, **44**, 145–60 (1981).
37. A. Dkhissi, N. Lacome and A. Perrin, *J. Mol. Spec.*, **194**, 156–162 (1999).
38. A. L. L. East, A. R. W. McKellar and J. K. G. Watson, *J. Chem. Phys.*, **109**, 4378–4383 (1998).
39. A. Dkhissi, P. Soulard, A. Perrin and N. Lacome, *J. Mol. Spec.*, **183**, 12–17 (1997).
40. B. Urban, A. Strobel and V. E. Bondybey, *J. Chem. Phys.*, **111**, 8939–8949 (1999).
41. A. L. L. East, *J. Chem. Phys.*, **109**, 2185–2193 (1998).
42. T. B. Reed and W. N. Lipscomb, *Acta. Cryst.*, **6**, 781–783 (1953).
43. R. L. Kuczkowski, *J. Am. Chem. Soc.*, **87**, 5259–5260 (1965).
44. J. S. Broadley and J. M. Robertson, *Nature*, **164**, 915–916 (1949).
45. B. S. Cartwright and J. M. Robertson, *Chem. Commun.*, 82–83 (1966).
46. P. Groth, *Nature*, **198**, 1081 (1963).
47. A. Kvick, R. K. McMullan and M. D. Newton, *J. Chem. Phys.*, **76**, 3754–3760 (1982).
48. D. W. Smith and K. Hedberg, *J. Chem. Phys.*, **25**, 1282–1283 (1956).
49. B. W. McClelland, G. Gundersen and K. Hedberg, *J. Chem. Phys.*, **56**, 4541–4545 (1972).
50. K. B. Borisenko, M. Kolonits, B. Rozsondai and I. Hargittai, *J. Mol. Struct.*, **413-414**, 121–131 (1997).
51. G. Wojcik and J. Holband, *Acta. Cryst.*, **B58**, 684–689 (2002).
52. A. Forni, P. Metragolo, T. Pilati and G. Resnati, *Cryst. Growth Des.*, **4**, 291–295 (2004).
53. I. Pastirk, M. Comstock and M. Dantus, *Chem. Phys. Lett.*, **349**, 71–78 (2001).
54. W. L. Jolly, *The Inorganic Chemistry of Nitrogen*, Benjamin, New York, 1964.
55. L. Pauling, *The Nature of the Chemical Bond*, Cornell University Press, New York, 1960.
56. F. Bitter, *Proc. Nat. Acad. Sci. USA.*, **15**, 638–642 (1929).
57. J. Aharoni and P. Scherrer, *Zeits. f. Phys.*, **58**, 749–765 (1929).

58. J. H. Van Vleck, *Electric and Magnetic Susceptibilities*, Oxford University Press, New York, 1932.
59. A. Volodin, D. Biglino, Y. Itagaki, *et al.*, *Chem. Phys. Lett.*, **327**, 165–170 (2000).
60. R. D. Feltham and J. Enemark, *Coord. Chem. Rev.*, **13**, 339 (1974).
61. P. W. Atkins and M. C. R. Symons, Elsevier, Amsterdam, 1967.
62. M. Al-Joboury, I., D. P. May and D. W. Turner, *J. Chem. Soc.*, 616–622 (1965).
63. H. Fukuzawa, X. J. Liu, *et al.*, *J. Phys. B*, **41**, 045102/1–045102/6 (2008).
64. J. C. Miller and R. N. Compton, *Chem. Phys. Lett.*, **93**, 453–9 (1982).
65. M. H. Kibel and G. L. Nyberg, *J. Elec. Spec. Rel. Phen.*, **17**, 1–13 (1979).
66. G. L. Price and B. G. Baker, *Surf. Sci.*, **68**, 507–15 (1977).
67. T. Darko, I. H. Hillier and J. Kendrick, *Chem. Phys. Lett.*, **45**, 188–90 (1977).
68. J. L. Gardner and J. A. R. Samson, *J. Elec. Spec. Rel. Phen.*, **2**, 153–60 (1973).
69. V. I. Kleimenov, Y. V. Chizhov and F. I. Vilesov, *Optika i Spektroskopiya*, **32**, 702–8 (1972).
70. O. Edqvist, L. Asbrink and E. Lindholm, *Zeits, Naturforsch. A*, **26**, 1407–10 (1971).
71. H. Lefebvre-Brion, *Chem. Phys. Let.* **9**, 463–4 (1971).
72. T. B. Borne, *Observation of excited molecular ions. Resonance photoelectron spectroscopy of nitric oxide and benzene*, Ph.D. Thesis, Northwestern University, Evanston, IL, 1969.
73. N. Form, T., B. Whitaker, J., L. Poisson and B. Soep, *Phys. Chem. Chem. Phys.*, **8**, 2925–32 (2006).
74. M. E. Jacox, *J. Phys. Chem. Ref. Data*, **27**, 115–393 (1998).
75. A. L. L. East and J. K. G. Watson, *J. Chem. Phys.*, **110**, 6099–6102 (1999).
76. V. Dribinski, A. B. Potter, I. Fedorov and H. Reisler, *Chem. Phys. Lett.*, **385**, 233–238 (2004).
77. E. Miescher and K. P. Huber, *Int. Rev. Sci.: Phys. Chem., Ser. 2*, **3**, 37–73 (1976).
78. K. P. Huber and M. Vervloet, *J. Mol. Spect.*, **129**, 1 (1988).
79. L. Burnelle, P. Beaudouin and L. J. Schaad, *J. Phys. Chem.*, **71**, 2240–7 (1967).
80. G. G. Havens, *Phys. Rev.*, **41**, 337–44 (1932).
81. G. G. Havens, *Phys. Rev.*, **43**, 992–1000 (1933).
82. T. J. Schaafsma, *Chem. Phys. Lett.*, **1**, 16–18 (1967).
83. J. A. Castle and R. Beringer, *Phys. Rev.*, **80**, 114 (1950).
84. F. J. Adrian, *J. Chem. Phys.*, **36**, 1692–3 (1962).
85. H. Seldes and R. Livingston, *J. Chem. Phys.*, **35**, 563 (1961).
86. B. H. J. Bielski, J. J. Freeman and J. M. Gebicki, *J. Phys. Chem.*, **72**, 1721–5 (1968).
87. D. S. Burch, W. H. Tanttila and M. Mizushima, *J. Chem. Phys.*, **61**, 1607–12 (1974).
88. S. Katsumata, H. Shiromaru, K. Mitani, S. Iwata and K. Kimura, *Chem. Phys.*, **69**, 423–31 (1982).
89. P. Warneck, *Chem. Phys. Lett.*, **3**, 532–3 (1969).
90. P. Natalis and J. E. Collin, *Chem. Phys. Lett.*, **2**, 79–82 (1968).
91. D. P. Chong, D. C. Frost, W. M. Lau and C. A. McDowell, *Chem. Phys. Lett.*, **90**, 332–6 (1982).
92. M. J. Campbell, J. Liesegang, J. D. Riley and J. G. Jenkin, *J. Phys. C*, **15**, 2549–58 (1982).
93. K. Nomoto, Y. Achiba and K. Kimura, *Bull. Chem. Soc. Jap.*, **52**, 1614–18 (1979).
94. W. Von Niessen, W. Domcke, L. S. Cederbaum and J. Schirmer, *J. Chem. Soc., Farad. Trans. 2*, **74**, 1550–8 (1978).
95. T. H. Gan, J. B. Peel and G. D. Willett, *J. Chem. Soc., Farad. Trans. 2*, **73**, 1459–63 (1977).
96. D. C. Frost, C. A. McDowell and N. P. C. Westwood, *J. Elect. Spect. Rel. Phen.*, **10**, 293–303 (1977).
97. D. L. Ames and D. W. Turner, *Proc. Roy. Soc. London A*, **348**, 175–86 (1976).
98. P. Baltzer, L. Karlsson, B. Wannberg, *et al.*, *Chem. Phys.*, **237**, 451–470 (1998).
99. J. E. Williams and J. N. Murrel, *J. Am. Chem. Soc.*, **93**, 7149 (1971).
100. T. Vladimiroff, *J. Am. Chem. Soc.*, **94**, 8250 (1972).
101. S. Skaarup, P. N. Skancke and J. E. Boggs, *J. Am. Chem. Soc.*, **98**, 6106 (1976).
102. K. A. Nguyen, M. S. Gordon, J. S. Montgomery and H. H. Michels, *J. Phys. Chem.*, **98**, 10072 (1994).
103. R. Gonzalez-Luque, M. Merchan and B. O. Roos, *Theor. Chim. Acta*, **88**, 425 (1994).
104. Y. Xie, H. F. Schaefer, III, F. Xiao-Yuan and R.-Z. Liu, *J. Chem. Phys.*, **111**, 2532 (1999).
105. H. L. Johnston and H. R. Weimer, *J. Am. Chem. Soc.*, **56**, 625–628 (1934).
106. M. L. McGlashan and D. J. B. Potter, *Proc. Roy. Soc. A.*, **267**, 478–482 (1962).

107. E. A. Guggenheim, *Mol. Phys.*, **10**, 401 (1966).
108. R. L. Scott, *Mol. Phys.*, **11**, 399–401 (1966).
109. I. C. Hisatsune and L. Zafonte, *J. Phys. Chem.*, **73**, 2980–2989 (1969).
110. H. Gershinowitz and H. Eyring, *J. Am. Chem. Soc.*, **57**, 985–991 (1935).
111. J. Heicklen and N. Cohen, *Adv. Photochem.*, **5**, 157–328 (1968).
112. B. Galliker, R. Kissner, T. Nauser, and W. H. Koppenol, *Chem. Eur. J.*, **15**, 6161–6168 (2009).
113. W. A. Guillory and H. S. Johnston, *J. Phys. Chem.*, **42**, 2457–2461 (1965).
114. E. D. J. Morris and H. S. Johnston, *J. Phys. Chem.*, **47**, 4282 (1967).
115. Y.-L. Zhao, M. D. Bartberger, K. Goto, *et al.*, *J. Am. Chem. Soc.*, **127**, 7964–7965 (2005).
116. D. S. Bohle and K. L. Smith, *Inorg. Chem.*, **47**, 3925–7 (2008).
117. M. D. Lim, I. M. Lorkovic and P. C. Ford, *Inorg. Chem.*, **41**, 1026–1028 (2002).
118. L. P. Kuhn, J. O. Doali and C. Wellman, *J. Am. Chem. Soc.*, **82**, 4792–4795 (1960).
119. K. McNeill and R. G. Bergman, *J. Am. Chem. Soc.*, **121**, 8260–8269 (1999).
120. D. Littlejohn, K. Y. Hu and S. G. Chang, *Inorg. Chem.*, **25**, 3131–3135 (1986).
121. T. L. Nunes and R. E. Powell, *Inorg. Chem.*, **9**, 511–512 (1970).
122. M. L. McKee, *J. Am. Chem. Soc.*, **117**, 1629–1637 (1995).
123. H. Tsukahara, T. Ishida and M. Mayumi, *Nitric Oxide*, **3**, 191–198 (1999).
124. F. Murad, C. K. Mittal, W. P. Arnold, *et al.*, *Adv. Cyclic Nucleotide Res.*, **9**, 145–58 (1978).
125. L. J. Ignarro, *Angew. Chem. Int. Ed.*, **38**, 1882–1892 (1999).
126. R. F. Furchgott, *Angew. Chem. Int. Ed.*, **38**, 1871–1880 (1999).
127. R. F. Furchgott, P. D. Cherry, J. V. Zawadzki and D. Jothianandan, *J. Cardiovasc. Pharmacol.*, **6** (Suppl. 2), S336–43 (1984).
128. L. J. Ignarro, J. N. Degnan, W. H. Baricos, P. J. Kadowitz and M. S. Wolin, *Biochim Biophys Acta* **718**, 49–59 (1982).
129. J. Stamler, M. E. Mendelsohn, P. Amarante, *et al.*, *Circul. Res.*, **65**, 789–95 (1989).
130. C. A. Gruetter, D. Y. Gruetter, J. E. Lyon, *et al.*, *J. Pharmacol. Exp. Therp.*, **219**, 181–6 (1981).
131. D. L. Granger, J. B. Hibbs, Jr., J. R. Perfect and D. T. Durack, *J. Clin. Invest.*, **81**, 1129–36 (1988).
132. J. B. Hibbs, Jr., R. R. Taintor and Z. Vavrin, *Science* **235**, 473–6 (1987).
133. J. B. Hibbs, Jr., Z. Vavrin and R. R. Taintor, *J. Immunol.*, **138**, 550–65 (1987).
134. M. Miwa, D. J. Stuehr, M. A. Marletta, *et al.*, *Carcinogenesis*, **8**, 955–8 (1987).
135. S. R. Tannenbaum, *Science*, **205**, 1332, 1334–1337 (1979).
136. A. M. Leone, R. M. Palmer, R. G. Knowles, *et al.*, *J. Biol. Chem.*, **266**, 23790–5 (1991).
137. N. S. Kwon, C. F. Nathan, C. Gilker, *et al.*, *J. Biol. Chem.*, **265**, 13442–5 (1990).
138. G. Richter-Addo, P. Legzdins and J. Burstyn, *Chem. Rev.*, **102**, 857–1270 (2002).
139. M. R. Gunther, B. E. Sturgeon and R. P. Mason, *Toxicology*, **177**, 1–9 (2002).
140. W. H. Koppenol, *Met. Ions Biol. Syst.*, **36**, 597–619 (1999).
141. S. Herold, *FEBS Lett.*, **439**, 85–88 (1998).
142. N. Arulsamy, D. S. Bohle, J. A. Butt, *et al.*, *J. Am. Chem. Soc.*, **121**, 7115–7123 (1999).
143. M. T. Mocella, M. S. Okamoto and E. K. Barefield, *Synth. Reac. Inorg. Met.-Org. Chem.*, **4**, 69–90 (1974).
144. A. P. Gaughan, Jr., B. J. Corden, R. Eisenberg and J. A. Ibers, *Inorg. Chem.*, **13**, 786–91 (1974).
145. R. O. Ragsdale, *Dev. Inorg. Nitrogen Chem.* (Ed. C. B. Colburn), Elsevier, New York, 1973.
146. G. A. Poskrebyshev, V. Shafirovich and S. V. Lymar, *J. Phys. Chem. A*, **112**, 8295–8302 (2008).
147. S. V. Lymar, V. Shafirovich and G. A. Poskrebyshev, *Inorg. Chem.*, **44**, 5212–5221 (2005).
148. G. A. Poskrebyshev, V. Shafirovich and S. V. Lymar, *J. Am. Chem. Soc.*, **126**, 891–899 (2004).
149. V. Shafirovich and S. V. Lymar, *J. Am.Chem. Soc.*, **125**, 6547–6552 (2003).
150. V. Shafirovich and S. V. Lymar, *Proc. Nat. Acad. Sci.*, **99**, 7340–7345 (2002).
151. M. J. Akhtar, F. T. Bonner, M. N. Hughes, *et al.*, *Inorg. Chem.*, **25**, 4635–9 (1986).
152. C. E. Donald, M. N. Hughes, J. M. Thompson and F. T. Bonner, *Inorg. Chem.*, **25**, 2676–7 (1986).
153. M. J. Akhtar, C. A. Lutz and F. T. Bonner, *Inorg. Chem.*, **18**, 2369–75 (1979).
154. F. T. Bonner and B. Ravid, *Inorg. Chem.*, **14**, 558–63 (1975).
155. F. T. Bonner and Y. Ko, *Inorg. Chem.*, **31**, 2514–19 (1992).

156. F. T. Bonner and M. N. Hughes, *Comm. Inorg. Chem.*, **7**, 215–34 (1988).
157. N. Arulsamy, D. S. Bohle, J. A. Imonigie and E. S. Sagan, *Inorg. Chem.*, **38**, 2716–2725 (1999).
158. G. B. Richter-Addo and P. Legzdins, *Metal Nitrosyls*, Oxford University Press, New York, 1992.
159. N. Arulsamy, D. S. Bohle, J. A. Imonigie and R. C. Moore, *Polyhedron*, **26**, 4737–4745 (2007).
160. W. Hieber and H. Tengler, *Z. Anorg.Allg. Chem.*, **318**, 136–54 (1962).
161. M. Herberhold and A. Razavi, *Angew. Chem., Int. Ed. Engl.*, **11**, 1092–4 (1972).
162. J. P. Collman, L. S. Hegedus, J. R. Norton and R. G. Finke, *Principles and Applications of Organotransition Metal Chemistry*, University Science Books, Mill Valley, 1987.
163. C. P. Brock, J. P. Collman, G. Dolcetti, *et al.*, *Inorg. Chem.*, **12**, 1304–13 (1973).
164. B. L. Haymore and J. A. Ibers, *J. Am. Chem. Soc.*, **96**, 3325–7 (1974).
165. B. Haymore and J. A. Ibers, *Inorg. Chem.*, **14**, 2610–2617 (1975).
166. J. Wang, M. P. Schopfer, A. A. N. Sarjeant and K. D. Karlin, *J. Am. Chem. Soc.*, **131**, 450–451 (2009).
167. P. S. Braterman (Ed.), *Reactions of Coordinated Ligands*, Vol. 2, Plenum, New York (1989).
168. E. M. R. Kiremire, *Orient. J. Chem.*, **21**, 161–182 (2005).
169. R. B. Woodward and C. Wintner, *Tet. Lett.*, 2689–2692 (1969).
170. L. K. Keefer, J. L. Flippen-Anderson, C. George, *et al.*, *Nitric Oxide*, **5**, 377–394 (2001).
171. J. A. Hrabie and L. K. Keefer, *Chem. Rev.*, **102**, 1135–1154 (2002).
172. H. Davy, *Bibl. Br. Sci. Arts.*, **20**, 350–367 (1802).
173. C. S. Marvel, *Organic Syntheses*, **4** (1925).
174. I. M. Kolthoff and A. Liberti, *Analyst*, **74**, 635–41 (1949).
175. R. S. Drago and F. E. Paulik, *J. Am. Chem. Soc.*, **82**, 96–98 (1960).
176. C. M. Maragos, D. Morley, D. A. Wink, *et al.*, *J. Med. Chem.*, **34**, 3242–3247 (1991).
177. J. E. Saavedra, T. M. Dunams, J. L. Flippen-Anderson and L. K. Keefer, *J. Org. Chem.*, **57**, 6134 (1992).
178. D. S. Bohle and I. Perepichka, *J. Org. Chem.*, **74**, 1621–1626 (2009).
179. Y.-N. Wang, D. S. Bohle, C. L. Bonifant, *et al.*, *J. Am. Chem. Soc.*, **127**, 5388–5395 (2005).
180. N. Arulsamy and D. S. Bohle, *Ang. Chem. Int. Ed. Engl.*, **41**, 2089–2091 (2002).
181. M. Ziche, S. Donnini, L. Morbidelli, *et al.*, *Chem. Med. Chem.*, **3**, 1039–1047 (2008).
182. N. Arulsamy and D. S. Bohle, *J. Org. Chem.*, **71**, 572–581 (2006).
183. E. V. Arnold, L. K. Keefer and J. A. Hrabie, *Tet. Lett.*, **41**, 8421–8424 (2000).
184. W. Traube, *Annalen* **300**, 81 (1898).
185. J. Hartung, *Chem. Rev.*, **109**, 4500–4517 (2009).
186. F. Seel, *Chem. Soc. (London) Spec. Publ. (Discussions)*, **10**, 7–20 (1957).
187. J. R. Ohlsen and J. Laane, *J. Am. Chem. Soc.*, **100**, 6948–6955. (1978).
188. G. Lunge and B. E., *Z. Angew. Chem.*, **19**, 881–894 (1906).
189. F. Raschig, *Z. Angew. Chem.*, **18**, 695–206 (1905).
190. W. H. Koppenol and R. Kissner, *Chem. Res. Tox.*, **11**, 87–90 (1998).
191. S. Goldstein, G. Czapski, J. Lind and G. Merenyi, *Chem. Res. Tox.*, **14**, 657–660 (2001).
192. T. Nauser, M. Merkofer, R. Kissner and W. H. Koppenol, *Chem. Res. Tox.*, **14**, 348–350 (2001).
193. N. Arulsamy and D. S. Bohle, *J. Am. Chem. Soc.*, **123**, 10860–10869 (2001).
194. S. R. Fletcher, A. Shortland, A. C. Skapski and G. Wilkinson, *J. Chem. Soc., Chem. Commun.*, 922–3 (1972).
195. H. Iinuma, T. Takeuchi, S. Kondo, *et al.*, *J. Antibiot.*, **25**, 497–500 (1972).
196. M. A. F. Jalal, M. B. Hossain and D. Van der Helm, *Acta Cryst.*, **C42**, 733–738 (1986).
197. T. A. Alston, D. J. T. Porter and H. J. J. Bright, *Biol. Chem.*, **260**, 4069–4074 (1985).
198. T. A. Alston, D. J. T. Porter and H. J. Bright, *Accts. Chem. Res.*, **16**, 418–424 (1983).
199. G. Powis and J. S. Kovach, *Biochem. Pharmacol.*, **30**, 771–776 (1981).
200. Y. K. S. Murthy, J. E. Thiemann, C. Coronelli and P. Sensi, *Nature (London)*, **211**, 1198–1199 (1966).
201. A. Murayama and S. Tamura, *Ag. Bio. Chem.*, **34**, 122–9 (1970).
202. A. Murayama and S. Tamura, *Ag. Bio. Chem.*, **34**, 130–4 (1970).
203. A. Murayama, K. Hata and S. Tamura, *Ag. Bio. Chem.*, **33**, 1599–1605 (1969).
204. S. Tamura, A. Murayama and K. Kagei, *Ag. Bio. Chem.*, **31**, 996–7 (1967).
205. S. Tamura, A. Murayama and K. Hata, *Ag. Bio. Chem.*, **31**, 758–9 (1967).

206. M. Nishio, M. Hasegawa, K. Suzuki, *et al.*, *J. Antibiotics*, **46**, 193–195 (1993).
207. M. Xian and B. J. Shuhler, *Tet. Lett.*, **48**, 1209–1212 (2007).
208. T. Natori, Y. Kataoka, S. Kato, H. Kawai and N. Fusetani, *Tet. Lett.*, **38**, 8349–8350 (1997).
209. D. M. Yost and R. J. Russell, *Systematic Inorganic Chemistry*, Prentice-Hall, New York, 1944.
210. R. Eisenberg and C. D. Meyer, *Acc. Chem. Res.*, **8**, 26–34 (1975).
211. J. A. McCleverty, *Chem. Rev.*, **79**, 53–76 (1979).
212. L. K. Keefer, *Ann. Rev. Pharm. Tox.*, **43**, 585–607, (2003).
213. R. E. Huie, *Toxicology*, **89**, 193–216 (1994).
214. C. C. Addison, *Chem. Rev.*, **80**, 21–39 (1980).
215. Y. Rees and G. H. Williams, *Advances in Free-Radical Chemistry* (Ed. G. H. Williams), Logos Press Ltd, New York, 1968.
216. L. L. Sloss, A.-K. Hjalmarsson, H. N. Soud, *et al.*, *Nitrogen Oxides Control Technology Fact Book*, Noyes Data Corp., Park Ridge, N.J., 1992.
217. D. Menzel and D. M. Meacher, *Reactive Oxygen Species in Biological Systems* (Eds D. L. Gilbert and C. A. Colton), Kluwer Academic, New York, 1999.
218. A. R. Butler and I. L. Megson, *Chem. Rev.*, **102**, 1155–1166 (2002).
219. H. H. Sisler, *Comprehensive Inorganic Chemistry*, Van Nostrand, Princeton, N.J., 1956.
220. T. Moeller, *Inorganic Chemistry*, John Wiley & Sons, Inc., New York, 1952.
221. J. Billingsley and A. B. Callear, *Trans. Faraday Soc.*, **67**, 589 (1971).
222. J. Mason, *J. Chem. Soc.*, 1288–95 (1959).
223. A. W. Shaw and A. J. Vosper, *J. Chem. Soc., Dalton Trans.*, 961–964 (1972).
224. T. C. Hall and F. E. Blacet, *J. Am. Chem. Soc.*, **20**, 1745 (1952).
225. A. H. Nielsen and W. Gordy, *Phys. Rev.*, **56**, 781–784 (1939).
226. R. H. Gillette and E. H. Eyster, *Phys. Rev.*, **56**, 1113–1119 (1939).
227. N. L. Nichols, C. D. Hause and R. H. Noble, *J. Chem. Phys.*, **23**, 57–61 (1955).
228. C. A. Burrus and W. Gordy, *Phys. Rev.*, **92**, 1437–1439 (1953).
229. J. J. Gallagher, F. D. Bedard and C. M. Johnson, *Phys. Rev.*, **93**, 729–733 (1954).
230. C. E. Dinerman and G. E. Ewing, *J. Chem. Phys.*, **53**, 626 (1970).
231. L. R. Maxwell and V. M. Mosley, *J.Chem Phys.*, **8**, 738–742 (1940).
232. S. Claesson, J. Donohue and V. Schomaker, *J. Chem. Phys.*, **16**, 207–210 (1948).
233. G. E. Moore, *J. Opt. Soc. Am*, **43**, 1045–1050 (1953).
234. P. Groth, *Acta Chem. Scand*, **17**, 2419–2422 (1963).
235. J. L. Domenech, A. M. Andrews, S. P. Belov, *et al.*, *J. Chem. Phys.*, **100**, 6993–6999 (1994).
236. H. W. Buschmann, *Zeits. f. Phys.*, **92**, 11–22 (1974).
237. L. E. Borrell and P. Warlop, *J. Phys. Ch.*, **92**, 4377–4383 (1990).
238. M. Cher, *J. Chem. Phys.*, **37**, 2564–2570 (1962).
239. T. Carrington and N. Davidson, *J. Phys. Chem.*, **57**, 418–427 (1953).
240. J. C. Drapier and J. B. Hibbs, *Meth. Enzym.*, **269**, 26–36 (1996).
241. R. Welter, L. Yu and C. A. Yu, *Arch. Biochem. Biophys.*, **331**, 9–16 (1996).
242. T. Jackson, A., E. Yikilmaz, A.-F. Miller and T. Brunold, C., *J. Am. Chem. Soc.*, **125**, 8348–63 (2003).
243. D. E. Coufal, P. Tavares, A. S. Pererira, *et al.*, *Biochem.*, **38**, 4504–453 (1999).
244. K. Pantopoulos, S. Mueller, A. Atzberger, *et al.*, *J. Biol. Chem.*, **272**, 9802–9808 (1997).
245. M. Boese, M. A. Keese, K. Becker, *et al.*, *J. Biol. Chem.*, **272**, 21767–21773 (1997).
246. N. E. Le Brun, S. C. Andrews, G. R. Moore and A. J. Thomson, *Biochem. J.*, **326** (Pt 1), 173–9 (1997).
247. I. Tocheva Elitza, I. Rosell Federico, A. G. Mauk and E. P. Murphy Michael, *Science*, **304**, 867–70 (2004).

5

Nitroxide Radicals: Properties, Synthesis and Applications

Hakim Karoui, François Le Moigne, Olivier Ouari and Paul Tordo
Laboratoire Chimie Provence, Aix-Marseille Université and CNRS, Marseille, France

5.1 Introduction

Since 1990, more than 4000 papers have appeared dealing with the synthesis, the physicochemical properties, and the applications of stable nitroxides and molecules composed of two or more nitroxide moieties. Within the limited pages of this chapter, only some relevant contributions that are not tackled in the other parts of this book are commented on or mentioned. Although the term "stable" nitroxide is used here in the Ingold sense,[1] the review also mentions results from the interesting branch of nitroxide chemistry relating to the "spin trapping" method[2] that involves persistent nitroxides.

The first organic nitroxide, porphyrexide **1** (Scheme 5.1), was prepared and named by Piloty and Schwerin in 1901,[3] and half a century after its radical character was elucidated by Holden *et al.*[4] by means of Electron Paramagnetic Resonance (EPR) spectroscopy. The next important contribution to the development of nitroxide chemistry came from Wieland *et al.*[5] and Meyer *et al.*,[6] who prepared diarylnitroxides as shown in Scheme 5.1. The first synthesis of 2,2,6,6-tetramethyl-4-piperidone-1-oxyl (4-oxo-TEMPO), **2**, by Lebedev *et al.*[7] in 1959 marked the beginning of the development of TEMPO derivatives a widely used class of nitroxides.

Rozantsev *et al.*[8] greatly contributed to this development showing that nitroxides can be involved in various organic reactions without direct involvement of the aminoxyl group. Over the period 1960–1980, the fundamental chemical and physical properties of stable nitroxides were established, largely through the pioneering research of the groups of Rozantsev[9] and Rassat.[10] Since then, the interest raised by nitroxides has continued to grow and now they find applications in many important fields of chemical and biological research, as illustrated by the number of books[11] summarizing the many contributions in this area. Together with the development of nitroxides, the interest of the scientific community in molecular magnetism and organic magnets favored the development of nitroxide multiradicals (mainly dinitroxides) and nitronyl nitroxides[12] (Scheme 5.2).

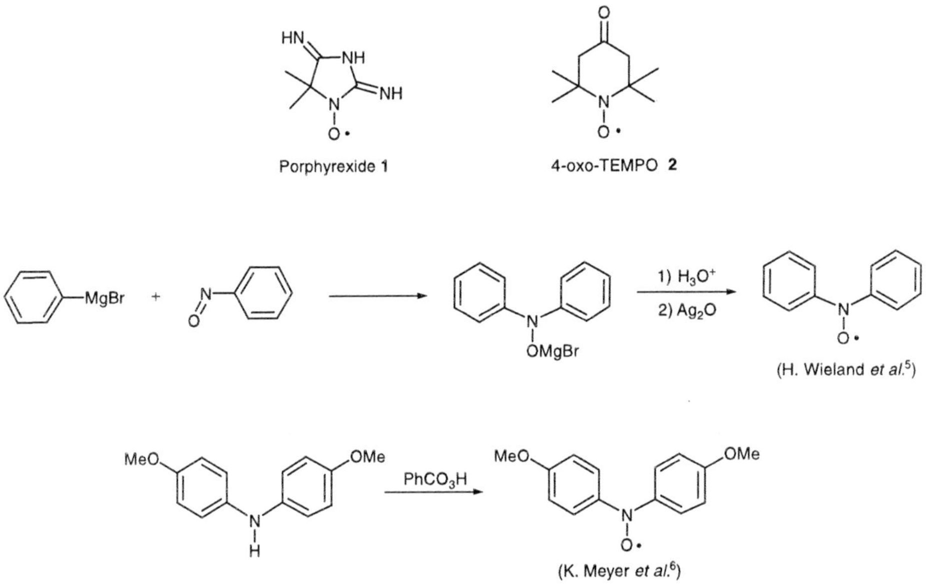

Scheme 5.1

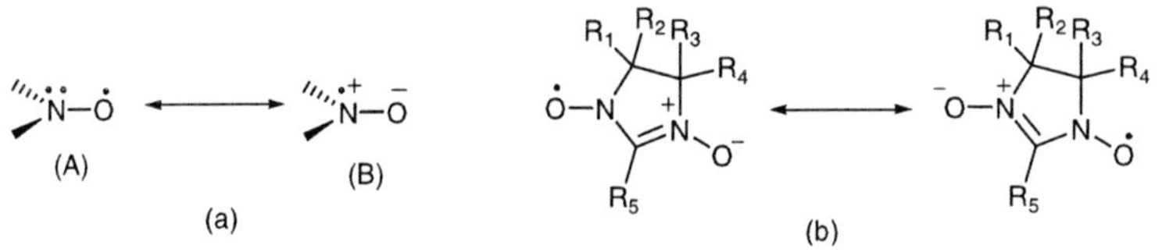

Scheme 5.2 Principal mesomeric structures of a nitroxide (left) and a nitronyl nitroxide (right)

5.2 Nitroxide structure

5.2.1 Characteristics of the aminoxyl group

The aminoxyl group of nitroxides is characterized by a π_{N-O} three-electron bond resulting from the overlap of the $2p_z$ orbitals of the nitrogen and oxygen atoms, and in terms of valence bond theory the contribution of the two main mesomeric structures, (A) and (B), to the actual electronic structure must be considered (Scheme 5.2). As a result of this three-electron N–O π system, the N–O bond of an aminoxyl group has a bond order of 1.5, as indicated by the bond energy of about 100 kcal/mol, and the bond length d_{NO} (1.25 Å < d_{NO} < 1.30 Å), which is midway between the energy and the bond length of a N–OH single bond (53 kcal/mol; ~ 1.43 Å) and a N=O double bond (145 kcal/mol; ~ 1.20 Å)[11d,13].

The spin density is mainly distributed on the nitrogen and oxygen atoms, being slightly higher on the latter. The spin distribution depends on the degree of pyramidalization of the nitrogen atom and on the polarity of the embedding medium.

From O–H bond dissociation energies in hydroxylamines, the gain in energy from the delocalization of the unpaired electron between the nitrogen and oxygen atoms amounts to 30 kcal/mol.[11d,14] A lower value

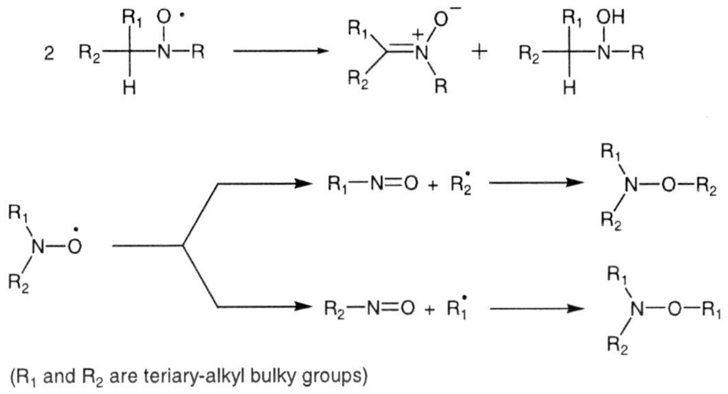

Scheme 5.3 Disproportionation of a nitroxide bearing hydrogen atoms on a carbon of the C–N(O) bonds and fragmentation of bulky nitroxides

(23 kcal/mol) obtained from HeI/HeII photoelectron spectroscopy has been recently proposed.[15] The gain in energy from the formation of an oxygen–oxygen bond has been calculated to be about 35 kcal/mol. Thus, for nitroxides the gain in energy from O–O dimerization is lower than the loss of the resonance energy of the two aminoxyl groups, making O–O dimerization thermodynamically unfavorable.

In general, nitroxides with a hydrogen atom α to the nitrogen atom are not stable and readily undergo disproportionation to form the corresponding hydroxylamine and nitrone (Scheme 5.3).[16] Thus, for most of the stable nitroxides, the hydrogen atoms α to the nitrogen atom have been replaced by alkyl or aryl groups. However, if the steric strain induced by these groups becomes too important, unimolecular decay of the nitroxide can occur through the homolytic cleavage of a C–N(O) bond leading to a nitroso compound and alkyl radical[16d,e] (Scheme 5.3).

5.2.2 X-ray structures of nitroxides

X-ray diffraction studies of some representative cyclic nitroxides, (five-, six- and seven-membered rings) accompanied the development of the nascent nitroxide chemistry. Values of important geometrical parameters are listed in Table 5.1. The N–O bond length is not significantly influenced by the ring size. However, while the aminoxyl group is planar for five-membered ring nitroxides (α close to 0°), it is clearly pyramidal for six- and seven-membered ring nitroxides (Scheme 5.4). In the solid state, nitroxides are nearly always monomeric radicals. However, in a few cases head-to-tail intermolecular interactions of the aminoxyl groups with N. . .N and O. . .O contacts shorter than the sum of the van der Waals radii can occur, resulting in a diamagnetic dimer (nitroxide **3** in Scheme 5.4).[17]

Table 5.1 *Geometrical parameters of cyclic nitroxides*

	N–O (Å)	C–N–C (°)	α (°)[a]
5-membered ring[18]	1.27	112–117	0–5
6-membered ring[19]	1.27–1.31	123–126	15–20
7-membered ring[20]	1.29	130	21

[a]Out-of-plane angle.

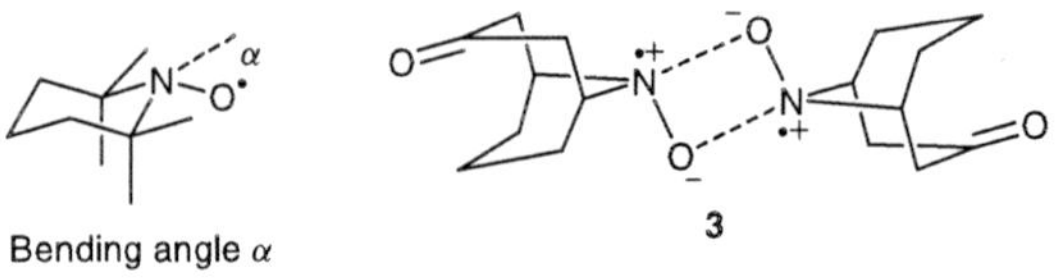

Scheme 5.4

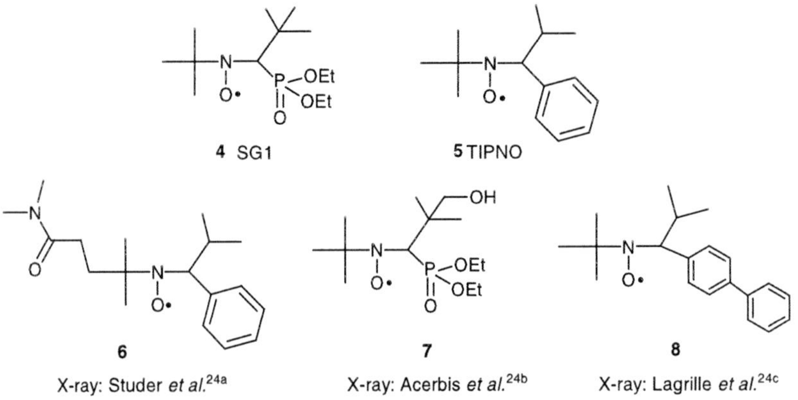

Scheme 5.5

In the last twenty years, the huge development in the use of nitroxides to control radical polymerization processes has led to the preparation of various new nitroxides.[21] Tordo *et al.* first showed that two α-hydrogen-bearing stable nitroxides (SG1 and TIPNO, Scheme 5.5) are particularly efficient for this purpose.[22a] Many stable analogs of SG1[22b] and TIPNO[23] have subsequently been prepared. X-ray structures of some of these[24] (Scheme 5.5, Figure 5.1) reveal that the α-hydrogen atom is located anti to the aminoxyl oxygen atom and close to the nodal plane of the π_{N-O} system, a geometry which is unfavorable to disproportionation, according to the mechanism proposed by Ingold.[16] Braslau *et al.*[25] characterized the compounds resulting from the thermal decomposition of TIPNO and proposed that TIPNO could disproportionate through a Single Electron Transfer (SET) mechanism involving a head-to-tail dimer.

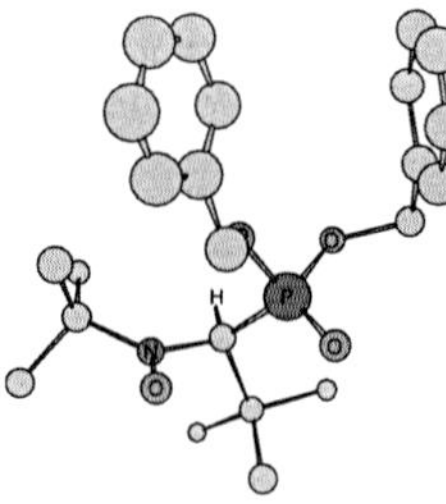

Figure 5.1 *X-ray structure of N-tert-butyl-N-(1-dibenzylphosphono-2,2-dimethylpropyl)nitroxide (only the α-H atom located anti to the aminoxyl O atom is shown).*[24d]

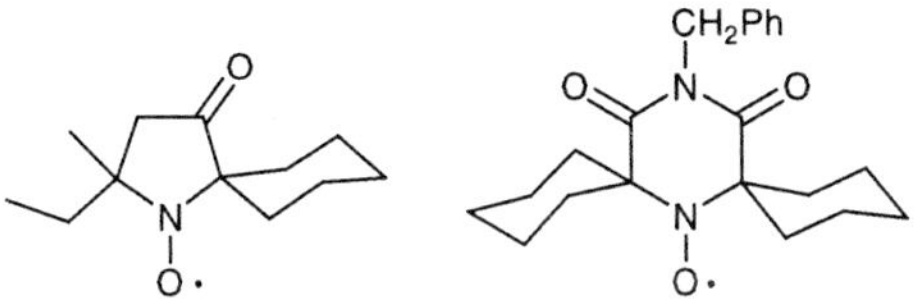

9 X-ray: Aldabbagh[26] **10** X-ray: Aldabbagh[26]

Scheme 5.6

Aldabagh *et al.*[26] used nitroxides **9** and **10** (Scheme 5.6) to control the free radical polymerization of styrene and discussed the results in terms of structural parameters. According to X-ray crystal structures, the six-membered ring of **10** deviates significantly from planarity as compared to the planar five-membered ring of **9**. Thus, **10** possesses a more exposed oxyl group leading to a higher rate of radical trapping (k_c) and a lower rate of NO–C bond homolytic cleavage in the corresponding alkoxyamine (Section 5.6.1, Scheme 5.31).

Nitronyl nitroxides, imino-nitroxides, and, to a lesser extent, nitroxides and dinitroxides are frequently used as stable spin-building blocks to prepare magnetic materials.[27] The X-ray structure and magnetic behavior of numerous crystalline materials formed with these building blocks have been reported.[28] Discussing the results of these numerous studies is beyond the scope of this chapter and only a few reports describing the X-ray structure of non-coordinated new nitroxides or dinitroxides (Scheme 5.7) are mentioned.

All-organic liquid crystals containing a chiral five-membered cyclic nitroxide unit within the rigid core (Scheme 5.8) have been prepared and their X-ray structures and magnetic properties have been studied.[29]

The X-ray structures of other nitroxides used in miscellaneous applications have also been determined; some[30] are shown in Scheme 5.9.

5.2.3 Quantum mechanical (QM), molecular dynamics (MD) and molecular mechanics (MM) calculations

The EPR parameters of a nitroxide are strongly dependent on its geometry and the nature of the embedding environment, particularly the polarity and the presence or absence of molecular oxygen dissolved in the solution. Nitroxides are widely used as reporter molecules to gain different kinds of information: motion of biomolecules (spin labeling), oxygen content of a medium (oximetry), pH values, formation of inclusion complexes,[31] and so on. As a result, it is important to fully understand how the molecular geometry and the solvent influence the magnitude of the hyperfine coupling constants (hccs) of a nitroxide. Kikuchi *et al.*[32] explored the conformational space of small nitroxides by means of Monte Carlo (MC) simulations, the hccs being then obtained by averaging results of quantum calculations on MC configurations. Barone *et al.*[33] developed various approaches to study the magnetic properties of large nitroxide systems *in vacuo* and in condensed phases. They showed that within the density functional theory (DFT) approach,[33e] the popular B3LYP hybrid functional coupled with the standard basis 6–31+G(d,p) was appropriate to reproduce the experimental geometries and hccs of nitroxides accurately. However, they also developed some special basis sets, like the EPRII, EPRIII. and N07D.[33g]

With respect to environmental effects on the magnitude of hccs, a supermolecule composed of the solute and the solvent molecules strongly coordinated to the aminoxyl group must be considered. Then, adding bulk solvent effects in a suitable way, for instance, by the Polarizable Continuum Model (PCM),

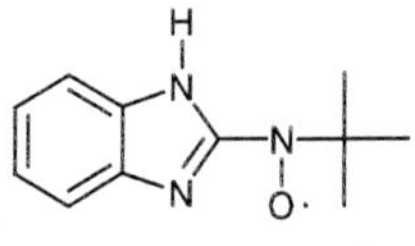

11, X-ray: Lahti *et al.*[28a]

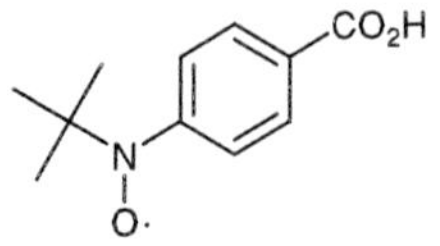

12, X-ray: Veciana *et al.*[28b]

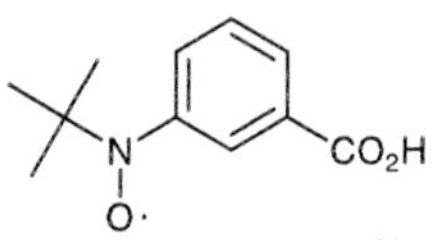

13, X-ray: Lahti *et al.*[28c]

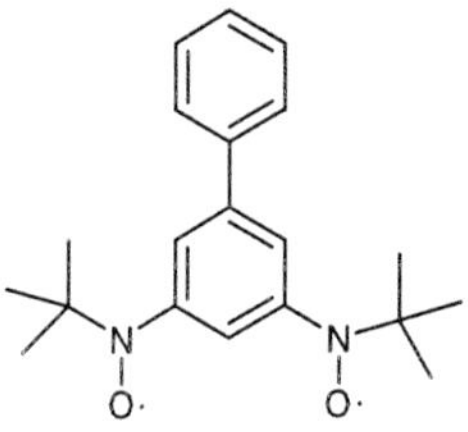

14, X-ray: Ishida *et al.*[28d]

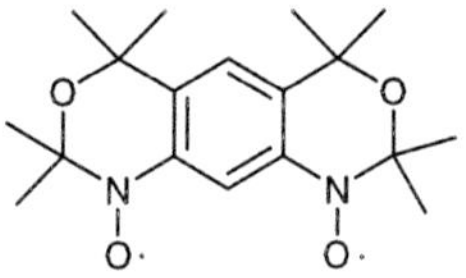

15, X-ray: Rajca *et al.*[28e]

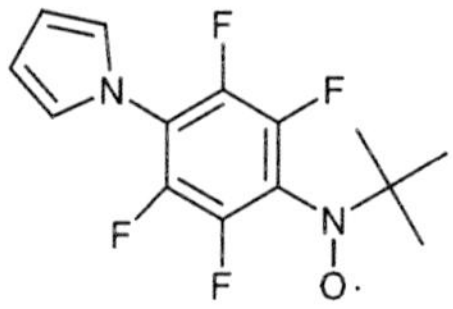

16, X-ray: Lahti *et al.*[28f]

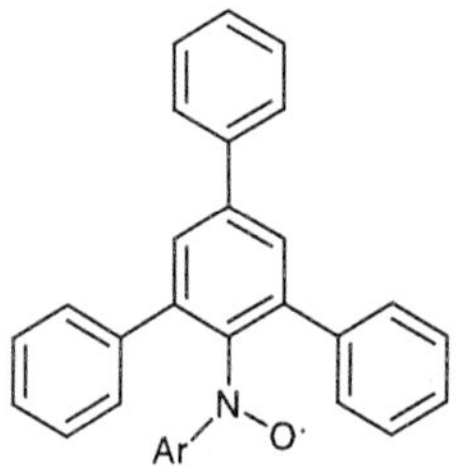

17, **18** Ar = diphenyl; 4-Br-phenyl
X-ray: Okada *et al.*[28g]

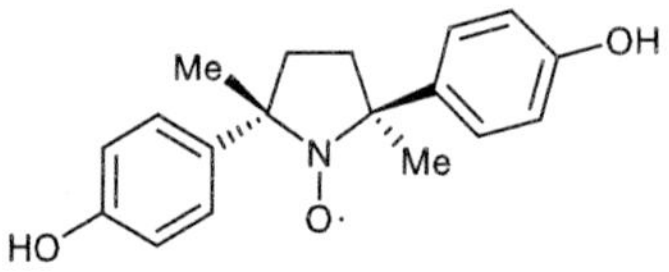

19, X-ray: Tamura *et al.*[28h]

Scheme 5.7

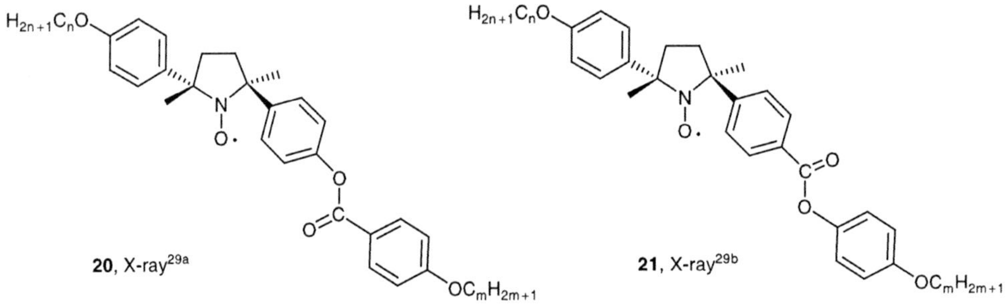

20, X-ray[29a]

21, X-ray[29b]

Scheme 5.8

can allow good agreement to be obtained between calculated and experimental hccs values. However, the number and the position of solvent molecules in the first solvation shell of any solute are intrinsically dynamic. In this context, Barone *et al.*[33f] carried out molecular dynamics (MD) simulations of nitroxides *in vacuo* and in aqueous solution within the Car–Parrinello (CP) framework. The hccs values were then computed at a DFT level and were shown to fit accurately with the experimental ones. This approach, although presently among the "state-of-the-art" molecular simulation techniques, is particularly

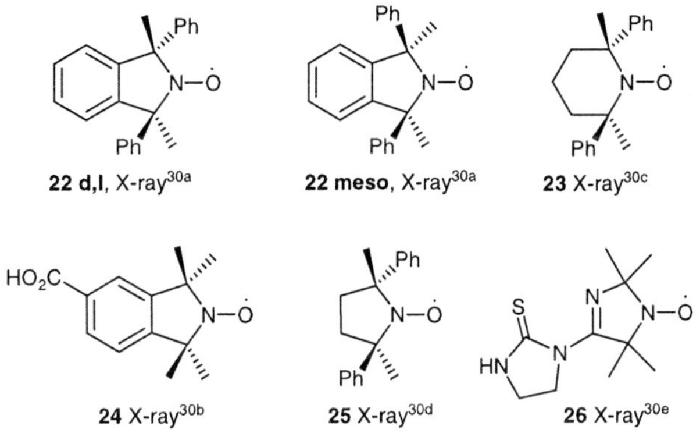

Scheme 5.9

resource demanding, and its application to large systems can be very tedious. Ferré *et al.*[34] developed an approach combining the analysis of MD trajectories generated by use of a classical polarizable force field, and quantum calculations coupled to the solvent instantaneous electrostatic potential, to compute averaged hyperfine coupling constants. This approach can be used readily to compute hccs of large systems.

The existing literature on the vibrational spectra of stable nitroxides has been critically reviewed and *ab initio* DFT calculations using UB3LYP at the 6–311++G(d,p) level were performed to obtain a theoretical band position of $\nu_{N-O^\bullet}$ (in the range 1450–1420 cm^{-1} for five-membered cyclic nitroxides and 1395–1340 cm^{-1} for six-membered cyclic and acyclic nitroxides).[35]

The nitroxide spin label MTSSL (Scheme 5.10), widely used in site-directed spin labeling of membrane proteins, has been studied with MD simulations.[36] MTSSL attached to a polyalanine α helix in explicit solvent was simulated to elucidate the factors affecting its conformational dynamics.

A simple MM approach based on the Cremer and Pople description of five-membered cycles has been applied to the conformational analysis of five-membered cyclic nitroxides.[37] The method has been implemented in the GenMol[38] software, and it was applied to determine the major conformers of β-phosphorylated nitroxides.

Dramatic line broadening resulting from chemical exchange was observed in the EPR spectra of either cyclic[39] (**27**, **28**, Scheme 5.10) or acyclic[40] β-diphosphorylated nitroxides. For the major conformer of the *trans* isomer **27t**, the two phosphoryl groups are axial and there is no exchange with the diequatorial conformer. However, for the *cis* isomer a fast exchange occurs between the axial–equatorial and equatorial–axial conformers, and the large difference between a_P(ax) and a_P(eq) ($\Delta a_P \approx 20$ Gauss) explains the large broadening effect observed. Calculations of the EPR spectra at different temperatures yielded the potential barriers for ring interconversion.

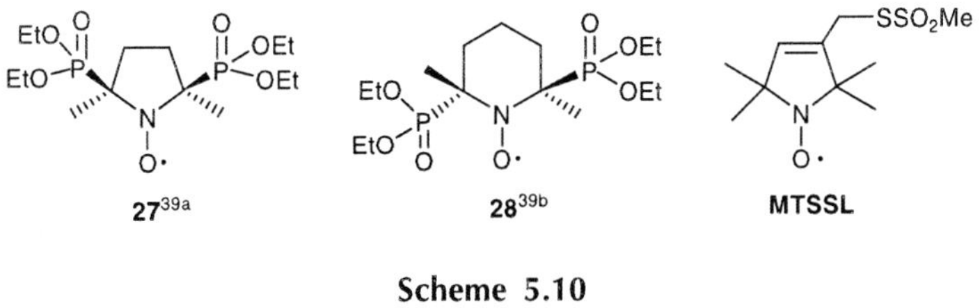

Scheme 5.10

5.2.4 Influence of solvent polarity on the EPR parameters of nitroxides

For many applications the usefulness of nitroxides as paramagnetic probes stems from the sensitivity of their EPR parameters to the polarity of the embedding medium. For example, the changes in a_N, $a_{O(17)}$ and g values for 4-oxo-TEMPO (Scheme 5.1) in solvents of increasing polarity are shown in Table 5.2.[41] The trend of a_N and $a_{O(17)}$ can be explained in terms of relative weights of the two most important resonance structures of the aminoxyl group; a_N increases and $a_{O(17)}$ decreases due to the stabilization of the resonance structure (B) with respect to (A) (Scheme 5.2). More generally, polar solvents increase a_N through the relative stabilization of the zwitterionic resonance structure (B).

Barone *et al.*[32d] investigated solvent polarity effects on a_N for 2,2,5,5-tetramethyl-3-carboxy-pyrrolidinoxyl **29**, and 2,2,6,6-tetramethyl-4-carboxy-piperidinoxyl **30** (Scheme 5.11). Part of their results are listed in Table 5.3. From these results, the authors concluded that the trends relative to aprotic and protic solvents are different, suggesting that not only the polarity of the solvent but also the possibility of explicit hydrogen bonds influence the value of a_N. With solvents that cannot form hydrogen bonds, inclusion of the bulk solvent effect by means of single-point PCM calculations on the gas phase geometries is sufficient to provide a_N values in good agreement with the experiments. However, with hydrogen bond forming solvents, the authors considered the formation of supermolecules containing either one (chloroform, methanol) or two (water) solvent molecules coordinated to the aminoxyl group. The use of these supermolecules for the calculations restore the agreement between experimental and calculated a_N values.

According to the Stone theory[42] of the g factor in organic free radicals, the most important contribution to Δg for nitroxides comes from an electronic excitation from an in-plane oxygen lone pair n_O to the π^* SOMO. When the lone pairs are stabilized by polar solvents the $n_O \rightarrow \pi^*$ gap increases with the consequent reduction of g value.

An experimental and computational study of the solvent dependence of the hyperfine couplings for the di-*tert*-butyl nitroxide (DTBN) has been carried out by Mattar *et al.*[43] The DTBN a^{iso} values were directly calculated in six different solvents, using the unrestricted UB1LYP HDF and UAHF methods. The trend of increasing a_N in going from non-polar to polar solvents is very well reproduced by the computations. The most difficult case was reproducing the a_N value in water. Nevertheless, the a_N values of DTBN-1H_2O supermolecule come within 0.38 G of the experimental value.

Table 5.2 *g-Factors and hccs of 4-oxo-TEMPO in solvents of increasing polarity*

Solvents	g-Factor	a_N (Gauss)	$a_{O(17)}$ (Gauss)
Benzene	2.0062	14.45	*19.29*
1-Butanol	2.0060	15.01	*19.11*
Water	*2.0058*	*16.01*	*17.86*

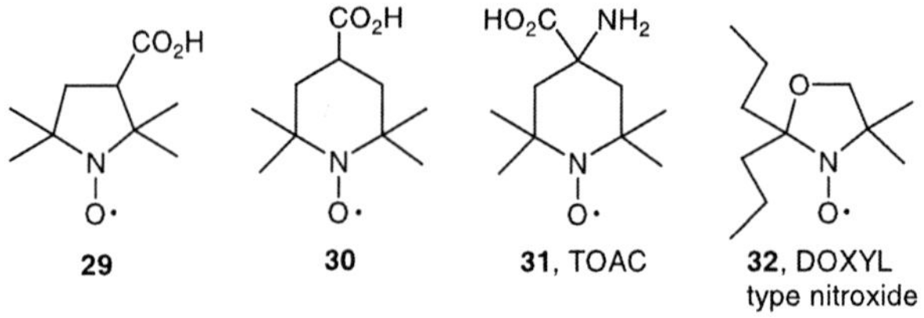

Scheme 5.11

Table 5.3 *Dielectric constant (ε) for different solvents and corresponding experimental and computed values of A_N (Gauss) for* ***29*** *and* ***30***

Solvent	**29**[c]		**30**[c]		ε
	Exp.	Calc.	Exp.	Calc.	
CCl_4	14.081	13.80	15.450	15.09	2.24
$CHCl_3$	14.697	14.05	15.947	15.31	4.90
		14.61[a]		15.71[a]	
		15.25[b]			
THF	14.252	14.14	15.528	15.40	7.58
CH_2Cl_2	14.363	14.17	15.749	15.42	8.93
CH_3COCH_3	14.287	14.27	15.601	15.52	20.70
CH_3OH	15.198	14.30	16.252	15.54	32.60
		15.29[a]		16.17[a]	
Ph-NO_2	14.471	14.30	15.731	15.54	34.85
H_2O	16.050	14.32	17.060	15.54	78.40
		15.25[a]		15.54[a]	
		16.12[b]		16.74[b]	

[a]Calculations including one explicit solvent molecule.
[b]Calculations including two explicit solvent molecules.
[c]Gas phase results: 13.35 G (**29**), 14.80 G (**30**).

Polarity contributions to a_N values of nitroxides have also been calculated[44] within the Onsager model. Different reaction fields have been proposed[45] as improvements of the original model and it has been shown[46] that polarizability of the nitroxide must be taken into account to obtain the best description of the solvent dependence of the isotropic ^{14}N-hyperfine couplings.

Marsh *et al.*[47] determined the polarity dependence of the isotropic ^{14}N-hyperfine couplings, a_N, and g-values in a wide range of protic and aprotic media, for a TOAC-containing dipeptide **31** and for a DOXYL-containing fatty acid **32** (Scheme 5.11). The correlation between data sets for TOAC and DOXYL nitroxides in various solvents was used to establish the polarity profile for isotropic hyperfine couplings of TOAC in a transmembrane peptide. This calibration can then be used to determine the location of TOAC at selected residue positions in a transmembrane or surface active peptide. A similar calibration procedure was also applied to a_N and g for MTSSL (Scheme 5.10) that is widely used in site-directed spin labeling studies of membrane proteins.

Nitroxide molecules are excellent probes of intermolecular hydrogen bonding interactions in polar solvents[48]; they interact with weak protic solvent molecules providing insight into the short-lived complexes formed. The distance and magnitude of the scalar interaction for the nitroxide hydrogen bonding at the methyl group in acetonitrile and the amino group in acetamide were computed. Scalar interactions were accurately modeled at the UB3LYP level in terms of the magnitude and attitude for substrate/nitroxide complexes.

5.3 Nitroxide multiradicals

Organic di- and polyradicals are especially relevant sensors to study weak interatomic/intermolecular interactions in large systems. Their spectroscopic and magnetic properties[49] depend on the electron spin–spin exchange coupling between unpaired electrons localized on different centers. The synthesis and handling of stable nitroxide multiradicals are relatively well controlled and nowadays these species are used in many applications.

5.3.1 Electron spin–spin exchange coupling

Nitroxide multiradicals, particularly dinitroxides, have been extensively used to study the dependence of exchange coupling (J) upon structural parameters. A few representative examples of the many recently studied species are shown in Scheme 5.12. The diazaadamantane dinitroxide **33** currently holds the transition temperature record ($T_c = 1.48$ K) for a fully organic nitroxide-based ferromagnet.[50] As expected based on symmetry considerations, the intramolecular interaction between the two orthogonal N–O groups is ferromagnetic. Intermolecular interactions are also ferromagnetic and a truly three-dimensional ferromagnetic order is observed at a Curie temperature of 1.48 K.

A series of six trimethylenemethane-type dinitroxides which differ in (*tert*-butylaminoxyl)phenyl ring torsion angles by virtue of different steric demands of the spacer groups was prepared.[51a] Electron spin–spin exchange values were obtained from solid state magnetic susceptibility measurements.[51b] The

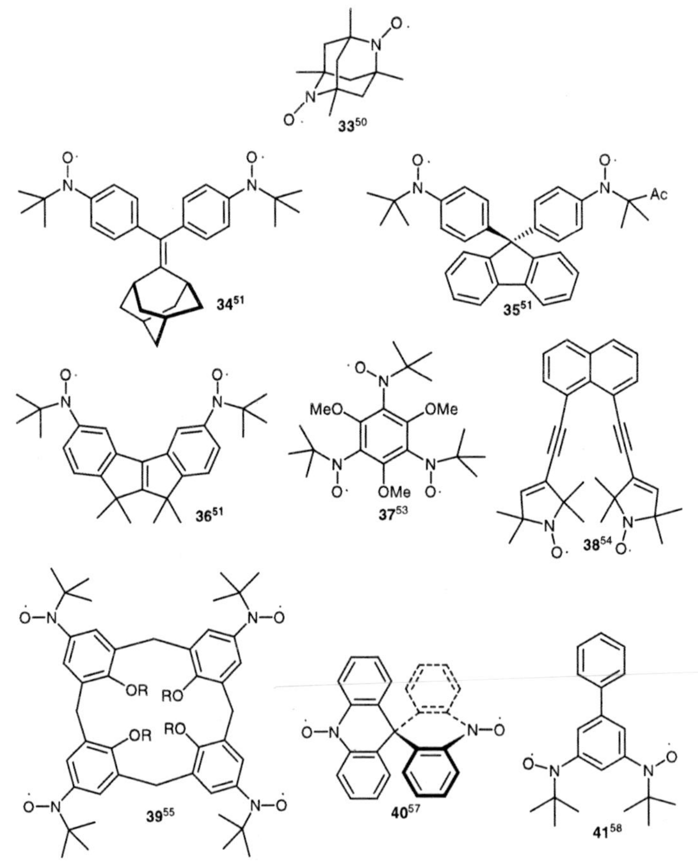

Scheme 5.12

exchange parameter is antiferromagnetic for **34** ($J = -7.30 \pm 0.34$ K), which has an average phenyl torsion $\phi = 54.1 \pm 0.6°$, while the strongest net ferromagnetic coupling ($J = +36.4 \pm 2.3$ K) is observed in **36**, in which the coupler is part of an essentially planar π system. The weakest net intramolecular exchange coupling is for **35**, which lacks a conjugated coupling unit. Using the Broken Symmetry (BS) approach, Datta *et al.*[52] performed calculations of J (at the UB3LYP/6-311+G(d,p) level) for several trimethylenemethane-based nitroxide diradicals including **36**. They showed that increasing both conjugation and planarity results in significant increase of the intramolecular ferromagnetic exchange interaction.

Triradical **37** has been prepared[53] and its X-ray crystallographic analysis revealed an all-*syn* conformation of the three aminoxyl groups, with large dihedral angles (90° and 70°) between the N–O bonds and the benzene ring plane. In the EPR spectrum of **37** in frozen solution at 6.5 K a fine structure characteristic of a randomly orientated quartet species was observed. Temperature dependence of the molar susceptibility of **37** on a microcrystalline sample suggested that a global antiferromagnetic interaction produces a doublet ground state.

Dinitroxide **38** was prepared by Shiemann *et al.*[54] from 3-ethynyl-2,2,5,5-tetramethyl-3-pyrroline-1-oxyl and 1,8-di-iodonaphtalene via a Sonogashira cross-coupling reaction. According to X-ray crystallographic analysis, the through-space distance between the oxygen atoms of the two aminoxyl groups is 8.9 Å and the distance between the nitrogen atoms is 7.2 Å. The CW X-band isotropic EPR spectrum of **38** clearly shows the five lines due to two strongly exchange-coupled nitroxide moieties. The plot of the doubly-integrated half-field signal intensity against temperature between 4.1 and 90 K leads to $2J = -4.1$ K, corresponding to a singlet ground state. From the temperature dependence of the magnetic susceptibility studied with a SQUID, one obtains $2J = -3.54$ K. UB3LYP/6-311G(df,pd) Broken Symmetry calculations revealed a strong ferromagnetic through-bond interaction via the 1,8-substituted naphthalene bridge and a competing strong antiferromagnetic through-space interaction via the acetylene groups. Both interactions are of the same order of magnitude, leading to the weak overall antiferromagnetic coupling observed.

Rajca *et al.*[55] described the synthesis, crystallography, and magnetic characterization of **39**, a stable macrocyclic tetranitroxide composed of a calix[4]arene which is functionalized with four *tert*-butylnitroxides at the upper rim. In the solid state, dimerization of one diagonal pair of nitroxides leads to a pinched cone conformation for **39** with strong intradimer antiferromagnetic coupling (singlet–triplet energy gap, $\Delta E_{ST} \approx 1$ kcal/mol).

Several calix[4]arenes with two opposite nitroxide moieties on the upper rims were prepared, and the influence of different factors on the through-space spin–spin exchange interactions in solution was investigated by EPR.[56]

Ishida *et al.* obtained dinitroxide **40** by oxidation of 9,9'(10*H*, 10'*H*)-spiroacridine with *m*-chloroperbenzoic acid, and determined its structure by X-ray crystallographic analysis.[57] Magnetic measurements on a polycrystalline sample of **40** revealed the presence of a dominant antiferromagnetic ($J \approx -69.9$ K) interaction, which was ascribed mainly to intermolecular interactions due to close contacts between two aminoxyl groups. However, the temperature dependence of the EPR peak heights approximately obeyed the Curie law, suggesting a triplet ground state for a free **40** molecule. UB3LYP DFT BS calculations supported this result with a calculated J value of +12.0 K. In the solid state, the ferromagnetic intramolecular spin–spin coupling is hidden by the stronger antiferromagnetic intermolecular coupling.

Remarkably strong intermolecular antiferromagnetic couplings were also observed by Ishida *et al.*[28d] in the usual crystallographic allotrope of biphenyl-3,5-diyl bis(*tert*-butyl nitroxide) **41** (α-**41**). X-ray crystallography revealed strong head-to-tail intermolecular dimeric interactions of the aminoxyl groups, with N...N and O...O contacts shorter than the sum of the van der Waals radii. These interactions can be considered as partly bonding, however the contribution of mesomeric form C (Scheme 5.13) is very small, since **41** has still a radical character in the solid state. As a result of these intermolecular interactions

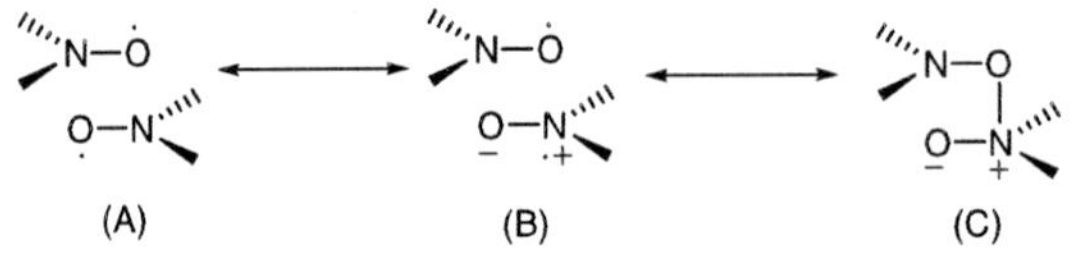

Scheme 5.13

the N–O nitrogen atoms are significantly pyramidalized and the N–O bonds are elongated. More recently Ishida *et al.*[58] succeeded to obtain a new crystallographic allotrope of **41** (β-**41**). X-ray data show that for β-**41** the intermolecular N–O...N–O distances are very long, and, as expected, magnetic susceptibility measurements revealed strong intramolecular ferromagnetic coupling.

5.3.2 Miscellaneous aspects of di- and polynitroxides

EPR studies have been devoted to numerous dinitroxides[59]; some of the studied species are shown in Scheme 5.14.

The impact of electron–electron spin interaction on electron spin relaxation of a series of nitroxide diradicals and a nitroxide tetraradical in glassy solvents between 10 and 300 K has been studied.[60] Relaxivity studies on dinitroxides and dendritic nitroxide multiradicals (**46**, Scheme 5.15) have been performed.[61] Dendritic polynitroxides exhibit higher per-nitroxide-based water relaxivity compared to a mononitroxide model. The higher generation polynitroxide dendrimers exhibit relaxivities exceeding those for gadolinium(III) chloride.

Prefluorescent nitroxides possess a fluorophore tethered by a short covalent link to a nitroxide moiety. The fluorophores exhibits low fluorescence due to enhanced intersystem crossing from the first excited singlet state of the fluorophore to its triplet state via electron exchange interactions with the nitroxide moiety. Upon radical trapping to form a diamagnetic alkoxyamine normal fluorophore emission is restored, and thus these prefluorescent nitroxides are extremely responsive sensors for the detection of free radical species. Recently Bottle *et al.*[62] prepared a new prefluorescent dinitroxide based on a 9,10 diphenylanthracene core

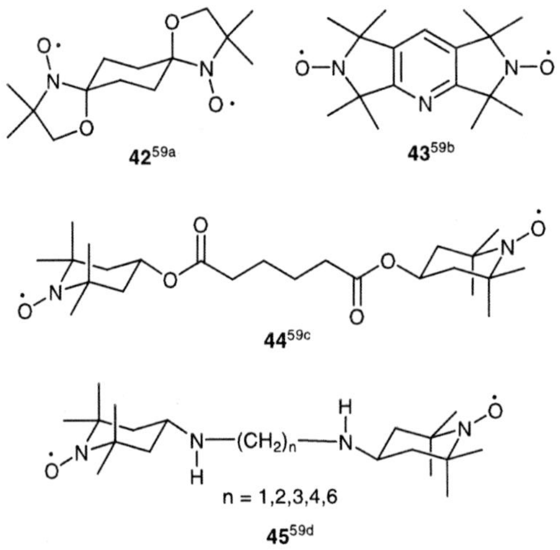

Scheme 5.14

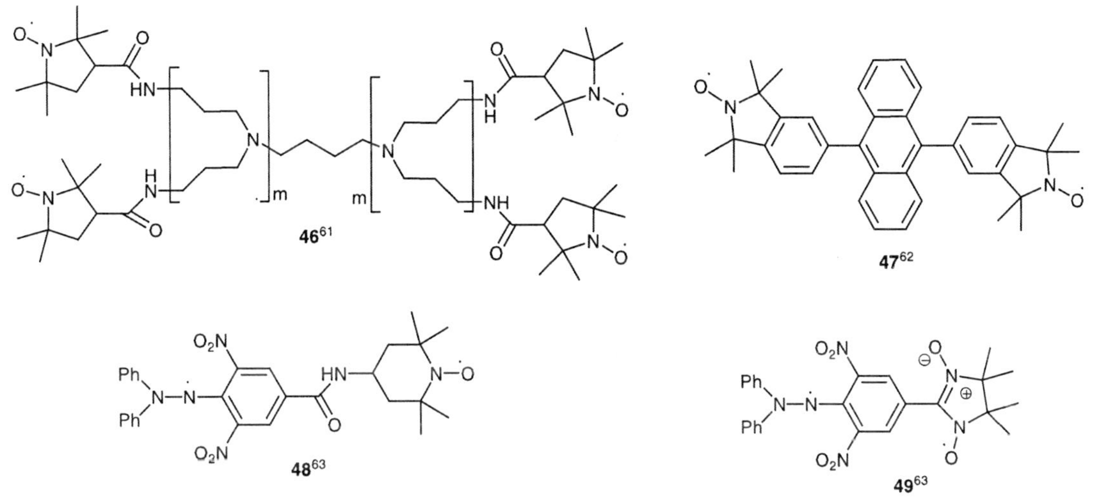

Scheme 5.15

(**47**, Scheme 5.15). Using **47**, the authors were able to image by fluorescence spectroscopy free radical damage in polypropylene.

The synthesis and characterization of a series of hetero-diradicals containing linked hydrazyl and nitroxide moieties have been reported (**48**, **49**, Scheme 5.15) and their EPR spectra have been discussed.[63]

5.4 Nitronyl nitroxides (NNOs)

Nitronyl nitroxides (**NNOs**) are stable paramagnetic species bearing interacting nitronyl and aminoxyl groups. The unpaired electron is equally delocalized on both N–O groups as demonstrated by theoretical calculations.[64] **NNOs** are characterized by a five line EPR signal with an intensity ratio 1 : 2 : 3 : 2 : 1 due to the coupling of the single electron with two equivalent nitrogen atoms (Figure 5.2); the nitrogen coupling

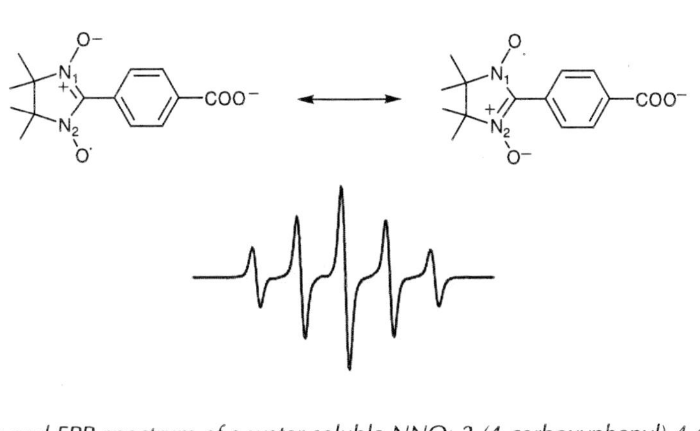

Figure 5.2 *Structure and EPR spectrum of a water soluble NNO: 2-(4-carboxyphenyl)-4,4,5,5-tetramethyl-4,5-dihydro-1H-imidazole-3-oxide-1-oxyl (**2-carboxyNNO**).*

constant is about half the value of a dialkyl nitroxide such as TEMPO. **NNOs** are mainly used for building magnetic materials and for trapping nitric oxide (to form imino-nitroxide, **INO**), a ubiquitous molecule involved in many biological processes.

5.4.1 Synthesis of nitronyl nitroxides

Generally, nitronyl nitroxide radicals are synthesized following a procedure based on Ullman's work (Scheme 5.16).[12] Usually, a precursor aldehyde is condensed with a bis-hydroxylamine in methanol. The resulting bis-hydroxyimidazolidine is then converted to the nitronyl nitroxide radical by oxidation using sodium periodate ($NaIO_4$) in either a two-phase mixture of water/dichloromethane or manganese dioxide in dichloromethane. Some new NNOs recently synthesized are reported in Scheme 5.16.[65–69]

In 2007, Dooley *et al.*[70] reported a new synthetic route leading to various substituted spin delocalized benzonitronyl nitroxides (Scheme 5.17).

5.4.2 Nitronyl nitroxide as a nitric oxide trap

Furchgott, Ignaro and Murad received the Nobel Prize in Medicine in 1998 for their "discoveries concerning nitric oxide as a signaling molecule in the cardiovascular system".* Apart from being responsible for the vascular smooth muscle vasorelaxation,[71] nitric oxide has other important physiological roles, such as a cytotoxic mediator of the immune system[72] or as a neurotransmitter in the central nervous system.[73] Nitric oxide is stable in oxygen-free solutions but it reacts with superoxide radical to form peroxynitrite

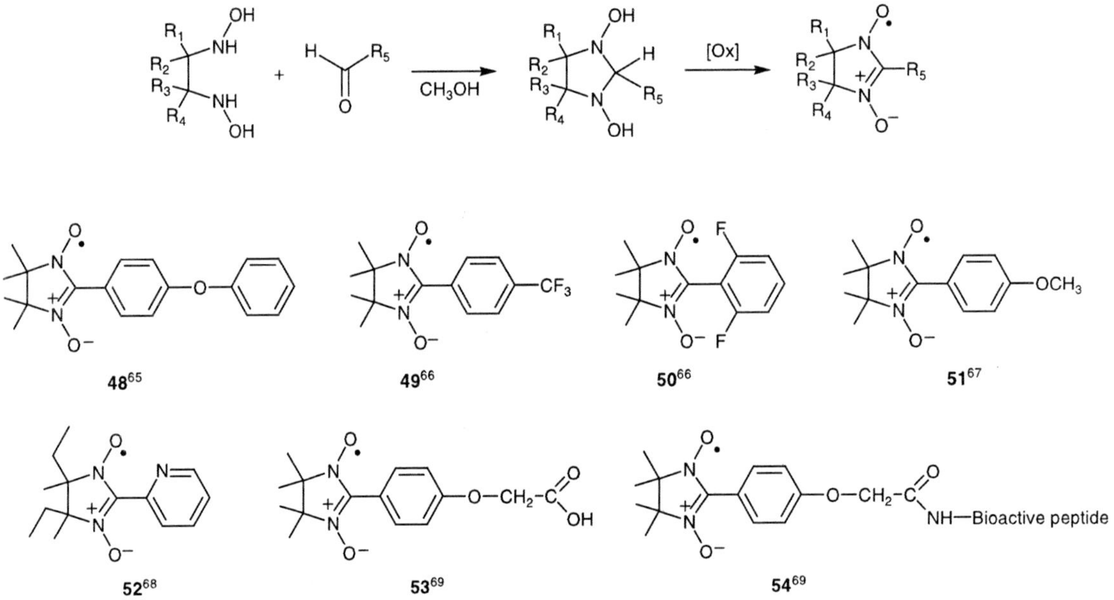

Scheme 5.16 Synthetic pathway for nitronyl nitroxide preparation

*12 October 1998. The Nobel Assembly at Karolinska Institutet has decided to award the Nobel Prize in Physiology or Medicine for 1998 jointly to Robert F. Furchgott, Louis J. Ignarro and Ferid Murad for their discoveries concerning "nitric oxide as a signalling molecule in the cardiovascular system".

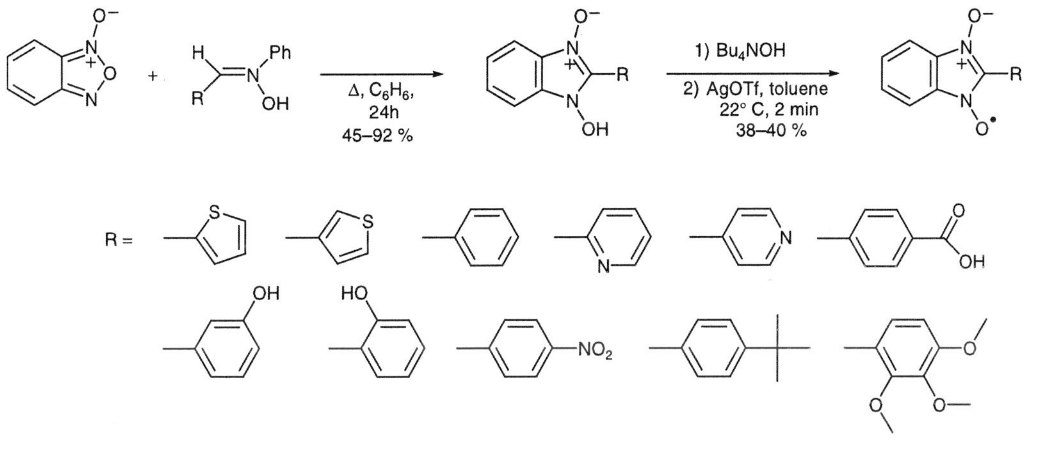

Scheme 5.17 Synthetic pathway for benzonitronyl nitroxide preparation

($ONOO^-$) at a nearly diffusion controlled rate.[74] Peroxynitrite is a cytotoxic species implicated in several pathophysiological conditions like atherosclerosis[75] and neurodegenerative diseases.[76]

To characterize nitric oxide production in living systems, usual EPR spin traps have been evaluated. Nitrone and nitroso compounds are not suitable because of the instability of the resulting spin adduct.[77] Other methods, such as trapping using cheletropic traps or exogenous iron chelates, were successfully developed. Here, after a brief discussion of these two approaches, the approach using NNOs as a nitric oxide scavenger is introduced.

The concept of a nitric oxide cheletropic trap (NOCT) is based on the addition reaction of nitric oxide to form a stable indolinoxyl type nitroxide (Scheme 5.18) with a simple and characteristic three line EPR spectrum. This approach was developed in the 1990s by Korth, Sustmann and Ingold.[78] However, the poor solubility and thermal sensitivity of the NOCT compounds, as well as the bioinstability of the resulting nitroxides, limited their use to monitor nitric oxide production in cellular systems.

Nitric oxide can bind very easily with Fe^{2+} chelates, and the diethyldithiocarbamate ferrous complex [$(DETC)_2Fe^{2+}$] (**55**, Scheme 5.19) is commonly used to trap nitric oxide produced in hydrophobic conditions.[79] The resulting $(DETC)_2$-Fe^{2+}@NO complex is detected as a three line EPR spectrum. Water soluble Fe^{2+}-dithiocarbamate complexes (*N*-methyl-D-glucamine (MGD) **56**, *N*-(dithiocarboxy) sarcosine **57**, and 2-hydroxyethyl dithiocarbamate **58**) were developed and successfully used to obtain evidence of real time nitric oxide production in septic shock-mice[80] or from purified neuronal Nitric Oxide Synthase (*n*NOS).[81] However, the trapping of nitric oxide by $(MGD)_2.Fe^{2+}$ is not selective; Mason showed that nitrite, an oxidation product of nitric oxide, can react with $(MGD)_2.Fe^{2+}$ to produce nitric oxide.[82]

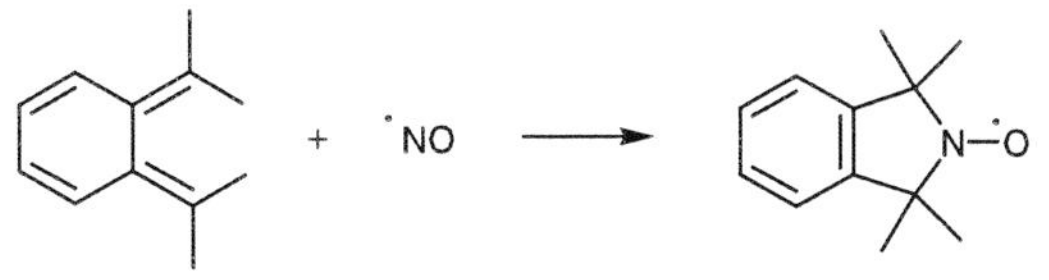

NOCT

Scheme 5.18 Trapping of NO with a NOCT

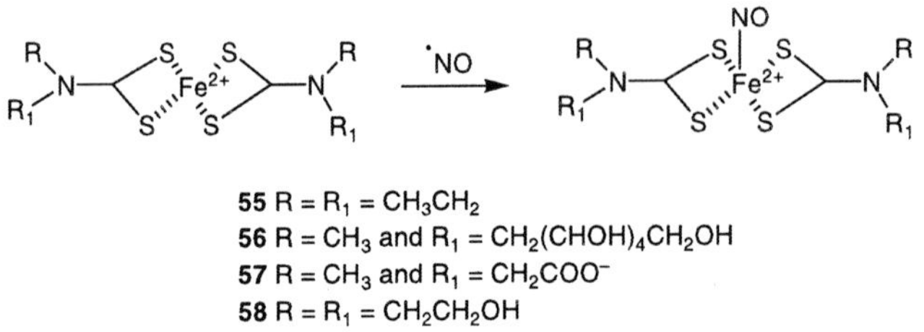

Scheme 5.19 Structure of Fe^{2+}-dithiocarbamate complexes

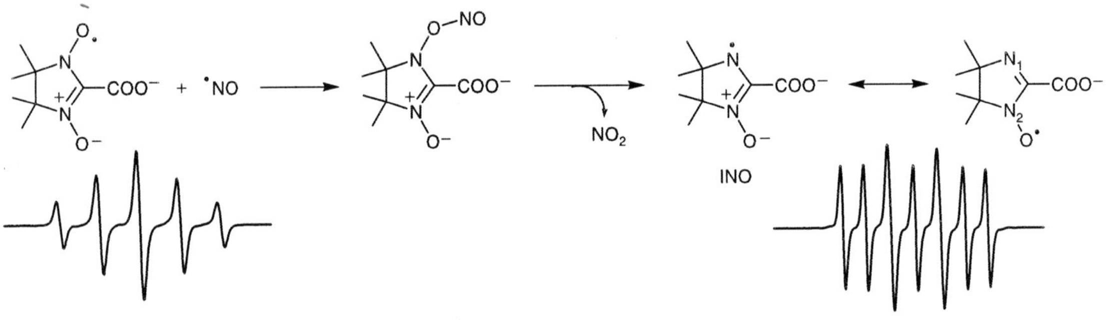

Figure 5.3 Trapping of nitric oxide with NNO and EPR signal of the resulting iminyl nitroxide (INO).

Nevertheless, dithiocarbamate complexes are widely used, even though the high quantities of added Fe^{2+} and dithiocarbamate ligands can initiate unwanted reactions and a high toxicity.

Kalyanaraman *et al*. were the first to suggest that nitronyl nitroxides could be a viable alternative to iron(II)-dithiocarbamate complexes[83] to characterize nitric oxide. They showed that 2-carboxyNNO reacts specifically with nitric oxide giving rise to an imino nitroxide (INO) which shows a characteristic EPR signal[83] composed of 7 lines with an intensity close to 1 : 1 : 2 : 1 : 2 : 1 : 1 ($a_{N1} = 4.5$ and $a_{N2} = 9.0$ G (Figure 5.3)). The rate constant of the reaction between NNOs and NO in aqueous solutions is in the region of $10^4\ M^{-1}s^{-1}$.[84]

As shown by Peng *et al*., who reported the synthesis of 30 different NNO labeled with amino acid fragments,[85] the possibility to modulate NNO solubility and specificity makes possible the detection of nitric oxide at different tissue sites. To improve nitric oxide rate trapping, Rosen *et al*. reported the first synthesis of dendrimer linked NNOs (from two to eight units of NNO).[86] Unfortunately, the EPR spectrum of the dendrimer-linked NNOs is broadened by spin exchange interactions, and the rate constant of trapping is similar to that observed with nitronyl nitroxides.

Despite the easy access to various NNOs, their use for specific EPR nitric oxide detection is not without limitation; it has been reported that NNOs can undergo fast reduction into EPR silent diamagnetic products. Indeed, Blasig *et al*. reported that NNOs can react with superoxide anion radical ($O_2^{\bullet -}$) with a rate constant of $8.8 \times 10^5\ M^{-1}s^{-1}$, which is more than two orders of magnitude higher than the value reported previously for the NNO's reaction with nitric oxide.[87]

5.4.3 Nitronyl nitroxides as building blocks for magnetic materials

Since the pioneering work of Cambi and coworkers on spin-crossover magnetic materials in the 1930s,[88] it has been a holy grail for chemists and physicists to create pure organic magnetic materials. In 1991,

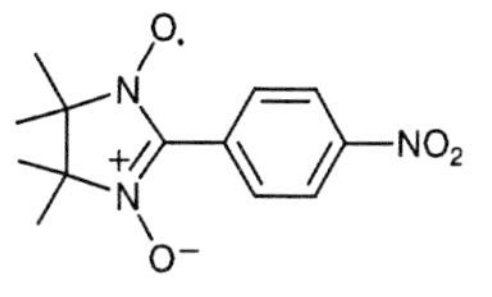

Scheme 5.20 p-Nitrophenyl nitronyl nitroxide

Tamura *et al*. reported the existence of ferromagnetic intermolecular interaction in the crystal structure (β phase) of *p*-nitrophenyl nitronyl nitroxide (Scheme 5.20), with a transition temperature to a ferromagnetically ordered state of 0.65 K.[89]

Since then, different strategies have been applied to design single molecules or organic free radicals based metallic complexes to get ferromagnetic interactions at the highest temperature possible. However, the best result ever achieved was to observe ferromagnetic properties at 35K with a sulfur–nitrogen radical.[90] Much work has been devoted to the pursuit of organic magnetic materials, and relevant reviews are available.[27d,49b,91–94] Herein, some NNOs which are important building blocks for the design of magnetic materials are mentioned.

Nitronyl nitroxides (NNOs), and to a lesser extent imino nitroxides (INOs), represent the most promising compounds to be used as bridging ligands for the construction of metal ion-based organic magnets. It is the approach chosen by Kahn, Caneschi, Rey, and Sessoli who pioneered in this field using nitronyl nitroxide radicals to bind transition metal ions ($M(hfac)_2$) (Scheme 5.21). They reported the magnetic behavior of one-dimensional chain metal–nitronyl nitroxide complexes showing the efficiency of such coordinating system.[95–101] The possibility of modeling the architecture of the whole molecule by using different substituents (R) can modify the arrangement of the molecules in the complex and influence the ferromagnetic properties.

To observe magnetization at high temperature will require increasing the dimensionality of the metal–nitroxide network. Two possibilities are currently developed: (1) increase the number of coordination sites on the nitronyl nitroxide radical by choosing the right substituents and (2) increase the number of organic free radicals around the metal. In order to add coordination sites on the NNO radicals, different groups developed the synthesis of multidentate NNOs (Figure 5.4 shows examples) and the metal–free radical complexes obtained were crystallized.[27,102–109]

To increase the number of radicals surrounding the transition metal ion, several groups developed the synthesis of bis-nitronyl nitroxides-based tetradentate ligands, such as 2,2'-bipyridine and phenanthroline *N*-oxide *N*-oxyl biradicals used with copper(II) and nickel(II) (Scheme 5.22, Figure 5.4).[110–112] Some other examples of stable chelate multiradicals used as building blocks for the preparation of molecule based magnets are reported in Figure 5.4.[113–115]

Novoa *et al*. reported recently the analysis of 47 different crystal structures of α-nitronyl nitroxide radicals and the influence of the relative disposition of the radical centers in the crystals on the magnetic properties.[116] Such theoretical study coupled with experimental data should allow the design of more

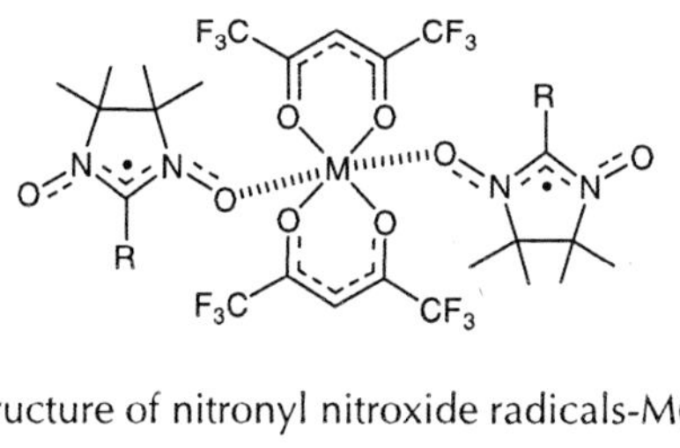

Scheme 5.21 Structure of nitronyl nitroxide radicals-$M(hfac)_2$ complexes

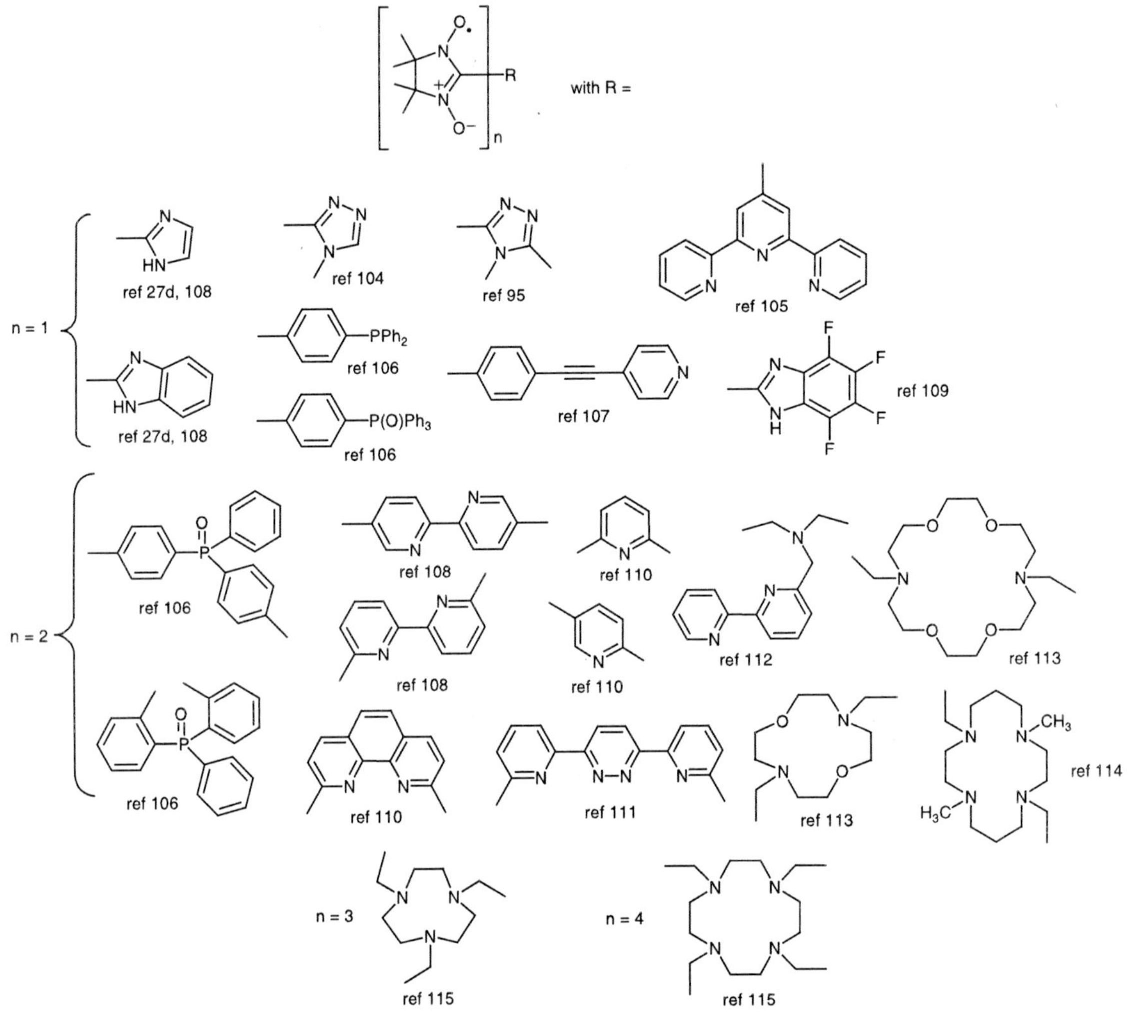

Figure 5.4 *Structure of nitronyl nitroxide-based chelates.*

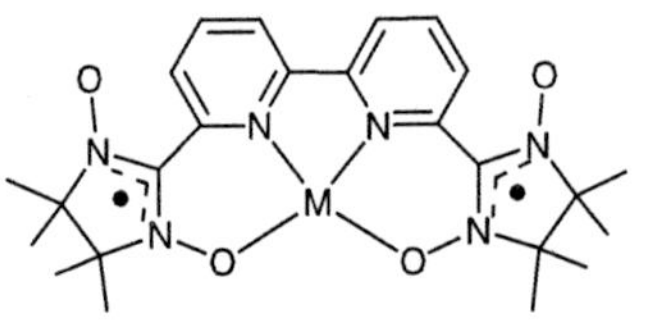

Scheme 5.22 Structure of bis(nitronyl nitroxide) radicals-M$(ClO_4)_2$ complexes

efficient nitronyl nitroxide radicals to prepare metal–organic free radical complexes exhibiting interesting ferromagnetic properties.

5.5 Synthesis of nitroxides

An impressive number of different stable nitroxides have been reported in the literature, and there are several books and reviews which cover early as well as more recent progress of the synthetic aspect of nitroxide chemistry.[9,11a–d,117] The synthesis of nitroxides $R_1N(^{\bullet}O)R_2$ are carried out through different synthetic routes that depend on the nature of the targeted R_1 and R_2 groups. However, in almost all the reported preparation of nitroxides, the aminoxyl group is introduced through an oxidation step involving either an amine or a hydroxylamine.

5.5.1 Oxidation of amines

Peracids,[117,118,9] dimethyldioxirane,[119] oxone,[120] and hydrogen peroxide[121] are widely used to oxidize secondary amines to nitroxides. Direct oxidation of tertiary amines to nitroxides[9,30a,122] is generally carried out with mCPBA or H_2O_2/sodium tungstate (Na_2WO_4) (Scheme 5.23).

5.5.2 Oxidation of hydroxylamines

Hydroxylamines can be easily oxidized to nitroxides. The main strategies to form an aminoxyl group through the oxidation of an N-hydroxy intermediate are shown in Scheme 5.24.[123–125] When different reactions must be carried out before the oxidation step, the hydroxylamine can be protected either by O-acetylation[126] or O-silylation with *tert*-butylchlorodimethylsilane.[28c,127]

5.5.3 Chiral nitroxides

Optically active[30,128] nitroxides and nitroxides for the control of free radical polymerization[22,23,129] have been also intensively studied and some results concerning their synthesis are presented here.

Tamura *et al.*[130] developed a new synthesis of α-asymmetric bicyclic nitroxide radicals, **64**, via the reduction with samarium diiodide (SmI_2) of homoallylic nitroenones and the subsequent addition of electrophiles (Scheme 5.25). Rychnovsky *et al.*[131] prepared several chiral piperazine and morpholine

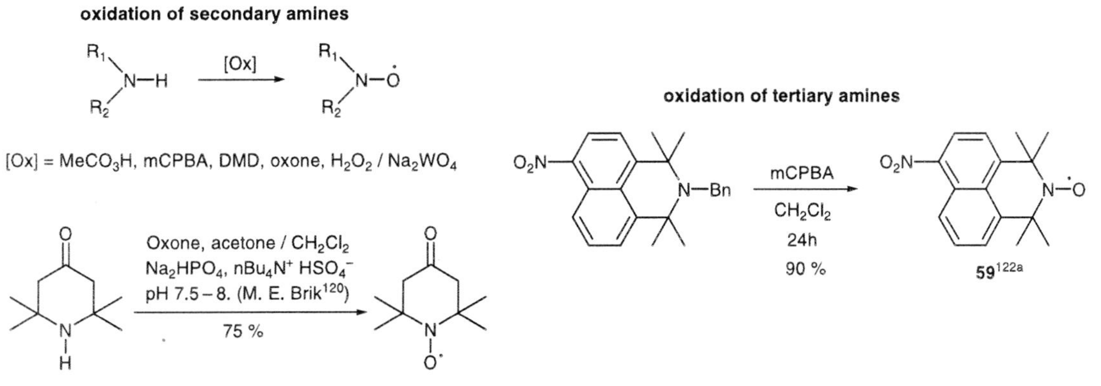

Scheme 5.23 Oxidation of amines to nitroxides

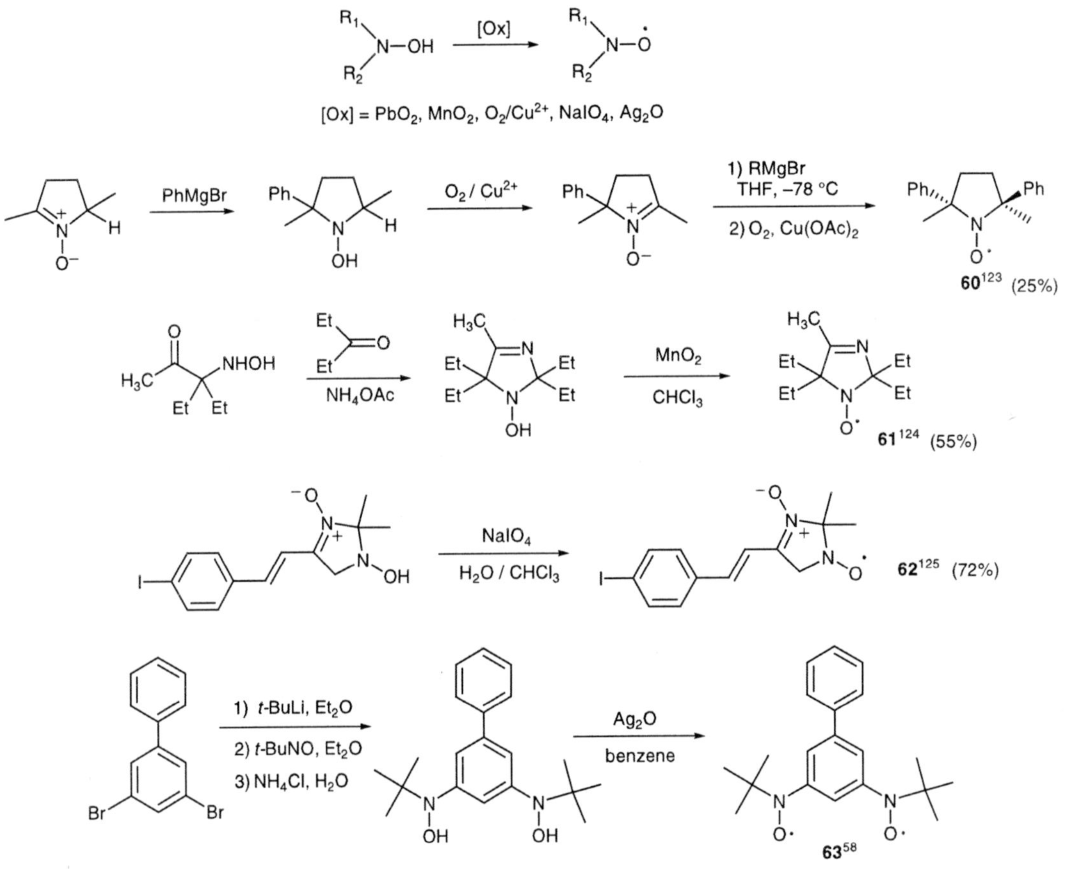

Scheme 5.24 Oxidation of hydroxylamines to nitroxides

nitroxides (**65**, **66**) using Lai's protocol to prepare hindered morpholinones and piperazinones. Greci *et al.*[132] obtained a series of 2-alkyl-2-phenyl-3,3-dimethylindolines by 1,2-organolithium addition to 2-phenyl-3,3-dimethyl-3*H*-indole. Oxidation of these indolines with mCPBA yielded a series of chiral indolinic nitroxides (**67**).

Several C_2-symmetrical chiral nitroxides have been prepared (Scheme 5.26), in which the asymmetric substituents are either immediately adjacent to the nitrogen on a pyrrolidine or piperidine ring.[30a,c,d] The asymmetric substituents can also be remote from the nitrogen on an azepine[133] or pyrrolidine ring.[134] A 1 : 2 mixture of meso/(±) diastereomers of *N*-benzyl-2,5-dicarboxymethylpyrrolidine treated with lithium diisopropylamide (LDA) and then with iodomethane yields **68** as a 1 : 6 mixture of meso/(±) diastereomers, which are chromatographically separated. Deprotection furnishes the pure amines, which after oxidation give meso- and (±)**69**.[135] Nitrone **70** is obtained by oxidation of (2R,5R)-2,5-bis(methoxymethyl)pyrrolidine with $MeReO_3$/urea.H_2O_2. Without purification **70** is converted to hydroxylamine **71** by the action of phenylmagnesium bromide in tetrahydrofuran. Oxidation of **71** to a nitrone followed by treatment with two equivalents of phenyllithium in tetrahydrofuran yields a hydroxylamine which is easily oxidized to **72**.[136] The total yield of **72** amounts to 20 %, the enantiomeric excess is higher than 98 %, the stereochemistry of the addition of phenylmagnesium bromide and phenyllithium agrees with the stereo-course proposed by Keana.[137]

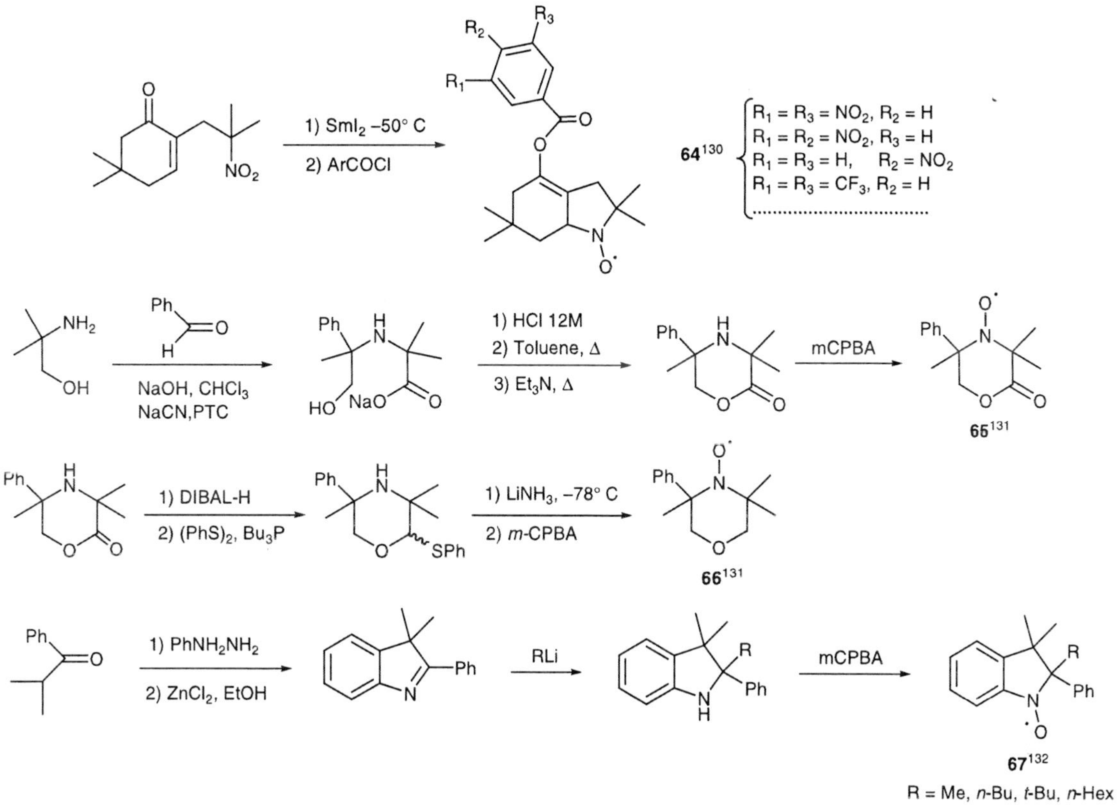

Scheme 5.25 Preparation of chiral nitroxides

The bis-phosphorylated pyrrolidine, (±)-**73**, was obtained in a one-pot reaction (65 % after distillation) by bubbling ammonia into a mixture of hexane-2,5-dione and diethylphosphite.[138] If the reaction mixture is left for three months at room temperature a mixture (61 : 39) of (±)-**73** and *meso*-**73** was formed. Crystals of *meso*-**73** were obtained and its stereochemistry established by X-ray diffraction. The corresponding pure nitroxides (±)-**27** and *meso*-**27** were obtained by oxidation with mCPBA in dichloromethane.

Tamura *et al.*[29a–d,139] have intensively studied all-organic liquid crystals containing chiral nitroxides within the rigid core (Scheme 5.27). Like **74** and **20**, most of the studied nitroxides were prepared from the 2,5-dimethylpyrroline-*N*-oxide using Keana's protocol.[140]

5.5.4 Nitroxide design for nitroxide mediated polymerization (NMP)

The Persistent Radical Effect (PRE)[129] (Section 5.6.1) established in the presence of a nitroxide has been widely investigated during the last fifteen years to prepare living polymers with narrow molar mass distribution, through radical polymerization. The efforts of several groups resulted in the development of the synthesis of many new nitroxides. Some of these results are illustrated here.

Tordo *et al.*, using straightforward chemistry, developed the synthesis of a series of original acyclic *β*-phosphorylated stable nitroxides (Scheme 5.28).[22,141] The synthetic route to SG1 allows the R_1,R_2,R_3 and R_4 groups to be changed, thus giving access to a versatile series of SG1 analogs. In order to

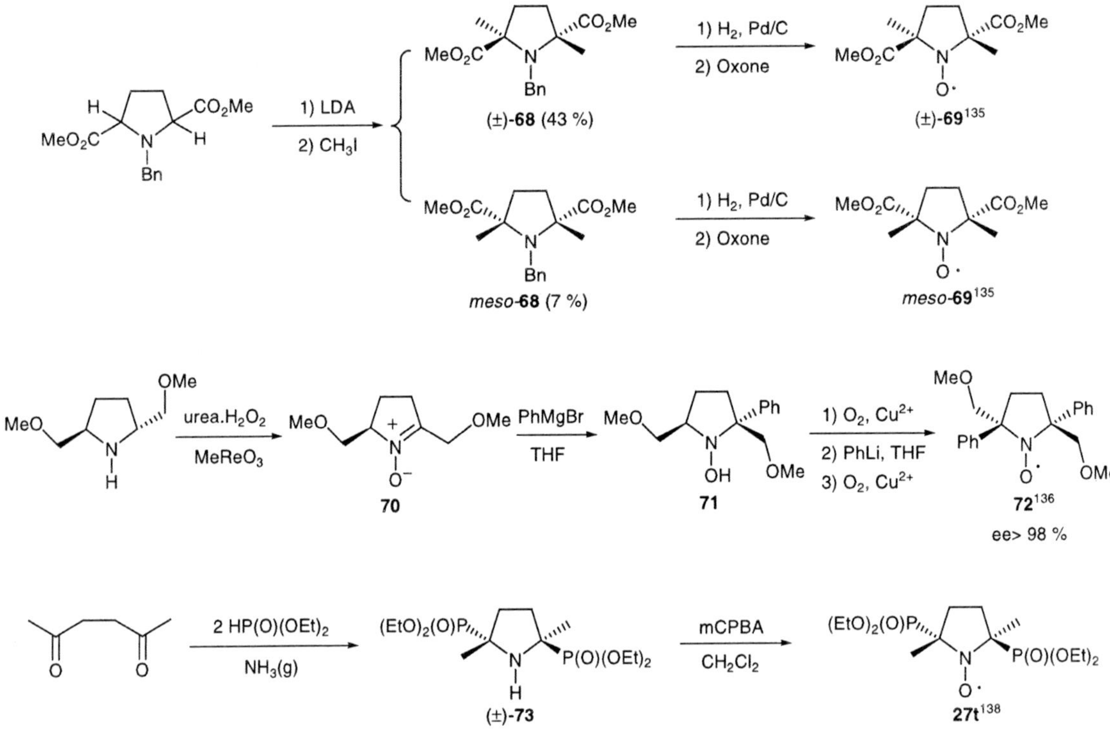

Scheme 5.26 Synthesis of C_2-symmetrical chiral nitroxides

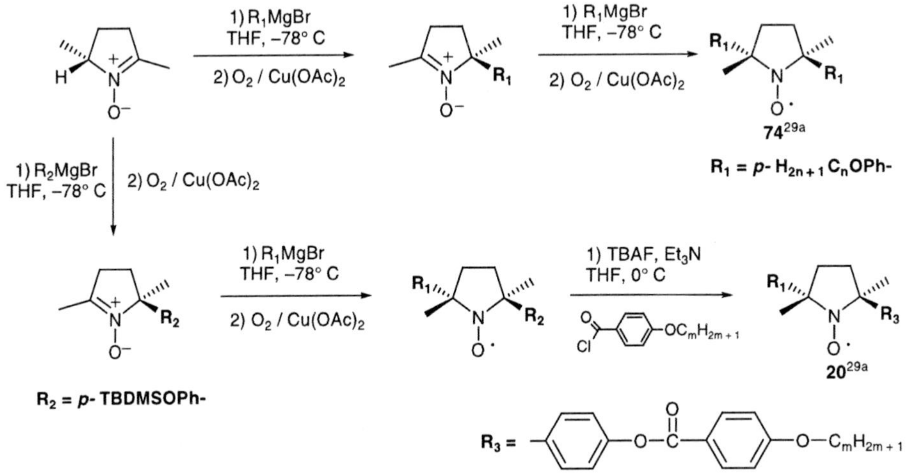

Scheme 5.27 Synthesis of chiral nitroxides for paramagnetic liquid crystals

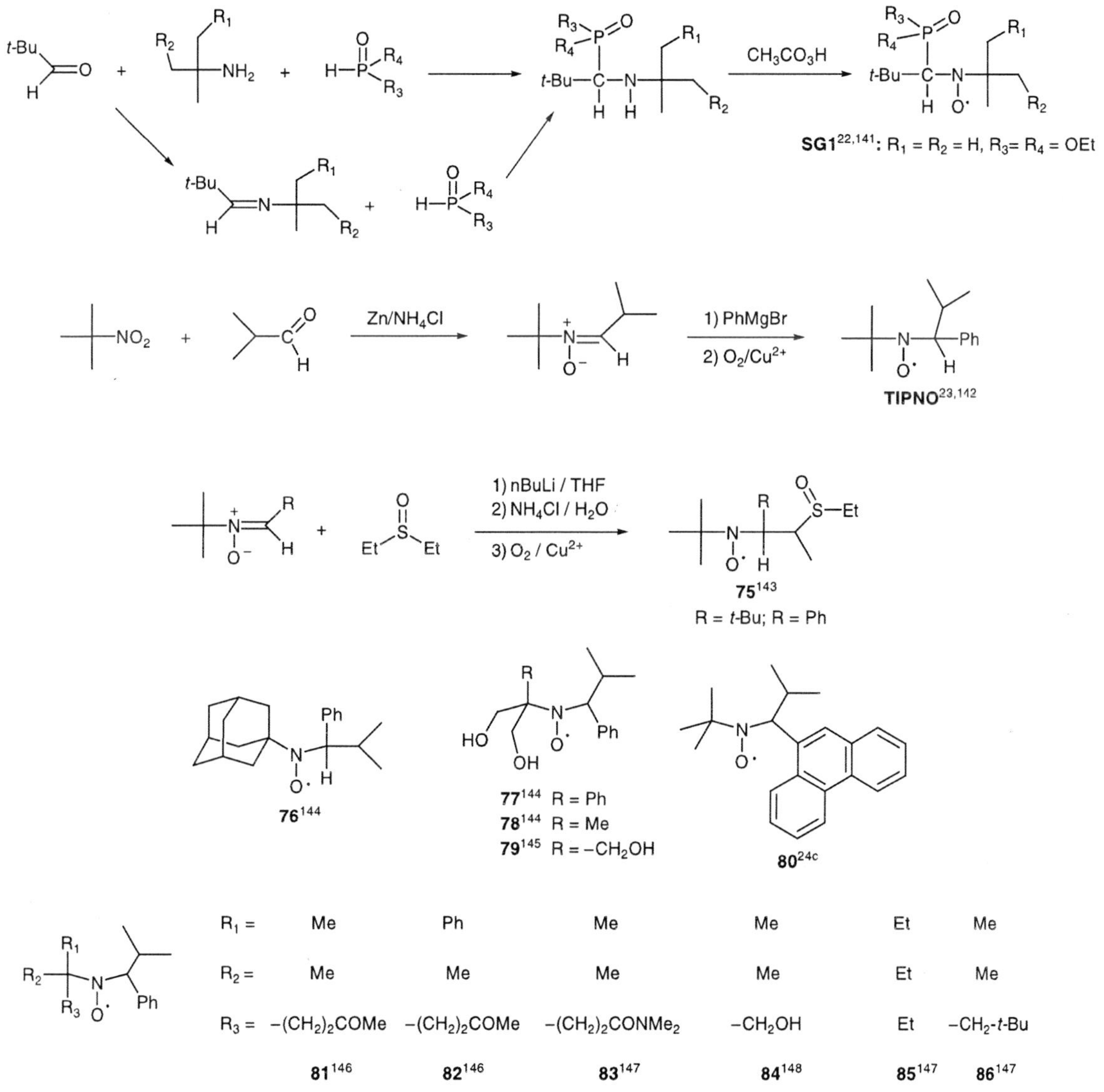

Scheme 5.28 Synthesis of SG1, TIPNO and some derivatives

understand the factors sustaining the very good efficiency of SG1 in the control of radical polymerization of styrenic and acrylic monomers, Tordo *et al.*[22a] tested different α-hydrogen bearing nitroxides and they found that TIPNO is also an efficient controller for nitroxide mediated polymerization. The chemistry of TIPNO was subsequently intensively developed,[23,142] mainly by Hawker, Braslau and their coworkers, and an impressive number of interesting derivatives were reported (Scheme 5.28: **75**,[143] **76–78**,[144] **79–80**,[145] **81–82**,[146] **83** and **85–86**,[147] **84**[148]).

The Bond Dissociation Energy (BDE) of the NO–C bond of the dormant alkoxyamine is a key factor to reach a successful controlled living polymerization in the course of nitroxide mediated polymerization. If the bond dissociation energy is too high (around 540 kJ/mol for many monomers), at usual temperatures

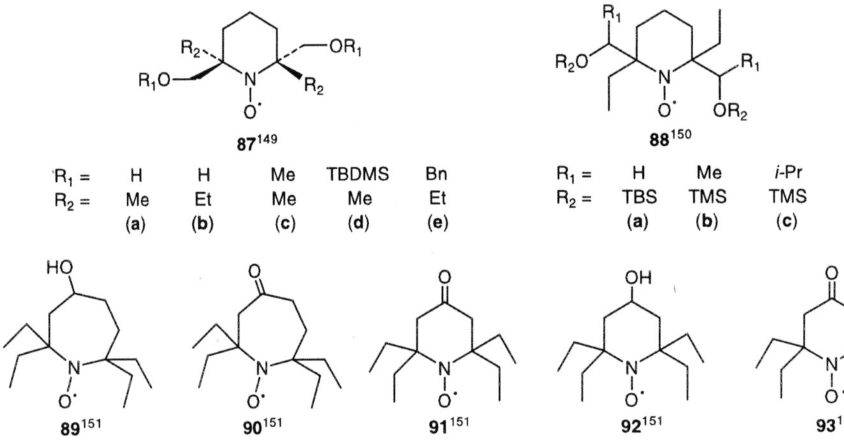

Scheme 5.29 Sterically crowded TEMPO and azepanoxyl derivatives

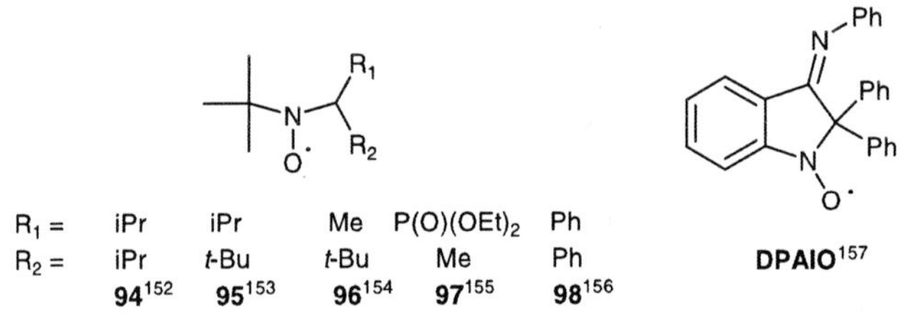

Scheme 5.30

of polymerization (100–130 °C), the nitroxide will behave as an inhibitor of polymerization. The bond dissociation energy of the NO–C bond can be decreased by increasing the steric hindrance around the nitrogen atom of the nitroxide moiety. Studer *et al.* prepared a series of TEMPO and seven-membered ring nitroxides, introducing a high steric crowding by the substitution of methyl groups with bulkier substituents. (Scheme 5.29: **87a–e**,[149] **88a–c**,[150] **89–93**[151]).

Many other new nitroxides have been tested to control the radical polymerization of different vinylic monomers. Some examples (**94**,[152] **95**,[153] **96**,[154] **97**,[155] **98**[156]) are shown in Scheme 5.30. Before closing this partial survey of recent developments in the synthesis of nitroxides, it is worth mentioning DPAIO (Scheme 5.30), the only nitroxide capable of controlling the radical polymerization of methyl methacrylate (MMA).[157] During MMA polymerization in the presence of TEMPO, TEMPO abstracts a hydrogen atom from MMA polymeric radicals to form a dead chain with a terminal double bond. In the case of DPAIO, due to the delocalization of the unpaired electron, hydrogen atom abstraction is disfavored and DPAIO is able to partially control the MMA polymerization.

5.6 Chemical properties of nitroxides

Nitroxides possess an interesting chemical reactivity that to some extent is characteristic of free radicals, with the exception of dimerization (homocoupling) and reactions with oxygen. The stability of nitroxide radicals makes it possible to carry out reactions selectively on functional groups not involving the unpaired electron. This topic has been thoroughly reviewed and further information can be found in the articles of Volodarsky,[158,11c] Rozantsev,[9] Keana,[117a] and Hideg.[159]

5.6.1 The Persistent Radical Effect

Another interesting feature of nitroxides is their involvement in kinetic situations controlled by the Persistent Radical Effect (Scheme 5.31).

The transient carbon-centered radical generated by the thermal decomposition of a sterically crowded alkoxyamine decays rapidly to yield various products through self reactions. However, the stable nitroxide accumulates in the medium and after a very short time (depending on the values of k_d and k_c) its concentration is much higher than that of the transient radical. At this point, the self-reactions of the transient radicals are stopped and if they do not participate to other fast chemical reactions, for example, cyclizations, addition to olefins, and so on, the only occurring reaction is the cross-coupling to regenerate the starting alkoxyamine. This phenomenon, also observed with other persistent free radicals, was explained by Fischer and Ingold in the 1980s.[129,160]

5.6.2 Redox reactions

Nitroxides are redox active species and they can undergo one-electron reduction or oxidation (Scheme 5.32). Their interesting redox properties have led to a wide range of applications extending their use in biology as SOD mimics (Chapter 17) and redox probes, and to materials sciences with the development of organic batteries (Chapter 14).

The redox potentials of nitroxides have been studied (bio)chemically,[161–164] electrochemically,[165–167] and theoretically.[168,169] Nitroxides can be reduced to the corresponding hydroxylamines and in some cases the reaction is irreversible. For cyclic nitroxides, the rate of reduction by ascorbate depends mainly on the ring size and correlates with reduction potentials.[170,171] The order of increasing favorability[172] for the

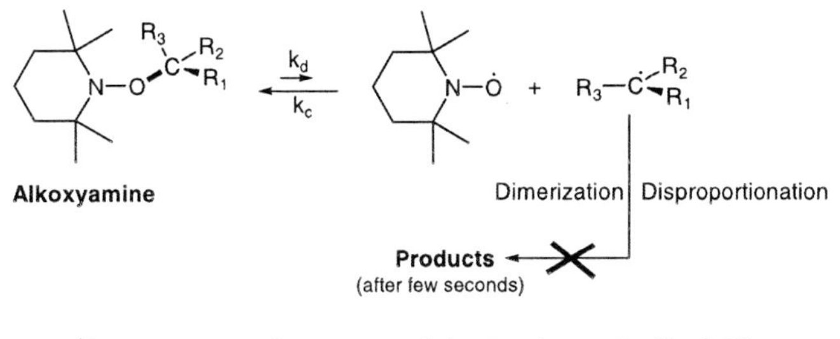

Scheme 5.31 Illustration of the Persistent Radical Effect

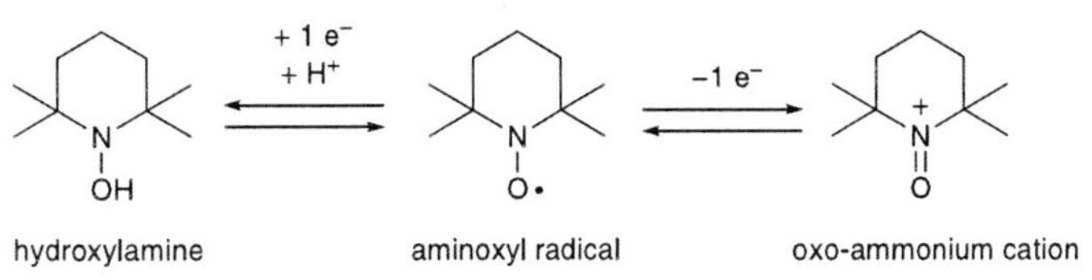

Scheme 5.32 Reversible redox properties of nitroxides

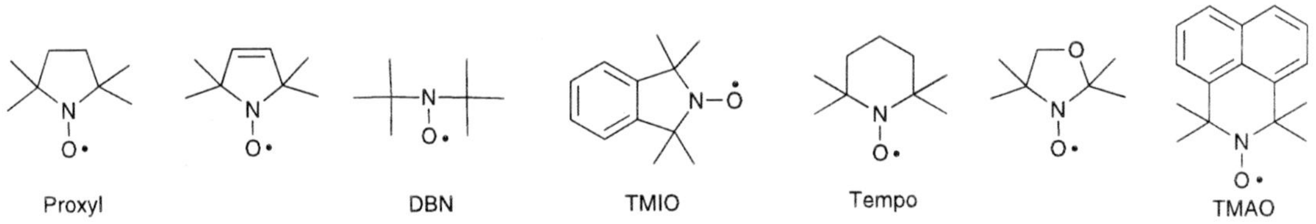

Scheme 5.33 Nitroxides classified by increasing ease of reduction

reduction is the following: pyrrolidine (PROXYL) < pyrroline < acyclic (DBN) < isoindoline (TMIO) ≪ piperidine (TEMPO) < oxazolidine < azaphenalene (TMAO) (Scheme 5.33). The rate constant for the pyrrolidine nitroxide radicals may be 30 to 40 times less than for similar piperidine nitroxide radicals.

Most nitroxides undergo reversible oxidation to the corresponding oxo-ammonium cation. The trend of oxidation potentials is not the inverse of the trend for reduction potentials. Generally, the lowest oxidation potentials are observed for the azaphenalene derivatives and the highest for the isoindoline derivatives. Piperidine and pyrrolidine derivatives have intermediate oxidation potentials. Oxazolidine derivatives are not easily oxidized due to the electro-withdrawing effect of the oxygen atom in the ring that destabilizes the positive charge of the resulting oxo-ammonium cation.

As a general rule, the principal factor affecting the redox potentials of a nitroxide is the nature of the ring. The substituents on the ring have a relatively small effect, except for azaphenalene-type nitroxides. It can be assumed that the ability of nitroxides to be oxidized or reduced can be related to the flexibility of the molecule, that is, how easily the nitrogen atom can be pyramidalized in the hydroxylamine and planarized in the oxo-ammonium cation. It is also observed that changing the groups surrounding the nitroxide moiety can have an effect on the oxidation and the reduction potentials. For instance, replacing the methyl groups by ethyl or cyclohexyl groups in the isoindoline, imidazoline, or piperidine series led to a decrease of the reduction rate by weak reductants.[173–175]

5.6.3 Approaches to improve the resistance of nitroxides toward bioreduction

Nitroxides are interesting probes and reporters in biological systems. However, their reduction by bioreductants, such as ascorbate anion, thiols and enzymatic pools, has limited their use *in vivo*. To circumvent this limitation, two approaches have been developed that are based on the synthesis of sterically hindered nitroxides and on the inclusion of nitroxides into host molecules.

Replacing the methyl groups by ethyl or cyclohexyl groups in the isoindoline,[176] imidazoline,[177] or piperidine[178] (Schemes 5.34 and 5.35) series led to a significant decrease of the reduction rate in *in vitro* experiments, in blood or in liver homogenates.

In the late 1970s, Rassat *et al.*[179] reported the partial protection of TEMPO to ascorbate reduction by adding β-cyclodextrin in the media. This approach has been successfully expanded to the spin trapping

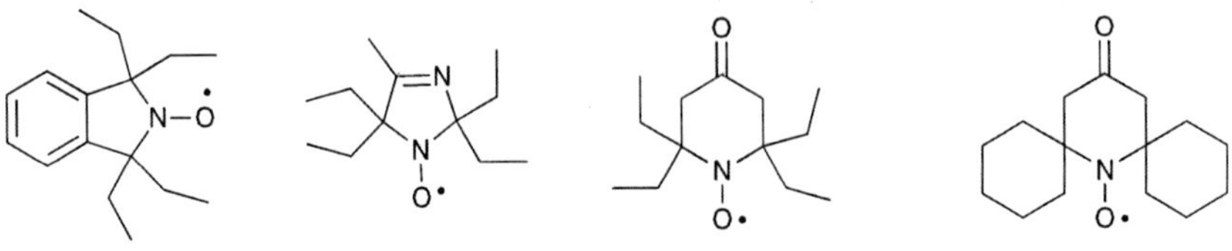

Scheme 5.34 Sterically hindered nitroxides with improved resistance to bioreduction

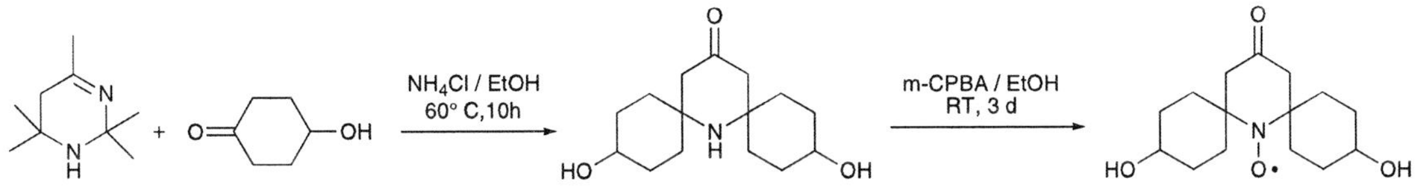

Scheme 5.35 Synthesis of a spirocyclohexyl TEMPO derivative

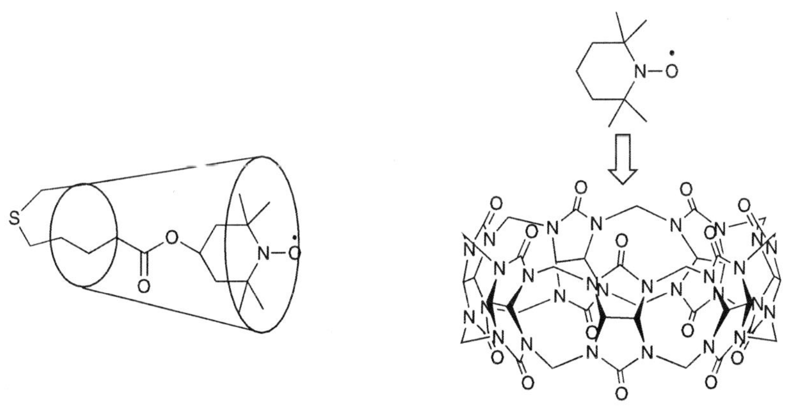

Scheme 5.36 TEMPO based [1]rotaxane (left) and TEMPO@CB[7] (right)

technique[180] and recently a [1]rotaxane based on a TEMPO unit linked to a β-cyclodextrin moiety has been synthesized and exhibited significant resistance to glutathione (GSH) reduction (Scheme 5.36).[246] The strong association constant ($K > 10^4\,M^{-1}$ at 298 K) and the perpendicular orientation of TEMPO in the TEMPO@CB[7] inclusion complex have shown to be interesting features to protect the paramagnetic probe against ascorbate reduction, with a half-life time increased by a factor of 60 (Scheme 5.36).[181]

5.6.4 Hydrogen abstraction reactions

Nitroxide radicals such as TEMPO and PROXYL derivatives are able to abstract hydrogen atoms only from weak hydrogen bonds, including thiols, phenols, allylic positions, and metal hydrides.[182] Electron-poor nitroxides, such as bis(trifluoromethyl)nitroxide, Fremy's salt,[183] acyl nitroxides, and phthalimide nitroxides (PINO), are more reactive as hydrogen abstractors and can abstract inactivated hydrogen atoms under ambient conditions. This reactivity is due to the higher oxygen–hydrogen bond dissociation energies of their corresponding hydroxylamines. Interesting discussions on these reactions can be found in recent review articles.[184–186]

TEMPO was shown to react at 80 °C with tributyltin hydride (Bu_3SnH) and triphenylgermanium hydride (Ph_3GeH) by hydrogen atom abstraction to yield the corresponding hydroxylamine and the stannyl or germyl radical, respectively. No reaction was observed with Et_3SiH or Ph_3SiH under the same conditions. Using $(Me_3Si)_3SiH$ (tris(trimethylsilyl)silane), the reaction yields the corresponding hydroxylamine and amine in a 1 : 1 ratio.[187]

Using ^2H isotope labeling, a study for distinguishing between the allylic hydrogen atom abstraction/addition and addition/abstraction mechanisms in the reaction of TEMPO with cyclohexene

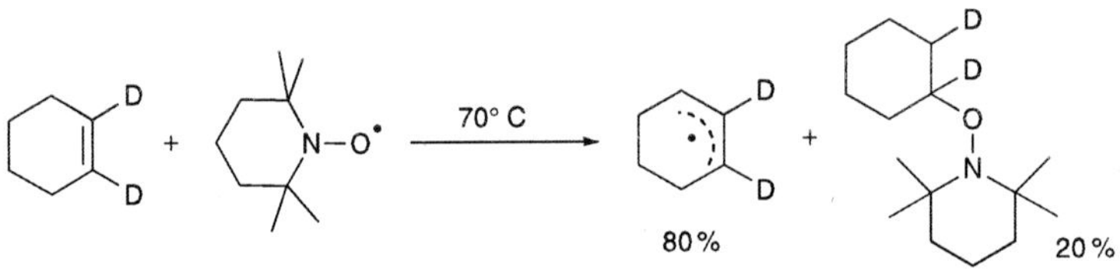

Scheme 5.37 First step in the reaction of TEMPO with cyclohexene

at 70 °C demonstrated that both pathways occur, the more important being the allylic hydrogen atom abstraction/addition process (80 %).[188] This work was undertaken to sort out the different mechanisms reported in the literature concerning the reaction of nitroxide with cyclohexene (Scheme 5.37).

Under photoirradiation, the surprising hydrogen atom abstraction by TEMPO from acetonitrile and toluene was reported with nearly quantitative yield.[189] Because the O–H bond energy in TEMPO hydroxylamine is known, TEMPO has been also used to confirm the Bond Dissociation Energy (BDE) of O–H bonds in non-heme iron and manganese complexes as model of metalloproteins.[190]

5.6.5 Cross-coupling reactions

Nitroxides have often been used as radical scavenging agents, due to their fast reaction with other free radicals, such as carbon- and nitrogen-centered radicals[191–194] and protein radicals.[195,196] Due to their efficient radical trapping behavior, nitroxides have been used for many years as stabilizers and inhibitors in plastics and rubber manufacture, as regulators in controlled/living radical polymerization, and as antioxidants in the biomedical field.[197,198]

Absolute rate constants for the cross-coupling of various nitroxides with several carbon-centered radicals have been determined and are in the range 5×10^7 to $2 \times 10^9\,M^{-1}s^{-1}$ at room temperature, the nature of the trapped radicals and of the nitroxide affecting mainly the rate value.[199] As a result, trapping of carbon-centered radicals with nitroxides is an important and effective method for the synthesis of alkoxyamines. When stabilized carbon-centered radicals are formed, the alkoxyamine O–C bond undergoes facile and clean thermal homolysis (Scheme 5.31).

Recently, it has been reported that photoinduced homolysis of alkoxyamines is possible, opening up possible applications of alkoxyamine free radical initiation at low temperature.[200] Indeed, the photolysis of an alkoxyamine bearing a quinoline antenna proceeds via intramolecular energy transfer from the quinoline moiety leading to the homolysis of the C–O alkoxyamine bond.

5.6.6 Nitroxides in synthetic sequences

The use of nitroxides in organic synthesis has been more and more important in the last decade, as illustrated by the growing number of review articles[201–204] and book chapters.[205] Nitroxides have often been used for the oxidation of primary and secondary alcohols as well as sulfides to the corresponding aldehydes and ketones, and sulfoxides respectively. A chapter of the book is devoted to this topic.

5.6.6.1 Trapping of carbon-centered radicals

In various radical initiation methodology, radical carbon–carbon bond formation has been successfully combined with subsequent nitroxide trapping reactions,[206–212] allowing the introduction of a functional group after removal of the alkoxyamine. For instance, Scheme 5.38 describes the reaction of an aryl iodide

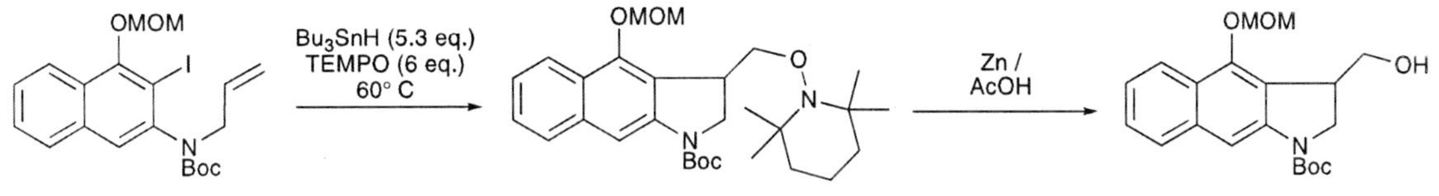

Scheme 5.38 Aryl radical cyclization and subsequent trapping by TEMPO

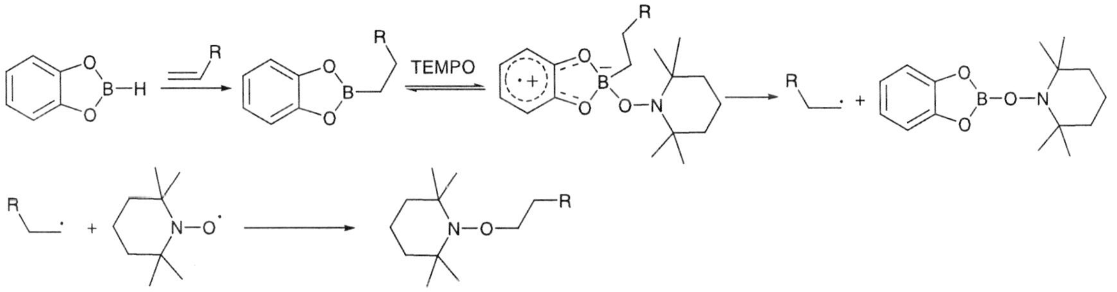

Scheme 5.39

derivative with tributyltin hydride in the presence of TEMPO leading, after fast 5-exo-cyclization, to an alkoxyamine that can be cleaved to afford an alcohol in a high overall yield.[213]

Addition of an olefin to catecholborane followed by the reaction of the resulting B-alkylcatecholborane with TEMPO is a very interesting method to generate selectively alkyl radicals. B-alkylboranes (and not trialkylboranes) react with two equivalents of TEMPO to afford in good yield the corresponding alkoxyamine (Scheme 5.39).[214]

Radical hydroxylation reactions of alkenes via the corresponding *B*-alkylcatecholboranes in the presence of TEMPO are an interesting alternative to the classical hydroboration followed by oxidative work-up procedure (Scheme 5.40).[215] For instance, the hydroboration of α-pinene affords with high diastereoselectivity the alkoxyamine, which can then be cleaved to afford the alcohol.

With specific substrates, the radical-mediated oxidation of organoboranes leads to products differing from those obtained using the classical oxidation reaction due to rearrangement of the intermediate radical. This strategy was also used for tandem process; however, this reaction is limited to radical addition to highly reactive olefins to avoid the TEMPO trapping of the initial radical.

The reaction of TEMPO with ketenes to give the 1,2-bis adducts (Scheme 5.41) has been extensively studied on a variety of ketenes and by theoretical calculations.[216–218] It was found that the first TEMPO addition occurs at the carbon–carbon double bond, leading to the most stabilized radical.

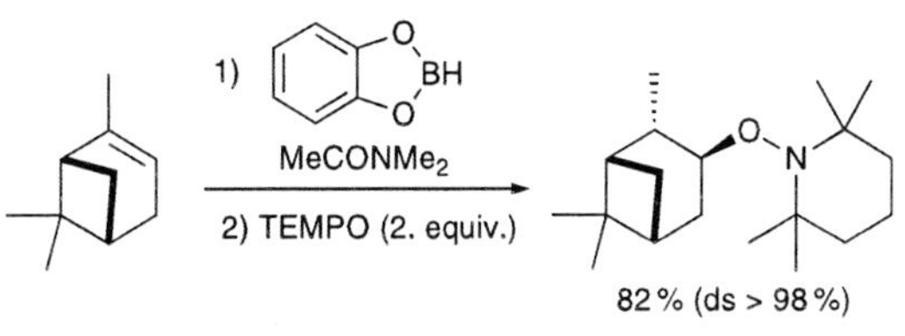

Scheme 5.40 TEMPO-mediated oxidation of α-pinene

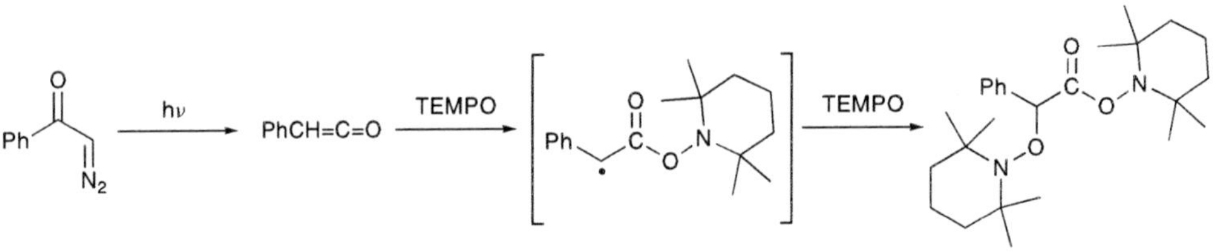

Scheme 5.41 Reaction of TEMPO with phenylketene

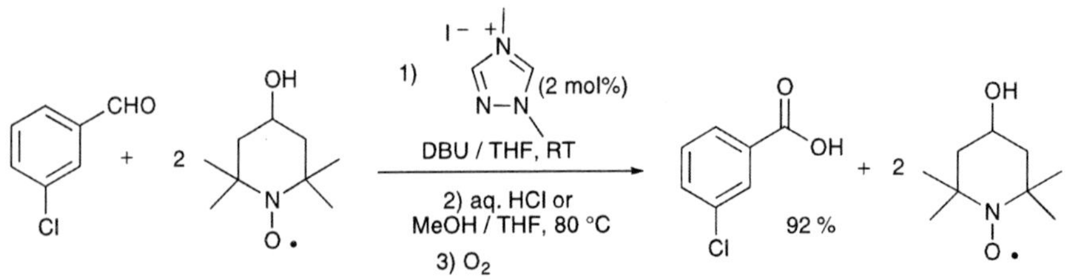

Scheme 5.42 N-Heterocyclic carbene-catalyzed oxidation of aldehydes by TEMPO

Diphenylcarbene reacts with TEMPO with a rate constant of $2.7 \times 10^8\, M^{-1}s^{-1}$ at room temperature leading to benzophenone and 2,2,6,6-tetramethylpiperidinyl radical in nearly quantitative yields.[219]

N-Heterocyclic carbene-catalyzed oxidation of aldehydes by using TEMPO as oxidant affords carboxylic acids in good to very good yields, even with sensitive aldehydes such as 2-thiophenecarboxaldehyde, 2-pyridinecarboxaldehyde, phenylglyoxal monohydrate or crotonaldehyde (Scheme 5.42).[220] The suggested mechanism implies the reaction of the N-heterocyclic carbene catalyst with an aldehyde to give the corresponding enamine that is oxidized by SET by two equivalents of TEMPO, followed by further reaction of the TEMPO hydroxylamine as a nucleophile to yield the corresponding ester, leading finally to the acid after hydrolysis.

Reactions of chiral nitroxides with prochiral carbon-centered radicals can yield two diastereomeric alkoxyamines (Scheme 5.43). Rigid C_2-symmetric nitroxides have shown to be the most efficient stereoselective trapping agents; however, low diastereoselectivities were observed.[221]

A tandem anionic/oxidative reaction sequence terminated by a TEMPO trapping reaction was developed and has shown to be an interesting strategy for the efficient preparation of functionalized chiral pyrrolidines in a one-pot sequence[222] and for the synthesis of 15-F_{2t}- isoprostane[223] (Scheme 5.44).

Recently, an enantioselective radical-mediated α-oxygenation of aldehydes using organocatalysts has been successfully developed (Scheme 5.45).[224] The methodology proceeds by oxidation of a chiral enamine

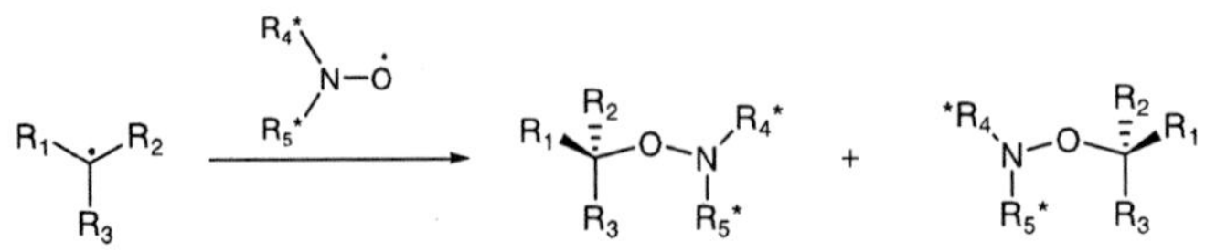

Scheme 5.43 Trapping of a prochiral radical with a chiral nitroxide

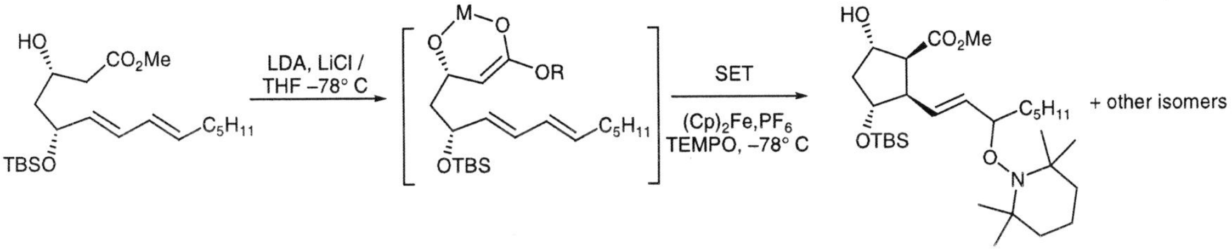

Scheme 5.44 Tandem anionic/oxidative reaction using TEMPO as trapping agent

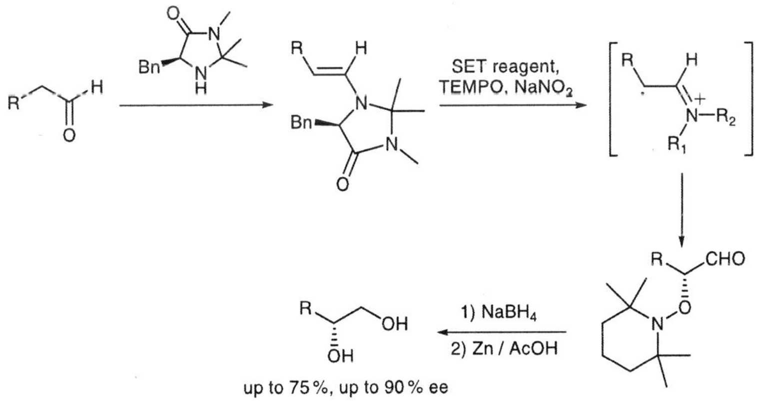

Scheme 5.45 Enantioselective radical-mediated α-oxygenation of aldehydes

by a SET reagent (such as Cp_2Fe, BF_4 or $FeCl_3$) and subsequent stereoselective trapping of the intermediate radical by TEMPO with good enantioselectivity.

5.6.6.2 Persistent Radical Effect in alkoxyamine addition reactions

Alkoxyamines have been also prepared as carbon-centered radical precursors in tin-free radical reactions. This approach is made possible and efficient because of the Persistent Radical Effect-mediated reversible alkoxyamine homolysis.[225] Isomerization, cyclization, and intermolecular additions (e.g., Scheme 5.46) and library expansion[226] have been performed successfully using this strategy.

Indeed, the use of the thermally-induced homolysis of activated alkoxyamines has been shown to be an easy and efficient procedure for the generation of carbon-centered radicals, leading to various clean reactions in good yields (Scheme 5.47).[227–232]

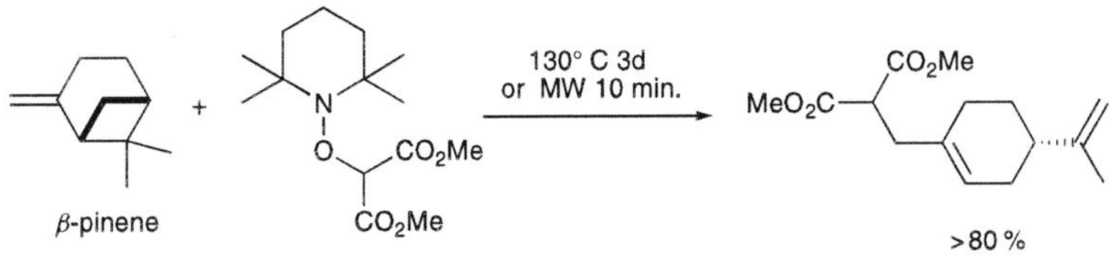

Scheme 5.46 Alkoxyamine addition on β-pinene

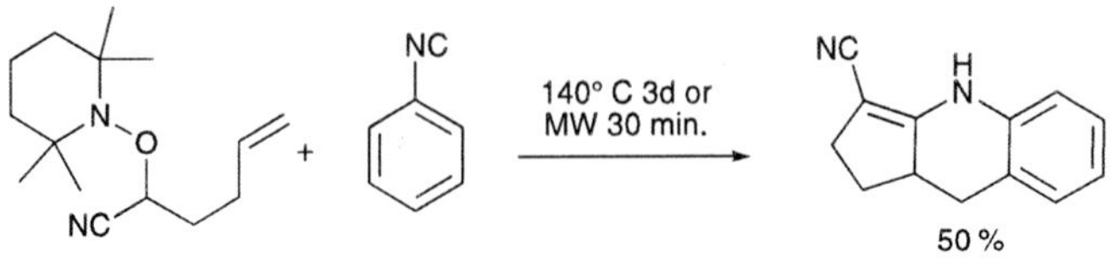

Scheme 5.47 Alkoxyamine addition to phenyl isonitrile

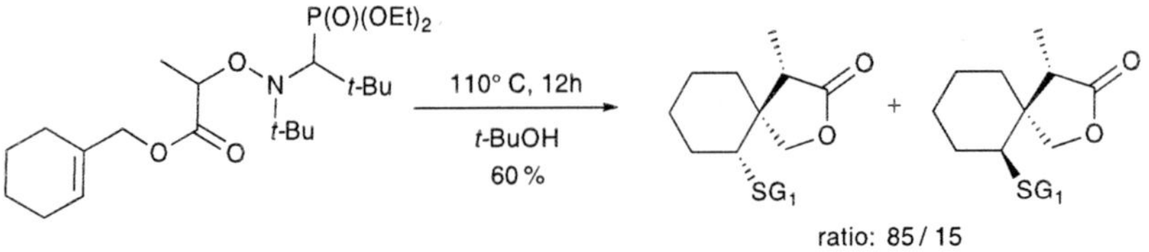

Scheme 5.48 Thermal homolysis of SG1-based alkoxyamine affording spiro-lactone with high stereoselectivity

However, TEMPO-based alkoxyamines present some drawbacks, such as high temperature of cleavage, long reaction times, and hydrogen abstraction side reactions. To overcome these limitations, it has been shown that steric effects from the nitroxide part play an important role on the reaction outcome and that by using hindered nitroxides, such as SG1 (Scheme 5.48), such drawbacks can be overcome.[233,234]

5.6.6.3 *Reactions with organometallic compounds*

The reactions of several classes of organometallic species with two equivalents of nitroxides have been reported since the 1970s.[235] The treatment of various organometallic compounds RM (with M = Li, Mg, Zn, Cu, Sm, Ti, Zr) with TEMPO was demonstrated to proceed through the initial attack of TEMPO on the metal atom to generate a carbon-centered radical that is subsequently trapped by the second equivalent of TEMPO (Scheme 5.49).

TEMPO was also used as oxidant in transition metal-catalyzed reactions. Copper-catalyzed C–N and C–O bond cross-couplings of secondary amines and phenols with arylboronic acid using TEMPO as oxidant were successfully developed (Scheme 5.50).[236]

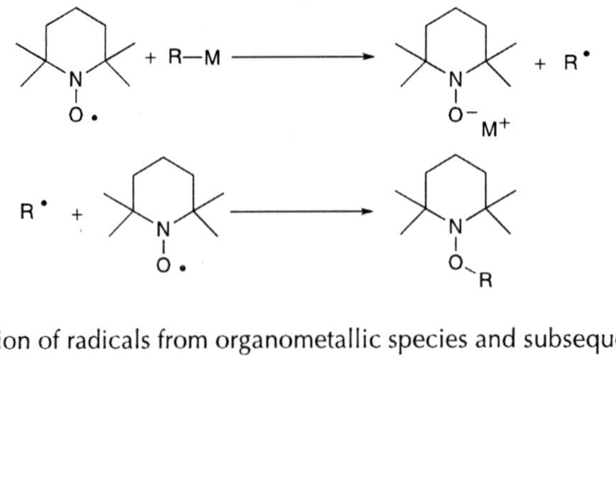

Scheme 5.49 Generation of radicals from organometallic species and subsequent trapping by TEMPO

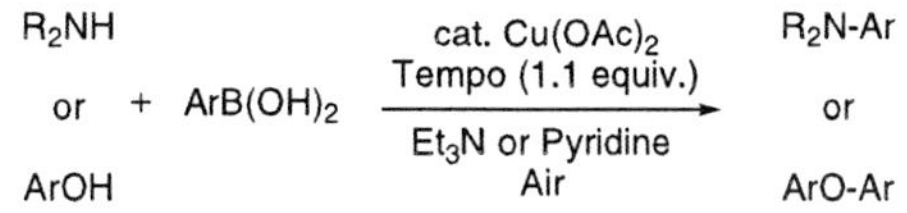

Scheme 5.50 Copper-catalyzed cross-coupling reactions using TEMPO as oxidant

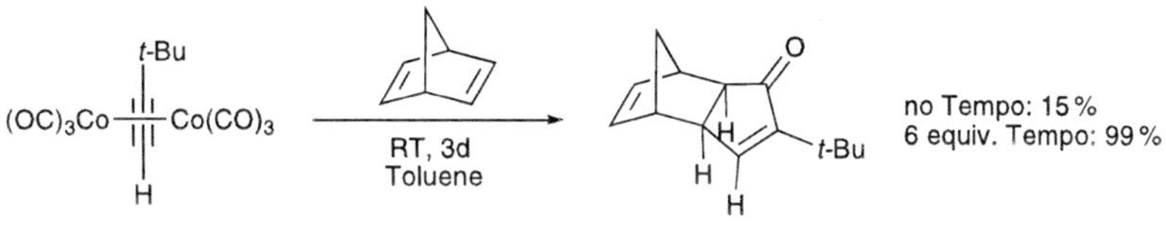

Scheme 5.51 TEMPO-catalyzed Pauson–Khan reaction

An acceleration of the Pauson–Khand reaction was found by using TEMPO, especially with sterically demanding alkynes (Scheme 5.51). DFT calculations supported the hypothesis that TEMPO promotes the decarbonylation of the cobalt complex through single-electron activation of the cobalt–carbon bonds, thus providing a low energy, radical pathway.[237]

For many years, TEMPO has also been used as ligand for transition metals and main group elements. It has been shown that Ti–O bonds in titanium nitroxide complexes are weaker than those of titanium alkoxides, and they can be cleaved at moderate temperature (60 °C) to generate titanium(III) complexes and aminoxyl radicals (Scheme 5.52). This result and DFT calculations have demonstrated also that the strength of these Ti–O bonds can be tuned by ancillary ligand modifications on titanium.[238]

Recently, "transition metal free" oxidative homocoupling reactions of aryl, alkenyl, and alkynyl based Grignard reagents have been reported using TEMPO as the stoichiometric oxidant.[239] The method appears to be versatile and very efficient. Reaction of TEMPO with Grignard reagents usually affords the corresponding alkoxyamines, as illustrated in Scheme 5.53. However, due to the destabilized character of aryl, alkenyl, and alkynyl radicals, the reaction outcome is different, and this is presumably due to the non-generation of free aryl, alkenyl, and alkynyl radical intermediates.

The oxidative rhodium-catalyzed coupling reaction of arenes and heteroarenes with various arylboronic acids via direct C–H arylation and using TEMPO as the oxidant has been reported recently (Scheme 5.54).[240] In the proposed mechanism, it is assumed that TEMPO acts as an oxidant (two

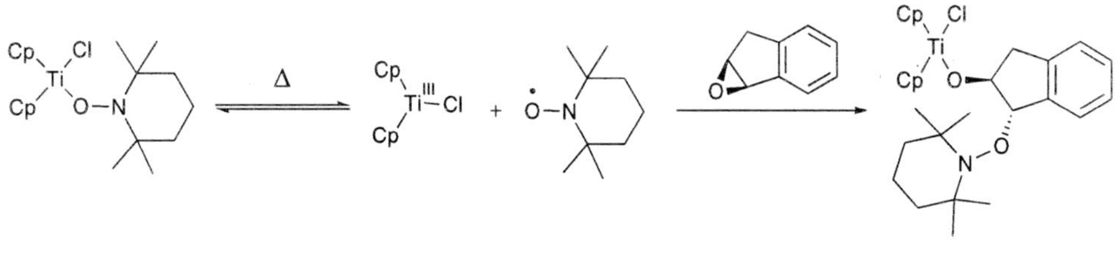

Scheme 5.52 Thermal homolysis of Ti–O bonds derived from titanium–TEMPO complex and subsequent ring opening mediated by the $Cp_2Ti(III)Cl$ complex followed by TEMPO trapping

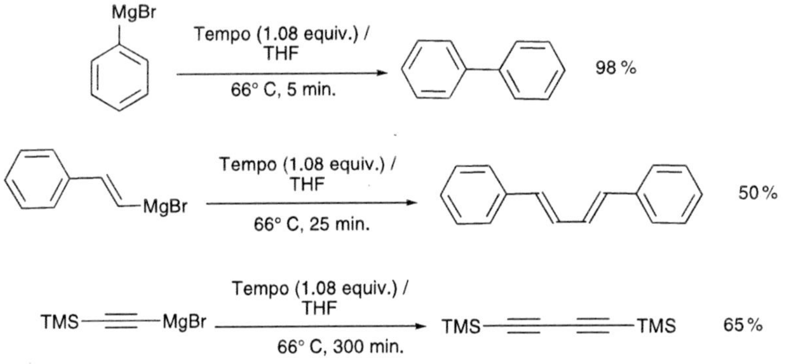

Scheme 5.53 Oxidative homocoupling of aryl, alkenyl, and alkynyl Grignard reagents using TEMPO

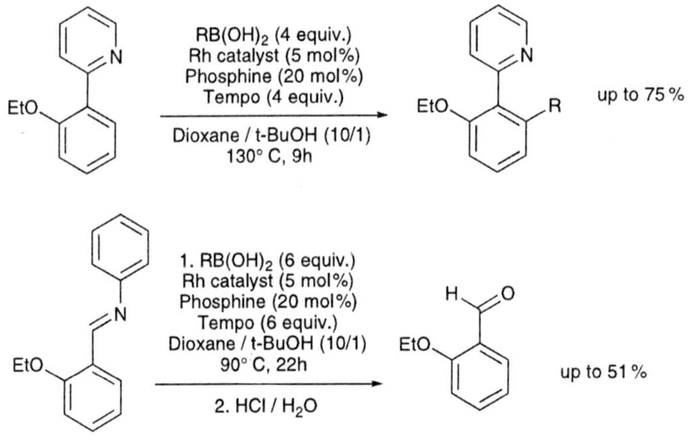

Scheme 5.54 Oxidative coupling of arenes with various arylboronic acids

equivalents) of the rhodium(I) complex to the rhodium(III) complex, and can also act as a base through the deprotonated (anionic) form of the corresponding hydroxylamine.

5.7 Nitroxides in supramolecular entities

The component structural units of supramolecular entities are typically held together by a variety of weak (non-covalent) interactions, including hydrogen bonding, $\pi-\pi$ stacking, dipolar interactions, London and van der Waals forces. The resulting host–guest complexes are thermodynamically weak and kinetically labile, making the kinetics of association and dissociation processes, which occur in the microsecond and submicrosecond time range, difficult to measure. In most cases the experimental NMR spectra appear as averages of those from the free and complexed species and only the equilibrium of association can be reached. However, EPR spectroscopy is characterized by a shorter time scale than NMR, so if a free radical is used as guest and the radical shows sensitivity to the embedding medium, different EPR signals can be detected for its complexed and uncomplexed forms. The stability of many nitroxides and the sensitivity of their EPR parameters to the polarity of the embedding medium[41,46,47] make them ideal guests to investigate various supramolecular assemblies.[31,241]

5.7.1 Interaction of nitroxides with cyclodextrins

Rassat[242] *et al*. were the first to show that EPR spectroscopy is well suited to study the association of nitroxides with cyclodextrins. When nitroxide **99** (Scheme 5.55) was used as a guest for β-cyclodextrin (β-CD) in water, the free and included (**99**@β-CD) species showed separated high field EPR lines. The spectrum with the smaller nitrogen hyperfine splitting (a_N) was assigned to the complexed nitroxide, which inside the β-CD cavity resides in a less polar environment than pure water. On the other hand, on using TEMPO as guest, no clear distinction between the free and included species can be observed, the formation of the host–guest being evidenced only by an increase of the EPR high-field line width.

Kotake and Janzen studied a series of nitroxide@β-CD complexes.[243] They were the first to show that a nitroxide radical with more than one relatively bulky functional group can form two distinct host–guest complexes (Scheme 5.55). Using a high-pressure EPR system, Sueishi *et al*.[243f] established that the external pressure either increases or decreases the equilibrium constant of group-in complexes, depending on the sizes of the included group relative to that of the cyclodextrin cavity.

Lucarini *et al*.[31,244] studied the association of benzyl *tert*-butyl nitroxide, **100** (Scheme 5.56), with α-, β-, and γ-cyclodextrin. In the presence of cyclodextrins they observed additional lines on the EPR spectrum of **100** that were assigned to the radical included in the host, in equilibrium with the free species. The significantly smaller nitrogen hyperfine coupling (Table 5.4) found with β- and γ-CD with respect to water suggests that the N–O group is deeply included in the cavity of the CDs. On the other hand, with α-CD, which has a smaller (≈ 5.2 Å) internal diameter than β-CD (≈ 6.6 Å) and γ-CD (≈ 8.4 Å), the very similar

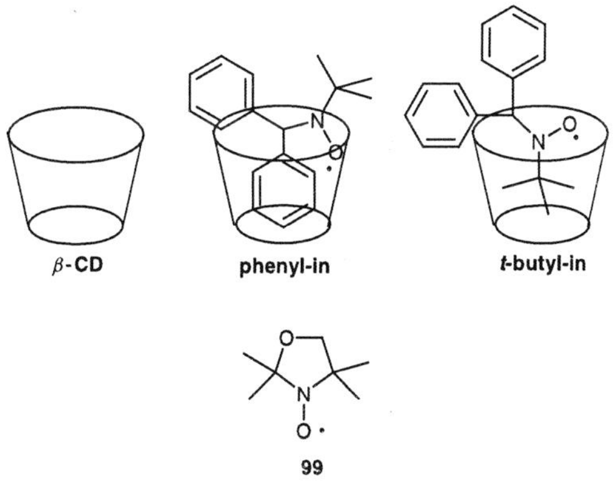

Scheme 5.55

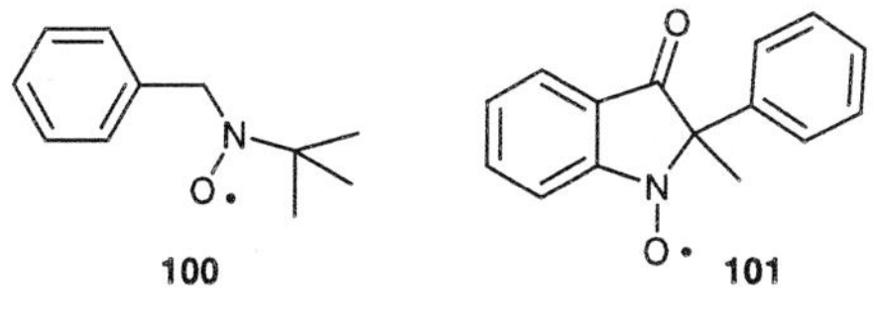

Scheme 5.56

Table 5.4 EPR parameters and binding constants for the inclusion complexes of **100** with cyclodextrins

Host	a_N/G	$a_{H\beta}$(2H)/G	g	K/M^{-1} (294 K)
none	16.69	10.57	2.0056	
α-CD	16.56	9.44	2.0058	14.3
β-CD	15.74	7.88	2.0058	1281
γ-CD	15.97	8.02	2.0058	50.7

value of a_N observed for the free and bound **100** indicates that also in the loose **100**@γ-CD complex the N–O group is exposed to the bulk water. Simulation of experimental EPR spectra recorded at different temperatures and host concentrations allowed the authors to determine the kinetic and thermodynamic parameters for association and dissociation of the inclusion complexes (Table 5.4).

Thanks to the significant difference of the EPR parameters of **100** in water, β-CD (Table 5.4) and SDS micelles ($a_N = 16.04$ G, $a_{H\beta} = 8.84$ G, $g = 2.0057$), the partitioning rate of **100** in β-CD/SDS micelles systems was determined.[245]

EPR spectroscopy has been employed[244c] to investigate the formation of complexes between heptakis-(2,6-O-dimethyl-β-cyclodextrin) (DM-β-CD) and different enantiomeric pairs of chiral nitroxides of general structure $PhCH_2N(O^{\bullet})CH(R)R'$. Inclusion by the DM-β-CD chiral cavity of the (R) and (S) enantiomers gave rise to diastereomeric inclusion complexes characterized by different EPR spectroscopic parameters. The corresponding complex stability constants were obtained by plotting the ratio between the concentrations of the complexed and free species as function of the DM-β-CD concentration in water. With the help of NMR studies on the complexes between DM-β-CD and the structurally related amine precursors of nitroxide, together with molecular dynamics, the factors responsible for chiral recognition by DM-β-CD were identified.

The binding behaviour of β-CD and 2,6-*O*-dimethyl-β-CD toward **100** in the presence of different alcohols has been studied by EPR.[244d] Data were found consistent with the formation of a binary complex alcohol@CD competing with the EPR monitored complex **100**@CD.

Lucarini *et al*. prepared[246] the first nitroxide-based [1]rotaxane by reacting a TEMPO derivative with 6-mercapto-βcyclodextrin (β-CD-SH) using alkaline water as reaction medium (Scheme 5.57). They demonstrated that the TEMPO moiety is irreversibly trapped inside the β-CD cavity, resulting in the efficient protection of the reduction of the aminoxyl group by GSH.

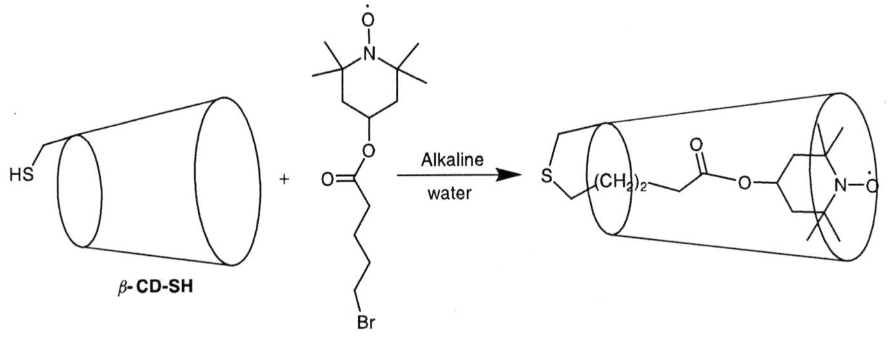

Scheme 5.57

Formation of inclusion complexes between several cyclodextrin derivatives and TEMPO and DOXYL-based spin probes was studied by EPR spectroscopy.[247] Competition between alkyl chains and nitroxide functionalities for cyclodextrin cavities leads to different types of complexation.

Complexation of β-cyclodextrin with flexible nitroxide diradicals linked by a polyethylene glycol chain was monitored by EPR spectroscopy.[248] Complexation with cyclodextrin reduces dramatically the flexibility of the biradicals and leads to the disappearance of the lines due to electron spin–spin exchange in their EPR spectra. In a different kind of spin exchange experiment, two TEMPO moieties were attached on the same rim of the cavity of two isomers of partially methylated β-cyclodextrins. The host–guest properties of these bis spin labeled cyclodextrins were investigated.[249] The nitroxides form inclusion complexes in the presence of unlabelled β-CD.

Greci *et al.*[250] have shown that the water solubility of 1,2-dihydro-2-methyl-2-phenyl-3H-indole-3-one-1-oxyl **101** (Scheme 5.56) is dramatically increased (1312-fold) in the presence of randomly methylated β-cyclodextrin (RM-β-CD). The decrease of a_N for the included **101** is particularly high ($\Delta a_N = 1.66\,G$) and, when included in the cavity of a RM-β-CD, radical **101** keeps its antioxidant activity towards the peroxidation of methyl linoleate and it is protected against photodegradation.

The formation of β-CD self-assembled structures in water above a critical aggregation concentration (3 mM) was established[251] from the EPR study of the interaction of β-CD with amphiphilic spin probes which contain a cyclic nitroxide moiety linked to different positions of their long aliphatic chain.

Tordo *et al.*[252] investigated the association of *tert*-butyl 2-methyl-1-phenylpropyl nitroxide (TIPNO, Scheme 5.5) with either 2,6-*O*-dimethyl-β-CD or permethylated-β-cyclodextrin (TRIMEB). They also prepared 236CDTIPNO composed of a TIPNO moiety covalently bound to TRIMEB and studied its self-association in water (Scheme 5.58). The results are consistent with a weak complex showing the TIPNO moiety capping the small cavity entrance.

5.7.2 Interaction of nitroxides with calix[4]arenes

The benzyl *tert*-butyl nitroxide probe **100** was also used to study inclusion phenomena in aqueous solution by two water soluble calix[4]arenes, **102** and **103**[253] (Scheme 5.59). With **102**, the value of the affinity constant is small ($K \approx 12.5\,M^{-1}$ at 298 K). Moreover, the hyperfine splitting at nitrogen remains essentially unaffected by complexation, indicating that the aminoxyl group in the complexed radical is exposed to

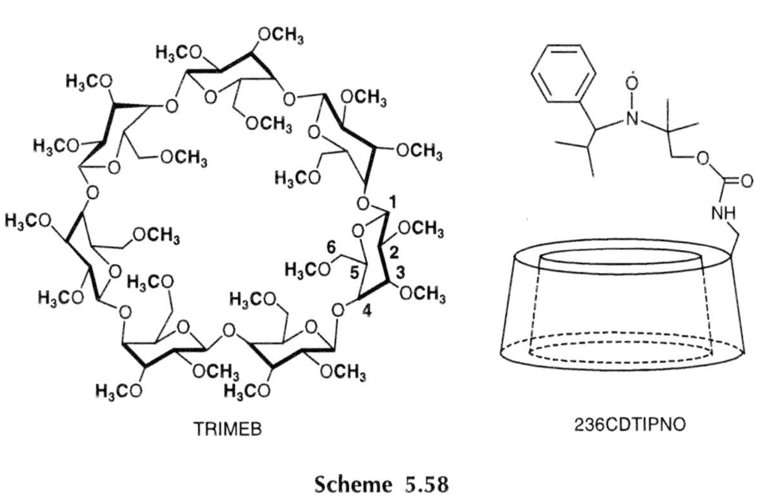

Scheme 5.58

bulk water. In the case of **103** it was not possible to detect separate signals from the free and bound nitroxide, thus suggesting that a time-averaged spectrum is observed; this could be due either to fast rates of association and dissociation or to small differences in the EPR parameters of the free and bound species.

The capsular inclusion complex of *para*-hexanoyl calix[4]arene, **104** (Scheme 5.59), with N-*tert*-butyl-N-(1-diethylphosphono-2,2-dimethylpropyl)nitroxide (SG1, Scheme 5.5) was shown to precipitate as yellow, large, square-like crystals. Single crystal X-ray diffraction showed how SG1 is inserted into the calixarene pocket.[254] EPR studies revealed a strong electron dipole–dipole interaction between guests entrapped in neighbouring capsules.

An inclusion complex of **104** with 4-methoxy-2,2,6,6-tetramethylpiperidine-*N*-oxyl, **105** (Scheme 5.59), was also crystallized and studied by X-ray diffraction and multi frequency EPR.[255]

5.7.3 Interaction of nitroxides with curcubiturils

Cucurbit[n]urils (CBn, n = 5–8, 10) are cyclic molecules consisting of n glycouril units and characterized by the presence of a hydrophobic cavity that is accessible through two identical carbonyl fringed portals.[256] Benzyl *tert*-butyl nitroxide **100** and TEMPO were used as probes to investigate host–guest interactions with cucurbit[7]uril, (CB[7]).[257a,183] With **100**, the EPR results demonstrate the formation of a stable inclusion complex **100**@CB[7], but no more information was accessible due to the interaction with CB[7] of the chemicals (parent amine and magnesium salt of monoperoxyphthalic acid) used to generate water solutions of **100**.

TEMPO was shown to form a strong inclusion complex with CB[7] ($K = 25000 \pm 2000\,M^{-1}$), the aminoxyl group being deeply immersed in the CB[7] cavity. The binding constant is one order of magnitude larger than that measured with β-CD; this huge enhancement was attributed to the larger equatorial width of CB[7] (7.3 Å against 6.5 Å for β-CD), which allows total inclusion of TEMPO inside the cavity of the host. In the presence of cations M^+, the formation of CB[7]@M^+ decreases the concentration of free CB[7], giving rise to a reduction of the apparent binding constant K_{EPR} for the host@nitroxide complex. With both nitroxides, EPR showed the formation of a ternary complex nitroxide@CB[7] (M^+). Also, it was shown that the combination of nitroxide radicals and EPR spectroscopy can be used to directly observe the concentration-dependent aggregation of CB[7] and CB[8] in water.[181,257b]

Crystals were obtained from a solution of **105** and CB[8].[183] The asymmetric unit of **105**@CB[8] contains six cucurbiturils, and the nitroxide inside each appeared to be disordered. The host–guest couples are

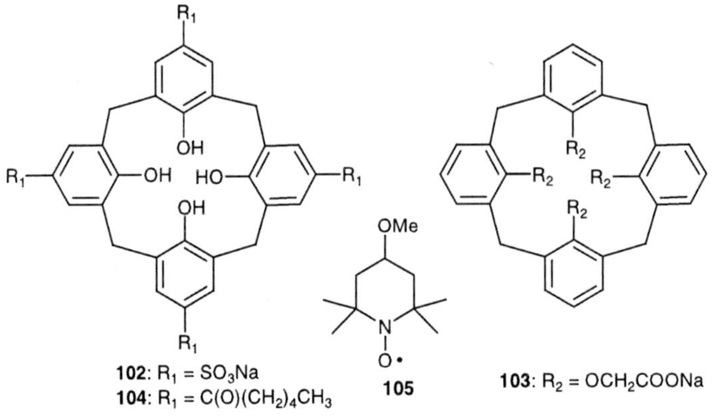

102: $R_1 = SO_3Na$
104: $R_1 = C(O)(CH_2)_4CH_3$

105

103: $R_2 = OCH_2COONa$

Scheme 5.59

arranged in supramolecular equilateral triangles. In each triangle, the distances between the oxygen atoms of the three nitroxides are in the range 8.4–9.0 Å, with an average value of 8.7 Å. When **105**@CB[8] crystals were dissolved in pure water, an X-band EPR signal composed of the superposition of the expected three-line spectrum from free **105** and a seven-line spectrum were observed. This additional spectrum exhibited a 1 : 3 : 6 : 7 : 6 : 3 : 1 seven-line pattern with a hyperfine coupling constant (a_N) of 5.13 G, corresponding to one-third of the value observed for an included TEMPO-like nitroxide ($a_N \approx 15.5$ G). This spectrum is characteristic of a system composed of three identical nitroxide moieties coupled through spin exchange with $J_{12}, J_{23}, J_{13} \gg a_N$ in agreement with the "trinitroxide" supraradical {**105**@CB[8]}$_3$ crystal structure.

5.7.4 Interaction of nitroxides with micelles

Two dynamic processes take place in micellar solutions: the faster process is the association–dissociation process involving the exchange of individual surfactant molecules between the micelles and the water phase, while the slower process is identified as the rearrangement of the system involving the creation and destruction of micelles.[258] When the micelles contain hydrophobic solutes, another type of dynamic process takes place, namely, the exchange of solute molecules between the micellar and the water phases. Monitoring of nitroxide probes with EPR has been used to obtain useful information on their interaction in micellar environments.[259]

Tordo *et al.*[260] investigated the interaction of a series of stable β-phosphorylated cyclic aminoxyl radicals with SDS micelles by EPR spectroscopy. The hyperfine coupling constant with the phosphorus atom (a_P) proved to be very sensitive to the environment; as a result, two distinct spectra corresponding to the nitroxide in bulk water and in micelles were observed (Scheme 5.60).

EPR investigation of the kinetics of the exit and re-entry processes in SDS micelles of **100** and various *para*-substituted benzyl *tert*-butyl nitroxides and *para*-substituted benzyl hydroxyalkyl nitroxides was also reported.[261] Micellar aggregation of sulfonate surfactants and alkyltrimethylammonium bromide surfactants [$CH_3(CH_2)_{n-1}N(CH_3)_3Br$; n = 6, 8, 12, 16] was investigated by EPR of TEMPO-choline and the sodium salt of 3-carboxy-PROXYL, respectively.[262]

There is limited information related to the participation of free radical processes and the antioxidant mechanism involved in biological membranes. A prefluorescent hydrophobic TEMPO derivative has been used to evaluate local reactivity of phenolic antioxidant in micellar systems[263] (Scheme 5.61). An apparent rate constant that directly reflects the relevance of antioxidant hydrophobicity on the reaction toward the nitroxide has been defined. Dramatic differences in the apparent rate constant were found on moving from methanol to micellar media (90- and 230-fold enhancements for SDS and Triton X100 micelles respectively).

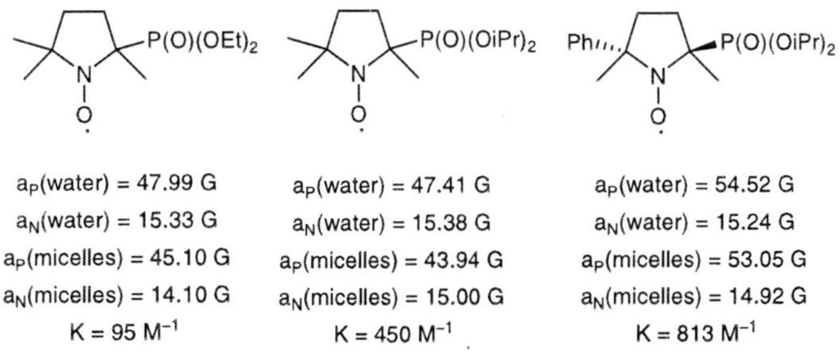

Scheme 5.60

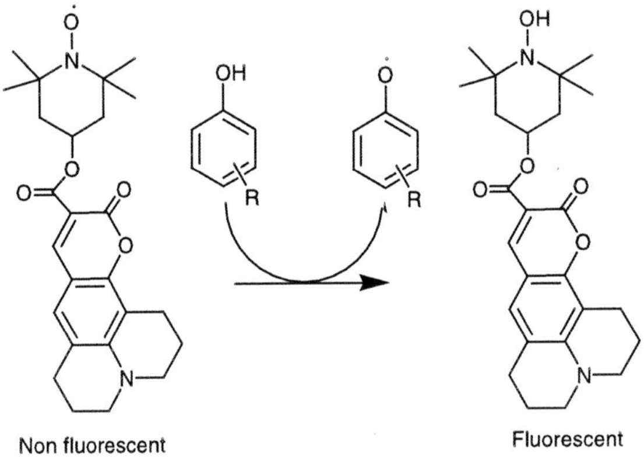

Scheme 5.61

An interesting analysis tool for the selective detection of local water inside soft molecular assemblies suspended in bulk water has been developed.[264] The ^{1}H NMR signal of water interacting with specifically localized stable nitroxides is amplified through Dynamic Nuclear Polarization.

The quenching of fluorescence of pyrene derivatives by nitroxides has been used to investigate the properties of different micelles.[265]

The partition isotherms and exchange rates of *para*-substituted benzyl hydroxyalkyl nitroxides between an aqueous solution and the monolayer of water soluble protected gold nanoparticles have been determined by EPR (Scheme 5.62).[266]

5.7.5 Fullerene-linked nitroxides

The synthesis and properties of fullerenes bearing nitroxide moiety(ies) have been reported in the last decade. Electron transfer, spin coupling in di-, trinitroxide–fullerene adducts, and interactions between the nitroxide unpaired electron and the fullerene excited triplet state have been studied (Scheme 5.63).[267–269]

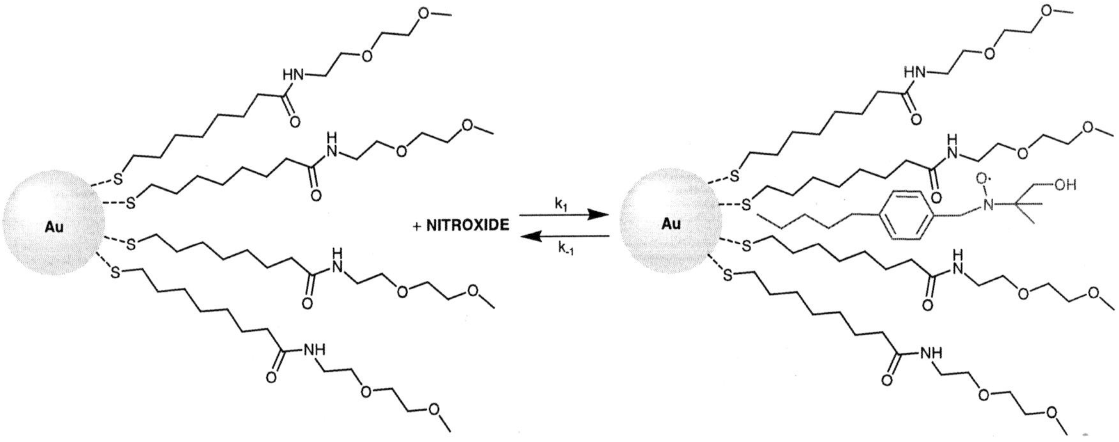

Scheme 5.62

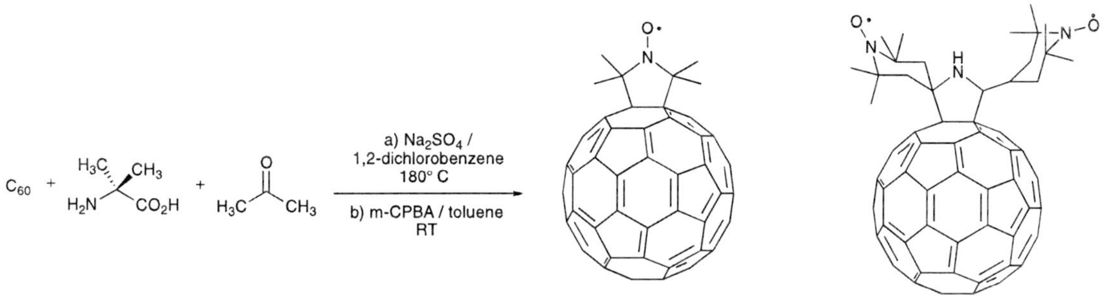

Scheme 5.63 A fulleropyrrolidine nitroxide and a fulleropyrrolidine dinitroxide

The inner environment of single-walled carbon nanotubes has been investigated by EPR using a fullerene nitroxide probe that was inserted using supercritical carbon dioxide.[270]

5.8 Nitroxides for dynamic nuclear polarization (DNP) enhanced NMR

NMR spectroscopy is a widely used analytical technique, and Magnetic Resonance Imaging (MRI) is one of the most powerful clinical imaging methods, due to a number of advantages including excellent resolution, versatility, non-invasive nature, and the use of low-energy frequency irradiation. However, the inherently low sensitivity of NMR is a major limitation and hampers the extension of its applications. The lack of sensitivity results from the low magnetic energy of nuclear spins compared to thermal energy at room temperature.

Improvements in instrumentation by the development of magnets operating at very high fields and of cryogenic probes have led to significant but still limited enhancement of sensitivity. In recent years, large NMR signal enhancements were obtained through different approaches based on spin polarization transfer. Among these approaches, experiments involving hyperpolarized noble gases,[271,272] parahydrogen-induced polarization (PHIP),[273,274] photochemical reaction centers (photoCIDNP),[275,276] or stable free radicals (Dynamic Nuclear Polarization, DNP)[277] are very appealing. From this set of methods, DNP appears to be the most versatile and permits a large number of applications both in solid and liquid states.

5.8.1 DNP for biological NMR and real-time metabolic imaging

In the last decade, the achievements of Griffin's group gave a strong emphasis to the development of DNP for enhancing the spin polarization in high field MAS NMR experiments,[278–280] and signal enhancement values up to 330 were reported.

Figure 5.5 shows the clear improvement in signal-to-noise ratio obtained using DNP SSNMR at 90 K and 5 T in the active site of bacteriorhodopsin (bR), a membrane protein acting as a light-driven ion pump.[281] The obtained enhancement of 50 provided sufficient resolution and shorter data acquisition to gain 2-D spectra with high quality.

More recently, the generation of hyperpolarized spin systems in the solid state at very low temperatures, followed by a rapid dissolution process and sample transfer to the NMR/MRI apparatus, has been developed by the Ardenkjær-Larsen and Golman groups. This technique gained a tremendous interest in the MRI field allowing real-time metabolic studies and improved spatial distribution of hyperpolarized molecules.[282–284] In Figure 5.6 (left part) are shown two ^{13}C NMR spectra of urea at 9.4 T, in water at room temperature. Spectrum (a) (1 scan), was obtained after DNP (20 % polarization), spectrum (b) was obtained at the

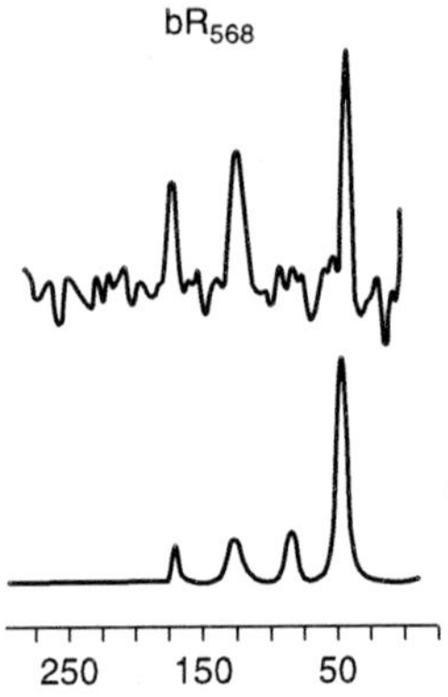

Figure 5.5 *DNP improvement of the 1D ^1H decoupled ^{15}N MAS spectrum of light adapted ζ-^{15}N-Lys-bR. (Reprinted from [281], Copyright 2001, with permission from Elsevier.)*

thermal equilibrium after 65 hours of acquisition. After the injection of $^{13}C_1$-enriched pyruvate in a rat (Figure 5.6, right), ^{13}C NMR spectra were acquired with a time interval of three seconds at 1.5 T on a clinical MRI. This experiment showed the capability of *in vivo* monitoring the fate of pyruvate in a rat using clinical MRI and opens up the way to clinical molecular imaging with MRI.

5.8.2 Nitroxides as polarizing agents for DNP

In 1953, Overhauser proposed an experiment where the large Boltzmann polarization of unpaired electrons could be transferred to surrounding nuclei by saturating the corresponding electron paramagnetic resonance (EPR) transition.[287] DNP is the extension of the original Overhauser effect from metals to frozen liquids[288] and liquids. A prerequisite for DNP is the presence of paramagnetic centers in the sample at concentration low enough to limit the paramagnetic broadening of the NMR spectrum. On irradiation close to the frequency of the EPR transition, the large polarization of the electron spin system can be transferred to the nearby nuclei through the dipolar interaction between the electron and nuclear spins.

Since the early 1960s, the DNP process was first demonstrated experimentally and intensively studied at low magnetic field.[289–294] The maximum theoretical enhancement achievable is given by the gyromagnetic ratio (γ_e/γ_i), being $\sim$ 660 for ^1H nuclei and $\sim$ 2600 for ^{13}C nuclei. However, extending DNP to higher fields proved to be challenging. Hyperpolarization processes can be performed in the liquid or in the solid states, at low or very low temperatures, and at 0.34 T or up to 5 T. Several different and even potentially multiple processes can then be involved in DNP experiments. To date, this has prevented the complete DNP mechanistic understanding for paramagnetic centers with broader EPR spectrum than the Larmor frequency of the nuclei that are polarized.

Different kinds of paramagnetic species (chromium(V) complexes, char, BDPA, trityls, nitroxides)[295–298] have been employed, and it appears that TEMPO-like nitroxides are among the most promising polarizing agents.[299,300] Nitroxides have been shown to have strong dipolar coupling to water, which can result in large DNP enhancement. Nitroxides have the additional advantages of being easily available and non-toxic at the concentration used for *in vivo* applications. It has been shown in ^1H-DNP experiments in aqueous solution and pumping at 9.7 GHz microwave frequency, that using ^2H and ^{15}N isotope-labeled nitroxides the effective saturation factor can be increased.[301]

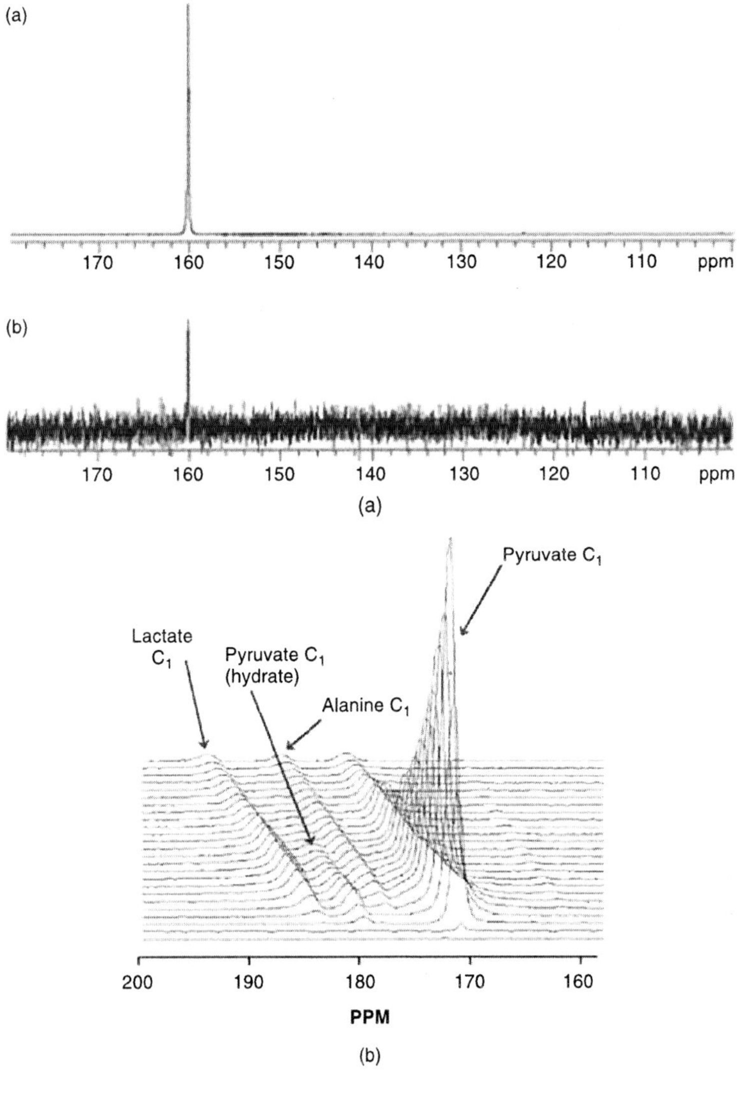

Figure 5.6 *(Left)* ^{13}C *NMR spectra at 9.4 T and room temperature of (A) urea (59.6 mM in water) hyperpolarized by the DNP-NMR method and (B) urea (59.6 mM in water) at thermal equilibrium (acquisition time: 65 h) (Reprinted from [285], Copyright 2003, with permission from National Academy of Sciences, U.S.A.)*[285] *(Right) In vivo metabolic production of lactate and alanine after the injection of hyperpolarized* $^{13}C_1$ *pyruvate in a rat.* ^{13}C *NMR spectra were acquired with a time interval of 3 s at 1.5 T on a clinical MRI. (Reprinted from [286], Copyright 2003, with permission from National Academy of Sciences, U.S.A.)*[286]

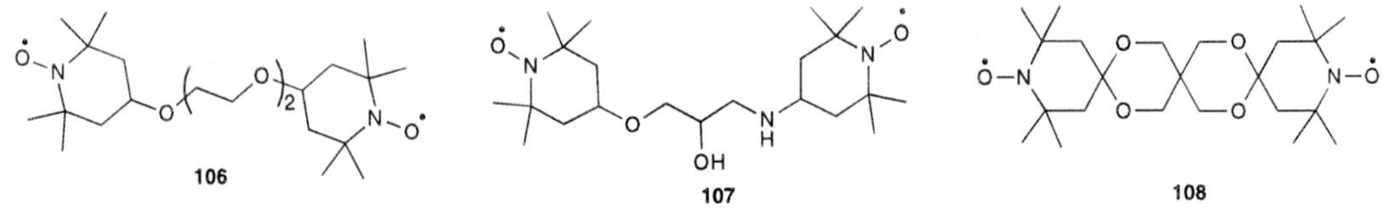

Scheme 5.64

At high magnetic field in the solid state (90 K) and pumping at 260 GHz microwave frequency, Griffin reported that using biradicals as polarizing agents, the Cross Effect (CE) mechanism leads to improved polarization enhancements when comparing with Solid Effect (SE) and Thermal Mixing (TM) mechanisms. The more efficient biradicals (Scheme 5.64) consist of two TEMPO nitroxides tethered by an ethylene glycol chain (**106**),[302] a more water soluble TEMPO-based dinitroxide bearing a propan-2-ol tethering chain (Totapol, **107**),[303] and a rigid binitroxide bearing orthogonal TEMPO moieties (bTbK, **108**).[304] Compared with that obtained with TEMPO, the signal enhancement observed with biradical dinitroxides **106**, **107** and **108** increased by factors of about 4, 5.5 and 7.5 respectively.

Immobilized TEMPO-based nitroxides on agarose gel were prepared and studied as possible polarizing medium for liquid DNP experiments at 0.35 T and room temperature for the preparation of radical-free steady state injection solutions.[305] However, the signal enhancement (−38) obtained using the TEMPO agarose gel is lower than with TEMPO (−110) dissolved in water at the same concentration. This effect was attributed to a lower coupling factor for the agarose-bound TEMPO.

5.9 Nitroxides as pH-sensitive spin probes

Local pH value is one of the most important parameters in the biochemistry of living systems and pH changes have been associated with many physiological and pathological processes. Current methods for assessing local pH involve invasive procedures or transparent samples, and typically can assess pH values from a limited numbers of locations. *In vitro* and *in vivo* measurements of pH in a non-invasive way would provide useful information in clinical diagnostic applications. In the last two decades, imidazoline and imidazolidine-based stable nitroxides[306–308] have shown to be interesting candidates as pH spin probes due to the effect of pH changes on their EPR spectra (Scheme 5.65).[309] The change in A_N between the protonated and the unprotonated forms can reach values close to 0.1 mT.

Along with progress in low-frequency EPR, such as Longitudinally-Detected EPR (LODEPR), Proton Electron Double-Resonance Imaging (PEDRI), and Field-Cycled Dynamic Nuclear Polarization (FC-DNP), the design of efficient pH-sensitive nitroxide spin probes is a key goal in the successful development of such diagnostic methods.

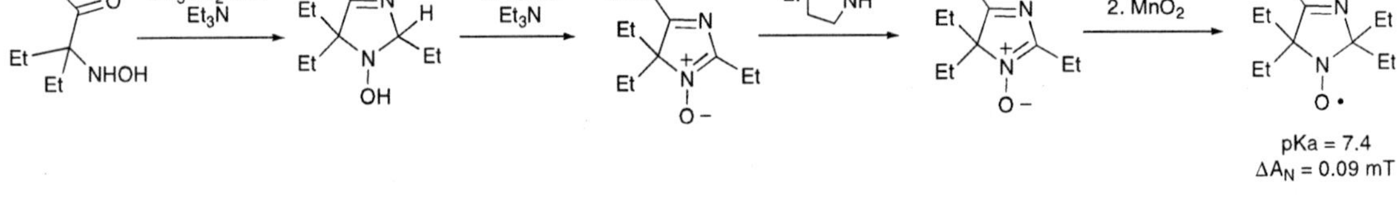

Scheme 5.65 Synthesis of a pH sensitive nitroxide

5.10 Nitroxides as prefluorescent probes

Nitroxides are efficient excited state quenchers, and thus they have been associated to dyes in inter- and intramolecular systems to form responsive fluorescent probes. Several mechanisms have been proposed for quenching by free radicals, for example, exchange induced relaxation processes (intersystem crossing and internal conversion), electron transfer, and energy transfer interactions.[310]

The nitroxide–fluorophore systems have been mostly employed to investigate dynamic processes in biochemical systems,[311] to evaluate the reactivity or the concentration of antioxidants in homogeneous or micellar systems (Scheme 5.66),[312,313] for the detection of free radicals,[314] and to study polymer degradation processes due to oxidative damages such as observed in polypropylene for instance.[315] Recently, the development of free radical sensors based on TEMPO functionalized quantum dots has also been reported.[316,317]

5.11 EPR-spin trapping technique

A chapter dealing with stable nitroxides and their applications cannot avoid mentioning EPR-spin trapping (EPR-ST), a technique which was first introduced in the 1970s by Janzen and Blackburn.[318] In the last 20 years, oxygen-centered free radicals have been recognized to be of particular biomedical importance, due to their role as critical mediators in various physiological and pathophysiological processes.[319] However, the combination of rapid decay and low steady state concentration prevents the direct EPR detection of most biological relevant radicals under physiological conditions. The use of EPR-ST allows detecting low concentrations of short-lived radicals. Indeed, a short-lived radical (**Y$^{\bullet}$**) reacts specifically with a diamagnetic molecule (mainly a nitrone or a nitroso compound) to produce a nitroxide spin adduct persistent enough to be detected by EPR spectroscopy (Scheme 5.67). The study of the spin adduct EPR signal, characterized by its spectral pattern and its hyperfine coupling constant(s), brings information about the nature and the structure of the radical species trapped.

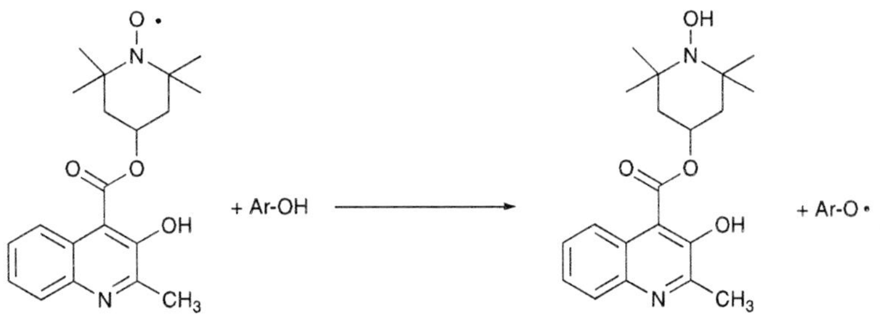

Scheme 5.66 Hydrogen transfer from phenols to a prefluorescent nitroxide

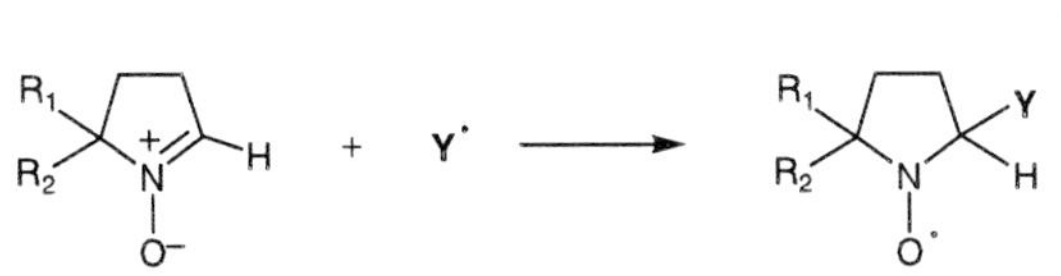

Scheme 5.67 Principle of EPR-spin trapping technique

This technique has been used to trap and study short-lived free radicals, either in organic solvents for mechanistic purposes[320] or in aqueous media, using nitrones as traps, to study the oxidative stress and, more specifically, the superoxide anion radical.[321] Superoxide is produced by one-electron reduction of molecular oxygen during mitochondrial respiration. It constitutes the main source of various reactive oxygen species *in vivo*, like peroxynitrite ($ONOO^-$), hydrogen peroxide (H_2O_2) and hydroxyl radical ($HO^\bullet$).

The study of transient free radicals in biological systems with EPR-ST is still limited by various drawbacks: (i) the reduction of nitroxide spin adducts to EPR silent compounds,[322] (ii) the small rate constants observed for the trapping of superoxide compared to the rates of its reactions with various endogenous biocomponents (superoxide dismutase (SOD), ferricytochrome c, thiols, nitric oxide), and (iii) the very low steady state concentration of radicals.

PBN (α-phenyl *tert*-butylnitrone) (**109**) and DMPO (5,5-dimethyl 1-pyrroline N-oxide) (**110**) were the first nitrones to be used as spin traps but their superoxide spin adducts exhibit a poor half-life (Scheme 5.68). To improve the efficiency of the trapping of superoxide, numerous spin traps were prepared; DEPMPO (**111**),[323] EMPO (**112**),[324] and numerous analogues reported by Nohl's group (**115**),[325] BMPO (**113**),[326] AMPO (**114**),[327] and NHS-DEPMPO (**116**),[328] from which different DEPMPO derivatives like Mito-DEPMPO (**117**)[329] and β-CD-DEPMPO (**119**)[330] were synthesized.

The half-lives of the superoxide spin adducts of these new nitrone spin traps ranges from 7 to 45 minutes, while for PBN and DMPO it barely reaches one minute (Scheme 5.68).

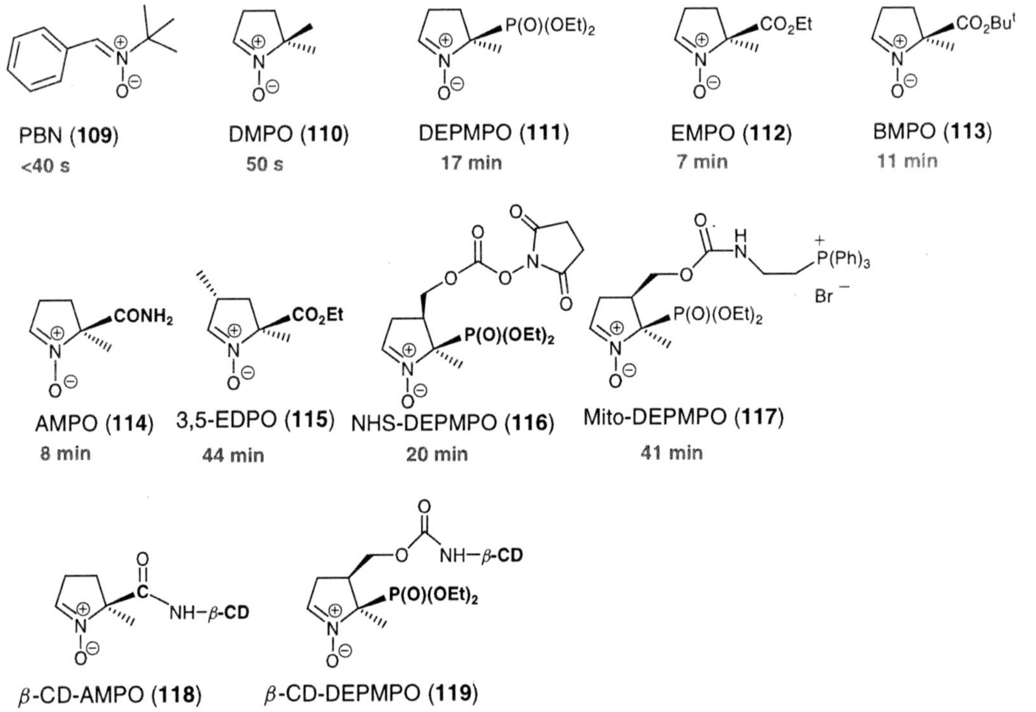

PBN (**109**) — $^{(1)}t_{1/2}$: <40 s; DMPO (**110**) 50 s; DEPMPO (**111**) 17 min; EMPO (**112**) 7 min; BMPO (**113**) 11 min

AMPO (**114**) 8 min; 3,5-EDPO (**115**) 44 min; NHS-DEPMPO (**116**) 20 min; Mito-DEPMPO (**117**) 41 min

β-CD-AMPO (**118**) 27.5 min; β-CD-DEPMPO (**119**) 36 min

Scheme 5.68 Structure of nitrone spin traps and half-life values of their respective superoxide spin adduct measured in buffer solutions at physiological pH

In vivo, the nitroxide spin adducts can be rapidly reduced to the corresponding EPR silent hydroxylamines, due to the presence of relatively high concentrations of reducing agents like reduced glutathione (GSH), L-ascorbate anion, and so on. The protection of spin adducts toward bioreduction is one of the most important current challenges for the EPR-ST study of free radicals released in biological systems and also to other important biological applications of nitroxides.

As already mentioned in Section 5.7, inclusion of nitroxides in the cavity of various hosts appears as a promising approach to achieve their protection in biological milieu. When cyclodextrin is used in buffer solutions to encapsulate superoxide adducts of various pyrroline-*N*-oxides, a sevenfold enhancement in stability and a partial protection against glutathione peroxidase- and L-ascorbate anion-induced reduction were reported.[331] The same enhancement was observed during the trapping of glutathiyl radical with PBN.[332] Bardelang reported the first synthesis of a nitrone grafted to a permethylated β-cyclodextrin.[333] Villamena published the synthesis of β-CD-AMPO, a cyclic nitrone grafted on a cyclodextrin (Scheme 5.68).[334] A DEPMPO-appended β-cyclodextrin (β-CD-DEPMPO **119**,[333] Scheme 5.68) was synthesized using NHS-DEPMPO (**116**) as versatile building block. Very promising preliminary *in vitro* results have been obtained with **119**: its superoxide adduct has a half-life of about 36 minutes and showed a greatly improved resistance to bioreduction processes compared to what was observed with DEPMPO in the presence of cyclodextrins.[333]

5.11.1 Immuno spin trapping

Mason and coworkers have developed a new highly specific and sensitive technique capable of detecting protein[335] and DNA radicals.[336] This technique was named "immuno spin trapping" and associates EPR spin trapping with immuno assays. DMPO (**110**) is used as the spin trap (Scheme 5.69). After a protein radical has been trapped by DMPO, the nitroxide spin adduct goes to one-electron oxidation, converting it to an EPR silent DMPO-protein adduct.

Then, this DMPO-protein adduct can be detected by polyclonal antibodies (anti-DMPO). This immuno technique was initially reported for the study and the identification of protein-derived free radical species generated during the reaction of myo-[338a] and hemoglobin[338b,337] with hydrogen peroxide. Since the first paper by Mason *et al.* in 2002 reporting the advantages of this technique, more than 20 papers have been published investigating free radical intermediate production during cell protein damaging processes.[338,339]

5.11.2 Conclusion

The goal of getting spin traps that afford nitroxide spin adducts exhibiting half-lives from 15 to 40 minutes has actually been achieved. However, combining a higher persistency for the nitroxide spin adducts and a faster addition of the free radical species on the nitrone should lead to higher steady state concentrations in solution, facilitating the free radical detection. Associated with an improved protection against bioreduction and the high performance EPR equipments available today, EPR-spin trapping should represent a unique tool to investigate the fate of free radicals in biological systems.

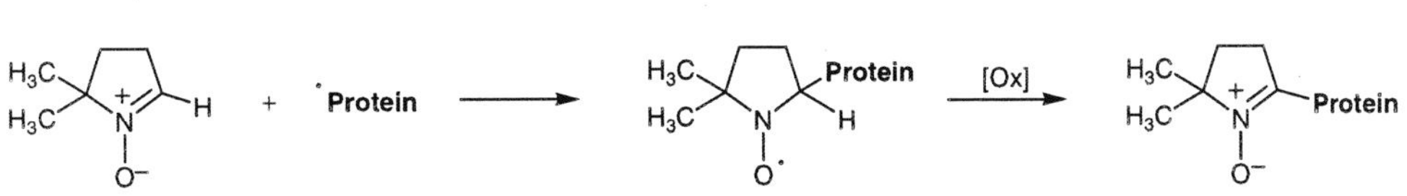

Scheme 5.69 Trapping of protein radical by DMPO

5.12 Conclusions

Nitroxides and di- or polynitroxides continue to arouse great interest in a large number of research topics: organic synthesis, Nitroxide Mediated Polymerization (NMP), Magnetic Resonance Imaging (MRI), free radical biology, supramolecular assemblies, Dynamic Nuclear Polarization (DNP), organic batteries, antioxidants, organic magnets, and so on. These different domains of research need suited nitroxides or polynitroxides exhibiting appropriate characteristics (redox potential, rate of trapping of free radicals, relaxivity, biocompatibility, ferromagnetic interactions, etc.). The unique advantage of nitroxides compared to other classes of stable free radicals is that the aminoxyl group can resist the experimental conditions needed to perform various organic reactions. As a consequence, the chemistry of nitroxides is very rich, offering access to an almost unlimited number of molecules suited for specific applications.

References

1. D. Griller and K. U. Ingold, *Acc. Chem. Res.*, **9**, 13–19 (1976).
2. E. G. Janzen, *Acc. Chem. Res.*, **4**, 31–40 (1971).
3. O. Piloty and B. G. Schwerin, *Berichte der Deutschen Chemischen Gesellschaft*, **34**, 1870–1887 (1901).
4. A. N. Holden, W. A. Yager and F. R. Merity, *J. Chem. Phys.*, **19**, 1319 (1951).
5. (a) H. Wieland and M. Offenbächer, *Berichte der Deutschen Chemischen Gesellschaft*, **47**, 2111–2115 (1914). (b) H. Wieland and K. Roth, *Berichte der Deutschen Chemischen Gesellschaft*, **53**, 210–230 (1920).
6. K. H. Meyer and W. Reppe, *Berichte der Deutschen Chemischen Gesellschaft*, **54**, 327–337 (1921).
7. O. L. Lebedev, M. I. Khidekel and G. A. Razuvaev, *Doklady Akademii Nauk SSSR*, **140**, 1327–1329 (1961).
8. M. B. Neiman, E. G. Rozantsev and Yu G. Mamedova, *Nature*, **196**, 472–474 (1962).
9. (a) E. G. Rozantsev and V. D. Sholle, *Synthesis*, 190–202 (1971). (b) E. G. Rozantsev and V. D. Sholle, *Synthesis*, 410–414 (1971). (c) E. G. Rozantsev and V. D. Sholle, *Synthesis*, 895–916 (1984).
10. (a) Y. Brunel, H. Lemaire, and A. Rassat, *Bull. Soc. Chim. Fr.*, **8**, 1895–1899 (1964). (b) R. M. Dupeyre and A. Rassat, *J. Am. Chem. Soc.*, **88**, 3180–3181 (1966). (c) R. Ramasseul and A. Rassat, *Tet. Lett.*, **48**, 4623–4624 (1971). (d) R. M. Dupeyre, A. Rassat, and J. Ronzaud, *J. Am. Chem. Soc.*, **96**, 6559–6568 (1974). (e) P. Michon and A. Rassat, *J. Am. Chem. Soc.*, **97**, 696–700 (1975). (f) R. Chiarelli, A. Jeunet, J. Michon, *et al.*, *Organic Magnetic Resonance*, **13**, 216–217 (1980).
11. (a) A. R. Forrester, J. M. Hay and R. H. Thomson, *Organic Chemistry of Stable Free Radicals*, Academic Press, New York, 1968. (b) E. G. Rozantsev, *Free Nitroxyl Radicals*, Plenum Press, New York, 1970. (c) L. B. Volodarsky, *Imidazoline Nitroxides*, CRC Press Inc., Boca Raton, 1988. (d) E. Breuer, H. G. Aurich and A. Nielsen, *Nitrones, Nitronates, and Nitroxides*, John Wiley and Sons Ltd, Chichester, 1989. (e) L. B. Volodarsky, V. A. Reznikov and V. I. Ovcharenko, *Synthetic Chemistry of Stable Nitroxides*, CRC Press Inc., Boca Raton, 1994. (f) N. Kocherginsky and H. M. Swartz, *Nitroxide Spin Labels: Reactions in Biology and Chemistry*, CRC Press Inc., Boca Raton, 1995. (g) A. Alberti (ed.), *Nitroxide Radicals and Nitroxide Based High-Spin Systems* Series: Landolt-Boernstein, Group 2, Subvolume D, Springer Verlag, New York (2005). (h) G. I. Likhtenshtein, J. Yamauchi, S. Nakatsuji, *et al.*, *Nitroxides: Applications in Chemistry, Biomedicine, and Materials Science*, Wiley-VCH Verlag GmbH, Weinheim, 2008.
12. E. F. Ullman, J. H. Osiecki, D. G. B. Boocock and R. Darcy, *J. Am. Chem. Soc.*, **94**, 7049–7059 (1972).
13. H. B. Hass and E. F. Riley, *Chem. Rev.*, **32**, 373–430 (1943).
14. L. R. Mahoney, G. D. Mendenhall and K. U. Ingold, *J. Am. Chem. Soc.*, **95**, 8610–8614 (1973).
15. I. Novak and L. J. Harrison, *J. Org. Chem.*, **69**, 7628–7634 (2004).
16. (a) D. F. Bowman, T. Gillan and K. U. Ingold, *J. Am. Chem. Soc.*, **93**, 6555–6561 (1971). (b) K. Adamic, D. F. Bowman, T. Gillan and K. U. Ingold, *J. Am. Chem. Soc.*, **93**, 902–908 (1971). (c) D. F. Bowman, J. L. Brokenshire, T. Gillan and K. U. Ingold, *J. Am. Chem. Soc.*, **93**, 6551–6555 (1971). (d) E. A. Lissa, M. A. Rubio, D. Araya, and G., Zanocco, *Int. J. Chem. Kinet*, **12**, 871–881 (1980). (e) S. Jousset and J. M. Catala, *Macromolecules*, **33**, 4705–4710 (2000).
17. A. Capiomont, B. Chion, and J. Lajzérowicz, *Acta Cryst. Sect. B*, **27**, 322–326 (1971)

18. (a) J. W. Turley and F. P. Boer, *Acta Cryst. Sect. B*, **28**, 1641–1644 (1972). (b) S. S. Ament, J. B. Wetherington, J. W. Moncrief, *et al.*, *J. Am. Chem. Soc.*, **95**, 7896–7897 (1973). (c) J. C. A. Boeyens and G. J. Kruger, *Acta Cryst. Sect. B*, **26**, 668–672 (1970). (d) B. Chion, A. Capiomont and J. Lajzérowicz, *Acta Cryst. Sect. B*, **28**, 618–619 (1972). (e) D. Bordeaux and J. Lajzérowicz, *Acta Cryst. Sect. B*, **33**, 1837–1840 (1977). (f) J. B., Wetherington, S. S. Ament, J. W. Moncrief, *Acta Cryst. Sect. B*, **30**, 568–573 (1974).
19. (a) A. Capiomont, D. Bordeaux and J. Lajzérowicz, *C. R. Acad. Sci. Sér. C*, **275**, 317–320 (1972). (b) L. J. Berliner, *Acta Cryst. Sect. B*, **26**, 1198–1202 (1970). (c) J. Lajzérowicz-Bonneteau *Acta Cryst. Sect. B*, **24**, 196–199 (1968). (e) D. Bordeaux and J. Lajzérowicz, *Acta Cryst. Sect. B*, **30**, 790–792 (1974). (f) R. P. Shibaeva, L. O. Atovmyan, M. G. Neigauv, *et al.*, *Zh. Strukt. Khim.*, **13**, 887–892 (1972). (g) A. Capiomont, *Acta Cryst. Sect. B*, **28**, 2298–2301 (1972).
20. A. Grand and P. Rey, *Acta Cryst. Sect. B*, **35**, 2149–2153 (1979).
21. See chapter devoted to "Living Radical Polymerization" (Chapter 11).
22. (a) S. Grimaldi, J. P. Finet, A. Zeghdaoui, *et al.*, *ACS Polym. Prepr.*, **38**, 651–652 (1997). (b) S. Grimaldi, J. P. Finet, F. Le Moigne, *et al.*, *Macromolecules*, **33**, 1141–1147 (2000).
23. C. J. Hawker, A. W. Bosman and E. Harth, *Chem. Rev.*, **101**, 3361–3388 (2001).
24. (a) A. Studer, K. Harms, C. Knoops, *et al.*, *Macromolecules*, **37**, 27–34 (2004). (b) S. Acerbis, D. Bertin, B. Boutevin, *et al.*, *Helv. Chim. Act.*, **89**, 2119–2132 (2006). (c) O. Lagrille, N. R. Cameron, P. A. Lovell, *et al.*, *J. Polym. Sc. A*, **44**, 1926–1940 (2006). (d) S. Grimaldi, J. P. Finet, D. Siri and P. Tordo, *Acta Cryst. Sect. C*, **54**, 1712–1714 (1998).
25. A. Nilsen and R. Braslau, *J. Polym. Sc. A*, **44**, 697–717 (2006).
26. F. Aldabbagh, P. Dervan, M. Phelan, *et al.*, *J. Polym. Sc. A*, **44**, 3892–3900 (2003).
27. (a) O. Kahn, *Magnetism: A Supramolecular Function*, Kluwer, Dordrecht, 1996 (b) P. M. Lahti, *Magnetic Properties of Organic Materials*, Marcel Dekker, New York, 1999. (c) J. S. Miller, M. Drillon, *Magnetism: Molecules to Materials*, vol. I-IV, Wiley-VCH Verlag GmbH, Weinheim, 2001-2003. (d). D. Luneau and P. Rey, *Coord. Chem. Rev.*, **249**, 2591–2611 (2005).
28. (a) Y. Miyazaki, A. Inaba, M. Sorai, *et al.*, *J. Phys. Chem. B*, **112**, 8144–8150 (2008). (b) D. Maspoch, L. Catala, P. Gerbier, *et al.*, *Chem. Eur. J.*, **8**, 3635–3645. (c) M. Baskett and P. M. Lahti, *Polyhedron*, **24**, 2645–2652 (2005). (d) G. Kurokawa, T. Ishida and T. Nogami, *Chem. Phys. Lett.*, **392**, 74–79 (2004). (e) A. Rajca, M. Takahashi, M. Pink, *et al.*, *J. Am. Chem. Soc.*, **129**, 10159–10170 (2007). (f) Z. Delen and P. M. Lahti, *Polyhedron*, **26**, 2031–2036 (2007). (g) S. Kanaya, M. Kozaki, D. Shiomi, *et al.*, *Synth. Met.*, **121**, 1808–1809 (2001). (h) N. Ikuma, R. Tamura, S. Shimono, N. Kawame, O. Tamada, J. *et al.*, *Mendeleev Commun.*, **13**, 109–111 (2003).
29. (a) N. Ikuma, R. Tamura, S. Shimono, *et al.*, *Angew. Chem. Int. Ed.*, **43**, 3677–3682 (2004). (b) Y. Uchida, R. Tamura, N. Ikuma, *et al.*, *Mol. Cryst. Liq. Cryst.*, **479**, 213/[1251]–221/[1259] (2007). (c) R. Tamura, Y. Uchida and N. Ikuma, *J. Mater. Chem.*, **18**, 2872–2876 (2008). (d) N. A. Chumakova, A. Kh. Vorobiev, N. Ikuma, *et al.*, *Mendeleev Commun.* **18**, 21–23 (2008).
30. (a) R. Braslau, V. Chaplinski and P. Goodson, *J. Org. Chem.*, **63**, 9857–9864 (1998). (b) S. E. Bottle, D. G. Gillies, D. L. Hughes, *et al.*, *J. Chem. Soc., Perkin Trans 2*, 1285–1291 (2000). (c) J. Einhorn, C. Einhorn, F. Ratajczak, *et al.*, *Tetrahedron Lett.*, **39**, 2565–2568 (1998). (d) N. Benfaremo, M. Steenbock, M. Klapper, *et al.*, *Liebigs Ann.*, 1413–1415 (1996). (e) J. F. Polienko, Th. Schanding, Y. V. Gatilov, *et al.*, *J. Org. Chem.*, **73**, 502–510 (2008).
31. P. Franchi, M. Lucarini and G. F. Pedulli, *Curr. Org. Chem.*, **8**, 1831–1849 (2004).
32. T. Yagi and O. Kikuchi, *J. Phys. Chem. A*, **103**, 9132–9137 (1999).
33. (a) V. Barone, A. Bencini and A. di Matteo, *J. Am. Chem. Soc.*, **119**, 10831–10837 (1997). (b) V. Barone, A. Bencini, M. Cossi, *et al.*, *J. Am. Chem. Soc.*, **120**, 7069–7078 (1998). (c) R. Improta, G. Scalmani and V. Barone, *Chem. Phys. Lett.*, **336**, 349–356 (2001). (d) G. A. A. Saracino, A. Tedeschi, G. D'Errico, *et al.*, *J. Phys. Chem. A*, **106**, 10700–10706 (1997). (d) M. Pavone, C. Benzi, F. De Angelis and V. Barone, *Chem. Phys. Lett.*, **395**, 120–126 (2004). (e) R. Improta and V. Barone, *Chem. Rev.*, **104**, 1231–1253 (2004). (f) M. Pavone, P. Cimino, O. Crescenzi, *et al.*, *J. Phys. Chem. B*, **111**, 8928–8939 (2007). (g) V. Barone, P. Cimino and E. Stendardo, *J. Chem. Theory and Comput.*, **4**, 751–764 (2008).
34. C. Houriez, N. Ferré, M. Masella and D. Siri, *J. Phys. Chem.*, **128**, 24504/1–24504/13 (2008).

35. L. Rintoul, A. S. Micallef and S. E. Bottle, *Spectrochimica Acta Part A*, **70**, 713–717 (2008).
36. D. Sezer, J. H. Freed and B. Roux, *J. Phys. Chem. B*, **112**, 5755–5767 (2008).
37. D. Siri, A. Gaudel-Siri and P. Tordo, *J. Mol. Struct. (Theochem)*, **582**, 171–185 (2002).
38. G. Pèpe and D. Siri, *Stud. Phys. Theor. Chem*,. **11**, 411 (1990).
39. (a) A. Rockenbauer, A. Gaudel-Siri, D. Siri, *et al.*, *J. Phys. Chem. A*, **107**, 3851–3857 (2003). (b) A. Rockenbauer, N. V. Nagy, F. Le Moigne, *et al.*, *J. Phys. Chem. A*, **108**, 5484–5489 (2004).
40. A. Rockenbauer, G. Olive, X. Rozanska, *et al.*, *J. Phys. Chem. A*, **107**, 3851–3857 (2003).
41. H. Hayat and B. L. Silver, *J. Phys. Chem.*, **77**, 72–78 (1973).
42. (a) A. J. Stone, *Mol. Phys.*, **6**, 509–515 (1963). (b) A. J. Stone, *Mol. Phys.*, **7**, 311–316 (1964).
43. S. M. Mattar and A. D. Stephens, *Chem. Phys. Lett.*, **347**, 189–198 (2001).
44. D. Marsh, *J. Magn. Reson.*, **157**, 114–118 (2002).
45. (a) H. Block and Walker, S. M., *Chem. Phys. Lett.*, **19**, 363–364 (1973). (b) S. Ehrenson, *J. Comput. Chem.*, **2**, 41–52 (1981). (c) M. S. Wertheim, *Mol. Phys.*, 25, 211–223 (1973). (d) M. S. Wertheim, **26**, 1425–1444 (1973).
46. D. Marsh, *J. Magn. Reson.*, **190**, 60–67 (2008).
47. D. Marsh and C. Toniolo *J. Magn. Reson.*, **190**, 211–221 (2008).
48. J. L. Russ, J. G. Kun-Hsiang Tsai, T. Glass, *et al.*, *J. Am. Chem. Soc.*, 129, 7018–7027 (2007), and references cited therein.
49. (a) A. Rajca, *Chem. Rev.*, **94**, 871–893 (1994). (b) S. J. Blundell and F. L. Pratt, *J. Phys: Condens. Matter*, 16, R771–R828 (2004). (c) A. Rassat, *Pure & Appl. Chem.*, **62**, 223–227 (1990).
50. R. Chiarelli, M. A. Novak, A. Rassat and J. L. Tholence, *Nature*, **363**, 147–149 (1993).
51. (a) D. A. Shultz, A. K. Boal, H. Lee and G. T. Farmer, *J. Org. Chem.*, **64**, 4386–4396 (1999). (b) D. A. Shultz, R. M. Fico Jr., H. Lee, *et al.*, *J. Am. Chem. Soc.*, **125**, 15426–15432 (2003)
52. Md. E. Ali, A. S. Roy and S. N. Datta, *J. Phys. Chem. A*, **111**, 5523–5527 (2007).
53. J. Fujita, M. Tanaka, H. Suemune, *et al.*, *J. Am. Chem. Soc.*, **118**, 9347–9351 (1996).
54. J. Fritscher, M. Beyer and O. Schiemann, *Chem. Phys. Lett.*, **364**, 393–401 (2002).
55. A. Rajca, M. Pink, T. Rojsajjakul, *et al.*, *J. Am. Chem. Soc.*, **125**, 8534–8538 (2003). See also, A. Rajca, S. Mukherjee, M. Pinkaud and S. Rajca, *J. Am. Chem. Soc.*, **128**, 13497–13507 (2006); A. Rajca, M. Pink, S. Mukherjee, *et al.*, *Tetrahedron*, **63**, 10731–10742 (2007); Q. Wang, Y. Li and G. Wu, *Chem. Commun.*, 1268–1269 (2002).
56. X. Hu, H. Yang and Y. Li, *Spectrochimica Acta Part A*, **70**, 439–444 (2008).
57. T. Ishida, M. Ooishi, N. Ishii, H. Mori and T. Nogami, *Polyhedron*, **26**, 1793–1799 (2007).
58. H. Nishimaki, S. Mashiyama, M. Yasui, *et al.*, *Chem. Mater.*, **18**, 3602–3604 (2006). See also, G. Spagnol, K. Shiraishi, S. Rajca and A. Rajca, *Chem. Commun.*, 5047–5049 (2005).
59. (a) S. Gambarelli, D. Jaouen, A. Rassat, *et al.*, *J. Phys. Chem.*, **100**, 9605–9609 (1996). (b) C. Fabre, S. Gambarelli, D. Jaouen, *et al.*, *Mag. Res. Chem.*, **33**, S129–S133 (1995). (c) C. Chachaty, S. Gambarelli and A. Rassat, *Mag. Res. Chem.*, **33**, S174–S177 (1995). (d) J. Szydlowska, K. Pietrasik, L. Glaz and A. Kaim, *Chem. Phys. Lett.*, **460**, 245–252 (2008).
60. H. Sato, V. Kathirvelu, G. Spagnol, *et al.*, *J. Phys. Chem. B*, **112**, 2818–2828 (2008).
61. A. J. Maliakal, N. J. Turro, A. W. Bosman, *et al.*, *J. Phys. Chem. B*, **107**, 8467–8475 (2008).
62. K. E. Fairfull-Smith, J. P. Blinco, D. J. Keddie, *et al.*, *Macromolecules*, **41**, 1577–1580 (2008), and references cited therein.
63. (a) P. Ionita, A. C. Whitwood and B. C. Gilbert, *J. Chem. Soc., Perkin Trans. 2*, 1453–1462 (2001). (b) P. Ionita. *Free Rad. Res.*, **40**, 59–65 (2006). (c) P. Ionita. *Lett. Org. Chem.*, **5**, 42–46 (2008).
64. (a) E. Ressouche, J. X. Boucherle, B. Gillon, *et al.*, *J. Am. Chem. Soc.*, **115**, 3610–3617 (1993). (b) A. Zheludev, V. Barone, M. Bonnet, *et al.*, *J. Am. Chem. Soc.*, **116**, 2019–2027 (1994).
65. A. Caneschi, D. Gatteschi, N. Lalioti, *et al.*, *Angew. Chem. Int. Ed.*, **40**, 1760–1763 (2001).
66. A. A. Bobko, E. G. Bagryanskaya, V. A. Reznikov, *et al.*, *Free. Radic. Biol. Med.*, **36**, 248–258 (2004).
67. L. Bogani, C. Sangregorio, R. Sessoli, and G. Gatteschi. *Angew. Chem. Int. Ed*, **44**, 5817–2821 (2005).
68. C. Hirel, L. Li, P. Brough, *et al.*, *Inorg. Chem.*, **46**, 7545–7552 (2007).
69. J. Zhang, M. Zhao, G. Cui, and S. Peng, *Bioorgan. Med. Chem.*, **16**, 4019–4028 (2008).

70. B. M. Dooley, S. E. Bowles, T. Storr, and N. L. Frank, *Org. Lett.* **9**, 4781–4783 (2007)
71. (a) L. J. Ignarro, G. M. Buga, K. S. Wood, *et al.*, *Proc. Natl. Acad. Sci. USA*, **84**, 9265–9269 (1987). (b) R. F. Furchgott, in *Vasodilation: Vascular Smooth Muscle, Peptides, Autonomic Nerves and Endothelium* (ed. Vanhoutte, P. M.), Raven, New York, 401–414 (1988).
72. R. M. J. Palmer, A. G. Ferrige, and S. Moncada, S. *Nature* **327**, 524–526 (1987).
73. J. Collier, and P. Vallance, *Trends Pharmacol. Sci*,. **10**, 427–431 (1989).
74. R. E. Huie and S. Padmaja, *Free Rad. Res. Commun.*, **18**, 195–199 (1993).
75. V. M. Darley-Usmar, N. Hogg, V. J. O'Leary, *et al.*, *Free Rad. Res. Comms.*, **17**, 9–20 (1992).
76. J. S. Beckman, *J. Dev. Physiol.* **15**, 53–59 (1991).
77. S. Pou, L. Keaton, W. Surichamorn, *et al.*, *Biochim. Biophys. Acta*, **1201**, 118–124 (1994).
78. (a) H.-G. Korth, K. U. Ingold, R. Sustmann, *et al.*, *Angew. Chem. Int. Ed.*, **31**, 891–893 (1992). (b) H.-G. Korth, R. Sustmann, P. Lommes, *et al.*, *J. Am. Chem. Soc.*, **116**, 2767–2777 (1994). (c) T. Paul, M. A. Hassan, H.-G. Korth, *et al.*, *J. Org. Chem.*, **61**, 6835–6848 (1996). (d) M. Bätz, H.-G. Korth, and R. Sustmann, *Angew. Chem. Int. Ed.*, **36**, 1501–1503 (1997). (e) P. Meineke, U. Rauen, H. de Groot, *et al.*, *Chem-Eur. J.*, **5**, 1738–1747 (1999). (f) P. Meineke, U. Rauen, H. de Groot, *et al.*, *Biol. Chem.*, **381**, 575–581 (2000). (g) F. S. Hornig, H.-G. Korth, U. Rauen, *et al.*, *Helv. Chim. Acta*, **89**, 2281–2296 (2006).
79. K. Tsuchiya, M. Takasugi, K. Minakuchi, and K. Fukuzawa, *Free Rad. Biol. Med.*, **21**, 733–737 (1996).
80. (a) C. S. Lai and A. M. Komarov, *FEBS Lett.*, **345**, 120–124 (1994). (b) H. Fujii, J. Koscielniak, and L. J. Berliner, *Magn. Reson. Med.*, **38**, 565–568 (1997).
81. Y. Xia and J. L. Zweier, *Proc. Natl. Acad. Sci. USA*, **94**, 12705–12710 (1997).
82. K. Tsuchiya, M. Yoshizumi, H. Houchi, and R. P. Mason, *J. Biol. Chem.*, **275**, 1551–1556 (2000).
83. (a) J. Joseph, B. Kalyanaraman, and J. S. Hyde, *Biochem. Biophys. Res. Co.* **192**, 926–934 (1993). (b) E. A. Konorev, M. M. Tarpey, J. Joseph, *et al.*, *Free Radical Biol. Med.*, **18**, 169–177 (1995).
84. T. Akaike, M. Yoshida, Y. Miyamoto, *et al.*, *Biochemistry*, **32**, 827–832 (1993).
85. J. Zhang, M. Zhao, G. Cui and S. Peng. *Bioorg. Med. Chem*, **16**, 4019–4028 (2008).
86. G. M. Rosen, S. Porasuphatana, P. Tsai, *et al.*, *Macromolecules*, **36**, 1021–1027 (2003).
87. R. F. Haseloff, S. Zollner, I. A. Kirilyuk, *et al.*, *Free Radic. Res.*, **26**, 7–17 (1997).
88. L. Cambi and L. Szegö, *Ber. Dtsch. Chem. Ges. B*, **64**, 2591–2598 (1931).
89. M. Tamura, Y. Nakazawa, D. Shiomi, *et al.*, *Chem. Phys. Lett.*, **186**, 401–404 (1991).
90. A. J. Banister, N. Bricklebank, I. Lavender, *et al.*, *Angew. Chem. Int. Ed.*, **35**, 2533–2535 (1996)
91. C. Benelli and D. Gatteschi, *Chem. Rev.*, **102**, 2369–2388 (2002).
92. S. J. Blundell and F. L. Pratt, *J. Phys.: Condens. Matter*, **16**, 771–828 (2004).
93. M. T. Lemaire, *Pure Appl. Chem.*, **76**, 277–293 (2004).
94. D. Gatteschi, L. Bogani, A. Cornia, *et al.*, *Solid State Sci.*, **10**, 1701–1709 (2008).
95. J. P. Sutter, A. Lang, O. Khan, *et al.*, *J. Magn. Magn. Mater.*, **171**, 147–152 (1997).
96. A. Caneschi, D. Gatteschi, J. Laugier and P. Rey, *J. Am. Chem. Soc.*, **109**, 2191–2192 (1987).
97. A. Caneschi, D. Gatteschi, P. Rey and R. Sessoli, *Inorg. Chem.*, **27**, 1756–1761 (1988).
98. A. Caneschi, D. Gatteschi, J. P. Renard, *et al.*, *Inorg. Chem.*, **28**, 3314–3319 (1989).
99. A. Caneschi, D. Gatteschi, P. Rey and R. Sessoli, *Acc. Chem. Res.*, **22**, 392–398 (1989).
100. A. Caneschi, D. Gatteschi, P. Rey and R. Sessoli, *Inorg. Chem.*, **30**, 3936–3941 (1991).
101. N. Ishii, Y. Okamura, S. Chiba, *et al.*, *J. Am. Chem. Soc.*, **130**, 24–25 (2008).
102. A. Caneschi, F. Ferraro, D. Gatteschi, *et al.*, *Inorg. Chem.*, **30**, 3162–3166 (1991).
103. A. Caneschi, P. Chiesi, L. David, *et al.*, *Inorg. Chem.*, **32**, 1445–1453 (1993).
104. J. P. Sutter, A. Lang, O. Khan, *et al.*, *J. Magn. Magn. Mater.*, **171**, 147–152 (1997).
105. C. Stroh, P. Turek, P. Rabu and R. Zeissel. *Inorg. Chem.*, **40**, 5334–5342 (2001).
106. C. Rancurel, H. Heise, F. H. Kohler, *et al.*, *J. Phys. Chem. A*, **108**, 5903–5914 (2004)
107. C. Stroh, M. Mayor and C. Von Hanisch. *Tet. Lett.*, **45**, 9623–9626 (2004)
108. D. Luneau, A. Borta, Y. Chumakov, *et al.*, *Inorg. Chim. Acta*, **361**, 3669–3676 (2008)
109. H. Murata, Y. Miyazaki, A. Inaba, *et al.*, *J. Am. Chem. Soc.*, **130**, 186–194 (2008).
110. D. Luneau, J. Laugier, P. Rey, *et al.*, *J. Chem. Soc.,Chem. Commun.*, 741–742 (1994).
111. R. Ziessel, *Mol. Cryst. Liq. Cryst.*, **273**, 101–110 (1995).

112. C. Stroh and R. Ziessel., *Chem. Commun.*, 1916–1917 (2002).
113. G. Ulrich, P. Turek, R. Ziessel, A. De Cian and J. Fischer, *Chem. Commun.*, 2461–2462 (1996).
114. K. Igarashi, T. Nogami and T. Ishida, *Polyhedron*, **9-10**, 1672–1677 (2009).
115. K. Igarashi, T. Nogami and T. Ishida, *Chem. Commun.*, 501–503 (2007).
116. M. Deumal, J. Cirujeda, J. Veciana and J. J. Novoa, *Chem. Eur. J.*, **5**, 1631–1642 (1999).
117. (a) J. F. W. Keana, *Chem. Rev.*, **78**, 37–64 (1978). (b) K. Hideg, T. Kálai and C. P. Sár, *J. Het. Chem.*, 42, 437–450 (2005).
118. (a) J. P. Gillet, O. Guerret and P. Tordo, *PCT Int. Appl.*, 2000, WO 0049027. P. Nesvadba, L. Bugnon and M. Von Bueren, *PCT Int. Appl.*, 2004, WO 2004085397.
119. (a) R. W. Murray and M. Singh, *Tetrahedron Lett.*, **29**, 4677–4680 (1988). (b) R. W. Murray and R. W. Jeyarannan, *J. Org. Chem.*, **50**, 2847–2853 51985). (c) R. W. Murray, *Chem. Rev.*, **89**, 1187–1201 (1989). (c) A. Rajca, M. Vale and S. Rajca, *J. Am. Chem. Soc.*, **130**, 9099–9105 (2008). C. Bonvalet, F. Bourelle, D. Scholler and A. Fegenbaum, *J. Chem. Res.*, **12**, 348–349 (1991).
120. M. E. Brik, *Tetrahedron Lett.*, **36**, 5519–5522 (1995).
121. R. M. Dupeyre, A. Rassat and J. Ronzaud, *J. Am. Chem. Soc.*, **96**, 6559–6568 (1974).
122. (a) J. P. Blinco, J. L. Hodgson, B. J. Morrow, *et al.*, *J. Org. Chem.*, **73**, 6763–6771 (2008). (b) M. A. Voinov, J. F. Polienko, T. Schanding, *et al.*, *J. Org. Chem.*, **73**, 6763–6771 (2008).
123. J. Einhorn, F. Ratajczak, I. Gautier-Luneau and J. L. Pierre, *J. Org. Chem.*, **62**, 9385–9388 (2008).
124. I. A. Kirilyuk, A. A. Bobko, I. A. Grigor'ev and V. V. Kramtsov, *Org. Biomol. Chem.*, **2**, 1025–1030 (2004).
125. S. F. Vasilevsky, S. V. Klyatskaya, O. L. Korovnikova, *et al.*, *Tetrahedron*, **62**, 4597–4602 (2006).
126. J. F. W. Keana, G. S. Heo and G. T. Gaughan, *J. Org. Chem.*, **50**, 2346–2351 (1985).
127. L. M. Field and P. M. Lahti, *Chem. Mater.*, **15**, 2861–2863 (2003).
128. N. Naik and R. Braslau, *Tetrahedron*, **54**, 667–696 (1998).
129. H. Fischer, *Chem. Rev.*, 101, 3581–3610 (2001).
130. R. Tamura, S. Susuki, N. Azuma, *et al.*, *J. Org. Chem.*, **60**, 6820–6825 (1995).
131. S. D. Rychnovsky, T. Beauchamp, R. Vaidyanathan and T. Kwan, *J. Org. Chem.*, **63**, 6363–6374 (1998).
132. G. Tommasi, P. Bruni, L. Greci, *et al.*, *J. Chem. Soc., Perkin Trans 2*, 2123–2128 (1999).
133. S. D. Rychnovsky, T. L. McLernon and H. Rajapakse, *J. Org. Chem.*, **61**, 1194–1195 (1996).
134. J. F. W. Keana, K. Hideg, G. B. Birrell, *et al.*, *Can. J. Chem.*, **60**, 1439–1447 (1982).
135. J. Einhorn, C. Einhorn, F. Ratajczak and J. L. Pierre, *Synth. Comm.*, **30**, 1837–1848 (2000).
136. T. Shibata, K. Uemae and Y. Yamamoto, *Tetrahedron: Asymmetry*, **11**, 2339–2346 (2000).
137. T. D. Lee, G. B. Keana and J. F. W. Keana, *J. Am. Chem. Soc.* **100**, 1618–1619 (1978).
138. F. Le Moigne and P. Tordo, *C. R. Acad. Sci. Paris, Chemistry*, **4**, 585–590 (2001).
139. (a) Y. Noda, S. Shimono, M. Baba, *et al.*, *J. Phys. Chem. B*, **110**, 23683–23687 (2006). (b) Y. Uchida, T. Uematsu, Y. Nakayama, *et al.*, *Chirality*, **20**, 282–287 (2008).
140. J. F. W. Keana, T. D. Lee and E. M. Bernard, *J. Am. Chem. Soc.*, **98**, 3052–3053 (1976).
141. C. Le Mercier, J. F. Lutz, S. Marque, *et al.*, in *Controlled/Living Radical Polymerization* (ed. K. Matyjaszewski), ACS Symposium Series, 768, American Chemical Society, Washington, DC, 2000.
142. D. Benoit, V. Chaplinski, R. Braslau and C. Hawker, *J. Am. Chem. Soc.*, **121**, 3904–3920 (1999).
143. (a) E. Drockenmuller and J. M. Catala, *Tet. Lett.*, **42**, 9011–9013 (2001). (b) E. Drockenmuller and J. M. Catala, *Macromolecules*, **35**, 2461–2466 (2002).
144. R. Braslau, G. O'Bryan, A. Nilsen, *et al.*, *Synthesis*, **9**, 1496–1506 (2005).
145. S. Marque, H. Fischer, E. Baier and A. Studer, *J. Org. Chem.*, **66**, 1146–1156 (2001).
146. G. O'Bryan, A. Nilsen and R. Braslau, *Macromolecules*, **40**, 7848–7854 (2007).
147. A. Studer, K. Harms, C. Knoop, *et al.*, *Macromolecules*, **37**, 27–34 (2004).
148. J. Ruehl and R. Braslau, *J. Polymer. Sci. Part A: Polym. Chem.*, **45**, 2015 (2007).
149. C. A. Knoop and A. Studer, *J. Am. Chem. Soc.*, **125**, 16327–16333 (2003).
150. K. O. Siegenthaler and A. Studer, *Macromolecules*, **39**, 1347–1352 (2006).
151. C. Wetter, J. Gierlich, C. A. Knoop, *et al.*, *Chem. Eur. J.*, **10**, 1156–1166 (2004).
152. S. Flakus, K. Mandel, M. Bartsch and G. Schmidt-Naake, *Macromol. Rapid Commun.*, **26**, 1698–1703 (2005).
153. N. Cameron and O. Lagrille, *Brit. UK Pat. Appl.*, GB 2442952 (2008).

154. E. Drockenmuller and J. M. Catala, *ACS Polym. Prepr.*, **44**, 677–678 (1997).
155. S. Abraham, J. H. Choi and I. Kim, *J. Polymer. Sci. Part A: Polym. Chem.*, **45**, 5559–5572 (2007).
156. R. Cuatepotzo-Diaz, M. Albores-Velasco, E. Saldívar-Guerra and F. B; Jiménez, *Polymer*, **45**, 815–824 (2004).
157. Y. Guillaneuf, D. Gigmes, S. R. A. Marque, *et al.*, *Macromolecules*, **40**, 3108–3114 (2007).
158. L. B. Volodarsky, *Pure Appl. Chem.*, **62**, 177–181 (1990).
159. K. Hideg, *Pure Appl. Chem.*, **62**, 207–212 (1990).
160. (a) H. Fischer, *J. Am. Chem. Soc.*, **108**, 3925–3927 (1986). (b) K. S. Focsaneamu and J. C. Scaiano, *Helv. Chim. Acta*, **89**, 2473–2482 (2006).
161. Y. Samuni, J. Gamso, A. Samuni, *et al.*, *Antioxid. Redox Signal.*, **6**, 587–595 (2004).
162. L. Marx and B. Schöllhorn, *New J. Chem.*, **30**, 430–434 (2006).
163. C. Mathieu, A. Mercier, D. Witt, *et al.*, *Free Radic. Biol. Med.*, **22**, 803–806 (1997).
164. W. R. Couet, R. C. Brasch, G. Sosnovsky, *et al.*, *Tetrahedron*, **41**, 1165–1172 (1985).
165. S. Manda, I. Nakanishi, K. Ohkubo, *et al.*, *Org. Biomol. Chem.*, **5**, 3951–3955 (2007).
166. J. L. Hodgson, M. Namazian, S. E. Bottle and M. L. Coote, *J. Phys. Chem. A*, **111**, 13595–13605 (2007).
167. Y. Kashiwagi, T. Nishimura and J. Anzai, *Electroch. Acta*, **47**, 1317–1320 (2002).
168. S. D. Rychnosky, R. Vaidyanathan, T. Beauchamp, *et al.*, *J. Org. Chem.*, **64**, 6745–6749 (1999).
169. J. P. Blinco, J. L. Hodgson, B. J. Morrow, *et al.*, *J. Org. Chem.*, **73**, 6763–6771 (2008).
170. N. Kocherginsky and H. M. Swartz, *Nitroxide spin labels, reactions in biology and chemistry*, CRC Press, Boca Raton, 1–265 (1995).
171. A. A. Bobko, I. A. Kirilyuk, I. A. Grigor'ev, *et al.*, *Free Radic. Biol. Med.*, **42**, 404–412 (2007).
172. J. F. Keana, S. Pou and G. M. Rosen, *Magn. Reson. Med.*, **5**, 525–536 (1987).
173. L. Marx, R. Chiarelli, T. Guiberteau and A. Rassat, *J. Chem. Soc. Perkin Trans 1*, 1181–1182 (2000).
174. I. A. Kirilyuk, A. A. Bobko, I. A. Grigor'ev and V. V. Khramtsov, *Org. Biomol. Chem.*, **2**, 1025–1030 (2004).
175. S. Okazaki, M. A. Mannan, K. Sawai, *et al.*, *Free Rad. Res.*, **41**, 1069–1077 (2007).
176. L. Marx, R. Chiarelli, T. Guiberteau and A. Rassat, *J. Chem. Soc. Perkin Trans 1*, 1181–1182 (2000).
177. I. A. Kirilyuk, A. A. Bobko, I. A. Grigor'ev and V. V. Khramtsov, *Org. Biomol. Chem.*, **2**, 1025–1030 (2004).
178. S. Okazaki, M. A. Mannan, K. Sawai, *et al.*, *Free Rad. Res.*, **41**, 1069–1077 (2007).
179. C. Ebel, K. U. Ingold, J. Michon and A. Rassat, *Tetrahedron Lett.*, **26**, 741–744 (1985).
180. H. Karoui, A. Rockenbauer, S. Pietri and P. Tordo, *Chem. Commun.*, **24**, 3030–3031 (2002).
181. D. Bardelang, K. Banaszak, H. Karoui, *et al.*, *J. Am. Chem. Soc.*, **131** (5), 5402–5404 (2009).
182. C. Jones and R. P. Rose, *New. J. Chem.*, **31**, 1484–1487 (2007).
183. K. A. Parker and D. S. Su, in *Encyclopedia of Reagents for Organic Synthesis* (ed. L. A. Paquette), Vol. 6, John Wiley & Sons, Inc., New York, 4271–4272 (1995).
184. F. Recupero and C. Punta, *Chem. Rev.*, **107**, 3800–3842 (2007).
185. C. Galli, P. Gentili and O. Lanzalunga, *Angew. Chem. Int. Ed.*, **47**, 4790–4796 (2008).
186. Y. Ishii, S. Sakaguchi and T. Iwahamab, *Adv. Synth. Catal.*, **343**, 393–427 (2001).
187. M. Lucarini, E. Marchesi, G. F. Pedulli and C. Chatgilialoglu, *J. Org. Chem.*, **63**, 1687–1683 (1998).
188. S. Coseri and K. U. Ingold, *Org. Lett.*, **6**, 1641–1643 (2004).
189. L. J. Johnston, M. Tencer and J. C. Scaiano, *J. Org. Chem.*, **51**, 2806–2608 (1986).
190. R. Gupta and A. S. Borovik, *J. Am. Chem. Soc.*, **125**, 13234–13242 (2003).
191. A. L. J. Beckwith, V. W. Bowry and K. U. Ingold, *J. Am. Chem. Soc.*, **114**, 4983–4992 (1992).
192. V. W. Bowry and K. U. Ingold, *J. Am. Chem. Soc.*, **114**, 4992–4996 (1992).
193. (a) A. Samuni, S. Goldstein, A. Russo, *et al.*, *J. Am. Chem. Soc.*, **124**, 8719–8724 (2002). (b) S. Goldstein and A. Samuni, *J. Phys. Chem. A.*, **111**, 1066–1072 (2007).
194. E. Amiani, R. Castagna, P. Astolfi and L. Greci, *Free Rad. Res.*, **39**, 325–336 (2005).
195. P. J. Wright, and A. M. English, *J. Am. Chem. Soc.*, **125**, 8655–8665 (2003).
196. M. A. Lam, D. I. Pattison, S. E. Bottle, *et al.*, *Chem. Res. Toxicol.*, **21**, 2111–2119 (2008).
197. A. T. Hoye, J. E. Davoren, P. Wipf, *et al.*, *Acc. Chem. Res.*, **41**, 87–97 (2008).
198. M. C. Krishna, W. DeGraff, O. H. Hankovszky, *et al.*, *J. Med. Chem.*, **41**, 3477–3492 (1998).
199. J. Sobek, R. Martschke and H. Fischer, *J. Am. Chem. Soc.*, **123**, 2849–2857 (2001).
200. A. Goto, J. C. Scaiano and L. Maretti, *Photochem. Photobiol. Sci.*, **6**, 833–835 (2007).

201. N. Naik and R. Braslau, *Tetrahedron*, **54**, 667–696 (1998).
202. W. Adam, C. R. Saha-Möller and P. A. Ganeshpure, *Chem. Rev*., **101**, 3499–3548 (2001).
203. A. Studer, *Chem. Soc. Rev*., **33**, 267–273 (2004).
204. T. Vogler and A. Studer, *Synthesis*, **13**, 1979–1993 (2008).
205. R. Braslau and M. O. Anderson, in *Radicals in Organic Synthesis* (eds P. Renaud and M. P. Sibi), Wiley-VCH Verlag GmbH, 127–146 (2001).
206. E. J. Corey and A. W. Gross, *J. Org. Chem*., **50**, 5391–5393 (1985).
207. T. N. Nagashima and D. P. Curran, *Synlett*, 330–332 (1996).
208. C. B. Gill, G. Pattenden and S. J. Reynolds, *Tetrahedron Lett*., **30**, 3229–3232 (1998).
209. C. Ollivier, R. Chuard and P. Renaud, *Synlett*, 807–809 (1999).
210. M. R. Heinrich, *Chem. Eur. J*., **15**, 820–833 (2009).
211. J. P. Goddard, C. Gomez, F. Brebion, *et al*., *Chem. Commun*., 2929–2931 (2007).
212. D. L. Boger and C. W. Boyce, *J. Org. Chem*., **65**, 4088–4100 (2000).
213. D. L. Boger, R. M. Garbaccio, and Q. Jin, *J. Org. Chem*., **62**, 8875–8891 (1997).
214. C. Cadot, P. I. Dalko, J. Cossy, *et al*., *J. Org. Chem*., **67**, 7193–7202 (2002).
215. A. P. Schaffner and P. Renaud, *Eur. J. Org. Chem*., **11**, 2291–2298 (2004).
216. W. Huang, H. Henry-Riyad and T. T. Tidwell, *J. Am. Chem. Soc*., **121**, 3939–3943 (1999).
217. A. D. Allen, B. Cheng, M. H. Fenwick, *et al*., *J. Org. Chem*., **66**, 2611–2617 (2001).
218. T. T. Tidwell and M. H. Fenwick, *Eur. J. Org. Chem*., **18**, 3415–3419 (2001).
219. H. L. Casal, N. H. Werstiuk, and J. C. Scaiano, *J. Org. Chem*., **49**, 5214–5217 (1984).
220. J. Guin, S. De Sarkar, S. Grimme and A. Studer, *Angew. Chem. Int. Ed*., **47**, 8727–8730 (2008).
221. (a) J. Einhorn, C. Einhorn, F. Ratajczak, *et al*., *J. Org. Chem*., **62**, 9385–9388 (1997). (b) R. Braslau, N. Naik and H. Zipse, *J. Am. Chem. Soc*., **122**, 8421–8434 (2000). (c) P. I. Dalko, *Tetrahedron Lett*., **40**, 4035–4036 (1999).
222. U. Jahn, M. Müller and S. Ausieker, *J. Am. Chem. Soc*., **122**, 5212–5213 (2000).
223. U. Jahn and E. Dinca, *Chem. Eur. J*., **15**, 58–62 (2009).
224. M. P. Sibi and M. Hasegawa, *J. Am. Chem. Soc*., **129**, 4124–4125 (2007).
225. H. Fischer, *Chem. Rev*., **101**, 3581–3610 (2001), see also the chapter on controlled/living radical polymerization.
226. D. Crich, D. Grant and A. A. Bowers, *J. Org. Chem*., **129**, 12106–12107 (2007).
227. (a) C. Leroi, B. Fenet, J. L. Couturier, *et al*., *Org. Lett*., **5**, 1079–1081 (2003). (b) C. Leroi, D. Bertin, P. E. Dufils, *et al*., *Org. Lett*., **5**, 4943–4945 (2003).
228. (a) A. Studer, *Angew. Chem. Int. Ed*., **39**, 1108–1111 (2000), (b) A. Studer, *Chem. Soc. Rev*., **33**, 267–273 (2004).
229. A. Teichert, K. Jantos, K. Harms and A. Studer, *Org. Lett*., **6**, 3477–3480 (2004).
230. Y. Uenoyama, M. Tsukida, T. Doi, *et al*., *Org. Lett*., **7**, 2985–2988 (2005).
231. B. Janza and A. Studer, *Org. Lett*., **8**, 1875–1878 (2005).
232. C. Wetter, K. Jantos, K. Woithe and A. Studer, *Org. Lett*., **5**, 2899–2902 (2003).
233. K. Molawi, T. Schulte, K. O. Siegenthaler, *et al*., *Chem. Eur. J*., **11**, 2335–2350 (2005).
234. D. Bertin, D. Gigmes, S. R. A. Marque and P. Tordo, *Tetrahedron*, **61**, 8752–8761 (2005).
235. (a) V. D. Sholle, V. A. Golubev and E. G. Rozantsev, *Dokl. Akad. Nauk. SSSR*, **200**, 137 (1971). (b) K. S. Root, C. L. Hill, L. M. Lawrence and G. M. Whitesides, *J. Am. Chem. Soc*., **111**, 5405–5412 (1989). (c) T. N. Nagashima and D. P. Curran, *Synlett*, 330–332 (1996). (d) P. I. Dako, *Tetrahedron Lett*., **40**, 4035–4036 (1999).
236. P. Y. S. Lam, G. Vincent, C. G. Clark, *et al*., *Tetrahedron Lett*., **42**, 3415–3418 (2001).
237. A. Lagunas, A. Mairata i Payeras, C. Jimeno and M. A. Pericàs, *Org. Lett*., **7**, 3033–3036 (2005).
238. (a) K. W. Huang, J. H. Han, A. P. Cole, *et al*., *J. Am. Chem. Soc*., **127**, 3807–3816 (2005). (b) S. DeBeer George, K. W. Huang, R. M. Waymouth and E. I. Solomon, *Inorg. Chem*., **45**, 4468–4477 (2006). (c) K. W. Huang, J. H. Han, C. B. Musgrave and R. M. Waymouth, *Organometallics*., **25**, 3317–3323 (2006).
239. M. S. Maji, T. Pfeifer and A. Studer, *Angew. Chem. Int. Ed*., **47**, 9547–9550 (2008).
240. T. Vogler and A. Studer, *Org. Lett*., **10**, 129–131 (2008).
241. J. Y.-C. Chen, N. Jayaraj, S. Jockusch, *et al*., *J. Am. Chem. Soc*., **130**, 7206–7207 (2008).

242. J. Martinie, J. Michon and A. Rassat, *J. Am. Chem. Soc.*, **97**, 1818–1823 (1975).
243. (a) Y. Kotake and E. G. Janzen, *Chem. Phys. Lett.*, **150**, 199 (1988). (b) Y. Kotake and E. G. Janzen, *J. Am. Chem. Soc.*, **110**, 3699–3701 (1988). (c) Y. Kotake and E. G. Janzen, *J. Am. Chem. Soc.*, **111**, 2066–2070 (1989). (d) Y. Kotake and E. G. Janzen, *J. Am. Chem. Soc.*, **111**, 5138–5140 (1989). (e) Y. Kotake and E. G. Janzen *J. Am. Chem. Soc.*, **114**, 2872–2874 (1992); (f) Y. Sueishi, H. Tobisako and Y. Kotake, *J. Phys. Chem. B*, **108**, 12623–12627 (2004).
244. (a) M. Lucarini, B. Luppi, G. F. Pedulli and B. P. Roberts, *Chem. Eur. J.*, **5**, 2048 (1999). (b) P. Franchi, M. Lucarini, G. F. Pedulli, *Angew. Chem. Int. Edn.*, **42**, 1842–1845 (2003). (c) P. Franchi, M. Lucarini, E. Mezzina, and G. F. Pedulli, *J. Am. Chem. Soc.*, **126**, 4343–4354 (2004). (d) P. Franchi, G. F. Pedulli and M. Lucarini, *J. Phys. Chem.*, **112**, 8706–8714 (2008).
245. E. Mileo, P. Franchi, R. Gotti, *et al.*, *Chem. Commun.*, 1311–1313 (2008).
246. P. Franchi, M. Fani, E. Mezzina and M. Lucarini, *Org. Lett.*, **10**, 1901–1904 (2008).
247. G. Ionita, A. Caragheorgheopol, H. Caldararu, *et al.*, *Org. Biomol. Chem.*, **7**, 598–602 (2009).
248. G. Ionita, V. Meltzer, E. Pincu and V. Chechik, *Org. Biomol. Chem.*, **5**, 1910–1914 (2007).
249. V. Chechik and G. Ionita, *New J. Chem.*, **31**, 1726–1729 (2007).
250. E. Damiani, R. Tursilli, A. Casolari, *et al.*, *Free Rad. Res.*, **39**, 41–49 (2005).
251. S. Rossi, M. Bonini, P. Lo Nostro and P. Bagioni, *Langmuir*, **23**, 10959–10967 (2007).
252. D. Bardelang, A. Rockenbauer, L. Jicsinszky, *et al.*, *J. Org. Chem*, **71**, 7657–7667 (2006).
253. P. Franchi, M. Lucarini, G. F. Pedulli and D. Sciotto, *Angew. Chem. Int. Edn.*, **39**, 263–265 (2000).
254. G. S. Ananchenko, K. A. Udachin, A. W. Coleman, *et al.*, *Chem. Commun.*, 223–225 (2008).
255. D. M. Polovyanenko, E. G. Bagryanskaya, A. Schnegg, *et al.*, *Phys. Chem. Chem. Phys.*, **10**, 5299–5307 (2008).
256. (a) J. W. Lee, S. Samal, N. Selvapalam, *et al.*, *Acc. Chem. Res.*, **36**, 621–630 (2003). (b) J. Lagona, P. Mukhopadhyay, S. Chakrabarti, and L. Isaacs, *Angew. Chem., Int. Edn.*, **44**, 4844–4870 (2005).
257. (a) E. Mezzina, F. Cruciani, G. F. Pedulli and M. Lucarini, *Chem. Eur. J.*, **13**, 7223–7233 (2007). (b) E. Mileo, E. Mezzina, F. Grepioni, *et al.*, *Chem. Eur. J.*, **15**, 7859–7862 (2009).
258. (a) N. Muller, in *Solution Chemistry of Surfactants* (ed. K. L. Mittal), Vol. 1, Plenum, New York, 1979. (b) J. Lang and R. Zana, in *Surfactant in Solutions: New Methods of Investigation* (ed. R. Zana), Marcel Dekker, New York, 1987. (c) P. D. T. Huibers, S. G. Oh and D. O. Shah in *Surfactants in Solutions* (eds A. K. Chattopadhay and K. L. Mittal), Vol. 64, Marcel Dekker, New York, 1995.
259. For pioneering work see: (a) A. S. Wagoner, O. H. Griffith and C. R. Christensen, *Proc. Nat. Acad. Sci. USA*, **57**, 1198–1205 (1967). (b) J. Oakes, *Nature*, **231**, 38–39 (1971). (c) K. K. Fox, *J. Chem. Soc., Faraday Trans*, **67**, 2802–2809 (1971). (d) N. M. Atherton and S. Strach, *J. Chem. Soc., Faraday Tran s 2*, **68**, 374–381 (1972).
260. C. Rizzi, C. Mathieu, B. Tuccio, *et al.*, *J. Chem. Soc., Perkin Trans. 2*, 2777–2781 (1999).
261. G. Brigati, P. Franchi, M. Lucarini, *et al.*, *Res. Chem. Intermed.*, **28**, 131–141 (2002).
262. (a) A. M. Tedeschi, G. D'Errico, E. Busi, *et al.*, *Phys. Chem. Chem. Phys.*, **4**, 2180–2188 (2002). (b) A. M. Tedeschi, L. Franco, M. Ruzzi, *et al.*, *Phys. Chem.Chem.Phys.*, **5**, 4204–4209 (2003).
263. C. Aliaga, J. M. Juárez, J. C. Scaiano and A. Aspée, *Org. Lett.*, **10**, 2147–2150 (2008).
264. E. R. McCarney, B. D. Armstrong, R. Kausik and S. Han, *Langmuir*, **24**, 10062–10072 (2008).
265. (a) J. Alvarez, E. A. Lissi and M. V. Encinas, *Langmuir*, **12**, 1738–1743 (1996). (b) D. Angelescu and M. Vasilescu, *J. Coll. Int. Sc.*, **244**, 139–144 (2001). (c) E. Szajdzinska-Pietek and M. Wolszcak, *J. Photochem. Photobiol. A: Chem.*, **112**, 245–249 (1998).
266. M. Lucarini, P. Franchi, G. F. Pedulli, *et al.*, *J. Am.Chem. Soc.*, **126**, 9326–9329 (2004).
267. (a) M. Mazzoni, L. Franco, C. Corvaja, *et al.*, *Chem. Phys. Chem.*, **3**, 527–531 (2002). (b) A. Polimeno, M. Zerbetto, L. Franco, *et al.*, *J. Am. Chem. Soc.*, **128**, 4734–4741 (2006). (c) C. Corvaja, F. Conti, L. Franco and M. Maggini, *C. R. Chimie*, **9**, 909–915 (2006).
268. F. Conti, C. Corvaja, A. Toffoletti, *et al.*, *J. Phys. Chem. A*, **104**, 4962–4967 (2000).
269. F. Conti, C. Corvaja, F. Busolo, *et al.*, *Phys. Chem. Chem. Phys.*, **11**, 495–502 (2009).
270. S. Campestrini, C. Corvaja, M. De Nardi, *et al.*, *Small*, **4**, 350–356 (2008).
271. B. Goodson, *J. Magn. Reson.*, **155**, 157–216 (2002).
272. L. Schröder, T. J. Lowery, C. Hilty, *et al.*, *Science*, **314**, 446–449 (2006).
273. C. R. Bowers and D. P. Weitekamp, *Phys. Rev. Lett.*, **57**, 2645–2648 (1986).

274. F. Reineri, A. Vial, G. Giovenzana, *et al.*, *J. Am. Chem. Soc.*, **130**, 15047–15053 (2008).
275. M. G. Zysmilich, *J. Am. Chem. Soc.*, **118**, 5867–5873 (1996).
276. S. Prakash, A. Alia, P. Gast, *et al.*, *Biochemistry*, **46**, 8953–8960 (2007).
277. T. Prisner and W. Köckenberger, *Appl. Magn. Reson.*, **34**, 213–544 (2008).
278. D. A. Hall, D. C. Maus, G. J. Gerfen, *et al.*, *Science*, **276**, 930–932 (1997).
279. T. Maly, G. T. Debelouchina, V. S. Bajaj, *et al.*, *J. Chem. Phys.*, **128**, 052211–19 (2008).
280. C. G. Joo, A. Casey, C. J. Turner and R. G. Griffin, *J. Am. Chem. Soc.*, **131**, 12–13 (2009).
281. V. S. Bajaj, M. K. Hornstein, K. E. Kreischer, *et al.*, *J. Magn. Reson.*, **189**, 251–279 (2007).
282. K. Golman, J. S. Petersson, P. Magnusson, *et al.*, *Magn. Reson. Med.*, **59**, 1005–13 (2008).
283. K. Golman, R. I. Zandt, M. Lerche, *et al.*, *Cancer Res.*, **66**, 10855–60 (2006).
284. F. A Gallagher, M. I. Kettunen, S. E. Day, *et al.*, *Nature*, **453**, 940–943 (2008).
285. J. H. Ardenkjaer-Larsen, B. Fridlund, A. Gram, *et al.*, *PNAS*, **100**, 10158–10163 (2003).
286. K. Golman, R. I. Zandt and M. Thaning, *Proc. Natl. Acad. Sci. USA*, **103**, 11270–11275 (2006).
287. A. W. Overhauser, *Phys. Rev.*, **92**, 411–415 (1953).
288. A. Abragam and M. Goldman, *Rep. Prog. Phys.*, **41**, 395–467 (1978).
289. K. H. Hausser and D. Stehlik, *Adv. Magn. Reson.*, **3**, 79–139 (1968).
290. R. A. Dweck, R. E. Richards and D. Taylor, *Ann. Rev. NMR Spectrosc.*, **2**, 293 (1969).
291. W. Müller-Warmuth and K. Meise-Gresch, *Adv. Magn. Reson.*, **11**, 1–45 (1983).
292. D. J. Lurie, D. M. Bussell, L. H. Bell and J. R. Mallard, *J. Magn. Reson.*, **76**, 366–370 (1988).
293. R. D. Bates, *Magn. Reson. Rev.*, **16**, 237 (1993).
294. G. Planinsic, T. Guiberteau and D. Grucker, *J. Magn. Reson. B*, **110**, 205–209 (1996).
295. A. Comment, B. van den Brandt, K. Uffmann, *et al.*, *Conc. Magn. Reson. B*, **31**, 255–269 (2007).
296. N. M. Loening, M. Rosay, V. Weis and R. G. Griffin, *J. Am. Chem. Soc.*, **124**, 8808–8809 (2002).
297. B. M. Odintsov, R. L. Belford, Z. Sh. Idiyatullin, *et al.*, *J. Magn. Reson.*, **135**, 435–443 (1998).
298. M. E. Merritt, C. Harrison, C. Storey, *et al.*, *PNAS*, **104**, 19773–19777 (2007).
299. P. Höfer, G. Parigi, C. Luchinat, *et al.*, *J. Am. Chem. Soc.*, **130**, 3254–3255 (2008).
300. A. Comment, J. Rentsch, F. Kurdzesau, *et al.*, *J. Magn. Reson.*, **194**, 152–155 (2008).
301. P. Höfer, P. Carl, G. Guthausen, *et al.*, *Appl. Magn. Reson.*, **34**, 393–398 (2008).
302. K. N. Hu, H. Yu, T. M. Swager and R. G. Griffin, *J. Am. Chem. Soc.*, **126**, 10844–10845 (2004).
303. C. Song, K. N. Hu, C. G. Joo, *et al.*, *J. Am. Chem. Soc.*, **128**, 11385–11390 (2006).
304. Y. Matsuki, T. Maly, O. Ouari, *et al.*, *Angew. Chem. Int. Ed.*, **48**, 4996–5000 (2009).
305. E. R. McCarney and S. Han, *J. Magn. Reson.*, **190**, 307–315 (2008).
306. J. F. W. Keana, M. J. Acarregui and S. L. M. Boyle, *J. Am. Chem. Soc.*, **104**, 827–830 (1982).
307. I. A. Kirilyuk, A. A. Boko, I. A. Grigor'ev and V. V. Khramtsov, *Org. Biol. Chem.*, **2**, 1025–1030 (2004).
308. V. V. Khramtsov, I. A. Grigor'ev, M. A. Foster and D. J. Lurie, *Antioxid. Redox Signal*, **6**, 667–676 (2004).
309. V. K. Khlestkin, V. V. Butakov, I. A. Grigor'ev, *et al.*, *Synthesis*, **20**, 3649–3653 (2005).
310. S. E. Herbelin and N. V. Blough, *J. Phys. Chem. B*, **102**, 8170–8166 (1998).
311. M. Kondo, M. Mehiri and S. L. Regen, *J. Am. Chem. Soc.*, **130**, 13771–13777 (2008).
312. E. Lozinsky, V. V. Martin, T. A. Berezina, *et al.*, *J. Biochem. Biophys. Methods*, **38**, 29–42 (1999).
313. C. Aliaga, J. M. Juárez-Ruiz, J. C. Scaiano and A. Aspée, *Org. Lett.*, **10**, 2147–2150 (2008).
314. N. V. Blough and D. J. Simpson, *J. Am. Chem. Soc.*, **110**, 1915–1917 (1988).
315. K. E. Fairfull-Smith, J. P. Blinco, D. J. Keddie, *et al.*, *Macromolecules*, **41**, 1577–1580 (2008).
316. (a) V. Maurel, M. Laferrière, P. Billone, *et al.*, *J. Phys. Chem. B*, **110**, 16353–16358 (2006). (b) C. Coenjarts, O. García, L. Llauger, *et al.*, *J. Am. Chem. Soc.*, **125**, 620–621 (2003).
317. D. Bardelang, M. B. Zaman, I. L. Moudrakovski, *et al.*, *Adv. Mater.*, **20**, 4517–4520 (2008).
318. (a) E. G. Janzen and B. J. Blackbum, *J. Am. Chem. Soc.*, **90**, 5909–5910 (1968). (b) E. G. Janzen, *Acc. Chem. Res.*, **4**, 31–40 (1971).
319. B. Halliwell and J. M. Gutteridge, *Free Radicals in Biology and Medicine*, Oxford University Press, 4th edn, 2007.
320. (a) P. G. Mekarbane and B. J. Tabner, *Magn. Reson. Chem.*, **36**, 826–832 (1998). (b) D. Gigmes, A. Gaudel-Siri, S. R. A. Marque, *et al.*, *Helv. Chim. Acta*, **89**, 2312–2326 (2006).

321. (a) P. Tordo, *Electron Paramagnetic Resonance*, **16**, 116–144 (1998). (b) J.-L. Clement and P. Tordo, *Electron Paramagnetic Resonance*, **20**, 29–49 (2007).
322. (a) G. M. Rosen, M. S. Cohen, B. E. Britigan and S. Pou, *Free Rad. Res.Comm.*, **9**, 187–195 (1990). (b) N. Kocherginsky and H. M. Swartz, *Nitroxide Spin Labels. Reactions in Biology and Chemistry*, CRC Press, Boca Raton, 32–48, 1995.
323. (a) C. Fréjaville, H. Karoui, F. Le Moigne, *et al.*, *J. Chem. Soc., Chem. Commun.*, 1793–1794 (1994). (b) C. Frejaville, H. Karoui, B. Tuccio, *et al.*, *J. Med. Chem.*, **38**, 258–265 (1995).
324. G. Olive, A. Mercier, F. Le Moigne, *et al.*, *Free Radical Biol. Med.*, **28**, 403–408 (2000).
325. (a) K. Stolze, N. Rohr-Udilova, T. Rosenau, *et al.*, *Bioorg. Med. Chem.*, **15**, 2827–2836 (2007). (b) K. Stolze, N. Udilova, T. Rosenau, *et al.*, *Bioorg. Med. Chem.*, **14**, 3368–3376 (2006). (c) K. Stolze, N. Udilova, T. Rosenau, *et al.*, *Biochem. Pharmacol.*, **69**, 1351–1361 (2005). (d) K. Stolze, N. Udilova, T. Rosenau, *et al.*, *Biochem. Pharmacol.*, **69**, 297–305 (2005).
326. (a) H. Zhao, J. Joseph, H. Zhang, *et al.*, *Free Radical Biol. Med.*, **31**, 599–606 (2001). (b) F. A. Villamena and J. L. Zweier, *J. Chem. Soc., Perkin Trans. 2*, 1340–1344 (2002).
327. (a) F. A. Villamena, A. Rockenbauer, J. Gallucci, *et al.*, *J. Org. Chem.*, **96**, 7994–8004 (2004). (b) K. Stolze, N. Rohr-Udilova, A. Hofinger and T. Rosenau, *Bioorg. Med. Chem.*, **16**, 8082–8089 (2008).
328. F. Chalier, M. Hardy, O. Ouari, *et al.*, *J. Org. Chem.*, **72**, 7886–7892 (2007).
329. M. Hardy, F. Chalier, O. Ouari, *et al.*, *Chem. Commun.*, 1083–1085 (2007).
330. M. Hardy, D. Bardelang, H. Karoui, *et al.*, *Chem. Eur. J.*, **15**, 11114–11118 (2009).
331. (a) H. Karoui, S. Pietri, A. Rockenbauer and P. Tordo, *Chem. Commun.*, **2002**, 3030–3031. (b) H. Karoui and P. Tordo, *Tetrahedron Lett.*, **45**, 1043–1045 (2004). (c) D. Bardelang, A. Rockenbauer, H. Karoui, *et al.*, *Org. Biomol. Chem.*, **4**, 2874–2882 (2006).
332. D. N. Polovyanenko, S. R. A. Marque, S. Lambert, *et al.*, *J. Phys. Chem. B*, **112**, 13157–13162 (2008).
333. D. Bardelang, J.-P. Finet, L. Jicsinszky, *et al.*, *Chem. Eur. J.*, **13**, 9344–9354 (2007).
334. Y. Han, B. Tuccio, R. Lauricella and F. A. Villamena, *J. Org. Chem.*, **73**, 7108–7117 (2008).
335. (a) C. D. Detweiler, L. J. Deterding, K. B. Tomer, *et al.*, *Free Radical Biol. Med.*, **33**, 364–369 (2002). (b) D. C. Ramirez, Y. R. Chen and R. P. Mason, *Free Radical Biol. Med.*, **34**, 830–839 (2003). (c) R. P. Mason, *Free Radical Biol. Med.*, **36**, 1214–1223 (2004). (d) K. Nakai and R. P. Mason, *Free Radical Biol. Med.*, **39**, 1050–1058 (2005). (e) A. Kesler, R. P. Mason and N. Hogg, *Free Radical Biol. Med.*, **40**, 507–515 (2006).
336. (a) D. C. Ramirez, S. E. Gomez Mejiba and R. P. Mason, *Nat. Methods*, **3**, 123–127 (2006). (b) D. C. Ramirez, S. E. Gomez Mejiba and R. P. Mason, *Nat. Protoc.*, **2**, 512–522 (2007).
337. (a) Y. Y. He, D. C. Ramirez, C. D. Detweiler, *et al.*, *Photochem. Photobiol.*, **77**, 585–591 (2003). (b) L. J. Deterding, D. C. Ramirez, J. R. Dubin, *et al.*, *J. Biol. Chem.*, **279**, 11600–11607 (2004). (c) C. D. Detweiler, O. M. Lardinois, L. J. Deterding, *et al.*, *Free Radical Biol. Med.*, **38**, 969–976 (2005).
338. D. C. Ramirez, S. E. Gomez Mejiba and R. P. Mason, *Free Radical Biol. Med.*, **38**, 201–214 (2005).
339. R. Sengupta, T. R. Billiar and D. A. Stoyanovsky, *Org. Biomol. Chem.*, **7**, 232–234 (2009).

6

The Only Stable Organic Sigma Radicals: Di-*tert*-Alkyliminoxyls

Keith U. Ingold

National Research Council, Ottawa, Canada

6.1 Introduction

Radicals can be divided into classes both on the basis of the kinetics of their bimolecular self-reactions and on the basis of their structural/electronic properties. Three kinetic classes are recognized:[1] (i) *Transient* radicals undergo their bimolecular self-reactions at, or close to, the diffusion-controlled limit; (ii) *Persistent* radicals undergo much slower bimolecular self-reactions and slow, or no, unimolecular decay reactions (such as β-scissions); (iii) *Stable* radicals do not undergo such reactions at ambient temperatures (or, if they do, the rates are negligible). The present author applies the following pragmatic test as to whether a radical should be described as *stable* or *persistent*. Only if the radical can be isolated, handled, and stored (for prolonged periods) as the pure compound under normal laboratory conditions, should the radical be described as *stable*. This means that the adjective *stable* should be confined to those radicals that do not react with oxygen or moisture and, from a practical perspective, can be "put in a bottle" and kept on a lab bench.[1]

There are two main structural/electronic classes of radicals. The vast majority of organic radicals have their unpaired electron occupying an orbital perpendicular to the local molecular framework. Such species are classified as π *radicals*. A few families of organic radicals have their unpaired electron occupying an orbital lying in the plane of the local molecular framework. Such species are classified as σ *radicals*.

Not surprisingly, most stable radicals are π radicals. Examples include, amongst many others[2]: perchlorotriphenylmethyl; Koelsch's radical; *N,N*-diphenyl-*N'*-picrylhydrazyl (DPPH$^\bullet$); di-*tert*-alkyl nitroxides (e.g., TEMPO, 2,2,6,6-tetramethylpiperidin-N-oxy); and pyridinyl.[3] A few types of σ radicals have been rendered *persistent* by steric protection of the radical center. Examples include: various phenyls[1,4]; various vinyls[5]; a cyclopropyl[6]; and various iminyls[7]; but none of these radicals could be rendered *stable*. The *only* stable organic σ radicals are certain iminoxyls, R,R'C=NO$^\bullet$, the subject of this chapter.

6.2 The discovery of stable iminoxyls

As Hicks[2] has pointed out: "From Gomberg's seminal report of the triphenylmethyl radical at the turn of the twentieth Century up to relatively recent times, essentially every new stable/persistent radical class was an accidental discovery". This, however, was not the case for the discovery of *stable* iminoxyl radicals. Their discovery was the goal of a 1972 comprehensive investigation of structural effects on the kinetics, mechanisms, and products of decay of a wide variety of alkyl- and aryl-substituted iminoxyls.[8] This particular study was one part of a much more general survey of structural effects on the kinetics and mechanisms of decay of all the more important classes of free radicals (such as peroxyls,[9] nitroxides,[10] alkyls,[11] etc.) that was undertaken in the author's laboratory in the 1960s and 1970s.

Iminoxyls were first identified in solution by Thomas[12] in 1964 using electron paramagnetic resonance (EPR) spectroscopy. He also described the Z–E isomerization of non-symmetrically substituted iminoxyls (Scheme 6.1).

On this basis, and on the basis of their large nitrogen hyperfine splittings ($a^N \sim 28–33$ G; cf., acyclic dialkyl nitroxides, $a^N \sim 14.0–15.5$ G), which indicates the presence of significant spin density on nitrogen in an orbital with considerable *s*-character, Thomas concluded that in iminoxyls the unpaired electron resides in a π-type orbital lying in the plane of the local molecular framework. That is, Thomas[12] concluded that iminoxyls are σ radicals. They can be represented by the canonical structures shown in Scheme 6.2.

The structural/electronic conclusions of Thomas[12] were quickly confirmed,[13,14] particularly by Symons[13] who estimated that 41 ± 5 % of the spin density in $R_2C{=}NO^\bullet$ was on the nitrogen atom in an orbital with a p/s ratio of 6.1 ± 0.7, independent of the nature of the R groups. More recently, the Mulliken spin densities on the oxygen and nitrogen have been calculated to be: O/N = +0.57/+0.45 for $H_2C{=}NO^\bullet$, +0.57/+0.46 for $Me_2C{=}NO^\bullet$, and +0.53/+0.48 for $(Me_3C)_2C{=}NO^\bullet$.[15] The oxygen and nitrogen spin densities for aromatic iminoxyls have been computed to be very similar,[16] for example, +0.58/+0.46 for $PhCH{=}NO^\bullet$ and +0.57/+0.46 for $Ph_2C{=}NO^\bullet$.

For the 1972 survey of iminoxyl radicals' decay kinetics,[8] the radicals were generated by oxidation of their parent oximes. Most of the kinetic work involved generation of the iminoxyls, usually at initial concentrations in the range $10^{-5}–10^{-6}$ M, using photochemically produced *tert*-butoxyl radicals (Equations 6.1 and 6.2):

$$Me_3COOCMe_3 \xrightarrow{h\nu} 2\ Me_3CO^\bullet \quad (6.1)$$

$$Me_3CO^\bullet + R,R'C{=}NOH \rightarrow Me_3COH + R,R'C{=}NO^\bullet \quad (6.2)$$

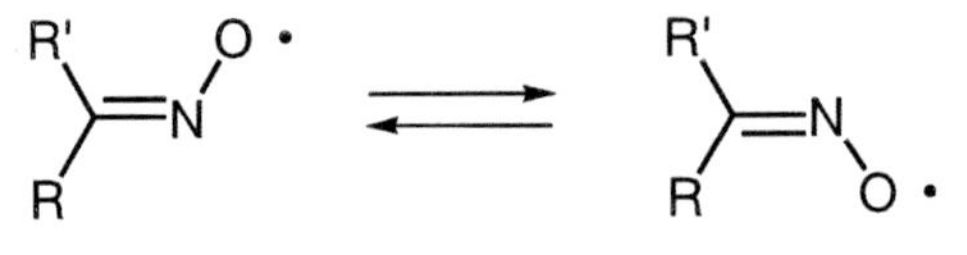

Scheme 6.1

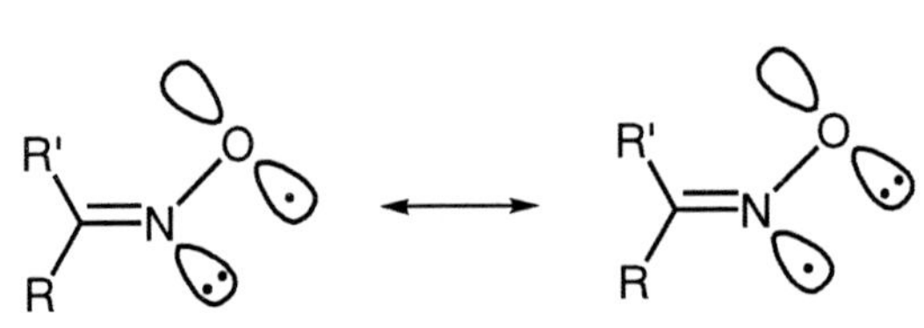

Scheme 6.2

The solvent was generally benzene and the temperature was normally 25 °C. The radicals were generated in the cavity of an EPR spectrometer and decay was monitored after shuttering the light. Diaryliminoxyls and alkylaryliminoxyls decayed with the expected second order kinetics, for example, Equation 6.3:

$$2Ph_2\ C{=}NO^\bullet \rightarrow \text{products} \qquad (2\ k_3{}^{298K} = 2.5 \times 10^3\ M^{-1}s^{-1}) \tag{6.3}$$

Surprisingly, most of the dialkyliminoxyls decayed with first order kinetics and with remarkably similar decay rate constants (Equation 6.4), for example,[8] $k_4{}^{298K} = 4 \times 10^{-2}\ s^{-1}$ for R,R′ = $(CH_3)_2$; $(CD_3)_2$; $(PhCH_2)_2$; CH_3,$PhCH_2$; c-C_5H_{10}; and others.

$$R,R'C{=}NO^\bullet \rightarrow \text{products} \tag{6.4}$$

However, the concentrations of these dialkyliminoxyls under steady UV irradiation was proportional to the *square root* of the light intensity. The first order decay kinetics arose because the iminoxyls were in equilibrium with their dimers and (most probably)[8] the decay of the dimers was rate controlling, Equation 6.5:

$$2\ R,R'C{=}NO^\bullet \rightleftarrows \{R,R'C{=}NO\}_2 \rightarrow \text{products} \tag{6.5}$$

The existence of these radical/dimer equilibria was demonstrated by showing that during the decay of an iminoxyl radical, the iminoxyl concentration could be temporarily increased by rapidly raising the temperature of the sample by 10 °C. Following this temperature-induced increase in concentration, the subsequent decay rate of the iminoxyl was faster than at the initial, lower temperature.[8] Di-isopropyliminoxyl decayed with first order kinetics ($k_4 = 7 \times 10^{-3}\ s^{-1}$) at low concentrations (2×10^{-7}–1×10^{-5} M) but with second order kinetics ($2\ k^{298K} = 1.9 \times 10^2\ M^{-1}s^{-1}$) at high concentrations (1×10^{-4}–1.3×10^{-3} M). A blue "head-to-tail" dimer of this iminoxyl was isolated (together with other products) by oxidation of di-isopropyl ketoxime with silver(I) oxide (Ag_2O), Equation 6.6[8]:

$$2\ \{(CH_3)_2CH\}_2C{=}NOH + Ag_2O \rightarrow \{(CH_3)_2CH\}_2C{=}NOC\{(CH_3)_2CH\}_2N{=}O + H_2O + 2Ag \tag{6.6}$$

Similar kinetic behavior was found for the decay of *tert*-butylmethyliminoxyl.[8]

The foregoing kinetic results encouraged the synthesis of di-*tert*-butyl ketoxime. This was no trivial task because the only reported synthesis[17] required extremely high pressures, 125 000 psi (8300 atm) at room temperature, which increased to 136 000 psi (9000 atm) at the reaction temperature of 75 °C. Fortunately, the equipment required for such high pressure chemistry was available at the National Research Council, though the amount of oxime that could be synthesized was very limited owing to the small size of very high pressure reaction vessels. Di-*tert*-butyliminoxyl, generated from the ketoxime (Equations 6.1 and 6.2) in benzene in an EPR spectrometer at concentrations from 1×10^{-6}–1×10^{-2} M, underwent no measurable decay over the course of several days.[8] We therefore set out to synthesize a much larger quantity of the ketoxime in order to prepare sufficient di-*tert*-butyliminoxyl to determine whether or not it could be isolated, purified, and "put in a bottle". This was made practicable by Hartzler's timely demonstration that hindered nitriles readily undergo nucleophilic attack.[18] The following synthetic procedure was developed.[19]

6.2.1 Synthesis of di-*tert*-butyl ketoxime

Into 40 ml of 1.25 M *tert*-butyl lithium in *n*-pentane under nitrogen was added, drop-wise and with magnetic stirring, 4.1 g (50 mmol) of pivalonitrile. After standing for one hour at 25 °C, there was added, successively: 5 ml absolute ethanol, 3 ml acetic acid, and 3.5 g hydroxylamine hydrochloride. An additional 65 ml of absolute ethanol was used to transfer the mixture to a flask with an attached condenser, and the mixture was refluxed for five hours. About 25 ml of water was then added and the mixture was cooled in ice. Colorless crystals formed and were filtered off, sucked partially dry, and sublimed at about one Torr. Yield of di-*tert*-butyl ketoxime: 3.3 g (46 %).[19]

Some 24 years later, a simpler and more elegant synthesis of this ketoxime (and other symmetric di-*tert*-alkyl ketoximes) was developed by Mendenhall and coworkers (Equation 6.7).[20] In their new procedure the alkylating agent was generated *in situ* from sodium and the tertiary chloride, thus conveniently avoiding expensive, pyrophoric reagents.

$$2\ \mathrm{Na} + t\text{-}\mathrm{RCl} + t\text{-}\mathrm{RCN} \rightarrow \mathrm{NaCl} + t\text{-}\mathrm{R_2C{=}NNa} \rightarrow\ \rightarrow t\text{-}\mathrm{R_2C{=}NOH} \tag{6.7}$$

6.2.2 Synthesis of di-*tert*-butyliminoxyl[19]

The ketoxime (0.3 g) in 30 ml benzene was shaken 90 min with 0.7 g silver oxide. The resulting blue solution was filtered and the filtrate was concentrated at 25 °C to a blue oil (0.22 g, 74 %) which was distilled twice *in vacuo* by the bulb-to-bulb procedure with the receiver immersed in liquid nitrogen. Di-*tert*-Butyliminoxyl is a blue liquid, freezing point −21 °C, $n^{294\mathrm{K}}\mathrm{D}$ 1.4452, $d^{295\mathrm{K}}$ 0.824, with a weak absorption extending from about 530 nm to beyond 800 nm, $\varepsilon(800\ \mathrm{nm}) = 5.1$ in benzene, 5.0 in 95 % ethanol, and 4.7 in cyclohexane.[19]

The IR spectrum shows an intense absorption at 1610 $\mathrm{cm^{-1}}$ from C=N and/or N–O stretching vibrations. A peak at 1650 $\mathrm{cm^{-1}}$ in the parent ketoxime (C=N vibration) does not appear in the spectrum of the radical. However, peaks at 2960, 1480, 1390, and 1366 $\mathrm{cm^{-1}}$ are also present in the IR spectrum of di-*tert*-butyl ketone and can, therefore, be assigned to C–H vibrations.

The EPR spectrum of di-*tert*-butyliminoxyl in dilute solutions is split into a triplet due to interaction of the unpaired electron with the ^{14}N nucleus. The magnitude of the ^{14}N splitting shows a very much smaller dependence on the solvent than is the case for the ^{14}N splitting of di-*tert*-butyl nitroxide. For example, at 23 °C $(\mathrm{Me_3C})_2\mathrm{C{=}NO^\bullet}$ has $a^{\mathrm{N}} = 31.32$ G in benzene and 31.55 G in acetic acid, whereas $(\mathrm{Me_3C})_2\mathrm{NO^\bullet}$ has $a^{\mathrm{N}} = 15.44$ G in benzene and 16.46 in acetic acid.[19] The lack of a significant solvent effects on both the EPR and visible spectra can be rationalized by noting that the nitrogen atom in the iminoxyl is approximately sp^2 hybridized and is, therefore, less polarizable and less basic than the sp^3 hybridized nitrogen atom in the nitroxide. The orbital energies and electron distribution in the iminoxyl are, therefore, less affected by a change in solvent polarity than is the case for the nitroxide.

The three main lines in the EPR spectrum of $(\mathrm{Me_3C})_2\mathrm{C{=}NO^\bullet}$ show additional hyperfine splittings by the hydrogen atoms. At 23 °C, the relative intensities of these lines are not consistent with 18 equivalent protons. However, the multiplet structure at −50 °C was simulated using two groups of nine protons with a^{H} (9H) = 0.481 and 0.772 G; these correspond to the protons on the anti and syn *tert*-butyl groups, respectively.[21] As the temperature was increased above −50 °C, the multiplet structure changes because inversion of the C=NO$^\bullet$ moiety becomes faster and the distinction between the anti and syn *tert*-butyl protons becomes blurred. At 72 °C, inversion is rapid on the EPR time scale ($k_{\mathrm{inv}} = 1.3 \times 10^6\ \mathrm{s^{-1}}$) and all eighteen protons become equivalent. Under these conditions, $a^{\mathrm{H}} = 0.627$ G, which is the mathematical average of the anti and syn proton hyperfine splittings given above. This proves that the spin on both sets of protons has the same sign. This sign was determined by taking the ^{1}H NMR spectrum of neat $(\mathrm{Me_3C})_2\mathrm{C{=}NO^\bullet}$.[21] At room temperature this spectrum shows a *single* broad peak ($\Delta H \sim 1600$ Hz) shifted to *low* field.[21] The average spin density on the *tert*-butyl protons is therefore *positive* in sign. The paramagnetic shift relative to $(\mathrm{Me_3C})_2\mathrm{C{=}NOH}$ is −45.7 ppm, which yields $a^{\mathrm{H}}(\mathrm{av}) = +0.61$ G, in excellent agreement with the EPR results. This positive spin on the *tert*-butyl protons in $(\mathrm{Me_3C})_2\mathrm{C{=}NO^\bullet}$ contrasts with the π radical, $(\mathrm{Me_3C})_2\mathrm{NO^\bullet}$, where the spin on the protons is negative, $a^{\mathrm{H}} = 0.11$ G.[21]

The following experiments demonstrated that di-*tert*-butyliminoxyl is monomeric in solution[19]:

(i) The molecular weight, measured by vapor pressure osmometry in benzene at 37 °C, was 160, a value in satisfactory agreement with the calculated molecular weight of 156.
(ii) A plot of the doubly integrated EPR signal intensity in cyclohexane at 25 °C against [$(Me_3C)_2C{=}NO^\bullet$] was linear from 0.01 to 0.29 M.
(iii) A plot of optical density (800 nm) in benzene at 25 °C against [$(Me_3C)_2C{=}NO^\bullet$] was linear from 0.015 to 0.16 M.
(iv) The EPR signal intensity increased by a factor of 2.70 on cooling a 0.2 M solution of $(Me_3C)_2C{=}NO^\bullet$ in isopentane from +25 to −150 °C. This increase is due to the Boltzmann factor plus solvent contraction and, within experimental error, it corresponds to the intensity increase of 2.84 found for a 10^{-4} M solution of $DPPH^\bullet$ under the same conditions.

The results described above all show that di-*tert*-butyliminoxyl does not undergo dimerization in solution, even at low temperatures. This behavior contrasts with the behavior of less sterically crowded dialkyliminoxyls for which there is abundant kinetic evidence for reversible dimerization at room temperature.[8]

Magnetic susceptibility measurements on di-*tert*-butyliminoxyl down to 5 K showed no sign of a paramagnetic to ferromagnetic transition.[22]

6.2.3 Stability of di-*tert*-butyliminoxyl

Disappointingly, *pure* $(Me_3C)_2C{=}NO^\bullet$ undergoes significant decomposition in a few hours at room temperature[19] (at 100 °C it detonates!). However, the pure radical can be stored indefinitely at −78 °C, at which temperature it is a solid. (This observation lead us to synthesize di-(1-adamantyl)iminoxyl in the expectation, fulfilled, that it would be a solid at room temperature and in the further expectation, also fulfilled, that it would therefore be a *genuinely* stable σ radical at ambient temperatures, see below.) At −20 °C, pure (liquid) $(Me_3C)_2C{=}NO^\bullet$ slowly developed several new IR bands at the expense of its strong 1610 cm^{-1} band.[19] This sample had partially solidified after nine weeks and was only weakly paramagnetic, though it remained blue. A diamagnetic, unstable dimer was isolated as a blue solid, melting point 48–53 °C, λ_{max} (benzene) 698 nm (ε 9.5). This dimer was assigned the same "head-to-tail" structure as the dimer obtained from di-isopropyliminoxyl (see above), that is: $(Me_3C)_2C{=}NOC(Me_3C)_2N{=}O$. Surprisingly, heating an air-free solution of this dimer to 50 °C does not yield the EPR signal of di-*tert*-butyliminoxyl but it does produce a small yield (1–3 %) of a nitroxide radical, most probably di-*tert*-butyl nitroxide. Much less of this nitroxide is formed if the reaction is carried out under air. The molecular products identified after $(Me_3C)_2C{=}NO^\bullet$ was allowed to decompose at room temperature for a week included di-*tert*-butyl ketone (42 %), di-*tert*-butyl nitrimine (20 %), and pivalonitrle (4 %). These products are consistent with the decomposition mechanism shown in Equations 6.8–6.12 (products identified are given in boldface).[19]

$$2\,(Me_3C)_2C{=}NO^\bullet \rightarrow (Me_3C)_2C{=}NOC(Me_3C)_2N{=}O \rightarrow (Me_3C)_2C{=}N^\bullet + \mathbf{(Me_3C)_2C{=}O} + {}^\bullet NO \quad (6.8)$$

$$(Me_3C)_2C{=}NO^\bullet + {}^\bullet NO \rightarrow (Me_3C)_2C{=}NONO \rightarrow (Me_3C)_2C{=}N^\bullet + {}^\bullet NO_2 \rightarrow \mathbf{(Me_3C)_2C{=}NNO_2} \quad (6.9)$$

$$(Me_3C)_2C{=}N^\bullet \rightarrow Me_3C^\bullet + \mathbf{Me_3CCN} \quad (6.10)$$

$$\text{Products} \leftarrow Me_3COO^\bullet \leftarrow O_2 + Me_3C^\bullet + {}^\bullet NO \rightarrow Me_3CN{=}O \quad (6.11)$$

$$Me_3C^\bullet + Me_3CN{=}O \rightarrow \mathbf{(Me_3C)_2NO^\bullet} \quad (6.12)$$

Table 6.1 *Values of $2k_{13}$ for iminoxyls $RR'C{=}NO^\bullet$ at 24 °C*

R	R′	$2k_{13}$ ($M^{-1}s^{-1}$)
Me_3C	Me_3C	2.1×10^{-5}
Me_3C	Me_2CEt	6.3×10^{-5}
Me_2CEt	Me_2CEt	6.9×10^{-5}
Me_3C	Et_3C	1.4×10^{-5}
Me_3C	Me_2CH	5.0×10^{-2}
Me_2CH	Me_2CEt	0.17
Me_3C	$PhCH_2CMe_2$	0.56
Me_3C	$PhCMe_2$	0.65
Me_2CH	Me_2CH	1.9×10^{2}
Me_3C	Me	2.0×10^{3}

The decay of $(Me_3C)_2C{=}NO^\bullet$ at 24 °C, both neat and in benzene, follows second order kinetics. This is also true for other hindered dialkyliminoxyls (Table 6.1).[8,19,23]

$$2\ R,R'C{=}NO^\bullet \rightarrow \text{products} \tag{6.13}$$

However, in dilute solution in degassed *n*-hexadecane the EPR spectrum of $(Me_3C)_2C{=}NO^\bullet$ could easily be recorded at 150 °C.[19]

In view of the foregoing, and until precise definitions are available, it is a matter of personal choice as to whether di-*tert*-butyliminoxyl should be classified as a stable, or as an extremely persistent, free radical.

6.3 Hydrogen atom abstraction by di-*tert*-butyliminoxyl

6.3.1 The O–H bond dissociation enthalpy (BDE) in $(Me_3C)_2C{=}NOH$

The rates of hydrogen atom abstraction by $(Me_3C)_2C{=}NO^\bullet$ are largely determined by thermodynamic factors and these are, of course, related to the O–H bond dissociation energy (BDE) in $(Me_3C)_2C{=}NOH$. This bond dissociation energy was determined calorimetrically in 1973 by measuring the heat evolved in the rapid, exothermic reaction that occurs upon the addition of a small quantity of solid hydrazobenzene to a large excess of di-*tert*-butyliminoxyl dissolved in 25 ml of well stirred benzene at room temperature (Equation 6.14).[24] The hydrazobenzene was quantitatively converted to *E*-azobenzene with a measured heat of reaction of -6.5 ± 0.5 kcal/mol.[24] This heat of reaction, combined with the heats of formation of *E*-azobenzene and hydrazobenzene, and with the heat of solution of azobenzene, yielded: BDE $[(Me_3C)_2C{=}NO{-}H] = 80.9$ kcal/mol in benzene.[24]

$$2\ (Me_3C)_2C{=}NO^\bullet + PhNHNHPh \rightarrow 2\ (Me_3C)_2C{=}NO{-}H + E\text{-}PhN{=}NPh \tag{6.14}$$

In 2004 the O–H bond dissociation energy in $(Me_3C)_2C{=}NOH$ was revised[25] downward by 1 kcal/mol to take account of a revision in the heat of formation of *E*-azobenzene[26] and by an additional 0.7 kcal/mol to correct for hydrogen bond formation between this oxime and a molecule of the benzene solvent. That is, the "best" gas phase value for BDE $[(Me_3C)_2C{=}NO{-}H]$ is $80.9 - 1.7 = 79.2$ kcal/mol.[25]

During the 1973 work, some oxime O–H bond dissociation energies were also determined in benzene by the EPR radical-equilibrium technique, rather than by calorimetry.[24] For $Me_3C(Me_2CH)C{=}NOH$ and

Me_3C(1-adamantyl)C=NOH, the O–H bond dissociation energies were found to be 84.3 and 80.0 kcal/mol, respectively, (which today would be corrected to 82.6 and 78.3 kcal/mol.)[25] Space filling molecular models suggested that in $Me_3C(Me_2CH)C{=}NOH$ there would be little steric repulsion between the hydroxyl group and a suitably orientated syn isopropyl group, an observation that lead to the prediction that simple, unhindered oximes, R,R′C=NOH with R,R′ = alkyl, hydrogen, *or aryl*, would have O–H bond dissociation energies of about 86 kcal/mol.[24] This conclusion was later disputed by Bordwell on the basis of oxime O–H bond dissociation energies determined by an electrochemical procedure.[27] However, Bordwell failed to appreciate the fact that *all known* iminoxyls are σ radicals and that, therefore, their unpaired electron cannot delocalize into the π electron systems of neighboring aromatic rings, that is, even $Ph_2C{=}NO^\bullet$ and other aryliminoxyls are σ radicals.[8,12–14,16,25] As a consequence, the fact that Bordwell's experimental technique yielded considerably lower O–H bond dissociation energies for aryliminoxyls than for unhindered alkyliminoxyls did not cause Bordwell concern, instead it was used to argue that aryliminoxyls had π-radical structures![27] However, in 2004 "order was restored" and the correctness of the 1973 prediction[24] (revised downward by about 2 kcal/mol, *vide supra*) was confirmed both by experiment and by theory.[25,16] For example, the following R,R′C=NO–H bond dissociation energies have been recommended[25]: 83.4, 84.6, and 81.6 kcal/mol for R = R′ = hydrogen, methyl, and phenyl, respectively.

The hydroxylamines derived by reduction of stable nitroxides in the TEMPO class have O–H bond dissociation energies of about 70 kcal/mol.[24] The stable (or highly persistent) di-*tert*-butyliminoxyl radical is, therefore, a considerably stronger oxidizing agent than stable dialkyl nitroxides and it has seen some limited use in synthesis, *vide infra*.

6.3.2 Oxidation of hydrocarbons with di-*tert*-butyliminoxyl

Benzylic hydrogen atoms are abstracted quite readily at elevated temperatures. For example,[19] $10^4 \times k_{15}/M^{-1}s^{-1} = 0.75$, 4.7, and 11.5 at 120 °C for toluene, ethylbenzene, and isopropylbenzene, respectively (Equations 6.15 and 6.16).

$$(Me_3C)_2C{=}NO^\bullet + RH \rightarrow (Me_3C)_2C{=}NOH + R^\bullet \qquad (6.15)$$

$$(Me_3C)_2C{=}NO^\bullet + R^\bullet \rightarrow (Me_3C)_2C{=}NOR \qquad (6.16)$$

The rate constant for reaction of this radical with diphenylmethane has also been reported ($9 \times 10^{-6}\,M^{-1}s^{-1}$ at 28 °C).[23] These reactions have a large deuterium kinetic isotope effect, for example, replacement of the tertiary (benzylic) hydrogen atom in isopropylbenzene by a deuterium atom yielded: $k_{15}^{H}/k_{15}^{D} = 14.4$ (±1.8) at 120 °C.[19] The temperature dependence of hydrogen atom abstraction from isopropylbenzene can be described by Equation 6.17[19]:

$$k_{15}^{H}/M^{-1}s^{-1} = 10^{6.3} \times \exp(-16.8\ \text{kcal/mol}/RT) \qquad (6.17)$$

The reactions of di-*tert*-butyliminoxyl with alkenes are much faster than with alkyl aromatic hydrocarbons, for example,[23] $10^5 \times k_{obs}/M^{-1}s^{-1} = 2.0$, 3.3, 4.0, and 5.1, at 28 °C for 2-methylpent-2-ene, 2,3-dimethylbut-2-ene, cyclohexene, and cyclopentadiene, respectively. These reactions are complete in about 90 minutes at room temperature.[28] With 1,4-cyclohexadiene there is a rapid, exothermic reaction at room temperature that yields stoichiometric amounts of the oxime and benzene according to Equation 6.18[19]:

$$2\ (Me_3C)_2C{=}NO^\bullet + c\text{-}C_6H_8 \rightarrow 2(Me_3C)_2C{=}NOH + C_6H_6 \qquad (6.18)$$

The reactions of this iminoxyl with mono-alkenes are of particular interest because they occur by two parallel, but mechanistically distinct, processes that give the same final products (!), the $(Me_3C)_2C{=}NO$– alkene oximes.[28] These two processes are:

(i) Initial abstraction of an allylic hydrogen atom followed by $(Me_3C)_2C{=}NO^{\bullet}$ addition to either end of the resulting allylic radical.
(ii) Initial addition of $(Me_3C)_2C{=}NO^{\bullet}$ to the double bond, followed by hydrogen atom abstraction from the resulting alkyl radical.

These two reaction mechanisms were identified by selectively deuterating the alkenes at the two vinylic positions and then using ^2H NMR spectroscopy to identify the position of the double bond and of the added $(Me_3C)_2C{=}NO$– moiety.[28] As is shown for cyclohexene in Scheme 6.3, the addition + abstraction mechanism (Equation 6.20) yields a single deuterated product, whereas the abstraction + addition mechanism (Equation 6.19) yields *both* this same product *and* a product unique to this mechanism (the product containing D_d and D_e).

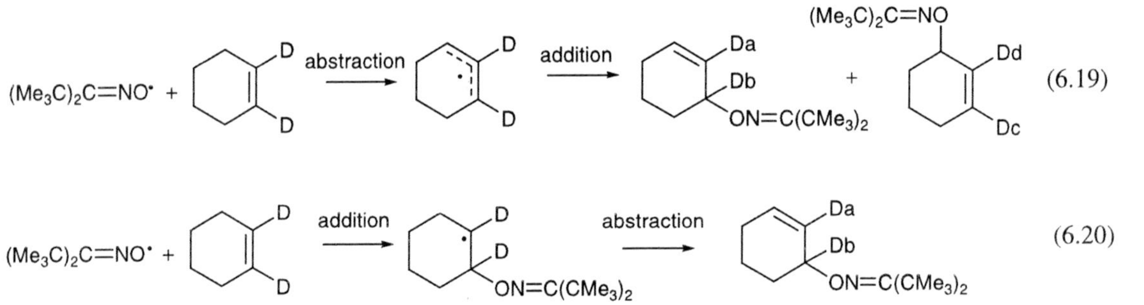

(6.19)

(6.20)

Scheme 6.3

The abstraction + addition process (Equation 6.19) is dominant in the three alkenes examined, cyclohexene, cyclooctene, and 3-hexene, with 90–92 % of the overall reaction occurring by this mechanism.[28] (As an aside, both mechanisms were also identified in the reactions of these three di-deuterated alkenes with 4-hydroxyTEMPO and with phthalimide-*N*-oxyl and, for all six of these reactions, about 90 % of the overall process occurs by the abstraction + addition mechanism.)[28]

6.3.3 Oxidation of phenols with di-*tert*-butyliminoxyl

These oxidations are faster than the hydrocarbon oxidations described above. The observed rate constants for the loss of $(Me_3C)_2C{=}NO^{\bullet}$ (measured by EPR) in the presence of $4\text{-}YC_6H_4OH$ in acetonitrile at 22 °C increase dramatically as the Y substituent is changed from an electron withdrawing group to an electron donating group: for example,[29] $k_{obs}/M^{-1}s^{-1}$ $(Y) = <5 \times 10^{-4}$ (NO_2), 0.03 (H), 0.2 (methyl), > 0.5 (MeO). This rate accelerating effect of electron donating substituents is normal for hydrogen atom abstractions from $4\text{-}YC_6H_4OH$ because $4\text{-}YC_6H_4O\text{–}H$ bond dissociation energies decrease dramatically as Y becomes more strongly electron donating,[30] for example, a decrease of about 10 kcal/mol on changing from a 4-NO_2 group to a 4-MeO group.[30]

The products of these reactions are highly dependent on the phenol's structure. For example,[29] *para*-hydroquinones and catechol were oxidized to the corresponding quinones in good yields (70–99 %);

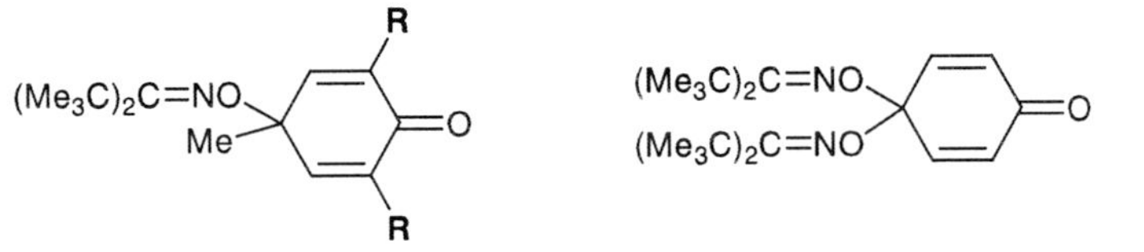

A (R = H, Me3C) *B*

Figure 6.1 *4-methyl-4-iminoxycyclohexadiones,* ***A****, and 4,4-bis-oximes,* ***B****.*

4-methyl-4-iminoxycyclohexadiones, ***A*** (Figure 6.1), were formed with *para*-cresol and 2,6-di-*tert*-butyl-4-methylphenol (BHT) in yields of 43 and 78 %, respectively; phenol and 1-naphthol gave the 4,4-bis-oximes, for example, ***B*** (Figure 6.1), in 58 and 87 % yields, respectively; whereas 2-naphthol gave 1,2-naphthoquinone in 82 % yield.

6.3.4 Oxidation of amines with di-*tert*-butyliminoxyl

Primary and secondary amines were converted into imines in good yields in a few hours at room temperature in pentane or hexane.[31] Due to the high inherent reactivity of most imines, the majority of the imine products were transformed *in situ* into the 2.4-dinitrophenylhydrazine derivatives of the corresponding carbonyl compounds. For example,[31] the isolated yields of 2,4-dinitrophenylhydrazones from Ph_2CHNH_2, $PhCH_2NH_2$, and $(PhCH_2)_2NH$ were 79, 68, and 78 %, respectively. However, *N*-benzylidinemethylamine, PhCH=NMe, was actually isolated from the reaction of $PhCH_2NHMe$ with $(Me_3C)_2C{=}NO^\bullet$ in yields (reaction time) of 49 % (30 min), 64 % (4 hours), and 76 % (24 hours) from reactions carried out at 69, 25, and $-5\,^\circ$C, respectively.[31] In retrospect, it seems likely that other imines could probably also have been isolated by a careful choice of reaction conditions and work-up procedures.

6.3.5 Oxidation of di-*tert*-butylketoxime with di-*tert*-butyliminoxyl

This hydrogen atom abstraction reaction is a *self-exchange*, or *identity*, reaction. Self-exchange reactions are precisely thermoneutral and their study has provided a wealth of understanding about reaction mechanisms, "intrinsic" activation energies, and so on. One remarkable series of observations has been that the self-exchange reactions of oxygen-centered π radicals (Equation 6.21), and other thermoneutral (or near thermoneutral) reactions involving the destruction and formation of oxygen-centered π radicals (Equation 6.22), have remarkably low activation energies (typically $\sim 2 \pm 1$ kcal/mol)[15] in comparison with the activation energies for self-exchange reactions involving carbon-centered π radicals, for example, 14.6 kcal/mol for the $CH_3^\bullet + CH_4 \rightarrow CH_4 + CH_3^\bullet$ reaction.[32]

$$\pi\text{-RO}^\bullet + \text{ROH} \rightarrow \text{ROH} + \pi\text{-RO}^\bullet \qquad (6.21)$$

$$\pi\text{-RO}^\bullet + \text{R}'\text{OH} \rightarrow \text{ROH} + \pi\text{-R}'\text{O}^\bullet \qquad (6.22)$$

A detailed theoretical study of the π-phenoxyl + phenol self-exchange by Mayer *et al.*[33] revealed an exciting fact. This self-exchange is *not* a 'simple' *hydrogen atom transfer* (HAT) with the proton being transferred together with one of its bonding electrons. Instead, the $PhO^\bullet$ + PhOH self-exchange is a *proton-coupled electron transfer* (PCET), with the proton and electron being transferred between different sets of orbitals. There is an initial formation of a $PhO^\bullet$---HOPh complex in which the phenol forms a hydrogen bond with a lone pair on the phenoxyl oxygen atom, $O^{\bullet}$. This complex was computed to be 9.9 kcal/mol lower in energy than the separated reactants.[33] The transition state for the self-exchange was also computed to be lower in energy by 1.3 kcal/mol than the separated reactants. These computational results imply that the $PhO^{\bullet}$ + PhOH self-exchange is essentially a diffusion-controlled process. This accounts for the very low activation energy found for $PhO^{\bullet}$ + PhOH and for related self-exchange and near thermoneutral π-$RO^{\bullet}$ + R′OH reactions. In the $PhO^{\bullet}$ + PhOH reaction, the proton was involved in a four electron, three center hydrogen bond and was computed to be transferred nearly in the plane of the two phenoxyls.[33] Accompanying this proton transfer between σ lone pairs is an electron transfer from the doubly occupied 2p-π atomic orbital on the oxygen in PhOH to the singly occupied 2p-π atomic orbital of $PhO^{\bullet}$. This intriguing mechanistic insight, which provides such a simple explanation for the much lower activation energies for the self-exchange reactions of oxygen-centered π radicals compared with the self- exchange reactions of carbon-centered π radicals, has been supported and expanded in subsequent computational work.[34]

An experimental study, using EPR spectroscopy, of the reaction of $(Me_3C)_2C{=}NO^{\bullet}$ with three oximes, RR′C=NOH (R = R′ = Me; R = Ph, R′ = H; R = Me_3C, R′ = Me_2CH) revealed that all three reactions were surprisingly slow.[35] The self-exchange reaction was therefore investigated, also by EPR, using a ^{17}O-labeled iminoxyl generated from di-*tert*-butyliminyl, $(Me_3C)_2C = N^{\bullet}$, as shown in Equations 6.1 and 6.23–6.25 below.

$$Me_3COOCMe_3 \xrightarrow{hv} 2\ Me_3CO^{\bullet} \tag{6.1}$$

$$Me_3CO^{\bullet} + (Me_3C)_2C{=}NH \rightarrow Me_3COH + (Me_3C)_2C{=}N^{\bullet} \tag{6.23}$$

$$(Me_3C)_2C{=}N^{\bullet} + {}^{34}O_2 \rightarrow \rightarrow (Me_3C)_2C{=}N{-}{}^{17}O^{\bullet} \tag{6.24}$$

$$(Me_3C)_2C{=}N{-}{}^{17}O^{\bullet} + (Me_3C)_2C{=}NOH \rightarrow (Me_3C)_2C{=}N{-}{}^{17}OH + (Me_3C)_2C{=}NO^{\bullet} \tag{6.25}$$

At 25 °C, k_{25} was only 1.3 (±0.3) $M^{-1}s^{-1}$ and the activation energy (measured only from 7 to 37 °C) was 7 ± 2 kcal/mol.[35] This rate constant was orders of magnitude smaller and the activation energy was considerably greater than for all known self-exchange reactions involving oxygen-centered π radicals.[15]

Because iminoxyls are σ radicals with their unpaired electron residing in an orbital coplanar with the local molecular framework, their self-exchange reactions (Equation 6.25) cannot occur by Mayer *et al.*'s[33] "classic" ($PhO^{\bullet}$/PhOH) PCET mechanism with the proton moving "in-plane" between two electron pairs and its accompanying electron moving between orbitals perpendicular to the molecular plane, *vide supra*. It was, therefore, a genuine surprise when a theoretical investigation of Equation 6.26 for R = hydrogen, methyl and Me_3C, revealed that these hydrogen atom transfers also occur by a PCET mechanism.[15]

$$R_2C{=}NO^{\bullet} + R_2C{=}NOH \rightarrow R_2C{=}NOH + R_2C{=}NO^{\bullet} \tag{6.26}$$

These reactions occur via a counter-intuitive cisoid transition state in which the seven atoms, >C=NO---H---ON=C<, lie in one plane (R = H and methyl), while for R = Me_3C these atoms lie in two planes twisted at 45 ° to relieve steric compression forces.[15] The planar transition state has the two N–O dipoles close to each other and pointing in almost the same direction with an O---H---O angle of 165°. A transoid transition state was also found but it lies several kcal/mol higher in energy than the cisoid despite a more favorable arrangement of the two N–O dipoles and the O---H---O atoms being almost linear. It was concluded[15] that the iminoxyl/oxime self-exchange occurs by a five center, cyclic PCET mechanism with the proton being transferred between electron pairs on the two oxygen atoms and the electron migrating between in-plane orbitals on the two nitrogen atoms, (Scheme 6.4) (N–N distance = 0.265 nm).

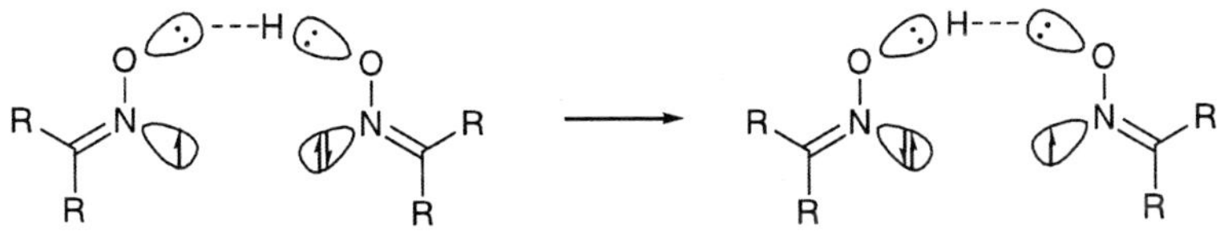

Scheme 6.4

6.4 Other reactions and non-reactions of di-*tert*-butyliminoxyl

The reaction of this iminoxyl with nitrogen monoxide, $^{\bullet}NO$, goes through the colorful sequence: blue → green → yellowish brown → colorless, and the final product is the nitrimine, $(Me_3C)_2C{=}NNO_2$.[19] There is, presumably, an initial coupling to form $(Me_3C)_2C{=}NONO$, and this is followed by decomposition to $^{\bullet}NO_2$ and the iminyl, $(Me_3C)_2C{=}N^{\bullet}$, which then couple to the nitrimine (Equation 6.9). In contrast to this facile radical + radical reaction, $(Me_3C)_2C{=}NO^{\bullet}$ does not react with the DPPH$^{\bullet}$ radical, nor with stable nitroxides.[19] This iminoxyl is also unreactive towards styrene, vinyl acetate, methyl iodide and triphenyl phosphine, but it does react with triethyl phosphate, potassium iodide-starch paper, bromine vapor, and ceric ammonium nitrate.[19] None of these last reactions has been examined in any detail.

6.5 Di-*tert*-alkyliminoxyls more sterically crowded than di-*tert*-butyliminoxyl

Since $(Me_3C)_2C{=}NO^{\bullet}$ decays via an irreversible head-to-tail dimerization, it was hypothesized that the replacement of one or both *tert*-butyl groups by sterically more demanding substituents would hinder dimer formation and, hence, yield much more persistent, and possibly stable, iminoxyls.[20] Unfortunately, this was not the case.[20]

The total steric effect associated with a substituent, relative to the steric effect of a methyl group, is best quantified by Taft's (logarithmic) steric substituents constants, E_s.[36] Values of E_s for some of the alkyl substituents of the iminoxyls listed in Table 6.1 (which gives the rate constants for their second order decay) are: Me, 0.00; Me_2CH, −0.47; Me_3C, −1.54; $EtCMe_2$, −1.95[37]; $PhCMe_2$, −2.06[37]; Et_3C, −3.8. The following sterically-protected iminoxyls have been isolated as blue liquids: $Me_3C(EtCMe_2)C{=}NO^{\bullet}$, $(EtCMe_2)_2C{=}NO^{\bullet}$, and $Me_3C(Et_3C)C{=}NO^{\bullet}$.[20] Although they are more sterically congested than $(Me_3C)_2C{=}NO^{\bullet}$, the second order rate constants ($2k_{13}$) for the decay of two of them, $Me_3C(EtCMe_2)C{=}NO^{\bullet}$ and $(EtCMe_2)_2C{=}NO^{\bullet}$, are about three times as great as $2k_{13}$ for di-*tert*-butyliminoxyl[23] (Table 6.1). Only $Me_3C(Et_3C)C{=}NO^{\bullet}$ decayed at a slightly lower rate than $(Me_3C)_2C{=}NO^{\bullet}$ (Table 6.1). However, the kinetics of decay of this iminoxyl did not correspond to a simple second order process and gave between 9 and 11 products (HPLC) with traces of pivalonitrile and triethylacetonitrile, but no detectable (<1 %) 4,4-diethyl-2,2-dimethylhexan-3-one.[23] In contrast, the decay of $(Me_3C)_2C{=}NO^{\bullet}$ gave 4 % pivalonitrile and 40 % di-*tert*-butyl ketone (*vide supra*).[19] At high concentrations of $Me_3C(Et_3C)C{=}NO^{\bullet}$ (0.2 M) decay followed reasonable first order kinetics ($k^{obs} = 2.3 \times 10^{-5}\ s^{-1}$ in acetonitrile at 23 °C), probably via an intramolecular hydrogen atom abstraction (Equations 6.27 and 6.28):

$$Me_3C(Et_3C)C{=}NO^{\bullet} \rightarrow Me_3CC({=}NOH)CEt_2(C^{\bullet}HCH_3) \quad (6.27)$$

$$Me_3CC({=}NOH)CEt_2(C^{\bullet}HCH_3) + Me_3C(Et_3C)C{=}NO^{\bullet} \rightarrow$$
$$Me_3CC({=}NOH)CEt_2(CHCH_3ON{=}C(Me_3C)CEt_3) \rightarrow \rightarrow \rightarrow \text{9 to 11 products} \quad (6.28)$$

This interesting attempt to prepare kinetically more stable di-*tert*-alkyliminoxyls radicals by increasing the steric bulk of the alkyl groups failed because of the appearance of a new decomposition mechanism, intramolecular hydrogen atom abstraction.[38]

6.6 Di-(1-Adamantyl)iminoxyl: a truly stable σ radical

This radical was synthesized[39] by the (atmospheric pressure) method used to prepare $(Me_3C)_2C{=}NO^{\bullet}$, and with the hope that it would be crystalline and would fit the present author's definition of a "stable" radical, *vide supra*. This hope was realized, pale blue crystals (Figure 6.2), melting point 121–122.5 °C.[39] In a potassium bromide pellet, this iminoxyl showed a strong, sharp IR band at 1595 cm^{-1} (C=N stretch), its EPR spectral parameters in benzene were: $a^{14N} = 31.14$ G, g = 2.00481 and, at natural abundance, $a^{15N} = 43.3$ G, $a^{13C} = 8.0$ and 13.6 G, and its extinction coefficient was 5.7 (800 nm).[39] The ^{1}H NMR of a 0.8 M solution of $(1\text{-Ad})_2C{=}NO^{\bullet}$ in C_6D_6 at 23 °C showed *three* peaks, all at low field, which indicates, surprisingly, that there was positive spin density on all three groups of protons.[21] The paramagnetic shifts relative to $(1\text{-Ad})_2C{=}NOCH_3$ in C_6D_6, were: −1.6 (width ~120 Hz), −14.2 (width ~650 Hz), and about −41 ppm (very broad), shifts that correspond to the EPR parameters $a^H = +0.02, +0.19$, and +0.55 (±04) G, respectively.[21] The largest a^H was assigned to protons at the 2-position of the adamantine skeleton and the others to protons at the 3- and 4-positions.[21] For comparison, in the nitroxide, $(1\text{-Ad})_2NO^{\bullet}$, $a^H = -0.405$ (2-CH_2), +0.54 (3-CH), and −0.035 and −0.015 G (axial and equatorial 4-CH_2). In benzene at 30 °C, the dipole moment of $(1\text{-Ad})_2C{=}NO^{\bullet}$ is 2.90 D, while those for $(1\text{-Ad})_2C{=}NH$ and $(1\text{-Ad})_2C{=}NOCH_3$ are 2.49 and 0.79 D, respectively.[39]

One reason for synthesizing a solid iminoxyl in 1974[39] was to determine its structure by X-ray crystallography. Disappointingly, $(1\text{-Ad})_2C{=}NO^{\bullet}$ always crystallized in extremely thin plates that were unsuitable for X-ray analysis. Fortunately, structural determinations by X-ray analysis are, today, much less important thanks to modern computational procedures which have produced accurate structural information for several iminoxyl radicals.[25]

Figure 6.2 The Di(1-Adamantyl)iminoxyl Radical. A full-colour version of this figure appears in the Colour Plate section of this book.

References

1. D. Griller and K. U. Ingold, *Acc. Chem. Res.*, **9**, 13–19 (1976).
2. R. G. Hicks, *Org. Biomol. Chem.*, **5**, 1321–1338 (2007).
3. E. M. Kosower, *Topics Curr. Chem.*, **112**, 117–162 (1983).
4. (a) L. R. C. Barclay, D. Griller, and K. U. Ingold, *J. Am. Chem. Soc.*, **96**, 3011–3012 (1974). (b) G. Brunton, D. Griller, L. R. C. Barclay, and K. U. Ingold, *J. Am. Chem. Soc.*, **98**, 6803–6811 (1976). (c) G. Brunton, J. A. Gray, D. Griller, *et al.*, *J. Am. Chem. Soc.*, **100**, 4197–4200 (1978). (d) K. U. Ingold, in *Hydrogen-Transfer Reactions* (Eds J. T. Hynes, J. P. Klinman, H.-H. Limbach, and R. L. Schowen), Vol. 2, John Wiley & Sons, Inc., New York, 875–893 (2007).
5. D. Griller, J. W. Cooper, and K. U. Ingold, *J. Am. Chem. Soc.*, **97**, 4269–4275 (1975).
6. V. Malatesta, D. Forest, and K. U. Ingold, *J. Am. Chem. Soc.*, **100**, 7073–7074 (1978).
7. D. Griller, G. D. Mendenhall, W. Van Hoof, and K. U. Ingold, *J. Am. Chem. Soc.*, **96**, 6068–6070 (1974).
8. J. L. Brokenshire, J. R. Roberts, and K. U. Ingold, *J. Am. Chem. Soc.*, **94**, 7040–7049 (1972).
9. See, e. g., (a) J. A. Howard and K. U. Ingold, *Can. J. Chem.*, **45**, 793–802 (1966). (b) K. U. Ingold, *Acc. Chem. Res.*, **2**, 1–9 (1969). (c) J. A. Howard, *Adv. Free Radical Chem.*, **4**, 49–173 (1972).
10. See, e. g., (a) K. Adamic, D. F. Bowman, T. Gillan, and K. U. Ingold, *J. Am. Chem. Soc.*, **93**, 902–908 (1971). (b) D. F. Bowman, J. L. Brokenshire, T. Gillan, and K. U. Ingold, *J. Am. Chem. Soc.*, **93**, 6551–6555 (1971). (c) D. F. Bowman, T. Gillan, and K. U. Ingold, *J. Am. Chem. Soc.*, **93**, 6555–6561 (1971). (d) G. D. Mendenhall and K. U. Ingold, *J. Am. Chem. Soc.*, **94**, 7166–7167 (1972).
11. See, e. g., (a) D. J. Carlsson and K. U. Ingold, *J. Am. Chem. Soc.*, **90**, 7047–7055 (1968). (b) G. B. Watts and K. U. Ingold, *J. Am. Chem. Soc.*, **94**, 491–494 (1972). (c) G. D. Mendenhall and K. U. Ingold, *J. Am. Chem. Soc.*, **95**, 3422 (1973).
12. J. R. Thomas, *J. Am. Chem. Soc.*, **86**, 1446–1447 (1964).
13. (a) M. C. R. Symons, *J. Chem. Soc.*, 2276–2277 (1965). (b) W. M. Fox and M. C. R. Symons, *J Chem. Soc. A*, 1503–1507 (1966).
14. See, e. g., (a) B. C. Gilbert, R. O. C. Norman, and D. C. Price, *Proc. Chem. Soc.*, 234 (1964). (b) H. Lemaire and A. Rassat, *Tetrahedron Lett.*, 3245–3248 (1964). (c) B. C. Gilbert and R. O. C. Norman, *J. Chem. Soc. B*, 86–91 (1966).
15. G. A. Dilabio and K. U. Ingold, *J. Am. Chem. Soc.*, **127**, 6693–6699 (2005).
16. S.-S. Chong, Y. Fu, L. Liu, and Q.-X. Guo, *J. Phys. Chem. A*, **111**, 13112–13125 (2007).
17. (a) W. H. Jones, E. W. Tristram, and W. F. Benning, *J. Am. Chem. Soc.*, **81**, 2151–2154 (1959). (b) W. H. Jones and E. W. Tristram, *US Patent* # 3,256,331 (1966).
18. H. D. Hartzler, *J. Am. Chem. Soc.*, **93**, 4527–4531 (1971).
19. G. D. Mendenhall and K. U. Ingold, *J. Am. Chem. Soc.*, **95**, 2963–2971 (1973).
20. B. M. Eisenhauer, M. Wang, H. Labaziewicz, *et al.*, *J. Org. Chem.*, **62**, 2050–2053 (1997).
21. K. U. Ingold and S. Brownstein, *J. Am. Chem. Soc.*, **97**, 1817–1818 (1975).
22. G. D. Mendenhall, private communication.
23. B. M. Eisenhauer, M. Wang, R. E. Brown, *et al.*, *J. Phys. Org. Chem.*, **10**, 737–746 (1997).
24. L. R. Mahoney, G. D. Mendenhall, and K. U. Ingold, *J. Am. Chem. Soc.*, **95**, 8610–8614 (1973).
25. D. A. Pratt, J. A. Blake, P. Mulder, *et al.*, *J. Am. Chem. Soc.*, **126**, 10667–10675 (2004).
26. P. Mulder, H.-G. Korth, D. A. Pratt, *et al.*, *J. Phys. Chem. A*, **109**, 2647–2655 (2005).
27. (a) F. G. Bordwell and G.-Z. Ji, *J. Org. Chem.*, **57**, 3019–3025 (1992). (b) F. G. Bordwell and S. Zhang, *J. Am. Chem. Soc.*, **117**, 4858–4861 (1995).
28. S. Coseri, G. D. Mendenhall, and K. U. Ingold, *J. Org. Chem.*, **70**, 4629–4636 (2005).
29. M. Ngo, K. R. Larson, and G. D. Mendenhall, *J. Org. Chem.*, **51**, 5390–5393 (1986).
30. See, e. g., (a) L. R. Mahoney and M. A. DaRooge, *J. Am. Chem. Soc.*, **92**, 890–899 (1970). (b) P. Mulder, O. W. Saastad, and D. Griller, *J. Am. Chem. Soc.*, **110**, 4090–4092 (1988). (c) M. Jonsson, J. Lind, T. E. Eriksen, and G. Merenyi, *J. Chem. Soc., Perkin Trans. 2*, 1567–1568 (1993). (d) D. D. M. Wayner, E. Lusztyk, and K. U. Ingold, *J. Org. Chem.*, **61**, 6430–6433 (1996). (e) D. A. Pratt, M. deHeer, P. Mulder, and K. U. Ingold, *J. Am. Chem. Soc.*, **124**, 5518–5526 (2001). (f) D. A. Pratt, G. A. DiLabio, P. Mulder, and K. U. Ingold, *Acc. Chem. Res.*, **37**, 334–340 (2004).

31. J. J. Cornejo, K. D. Larson, and G. D. Mendenhall, *J. Org. Chem.*, **50**, 5382–5383 (1985).
32. S. W. Benson, *Thermochemical Kinetics*, 2nd edn, John Wiley & Sons, Inc., New York (1976).
33. J. M. Mayer, D. A. Hrovat, J. L. Thomas, and W. T. Borden, *J. Am. Chem. Soc.*, **124**, 11142–11147 (2002).
34. G. A. DiLabio and E. R. Johnson, *J. Am. Chem. Soc.*, **129**, 6199–6203 (2007).
35. G. D. Mendenhall and K. U. Ingold, *J. Am. Chem. Soc.*, **95**, 627–628 (1973).
36. R. W. Taft, Jr., in *Steric Effects in Organic Chemistry* (ed. M. S. Newman), John Wiley & Sons, Inc., New York, Chapter 13 (1956).
37. K. U. Ingold, *J. Phys. Chem.*, **64**, 1636–1642 (1960).
38. In this connection, it is interesting to note that acyclic di-*tert*-alkyl nitroxides that are more sterically crowded than $(Me_3C)_2NO^\bullet$.also have a reduced stability due to intramolecular hydrogen atom abstractions. Private communication from the late Professor Andre Rassat.
39. D. Lindsay, E. C. Horswill, D. W. Davidson, and K. U. Ingold, *Can. J. Chem.*, **52**, 3554–3556 (1974).

7

Verdazyls and Related Radicals Containing the Hydrazyl [R_2N–NR] Group

Robin G. Hicks

Department of Chemistry, University of Victoria, Victoria, Canada

7.1 Introduction

The three most famous organic stable[1] radicals are triphenylmethyl (trityl, **1**), 2,2,6,6-tetramethylpiperidine-*N*-oxyl (TEMPO, **2**), and *N,N*-diphenyl-*N'*-picrylhydrazyl (DPPH, **3**). The first two of these are the quintessential examples of triarylmethyl and nitroxyl (nitroxide) radicals, respectively – two classes of stable radicals which continue to receive intense interest (and not coincidentally which are the subject of two other chapters in this book). Similarly, DPPH is by far the best known derivative of a hydrazyl radical **4**. In fact, DPPH actually stands out as an exception among hydrazyls in terms of its practical stability; while most generic hydrazyls are *persistent* (lifetimes ranging from minutes to days), only a very small number of these radicals qualify as stable (i.e. isolable), and most of these tend to closely resemble the structure of DPPH itself.

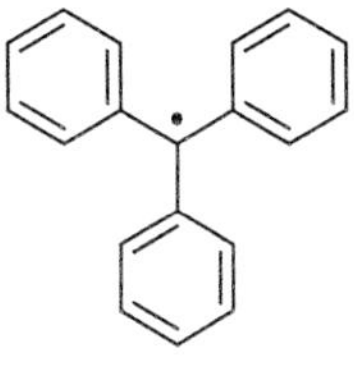

1

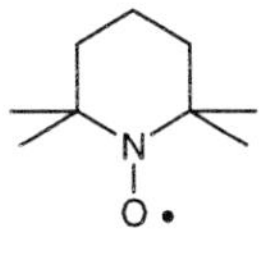

2

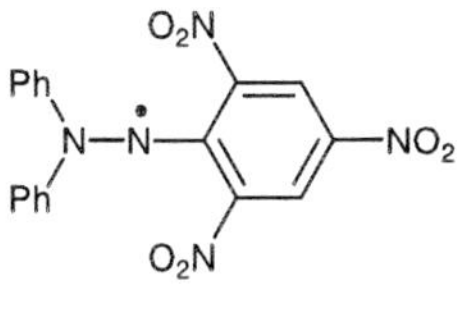

3

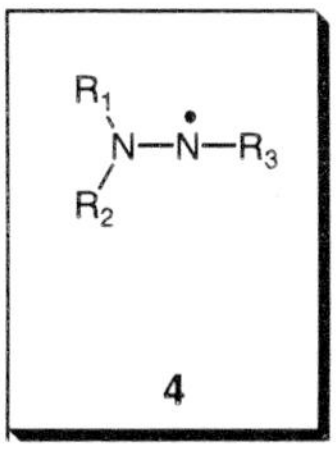

4

Hydrazyl radicals,[2] including the ubiquitous DPPH,[3] were discovered in the 1920s by Goldschmidt; hydrazyl radical chemistry was comprehensively reviewed in Forrester *et al*. 1968 book on stable free radicals.[4] The scientific activity associated with hydrazyl radicals has become completely dominated by

DPPH itself, mainly owing to its widespread use as (i) an Electron Paramagnetic Resonance (EPR) standard, (ii) a radical scavenger in polymer chemistry, and (iii) an indicator for antioxidant chemistry. However, outside of these areas, in the intervening 40+ years developments in the fundamental aspects of hydrazyl radical chemistry – particularly in hydrazyls other than DPPH – have been rare.

In contrast to the relatively poor stability (compared to DPPH) of most hydrazyls, there are several kinds of stable radicals in which the hydrazyl subunit (R_2NNR') is incorporated into a delocalized (and often cyclic) π system. Neugebauer reviewed the field of so-called "hydrazinyl" radicals (nearly all of which was his own work) in 1973; in the approximately 35 years since there have been significant developments in the synthesis, chemical and physical properties of hydrazyl-containing radicals, particularly verdazyls; these topics are the principal foci of this chapter.[5]

7.2 Verdazyl radicals

Verdazyl radicals (general structure **5**) were discovered in the 1960s by Kuhn and Trischmann.[6] Their chemistry has grown steadily – largely enabled by their outstanding chemical stability – to the point where they are now one of the larger families of stable radicals. Verdazyls can be categorized according to the nature of the C6 ring carbon atom. Verdazyls with structure **6** (hereafter referred to as "Kuhn verdazyls") have a saturated carbon at C6 and nearly always have aromatic substituents on each of the nitrogen atoms (R′ and R″). Verdazyls containing a carbonyl (**7**) or thiocarbonyl (**8**) group at C6 were developed in the 1980s by Neugebauer. General aspects of verdazyl radical chemistry have not been reviewed in over 30 years,[7] although two reviews of the magnetic properties of verdazyls have appeared recently.[8,9] As such, verdazyl-based magnetism will not be discussed in detail here.

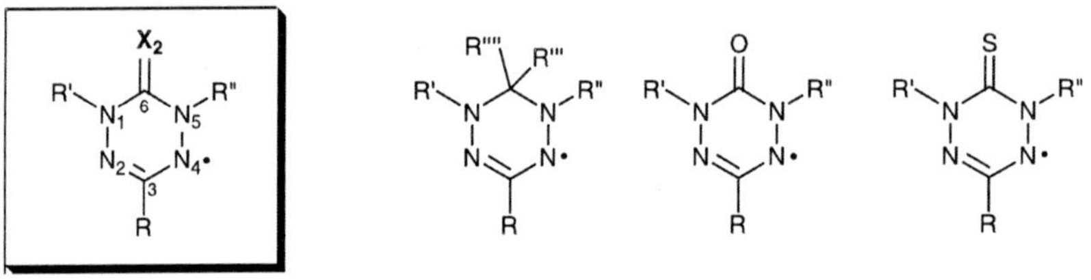

5 **6** **7** **8**

7.2.1 Synthesis of verdazyls

7.2.1.1 Verdazyls from formazans

Formazans **9** are a venerable class of compounds[10] which are used principally as redox indicators in cell chemical biology.[11] In 1963 Kuhn and Trischmann reported that the reaction of triphenylformazan (**9**, R = R′ = R″ = phenyl) with methyl halides yields the corresponding 1,3,5-triphenylverdazyl radical **6** (R = R′ = R″ = phenyl; R‴ = R⁗ = H) (Scheme 7.1) instead of the alkylated formazan **10**. This reaction proceeds via cyclization of **10** to give the so-called "leuco" verdazyl **11**, which is subsequently oxidized to give the radical; the isolation of the leuco compound is not necessary, and the verdazyl is often obtained directly from the formazan reaction using air as the oxidizing agent (though other oxidants such as $Fe(CN)_6^{3-}$ and $FeCl_3$ have been used[12,13]). The most common alkylating agents are methyl bromide or iodide, although secondary alkyl halides have been successfully used as well; there is one example of a verdazyl with two C6 substituents derived from a tertiary alkyl bromide. The N-alkylated formazan **10** can

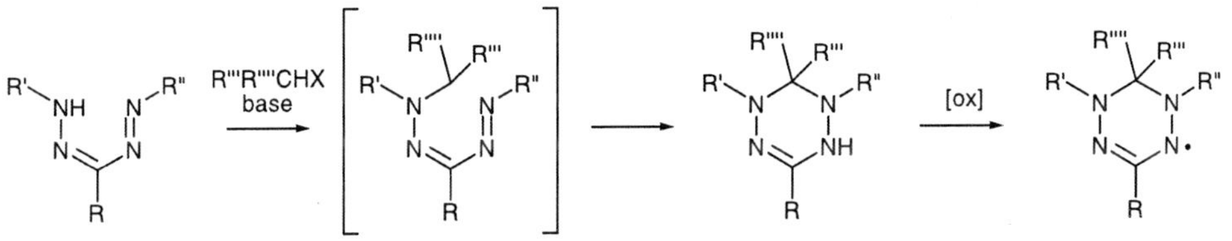

9 10 11 6

Scheme 7.1

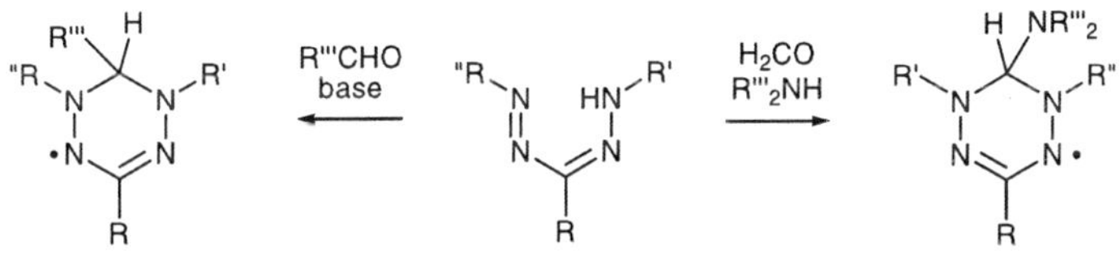

5 9 5

Scheme 7.2

occasionally be isolated, and then heated to proceed to the leuco verdazyl and, ultimately (in the presence of an oxidant), the verdazyl.[12,14,15]

A variation on the formazan synthesis involves their reactions with formaldehyde or aliphatic aldehydes under basic conditions (Scheme 7.2). This reaction presumably also proceeds via a leuco intermediate, based on the observation that oxygen is required for radical formation. If formaldehyde is used with an equivalent amount of a secondary amine (e.g., pyrrolidine, piperidine) then 6-aminoverdazyls are produced (Scheme 7.2).[16,17] Reactions of formazans with formaldehyde under *acidic* conditions give cationic verdazylium salts **12**, which can then be reduced to the radicals (Scheme 7.3).[15,18–20] The verdazylium salts can be generated *in situ*[12,21] and reduced or they can be isolated.[22] The reduction step can be carried out with formaldehyde and base,[21,23] or with ascorbic acid[22]; the latter reagent can, depending on the solvent system, reduce the cation either to the radical **5** or to the leuco compound **11**, which can then be air-oxidized to the radical.

An alternative route to verdazyls from formazans was reported by Kuhn, Neugebauer *et al.* in which triphenyltetrazolium cation **13** (R = R′ = R″ = phenyl; the product of oxidation of the corresponding formazan[10]) was converted to triphenylverdazyl by reaction with diazomethane (Scheme 7.4).[16] This interesting transformation has not been further exploited.

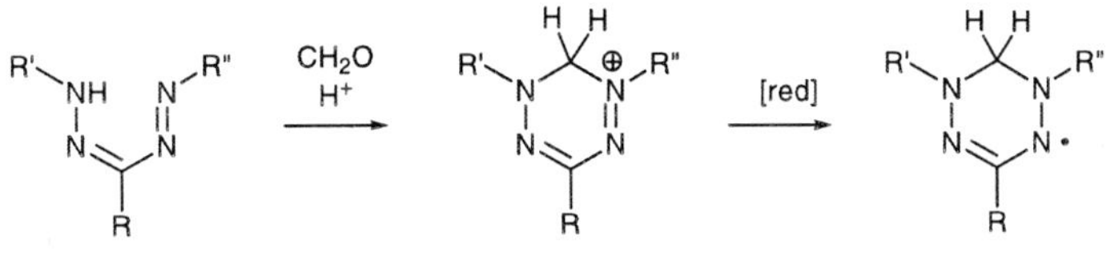

9 12 5

Scheme 7.3

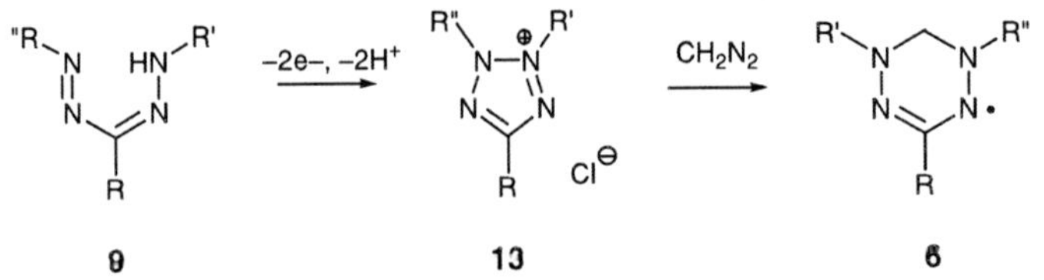

Scheme 7.4

Collectively, the methods for making Kuhn verdazyls from formazans are broadly applicable, facilitated in large part by the availability of a large number of the precursor formazans.[10] The R group in the 3-position of the radical can be hydrogen, alkyl, aryl, cyano, nitro, sugars, carbonyl compounds, and so on. The nitrogen substituents R′ and R″ are usually aromatic. A variety of Kuhn verdazyl-based di-, tri-, and tetra-radicals have been prepared (Table 7.1) wherein the point of verdazyl attachment can be C3, C6 or one of the nitrogen atoms. The formazan methodologies have also been used in the synthesis of polymers containing verdazyls, either as pendant groups or incorporated into the polymer backbone (Figure 7.1).

7.2.1.2 *Verdazyls from hydrazides and bis-hydrazides: 6-oxoverdazyls*

Neugebauer developed the synthesis of so-called "6-oxoverdazyls" **7** in which the C6 carbon is part of a carbonyl group (Scheme 7.5). 2,4-Disubstituted bis(carbohydrazides) **16**[40] condense with aldehydes to give 1,2,4,5-tetrazane-6-ones **17** which can then be oxidized to the 6-oxoverdazyls **7**.[41,42] Several oxidants can be used in the final step, including silver(I) oxide (Ag_2O), tripotassium hexacyanoferrate ($K_3Fe(CN)_6$), lead(IV) oxide (PbO_2), periodate, and benzoquinone.

Certain bis(hydrazides) **16** (R = methyl, CH_2Ph) were later found to be accessible directly from the monoalkylhydrazines (eliminating the need to prepare the hydrazone **14** and subsequent NH_2 deprotection of bis-benzylidene intermediate **15**) either by performing the reaction at low temperatures[43] and/or by using triphosgene (bis(trichloromethyl)carbonate) as a phosgene alternative (Scheme 7.6).[44]

Neugebauer has made 6-thioxoverdazyls **8** by using thiophosgene in place of phosgene (Scheme 7.7).[43] Thus, dehydrogenation of the 6-thioxotetrazanes **18** gives 6-thioxoverdazyls **8**.[42] An additional feature of the thioxotetrazanes **18** consists of their desulfurization to the corresponding tetrazanes **19** with a saturated carbon at C6. Subsequent oxidation gives the only example of verdazyls of structure **6** in which the two substituents on the nitrogen atoms are alkyl groups. Although radicals **6** are persistent they cannot be isolated.[45]

The main limitation of the syntheses described in Schemes 7.5–7.7 is the range of bis-hydrazide reagents, which until recently was confined to R′ = methyl or benzyl. Brook has recently developed a method for making 2,4-bis(isopropyl)hydrazide **23** as shown in Scheme 7.8, which provides entry into N,N-diisopropyl substituted 6-oxoverdazyls.[46]

Neugebauer[47] and Milcent[48] have expanded the repertoire of verdazyl N-substituents by constructing the tetrazane ring in a step-wise manner (Scheme 7.9). Careful reaction of a hydrazone **24** with phosgene or thiophosgene gives the N-chloroformylhydrazone **25** (X = O, S), which is subsequently treated with a monosubstituted hydrazine to give tetrazane **26**; oxidation using standard protocols gives 6-oxo or 6-thioxoverdazyls. This synthetic route considerably expands the possible derivatives of **7** and **8** by allowing for (1) aromatic substituents and (2) differential substitution on the two nitrogen atoms.

Overall, a wide range of 6-oxo (and thioxo)verdazyl derivatives has been prepared. All three substituents (R, R′, R″) can be alkyl or aryl and the Milcent procedure offers the opportunity to prepare radicals with different N1/N5 substituents. Some diradicals based on 6-oxoverdazyls have been studied in which the

Table 7.1 *Di-, tri-, and tetraradicals based on verdazyls* ***6***

Structure type	Linker 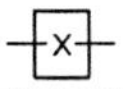	Reference
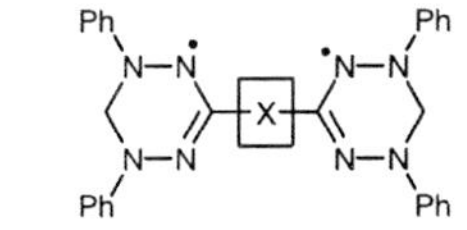	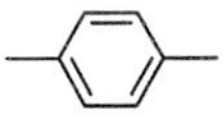	19, 24
	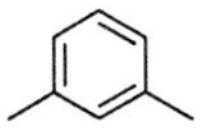	19, 24
	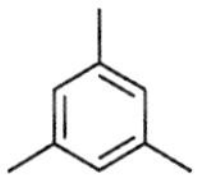	25
	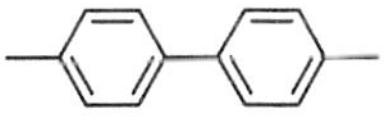	19
	$(CH_2)_4$	19
	nothing	26
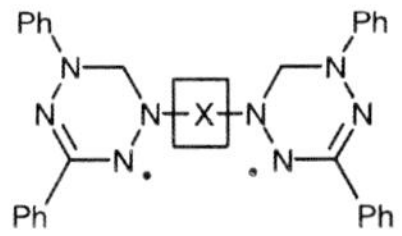	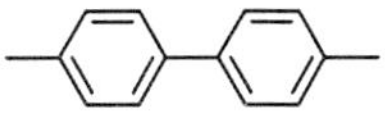	19, 27
	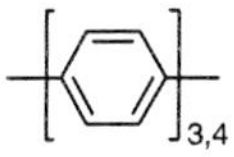	19
	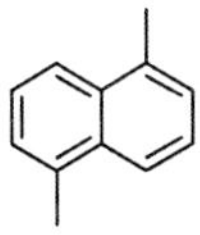	27
	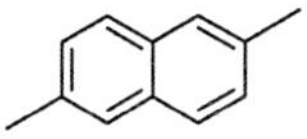	27
	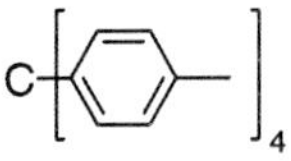	28
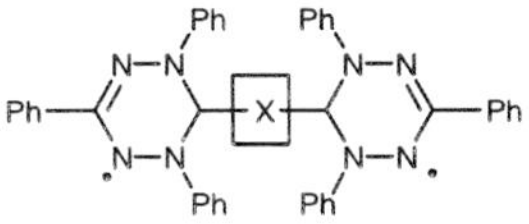	$(CH_2)_4$	19
	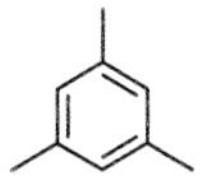	29

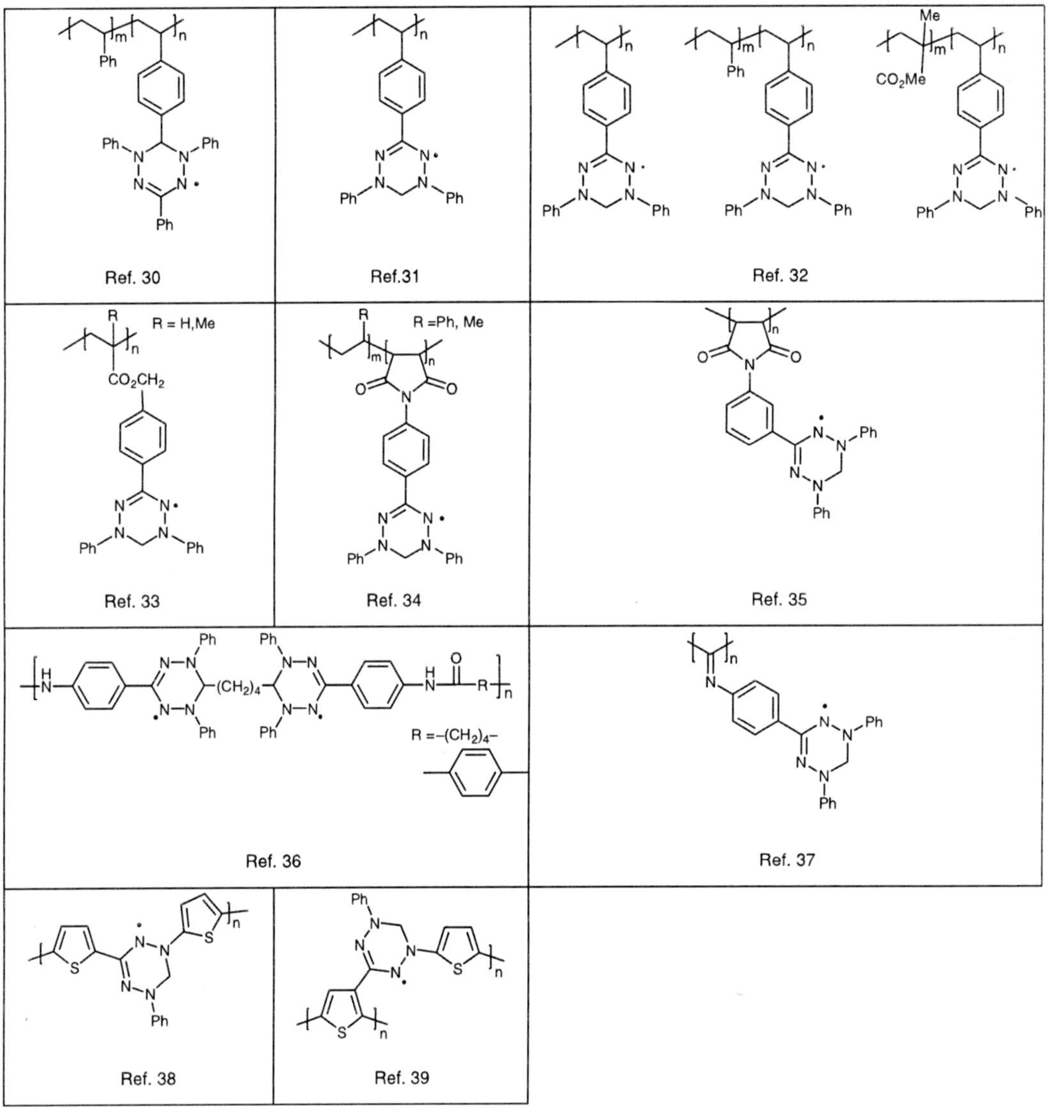

Figure 7.1 Polymers containing verdazyls **6**.

two radicals are linked (via the C3 carbon) by alkyl[42,49] or aromatic spacers (e.g., 1,3- or 1,4-phenylene or 2,5-thienyl)[50,51], or by nothing at all.[42,50,52]

7.2.2 Stability, physical properties and electronic structure of verdazyls

As a general class, verdazyls are among the most robust of radicals. Nearly all verdazyls that have been synthesized are isolable and can be handled and stored indefinitely without decomposition; derivatives with very small groups at C3 (e.g. R = H, methyl) gradually decompose, but this occurs over a period

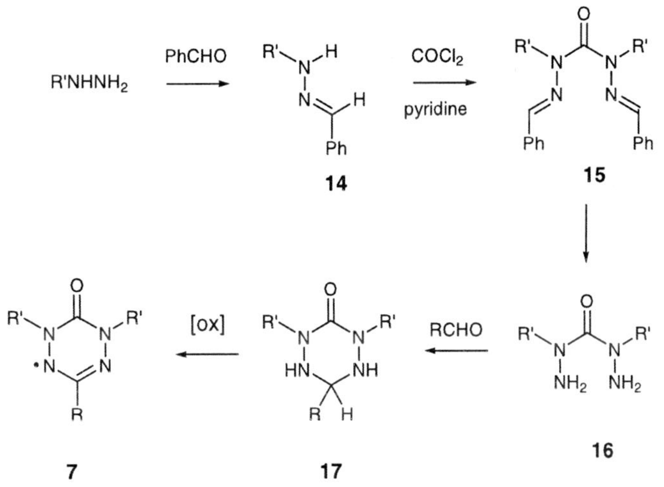

Scheme 7.5

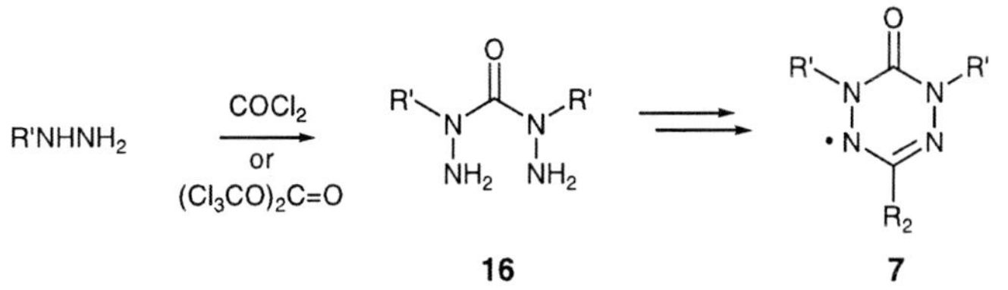

Scheme 7.6

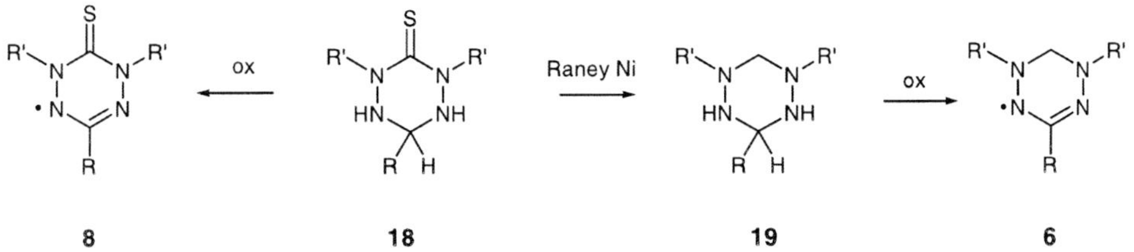

Scheme 7.7

of weeks. The lone general exception is the family of 6-oxoverdazyls **7** with methyl substituents on each of the two nitrogen atoms, which disproportionate (Section 7.2.3.3) – although this process can be sufficiently slow to render some derivatives stable as opposed to long lived/persistent. Verdazyls are air and water stable and remain monomeric in solution and the solid state; there are no examples of verdazyl dimerization in solution (the electronic spectrum of verdazyl **27** is concentration dependent, but in this instance the association-driven spectral changes arise from the amphiphilic nature of the molecule rather than radical–radical association[53]). The lone solid state example of a verdazyl dimer is the intramolecularly associated ferrocene-linked diradical **28**.[54] In the solid state the two verdazyl rings are eclipsed and have a centroid-to-centroid distance (~3.18 Å) which is well within van der Waals separation (and significantly

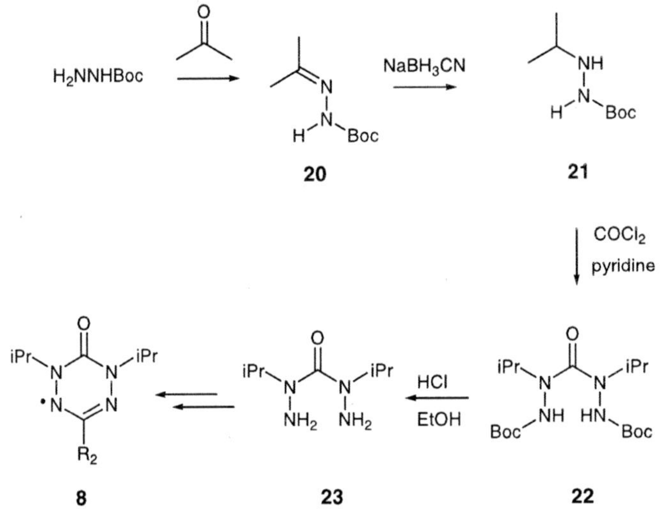

Scheme 7.8

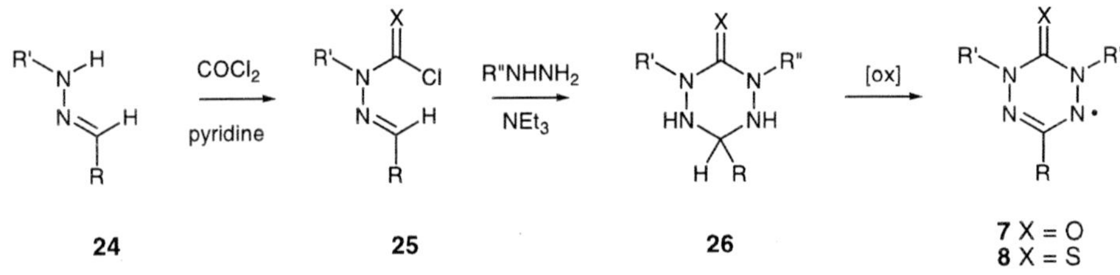

Scheme 7.9

shorter than the Cp-to-Cp distance of 3.30 Å). The molecule is diamagnetic, indicating that spin pairing is complete via π dimerization. Notably, the solution characterization data indicate that the two verdazyls are very weakly interacting, suggesting that in solution the diradical does *not* adopt the π-dimer structure.

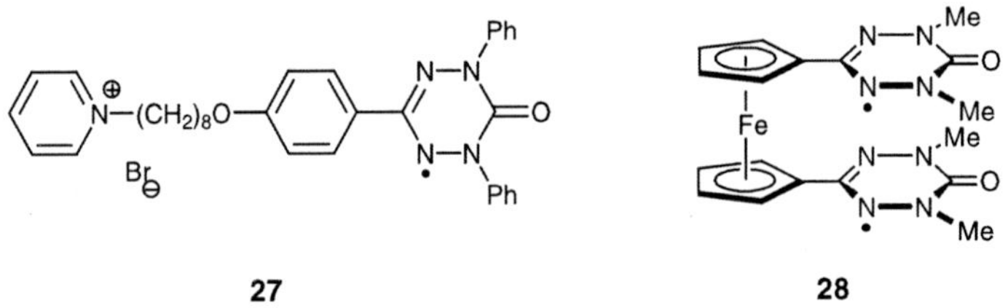

27 **28**

Not surprisingly, verdazyl radicals are highly coloured compounds. Kuhn verdazyls **6** are typically green, with low energy absorptions between 700–750 nm (extinction coefficients $\sim 10^3\ M^{-1}\ cm^{-1}$).[12] In comparison, 6-oxoverdazyls have their longest wavelength absorption maximum (λ_{max}) between 400 and 560 nm and thus are yellow to red in solution.[42,47] For both of these verdazyl types the C3 substituent has a significant influence on the absorption wavelength: radicals with R = hydrogen or alkyl generally have

λ_{max} values about 50 nm lower than the corresponding species with a C3 phenyl group. Electron donating groups on the C3 phenyl cause further red shifts,[55] suggesting that the lowest energy absorption involves a transition from an occupied orbital to the singly occupied molecular orbital (SOMO). The substituents on the nitrogen atoms are also important: 6-oxoverdazyls with N-alkyl groups (e.g., R′, R″ = methyl, isopropyl) have absorption bands near 400–450 nm and appear yellow to dark orange, whereas N-aryl substituted 6-oxoverdazyls are deep red with λ_{max} ranging from 500 to 550 nm.

X-ray crystal structures of many derivatives of verdazyls have been reported. Structurally characterized 6-oxoverdazyls **7** include *N,N′*-dialkyl[49,51,52,56–59] and *N,N′*-diaryl[47,60] derivatives: In all cases the N4C2 ring is planar, with N–N and C3–N bond lengths (1.34–1.37 Å and 1.32–1.35 Å, respectively) indicative of a delocalized π system. In *N,N′*-diaryl substituted oxoverdazyls (Figure 7.2a) the aromatic substituents are moderately twisted with respect to the verdazyl plane (torsion angles ~30°). In the lone example of a crystallographically characterized 6-thioxoverdazyl (Figure 7.2b) the N-phenyl rings are twisted to become nearly perpendicular to the verdazyl.[47] Kuhn verdazyls **6** in which C6 is a methylene (CH_2) group adopt a half-chair conformation; C6 is displaced from the verdazyl plane by ~0.6 Å (Figure 7.2c).[61,62] In cases in which one of the two C6 substituents (R‴, R⁗) is *not* hydrogen, the larger group occupies the axial position.[62,63]

EPR spectroscopy has provided many insights into the electronic structure of these radicals. The majority of these studies are solution EPR experiments, but the spectra are typically complex because of the large number of spin active nuclei to which the unpaired electron can couple. Isotopic labeling studies (^{2}H, ^{15}N) and related spectroscopic techniques, such as ENDOR,[42,47,65] ELDOR,[66] and NMR,[13,67,68] have been commonly employed to obtain more precise/complete pictures of hyperfine coupling (magnitude and sign) to all nuclei, including the protons on all substituents and the skeletal methylene group of the Kuhn verdazyls.

The EPR spectra of Kuhn verdazyls **6** are typically dominated by a nine-line pattern, which arises from hyperfine coupling to the four heterocyclic nitrogen atoms. Although there are two chemically inequivalent pairs of nitrogen atoms (N1/N5 and N2/N4), the hyperfine coupling constants for the two pairs are usually close in magnitude, and as such the four nitrogen atoms can occasionally appear to be "equivalent" within the resolution limits of the spectrum. The resonance structures shown in Figure 7.3 can be invoked to rationalize the presence of spin density on all four nitrogen atoms; structures **a** and **b** are diazaallyl-type resonance forms, whereas **c** and **d** are zwitterionic contributors. EPR data for verdazyls with electron

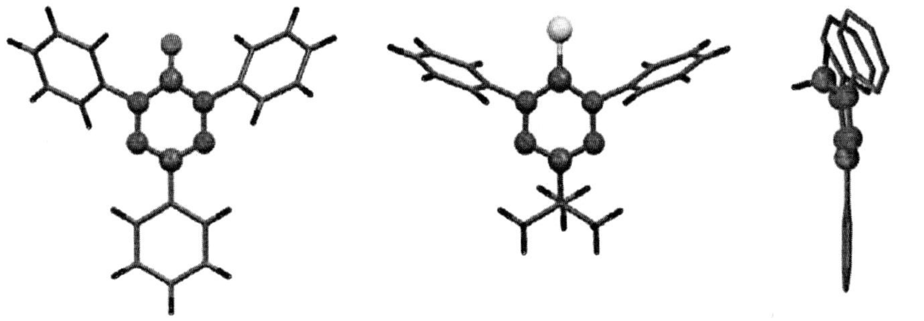

(a) (b) (c)

Figure 7.2 *X-ray crystal structures of (a) 1,3,5-triphenyl-6-oxoverdazyl,[47] (b) 3-t-butyl-1,5-diphenyl-6-thioxoverdazyl,[47] and (c) 1,3,5-triphenylverdazyl[64] (H atoms except for the ring CH_2 protons omitted for clarity).*

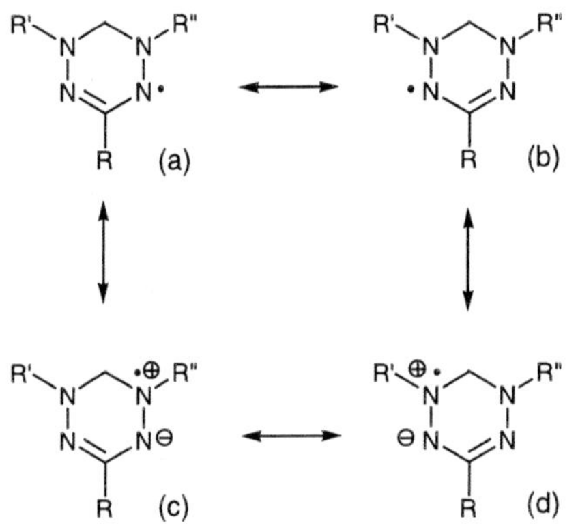

Figure 7.3 Resonance structures for Kuhn verdazyls **6**.

donating or electron withdrawing aromatic substituents on the nitrogen atoms show predictable changes in the hyperfine couplings: more electron poor groups (e.g., nitrophenyl) have relatively smaller hyperfine couplings to the N1/N5 pair as the electron withdrawing groups compete for the nitrogen's lone pair.[12,15,20]

The EPR spectra of 6-oxoverdazyls **7** and 6-thioxoverdazyls **8** differ in subtle but systematic ways from those of the Kuhn verdazyls **6**. For **7** and **8** the N1/N5 constants are consistently smaller than N2/N4; typical values for N1/N5 are 5.1–5.4 G if $R'/R'' =$ alkyl and ~4.5 G if the $R'/R'' =$ aryl; the N2/N4 constants are consistently 6.3–6.6 G.[42,47] The differences between these and the Kuhn radicals can be understood in the context of the resonance structures of Figure 7.3 as applied to the 6-oxoverdazyl framework: the electron withdrawing carbonyl group adjacent to N1/N5 competes for electron density from these atoms, thereby decreasing the relative importance of resonance structures **c** and **d** (compared to **a** and **b**). From a practical perspective, the EPR spectra of 1,5-dialkyl-6-oxoverdazyls are rendered more complex by additional hyperfine coupling: alkyl group α-protons hfc's of between 5.2–5.8 G, 2.9–3.0 G, and 1.2–1.5 G are typical ranges for $R' = R'' =$ methyl, CH_2Ph, and isopropyl, respectively.

Verdazyl radicals have been subjected to a large number of computational investigations. Kuhn verdazyls **6** have been studied computationally, mainly with semi-empirical (McLcachlan,[15,67,69] INDO/PNDO[70]) methods, although a more recent density functional theory (DFT) study has been reported.[71] In contrast, calculations on 6-oxoverdazyl derivatives have been principally based on DFT or hybrid (HF/DFT) methods,[51,56,72–75] and occasionally multiconfigurational methods (CASSCF) have been employed.[75,76] In general terms, the unpaired electron in all verdazyl types is part of a 7π-electron system delocalized over the N1-N2-C3-N4-C5 portion of the heterocycle; C6 is not formally conjugated because it is either a saturated carbon (in **6**) or part of an exocyclic double bond (**7**, **8**). The singly occupied molecular orbital (SOMO) in all cases is a π^* orbital based on the four nitrogen atoms (Figure 7.4a shows the SOMO for a derivative of **6** with R = phenyl and $R' = R'' =$ isopropyl)[51]. One of the nodal planes of this orbital passes through both C3 and C6, which consequently prevents direct conjugative overlap with any π framework which may exist on the C3 substituent. However, spin polarization effects (i.e., electron correlation) produce a small amount of spin density on C3 (as well as N1 and N5 when appropriate) substituents. The trends in calculated spin density distributions for the Kuhn and 6-oxoverdazyls corroborate the experimental nitrogen hyperfine coupling constants (Table 7.2). Overall, the spin density is more or less equally shared by the four nitrogen atoms in the Kuhn verdazyls, with slightly higher spin found on N1/N5. In contrast, for the oxoverdazyls the N2/N4 pair carries substantially more spin than N1/N5. In all cases the four ring nitrogen atoms carry the majority of the spin density.

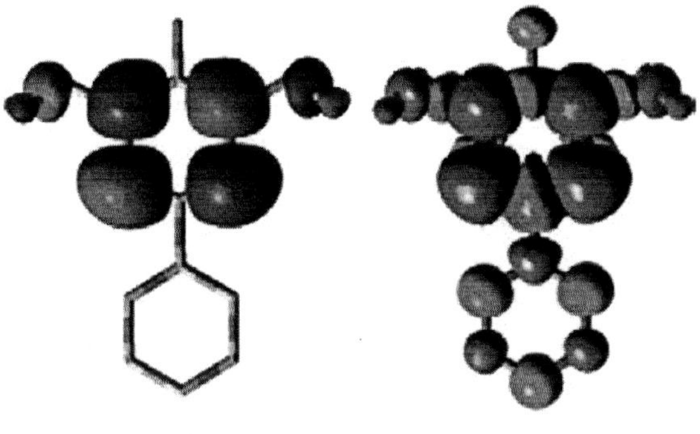

(a) (b)

Figure 7.4 *(a) SOMO and (b) spin density plot for 1,5-diisopropyl-3-phenyl-6-oxoverdazyl (blue = positive spin density, green = negative spin density).[51] (Reprinted with permission from [51]. Copyright 2007 American Chemical Society.) A full-colour version of this figure appears in the Colour Plate section of this book.*

Table 7.2 *Summary of ranges of EPR nitrogen hyperfine coupling constants and DFT calculated spin densities for verdazyls*

Verdazyl general structure	N1/N5 hyperfine coupling (G) *N1/N5 calculated spin density*	N2/N4 hyperfine coupling (G) *N2/N4 calculated spin density*
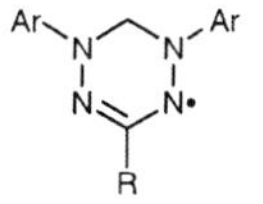	5.7–6.2 *0.24*	5.4–6.1 *0.16*
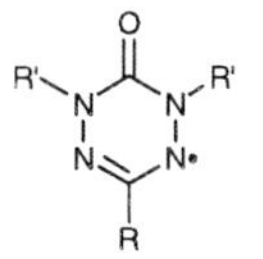	5.1–5.4 *0.18–0.21*	6.4–6.6 *0.34–0.40*
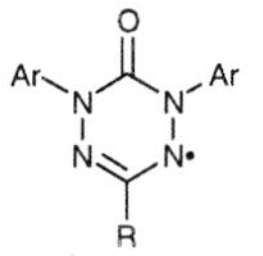	4.4–4.5G —	6.3–6.5G —

To summarize, verdazyls are delocalized radicals in terms of the unpaired electron distribution *within* the verdazyl ring, but delocalization onto the ring *substituents* is relatively ineffective. This feature of their electronic structure is manifested in a number of different ways. For example, verdazyl-substituted cyclophanes **29** and **30** were designed to probe the extent of through-space spin perturbation. Detailed EPR and NMR analyses indicate no through space (transannular) spin transfer in **29** to the "remote" aromatic group, a consequence of the small amount of spin density on the N-aromatic substituent comprising the other half of the cylophane.[77] Analogous cyclophane-bridged diradicals **30** have correspondingly very weak intramolecular spin–spin interactions,[78] which contrasts the relatively strong through-space interactions in cyclophane-based triplet dicarbenes, such as **31**.[79]

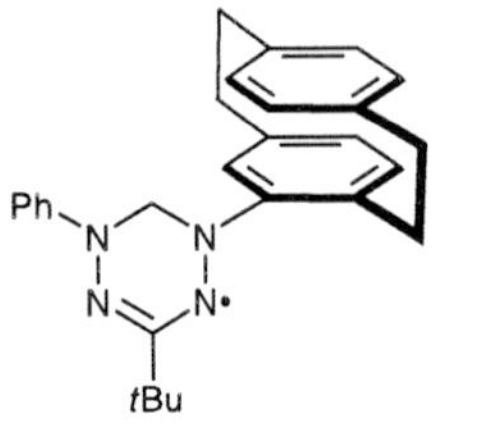

29

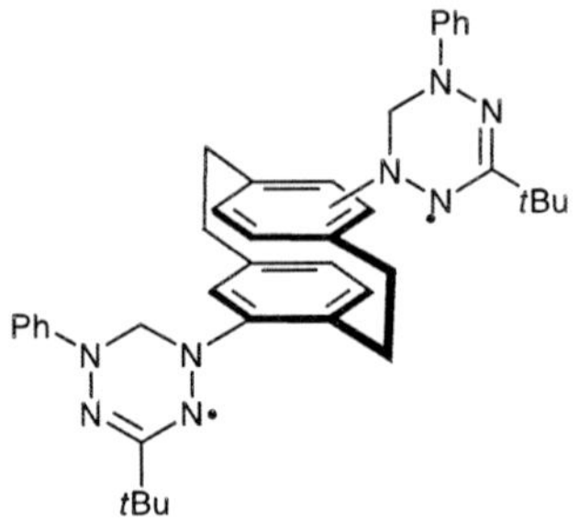

30

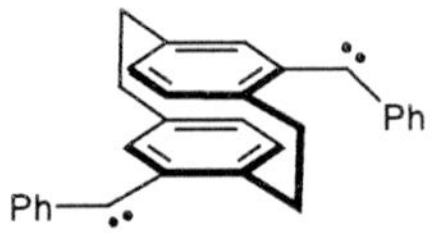

31

The localized electronic structure attenuates the "communication" of verdazyl radicals with other covalently-linked chromophores. Experimental and computational[52,73,80–82] studies on a variety of C3-linked verdazyl di- (or tri- or tetra-) radicals **32** or their N-linked[19,27,83] counterparts **33** show the two radicals to invariably be fairly weakly electronically coupled, irrespective of whether the diradical in question is based on a Kuhn[19,24–26,28,84] or 6-oxoverdazyl[42,49–52] structure. Even the 6-oxoverdazyl diradical, in which the two radicals are *directly* linked, remains an open shell singlet diradical in its ground state, with the triplet excited state some 760 cm^{-1} higher in energy.[52] Electronic and magnetic communication has been investigated in polyradicals of general structure **34**, in which a verdazyl is linked by a π-conjugated spacer "X" to a different open shell chromophore, for example, nitronyl nitroxide,[82] tetrathiafulvalene radical cation,[81] triplet nitrene (generated by photolysis of the corresponding azide),[63,85] or triplet *excited* states of organic compounds, such as pyrene or anthracene.[74,86] In all of these cases, where spin–spin interactions were probed quantitatively (using EPR spectroscopy and/or computational studies) the exchange coupling energies remain on the order of a few hundred wavenumbers.

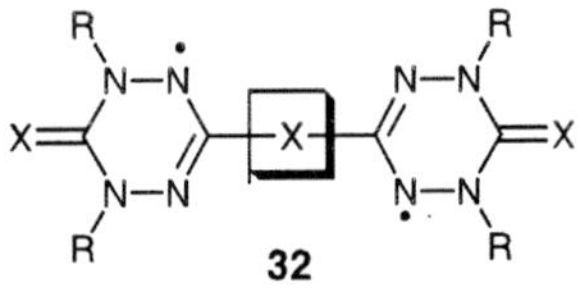

32

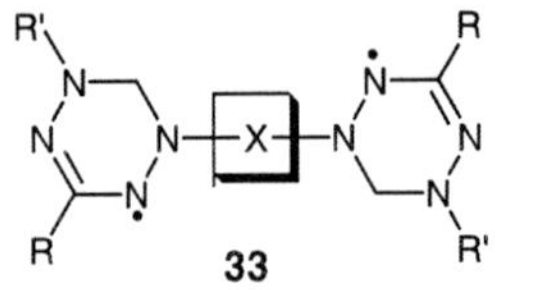

33

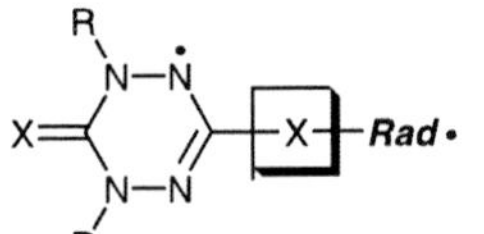

34

7.2.3 Verdazyl radical reactivity

Verdazyls are stable radicals but are also species with a substantial and varied chemistry. The general stability of verdazyls is sufficient for chemical transformations on other parts of the molecule to take place without affecting the radical itself. For example, amino-substituted verdazyl **35** can be converted into amide **36** or imine **37** derivatives (Scheme 7.10),[21] and 1,3,5,-triphenylverdazyl **38** can be sulfonated to give a water soluble derivative **39** (Scheme 7.11).[87] Many of the verdazyl-containing polymers shown in Figure 7.1 were made via multistep syntheses in which the pre-assembled verdazyl is carried through reactions which do not affect the radical center.

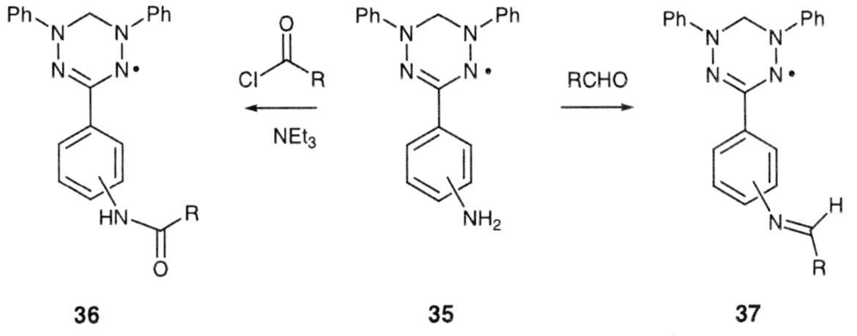

Scheme 7.10

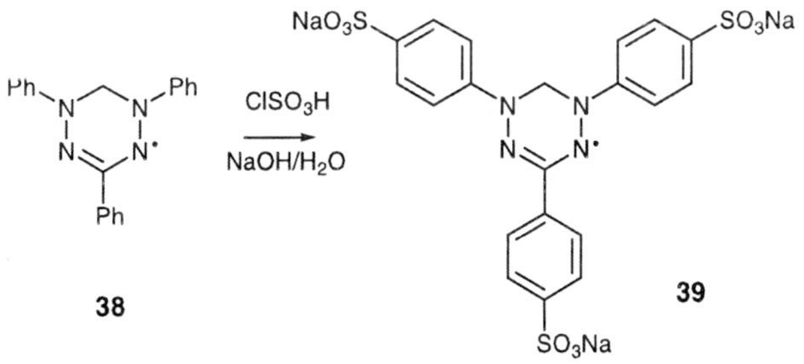

Scheme 7.11

7.2.3.1 *Electrochemical properties and electron transfer reactions*

There have been several reports on the electrochemical characterization of specific verdazyl radicals,[44,50,51,88,89] as well as more systematic studies of structure/property relationships.[90,91] Verdazyl radicals undergo reversible one-electron oxidation and one-electron reduction processes, thereby providing a triad of stable oxidation states (Scheme 7.12). The oxidation and reduction potentials are strongly influenced by the verdazyl structure and substituents. For example, Kuhn verdazyls **6** are oxidized in the range of −0.2 to −0.4V vs. ferrocene in acetonitrile, whereas introduction of a carbonyl group at C6 (i.e., 6-oxoverdazyls **7**) raises the oxidation potential by approximately half a volt (for equivalent R groups).[91] Reduction potentials for verdazyls fall in the range of −1.0 to −1.3 V. Electron withdrawing and donating substituents have predictable effects on the oxidation and reduction potentials – but the effects of the C3 substituent (R) are somewhat attenuated compared to the effects of the N1 and N5 substituents (R′, R″), due to the former substituent being attached to the verdazyl ring on a SOMO nodal plane.

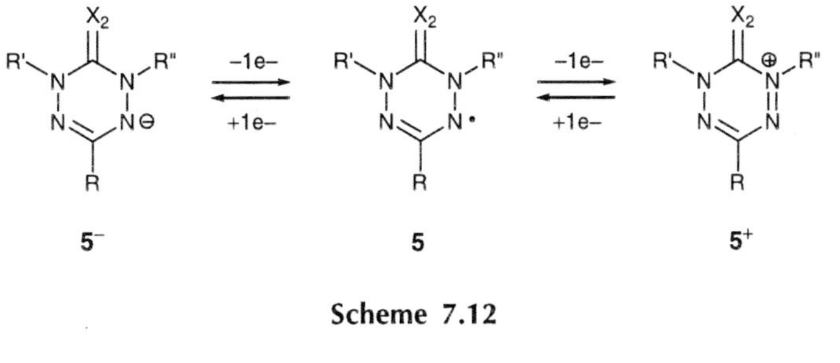

Scheme 7.12

While the redox chemistry of verdazyls is fairly extensive, there are only a few reports in which the redox activity of verdazyls is exploited in electron transfer reactions (i.e., without bond making/breaking). Nakatsuji prepared charge transfer salts of verdazylium cations (**5**$^+$) and radical anions of the acceptor molecules TCNQ **40**, tetrafluoroTCNQ **41**, and DDQ **42** by reaction of the two neutral reagents.[89] Hynes has studied the kinetics of outer sphere electron transfer reactions between Kuhn verdazyl radicals and coordination complexes of Co^{3+}, Ru^{3+}, Fe^{3+} and Fe_3 clusters.[92]

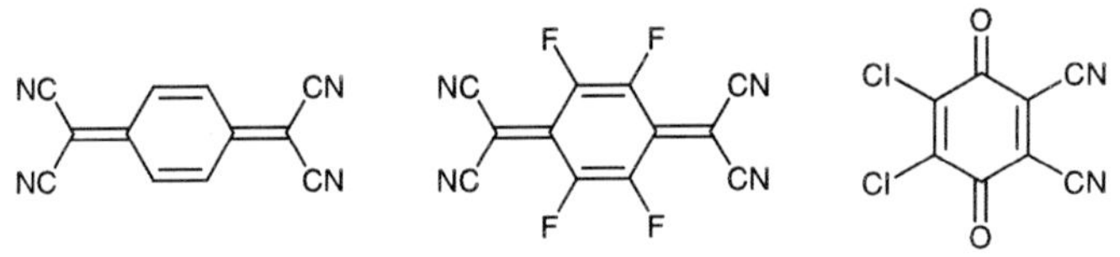

40 **41** **42**

Reactions of verdazyls **6** with oxidizing agents (halogens,[12] metal-based reagents such as $FeCl_3$, $AuCl_3$, $PtCl_4$,[93] tetranitromethane[93]) normally give the corresponding verdazylium cations **12**. Oxidations with peracids[94] or peroxides,[95] or heating verdazyls in oxygenated solutions containing activated charcoal,[96] induce ring opening to form either 1,2,4-triazoles **43** or N-formylformazans **44** (Scheme 7.13). 1,2,4-Triazoles are also produced by simple thermolysis of verdazyls, either by heating the radical as a solid to 200 °C[12] or in solution at 80 °C.[97]

There have been a few studies on the chemistry of verdazylium cations. They undergo a complex reaction with hydroxide to ultimately give the corresponding radicals and N-formyl formazans **44** (Scheme 7.14).[23] Heating mixtures of verdazylium cations and verdazyl radicals produces C6–N2 coupled products **45** along with one equivalent of leuco verdazyl (Scheme 7.15).[98]

7.2.3.2 Hydrogenation and related reductions

Leuco verdazyls **11**, the initial cyclic product from formazan alkylation reactions (Schemes 7.1 and 7.2) are oxidized (most commonly by atmospheric oxygen) to produce radicals **6**. The reverse

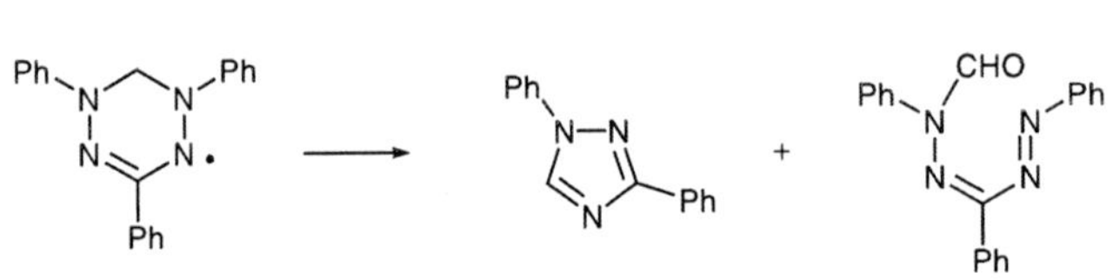

38 **43** **44**

Scheme 7.13

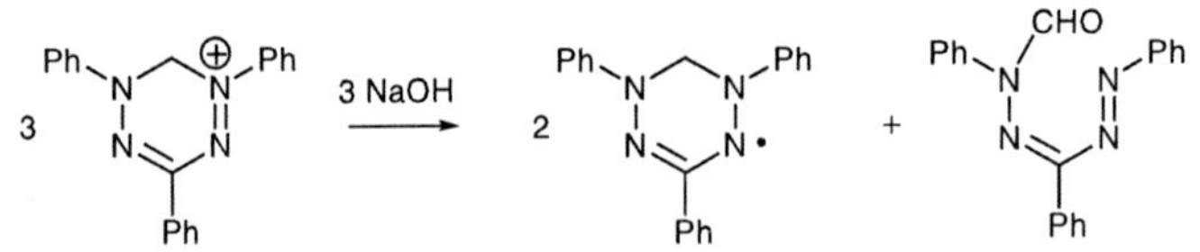

38+ **38** **44**

Scheme 7.14

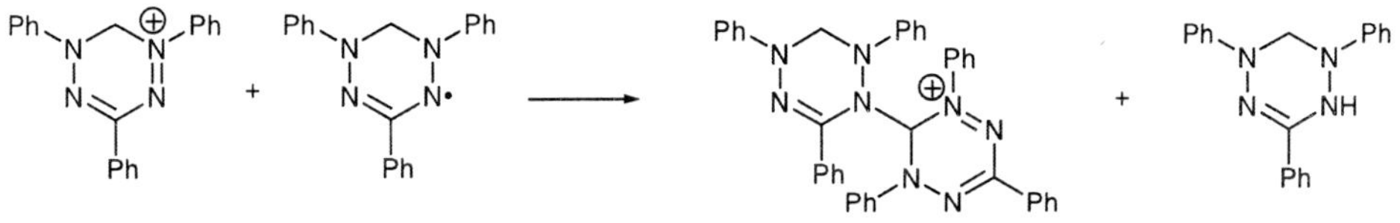

38+ 38 45 46

Scheme 7.15

reaction – reduction of verdazyls to leuco compounds – has been documented extensively as well. The N–H bond in the leuco compounds is relatively weak (~70 kcal/mol)[99]; accordingly, many of the hydrogenating reagents – hydrogen/palladium,[12] sodium dithionite,[12] thiols,[12,100] 1,4-dihydropyridines,[101] hydrazines,[102] secondary phosphines,[103] and ascorbic acid[104] – are good hydrogen atom donors/reducing agents.

Leuco versions of 6-oxoverdazyls (**47**) are relatively rare. They have been made by hydrogen reduction of the radicals,[41,42] and there is one report of oxidation of a tetrazane **17** to give a leuco compound (Scheme 7.16).[44] As is the case with the leuco compounds **11, 47** are readily converted to radicals by atmospheric oxygen. 6-Oxoverdazyls can also be reduced to the leuco compounds with ascorbic acid, although the reaction is rather sluggish compared to the reduction of nitroxides by this method.[105]

1,3,5-Triphenylverdazyl **38** undergoes redox reactions with a variety of organometallic reagents (Scheme 7.17; M = Mg, Li, Al).[106] The verdazyl is initially reduced by the carbanion, producing a

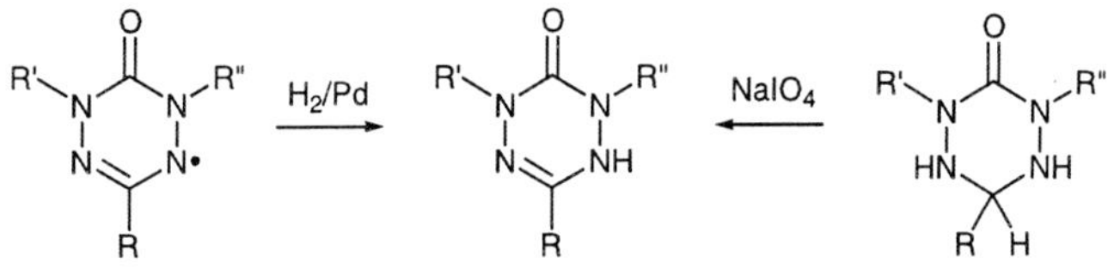

7 47 17

Scheme 7.16

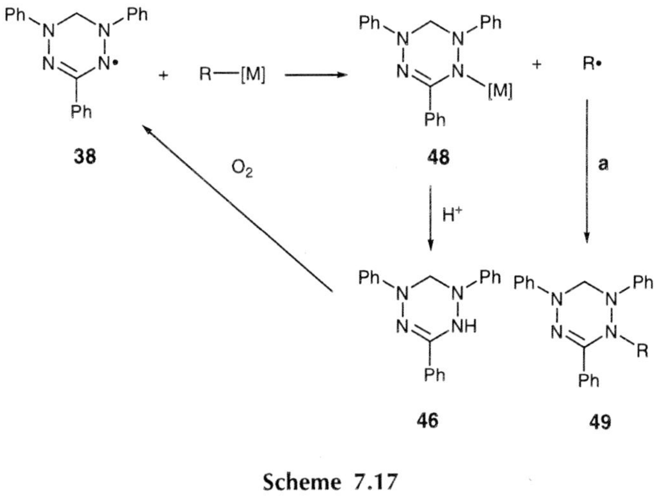

Scheme 7.17

metallated (reduced) radical **48** and an alkyl radical. The latter couples with more verdazyl to give the N-alkylated species **49**, while **48** is protonated upon work-up to initially give the leuco derivative **46**, which is subsequently oxidized in the presence of air (also introduced during work-up) back to the radical **38**.

7.2.3.3 Disproportionation reactions

One of the very first reactions reported for triphenylverdazyl **38** was its reaction with HCl, which leads to verdazylium cation **38** and leuco verdazyl **46** as the two products (Scheme 7.18).[6,12] This redox disproportionation reaction can be induced by other protic acids[107] as well as Lewis acidic metal ions such as Zn^{2+} and Cd^{2+}.[108,109]

Dvorko and Ponomareva have developed a method for studying heterolysis kinetics of alkyl halides based on their reactions with verdazyls. According to their work, the initial reaction between the solvent-separated ion pair of, for example, *t*-butyl chloride with 1,3,5-triphenyl verdazyl produces an N-alkylated verdazyl **50** and a verdazylium halide salt **38**$^+$ (Scheme 7.19). The alkylated verdazyl can fragment to give a leuco verdazyl **46** and isobutene via hydrogen atom transfer, or alternatively undergo a solvolysis reaction (which produces the same leuco product). The verdazyl is employed as an indicator for kinetic purposes: small concentrations of verdazyl ($\sim 10^{-4}$ M) are employed compared to substrate (often) and the consumption of radical or the appearance of verdazylium cation can be monitored by UV-Visible spectroscopy. This reaction is quite general and a massive amount of data has been compiled on substrate, solvent, and salt effects on alkyl halide heterolysis chemistry based on this scheme.[110] However, the validity of this reaction scheme has been seriously challenged by Serebryakov *et al.*[111] The basis for their objection is that, under typical reaction conditions in which the verdazyl concentration is low and the alkyl halide is in a large excess (often $10^3 - 10^4$-fold or more), traces of water present can affect the same reaction with verdazyls – that is, formation of a verdazylium cation and a leuco verdazyl (see above).

The redox disproportionation of 6-oxoverdazyls has yet to be reported. However, 1,5-dimethyl-6-oxoverdazyls (e.g., **51**) undergo a disproportionation reaction involving hydrogen atom transfer from an N-methyl proton, which initially produces a leuco compound **52** and an azomethine imine intermediate **53**. The latter is not observed but dimerizes in a [3+3] cycloaddition reaction to give the novel tricyclic structure **54** (Scheme 7.20).[42] The leuco compound is ultimately air-oxidized to regenerate the radical.

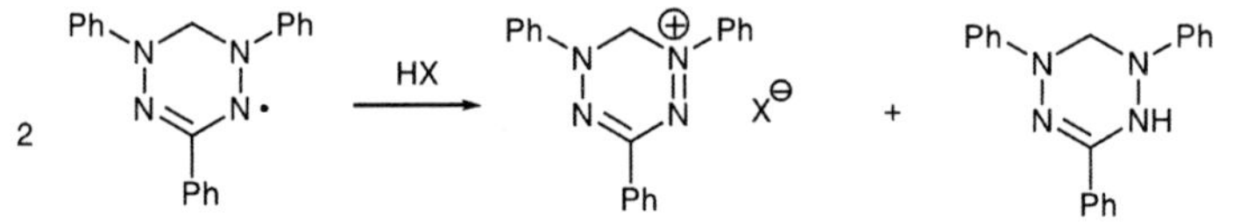

Scheme 7.18

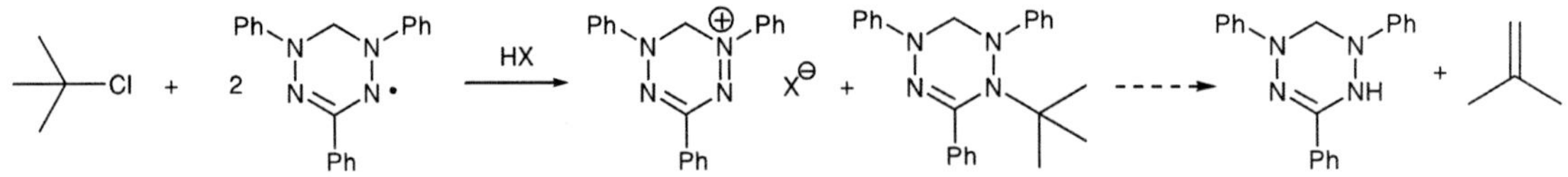

Scheme 7.19

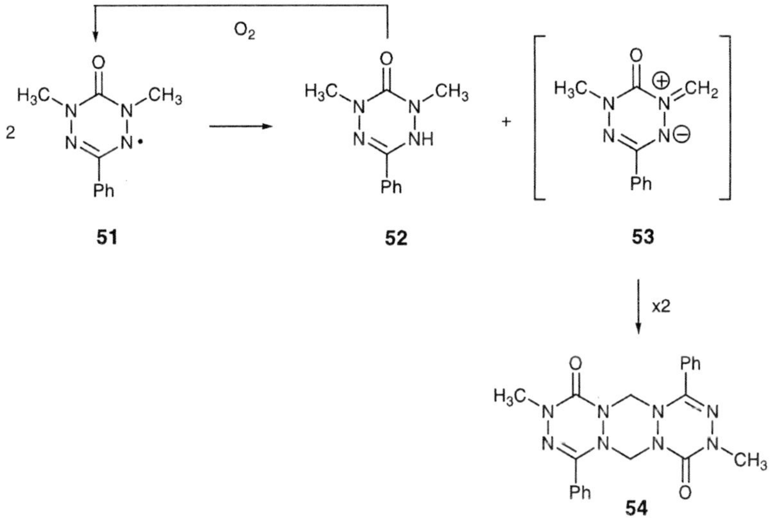

51 52 53 54

Scheme 7.20

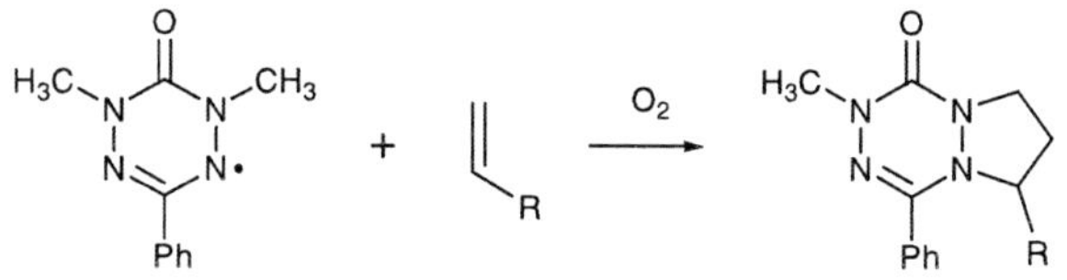

51 55

Scheme 7.21

This reaction lies at the heart of the lack of long-term stability of many 1,5-dimethyl-6-oxoverdazyls. Interestingly, the recently synthesized 1,5-diisopropyl 6-oxoverdazyl analogues (Scheme 7.8) appear to be indefinitely stable with respect to disproportionation even though they still possess an α-CH proton.[46]

Georges has harnessed this disproportionation reaction of 1,5-dimethyl-6-oxoverdazyls for synthetic purposes. Alkenes serve as effective trapping agents for the azomethine imine intermediates **53**, leading to bicyclic structures **55** via a 3+2 cycloaddition; running the reactions in the presence of oxygen significantly improves the yields of **55** by converting the leucoverdazyl byproduct **52** back to a radical, whereupon the radical can react with more alkene (Scheme 7.21).[112]

7.2.3.4 Radical coupling reactions of verdazyls

The reactions of verdazyls with Grignards, organolithiums, and so on (Scheme 7.17) includes a coupling reaction between a verdazyl radical and an alkyl radical, the latter of which is generated by electron transfer from the initial alkyl anion. Alkylated verdazyls can be made by treatment of the corresponding leuco compound with an alkyl halide and base,[19,42] but verdazyls also react cleanly with carbon-centered radicals to give tetrazine compounds of general structure **49** (cf. Scheme 7.17). For example, heating verdazyls in the presence of AIBN (2,2-azobisisobutyronitrile) generates the 2-cyanopropyl-substituted tetrazines **56** – which can be reverted back to radicals in high yield upon heating to 150 °C (Scheme 7.22).[42,83,113]

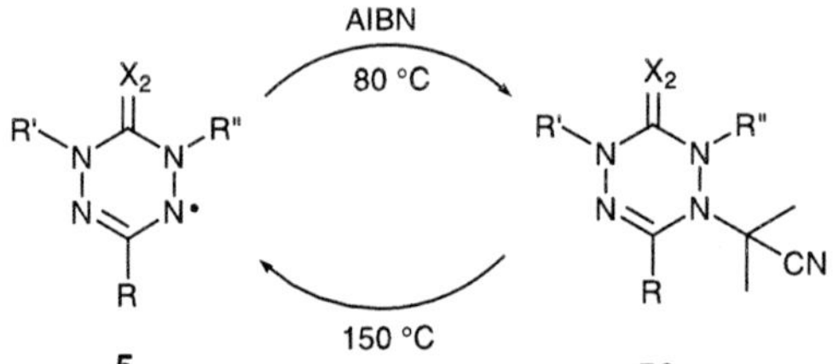

5 56

Scheme 7.22

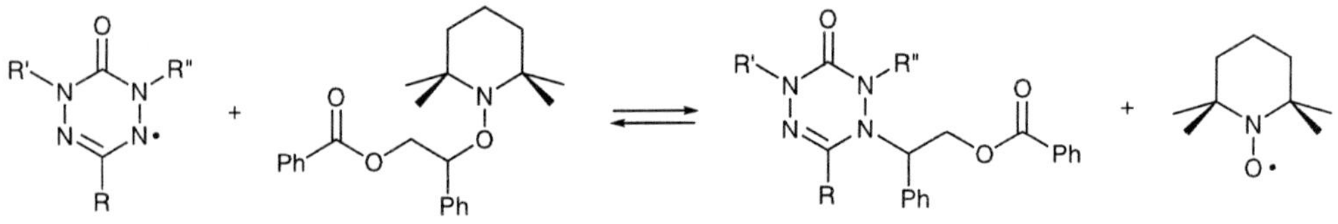

7 57 58 59

Scheme 7.23

N-alkyl tetrazines can also be made using another radical–alkyl adduct as the alkyl radical source. Thus, heating mixtures of 6-oxoverdazyls **7** and the alkoxyamine **57** leads to transfer of the substituted benzyl radical from the TEMPO radical **59** to the verdazyl (Scheme 7.23).[105] The thermal lability of both the C–N bond in **58** as well as the C–O bond in **57** sets up an equilibrium situation, but in some examples the N-alkyl verdazyls can be isolated in high yield. The addition of ascorbic acid also pushes the equilibrium towards the N-alkylverdazyl as it was shown that ascorbic acid reduces the TEMPO radical more rapidly than verdazyls.

The rapid reaction of verdazyls with reactive carbon-centered radicals was exploited early in the development of verdazyl chemistry as polymerization inhibitors. Much like DPPH, the addition of verdazyls to free radical polymerization processes by reacting with the polymer radical chain.[114] The fact that the tetrazine N–C bond can be homolyzed at high temperatures has led to investigations of verdazyl radicals as reagents for the *control* of free radical polymerization – a field of study which has been dominated by nitroxide radicals (and which is discussed in detail in Chapter 11 of this book). Yamada studied the ability of 1,3,5-triphenylverdazyl to control the radical polymerization of styrene. Although some degree of molecular weight control was realized, the process could not be considered living due to relatively high polydispersities.[115] More recently, Georges and Hicks employed 1,5-dimethyl-6-oxoverdazyl as the radical agent (as well as the alkylated species **58** (Scheme 7.23) as a 'masked' radical) and showed that these verdazyls can control the polymerization of styrene and *n*-butyl acrylate in a pseudo-living manner.[105,112] The difference in behavior between 1,3,5-triphenylvedazyl and the 6-oxoverdazyls has been attributed to the gradual decomposition of the latter radicals (see above), which keeps radical concentration from building up in too much excess as inevitable termination steps of polymerization lower chain concentrations.

7.2.3.5 *Verdazyl radical coordination chemistry*

The first reports of the interaction of verdazyl radicals with metals focused on electron transfer or decomposition processes of the radicals, although metal complexes of the radicals were not isolated or observed directly.[108,116] The first genuine coordination complex containing a verdazyl ligand was Brook's one-dimensional coordination polymer **60** (X = Cl, Br, I) based on a bis(1,5-dimethyl-6-oxoverdazyl) diradical coordinated to copper(I) halide salts.[117] In this structure the two verdazyls create a chelating environment for the metal ions – a structural feature which has dominated the rapid growth of verdazyl coordination chemistry in the past decade. Structures **61–65**, all of which have bidentate or tridentate binding sites for metals, have been the most commonly used verdazyl ligands. Many complexes of first row d-block metals have been made including copper(I),[118] the divalent ions of manganese,[59,119,120] cobalt,[121,122] nickel,[59,119–121,123] copper,[57,120] and zinc,[120,124] as well as cadmium(II),[124] mercury(II),[124] ruthenium(II),[125] silver(I),[126] and lanthanide (III) ions.[127] In all of these complexes the radicals remain as such upon coordination, that is, no electron transfer takes place. The magnetic properties of metal–verdazyls has been the biggest focus of research in this area, and has been discussed elsewhere.[8,128] Reaction chemistry and other uses of metal–verdazyl complexes remain wholly uncharted territory.

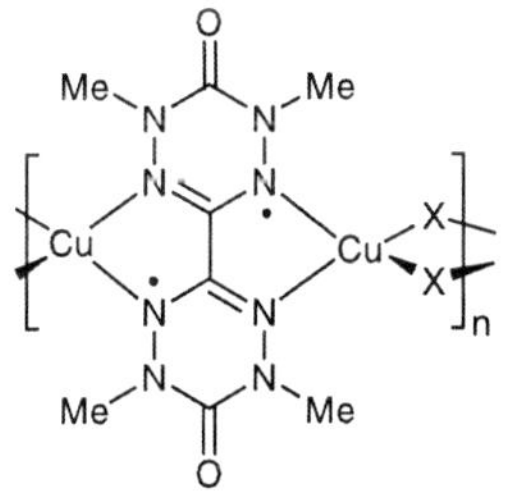

60

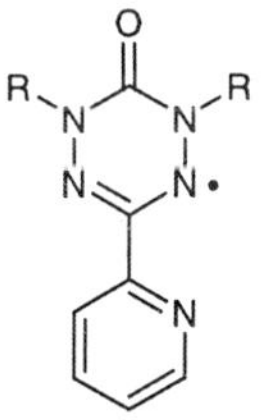

61

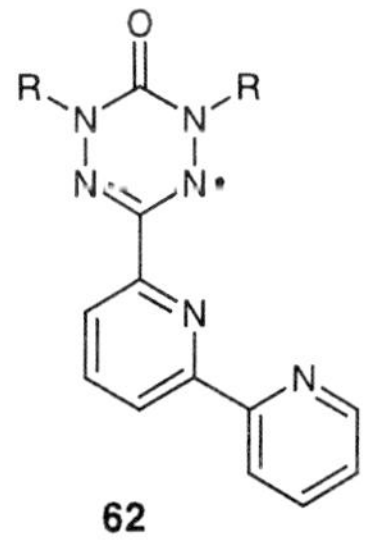

62

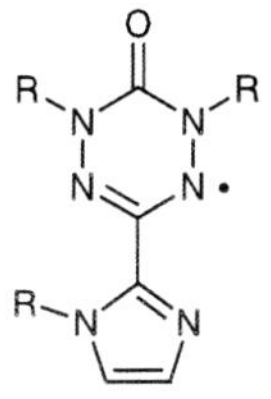

63

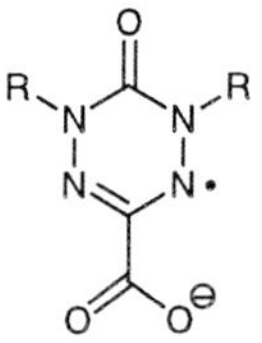

64

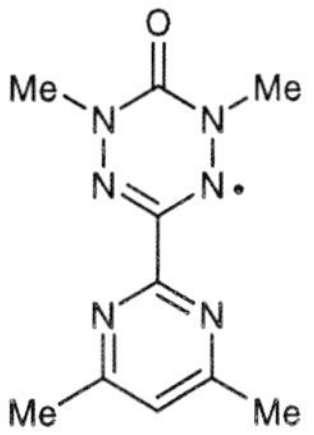

65

A subset of verdazyl coordination chemistry consists of complexes in which the verdazyl ring is not directly bound to the metal. This includes ferrocene-based bis-verdazyl **28** and the corresponding ferrocene-*mono*verdazyl, where the verdazyl is simply a substituent on the η^5-cyclopentadienide ring(s).[54] The verdazyl–lanthanide complexes mentioned above consist of imidazole-verdazyl **63** (R=H) asymmetrically bridging two lanthanide ions. One metal binds to the conventional bidentate site while the other is attached at the carbonyl oxygen.[127] Finally, verdazyls have been bound to metals using deliberately designed remote donor sites, for example, 4-pyridyl substituted radical **66**.[129] In this vein Fujita has employed 1,3,5-tris(4-pyridyl)verdazyl **67** as an *exotridentate* ligand building block for so-called "spin cages" such as the hexanuclear structure **68** (Scheme 7.24).[130,131] Self-assembled metal–verdazyl based structures based on *direct* verdazyl–metal coordination have yet to appear, although several key ligand building blocks have been described.[132]

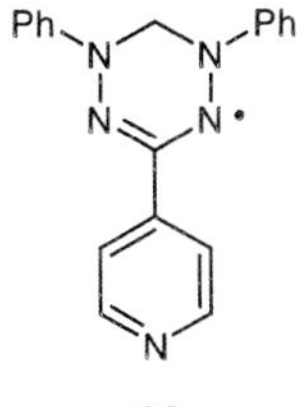

66

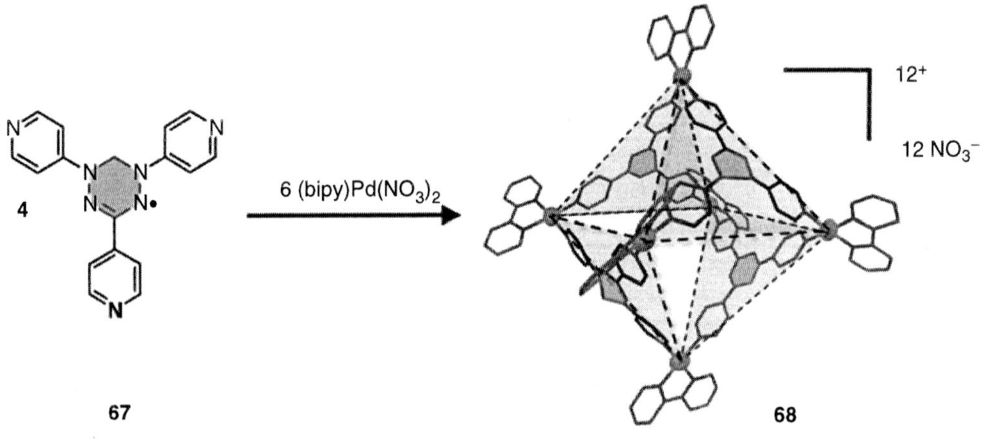

Scheme 7.24

7.2.4 Inorganic verdazyl analogues

There are a few examples of radicals based on the verdazyl template in which one of the two ring carbon atoms (C3 or C6) is replaced by another element. 6-Phosphaverdazyls **71**[133–135] have been made based on adaptation of the "bis-hydrazide" route (Scheme 7.25): phosphonic acid bis(1-methyl hydrazide)s **69** can be made from the corresponding phosphonic acid dichloride and methylhydrazine. The bis-hydrazides condense with orthoesters to give phosphatetrazines **70** (i.e., phosphorus-containing "leuco" verdazyls compounds), which, in contrast to the all-organic leuco verdazyls, are air stable. Oxidation to the phosphaverdazyl is achieved with lead dioxide or periodate. In an analogous manner, reaction of 2,4-dimethylcarbohydrazide **72** with trichlorophosphoranes gives 3-phosphatetrazines **73**, which can be oxidized (I_2) to yield 3-phosphaverdazyls **74** (Scheme 7.26).[134]

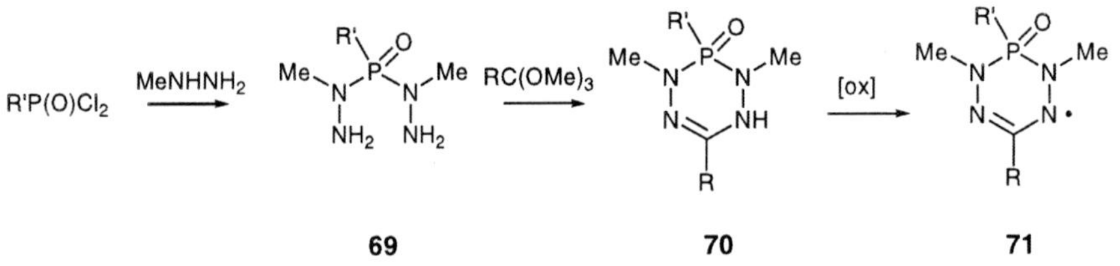

Scheme 7.25

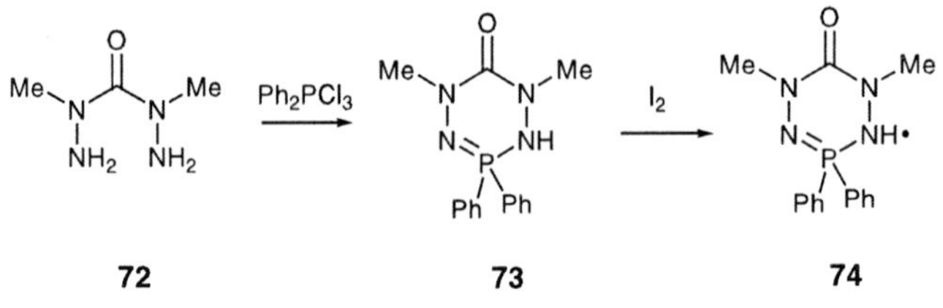

Scheme 7.26

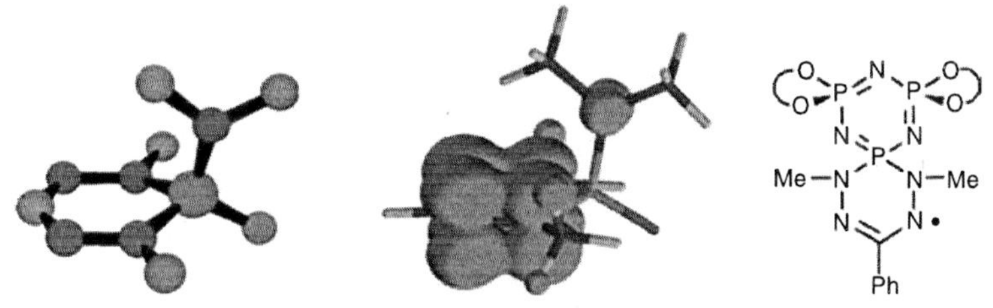

75

(a) (b) (c)

Figure 7.5 *Optimized structure (a) and positive spin density plot (b) for 6-phosphaverdazyl* ***x*** *(R = Ph, R′ = NMe_2).*[135] *(c) Structure of 6-phosphaverdazyl/cyclotriphosphazene spirocycle* ***75***.[136]

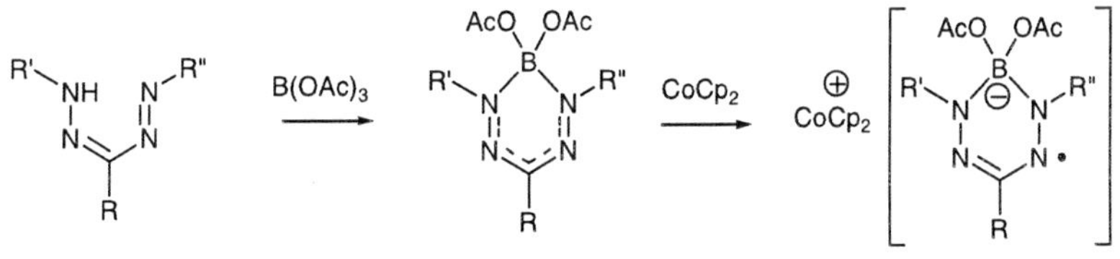

9 76 77

Scheme 7.27

Phosphaverdazyl radicals are persistent in solution but slowly decompose, most likely due *not* to the presence of phosphorus but due to disproportionation via hydrogen atom transfer from the N-methyl groups (Scheme 7.20). The electronic structures of these radicals have been probed through EPR[133–136] and computational[135,137] studies and fundamentally resemble verdazyls in their electronic structure. The small ^{31}P hyperfine couplings (<5 G) are indicative of small spin density at the heteroatom. Some 6-phosphaverdazyl derivatives have a nitrogen atom attached directly to the 6-phosphaverdazyl phosphorus (e.g., **71**, R′ = NMe_2 and phosphazene-verdazyl spirocycle **75** (Figure 7.5c)), and the non-negligible spin density on these nitrogen atoms has been suggested to arise from a through-space, spiroconjugation mechanism (Figure 7.5a, 7.5b for **71**, R′ = NMe_2).[135,136]

The only other inorganic verdazyls are boron-containing verdazyl radical anions. Reactions of formazans with "diboron tetraacetate"[138] (more likely boron triacetate; generated *in situ* from boric acid ($B(OH)_3$) and acetic anhydride) gives formazan/$B(OAc)_2$ complexes **76** ("boratatetrazine") (Scheme 7.27), which can be reduced with cobaltocene to give highly reactive, but persistent borataverdazyl radical anions.[139]

7.3 Tetraazapentenyl radicals

Reactions of triethyl orthoformate with *N,N*-diarylhydrazines (Ar = substituted phenyl or carbazolyl) produces 1,2,4,5-tetrazapentenes **78**, which can be oxidized (PbO_2) to the corresponding tetraazapentenyl radicals **79** (Scheme 7.28).[140] These radicals can be considered as hydrazyls in which a hydrazyl-type radical center is conjugated to a hydrazone (R_2N–N=C) moiety, permitting the unpaired electron to delocalize between two structurally equivalent fragments. Radicals **79** are persistent enough to be isolated as crystalline, monomeric radicals, although in solution they gradually decompose. Heating solutions of

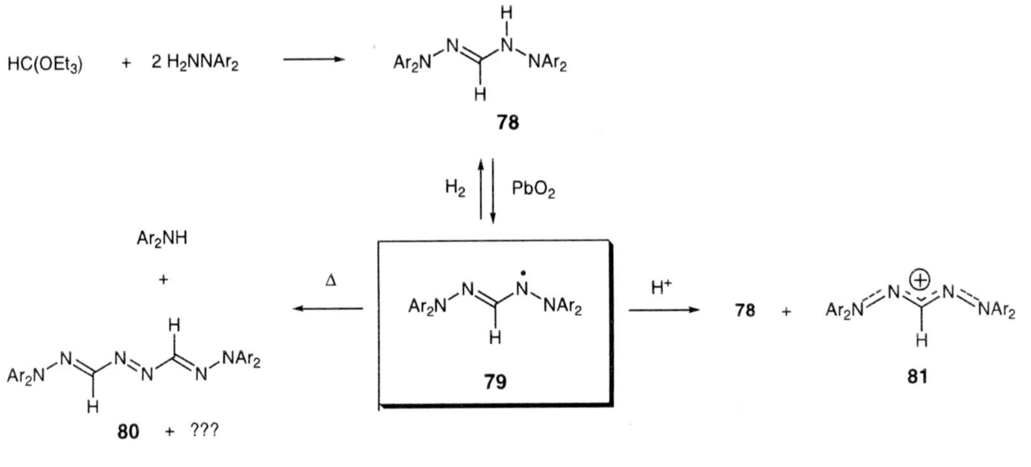

Scheme 7.28

these radicals accelerates the decomposition process, affording diarylamines, azo compounds **80**, and other unidentifiable products. Dissolving the radicals in formic acid causes disproproportionation to the parent tetrazapentene **78** and a cyanine-like cation **81** (Scheme 7.28).

The EPR spectra of radicals **79** are quite complex; assignment of hyperfine coupling constants required isotopically enriched (^{2}H on some/all of the aryl positions, ^{15}N on the "inner" nitrogen atoms) samples.[141] These labeling studies, coupled with McLachlan molecular orbital calculations, permitted assignment of the larger spin density values to the "inner" nitrogen atoms ($\rho \sim 0.26$ each) and the smaller spin density to the "outer" nitrogen atoms ($\rho \sim 0.14$ each); the outer nitrogen substituents possess a small amount of spin and the central methine carbon carries a small *negative* spin density ($\rho \sim -0.08$) which arises from spin polarization effects.

7.4 Tetrazolinyl radicals

Photolysis of 2,3-diaryltetrazolium cations (**13**) causes an intramolecular aryl–aryl coupling reaction to occur, resulting in the fused "phototetrazolium" cations **82** in high yield (Scheme 7.29).[142] The aromatic groups include substituted phenyl[143,144] and pyridyl,[145] the latter leading to azabiphenylene–tetrazolium salts. These cations can be reduced to stable radicals **83** ("Kuhn–Jerchel" radicals[146]) with sodium/ammonia, sodium amalgam, dithionite, hypophosphite, or tin(II) chloride ($SnCl_2$).[143–145] Harsher reducing agents (e.g., Raney nickel) or acid fragments the radical to a diazaphenanthrene **84**, but otherwise many of these fused tetrazolinyl radicals are stable and isolable as crystalline compounds.

Kuhn–Jerchel radicals **83** are a subset of the more general class of tetrazolinyl radicals **85** whose observation and elucidation were developed independently. The formazan–tetrazolium redox couple[10] has found widespread use as a redox indicator in cell biology.[11] Both species can give rise to unusually strong EPR signals, in solid form as well as in solution, which were shown to arise from a paramagnetic impurity.[147] The EPR active species was later identified as a tetrazolinyl radical **85**, essentially a redox form intermediate between formazan and tetrazolium salt.[148] Accordingly, tetrazolinyls **85** can be made by (i) reduction of the corresponding tetrazolium salts (by analogy to Kuhn–Jerchel radicals **83**) (Scheme 7.30),[148–150] (ii) oxidation of formazans (oxygen and base; diarylaminyl radicals generated from a tetraarylhydrazine;

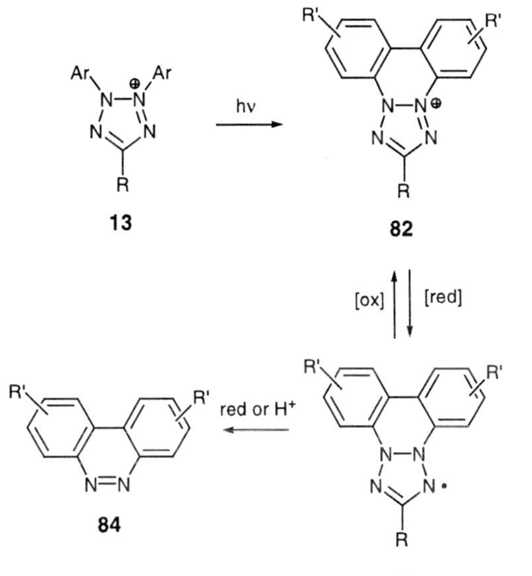

Scheme 7.29

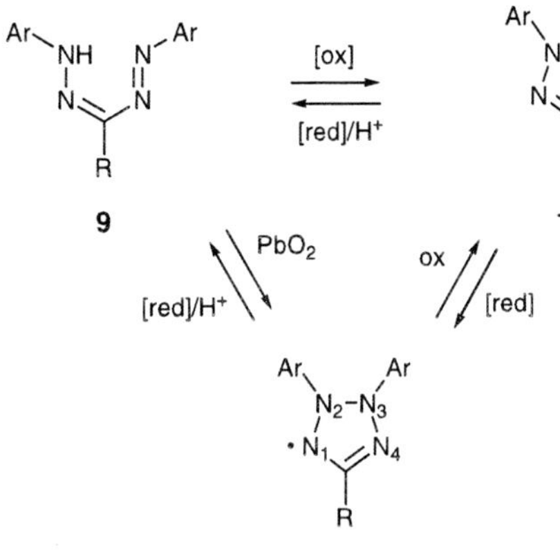

Scheme 7.30

PbO_2) of a formazan,[148,149,151–153] or (iii) a comproportionation reaction between a formazan and tetrazolium salt.[149,151] Electrochemical studies of tetrazolium salts show that the initial one-electron reduction to the radical is reversible, but addition of a second electron leads to irreversible (electrochemically) N–N bond cleavage and formation of the formazan anion.[154]

Most tetrazolinyl radicals **85** are long-lived in solution, but only one derivative (R = *t*Bu, Ar = *p*-nitrophenyl) has been isolated as a pure compound.[152] There appears to be a general correlation between the persistence of these radicals and the presence of electron withdrawing groups on the N-aromatic rings. The electron withdrawing/donating capabilities of Ar also strongly influence the N2/N3 EPR hyperfine coupling constants, which range from 8.3 G for electron donating groups (e.g., *p*-methoxyphenyl) to 6.1 G for *p*-nitrophenyl.[148,151,155] The N1/N4 hyperfine coupling constants remain constant (5.6–5.7 G) for all derivatives. Kuhn–Jerchel radicals **83** appear to be somewhat more stable as several of these are pure crystalline solids.[144] The nitrogen hfcs for **83** are significantly smaller for these radicals compared to **85** (a(N1/N4) = 3.8 G, a(N2/N3) = 5.6–7.5 G).[146]

7.5 1,2,4-Triazolinyl radicals

In 1989, Neugebauer discovered that leuco verdazyls **11** undergo ring contraction upon either oxidation or thermolysis to give a triazoline ring.[156] Oxidation (TCNE (tetracyanoethylene) or iron(III) chloride) affords the 1,2,4-triazolinyl radical **86** directly (and not the radical cation of **11** as was initially proposed[157]), whereas the thermolysis product (made using phenylhydrazine as a reducing agent) is the parent triazoline **87**. The two five-membered rings can be interconverted by redox reactions shown in Scheme 7.31. Alternative methods for making the triazolines **87** are based on condensation reactions of amidrazones **88** or hydrazones **89**.[158]

Triazolinyls **86** with N-aryl substituents *and* with R‴ = Aryl are stable, isolable species, one of which has been crystallographically characterized[156]; derivatives in which either R′ or R‴ are methyl groups can be spectroscopically observed but not isolated as pure compounds.[158] Extensive EPR studies – including ENDOR and complemented by NMR – have facilitated the assignment of hyperfine coupling to the three distinct nitrogen atoms: the hyperfine coupling constant ranges for N1, N2, and N4 are 7.4–7.6 G, 6.2–6.5 G, and 3.8–4.0 G, respectively. Very small proton hyperfine coupling constants are observed for C3-phenyl substituents, implying very small spin density at the C3 ring carbon of the radical SOMO.

The only reactivity studies involving these radicals involves their use in controlled/living radical polymerization (cf. verdazyls, Section 7.2.3.4; see also Chapter 11).[159] One interesting aspect of their chemistry involves the thermal decomposition of the triazolinyl to a triazole and a phenyl radical (Scheme 7.32); as was the case for verdazyls, this radical decomposition is believed to have beneficial implications for controlling polymerization.

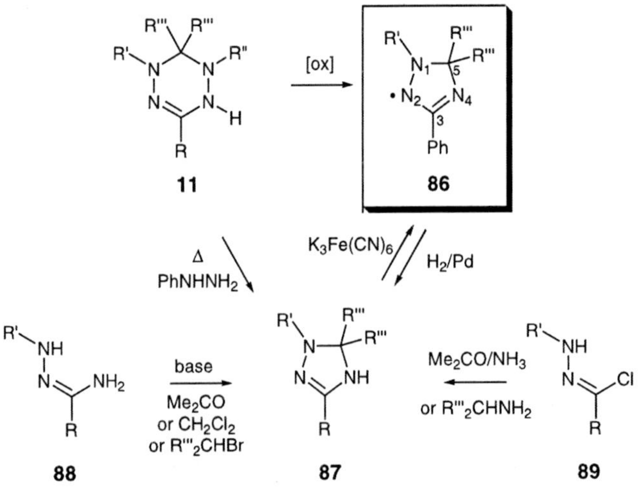

Scheme 7.31

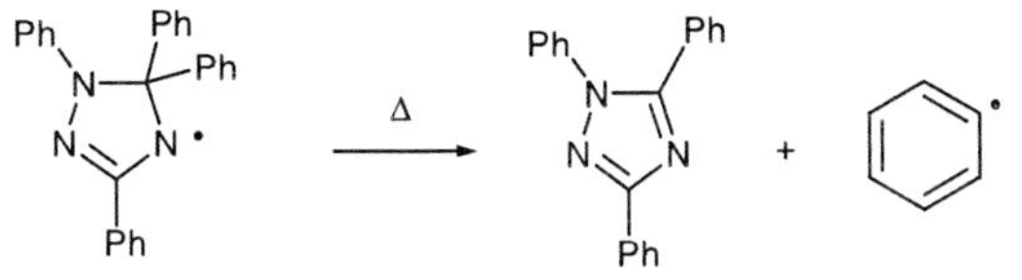

90 91

Scheme 7.32

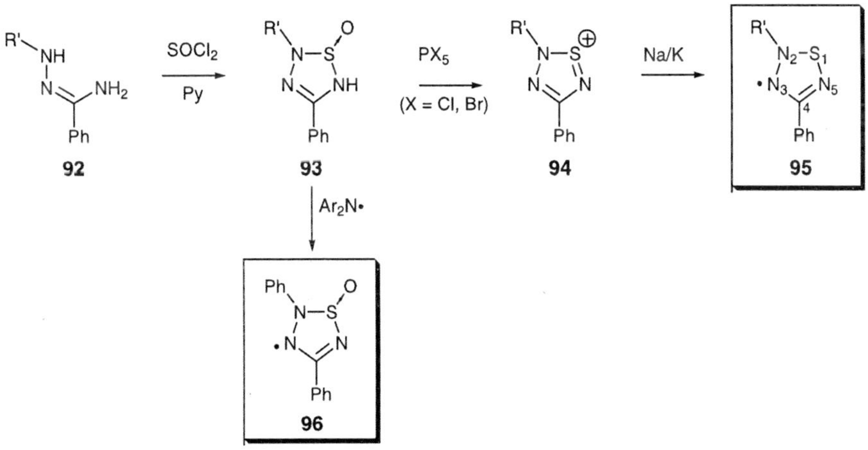

92 93 94 95 96

Scheme 7.33

Closely related to the triazolinyls are 1,2,3,5-thiatriazolinyls **95** and 1-oxo-1,2,3,5-thiatriazolinyls **96**, in which the saturated ring carbon (C6) of **86** is replaced by a divalent or tetravalent sulfur respectively. Both radicals are accessed from 1,2,3,5-thiatroazoline-1-oxides **93**, which can be made from amidrazones **92** and thionyl chloride (Scheme 7.33). Oxidation of **93** with diarylaminyl radicals gives the oxothiatriazolinyl **96**,[160] while reaction with phosphorus pentahalides gives the thiatriazolium cations **94**, which can be reduced to the corresponding radicals **95**.[161] These radicals are reported to be monomeric in solution down to 230 K, but none were isolated. EPR parameters (deduced from a combination of ENDOR studies and isotopic (^{2}H, ^{15}N) labeling) suggest somewhat different spin distributions between the two types of radicals, that is, for **95** (R′ = phenyl) a(N2) = 7.85 G, a(N3) = 4.15 G, a(N5) = 5.10 G whereas for **96** a(N2) = 8.19 G, a(N3) = 4.03 G, a(N5) = 2.35 G. These differences have been interpreted in the context of the involvement of the sulfur atom in the π framework of the radical: in **95** the sulfur atom is a component of a cyclic 7π-electron radical, whereas the S-oxide unit interrupts conjugation in **96** and produces a non-cyclic, 5π-electron system more akin to the 1,2,4-triazolinyls **86**.[160,161]

7.6 1,2,4,5-Tetrazinyl radicals

Oxidation of 1,4-dihydrotetrazines **97** with diarylaminyl radicals gives solutions of the persistent 1,2,4,5-tetrazinyl radicals **98** (R′ = methyl, CH_2Ph; R = R′ = H, phenyl) (Scheme 7.34).[162] The low symmetry of their structure leads to rather complex EPR spectra, but a combination of isotopically enriched derivatives

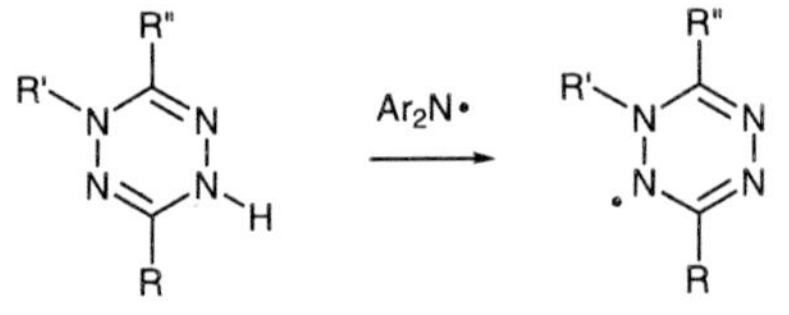

97 **98**

Scheme 7.34

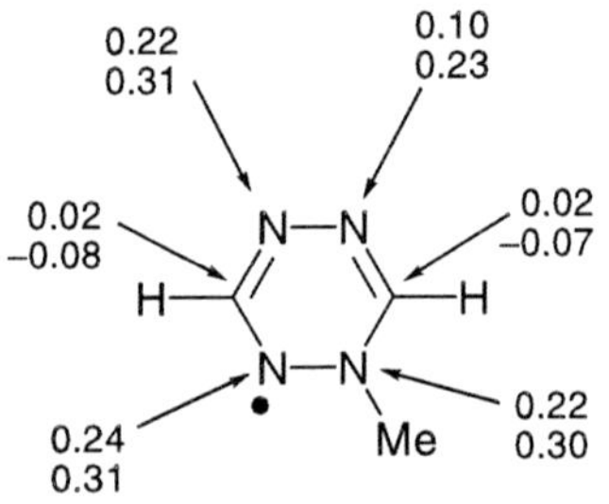

*Figure 7.6 Experimental (EPR; top) and computational (McLachlan; bottom) spin densities in a 1,2,4,5-tetrazinyl model radical **98**.*

(^{2}H, ^{15}N) permitted full assignment of all of the nitrogen and proton hyperfine coupling constants. The spin density distribution picture which emerges (Figure 7.6) is qualitatively not unlike that of the verdazyl radical system. These radicals were not isolated.

7.7 Benzo-1,2,4-triazinyl radicals

The benzo-1,2,4-triazinyl radical system **99** was first reported by Blatter *et al.*[163] in 1968 and was subsequently developed principally by Neugebauer.[164] The most effective synthetic route to these radicals starts from diaryl amidrazones **100** (Scheme 7.35). Oxidation with mercuric oxide affords azo compounds **101**, which can be ring-closed a number of different ways to give either the radical **99e** directly (heat, mercury(II) oxide), the corresponding benzo-1,4-dihydro-1,2,4-triaz*ine* **103** (thermolysis) or its hydrobromide salt (HBr) **104**. The latter two species can be interconverted by acid/base chemistry; both can be converted to the corresponding radical with strong base – a reaction which presumably initially yields the deprotonated triazine anion which is air (O_2) oxidized to give the radical.

A few other methods are known to produce benzo-1,2,4-triazinyl radicals (Scheme 7.36). Blatter *et al.* showed that heating hydrazide derivatives **104** followed by treatment with base (and presumably air) gives the corresponding radical. Low yields of radicals have been obtained from the reactions of nitrilimines **105** with reagents of general structure **106** (e.g., iminophosphoranes (E = PPh_3),[165] sulfimides (E = SMe_2),[166] nitrosobenzene (E = O)).[167]

Benzo-1,2,4-triazinyl radicals are extremely stable – most of the derivatives prepared to date are isolable and do not decompose over time. The radicals possess substantial hyperfine coupling constants to all three nitrogen atoms (a(N1) = 7.0–7.5 G, a(N2) = 5.1 G, a(N4) = 5.1 G) and smaller ^{1}H hyperfine coupling can be observed (by EPR or NMR) to aromatic protons on the N1 and C3 substituents as well as the benzannulated ring protons (C5–C8).[164] The EPR data, in conjunction with ENDOR[168] and semi-empirical

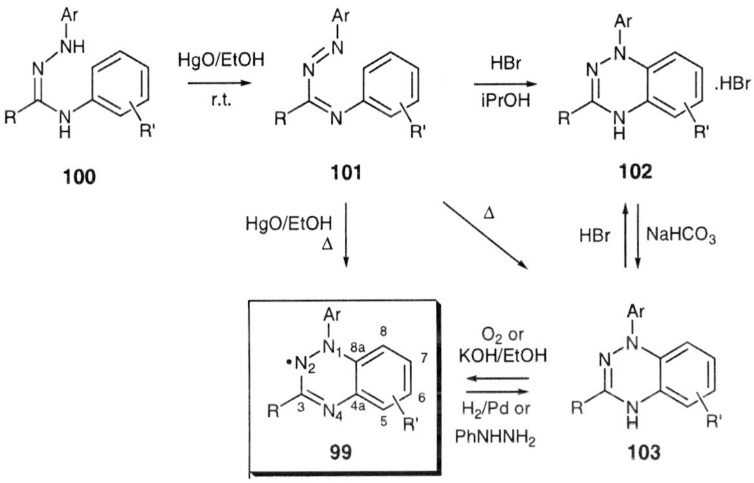

Scheme 7.35

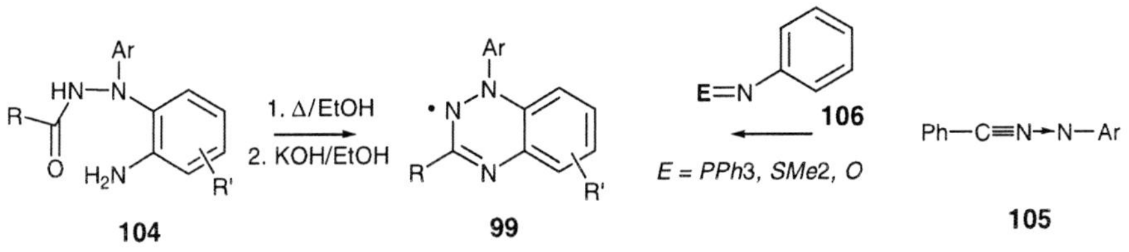

Scheme 7.36

molecular orbital calculations, produce a spin distribution picture in which 60–70% of the spin density resides on the three nitrogen atoms; the rest is shared among the benzo ring carbons, that is, each position possesses a small proportion (4–7% per site) of spin density.

Neugebauer has explored some reaction chemistry of the benzotriazinyls (Scheme 7.37).[164] One of their more interesting reactions is with acid. Whereas other related radicals (verdazyls, tetraazapentenyls) undergo disproportionation upon protonation, the benzotriazinyls appear to be stable with respect to disproportionation. Stable solutions of benzotriazinyl radical cations **107** can be generated by treatment of the neutral radical with strong acid (trifluoroacetic acid CF_3CO_2H).[169] The identity of the radical cations was confirmed principally by EPR spectroscopy; radical protonation led to changes to the nitrogen hyperfine couplings as well as an additional large coupling (4.8–5.0 G) to the added proton. Other reactions of the radicals include oxidations either to cations **108** or to quinoidal type compound **109** (by boiling in charcoal in air). As mentioned above, the radicals can be hydrogenated to the leuco compounds **103**, which can be alkylated – though a mixture of N2 and N4 alkylated compounds **110** and **111** respectively are produced.

The outstanding stability of benzotriazyls has enabled a few investigations of their solid state (materials) properties. Intermolecular magnetic exchange interactions have been examined in crystalline derivatives of these radicals[170,171] Wudl *et al*. reacted a benzotriazinyl (**99**, Ar = R = Ph; R′ = H) with the good electron acceptor TCNQ (tetracyanoquinodimethane) **40** to give a pressure-sensitive semiconducting charge transfer complex $(\mathbf{99})_2(\mathrm{TCNQ})_5$.[172] Analysis of the crystal structure of the complex (and comparison with the structure of the neutral structure[171,173]) lead to the conclusion that the benzotriazine was oxidized to its

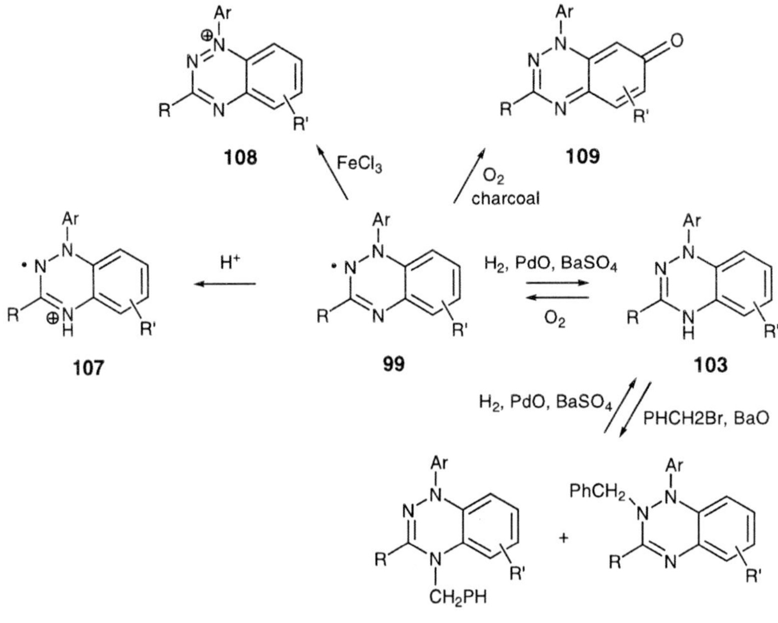

Scheme 7.37

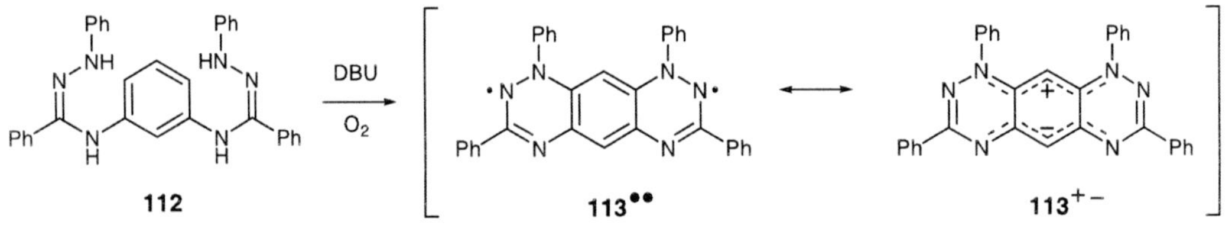

Scheme 7.38

corresponding cation (i.e., **108**) whereas the TCNQ is only partially reduced, in accord with the unusual stoichiometry.

An interesting class of putative benzotriazinyl-based diradicals was developed by Wudl *et al.*[174] Intramolecular cyclization of the bifunctional bis(amidrazone) **112** based on a modified version of the Blatter/Neugebauer synthesis affords the hexaazaanthracene compound **113**, which can formally be described as a diradical (**113••**) (Scheme 7.38).[174] However, many of the characteristics of this compound (for example it has a normal NMR spectrum, is EPR silent, and is strongly fluorescent) argue against a diradical formulation and in favour of a closed shell configuration. X-ray structural analysis and DFT calculations indicate that the zwitterionic formulation (**113**$^{+-}$), with two parallel cyanine-like chromophores, is the best description of the ground state. The diradical triplet state is calculated to be relatively low-energy excited state (18.9 kcal/mol higher than the ground state), and the strong fluorescence exhibited by this molecule has been attributed to an intramolecular electron transfer process to generate an excited state diradical species.[175]

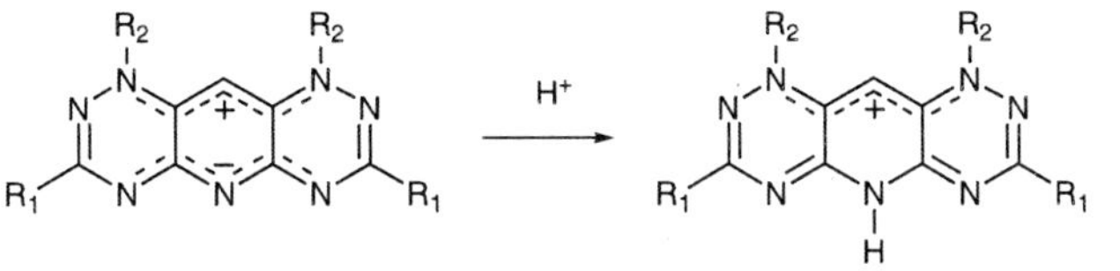

114 **115**

Scheme 7.39

Schreiner *et al.* have explored heptaaza-anthracenes **114**, which are pyridine-based analogues of **113**.[176,177] These compounds are synthesized in a manner analogous to the one used to make the hexaazaanthracene and have similar electronic structures (ground state zwitterions, low lying triplet diradical excited states). These derivatives are strongly solvatochromic and can be protonated (trifluoroacetic acid) to give stable cations **115** in which the central (pyridine) nitrogen is the site of protonation (Scheme 7.39). The influence of donor and acceptor groups on the aromatic substituents on the singlet–triplet energy separation was explored computationally. Although the changes are modest within a series where R_1 and R_2 are both substituted phenyl (15–20 kcal/mol), more dramatic changes in the electron configurational energetics are predicted for derivatives with donor ($R_1 = NMe_2$) and acceptor ($R_2 = NO_2$) groups attached *directly* to the polycyclic skeleton.[177]

7.8 Summary

Verdazyl radicals have enjoyed steady attention over the past four and a half decades, and stand out as the dominant species among hydrazyl-based radicals. A range of other species have received far less attention but add to the collection of stable radicals based on the hydrazyl ($R_2NNR^{\bullet}$) functional group. As such, hydrazyl deserves to be considered as another pre-eminent "building block" for stable radicals, both in terms of variety and stability, along with the nitroxyl ($R_2NO^{\bullet}$) moiety (Chapter 5) and the thiazyl ($NS^{\bullet}$) unit (Chapter 9).

References

1. In the context of radical chemistry the term "stable" has long been used subjectively. The term "stable radical" was originally and arbitrarily applied to any system with a sufficiently long lifetime to be observable by "conventional" spectroscopic methods. There have been attempts to specify a "stable" radical as one which is *isolable* (see, e.g., D. Griller and K. U. Ingold, *Acc. Chem. Res.* **9**, 13–19 (1976) and P. P. Power, *Chem. Rev.* **103**, 789–809 (2003)) and as such distinct from *persistent* radicals, which have some degree of kinetic stability but which cannot be isolated. Nonetheless, for the purposes of this review there is some utility in maintaining a degree of flexibility in distinguishing between stable and persistent radicals (for example, in a strict sense trityl would fall firmly within the "persistent" category, being a long lived but reactive radical and one which can not be isolated as a pure compound).
2. (a) S. Goldschmidt, *Ber. Deutsch. Chem. Gesel.* **B53**, 44–62 (1920). (b) S. Goldschmidt and K. Euler, *Ber. Deutsch. Chem. Gesel.* **B55**, 616–628 (1922). (c) S. Goldschmidt, A. Wolf, E. Wolffhardt, *et al.*, *Ann. Chem.* **427**, 194–226 (1924). (d) S. Goldschmidt and J. Bader, *Ann. Chem.* **473**, 137–162 (1929).
3. S. Goldschmidt and K. Renn, *Ber. Deutsch. Chem. Gesel.* **B55**, 628–643 (1922).
4. A. R. Forrester, J. M. Hay and R. H. Thomson, *Organic Chemistry of Stable Free Radicals*, Academic Press, New York, 1968, 137–179.

5. The hydrazyl-based radical literature includes a large number of papers from the Soviet/Russian literature, which have proven extremely difficult to obtain and, unfortunately, have not been translated into English. As such, the *accessible* content of many of these papers is limited to their abstracts. The available information in these abstracts has been incorporated into this review with the relevant papers cited. This approach was taken out of practical considerations and is done with apologies to the many scientists who have made significant contributions to the chemistry of verdazyls and related stable radicals.
6. R. Kuhn and H. Trischmann, *Angew. Chem. Int. Ed. Engl.* **2**, 155 (1963).
7. (a) F. A. Neugebauer, *Angew. Chem. Int. Ed. Engl.* **12**, 455–464 (1973). (b) O. M. Polumbrik, *Russ. Chem. Rev.* **47**, 767–785 (1978).
8. B. D. Koivisto and R. G. Hicks, *Coord. Chem. Rev.* **249**, 2612–2630 (2005).
9. K. Mukai, in *Carbon-Based Magnetism* (eds T. Makarova and F. Palacio), Elsevier, Amsterdam, 2006, 75–106.
10. A. W. Nineham, *Chem. Rev.* **55**, 355–483 (1955).
11. E. Seidler, *Progr. Histochem. Cytochem.* **24**, 1–86 (1991).
12. R. Kuhn and H. Trischmann, *Monatsh. Chem.* **95**, 457–479 (1964).
13. H. Brunner, *Tetrahedron*, **27**, 3611–3621 (1971).
14. (a) G. McConnachie and F. A. Neugebauer, *Tetrahedron*, **31**, 555–560 (1975). (b) F. A. Neugebauer, and M. Jenne, *Chem. Ber.* **102**, 3082–3087 (1969). (c) F. A. Neugebauer and M. Jenne, *Tet. Lett.* 791–& (1969).
15. F. A. Neugebauer, *Monatsh. Chem.* **98**, 231–244 (1967).
16. R. Kuhn, F. A. Neugebauer, and H. Trischmann, *Monatsh. Chem.* **97**, 846–852 (1966).
17. A. R. Katritzky, S. A. Belyakov, H. D. Durst, *et al.*, *Can. J. Chem.* **72**, 1849–1856 (1994).
18. R. Kuhn and G. Fischers, *Monatsh. Chem.* **97**, 517–524 (1966).
19. R. Kuhn, F. A. Neugebauer and H. Trischmann, *Monatsh. Chem.* **97**, 525–538 (1966).
20. F. A. Neugebauer, *Tetrahedron*, **26**, 4853–4857 (1970).
21. F. A. Neugebauer and R. Bernhard, *Chem. Ber.* **107**, 529–536 (1974).
22. A. R. Katritzky and S. A. Belyakov, *Synthesis*, 17–19 (1997).
23. E. A. Ponomareva, P. V. Tarassenko and G. F. Dvorko, *Angew. Chem. Int. Ed. Eng.* **14**, 431–432 (1975).
24. R. Kuhn, *Angew. Chem. Int. Ed. Eng.* **3**, 762–763 (1964).
25. R. Kuhn, F. A. Neugebauer, and H. Trischmann, *Angew. Chem. Int. Ed. Eng.* **4**, 72 (1965). (b) G. Kothe, H. Zimmermann and F. A. Neugebauer, *Angew. Chem. Int. Ed. Eng.* **11**, 830–832 (1972).
26. F. A. Neugebauer, H. Fischer and P. Meier, *Chem. Ber.* **113**, 2049–2051 (1980).
27. F. A. Neugebauer, *Angew. Chem. Int. Ed. Eng.* **6**, 362–363 (1967).
28. F. A. Neugebauer, H. Fischer and R. Bernhardt, *Chem. Ber.* **109**, 2389–2394 (1976).
29. A. Lang, H. Naarmann, G. Rosler, *et al.*, *Mol. Phys.* **79**, 1051–1062 (1993).
30. Y. Kurusu, H. Yoshida and M. Okawara, *Tet. Lett.* 3595–3597 (1967).
31. M. Kinoshita and R. C. Schulz, *Makromol. Chem. Phys.* **111**, 137–145 (1968).
32. Y. Miura, M. Kinoshita, and M. Imoto, *Makromol. Chem.* **146**, 69–77 (1971).
33. Y. Miura, M. Kinoshita, and M. Imoto, *Makromol. Chem.* **157**, 51–61 (1972). (b) M. Kamachi, H. Enomoto, M. Shibasaka, *et al.*, *Polym. J.* **18**, 439–441 (1986).
34. Y. Miura, N. Makita and M. Kinoshita, *J. Macromol. Sci. Chem.* **A7**, 1007–1012 (1973).
35. Y. Miura and M. Kinoshita, *Makromol. Chem. Phys.* **175**, 23–29 (1974).
36. F. A. Neugebauer and H. Trischmann, *J. Polym. Sci B Polym. Lett.* **6**, 255–256 (1968).
37. J. Bosch, C. Rovira, J. Veciana, *et al.*, *Synth. Met.* **55**, 1141–1146 (1993).
38. M. Hmyene, H. Naarmann, H. Winter, *et al.*, *J. Phys. Cond. Matt.* **6**, L511–L515 (1994).
39. M. Hmyene, A. Hakam, H. Naarmann and E. Dormann, *Ann. Chim. Sci. Mater.* **29**, 55–61 (2004).
40. L. Raphaelian, H. Hooks and G. Ottmann, *Angew. Chem. Int. Ed. Engl.* **6**, 363–364 (1967).
41. F. A. Neugebauer and H. Fischer, *Angew. Chem. Int. Ed. Engl.* **19**, 724–725 (1980).
42. F. A. Neugebauer, H. Fischer and R. Siegel, *Chem. Ber.* **121**, 815–822 (1988).
43. F. A. Neugebauer, H. Fischer, R. Siegel and C. Krieger, *Chem. Ber.* **116**, 3461–3481 (1983).
44. C. L. Barr, P. A. Chase, R. G. Hicks, *et al.*, *J. Org. Chem.* **64**, 8893–8897 (1999).
45. F. A. Neugebauer and R. Siegel, *Angew. Chem. Int. Ed. Eng.* **22**, 320–321 (1983).
46. E. C. Pare, D. J. R. Brook, A. Brieger, *et al.*, *Org. Biomol. Chem.* **3**, 4258–4261 (2005).

47. F. A. Neugebauer, H. Fischer and C. Krieger, *J. Chem. Soc. Perkin Trans. 2*, 535–544 (1993).
48. R. Milcent, G. Barbier, S. Capelle and J. P. Catteau, *J. Heterocycl. Chem.* **31**, 319–324 (1994).
49. D. J. R. Brook and G. T. Yee, *J. Org. Chem.* **71**, 4889–4895 (2006).
50. R. M. Fico, M. F. Hay, S. Reese, *et al.*, *J. Org. Chem.* **64**, 9386–9392 (1999).
51. J. B. Gilroy, S. D. J. McKinnon, P. Kennepohl, *et al.*, *J. Org. Chem.* **72**, 8062–8069 (2007).
52. D. J. R. Brook, H. H. Fox, V. Lynch and M. A. Fox, *J. Phys. Chem.* **100**, 2066–2071 (1996).
53. K. Suzuki, M. M. Matsushita, H. Hayashi, *et al.*, *New J. Chem.* **32**, 2201–2208 (2008).
54. B. D. Koivisto, A. S. Ichimura, R. McDonald, *et al.*, *J. Am. Chem. Soc.* **128**, 690–691 (2006).
55. V. Chemistruck, D. Chambers and D. J. R. Brook, *J. Org. Chem.* **74**, 1850–1857 (2009).
56. R. G. Hicks, M. T. Lemaire, L. Ohrstrom, *et al.*, *J. Am. Chem. Soc.* **123**, 7154–7159 (2001).
57. J. B. Gilroy, B. D. Koivisto, R. McDonald, *et al.*, *J. Mater. Chem.* **16**, 2618–2624 (2006).
58. M. J. Plater, S. Kemp, E. Coronado, *et al.*, *Polyhedron*, **25**, 2433–2438 (2006).
59. L. Norel, F. Pointillart, C. Train, *et al.*, *Inorg. Chem.* **47**, 2396–2403 (2008).
60. (a) K. Mukai, T. Hatanaka, N. Senba, *et al.*, *Inorg. Chem.* **41**, 5066–5074 (2002). (b) K. Mukai, N. Senba, T. Hatanaka, *et al.*, *Inorg. Chem.* **43**, 566–576 (2004).
61. (a) N. Azuma, *Bull. Chem. Soc. Jpn.* **53**, 2671–2672 (1980). (b) N. Azuma, K. Tsutsui, Y. Miura and T. Higuchi, *Bull. Chem. Soc. Jpn.* **54**, 3274–3278 (1981). (c) A. A. Dvorkin, O. M. Polumbrik, I. G. Riabokon, *et al.*, *Dokl. Akad. Nauk SSSR*, **277**, 1134–1138 (1984). (d) L. Parkanyi, C. Kertesz, A. J. Mayr and M. P. Eastman, *J. Mol. Struct.* **447**, 141–150 (1998). (e) K. Mukai, S. Jinno, Y. Shimobe, *et al.*, *Polyhedron*, **20**, 1537–1544 (2001). (f) T. Ojima, H. Akutsu, J. Yamada and S. Nakatsuji, *Polyhedron*, **20**, 1335–1338 (2001). (g) K. Mukai, M. Matsubara, H. Hisatou, *et al.*, *J. Phys. Chem. B*, **106**, 8632–8638 (2002). (h) K. Mukai, S. Jinno, Y. Shimobe, *et al.*, *J. Mater. Chem.* **13**, 1614–1621 (2003). (i) K. Mukai, D. Shiba, K. Mukai, *et al.*, *Polyhedron*, **24**, 2513–2521 (2005). (j) K. Mukai, D. Shiba, K. Yoshida, *et al.*, *Bull. Chem. Soc. Jpn.* **78**, 2114–2123 (2005).
62. (a) P. M. Allemand, G. Srdanov and F. Wudl, *J. Am. Chem. Soc.* **112**, 9391–9392 (1990). (b) M. Mito, K. Takeda, K. Mukai, *et al.*, *J. Phys. Chem. B*, **101**, 9517–9524 (1997).
63. P. R. Serwinski, B. Esat, P. M. Lahti, *et al.*, *J. Org. Chem.* **69**, 5247–5260 (2004).
64. D. E. Williams, *Acta Cryst.* **B29**, 96–102 (1973).
65. K. Mukai, T. Yamamoto, M. Kohno, *et al.* u, *Bull. Chem. Soc. Jpn.* **47**, 1797–1798 (1974).
66. K. Mukai, H. Shikata, N. Azuma and K. Kuwata, *J. Mag. Reson.* **35**, 133–137 (1979).
67. (a) F. A. Neugebauer, H. Trischmann and G. Taigel, *Monatsh. Chem.* **98**, 713–725 (1967). (b) F. A. Neugebauer, H. Brunner and K. H. Hausser, *Tetrahedron*, **27**, 3623–3628 (1971).
68. F. A. Neugebauer and H. Brunner, *Tetrahedron*, **30**, 2841–2850 (1974).
69. P. H. H. Fischer, *Tetrahedron*, **23**, 1939–1952 (1967).
70. (a) L. S. Degtyarev and Y. I. Gorlov, *J. Struct. Chem.* **16**, 715–718 (1975). (b) O. M. Polumbrik, A. M. Nesterenko and L. N. Markovskij, *Dop. Akad. Nauk Ukr. Ser. B*, 521–524 (1977). (c) O. M. Polumbrik, A. M. Nesterenko and L. N. Markovskii, *Dokl. Akad. Nauk SSSR*, **240**, 640–643 (1978). (d) L. N. Markovsky, O. M. Polumbrik and A. M. Nesterenko, *Int. J. Quant. Chem.* **16**, 891–895 (1979). (e) O. M. Polumbrik, A. M. Nesterenko and L. N. Markovskii, *J. Struct. Chem.* **20**, 487–490 (1979). (f) L. S. Degtyarev, A. A. Stetsenko and Y. I. Gorlov, *Chem. Phys. Lett.* **69**, 323–326 (1980).
71. L. Ohrstrom, A. Grand and B. Pilawa, *Acta Chem. Scand.* **50**, 458–461 (1996).
72. K. Mukai, M. Nuwa, K. Suzuki, *et al.*, *J. Phys. Chem. B*, **102**, 782–787 (1998).
73. V. Barone, A. Bencini, I. Ciofini and C. Daul, *J. Phys. Chem. A*, **103**, 4275–4282 (1999).
74. (a) Y. Teki, M. Kimura, S. Narimatsu, *et al.*, *Bull. Chem. Soc. Jpn.* **77**, 95–99 (2004). (b) Y. Teki, H. Tamekuni, K. Haruta, *et al.*, *J. Mater. Chem.* **18**, 381–391 (2008).
75. J. Jornet, M. Deumal, J. Ribas-Arino, *et al.*, *Chem. Eur. J.*, **12**, 3995–4005 (2006).
76. J. B. Rota, L. Norel, C. Train, *et al.*, *J. Am. Chem. Soc.* **130**, 10380–10385 (2008).
77. (a) F. A. Neugebauer and H. Fischer, *Tet. Lett.* 3345–3348 (1977). (b) F. A. Neugebauer, H. Fischer and H. Brunner, *Tetrahedron*, **37**, 1391–1396 (1981).
78. F. A. Neugebauer and H. Fischer, *J. Chem. Soc. Perkin Trans. 2*, 896–900 (1981).
79. A. Izuoka, S. Murata, T. Sugawara and H. Iwamura, *J. Am. Chem. Soc.* **109**, 2631–2639 (1987).

80. (a) M. T. Green and T. A. McCormick, *Inorg. Chem.* **38**, 3061–3065 (1999). (b) G. Chung and D. Lee, *Chem. Phys. Lett.* **350**, 339–344 (2001). (c) D. Bhattacharya and A. Misra, *J. Phys. Chem. A*, **113**, 5470–5475 (2009). (d) Y. Masuda, M. Kuratsu, S. Suzuki, *et al.*, *Polyhedron*, **28**, 1950–1954 (2009).
81. V. Polo, A. Alberola, J. Andres, *et al.*, *Phys. Chem. Chem. Phys.* **10**, 857–864 (2008).
82. I. A. Latif, A. Panda and S. N. Datta, *J. Phys. Chem. A*, **113**, 1595–1600 (2009).
83. F. A. Neugebauer, R. Bernhardt and H. Fischer, *Chem. Ber.* **110**, 2254–2275 (1977).
84. (a) K. Mukai, N. Azuma, H. Shikata and K. Ishizu, *Bull. Chem. Soc. Jpn.* **43**, 3958–3960 (1970). (b) P. Kopf, K. Morokuma and R. Kreilick, *J. Chem. Phys.* **54**, 105–110 (1971). (c) N. Azuma, K. Ishizu and K. Mukai, *J. Chem. Phys.* **61**, 2294–2296 (1974).
85. P. M. Lahti, B. Esat, Y. Liao, *et al.*, *Polyhedron*, **20**, 1647–1652 (2001).
86. (a) Y. Teki, M. Nakatsuji and Y. Miura, *Int. J. Mod. Phys. B*, **15**, 4029–4031 (2001). (b) Y. Teki, M. Nakatsuji and Y. Miura, *Mol. Phys.* **100**, 1385–1394 (2002). (c) T. Toichi and Y. Teki, *Polyhedron*, **24**, 2337–2340 (2005). (d) I. Ciofini, C. Adamo, Y. Teki, *et al.*, *Chem. Eur. J.* **14**, 11385–11405 (2008). (e) N. Mihara and Y. Teki, *Inorg. Chim. Acta*, **361**, 3891–3894 (2008). (f) I. Matsumoto, I. Ciofini, P. P. Laine and Y. Teki, *Chem. Eur. J.* **15**, 11210–11220 (2009).
87. I. A. Bezvershenko and V. K. Premyslov, *Khim. Geter. Soedin.* 1129–1130 (1985).
88. (a) J. S. Jaworski, *J. Electroanal. Chem.* **300**, 167–174 (1991). (b) J. S. Jaworski and I. Krawczyk, *Monatsh. Chem.* **123**, 43–50 (1992). (c) M. Chahma, X. S. Wang, A. van der Est and M. Pilkington, *J. Org. Chem.* **71**, 2750–2755 (2006). (d) M. Chahma, K. Macnamara, A. Van der Est, *et al.*, *New J. Chem.* **31**, 1973–1978 (2007).
89. S. Nakatsuji, A. Kitamura, A. Takai, *et al.*, *Z. Naturforsch.* **B53**, 495–502 (1998).
90. (a) O. M. Polumbrik, A. V. Misyura, N. G. Vasilkevich and L. N. Markovskii, *Zh. Org. Khim.* **21**, 871–875 (1985). (b) O. M. Polumbrik, I. G. Ryabokon, N. G. Vasilkevich, *et al.*, *Zh. Org. Khim.* **21**, 185–193 (1985).
91. J. B. Gilroy, S. D. J. McKinnon, B. D. Koivisto and R. G. Hicks, *Org. Lett.* **9**, 4837–4840 (2007).
92. (a) L. Keeney and M. J. Hynes, *Dalton Trans.* 1524–1531 (2005). (b) L. Keeney and M. J. Hynes, *Dalton Trans.* 133–138 (2005). (c) S. Conway, L. Keeney and M. J. Hynes, *Dalton Trans.* 2878–2882 (2009).
93. R. Kuhn, F. A. Neugebauer, and H. Trischmann, *Monatsh. Chem.* **97**, 1280–1289 (1966).
94. E. A. Ponomareva, P. V. Tarasenko and G. F. Dvorko, *Dop. Akad. Nauk Ukr. Ser. B*, 142–146 (1977).
95. (a) O. M. Polumbrik, V. I. Staninets and B. M. Yarmolyuk, *Reakt. Spos. Org. Soedin.* **12**, 161–166 (1975). (b) O. M. Polumbrik, B. M. Yarmolyuk and E. I. Zaika, *Zh. Fiz. Khim.* **49**, 742–744 (1975). (c) M. S. Sytilin, *Zh. Fiz. Khim.* **51**, 235–236 (1977). (d) I. G. Ryabokon, O. M. Polumbrik and L. N. Markovskii, *Dop. Akad. Nauk Ukr. Ser. B*, 47–50 (1984).
96. R. Kuhn, F. A. Neugebauer and H. Trischmann, *Monatsh. Chem.* **98**, 726–730 (1967).
97. F. A. Neugebauer, W. Otting, H. Trischmann and H. O. Smith, *Chem. Ber.* **105**, 549–553 (1972).
98. P. V. Tarasenko, E. A. Ponomaryova and G. F. Dvorko, *Khim. Geterot. Soedin.* 270–271 (1983).
99. (a) O. M. Polumbrik and I. G. Ryabokon, *Zh. Org. Khim.* **14**, 1332–1335 (1978). (b) A. V. Misyura, O. M. Polumbrik and L. N. Markovskii, *Zh. Org. Khim.* **25**, 424–431 (1989).
100. (a) O. M. Polumbrik and O. I. Zaika, *Dop. Akad. Nauk Ukr. Ser. B*, 338–341 (1975). (b) O. Polumbrik and E. Zaika, *Org. React.* **14**, 388–403 (1977). (c) O. M. Polumbrik and I. G. Ryabokon, *Zh. Org. Khim.* **21**, 1975–1980 (1985).
101. O. M. Polumbrik, G. F. Dvorko, E. A. Ponomareva, and E. I. Zaika, *Zh. Org. Khim.* **8**, 2417–2422 (1972). (b) O. M. Polumbrik, E. I. Zaika and G. F. Dvorko, *Zh. Org. Khim.* **10**, 1953–1955 (1974). (c) O. M. Polumbrik, E. J. Zaika and G. F. Dvorko, *Ukr. Khim. Zh.* **40**, 963–966 (1975).
102. (a) O. M. Polumbrik, N. G. Vasilkevich, and G. F. Dvorko, *Ukr. Khim. Zh.* **40**, 1161–1166 (1974). (b) O. M. Polumbrik, N. G. Vasilkevich and G. F. Dvorko, *Zh. Org. Khim.* **11**, 770–773 (1975). (c) P. V. Tarasenko and E. A. Ponomareva, *Vest. Kiev. Polit. Inst. Khim. Mashin. Tekhnol.* **14**, 25–27 (1977). (d) O. M. Polumbrik, I. G. Ryabokon and L. N. Markovskii, *Zh. Org. Khim.* **18**, 1060–1065 (1982).
103. A. V. Misyura, O. M. Polumbrik and L. N. Markovskii, *Dokl. Akad. Nauk Ukr. Ser. B*, 58–61 (1988).
104. N. I. Buryak, O. M. Polumbrik and A. A. Yasnikov, *Ukr. Khim. Zh.* **51**, 1111–1112 (1985).
105. E. K. Y. Chen, S. J. Teertstra, D. Chan-Seng, *et al.*, *Macromolecules*, **40**, 8609–8616 (2007).

106. (a) Y. Miura, Y. Morimoto and M. Kinoshita, *Bull. Chem. Soc. Jpn.* **48**, 3765–3766 (1975). (b) Y. Miura, Y. Morimoto and M. Kinoshita, *Chem. Soc. Jpn.* **49**, 253–255 (1976). (c) Y. Miura, Y. Morimoto and M. Kinoshita, *Chem. Soc. Jpn.* **49**, 1715–1716 (1976).
107. (a) O. M. Polumbrik, E. A. Ponomareva, E. I. Zaika and G. F. Dvorko, *Zh. Org. Khim.* **8**, 1925–& (1972). (b) O. M. Polumbrik and E. I. Zaika, *Org. React.* **14**, 375–387 (1977).
108. L. S. Degtyarev, Y. A. Maletin and A. A. Stetsenko, *Zh. Obshch. Khim.* **51**, 2387–2388 (1981).
109. N. G. Strizhakova, Y. A. Maletin and I. A. Sheka, *Zh. Obshch. Khim.* **53**, 498–502 (1983).
110. (a) E. A. Ponomareva, T. L. Pervishko and G. F. Dvorko, *Org. React.* **16**, 113–128 (1979). (b) T. L. Pervishko, E. A. Ponomareva and G. F. Dvorko, *Org. React.* **24**, 199–206 (1987). (c) G. F. Dvorko and T. V. Cherevach, *Zh. Obshch. Khim.* **60**, 2151–2160 (1990). (d) T. L. Pervishko, E. A. Ponomareva and G. F. Dvorko, *Org. React.* **27**, 34–41 (1990). (e) G. F. Dvorko and E. A. Ponomareva, *Usp. Khim.* **60**, 2089–2112 (1991). (f) E. A. Ponomareva, I. V. Koshchii, T. L. Pervishko and G. F. Dvorko, *Russ. J. Gen. Chem.* **70**, 907–916 (2000). (g) G. F. Dvorko, I. V. Koshchii, A. M. Prokopets and E. A. Ponomareva, *Russ. J. Gen. Chem.* **72**, 1882–1893 (2002). (h) G. F. Dvorko, V. V. Zaliznyi and N. E. Ponomarev, *Russ. J. Gen. Chem.* **72**, 1414–1428 (2002). (i) N. E. Ponomarev, A. V. Stambirskii and G. F. Dvorko, *Russ. J. Gen. Chem.* **72**, 79–85 (2002). (j) G. F. Dvorko, I. V. Koshchii and E. A. Ponomareva, *Russ. J. Gen. Chem.* **73**, 204–212 (2003). (k) G. F. Dvorko, I. V. Koshchii and E. A. Ponomareva, *Russ. J. Gen. Chem.* **73**, 569–574 (2003). (l) G. F. Dvorko, E. A. Ponomareva and M. E. Ponomarev, *J. Phys. Org. Chem.* **17**, 825–836 (2004). (m) N. E. Ponomarev, M. V. Stambirskii, G. F. Dvorko and A. V. Bazil'chuk, *Russ. J. Org. Chem.* **40**, 489–496 (2004). (o) G. F. Dvorko, E. A. Ponomareva, N. N. Golovko and T. L. Pervishko, *Russ. J. Gen. Chem.* **75**, 94–99 (2005). (p) N. E. Ponomarev, M. V. Stambirskii and G. F. Dvorko, *Russ. J. Gen. Chem.* **75**, 883–890 (2005). (q) N. E. Ponomarev, V. V. Zaliznyi and G. F. Dvorko, *Russ. J. Gen. Chem.* **75**, 1430–1436 (2005). (r) G. F. Dvorko, E. A. Ponomareva, N. E. Ponomarev, *et al.*, *Russ. J. Gen. Chem.* **77**, 1535–1558 (2007). (s) G. F. Dvorko, E. A. Ponornareva, M. E. Ponomarev and M. V. Stambirsky, *Progr. React. Kin. Mech.* **32**, 73–118 (2007). (t) N. E. Ponomarev, V. V. Zaliznyi and G. F. Dvorko, *Russ. J. Gen. Chem.* **77**, 1204–1214 (2007).
111. (a) I. M. Serebryakov, E. B. Kryzhanovskaya, S. S. Dzhurinskaya, *et al.*, *Russ. J. Gen. Chem.* **75**, 1331–1335 (2005). (b) I. M. Serebryakov, E. B. Kryzhanovskaya and S. S. Dzhurinskaya, *Russ. J. Gen. Chem.* **76**, 798–800 (2006).
112. A. Yang, T. Kasahara, E. K. Y. Chen, *et al.*, *Eur. J. Org. Chem.* 4571–4574 (2008).
113. F. A. Neugebauer, *Monatsh. Chem.* **97**, 853–856 (1966).
114. M. Kinoshita and Y. Miura, *Makromol. Chem.* **124**, 211–& (1969).
115. (a) B. Yamada, H. Tanaka, K. Konishi and T. Otsu, *J. Macromol. Sci.* **A31**, 351–366 (1994). (b) B. Yamada, Y. Nobukane and Y. Miura, *Polym. Bull.* **41**, 539–544 (1998).
116. (a) G. F. Dvorko, L. S. Degtyarev and A. K. Tomashchik, *Dokl. Akad. Nauk SSSR*, **202**, 1073–1075 (1972). (b) G. F. Dvorko and L. S. Degtyarev, *Zh. Org. Khim.* **10**, 1554–1555 (1974). (c) N. G. Strizhakova, Y. A. Maletin and I. A. Sheka, *Ukr. Khim. Zh.* **45**, 1039–1041 (1979). (d) Y. A. Maletin, A. S. Grigor'eva, E. E. Kriss and N. F. Konakhovich, *Koord. Khim.* **7**, 1464–1470 (1981). (e) Y. A. Maletin, N. G. Strizhakova and I. A. Sheka, *Zh. Obshch. Khim.* **51**, 1119–1126 (1981). (f) T. G. Samarskaya, L. N. Ganyuk and V. M. Ogenko, *Mendel. Commun.* 19–20 (1994).
117. D. J. R. Brook, V. Lynch, B. Conklin and M. A. Fox, *J. Am. Chem. Soc.* **119**, 5155–5162 (1997).
118. (a) D. J. R. Brook, S. Fornell, B. Noll, *et al.*, *J. Chem. Soc. Dalton Trans.* 2019–2022 (2000). (b) T. M. Barclay, R. G. Hicks, M. T. Lemaire and L. K. Thompson, *Inorg. Chem.* **40**, 6521–6524 (2001). (c) D. J. R. Brook and V. Abeyta, *J. Chem. Soc. Dalton Trans.*, 4219–4223 (2002). (d) J. E. Stevens, D. J. R. Brook and V. W. Abeyta, *Polyhedron*, **22**, 2241–2247 (2003).
119. (a) R. G. Hicks, M. T. Lemaire, L. K. Thompson and T. M. Barclay, *J. Am. Chem. Soc.* **122**, 8077–8078 (2000). (b) T. M. Barclay, R. G. Hicks, M. T. Lemaire and L. K. Thompson, *Inorg. Chem.* **40**, 5581–5584 (2001).
120. T. M. Barclay, R. G. Hicks, M. T. Lemaire and L. K. Thompson, *Inorg. Chem.* **42**, 2261–2267 (2003).
121. T. M. Barclay, R. G. Hicks, M. T. Lemaire, *et al.*, *Chem. Commun.* 1688–1689 (2002).
122. M. T. Lemaire, T. M. Barclay, L. K. Thompson and R. G. Hicks, *Inorg. Chim. Acta*, **359**, 2616–2621 (2006).
123. T. M. Barclay, R. G. Hicks, M. T. Lemaire and L. K. Thompson, *Chem. Commun.* 2141–2142 (2000).

124. D. J. R. Brook, S. Fornell, J. E. Stevens, *et al.*, *Inorg. Chem.* **39**, 562–567 (2000).
125. J. Z. Wu, E. Bouwman, J. Reedijk, *et al.*, *Inorg. Chim. Acta*, **351**, 326–330 (2003).
126. F. Pointillart, C. Train, P. Herson, *et al.*, *New J. Chem.* **31**, 1001–1006 (2007).
127. L. Norel, L. M. Chamoreau, Y. Journaux, *et al.*, *Chem. Commun.* 2381–2383 (2009).
128. R. G. Hicks, *Aust. J. Chem.* **54**, 597–600 (2001).
129. A. V. Yakovenko, S. V. Kolotilov, O. Cador, *et al.*, *Eur. J. Inorg. Chem.* 2354–2361 (2009).
130. K. Nakabayashi, Y. Ozaki, M. Kawano and M. Fujita, *Angew. Chem. Int. Ed.* **47**, 2046–2048 (2008).
131. Y. Ozaki, M. Kawano and M. Fujita, *Chem. Commun.* 4245–4247 (2009).
132. R. G. Hicks, B. D. Koivisto and M. T. Lemaire, *Org. Lett.* **6**, 1887–1890 (2004).
133. (a) P. P. Kornuta, V. N. Bobkov, O. M. Polumbrik and L. N. Markovskii, *Zh. Obshch. Khim.* **48**, 697–698 (1978). (b) F. A. Neugebauer and H. Fischer, *Z. Naturforsch.* **B35**, 250–251 (1980). (c) P. P. Kornuta, V. N. Bobkov, O. M. Polumbrik and L. N. Markovskii, *Zh. Obshch. Khim.* **51**, 2449–2456 (1981).
134. R. G. Hicks and R. Hooper, *Inorg. Chem.* **38**, 284–286 (1999).
135. R. G. Hicks, L. Ohrstrom and G. W. Patenaude, *Inorg. Chem.* **40**, 1865–1870 (2001).
136. T. M. Barclay, R. G. Hicks, A. S. Ichimura and G. W. Patenaude, *Can. J. Chem.* **80**, 1501–1506 (2002).
137. A. M. Nesterenko, O. M. Polumbrik and L. N. Markovskii, *J. Struct. Chem.* **25**, 209–214 (1984).
138. (a) B. I. Stepanov and G. V. Avramenko, *Zh. Obshch. Khim.* **50**, 358–362 (1980). (b) G. V. Avramenko, T. S. Kolodina and B. I. Stepanov, *Zh. Obshch. Khim.* **53**, 2286–2291 (1983). (c) T. S. Kolodina, G. V. Avramenko and B. I. Stepanov, *Zh. Obshch. Khim.* **53**, 1340–1343 (1983). (d) B. E. Zaitsev, Z. V. Bezuglaya, G. V. Avramenko, *et al.*, *Zh. Obshch. Khim.* **55**, 1398–1403 (1985). (e) B. I. Stepanov, G. V. Avramenko, S. Khamud and S. I. Mustafaeva, *Zh. Obshch. Khim.* **56**, 390–392 (1986). (f) Z. V. Bezuglaya, M. Naser, G. V. Avramenko and B. I. Stepanov, *Zh. Obshch. Khim.* **61**, 1902–1903 (1991).
139. J. B. Gilroy, M. J. Ferguson, R. McDonald, *et al.*, *Chem. Commun.* 126–128 (2007).
140. (a) R. Kuhn, H. Trischmann and F. A. Neugebauer, *Angew. Chem. Int. Ed.* **3**, 232 (1964). (b) F. A. Neugebauer and H. Trischmann, *Monatsh. Chem.* **97**, 554–569 (1966).
141. F. A. Neugebauer, *Chem. Ber.* **106**, 1716–1723 (1973).
142. I. Hausser, D. Jerchel and R. Kuhn, *Chem. Ber.* **82**, 195–199 (1949).
143. R. Kuhn and D. Jerchel, *Liebigs Ann. Chem.* **578**, 1–5 (1952).
144. D. Jerchel and H. Fischer, *Liebigs Ann. Chem.* **590**, 216–231 (1954).
145. D. Jerchel and H. Fischer, *Chem. Ber.* **89**, 563–570 (1956).
146. F. A. Neugebauer, *Chem. Ber.* **102**, 1339–1346 (1969).
147. R. T. Lofberg, *Nature*, **206**, 503–505 (1965).
148. F. A. Neugebauer, *Tet. Lett.* 2129–2132 (1968).
149. O. W. Maender and G. A. Russell, *J. Org. Chem.* **31**, 442–446 (1966).
150. Y. Deguchi and Y. Takagi, *Tet. Lett.* 3179–3180 (1967).
151. F. A. Neugebauer and G. A. Russell, *J. Org. Chem.* **33**, 2744–2746 (1968).
152. F. A. Neugebauer, *Angew. Chem. Int. Ed. Engl.* **8**, 520 (1969).
153. (a) N. Azuma, K. Mukai and K. Ishizu, *Bull. Chem. Soc. Jpn.* **43**, 3960A–3962 (1970). (b) R. M. Zaripova, A. A. Vafina, B. G. Liorber and V. N. Ivanova, *Bull. Acad. Sci. USSR Div. Chem. Sci.* **36**, 1081–1082 (1987).
154. K. B. Umemoto, *Bull. Chem. Soc. Jpn.* **62**, 3783–3789 (1989).
155. F. A. Neugebauer, *Tetrahedron*, **26**, 4843–4851 (1970).
156. F. A. Neugebauer, H. Fischer and C. Krieger, *Angew. Chem. Int. Ed. Engl.* **28**, 491–492 (1989).
157. A. M. Nesterenko, O. M. Polumbrik and L. N. Markovskii, *Zh. Org. Khim.* **19**, 1961–1964 (1983).
158. F. A. Neugebauer and H. Fischer, *Tetrahedron*, **51**, 12883–12898 (1995).
159. (a) D. Colombani, M. Steenbock, M. Klapper and K. Mullen, *Macromol. Rapid Commun.* **18**, 243–251 (1997). (b) M. Steenbock, M. Klapper and K. Mullen, *Acta Polym.* **49**, 376–378 (1998). (c) M. Steenbock, M. Klapper and K. Mullen, *Macromol. Chem. Phys.* **199**, 763–769 (1998). (d) M. Steenbock, M. Klapper, K. Mullen, *et al.*, *Macromolecules*, **31**, 5223–5228 (1998). (e) A. Dasgupta, T. Brand, M. Klapper and K. R. Mullen, *Polym. Bull.* **46**, 131–138 (2001). (f) M. Klapper and K. Mullen, *Macromol. Symp.* **163**, 1–23 (2001). (g) N. S. Khelfallah, M. Peretolchin, M. Klapper and K. Mullen, *Polym. Bull.* **53**, 295–304 (2005).

160. F. A. Neugebauer, *Mag. Reson. Chem.* **30**, 84–86 (1992).
161. F. A. Neugebauer, H. Fischer, R. Crockett and C. Krieger, *J. Chem. Soc. Perkin Trans. 2*, 1619–1626 (1990).
162. F. A. Neugebauer and R. Siegel, *Chem. Ber.* **118**, 2157–2163 (1985).
163. H. M. Blatter and H. Lukaszewski, *Tet. Lett.* 2701–2705 (1968).
164. F. A. Neugebauer and I. Umminger, *Chem. Ber.* **113**, 1205–1225 (1980).
165. R. Huisgen and J. Wulff, *Chem. Ber.* **102**, 1848–1858 (1969).
166. T. L. Gilchrist, C. J. Harris and C. W. Rees, *J. Chem. Soc. Chem. Commun.* 485–486 (1974).
167. D. H. R. Barton, J. W. Ducker, W. A. Lord and P. D. Magnus, *J. Chem. Soc. Perkin Trans. 1*, 38–42 (1976).
168. (a) F. A. Neugebauer and G. Rimmler, *Mag. Reson. Chem.* **26**, 595–600 (1988). (b) M. K. Kadirov, B. I. Buzykin and N. G. Gazetdinova, *Russ. Chem. Bull.* **51**, 1796–1799 (2002).
169. F. A. Neugebauer and I. Umminger, *Chem. Ber. Rec.* **114**, 2423–2430 (1981).
170. K. Mukai, K. Inoue, N. Achiwa, *et al.*, *Chem. Phys. Lett.* **224**, 569–575 (1994).
171. C. Krieger and F. A. Neugebauer, *Acta Cryst.* **C52**, 3124–3126 (1996).
172. K. A. Hutchison, G. Srdanov, R. Menon, *et al.*, *J. Am. Chem. Soc.* **118**, 13081–13082 (1996).
173. A. T. Gubaidullin, B. I. Buzykin, I. A. Litvinov and N. G. Gazetdinova, *Russ. J. Gen. Chem.* **74**, 939–943 (2004).
174. K. Hutchison, G. Srdanov, R. Hicks, *et al.*, *J. Am. Chem. Soc.* **120**, 2989–2990 (1998).
175. K. A. Hutchison, K. Hasharoni, F. Wudl, *et al.*, *J. Am. Chem. Soc.* **120**, 6362–6365 (1998).
176. P. Langer, A. Bodtke, N. N. R. Saleh, *et al.*, *Angew. Chem. Int. Ed. Engl.* **44**, 5255–5259 (2005).
177. P. Langer, S. Amiri, A. Bodtke, *et al.*, *J. Org. Chem.* **73**, 5048–5063 (2008).

8

Metal Coordinated Phenoxyl Radicals

Fabrice Thomas

University of Grenoble, Department of Molecular Chemistry, Grenoble, France

8.1 Introduction

The term "phenoxyl" radical was first introduced in 1914 by Pummerer[1] to designate species involved in the oxidation of naphthols and phenanthroles. It was not until the 1960s that the existence of phenoxyls was demonstrated by Electron Paramagnetic Resonance (EPR). The term "stable" is commonly used to designate radicals such as nitroxides, trityls, and so on. Although the term "stable" had been associated with some phenoxyl radicals in 1967,[2] it must be realized that phenoxyl radicals exhibiting stabilities comparable to those of nitroxides were rather rare at this time. During the 1970s, Reichard *et al.*[3] demonstrated that radicals related to phenoxyls could also arise from mono-electronic oxidation of the phenolic side chain of tyrosines in some proteins, and thus could be involved in biological processes. The new term "tyrosyl" was then introduced to designate these residues, and many other biological systems involving such radicals have been described.[4,5] One of the more important recent advances in tyrosyl radical history was achieved in the 1990s with the characterization of the galactose oxidase (GO) active site.[6–8] For the first time it was demonstrated that tyrosyls could exist coordinated to a metal ion. To better understand this association, chemists have subsequently developed many complexes involving coordinated phenoxyl radicals.[9–13] Elucidation of the properties of coordination compounds involving redox active ligands, especially those involving sterically hindered phenolates, is one of the most recent and fascinating topics of interest in bioinorganic chemistry. The challenge in this kind of chemistry is the right description of the electronic structure of the M^{n+}–OPh entity (where OPh represents a phenolate group) once it has been oxidized by one electron. In principle, either the $M^{(n+1)+}$–OPh, that is the oxidation is metal based, or the M^{n+}–$^{\bullet}$OPh form, that is the oxidation is ligand based, could be obtained. The right description is thus not obvious and, as will be shown below, it strongly depends on the nature of the metal ion, the denticity of the ligand, the substituents of the phenolate precursor and even the temperature. Coordination is also found to improve the radical stability, and exerts an extraordinary control on their magnetic properties and reactivity (regio- and stereoselective oxidations are promoted by a radical).

An exhaustive description of all the coordinated phenoxyl radicals reported so far is a very difficult task, as this chemistry has bloomed during the last decade. Historically, complexes involving first row d transition metals were the first to be described, and they are certainly the most well known. For this reason, this chapter is limited to these complexes. It is also important to mention that semiquinonate radicals can be related to phenoxyls. This class of compounds was intensively investigated in the 1980–1990s and excellent reviews are available for readers interested in these systems.[14–16] Only the most recent advances in metal–semiquinonate complexes that deal with the modeling of the GO active site are covered here.

The general properties of phenoxyl radicals (coordinated or not) are briefly presented prior to the description of the structural, spectroscopic and electrochemical properties of their complexes. For the sake of clarity, complex chemistry will be classified by metal, from the left to the right of the periodic table. Vanadium complexes, for which the redox chemistry is rather metal-centered, and chromium complexes, which exhibit pure ligand-centered redox chemistry, are presented first. Subsequent sections are devoted to manganese, iron, cobalt, nickel and, finally, of course, copper complexes, that is those for which the electronic structure is the most difficult to describe, as both the metal and the ligand(s) are redox active. A special emphasis is given to copper and nickel complexes that have created a great interest recently. The last section deals with complexes of the redox-innocent zinc ion. Scandium and titanium ions are omitted, since very few results concerning them have been reported.[17,18]

8.2 General properties of phenoxyl radicals

8.2.1 Electronic structure and stabilization

Phenoxyl radicals are electron deficient, making them highly reactive.[2,19] Even though specific techniques such as flash photolysis or pulsed radiolysis (and UV-Vis or EPR monitoring) have been successfully used for their detection, simple phenoxyls are rather uncommon. Chemists have thus developed strategies in order to stabilize them enough to allow their handling and characterization by conventional techniques. This has been achieved by following rules that are inherent to the specific electronic distributions in these radicals.

Phenoxyl radicals are usually viewed with an electron localized at the oxygen atom. This representation is not entirely true since the spin density is not exclusively located at this atom. As a matter of fact they are better described in terms of the canonical structures depicted in Figure 8.1. From this representation it is evident that a significant amount of spin density is located at the *ortho* and *para* positions, thus enhancing their reactivity. Stabilization thus implies substitution at these positions. Some examples of stable phenoxyl radicals that do not have any *para* substituent exist but they are very rare.[20,21]

To enhance stabilization, the substituents must be also able to compensate efficiently for electronic deficiency by resonance or inductive effects. To date, the most stable phenoxyl radicals are those substituted at their *ortho* and *para* positions by aromatic, O,N,S-alkyl and sterically hindered alkyl groups. Among them, *tert*-butyl is the most common substituent, probably because of the commercial availability of many precursors for organic synthesis. Figure 8.2 shows selected copper(II) radical species classified by decreasing stability.[22,23] The increased lifetime of (a) relative to (b) is explained by the more electron

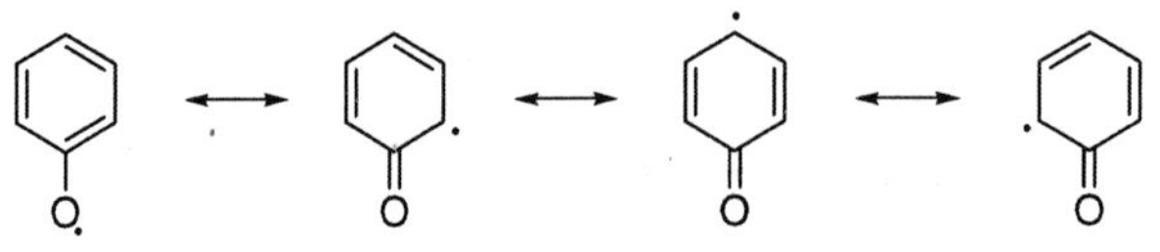

Figure 8.1 Canonical forms of phenoxyl radicals.

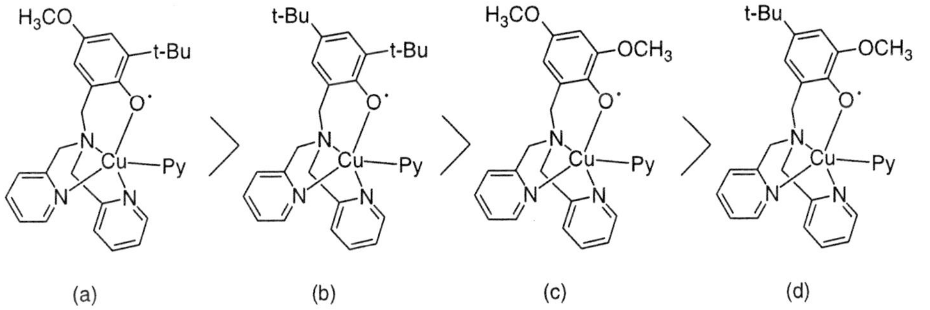

Figure 8.2 Relative stabilities of copper(II)–radical complexes (Py denotes pyridine). (See Table 8.6 for references and nomenclature).

donating properties of the methoxy compared to the *tert*-butyl group (σ^+ Hammet = −0.778 vs −0.256). The same would hold for (b) relative to (d), but the opposite is experimentally observed. This paradox is explained by the increased bulk provided by the *tert*-butyl group around the oxygen atom, thus decreasing its reactivity, and enhancing the overall stability of the molecule.

8.2.2 Electrochemistry of phenoxyl radicals

Phenoxyl radicals are usually obtained by oxidation of a phenol precursor. In the absence of a metal, the oxidation has to be performed under alkaline conditions for two reasons. The first reason is that the oxidation potential of phenols is high. For example, the oxidation potential of one of the most easily oxidizable phenols, the tri-*tert*-butylphenol, is +1.07 V vs Fc^+/Fc.[24,25] When electrochemical measurements are performed in the presence of a strong base, such as sodium hydroxide, this value falls to −0.68 V[26]; the base deprotonates the phenol (pKa ≈ 12.2 for the tri-*tert*-butylphenol in water),[27] thus affording a phenolate that is electron enriched and, consequently, easier to oxidize.

The second reason is mechanistic. Oxidation of a phenolate is a very simple process from an electrochemical point of view, as it is a simple electron transfer. In contrast, oxidation of phenols involves several elemental steps (except under certain particular conditions, as noted below).[19] The first step is electron removal, which affords a phenoxyl radical cation. Its pKa has been estimated to be −9.5 in acetonitrile for the tri-*tert*-butylphenoxyl,[24] meaning that such species could only be detected in highly acidic media (H_2SO_4 12 M). In moderate acids, the radical cation is deprotonated to give a phenoxyl radical (proton transfer coupled to electron transfer). Since the phenoxonium/phenoxyl redox potential is lower than the oxidation potential of the initial phenol, the phenoxyl is again one-electron oxidized giving rise to a phenoxonium ion. The overall mechanism is thus a two-electron oxidation, affording a phenoxonium and a released proton. A particular case often encountered in biological systems concerns phenols that are hydrogen bonded to an adjacent weak base (amine for instance).[28–32] The overall mechanism also implies a proton coupled electron transfer but it is, in these cases, a single kinetic step. As a result the oxidation potential is much lower and the oxidation product in this case is a phenoxyl radical hydrogen bonded to a neighboring ammonium proton and not a phenoxonium ion.

The electrochemistry of coordinated phenols has not been well studied, presumably because of their high oxidation potentials due to the positive charge of the metal located in the vicinity of the oxygen atom. Oxidation of coordinated phenolates is more interesting, as this takes place at potentials intermediate between those of coordinated (or not) phenols and free phenolates, and the mechanism involves a single-electron transfer process, affording phenoxyl radicals.

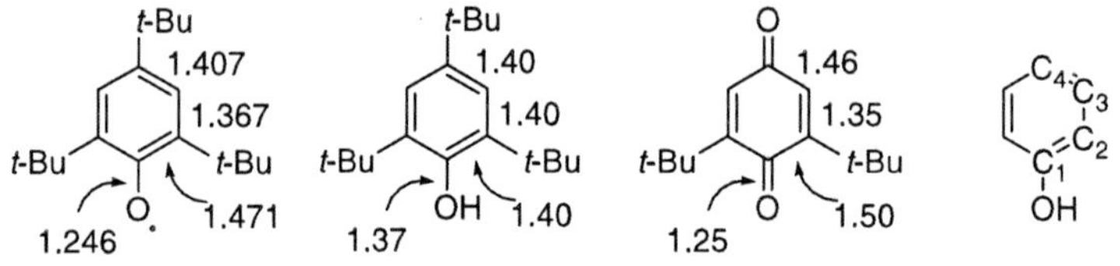

(a) (b) (c) (d)

Figure 8.3 *Bond distances in: (a) the tri-*tert*-butylphenoxyl radical; (b) the tri-*tert*-butylphenol; and (c) the di-*tert*-butylquinone. The numbering used in the text is indicated in (d). ([34] Reproduced by permission of the Royal Society of Chemistry.)*

8.2.3 Structure of non-coordinated phenoxyl radicals

Although phenoxyl radicals have been known for over six decades, very few X-ray crystal structures have been reported to date.[33,34] This is the consequence of the low stability of these molecules, combined with their high solubility in most conventional solvents. Until very recently, structural aspects were only investigated by computational methods. The first high resolution X-ray crystal structure was reported in 2008 for the well known tri-*tert*-butylphenoxyl radical. The bond lengths observed in the tri-*tert*-butylphenoxyl radical are compared to those of the tri-*tert*-butylphenol precursor in Figure 8.3. The C1–O bond length in the radical is significantly shorter than in the phenol, and is quite similar to the C1=O bond length in the 2,6-di-*tert*-butyl-1,4-benzoquinone.[35] The carbon–carbon bond lengths in the phenoxyl ring are inequivalent, with a C2–C3 bond length that is shorter than the C3–C4, itself shorter than the C1–C2 bond length. This alternance compares well with that found in asymetrical quinones such as 2,6-di-*tert*-butyl-1,4-benzoquinone and reflects the significant double bond character of the C1–O, C2–C3 and C3–C4 bonds, as expected for the canonical forms depicted in Figure 8.1.

8.2.4 UV-Vis spectroscopy

As long as 80 years ago it was reported that phenoxyl radicals exhibit an intense color, in contrast with the precursor phenols that are usually colorless.[36,37] The first visible spectra of phenoxyl radicals were reported in the 1950s. A feature common to most of these species is the intense bands in the 370–440 nm range along with a less intense transition at higher wavelengths (600–800 nm), as illustrated in Figure 8.4. The position of these bands is slightly (but significantly) dependent on the substituent's properties, as well as the protonation state of the radical. As an example, the λ_{max} for the tri-*tert*-butylphenoxyl radical at ≈400 nm is shifted to 419 nm in 12 M sulphuric acid (protonation of the oxygen).[38] As shown below, metal coordination can also shift these transitions and enhance their intensities. These bands originate from $\pi-\pi^*$ transitions, although their exact attribution has been debated, sometimes controversially, for several years.[39] In the presence of a metal ion, complication of the UV-Vis spectra is often observed. This is mainly due to the presence of additional bands corresponding to charge transfer (CT) and d–d transitions.

8.2.5 EPR spectroscopy

The free phenoxyl radicals are paramagnetic ($S = 1/2$) species that could be easily characterized by EPR and related magnetic spectroscopies. From the g-tensor and the hyperfine coupling constants the electronic distribution in the radical can be elucidated. Most of the non-coordinating phenoxyl radicals exhibit isotropic g-tensor values ranging between 2.003 and 2.007.[40] Resolution of the anisotropy can be

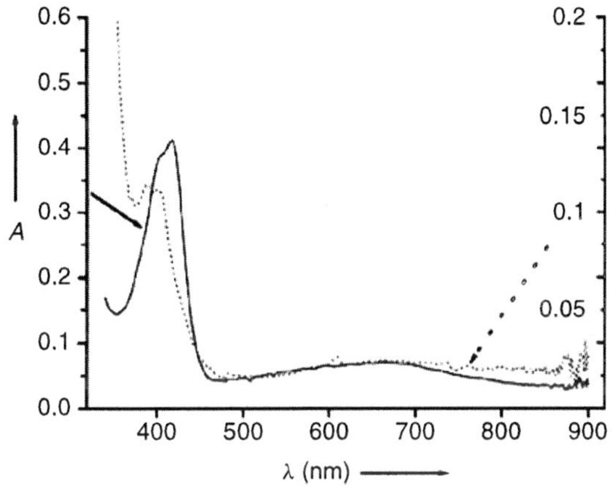

Figure 8.4 *UV-Vis spectra of: solid lines: complex (b) depicted in Figure 8.2 (0.11 mM in acetonitrile) (see Table 8.6 for references and nomenclature). Dotted lines: the radical of the free ligand (0.24 mM in dichloromethane). For this latter species the absorbance is underestimated because of its instability (unpublished results).*

achieved by increasing the frequency (higher than 95 GHz).[41,42] As an example, the three principal g-tensor components of the tri-*tert*-butylphenoxyl radical are 2.0072, 2.0043 and 2.0024 ($g_{average}$ = 2.0046). While the g_y and g_z values are usually similar in all the phenoxyls (and close to the free electron g_e value of 2.0023 for g_z), the g_x values are strongly dependent on the electrostatic environment around the oxygen atom; when it is hydrogen bonded to a positively charged ammonium group, the g_x component can be as low as 2.0064.[43,44]

Coordination to a metal ion strongly affects the EPR spectrum of phenoxyls. Diamagnetic metals lead to ($S = 1/2$) systems. For phenoxyl radicals containing diastereotopic benzyl hydrogen atoms, strong hyperfine coupling to one hydrogen atom is usually observed, together with an additional interaction with the metal nuclear spin if this latter is different from zero. In the presence of a paramagnetic metal ion, the system has to be described by a total spin state, S_t, arising from exchange and dipolar interactions between the metal and radical electron spins. That results in profound changes in the EPR spectra, as shown below.

8.3 Occurrence of tyrosyl radicals in proteins

Since the 1970s it has been recognized that stable radicals exist in enzymes and that they are involved in numbers of biological processes. Cysteinyl (RNR class I and II), glycyl (anaerobic RNR class III) and tryptophanyl (cyt c peroxydases) radicals are all known, but the most representative family is that of the tyrosyl radicals (photosystem II, Prostaglandine synthase, galactose oxidase, glyoxal oxidase etc.).[45] It is noteworthy that nearly all of these proteins are metalloenzymes, and so the organic cofactor is believed to complete or assist the metal-driven electron transfer. While most of the above mentioned enzymes involve a free or hydrogen bonded radical, galactose oxidase (GO) and glyoxal oxidase are unique in that the radical is coordinated to a metal ion.[46]

Galactose oxidase catalyses the oxidation of a wide range of primary alcohols to their corresponding aldehydes, with concomitant reduction of molecular oxygen to hydrogen peroxide. The active site of GO comprises a copper atom in a distorted square pyramidal geometry with Tyr495 occupying the axial

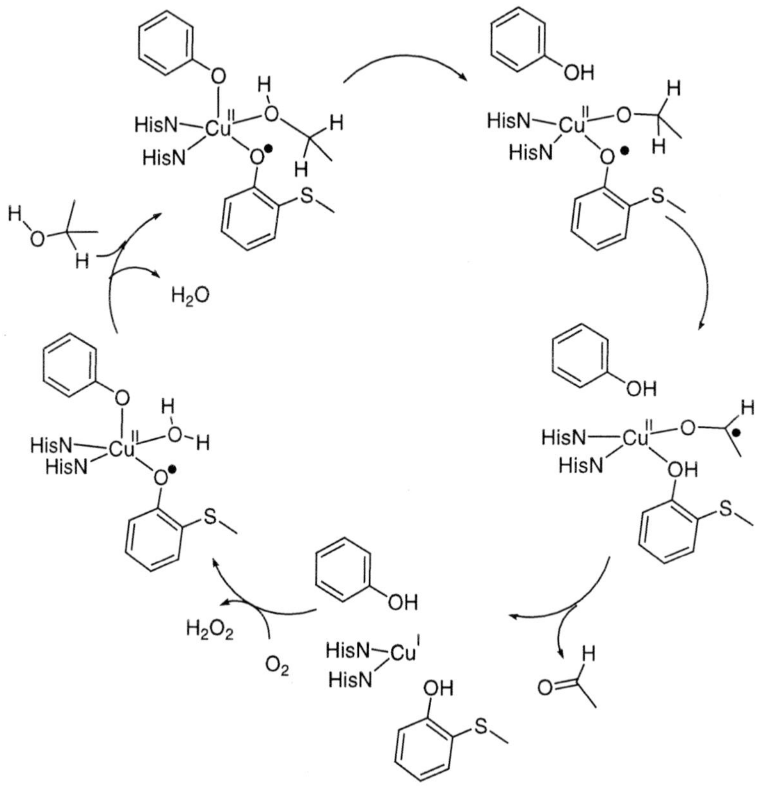

Figure 8.5 *Active site and free radical catalysis by GO. (Adapted with permission from: [53] Copyright 1993 American Chemical Society; [54] Copyright 1998 American Chemical Society; [55] Copyright 1998 Elsevier.)*

position (Figure 8.5). His581, His496, Tyr272 and an exogenous ligand (H_2O or acetate, replacing the substrate) coordinate to the copper in equatorial positions.[47,48] A striking feature of this complex is the cross-linking of Tyr272 to Cys228 through a thioether bond at the *ortho* position of the hydroxly group, presumably to lower the tyrosyl/tyrosine redox potential. The enzyme can exist in three well defined oxidation levels: the copper(II)–tyrosyl radical oxidized form, an intermediate copper(II)–tyrosinate form and the reduced copper(I)–tyrosine form, with only the former and the latter being catalytically active. The tyrosyl/tyrosinate redox couple has been estimated at +0.01 V vs Fc^+/Fc at pH 7 and extensive spectroscopic studies on the active oxidized form have shown that the radical is mainly harbored by the equatorially-bound Tyr272 residue. The enzyme is found to be relatively stable in this radical form with a $t_{1/2}$ of 7.2 days for its self-decomposition in absence of substrate.[49] The radical leads to an unusual optical spectrum for GO, with intense bands at 445 nm and 800 nm[50] mainly attributed to charge transfer and π–π^* transitions of the tyrosyl group. Strong antiferromagnetic exchange coupling between the tyrosyl radical and the paramagnetic copper(II) ion induces a diamagnetic ground state, with a singlet–triplet splitting greater than 200 cm^{-1}.[51,52]

Catalysis of galactose oxidase proceeds in a ping-pong turnover reaction (Figure 8.5).[53–55] In the first half-reaction the alcohol binds to the active site and the tyrosyl radical abstracts a hydrogen atom from the substrate (the thermodynamic affinity of a tyrosyl radical for a hydrogen atom is high).[56] The metal ion then reduces the transient ketyl radical in a single electron transfer mechanism.[57,58] Large non-classical kinetic isotope effects have been reported, as expected for the tunneling of a hydrogen atom in the rate limiting step.[50] In the second half-reaction, the aldehyde is released into the medium and the copper(I)–tyrosine site binds oxygen and reduces it very quickly to H_2O_2, thus regenerating the initial copper(II)–tyrosyl radical state.

8.4 Complexes with coordinated phenoxyl radicals

To create stable metal–phenoxyl complexes, each of the phenol(ate)s has either to be incorporated into elaborate polydentate ligands, or the metal:ligand stoichiometry has to be strictly controlled. These are important aspects to consider as each metal ion exhibits specific preferences in the terms of geometry and/or kind of coordinating atom. Solvent molecules can also be involved in coordination if the donor set of the ligand is not sufficient for a given metal ion. Most of the ligands studied belong to the four main classes briefly described in the next section. As all the ligands having at least one coordinating phenol, that is a huge number of molecules, are a priori precursors of phenoxyl radicals, the following is limited to ligands specifically designed to stabilize coordinated phenoxyl radicals, that is for which the phenols are *ortho* and *para* protected by electron donating and/or bulky groups. The nomenclature used to present the complexes is: H_xL_y for a ligand y possessing x acido-basic sites and $[M(H_xL_y)(X)_z]$ for the M complex of H_xL_y bearing z coordinated exogenous ligands X. Following the general discussion on the architecture of the ligands, the radical complexes will be presented in distinct parts, each being devoted to a specific first row d transition metal. Reported electrochemical potentials for complexes of each metal are tabulated at the beginning of each metal-specific section.

8.4.1 General ligand structures

The macrocyle 1,4,7-triazacyclononane (TACN) is a modulable scaffold with high complexing ability. TACN coordinates to a metal ion by three amine groups, and can be further derivatized to incorporate one, two or three additional coordinating groups such as phenols. This modularity makes them good chelators for metals requiring four, five or six ligands, that is metals of the first row. Some representative structures are shown in Figure 8.6.

Tripodal ligands possess a pivotal nitrogen on which one phenol and two pyridines, or two phenols and one pyridine, can be introduced (Figure 8.7). Their copper complexes are excellent structural and sometime functional models of the GO active site. The phenols have been diversely substituted, as well as the pyridine, and the length of the linker can also be modulated. The solution chemistry of their copper complexes is the most extensively studied of the four classes of ligands.

Tetradentate Schiff bases are famous catalysts for oxidation (Jacobsen's catalyst for example) that involve two salicylidene moieties joined together with a linker that modulates the geometry around the metal ion (Figure 8.8): binaphthyl linkers constrain the metal ion in a tetrahedral environment whereas *o*-phenylenediamines make the ligand almost planar and highly conjugated, thus favoring a square planar environment around the metal ion. As will be shown, the phenoxyl/phenolate redox potentials in this series are usually high, due to the conjugation of the imine with the phenolate, but the radical species are rather stable again due to conjugation. Phenol-benzimidazole are not strictly speaking Schiff bases but they will be included in this section for clarity, as they coordinate in a six-membered chelation ring with one oxygen and one sp2 nitrogen.

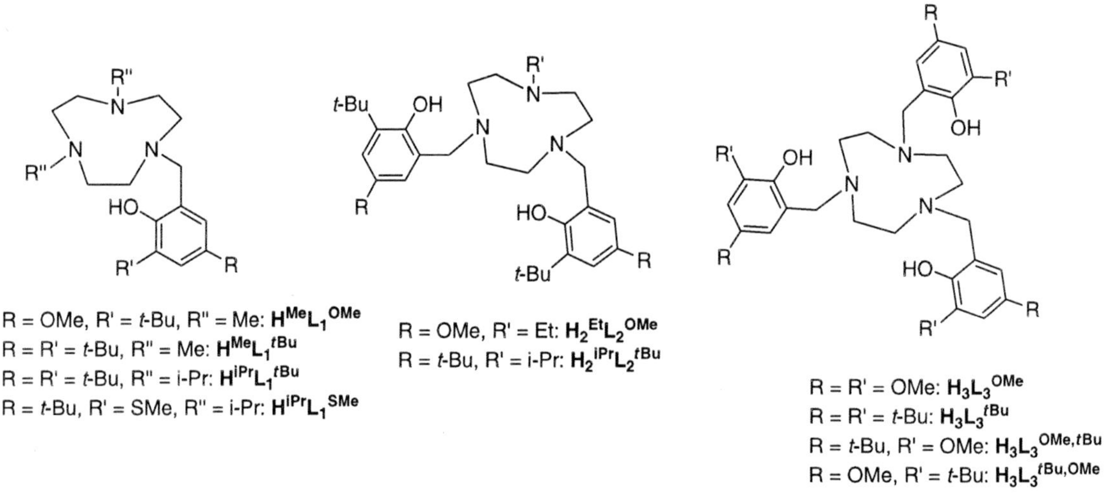

Figure 8.6 *Representative TACN ligands.*

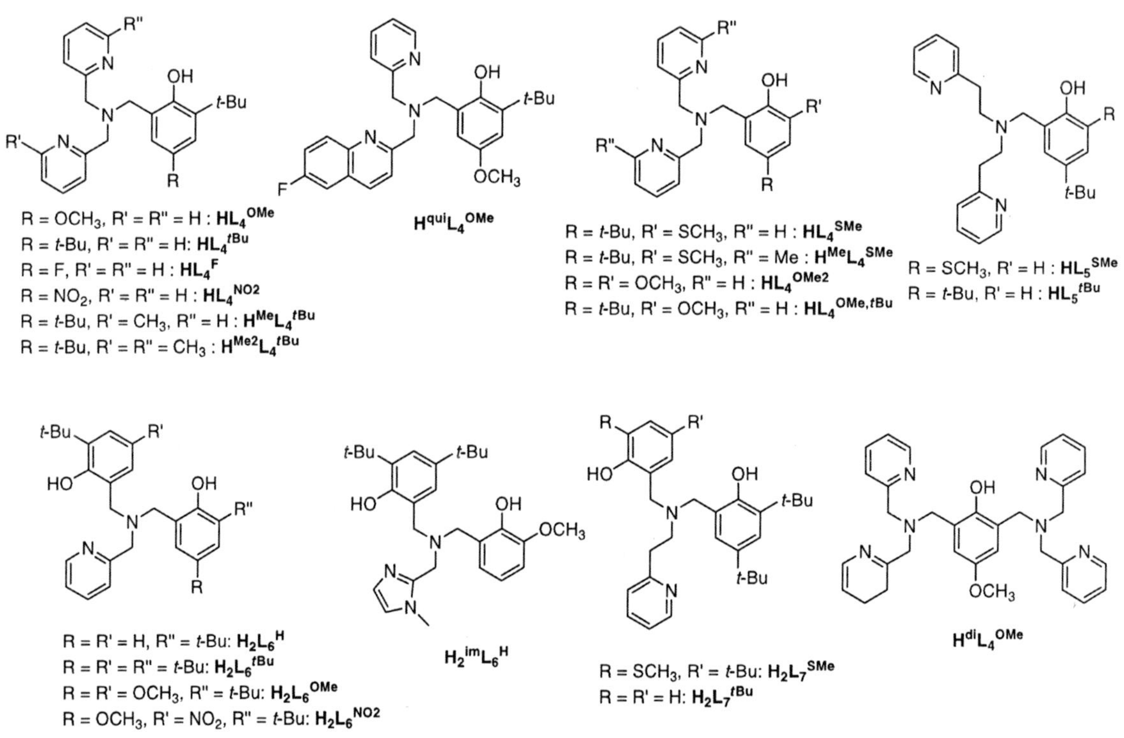

Figure 8.7 *Representative tripodal ligands.*

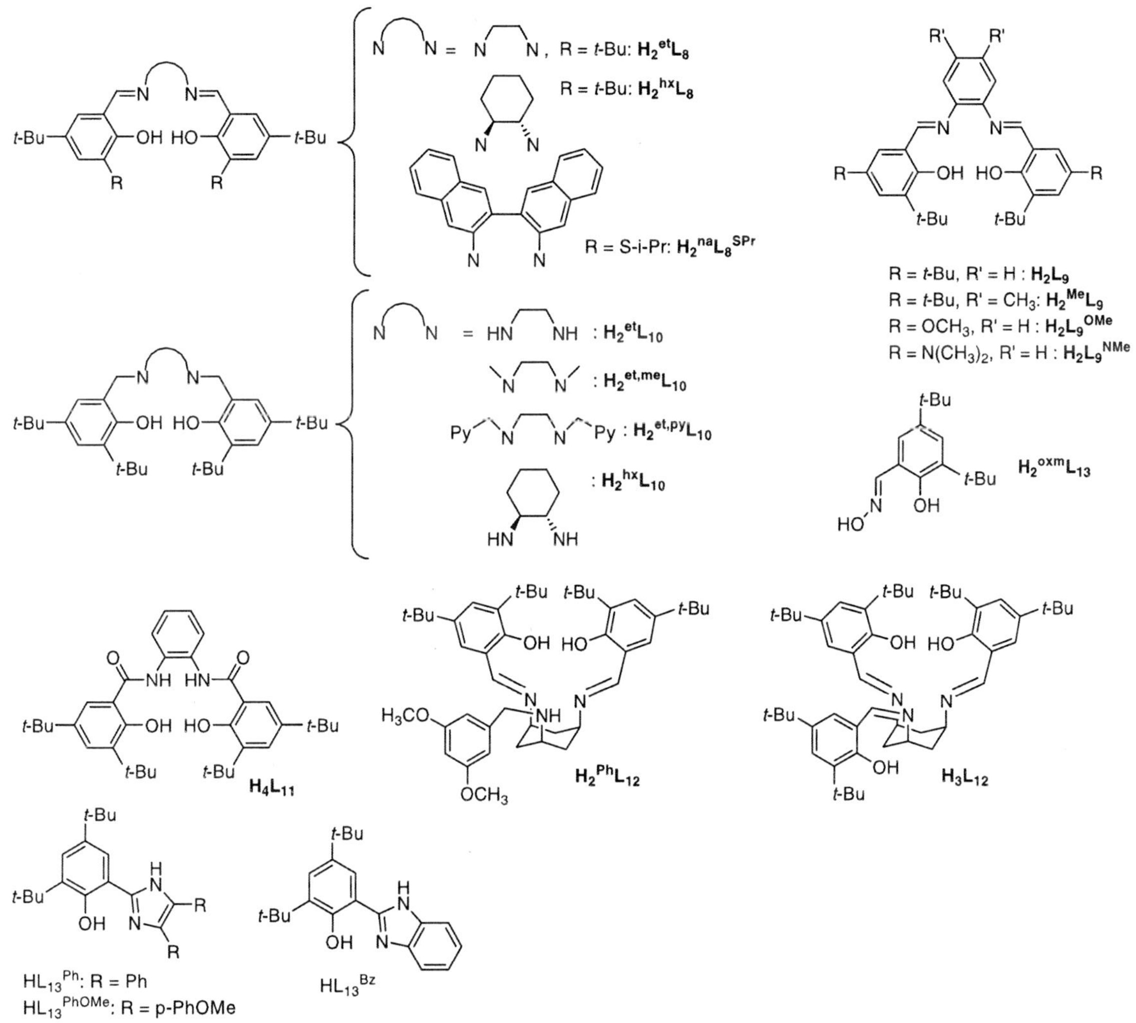

Figure 8.8 *Representative salen and reduced salen ligands.*

The last ligand family is that of sterically hindered *o*-iminobenzosemiquinones. For synthetic reasons mainly the di-*tert*-butylated form of the *o*-iminobenzosemiquinone is used. Although this bidentate elemental unit has been used to get number of metal radical complexes simply by controlling judiciously the ligand:metal stoichiometry, it has been also incorporated into tri- and tetradentate architectures in order to make efficient catalysts. Although phenoxyl and *o*-iminobenzosemiquinonate radicals exhibit closely related structures, the former are more intimately related to GO from a reactivity point of view. Nevertheless, phenoxyl radical stability is not always high, and most of the time the compounds are solely isolated on their metal phenolate form (and then oxidized into radicals species), meaning that structural data are generally missing. The *o*-iminobenzosemiquinonate radicals offer the advantage of very high stability of

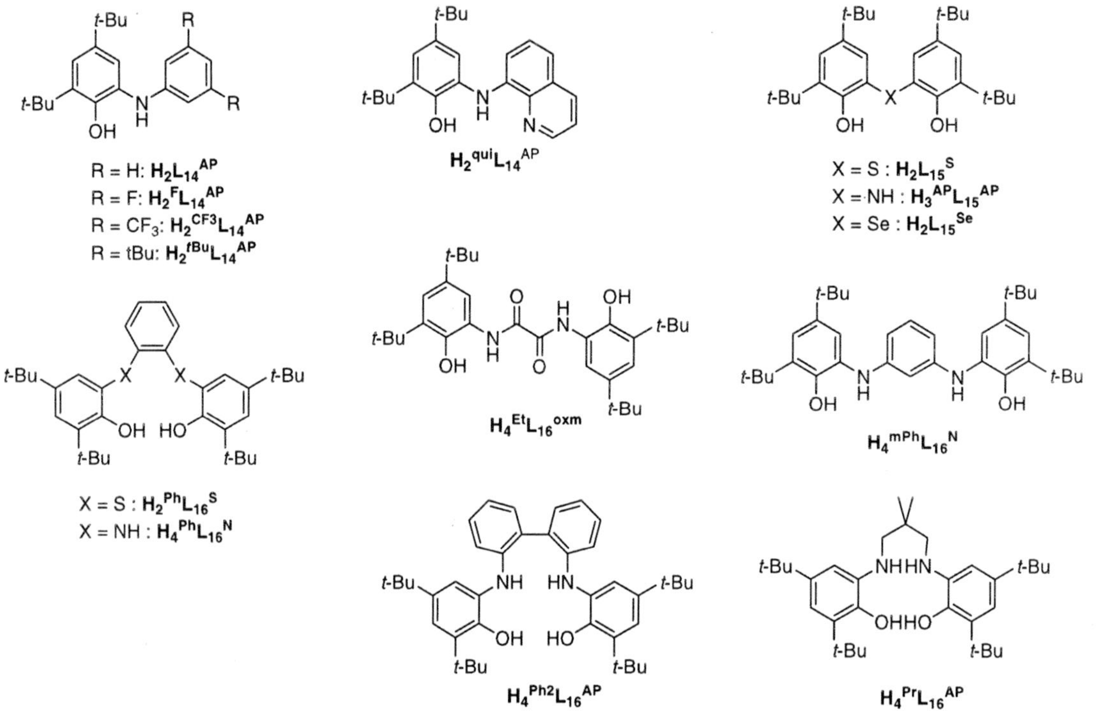

Figure 8.9 *Representative aminophenol ligands.*

the radical species. Many of them can be isolated and crystallized in their radical form thus contributing widely to their success. Some representative ligands are depicted in Figure 8.9.

8.4.2 Vanadium complexes

Among the metals presented in this section, the vanadium ion is the most subject to a metal-centered redox activity in the usual potential range. As an example, the oxidative chemistry of vanadium–phenolate complexes of the TACN ligands $H_3L_3{}^{tBu}$ or $H_3L_3{}^{tBu,OMe}$ involves exclusively the V^{III}/V^{II}, V^{IV}/V^{III}, V^{V}/V^{IV} redox couples.[59,60] The electrochemical behavior of complexes of ligands involving aminophenolate moieties (i.e. moieties that could be oxidized much more easily than other phenolates into radicals), such as $H_2L_{15}{}^{S}$ and $H_2L_{15}{}^{Se}$, also suggests metal-centered oxidations to V^V or $V^{IV} = O$.[61] $H_2L_{14}{}^{AP}$ is an exception, since the radical species $[V^V(L_{14}{}^{AP})_2(L_{14}{}^{ISQ})^{\bullet}]$ has been evidenced by X-ray diffraction.[62] Although the classical phenoxyl radical band at ≈400 nm (see above) is absent in the visible spectrum of this species, the EPR spectrum is classical for an $S = 1/2$ system with a g-value of 2.0045 (the V^V ion is diamagnetic). The absence of hyperfine splitting with the ^{51}V ion further precludes the formulation as a paramagnetic d^1 V^{IV} ion. Oxidation at −0.10 V affords the diradical $[V^V(L_{14}{}^{AP})(L_{14}{}^{ISQ})^{\bullet\bullet}{}_2]^+$, whereas at lower potentials reduction of the *o*-iminobenzosemiquinone radical followed by reduction of the metal ion is observed according to Equation 8.1:

$$[V^{III}(L_{14}{}^{AP})_3]^{3-} \underset{+e^-}{\overset{-e^-}{\rightleftharpoons}} [V^{IV}(L_{14}{}^{AP})_3]^{2-} \underset{+e^-}{\overset{-e^-}{\rightleftharpoons}} [V^{V}(L_{14}{}^{AP})_3]^{-} \underset{+e^-}{\overset{-e^-}{\rightleftharpoons}} [V^{V}(L_{14}{}^{AP})_2(L_{14}{}^{ISQ})^{\bullet}] \underset{+e^-}{\overset{-e^-}{\rightleftharpoons}} [V^{V}(L_{14}{}^{AP})(L^{ISQ})^{\bullet\bullet}{}_2]^{+} \quad (8.1)$$

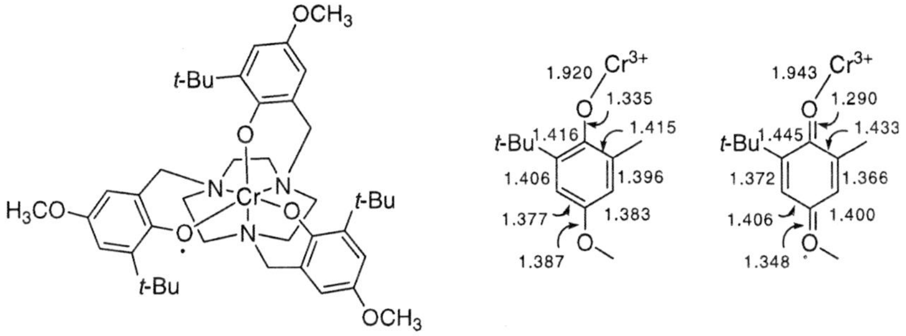

Figure 8.10 *Structure and selected bond distance in the first metal–phenoxyl radical complex crystallized ($[Cr^{III}(L_3{}^{tBuOMe})^\bullet]^+$). ([63] Reproduced by permission of the Royal Society of Chemistry.)*

8.4.3 Chromium complexes

The chromium(III) ion is redox innocent in the classical potential range. It is paramagnetic ($S_{Cr} = {}^3\!/_2$) and favors an octahedral geometry. Historically, the chromium complex $[Cr^{III}(L_3{}^{tBu,OMe})^\bullet]^+$ was the first metal–radical compound to have its X-ray crystal structure solved (1996).[63] Its structure (Figure 8.10) shows a quinoid distribution of bond lengths in the phenoxyl ring, as observed in the structure of the tri-*tert*-butylphenoxyl radical. As expected the metal–oxygen bond in the phenoxyl ring is weakened (longer) compared to the free phenolate. It is noteworthy that the presence of inequivalent metal–oxygen bonds suggests that the electronic hole is located on a single ring. Cyclic voltammetry exhibits one cathodic wave (reduction of the phenoxyl) at −0.03 V vs Fc^+/Fc, as well as two anodic waves showing that the di- and triphenoxyl radical species are also accessible (Equation 8.2):

$$[Cr^{III}(L_3{}^{tBu,OMe})] \rightleftharpoons [Cr^{III}(L_3{}^{tBu,OMe})^\bullet]^+ \rightleftharpoons [Cr^{III}(L_3{}^{tBu,OMe})^{\bullet\bullet}]^{2+} \rightleftharpoons [Cr^{III}(L_3{}^{tBu,OMe})^{\bullet\bullet\bullet}]^{3+} \quad (8.2)$$

The ground state of $[Cr^{III}(L_3{}^{tBu,OMe})^\bullet]^+$ is ($S_t = 1$) (but it is X-Band EPR silent), whereas the diradical $[Cr^{III}(L_3{}^{tBu,OMe})^{\bullet\bullet}]^{2+}$ exhibits an ($S_t = {}^1\!/_2$) signal at g = 1.97, and the triradical $[Cr^{III}(L_3{}^{tBu,OMe})^{\bullet\bullet\bullet}]^{3+}$ is diamagnetic. This indicates that the coupling between the chromium(III) ion ($S_{Cr} = {}^3\!/_2$) and each ($S_{rad} = {}^1\!/_2$) radical is antiferromagnetic. Interestingly, the UV-Vis spectra of chromium(III)–phenoxyl radicals are, in general, rich and particular, due to an admixture of the SOMO of the ligand radical and the d orbitals of the metal ion. As a consequence, the expected phenoxyl bands at ≈400 nm could be distinguished only for the mono- and dications $[Cr^{III}(L_3{}^{tBu,OMe})^\bullet]^+$ and $[Cr^{III}(L_3{}^{tBu,OMe})^{\bullet\bullet}]^{2+}$. The $[Cr^{III}({}^{Me}L_1{}^{OMe})(Bu_2acac)]^+$ complex formed with a ligand involving a single phenolate (Table 8.1)[64] is comparatively harder to oxidize into a monoradical than $[Cr^{III}(L_3{}^{tBu,OMe})]$, likely due to its positive charge. The coupling between the metal and the radical in $[Cr^{III}({}^{Me}L_1{}^{OMe})^\bullet(Bu_2acac)]^{2+}$ is antiferromagnetic as in $[Cr^{III}(L_3{}^{tBu,OMe})^\bullet]^+$. Q-Band

Table 8.1 *Redox potentials of representative chromium complexes*

Complex	$E_{1/2}$ [V vs Fc^+/Fc]	Ref.
$[Cr^{III}({}^{Me}L_1{}^{OMe})(Bu_2acac)]^+$	0.31	[64]
$[Cr^{III}({}^{iPr}L_1{}^{tBu})(acac)]^+$	0.51	[64]
$[Cr^{III}(L_3{}^{tBu})]$	0.09, 0.55, 0.77	[63]
$[Cr^{III}(L_3{}^{tBu,OMe})]$	−0.03, 0.24, 0.45	[63]
$[Cr^{III}(L_{14}{}^{ISQ})^{\bullet\bullet\bullet}{}_3]$	−1.77, −1.26, −0.10, 0.64, 1.10	[62]

EPR data are available for the former, showing that the ($S_t = 1$) system is in the limit of strong coupling ($|D| = 0.52\,cm^{-1}$, $E/D = 0.05$).

As pointed out in the introduction, amino substituents are known to greatly stabilize phenoxyl radicals. For instance the *triradical* $[Cr^{III}(L_{14}{}^{ISQ})^{\bullet\bullet\bullet}{}_3]$ could be prepared from $H_2L_{14}{}^{AP}$ and chromium without specific experimental conditions. This is due to the very low redox potentials of this class of ligands. The first iminobenzosemiquinone/aminophenolate redox couple is, for example, too low to be measured in dichloromethane, whereas the di and tri-radicals are successively formed at −1.77 and −1.26 V (remember that the first phenoxyl/phenolate redox wave of $[Cr^{III}(L_3{}^{tBu,OMe})]$ is observed at −0.03 V).[62] The redox activity of $[Cr^{III}(L_{14}{}^{ISQ})^{\bullet\bullet\bullet}{}_3]$ is summarized below (Equation 8.3):

$$[Cr^{III}(L_{14}{}^{AP})_2(L_{14}{}^{ISQ})^{\bullet}]^{2-} \underset{+e^-}{\overset{-e^-}{\rightleftharpoons}} [Cr^{III}(L_{14}{}^{AP})(L_{14}{}^{ISQ})^{\bullet\bullet}{}_2]^{-} \underset{+e^-}{\overset{-e^-}{\rightleftharpoons}} [Cr^{III}(L_{14}{}^{ISQ})^{\bullet\bullet\bullet}{}_3]$$

$$\underset{+e^-}{\overset{-e^-}{\rightleftharpoons}} [Cr^{III}(L_{14}{}^{ISQ})^{\bullet\bullet}{}_2(L_{14}{}^{BQ})]^{+} \underset{+e^-}{\overset{-e^-}{\rightleftharpoons}} [Cr^{III}(L_{14}{}^{ISQ})^{\bullet}(L_{14}{}^{BQ})_2]^{2+} \underset{+e^-}{\overset{-e^-}{\rightleftharpoons}} [Cr^{III}(L_{14}{}^{BQ})_3]^{3+} \quad (8.3)$$

The magnetic orbitals of the radical ligands in $[Cr^{III}(L_{14}{}^{ISQ})^{\bullet\bullet\bullet}{}_3]$ adopt an arrangement relatively similar to those found in $[Cr^{III}(L_3{}^{tBu,OMe})^{\bullet\bullet\bullet}]^{3+}$. As a consequence, antiferromagnetic coupling between the radicals and the metal ion also exists in $[Cr^{III}(L_{14}{}^{ISQ})^{\bullet\bullet\bullet}{}_3]$, which is consequently diamagnetic ($J = -436\,cm^{-1}$). The X-ray crystal structure of $[Cr^{III}(L_{14}{}^{ISQ})^{\bullet\bullet\bullet}{}_3]$, as well as those of many other complexes involving *o*-iminobenzosemiquinonate ligands, has been solved. A quinoid distribution of bond lengths is observed in the *o*-iminobenzosemiquinonate ring (as for the tri-*tert*-butylphenoxyl) with shortening of the carbon–oxygen bond, thus emphasizing its double bond character, and elongation of the adjacent carbon–carbon bonds. Interestingly no substantial change in these bond lengths is observed upon substitution of the chromium by other metal ions. The mean bond distances given in Figure 8.11 are thus valid for most of the complexes described below.

8.4.4 Manganese complexes

The manganese ion normally adopts an octahedral geometry, as chromium, but in contrast it is redox active, and can exist at the (+II, +III and +IV) redox states in the usual potential range. The study of manganese complexes is, therefore, much more complicated than that of the chromium analogues.

The complex $[Mn^{III}({}^{Me}L_1{}^{OMe})(Bu_2acac)]^+$ has been prepared from the monophenolate TACN ligand $H^{Me}L_1{}^{OMe}$.[64] It exhibits both a metal-centered reduction wave at −0.61 V and a ligand-centered oxidation one at +0.42 V (Table 8.2) according to Equation 8.4:

$$[Mn^{II}({}^{Me}L_1{}^{OMe})(Bu_2acac)] \underset{+e^-}{\overset{-e^-}{\rightleftharpoons}} [Mn^{III}({}^{Me}L_1{}^{OMe})(Bu_2acac)]^{+} \underset{+e^-}{\overset{-e^-}{\rightleftharpoons}} [Mn^{III}({}^{Me}L_1{}^{OMe})^{\bullet}(Bu_2acac)]^{2+} \quad (8.4)$$

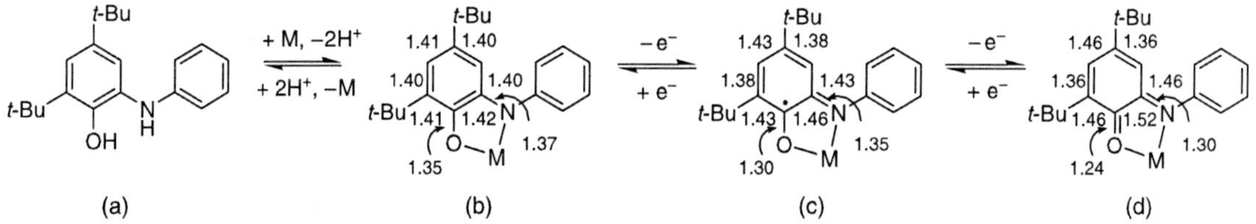

Figure 8.11 Schematic representation and metrical parameters for the coordinated o-iminobenzosemiquinone unit under different redox states: (a) aminophenol ($H_2L_{14}{}^{AP}$); (b) deprotonated aminophenolate ($L_{14}{}^{AP}$) complex; (c) o-iminobenzosemiquinonate ($L_{14}{}^{ISQ}$) complex; and (d)o-iminobenzoquinone ($L_{14}{}^{BQ}$) complex. ([67] Reproduced by permission of the Royal Society of Chemistry.)

Table 8.2 *Redox potentials of representative manganese complexes*

Complex	$E_{1/2}$ [V vs Fc^+/Fc]	Ref.
$[Mn^{III}(^{Me}L_1{}^{OMe})(Bu_2acac)]^+$	−0.61,* 0.42, 0.89(irr)	[64]
$[Mn^{IV}(L_3{}^{tBu,OMe})]^+$	−0.92,* −0.55,* 0.56, 0.74 (irr), 0.89 (irr)	[17]
$[Mn^{III}(^{et,py}L_{10})]^+$	−0.65, 0.36, 1.15	[65]
$[Mn^{IV}(^{qui}L_{14}{}^{AP})_2]$	−1.10, −0.33, 0.09	[67]
$[Mn^{IV}(^{mPh}L_{16}{}^{ISQ})^{\bullet\bullet}(^{mPh}L_{16}{}^{N,ISQ})^{\bullet\bullet}{}_2]$	−1.23, −1.02 (red), −0.34, −0.12, 0.10 (irr)	[70]
$[Mn^{III}(L_{15}{}^{Se})_2]^-$	−0.05#	[68]

*: metal-centered process; irr: irreversible couple, $E_p{}^a$ is given #: E° value attributed to the phenoxyl/phenolate redox wave

The manganese(III)–radical – rather than manganese(IV)–phenolate – character of the dication is evident from UV-Vis spectroscopy (bands at 428 and 1015 nm) and resonance Raman (rR) measurements. Upon excitation in resonance with the $\pi-\pi^*$ transition of the phenoxyl (380–430 nm), the rR band originating from the mode ν_{8a} (C=C stretching) is observed at the expected value (1619 cm^{-1}) for phenoxyl radicals. The ground state is $S_t = 3/2$ as a result of a strong antiferromagnetic coupling between the radical and the octahedral paramagnetic metal ion ($S_{Mn} = 2$). The zero field splitting parameters are E/D $= 0.33\,cm^{-1}$, $D_t = 5\,cm^{-1}$, with $g_x = g_y = 2.05$ and $g_z = 1.85$. It is instructive to compare the properties of $[Mn^{III}(^{Me}L_1{}^{OMe})(Bu_2acac)]^+$ with those of $[Mn^{III}(L_3{}^{tBu,OMe})]$. In both cases the first electron removal is reversible; however, while the former complex affords a manganese(III)–phenoxyl radical species, oxidation of the latter affords a manganese(IV)–phenolate species. An increasing number of phenolate oxygens in the metal coordination sphere thus favors metal-centered rather than ligand-centered redox process, as also observed for $[Mn^{III}(^{et,py}L_{10})]$.[65] The additional redox wave observed at +0.56 V for $[Mn^{III}(L_3{}^{tBu,OMe})]$ is attributed to the formation of the Mn(IV)–phenoxyl species $[Mn^{IV}(L_3{}^{tBu,OMe})^{\bullet}]^{2+}$ according to Equation 8.5:

$$[Mn^{II}(L_3{}^{tBu,OMe})]^- \underset{+e^-}{\overset{-e^-}{\rightleftharpoons}} [Mn^{III}(L_3{}^{tBu,OMe})] \underset{+e^-}{\overset{-e^-}{\rightleftharpoons}} [Mn^{IV}(L_3{}^{tBu,OMe})]^+ \underset{+e^-}{\overset{-e^-}{\rightleftharpoons}} [Mn^{IV}(L_3{}^{tBu,OMe})^{\bullet}]^{2+} \quad (8.5)$$

From the aminophenolate ligand $H_2L_{14}{}^{AP}$, two diradical complexes, namely $[Mn^{III}(L_{14}{}^{ISQ})^{\bullet\bullet}{}_2(L_{14}{}^{AP})]$ and $[Mn^{IV}(L_{14}{}^{ISQ})^{\bullet\bullet}{}_2(HL_{14}{}^{AP})]$ can be isolated.[66] Interestingly, the redox state of the metal center is evident in the structure, not only from the shortening of the mean Mn–N/O bond lengths (2.017 and 1.934 Å in the manganese(III) and manganese(IV) complexes, respectively) but also from the deformation of the coordination polyhedron, because high spin d^4 ions (such manganese(III)) are known to exhibit a strong Jahn–Teller distortion, whereas the d^3 (manganese(IV)) configuration does not exhibit such distortion. The coupling between the radical spins and the manganese ion is, as expected, antiferromagnetic, due to

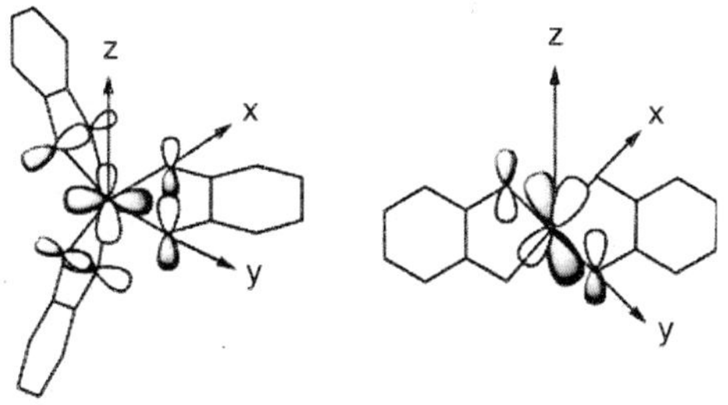

Figure 8.12 *Overlap between the π and selected d orbitals in octahedral and square planar complexes.*

an optimal orbital overlap in these octahedral complexes (Figure 8.12). The ground state is consequently ($S_t = 1$) for $[Mn^{III}(L_{14}{}^{ISQ})^{\bullet\bullet}{}_2(L_{14}{}^{AP})]$ ($J = -300\,cm^{-1}$ and $D_t = 3.4\,cm^{-1}$), whereas it is ($S_t = 1/2$) with a stronger J of $-470\,cm^{-1}$ in the case of $[Mn^{IV}(L_{14}{}^{ISQ})^{\bullet\bullet}{}_2(HL_{14}{}^{AP})]$.

Other octahedral manganese complexes have been obtained from the two tridentate ligands $H_2{}^{qui}L_{14}$ and $H_2L_{15}{}^{Se/S}$ in the presence of one metal,[67,68] or three tetradentate ligands $H_4{}^{mPh}L_{16}{}^{N}$ in the presence of two equivalents of manganese.[69] One-electron oxidation of $[Mn^{III}(L_{15}{}^{Se})_2]^-$ affords the corresponding *o*-iminobenzosemiquinonate radical, but only as a transient species (its existence has been evidenced indirectly by cyclic voltammetry measurements) in the course of formation of the thermodynamically stable manganese(IV) complex $[Mn^{IV}(L_{15}{}^{Se})_2]$. In contrast, a manganese–radical complex could be isolated from $H_2{}^{qui}L_{14}{}^{AP}$, in particular as crystals. The structures of $[Mn^{IV}({}^{qui}L_{14}{}^{ISQ})^{\bullet}({}^{qui}L_{14}{}^{AP})]^+$ and $[Mn^{IV}({}^{qui}L_{14}{}^{AP})_2]$ are very interesting, as no significant change in the bond distances is observed in spite of changes in oxidation state of one subunit. This shows that the unpaired π-radical electron is *delocalized* over both ligands, in contrast with $[Cr^{III}(L_3{}^{tBuOMe})^{\bullet}]^+$ for which the unpaired electron is *localized* on a single ring.

$[Mn^{IV}({}^{mPh}L_{16}{}^{ISQ})^{\bullet\bullet}({}^{mPh}L_{16}{}^{N,ISQ})^{\bullet\bullet}{}_2]$ is a dimetallic complex in which each manganese ion is coordinated to two *o*-iminobenzosemiquinonate radical moieties. Its electronic structure is complicated and can be summarized by strong antiferromagnetic interactions between the radicals and each metal, which in turn coupled weakly antiferromagnetically (intermetallic distance of 6.7 Å).[70] It is noticeable that $[Mn^{IV}({}^{mPh}L_{16}{}^{ISQ})^{\bullet\bullet}({}^{mPh}L_{16}{}^{N,ISQ})^{\bullet\bullet}{}_2]$, as $[Mn^{IV}(L_{14}{}^{ISQ})^{\bullet\bullet}{}_2(L_{14}{}^{AP})]$, exhibits a significant catalytic activity for the aerobic oxidation of catechols into quinones, with 500 turnovers achieved within 24 hours for the former. $[Mn^{IV}({}^{mPh}L_{16}{}^{ISQ})^{\bullet\bullet}({}^{mPh}L_{16}{}^{N,ISQ})^{\bullet\bullet}{}_2]$ also catalyses the aerobic oxidation of hindered phenols to diphenoquinones in a mechanism that involves an exclusively ligand-based redox chemistry.

8.4.5 Iron complexes

In the usual potential range, iron can exist both at the +II and +III redox state (and even redox states as high as +VI have been described).[71] Hence, in iron phenolate complexes, the redox processes could be either metal-centered or ligand-centered according to Equation 8.6:

$$Fe^{II}(\text{phenolate}) \rightleftharpoons Fe^{III}(\text{phenolate}) \text{ or } Fe^{II}(\text{phenolate}) \rightleftharpoons Fe^{II}(\text{phenoxyl}) \quad (8.6)$$

In fact, coordination of the phenolate moieties to the iron center lowers the iron(III)/iron(II) redox potential, and thus stabilizes the (+III) redox state. Consequently, the iron(II) ion is oxidized prior to the phenolate, and all the phenoxyl radical complexes of iron involve a metal at the +III redox state at least.

Depending on the ligand properties, different spin states can be attained for the iron(III) ion: $S_{Fe} = 1/2$, $3/2$ or $5/2$. Furthermore, because the radical and the metal ion spins are magnetically interacting, the number of possible ground states is even higher. This versatility, combined with the fact that only the total (and not local) spin state is accessible by techniques such as EPR or magnetic measurements, creates challenges in obtaining the correct description of the electronic structure of these complexes. As it turns out, determination of the local spin state is much easier than expected, since iron exhibits a Mössbauer effect. The local state of the iron can, therefore, be easily and specifically probed with this technique, thus greatly facilitating the chemist's task.

Historically, the first coordinated phenoxyl radical described was generated from an iron complex of a tris-phenolate TACN ligand in 1993.[72] The expected spin state for a iron(III) ion in the N_3O_3 octahedral environment provided by the tris-phenolate TACN ligand is ($S_{Fe} = 5/2$). This can be confirmed by the X-ray crystal structures of some complexes like $[Fe^{III}(L_3{}^{tBu,OMe})]$, which show the typical Fe–O and Fe–N bond lengths of 1.92 and 2.22 Å, respectively. In addition, EPR reveals a typical signal at $g = 4.3$ and isomer shifts of 0.49–0.54 mm s^{-1} are observed in the Mössbauer spectra. $[Fe^{III}(L_3^{tBu,OMe})]$, as well as its derivatives, exhibits ligand-centered electrochemical activity, as summarized in Equation 8.7 (L denotes any ligand of this family). The mono- and diradicals are usually stable enough to be characterized, whereas the triradicals are found to be poorly stable in solution.[17,73]

$$[Fe^{III}(L)] \underset{+e^-}{\overset{-e^-}{\rightleftharpoons}} [Fe^{III}(L)^\bullet]^+ \underset{+e^-}{\overset{-e^-}{\rightleftharpoons}} [Fe^{III}(L)^{\bullet\bullet}]^{2+} \underset{+e^-}{\overset{-e^-}{\rightleftharpoons}} [Fe^{III}(L)^{\bullet\bullet\bullet}]^{3+} \quad (8.7)$$

All of the mono- and diradical complexes display the typical π–π* phenoxyl band at ≈400 nm and an additional phenolate-to-iron charge transfer transition at 500–600 nm. As with the chromium complexes, a strong antiferromagnetic coupling is observed between the metal and the radical spins ($S_{rad} = 1/2$). The ground state is consequently ($S_t = 2$) for $[Fe^{III}(L_3^{tBu,OMe})^\bullet]^+$ ($J = -80\,cm^{-1}$, $E/D_t = 0.28$ and $D_t = +0.46\,cm^{-1}$) and ($S_t = 3/2$) for $[Fe^{III}(L_3^{tBu,OMe})^{\bullet\bullet}]^{2+}$ with $E/D_t = 0.07$ and $D_t = +0.51\,cm^{-1}$.

Iron radical complexes are rare in the Schiff base series.[74] Nevertheless, the tris(phenolate) complex $[Fe(L_{12})]$ has been described. This complex exhibits two reversible oxidation waves at +0.20 and +0.55 V (Table 8.3), corresponding to the successive oxidation of coordinating phenolates into phenoxyl radicals (the trication is unstable). Oxidation of $[Fe(L_{12})]$ to $[Fe(L_{12})^\bullet]^+$ and $[Fe(L_{12})^{\bullet\bullet}]^{2+}$ is accompanied by the appearance of the π–π* band of the phenoxyl radicals at 405 nm and a loss of intensity of the phenolate to iron charge transfer transition as above.

Several X-ray crystal structures of iron complexes (with $S_{Fe} = 5/2$, $3/2$ and $1/2$) involving *o*-iminobenzosemiquinones are available. While they all exhibit similar C–C, C–O and C–N bond lengths (Figure 8.11), the metal–oxygen and metal–nitrogen bond lengths were found to be significantly affected by the spin state of the iron. As an example, the metal–oxygen and metal–nitrogen bond distances in $[Fe^{III}(L_{14}^{ISQ})^{\bullet\bullet\bullet}{}_3]$ (high spin metal ion) are 2.014 and 2.057–2.149 Å, respectively,[62] whereas they are only 1.892(2) and 1.920(3) Å in $[Fe^{III}(L_{14}^{ISQ})^\bullet(cyclam)]^{2+}$, in which the increased number of coordinating nitrogen atoms favors a low spin state (doublet at $\delta = 0.33$ mm s^{-1} in the Mössbauer spectra).

The $[Fe^{III}(L_{14}^{ISQ})^{\bullet\bullet\bullet}{}_3]$ complex exhibits redox potentials much lower than those of tris-phenolate TACN complex $[Fe^{III}(L_3^{tBu,OMe})]$, thus attesting that the triradical species formed from aminophenolates are more stable (thermodynamically). Nevertheless, despite that its cyclic voltammetry curve exhibits four reversible

Table 8.3 *Redox potentials of representative iron complexes*

Complex	$E_{1/2}$ [V vs Fc^+/Fc]	Ref.
$[Fe^{III}(L_3^{tBu})]$	−1.78,* 0.38, 0.65, 0.96(irr)	[73]
$[Fe^{III}(L_3^{tBu,OMe})]$	−1.81,* 0.14, 0.38, 0.64	[73]
$[Fe^{III}(L_3^{OMe,tBu})]$	−1.79,* 0.30, 0.50, 0.71	[73]
$[Fe^{III}(L_3^{OMe})]$	−1.77,* 0.11, 0.28, 0.46	[73]
$[Fe^{III}(L_{12})]$	−1.88,* 0.20, 0.55, 1.07(irr)	[74]
$[(tmtacn)Fe^{III}(^{oxm}L_{13}^{tBu})_3Fe^{III}]$	0.31, 0.47, 0.89	[80]
$[Fe^{III}(^{et,py}L^{10})]^+$	−1.2,* 0.73, 1.07	[65]
$[Fe^{III}(L_{14}^{ISQ})^\bullet(cyclam]^{2+}$	−0.42, 0.80	[75]
$[Fe^{III}(L_{14}^{ISQ})^\bullet(tren]^{2+}$	−1.33 (irr), −0.52, 0.67	[76]
$[Fe^{III}(L_{14}^{ISQ})^{\bullet\bullet\bullet}{}_3]$	−1.51 (irr), −1.31, −1.12, −0.35, 0.27	[62]
$[Fe^{III}(^{tBu}L_{14}^{ISQ})^{\bullet\bullet\bullet}{}_3]$	−1.29, −0.42, 0.29	[79]
$[Fe^{III}(^{F}L_{14}^{ISQ})^{\bullet\bullet\bullet}{}_3]$	−1.57,* −1.27, −0.87, −0.10, 0.53	[79]
$[Fe^{III}{}_2(^{Et}L_{16}^{oxm})Cl_4]2-$	0.16, 0.45	[81]

*: metal-centered process; irr: irreversible couple, E_p^a is given.

redox waves attributed as depicted below in Equation 8.8, the fully oxidized and reduced forms are rather unstable due to the lability of the ligands.

$$[\mathrm{Fe^{III}(L_{14}{}^{AP})(L_{14}{}^{ISQ})^{\bullet\bullet}{}_2}]^- \underset{+e^-}{\overset{-e^-}{\rightleftharpoons}} [\mathrm{Fe^{III}(L_{14}{}^{ISQ})^{\bullet\bullet\bullet}{}_3}] \underset{+e^-}{\overset{-e^-}{\rightleftharpoons}} [\mathrm{Fe^{III}(L_{14}{}^{ISQ})^{\bullet\bullet}{}_2(L_{14}{}^{BQ})}]^+ \underset{+e^-}{\overset{-e^-}{\rightleftharpoons}} [\mathrm{Fe^{III}(L_{14}{}^{ISQ})^{\bullet}(L_{14}{}^{BQ})_2}]^{2+} \quad (8.8)$$

The three *o*-iminobenzosemiquinonate ligands of $[Fe^{III}(L_{14}{}^{ISQ})^{\bullet\bullet\bullet}{}_3]$ couple antiferromagnetically ($J = -120\,cm^{-1}$) to the high spin metal center as in $[Cr^{III}(L_{14}{}^{ISQ})^{\bullet\bullet\bullet}{}_3]$ due to geometrical analogies.[62] $[Fe^{III}(L_{14}{}^{ISQ})^{\bullet\bullet\bullet}{}_3]$ also exhibits the typical bands for the *o*-iminobenzosemiquinonate radical at 423 nm (6000) and 746 nm ($8400\,M^{-1}\,cm^{-1}$). $[Fe^{III}(L_{14}{}^{ISQ})^{\bullet}(cyclam)]^{2+}$[75] as well as $[Fe^{III}(L_{14}{}^{ISQ})^{\bullet}(tren)]^{2+}$[76] are diamagnetic due to the antiferromagnetic coupling between the ($S_{Fe} = 1/2$) metal spin and the ($S_{rad} = 1/2$) radical spin. These complexes exhibit the typical bands for the *o*-iminobenzosemiquinonate radical at ≈420 and 650 nm with somewhat weaker intensities compared to $[Fe^{III}(L_{14}{}^{ISQ})^{\bullet\bullet\bullet}{}_3]$ because of the decreased number of *o*-iminobenzosemiquinonate subunits. It can be noticed that the *o*-iminobenzosemiquinonate moiety of $[Fe^{III}(L_{14}{}^{ISQ})^{\bullet}(cyclam)]^{2+}$ and $[Fe^{III}(L_{14}{}^{ISQ})^{\bullet}(tren)]^{2+}$ could be either oxidized into a benzoquinone or reduced into an aminophenolate affording species with closed shell ligands that are consequently paramagnetic ($S_t = S_{Fe} = 1/2$).

In the diradical complex $[Fe^{III}(L_{14}{}^{ISQ})^{\bullet\bullet}{}_2(X)]$ the iron atom exhibits a square pyramidal geometry, where the axial coordination site is occupied by an exogenous anion (halogen X, SCN^-, $N_3{}^-$).[77–79] This fifth ligand is of prime importance in the stabilization of a particular metal spin state rather than another. When X = SCN^- or Cl^- the iron(III) ion is high spin ($S_{Fe} = 5/2$), and consequently the total spin state is ($S_t = 3/2$) (antiferromagnetic coupling with the two *o*-iminobenzosemiquinonate radicals). When X = Br^-, I^- or $N_3{}^-$ the iron can exist either at the low ($S_{Fe} = 1/2$) or intermediate ($S_{Fe} = 3/2$) spin state in a temperature-dependent spin crossover behavior. It is noticeable that a ($S_t = 1/2$) ground state is attained in both cases. The EPR spectrum of $[Fe^{III}(L_{14}{}^{ISQ})^{\bullet\bullet}{}_2(I)]$ is characterized by signals at g = 2.189 and 2.134 ≫ g_e, confirming that the unpaired electron resides in a d orbital, thus ($S_{Fe} = 3/2$).

In the diradical complex $[Fe^{III}({}^{qui}L_{14}{}^{ISQ})^{\bullet\bullet}{}_2]^+$ the iron is octahedral and low spin.[67] Remarkably, despite that the ground state is ($S_t = 1/2$), that is similar to that of $[Fe^{III}(L_{14}{}^{ISQ})^{\bullet\bullet}{}_2(I)]$, its EPR spectrum differs significantly with a rather isotropic signal at g = 1.9905. Therefore, the SOMO of $[Fe^{III}({}^{qui}L_{14}{}^{ISQ})^{\bullet\bullet}{}_2]^+$ has a marked organic radical character, which can only be explained by considering that the iron is ($S_{Fe} = 1/2$) and not ($S_{Fe} = 3/2$).

Homobimetallic complexes were isolated from several kinds of ligands. In the $[(tmtacn)Fe^{III}({}^{oxm}L_{13}{}^{tBu})_3Fe^{III}]$ complex, the imine nitrogen and phenolic oxygen of the tridentate Schiff base ligand bind one metal ion, whereas the terminal oxamate oxygen binds another metal (the coordination sphere of this latter is completed by the tmtacn ligand).[80] Three oxidation waves attributed to the successive oxidation of the phenolates into phenoxyl radicals were observed but the radicals could not be quantitatively generated. In $[Fe^{III}{}_2({}^{Et}L_{16}{}^{oxm})^{\bullet}(Cl)_4]^-$, the iron atom does not exhibit an octahedral but rather a square pyramidal geometry in which three coordinating atoms are provided by the bis(tridentate) ligand and two other from chloride anions.[81] The EPR spectrum of this complex displays a rhombic ($S_t = 1/2$) signal at $g_x = 1.998$, $g_y = 1.944$, and $g_z = 1.825$. This ground state is attained via an intramolecular antiferromagnetic coupling between an *o*-iminobenzosemiquinonate radical and a high spin ferric ion (J of $\approx -100\,cm^{-1}$), yielding a ($S^* = 2$), which in turn couples weakly antiferromagnetically to the second high spin ferric ion ($J \approx -10\,cm^{-1}$). $[Fe^{III}{}_2({}^{mPh}L_{16}{}^{N})_3{}^{6\bullet}]$ has been obtained by two independent teams from the same ligand.[69,82] Compared to $[Mn^{IV}({}^{mPh}L_{16}{}^{ISQ})^{\bullet\bullet}({}^{mPh}L_{16}{}^{N,ISQ})^{\bullet\bullet}{}_2]$ it is a hexaradical complex with a singlet ground state resulting from both ferro- and antiferromagnetic exchanges that will not be detailed here.

8.4.6 Cobalt complexes

The cobalt ion is commonly encountered at its paramagnetic (+II) and (often) diamagnetic (+III) redox state in the usual potential range. Depending on the geometry (tetra or hexacoordinated), and the ligands, either one or the other redox state could be stabilized.

Cobalt(III) complexes have been synthesized from mono- and tris-phenolate TACN ligands.[60,64] Interestingly, they are mainly obtained by reacting the cobalt(II) and not the cobalt(III) salt with the ligand in the presence of air. The cobalt complex $[Co^{III}(^{Me}L_1{}^{OMe})(Ph_2acac)]^+$ exhibits strong structural and electrochemical analogies with the chromium complex $[Cr^{III}(^{Me}L_1{}^{OMe})(Ph_2acac)]^+$ described above. The phenolate moiety is, for example, oxidized into a phenoxyl radical at roughly the same potential (0.24 vs 0.30 V, Table 8.4) and its UV-Vis spectrum exhibits the typical absorption band at 405 nm (Figure 8.13). The phenoxyl species is, nevertheless, paramagnetic ($S = 1/2$), since the cobalt(III) ion is diamagnetic. Hyperfine interactions with a benzylic hydrogen and also the cobalt nucleus ($I = 7/2$) unambiguously show that the metal remains coordinated upon oxidation in $[Co^{III}(^{iPr}L_1{}^{tBu})^\bullet(acac)]^{2+}$. Interestingly, $[Co^{III}(^{iPr}L_1{}^{tBu})^\bullet(Cl_4\text{-cat})^\bullet]$, a complex involving two coordinated radicals ($(Cl_4\text{-cat})^\bullet$ denotes the tetrachlorosemiquinonate radical), could be prepared. The two paramagnetic centers interact ferromagnetically with each other with $J = +12\,cm^{-1}$.

The complexes $[Co^{II}(L_{13}{}^{Ph})_2]$ and $[Co^{II}(L_{13}{}^{Bz})_2]$ contain two bidentate *N,O*-ligands (donor set similar to that of $H_2L_{14}{}^{AP}$) coordinated to a *tetrahedral* cobalt(II) ion (and *not* to an *octahedral* cobalt(III) ion).[83] $[Co^{II}(L_{13}{}^{Ph})_2]$ can be reversibly oxidized at +0.16 V to the monoradical species $[Co(L_{13}{}^{Ph})(L_{13}{}^{Ph})^\bullet]^+$, whose ground state is ($S_t = 1$) as a result of antiferromagnetic coupling ($J < -250\,cm^{-1}$, $D_{Co} = -24.5 \pm 2\,cm^{-1}$) between the radical and the paramagnetic cobalt(II) ion ($S_{Co} = 3/2$). Surprisingly, cyclic voltammetry of a closely related complex, $[Co^{II}(L_{13}{}^{Bz})_2]$, revealed a chemically *irreversible* process at $E_p^a = 0.35$ V. Interestingly, another cobalt complex, $[Co^{III}(L_{13}{}^{Bz})_3]$, which contains an octahedral central cobalt(III) ion, could be isolated from $HL_{13}{}^{Bz}$. In contrast to $[Co^{II}(L_{13}{}^{Bz})_2]$ its cyclic voltammetry exhibits a *reversible*

Table 8.4 *Redox potentials of representative cobalt complexes*

Complex	$E_{1/2}$ [V vs Fc^+/Fc]	Ref.
$[Co^{III}(^{Me}L_1{}^{OMe})(Ph_2acac)]^+$	−1.06 (irr), 0.24	[64]
$[Co^{III}(^{iPr}L_1{}^{tBu})(acac)]^+$	0.46	[60]
$[Co^{III}(^{iPr}L_1{}^{tBu})(Cl_4\text{-cat})]$	0.16, 0.62	[60]
$[Co^{III}(L_3{}^{tBu})]$	0.01, 0.41, 0.75	[60]
$[Co^{III}(L_4{}^{tBu})(acac)]^+$	0.51	[84]
mer-$[Co^{III}(L_4{}^{tBu})(SQ)]^+$	0.50	[84]
fac-$[Co^{III}(^{Me2}L_4{}^{tBu})(SQ)]^+$	0.27	[84]
$[Co^{II}(L_{13}{}^{Ph})_2]$	0.16, 0.51	[83]
$[Co^{II}(L_{13}{}^{PhOMe})_2]$	0.12, 0.49	[83]
$[Co^{II}(L_{13}{}^{Bz})_2]$	0.35(irr)	[83]
$[Co^{III}(L_{13}{}^{Bz})_3]$	0.08, 0.52	[83]
$[Co^{III}(L_{14}{}^{ISQ})^\bullet(tren)]^{2+}$	−1.36 (irr), −0.58, 0.27	[76]
$[Co^{III}(L_{14}{}^{ISQ})^{\bullet\bullet\bullet}]$	−1.31, −0.99, −0.34, 0.20	[62,86]
$[Co^{II}(^{qui}L_{14}{}^{ISQ})^\bullet(^{qui}L_{14}{}^{BQ})]^+$	−0.97, −0.53, 0.31, 0.64	[67]

irr: irreversible couple, $E_p{}^a$ is given.

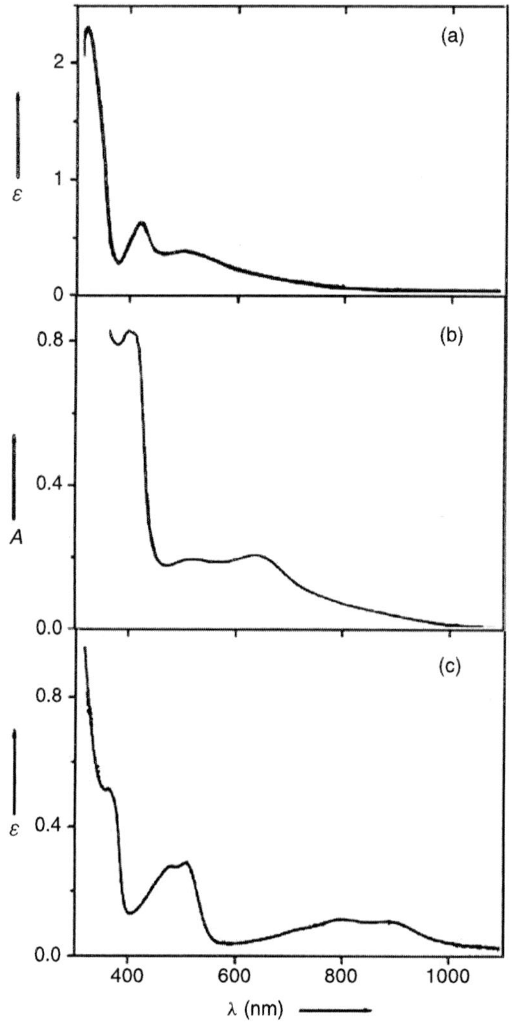

Figure 8.13 UV-Vis spectra of: (a) $[Co^{III}(^{Me}L_1{}^{OMe})^{\bullet}(acac)]^{2+}$ in dichloromethane; (b) $[Co^{III}(L_4{}^{tBu})^{\bullet}(acac)]^{2+}$ in acetonitrile; and (c) $[Co^{III}(L_{14}{}^{ISQ})^{\bullet}(tren)]^{2+}$ in dichloromethane. (Adapted from [64], [84] and [76].) (a) and (c): ε value in 10^4 M^{-1} cm^{-1} (b): Absorbance of a 0.5 mM solution. (Reprinted from [64], copyright 2000, with permission from Elsevier.)

redox couple at the lower value of 0.08 V. Bulk electrolysis of $[Co^{II}(L_{13}{}^{Bz})_2]$ at 0.35 V or $[Co^{III}(L_{13}{}^{Bz})_3]$ at 0.08 V affords a similar oxidized species that has been identified as $[Co^{III}(L_{13}{}^{Bz})_2(L_{13}{}^{Bz})^{\bullet}]^+$. Therefore, $[Co^{II}(L_{13}{}^{Bz})(L_{13}{}^{Bz})^{\bullet}]^+$ is likely formed as a transient species upon oxidation of $[Co^{II}(L_{13}{}^{Bz})_2]$, and rearranges into $[Co^{III}(L_{13}{}^{Bz})_2(L_{13}{}^{Bz})^{\bullet}]^+$. The increased steric bulk of the diphenylimidazole relative to the benzimidazole moiety has been proposed to account for the different behavior observed for $[Co^{II}(L_{13}{}^{Ph})_2]$ and $[Co^{II}(L_{13}{}^{Bz})_2]$.

A series of octahedral cobalt(III) complexes has been prepared from tetradentate N_3O tripodal ligands.[84] The octahedral coordination sphere of the metal is completed with an additional bidentate ligand, either acetylacetonate (acac) or the semiquinone (SQ) radical. The phenolate complex $[Co^{III}(L_4{}^{tBu})(acac)]^+$ can be oxidized to the corresponding phenoxyl radical species $[Co^{III}(L_4{}^{tBu})^{\bullet}(acac)]^{2+}$, which is characterized by an absorbance maximum at 412 nm and an ($S = 1/2$) EPR signal at $g \approx 2.00$. Oxidation of the paramagnetic *mer*-$[Co^{III}(L_4{}^{tBu})(SQ)]^+$ complex occurs at $E_p^a = 0.50$ V, and also affords a radical species, namely *mer*-$[Co^{III}(L_4{}^{tBu})^{\bullet}(SQ)]^{2+}$. In contrast, oxidation of *fac*-$[Co^{III}({}^{Me2}L_4{}^{tBu})(SQ)]^+$ does not afford a phenoxyl radical, but instead a *fac*-$[Co^{III}({}^{Me2}L_4{}^{tBu})(quinone)]^{2+}$ species with two closed shell ligands, illustrating that the isomerism and the steric bulk around the pyridine nitrogen are non-innocent in the redox process.

Complexes isolated from aminophenolate ligands are mainly $[Co^{III}(L_{14}{}^{ISQ})^{\bullet}(tren)]^{2+}$ (monoradical),[76] $[Co^{III}(L_{14}{}^{ISQ})^{\bullet\bullet}{}_2X]$ (diradical)[85] and $[Co^{III}(L_{14}{}^{ISQ})^{\bullet\bullet\bullet}{}_3]$ (triradical).[62,86] Of these, $[Co^{III}(L_{14}{}^{ISQ})^{\bullet}(tren)]^{2+}$ exhibits in its EPR spectrum an ($S = 1/2$) signal at $g = 1.997$ with hyperfine splitting due to the interaction of the electron spin with the diamagnetic ^{59}Co (14 G) and the ^{14}N (8.8 G) nuclei. Its UV-Vis spectrum (Figure 8.13) reveals *o*-iminobenzosemiquinonate bands at 476 (3100), 497 (3200) and 851 nm (980 M^{-1} cm^{-1}). As $[Fe^{III}(L_{14}{}^{ISQ})^{\bullet}(tren)]^{2+}$ it could be one-electron oxidized or reduced according to Equation 8.9:

$$[Co^{III}(L_{14}{}^{AP})(tren)]^+ \underset{+e^-}{\overset{-e^-}{\rightleftharpoons}} [Co^{III}(L_{14}{}^{ISQ})^{\bullet}(tren)]^{2+} \underset{+e^-}{\overset{-e^-}{\rightleftharpoons}} [Co^{III}(L_{14}{}^{BQ})(tren)]^{3+} \tag{8.9}$$

In $[Co^{III}(L_{14}{}^{ISQ})^{\bullet\bullet}{}_2X]$ (where X = iodine or chloine) the metal ion adopts a square pyramidal geometry, similarly to $[Fe^{III}(L_{14}{}^{ISQ})^{\bullet\bullet}{}_2X]$. The UV-Vis features of these two complexes are also similar, with a strong band at ≈670 nm.[85] Nevertheless, the ground state remains ($S_t = 0$) (antiferromagnetic coupling between the radicals) regardless of the identity of X and no temperature dependence of S_t has been reported, unlike $[Fe^{III}(L_{14}{}^{ISQ})^{\bullet\bullet}{}_2X]$.

In contrast to $[Co^{III}(L_{14}{}^{ISQ})^{\bullet\bullet}{}_2X]$, the ground state of the triradical $[Co^{III}(L_{14}{}^{ISQ})^{\bullet\bullet\bullet}{}_3]$ is ($S_t = 3/2$), that is the radicals couple *ferromagnetically* rather than *antiferromagnetically* in the former case.[86] What is the reason for the difference in magnetic coupling? The three π^* magnetic orbitals of the radical ligands in $[Co^{III}(L_{14}{}^{ISQ})^{\bullet\bullet\bullet}{}_3]$ are orthogonal to one each other (octahedral metal) and consequently cannot couple antiferromagnetically one to each other (Figure 8.12). A new question then arises: Why does the isostructural $[Fe^{III}(L_{14}{}^{ISQ})^{\bullet\bullet\bullet}{}_3]$ exhibit a ($S_t = 1$) ground state resulting from strong *antiferromagnetic* coupling? The reason is obvious, as the interactions involve both the ligands and the paramagnetic iron in this complex, and not solely the ligand radical spins. The subsequent orbital overlap is higher, resulting in a stronger and antiferromagnetic coupling. The redox chemistry of $[Co^{III}(L_{14}{}^{ISQ})^{\bullet\bullet\bullet}{}_3]$ is summarized by Equation 8.10:

$$\begin{aligned}&[Co^{III}(L_{14}{}^{AP})_2(L_{14}{}^{ISQ})^{\bullet}]^{2-} \underset{+e^-}{\overset{-e^-}{\rightleftharpoons}} [Co^{III}(L_{14}{}^{AP})(L_{14}{}^{ISQ})^{\bullet\bullet}{}_2]^- \underset{+e^-}{\overset{-e^-}{\rightleftharpoons}} [Co^{III}(L_{14}{}^{ISQ})^{\bullet\bullet\bullet}{}_3]\\ &\underset{+e^-}{\overset{-e^-}{\rightleftharpoons}} [Co^{III}(L_{14}{}^{ISQ})^{\bullet\bullet}{}_2(L_{14}{}^{BQ})]^+ \underset{+e^-}{\overset{-e^-}{\rightleftharpoons}} [Co^{III}(L_{14}{}^{ISQ})^{\bullet}(L_{14}{}^{BQ})_2]^{2+}\end{aligned} \tag{8.10}$$

A surprising example in this series is the complex formed with the $H_2{}^{qui}L_{14}{}^{AP}$ ligand.[68] No X-ray crystal structure could be obtained, but its magnetic properties suggest the presence of a cobalt(II) ion rather than cobalt(III) antiferromagnetically coupled to a ligand radical ($|J| > 200$ cm^{-1}). This complex, namely $[Co^{II}({}^{qui}L_{14}{}^{ISQ})^{\bullet}({}^{qui}L_{14}{}^{BQ})]^+$ shows a rich redox chemistry, as does $[Co^{III}(L_{14}{}^{ISQ})^{\bullet\bullet\bullet}{}_3]$ but attribution of the oxidation/reduction sites is less detailed.

8.4.7 Nickel complexes

The nickel ion is a redox active metal that can exist either in the (+II) or (+III) redox state in the usual potential range. In the former state the nickel can be either paramagnetic (octahedral) or diamagnetic (square planar).

Until the beginning of 2000, only a few nickel–radical complexes had been reported. The TACN complex $[Ni^{II}(^{Me}L_1{}^{OMe})^{\bullet}(Ph_2acac)]^+$ is one such example, in which the octahedral ligand field stabilizes the high spin state of the nickel(II) ion.[65] Its ground state is ($S_t = 3/2$) as a result of a ferromagnetic coupling between the high spin nickel(II) ($S_{Ni} = 1$) and the phenoxyl radical ($S_{rad} = 1/2$). The zero field splitting parameters obtained by EPR are $|D| > 1\,cm^{-1}$ and $E/D = 0.13$ (only a lower limit is obtained for D since its absolute value is higher that the quantum of the X-Band EPR), with a low g tensor anisotropy ($g_x = g_y = 2.10$, $g_z = 2.12$).

Nickel complexes of tripodal ligands involving a single di-*tert*-butylphenol were described four years later. These complexes consist of a high spin nickel ion within a square pyramidal geometry, the coordination sphere being completed to five with an exogenous ligand.[87] The cyclic voltammetry curves of $[Ni^{II}(L_4{}^{tBu})(Cl)]$ and its derivatives exhibit a single reversible ligand-centered redox couple attributed to the formation of nickel(II)–radical species (Table 8.5). These species are characterized by strong absorption bands at ≈405 nm, 700 and 1100 nm, whereas their EPR spectra display isotropic signals at $g = 2.23$. This value is rather unusual and corresponds to an ($S_t = 1/2$) system arising from antiferromagnetic coupling (>100 cm^{-1}) between the ($S_{Ni} = 1$) and ($S_{rad} = 1/2$) spins. Interestingly, the N-donor character of the pyridines has been shown to significantly influences the stability of the radical species, with a $t_{1/2}$ of 45 minutes at 20 °C for $[Ni^{II}(L_4{}^{tBu})(Cl)]$, whereas it is only 5.9 minutes at −20 °C for $[Ni^{II}(^{Me2}L_4{}^{tBu})(Cl)]$.

In 2003, Shimazaki *et al*. published a key article concerning the oxidative behavior of diamagnetic pseudo square planar nickel salen complexes.[88] They showed that the nickel(II)–phenolate complex $[Ni^{II}(^{hx}L_8)]$ could be oxidized into a paramagnetic ($S = 1/2$) species, and that the oxidation site could be shifted by solvent and/or temperature changes. EPR measurements revealed that a metal-centered process is achieved in DMF, whereas a thermal equilibrium between the nickel(II)–radical and the nickel(III)–phenolate form is observed in dichloromethane. For instance, the EPR spectrum recorded in dichloromethane at −40 °C exhibits a radical signal at $g \approx 2.04$ that splits below the solvent freezing point into $g_1 = 2.30$, $g_2 = 2.23$, $g_3 = 2.02$, that is an anisotropy that is typical for nickel(III)–phenolate complexes.

Table 8.5 *Redox potentials of representative nickel complexes*

Complex	$E_{1/2}$ [V vs Fc$^+$/Fc]	Ref.
$[Ni^{II}(^{Me}L_1{}^{OMe})(Ph_2acac)]$	−0.36	[64]
$[Ni^{II}(L_4{}^{tBu})(Cl)]$	0.04	[87]
$[Ni^{II}(^{Me2}L_4{}^{tBu})(Cl)]$	0.14	[87]
$[Ni^{II}(^{hx}L_8)]$	0.46, 0.80	[88]
$[Ni^{II}(^{et}L_8)]$	0.59, 1.05	[95]
$[Ni^{II}(L_9)]$	0.58, 0.80	[92]
$[Ni^{II}(^{Me}L_9)]$	0.49, 0.78	[91]
$[Ni^{II}(L_9{}^{OMe})]$	0.36	[90]
$[Ni^{II}(L_9{}^{NMe})]$	−0.14, 0.00	[90]
$[Ni^{II}(^{et,py}L_{10})]$	0.09, 0.36	[95]
$[Ni^{II}(^{et,me}L_{10})]$	0.41*	[95]
$[Ni^{II}(L_{11})]2-$	−0.25, 0.31, 0.71	[92]
$[Ni^{II}(L_{14}{}^{ISQ})^{\bullet\bullet}{}_2]$	−1.64, −1.07, 0.04*	[96]
$[Ni^{II}(^{Pr}L_{16}{}^{ISQ})^{\bullet\bullet}]$	−1.84, −1.18, 0.00, 0.26	[97]
$[Ni^{II}(L_{14}{}^{ISQ})^{\bullet}(tren)]^+$	−1.28, −0.36, 0.97	[76]

*: two-electron process.

The metal-centered oxidation process in coordinating solvents is not surprising: The nickel(III) ion is known to adopt an octahedral geometry, thus solvent molecules likely complete the metal coordination sphere from four to six upon oxidation. The "valence tautomerism" observed in dichloromethane is more interesting, as it implies an electronic redistribution induced by temperature changes, despite the fact that the nickel(III) ion could not be, in principle, tetracoordinated. Interestingly, one-electron oxidized triple-salen complexes also exhibit a temperature dependent EPR spectrum.[89] Lately it has been reported in some related salen complexes that although the nature of the spacer (phenyl, cyclohexyl, ethyl)[90,91] does not significant influence the phenomenon, changes in the solvent quality could dramatically affect it. For instance, replacing the conventional dichloromethane (distilled prior to measurements) by high quality anhydrous >99.8% dichloromethane results in the formation of solely the nickel(II)–radical species, irrespective of the temperature. In order to explain these results, the binding constants of an exogenous ligand, pyridine, were determined for a series of nickel(II)–radical complexes. An important conclusion is that pyridine binds tightly to the Schiff base complex, and that the affinity increases by lowering the temperature.[92] For the equilibrium $[Ni^{II}(L_9)^{\bullet}]^+ + 2\ Py \rightleftharpoons [Ni^{III}(L_9)(Py)_2]^+$, the K values are $10^{8.8}$ at 243 K and $10^{6.7}$ at 298 K. It has thus been proposed that weak exogenous ligands present in dichloromethane (impurities or supporting electrolyte) could coordinate sufficiently at low temperature to promote the pseudo-square planar Ni(II)–radical $\leftrightarrow$ octahedral Ni(III)–phenolate conversion. This hypothesis has been further confirmed by three facts. Firstly, the nickel(II)–radical complex of a salen derived ligand involving amidates instead of imines $[Ni^{II}(L_{11})^{\bullet}]^-$, which presents a much lower binding constant for pyridine ($10^{5.4}$ at 238 K), does not exhibits any temperature-dependent valence tautomerism. It is interesting to note that usually amidates stabilize the nickel(III) ion, which were not observed in that case. Secondly, variable temperature EPR measurements on a powder sample of the nickel(II)–radical species $[Ni^{II}(^{hx}L_8)^{\bullet}]^+$ revealed that this species remains a ligand-centered radical in the solid state regardless of the temperature.[93] Finally, the X-ray crystal structure of $[Ni^{II}(^{hx}L_8)^{\bullet}]^+$ was solved in 2007,[94] definitely confirming the radical state of the complex.

A remarkable feature in the structure of $[Ni^{II}(^{hx}L_8)^{\bullet}]^+$ is the contraction in the metal–oxygen and metal–nitrogen bonds upon oxidation (in other phenoxyl radical complexes these bond distances increase): the Ni–O bonds shorten from 1.856 to 1.827 and 1.830 Å, whereas the M–N bonds decrease from 1.861 to 1.825 and 1.843 Å. An explanation is the removal of an electron from a primarily ligand-based antibonding orbital that results in an increased M $\rightarrow$ L back donation. Another peculiarity in this structure is the absence of a quinoid distribution of bond lengths, likely because of delocalization of the radical over both ring systems. The same year, a series of nickel(II)–phenoxyl radical complexes was obtained from salen ligands involving phenolate *p*-substituents with different electron donating properties (Figure 8.14).[91] Of course the more electron donating substituents give rise to complexes that are easier to oxidize. More interestingly, solution EPR measurements revealed a shift of g_{iso} from 2.006 (typical value for zinc(II)–phenoxyl radical complexes of salen ligands) for the strong electron donating $N(CH_3)_2$ substituant to 2.017 for the intermediate methoxy substituent and 2.034 for the weaker electron donating *tert*-butyl group. The $N(CH_3)_2$ group is so electron donating that it could easily compensate for the electron deficiency on the phenoxyl ring. The SOMO thus has a strong radical character. In contrast, the less electron donating *tert*-butyl group does not compensate so easily the electron hole and the metal orbitals take over. The SOMO thus exhibits a more marked metal character. It is noteworthy that all the complexes exhibit several absorption bands in the visible spectrum, thus making attribution of the oxidation site ambiguous solely on the basis of this spectroscopy.

Although an octahedral environment around the metal ion in salen-derived complexes seems to stabilize the nickel(III)–phenolate valence tautomer, octahedral radical complexes of nickel(II) have also been reported. They are, however, only obtained by oxidation of complexes of *reduced* hexadentate salen ligands in which the nickel(II) ion is octahedral. $[Ni^{II}(^{et,Py}L_{10})^{\bullet}]^+$ exhibits the intense radical absorption bands at 391 (3160), 405 (3140) and 611 nm ($2160\,M^{-1}\,cm^{-1}$), and an isotropic EPR signal at $g_{iso} = 2.22$, suggesting an antiferromagnetic coupling between the nickel ion ($S_{Ni} = 1$) and the radical spin. It is

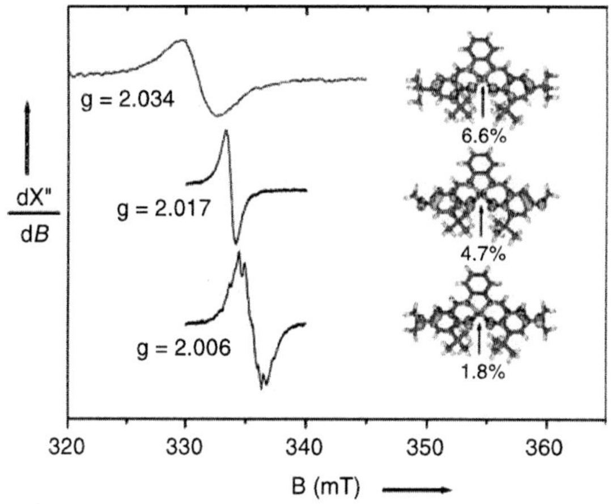

Figure 8.14 *Isotropic EPR spectra of 1 mM dichloromethane solutions of: (a)* $[Ni^{II}(L_9)^{\bullet}]^+$*; (b)* $[Ni^{II}(L_9^{OMe})^{\bullet}]^+$*; and (c)* $[Ni^{II}(L_9^{NMe})^{\bullet}]^+$*. ([90] Reproduced by permission of the Royal Society of Chemistry.) A full-colour version of this figure appears in the Colour Plate section of this book.*

noteworthy that the kind of coupling depends on geometrical parameters, as the octahedral nickel complex of the tetradentate ligand $H_2{}^{et,Me}L_{10}$ in the presence of pyridine, namely $[Ni^{II}({}^{et,Me}L_{10})^{\bullet}(Py)_2]^+$, is a ($S_t = 3/2$) system.[95] An important conclusion in this series of octahedral complexes is that the imine nitrogens of classical Schiff bases tend to favor a metal-centered oxidation process, whereas amines favour a ligand-centered process.

o-Iminobenzosemiquinone radical ligands have been incorporated both in square planar and octahedral nickel(II) complexes. The *trans*-$[Ni^{II}(L_{14}{}^{ISQ})^{\bullet\bullet}{}_2]$ complex is planar within experimental error, and the metal ion is consequently diamagnetic.[96] In this geometry the overlap between the π orbitals of the radicals and one of the filled t_{2g} metal orbital is significant (Figure 8.12), leading to a strong antiferromagnetic coupling. $[Ni^{II}(L_{14}{}^{ISQ})^{\bullet\bullet}{}_2]$ is therefore a singlet diradical. Its cyclic voltammetry curve exhibits two cathodic waves attributed to the reduction of each of the *o*-iminobenzosemiquinone moieties, and a single bielectronic anodic wave that corresponds to the simultaneous one-electron oxidation of the two *o*-iminobenzosemiquinone moieties according to Equation 8.11:

$$[Ni^{II}(L_{14}{}^{AP})_2]^{2-} \underset{+e^-}{\overset{-e^-}{\rightleftharpoons}} [Ni^{II}(L_{14}{}^{AP})(L_{14}{}^{ISQ})^{\bullet}]^- \underset{+e^-}{\overset{-e^-}{\rightleftharpoons}} [Ni^{II}(L_{14}{}^{ISQ})^{\bullet\bullet}{}_2] \underset{+2e^-}{\overset{-2e^-}{\rightleftharpoons}} [Ni^{II}(L_{14}{}^{BQ})_2]^{2+} \quad (8.11)$$

Trans-$[Ni^{II}(L_{14}{}^{ISQ})^{\bullet\bullet}{}_2]$ exhibits a remarkable feature in its UV-Vis spectrum, a very intense absorption at 890 nm (37 000 $M^{-1}\,cm^{-1}$) assigned to the spin- and dipole-allowed ligand-to-ligand transition shown in Figure 8.15. Similar bands have been also reported for square planar complexes of other metals (copper, palladium). As expected the band loses most of its intensity in the monoradical $[Ni^{II}(L_{14}{}^{AP})(L_{14}{}^{ISQ})]^-$ with an ε value of only 7500 $M^{-1}\,cm^{-1}$ at 880 nm. Interestingly no redox process leading to a nickel(III) ion has been reported in this series.

In the diradical complex $[Ni^{II}({}^{Pr}L_{16}{}^{ISQ})^{\bullet\bullet}]$, the *O,N* *o*-iminobenzosemiquinone moieties are forced to be *cis* one to each other.[97] Like $[Ni^{II}(L_{14}{}^{ISQ})^{\bullet\bullet}{}_2]$, this complex is a singlet diradical characterized by an intense ligand-to-ligand transition at 854 nm (30 000 $M^{-1}\,cm^{-1}$). The one-electron reduced form of $[Ni^{II}({}^{Pr}L_{16}{}^{ISQ})^{\bullet\bullet})]$ has been characterized by EPR, revealing that it is a mixed-valent ($S = 1/2$) species where the unpaired electron is delocalized over both ligands.

The $[Ni^{II}(L_{14}{}^{ISQ})^{\bullet}(tren)]^-$ complex is different from the others of the series in the sense that the nickel ion is octahedral, and thus paramagnetic. Consequently, the e_g orbitals that host the metal unpaired electrons

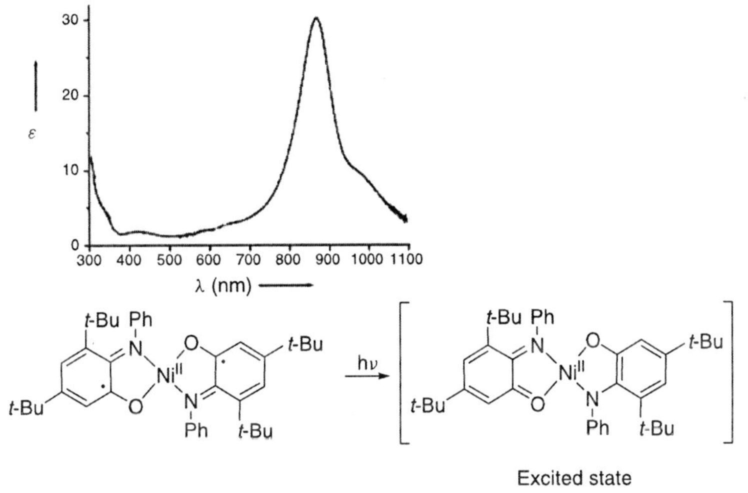

Figure 8.15 Intraligand charge transfer transition in $[Ni^{II}(^{14}L^{ISQ})^{\bullet\bullet}{}_2]$. ε in mM cm^{-1}. (Adapted with permission from [96]. Copyright 2001 American Chemical Society.)

are orthogonal to the half-filled ligand π orbitals. Then, the coupling between the *o*-iminobenzosemiquinone moieties and the metal is expected to be strongly ferromagnetic. EPR spectroscopy confirms this assignment, an ($S_t = 3/2$) signal being detected at $g_x = g_y = 2.10$ and $g_z = 2.06$ with $|D|_{3/2} = 10 \pm 5\,cm^{-1}$ and $E/D = 0.21$. Interestingly, this complex is the only one in the $H_2L_{14}{}^{AP}$ series that could be finally oxidized into a ($S = 1/2$) nickel(III) complex, perhaps due to its octahedral preorganization (Equation 8.12):

$$[Ni^{II}(L_{14}{}^{AP})(tren)] \underset{+e^-}{\overset{-e^-}{\rightleftharpoons}} [Ni^{II}(L_{14}{}^{ISQ})(tren)]^+ \underset{+e^-}{\overset{-e^-}{\rightleftharpoons}} [Ni^{II}(L_{14}{}^{BQ})(tren)]^{2+} \underset{+e^-}{\overset{-e^-}{\rightleftharpoons}} [Ni^{III}(L_{14}{}^{BQ})(tren)]^{3+} \quad (8.12)$$

8.4.8 Copper complexes

The copper complexes were those that attracted the greatest and earliest interest due to the occurence of the copper-radical entity in Galactose Oxidase. The copper(II)-phenolate complex $[Cu^{II}(^{iPr}L_1{}^{tBu})(CH_3CN)]^+$ [98,99] exhibits in its cyclic voltammetry curve a reversible signal at 0.18 V attributed to the phenoxyl/phenolate redox couple (Table 8.6). The electrogenerated radical species is relatively stable and EPR silent, with absorption bands at 410 (4000) and 672 nm (1000 $M^{-1}\,cm^{-1}$) typical for phenoxyl radicals. Interestingly, the copper(II)–phenolate complex could also be isolated as a benzyl alcoholate adduct $[Cu^{II}(^{iPr}L_1{}^{tBu})(OCH_2Ph)]$. The electrochemical behavior of this complex is characterized by a total loss of reversibility of the redox wave corresponding to the phenoxyl/phenolate couple, as expected for an electrocatalytic process. In addition, bulk electrolysis of $[Cu^{II}(^{iPr}L_1{}^{tBu})$ (benzyl alcoholate)] affords benzaldehyde in a 46% yield showing that the catalyst is only active as its Cu^{II}-phenoxyl radical form as galactose oxidase. The TACN scaffold has been developed concomitantly by Wieghardt *et al.*[100]: from $H^{Me}L_1{}^{OMe}$ and in the presence of acetylacetonate, a copper(II)–phenolate complex in which the metal is six-coordinate was isolated.[101] Interestingly, the Cu^{II}–O–C bond angle (α) and the dihedral angle (β) between the x,y plane at the copper(II) ion and the phenyl ring of the phenolate ligand differ significantly when the coordination number increases to six (Figure 8.16). This results in changes in orbital overlap that are sufficient to invert the sign of the magnetic coupling between the metal and the radical. For instance, it is antiferromagnetic in $[Cu^{II}(^{iPr}L_1{}^{tBu})^{\bullet}(Cl)]^+$ whereas it is ferromagnetic in $[Cu^{II}(^{Me}L_1{}^{OMe})^{\bullet}(Ph_2acac)]$.

Table 8.6 *Redox potentials of representative copper complexes*

Complex	$E_{1/2}$ [V vs Fc^+/Fc]	Ref.
$[Cu^{II}(^{Me}L_1{}^{OMe})(Ph_2acac)]$	−0.33	[100,101]
$[Cu^{II}(^{iPr}L_1{}^{tBu})(CH_3CN)]^+$	0.31	[98,99]
$[Cu^{II}(^{Et}L_2{}^{OMe})]$	−0.10, 0.14	[102]
$[Cu^{II}(H^{Me}L_2{}^{OMe})]^+$	0.14	[102]
$[Cu^{II}(HL_3{}^{tBu,OMe})]$	−0.06, 0.12	[101]
$[Cu^{II}(H_2L_3{}^{tBu,OMe})]^+$	0.26	[100]
$[Cu^{II}(HL_3{}^{tBu})]$	0.15, 0.38	[100]
$[Cu^{II}(H_2L_3{}^{tBu})]^+$	0.44	[100]
$[Cu^{II}(L_4{}^{OMe})(CH_3CN)]^+$	0.02	[22]
$[Cu^{II}(L_4{}^{tBu})(CH_3CN)]^+$	0.15	[22]
$[Cu^{II}(L_4{}^{F})(CH_3CN)]^+$	0.2	[22]
$[Cu^{II}(L_4^{NO2})(CH_3CN)]^+$	0.7	[22]
$[Cu^{II}(^{Me}L_4{}^{tBu})(Cl)]$	0.14	[103]
$[Cu^{II}(^{Me2}L_4{}^{tBu})(Cl)]$	0.19	[103]
$[Cu^{II}(L_4{}^{SMe})(OAc)]$	0.01	[105]
$[Cu^{II}(^{Me}L_4{}^{SMe})(OAc)]$	0.03	[105]
$[Cu^{II}{}_2(L_4^{OMe2})_2]^{2+}$	0.45	[23]
$[Cu^{II}(L_4^{OMe2})(Pyr)]^+$	0	[23]
$[Cu^{II}{}_2(L_4{}^{OMe,tBu})_2]^{2+}$	0.66	[23]
$[Cu^{II}(L_4{}^{OMe,tBu})(Pyr)]^+$	0.14	[23]
$[Cu^{II}(L_5{}^{SMe})(OAc)]$	0.05	[104]
$[Cu^{II}(L_5{}^{tBu})(CH_3CN)]^+$	0.35	[104]
$[Cu^{II}{}_2(L_6{}^{H})_2]$	0.70(irr)	[20]
$[Cu^{II}(L_6{}^{H})(CH_3CN)]$	0.11, 0.49(irr)	[20]
$[Cu^{II}(HL_6{}^{tBu})(OAc)]$	0.2	[107]
$[Cu^{II}(L_6{}^{OMe})]_2$	−0.14, 0.12	[22]
$[Cu^{II}(L_6{}^{OMe})(Pyr)]$	−0.16	[22]
$[Cu^{II}(HL_6^{NO2})(OAc)]$	0.08	[22,108]
$[Cu^{II}(L_6{}^{NO2})(Pyr)]$	−0.08	[108]
$[Cu^{II}(^{im}L_6{}^{H})(CH_3CN)]$	−0.07	[21]
$[Cu^{II}(L_7{}^{SMe})(Pyr)]$	0.10, 0.33	[109]
$[Cu^{II}{}_2(L_7{}^{tBu})_2]$	0.12, 0.26	[112]
$[Cu^{II}(^{na}L_8{}^{SPr})]$	0.62	[113]
$[Cu^{II}(^{et}L_8)]$	0.31, 0.79	[116]
$[Cu^{II}(^{hx}L_8)]$	0.45, 0.65	[115,116,118]
$[Cu^{II}(L_9)]$	0.65, 0.83	[116]
$[Cu^{II}(^{hx}L_{10})]$	0.08, 0.21	[115,118]
$[Cu^{II}(^{et}L_{10})]$	0.11, 0.45	[116,119]
$[Cu^{II}(^{Ph}L_{12})]$	0.34, 0.58	[124]
$[Cu^{II}(L_{13}{}^{Ph})_2]$	0.17, 0.51	[126,128]
$[Cu^{II}(L_{13}{}^{Bz})_2]$	0.36, 0.58	[128]
$[Cu^{II}(L_{13}{}^{PhOMe})_2]$	0.11, 0.44	[128]
$[Cu^{II}(L_{14}{}^{ISQ})^{\bullet\bullet}{}_2]$	−1.32, −1.02, −0.26, 0.37	[96]
$[Cu^{II}(L_{14}{}^{ISQ})^{\bullet}(tmtacn)]^+$	−0.06, −0.77	[87]
$[Cu^{II}{}_2(L_{15}{}^{S})^{\bullet\bullet}{}_2]$	−1.26, −0.29	[129]
$[Cu^{II}(^{AP}L_{15}{}^{ISQ})^{\bullet}(NEt_3)]$	−1.06, −0.14	[130]
$[Cu^{II}(^{Ph}L_{16}{}^{AP})]2-$	−1.42, −0.66, −0.06, 0.41	[133]
$[Cu^{II}(^{Ph}L_{16}{}^{S})]$	0.62(irr), 0.99(irr)	[135]
$[Cu^{II}(^{Ph2}L_{16}{}^{AP})]^{2-}$	−0.98, −0.61, −0.16, 0.35	[134]

irr: irreversible couple, $E_p{}^a$ is given.

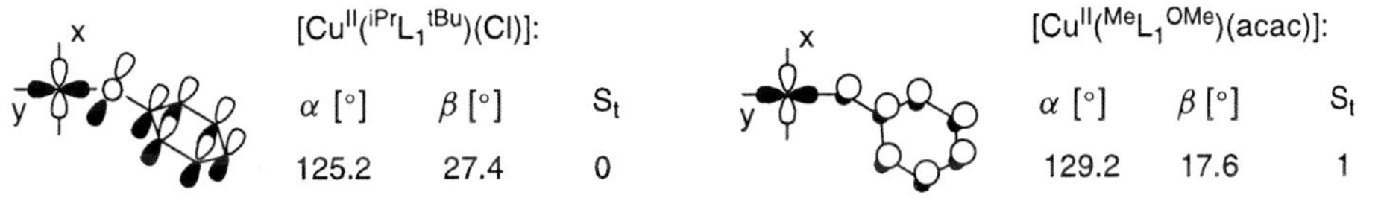

Figure 8.16 *Ground state modulation by changing the Cu^{II}–O–C angle. (Reproduced with permission from [101]. Copyright Wiley-VCH Verlag GmbH & Co. KGaA.)*

Using tris-phenol ligands, copper(II) complexes can be isolated with different protonation states, that is 1 phenolates + 2 phenols or 2 phenolates + 1 phenol. The cyclic voltammogram of the former species exhibits a single reversible redox wave whereas two waves are observed in the latter compound, both of which are attributed to the phenoxyl/phenolate couples. It is notable that the potential values associated to the first electron transfer are dependent on the protonation state, as expected for changes in the global charge of the complexes. The copper complex of $H_2{}^{Et}L_2{}^{OMe}$ was also found to be the precursor of stable bis(phenoxyl) radicals.[102] Formation of such species can be easily monitored by UV-Vis spectroscopy: $[Cu^{II}(^{Et}L_2{}^{OMe})^{\bullet\bullet}]^{2+}$ exhibits a band at 410 nm whose intensity is about twice that of the monoradical $[Cu^{II}(^{Et}L_2{}^{OMe})^{\bullet}]^{+}$. Its 4 K EPR spectrum is characterized by signals distributed over the entire spectral width indicating a dominant ferromagnetic coupling between the Cu^{II} and each of the radicals yielding a ($S_t = 3/2$) system with large D values.

The properties of copper(II) complexes of tripodal ligands involving either a bis(pyridyl)alkylamine connected to a single phenol (N_3O donor set), or a mono(pyridyl)alkylamine attached to two phenols (N_2O_2 donor set) have been extensively studied. For instance, the ligand properties have been modulated by modifying (i) the length of the alkyl chain linking the pyridine and the pivotal nitrogen, (ii) the steric bulk provided by the *ortho* substituent of the phenol, (iii) the electronic properties of the *para* substituent of the phenol, and (iv) the basicity of the pyridine donors. Using different copper(II) salts, a large number of phenolate complexes, monomeric or dimeric, have been prepared (Figure 8.17). Many of them have been structurally characterized by X-ray crystallography, but structural data concerning the copper(II)–phenoxyl species are still missing.[22,23,103–111]

Mononuclear copper(II)–phenolate complexes of general formula [Cu(L)(X)] are obtained from N_3O ligands in the presence of triethylamine if steric bulk (e.g. *tert*-butyl group) is provided by the *ortho* substituent of the phenol. The copper(II) ion in these complexes is square pyramidal, with a solvent or exogenous anion that completes the N_3O donor set of the ligand. The cyclic voltammograms of these complexes exhibit single one-electron redox waves that correspond to the oxidation of the phenolate moiety to a phenoxyl radical. The oxidized species exhibit the typical π–π^* transitions of phenoxyls at ≈410 and 600 nm, and their ground state is either ($S_t = 0$) or ($S_t = 1$) depending on the ligand (and the axial or equatorial positioning of the phenolate). Both the oxidation potentials and the chemical stability of the copper(II)–phenoxyl radical species correlate with the $\sigma^{+}{}_{Hammet}$ of the phenolate *p*-substituent, with the most electron donating substituent affording the most stable radical.[22] In addition, a dependence of the radical stability on the N-donor properties was observed: The weaker N-donor ability of the 2-methylpyridine or quinoline group compared to pyridine strongly destabilizes the radical complex. As an example, the half-life of $[Cu^{II}(^{Me}L_4{}^{tBu})^{\bullet}(Cl)]^{2+}$ is 65 minutes at −20 °C whereas it is only 9 minutes at −40 °C for $[Cu^{II}(^{Me\ 2}L_4{}^{tBu})^{\bullet}(Cl)]^{2+}$.[103] Interestingly, it has been shown for $[Cu^{II}(L_4{}^{tBu})^{\bullet}(CH_3CN)]^{2+}$ and some other complexes that the radical quickly comproportionates in the presence of Cu^{+} and an excess of ligand, according to Equation 8.13[23]:

$$[Cu^{II}(L_4{}^{tBu})^{\bullet}(CH_3CN)]^{2+} + Cu^{+} + HL_4{}^{tBu} \xrightarrow{NEt_3} 2[Cu^{II}(L_4{}^{tBu})(CH_3CN)]^{+} \quad (8.13)$$

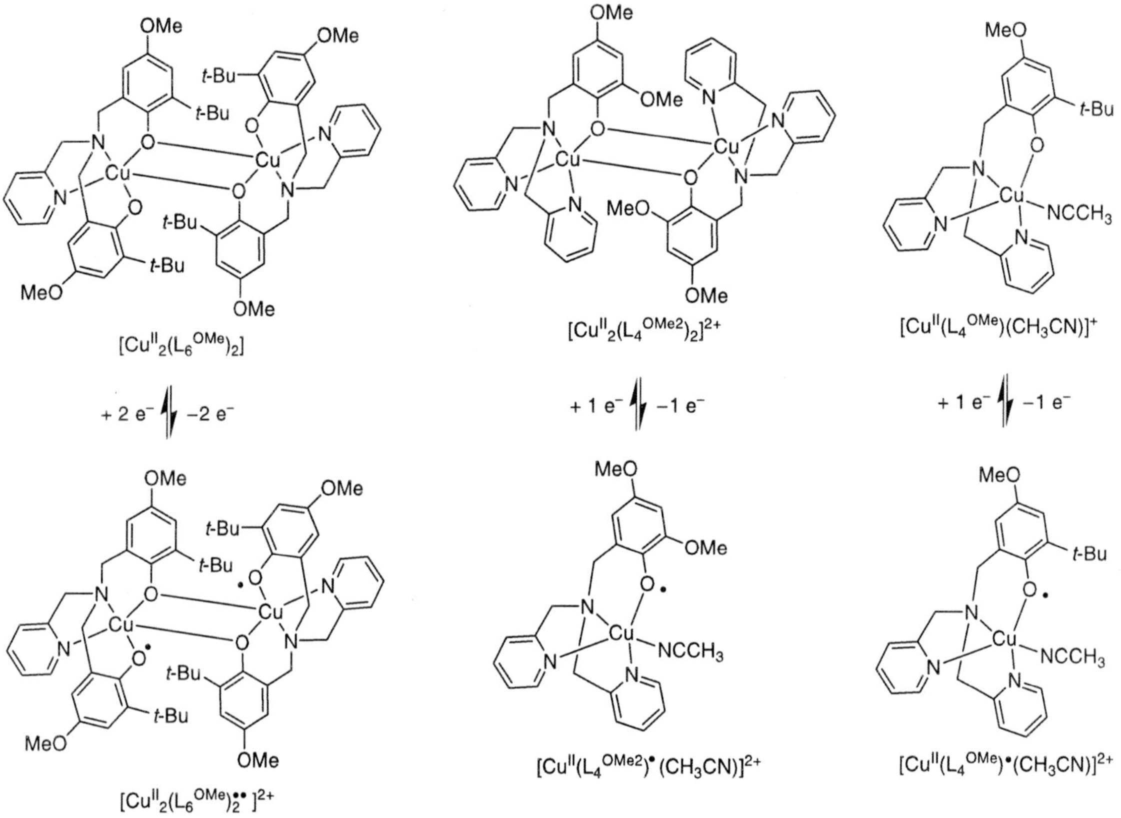

Figure 8.17 *Solution and oxidative chemistry of representative tripodal ligands.*

The catalytic activity of some of these complexes has been tested using benzyl alcohol as substrate. For example, $[Cu^{II}(L_5{}^{SMe})^{\bullet}(NO_3)]^+$ oxidizes stoichiometrically benzyl alcohol into benzaldehyde in a mechanism similar to that observed in galactose oxidase (GO): the substrate binds to the metal on its alcoholate form, then the radical performs the rate limiting hydrogen abstraction and the transient ketyl radical reduces the copper(II) center according to Equations 8.14 and 8.15[111]:

$$[Cu^{II}(L_5{}^{SMe})^{\bullet}]^+ + PhCH_2OH \rightleftharpoons [Cu^{II}(L_5{}^{SMe})^{\bullet}(OCH_2Ph)]^+ + H^+ \tag{8.14}$$

$$[Cu^{II}(L_5{}^{SMe})^{\bullet}(OCH_2Ph)]^+ \rightarrow [Cu^{I}(HL_5{}^{SMe})]^{2+} + PhCHO \tag{8.15}$$

Dimeric species of general formula $[Cu^{II}{}_2(L)_2]$ are obtained from N_3O ligands bearing a phenol without steric bulk at the ortho positions, or N_2O_2 ligands irrespective of the bulkiness of the phenolate *o*-substituent, and a copper(II) salt that does not contain a strong monoanionic donor. Nevertheless, addition of donors such as pyridine or acetate in the medium easily promotes the dimer to monomer conversion according to Equation 8.16:

$$[Cu^{II}{}_2(L)_2] + 2Py \rightleftharpoons 2[Cu^{II}(L)(Py)] \tag{8.16}$$

In the dimers formed with N_3O ligands, the copper ions are bridged by one phenolate oxygen of each ligand. The prototypical complex $[Cu^{II}{}_2(L_4^{OMe2})_2]^{2+}$ exhibits an irreversible anodic peak E_p^a in its cyclic voltammetry curve at +0.45 V, associated with a cathodic peak E_p^c at the much lower potential values +0.03 V. E_p^a is found to be higher than $E_{1/2}$ of the corresponding mononuclear complex obtained in pyridine $[Cu^{II}(L_4^{OMe2})(py)]^+$ ($E_{1/2}$ = 0.00 V). This is not surprising as coordination to two instead of one positively charged metal ion makes the oxygen less electronegative. In contrast, E_p^c is close to $E_{1/2}$ of $[Cu^{II}(L_4^{OMe2})(py)]^+$, thus showing that $[Cu^{II}_2(L_4^{OMe2})_2]$ is likely converted into a monomer upon oxidation. As bulk electrolysis of $[Cu^{II}_2(L_4{}^{OMe,tBu})_2]$ produces species that are characterized by strong absorption bands at ≈410 and 600 nm, it has been proposed that the monomer is the radical $[Cu^{II}(L_4^{OMe2})^\bullet(CH_3CN)]^{2+}$.[23]

The dimers formed with N_2O_2 ligands are very interesting as the ligand exhibits two distinct phenolates, one bridging and one (presumably easier to oxidize) non-bridging. The radical species remain dimeric in this case as the phenoxyl is generated on the non-bridging phenolate. The protypical complex $[Cu^{II}_2(L_6{}^{OMe})^{\bullet\bullet}{}_2]^{2+}$ (Figure 8.17) exhibits the expected phenoxyl band at 416 nm (3370 M^{-1} cm^{-1}) and is found to be EPR silent as the result of antiferromagnetic exchange coupling between the two copper ions (separated by 2.8–3 Å) mediated by the μ-phenolato bridges. It is noteworthy that the stability of these dimeric radicals is low compared to the monomers. For instance the half-life of $[Cu^{II}_2(L_6{}^{OMe})^{\bullet\bullet}{}_2]^{2+}$ is five minutes at 298 K, that is much lower than that of the monomer $[Cu(L_4{}^{OMe})^\bullet(Py)]^{2+}$ (154 minutes at 298 K) although the former possesses a similar 2-*tert*-butyl-4-methoxyphenoxyl moiety.[22] The sole exception is $[Cu^{II}_2({}^{im}L_6{}^{H})^{\bullet\bullet}{}_2]^{2+}$, which was found to be exceptionally stable.[21] In some cases, dissolution of the dimers in coordinating solvents results in the formation of monomers, as observed for $[Cu^{II}_2(L_6{}^{H})_2]$. The oxidation product is then the radical monomer $[Cu(L_6{}^{H})^\bullet(CH_3CN)]^+$ characterized by an unusual broad EPR signal at g ≈ 2. This signal likely arises from a weak ferromagnetic or a dipole–dipole interaction between the copper(II) and the phenoxyl radical, suggesting an axial positioning of the phenoxyl radical.[20] $[Cu^{II}_2(L_7{}^{tBu})_2]$ was also shown to dissociate upon oxidation into the radical monomer $[Cu^{II}(L_7{}^{tBu})^\bullet(H_2O)]^+$ that quickly decomposes.[112]

There are only two examples of dicopper(II) complexes in which the bridging atom is a phenoxyl oxygen (only one being stable enough to be characterized).[110] Both examples were isolated from the dinucleating ligand $H^{di}L_4{}^{OMe}$, depending on the amount of triethylamine added during the synthesis. $[Cu^{II}_2({}^{di}L_4{}^{OMe})(CH_3CN)_2]^{3+}$, which is the precursor of the most stable radical species, contains two metal ions that are separated by 4.06 Å (an exogenous acetonitrile completes to five the metal coordination sphere). The radical $[Cu^{II}_2({}^{di}L_4{}^{OMe})^\bullet(CH_3CN)_2]^{4+}$ has a (S_t = 3/2) ground state, with a D value of −0.056 cm^{-1} intermediate between that of triradicals and that of copper(II) coordinated bis-phenoxyl radicals.

Radical complexes of salen ligands were first described in 1996. The phenolate moieties of $[Cu^{II}({}^{na}L_8{}^{SPr})]$ are *ortho* substituted by the electron donating SCH_3 group, thus mimicking the thioether bond in GO.[113] Oxidation at +0.62 V results in the formation of $[Cu^{II}({}^{na}L_8{}^{SPr})^\bullet]^+$. Although the expected π–π* transitions of the phenoxyl radicals were obscured by intense charge transfer transitions, the copper(II)–radical rather copper(III)–phenolate redox state of the one-electron oxidized complex was unambiguously established by EXAFS and X-ray absorption spectroscopy. $[Cu^{II}({}^{na}L_8{}^{SPr})^\bullet]^+$ is EPR silent as the result of antiferromagnetic coupling between the coordinated radical and the metal spins. Interestingly, this complex exhibits a significant GO-like catalytic activity, with 1300 turnovers in 20 hours for the aerobic oxidation of benzyl alcohol to benzaldehyde (at 295 K).[114] This may be explained by the significant tetrahedral distortion induced around the copper atom by the binaphthyl linker, thus stabilizing the reduced copper(I) form of the catalyst. Other salen complexes were synthesized in order to modulate the flexibility of the linker, and the geometry around the metal ion. These complexes usually exhibit two reversible oxidation waves on their cyclic voltammetry curves that are attributed to the successive formation of mono- and diradical species (Equation 8.17).[115–117] Interestingly, the rigid scaffold of $[Cu^{II}(L^9)]$ enhances the electronic communication between the two phenolate moieties. As a consequence the $\Delta E_{1/2} (= E^2{}_{1/2} - E^1{}_{1/2})$ is the highest in the series.

$$[Cu^{II}(L^9)] \rightleftharpoons [Cu^{II}(L^9)^\bullet]^+ \rightleftharpoons [Cu^{II}(L^9)^{\bullet\bullet}]^{2+} \quad (8.17)$$

NIR spectroscopy reveals a remarkable feature for $[Cu^{II}(^{Hx}L_8)^\bullet]^+$: it exhibits a high intensity band at 1750 nm attributed to a phenolate–phenoxyl charge transfer.[115] Recently, the X-ray crystal structure of this complex was solved.[118] In contrast to the results of X-ray absorption spectroscopy, the copper atom was found to be in the (+III) oxidation state, whereas the ligand remains on its closed shell phenolate form. These results have been rationalized by considering the existence of the following temperature dependent equilibrium in solution (Equation 8.18):

$$[Cu^{III}(^{Hx}L_8)]^+ \overset{\Delta}{\rightleftharpoons} [Cu^{II}(^{Hx}L_8)^\bullet]^+ \quad (8.18)$$

Several diradical complexes were obtained from salen ligands. The electronic structures of $[Cu^{II}(^{Hx}L_8)^{\bullet\bullet}]^{2+}$ and $[Cu^{II}(L_9)^{\bullet\bullet}]^{2+}$ compare well with that of $[Cu^{II}(^{Et}L_2{}^{OMe})^{\bullet\bullet}]^{2+}$. In contrast, the EPR spectrum of $[Cu^{II}(^{Et}L_8)^{\bullet\bullet}]^{2+}$ is the superposition of two spectra, one typical of an organic radical (narrow signal at g = 2.005), and the other corresponding to a mononuclear copper(II) complex, thus showing that the ethyl linker is too flexible to maintain both phenoxyl moieties coordinated to the metal ion.[116]

Copper(II) complexes of reduced salen ligands are much more readily oxidizable than the corresponding "true" salen, the first electron transfer being for example observed at 0.02 V for $[Cu^{II}(^{Et}L_{10})]$,[119] and their copper(II)–phenoxyl complexes exhibit UV-Vis spectra more typical with bands at ≈410 and 600 nm.[119–122] From a reactivity point of view, $[Cu^{II}(^{Hx}L_{10})^\bullet]^+$ stoichiometrically oxidizes benzyl alcohol in the absence of base. Despite the fact that this complex is a weaker oxidizer than $[Cu^{II}(^{Hx}L_8)^\bullet]^+$, an acceleration of the rate of reaction is observed and attributed to substrate preorganization upon binding.[123] On the other hand, $[Cu^{II}(^{Et}L_{10})^\bullet]^+$ oxidizes several unactivated primary alcohols into aldehydes in the presence of base with turnover numbers up to 32.[119]

$[Cu^{II}(^{Ph}L_{12})]$ is a particular salen complex in the sense that the copper resides in a square pyramidal geometry as in GO. Unlike GO, which has a four-atom donor set, leaving the fifth position vacant for substrate binding, no position is available in $[Cu^{II}(^{Ph}L_{12})]$ for substrate binding and subsequent catalytic activity. Nevertheless, this complex oxidizes benzyl alcoholate in the presence of base.[124,125] This paradox is explained by a pentacoordinated copper(II)–radical → hexacoordinated copper(III)–phenolate conversion induced by coordination of the alcoholate substrate. Catalysis then proceeds by subsequent electron transfer to the metal.

A series of phenol–imidazole copper complexes has attracted a lot of interest,[126–128] as $[Cu^{II}(L_{13}{}^{Ph})(L_{13}{}^{Ph})^\bullet]^+$ was the first phenoxyl radical coordinated to a copper(II) ion that could be crystallized. In its X-ray crystal structure the C–O distance of 1.264(5) Å emphasizes the C=O character of this bond (it is 1.322(5) Å in the corresponding bis-phenolate complex $[Cu^{II}(L_{13}{}^{Ph})_2]$). As expected, the metal–ligand bonds in $[Cu^{II}(L_{13}{}^{Ph})(L_{13}{}^{Ph})^\bullet]^+$ are significantly longer than in $[Cu^{II}(L_{13}{}^{Ph})_2]$. All these bond lengths compared well with those obtained in the chromium(III)–radical complex $[Cr^{III}(L_3{}^{tBuOMe})^\bullet]^+$.

The first functional model of GO involving an *o*-iminobenzosemiquinone was obtained from $H_2L_{15}{}^S$.[129] Mixing this tridentate ligand with copper(II) in the presence of triethylamine and air affords the dinuclear copper(II)–phenoxyl complex $[Cu^{II}{}_2(L_{15}{}^S)^{\bullet\bullet}{}_2]$, which is characterized by an absorption band at 404 nm ($8000\,M^{-1}\,cm^{-1}$) and no signal in its EPR spectrum. Under similar conditions, by replacing $H_2L_{15}{}^S$ by $H_3{}^{AP}L_{15}{}^{AP}$, the diamagnetic mononuclear copper(II)–phenoxyl complex $[Cu^{II}(^{ISQ}L_{15}{}^{AP})^\bullet(NEt_3)]$ was obtained and structurally characterized[130] (Figure 8.18; a similar complex involving two pyridines instead of the triethylamine ligand has been reported previously).[131] The magnetic coupling between the metal and

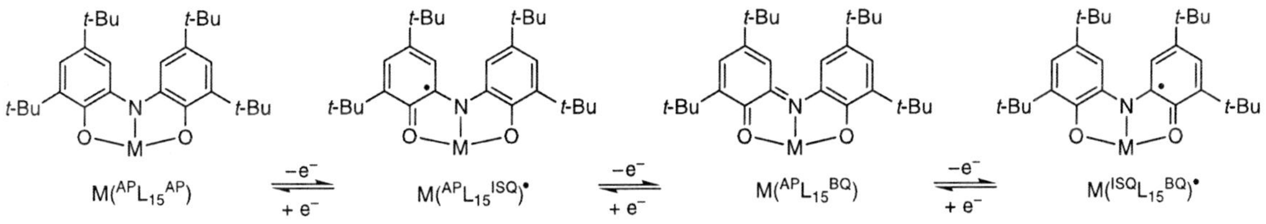

Figure 8.18 *Oxidative behavior of* $H_3{}^{AP}L_{15}{}^{AP}$ *in the presence of metal.*

the radical spin has been estimated to be $-137\,cm^{-1}$. It could be reduced into $[Cu^{II}(^{AP}L_{15}{}^{AP})(NEt_3)]^-$ at -1.06 or oxidized at -0.14 V into $[Cu^{II}(^{AP}L_{15}{}^{BQ})(NEt_3)]^+$.

Both $[Cu^{II}{}_2(L_{15}{}^{S})^{\bullet\bullet}{}_2]$ and $[Cu^{II}(^{AP}L_{15}{}^{ISQ})^{\bullet}(NEt_3)]$, and $[Cu^{II}(L_{15}{}^{Se})^{\bullet}(NEt_3)]$ (the selenium analogue of the latter complex), are efficient catalysts for the aerobic oxidation of alcohols. For instance $[Cu^{II}(^{AP}L_{15}{}^{ISQ})^{\bullet}(NEt_3)]$ oxidizes aerobically a 0.125 M ethanol solution in an overall yield of 55% (2.65×10^{-5} M in catalyst) with release of H_2O_2. The mechanism is very similar to that performed by GO, the reduced catalyst being a copper(I)–phenol species. Interestingly, this is not its sole catalytic activity of the complex of $H_2{}^{Ph}L_{15}{}^{Se}$, as it also reacts with benzylamine to form benzylidinebenzylamine.[132] The reaction is believed to be initiated by aerobic oxidation of the dimer $[Cu^{II}{}_2(L_{15}{}^{Se})_2(NH_2CH_2Ph)_2]$ into a radical complex. A simplified mechanism is depicted below (Equations 8.19–8.21), but note that for clarity only one half of the dimeric complex is shown. The reaction product PhCHO reacts with $PhCH_2NH_2$ to form benzylidinebenzylamine, whereas the reduced form of the catalyst (copper(I) complex) is oxidized by air into the radical dimer.

$$[Cu^{II}(L_{15}{}^{Se})(NH_2CH_2Ph)] \rightarrow [Cu^{II}(L_{15}{}^{Se})^{\bullet}(NH_2CH_2Ph)_2] \text{ in the presence of } O_2 \text{ and } NH_2CH_2Ph \tag{8.19}$$

$$[Cu^{II}(L_{15}{}^{Se})^{\bullet}(NH_2CH_2Ph)_2] \rightarrow [Cu^{II}(HL_{15}{}^{Se})(NH_2CH^{\bullet}Ph)(NH_2CH_2Ph)](\text{H abstraction as GO}) \tag{8.20}$$

$$[Cu^{II}(HL_{15}{}^{Se})(NH_2CH^{\bullet}Ph)(NH_2CH_2Ph)] \rightarrow [Cu^{I}(HL_{15}{}^{Se})(NH{=}CHPh)(NH_2CH_2Ph)]$$
$$+ H_2O \rightarrow [Cu^{I}(HL_{15}{}^{Se})(NH_3)(NH_2CH_2Ph)] + PhCHO \tag{8.21}$$

The copper complex of the tetradentate ligand $H_4{}^{Ph}L_{16}{}^{N}$ could be obtained in five oxidation states (Figure 8.19).[133] $[Cu^{II}(^{ISQ}L_{16}{}^{N})^{\bullet}]^-$ and $[Cu^{II}(^{BQ}L_{16}{}^{N,ISQ})^{\bullet}]^+$ are diamagnetic species due to antiferromagnetic coupling of the copper unpaired electron in a d_{x2-y2} orbital and the radical, whereas the other compounds are paramagnetic with ($S_t = S_{Cu} = 1/2$). $[Cu^{II}(^{ISQ}L_{16}{}^{N})^{\bullet}]^-$ can be oxidized by oxygen to give $[Cu^{II}(^{BQ}L_{16}{}^{N,ISQ})^{\bullet}]^+$ and one equivalent of H_2O_2, whereas $[Cu^{II}(^{BQ}L_{16}{}^{N,ISQ})^{\bullet}]^+$ reacts in the presence of triethylamine with alcohols to give the corresponding aldehyde and $[Cu^{II}(^{ISQ}L_{16}{}^{N})^{\bullet}]^-$. Oxidation of alcohols by these complexes is thus catalytic in the presence of oxygen (and base), with 5000 turnovers achieved in 50 hours for the oxidation of ethanol (turnover frequency of $0.03\ s^{-1}$). Interestingly, the first half-reaction involves substrate binding and further rate limiting hydrogen abstraction (Equations 8.22–8.23), as observed in GO. Nevertheless, in contrast with GO the reduced form of the catalyst is a copper(II) coordinated radical and not a copper(I) complex.

$$[Cu^{II}(^{BQ}L_{15}{}^{N,ISQ})^{\bullet}]^+ + RCH_2OH \rightleftharpoons [Cu^{II}(^{BQ}L_{15}{}^{N,ISQ})^{\bullet}(RCH_2O^-)] + H^+$$
$$\rightarrow [Cu^{II}(^{ISQ}L_{15}{}^{N})^{\bullet}]^- + RCH_2O + 2H^+ \tag{8.22}$$

$$[Cu^{II}(^{ISQ}L_{15}{}^{N})^{\bullet}]^- + 2H^+ + O_2 \rightarrow [Cu^{II}(^{BQ}L_{15}{}^{N,ISQ})^{\bullet}]^+ + H_2O_2 \tag{8.23}$$

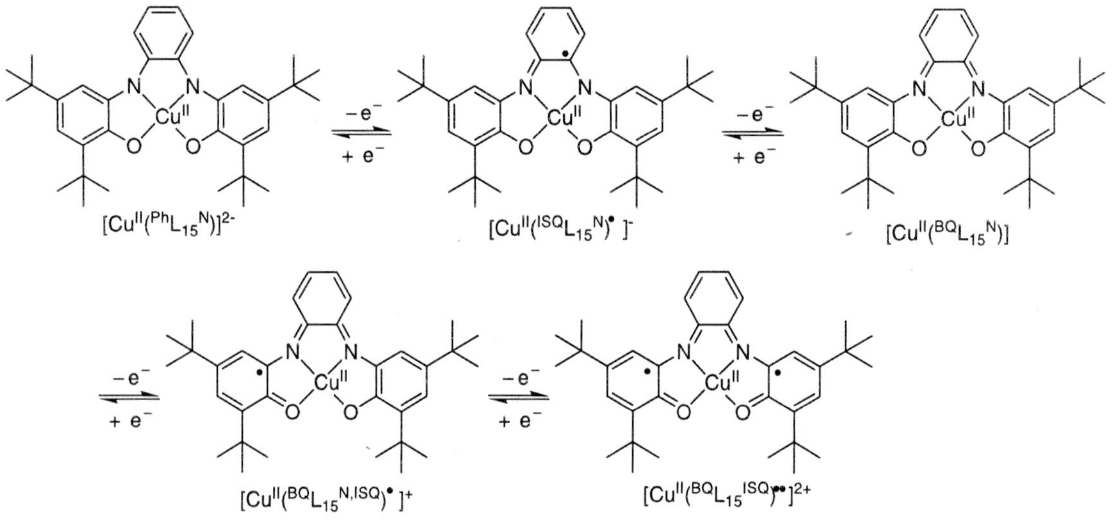

Figure 8.19 *Oxidative behavior of the metal complexes of $H_4{}^{Ph}L_{16}N$. (Adapted with permission from [133]. Copyright 1999 American Chemical Society.)*

Efficient oxidation of benzyl alcohol was also achieved under basic medium with the derivative diradical complex $[Cu^{II}({}^{Ph2}L_{16}{}^{ISQ})^{\bullet\bullet}]$,[134] whereas the radical form of $[Cu^{II}({}^{Ph}L_{16}{}^{S})]$, the sulfur analogue of $[Cu^{II}({}^{Ph}L_{16}{}^{AP})]$ could not be prepared due to its high unstability.[135]

From the prototypical ligand $H_2L_{14}{}^{AP}$ two complexes were isolated. In $[Cu^{II}(L_{14}{}^{ISQ})^{\bullet}(tmtacn)]^+$ the half-filled d_{x2-y2} metal orbital and the π orbital of the ligand radical are orthogonal: a strong *ferromagnetic* coupling is thus observed ($J = 195\,cm^{-1}$) and the ground state is ($S_t = 1$).[96] The diradical complex $[Cu^{II}(L_{14}{}^{ISQ})^{\bullet\bullet}{}_2]$ exhibits strong similarities with $[Ni^{II}(L_{14}{}^{ISQ})^{\bullet\bullet}{}_2]$ both from a structural and electrochemical point of view. A strong *antiferromagnetic* coupling ($\approx -400\,cm^{-1}$) is operative between the two ligand radicals as in $[Ni^{II}(L_{14}{}^{ISQ})^{\bullet\bullet}{}_2]$, and dominates the ferromagnetic interaction between the copper(II) and its adjacent radical ligand. The EPR spectrum of $[Cu^{II}(L_{14}{}^{ISQ})^{\bullet\bullet}{}_2]$ consequently exhibits parameters close to those of isolated ($S_{Cu} = 1/2$) copper(II) ions with a d_{x2-y2} magnetic orbital. From a catalytic point of view, some derivatives like $[Cu^{II}({}^{CF3}L_{14}{}^{ISQ})^{\bullet\bullet}{}_2]$ efficiently oxidize benzyl alcohol.

Many other complexes of interest involving none of the previous general structures have been reported as the tris(3-arylpyrazolyl)hydroborate copper(II) complexes involving 2-hydroxy-5-methyl-3-methylsulfanyl-benzaldehyde or 5-*tert*-butyl-2-hydroxy-3-methylsulfanyl-benzaldehyde as co-ligand. Surprisingly, a coordinated phenoxyl radical can be generated only by using the former co-ligand, thus showing that in some instances the methyl substituent can significantly stabilize radicals.[136,137] Some other scaffolds such as calix[6]arenes functionalized with one di-*tert*-butylphenol and three *N*-methylimidazole groups have been tested. They give rise to relatively stable copper(II)–phenoxyl radicals ($E_{1/2} = 0.32\,V$)[138] able to mediate the stoichiometric two-electron oxidation of benzyl alcohol.

8.4.9 Zinc complexes

The zinc ion has been widely used as structural analogue of copper in complexes mimicking the GO active site. The phenolate–zinc complexes are colorless (d^{10} configuration of the metal ion) and diamagnetic, whereas the monoradicals are paramagnetic ($S = 1/2$) and intensely colored.

Table 8.7 *Redox potentials of representative zinc complexes*

Complex	$E_{1/2}$ [V vs Fc^+/Fc]	Ref.
$[Zn^{II}(^{iPr}L_1{}^{tBu})(Cl)]$	0.21	[99]
$[Zn^{II}(^{iPr}L_1{}^{SMe})(Cl)]$	0.18	[99]
$[Zn^{II}(^{Me}L_1{}^{tBu})(Ph_2acac)]$	−0.09	[139]
$[Zn^{II}(^{Me}L_1{}^{OMe})(Ph_2acac)]$	−0.28	[139]
$[Zn^{II}(^{iPr}L_2{}^{tBu})]$	0.17, 0.50	[99]
$[Zn^{II}(^{Et}L_2{}^{OMe})]$	−0.10, 0.15	[102]
$[Zn^{II}(L_3{}^{tBu,OMe})]^-$	−0.63, −0.28, −0.06	[139]
$[Zn^{II}(HL_3{}^{tBu,OMe})]$	−0.28, −0.05, 0.80(irr)	[139]
$[Zn^{II}(H_2L_3{}^{tBu,OMe})]^+$	0.17, 0.94(irr)	[139]
$[Zn^{II}(H_2L_3{}^{tBu})]^+$	0.39, 1.40(irr)	[139]
$[Zn^{II}(L_5{}^{SMe})(CH_3CN)]^+$	0.42	[104]
$[Zn^{II}(L_5{}^{tBu})(CH_3CN)]^+$	0.41	[104]
$[Zn^{II}(L_7{}^{SMe})(Pyr)]$	0.41(irr)	[109]
$[Zn^{II}{}_2(L_7{}^{tBu})_2]$	0.20, 0.37	[112]
$[Zn^{II}(L^9)(EtOH)]$	0.45, 0.67	[140]
$[Zn^{II}(H^{et,py}L_{10})]^+$	0.39	[121]
$[Zn^{II}(L_{13}{}^{Ph})_2]$	0.23, 0.51	[127]
$[Zn^{II}(L_{13}{}^{Bz})_2]$	0.41, 0.58	[128]
$[Zn^{II}(L_{13}{}^{PhOMe})_2]$	0.18, 0.47	[128]
$[Zn^{II}(^{AP}L_{15}{}^{ISQ})^\bullet(NEt_3)]$	−1.28, −0.29	[141]
$[Zn^{II}(^{AP}L_{15}{}^{BQ})_2]$	−1.28, −1.06, 0.43, 0.83	[141]
$[Zn^{II}(^{Ph}L_{16}{}^{AP})]$	−1.29, −0.64, 0.03, 0.37	[133]

irr: irreversible couple, $E_p{}^a$ is given

The first zinc(II)–phenoxyl radical complexes to be described were those from TACN ligands.[99,102,139] As with their copper(II) analogues, the complexes can be isolated with different protonation states. The electrochemically generated zinc(II)–radical species (Table 8.7) usually exhibit a green-blue color, with the $\pi-\pi^*$ transitions of the phenoxyl radicals being observed at ≈400 nm (their intensities increase with the number of coordinated phenoxyl moieties). Interestingly, the λ_{max} is only slightly shifted when it is compared to that of the copper analogue. The monoradicals exhibit in their EPR spectra an ($S = 1/2$) signal at the expected g_{iso} value of ≈2.005. A strong hyperfine coupling with one of the benzylic hydrogens (up to 6 G) evidences a rigid six-membered coordination ring, thus demonstrating that the radical remains coordinated. The diradical $[Zn^{II}(^{Et}L_2{}^{OMe})^{\bullet\bullet}]^{2+}$ can be obtained, and the two unpaired electrons interact one with each other via exchange and weak dipolar coupling. The magnitudes are −3 and 0.01 cm^{-1} respectively (with $D = 0.014\,cm^{-1}$ and $E/D = 0$), thus the ground state is diamagnetic.[102]

Tripodal ligands involving a 2-methylthio-4-*tert*-butylphenolate (with a N_3O or N_2O_2 coordination sphere) have been developed in order to better understand the influence of the Cys-Tyr cross-link in GO.[104,109] While the monoradical species generated from the zinc(II) bis-phenolate complex $[Zn^{II}{}_2(L_7{}^{SMe})_2]^{2+}$ is poorly stable in solution, $[Zn^{II}(L_5{}^{SMe})(CH_3CN)]^+$ and $[Zn^{II}(L_5{}^{tBu})(CH_3CN)]^+$, which contain a single phenolate moiety, can be oxidized to stable monoradical species. Both radicals exhibit the typical $\pi-\pi^*$ phenoxyl bands at ≈ 410 nm, but there is a remarkable difference between $[Zn^{II}(L_5{}^{SMe})(CH_3CN)]^+$ and $[Zn^{II}(L_5{}^{tBu})(CH_3CN)]^+$ in that only the former exhibits a transition at 887 nm (510 $M^{-1}\,cm^{-1}$) similar to that reported in GO. This transition has been attributed to an intramolecular charge transfer from the benzene ring to the methylthio group. $[Zn^{II}(L_5{}^{SMe})^\bullet(NO_3)]^+$, as its copper analogue, mediates the two-electron oxidation of benzylalcohol. Although the reaction proceeds under first order conditions for the copper complex, it obeys second order kinetics with respect to substrate concentration for the zinc one. The active species is thus different, likely a dimer instead of a monomer for the zinc complex (Equations 8.24–8.25)[111]:

$$[Zn^{II}(L_5{}^{SMe})^{\bullet}]^{+} + PhCH_2OH \rightleftharpoons \tfrac{1}{2}[Zn^{II}(L_5{}^{SMe})^{\bullet}(OCH_2PhCH_2)]_2{}^{2+} + H^{+} \quad (8.24)$$

$$[Zn^{II}(L_5{}^{SMe})^{\bullet}(OCH_2Ph)]_2{}^{2+} \rightarrow 2[Zn^{II}(HL_5{}^{SMe})]^{2+} + PhCHO + PhCH_2OH \quad (8.25)$$

The diradical $[Zn^{II}{}_2(L_7{}^{tBu})^{\bullet\bullet}{}_2]^{2+}$ was also obtained. In this complex the radicals are not bridging (see copper complexes) and well separated, leading to very weak interactions between each of them. Its EPR spectrum thus contrasts sharply with that of $[Zn^{II}({}^{Me}L_2{}^{OMe})^{\bullet\bullet}]^{2+}$ as it is constituted by a single ($S = \tfrac{1}{2}$) signal at $g = 2.006$.[112]

Zinc(II)–phenoxyl radical complexes derived from salen ligands are rare.[140] Although the metal ion is tetracoordinated in the corresponding copper complexes, the zinc ion in $[Zn^{II}(L_9)(EtOH)]$ is pentacoordinated and square pyramidal, with a solvent molecule that occupies the apical position. This complex can be oxidized into the radical species $[Zn^{II}(L_9)^{\bullet}(EtOH)]^{+}$ characterized by an isotropic EPR signal at $g \approx 2.005$. No hyperfine splitting with the benzylic hydrogens could be resolved, as a consequence of their orientation parallel to the phenoxyl ring. The diradical $[Zn^{II}(L_9)^{\bullet\bullet}(EtOH)]^{2+}$ has a diamagnetic ground state, with an exchange interaction J of $-34\,cm^{-1}$ stronger than that observed for diradicals of TACN-type ligands. This highlights the contribution of the rigid planar scaffold in the exchange. The zero field splitting parameters for the excited ($S_t = 1$) state are $|D| = 0.0113\,cm^{-1}$ and $E = 0.11\,cm^{-1}$. The zinc(II)–phenoxyl radical complexes of penta- and hexadentate reduced Schiff bases such as $[Zn^{II}(H^{et,py}L_{10})^{\bullet}]^{2+}$ have been reported, but their low chemical stability at room temperature precluded detailed investigation on its properties.[120,121]

The complexes $[Zn^{II}(L_{13}{}^{Ph})_2]$, $[Zn^{II}(L_{13}{}^{PhOMe})_2]$ and $[Zn^{II}(L_{13}{}^{Bz})_2]$ are isostructural with their copper analogues, and exhibit a similar redox chemistry (although the monoradicals are paramagnetic). The X-ray crystal structure of the zinc(II) radical complex $[Zn^{II}(L_{13}{}^{Ph})(L_{13}{}^{Ph})^{\bullet}]^{+}$ is available, although it could be only partially determined, presumably due to disordered solvent molecules within the crystal lattice.[127] The main point is an elongation of the M–O and M–L bonds upon oxidation of the phenolate into the phenoxyl radical similar to that observed in the copper analogue. The large uncertainties in the C–C, C–N and C–O bonds preclude further detailed structural investigations.

The oxidative chemistry of the zinc complexes of *o*-iminobenzosemiquinonate ligands is illustrated below through two main studies. One study deals with the zinc complex of the tetradentate ligand $H_4{}^{Ph}L_{16}{}^{N}$.[133] Similar to its copper(II) analogue, the zinc compounds can be obtained in five different forms stable enough to be characterized, three of which are radicals. The monoradicals $[Zn^{II}({}^{ISQ}L_{16}{}^{N})^{\bullet}]^{-}$ and $[Zn^{II}({}^{BQ}L_{16}{}^{N,ISQ})^{\bullet}]^{+}$ exhibit the classical transitions at 385 (4200), 401 (4000) and 392 (2900), 411 nm ($3800\,M^{-1}\,cm^{-1}$) respectively. The EPR spectrum of the monoanion $[Zn^{II}({}^{ISQ}L_{16}{}^{N})^{\bullet}]^{-}$ displays a signal at $g = 2.0045$ with hyperfine splittings from only one nitrogen atom and one proton, as expected for a spin density mainly located at the phenylene diamine bridge. In contrast, the spectrum of $[Zn^{II}({}^{BQ}L_{16}{}^{N,ISQ})^{\bullet}]^{+}$ shows hyperfine splittings from two nitrogen atoms and four different protons, indicating delocalization of the unpaired electron over the iminoquinone and both phenolate rings. Finally, $[Zn^{II}({}^{BQ}L_{16}{}^{ISQ})^{\bullet\bullet}]^{2+}$ is EPR silent has a result of strong magnetic coupling between each of the *o*-iminobenzosemiquinonate rings. From a reactivity point of view, $[Zn^{II}({}^{BQ}L_{16}{}^{N,ISQ})^{\bullet}]^{+}$ reacts aerobically in the presence of triethylamine with alcohols similarly to its copper analogue (for which no metal-centered redox activity was observed), although it is less efficient (170 turnovers in 24 hours).

The paramagnetic ($S = \tfrac{1}{2}$) $[Zn^{II}({}^{AP}L_{15}{}^{ISQ})^{\bullet}(NEt_3)]$ complex was crystallized in 1999.[141] It is found to be isomorphous and isostructural with its copper(II) analogue $[Cu^{II}({}^{AP}L_{15}{}^{ISQ})^{\bullet}(NEt_3)]$. More interestingly, in the presence of two equivalents of ligand, two species in which the zinc ion is octahedral could be crystallized. One is the unstable diradical species $[Zn^{II}(^{AP}L_{15}{}^{ISQ})^{\bullet}(^{ISQ}L_{15}{}^{BQ})^{\bullet}]$, in which the spins are essentially uncoupled as in $[Zn^{II}{}_2(L_7{}^{tBu})^{\bullet\bullet}{}_2]^{2+}$. It converts into its valence tautomer that could be also crystallized, the stable diamagnetic $[Zn^{II}(^{AP}L_{15}{}^{BQ})_2]$ complex (Figure 8.18).

8.5 Conclusions

The coordination chemistry of phenoxyl radicals is a main topic in bioinorganic chemistry which emerged in the middle of the 1990s. Many complexes involving the M^{n+}–PhO^- entity (PhO^- denotes the phenolate precursor) have been developped and oxidized, giving insights onto the influence of the metal, ligand field, and so on, on the localization of the oxidation site. Which parameters favor a ligand-centered (affording M^{n+}–$PhO^{\bullet}$ species) rather than a metal-centered process (affording $M^{(n+1)+}$–PhO^-)? Clearly the nature of the metal has an influence as metal-centered processes are, for example, usually observed for the vanadium ion. The ligand has also a profound influence, as aminophenols are more prone than Mannich or Schiff bases to stabilize radicals. In some instances, the kind of coordinating atoms affects the oxidation locus, as observed in manganese complexes, whereas in some nickel and copper species the energy gap between the M^{n+}–$PhO^{\bullet}$ and $M^{(n+1)+}$–PhO^- forms is so low that external stimuli could promote shifts in the oxidation site.

Close inspection of the geometrical parameters in structurally characterized radicals or spectroscopic data also reveals that the delocalization of the spin density could differ notably with respect to the metal complex. For instance, while it is localized in $[Cr^{III}(L_3{}^{tBuOMe})^{\bullet}]$, it is fully delocalized in $[Ni^{II}(^{Hx}L_8)^{\bullet}]$. The geometry of the complex (dictated by the ligand and metal) also obviously affects the orbital overlap between the paramagnetic centers, and thus the magnitude of the magnetic coupling.

Clearly this metal–radical chemistry covers many different aspects of coordination and organic radical chemistry. The work described here shows how it is fascinating, and lots more remains to be done to better understand the fundamentals of this chemistry.

8.6 Abbreviations

Acac: acetylacetonate; Bu_2acac: dibutylacetylacetonate; Ph_2acac: diphenylacetonate; tacn: triaza-cyclononane; dmtacn: dimethyltriazacyclononane; tren: tris(2-aminoethyl)amine; GO: galactose oxidase; DTBC: Di-*tert*-butylcatechol; DTBQ: Di-*tert*-butylquinone; SQ: Semiquinone; Cl_4-cat$^{\bullet}$: Tetrachlorosemiquinonate radical; Py: pyridine. Ton: turnover number.

References

1. R. Pummerer, and F. Frankfurter, *Chem. Ber.*, **47**, 1479 (1914).
2. E. R. Altwicker, *Chem. Rev.*, **67**, 475–531 (1967).
3. A. Ehrenberg, and P. Reichard, *J. Biol. Chem.*, **247**, 3485–3488 (1972).
4. J. L. Pierre, *Chem. Soc. Rev.*, **29**, 251–257 (2000).
5. J. Stubbe, and W. Van Der Donk, *Chem. Rev.*, **98**, 705–762 (1998).
6. C. D. Borman, C. G. Saysell, A. Sokolowski, *et al.*, *Coord. Chem. Rev.*, **190–192**, 771–779 (1999).
7. M. J. McPherson, M. R. Parsons, R. K. Spooner, and C. M. Wilmot, Galactose Oxidase, in *Handbook of metalloproteins* (eds A. Messerschmidt, R. Huber, K. Wieghardt, and T. Poulos), John Wiley and Sons Ltd, Chichester, 2001.
8. J. W. Whittaker, *Chem. Rev.*, **103**, 2347–2363 (2003).

9. B. A. Jazdzewski, and W. B. Tolman, *Coord. Chem. Rev.*, **200–202**, 633–685 (2000).
10. H. J. Krüger, *Angew. Chem. Int. Ed.*, **38**, 627–631 (1999).
11. S. Itoh, M. Taki, and S. Fukuzumi, *Coord. Chem. Rev.*, **198**, 3–20 (2000).
12. P. Chaudhuri, and K. Wieghardt, *Prog. Inorg. Chem.*, **50**, 151–216 (2001).
13. F. Thomas, *Eur. J. Inorg. Chem.*, 2379–2404 (2007).
14. C. G. Pierpont, and R. M. Buchanon, *Coord. Chem. Rev.*, **38**, 45–87 (1981).
15. C. G. Pierpont, and C. W. Lange, *Prog. Inorg. Chem.*, **41**, 331–442 (1994).
16. C. G. Pierpont, *Coord. Chem. Rev.*, **216–217**, 99–125 (2001).
17. B. Adam, E. Bill, E. Bothe, *et al.*, *Chem. Eur. J.*, **3**, 308–319 (1997).
18. S. Bruni, A. Caneschi, F. Cariati, *et al.*, *J. Am. Chem. Soc.*, **116**, 1388–1394 (1994).
19. A. J. Bard, and H. Lund (Eds), Encyclopedia of Electrochemistry of the Elements, vol. XI, Organic section, Marcel Dekker, New York, 1976.
20. D. Zurita, I. Gautier-Luneau, S. Ménage, *et al.*, *J. Biol. Inorg. Chem.*, **2**, 46–55 (1997).
21. Y. Shimazaki, S. Huth, S. Hirota, and O. Yamauchi, *Inorg. Chim. Acta*, **331**, 168–177 (2002).
22. A. Philibert, F. Thomas, C. Philouze, *et al.*, *Chem. Eur. J.*, **9**, 3803–3812 (2003).
23. F. Michel, F. Thomas, S. Hamman, *et al.*, *Chem. Eur. J.*, **10**, 4115–4125 (2004).
24. F. G. Bordwell, and J. P. Cheng, *J. Am. Chem. Soc.*, **113**, 1736–1743 (1991).
25. In order to facilitate comparison between the systems, all electrochemical potentials are given relative to the Fc^+/Fc reference. When potentials are reported vs other standards, corrections are applied according to the following references: N. G. Connelly, and W. E. Geiger, *Chem. Rev.*, **96**, 877–910 (1996); V. V. Pavlishchuk, and A. W. Addison, *Inorg. Chim. Acta*, **298**, 97–102 (2000). Differences in solvent, electrolyte, ionic strength may, however, complicate the comparison.
26. R. D. Webster, *Electroch. Commun.*, **5**, 6–11 (2003).
27. I. M. Kolthoff, and M. K. Chantooni. *J. Am. Chem. Soc.*, **87**, 4428–4436 (1965).
28. T. Maki, Y. Araki, Y. Ishida, *et al.*, *J. Am. Chem. Soc.*, **123**, 3371–3372 (2001).
29. J. M. Mayer, and I. J. Rhile, *J. Am. Chem. Soc.*, **126**, 12718–12719 (2004).
30. I. J. Rhile, T. F. Markle, H. Nagao, *et al.*, *J. Am. Chem. Soc.*, **128**, 6075–6088 (2006).
31. M. Sjödin, T. Irebo, J. E. Utas, *et al.*, *J. Am. Chem. Soc.*, **128**, 13076–13083 (2006).
32. C. Constentin, M. Robert, and J. M. Savéant, *J. Am. Chem. Soc.*, **128**, 4552–4553 (2006).
33. D. E. Williams, *Mol. Phys.*, **16**, 145–151 (1969).
34. V. W. Manner, T. F. Markle, J. H. Freudenthal, *et al.*, *Chem. Commun.*, 256–258 (2008).
35. G. G. Aleksandrov, Y. T. Struchkov, D. I. Kalinin, and M. G. Neigauz, *Zh. Strukt. Khim.*, **14**, 852–858 (1973).
36. S. Goldschmidt and W. Schmidt, *Ber.*, **55**, 3197 (1922).
37. C. D. Cook, C. B. Depathi, and E. S. English, *J. Org. Chem.*, **24**, 1356–1358 (1959).
38. T. N. Das, *J. Phys. Chem. A*, **109**, 3344–3351 (2005).
39. J. G. Radziszewski, M. Gil, A. Gorski, *et al.*, *J. Chem. Phys.*, **115**, 9733–9738 (2001).
40. S. Itoh, S. Takayama, R. Arakawa, *et al.*, *Inorg. Chem.*, **36**, 1407–1416 (1997).
41. A. Y. Bresgunov, A. A. Dubinsky, O. G. Poluektov, *et al.*, *Mol. Phys.*, **75**, 1123–1131 (1992).
42. T. Yamaji, I. S. M. Saiful, M. Baba, *et al.*, *J. Phys. Chem. A*, **111**, 4612–4619 (2007).
43. F. Thomas, O. Jarjayes, H. Jamet, *et al.*, *Angew. Chem. Int. Ed.*, **43**, 594–597 (2004).
44. L. Benisvy, R. Bittl, E. Bothe, *et al.*, *Angew. Chem. Int. Ed.*, **44**, 5314–5317 (2005).
45. See the special issue of *Chem. Rev.* **103** (6) (2003) devoted to radical metalloenzymes.
46. M. M. Whittaker, P. J. Kersten, D. Cullen, and J. W. Whittaker, *J. Biol. Chem.*, **274**, 36226–36232 (1999).
47. N. Ito, S. E. V. Philips, C. Stevens, *et al.*, *Nature*, **350**, 87–90 (1991).
48. N. Ito, S. E. V. Philips, K. D. S. Yadav, and P. F. Knowles, *J. Mol. Biol.*, 794–814 (1994).
49. C. Wright, and A. G. Sykes, *J. Inorg. Biochem.*, **85**, 237–243 (2001).
50. M. M. Whittaker, and J. W. Whittaker, *J. Biol. Chem.*, **263**, 6074–6080 (1988).
51. M. M. Whittaker, C. A. Ekberg, J. Peterson, *et al.*, *J. Mol. Cat. B*, **8**, 3–15 (2000).
52. D. Rokhsana, D. M. Dooley, and R. K. Szilagyi, *J. Am. Chem. Soc.*, **128**, 15550–15551 (2006).
53. M. M. Whittaker, D. P. Ballou, and J. W. Whittaker, *Biochemistry*, **37**, 8426–8436 (1998).

54. B. P. Branchaud, M. P. Montague-Smith, D. J. Kosman, and F. R. Mc Laren, *J. Am. Chem. Soc.*, **64**, 798–800 (1993).
55. R. M. Wachter, and B. P. Branchaud, *Biochim. Biophys. Acta*, **1384**, 43–54 (1998).
56. M. Fontecave, and J. L. Pierre, *Bull. Soc. Chim. Fr.*, **133**, 653–660 (1996).
57. M. M. Whittaker, and J. W. Whittaker, *Biochemistry*, **40**, 7140–7148 (2001).
58. S. G. Minasian, M. M. Whittaker, and J. W. Whittaker, *Biochemistry*, **43**, 13683–13693 (2004).
59. U. Auerbach, T. Weyhermüller, K. Wieghardt, *et al.*, *Inorg. Chem.*, **32**, 508–519 (1993).
60. A. Sokowski, B. Adam, T. Weyhermüller, *et al.*, *Inorg. Chem.*, **36**, 3702–3710 (1997).
61. T. K. Paine, T. Weyhermüller, L. D. Slep, *et al.*, *Inorg. Chem.*, **43**, 7324–7338 (2004).
62. H. Chun, C. N. Verani, P. Chaudhuri, *et al.*, *Inorg. Chem.*, **40**, 4157–4166 (2001).
63. A. Sokolowski, E. Bothe, E. Bill, *et al.*, *Chem. Commun.*, 1671–1672 (1996).
64. J. Müller, A. Kikuchi, E. Bill, *et al.*, *Inorg. Chim. Acta*, **297**, 265–277 (2000).
65. A. dos Anjos, A. J. Bortoluzzi, M. S. B. Caro, *et al.*, *J. Braz. Chem. Soc.*, **17**, 1540–1550 (2006).
66. H. Chun, P. Chaudhuri, T. Weyhermüller, and K. Wieghardt, *Inorg. Chem.*, **41**, 790–795 (2002).
67. K. S. Min, T. Weyhermüller, and K. Wieghardt, *Dalton Trans.*, 178–186 (2004).
68. T. K. Paine, T. Weyhermüller, E. Bothe, *et al.*, *Dalton Trans.*, 3136–3144 (2003).
69. S. Mukherjee, E. Rentschler, T. Weyhermüller, *et al.*, *Chem. Commun.*, 1828–1829 (2003).
70. S. Mukherjee, T. Weyhermüller, E. Bothe, *et al.*, *Dalton Trans.*, 3842–3853 (2004).
71. J. F. Berry, E. Bill, E. Bothe, *et al.*, *Science*, **312**, 1937–1941 (2006).
72. J. Hockertz, S. Steenken, K. Wieghardt, and P. Hildebrandt, *J. Am. Chem. Soc.*, **115**, 11222–11230 (1993).
73. M. D. Snodin, L. Ould-Moussa, U. Wallman, *et al.*, *Chem. Eur. J.*, **5**, 2554–2565 (1999).
74. A. K. Nairn, R. Bhalla, S. P. Foxon, *et al.*, *J. Chem. Soc., Dalton Trans.*, 1253–1255 (2002).
75. H. Chun, E. Bill, E. Bothe, *et al.*, *Inorg. Chem.*, **41**, 5091–5099 (2002).
76. K. S. Min, T. Weyhermüller, and K. Wieghardt, *Dalton Trans.*, 1126–1132 (2003).
77. H. Chun, T. Weyhermüller, E. Bill, and K. Wieghardt, *Angew. Chem. Int. Ed.*, **40**, 2489–2492 (2001).
78. H. Chun, E. Bill, T. Weyhermüller, and K. Wieghardt, *Inorg. Chem.*, **42**, 5612–5620 (2003).
79. S. Mukherjee, T. Weyhermüller, E. Bill, *et al.*, *Inorg. Chem.*, **44**, 7099–7108 (2005).
80. C. N. Verani, E. Bothe, D. Burdinski, *et al.*, *Eur. J. Inorg. Chem.*, 2161–2169 (2001).
81. U. Beckmann, E. Bill, T. Weyhermüller, and K. Wieghardt, *Eur. J. Inorg. Chem.*, 1768–1777 (2003).
82. A. Dei, D. Gatteschi, C. Sangregorio, *et al.*, *Inorg. Chem.*, **42**, 1701–1706 (2003).
83. L. Benisvy, E. Bill, A. J. Blake, *et al.*, *Dalton Trans.*, 3647–3653 (2004).
84. Y. Shimazaki, R. Kabe, S. Huth, *et al.*, *Inorg. Chem.*, **46**, 6083–6090 (2007).
85. D. Herebian, P. Ghosh, H. Chun, *et al.*, *Eur. J. Inorg. Chem.*, 1957–1967 (2002).
86. C. N. Verani, S. Gallert, E. Bill, *et al.*, *Chem. Commun.*, 1747–1748 (1999).
87. Y. Shimazaki, S. Huth, S. Karasawa, *et al.*, *Inorg. Chem.*, **43**, 7816–7822 (2004).
88. Y. Shimazaki, F. Tani, K. Fului, *et al.*, *J. Am. Chem. Soc.*, **125**, 10512–10513 (2003).
89. T. Glaser, M. Heidemeier, R. Frölich, *et al.*, *Inorg. Chem.*, **44**, 5467–5482 (2005).
90. O. Rotthaus, O. Jarjayes, C. Perez Del Valle, *et al.*, *Chem. Commun.*, 4462–4464 (2007).
91. L. Benisvy, R. Kannappan, Y. F. Song, *et al.*, *Eur. J. Inorg. Chem.*, 631–633 (2007).
92. O. Rotthaus, O. Jarjayes, F. Thomas, *et al.*, *Chem. Eur. J.*, **12**, 2293–2302 (2006).
93. Y. Shimazaki, T. Yajima, F. Tani, *et al.*, *J. Am. Chem. Soc.*, **129**, 2559–2568 (2007).
94. T. Storr, E. C. Wasinger, R. C. Pratt, and T. D. P. Stack, *Angew. Chem. Int. Ed.*, **46**, 5198–5201 (2007).
95. O. Rotthaus, F. Thomas, O. Jarjayes, *et al.*, *Chem. Eur. J.*, **12**, 6953–6962 (2006).
96. P. Chaudhuri, C. N. Verani, E. Bill, *et al.*, *J. Am. Chem. Soc.*, **123**, 2213–2223 (2001).
97. K. S. Min, T. Weyhermüller, E. Bothe, and K. Wieghardt, *Inorg. Chem.*, **43**, 2922–2931 (2004).
98. J. A. Halfen, V. G. Young Jr, and W. B. Tolman, *Angew. Chem. Int. Ed.*, **35**, 1687–1690 (1996).
99. J. A. Halfen, B. A. Jazdzewski, S. Mahapatra, *et al.*, *J. Am. Chem. Soc.*, **119**, 8217–8227 (1997).
100. A. Sokolowski, H. Leutbecher, T. Weyhermüller, *et al.*, *J. Biol. Inorg. Chem.*, **2**, 444–453 (1997).
101. J. Müller, T. Weyhermüller, E. Bill, *et al.*, *Angew. Chem. Int. Ed.*, **37**, 616–619 (1998).
102. E. Bill, J. Müller, T. Weyhermüller, and K. Wieghardt, *Inorg. Chem.*, **38**, 5795–5802 (1999).

103. Y. Shimazaki, S. Huth, S. Hirota, and O. Yamauchi, *Bull. Chem. Soc. Jpn.*, **73**, 1187–1195 (2000).
104. S. Itoh, M. Taki, H. Kumei, *et al.*, *Inorg. Chem.*, **39**, 3708–3711 (2000).
105. M. Taki, H. Hattori, T. Osako, *et al.*, I *norg. Chim. Acta*, **357**, 3369–3381 (2004).
106. F. Michel, S. Hamman, F. Thomas, *et al.*, *Chem. Commun.*, 4122–4124 (2006).
107. Y. Shimazaki, S. Huth, A. Odani, and O. Yamauchi, *Angew. Chem. Int. Ed.*, **39**, 1666–1669 (2000).
108. F. Thomas, G. Gellon, I. Gautier-Luneau, *et al.*, *Angew. Chem. Int. Ed.*, **41**, 3047–3050 (2002).
109. M. Taki, H. Kumei, S. Nagatomo, *et al.*, *Inorg. Chim. Acta*, **300-302**, 622–632 (2000).
110. F. Michel, S. Torelli, F. Thomas, *et al.*, *Angew. Chem. Int. Ed.*, **44**, 438–440 (2005).
111. S. Itoh, M. Taki, S. Takayama, *et al.*, *Angew. Chem. Int. Ed.*, **38**, 2774–2776 (1999).
112. A. Mukherjee, F. Lloret, and R. Mukherjee, *Inorg. Chem.*, **47**, 4471–4480 (2008).
113. Y. Wang, and T. D. P. Stack, *J. Am. Chem. Soc.*, **118**, 13097–13098 (1996).
114. Y. Wang, J. L. Dubois, B. Hedman, *et al.*, *Science*, **279**, 537–540 (1998).
115. R. C. Pratt, and T. D. P. Stack, *J. Am. Chem. Soc.*, **125**, 8716–8717 (2003).
116. F. Thomas, O. Jarjayes, C. Duboc, *et al.*, *Dalton Trans.*, 2662–2669 (2004).
117. I. Sylvestre, J. Wolowska, C. A. Kilner, *et al.*, *Dalton Trans.*, 3241–3249 (2005).
118. T. Storr, P. Verma, R. C. Pratt, *et al.* k, *J. Am. Chem. Soc.*, **130**, 15448–15459 (2008).
119. E. Saint-Aman, S. Ménage, J.-L. Pierre, *et al.*, *New J. Chem.*, 393–394 (1998).
120. M. Vaidyanathan, M. Palaniandavar, and R. S. Gopalan, *Ind. J. Chem.*, **42A**, 2210–2222 (2003).
121. A. dos Anjos, A. J. Bortoluzzi, B. Szpoganicz, *et al.*, *Inorg. Chim. Acta*, **358**, 3106–3114 (2005).
122. A. dos Anjos, A. J. Bortoluzzi, R. E. H. M. B. Osorio, *et al.*, *Inorg. Chem. Comm.*, **8**, 249–253 (2005).
123. R. C. Pratt, and T. D. P. Stack, *Inorg. Chem.*, **44**, 2367–2375 (2005).
124. A. K. Nairn, S. J. Archibald, R. Bhalla, *et al.*, *Dalton Trans.*, **1**, 172–176 (2006).
125. E. Zueva, P. H. Walton, and J. E. McGrady, *Dalton Trans.*, **1**, 159–167 (2006).
126. L. Benisvy, A. J. Blake, D. Collison, *et al.*, *Chem. Commun.*, 1824–1825 (2001).
127. L. Benisvy, A. J. Blake, D. Collison, *et al.*, *Dalton Trans.*, 1975–1985 (2003).
128. L. Benisvy, E. Bill, A. J. Blake, D. Collison, *et al.*, *Dalton Trans.*, 258–267 (2006).
129. P. Chaudhuri, M. Hess, U. Flörke, and K. Wieghardt, *Angew. Chem. Int. Ed.*, **37**, 2217–2220 (1998).
130. P. Chaudhuri, M. Hess, T. Weyhermüller, and K. Wieghardt, *Angew. Chem. Int. Ed.*, **38**, 1095–1098 (1999).
131. G. Speier, J. Csihony, A. M. Whalen, and C. G. Pierpont, *Inorg. Chem.*, **35**, 3519–3524 (1996).
132. T. K. Paine, T. Weyhermüller, K. Wieghardt, and P. Chaudhuri, *Dalton Trans.*, 2092–2101 (2004).
133. P. Chaudhuri, M. Hess, J. Müller, *et al.*, *J. Am. Chem. Soc.*, **121**, 9599–9610 (1999).
134. P. Chaudhuri, K. Wieghardt, T. Weyhermüller, *et al.*, *Biol. Chem.*, **386**, 1023–1033 (2005).
135. T. Kruse, and T. Weyhermüller, K. Wieghardt, *Inorg. Chim. Acta*, **331**, 81–89 (2002).
136. M. A. Halcrow, L. M. L. Chia, X. Liu, *et al.*, *Chem. Commun.*, 2465–2466 (1998).
137. I. Sylvestre, J. Wolowska, E. J. L. McInnes, *et al.*, *Inorg. Chim. Acta*, **358**, 1337–1341 (2005).
138. O. Sénèque, M. Campion, B. Douziech, *et al.*, *Dalton Trans.*, 4216–4218 (2003).
139. A. Sokolowski, J. Müller, T. Weyhermüller, *et al.*, *J. Am. Chem. Soc.*, **119**, 8889–8900 (1997).
140. O. Rotthaus, O. Jarjayes, F. Thomas, *et al.*, *Dalton Trans.*, 889–895 (2007).
141. P. Chaudhuri, M. Hess, K. Hildenbrand, *et al.*, *Inorg. Chem.*, **38**, 2781–2790 (1999).

9

The Synthesis and Characterization of Stable Radicals Containing the Thiazyl (SN) Fragment and Their Use as Building Blocks for Advanced Functional Materials

Robin G. Hicks
Department of Chemistry, University of Victoria, Victoria, Canada

9.1 Introduction

Many radicals which qualify as stable (isolable) or persistent (lifetime of hours or days in solution) are based on first row elements, particularly carbon, nitrogen, and oxygen. Examples include simple inorganic species (NO, NO_2, O_2) and a wide range of organic or "organomain group" radicals, such as nitroxides, hydrazyls, phenalenyls, and many others. The stability of these compounds arises from a combination of effects endemic to the chemistry of the first row p-block elements. Firstly, these elements are capable of engaging in effective π bonding, which can provide a framework for the stabilization of the unpaired electron in a delocalized π framework. In addition, the elements nitrogen and oxygen often have additional electron lone pairs. The well known "alpha effect" – the enhanced reactivity of nitrogen- and oxygen-based nucleophiles when catenated to a second lone-pair-containing heteroatom[1] – is believed to be based on repulsive interactions between lone pairs on adjacent atoms. This effect almost certainly also contributes to the inherently low strength of N–N, N–O, and O–O σ bonds. When the bond in question is between fragments which are themselves already part of another N–N, O–O, or N–O bond, the longer chain of catenated chain of lone-pair-rich heteroatoms becomes thermodynamically unstable. For example, the peroxide-type dimers R_2NOONR_2 of nitroxide radicals $R_2NO^\bullet$ are non-existent. In short, stable radicals based on the first row elements benefit from efficient π delocalization and the low (often negligible) σ bond strength in the putative dimers.

The stabilizing effects of delocalization and lone pair repulsion are significantly diminished beyond the first row of the periodic table. In the second row of the p-block and beyond, the larger size of the valence orbitals makes for more diffuse lone pairs; the alpha effect in heavier atom chemistry is not well documented but is anticipated to be less important than is known for nitrogen and oxygen. In addition, most heavy main group elements show a strong preference for σ over π bonding.[2] As a result, the many different examples of stable (isolable) radicals based on heavy p-block elements rely heavily on bulky substituents for their kinetic and thermodynamic stability.[3] In the absence of steric bulk, most radicals in which substantial spin density is located on a heavy main group element are not stable or even persistent.

The lone and dramatic exception is sulfur, specifically when linked to nitrogen. To be sure, the differences in atomic properties of sulfur and oxygen (atomic radius, electronegativity, σ/π energetic) do produce distinctive chemistries in many respects, including in the structures and stabilities of radical species. The most fundamental distinction can be found in the common forms of the two elements (O_2 vs S_8), and the stable open shell species O_2, NO, and NO_2 have no counterparts in sulfur analogues as stable entities. The dissimilarities between the two chalcogens is further demonstrated by the fact that while stable nitroxides $R_2NO^{\bullet}$ are ubiquitous, the corresponding thionitroxides $R_2NS^{\bullet}$ are not stable and dimerize to disulfides R_2NSSNR_2.[4]

However, the thiazyl (herein defined as an unsaturated SN unit in which both sulfur and nitrogen are two-coordinate) π bond is of considerable strength and it is instructive to compare this fundamental building block of thiazyl chemistry with a carbon–carbon π bond (Figure 9.1).[5] The –S=N–moiety has a three-electron π bond, with the third electron in a π^* orbital. Conjugated species containing multiple SN units have often been referred to as "electron rich" because as they have more π electrons than atoms.[6] However, the higher electronegativity of both sulfur and nitrogen relative to carbon provides some stability to these "electron rich" species. For example, the cyclic S_3N_3 anion (Section 9.2.2) is a perfectly stable,

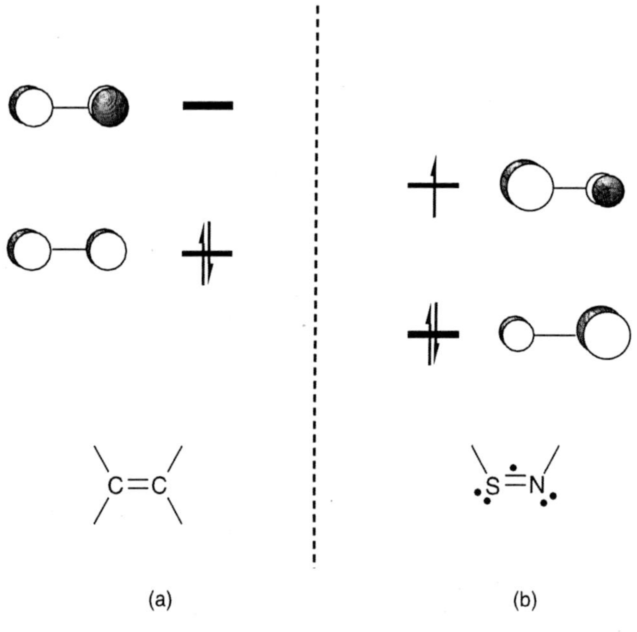

Figure 9.1 Qualitative π molecular orbital diagrams for C_2H_4 (left) and HSNH (right).

π-conjugated six-membered ring with ten π electrons; the corresponding isovalent (10π) hydrocarbon would be the exceedingly reactive benzene tetra-anion.

The collective properties of the thiazyl moiety are the foundation behind an incredibly diverse array of structure types which have open shell configurations and yet qualify as persistent or stable. Moreover, as a consequence of the differences between the elements noted above, NS- and NO-containing radical chemistries are entirely different from one another. Thiazyl radicals have been actively studied since the 1970s and are now well established from both fundamental and applied (materials) perspectives. This chapter has as its central focus the many classes of stable neutral radicals built from the thiazyl (SN) unit. The synthesis, electronic, molecular structures, and reactivity of different thiazyl radical types is the central focus; a separate section introducing the reader to the various uses of these radicals in advanced materials design is also presented.

The scope of thiazyl radical chemistry has grown tremendously over the past 30 years. The field is now very large and there are many general and specific reviews relevant to this chapter. Reviews on specific aspects of thiazyl radicals are cited in the appropriate sections. For general aspects of π-conjugated sulfur–nitrogen compounds, readers are directed to the older reviews of (among many others) Heal,[7] Gleiter,[8] Chivers,[6] and Oakley.[5] The recent book by Chivers provides the most up-to-date coverage of sulfur–nitrogen chemistry.[9] Among more topic-specific reviews, the electron paramagnetic resonance (EPR) spectroscopy of thiazyl radicals was reviewed some 20 years ago by Preston and Sutcliffe[10]; the electrochemical characterization of cyclic thiazyl radicals (among other species) was compiled in 2000 by Boeré,[11] and the relatively new area of thiazyl radical coordination chemistry has been recently reviewed by Preuss.[12] Kaszynski has compared the ability of *ab initio* and DFT computational methods to predict a variety of spectroscopic and structural parameters for thiazyls.[13]

9.2 Radicals based exclusively on sulfur and nitrogen

So-called "binary" sulfur–nitrogen compounds (those only containing sulfur and nitrogen) were historically one of the more structurally intriguing – and difficult to understand – classes of main group systems. Many compounds of this type are thermally unstable and shock sensitive, features which arise from the thermodynamic driving force of N_2 and S_8 formation. Furthermore, reactions of sulfur–nitrogen compounds often involve extensive skeletal rearrangements which proved challenging to understand from a mechanistic perspective. Nonetheless, this field has matured considerably in the past three decades, to the point where the geometric and electronic structures and reactivity patterns of sulfur–nitrogen compounds are now reasonably well understood.[9] Within the context of radical species, the few stable binary sulfur–nitrogen radicals are quite different in nature from their nitrogen oxide counterparts. The sections below cover the structurally important SN radicals, although some do not qualify as stable. Boeré and Chivers have compiled calculated and (where available) EPR spectroscopic characterization of a wide range of binary SN compounds, both neutral and charged.[14]

9.2.1 $NS^•$ and $SNS^•$

The $NS^•$ and $SNS^•$ radicals, sulfur analogues of the ubiquitous $NO^•$ and $^•NO_2$, are mentioned briefly here even though they are not stable radicals in any practical sense. The electrochemical reduction of the thionitrosonium cation $[NS]^+$ is irreversible, leading to the $S_3N_2{}^{•+}$ radical cation and other radical-containing species, indicating that radical associations and rearrangements take place upon reduction.[15] A variety of spectroscopic and computational studies have been reported for the $NS^•$[16] and $SNS^•$[17,18] radicals. Interestingly, condensed phase studies (Ar matrix, cryogenic temperatures) on $SNS^•$ indicate the presence of an asymmetric isomer, $NSS^•$.[17] Thionitrosyl complexes [M]–NS, though less well developed than the corresponding nitrosyl metal chemistry,[19] have received some attention,[20] although the source of the thionitrosyl ligand is invariably something other than NS$^•$ itself.

9.2.2 $S_3N_3^•$

The cyclic radical $S_3N_3^•$ (**1**) has long been recognized as an important, but elusive, binary species. This radical is one of the main gas phase species (detected using photoelectron spectroscopy) from the vaporization of poly(thiazyl), $(SN)_x$ (see below) or from the pyrolysis of the cage molecule S_4N_4 (**2**).[21,22] The prediction of a planar, cyclic structure for **1** was supported by calculations using varying levels of theory (*ab initio* HF,[22] DFT[14] and multiconfigurational[23] studies), although the ring structure is predicted to be slightly distorted from the ideal D_{3h} structure; the incomplete occupancy of the upper degenerate π molecular orbitals is predicted to lead to a Jahn–Teller distortion to a C_{2v} symmetry structure (Figure 9.2). The corresponding closed shell anion $S_3N_3^-$ is a stable species whose structure and reaction chemistry have been explored,[24,25] but the seemingly facile entry into the neutral radical via one-electron oxidation of $S_3N_3^-$ – either chemically or electrochemically – has failed to lead to detectable quantities of the neutral radical in solution.[14,15,25,23]

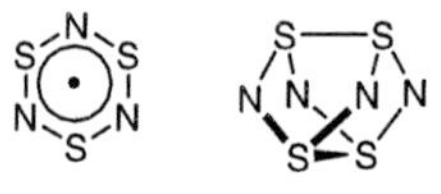

1 2

9.2.3 $S_3N_2^{•+}$ and related radical cations

Although not a neutral species, the S_3N_2 radical cation **3** deserves mention because it is the chronological and conceptual forerunner to a large number of 7π electron heterocyclic (carbon-containing) thiazyl radicals

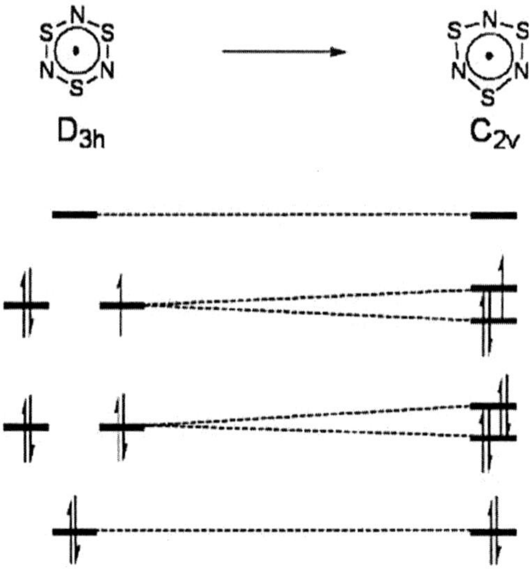

Figure 9.2 *Qualitative π-molecular orbital diagram illustrating the first order Jahn–Teller distortion for $S_3N_3^•$. The magnitude of the structural change indicated is exaggerated.*

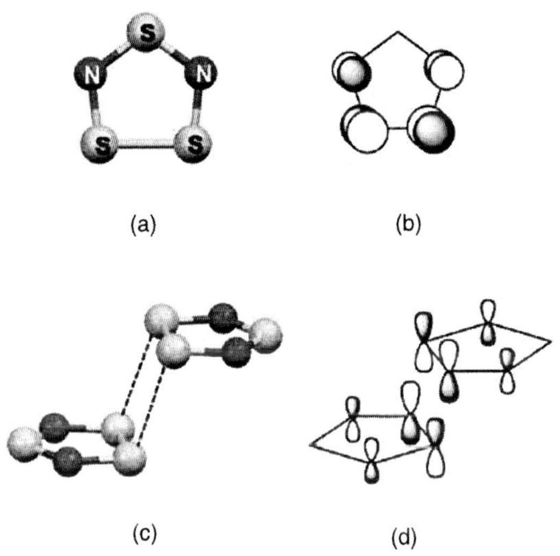

Figure 9.3 *(a) structure of $S_3N_2{}^{\bullet+}$ **3**; (b) π SOMO of **3**; (c) π dimer of **3**; (d) HOMO of the π dimer.*

discussed in subsequent sections. Salts of this cation are produced from the reaction of S_4N_4 with a variety of reagents, including oxidants,[26,27] protic acids,[27,28] and Lewis acids.[29,30] Several X-ray structures of S_3N_2 radical ion salts have been reported,[26,27,29,31,32] all of which consist of five-membered rings associated into dimers (Figure 9.3a,c). The S–N bond lengths within the rings (typically ~1.6 Å for the disulfide S–N bonds and ~1.56 Å for the SN bond to the unique S) are consistent with a delocalized π system, for which the singly occupied molecular orbital (SOMO) is a π^* orbital with a_2 symmetry (Figure 9.3b). EPR[28,33,34] and computational[35] studies corroborate this electronic structure description.

The S_3N_2 radical cations associate reversibly in solution; the dimerization enthalpy has been measured to be −11 kcal/mol.[33] In the solid state the rings associate into centrosymmetric dimers (Figure 9.3c). The closest contacts within the dimer are between the disulfide sulfur atoms, which are approximately 3 Å apart – a distance which is intermediate between a typical S–S covalent bond (~2.1 Å) and a van der Waals contact distance (~3.6 Å). The nature of the interaction between the S_3N_2 rings was discussed initially in terms of a four-center, two-electron bond[29,31]; these can now be regarded as the first structurally characterized examples of "π dimers", which are now known to be pervasive in the structural chemistry of many different kinds of stable, flat, delocalized radicals. The radical–radical interaction involves σ-type overlap between two π SOMOs (the bonding combination of SOMOs, i.e., the HOMO of the dimer, is depicted in Figure 9.3d) but noncovalent interactions between the S_3N_2 cations and counter-anions (Cl^-, $AsF_6{}^-$, etc.) are also believed to contribute to the stability of the dimers of **3** in the solid state.

Several selenium-containing variants of $S_3N_2{}^{\bullet+}$ (**3b–3e**) are also known.[36] The structures of all of these radical cations are isostructural to that of $S_3N_2{}^{\bullet+}$, that is, centrosymmetric π dimers. The Se–Se and Se–S inter-ring "bonds" within the dimers for **3c, 3d** and **3e** are all approximately 3.1 Å, that is, much longer than normal E–E′ σ bonds but much shorter than non-bonding contact distances.

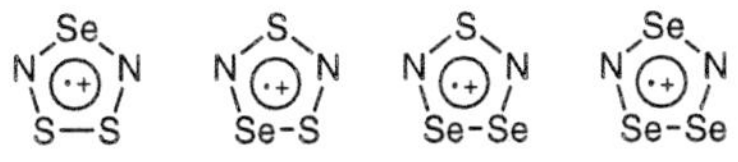

3b **3c** **3d** **3e**

9.2.4 Poly(thiazyl), $(SN)_x$

The discovery in the 1970s of the metallic and superconducting properties of poly(thiazyl), $(SN)_x$, is often cited as the catalyst behind the resurgence of interest in sulfur–nitrogen chemistry, particularly radical-based compounds. The preparation and properties of $(SN)_x$ have been comprehensively reviewed first by Labes *et al.*[37] and subsequently by Banister and Gorrell[38]; an overview of the chemical and structural aspects are presented here, while the physical properties are discussed in Section 9.4.

$(SN)_x$ was originally made by the solid state polymerization of S_2N_2 **4** (the mechanism of which has received significant attention[39]), which in turn is synthesized by heating S_4N_4 over silver wool (Scheme 9.1). This method produces the highest quality polymeric material, though the S_2N_2 polymerization can also be triggered photochemically in solution.[40] These methods suffer from the shock sensitivity of both S_4N_4 and S_2N_2. Although some improvements have been made to this procedure,[41] effort has gone into the development of alternative synthetic routes to the polymer which bypass these capricious compounds. The most effective methods (Scheme 9.1) involve reactions of thiazyl halides with azides,[42] or the reduction of sulfur–nitrogen cations.[43]

The polymer is, in fact, a highly crystalline substance, forming (from the solid state polymerization of S_2N_2) as fibrous golden needles.[44] The structure of $(SN)_x$ (Figure 9.4) is comprised of planar, puckered (*cis–trans*) SN chains. The two distinct SN bonds are very close in length (1.59 and 1.63 Å) and are indicative of significant and delocalized π bonding. There are also contacts *between* chains which are substantially below van der Waals contact distances (~3.1 Å from the originally published X-ray diffraction study[45]; a subsequent neutron diffraction study produced somewhat longer interchain contacts[46]). $(SN)_x$, like its precursors S_4N_4 and S_2N_2, is not reactive to water but, *unlike* its molecular precursors, the polymer is not shock sensitive. Perhaps mot interestingly the polymer can be vaporized (depolymerized) and re-condensed; in the gas phase several small molecules have been detected, including the S_3N_3 radical (see above), S_2N_2 and S_4N_4.

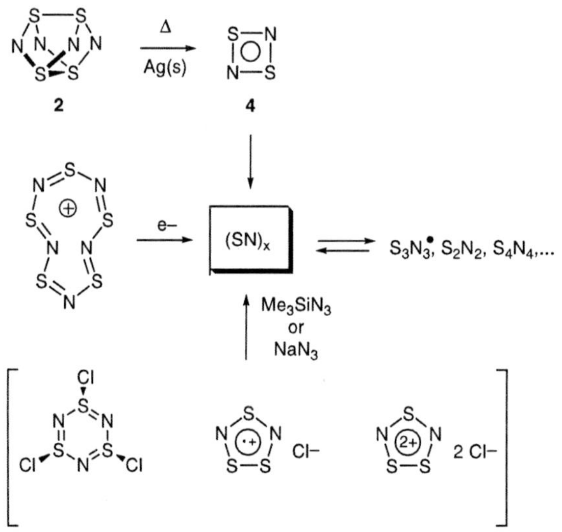

Scheme 9.1

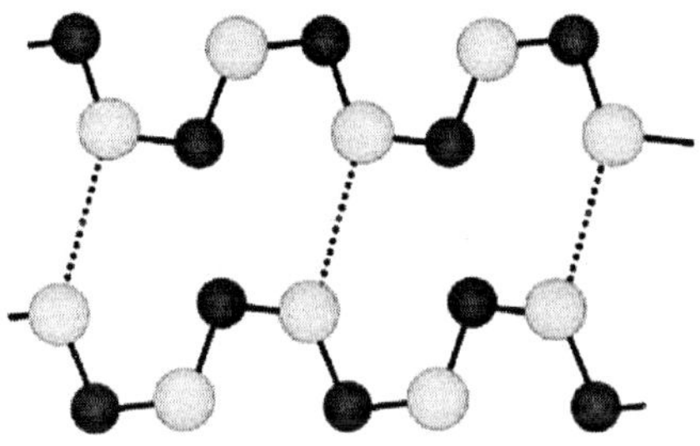

Figure 9.4 Schematic of the structure of $(SN)_x$. Dashed lines indicate interchain S–S contact; see text for details.[45]

9.3 "Organothiazyl" radicals

Many different classes of thiazyl radicals exist in which "SN" units are combined with carbon – often (but not always) in cyclic structures – and, as such, lie at the interface of organic and inorganic chemistry. These "organothiazyl" radicals generally do not suffer from the shock sensitivity which plagues many binary sulfur–nitrogen compounds, and the synthetic routes tend to be less mechanistically complex. Moreover, the organic components of many of these radicals – either carbon as a *con*stituent of a π conjugated radical or as a *sub*stituent on nitrogen, sulfur or carbon – open up the possibility of making several derivatives of a particular radical class, and therefore of controlling the chemical and physical properties of these radical classes.

9.3.1 Thioaminyl radicals

Thioaminyl radicals $[RNSR']^{\bullet}$ **5** are the simplest thiazyl-based "organic" radicals. First made in the 1970s, these radicals have been studied predominantly by Miura. Thioaminyls are normally made by oxidation of sulfenamides **6**, which are in turn made from the condensation of amines (RNH_2) and sulfenyl chlorides (ClSR′) (Scheme 9.2). The radicals can also be made by photolysis of dimercaptoamines **8** $RN(SR')_2$.[47]

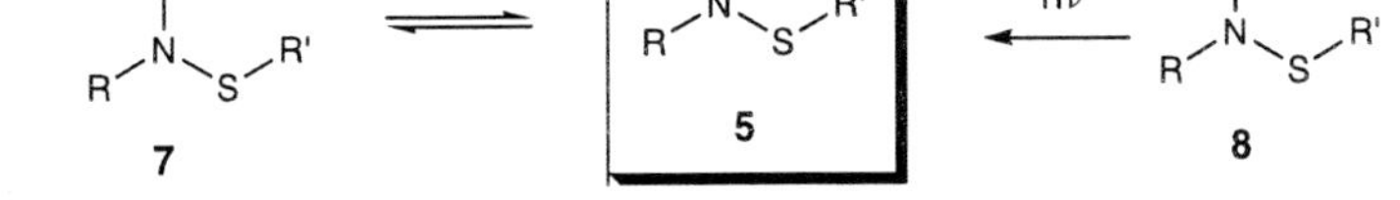

Scheme 9.2

Most derivatives of **5** have alkyl or aryl substituents, though other R groups have also been pursued (see below). In general, these radicals qualify as persistent, having lifetimes of at least a few hours, and many persist for days or longer. Many, but not all, thioaminyls appear to be unreactive to oxygen. The longest lived of these radicals have relatively large aromatic groups on nitrogen and electron withdrawing aromatic groups on sulfur. In particular, N-arylthioaminyls, which have *either* an N-aryl group with two large *ortho* substituents such as *t*-butyl[48] or phenyl,[49] *or* is a substituted pyrene,[50] can be isolated as pure materials, and several of these have been structurally characterized (see below). Derivatives with smaller substituents reversibly associate in solution to dimers which can in some cases be isolated. The dimers are most likely (based on characterization data) NN bonded, hydrazine-type structures **7** (Scheme 9.2). The dimerization enthalpies (ΔH_{dim}) for a few derivatives have been determined through variable temperature EPR studies. When R = alkyl, ΔH_{dim} is approximately 20 kcal/mol,[47] but when R = aryl the delocalization of spin density into the aromatic substituent (see below) weakens the dimer bond to 13–14 kcal/mol.[51]

Among the radicals labeled as 'persistent', detailed studies of the decomposition products of two thioaminyl derivatives have been reported. A deoxygenated benzene solution of thioaminyl **5a** (in equilibrium with its hydrazine-like dimer) was left for three months, after which no radical remained.[52] The major products (Scheme 9.3) were the corresponding sulfenamide **6a**, bis(arylthio)amine **8a**, and other products (**9, 10**) lacking sulfur altogether. Collectively the array of products implicates S–N bond scission (among other things) in the decomposition process. In contrast, heating a solution of the nominally stable thioaminyl radicals **5b** and **5c** gives in high yield (Scheme 9.4) two products – sulfenamide **6b** and the *ortho*-phenyl cyclized product **11** – which strongly suggest a radical disproportionation (hydrogen atom transfer).

EPR spectroscopy has been employed extensively to probe the electronic structure of thioaminyls (many of the early thioaminyls were generated exclusively for EPR studies and were not explicitly synthesized or isolated). From the large collection of available spectroscopic data the following trends can be identified: (1) the nitrogen hyperfine coupling constant a(N) is sensitive to the nitrogen substituent R and, to a much smaller extent, the sulfur substituent R′; (2) The largest a(N) values – between 12.0 and 12.5 G – are achieved when there is no possibility of delocalization onto the N–R group, for example, R = alkyl[47,53] or aromatic derivative such as 2,4,6-tri-*t*-butylphenyl, in which the bulk of the *t*-butyl groups forces the N-phenyl ring to be nearly perpendicular to the N–S bond[48,54]; (3) N-aryl substituents, which are able to maintain some degree of coplanarity with the thioaminyl unit, permit spin density to delocalize onto the substituent, as evidenced by smaller a(N) values (8–10 G) and often measurable, even substantial,

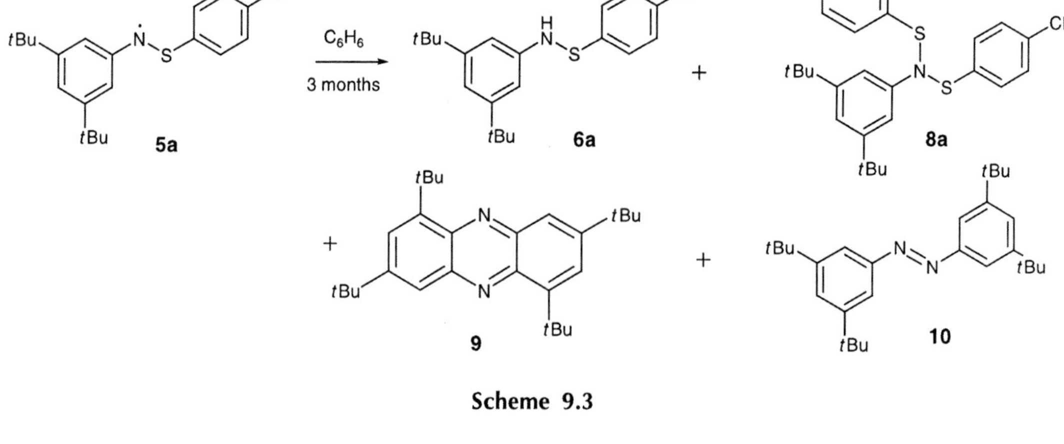

Scheme 9.3

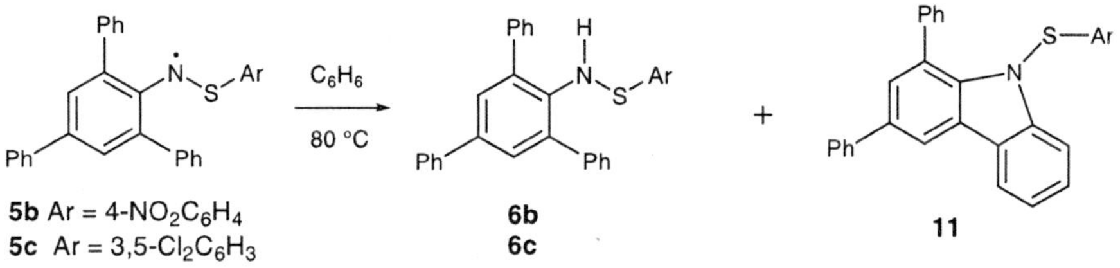

Scheme 9.4

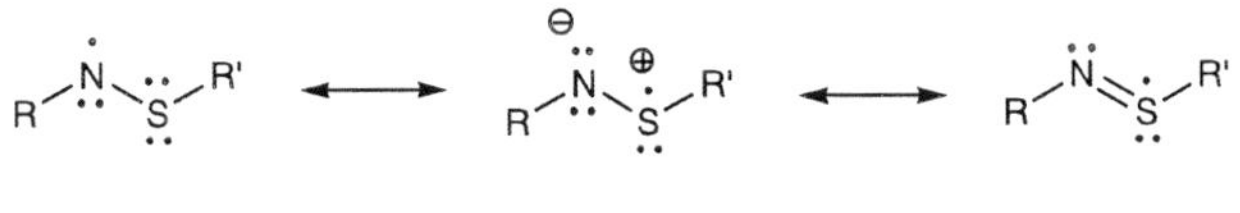

Scheme 9.5

hyperfine coupling to *ortho* and *para* N-phenyl protons[49,55]; (4) semi-empirical calculations, in conjunction with EPR spectral analyses, lead to a picture of the spin distribution which places the highest spin density on nitrogen (~0.50–0.60 for R = alkyl, ~0.4 for R = aryl) and relatively small but still substantial spin on sulfur (~0.25–0.30 when the NR substituent is alkyl, ~0.15–0.20 when NR = aryl).[47,51] The first two resonance structures in Scheme 9.5 are typically invoked to describe the delocalization of spin onto sulfur; structure **5c** is less often depicted, but reflects the (partial) S–N π bond (with the odd electron in an S–N π^* orbital, the overall S–N bond order is 1.5).

The bond order formulation described above is supported by several thioaminyl radical X-ray crystal structures, for which the S–N bond length is consistently between 1.60 and 1.63 Å – significantly shorter than S–N single bonds, for example, in sulfonamides for which bond length is closer to 1.70 Å. All of the crystallographically characterized thioaminyls have electron-deficient aromatic substituents on sulfur (nitrophenyl, chlorophenyl, pyridyl, etc.) and bulky N-aromatic groups containing two *ortho*-phenyl substituents.[54,56,57] All of these structures have similar structural features, namely: (i) a *trans*-coplanar disposition of the nitrogen and sulfur substituents (the CNSC torsion angle is close to 180°); (ii) an N-phenyl ring which is essentially coplanar with the N–S bond; and (iii) flanking *ortho* phenyl groups which are twisted with respect to the central N-phenyl ring. Figure 9.5 shows the structure of one representative example.

The reactivity of thioaminyl radicals has received sporadic attention. The reactions of one derivative with several phenols have been reported.[58] These reactions are fast (10–30 minutes) and a number of

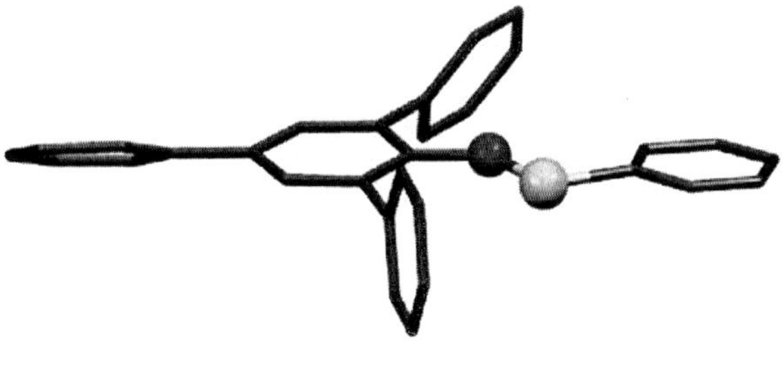

Figure 9.5 X-ray structure of N-(2-pyridylthio)-(2,4,6-triphenylphenyl)aminyl.[57]

different reaction paths are evident, including hydrogen abstraction from the phenol, radical coupling between the thioaminyl and the resulting phenoxyl, and S–N bond scission. In contrast, thioaminyls react relatively slowly (5–14 hours at 50 °C) with olefins to give 1,2-addition products.[59] Finally, the electron transfer properties of a few closely related derivatives (R = 2,4,6-triphenyl- or 2,4,6-tri-*t*-butylphenyl, R′ = chloro- or nitrophenyl) have been examined using cyclic voltammetry.[60] The radicals undergo reduction and oxidation processes, both of which are reversible – although the reversibility of the oxidation process is scan rate dependent, implying a reactive cation species. The reductions to give sulfenamide anions occur in the potential range −0.75 to −0.95 V vs SCE in acetonitrile, while the oxidation potentials lie between +0.42 and 0.45 V.

There are a handful of thioaminyl radicals with substituents other than simple alkyl or aryl groups. N-acylthioaminyl radicals **12** can be obtained by oxidation of the corresponding N-acylsulfenamides either by photolysis or by reaction with *t*-butoxyl radicals obtained from the thermolysis of di-*t*-butyl-diperoxyoxalate. These radicals are strongly persistent, with minimal decay over a period of a few days, although the radicals were not isolated. The carbonyl group adjacent to nitrogen opens up the possibility of delocalization of spin and/or charge, that is, through resonance structures **12c** and **12d** in addition to the conventional thioaminyl resonance structures **12a** and **12b** (Scheme 9.6). Analysis of EPR spectra of derivatives of **12**, including studies using ^{33}S and ^{17}O enriched samples,[61] indicate that the sulfur atom in these radicals carries substantially more spin density (~0.47) than in N-alkyl or N-aryl thioaminyls, and the nitrogen atom has correspondingly smaller amount of spin as evidenced by smaller $a(N)$ values (6.8–8.5 G). However, the very small $a(^{17}O)$ values obtained indicate that there is negligible spin density on the oxygen, that is, resonance structure **12d** is not an important contributor. Cyclic analogue **13** has a much larger $a(^{17}O)$ value, indicating that the corresponding resonance structure **d** for radical **13** is now significant, that is, tying the radical into a cyclic, conjugated framework enables π delocalization into the carbonyl group.[62]

There have been a few reports of sulfinyl- and sulfonylaminyls **14** and **15**, respectively, once again generated from the corresponding amides.[63] The g-values for these radicals (~2.003) are significantly lower than for thioaminyls (~2.006), suggesting much less spin density on the sulfur atom in derivatives of **14** and **15**. This can be understood based on the much more electron poor nature of the sulfur atom in these radicals, which attenuates the sulfur's ability to carry spin and charge (cf. resonance structures in Scheme 9.6). These radicals also dimerize to hydrazines more readily than thioaminyls, and in fact solutions of the dimers of **15** are EPR silent at room temperature; only on heating to 60 °C can the radical

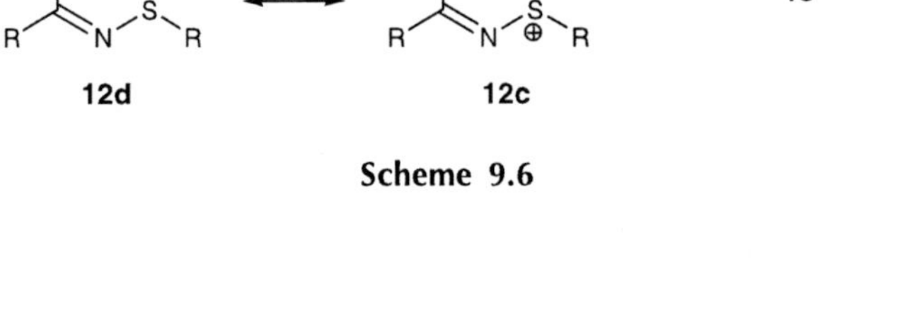

Scheme 9.6

be detected.[64] One selenoaminyl radical **16** has been spectroscopically (EPR) characterized; analysis of the hyperfine coupling suggest an electronic structure qualitatively similar to the thioaminyls.[65]

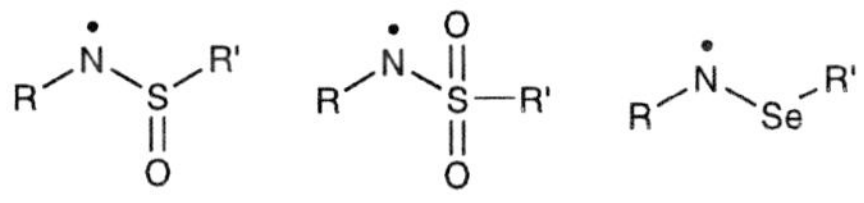

14 **15** **16**

Aside from the aforementioned thiocarboxamidyl radicals **13**, other examples of cyclic radicals containing a lone thiazyl unit are rare. Kaszynski has described the synthesis and spectroscopic characterization of polycyclic systems **17** and **18**.[66] The EPR parameters (a(N) = 8.5 and 8.7 G, respectively) are qualitatively similar to acyclic thioaminyls with N-aryl groups, although the general stability of **17** and **18** (half lives of a few days in solution) are substantially improved compared to "simple" unsubstituted thioaminyls **5** (R = R′ = phenyl).[50] Oakley has spectroscopically characterized benzene-fused cyclic thioaminyl and selenoaminyl radicals **19** (E = S, Se) by reduction of the corresponding cations, which were accessed by alkylation of the benzochalcogenadiazole.[67] EPR spectroscopic and DFT computational studies reveal substantial spin density on both the di- and tricoordinate nitrogens (spin densities ~ 0.42 and 0.14 respectively for both E = S and Se) as well as on the chalcogen (~0.2 for E = S and Se).

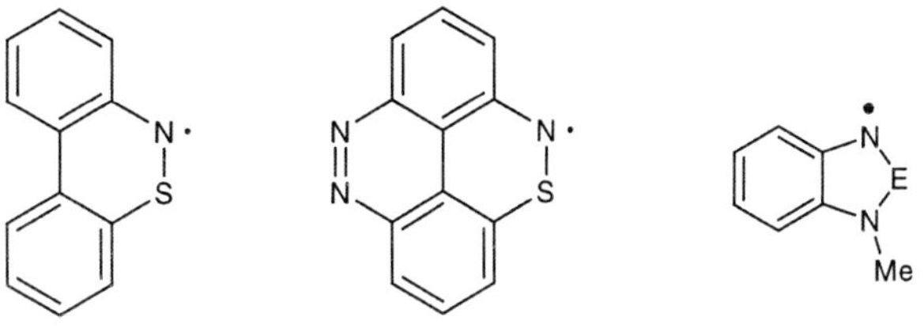

17 **18** **19**

9.3.1.1 Dithioaminyl radicals

Dithioaminyl radicals [RSNSR]$^{\bullet}$ **20** are a subclass of thioaminyls **5** in which the substituent (R) on nitrogen is also a sulfenyl (SR) group. These radicals can be made using methods analogous to those for simple thioaminyls, that is, oxidation of the corresponding bis(sulfenyl)amines **21** (Scheme 9.7).[68–70] EPR characterization and semi-empirical calculations indicate that the nitrogen atom is the predominant site of spin density with smaller amounts of spin on the two flanking sulfur atoms. Dithioaminyls with trifluoromethyl groups on sulfur,[71] and a related derivative $(F_5S)_2N^{\bullet}$[72] have been generated by heating solutions of the corresponding hydrazines **22**. These radicals have slightly higher a(N) values (13.2 and 13.5 G) than the bis(arylthio)aminyls (a(N) 11.3–11.5 G). The hydrazine N–N bond dissociation energy (BDE) of $(F_3CS)_2NN(SCF_3)_2$ was determined to be only 7.7 kcal/mol, that is, somewhat lower than the corresponding BDE for thioaminyls **5**.

S_4N_4 **1** and (bi)cyclic olefins, for example, norbornene, react to give (among other products) 2:1 adducts of structure **23** (Scheme 9.8).[73] Photolysis of these adducts (which can be generated *in situ*) produce solutions with strong and long-lived EPR spectra. The radicals responsible for the EPR signal were originally assigned the aziridylthyl (thionitroxide) structure **25**.[74] Subsequent studies showed the

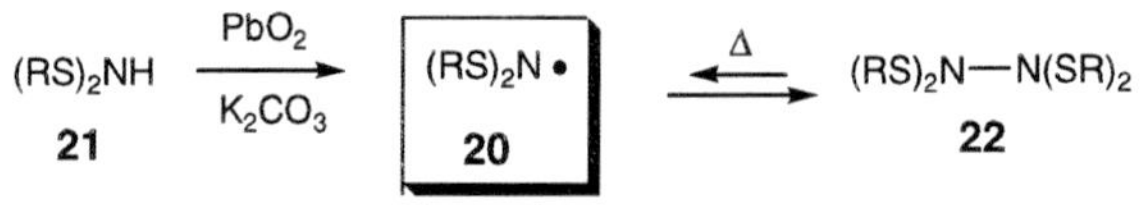

Scheme 9.7

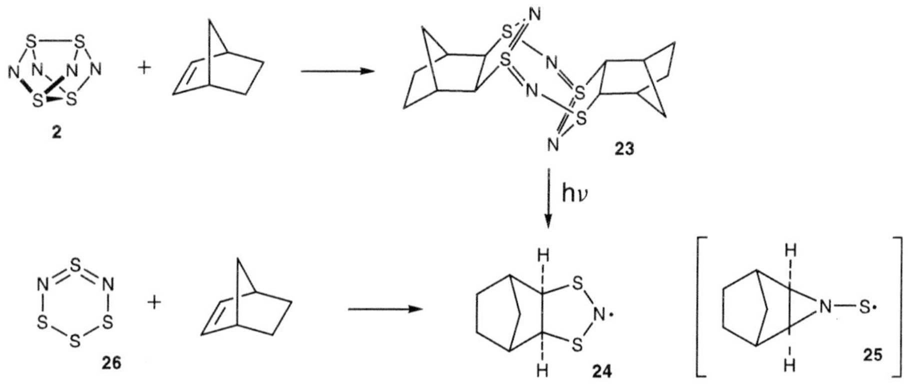

Scheme 9.8

correct structural assignment to be 1,3,2-dithiazolidinyl radicals **24**, which are cyclic analogues of **20**.[75] EPR-active solutions of these radicals can be generated made more directly by the reaction of S_4N_2 **26** with (bi)cyclic alkenes.[76] Several different derivatives of **24** have been characterized by EPR spectroscopy (a(N) is consistently ~13 G), but no derivatives have been isolated. These radicals reversibly dimerize in solution; dimerization enthalpies are in the range 7–11 kcal/mol.[77]

9.3.1.2 *Delocalized thioaminyl radicals*

The reaction of benzamidine with arylsulfenyl chlorides gives, depending on stoichiometry, bis- or tris(arylthio)benzamidines **27** and **28** respectively (Scheme 9.9). Both of these species can be converted into thioaminyl radicals **29** in which a sulfen*imide* fragment (=NSAr) is conjugated directly to the thioaminyl radical (–N•–S–Ar)component.[78] These radicals are highly persistent and insensitive to oxygen, and dimerize reversibly to give N–N bonded dimers **30** which are even more weakly bound (BDE ~ 6–7 kcal/mol) than simple thioaminyl radical derivatives **5**. The increase in radical stability (with respect to dimerization) can be ascribed to delocalization over two structurally equivalent RSNR units. The delocalization is supported by the EPR spectra of **29** which consist of a dominant five-line pattern arising from hyperfine coupling to two equivalent nitrogen atoms (a(N) = 6.0 – 6.1G).

In related work, Miura reported the reaction of arenesulfenyl chlorides with imidoylamidines **31** to give bis- and tris-arylsulfenyl compounds **32** and **33**, respectively.[79] Oxidation (of **32**) or photolysis (of **33**) produces radicals **34** as long lived and oxygen-insensitive species (Scheme 9.10). EPR characterization of the radicals, coupled with semi-empirical molecular orbital calculations, indicate that the spin is highly delocalized, shared predominantly between the two equivalent nitrogens (~19 % each), the central nitrogen (~17 %), and both sulfur atoms (~15 % each). The radicals associate reversibly into diamagnetic dimers in solution. Interestingly, one of the dimers was reported to have a genuine N–N bond at the terminal nitrogens based on X-ray crystallographic studies, but no crystallographic information was given.[79]

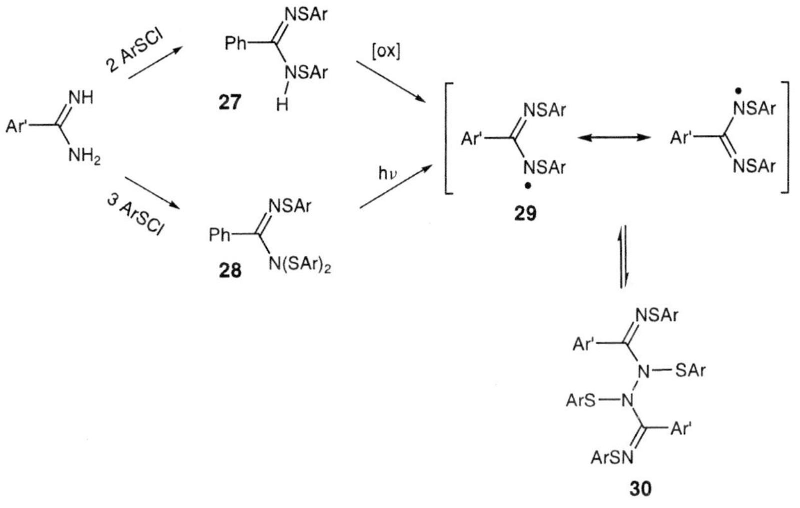

Scheme 9.9

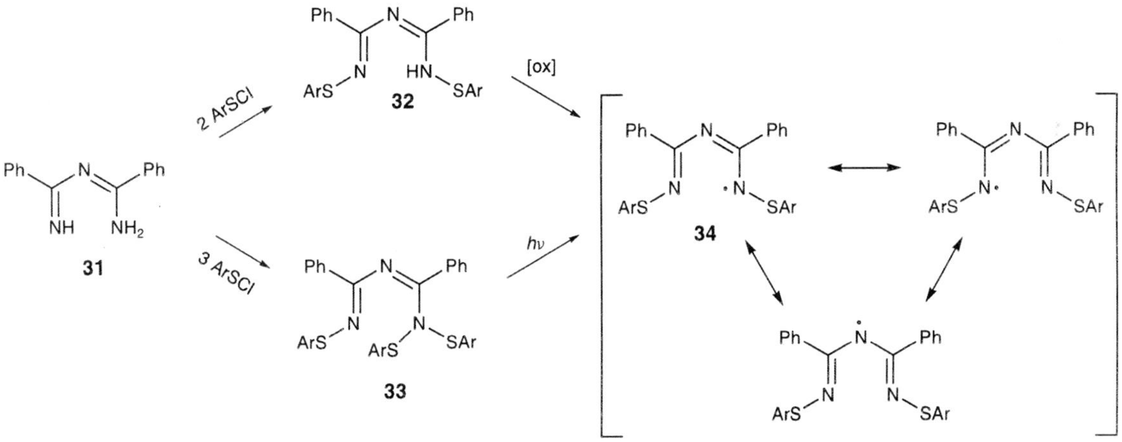

Scheme 9.10

9.3.2 1,2,3,5-Dithiadiazolyl radicals

First discovered in the 1970s, the 1,2,3,5-dithiadiazolyl radical **35** is one of the most heavily studied thiazyl radical systems; Rawson and Banister comprehensively reviewed their chemistry in 1995.[80] Synthetic routes to 1,2,3,5-dithiadiazolyl radicals nearly always proceed through the corresponding dithiadiazolium cation **36** (Scheme 9.11). The lone alternative route is the rearrangement of isomeric 1,3,2,4-dithiadiazolyl radicals **37** (this reaction is discussed in Section 9.3.5). The cations were first made from the reactions of $S_3N_3Cl_3$ with nitriles[81] or diazenes,[82] or from the reactions of nitriles with a mixture of sulfur dichloride, ammonium chloride, and chlorine.[83] A more recent use of $S_3N_3Cl_3$ in the synthesis of dithiadiazolium chlorides involves the cycloaddition reaction of the NSCl monomer (obtained by heating $S_3N_3Cl_3$) with *in situ* generated nitrile sulfides.[84] The reaction of amidines or their hydrochloride salts with sulfur dichloride (SCl_2)

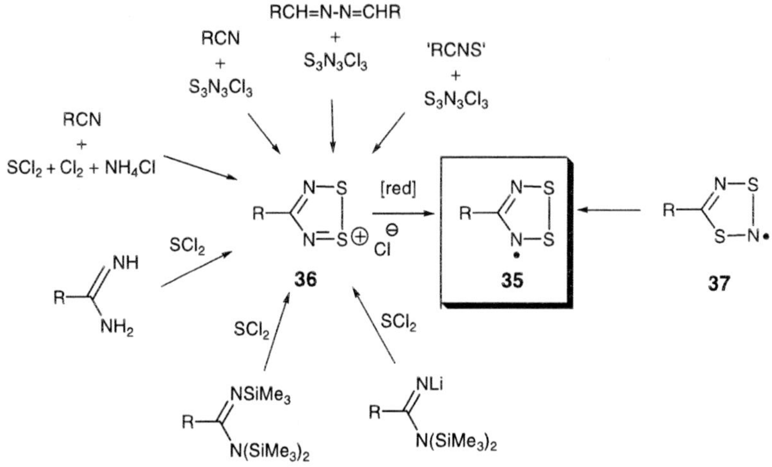

Scheme 9.11

produces dithiadiazolium chlorides in low yields[85]; the use of *N,N,N'*-tris(trimethylsilyl) derivatives,[86] or *N*-lithio-*N',N'*-bis(trimethylsilyl)amidines[87] improves yields significantly and permits the synthesis of selenium analogues (Section 9.3.2.1). Other routes have been reported but tend to be substrate-specific (for example Mews has developed routes to the 4-halogeno-1,2,3,5-dithiadiazolium cations starting from *N,N'*-bis(trimethylsilyl)carbodiimide).[88] Reduction of the dithiadiazolium cations **36** to the radicals **35** can be carried out with metals (sodium,[89] zinc, mercury, potassium, zinc/copper couple[90]) organic/organometallic reagents (tetramethyl-*p*-phenylenediamine, triphenylverdazyl,[89] triphenylantimony[91,92]), or reducing anions (SCN^-, I^-, Grignard reagents, CN^-, N_3^-).[90]

The electronic structure of 1,2,3,5-dithiadiazolyl radicals **35** has been thoroughly probed. The π^* SOMO (**38**) has a nodal plane at the ring carbon atom, which prevents direct electronic conjugation of the substituent R with this orbital. Consequently, many of the spectroscopic methods used to probe these radicals reveal minimal sensitivity of spectral parameters to substituent effects: (1) The EPR spectra of 1,2,3,5-dithiadiazolyls consist of a 1 : 2 : 3 : 2 : 1 pentet arising from coupling to two equivalent nitrogen atoms with a(N) values lying within a very small range (4.9–5.3 G); hyperfine coupling to ^{33}S, when observed, is also relatively insensitive (6.1–6.8 G) to the R group.[80] (2) The ionization potentials of several derivatives of **35** are substantially less sensitive to R than for, for example, the 1,2,4,6-thiatriazinyl radicals (Section 9.3.8), for which the substituents' π system can overlap directly with the SOMO.[91,93] (3) Cyclic voltammetry data for a wide range of radicals (or their corresponding cations) have been collected; the radicals can be reversibly oxidized to the known cations and reduced to the corresponding anions.[11] When R is a *para*- or *meta*-substituted phenyl group, the range of potentials for both oxidation and reduction fall within a narrow potential window: oxidation potentials of these radicals vary from +0.57 to +0.68 V vs SCE in acetonitrile.[87,94,95] Radicals with electron donating or withdrawing groups *directly* attached to the ring show a stronger substituent dependence.[96] For example, radical **35** with R = NMe_2 is oxidized at +0.35 V and reduced at −0.96 V, while the derivative with R = CF_3 has oxidation and reduction potentials of +0.91 and −0.42 V respectively. The gas phase electronic spectrum of one derivative (R = H) has been recorded; the transition centered at 617 nm corresponds to excitation into what is essentially an S–S σ^* orbital.[97]

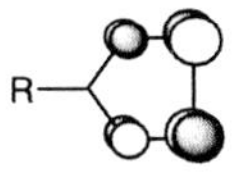

38

A large number of 1,2,3,5-dithiadiazolyl radicals have been structurally characterized by X-ray crystallography. Several different structure types involving radical dimers have been noted (Figure 9.6). The most common motif is the *cis*-cofacial dimer structure (Figure 9.6a), in which the two rings are directly superimposed and nearly parallel. The sulfur–sulfur distances between rings within the dimer (dashed lines in Figure 9.6a) are typically 3.0–3.1 Å. There are approximately 20 examples of dithiadiazolyl dimers of this type; nearly all are aromatic[92,94,98–101] or heteroaromatic[102,103] derivatives, or have R groups which are small (e.g., R = H,[104,105] Cl,[106] F,[107] CN[108]). Several derivatives adopt a twisted structure (Figure 9.6b) in which two rings are rotated by approximately 90° with respect to each other. The lone S–S contact is the shortest one between rings and is typically between 2.8 and 3.1 Å. This structure is adopted by radicals for which the size of R prohibits the *cis*-cofacial structure from forming for steric reasons (e.g., R = CH_3,[109] CF_3,[88] Me_2N,[93] adamantyl[84]) but also the halogeno (R = Br, F, Cl)[106,107,110] and one derivative with an aromatic substituent.[111] There are two possible *trans*-type dimer structures. One of these has the two rings slipped along their molecular long axis relative to one another, so that they only interact via the sulfur atoms (*trans*-antarafacial, Figure 9.6c), analogous to the structure of the S_3N_2 radical cation dimers (Section 9.2.3). There are only two examples of this dimer type; one (R = 2,4,6-$(CF_3)_3C_6H_2$-)[112] has considerable steric bulk while the other (R = 3-cyanophenyl)[92] does not. The other *trans* associative mode, also relatively rare, is *trans*-cofacial (Figure 9.6d), in which the two rings overlap exclusively via S–N contacts which range from 3.09 to 3.26 Å.[113–116] Finally, there are a small number of derivatives

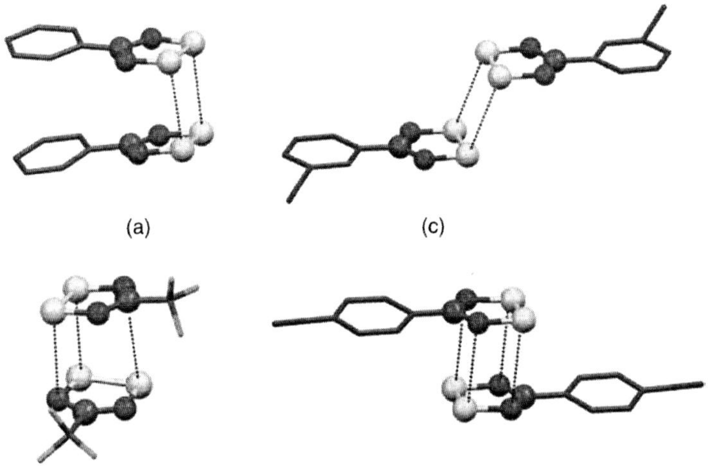

Figure 9.6 Examples of structural classes of π dimers of 1,2,3,5-dithiadiazolyl radicals **35**. (a) cis-cofacial (R = phenyl),[98] (b) twisted (R = CF_3),[88] (c) trans-antarafacial (R = 3-cyanophenyl),[92] (d) trans-cofacial (R = 4-iodophenyl).[113]

which are monomeric in the solid state.[117,118] The variety of structure types, and lack of correlation with the nature of the substituent R (aside from steric barriers to adopting the *cis*-cofacial structure), have been ascribed to the small differences in dimerization enthalpies between the various isomers,[88] coupled with the fact that the dimers are weakly bound to start with (see below). It is generally understood that intermolecular packing effects play a strong role in determining what (if any) dimeric structure is obtained for a particular derivative. In fact, some derivatives (R = 3-cyanopenyl,[92] Cl,[106,107,110] H[104,105]) adopt more than one kind of structure (i.e., polymorphism), and one compound – a diradical (see below) – has two different dimerization modes within the same single crystalline phase.[119]

The *solution* association thermodynamics of two dithiadiazolyl radicals has been studied quantitatively using variable temperature EPR. The radical dimerization enthalpies for R = phenyl[33] and R = *t*-butyl[120] are quite similar (−8.4 kcal/mol and −7.4 kcal/mol, respectively) despite the fact that they likely adopt different dimeric structures: the probable *cis*-cofacial geometry known for the phenyl derivative in the solid state[98] cannot be adopted by the *t*-butyl compound because of the steric bulk of the butyl group.

Several 1,2,3,5-dithiadiazolyl-based di- and triradicals have also been prepared. The very low solubility of these materials, coupled with their tendency to dimerize in the solid state and solution, has prevented examination of the intramolecular spin alignment in these radicals. The electronic state energetics of diradical **39**,[121] in which the two CN_2S_2 rings are directly attached to one another at the carbon atoms, have been investigated using computational methods.[122] The aforementioned nodal plane present in the dithiadiazolyl SOMO (**38**) prevents direct (conjugative) coupling of the spins. As such, **39** has two half-filled molecular orbitals **40a** and **40b**, which are simply the linear combinations of the SOMOs on each ring. Calculations on this diradical indicate that it has an open shell singlet ground state, with a triplet state only 0.5–1.0 kcal/mol higher in energy, that is, the unpaired electrons are very weakly interacting. These computational studies, coupled with the experimental carbon–carbon bond length of 1.49 Å, indicate that putative resonance structure **41** (a possible representation of a closed shell formulation) is not an accurate representation of the structure of this molecule. All of the other dithiadiazolyl diradicals have a 'spacer' group between the two rings (i.e., **42**), and intramolecular spin interactions are expected to be even weaker than for **39**.

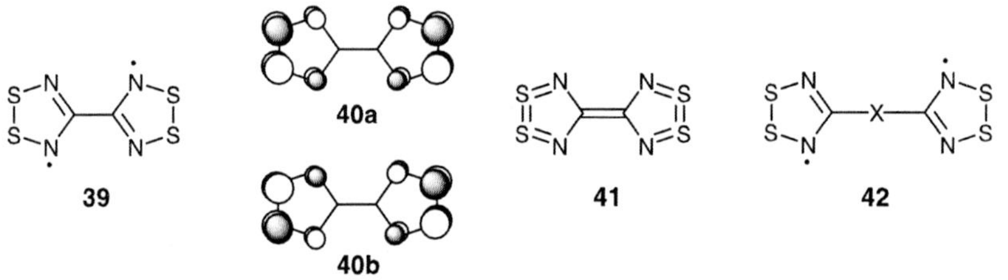

39 40a 40b 41 42

Solid state structures have been reported for nearly a dozen 1,2,3,5-dithiadiazolyl-based di/triradicals. Most adopt *cis*-cofacial dimeric arrangements at each ring, but the presence of two (or three) radicals on each molecule has consequences for the solid state assemblies. Discrete dimers are often obtained, as in, for example, the structure of the 1,4-phenylene-bridged diradical[123] (Figure 9.7a) among others.[102,121,124,125] However, for a few derivatives (the 1,3-phenylene-[126] and 5-cyano-1,3-phenylene-[127] bridged diradicals and the 1,3,5-phenylene-[128] bridged triradical) each of the radicals within a molecule associate with *different* neighboring molecules, creating polymeric chains (Figure 9.7b). Two diradical derivatives remain unassociated in the solid state. The 5-*t*-butyl-1,3-phenylene-bridged bis(dithiadiazolyl) cannot adopt a cofacial structure because of the *t*-butyl substituent; instead an antiparallel, but more regularly spaced, stacked

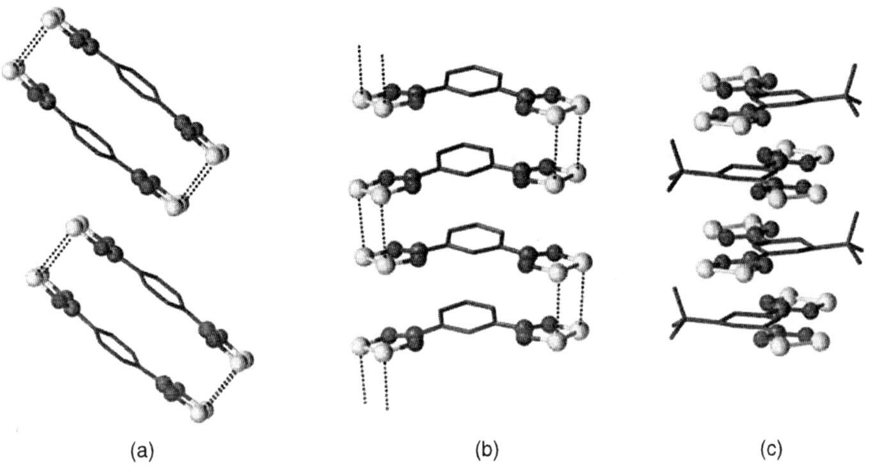

Figure 9.7 *Solid state association modes for 1,2,3,5-dithiadiazolyl diradicals: (a) discrete dimers for 1,4-phenylenebis(1,2,3,5-dithiadiazolyl)*[123]*; (b) polymeric stacking in 1,3-phenylenebis(1,2,3,5-dithiadiazolyl),*[126] *(c) unassociated diradical stacks in 5-t-butyl-1,3-phenylenebis(1,2,3,5)-dithiadiazolyl.*[129]

array is adopted (Figure 9.7c).[129] The other non-*cis*-cofacially bound diradical is **43**, in which one of the two CN_2S_2 rings associates in a *trans*-cofacial manner while the other remains monomeric.[119]

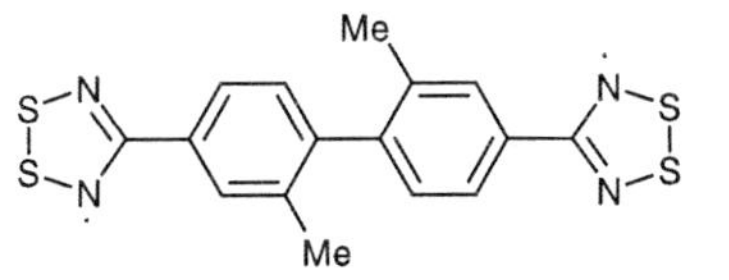

43

The dominant thrust in terms of reactivity studies of 1,2,3,5-dithiadiazolyl radicals (aside from their coordination chemistry[12]) has been redox-based chemistry. The radicals **35** can be oxidized to dithiadiazolium cations **36** with chlorine, bromine, sulfuryl chloride (SO_2Cl_2), thionyl chloride ($SOCl_2$), and other oxidizing agents.[90] In a similar vein dithiadiazolyls can perform reductive coupling chemistry, converting E–X bonds (E = P, Si, B substrates; X = Cl, Br) into E–E dimers and giving the dithiadiazolium halide salts as byproducts.[130] The redox chemistry between dithiadiazolyls and iodine is more complex. Dithiadiazolium iodide salts (made by metathesis reactions of corresponding chloride salts with I^-) were originally reported to spontaneously convert to iodine and the radical.[90] Oakley later showed that the gas phase reactions of dithiadiazolyls with iodine lead to partially oxidized, π-stacked species with very high electrical conductivity (Section 9.4.2).[131] Solution or gas phase reactions of the 4-phenyl-1,2,3,5-dithiadiazolyl radical with an excess of iodine give the corresponding dithiadiazolium cation as its triiodide salt; however, stoichiometric reactions with iodine produce material of composition [$PhCN_2S_2I$] but whose structure is trimeric – that is, the compound consists of a triple-decker array of dithiadiazole rings and an associated triiodide anion (Figure 9.8a).[131] An analogous structure based on a chloride salt, [$p\text{-}ClC_6H_4CN_2S_2]_3Cl$, has also been reported but its synthesis was not described.[132] Based on the tri-iodide anion, the triple-decker stack is

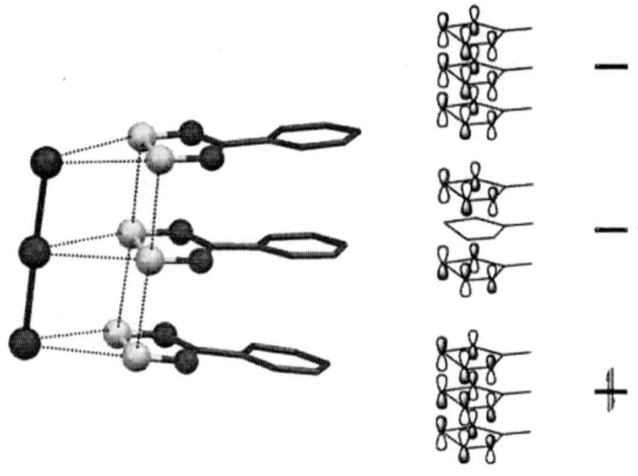

(a) (b)

Figure 9.8 X-ray structure of [PhCN$_2$S$_2$]$_3$I$_3$ (left) and frontier orbital diagram for the [PhCN$_2$S$_2$]$_3$ cation (right).[131]

cationic and therefore contains (on average) mixed-valent dithiadiazole rings. (i.e., $RCN_2S_2^{+0.33}$). This formulation is supported by the bonding metrics within the trimeric stacks: *neutral* 1,2,3,5-dithiadiazolyls have disulfide bond lengths of ~2.09 Å, whereas the S–S bond in the corresponding dithiadiazolium cations is shorter, typically ~2.00 Å. These structural differences are based on the population of orbital **38**, which is S–S antibonding in nature; in the radical this orbital is singly occupied whereas in the cation it is empty. The corresponding bonds in the triple-decker structure are intermediate in length but on average closer to the bond lengths for neutral rather than cationic dithiadiazole rings. (2.06 for the outer rings and 2.08 for the inner one). The S–S distances *between* rings are on average (3.18 Å) longer than the inter-radical distances in the neutral dimers (~3.1 Å), which can be rationalized by considering the frontier molecular orbital diagram in Figure 9.8b. The trimers are held together by a two-electron, three-ring "bond" (i.e., linear combination of three radical SOMOs), as compared to the neutral dimers wherein two electrons are shared by only two rings.

Reactions of the $PhCN_2S_2$ radical with cationic binary sulfur–nitrogen species lead to unusual ring transformations of the latter (Equations 9.1–9.4) and produce *mixed* π dimer structures in which one half

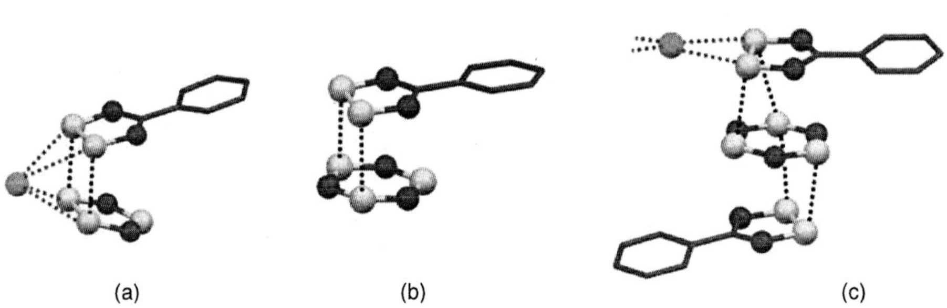

(a) (b) (c)

Figure 9.9 X-ray structure of (a) [PhCN$_2$S$_2$][S$_3$N$_2$]Cl, (b) [PhCN$_2$S$_2$][S$_3$N$_3$] and (c) [PhCN$_2$S$_2$]$_2$[S$_3$N$_3$]Cl.[133] *Black dashed lines indicate short intermolecular S–S contacts and green dashed lines indicate Cl–S contacts.*

of the dimer is a dithiadiazole and the other is a different thiazyl ring, specifically S_3N_2 or S_3N_3.[133] The S_3N_2-containing product (Equations 9.3 and 9.4; Figure 9.9a) is a π dimer between a neutral dithiadiazolyl radical and trithiadiazolium radical cation. The chloride is loosely associated with the heterodimeric unit via bifurcated Cl...S contacts, which are quite common in chloride salts of dithiadiazolium cations.[80] The S_3N_3 containing structure $[PhCN_2S_2][S_3N_3]$ (Equations 9.1 and 9.2; Figure 9.9b) is more intriguing as it harkens to the possible existence of the elusive S_3N_3 radical (Section 9.2.2) "trapped" as part of a hetero-π-dimer, although the ionic formulation of $PhCN_2S_2^+S_3N_3^-$ cannot be ruled out. Analysis of the bonding in the CN_2S_2 ring suggests that dithiadiazole ring is closer to a neutral than a cationic ring (see analysis above for $[(PhCN_2S_2)_3]I_3$) and (gas phase) calculations also suggest little charge transfer between the dithiadiazole and S_3N_3 species. The cationic triple-decker structure $[PhCN_2S_2]_2[S_3N_3]^+$ (Figure 9.9c) could be described as an S_3N_3 anion sandwiched between two $PhCN_2S_2$ cations, or alternatively a partial charge transfer system. The chloride anion is strongly associated with two $PhCN_2S_2$ rings, a finding which may imply a chemical role of the anion, given that the dithiadiazolyl radical reacts with the $AlCl_4^-$ salt of $S_5N_5^+$ (Equation 9.2) to give exclusively the hetero-π-dimer (Figure 9.9b) instead of the hetero-π-trimer.

$$PhCN_2S_2^{\bullet} + [S_5N_5^+]Cl^- \rightarrow [PhCN_2S_2]_2[S_3N_3]Cl + [PhCN_2S_2][S_3N_3] \quad (9.1)$$

$$PhCN_2S_2^{\bullet} + [S_5N_5^+]AlCl_4^- \rightarrow [PhCN_2S_2][S_3N_3] \quad (9.2)$$

$$PhCN_2S_2^{\bullet} + [S_4N_3^+]Cl^- \rightarrow [PhCN_2S_2][S_3N_2]Cl \quad (9.3)$$

$$PhCN_2S_2^{\bullet} + [S_3N_2Cl^+]FeCl_4^- \rightarrow [PhCN_2S_2][S_3N_2]Cl \quad (9.4)$$

9.3.2.1 1,2,3,5-Diselenadiazolyls

Selenium analogues of the 1,2,3,5-dithiadiazole ring (**45**) were first described by Oakley in 1989.[86] These radicals are made by reduction (triphenylantimony) of the corresponding 1,2,3,5-diselenadiazolium cations **44**, which are in turn prepared exclusively from the reaction of silylated amidines (either *N,N,N'*-tris(trimethylsilyl)[86] or *N*-lithio-*N',N'*-bis(trimethylsilyl)[92]) with selenium dichloride ($SeCl_2$) (Scheme 9.12). The selenium dichloride is generated *in situ* from selenium tetrachloride ($SeCl_4$) and a reducing agent, either triphenylantimony[86,123,126] or elemental selenium.[92,134]

Studies on diselenadiazolyl radicals **45** are more limited compared to their sulfur counterparts **35** owing, in part, to the lower solubility and thermal stability of the former. The few EPR spectra which have been reported[86,134] consist of a broad featureless singlet; g-values for **45** are consistently higher than for the sulfur analogues **35**, as anticipated for radicals with substantial spin density on selenium. Computational and experimental studies on the prototypical diselenadiazolyl radical (R = H) point to an electronic structure analogous to that for the corresponding sulfur-based radical.[104] For example the SOMO for the CN_2Se_2 ring is similar in symmetry and composition to the CN_2S_2 SOMO **38**, and the (computed) spin distributions are quite similar in the two species.

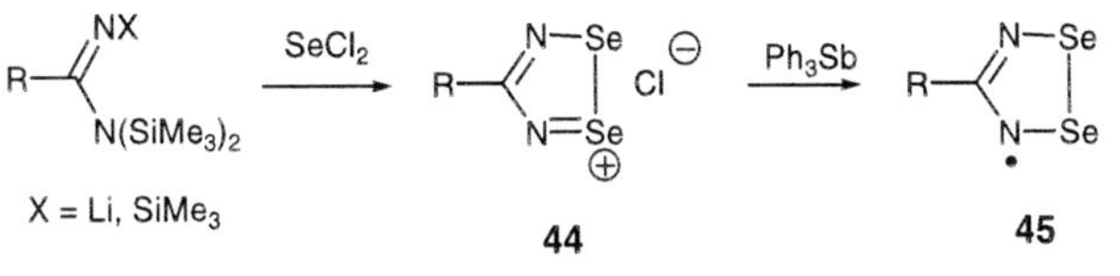

Scheme 9.12

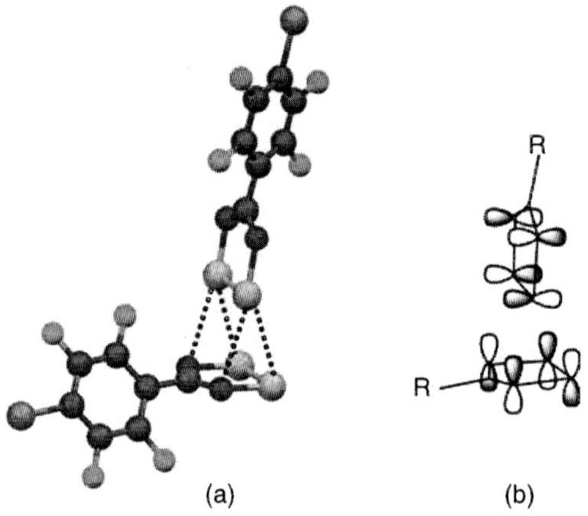

Figure 9.10 (a) X-ray structure of 4-(p-chloro-tetrafluorophenyl)-1,2,3,5-diselenadiazolyl. (b) Bonding combination of two SOMOs.[139]

Nearly all of the structurally characterized diselenadiazolyl radicals adopt the cofacial π dimer structure, including R = phenyl,[86] cyanophenyl,[92,135] 2,5-diflurophenyl,[99] 5-cyano-2-furyl,[136] and R = hydrogen.[134] The methyl derivative adopts a twisted structure analogous to its sulfur analogue.[137] Several diselenadiazolyl diradicals also adopt the superimposed π dimer structure; the 1,4-,[123] 1,3-,[126] and 5-cyano-1,3-phenylene-bridged[127] diradicals are all isostructural with their dithiadiazolyl counterparts (see above), and the 1,3-phenylene-bridged diradical also exists as a second polymorph in which the diradicals exist as discrete dimers instead of the zigzag extended structure.[138] Two derivatives (R = p-chloro- and p-bromo-tetrafluorophenyl) adopt an unusual edge-to-face dimeric structure (Figure 9.10).[139] The inter-radical distances between selenium atoms are 3.115 and 3.187 Å and there are also Se–N contacts (3.064/3.042 Å). The association of the radicals, and the observed diamagnetism of this material, can be rationalized by orbital coupling of the two radical SOMOs, as shown in Figure 9.10b.

9.3.3 1,3,2,4-Dithiadiazolyl radicals

1,3,2,4-Dithiadiazolyls **37** are made exclusively by reduction of the corresponding cations; cycloaddition reactions of the versatile reagent $[SNS^+]AsF_6^-$[140] **46** with a wide range of nitriles or polynitriles gives the 1,3,2,4-dithiadiazolium salts **47** or poly(dithiadiazolium) salts **48**, respectively (Scheme 9.13). The cations can be reduced (silver, sodium dithionite, triphenylantimony, ferrocene) to the (poly)radicals **37** (**49**). Alkyl,[120,141,142] aryl,[143,144] and halogenated[145,146] derivatives of **37** have been reported while the di- and triradicals **49** are mostly of the phenylene-bridge type. The diradical in which the two rings are directly bound (X = nothing) has received substantial attention (see below).

The EPR spectra of 1,3,2,4-dithiadiazolyl radicals consist of a dominant 1 : 1 : 1 triplet arising from coupling to the nitrogen atom flanked by both sulfur atoms, with hyperfine coupling constants generally ~11 G. Coupling to the nitrogen adjacent to the ring carbon is much smaller (0.5–0.6 G). The large differences in hyperfine coupling to nitrogen can be rationalized by considering the SOMO for these radicals (**50**), a π^* orbital which is concentrated heavily on the SNS fragment.[142] The small coefficient at the ring carbon attenuates significant substituent effects on the EPR spectra of these radicals.

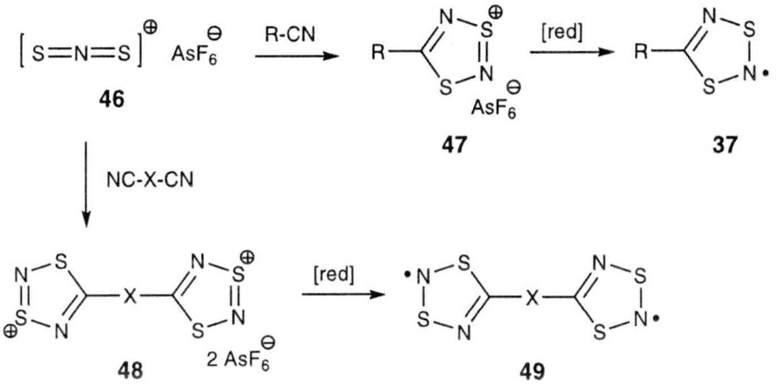

46 47 37 48 49

Scheme 9.13

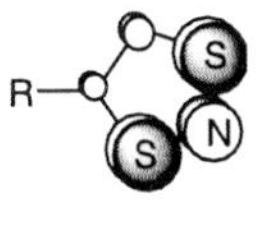

50

The feature of 1,3,2,4-dithiadiazolyl radicals **37** which has garnered the most attention is their thermal or photochemical rearrangement to the isomeric 1,2,3,5- radicals **35**. The thermal reaction occurs in solution[120,143,145,146] and the solid state,[147] and can also be induced photochemically.[143,144] This transformation is most conveniently monitored in solution using EPR spectroscopy, although UV-Visible spectroscopy has also been employed.[144] The substituent has a significant influence on the rate of this reaction: strongly electron withdrawing R groups tend to rearrange within minutes while more electron rich substituents can be isolated without rearrangement and persist for days or even weeks in solution in the dark. The proposed mechanism for the thermal rearrangement (Scheme 9.14) is based on the observation that the isomerization rate is concentration dependent, specifically second order in radical.[145] The process involves association (π dimerization) of the 1,3,2,4-radicals (the enthalpy of dimerization has been measured to be ~ -4.5 kcal/mol[144]) followed by a skeletal rearrangement of the dimers **51** to *trans*-type π dimers **52** of the 1,2,3,5-isomer. The lone X-ray structure of a 1,3,2,4-dithadiazoly (di)radical provides supporting evidence for this mechanism, as each radical associates in exactly the structure **51** proposed in the original mechanism (Figure 9.11).[148] Interestingly, studies on the *photochemical* rearrangement suggest that this process is unimolecular; a mechanism has been proposed involving initial fragmentation of the radical into a nitrile and the SNS$^{\bullet}$ radical.[143,144]

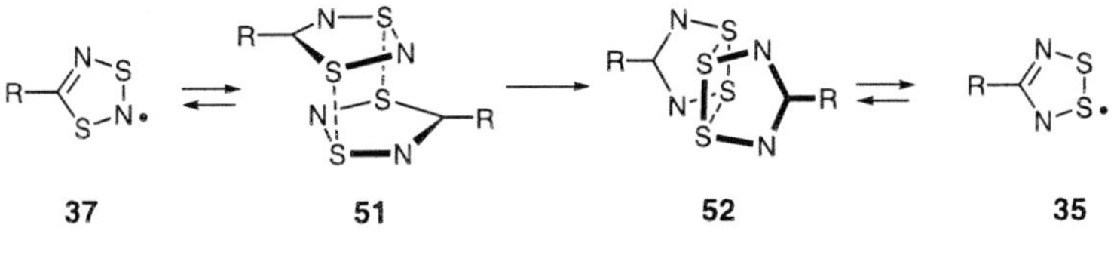

37 51 52 35

Scheme 9.14

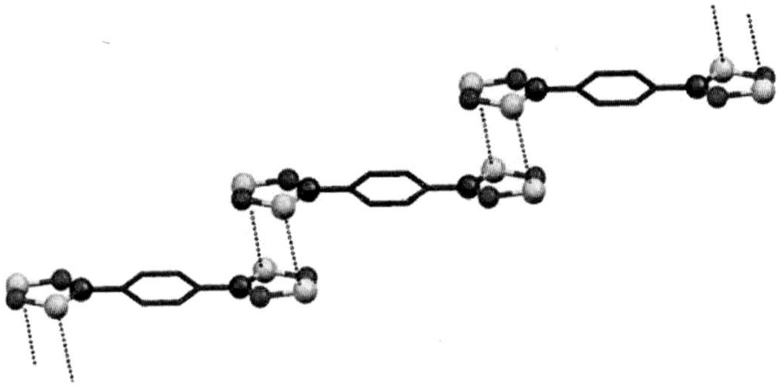

Figure 9.11 X-ray structure of 1,4-phenylenebis(1,3,2,4-dithiadiazolyl) diradical.[148]

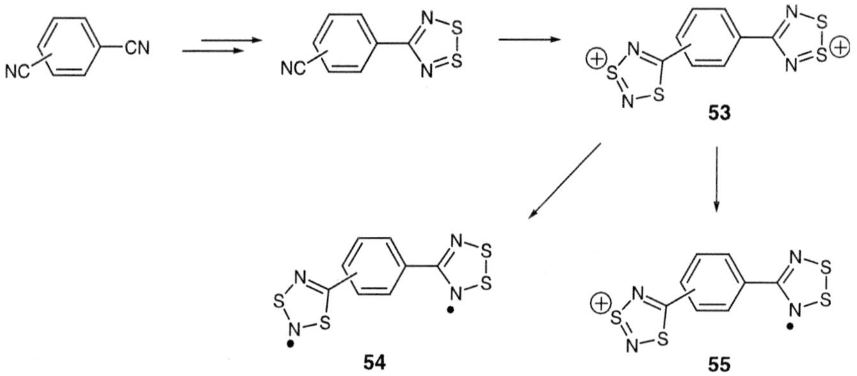

Scheme 9.15

Several poly(1,3,2,4-dithiadiazolyl) radicals **49** have been explored, with phenylene-based spacers (derived from di- and tricyanobenzenes) or single carbon spacers (from tricyanomethanide). Reduction of the polycations leads to the polyradicals, which rearrange slowly in the solid state (or more rapidly with heating) to the 1,2,3,5-isomers.[147,148] Banister has cleverly synthesized dithiadiazolium dications **53** – in which each of the two heterocycles is a different dithiadiazole isomer – by assembling the rings in stepwise fashion (Scheme 9.15; Ar = 1,4- and 1,3-phenylene).[149] Reduction of both rings leads to the mixed diradicals **54**. The EPR spectra of these consist of a superposition of the spectra of each type of dithiadiazolyl radical, suggesting very weak intramolecular radical–radical interactions. The fact that the 1,3,2,4- cations are reduced at significantly lower potentials (+0.3–0.4 V vs SCE for R = Ar)[87,95] than the corresponding 1,2,3,5- isomers (+0.6–0.7 V vs SCE) permits the selective reduction of the latter ring to afford radical cations **55**.

Attempts to prepare the 1,3,2,4-dithiadiazolyl diradical **59** with the two rings directly attached were ultimately successful, but *en route* produced a multitude of species of varying oxidation state and dithiadiazole isomer.[108,115,150,151] One-electron reduction of bis(1,3,2,4-dithiadiazolium) dication **56** gives radical cation **58**, in which the radical ring has rearranged to the 1,2,3,5- isomer; this product presumably proceeds via the bis(1,3,2,4) radical cation **57** (Scheme 9.16). The structure of **58** consists of *trans*-cofacial π dimers of the 1,2,3,5- radicals. Further reduction of **58** gives the 'mixed' diradical **61**, which adopts a cofacial but centrosymmetric dimer in which each π dimer is built from the two different CN_2S_2 rings (Figure 9.12).

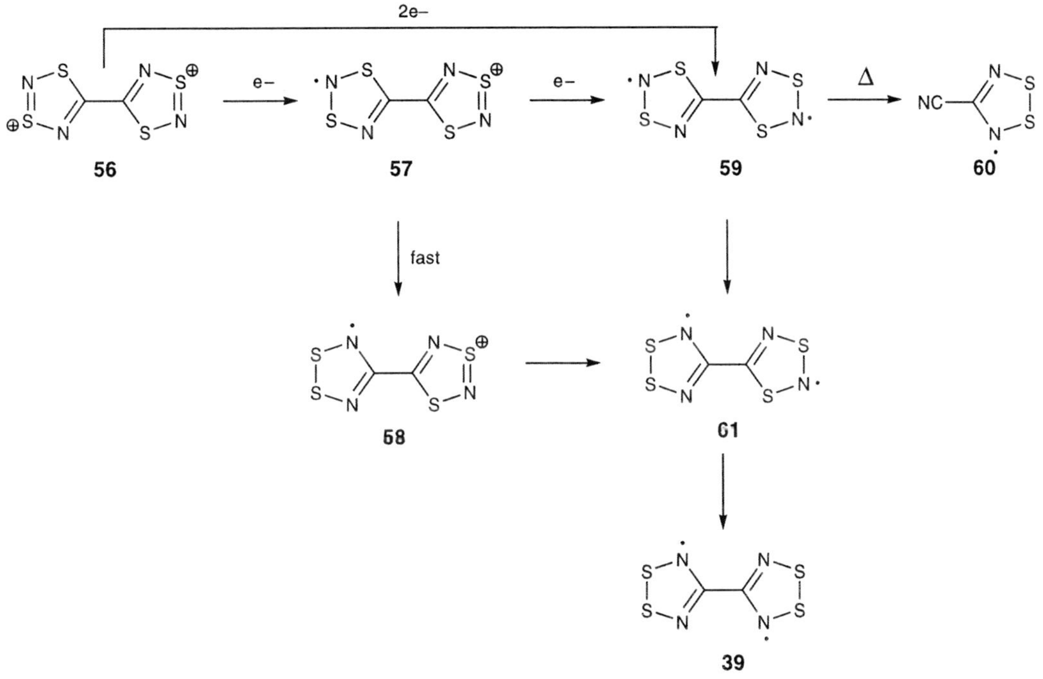

Scheme 9.16

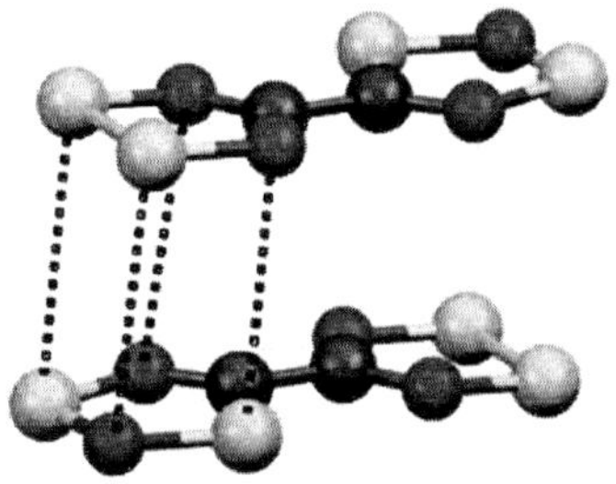

Figure 9.12 *X-ray crystal structure of (NSSNC)-(CNSNS)* ***61***.[115,151]

The mixed diradical ultimately isomerizes to the bis(1,2,3,5) diradical **39**, which had been independently synthesized.[121] Two-electron reduction of the dication **56** gives the bis(1,3,2,4) diradical **59**, which has not been structurally characterized. Attempted sublimation causes **59** to decompose to (among other things) the cyano-substituted-1,2,3,5-dithiadiazolyl **60**.[108] High level computational studies indicate that **59**[108] and **61**[115] are, like **39**,[104] biradical in nature with very weakly interacting spins.

9.3.4 1,3,2-Dithiazolyl radicals

The 1,3,2-dithiazolyl radical system also dates back to the 1970s; the benzannelated subset of these radicals **65** was reviewed by Rawson in 2000.[152] 1,3,2-Dithiazolyls were originally spectroscopically

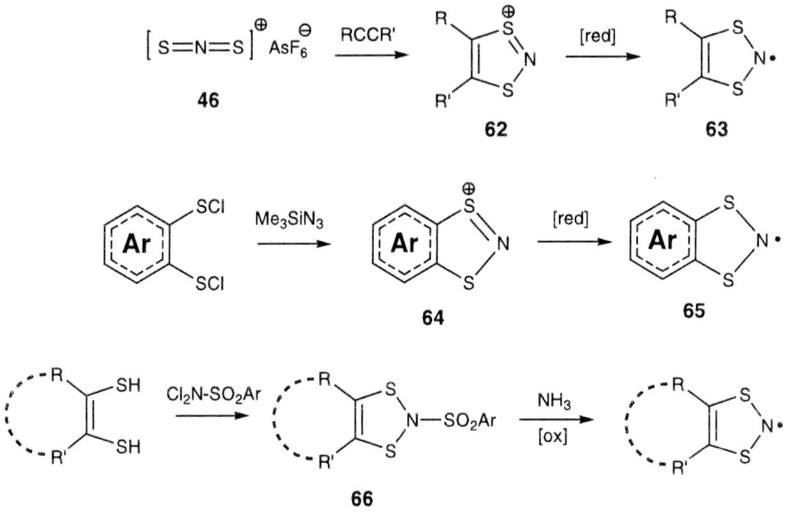

Scheme 9.17

(EPR) identified from the reactions of alkynes with either S_4N_4 or S_4N_2[75,153] (cf. the synthesis of 1,3,2-dithiazolidinyls, Section 9.3.1.1). More rational and general synthetic routes have since been developed. The cycloaddition of SNS^+ **46** with alkynes gives the 1,3,2-dithiazolium cations **62**, which can subsequently be reduced to the radicals **63** (Scheme 9.17).[141,142,154] Aromatic ring-fused derivatives **65** can be accessed by reduction of the cations **64**, which themselves are made by the condensation ortho-bis-sulfenyl chlorides with trimethylsilyl azide.[155,156] A third route involves reactions of N-arylsulfonyl dithiazoles **66** with ammonia[157–159]; this reaction initially produces the corresponding N–H heterocycle, which is subsequently oxidized (air, or ferricyanide) to give the radical.

The electronic structure of 1,3,2-dithiazolyls differ in substantial but predictable ways from the (non-conjugated) 1,3,2-dithiazolidinyl radicals **24**. The former, being part of a larger π conjugated skeleton (particularly if fused to an aromatic ring, i.e., **65**) have smaller hyperfine coupling constants to nitrogen (~11 G vs ~13 G for **24**) and (where detected) larger coupling to sulfur. The SOMOs **67** and **68** are both heavily concentrated on the SNS portion of the ring, but there are non-negligible contributions to the SOMO elsewhere on the molecule.

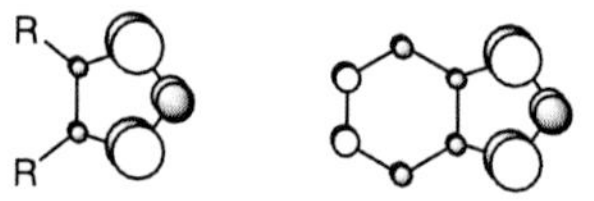

67 **68**

The electrochemical properties of several 1,3,2-dithiazolyl radicals have been reported. In general, these radicals can be reversibly oxidized to dithiazolium cations; their reduction to dithiazolide anions is, with few exceptions, irreversible. Unlike the 1,2,3,5-dithiadiazolyls, the redox potentials of these radicals are highly structure dependent (Table 9.1). Relatively electron rich fused derivatives (**69, 74, 76**) have very low oxidation potentials, but more electron poor derivatives (either with electron withdrawing substituents (**72, 73**) or incorporation of nitrogen into the annelated ring raise the oxidation potential, (**70, 71, 75, 77, 78, 79**) in some cases to over +1 V vs Fc/Fc^+. There also appears to be a correlation between redox properties and the amount of spin delocalization *away* from the dithiazole ring onto the other fused rings:

Table 9.1 *Cyclic voltammetry data for benzo-1,3,2-dithiazolyls. Values are in V vs SCE in acetonitrile*

Compound **(number)**	E^{red}	E^{ox}	a(N), G	Reference
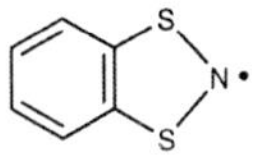 **69**	−1.20[a]	+0.15	11.27	160
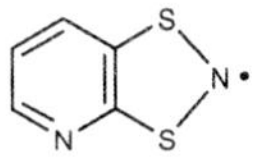 **70**	−1.09[a]	+0.30	11.38	161
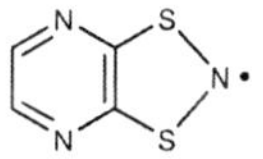 **71**	−0.88[a]	+0.53	11.25	160
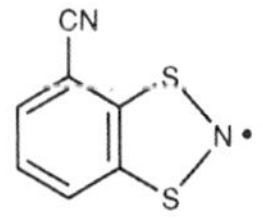 **72**	b	+0.49	11.5	162
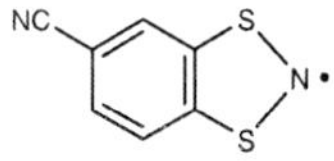 **73**	b	+0.45	11.1	162
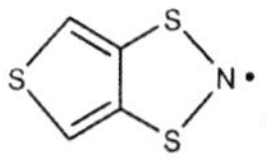 **74**	−0.73[a]	+0.25	11.6	163
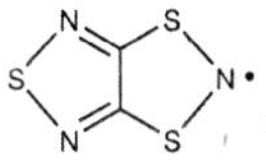 **75**	−0.66[a]	+0.65	11.15	160
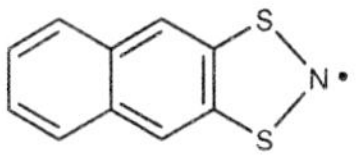 **76**	−1.08[a]	+0.27	11.40	160
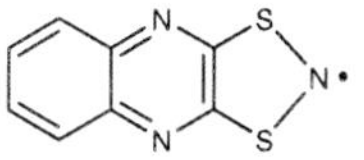 **77**	−0.73[a]	+0.62	10.89	160
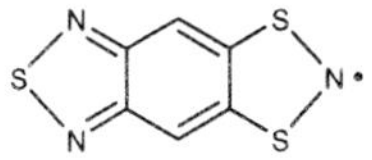 **78**	−0.76	+0.46	10.98	160
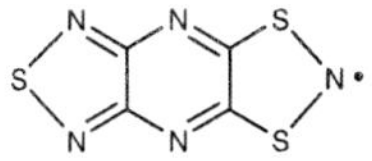 **79**	−0.06	+1.06	9.59	160
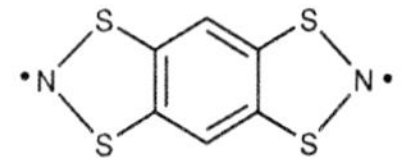 **80**	−1.3[a], −1.8[a]	+0.16, +0.74	b	164

[a]Irreversible process.
[b]Value not given.

EPR hyperfine coupling constants to the dithiazole nitrogen are typically in the range 11−11.5 G, but are noticeably smaller for the radicals which are more easily reduced (e.g., **77**, **78**, and especially **79**).

The solution EPR spectra of 1,3,2-dithiazolyls show no temperature dependence over a wide temperature range, indicating that the enthalpy of dimerization for these radicals is negligible (in contrast to, e.g., 1,2,3,5-dithiadiazolyls).[154,165] However, the solid state structures of these radicals reveal a number of different possibilities. Two types of dimer structures have been found. The two structurally characterized monocyclic derivatives (**63**, R/R′ = CN[158] or CF_3[166]) both adopt a cofacial arrangement (Figure 9.13a), as

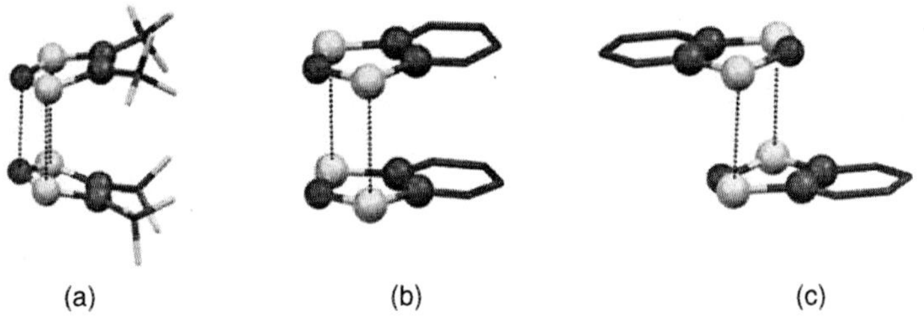

Figure 9.13 X-ray structures of 1,3,2-dithiazolyl radical dimers: (a) 4,5-bis(trifluoromethyl) derivative,[166] (b) **70**,[161] and (c) **69**.[165]

does pyridyl-fused radical **70**[161] (Figure 9.13b) and the radical cation of **80**.[167] There is substantial variation in the structural specifics of the dimers: the interplanar separation in the 4,5-dicyano derivative is 3.14 Å but substantially longer in **80**$^{+}$ (which exists as two different polymorphs, with average S–S distances of 3.35 and 3.26 Å). The two rings are substantially tilted in the bis(trifluoromethyl) structure, such that the N–N intradimer distance (2.86 Å) is shorter than the S–S distances (average 3.17 Å). The benzo-fused radical **69** is one of two derivatives to dimerize centrosymmetrically, with average d(S–S) of 3.17 Å (Figure 9.13c),[165] the other being **78** (the latter of which crystallizes with two radicals in the asymmetric unit – one of which adopts the centrosymmetric structure with d(S–S) ~3.23 Å, while the other remains unassociated).[160] Other derivatives remain monomeric in the solid state (**73, 76, 78, 80** and the methyl-substituted benzo-fused radical **81**[168]). Finally, several dithiazolyls have temperature-dependent solid state structures, consisting of a "high temperature" phase of unassociated radicals and a "low temperature" phase in which the radicals form cofacial π dimers. These temperature-dependent structures are discussed further in Section 9.4.3.1.

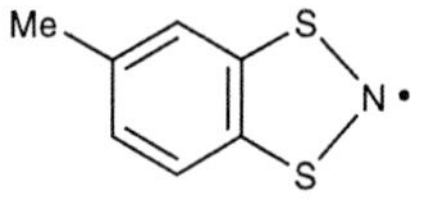

81

Selenium analogues of 1,3,2-dithiazolyls are all but unknown. Attempts to prepare the 1,3,2-diselenazolyl radicals **82** and **83** were explored by reduction of the corresponding diselenazolium cations, but the radicals were too unstable to isolate; the major species isolated in both instances was the tetraselenocine **84**.[169]

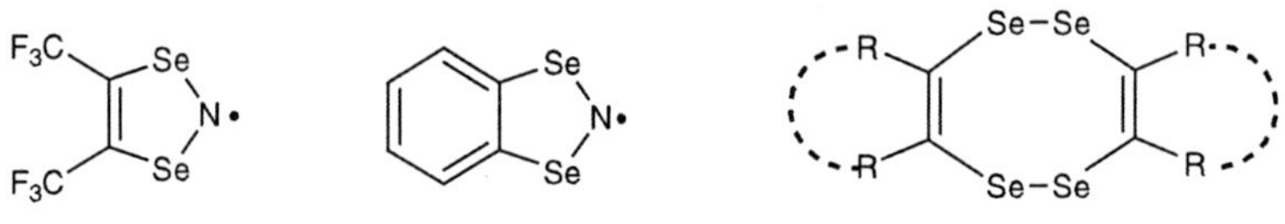

82 **83** **84**

9.3.5 1,2,3-Dithiazolyl radicals

Development of the 1,2,3-dithiazolyl radical system took place in two distinct stages. Mayer reported their generation and identification by EPR spectroscopy,[170,171] while in the past decade Oakley has devised new synthetic strategies and new derivatives which are amenable to more comprehensive characterization.

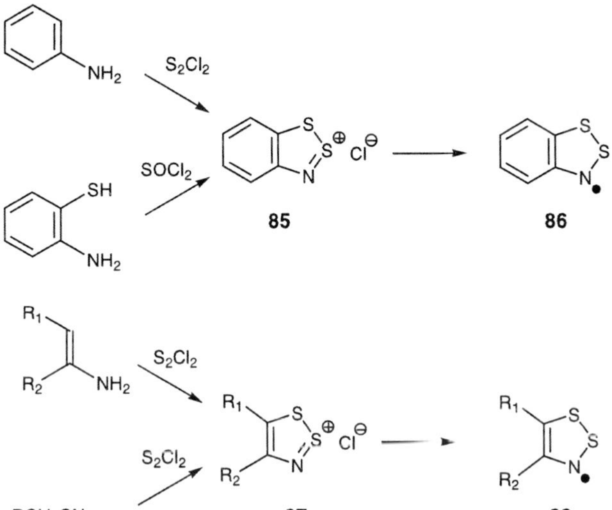

Scheme 9.18

As is the case for the isomeric 1,3,2-dithiazolyls, the 1,2,3-radicals can be monocyclic (**88**) or fused to another ring (**86**). Both types are made by reduction of the corresponding dithiazolium cations **85** and **87** respectively (Scheme 9.18). The polycyclic cations **85** can be made either by reactions of anilines with sulfur dichloride (the Herz reaction) or from *o*-aminothiophenols and thionyl chloride ($SOCl_2$); the former route usually leads to partial chlorination of the aromatic ring as a side reaction. The monocyclic species **87** are synthesized from reactions analogous to the Herz reaction, but using enamines[171] or primary nitriles[172–174] instead of anilines as the substrates. In the enamine reactions R_1 = a carbonyl group, while in the reaction with nitriles R_2 winds up being a chlorine atom.

The π SOMO of 1,2,3-dithiadiazolyl radicals **89** is antibonding with respect to the four ring atoms centered on the disulfide linkage. There is effectively no contribution from C4 (adjacent to nitrogen) by analogy to the 1,2,3,5-dithiadiazolyls, the SOMO of which (**38**) has a formal nodal plane passing through its C4 atom. The presence of substantial spin density on C5, however, permits spin delocalization onto R_1 (for **88**) or a fused ring (e.g., **86**), which has substantial consequences for the electronic structure and reactivity of these radicals. For example, the EPR nitrogen hyperfine coupling constant for these radicals are substituent dependent, spanning a relatively large range from 4.5 G up to over 8.0 G. Moreover, many derivatives cannot be isolated, and this may be due to decomposition pathways arising from substantial spin density at C5 (adjacent to sulfur). In fact, the reduction of the 4,5-dichloro cation **87** (Appel's salt) leads to the fulvalene-based structure **90**, most likely via C5–C5 bond formation from the radical **88** ($R_1 = R_2 =$ Cl) and subsequent reductive elimination of the "extra" chlorides to give the exocyclic doubly bonded dimer.[175]

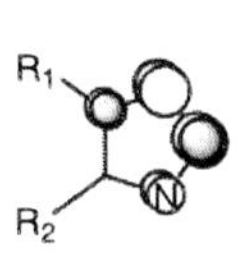

89

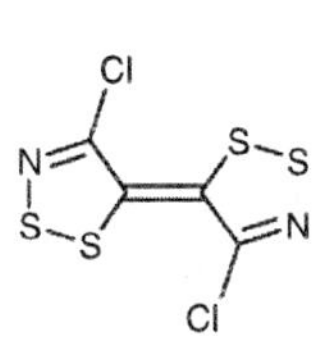

90

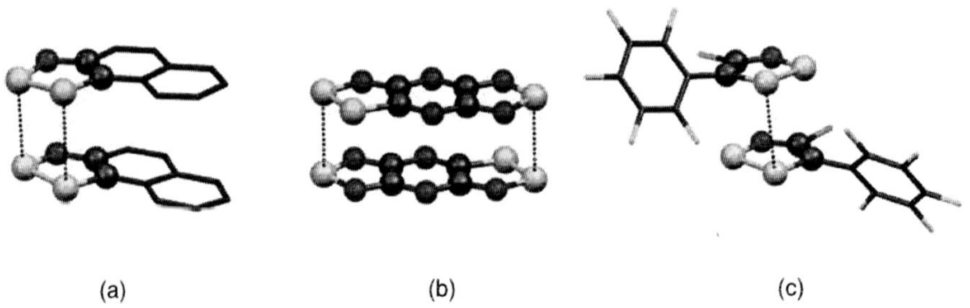

Figure 9.14 *X-ray structures of 1,2,3-dithiazolyl radicals (a)* ***91****, (b)* ***92****, and (c)* ***93****.*

A total of four 1,2,3-dithiazolyl radicals have been structurally characterized – each of which adopts a different structure type (Figure 9.14). 2,3-Naphtho-fused derivative **91** has a *cis*-cofacial dimer structure (Figure 9.14a) – albeit one in which the two rings are tilted significantly, as evidenced by the disparate S–S inter-radical distances of 3.053 and 3.309 Å[176] – while the tricyclic species **92** adopts a cofacial but *centrosymmetric* dimer structure (Figure 9.14b), in which each of the two inter-radical S–S contacts (3.23 Å) involve the dithiazole S2 and the *thiadiazole* S (Figure 9.13b).[177] The (pentafluorophenyl)(chloro)radical **93** has a twisted ("gauche") structure (Figure 9.14c) with an intradimer S–S contact distance of 3.299 Å,[173] and isothiazole-substituted radical **94** is monomeric and adopts a slipped π stacked structure.[174] In all of the dimerized derivatives, it is noteworthy that the S–S distances between rings are generally longer than the corresponding S–S interactions in other kinds of thiazyl radical dimers.

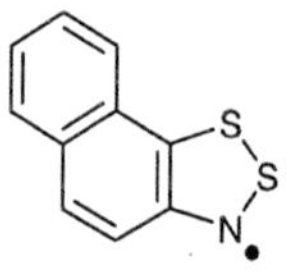

91

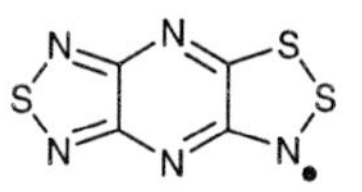

92

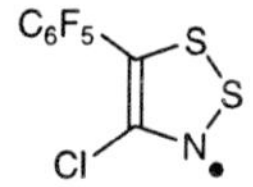

93

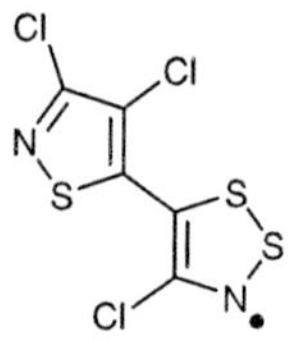

94

Oakley has attempted to incorporate selenium and even tellurium into the 1,2,3-dithiazolyl framework. The 1,2-naphtho-fused radicals **95a–c** – in which one or both of the sulfur atoms is replaced by selenium – could all be synthesized and spectroscopically characterized; the introduction of selenium in either or both chalcogen positions (E_1, E_2) appears to have minor consequences for the radicals' overall electronic structure, but isolation of the selenium radicals was thwarted by their thermal instability.[176] The same comparative features are true for the benzo fused selenium radical **96a**.[178] Attempts to generate tellurium-based radical **96b** failed, possibly because the corresponding tellurathiazolium *cation* is strongly associated in solid state and solution. Indeed, while the electrochemical reduction of the thiaselenazolium cation (**96a**$^+$) follows closely the behavior of the analogous benzo-1,2,3-dithiazolium cation, the corresponding thiatellurazolium cation **96b**$^+$ is irreversibly reduced at much higher potentials.[178]

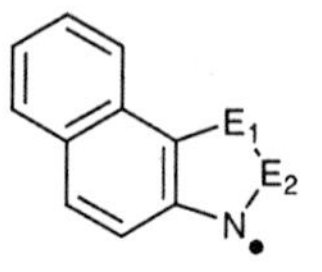

a E_1 = S, E_2 = Se
b E_1 = Se, E_2 = S
c E_1 = E_2 = Se

95

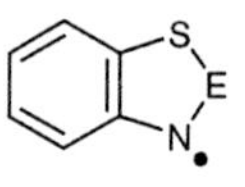

a E = Se
b E = Te

96

9.3.6 Bis(1,2,3-dithiazole) and related radicals

Oakley has developed an extensive series of bis(1,3,2-dithiazole) radicals **97** – including selenium variants – incorporated into a tricyclic system with mirror plane symmetry. There are over 20 derivatives with various substituent (R_1, R_2) and chalcogen atom (E_1/E_2 = S, Se) combinations. These have been called "resonance stabilized" bis(1,2,3-dithiazoles) because their general stability is greatly improved compared to the "parent" 1,2,3-dithiazolyl radical.

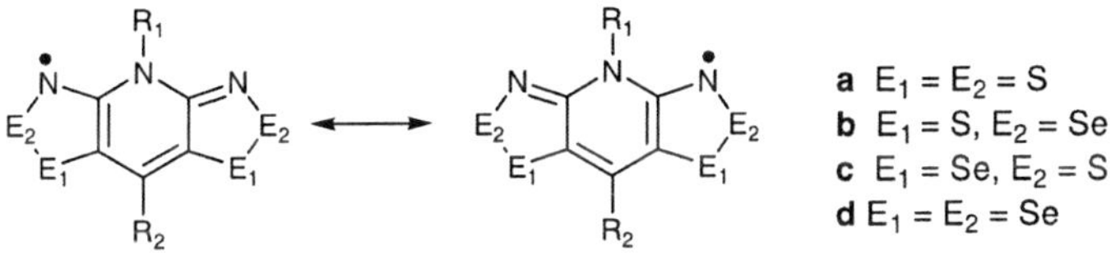

97

This work was prefaced by investigations of compound **99**, which was synthesized by deprotonation of pyridinium salt **98** using proton sponge.[179] Although a diradical resonance structure can in principle be depicted (**99••**) the experimental and computational data all point to a closed shell, zwitterionic electronic structure (Scheme 9.19).

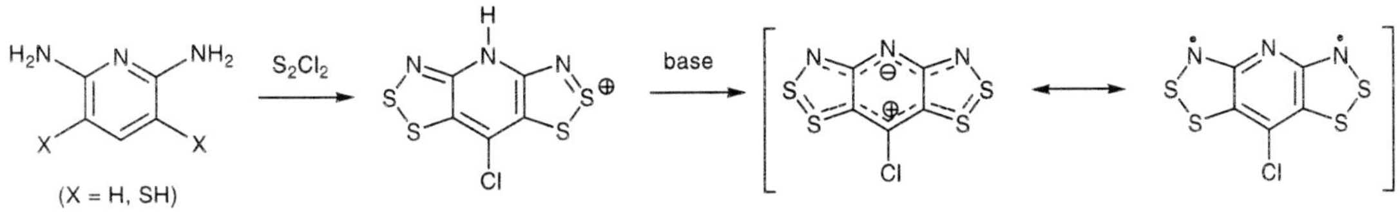

Scheme 9.19

Compound **99** can be alkylated at the pyridine nitrogen, affording N-alkylpyridinium cations **100** which can subsequently be reduced with decamethylferrocene to give the bis(1,2,3-dithiazole) radicals **97a** (Scheme 9.20).[180] A more efficient synthetic route involves alkylation of 2,6-diaminopyridine followed by a Herz-type reaction to give the N-alkylated cation **100** directly.[181] This route has the additional advantage of allowing for R_2 groups other than chlorine on C4. Pyrazine-based radicals **101**, in which the C4 atom is replaced by nitrogen, can be made by a modified sequence starting from 2,6-diaminopyrazine.[182,183]

Methods have been developed for site-specific replacement of either (or both) of the sulfur atom positions in **97a** by selenium. The 2-seleno derivatives **97b** were originally made by the reaction of S-protected

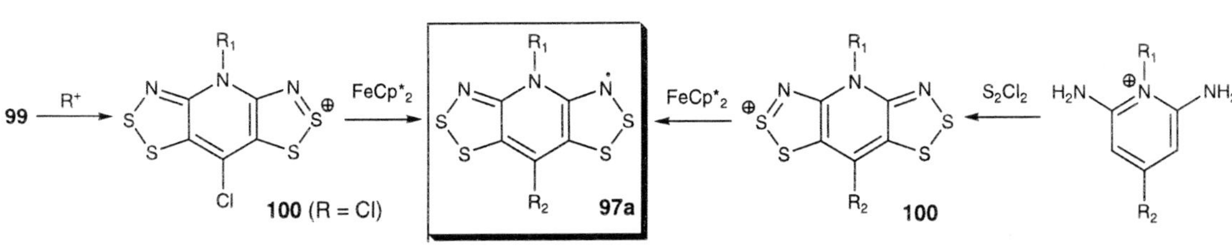

Scheme 9.20

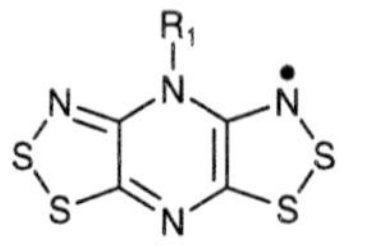

101

3,5-dithio-2,6-diaminopyridine with a mixture of selenium tetrachloride ($SeCl_4$) and selenium oxychloride ($SeOCl_2$) (Scheme 9.21).[184,185] The resulting NH pyridinium cations **102** can then be converted to N-alkyl derivatives **103** – through a series of steps analogous to the synthesis of N-alkyl bis(dithiazole) cations **100** – and then reduced to the radical. A more direct route to **103** involves direct substitution of the sulfur atoms of **100** in the 2-positions by selenium using selenium dioxide (SeO_2).[186,187]

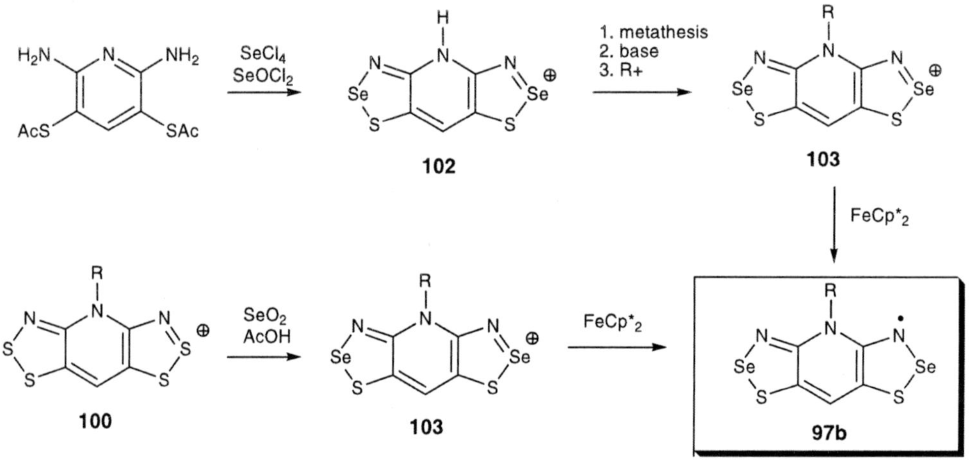

Scheme 9.21

The 1-selenathiazole-based radicals **97c** are made starting from 2,6-diamino-3,5-bis(methylselenyl) pyridine **104**, which reacts with thionyl chloride to give bis(1-selenathiazole) cations **105** (Scheme 9.22).[188,189] The metathesis/deprotonation/alkylation sequence gives the N-alkyl cations **106**, which are subsequently reduced to **97c**. The N-alkyl cations **106** can be converted to bis(1,2,3-diselenazole) cations **18** and subsequently reduced to the bis(diselenazole) radicals **97d**.[188]

All four sulfur/selenium variants of the bis(1,2,3-dithiazole) radicals **97a–d** have remarkably similar SOMOs **108**; the contributions from each of the two outer C_2NE_2 rings is reminiscent of the 1,2,3-dithiazolyl SOMO **89**.[184] Notably, there is a nodal plane passing through the pyridinium nitrogen and the carbon *trans* to it in the central ring, thereby limiting direct conjugative effects of the substituents R_1 and R_2. As such, the EPR spectra of these radicals are normally dominated by coupling to the two outer nitrogens (a(N) ~3.1–3.2 G), with slightly higher values for selenium-containing radicals. Coupling to the central nitrogen (a(N) ~0.6 G), and to protons on the R_1 and R_2 substituents, is occasionally also observable. These radicals undergo multiple electron transfer processes – they can be easily oxidized to cations (E_{ox}^1 between +0.08 and −0.21 V vs SCE) and dications ($+1.2 < E_{ox}^2 < +1.6$ V) and can be

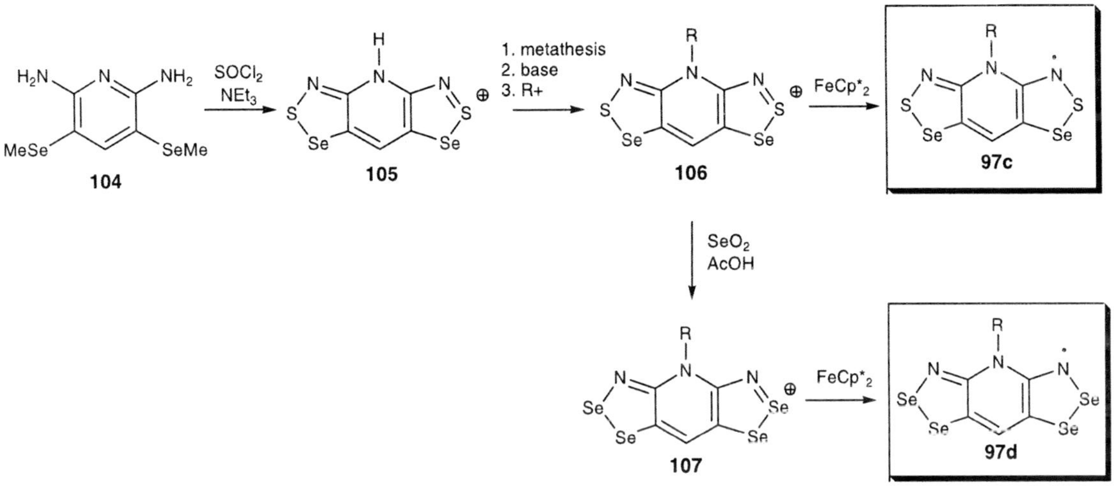

Scheme 9.22

reduced to anions ($-0.62 > E_{red}{}^1 > -1.06$ V) although the latter process is not always reversible; when R_1 = hydrogen, subsequent proton transfer events render the electron transfer processes irreversible. The effects of the substituents (R_1 and R_2) and chalcogen (E_1 and E_2) on the redox potentials are qualitatively predictable based on atom or group electronegativities.

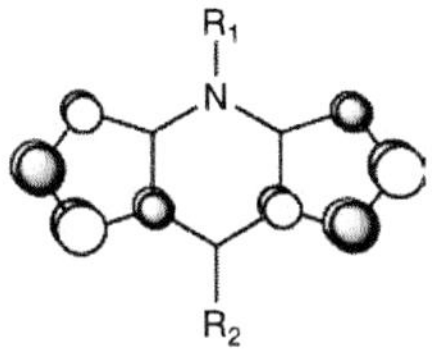

108

The majority of bis(dithiazole)s **97**, including selenium variants, exist as unassociated radicals in the solid state. There are a few exceptions whose associated structures are quite different from typical thiazyl radical dimers. Bis(selenathiazole) **97b** (R_1 = ethyl, R_2 = H) adopts a dimeric structure **109** with a Se–Se bond length (2.46 Å) close to that of a "normal" sigma bond.[185] The S–Se bond within the ring is essentially ruptured as a consequence of the electronic structure reorganization suggested by the resonance structure **109** (and supported by the bond metrics of the structure). Pyrazine-based bis(dithiazole) **101** (R_1 = ethyl) also adopts this dimeric structure, with the inter-radical S–S bond (2.17 Å) again being very close to that of a typical disulfide.[183] Interestingly, this molecule is bimoprhic; the second isomer is an unusual (for thiazyl radicals) carbon–carbon σ bonded dimer **110** with a relatively long (1.61 Å) C–C bond. Computational studies on dimerization processes on all four radical types (**98a–d**) suggest very low dimerization enthalpies to form chalcogen based dimers,[184] and the C–C dimer **111** is predicted to be weakly endothermic and as such was described as a kinetic product. Other derivatives of both of these radicals do not form these unusual dimers, providing further evidence of the subtleties of the dimerization process in these delocalized radicals.[182,186]

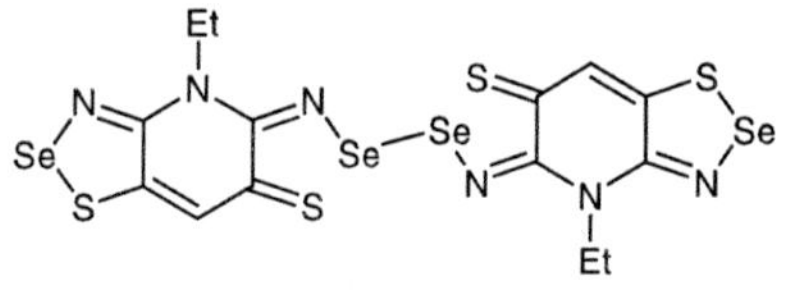

109

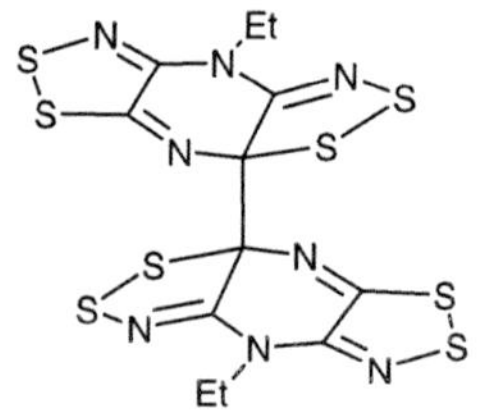

110

9.3.7 1,2,4-Thiadiazinyl radicals

Oxidation of annelated 4H-1,2,4-thiadiazines[190] **111** with lead oxide or silver oxide affords the corresponding 1,2,4-thiadiazinyl radicals **112**.[191] The simple benzo-, pyrido-, and pyrazino- fused radicals were not sufficiently stable to be isolated, but tetrafluorobenzo and tetrachlorobenzo-fused radicals have been structurally characterized (Figure 9.15). Both radicals are monomeric in the solid state. The tetrachloro-based radical does not adopt a stacked structure, but the tetrafluoro derivative consists of stacks of radicals in which the orientation of the radicals alternates up and down the stacks, leading to relatively short S–N contacts of ~3.2 Å. EPR and computational studies indicate that the dominant contributions to the 1,2,4-thiadiazine SOMO come from the thioaminyl (SN) unit and the other ring nitrogen atom; there is little spin delocalization onto the annelated portion of these radicals.

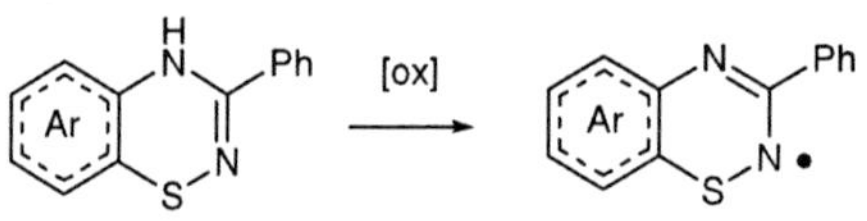

111 112

"Resonance-stabilized" bis(1,2,4-thiadiazine) radicals **115** have been developed by Oakley.[192] Condensation of bis-amidine **113** with disulfur dichloride (S_2Cl_2) affords the bis(thiadiazinium) cation **114**

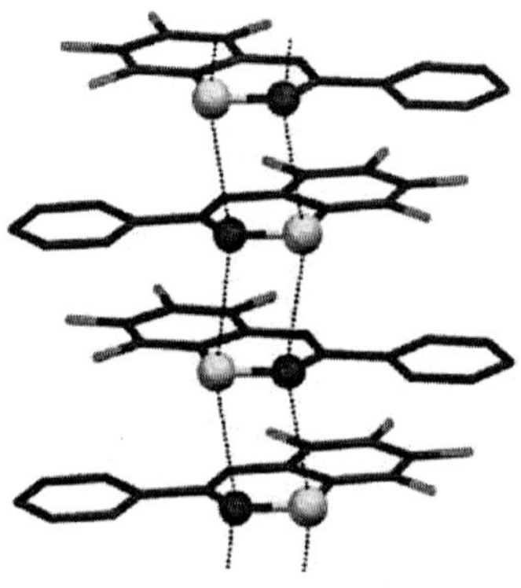

Figure 9.15 X-ray structure of tetrafluorobenzo-1,2,4-thiadiazinyl.[191]

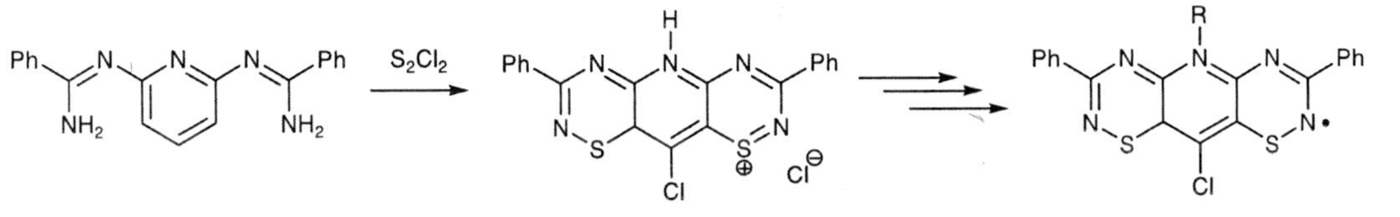

113 114 115

Scheme 9.23

which can be taken through a sequence of steps reminiscent of the bis(1,2,3-dithiazole) radicals **97** (i.e., metathesis/deprotonation/alkylation/reduction) to ultimately give the radicals **115** (R = methyl, ethyl) (Scheme 9.23). These radicals – which are monomeric in the solid state – are, similarly to the bis(1,2,3-dithiazoles), highly delocalized but with little spin density residing on the two central ring atoms which lie on the nodal plane of the SOMO (**116**).

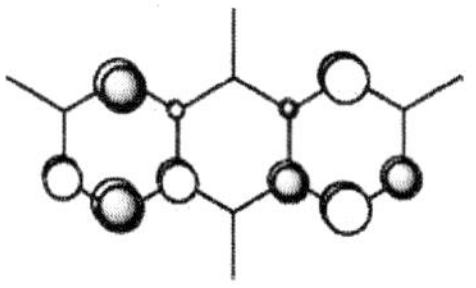

116

9.3.8 1,2,4,6-Thiatriazinyl and -selenatriazinyl radicals

Although not as frequently studied as some of the other thiazyl radical ring systems, the 1,2,4,6-thiatriazinyl skeleton **117** – along with its selenium and phosphorus-based variants discussed below – offers many key insights into relationships between molecular and electronic structure in thiazyl radicals. These radicals have been spectroscopically identified as decomposition products of dithiatetrazocines **119**,[193] but the synthetic potential of this route has not been explored. Thiatriazinyls can be synthesized by reduction of the corresponding S-chlorothiatriazines **118**. The first derivative of **118** to be made was the 1,3,5-trichloro derivative from dicyanamide salts and thionyl chloride[194]; this substrate can be converted to other thiatriazines with different substituents on carbon and sulfur through nucleophilic substitution chemistry.[195] The two general routes into the S-chlorothiatriazines involve the reaction of amidines with $S_3N_3Cl_3$,[91,196,197] or the condensation of imidoylamidines **120** with sulfur dichloride (Scheme 9.24).[198,199] The latter route permits the synthesis of unsymmetrically substituted derivatives. Reduction of the S-chlorothiatriazines to the radicals has been achieved with metals or triphenylantimony.

The 3,5-diphenyl derivative was the first 1,2,4,6-thiatriazinyl radical **117** to be studied in detail by EPR, structural, and computational studies.[196,197] The SOMO for this radical (**121**) has large contributions from the sulfur and all three nitrogen p-orbitals; the EPR spectrum of this radical has (nearly) equivalent hyperfine coupling to all three nitrogen atoms. Derivatives with more electron withdrawing groups on carbon (e.g., 4-$O_2NC_6H_4$, Cl, CF_3) have the effect of lowering spin density on N2 and N6 (the nitrogens flanking the sulfur atom) relative to N4, as assessed by the changes in EPR nitrogen hyperfine coupling constants.[91,200] The pronounced substituent effect on radical spin distributions and ionization potentials[91] contrasts the situation for, for example, 1,2,3,5-dithiadiazolyls for which the symmetry properties of the SOMO preclude direct conjugative substituent effects.

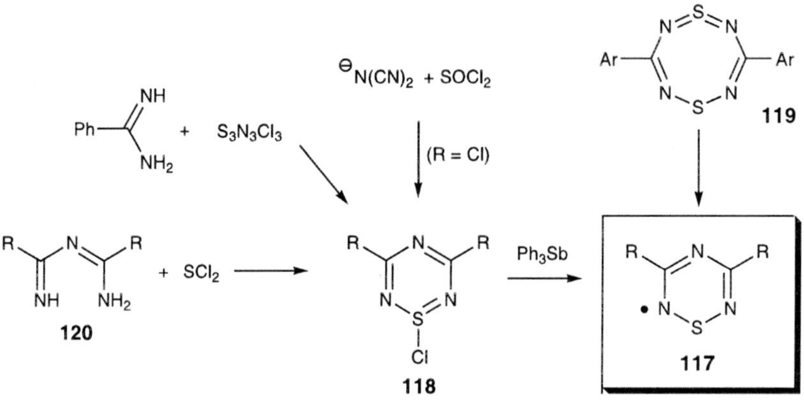

Scheme 9.24

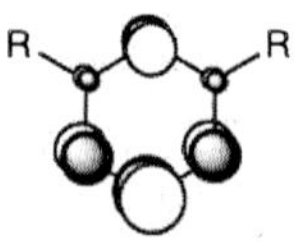

121

Thiatriazinyl radicals are in equilibrium with diamagnetic dimers in solution.[196] There is but one solid state structure of a thiatriazinyl, namely the diphenyl derivative (Figure 9.16).[197] The structure consists of a cofacial dimer with a S–S distance of 2.67 Å – considerably shorter than the S–S inter-ring distances in other thiazyl radicals. There is also a noticeable tilting of the two rings away from one another, which has lead to speculation as to whether this structure is best considered as having a weak S–S σ bond or is better regarded as a π dimer complex.[197]

The chemistry of thiatriazinyl radicals has been little explored aside from some fundamental redox chemistry (Scheme 9.25).[201] The radicals **117** can be re-oxidized to the S-chlorothiatriazines **118** with sulfuryl chloride; the use of nitrosonium salts of weakly coordinating anions (BF_4^-, PF_6^-) produces the thiatriazinium cations **122**. Reduction of the radicals with lithium in liquid ammonia, followed by quenching

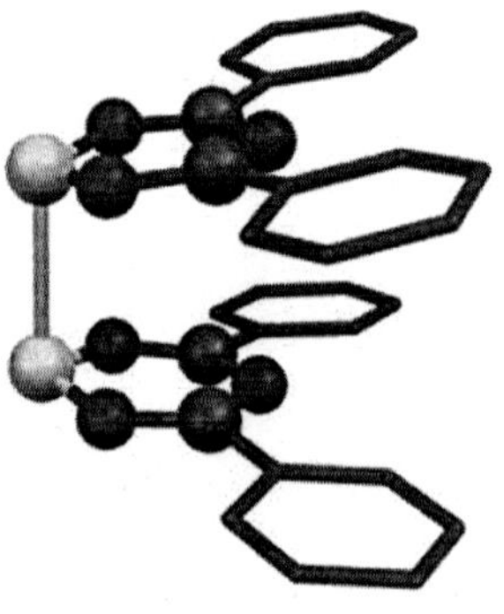

Figure 9.16 *Structure of the dimer of the 3,5-diphenyl-1,2,4,6-thiatriazinyl radical.*[197]

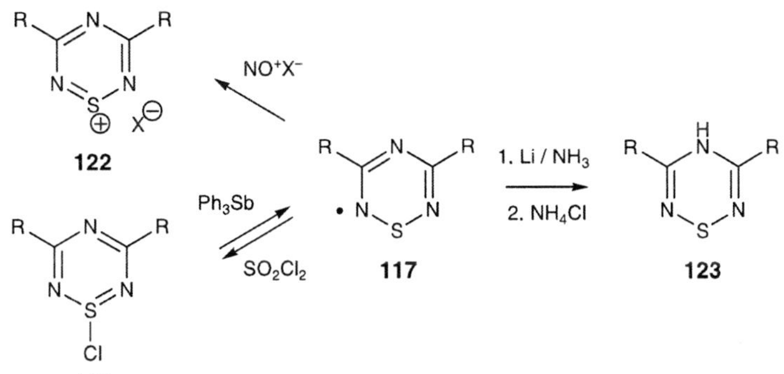

Scheme 9.25

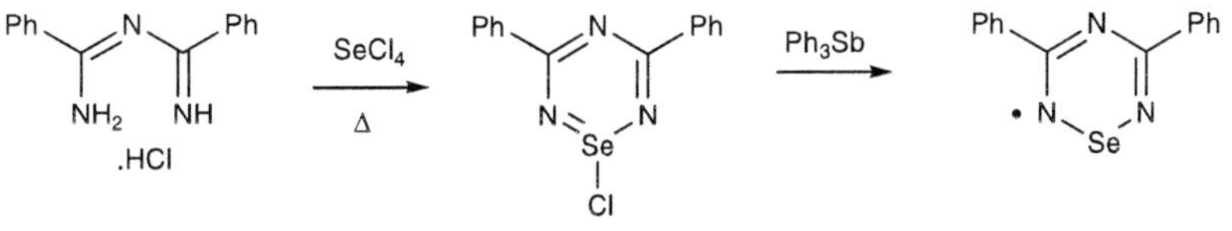

Scheme 9.26

with ammonium chloride, gives the imide **123**. The experimentally observed site of protonation (N4) in **123** is consistent with the calculated charge distribution in the intermediate thiatriazinide anion.

A single example of a 1,2,4,6-selenatriazinyl radical **124** has been prepared by the imidoylamidine route previously described for the thiatriazine (Scheme 9.26).[199] The ring-forming reaction passes through a series of isolable intermediates but ultimately gives the Se-chlorothiatriazine **125** upon prolonged heating. The electronic and molecular structural features of the radical (made by Ph_3Sb reduction of **125**) mirror those of the corresponding thiatriazinyl – the spin density is comparable on all three nitrogens, and the radical exists as a cofacial dimer analogous to the sulfur radical with an Se–Se "bond" length of 2.79 Å. Comparative solution EPR studies indicate that the selenatriazinyl radicals associate more strongly than the thiatriazinyls (dimer dissociation constants of 3×10^{-2} M for **117** vs 5×10^{-4} M for **124**).

9.3.8.1 Phosphorus-containing thia- and selenatriazinyl radicals

The incorporation of phosphorus into the thia (selena) triazinyl skeleton has significant consequences for the electronic structure of the resulting radicals. Condensations of the phosphorus-containing imidoylamidine **126** with sulfur dichloride or selenium tetrachloride gave the corresponding sulfur- or selenium-chlorotriazines **127** and **128** respectively, which were subsequently reduced (Ph_3Sb) to give phosphathiatriazinyl and phosphaselenatriazinyl radicals **129** and **130** (Scheme 9.27).[202]

Whereas the SOMO (**121**) of radicals **117** and **124** are fairly evenly distributed around the ring, analysis of the EPR spectra of the phosphorus containing radicals **129** and **130**, coupled with semi-empirical calculations, indicate that their SOMOs mainly reside on the SN unit adjacent to the phosphorus atom (Figure 9.17a). The phosphorus atom carries very little spin density but has a substantial positive charge in both radicals; in this context, the phosphorus centers are best regarded as phosphonium cations, with the

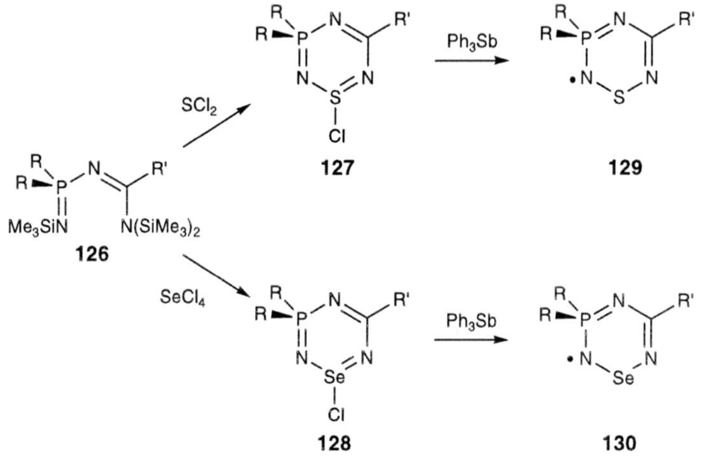

Scheme 9.27

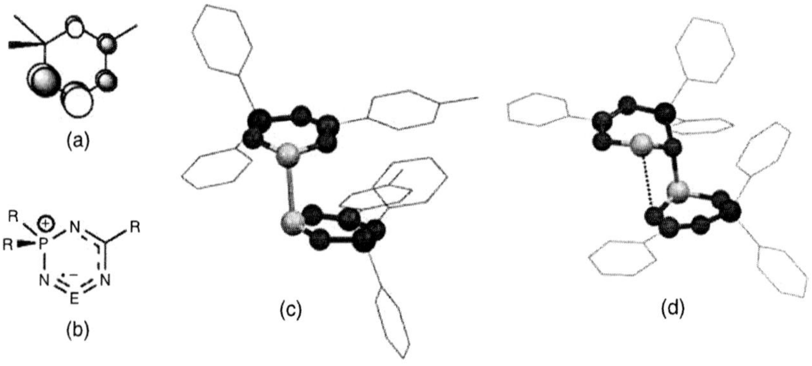

Figure 9.17 *(a) SOMO and (b) representation of charge distribution in phosphathiatriazinyl* ***129****. (c) X_ray structure of* ***129*** *(R = Ph, R′ = p-tol). (d) X-ray structure of* ***130*** *(R = R′ = phenyl).*[202]

negative charge shared among the remaining atoms in the ring (Figure 9.17b). The heteroatom perturbation also has ramifications for the nature of the radical dimer structures. Phosphathiatriazinyl **129** (R = R′ = phenyl) exists as an S–S bound dimer (d(SS) = 2.48 Å) but, unlike the thiatriazinyl **117**, has the two rings are twisted with respect to one another (Figure 9.17c) – probably due to the steric demands of the two phosphorus substituents. The dimer of phosphaselenatriazinyl **130** is based on a Se–N bond (1.99 Å), supported by a secondary Se–N contact of 3.11 Å (Figure 9.17d). Accompanying this unusual dimerization mode is the fact that the bond lengths within the two rings of the dimer are markedly different from one another. The implications of the bonding are that the dimer consists of anionic (nitrogen-bound) and cationic (selenium-bound) rings (**130**$^-$ and **130**$^+$ respectively) which result from electron transfer from one radical to another (Scheme 9.28) – in other words, while the inter-ring bond in **129** (and **117**) can be described as covalent, the Se–N interaction in **130** is better formulated as a donor-acceptor bond.

Replacement of both skeletal carbon atoms in **117/124** by phosphorus leads to the "all-inorganic" diphosphathia-(selena)triazinyl radicals **136/137**. S-Chloro-3,5,1,2,4,6-diphosphathiatriazines **131** were

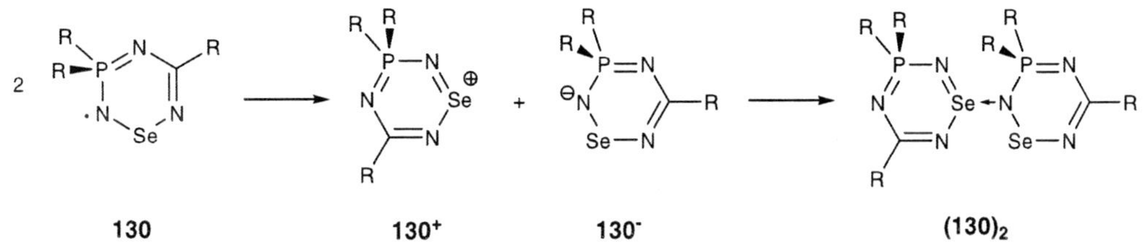

Scheme 9.28

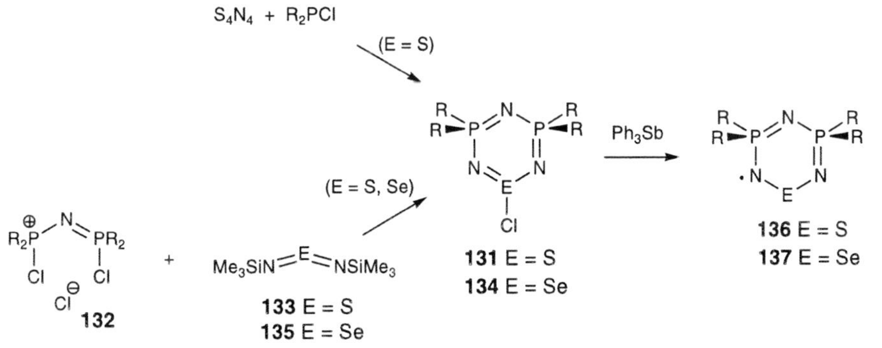

Scheme 9.29

originally obtained from the reactions of tetranitrogen tetrasulfide (S_4N_4) with chlorophosphines[203] but can also be made from the reaction of phosphonium salts **132** and bis(trimethylsilyl)sulfur diimide **133** (Scheme 9.29)[202]; the latter route allows entry into the corresponding Se-chlorodiphosphaselenatriazine **134** by using the analogous selenium diimide reagent **135**.[202] Reduction of **131** and **134** with triphenylantimony gives the radicals **136** and **137** respectively.

The electronic structure of the $P_2N_3E^{\bullet}$ radicals continues the trend established by the introduction of the first phosphorus (see above). EPR and computational studies indicate that the SOMO **138** is now totally confined to the NSN portion of the ring and the charge distribution strongly suggests an internal salt formulation **139**.

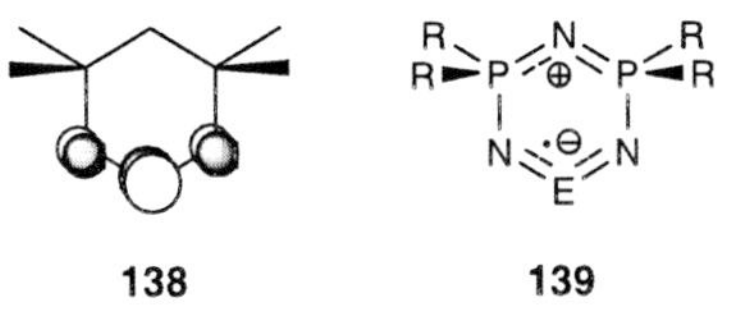

The long term instability of selenium-based radical **137** precluded its structural characterization. The structure of sulfur-based radical **136** consists of a very unusual dimeric structure **140** based on two fused seven-membered rings sharing a long (2.39 Å) disulfide bond (this structure has also been described as a 12 membered ring with a trans-annular S–S bond) (Figure 9.18).[204] The proposed mechanism of (reversible) formation of this dimer is shown in Scheme 9.30. Two radicals **136** associate and undergo S–N bond metathesis via transition state **141** to give a putative twelve membered ring "diradical" structure **142**, which

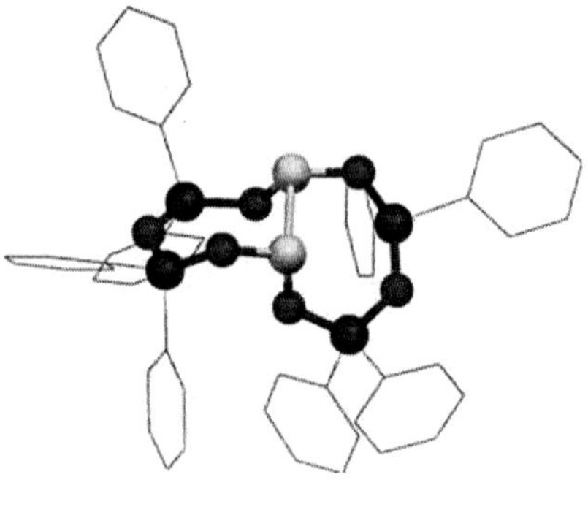

Figure 9.18 *Structure of the dimer **140** of the $Ph_4P_2N_3S^{\bullet}$ radical **136**.*[206]

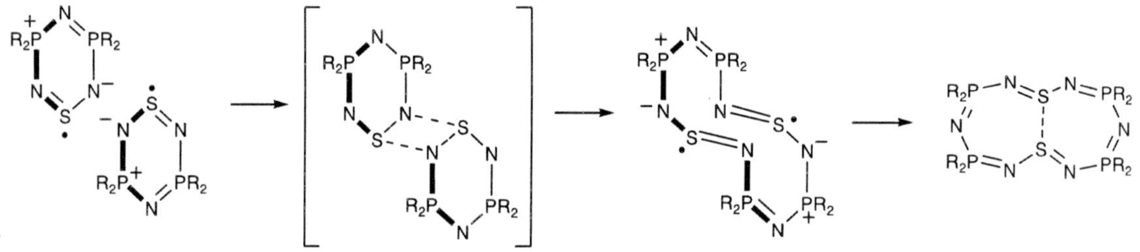

Scheme 9.30

undergoes intramolecular coupling to give the disulfide-bridged bicyclic structure **140**. This rearrangement process – essentially a S–N bond metathesis – is analogous to the scrambling reactions of sulfur diimide radical anions (Equation 9.5).[205]

$$[RN{=}S{=}NR]^{\bullet -} + [R'N{=}S{=}NR']^{\bullet -} \rightarrow 2[RN{=}S{=}NR']^{\bullet -} \quad (9.5)$$

All of the aforementioned studies on thiatriazinyls are based on the 1,2,4,6-thiatriazine topology. Kaszynski has performed DFT calculations on other possible six-membered ring radicals with the C_2N_3S skeleton. The 1,2,4,6-isomer **117** was found to be at least 20 kcal/mol more stable than the other thiatriazine isomers, for example, the 1,2,4,5-structure **143**. Attempts to make **143** – essentially a heterocyclic thioaminyl radical – were unsuccessful.[207]

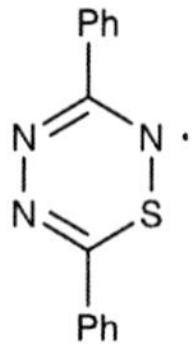

143

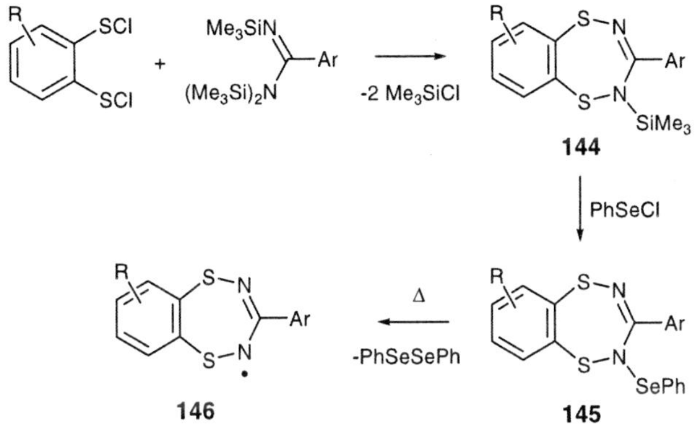

Scheme 9.31

9.3.9 Larger cyclic thiazyl radicals

Stable thiazyl radicals in seven- and larger-membered rings are rare. Chivers developed routes to benzodithiadiazepines **144** through reactions of benzenebis(sulfenyl chlorides) with silylated amidines (Scheme 9.31).[208] The remaining N–$SiMe_3$ group can be converted to a phenylselenyl substituent to give **145** which, upon heating, leads to N–Se homolysis and generation of the benzo-1,4,5,7-dithiadiazepinyl radicals **146**. However, these radicals proved to be too unstable to isolate. Attempts to access the radicals by chlorination of **144** lead to the benzo-1,3,2-dithiazolium cation (cf. **64**) which was postulated to form via elimination of a nitrile (ArCN) from the *in situ* generated radical **146**.

Oakley has reported the EPR characterization of an eight-membered ring radical **147** which was generated by reduction of the corresponding S-chloro compound.[209] The hyperfine coupling parameters suggest that this radical can be described as an "internal salt" analogous to diphosphathiatriazinyls **139**, that is, the NSN portion of **147** is formulated as a radical anion with the corresponding positive charge delocalized mainly on the NPN fragment opposite to the sulfur atom. This radical also exhibited unusually large hyperfine coupling constants to the two flanking phosphorus atoms (24.5 G each), indicative of a non-planar ring conformation which permits σ/π mixing and thereby increases the Fermi contact terms to these atoms. The non-planarity of **147**, and possible dynamic behavior thereof, is also inferred from its temperature dependent EPR spectrum.

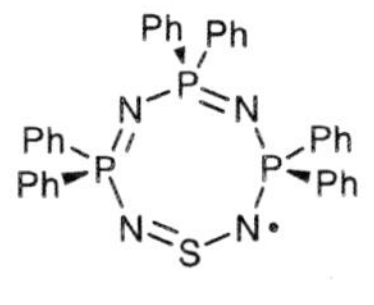

147

9.4 Thiazyl radicals as "advanced materials"

The discovery of the remarkable solid state electronic properties of poly(thiazyl) is generally regarded as the main catalyst for a resurgence of interest in the chemistry of so-called "sulfur–nitrogen" compounds, and particularly those based on radicals.[210] The properties of poly(thiazyl) have been comprehensively reviewed by Labes[37] and by Banister and Gorrell,[38] and only a short summary of the salient characteristics of this material is presented here. Poly(thiazyl) is a highly conducting material, with a room temperature conductivity of approximately 1000 S/cm. Upon lowering the temperature the conductivity increases by as much as 100-fold and becomes superconducting at 0.26 K (it should be emphasized here that the confirmation of superconductivity in $(SN)_x$ predated the discovery of the copper oxide "high T_C" superconductors by about 15 years; at the time of the $(SN)_x$ breakthrough, the record superconducting transition temperature (T_C) for any material was only 23 K). Extensive experimental and computational investigations reveal $(SN)_x$ to be an anisotropic, but genuine three-dimensional solid state material. This can be understood most simply by considering the electronic structure of a single $(SN)_x$ chain, which is predicted to undergo a Peierls distortion and hence be at best a semiconductor. The solid state structure of $(SN)_x$ (Figure 9.4) also provides compelling evidence for significant interactions between SN chains. The multidimensional electronic structure is crucial to understanding this unique material, and this (along with the benefit of hindsight) makes it possible to understand why attempts to prepare "hybrid" thiazyl conducting polymers combining SN units and organic components failed.[211]

In addition to inspiring number of efforts to make new sulfur–nitrogen-containing polymers, the transport properties of $(SN)_x$ had a large impact on the field of inorganic compounds of sulfur and nitrogen, particularly radical compounds. As a result of nearly three decades of sustained effort, the foundation of fundamental studies of thiazyl radicals has enabled more targeted studies of intriguing, and possibly technologically relevant, physical properties such as charge transport and magnetism. This evolution ("from molecules to materials") has positioned thiazyl radicals collectively as one of the more important classes of molecular building blocks for advanced materials. Detailed coverage of the literature in this area would require a separate (and very large) chapter as a companion to this one. Instead, the following sections provide a brief introduction to the main physical properties exhibited by thiazyl radicals; subsequent sections are organized phenomenologically rather than by radical type. Rawson provided a general overview of the materials properties of thiazyl radicals a few years ago,[212] and specific materials-oriented pursuits have been covered at various stages by Oakley,[210,213] Awaga,[214] and Rawson.[215]

9.4.1 Charge transport properties of thiazyl radicals

The pursuit of electrically conducting materials has been a major thrust of molecular thiazyl radical research for some time. In addition to the general context established by poly(thiazyl), a more specific rationale for the construction of conducting materials based on neutral radicals was independently developed by Haddon.[216,217] The basic premise was that a π stacked array of neutral radicals, with significant inter-radical orbital overlap, should give rise to a half-filled conduction band (Figure 9.19d). The presence of charge carriers (unpaired electrons) and a partly filled band fulfill two requirements for metallic type conduction. This approach contrasts the more established route to molecular metals in which unpaired electrons are generated by electron transfer from a donor to an acceptor molecule, that is, conducting π stacks of radical *ions* (Section 9.4.2). However, a uniformly spaced stack of uncharged radicals has some pitfalls. If intermolecular overlap within the stack is poor, the unpaired electrons can simply localize on each molecule and render the material a Mott insulator (Figure 9.19e). Even with substantial overlap (i.e., bandwidth in the solid state), a one-dimensional structure of this type is inherently unstable and can undergo a Peierls distortion, a structural distortion which opens up a band gap and renders the material a semiconductor at best. In chemical terms this can be described as radical dimerization, leading to small gap if there is still sufficient inter*dimer* overlap (Figure 9.19c) or, in extreme cases, localized dimers with little to no interaction with neighboring dimers (Figure 9.19b).

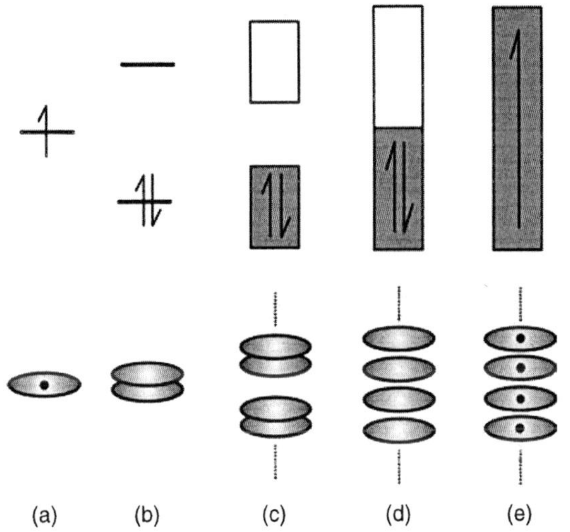

Figure 9.19 *Qualitative orbital energy diagrams for (a) the SOMO of a single π radical, (b) the HOMO and LUMO of a radical π dimer, (c) valence and conduction bands for a stack of radical π dimers, (d) a half-filled conduction band for a stack of uniformly spaced radicals, and (e) a Mott insulator for a stack of uniformly spaced radicals.*

At the time of Haddon's proposal, no such stacked radical structures existed; in fact, the precious few examples of stable neutral radicals in existence did not have the flat, delocalized, sterically unencumbered structure to form any sort of π stack at all (Haddon's proposal focused on the phenalenyl radical as the building block of choice for other reasons – see below – but it took over twenty years for stable versions of these radicals to be realized[218]). The development of several classes of thiazyl radicals in the 1970s and 1980s set the stage for some of these to be explored as candidates for neutral radical-based conducting materials. The 1,2,3,5-dithiadiazolyl (**35**) and diselenadiazolyl (**45**) radicals were the first to be seriously pursued in this regard, owing to their relative ease of synthesis, stability, and preponderance of heteroatoms (sulfur and nitrogen) available for intermolecular interactions. The latter feature was believed to be of possible use in providing additional orbital overlap which could aid in decreasing the band gap (Peierls distortions are prevalent for low dimensional (anisotropic) structures).[213,219]

Extensive investigations into the solid state structures and properties of a wide range of 1,2,3,5-dithiadiazolyl **35** and diselenadiazolyl radicals **45** produced several small band gap semiconductors (the sulfur-based radicals are insulators (conductivity $>10^{-8}$ S/cm) but several of the selenium variants have room temperature conductivities of up to $\sim 10^{-5}$ S/cm). These efforts also lead to a greater appreciation of molecular and solid state design principles which guided subsequent efforts. With few exceptions these radicals formed π dimer structures in the solid state, precluding the possibility of the half-filled band architecture of Figure 9.19d. However, many of these derivatives adopt π-stacked structures of *cis*-cofacial π dimers; the distance between dimers are typically close to the van der Waals contact distances between sulfur or selenium atoms. The interdimer interactions can, and often do, lead to appreciable band formation in the solid state. Band gaps for these materials range from values as high as 2.0 eV (in cases where intermolecular overlap is very poor) down to as little as ~0.5 eV. The 5-cyano-2-furyl substituted diselenadiazolyl radical (Figure 9.20) is one of the best examples – EHMO band calculations reveal that valence and conduction bands are due solely to π stacked structure (i.e., one-dimensional in nature), which

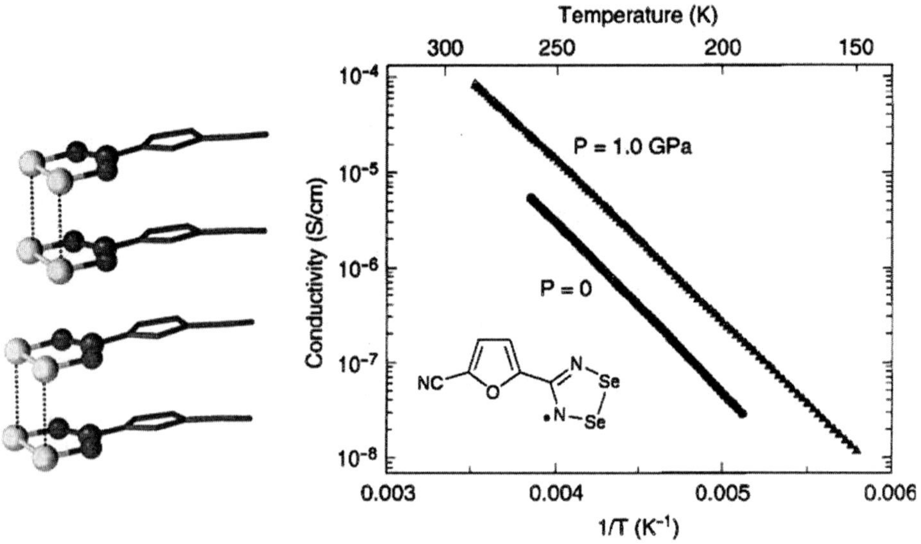

(a) (b)

Figure 9.20 *(a) π stacking of cis-cofacial π dimers and (b) variable temperature conductivity for 4-(5-cyano-2-furyl)-1,2,3,5-diselenadiazolyl.*[136] *(b Adapted with permission from [136]. Copyright 2001 American Chemical Society.)*

was cited as one of the most efficient of all the stacked structures for radicals of this type.[136] The room temperature conductivity is 10^{-5} S/cm – by far the highest value for a "mono" radical structure[92,135] – and the temperature dependence of conductivity is that of a classic intrinsic semiconductor, that is, linear log(σ) vs 1/T plot (Figure 9.20b). The application of external pressure leads to an enhancement of the conductivity, presumably by compressing the structure, which leads to stronger intermolecular overlap.

However, the one-dimensional stacking of dimers is not always the only contribution to the overall (three dimensional) solid state electronic structure and transport properties. The EHMO calculated band structures of the 1,4- and 1,3-benzene-bridged diradicals (whose structures are shown in Figures 9.6a and 9.6b respectively) both reveal significant isotropic band structures despite the stacking (1,3-diradical) or pseudostacking (1,4-diradical) motifs.[123,126] Although the sulfur-based diradicals are insulators, the selenium analogues show semiconductive behavior, and are superior in this respect to several diselenadiazolyl *mono*radicals.[92,135] Thus in the absence of very efficient π stacking, both *intra*- and *inter*stack interactions help to create bandwidth. In this respect, the fact that the dithia- and diselenadiazolyl di- and triradicals tend to have smaller band gaps and higher conductivities is probably a consequence of their simply having "more radical" and "less substituent" in their molecular structures, thereby facilitating more extensive intermolecular contact networks. This hypothesis is supported by, for example, the smallest possible diselenadiazolyl **45** (R = H), with just a proton as substituent. This compound is a markedly better conductor than most of the other monofunctional RCN_2Se_2 radicals.[134] Similarly, the dithiadiazolyl diradical **39,** which has "no" substituent, is one of the few dithiadiazolyl radicals with measurable conductivity.[121,122] Both of these radical derivatives have π stacks of cofacial π dimers as well as extensive interstack contacts.

The propensity of the 1,2,3,5-dithia- and diselenadiazolyl to dimerize in the solid state remains a major obstacle to the realization of the regular (monomeric) radical stacked structure (Figure 9.18d). Moreover, there is another aspect of these radicals which makes them less than ideal candidates for molecular conductors. Haddon discussed the importance of minimizing the ionic fluctuation energy (U_{eff}) of a molecular conductor, that is, energy associated with disproportionation of two radicals into a cation and anion.[216,217] This quantity – also referred to as the energy of disproprotionation, ΔH_{disp}, *an intrinsic molecular property* – is the difference between the ionization energy (IE) and electron affinity (EA) of a radical. Semi-empirical calculations suggested that, for the odd alternant hydrocarbon radicals initially under consideration, the U_{eff} quantities compared favorably to those determined for the radical ion constituents of the more well known charge transfer salts.[217]

Experimentally, U_{eff} can be determined if both the ionization energy and electron affinity or known, but more commonly this quantity is estimated from solution electrochemical methods, where the difference between a neutral radical's oxidation and reduction potential – referred to as the cell *potential*, "E_{cell}" – correlates with IP–EA (IP = Ionization Potential). The electrochemical method does not provide accurate absolute U_{eff} values, but the correlation between E_{cell} and either calculated or gas phase determined (IE–EA) U_{eff} is very good, as shown by Boeré who investigated the solution electrochemical properties of a large number of dithia- and diselenadiazolyl radicals.[96] These radicals turn out to have quite large E_{cell} values (1.43 ± 0.06 V for dithiadiazolyls and 1.25 ± 0.03 V for diselenadiazolyls) which are not very sensitive to substituent effects (Section 9.3.2). The implications are that, even if the desired uniform radical π stacked structure based on dithia- or diselenadiazolyl radicals could be realized, effective (metallic) charge transport would be unlikely because of the high ΔH_{disp} values inherent to these radicals.

Over the past decade Oakley has pursued other types of thiazyl radicals as building blocks for single-component conductors, using the cell potential (disproportionation energy) as a central consideration in molecular design. The 1,3,2-dithiazolyl (Section 9.3.4) was identified[164] as another candidate radical class; compared to the 1,2,3,5-dithiadiazolyls, the 1,3,2-dithiazolyls have a lower tendency to dimerize in the solid state and their redox properties (Table 9.1) are more substrate dependent, and therefore tunable. However, the E_{cell} values of most 1,3,2-dithiazolyls are ~1.2 V, although for one derivative (**80**) this value approaches 1 V. Conductivities of 1,3,2-dithiazolyls remain low, even for derivatives which do not dimerize and π stack in the solid state – that is, these materials are Mott insulators.[220,159]

The cell potentials of 1,2,3-dithiazolyl radicals (Section 9.3.5) tend to be somewhat smaller than their isomeric 1,3,2-derivatives. However, among the relatively small collection of structurally characterized derivatives only two have solid state packing patterns which could facilitate charge transport. The lone example of a π stack of *monomeric* 1,2,3-dithiadiazolyls (compound **94**) is, in fact, a Mott insulator.[174] Radical **92** assembles in the solid state into slipped π stacks of cofacial but antiparallel (centrosymmetric) π dimers (Figure 9.14b). This material has a room temperature conductivity of nearly 10^{-4} S/cm, at the time by far the highest conductivity for a sulfur-based radical.[177] The conductivity is thermally activated, with a band gap of ~0.4 eV (determined both experimentally and estimated computationally by EHMO methods). Interestingly, EHMO band calculations reveal that the orbital dispersion in this material is not significantly higher than that of other radicals with similar solid state architectures (i.e., stacked π dimers. Instead the low band gap is a direct consequence of the relatively small inherent HOMO–LUMO gap in the π dimmer, which arises from the head-to-tail mode of association. This is also reflected in the relatively long inter-radical S–S distances within the dimer.

The most promising thiazyl radical-based conductors to date are based on the resonance-stabilized bis(1,2,3-dithiazolyls) and their selenium analogues **97a–d** (Section 9.3.6). The cell potentials of these radicals are significantly smaller than all of the other thiazyl radical types; E_{cell} values are in the range 0.7–0.85 V, with the smallest values belonging to some of the selenium containing derivatives **97b**. These radicals offer the additional advantage of (with a few exceptions, noted in Section 9.3.6) remaining monomeric in the solid state. The bis(dithiazole) radicals **97a** show, with few key exceptions, remarkably consistent solid state properties across several derivatives,[180,181,187,221–223] namely: (i) relatively one-dimensional electronic structures arising from slipped π stacks of radicals; the overall bandwidths of approximately 0.5 eV (EHMO calculated) arise mostly from intrastack overlap; and (ii) intrinsic semiconductive behavior, with room temperature conductivities between 10^{-6}–10^{-5} S/cm and thermal activation energies of ~0.4 eV. A significant outlier among the bis(dithiazoles) in terms of properties is a derivative of pyrazine-containing radical **101** (R_1 = methyl).[182] This compound adopts a different type of stacking pattern, one in which the π stacks are not slipped (tilted) but also not cofacially superimposed, the result of which is a complex network of intra- and interstack overlap. This produces an unusually large bandwidth (1.5 eV) and exceptionally high conductivity for a thiazyl radical: the room temperature conductivity is nearly 10^{-3} S/cm and the activation energy is only 0.19 eV, or *half* the value of the other bis(dithiazole) materials. Incorporation of selenium into either (**97b, 97c**) or both (**97d**) chalcogen positions of these radicals does not dramatically perturb the molecular features (e.g., spin distributions, ion energetics), but does lead to substantially better intermolecular interactions, as manifested by larger bandwidths (up to 1 eV), higher room temperature conductivities (10^{-6} up to 10^{-3} S/cm) and smaller activation energies (generally ~0.3–0.17 eV).[184–189221–224]

9.4.2 Thiazyl radical-based charge transfer salts

The use of thiazyl radicals as single-component conductors represents an alternative to the more conventional charge transfer (CT) salt based materials. Charge transfer salts consist of donor (D) and acceptor (A) molecules between which there is *partial* charge transfer from D to A, that is, one or both components have non-integral charge states. The solid state structures of these materials consist of π stacks of all D and/or all A (as opposed to alternating stacks DADADA...), which gives rise to partially filled conduction bands. In molecular terms the π stacks are formally *mixed valent*, consisting of a mixture of neutral (closed shell) molecules and the corresponding radical ions. For example, the prototypical charge transfer salt TTF–TCNQ (tetrathiafulvalene–tetracyanoquinodimethane) has a degree of charge transfer of 0.59, that is, on average the TTF and TCNQ π stacks contain 59 % radical cations and radical anions, respectively.

Thiazyl radicals have also been explored as the donor component in charge transfer type conducting materials. Phenomenologically, the use of a neutral radical instead of a closed shell species as donor has the effect of changing the occupancy of the conduction band (Figure 9.21): the unoxidized stack of closed shell donors has a filled conduction band (Figure 9.21a), whereas the neutral radical stack has a half filled band (Figure 9.21c; cf. Figure 9.19d). Partial oxidation of the closed shell species creates π stacks consisting of a mixture of radical cations and neutral molecules, whereas the mixed valent stacks in the neutral radical case are combinations of *closed shell* cations and neutral radicals.

In 1984, Wolmershauser reported that the benzo-1,3,2-dithiazolyl radical (**69**) reacts with TCNQ to give an insoluble material with 1 : 1 stoichiometry and a room temperature conductivity of ~1 S/cm.[155] Based on the analogy to the TTF–TCNQ family of charge transfer salts, the authors speculated that the structure of their material consisted of segregated stacks of partially oxidized dithiazolyl radicals and partially reduced TCNQ stacks as well. This turned out to be the case, as reported by Awaga in 2008:[225] the X-ray structure indeed consists of the two components in separate π stacks (Figure 9.22), and analyses of the structural (S–N bond lengths) and spectroscopic (ν(CN) in the infrared spectrum) data suggest a degree of charge transfer of approximately 0.6. This value is remarkably close to the charge transfer in TTF–TCNQ and can be understood based on the fact that the oxidation potential of **69** is virtually identical to that of TTF. However, *unlike* TTF–TCNQ the conductivity of **69**:TCNQ is *not* metallic but instead thermally activated. This behavior is believed to be based on relatively poor ion energetics (high ionic fluctuation energy) associated with radical **69**.[225] A highly conducting charge transfer complex of **69** with an anionic nickel bis(dithiolene) has also been reported. In this material the 1,3,2-benzodithiazoles do not π stack but the metal dithiolene fragment does, implicating the latter as the major contributor to the charge transport properties.[226]

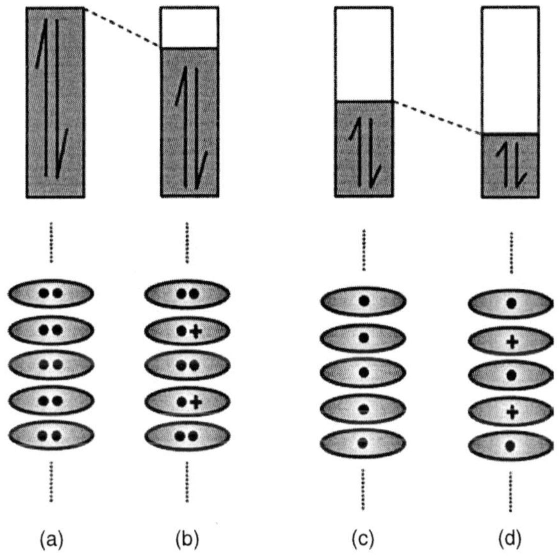

Figure 9.21 *Schematic conduction band occupancies for π stacks of (a) neutral closed shell donor, (b) partially oxidized closed shell donor, (c) neutral radical donor, and (d) partially oxidized radical donor.*

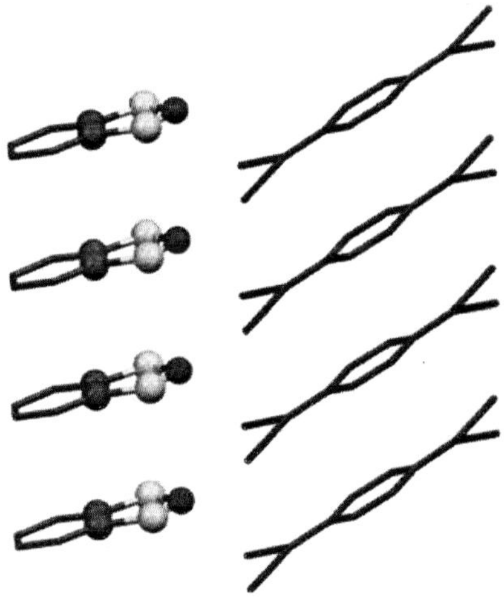

Figure 9.22 *X-ray structure of (benzo-1,3,2-dithiazolyl):TCNQ.*[225]

Several highly conducting charge transfer salts based on 1,2,3,5-dithiadiazolyl radicals and halogens have been studied. The first, and best studied, of these radicals to be studied in this context was the 1,4-benzene-bridged diradical (**148**). Cosublimation of this species with iodine leads to a crystalline material of composition [**148**][I].[227] The structure of this material (Figure 9.23) consists of diradicals assembled into perfectly aligned π stacks, with an interplanar distance of 3.415 Å – longer than typical π dimer distances for these radicals but still significantly shorter than a S–S van der Waals contact (Figure 9.23a).

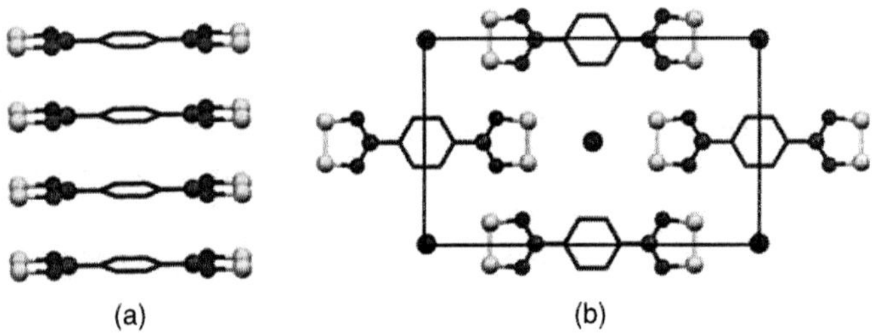

Figure 9.23 Structure of [148][I]. (a) perfectly aligned p stacks of $[148]^{\delta+}$. (b) view of the structure down the stacking axis. Purple spheres represent linear columns of disordered polyiodide.

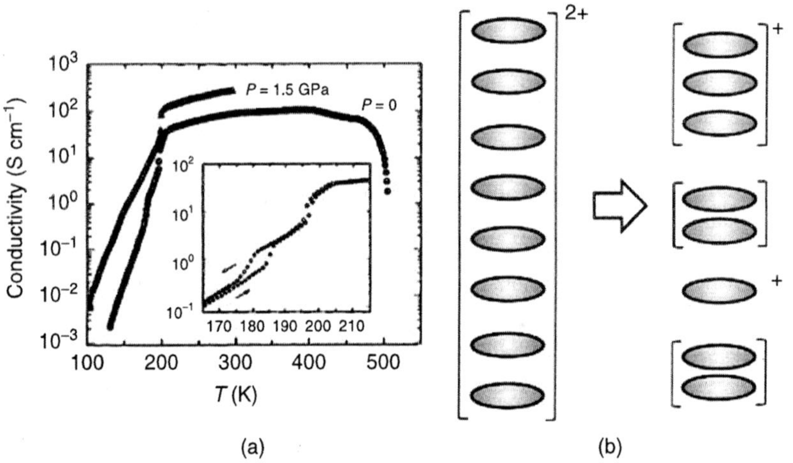

Figure 9.24 (a) Temperature dependence of conductivity and (b) structural model for the metal–insulator transition below 200 K in [**148**][I].[227,228] (a Reprinted by permission from Macmillan Publishers Ltd [228], copyright 1993.)

Iodine channels occupy channels between the diradical columns (Figure 9.23b) as (disordered) polyiodide chains; iodide (I^-) anions would be too large to fit into the lattice of the diradical columns. The S–S and S–N bonds within the rings are intermediate between cation and anion, suggesting that a partial positive charge is borne by the CN_2S_2 rings. This material has a room temperature conductivity of ~100 S/cm at room temperature; above 200 K the conductivity is metallic in nature (Figure 9.24a) Below 200 K the conductivity drops sharply, and the material becomes diamagnetic. Low temperature crystallography studies reveal the presence of a structural distortion – a charge density wave – from which band filling of the heterocyclic π stacks of 3/8 can be directly inferred.[227,228] This, in turn, means that each CN_2S_2 ring is one-quarter oxidized, leading to a total charge distribution of $[\mathbf{148}]^{+0.5}[I]^{-0.5}$. In the high temperature (metallic) regime, the charges are formalisms in a molecular sense; both the dithiazole and iodide chains are mixed valent in the solid state structure. Below the metal–insulator transition, however, the structure consists of irregular π stacks, which can be modeled as a combination of dithiazolium cations, two neutral π dimers, and one cationic π trimer (cf. the triiodide salt of $(PhCN_2S_2)_3{}^+$,[131] (Figure 9.24b).

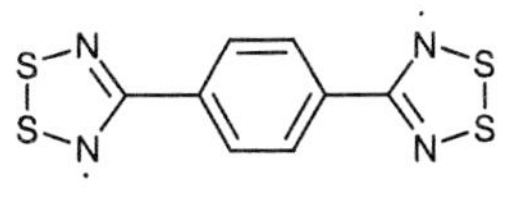

148

Iodine charge transfer salts of other dithia- and diselenadiazolyl diradicals include those based on the 1,4- (**148, 150**) and 1,3-phenylene (**149, 151**) linked diradicals,[228,229] heterocyclic (**152, 153, 154**) spacers,[230,231] and the diradical **39** with the two dithiadiazolyls directly attached to one another.[122] The selenium-based systems are made by electrochemical reduction of the corresponding dications in the presence of iodine, because the diselenadiazolyl diradicals are too involatile for the cosublimation chemistry to work. The structure and electronic properties of [**148**][I] are qualitatively reproduced in the iodide charge transfer salts of **149–155**, that is, regularly spaced π stacks alongside disordered polyiodide channels, high (>10 S/cm) room temperature conductivities, and metal–insulator transitions between 200 and 270 K. Band structure calculations suggest conduction bandwidths of 3–4 eV, based largely on the "perfect" π stacked structures (i.e., interstack interactions are, in a relative sense, far less significant in these materials). There appears to be some degree of flexibility in the degree of charge transfer from the diradicals, based on (i) analyses of the S–S and S–N bond lengths in the charge transfer salts and (ii) in a few instances, the observation of other stoichiometries: diradicals **152** and **154** afford iodide salts with a diradical:iodine ratio of 2 : 1 in addition to the more common 1 : 1 ratio.[230] The 2 : 1 salts have lower conductivities ($\sim 10^{-5}$ S/cm) than the 1 : 1 materials. Bromide-based charge transfer salts of **148** and **149** have also been made by a comproportionation reaction between the diradical and its corresponding dication (dibromide) salt.[228] These materials contain fully reduced bromide (Br^-), which means that the diradicals carry a full positive charge and are mixed valent. However the oxidized diradicals do not adopt the cofacially aligned π stacked structure and in fact the individual CN_2S_2 rings can be assigned as neutral or cationic based on analysis of their S–S and S–N bond lengths. Accordingly the bromide materials are poor conductors.

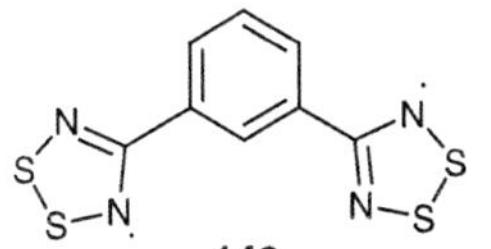

149

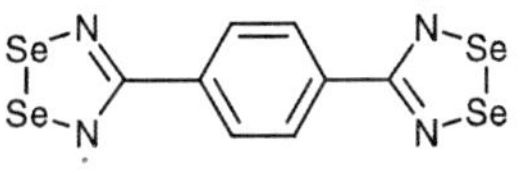

150

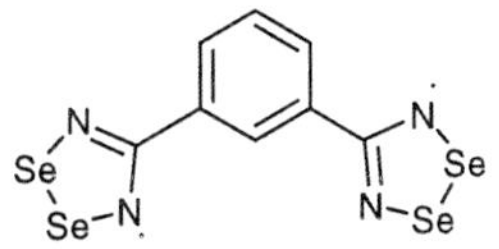

151

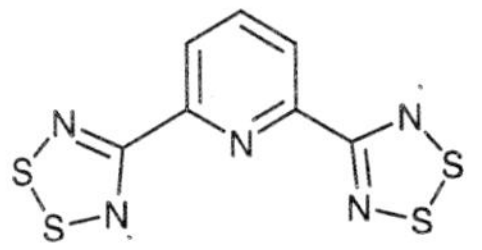

152

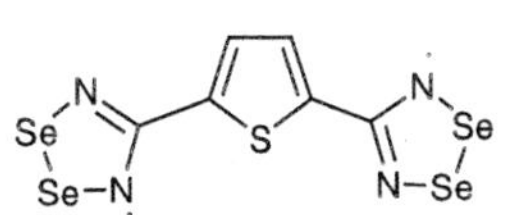

153

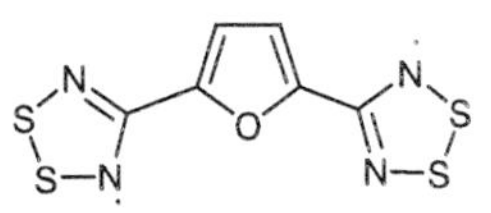

154

The only charge transfer salts based on dithiadiazolyl *mono*radicals are the (non-conducting) "triple-decker" structures based on the phenyl derivative (**35**, R = phenyl; Figure 9.8) and a conducting stacked structure based on the parent dithiadiazolyl HCN_2S_2.[105] This radical has two crystalline phases,[104] one of which consists of π-stacked π-dimers clustered into cyclic arrays with small void channels (Figure 9.25a). Cosublimation of this radical with iodine produces the same stacked arrangement, but now with undimerized radicals (Figure 9.25b,c) and disordered polyiodide in the channels. This material is noteworthy for two reasons: (i) it is the only dithiadiazolyl charge transfer salt in which the basic packing pattern of radicals is retained upon reaction with iodine; and (2) the stoichiometry, $(HCN_2S_2)_6I_{1.1}$, is more iodine-poor than

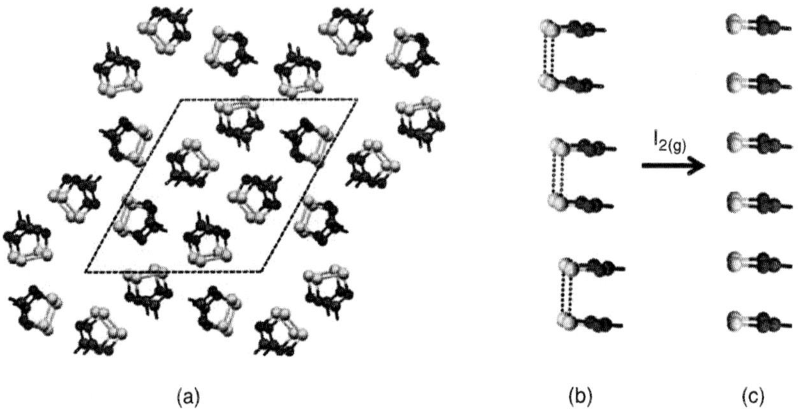

Figure 9.25 (a) Structure of β-HCN_2S_2 viewed down the π stacks. (b) Stacking of π dimers of β-HCN_2S_2. c) Structure of stacks in $(HCN_2S_2)_6I_{1.1}$.[105]

the diradical/iodide salts, suggesting a much smaller degree of charge transfer from the radical – which nonetheless still leads to high conductivity (15 S/cm at room temperature).

9.4.3 Magnetic properties of thiazyl radicals

The magnetic properties of stable radical-based materials have been enthusiastically explored for over two decades.[232] The goals of these pursuits work span fundamental science (the development of structure–magnetism relationships in radical-based compounds and materials) and more practical challenges (the creation of molecule-based magnetically ordered materials). The use of molecular or organic components in the design and construction of new magnetic materials – whether in radical-based molecular crystals,[233,234] incorporated into polymers,[235] or combined with metals as radical ligands[236] – offers a number of possible advantages including solubility, processibility, and multifunctionality.[237] Early efforts in this field were dominated by nitroxide radicals[234,238] owing to their excellent stability; triarylmethyl radicals[239] and verdazyl radicals[240] have also played significant roles.

Thiazyl radicals are relatively new additions to the chemists' toolkit of paramagnetic building blocks, largely because the synthesis and characterization of most thiazyl radicals was not fully developed until the 1980s and 1990s. However, interest in the magnetic properties of thiazyl radicals has steadily grown to the point where these radicals are now among the most important ones in molecular magnetism. Rawson reviewed the magnetic properties of thiazyls and outlined many of the general challenges of the field of designing molecule-based magnets.[212,215] In the past decade in particular, a very large number what could be called "case studies" – that is, papers which report the structural and magnetic characterization of a small number of closely related thiazyl radicals – have been published. Rather than provide a comprehensive account of this body of work, this section aims to distill some of the general magnetic trends and the more important developments in thiazyl radical magnetochemistry.

9.4.3.1 Magnetic properties of thiazyl radical-based π dimers and π stacks

In addition to playing a central role in charge transport properties (Sections 9.4.1 and 9.4.2), thiazyl radical π dimers and π stacks also lead to a rich array of magnetic properties. A generic orbital interaction diagram

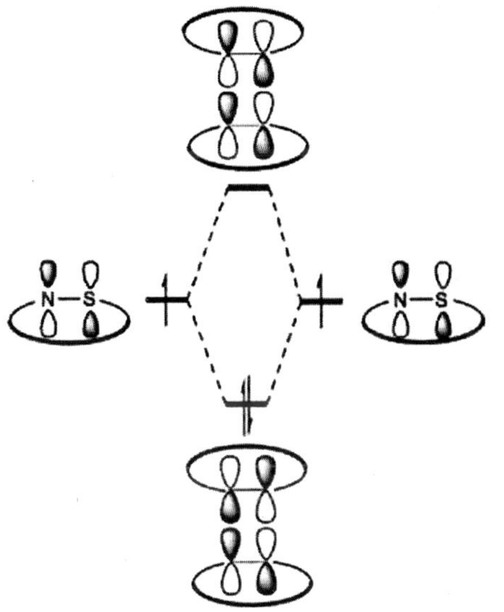

Figure 9.26 *Schematic interaction diagram for π dimer formation from two cyclic thiazyl radicals. The sulfur and nitrogen atoms are not indicated in the dimer orbitals because the intermolecular overlap can be between atoms of the same type or not.*

for thiazyl radical π dimerization is presented in Figure 9.26. The SOMO–SOMO interaction involves end-to-end interaction between the atomic p orbitals which constitute the radical SOMO. The weak, diffuse bonding nonetheless suggests that π dimers should be diamagnetic and, as such, relatively uninteresting from a magnetism perspective. In fact, this picture of radical association is far from complete. There are a few examples of thiazyl π dimer structures (e.g., a few dithiadiazolyls[115,116] and one thioaminyl radical[241]) in which the inter-radical distances are significantly longer (~3.25 Å). The magnetic susceptibility of these compounds exhibit notable paramagnetism at higher temperatures, which arises from thermal population of a triplet excited state which becomes accessible when the radical–radical interaction is sufficiently weak.[115,116]

Several π dimer structures based on 1,2,3,5-dithiadiazolyl polyradicals show paramagnetic behavior at high temperatures, though the origin of this effect is the thermally induced dissociation of the dimer "bonds". Figure 9.27a depicts the magnetic susceptibility for 1,3-phenylene-bridged diradical **149**, which adopts a π stacked, π dimer structure (Figure 9.7b).[126] This material is diamagnetic up to nearly 400 K, above which there is a significant increase in susceptibility. This behavior was ascribed to the extensive uncoupling of π dimers to generate radicals: The susceptibility near 500 K corresponds to approximately one spin per molecule. If the temperature is lowered prior to the onset of decomposition at 500 K the susceptibility increase is quasi-reversible – the temperature *decrease* profile does not follow the temperature *increase* profile, suggesting that not all spins re-align into π dimers (i.e., radical "defect" sites remain in the lattice upon re-cooling). Interestingly, this phenomenon occurs *only* for the dithiadiazolyl structures which adopt π stacks of π dimers[126–128]; radical dimer structures which are not π stacked (e.g., **148**,[123] see Figure 9.7a, and others[124]) only show irreversible susceptibility increases at the decomposition point of the material (Figure 9.27b for **148**). The dependence of the pseudo-reversible "spin breakout" on the presence of a π-stacked structure implies that cooperative (inter-dimer) effects between π dimers are a prerequisite for this process.

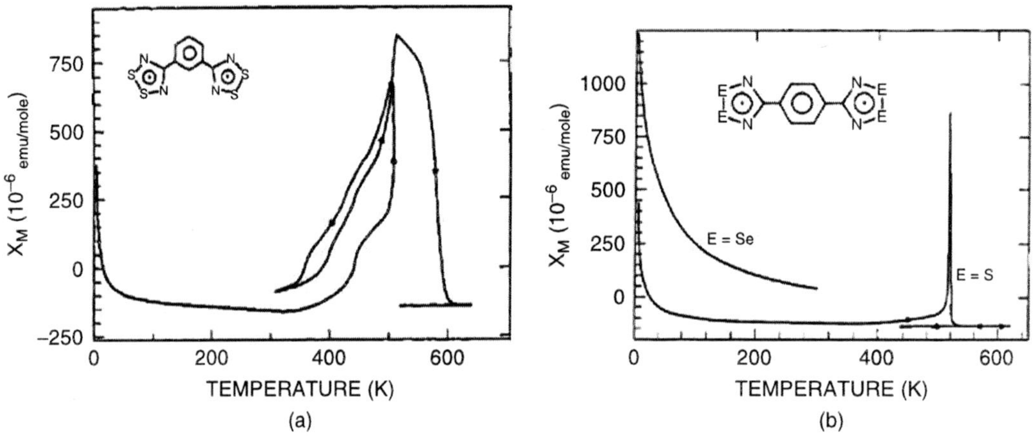

*Figure 9.27 Variable temperature magnetic susceptibility of (a) **149**[126] and (b) **148**.[123] (Reprinted with permission from [123]. Copyright 1991 American Chemical Society.)*

The reversible uncoupling of π dimers of 1,3,2-dithiazolyl radicals has been explored more extensively, facilitated in part by the much lower diamagnetic/paramagnetic transition regimes for these radicals compared to the 1,2,3,5-dithiadiazolyls (probably a reflection of the lower dimerization enthalpies for the former). The dithiazolyls typically have a relatively sharp transition between a low temperature diamagnetic state (based on a π-dimer π-stacked assembly) and a high temperature paramagnetic state (consisting of π stacks of unassociated radicals); in several instances both the diamagnetic and paramagnetic states have been crystallographically characterized. The most intriguing aspect of the monomer/dimer transition in 1,3,2-dithiazolyls is the fact that the transition temperature depends on whether the temperature is being *increased* or *decreased*. Thus, there are *two* transition temperatures (Table 9.2), and the regime in between the two is one in which the radical can be *either* dimeric or monomeric, depending on sample history – that is, these materials possess a form of *magnetic bistability*. This phenomenon was first observed by Oakley for radical **79**.[159] Shortly thereafter Awaga[242] – and later Rawson[101] – reported the first example (radical **75**) in which the bistable regime included ambient temperature, and several other derivatives exhibiting this unusual "spin transition" behavior have been discovered (Table 9.2).[160,162,243,244] Further investigations into **75** have demonstrated that the bistability can be induced by photolysis,[245] and that application of external pressure shifts both of the transitions to higher temperatures.[246]

As an illustrative example, Figure 9.28 shows the high and low temperature structures (both of which were solved *at the same temperature* (323 K), that is, within the bistable temperature window) and magnetic susceptibility for pyrazine-fused dithiazolyl radical **71**.[243] Upon raising the temperature from <250 K the material is diamagnetic until a sharp transition to a paramagnetic state at 343 K ($T_C\uparrow$). If the sample is subsequently cooled, the material retains its paramagnetism down to 297 K ($T_C\downarrow$). Thus, in the bistable window the radical can be either diamagnetic or paramagnetic depending on whether the sample was previously cooled below 297 K or heated above 347 K, respectively. The magnetic susceptibility in the high temperature regime is much lower than expected for a strict paramagnet, consistent with strong antiferromagnetic exchange between radicals in the π stacks. Analysis of the magnetism of another derivative (**75**) in the high temperature form reveals strong antiferromagnetic interactions within the π stacks, with $J = -224\,\mathrm{cm}^{-1}$ (in addition to significant *inter*stack magnetic interactions, $J' = -42\,\mathrm{cm}^{-1}$).[101,242] Other bistable dithiazolyls have qualitatively similar magnetic properties (that is, strongly antiferromagnetically coupled radicals)

Table 9.2 *π Dimer/monomer transition temperatures for 1,3,2-dithiazolyl radicals. Tc↓ and Tc↑ refer to the monomer/dimer transition temperatures whereupon temperature is decreasing or increasing, respectively, and ΔTc is the difference between the two transition temperatures*

Compound **(number)**	Tc↓	Tc↑	ΔTc	Reference
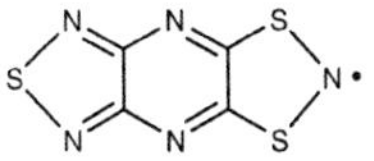 **79**	~125 K	~180 K	55 K	159
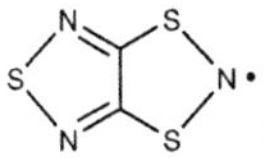 **75**	230 K	305 K	75 K	242
	234 K	317 K	83 K	101
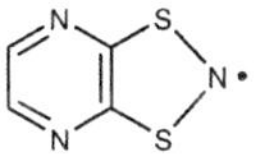 **71**	297 K	343 K	46 K	160, 243
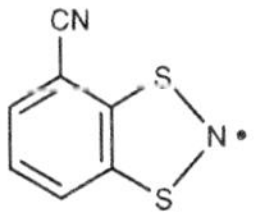 **72**	250 K	250 K	0 K	162, 244
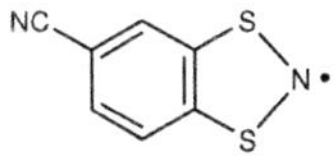 **73**	291 K	304 K	13 K	162

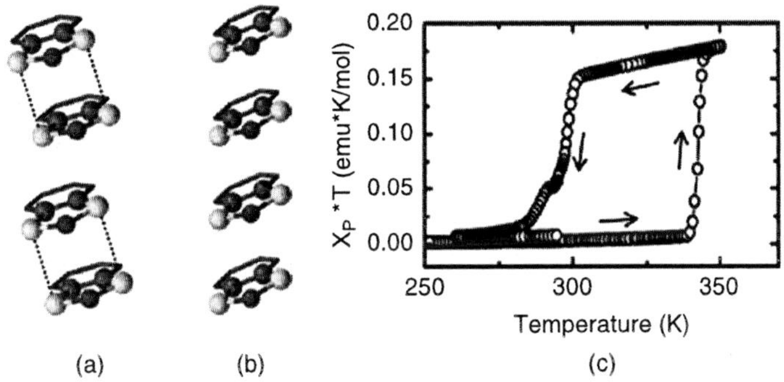

Figure 9.28 *(a) Low temperature X-ray structure of 1,3,2-dithiazolyl **71**. (b) High temperature X-ray structure. (c) Temperature dependence of χT for **71**.[243] (c Reprinted with permission from [243]. Copyright 2004 American Chemical Society.)*

in their high temperature form, although quantitative analyses of their temperature-dependent magnetic susceptibility were not reported.

Oakley has proposed a mechanism which accounts for the large bistability regime (large ΔTc) of some of these radicals (Figure 9.29).[243] The hysteretic nature of the phase changes require consideration of *interstack* interactions, both between sulfur and nitrogen atoms as well as sulfur and sulfur atoms. For radical **71** lateral (interstack) interactions must first be broken (Figure 9.29, step A) before the monomeric radicals can rotate (step B) to position themselves in optimal orientation to form the π dimer structure (step C). The reverse process (dimer disassociation) analogously requires a prior weakening of interactions between the π dimer stacks. Thus, while *intrastack* interactions are clearly a central feature of the hysteretic

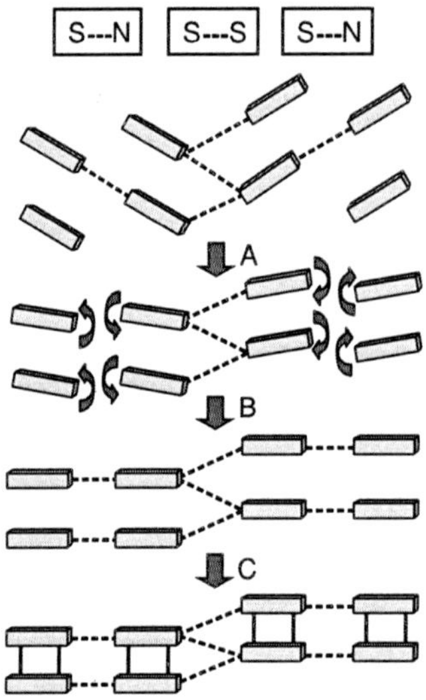

*Figure 9.29 Schematic illustration of the conversion of the high temperature (monomeric) to low temperature (π dimeric) forms of 1,3,2-dithiazolyl radical **71**.[243] (Reprinted with permission from [243]. Copyright 2004 American Chemical Society.) A full-colour version of this figure appears in the Colour Plate section of this book.*

monomer/dimer transitions, the subtleties of *interstack* interactions provide a means of rationalizing the large ΔT_C in these systems. This hypothesis is supported by the fact that extensive interstack contacts are present in the derivatives which have large ΔT_C (**71**, **75**, and **79**) whereas the cyanophenyl-fused dithiazolyls **72** and **73** – which do not have a preponderance of sulfur and nitrogen atoms on their molecular peripheries – have much smaller ΔT_C values (zero in once case).

Monomer–dimer transitions of a different nature have been reported by Awaga. The radical cation of the benzobis(1,3,2-dithiazolyl) diradical **80** forms one-dimensional coordination chains with InX_4^- anions (X = Br, Cl) (Figure 9.30a).[247] The radical cations are coordinated to *cis* coordination sites of the pseudo-octahedral-based indium-centered anions. The nitrogen atoms necessarily are in fairly close contact; at 270 K (X = Br) each of the N–N distances A and B (dashed lines in Figure 9.30a) are equivalent (3.168 Å). However, a phase transition at 250 K leads to subtle but important structural changes: the N–N distances "A" shorten to 2.899 Å while d(NN(B)) increases to 3.283 Å, and the coordinate bonds (In–N) change from all being the same (2.68 Å) in the high temperature regime to alternating in the low temperature structure, with d(In–N(A) = 2.455 Å and d(In–N(B)) = 2.887 Å. It appears that below the phase transition the electronic structure of the radical cation evolves to separates the spin and charge at opposite ends of the molecule, as represented below by $\mathbf{80^+}$. The structural features of one end (containing N(B)) resemble a dithiazolium cation, whereas the other end (containing N(A)) is best described a dithiazolyl radical which forms a dimer with the neighboring molecule bound to the same indium ion. Indeed, the local structure around the N(A)–N(A) interaction resembles that of the dimer of the bis(trifluoromethyl)dithiazolyl, (cf. Figure 9.13a). Modeling of the magnetic behavior above and below the phase transition suggests dramatically different magnetic interactions in the two forms (Figure 9.30b); in the low temperature regime the coupling is very strongly antiferromagnetic as expected for pseudo-π-dimers, while in the high temperature regime the coupling, though still substantial, is comparatively much weaker.

The majority of bis(1,2,3-dithiazole) and related radicals **97a–d** adopt *undimerized* π-stacked structures in the solid state. In these systems the stacks are usually "slipped" – that is, when viewed from above,

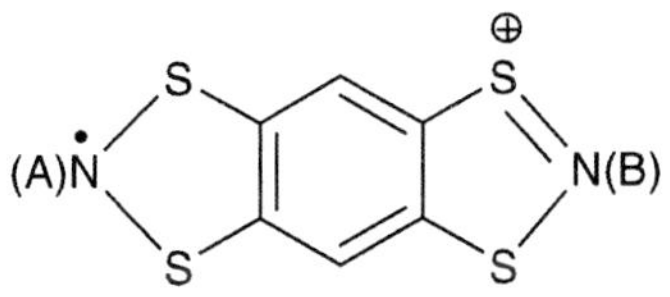

80⁺

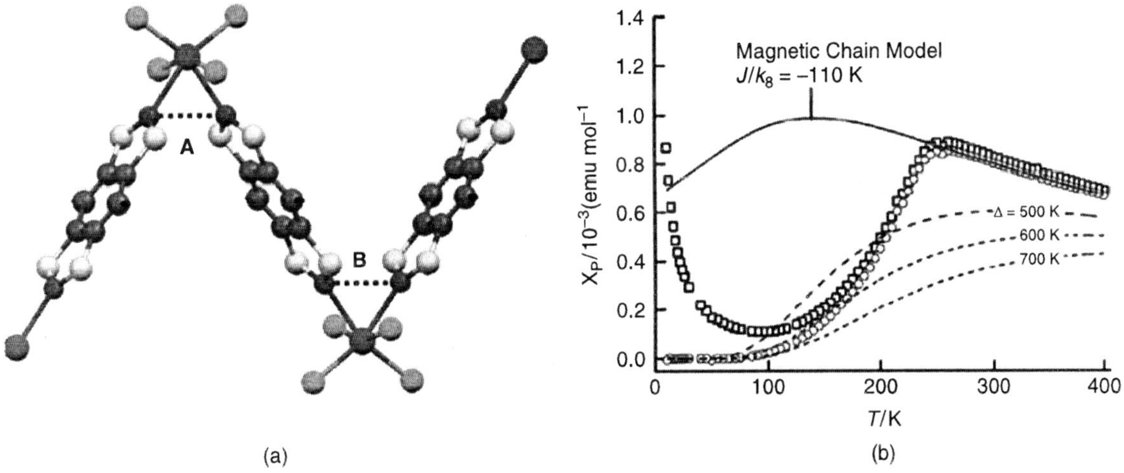

Figure 9.30 *(a) Structure and (b) temperature dependence of the magnetic susceptibility of $80^+.InBr_4^-$.*[247] *(Reproduced with permission from Wiley-VCH Verlag GmbH.)*

neighboring radicals within a stack are translated with respect to one another in the plane perpendicular to the stack (Figure 9.31b). Oakley has performed DFT calculations to predict the magnetic exchange interaction between π-stacked radical pairs for all four sulfur/selenium variations of the bis(dithiazole) radical as a function of the degree of slippage.[223] The energy surface plots generated (Figure 9.31a) provide qualitative, but effective predictive insight into the relationships between the structural nature of a particular radical π stack and the magnetic interactions (ferro- vs antiferromagnetic) observed within the stacks. This study is an extremely rare example of *predictive* magnetostructural correlations in molecular systems, in sharp contrast to the conventional "case by case" analysis approach; the latter approach is in part necessitated by the challenges in identifying meaningful structure/magnetism relationships in other radical classes.[248]

9.4.3.2 *Magnetically ordered thiazyl radicals*

The creation of magnetically ordered organic solids has long been one of the biggest challenges in the general field of molecule-based magnetism. Historically, nitroxide radicals have dominated these studies, and there are a few dozen examples of magnetically ordered materials based on these radicals.[234] The magnetic ordering temperatures (T_C) for these radicals are generally extremely low – the majority of nitroxide magnets have T_C's below 0.4 K and only a small number are above 0.5 K. Two important landmark examples are the *p*-nitrophenyl-substituted nitronyl nitroxide (T_C 0.60 K), the first purely organic magnet,[249]

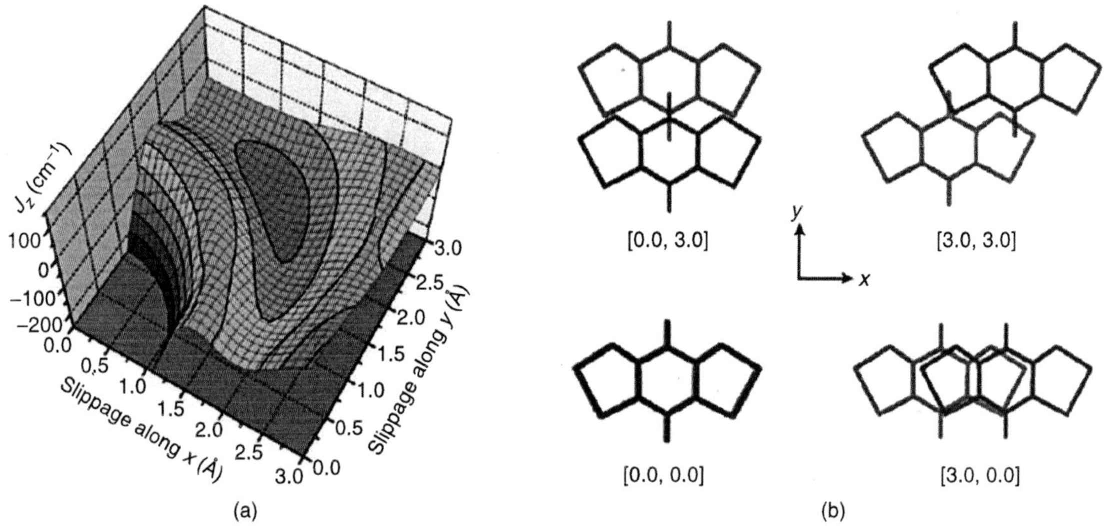

Figure 9.31 (a) Surface plot of calculated intermolecular magnetic exchange (J_π) between two π stacked bis(dithiazolyl) radicals **97a** *($R_1 = R_2 = H$) as a function of displacement of the two radicals along x and y vectors. (b) Definition of x and y displacement vectors.[223] (Reprinted with permission from [223]. Copyright 2009 American Chemical Society.) A full-colour version of this figure appears in the Colour Plate section of this book.*

and Rassat's adamantyl dinitroxide, which has the highest T_C among all nitroxides.[250] Among non-nitroxide radicals, the radical ion salt $C_{60}{}^{+\bullet}TDAE^-$ (TDAE = tetrakis(dimethylamino)ethylene) deserves mention as the material with the highest T_C (16 K) for a material based exclusively on first row elements.[251]

The magnetically ordered thiazyl-based radicals are listed in Table 9.3. A diverse range of molecular structure types are represented, and the magnetic ordering temperatures are generally *much* higher than is typically found for the nitroxide-based magnets. The first reported magnetically ordered thiazyl radical was Rawson's cyano-substituted perfluorphenyl 1,2,3,5-dithiadiazolyl **155**.[252] This discovery generated a great deal of interest, not only as the first example of a thiazyl radical to magnetically order, but also because the ordering temperature of 36 K was over an order of magnitude higher than the previous record for neutral radical magnets. The magnetically ordered state in this system is actually *antiferromagnetic*, but one in which the ordering is *canted* – a phenomenon which arises when the antiferromagnetic alignment of spins in two sublattices is not entirely antiparallel. As a result, the material does exhibit spontaneous magnetization below T_C but the magnetization is much smaller than would be expected for *ferromagnetic* ordering. The only other magnetically ordered dithiadiazolyl is the *p*-nitro-perfluorophenyl analogue **156** which orders ferromagnetically below 1.3 K[118]

There are a few 1,3,2-dithiazolyl radical based magnets. Benzo-1,3,2-dithiazolyl **69** was initially characterized as a (diamagnetic) π dimer at room temperature.[165] The discovery of magnetic bistability in other 1,3,2-dithiazolyls prompted a subsequent investigation of the high temperature magnetic behavior of **69**.[253] Indeed this radical undergoes a phase transition to a paramagnetic state at 346 K and reforms the diamagnetic state upon cooling to ~270 K. However, upon rapid cooling the paramagnetic state (of unknown solid state structure) remains stable all the way down to 11 K, whereupon antiferromagnetic ordering sets in.

Awaga has also investigated the magnetic properties of salts of the benzobis(1,3,2-dithiazole) radical cation, **80**$^+$. The $GaCl_4{}^-$[254,255] and $FeCl_4{}^-$[256] salts initially crystallize as solvates (acetonitrile and acetone,

Table 9.3 *Magnetically ordered thiazyl radicals*

Compound	Ordering type	Ordering temperature (K)	Reference
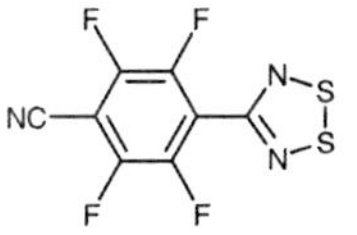 **155**	Canted AFM	36	252
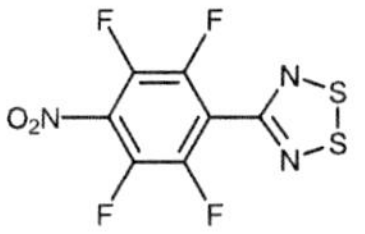 **156**	FM	1.3	118
69	AFM	11	253
80$^+$ $GaCl_4^-$	FM	6.7	254, 255
80$^+$ $FeCl_4^-$	FM	44	256
80$^+$ $GaBr_4^-$	FM	0.4	257
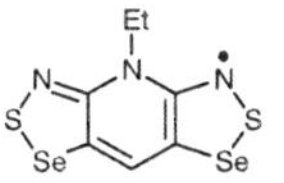 **157**	Canted AFM	18	189
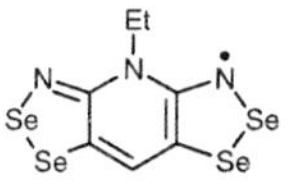 **158**	Canted AFM	27	189
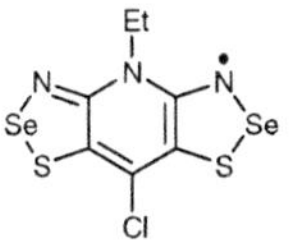 **159**	FM	12.8	221, 222, 224
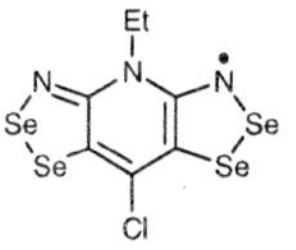 **160**	FM	17.0	222
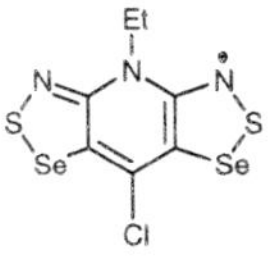 **161**	Canted AFM	14	222
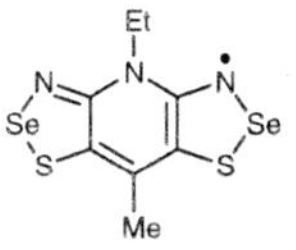 **162**	FM	13.6	224
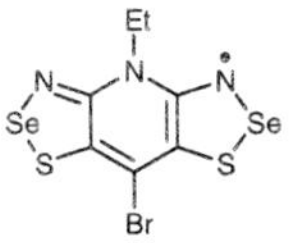 **163**	FM	14.1	224

respectively); both solvated structures contain cofacial π dimers of **80**$^+$ radical cation, and are thus diamagnetic; non-coordinated counterions and solvent molecules complete both lattices. However, the solvent molecules can be removed by applying a gentle vacuum, leading to materials which are paramagnetic and ferromagnetically order at the relatively high temperatures of 6.7 K ($GaCl_4$)[254] and 44 K ($FeCl_4$).[256] The structure of the desolvated $FeCl_4$ salt is not known but, based on the paramagnetism of the sample, must consist of monomeric **80**$^+$ molecules (the high spin iron(III) most likely makes an important contribution to the bulk magnetic properties as well). A structure of a desolvated $GaCl_4^-$ salt (prepared by crystallization from a solvent system not containing acetonitrile) indeed contains *monomeric* radicals and not π dimers.[255] This material has a T_C of 7.0 K and, as such, may well be the same structure as the one obtained by removal of acetonitrile from the solvated structure. The related $GaBr_4^-$ salt of **80**$^+$ (no solvent) also consists of monomeric radical cations; however, the solid state packing in this structure is different from that of the tetrachlorogallate, and the ferromagnetic ordering temperature is lower (0.4 K).[257]

Although the bis(1,2,3-dithiazole) radical family **97a–d** was initially designed and pursued for possible uses as *conducting* materials (Section 9.4.1), their magnetic properties have turned out to be among the more striking among all molecular radical systems. As can be seen from Table 9.3, several derivatives have magnetic ordering temperatures above 10 K. Three derivatives (**157**,[189] **158**,[189] **161**[222]) order antiferromagnetically but, like 1,2,3,5-dithiadiazolyl **155**, are spin-canted and, as such, are described as weak ferromagnets. Genuine ferromagnetic ordering has been established for four of these radicals (**159**,[221,222,224] **160**,[222] **162**,[224] **163**[224]); all four have ordering temperatures which rank among the very highest for molecular systems. Interestingly all of the magnetically ordered derivatives contain selenium (the corollary of which is that the all-sulfur radicals **97a** do not magnetically order). Heavy atom incorporation introduces significant spin orbit coupling effects, which are believed to contribute to both the spin canting[189] in **157**,

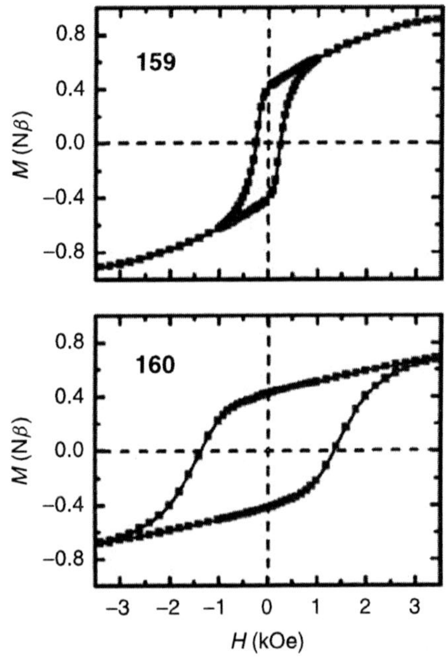

Figure 9.32 Magnetization vs magnetic field hysteresis loops for ***159*** *and* ***160***.[222] *(Reprinted with permission from [222]. Copyright 2008 American Chemical Society.)*

158 and **161** as well as the extremely high coercivities observed for **159, 162, 163**, and particularly the all-selenium derivative **160** (Figure 9.32). The coercivities of organic magnets constructed exclusively from lighter elements are normally at least two orders of magnitude smaller than in these systems.

9.5 Conclusions

The field of thiazyl radical chemistry has evolved significantly since its inception in the 1970s. Early stages necessarily focused on synthetic and electronic structure aspects; the latter aspect is clearly a fertile area owing to a number of phenomena, such as redox activity and the weak, reversible, intermolecular associations of many of these radicals. In more recent times the materials thrust has taken center stage as the fundamental chemistry has matured. Several different kinds of thiazyl radicals are now in play as molecular conductors, magnets, or switches, and their properties differ sharply from some of the other "more organic" radicals. Device applications of these systems have not yet been reported, for example, as components of field effect transistors, although Awaga has recently demonstrated the use of the 1,2,3,5-dithiadiazolyl diradical **39** as a photoconductor in a thin-film-based device.[258] In any event it is clear (to this author, at least) that thiazyl radical chemistry has flourished and will continue to do so.

References

1. M. B. Smith and J. March, *March's Advanced Organic Chemistry*, John Wiley & Sons, Inc., New York, 2001.
2. (a) W. Kutzelnigg, *Angew. Chem. Int. Ed.*, **23**, 272–295 (1984). (b) N. C. Norman, *Polyhedron*, **12**, 2431–2446 (1993).
3. P. P. Power, *Chem. Rev.*, **103**, 789–809 (2003).
4. W. C. Danen and D. D. Newkirk, *J. Am. Chem. Soc.*, **98**, 516–520 (1976).
5. R. T. Oakley, *Progr. Inorg. Chem.*, **36**, 299–391 (1988).
6. T. Chivers, *Chem. Rev.*, **85**, 341–365 (1985).
7. H. G. Heal, *The Inorganic Heterocyclic Chemistry of Sulfur, Nitrogen, and Phosphorus*, Academic Press, London, 1980.
8. R. Gleiter, *Angew. Chem. Int. Ed.*, **20**, 444–452 (1981).
9. T. Chivers, *A Guide to Chalcogen-Nitrogen Chemistry*, World Scientific, Singapore, 2005.
10. K. F. Preston and L. H. Sutcliffe, *Magn. Reson. Chem.*, **28**, 189–204 (1990).
11. R. T. Boere and T. L. Roemmele, *Coord. Chem. Rev.*, **210**, 369–445 (2000).
12. K. E. Preuss, *Dalton Trans.*, 2357–2369 (2007).
13. (a) P. Kaszynski, *J. Phys. Chem. A*, **105**, 7615–7625 (2001). (b) P. Kaszynski, *J. Phys. Chem A*, **105**, 7626–7633 (2001).
14. R. T. Boere, H. M. Tuononen, T. Chivers and T. L. Roemmele, *J. Organometal. Chem.*, **692**, 2683–2696 (2007).
15. H. P. Fritz, R. Bruchhaus, R. Mews and H. U. Hofs, *Z. Anorg. Allg. Chem.*, **525**, 214–220 (1985).
16. (a) L. R. Peebles and P. Marshall, *Chem. Phys. Lett.*, **366**, 520–524 (2002). (b) P. A. Denis, *J. Phys. Chem. A*, **108**, 11092–11100 (2004).
17. (a) R. D. Brown, P. S. Elmes and D. McNaughton, *J. Mol. Spec.*, **140**, 390–400 (1990). (b) P. Hassanzadeh and L. Andrews, *J. Am. Chem. Soc.*, **114**, 83–91 (1992).
18. (a) Y. Yamaguchi, Y. M. Xie, R. S. Grev and H. F. Schaefer, *J. Chem. Phys.*, **92**, 3683–3687 (1990). (b) C. L. Collins, Y. Yamaguchi and H. F. Schaefer, *J. Chem. Phys.*, **98**, 4777–4782 (1993). (c) M. Iraqi, N. Goldberg and H. Schwarz, *Chem. Ber.*, **127**, 1171–1173 (1994). (d) A. T. Wong and G. B. Bacskay, *Chem. Phys. Lett.*, **217**, 17–23 (1994). (e) M. T. Nguyen, R. Flammang, N. Goldberg and H. Schwarz, *Chem. Phys. Lett.*, **236**, 201–205 (1995).
19. T. W. Hayton, P. Legzdins and W. B. Sharp, *Chem. Rev.*, **102**, 935–991 (2002).

20. (a) H. W. Roesky and K. K. Pandey, *Adv. Inorg. Chem.*, **26**, 337–356 (1983). (b) K. K. Pandey, *Progr. Inorg. Chem*, **40**, 445 (1992).
21. W. M. Lau, N. P. C. Westwood and M. H. Palmer, *J. Chem. Soc. Chem. Commun.*, 752–753 (1985).
22. W. M. Lau, N. P. C. Westwood and M. H. Palmer, *J. Am. Chem. Soc.*, **108**, 3229–3237 (1986).
23. R. T. Boere, T. Chivers, T. L. Roemmele and H. M. Tuononen, *Inorg. Chem.*, **48**, 7294–7306 (2009).
24. (a) J. Bojes and T. Chivers, *Inorg. Chem.*, **17**, 318–321 (1978). (b) T. Chivers, W. G. Laidlaw, R. T. Oakley and M. Trsic, *J. Am. Chem. Soc.*, **102**, 5773–5781 (1980). (c) N. Burford, T. Chivers, A. W. Cordes, *et al.*, *Inorg. Chem.*, **20**, 4430–4432 (1981). (d) T. Chivers, R. T. Oakley, A. W. Cordes and W. T. Pennington, *J. Chem. Soc. Chem. Commun.*, 1214–1215 (1981). (e) T. Chivers, A. W. Cordes, R. T. Oakley and W. T. Pennington, *Inorg. Chem.*, **22**, 2429–2435 (1983). (f) C. G. Marcellus, R. T. Oakley, A. W. Cordes and W. T. Pennington, *Can. J. Chem.*, **62**, 1822–1827 (1984).
25. T. Chivers and M. N. S. Rao, *Can. J. Chem.*, **61**, 1957–1962 (1983).
26. R. J. Gillespie, P. R. Ireland and J. E. Vekris, *Can. J. Chem.*, **53**, 3147–3152 (1975).
27. R. J. Gillespie, J. P. Kent and J. F. Sawyer, *Inorg. Chem.*, **20**, 3784–3799 (1981).
28. S. A. Fairhurst, K. F. Preston and L. H. Sutcliffe, *Can. J. Chem.*, **62**, 1124–1126 (1984).
29. A. J. Banister, H. G. Clarke, I. Rayment and H. M. M. Shearer, *Inorg. Nucl. Chem. Lett.*, **10**, 647–654 (1974).
30. H. W. Roesky and A. Hamza, *Angew. Chem. Int. Ed.*, **15**, 226–226 (1976).
31. R. Gleiter, R. Bartetzko and P. Hofmann, *Z. Naturforsch. B*, **35**, 1166–1170 (1980).
32. (a) B. Krebs, G. Henkel, S. Pohl and H. W. Roesky, *Chem. Ber.*, **113**, 226–232 (1980). (b) U. Thewalt and M. Burger, *Z. Naturforsch. B*, **36**, 293–296 (1981). (c) H. W. Roesky, M. Witt, J. Schimkowiak, *et al.*, *Angew. Chem. Int. Ed.*, **21**, 538–539 (1982). (d) R. W. H. Small, A. J. Banister and Z. V. Hauptman, *J. Chem. Soc. Dalton Trans.*, 1377–1381 (1984). (e) B. Ayres, A. J. Banister, P. D. Coates, *et al.*, *J. Chem. Soc. Dalton Trans.*, 3097–3103 (1992).
33. S. A. Fairhurst, K. M. Johnson, L. H. Sutcliffe, *et al.*, *J. Chem. Soc. Dalton Trans.*, 1465–1472 (1986).
34. (a) E. Awere, J. Passmore, K. F. Preston and L. H. Sutcliffe, *Can. J. Chem.*, **66**, 1776–1780 (1988). (b) K. M. Johnson, K. F. Preston and L. H. Sutcliffe, *Magn. Reson. Chem.*, **26**, 1015–1019 (1988). (c) K. F. Preston, J. P. Charland and L. H. Sutcliffe, *Can. J. Chem.*, **66**, 1299–1303 (1988).
35. B. M. Gimarc and D. S. Warren, *Inorg. Chem.*, **30**, 3276–3280 (1991).
36. (a) G. Wolmershauser, C. R. Brulet and G. B. Street, *Inorg. Chem.*, **17**, 3586–3589 (1978). (b) R. J. Gillespie, J. P. Kent and J. F. Sawyer, *Inorg. Chem.*, **20**, 4053–4060 (1981). (c) E. G. Awere, J. Passmore, P. S. White and T. Klapotke, *J. Chem. Soc. Chem. Commun.*, 1415–1417 (1989). (d) A. Haas, J. Kasprowski, K. Angermund, *et al.*, *Chem. Ber.*, **124**, 1895–1906 (1991). (e) E. G. Awere, W. V. F. Brooks, J. Passmore, *et al.*, *J. Chem. Soc. Dalton Trans.*, 2439–2449 (1993). (f) A. Maaninen, J. Konu, R. S. Laitinen, *et al.*, *Inorg. Chem.*, **40**, 3539–3543 (2001).
37. M. M. Labes, P. Love and L. F. Nichols, *Chem. Rev.*, **79**, 1–15 (1979).
38. A. J. Banister and I. B. Gorrell, *Adv. Mater.*, **10**, 1415–1429 (1998).
39. R. C. Mawhinney and J. D. Goddard, *Inorg. Chem.*, **42**, 6323–6337 (2003).
40. P. Love, H. I. Kao, G. H. Myer and M. M. Labes, *J. Chem. Soc. Chem. Commun.*, 301–302 (1978).
41. M. W. R. Witt, W. I. Bailey and R. J. Lagow, *J. Am. Chem. Soc.*, **105**, 1668–1669 (1983).
42. (a) J. Passmore and M. N. S. Rao, *J. Chem. Soc. Chem. Commun.*, 1268–1269 (1980). (b) F. A. Kennett, G. K. Maclean, J. Passmore and M. N. S. Rao, *J. Chem. Soc. Dalton Trans.*, 851–857 (1982). (c) A. J. Banister, Z. V. Hauptman, J. Passmore, *et al.*, *J. Chem. Soc. Dalton Trans.*, 2371–2379 (1986).
43. (a) A. J. Banister, Z. V. Hauptman and A. G. Kendrick, *J. Chem. Soc. Chem. Commun.*, 1016–1018 (1983). (b) H. P. Fritz and R. Bruchhaus, *Z. Naturforsch. B*, **38**, 1375–1382 (1983). (c) A. J. Banister, Z. V. Hauptman, A. G. Kendrick and R. W. H. Small, *J. Chem. Soc. Dalton Trans.*, 915–924 (1987).
44. J. Stejny, J. Dlugosz and A. Keller, *J. Mater. Sci.*, **14**, 1291–1300 (1979).
45. M. J. Cohen, A. F. Garito, A. J. Heeger, *et al.*, *J. Am. Chem. Soc.*, **98**, 3844–3848 (1976).
46. G. Heger, S. Klein, L. Pintschovius and H. Kahlert, *J. Sol. St. Chem.*, **23**, 341–347 (1978).
47. Y. Miura and M. Kinoshita, *J. Org. Chem.*, **49**, 2724–2728 (1984).
48. Y. Miura, A. Yamamoto, Y. Katsura and M. Kinoshita, *J. Chem. Soc. Chem. Commun.*, 37–37 (1980).

49. (a) Y. Miura and A. Tanaka, *J. Chem. Soc. Chem. Commun.*, 441–442 (1990). (b) Y. Miura, Y. Kitagishi and S. Ueno, *Bull. Chem. Soc. Jpn.*, **67**, 3282–3288 (1994).
50. (a) Y. Miura, E. Yamano, A. Tanaka and Y. Ogo, *Chem. Lett.*, 1831–1834 (1992). (b) Y. Miura, E. Yamano, A. Tanaka and J. Yamauchi, *J. Org. Chem.*, **59**, 3294–3300 (1994). (c) Y. Miura, H. Oka, E. Yamano, *et al.*, *Bul. Chem. Soc. Jpn.*, **68**, 1187–1192 (1995). (d) Y. Miura and E. Yamano, *J. Org. Chem.*, **60**, 1070–1073 (1995).
51. Y. Miura, A. Yamamoto, Y. Katsura and M. Kinoshita, *J. Org. Chem.*, **45**, 3875–3880 (1980).
52. Y. Miura, A. Yamamoto and M. Kinoshita, *Bull. Chem. Soc. Jpn.*, **54**, 3215–3216 (1981).
53. (a) Y. Miura, H. Asada and M. Kinoshita, *Chem. Lett.*, 1085–1088 (1978). (b) Y. Miura, H. Asada, M. Kinoshita and K. Ohta, *J. Phys. Chem.*, **87**, 3450–3455 (1983). (c) Y. Miura, M. Isogai and M. Kinoshita, *Bull. Chem. Soc. Jpn.*, **58**, 751–752 (1985).
54. Y. Miura, A. Yamamoto, Y. Katsura, *et al.*, *J. Org. Chem.*, **47**, 2618–2622 (1982).
55. (a) Y. Miura, Y. Katsura and M. Kinoshita, *Chem. Lett.*, 409–412 (1977). (b) Y. Miura and M. Kinoshita, *Bull. Chem. Soc. Jpn.*, **50**, 1142–1146 (1977). (b) Y. Miura, Y. Katsura and M. Kinoshita, *Bull. Chem. Soc. Jpn.*, **52**, 1121–1125 (1979). (c) Y. Miura, H. Asada and M. Kinoshita, *Bull. Chem. Soc. Jpn.*, **53**, 720–725 (1980).
56. (a) Y. Miura, M. Momoki, T. Fuchikami, *et al.*, *J. Org. Chem.*, **61**, 4300–4308 (1996). (b) Y. Teki, K. Itoh, A. Okada, *et al.*, *Chem. Phys. Lett.*, **270**, 573–579 (1997). (c) Y. Miura, S. Kurokawa, M. Nakatsuji, *et al.*, *J. Org. Chem.*, **63**, 8295–8303 (1998). (d) Y. Miura, M. Momoki, M. Nakatsuji and Y. Teki, *J. Org. Chem.*, **63**, 1555–1565 (1998). (e) Y. Miura, S. Nakamura and Y. Teki, *J. Org. Chem.*, **68**, 8244–8247 (2003).
57. Y. Miura, Y. Oyama and Y. Teki, *J. Org. Chem.*, **68**, 1225–1234 (2003).
58. Y. Miura, A. Yamamoto and M. Kinoshita, *Bull. Chem. Soc. Jpn.*, **56**, 1476–1481 (1983).
59. Y. Miura, T. Kunishi and M. Kinoshita, *Bull. Chem. Soc. Jpn.*, **58**, 1696–1698 (1985).
60. (a) Y. Miura, A. Yamamoto and M. Kinoshita, *Electrochim. Acta*, **29**, 1731–1732 (1984). (b) Y. Miura and A. Tanaka, *Electrochim. Acta*, **37**, 2095–2098 (1992).
61. (a) Y. Miura, Y. Shibata and M. Kinoshita, *Bull. Chem. Soc. Jpn.*, **59**, 3291–3292 (1986). (b) Y. Miura, Y. Shibata and M. Kinoshita, *J. Org. Chem.*, **51**, 1239–1243 (1986).
62. Y. Miura, *J. Org. Chem.*, **53**, 2850–2852 (1988).
63. (a) Y. Miura, Y. Nakamura and M. Kinoshita, *Bull. Chem. Soc. Jpn.*, **51**, 947–948 (1978). (b) Y. Miura and Y. Nakamura, *Bull. Chem. Soc. Jpn.*, **63**, 1154–1159 (1990).
64. (a) Y. Miura, Y. Nakamura and M. Kinoshita, *Bull. Chem. Soc. Jpn.*, **54**, 3217–3218 (1981). (b) Y. Miura and T. Ohnishi, *J. Org. Chem.*, **53**, 3012–3016 (1988).
65. Y. Miura and O. Tsumori, *Bull. Chem. Soc. Jpn.*, **60**, 4154–4156 (1987).
66. V. Benin and P. Kaszynski, *J. Org. Chem.*, **65**, 8086–8088 (2000).
67. M. Risto, R. W. Reed, C. M. Robertson, *et al.*, *Chem. Commun.*, 3278–3280 (2008).
68. Y. Miura, N. Makita and M. Kinoshita, *Bull. Chem. Soc. Jpn.*, **50**, 482–486 (1977).
69. Y. Miura and M. Kinoshita, *Bull. Chem. Soc. Jpn.*, **53**, 2395–2396 (1980).
70. Y. Miura, T. Kunishi and M. Kinoshita, *J. Org. Chem.*, **50**, 5862–5865 (1985).
71. K. Schlosser and S. Steenken, *J. Am. Chem. Soc.*, **105**, 1504–1506 (1983).
72. J. S. Thrasher and J. B. Nielsen, *J. Am. Chem. Soc.*, **108**, 1108–1109 (1986).
73. (a) M. R. Brinkman and C. W. Allen, *J. Am. Chem. Soc.*, **94**, 1550–1553 (1972). (b) A. M. Griffin and G. M. Sheldrick, *Acta Cryst. B*, **B 31**, 895–896 (1975).
74. (a) M. R. Brinkman and L. H. Sutcliffe, *J. Magn. Reson.*, **28**, 263–270 (1977). (b) S. A. Fairhurst, W. R. McIlwaine and L. H. Sutcliffe, *J. Magn. Reson.*, **35**, 121–132 (1979).
75. S. Rolfe, D. Griller, K. U. Ingold and L. H. Sutcliffe, *J. Org. Chem.*, **44**, 3515–3519 (1979).
76. S. A. Fairhurst and L. H. Sutcliffe, *J. Chem. Soc. Faraday Trans. I*, **76**, 1490–1509 (1980).
77. S. A. Fairhurst, N. J. Stewart and L. H. Sutcliffe, *Magn. Reson. Chem.*, **25**, 60–64 (1987).
78. (a) Y. Miura, T. Kunishi and M. Kinoshita, *Chem. Lett.*, 885–888 (1983). (b) Y. Miura, T. Kunishi, M. Isogai and M. Kinoshita, *J. Org. Chem.*, **50**, 1627–1632 (1985).
79. Y. Miura and A. Tanaka, *J. Org. Chem.*, **56**, 3950–3954 (1991).
80. J. M. Rawson, A. J. Banister and I. Lavender, in *Advances in Heterocyclic Chemistry*, *62*, 137–247 (1995)

81. G. G. Alange, A. J. Banister, B. Bell and P. W. Millen, *Inorg. Nucl. Chem. Lett.*, **13**, 143–144 (1977).
82. H. W. Roesky and T. Muller, *Chem. Ber.*, **111**, 2960–2964 (1978).
83. G. G. Alange, A. J. Banister, B. Bell and P. W. Millen, *J. Chem. Soc. Perkin Trans. 1*, 1192–1194 (1979).
84. J. N. Bridson, S. B. Copp, M. J. Schriver, *et al.*, *Can. J. Chem.*, **72**, 1143–1153 (1994).
85. (a) M. Amin and C. W. Rees, *J. Chem. Soc. Chem. Commun.*, 1137–1138 (1989). (b) M. Amin and C. W. Rees, *J. Chem. Soc. Perkin Trans. 1*, 2495–2501 (1989).
86. P. D. Belluz, A. W. Cordes, E. M. Kristof, *et al.*, *J. Am. Chem. Soc.*, **111**, 9276–9278 (1989).
87. C. M. Aherne, A. J. Banister, I. B. Gorrell, *et al.*, *J. Chem. Soc. Dalton Trans.*, 967–972 (1993).
88. H. U. Hofs, J. W. Bats, R. Gleiter, *et al.*, *Chem. Ber.*, **118**, 3781–3804 (1985).
89. L. N. Markovski, O. M. Polumbrik, V. S. Talanov and Y. G. Shermolovich, *Tetrahedron Lett.*, **23**, 761–762 (1982).
90. A. J. Banister, N. R. M. Smith and R. G. Hey, *J. Chem. Soc. Perkin Trans.1*, 1181–1186 (1983).
91. R. T. Boere, R. T. Oakley, R. W. Reed and N. P. C. Westwood, *J. Am. Chem. Soc.*, **111**, 1180–1185 (1989).
92. A. W. Cordes, R. C. Haddon, R. G. Hicks, *et al.*, *Inorg. Chem.*, **31**, 1802–1808 (1992).
93. A. W. Cordes, J. D. Goddard, R. T. Oakley and N. P. C. Westwood, *J. Am. Chem. Soc.*, **111**, 6147–6154 (1989).
94. R. T. Boere, K. H. Moock and M. Parvez, *Z. Anorg. Allg. Chem.*, **620**, 1589–1598 (1994).
95. C. M. Aherne, A. J. Banister, T. G. Hibbert, *et al.*, *Polyhedron*, **16**, 4239–4245 (1997).
96. R. T. Boere and K. H. Moock, *J. Am. Chem. Soc.*, **117**, 4755–4760 (1995).
97. J. Campbell, D. Klapstein, P. F. Bernath, *et al.*, *Inorg. Chem.*, **35**, 4264–+ (1996).
98. A. Vegas, A. Perezsalazar, A. J. Banister and R. G. Hey, *J. Chem. Soc. Dalton Trans.*, 1812–1815 (1980).
99. L. Beer, A. W. Cordes, D. J. T. Myles, *et al.*, *CrystEngComm*, **2**, 109–114 (2000).
100. (a) A. J. Banister, A. S. Batsanov, O. G. Dawe, *et al.*, *Phosph. Sulf. Sil. Relat. El.*, **125**, 553–556 (1997). (b) A. M. T. Bell, J. N. B. Smith, J. P. Attfield, *et al.*, *New J. Chem.*, **23**, 565–567 (1999). (c) C. S. Clarke, D. A. Haynes, J. M. Rawson and A. D. Bond, *Chem. Commun.*, 2774–2775 (2003).
101. G. D. McManus, J. M. Rawson, N. Feeder, *et al.*, *J. Mater. Chem.*, **11**, 1992–2003 (2001).
102. A. W. Cordes, C. M. Chamchoumis, R. G. Hicks, *et al.*, *Can. J. Chem.*, **70**, 919–925 (1992).
103. (a) N. G. R. Hearns, K. E. Preuss, J. F. Richardson and S. Bin-Salamon, *J. Am. Chem. Soc.*, **126**, 9942–9943 (2004). (b) M. Jennings, K. E. Preuss and J. Wu, *Chem. Commun.*, 341–343 (2006).
104. A. W. Cordes, C. D. Bryan, W. M. Davis, *et al.*, *J. Am. Chem. Soc.*, **115**, 7232–7239 (1993).
105. C. D. Bryan, A. W. Cordes, R. C. Haddon, *et al.*, *J. Am. Chem. Soc.*, **116**, 1205–1210 (1994).
106. A. D. Bond, D. A. Haynes, C. M. Pask and J. M. Rawson, *J. Chem. Soc. Dalton Trans.*, 2522–2531 (2002).
107. C. Knapp, E. Lork, K. Gupta and R. Mews, *Z. Anorg. Allg. Chem.*, **631**, 1640–1644 (2005).
108. G. Antorrena, S. Brownridge, T. S. Cameron, *et al.*, *Can. J. Chem.*, **80**, 1568–1583 (2002).
109. A. J. Banister, M. I. Hansford, Z. V. Hauptman, *et al.*, *J. Chem. Soc. Dalton Trans.*, 1705–1713 (1989).
110. C. S. Clarke, S. I. Pascu and J. M. Rawson, *CrystEngComm*, **6**, 79–82 (2004).
111. A. J. Banister, A. S. Batsanov, O. G. Dawe, *et al.*, *J. Chem. Soc. Dalton Trans.*, 2539–2541 (1997).
112. A. Alberola, C. S. Clarke, D. A. Haynes, *et al.*, *Chem. Commun.*, 4726–4728 (2005).
113. N. Bricklebank, S. Hargreaves and S. E. Spey, *Polyhedron*, **19**, 1163–1166 (2000).
114. R. T. Boere, L. Y. Goh, C. Y. Ang, *et al.*, *J. Organometal. Chem.*, **692**, 2697–2704 (2007).
115. A. Decken, T. S. Cameron, J. Passmore, *et al.*, *Inorg. Chem.*, **46**, 7436–7457 (2007).
116. K. V. Shuvaev, A. Decken, F. Grein, *et al.*, *Dalton Trans.*, 4029–4037 (2008).
117. (a) A. J. Banister, N. Bricklebank, W. Clegg, *et al.*, *J. Chem. Soc. Chem. Commun.*, 679–680 (1995). (b) A. Alberola, R. J. Less, F. Palacio, *et al.*, *Molecules*, **9**, 771–781 (2004).
118. A. Alberola, R. J. Less, C. M. Pask, *et al.*, *Angew. Chem. Int. Ed.*, **42**, 4782–4785 (2003).
119. T. M. Barclay, A. W. Cordes, N. A. George, *et al.*, *Chem. Commun.*, 2269–2270 (1999).
120. W. V. F. Brooks, N. Burford, J. Passmore, *et al.*, *J. Chem. Soc. Chem. Commun.*, 69–71 (1987).
121. C. D. Bryan, A. W. Cordes, R. C. Haddon, *et al.*, *J. Chem. Soc. Chem. Commun.*, 1447–1448 (1994).
122. C. D. Bryan, A. W. Cordes, J. D. Goddard, *et al.*, *J. Am. Chem. Soc.*, **118**, 330–338 (1996).
123. A. W. Cordes, R. C. Haddon, R. T. Oakley, *et al.*, *J. Am. Chem. Soc.*, **113**, 582–588 (1991).

124. A. W. Cordes, R. C. Haddon, R. G. Hicks, *et al.*, *Inorg. Chem.*, **32**, 1554–1558 (1993).
125. A. W. Cordes, R. C. Haddon, C. D. MacKinnon, *et al.*, *Inorg. Chem.*, **35**, 7626–7632 (1996).
126. M. P. Andrews, A. W. Cordes, D. C. Douglass, *et al.*, *J. Am. Chem. Soc.*, **113**, 3559–3568 (1991).
127. A. W. Cordes, R. C. Haddon, R. G. Hicks, *et al.*, *Chem. Mater.*, **5**, 820–825 (1993).
128. A. W. Cordes, R. C. Haddon, R. G. Hicks, *et al.*, *J. Am. Chem. Soc.*, **114**, 5000–5004 (1992).
129. R. A. Beekman, R. T. Boere, K. H. Moock and M. Parvez, *Can. J. Chem.*, **76**, 85–93 (1998).
130. N. Adamson, A. J. Banister, I. B. Gorrell, *et al.*, *J. Chem. Soc. Chem. Commun.*, 919–921 (1993).
131. C. D. Bryan, A. W. Cordes, R. C. Haddon, *et al.*, *Chem. Mater.*, **6**, 508–515 (1994).
132. A. J. Banister, J. A. K. Howard and S. E. Lawrence, *Phosph. Sulf. Sil. Relat. El.*, **93**, 437–438 (1994).
133. A. J. Banister, M. I. Hansford, Z. V. Hauptman, *et al.*, *J. Chem. Soc. Dalton Trans.*, 2793–2802 (1990).
134. A. W. Cordes, S. H. Glarum, R. C. Haddon, *et al.*, *J. Chem. Soc. Chem. Commun.*, 1265–1266 (1992).
135. W. M. Davis, R. G. Hicks, R. T. Oakley, *et al.*, *Can. J. Chem.*, **71**, 180–185 (1993).
136. J. F. Britten, O. P. Clements, A. W. Cordes, *et al.*, *Inorg. Chem.*, **40**, 6820–6824 (2001).
137. C. D. Bryan, A. W. Cordes, R. T. Oakley and R. Spence, *Acta Cryst. C*, **C51**, 2402–2404 (1995).
138. A. W. Cordes, R. C. Haddon, R. G. Hicks, *et al.*, *J. Am. Chem. Soc.*, **114**, 1729–1732 (1992).
139. N. Feeder, R. J. Less, J. M. Rawson, *et al.*, *Chem. Commun.*, 2449–2450 (2000).
140. A. J. Banister, R. G. Hey, G. K. Maclean and J. Passmore, *Inorg. Chem.*, **21**, 1679–1680 (1982).
141. G. K. Maclean, J. Passmore, M. J. Schriver, *et al.*, *J. Chem. Soc. Chem. Commun.*, 807–808 (1983).
142. G. K. Maclean, J. Passmore, M. N. S. Rao, *et al.*, *J. Chem. Soc. Dalton Trans.*, 1405–1416 (1985).
143. J. Passmore, X. P. Sun and S. Parsons, *Can. J. Chem.*, **70**, 2972–2979 (1992).
144. J. Passmore and X. P. Sun, *Inorg. Chem.*, **35**, 1313–1320 (1996).
145. N. Burford, J. Passmore and M. J. Schriver, *J. Chem. Soc. Chem. Commun.*, 140–142 (1986).
146. N. Burford, J. Passmore, M. Schriver, *et al.*, *J. Fluor. Chem.*, **35**, 108–108 (1987).
147. C. Aherne, A. J. Banister, A. W. Luke, *et al.*, *J. Chem. Soc. Dalton Trans.*, 1277–1282 (1992).
148. A. J. Banister, J. M. Rawson, W. Clegg and S. L. Birkby, *J. Chem. Soc. Dalton Trans.*, 1099–1104 (1991).
149. (a) A. J. Banister, I. Lavender, J. M. Rawson and R. J. Whitehead, *J. Chem. Soc. Dalton Trans.*, 1449–1450 (1992). (b) A. J. Banister, I. Lavender, J. M. Rawson, *et al.*, *J. Chem. Soc. Dalton Trans.*, 1421–1429 (1993).
150. (a) S. Parsons, J. Passmore, M. J. Schriver and P. S. White, *J. Chem. Soc. Chem. Commun.*, 369–371 (1991). (b) S. Parsons, J. Passmore and P. S. White, *J. Chem. Soc. Dalton Trans.*, 1499–1507 (1993).
151. T. S. Cameron, M. T. Lemaire, J. Passmore, *et al.*, *Inorg. Chem.*, **44**, 2576–2578 (2005).
152. J. M. Rawson and G. D. McManus, *Coord. Chem. Rev.*, **189**, 135–168 (1999).
153. (a) S. A. Fairhurst, R. S. Pilkington and L. H. Sutcliffe, *J. Chem. Soc. Faraday Trans. I*, **79**, 925–940 (1983). (b) S. R. Harrison, R. S. Pilkington and L. H. Sutcliffe, *J. Chem. Soc. Faraday Trans. I*, **80**, 669–689 (1984).
154. E. G. Awere, N. Burford, C. Mailer, *et al.*, *J. Chem. Soc. Chem. Commun.*, 66–69 (1987).
155. G. Wolmershauser, M. Schnauber and T. Wilhelm, *J. Chem. Soc. Chem. Commun.*, 573–574 (1984).
156. G. Wolmershauser and R. Johann, *Angew. Chem. Int. Ed.*, **28**, 920–921 (1989).
157. G. Wolmershauser and G. Kraft, *Chem. Ber.*, **122**, 385–387 (1989).
158. G. Wolmershauser and G. Kraft, *Chem. Ber.*, **123**, 881–885 (1990).
159. T. M. Barclay, A. W. Cordes, N. A. George, *et al.*, *J. Am. Chem. Soc.*, **120**, 352–360 (1998).
160. J. L. Brusso, O. P. Clements, R. C. Haddon, *et al.*, *J. Am. Chem. Soc.*, **126**, 8256–8265 (2004).
161. A. Alberola, O. P. Clements, R. J. Collis, *et al.*, *Cryst. Gr. Des.*, **8**, 155–161 (2008).
162. A. Alberola, J. Burley, R. J. Collis, *et al.*, *J. Organometal. Chem.*, **692**, 2750–2760 (2007).
163. A. Alberola, R. D. Farley, S. M. Humphrey, *et al.*, *Dalton Trans.*, 3838–3845 (2005).
164. T. M. Barclay, A. W. Cordes, R. H. deLaat, *et al.*, *J. Am. Chem. Soc.*, **119**, 2633–2641 (1997).
165. E. G. Awere, N. Burford, R. C. Haddon, *et al.*, *Inorg. Chem.*, **29**, 4821–4830 (1990).
166. S. Brownridge, H. B. Du, S. A. Fairhurst, *et al.*, *J. Chem. Soc. Dalton Trans.*, 3365–3382 (2000).
167. G. Wolmershauser, G. Wortmann and M. Shnauber, *J. Chem. Res. - S*, 358 (1988).
168. G. D. McManus, J. M. Rawson, N. Feeder, *et al.*, *J. Mater. Chem.*, **10**, 2001–2003 (2000).
169. (a) G. Wolmershauser, W. Kaim, G. Heckmann and A. Lichtblau, *Z. Naturforsch. B*, **47**, 675–679 (1992). (b) K. B. Borisenko, M. Broschag, I. Hargittai, *et al.*, *J. Chem. Soc. Dalton Trans.*, 2705–2712 (1994).

170. (a) R. Mayer, S. Bleisch, G. Domschke, *et al*., *Z. Chem*., **21**, 147–148 (1981). (b) R. Mayer, G. Domschke, S. Bleisch, *et al*., *Z. Chem*., **21**, 264–265 (1981). (c) R. Mayer, G. Domschke, S. Bleisch, *et al*., *Coll. Czech. Chem. Commun*., **49**, 684–703 (1984).
171. R. Mayer, G. Domschke, S. Bleisch and A. Bartl, *Z. Chem*., **21**, 324–325 (1981).
172. R. Appel, H. Janssen, M. Siray and F. Knoch, *Chem. Ber*., **118**, 1632–1643 (1985).
173. T. M. Barclay, L. Beer, A. W. Cordes, *et al*., *Chem. Commun*., 531–532 (1999).
174. L. Beer, A. W. Cordes, R. C. Haddon, *et al*., *Chem. Commun*., 1872–1873 (2002).
175. T. M. Barclay, A. W. Cordes, R. T. Oakley, *et al*., *Chem. Commun*., 1039–1040 (1998).
176. R. T. Oakley, R. W. Reed, C. M. Robertson and J. F. Richardson, *Inorg. Chem*., **44**, 1837–1845 (2005).
177. T. M. Barclay, A. W. Cordes, R. C. Haddon, *et al*., *J. Am. Chem. Soc*., **121**, 969–976 (1999).
178. M. Risto, A. Assoud, S. M. Winter, *et al*., *Inorg. Chem*., **47**, 10100–10109 (2008).
179. L. Beer, R. T. Oakley, J. R. Mingie, *et al*., *J. Am. Chem. Soc*., **122**, 7602–7603 (2000).
180. L. Beer, J. L. Brusso, A. W. Cordes, *et al*., *J. Am. Chem. Soc*., **124**, 9498–9509 (2002).
181. (a) L. Beer, J. L. Brusso, A. W. Cordes, *et al*., *Chem. Commun*., 2562–2563 (2002). (b) L. Beer, J. F. Britten, J. L. Brusso, *et al*., *J. Am. Chem. Soc*., **125**, 14394–14403 (2003). (c) L. Beer, J. F. Britten, O. P. Clements, *et al*., *Chem. Mater*., **16**, 1564–1572 (2004).
182. A. A. Leitch, R. W. Reed, C. M. Robertson, *et al*., *J. Am. Chem. Soc*., **129**, 7903–7914 (2007).
183. A. A. Leitch, C. E. McKenzie, R. T. Oakley, *et al*., *Chem. Commun*., 1088–1090 (2006).
184. L. Beer, J. L. Brusso, R. C. Haddon, *et al*., *J. Am. Chem. Soc*., **127**, 18159–18170 (2005).
185. L. Beer, J. L. Brusso, R. C. Haddon, *et al*., *Chem. Commun*., 1543–1545 (2005).
186. L. Beer, J. L. Brusso, R. C. Haddon, *et al*., *Chem. Commun*., 5745–5747 (2005).
187. J. L. Brusso, S. Derakhshan, M. E. Itkis, *et al*., *Inorg. Chem*., **45**, 10958–10966 (2006).
188. J. L. Brusso, K. Cvrkalj, A. A. Leitch, *et al*., *J. Am. Chem. Soc*., **128**, 15080–15081 (2006).
189. A. A. Leitch, J. L. Brusso, K. Cvrkalj, *et al*., *Chem. Commun*., 3368–3370 (2007).
190. J. Zienkiewicz, P. Kaszynski and V. G. Young, *J. Org. Chem*., **69**, 2551–2561 (2004).
191. J. Zienkiewicz, P. Kaszynski and V. G. Young, *J. Org. Chem*., **69**, 7525–7536 (2004).
192. (a) L. Beer, R. C. Haddon, M. E. Itkis, *et al*., *Chem. Commun*., 1218–1220 (2005). (b) A. A. Leitch, R. T. Oakley, R. W. Reed and L. K. Thompson, *Inorg. Chem*., **46**, 6261–6270 (2007).
193. (a) N. P. Gritsan, K. V. Shuvaev, S. N. Kim, *et al*., *Mendeleev Commun*., **17**, 204–206 (2007). (b) C. Knapp, P. G. Watson, E. Lork, *et al*., *Inorg. Chem*., **47**, 10618–10625 (2008).
194. (a) J. Geevers, J. T. Hackmann and W. P. Trompen, *J. Chem. Soc. C*, 875–878 (1970). (b) W. Schramm, G. Voss, G. Rembarz and E. Fischer, *Z. Chem*., **14**, 471–472 (1974).
195. (a) E. Fischer, E. Jaudasprezel, R. Maggiulli, *et al*., *Chem. Ber*., **124**, 1347–1352 (1991). (b) T. V. V. Ramakrishna, A. J. Elias and A. Vij, *Inorg. Chem*., **38**, 3022–3026 (1999). (c) A. D. Stoller, *J. Heterocycl. Chem*., **37**, 583–595 (2000).
196. A. W. Cordes, P. J. Hayes, P. D. Josephy, *et al*., *J. Chem. Soc. Chem. Commun*., 1021–1022 (1984).
197. P. J. Hayes, R. T. Oakley, A. W. Cordes and W. T. Pennington, *J. Am. Chem. Soc*., **107**, 1346–1351 (1985).
198. (a) P. P. Kornuta, L. I. Derii and L. N. Markovskii, *Zh. Organich. Khim*., **16**, 1308–1313 (1980). (b) L. N. Markovski, P. P. Kornuta, L. S. Katschovskaya and O. M. Polumbrik, *Sulfur Lett*., **1**, 143 (1983).
199. R. T. Oakley, R. W. Reed, A. W. Cordes, *et al*., *J. Am. Chem. Soc*., **109**, 7745–7749 (1987).
200. R. T. Boere and T. L. Roemmele, *Phosph. Sulf. Sil. Relat. El*., **179**, 875–882 (2004).
201. R. T. Boere, A. W. Cordes, P. J. Hayes, *et al*., *Inorg. Chem*., **25**, 2445–2450 (1986).
202. K. Bestari, A. W. Cordes, R. T. Oakley and K. M. Young, *J. Am. Chem. Soc*., **112**, 2249–2255 (1990).
203. (a) T. Chivers, M. N. S. Rao and J. F. Richardson, *J. Chem. Soc. Chem. Commun*., 982–983 (1982). (b) T. Chivers and M. N. S. Rao, *Inorg. Chem*., **23**, 3605–3609 (1984).
204. T. Chivers, M. N. S. Rao and J. F. Richardson, *J. Chem. Soc. Chem. Commun*., 186–187 (1983).
205. K. Bestari, R. T. Oakley and A. W. Cordes, *Can. J. Chem*., **69**, 94–99 (1991).
206. T. Chivers, M. N. S. Rao and J. F. Richardson, *J. Chem. Soc. Chem. Commun*., 702–703 (1983).
207. J. M. Farrar, M. K. Patel, P. Kaszynski and V. G. Young, *J. Org. Chem*., **65**, 931–940 (2000).
208. V. Chandrasekhar, T. Chivers, *et al*., *Inorg. Chem*., **36**, 4772–4777 (1997).
209. K. T. Bestari, A. W. Cordes and R. T. Oakley, *J. Chem. Soc. Chem. Commun*., 1328–1330 (1988).

210. R. T. Oakley, *Can. J. Chem.*, **71**, 1775–1784 (1993).
211. (a) F. Wudl, E. T. Zellers and D. Nalewajek, *J. Org. Chem.*, **45**, 3211–3215 (1980). (b) I. Ernest, W. Holick, G. Rihs, *et al.*, *J. Am. Chem. Soc.*, **103**, 1540–1544 (1981). (c) O. J. Scherer, G. Wolmershauser and R. Jotter, *Z. Naturforsch. B*, **B37**, 433–436 (1982). (d) G. Wolmershauser, J. Fuhrmann, R. Jotter, *et al.*, *Mol. Cryst. Liq. Cryst.*, **118**, 435–438 (1985). (e) J. C. W. Chien and S. Ramakrishnan, *Macromolecules*, **21**, 2007–2010 (1988).
212. J. M. Rawson, A. Alberola and A. Whalley, *J. Mater. Chem.*, **16**, 2560–2575 (2006).
213. A. W. Cordes, R. C. Haddon and R. T. Oakley, *Adv. Mater.*, **6**, 798–802 (1994).
214. (a) K. Awaga, T. Tanaka, T. Shirai, *et al.*, *Bull. Chem. Soc. Jpn.*, **79**, 25–34 (2006). (b) K. Awaga, T. Tanaka, T. Shirai, *et al.*, *Compt. Rend. Chim.*, **10**, 52–59 (2007).
215. J. M. Rawson and F. Palacio, in *Pi-Electron Magnetism: From Molecules to Magnetic Materials* (ed. J. Veciana), Structure and Bonding Series, Vol. **100**, Springer, Berlin, 2001, 93–128.
216. R. C. Haddon, *Nature*, **256**, 394–396 (1975).
217. R. C. Haddon, *Aust. J. Chem.*, **28**, 2343–2351 (1975).
218. X. Chi, M. E. Itkis, B. O. Patrick, *et al.*, *J. Am. Chem. Soc.*, **121**, 10395–10402 (1999).
219. A. W. Cordes, R. C. Haddon and R. T. Oakley, in *The Chemistry of Inorganic Ring Systems* (ed. R. Steudel), Elsevier, Amsterdam, 1992, 295–321.
220. T. M. Barclay, A. W. Cordes, N. A. George, *et al.*, *Chem. Commun.*, 873–874 (1997).
221. C. M. Robertson, D. J. T. Myles, A. A. Leitch, *et al.*, *J. Am. Chem. Soc.*, **129**, 12688–12689 (2007).
222. C. M. Robertson, A. A. Leitch, K. Cvrkalj, *et al.*, *J. Am. Chem. Soc.*, **130**, 8414–8425 (2008).
223. A. A. Leitch, X. Y. Yu, S. M. Winter, *et al.*, *J. Am. Chem. Soc.*, **131**, 7112–7125 (2009).
224. C. M. Robertson, A. A. Leitch, K. Cvrkalj, *et al.*, *J. Am. Chem. Soc.*, **130**, 14791–14801 (2008).
225. K. Awaga, Y. Umezono, W. Fujita, *et al.*, *Inorg. Chim. Acta*, **361**, 3761–3770 (2008).
226. S. S. Staniland, W. Fujita, Y. Umezono, *et al.*, *Chem. Commun.*, 3204–3206 (2005).
227. C. D. Bryan, A. W. Cordes, R. M. Fleming, *et al.*, *Nature*, **365**, 821–823 (1993).
228. C. D. Bryan, A. W. Cordes, R. M. Fleming, *et al.*, *J. Am. Chem. Soc.*, **117**, 6880–6888 (1995).
229. C. D. Bryan, A. W. Cordes, N. A. George, *et al.*, *Chem. Mater.*, **8**, 762–768 (1996).
230. A. W. Cordes, N. A. George, R. C. Haddon, *et al.*, *Chem. Mater.*, **8**, 2774–2778 (1996).
231. A. W. Cordes, R. C. Haddon, R. G. Hicks, *et al.*, *Can. J. Chem.*, **76**, 307–312 (1998).
232. (a) P. M. Lahti (ed.), *Magnetic Properties of Organic Materials*, Marcel Dekker, New York, 1999. (b) T. L. Makarova and F. Palacio (eds), *Carbon-based Magnetism:An Overview of the magnetism of metal-free carbon-based compounds and materials* Elsevier, Amsterdam, 2006.
233. J. J. Novoa, P. Lafuente, M. Demual and F. Mota, in *Magnetism: Molecules to Materials IV* (eds J. S. Miller and M. Drillon), Wiley-VCH Verlag GmbH, Weinheim, 2003, 65–117.
234. D. B. Amabilino and J. Veciana, in *Magnetism: Molecules to Materials II* (eds J. S. Miller and M. Drillon), Wiley-VCH Verlag GmbH, Weinheim, 2003, 1–60.
235. (a) Y. Miura, in *Magnetic Properties of Organic Materials* (ed. P. M. Lahti), Marcel Dekker, New York, 1999, 267–284. (b) H. Nishide and T. Kaneko, in *Magnetic Properties of Organic Materials* (ed. P. M. Laht), Marcel Dekker, New York, 1999, 285–304.
236. A. Caneschi, D. Gatteschi, R. Sessoli and P. Rey, *Acc. Chem. Res.*, **22**, 392–398 (1989).
237. J. S. Miller, *Adv. Mater.*, **6**, 322–324 (1994).
238. A. Caneschi, D. Gatteschi and P. Rey, *Progr. Inorg. Chem.*, **39**, 331–429 (1991).
239. A. Rajca, *Chem. Eur. J.*, **8**, 4834–4841 (2002).
240. B. D. Koivisto and R. G. Hicks, *Coord. Chem. Rev.*, **249**, 2612–2630 (2005).
241. Y. Miura, T. Tomimura and Y. Teki, *J. Org. Chem.*, **65**, 7889–7895 (2000).
242. W. Fujita and K. Awaga, *Science*, **286**, 261–262 (1999).
243. J. L. Brusso, O. P. Clements, R. C. Haddon, *et al.*, *J. Am. Chem. Soc.*, **126**, 14692–14693 (2004).
244. A. Alberola, R. J. Collis, S. M. Humphrey, *et al.*, *Inorg. Chem.*, **45**, 1903–1905 (2006).
245. H. Matsuzaki, W. Fujita, K. Awaga and H. Okamoto, *Phys. Rev. Lett.*, **91**, 017403 (2003).
246. T. Tanaka, W. Fujita and K. Awaga, *Chem. Phys. Lett.*, **393**, 150–152 (2004).
247. W. Fujita, K. Kikuchi and K. Awaga, *Angew. Chem. Int. Ed.*, **47**, 9480–9483 (2008).

248. (a) M. Deumal, J. Cirujeda, J. Veciana and J. J. Novoa, *Adv. Mater.*, **10**, 1461–1466 (1998). (b) M. Deumal, J. Cirujeda, J. Veciana and J. J. Novoa, *Chem. Eur. J.*, **5**, 1631–1642 (1999).
249. (a) M. Kinoshita, P. Turek, M. Tamura, *et al.*, *Chem. Lett.*, 1225–1228 (1991). (b) M. Tamura, Y. Nakazawa, D. Shiomi, *et al.*, *Chem. Phys. Lett.*, **186**, 401–404 (1991).
250. R. Chiarelli, M. A. Novak, A. Rassat and J. L. Tholence, *Nature*, **363**, 147–149 (1993).
251. (a) P. M. Allemand, K. C. Khemani, A. Koch, *et al.*, *Science*, **253**, 301–303 (1991). (b) A. Omerzu, D. Arcon, R. Blinc and D. Mihailovic, in *Magnetism: Molecules to Materials II* (eds J. S. Miller and M. Drillon), Wiley-VCH Verlag GmbH, Weinheim, 2003, 123–147.
252. A. J. Banister, N. Bricklebank, I. Lavender, *et al.*, *Angew. Chem. Int. Ed.*, **35**, 2533–2535 (1996).
253. W. Fujita, K. Awaga, Y. Nakazawa, *et al.*, *Chem. Phys. Lett.*, **352**, 348–352 (2002).
254. W. Fujita and K. Awaga, *Chem. Phys. Lett.*, **357**, 385–388 (2002).
255. W. Fujita and K. Awaga, *Chem. Phys. Lett.*, **388**, 186–189 (2004).
256. W. Fujita, K. Awaga, M. Takahashi, *et al.*, *Chem. Phys. Lett.*, **362**, 97–102 (2002).
257. K. Shimizu, T. Gotohda, T. Matsushita, *et al.*, *Phys. Rev. B*, **74**, 172413 (2006).
258. A. Iwasaki, L. G. Hu, R. Suizu, *et al.*, *Angew. Chem. Int. Ed.*, **48**, 4022–4024 (2009).

10

Stable Radicals of the Heavy p-Block Elements

Jari Konu and Tristram Chivers

Department of Chemistry, University of Calgary, Calgary, Canada

10.1 Introduction

The primary focus of this chapter is on recent developments in the chemistry of stable radicals of the heavy p-block elements. In the context of the content of the other chapters in this book, the heavy p-block elements are taken to include all third row elements (i.e., aluminum, silicon, phosphorus, sulphur and chlorine) and their heavier congeners. However, sulfur–nitrogen and related chalcogen–nitrogen radicals are not included because they are the topics of Chapters 5 and 14.

The topic "Persistent and Stable Radicals of the Heavier Main Group Elements" has been covered in a comprehensive review by Power, which comprises the literature up to circa mid-2002.[1a] Consequently, the emphasis in the current account is on progress that has been made in the past 6–7 years, although comparisons with earlier seminal contributions will be made where appropriate. The choice of material to be discussed is made on the basis of the literature definitions of "stable" and "persistent". Both terms can be applied to long-lived radicals. However, the term "stable" is used to describe radicals that can be *isolated* and stored under an inert atmosphere for long periods, whereas the designation "persistent" refers to radicals that have relatively long lifetimes *under the conditions that they are generated*. For practical purposes, these definitions imply that the solid state structures of stable radicals can, in principle, be determined by X-ray crystallography, while the characterization of persistent radicals is limited to the application of spectroscopic methods (most commonly electron paramagnetic resonance (EPR) spectroscopy) either in solution or, infrequently, in the gas phase. Although the chemistry of stable radicals is emphasized in the current chapter, information on persistent radicals will be included when either it is pertinent to the discussion of the congeneric stable radicals or it introduces important insights into the behavior or properties of a particular type of radical for which stable analogues are unknown. While the emphasis is on neutral species, relevant examples of charged species, that is, anion or cation radicals, are also discussed in the context of isoelectronic relationships and in cases where they enhance our understanding of the electronic structures and bonding. In recent years there has been increasing interest in stable biradicals involving the p-block elements[2,3] and the intriguing chemistry associated with these species will also be covered in the appropriate sections of this chapter.

The principal strategy for conferring stability on heavy p-block element radicals is the use of extremely bulky substituents to provide a kinetic barrier to oligomerization processes, for example, dimerization. For cyclic systems, delocalization of the unpaired electron over the entire (or part of) the ring system may also contribute to stability. For heterocyclic radicals stability may also be enhanced by the presence of highly electronegative atoms. The primary interest in stable radicals of the heavy p-block elements has focused on the insights that structural studies, in conjunction with molecular orbital calculations, provide into the bonding in these paramagnetic systems. From the viewpoint of applications, the major advancements that have been made in the design and generation of chalcogen–nitrogen radicals with potentially useful conducting or magnetic properties are covered in Chapters 5 and 14.

The subsequent sections of this chapter are organized sequentially according to groups 13, 14, 15, 16 and 17 in the Periodic Table. For each group, radicals that incorporate the third-row elements will be discussed before their heavier analogues; boron-containing radicals are also considered in the section on group 13, since they are not covered elsewhere in this book. Neutral radicals are presented before charged species, that is, anion or cation radicals. In the case of ring systems, paramagnetic homocycles are described prior to heterocycles. For the major types of radical the methods of synthesis are illustrated along with information on their molecular and electronic structures, as revealed by X-ray structural determinations, molecular orbital calculations and spectroscopic techniques.

10.2 Group 13 element radicals

10.2.1 Boron

10.2.1.1 Radical anions/early studies

The early work on boron-centered radicals was stimulated by the isoelectronic relationship of boron triaryl radical anions $[BR_3]^{\bullet-}$ (**1**) (Scheme 10.1) and the neutral carbon-centered radicals $[CR_3]^{\bullet}$ (R = aryl).[4] EPR studies of $[BPh_3]^{\bullet-}$ and its deuterated derivatives, as well as derivatives of the type $[B(4\text{-}XC_6H_4)_3]^{\bullet-}$ (X = Cl, OMe), have revealed that the unpaired electron in **1** is located primarily at boron ($a(^{11}B) = 7.84$ G).[5] The relatively low value of the hyperfine coupling constant (hfcc) indicates that the unpaired electron density is mainly associated with the 2p orbital and that the geometry of the $[BR_3]^{\bullet-}$ radical anions is planar rather than pyramidal. This conclusion was subsequently corroborated by the crystal structure of the thermally stable, ion-separated salt, $[Li(12\text{-crown-}4)_2][BMes_3]$, which decomposes only at about 240 °C.[6] The trigonal planar boron center shows only small deviations from 120° in the C–B–C angles; the B–C bond lengths are elongated by about 0.02 Å from those observed in neutral $BMes_3$.

Another example of early seminal studies on boron-centered radicals is the one-electron reduction of diboron(4) compounds, R_2BBR_2 (R = alkyl, aryl). The resulting diborane(4) radical

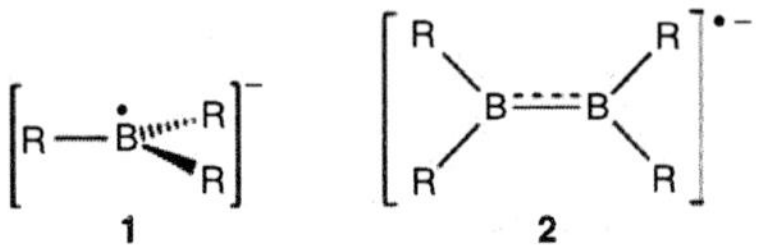

Scheme 10.1

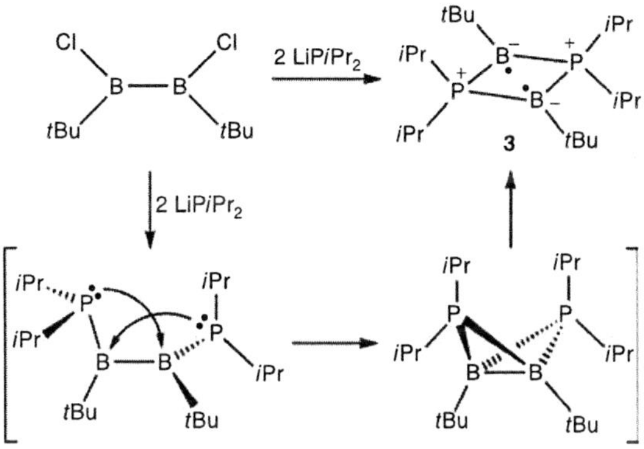

Scheme 10.2

anion, $[R_2BBR_2]^{\bullet-}$ (**2**) (Scheme 10.1), was first structurally characterized as the contact ion pair, $[Li(OEt_2)_2][MeO(Mes)BB(Mes)OMe]$, in which the Li^+ cation is bound to the anion by coordination to the methoxy oxygens.[7] The broadness of the electron paramagnetic resonance (EPR) signal of this paramagnetic species precluded the determination of hfcc values. However, the solvent-separated ion pair, $[K(18\text{-crown-}6)(THF)_2][Mes_2BB(Ph)Mes]$, which was obtained by one-electron reduction with potassium in tetrahydrofuran (THF) followed by the addition of 18-crown-6, exhibits a seven-line EPR pattern ($g = 2.0063$) with $a(^{11}B) = 13$ G due to the coupling to two boron centers (^{11}B, $I = 3/2$, 80.2 %) consistent with the formation of a π radical.[8] A π-bond order of 0.5 was indicated by comparison of the structural parameters of this radical with those of the neutral precursor, $Mes_2BB(Ph)Mes$.

10.2.1.2 Biradicaloids

A biradicaloid is a closed shell species derived from a singlet biradical by a weak interaction between the radical centers.[2,3] The term biradicaloid is applied to p-block element compounds in which the biradical character is substantially less than that of organic biradicals.[3] The study of such biradicaloids raises fundamental questions about the nature of chemical bonding, as well as the prospect of generating materials with unique magnetic properties, as indicated by the following examples involving cyclic B_2P_2 systems. The boron-centered biradicaloid, $(i Pr_2P)_2(Bt Bu)_2$ (**3**) is prepared by the reaction of the 1,2-dichloroborane, Cl(*t*Bu)BB(*t*Bu)Cl, with two equivalents of $LiPi Pr_2$.[9] The biradicaloid **3** is an isoelectronic analogue of the carbon-centered biradicaloids $(RP)_2(CR')_2$ (Scheme 10.16); it is presumably formed by a rearrangement of the initially formed acyclic P–B–B–P skeleton (Scheme 10.2).

The thermally stable (> 200 °C), air-sensitive yellow crystals of **3** display a perfectly planar P_2B_2 ring with a transannular B–B distance of 2.57 Å (cf. 1.76 Å for a B–B single bond), indicative of biradical character for **3**.[9] The absence of an EPR signal both in solution and in the solid state from −80 °C to room temperature suggests a singlet ground state for **3**. This inference is supported by DFT calculations, which predict that the singlet state is 17.2 kcal mol^{-1} lower in energy than the triplet state biradical, thus suggesting coupling between the two radical sites. The structures of four-membered rings $(R_2P)_2(BR')_2$ are markedly influenced by the nature of the substituents on boron and phosphorus. Replacement of the *tert*-butyl group on boron in **3** by a 2,3,5,6-tetramethylphenyl group generates a folded structure with a B–B distance of 2.24 Å, while changing the *iso*-propyl groups on phosphorus to phenyl substituents produces an even more folded structure with a B–B distance of 1.99 Å.[9c] Thus, it seems that the biradical nature of the P_2B_2 ring in **3** results from the combination of the steric strain imposed by the substituents on the phosphorus and boron and the σ donor property of the *tert*-butyl group at boron.

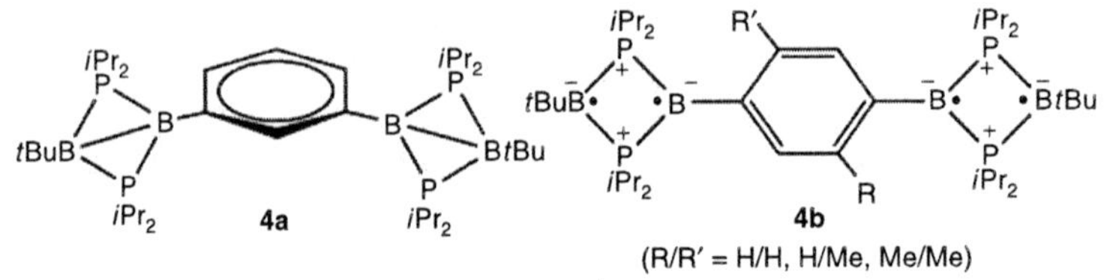

Scheme 10.3

In a fascinating recent development it has been shown that inverting the role played by the phosphorus and boron moieties, for example, in $(PhP)_2(BCy_2)_2$, produces a four-membered ring with a transannular P–P bond that is slightly longer than a typical single bond.[9d] The easily accessible biradical character of $(PhP)_2(BCy_2)_2$ is indicated by the facile reaction with tributyltin hydride to give a mixture of *cis* and *trans*-$(PhPH)_2(BCy_2)_2$.

In the context of producing antiferromagnetic low spin polymers, two four-membered P_2B_2 biradicaloids have been linked together via a phenylene spacer group in the derivatives 1,3-[B(iPr$_2$P)$_2$B(tBu)]$_2$C$_6$H$_4$ (**4a**) and 1,4-[B(iPr$_2$P)$_2$B(tBu)]$_2$-2,5-R,R′C$_6$H$_2$ (**4b**, R/R′ = H/H, H/methyl, methyl/methyl) (Scheme 10.3).[10] The tetraradical nature of **4a** and **4b** is dependent on the communication of two biradicaloid units through the *para*-phenylene linker. Crystal structure determinations reveal a coplanar arrangement of the C_6 and B_2P_2 rings in **4b** (R = R′ = H) with B–B distances of about 2.57 Å, whereas the steric strain in the *meta*-isomer **4a** and that imposed by the methyl substituents in **4b** (R = R′ = methyl) results in folding of the P_2B_2 ring to give bicyclic structures with transannular B–B bond lengths of about 1.88 Å. Similarly to **3**, the tetraradicaloid **4b** (R = R′ = H) shows no EPR signal indicating a singlet ground state.

10.2.2 Aluminum, gallium, and indium

10.2.2.1 Neutral radicals: acyclic and homocyclic systems

Investigations of both acyclic and cyclic radicals involving M–M bonds (M = aluminum or gallium) with bulky substituents on the group 13 metal center have provided insights into the influence of the metal center on relative stabilities as a result of differences in covalent radii and/or M–M bond strengths.

Thermolysis of the dimer $R'_2AlAlR'_2$ in heptane leads, via Al–Si bond cleavage, to a persistent acyclic radical, $[R'_2AlAlR']^\bullet$ (**5a**), that exists as an intermediate in the formation of the stable three-membered ring $[Al_3R'_4]^\bullet$ (R′ = SitBu$_3$) (**6a**) (Scheme 10.4).[11] The isolation of black-green crystals of this homocyclic neutral radical is a cogent illustration of the ability of bulky substituents to stabilize cyclic radicals of main group elements.

The planar, three-membered ring in **6a** forms an isosceles triangle in which two of the aluminums are three-coordinate and one is four-coordinate. The Al–Al bond between the two three-coordinate aluminums is somewhat shorter than those involving the four-coordinate aluminum atom (2.703 vs 2.756 Å, cf. 2.751 Å for the Al–Al single bond in $R'_2AlAlR'_2$). The poorly resolved EPR spectrum of **6a** in C_6D_{12} precludes a definitive identification of the radical present in solution. The observed hfcc values of about 3 G and two inequivalent couplings close to 13 G ($g = 2.0053$) may indicate cleavage of one of the elongated Al–Al bonds in solution leading to the acyclic radical $[R'_2Al\text{-}AlR'\text{-}AlR']^\bullet$, as suggested by DFT calculations.[11] Alternatively, the slight inequivalence of the two larger hfcc values may result from the disparity in the geometry about the two three-coordinate aluminum atoms in the solid state structure; one of the R′Al units in **6a** is planar while the second adopts a pyramidal arrangement.

The acyclic intermediate **5a** was identified only by its EPR spectrum in solution ($a(^{27}Al) = 21.8$ and 18.9 G).[11] In contrast, the heavier gallium congener, $[R'_2GaGaR']^\bullet$ (R′ = SitBu$_3$) (**5b**), has been isolated

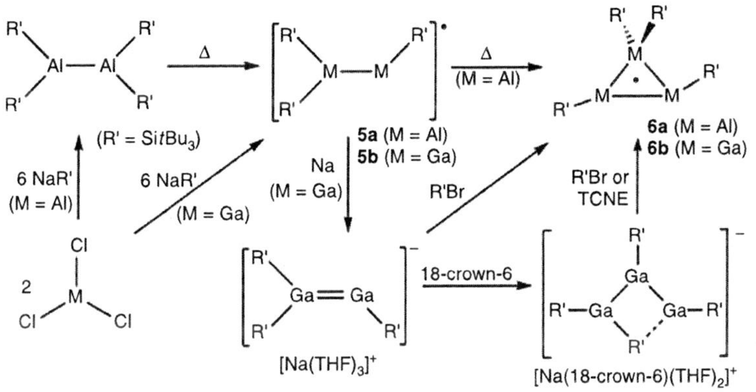

Scheme 10.4

as blue-black crystals from the reaction between NaR′ and gallium trichloride ($GaCl_3$) in 3 : 1 molar ratio (Scheme 10.4).[12a] The structure of **5b** exhibits a trigonal planar $Si_2Ga^{II}Ga^{I}$ unit and a slightly bent $Ga^{II}Ga^{I}Si$ chain ($\Sigma \angle Ga^{II}$ 359.6° and $\angle Ga^{II}–Ga^{I}–Si$ 170.3°) (the superscripts I and II refer to the formal oxidation states of the gallium centers). The Ga–Ga σ bond is formed by overlap of the sp^2- and sp-hybridized gallium atoms. The odd electron occupies a bonding π molecular orbital formed by overlap of the p_z orbitals on the two gallium atoms resulting in a formal bond order of 1.5 for the Ga–Ga bond. The EPR spectrum of **5b** shows a 64-line signal ($g = 1.9947$) arising from coupling of the unpaired electron with the two inequivalent gallium centers (^{69}Ga, $I = 3/2$, 60.1 %; ^{71}Ga, $I = 3/2$, 30.9 %).[12a] The two different hfcc values ($a_1(^{69/71}Ga) = 50/64$ G and $a_2(^{69/71}Ga) = 32/41$ G) could not be assigned to the two- and three-coordinate gallium atoms with certainty.

The three-membered ring $[Ga_3R'_4]^•$ (R′ = Si*t*Bu$_3$) (**6b**), the gallium analogue of **6a**, is synthesized either by oxidation of the acyclic anion $[R'_2Ga–GaR'–GaR']^-$ or by the reaction of $[R'_2GaGaR']^-$ with R′Br (Scheme 10.4).[12b] In contrast to the aluminum derivative **6a**, the neutral radical **6b** exhibits a Ga–Ga bond length between the two three-coordinate galliums that is about 0.35 Å longer than those involving the four-coordinate R'_2Ga unit (2.879(1) vs 2.527(1) Å), implying the onset of Ga–Ga bond cleavage. In solution, the three-membered ring **6b** decomposes to give the EPR signal of the acyclic radical $[R'_2GaGaR']^•$ (**5b**) either after prolonged standing or upon warming; the initial EPR signal arising from **6b** is not sufficiently resolved to be unambiguously analyzed.[12b]

The dark crimson neutral radical $[Al(AlR_3)_2]^•$ (R = N(SiMe$_2$Ph)$_2$) (**7**) is isolated in 15 % yield from the reaction of a metastable solution of aluminum chloride (AlCl) in toluene/diethyl ether with an equimolar amount of LiR at −78 °C.[13] The structure of the D_{3d}-symmetric Al_7 cluster is comprised of an aluminum atom linked symmetrically to two Al_3 rings (Figure 10.1). The Al–Al bonds within the three-membered rings are significantly shorter than those involving the central aluminum atom (2.61 vs 2.73 Å). On the basis of detailed bonding analysis of **7** and the corresponding monoanion $[Al(AlR_3)_2]^-$ (R = N(SiMe$_3$)$_2$) (**8**), this fascinating cluster radical is referred to as a "metalloid" system.[13]

10.2.2.2 Neutral radicals: heterocyclic systems

Two classes of heterocyclic radicals involving group 13 elements have been thoroughly investigated. The first involves a series of neutral bis-boraamidinate (*bam*) radicals, $\{M[PhB(\mu\text{-}N tBu)_2]_2\}^•$ (M = B, Al, Ga, In) (**9a–d**) (Scheme 10.5), which are obtained by one-electron oxidation of the corresponding monoanions. The aluminum (**9b**) and gallium (**9c**) derivatives form thermally stable dark red and green

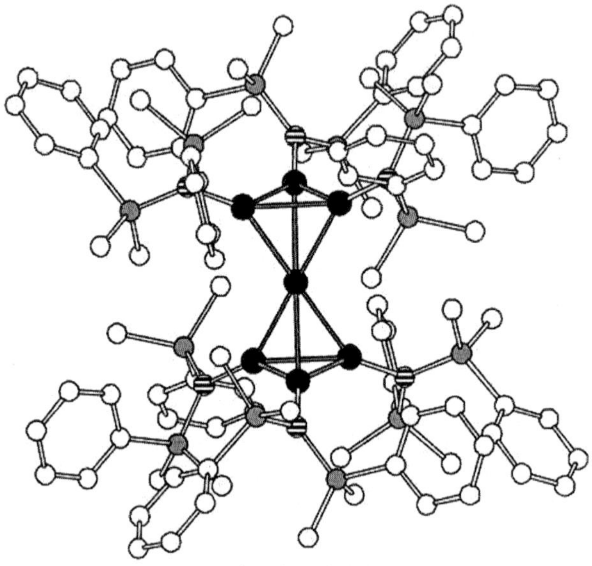

Figure 10.1 *Crystal structure of $[Al(Al_3R_3)_2]$ (7, R = $N(SiMe_2Ph)_2$). (Reproduced with permission from [13]. Copyright Wiley-VCH Verlag GmbH & Co. KGaA.)*

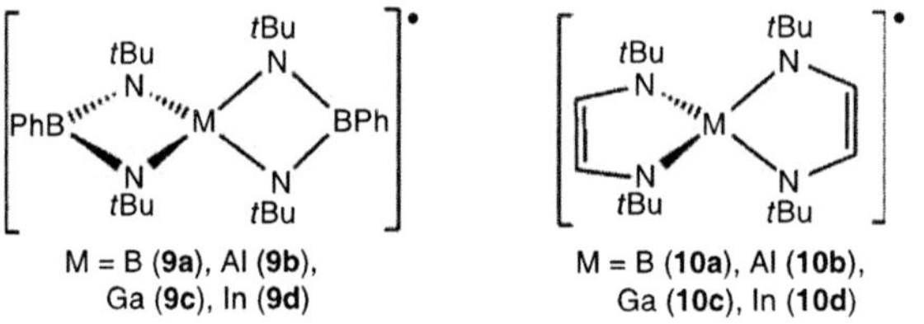

Scheme 10.5

crystals, respectively, which exhibit a spirocyclic arrangement of the two *bam* ligands.[14] The comparable N,N'-chelated bis-diazabutadiene system, $\{M[(CH)_2(\mu\text{-}NtBu)_2]_2\}^{\bullet}$ (M = B, Al, Ga, In) (**10a–d**) (Scheme 10.5), has been studied both experimentally (**10b–c**)[15] and computationally (**10a–d**).[16] Although the compounds **9** and **10** are not group 13-centered radicals, these stable, spirocyclic paramagnetic compounds serve as examples of the characterization of main group radicals containing multiple quadrupolar nuclei by EPR spectroscopy, supported by electronic structures obtained from DFT calculations.

DFT calculations of the model radicals $\{M[PhB(\mu\text{-}NMe)_2]_2\}^{\bullet}$ (M = B, Al, Ga, In) predict D_{2d}-symmetric geometries consistent with the observed solid state structures of **9b** and **9c**.[14] The single point calculations on the optimized geometries show that the SOMOs consist solely of nitrogen p-orbitals (Figure 10.2 for M = aluminum); they are equally delocalized over all four nitrogen atoms as revealed by the spin densities derived from Mulliken population analysis. The EPR spectra of **9a–d** show that the central group 13 metal atom and the NBN boron atoms have non-zero spin density values arising from polarization effects. The experimental spectra of these spirocyclic neutral radicals can, therefore, be simulated by assuming hyperfine interaction of the unpaired electron with four equivalent nitrogen centers, two boron atoms and the central group 13 element (^{27}Al, $I = 5/2$, 100 %; ^{69}Ga, $I = 3/2$, 60.1 %; ^{71}Ga, $I = 3/2$, 39.9 %; ^{113}In, $I = 9/2$, 4.3 %;

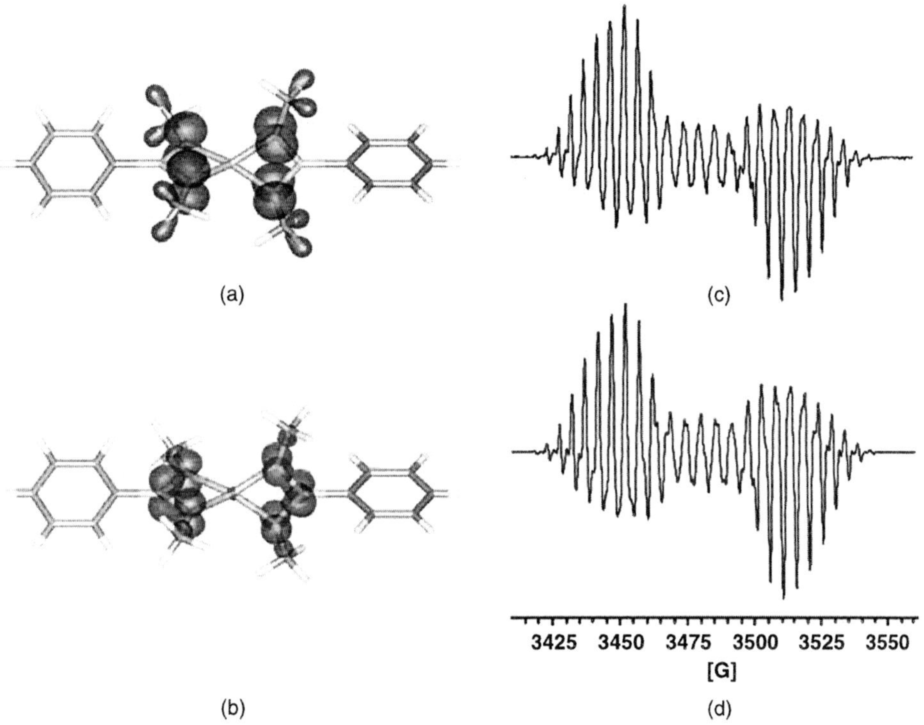

Figure 10.2 *(a) SOMO of* $\{Al[PhB(\mu\text{-}NMe)_2]_2\}^{\bullet}$ *drawn at isosurface level* ± 0.05*; (b) spin density map of* $\{Al[PhB(\mu\text{-}NMe)_2]_2\}^{\bullet}$ *drawn at isosurface level 0.02 (α-spin density) and −0.002 (β-spin density); (c) experimental and (d) simulated X-band EPR spectra of a diethyl ether solution of* $\{Al[PhB(\mu\text{-}NtBu)_2]_2\}^{\bullet}$ *(**9b**) at 298 K. ([14a] Reproduced by permission of the Royal Society of Chemistry.)*

^{115}In, $I = {}^{9}/_{2}$, 95.7 %) (Figure 10.2c,d for M = Al). Thus, the DFT calculations and spectral simulations confirm the retention of the spirocyclic structures in solution and indicate uniform spin delocalization throughout both ligands in **9a–d**.[14]

The *N*,*N*′-chelated bis-diazabutadiene aluminum and gallium radicals **10b** and **10c** are prepared by co-condensation of the metal vapor with an excess of 1,4-di-*tert*-butyl-1,4-diazabutadiene.[15] In contrast to the bis-*bam* systems **9**, both the solid state structures and the EPR spectra of **10b** and **10c** indicate that the unpaired electron is located on only one of the ligands with spin density contributions from two nitrogens and two hydrogen atoms in addition to a significantly smaller contribution from the group 13 center.[15d] In agreement with the experimental results, distorted tetrahedral C_s (M = B) and C_{2v} (M = Al, Ga) symmetries with the unpaired electron localized on one ligand were also predicted by computational methods.[16] Interestingly, however, the computations indicate that the experimentally inaccessible indium radical **10d** optimizes to a pyramidal geometry (C_2 symmetry) in which the spin density is equally delocalized over both ligands similarly to the bis-*bam* compounds **9**.

10.2.2.3 Radical anions

The stabilizing effect of the sterically bulky tBu$_2$(Me)Si group was also used in the production of trigonal planar radical anions $[MR_3]^{\bullet-}$ (M = Al (**11a**), Ga (**11b**); R = Si(Me)tBu$_2$) (Figure 10.3a). These

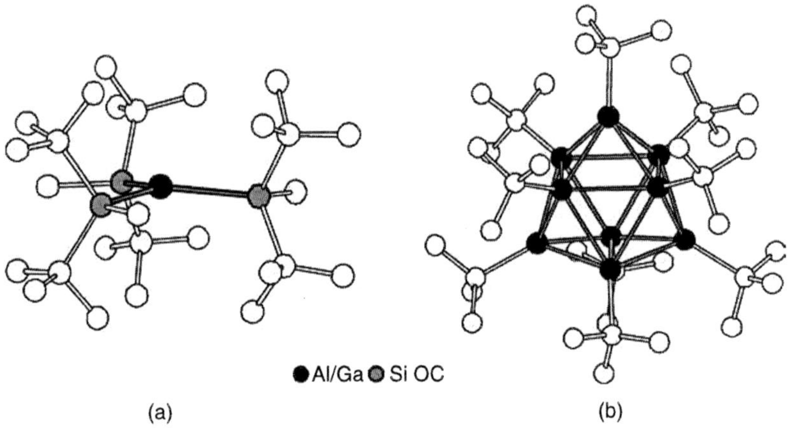

Figure 10.3 Crystal structures of (a) $[Al(SiMetBu_2)_3]^{\bullet-}$ (**11a**) and (b) $[Ga_9(tBu)_9]^{\bullet-}$ (**12**) radical anions. ((a) Reprinted with permission from [17]. Copyright 2005 American Chemical Society. (b) Reproduced with permission from [19]. Copyright Wiley-VCH Verlag GmbH & Co. KGaA.)

odd-electron compounds are prepared by direct metallation of the neutral precursors with alkali metals (lithium, sodium, potassium), and deep red crystals of the ion-separated potassium salts were isolated by recrystallization in the presence of [2.2.2]cryptand.[17] The Si–M bond lengths (M = Al, Ga) in the nearly planar radical anions **11a,b** are somewhat shorter than those in the neutral congeners suggesting that the unpaired electron predominantly occupies the $3p_z$ and $4p_z$ orbital of the central aluminum and gallium, respectively. By contrast, the prototypical pyramidal radical anions $[AlH_3]^{\bullet-}$ and $[GaH_3]^{\bullet-}$ have significant electron population in the corresponding s orbitals.[18] Localization of the spin density on the heavier group 13 center is also supported by the EPR spectra, which exhibit a well-resolved sextet ($a(^{27}Al) = 62\,G$, $g = 2.0050$) and two sets of quartets ($a(^{69}Ga) = 123\,G$, $a(^{71}Ga) = 157\,G$, $g = 2.0150$) for **11a** and **11b**, respectively. The relatively small hfcc values, as well as the observed crystal structures, are consistent with group 13 centered π radicals (cf. values of 154, 420 and 534 G for $a(^{27}Al)$, $a(^{69}Ga)$ and $a(^{71}Ga)$ in $[AlH_3]^{\bullet-}$ and $[GaH_3]^{\bullet-}$, respectively).

The gallium-containing radical anion cluster, $[Ga_9R_9]^{\bullet-}$ (R = *t*Bu) (**12**) (Figure 10.3b), is produced via one-electron reduction of the corresponding neutral compound with Cp^*_2Co.[19] The crystal structure reveals a molecular core comprised of nine gallium atoms arranged as a tricapped trigonal prism, analogous to the boron subhalide radical anions $[B_9X_9]^{\bullet-}$ (X = Cl, Br).[20] The Ga–Ga bonds parallel to the threefold rotational axis in **12** are significantly shorter in length (by about 17 pm) than those in the neutral precursor resulting in a more regular prism. The single electron is essentially localized in the p orbitals of the gallium atoms (0.993 e^-) with the capping galliums bearing most of the negative charge (80 %). Due to the numerous overlapping lines, only a poorly resolved EPR spectrum was obtained ($a(^{69/71}Ga) < 30\,G$).

10.3 Group 14 element radicals

The most significant recent developments in the chemistry of stable radicals of the heavy p-block elements have involved the group 14 elements. The primary contributor to this field is the group of Sekiguchi, whose findings have been discussed in detail in several recent reviews.[21–23] In this section the foremost aspects of this exciting work will be presented in the context of cognate organic radicals and related group 13 and group 15-centered radicals.

10.3.1 Cyclic group 14 radicals

10.3.1.1 Neutral radicals

The first homocyclic, neutral group 14 radical was reported by Power *et al*. in 1997.[24] This highly protected cyclotrigermenyl, $[(2,6\text{-}Mes_2C_6H_3)Ge]_3^{\bullet}$ (**13**), is obtained as dark blue crystals by the reduction of $(2,6\text{-}Mes_2C_6H_3)GeCl$ with KC_8 in THF (Scheme 10.6). The disordered triangular Ge_3 core in the crystal structure of **13** is consistent with a cyclogermenyl radical (germanium analogue of cyclopropenyl radical) in which one of the Ge–Ge bonds shows double bond character. The EPR spectrum of **13** displays a single resonance at $g = 2.0069$ with an hfc constant of $a(^{73}Ge) = 16\,G$ to one germanium nucleus (^{73}Ge, $I = {}^9/_2$, 7.8 %). The relatively low hyperfine coupling is consistent with a π radical in which the odd electron is localized on one of the planar sp^2-hybrized germanium centers.

A comparable, saturated Ge_3 triangle in the bicyclic, neutral radical **14** (Scheme 10.7) is obtained by one-electron oxidation of the parent anion.[25] The Ge=Ge double bond observed in **13** has been formally replaced by the bridging $-H_2C\text{–}C(Me){=}C(Me)\text{–}CH_2-$ unit in **14**. Similarly to **13**, the Ge_3 core in **14** is planar with an sp^2-hybridized germanium atom at the radical center. The EPR spectrum of **14** shows two independent sets of multiplets ($g = 2.0210$ and 2.0223) with $a_1(^{73}Ge) = 34\,G$ and $a_2(^{73}Ge) = 26\,G$ due

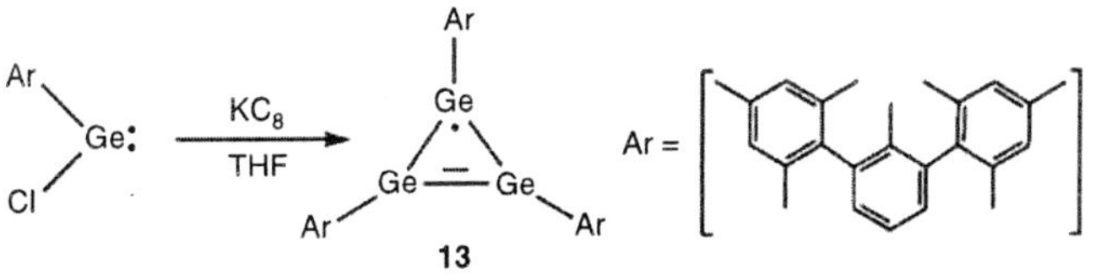

Scheme 10.6

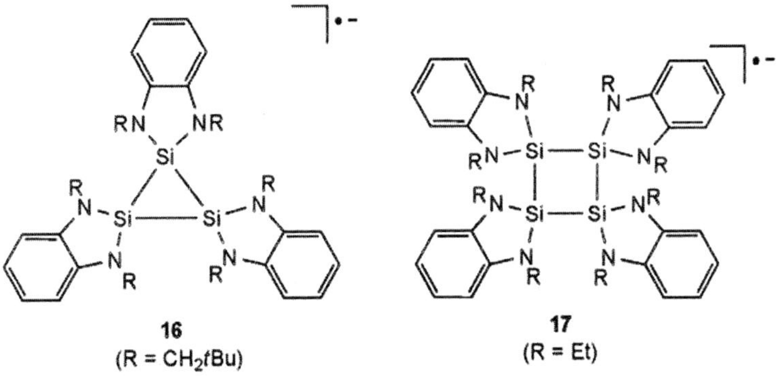

Scheme 10.7

to *endo,exo* isomerism; the small hfc constants indicate a π radical with the unpaired electron localized on the planar germanium center in solution.

An unusual example of a stable silyl radical, the red-purple cyclotetrasilenyl $[\{tBu_2(Me)Si\}_3\{Si(tBu)_2\}]^{\bullet}$ (**15**) (Scheme 10.7), is produced by one-electron reduction of the corresponding cation.[26] This all-silicon congener of the cyclobutenyl radical exhibits a nearly planar Si_4 ring with two Si–Si bond lengths that are intermediate between single and double bonds, while the other two Si–Si bonds are close to single bond values, thus suggesting an allylic structure. Consistently, the EPR spectrum ($g = 2.0058$) indicates additional spin density located on the terminal silicon atoms of the *pseudo*-allyl unit with $a_1(^{29}Si) = 40.7$ G and $a_2(^{29}Si) = 37.4$ G, while the hfcc value of the central silicon is expectedly somewhat smaller ($a_3(^{29}Si) = 15.5$ G; ^{29}Si, $I = 1/2$, 4.7 %) (Figure 10.4a).

10.3.1.2 Radical anions

The cyclic tri-and tetrasilane radical anions, $[Si\{1,2\text{-}(\mu\text{-}NCH_2tBu)_2C_6H_4\}]_3^{\bullet-}$ (**16**) and $[Si\{1,2\text{-}(\mu\text{-}NEt)_2C_6H_4\}]_4^{\bullet-}$ (**17**) (Scheme 10.7), are isolated as green crystals of ion-separated salts from the reduction of $Si[1,2\text{-}(\mu\text{-}NCH_2tBu)_2C_6H_4]$ and $Cl_2Si[1,2\text{-}(\mu\text{-}NEt)_2C_6H_4]$, respectively, with alkali metals in THF.[27,28] The radical anions **16** and **17** exhibit planar Si_3 and Si_4 rings, comparable to those of the cyclic trigermenyl and tetrasilenyl radicals, **13** and **15**, respectively. In marked contrast to **13** and **15**, however, the EPR spectra of **16** and **17** reveal significant delocalization of the spin density over the silane rings and onto the nitrogen atoms of the N,N'-chelated $(1,2\text{-}(\mu\text{-}NR)_2C_6H_4)$ substituents (R = CH_2tBu, ethyl) as indicated by the characteristic multiplets arising from coupling to six (**16**) or eight (**17**) equivalent ^{14}N nuclei ($a(^{14}N) = 4.6$ and 3.5 G for **16** and **17**, respectively; ^{14}N, $I = 1$, 99.6 %). The disparity in the distribution of the unpaired electron is clearly evident from a comparison of the EPR spectra of **15** and **17** (Figure 10.4). Coupling to the spin $I = 1/2$ ^{29}Si isotope with low natural abundance (4.7 %) in **15** and to the spin $I = 1$ ^{14}N isotope with high natural abundance in **17** results in a substantially different appearance of the spectra.

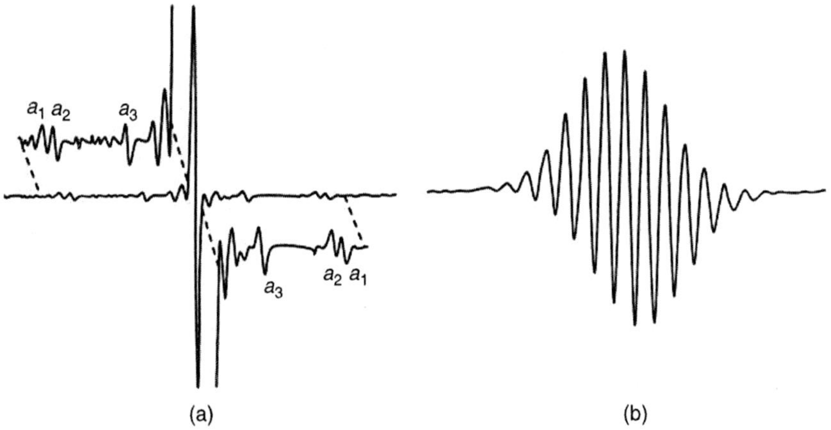

Figure 10.4 *EPR spectrum of (a)* ***15*** *($a_1(^{29}Si) = 40.7$ G, $a_2(^{29}Si) = 37.4$ G and $a_3(^{29}Si) = 15.5$ G; ^{29}Si, $I = 1/2$, 4.7 %), and (b)* ***17*** *($a(^{14}N) = 3.5$ G; ^{14}N, $I = 1$, 99.6 %). ((a) Reprinted with permission from [26]. Copyright 2001 American Chemical Society. (b) Reproduced with permission from [28]. Copyright Wiley-VCH Verlag GmbH & Co. KGaA.)*

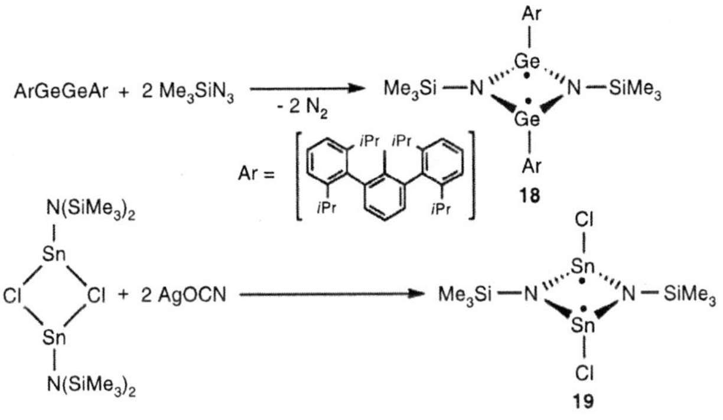

Scheme 10.8

10.3.1.3 Biradicaloids

Two examples of group 14-centered biradicaloids involving four-membered M_2N_2 rings (M = Ge, Sn) have been characterized, namely $[(ArGe)_2(\mu\text{-}NSiMe_3)_2]$ (**18**, Ar = 2,6-$Dipp_2C_6H_3$, Dipp = 2,6-$i Pr_2C_6H_3$) and $[(ClSn)_2(\mu\text{-}NSiMe_3)_2]$ (**19**). The Ge_2N_2 system **18** is obtained as dark violet crystals from the reaction of a "digermyne" ArGeGeAr with trimethylsilyl azide,[29] whereas the colorless, diamagnetic Sn_2N_2 derivative **19** was formed unexpectedly in the reaction between $(Me_3Si)_2NSn(\mu\text{-}Cl)_2SnN(SiMe_3)_2$ and silver isocyanate (Scheme 10.8).[30] In both compounds the four-membered M_2N_2 ring (M = Ge, Sn) is planar with pyramidal metal centers and trigonal planar nitrogen atoms. The substituents on the metal atoms in **18** and **19** assume a *trans* arrangement.

The transannular distance between the two metal atoms in the four-membered rings in **18** and **19** is substantially longer than the corresponding M–M single bond (by about 0.3 and about 0.6 Å, respectively) indicating some diradical character. As is typical for singlet state biradicaloids, no EPR signal was observed for either species. Consistently, DFT calculations reveal the absence of transannular M···M contacts and an energy gap of about 17 kcal mol^{-1} in favor of the singlet state over the triplet state for both heterocyclic systems (cf. 17.2 kcal mol^{-1} for the boron-centered biradicaloid, $(iPr_2P)_2(BtBu)_2$ (**3**)). The diradical character of **18** is also indicated by the facile addition of molecular hydrogen to give a compound tentatively identified as $Ar(H)Ge(\mu\text{-}NSiMe_3)_2Ge(H)Ar$.[29]

10.3.2 Acyclic group 14 radicals

10.3.2.1 Neutral radicals

The classic organic radicals involving three-coordinate carbon centers range from those with a fleeting existence, for example the methyl radical $[CH_3]^•$, to species that are sufficiently stable to be isolated in the solid state, for example the perchlorinated radical $[(C_6Cl_5)_3C]^•$ (Chapter 2). For the heavier group 14 elements, the simplest radicals $[EH_3]^•$ (E = Si, Ge, Sn, Pb) have very short lifetimes. However, by using the combination of steric protection and electronic properties provided by very bulky silyl groups, a remarkable class of neutral radicals of the type $[R_3E]^•$ have been isolated recently by Sekiguchi *et al.* (E = Si (**20**),[31] Ge (**21**),[31] Sn (**22**)[32]; R = $SiMetBu_2$) and by Klinkhammer *et al.* (E = Pb (**23**)[33]; R = $SiEt(SiMe_3)_2$). These stable, paramagnetic compounds are produced by a common synthetic approach in which the parent anion $[R_3E]^-$ is generated *in situ* and then subjected to mild, one-electron oxidation (Scheme 10.9).

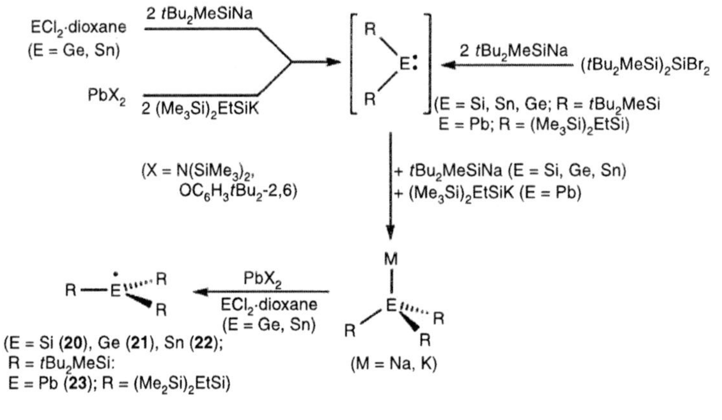

Scheme 10.9

The crystal structures of the silicon, germanium, and tin derivatives **20–22** show trigonal planar geometry with an sp^2-hybridized central atom, whereas the lead compound **23** displays a slight, albeit almost negligible, distortion towards a pyramidal arrangement ($\Sigma\ \angle$ Pb 355°), possibly owing to the bulkier substituent ($SiEt(SiMe_3)_2$ vs $SiMe\textit{t}Bu_2$). The stability of the acyclic radicals in this series is attributed primarily to the use of sterically demanding, electropositive silyl substituents. However, additional stability is achieved in **20–23** by delocalization of the unpaired electron via hyperconjugation between the p_z orbital of the central group 14 atom and the σ^* orbital of the Si–C(*t*Bu) (**20–22**) or Si–Si(Me) (**23**) bonds. This interaction is also evident from the structural parameters, which exhibit a slight shortening in the E–Si bonds (E = Sn, Pb) compared to the anionic precursors.[32,33] Interestingly, the Si–Si bond lengths in **20** are about 0.08 Å shorter than the Al–Si bonds observed in the isoelectronic radical anion $[(\textit{t}Bu_2MeSi)_3Al]^{\bullet-}$ (**11a**), consistent with the difference in the sum of covalent radii (Si–Si 2.22 Å vs Al–Si 2.32 Å).

The solution EPR spectra of **20** and **21** show the expected patterns at $g = 2.0056$ and 2.0229 with hfcc values of $a(^{29}Si) = 58$ G and $a(^{73}Ge) = 20$ G, respectively, in addition to smaller couplings to the silicon atoms of the $SiMe\textit{t}Bu_2$ substituents ($a(^{29}Si) = 7.9$ and 7.3 G).[31] Only one set of satellites is observed in the EPR spectrum of **22** in solution at $g = 2.0482$ (^{117}Sn, $I = 1/2$, 7.8 %; ^{119}Sn, $I = 1/2$, 8.6 %) owing to the overlap of the slightly broadened signals and an average coupling constant of 329 G was reported (cf. typical values of 1325–3426 G for $a(^{117/119}Sn)$ in tin-centered σ radicals).[32,34] For the lead congener **23**, a broad EPR signal is observed at room temperature ($g = 2.160$), but an approximate hfc constant of $a(^{207}Pb) = 520$ G was obtained from the frozen solution (^{207}Pb, $I = 1/2$, 22.1 %).[33] The EPR spectra of **20–23** indicate the retention of the planar geometry with sp^2-hybridized group 14 centers, and a true π radical character for these paramagnetic species in solution.

10.3.2.2 Charged radicals (anions and cations)

By taking advantage of the relatively low lying LUMOs and high lying HOMOs of disilenes compared to those of the C=C double bond of alkenes, either one-electron oxidation of the neutral disilene, $[R_2Si{=}SiR_2]$ (R = $SiMe\textit{t}Bu_2$), or one-electron reduction is easily achieved. Thus, the ion-separated disilene radical cation $[R_2Si{=}SiR_2]^{\bullet+}$ (**24**)[35] and the corresponding radical anion $[R_2Si{=}SiR_2]^{\bullet-}$ (**25**)[36] are formed by reactions of the neutral disilene with $[Ph_3C]^+[(C_6F_5)_4B]^-$ or *t*BuLi in THF, respectively (Scheme 10.10). In a similar vein, the tin-containing radical anion, $[R_2Sn{=}SnR_2]^{\bullet-}$ (**26**) (R = $SiMe\textit{t}Bu_2$), is produced from the neutral precursor by one-electron reduction with potassium in the presence of [2.2.2]cryptand (Scheme 10.10).[37]

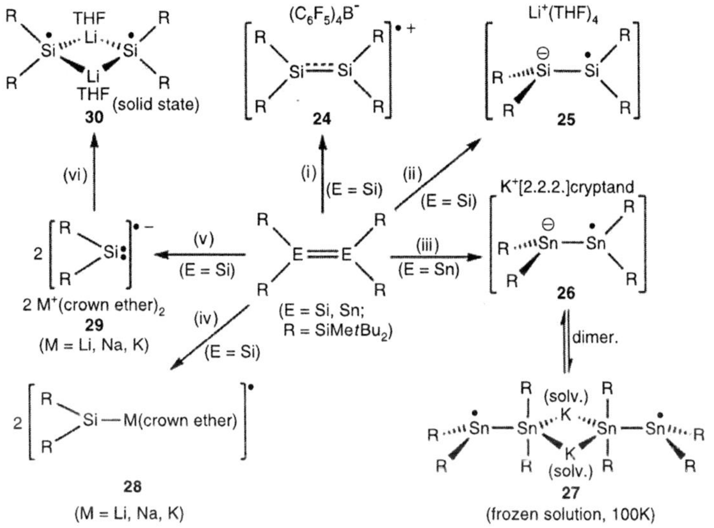

Scheme 10.10 Utilization of the neutral disilene and distannene, $R_2E{=}ER_2$ (E = Si, Sn; R = SiMe*t*Bu$_2$), in radical formation. (i) $[Ph_3C]^+[(C_6F_5)_4B]^-$ [35]; (ii) *t*BuLi[36]; (iii) K/[2.2.2]cryptand[37]; (iv) 2.2 equivalents M/Naphthalenide (M = Li, Na, K), 2 equivalents crown ether[38]; (v) 2.2 equivalents M/Naphthalenide (M = Li, Na, K), 4 equivalents crown ether; (vi) 10 equivalents LiBr, THF (M = Li)[38]

Crystal structure determinations of the radical anions **25** and **26** revealed one trigonal planar and one pyramidal metal center in both cases, whereas the paramagnetic cation **24** exhibits near planar geometry at both central silicon atoms. The E–E bond length in **25** (E = Si) and **26** (E = Sn) is elongated by about 0.08 (2 %) and 0.23 (9 %) Å, respectively, from that observed in the neutral parent molecule, while the Si–Si bond in the radical cation **24** is intermediate in length between the distances seen in the anion **25** and the neutral precursor. These structural features suggest that the radical anions **25** and **26** consist of an sp^3-hybridized silyl/stannyl anion and an sp^2-hybridized silyl/stannyl radical center, while the unpaired electron in the cation **24** is delocalized over the central Si–Si bond (Scheme 10.10). In solution, however, the EPR spectra of the paramagnetic disilene cation and anion, **24** and **25**, display virtually identical hfc constants at room temperature ($a(^{29}Si) = 23$ and 24.5 G; $g = 2.0049$ and 2.0061, respectively) that are approximately half of that observed for the structurally similar neutral silyl radical, $(tBu_2MeSi)_3Si^\bullet$ (**20**, $a(^{29}Si) = 58$ G), thus indicating delocalization of the unpaired electron in both of the charged disilene radicals. At low temperature (120 K), the delocalization in **25** is suppressed, as evinced by the increase in the hfc constant to 45 G. The tin analogue **26**, on the other hand, exhibits a central signal ($g = 2.0517$) with two pairs of satellites ($a(^{117/119}Sn_\alpha) = 340$ G, $a(^{117/119}Sn_\beta) = 187$ G, cf. 329 G for **22**) in the solution EPR spectrum at room temperature, consistent with the localization of the single electron on one of the tin centers as was also inferred from the solid state structure. It was suggested that the distannene radical anion **26** aggregates to a paramagnetic triplet state dimer **27** (Scheme 10.10) under glass matrix conditions (100 K) in 2-methyl-THF based on the observation of a signal corresponding to the forbidden $\Delta M_s = 2$ transition in the half-field region of the spectrum (1631 G).[37b]

While the mild, one-electron reduction of the disilene $R_2Si{=}SiR_2$ (R = SiMe*t*Bu$_2$) affords the corresponding radical anion **25**, the use of a stronger reducing agent, M/Naphthalenide (M = Li, Na, K), results in cleavage of the central Si=Si bond and formation of "monomeric" radicals, either as contact ion pairs $[R_2SiM(\text{crown-ether})]^\bullet$ (**28**) or as ion-separated salts $[M(\text{solv})_n]^+[R_2Si]^{\bullet-}$ (**29**) (M = Li, Na, K;

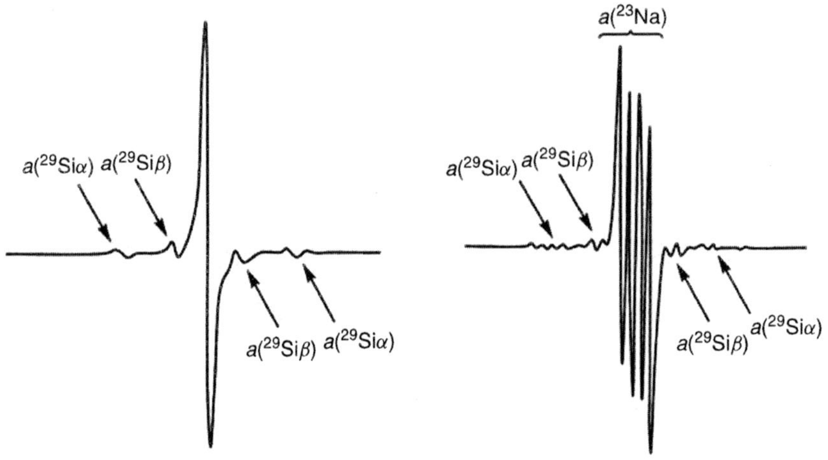

Figure 10.5 *EPR spectrum of (a) solvent-separated ion pair* $[R_2Si]^{\bullet-}[Na^+(THF)_4]$ ***(29)*** *in DME* ($a(^{29}Si_\alpha) = 29.1$ G *and* $a(^{29}Si_\beta) = 10.2$ G), *and (b) contact ion pair* $[R_2SiNa(15\text{-crown-}5)]^\bullet$ ***(28)*** *in toluene* ($a(^{29}Si_\alpha) = 29.1$ G, $a(^{29}Si_\beta) = 10.2$ G *and* $a(^{23}Na) = 1.9$ G); $g = 2.0074$, $R = SiMetBu_2$. *(Reprinted with permission from [38a]. Copyright 2007 American Chemical Society.)*

solv = crown ether (n = 2), THF, DME (n = 4); R = $SiMe\textit{t}Bu_2$), via a two-electron reduction process (Scheme 10.10).[38] The possibility of obtaining neutral radicals **28** or anionic paramagnetic species **29** (the anion radical of a silylene) was first noted when the EPR spectra of $[R_2Si]^\bullet Na(15\text{-crown-}5)_n$ were recorded both in polar (THF, DME) and non-polar (toluene) solvents.[38a] It was apparent from the EPR spectra (Figure 10.5) that in polar solvents the paramagnetic species exists as an ion-separated salt **29** with $a(^{29}Si_\alpha) = 29.1$ G and $a(^{29}Si_\beta) = 10.2$ G, whereas in the non-polar solvent the coupling to the Na^+ cation was also observable for the contact ion pair **28** ($a(^{23}Na) = 1.9$ G; ^{23}Na, $I = 3/2$, 100 %). Subsequently, the controlled formation of **28** and **29**, both in solid state and in solution, was achieved by using the appropriate amount of crown ether according to the reactions (iv) and (v) in Scheme 10.10.[38]

The solvent-separated ion pair **29** (M = Li, crown ether = 12-crown-4, Scheme 10) is converted to the contact ion pair **28** by removal of the crown ether with lithium bromide. The product exists as a monomer in solution, as shown by the EPR signal with hfc constants of $a(^{29}Si_\alpha) = 34.0$ G, $a(^{29}Si_\beta) = 9.7$ G and $a(^7Li) = 1.6$ G (^7Li, $I = 3/2$, 92.5 %), but dimerizes in the solid state to the diradical $[(\mu\text{-Li}\cdot\text{THF})_2(SiR_2)_2]^{\bullet\bullet}$ (**30**) (R = $SiMe\textit{t}Bu_2$) (Scheme 10.10), as proven by the X-ray crystal structure.[38] The diradical **30** was produced earlier by UV irradiation of the frozen hexane solution (150 K) of the *gem*-dilithiosilane, $[(R_2SiLi_2)(R_2HSiLi)_2]$ (R = $SiMe\textit{t}Bu_2$), and its triplet state nature was determined from the EPR spectrum, which consists of a signal with four ^{29}Si sidebands arising from $\Delta M_s = 1$ transitions and a signal at half-field (1675 G) resulting from the forbidden $\Delta M_s = 2$ transition.[39]

In contrast to linear alkynes RC≡CR, the heavier group 14 congeners, RE≡ER (E = Si, Ge, Sn), assume a *trans*-bent geometry with respect to the triple bond.[40,41] Similarly to the heavy alkene analogs $R_2E{=}ER_2$ (E = Si, Sn) described above, this results in a relatively low energy level for the LUMO in the heavy alkyne compounds. By exploiting the resulting lowered potential for reduction, the disilyne radical anion, $[RSi{\equiv}SiR]^{\bullet-}$ (R = $Si(\textit{i}Pr)[CH(SiMe_3)_2]_2$) (**31**), is obtained directly from the neutral precursor by one-electron reduction with KC_8 (Scheme 10.11).[40] The analogous germylene and stannylene radical anions, $[RE{\equiv}ER]^{\bullet-}$ (M = Ge, R = 2,6-$Tripp_2C_6H_3$ (**32a**) or 2,6-$Dipp_2C_6H_3$ (**32b**); E = Sn, R = 2,6-$Tripp_2C_6H_3$

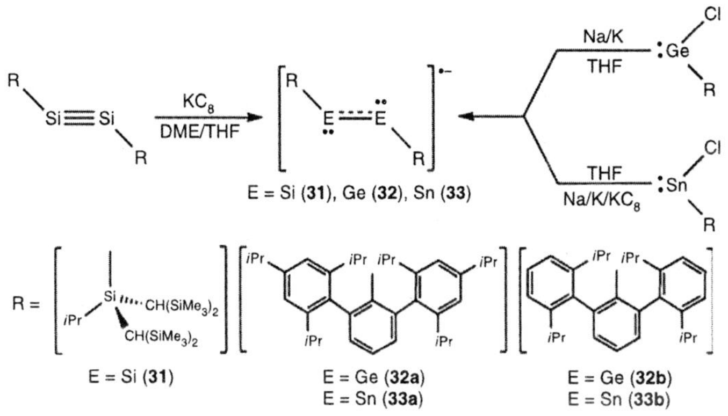

Scheme 10.11

(**33a**) or 2,6-$Dipp_2C_6H_3$ (**33b**); Tripp = 2,4,6-$i$$Pr_3C_6H_2$, Dipp = 2,6-$i$$Pr_2C_6H_3$), however, are produced by the reduction of RECl (E = Ge, Sn) (Scheme 10.11).[41]

The crystal structures of the ion-separated, paramagnetic species **31–33** show retention of the *trans*-geometry around the central RE–ER bond.[40,41] Most significantly, the central E–E bond is markedly elongated in the radical anions (by about 0.11, 0.03 and 0.13 Å in **31, 32b**, and **33a**, respectively) compared to the neutral molecules, even to the extent of approaching single bond values. This extensive lengthening of the E–E bond can be explained by a structure in which both group 14 centers bear a lone pair and the unpaired electron is delocalized over the E–E p_π orbital, thus giving a formal bond order of 1.5 (Scheme 10.11). In addition, narrowing of the E–E–R angles compared to the neutral, triply bonded species (by about 21, 25 and 31° in **31, 32b** and **33a**, respectively) contributes to a further elongation of the E–E distance.[41b]

The EPR spectrum of **31** shows a triplet arising from coupling to the α hydrogen of the *iso*-propyl groups ($a(^1H) = 2.3\,G$). This triplet is accompanied by two pairs of satellites with $a(^{29}Si_\alpha) = 39.2\,G$ and $a(^{29}Si_\beta) = 22.4\,G$. The magnitude of the hfc constants indicates the presence of a π radical and extensive delocalization of the odd electron over the Si–Si bond and onto the substituents.[40] Consistently, relatively small hfc constants of $a(^{73}Ge) = 7.5\,G$, $a(^{117}Sn) = 8.3\,G$ and $a(^{119}Sn) = 8.5\,G$ were obtained for **32b** and **33a**, respectively,[41] indicative of low unpaired electron density at the germanium and tin centers, and localization of the odd electron in an orbital of π symmetry.

10.4 Group 15 element radicals

Stable radicals involving group 15 elements are restricted primarily to phosphorus-containing systems. A review that emphasizes the diversity of both stable and persistent phosphorus radicals covers the literature up to 2004.[42]

10.4.1 Phosphorus

10.4.1.1 Neutral radicals

The structurally simplest phosphorus-centered radicals are the two-coordinate neutral phosphinyl radicals, $[R_2P]^\bullet$ (**34**). Numerous derivatives of this paramagnetic species have been produced since its first discovery 40 years ago,[43] most frequently by photolysis of the appropriate chlorophosphine in the presence of

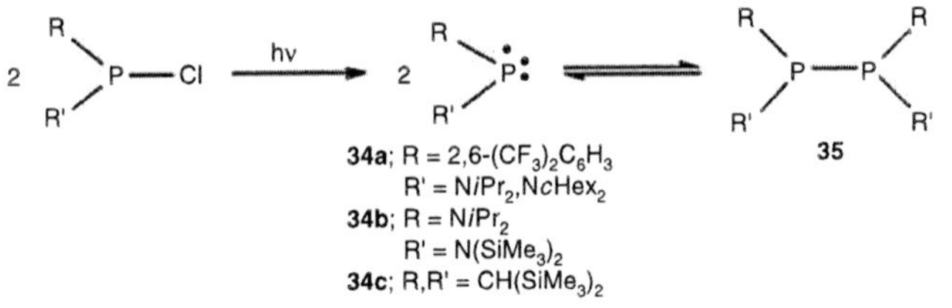

Scheme 10.12

an electron-rich alkene (Scheme 10.12).[44] The monomeric radicals **34** dimerize to the corresponding diamagnetic diphosphines, $[R_2P]_2$ (**35**), both in the solid state and upon cooling the solutions of these odd-electron species.[45,46] Remarkable stability has been achieved, however, for example with R = $CH(SiMe_3)_2$ substituents (**34c**), resulting in derivatives that are indefinitely stable both in solution and in the gas phase.[47]

The gas phase electron diffraction structures of **34b** (R/R′ = $N(SiMe_3)_2/N\mathit{i}Pr_2$)[46] and **34c** (R = R′ = $CH(SiMe_3)_2$)[47] reveal a V-shaped arrangement (∠NPN = 99.0° and ∠CPC = 104.0°, respectively) around the central phosphorus atom. Most significantly, the structure of the extremely stable radical **34c** shows a *syn,syn* conformation for the $CH(SiMe_3)_2$ groups, in contrast to the *syn,anti* arrangement observed for the dimeric congener **35c**. Based on DFT calculations, the disparity in the conformations of **34c** and **35c** is a significant contributor to the remarkable stability of the monomeric paramagnetic species **34c** in solution, that is the transformation between **34c** and **35c** not only involves P–P bond cleavage/formation, but also an isomerization of the substituents. Calculations predict an endothermic process of 95 kJ mol^{-1} for the P–P bond cleavage upon going from **35c** to **34c**, while the subsequent isomerization of the paramagnetic species **34c**, to give the *syn,syn* conformation, releases an energy of 67.5 kJ mol^{-1} for each of the monomeric units, thus giving an overall exothermic process by about 40 kJ mol^{-1} for the conversion of **35c** to **34c**.[47] Therefore, the absence of dimerization and the stability of **34c** in solution is attributed not to the kinetic stability imparted by the bulky ligands, but to the energy required for the conformational change prior to the dimerization.

A phosphinyl radical $[R_2P]^\bullet$ (R = NV[N(Np)Ar]$_3$; Np = neopentyl, Ar = 3,5-$Me_2C_6H_3$) (**34d**) that is stable both in solution and in the solid state is produced by one-electron reduction of the chlorine-containing precursor.[48] Similarly to **34b** and **34c**, the solid state structure of **34d** shows a V-shaped phosphorus center (∠NPN = 110.9°). The stability of the radical **34d** is achieved by delocalization of the unpaired electron onto the vanadium centers, which form a redox couple (IV/V) (Scheme 10.13). The delocalization is also evident from the EPR spectrum of **34d**. Whereas the "true" phosphinyl radicals **34a–c** show hfc constants of 75.9–96.3 G to the central phosphorus atom (^{31}P, I = ½, 100 %) with significantly smaller coupling to the substituents, the hfcc value of 42.5 G to the phosphorus atom in **34d** is substantially smaller and the coupling of $a(^{51}V)$ = 23.8 G to the two transition metal centers indicates extensive delocalization of

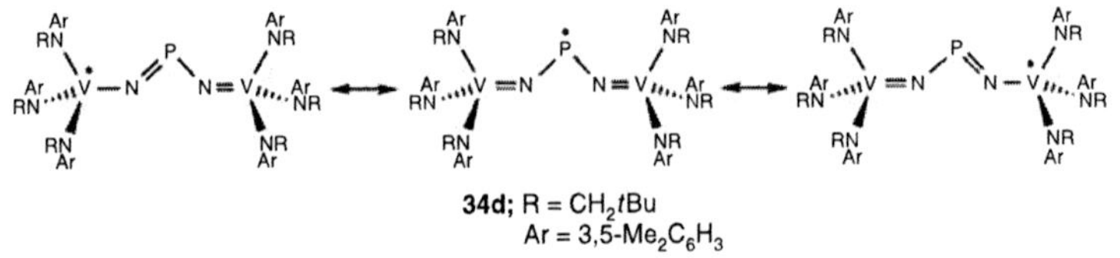

Scheme 10.13 Resonance-stabilized radical **34d** by the vanadium (IV/V) redox couple

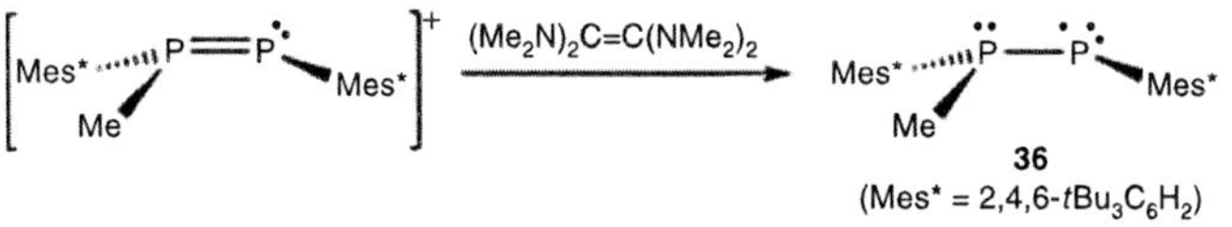

Scheme 10.14

the unpaired electron. The hfcc values of the paramagnetic phosphinyls **34a–c** are indicative of π radicals in which the odd electron is located predominantly in the phosphorus 3p orbital (**34a**: $a(^{31}P) = 87.0$ G, $a(^{14}N) = 5.56$ G, $6 \times a(^{19}F) = 10.1$ G, $2 \times a(^{1}H) = 1.0$ G;[45] **34b**: $a(^{31}P) = 75.9$ G, $a(^{14}N) = 5.95$ G;[44] **34c**: $a(^{31}P) = 96.3$ G, $2 \times a(^{1}H) = 6.40$G:[44] **34d**: $a(^{31}P) = 42.5$ G, $2 \times a(^{51}V) = 23.8$ G [48]).

The persistent diphosphanyl radical, [Mes*MePPMes*]$^{\bullet}$ (**36**), a close relative of the phosphinyl radicals **34**, was first observed in cyclic voltammetric studies of the diphosphene salt [Mes*MeP=PMes*][OSO_2CF_3].[49a] Subsequently, **36** was produced by reduction of the corresponding cation with tetrakis(dimethylamino)ethene (Scheme 10.14). In contrast to the phosphinyl radicals **34a–c**, the paramagnetic species **36** can be isolated as a solid, although magnetic measurements indicate a partial dimerization (10 %) during the liquid to solid transformation; no crystal structure has been reported.[49] Slow decomposition of **36** occurs in the solid state while in solution the half-life is about 90 minutes.

The EPR spectrum of **36** shows a four-line signal consistent with coupling to two inequivalent phosphorus atoms ($a(^{31}P) = 139.3$ and 89.3 G), thus confirming the identity of **36** as a diphosphanyl radical.[49] The larger of the two hfc constants is attributed to the two-coordinate phosphorus center, which is expected to carry the major part of the spin density, and the smaller coupling is due to the three-coordinate phosphorus. Consistently, DFT calculations disclose a small s contribution to the SOMO (10 %), while confirming that most of the spin density resides in a p orbital of the two-coordinate (74 %) and three-coordinate (15 %) phosphorus atoms. Since the calculations predict minimal delocalization of the unpaired electron, the stability of **36** is attributed mainly to the steric protection provided by the Mes* substituents.

Several short-lived 1,3-diphosphaallyl radicals have been suggested either as reaction intermediates or on the basis of EPR spin-trapping experiments.[50] However, the introduction of amino groups has a powerful stabilizing effect, and the stable diphosphaallyl radical [($i$$Pr_2NP)_2C(NiPr_2)$]$^{\bullet}$ (**37**) (Scheme 10.15) is produced from the cyclic cation [($i$$Pr_2N)_2P(\mu_3$-P)C(N$i$$Pr_2$)]$^+$ either by electrolysis or by reduction with lithium metal.[51] The red crystals of **37** were not suitable for X-ray analysis, but the FAB mass spectrum shows the expected molecular ion. The EPR spectrum of **37** in THF exhibits a significantly broadened five-line signal ($g = 2.0048$) with an intensity ratio of 1 : 3 : 4 : 3 : 1 that remains unchanged in the temperature range −60 to +25 °C. The spectrum can be interpreted by invoking hyperfine coupling to two equivalent phosphorus nuclei ($a(^{31}P) = 9.4$ G), two equivalent nitrogens ($a(^{14}N) = 1.5$ G) and a unique nitrogen atom ($a(^{14}N) = 9.9$ G), thus indicating significant stabilization of the radical by delocalization of the unpaired electron over a total of five atoms.

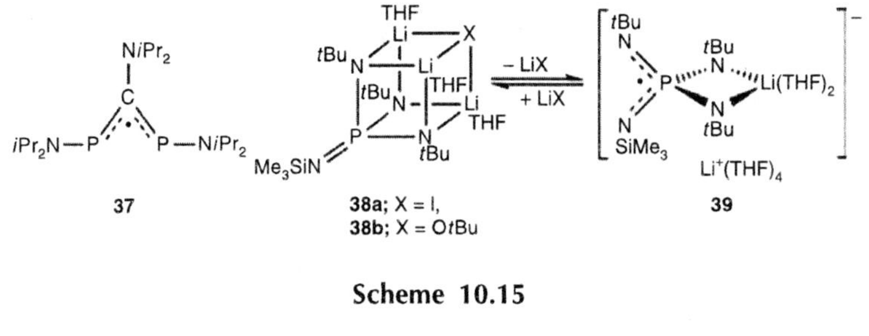

Scheme 10.15

The phosphorus(V)-containing dianion radical $[P(NtBu)_3(NSiMe_3)]^{2-\bullet}$ is generated, as the dilithium derivative, by one-electron oxidation of the trilithium salt of the corresponding tetrakisimidophosphate trianion $[P(NtBu)_3(NSiMe_3)]^{3-}$ (a tetraimido analogue of orthophosphate PO_4^{3-}).[52] The formally neutral radical $\{Li_2[P(NtBu)_3(NSiMe_3)]\}^\bullet$ exists as deep blue crystals that adopt highly distorted PN_3Li_3X cubic structures (**38a**, X = I; **38b**, X = O*t*Bu), in which an entrapped LiX molecule provides one edge of the cube (Scheme 10.15). In an extremely dilute THF solution, solvation of the Li^+ counterions causes cleavage of the cubic structure to give the solvent-separated ion pair **39** and a molecule of $(THF)_3LiI$. The EPR spectrum of **39** displays a complicated, approximately fifty-line signal ($g = 2.0063$) that is best simulated with hyperfine couplings to one phosphorus atom ($a(^{31}P) = 23.1$ G), two equivalent nitrogen centers ($a(^{14}N) = 5.38$ G), two unique nitrogen atoms ($a(^{14}N) = 7.38$ and 1.93 G), and a single lithium nucleus ($a(^{7}Li) = 0.30$ G). The relatively small ^{31}P hfc constant in **39**, compared to those in **34** and **36**, indicates a small amount of spin density located at the phosphorus center. This conclusion is supported by DFT calculations that predict a SOMO consisting predominantly of nitrogen-based 2p orbitals for **39**.[52]

The 1,3-diphosphacyclobuten-4-yl radical, $[(Mes^*C)_2(PtBu)P]^\bullet$ (**40**, Mes* = 2,4,6-$tBu_3C_6H_2$), is isolated as deep red crystals by treatment of the phosphaalkyne Mes*C≡P with *t*BuLi followed by one-electron oxidation with iodine (Scheme 10.16).[53] The structure of **40** shows a nearly planar four-membered ring with trigonal planar carbon atoms, and both pyramidal and two-coordinate phosphorus atoms. The bond parameters suggest a pure single bond for the P–C bonds involving the three-coordinate phosphorus atom and partial double bond character for the other two P–C bonds. The room temperature EPR spectrum of **40** in toluene exhibits a four-line central signal with small satellites ($g = 2.0025$), from which hfc constants of $a(^{31}P_\alpha) = 20.4$ G, $a(^{31}P_\beta) = 10.2$ G and $a(^{13}C) = 30.2$ G were obtained. The relatively low ^{31}P hfcc values compared to those observed in phosphinyl and diphosphanyl radicals (**34, 36**), as well as the metrical parameters, support an allylic radical in which the unpaired electron is delocalized mainly in the CPC unit as depicted in Scheme 10.16 (cf. cyclotetrasilenyl **15**, Scheme 10.7). *Ab initio* calculations, however, predict that 80–90 % of the spin density resides on the two carbon atoms.

10.4.1.2 Biradicaloids

The treatment of the cyclic anion formed from Mes*C≡P and *t*BuLi with methyl iodide, instead of iodine, provides a novel route to the carbon-centered biradicaloid $[(Mes^*C)_2(PtBu)(PMe)]^{\bullet\bullet}$ (**41**) with a four-membered P_2C_2 ring (Scheme 10.16).[54] A number of different derivatives of **41**, $[(RC)(R'C)(PMes^*)_2]^{\bullet\bullet}$

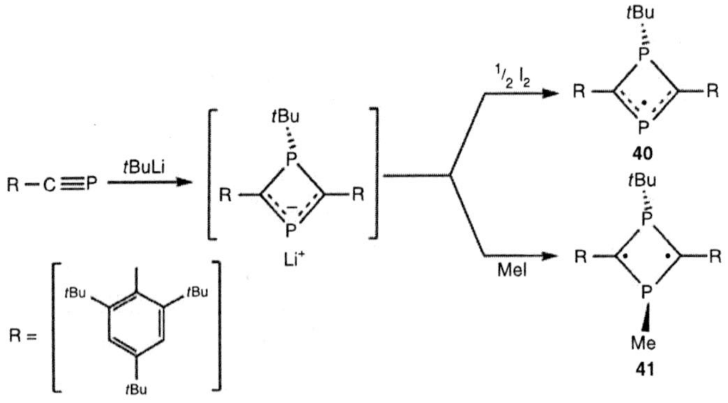

Scheme 10.16

(R, R′ = Cl; R = $SiMe_3$, R′ = H, $AlMe_3$), were originally reported by Niecke *et al.*,[55] and they all show similar structural features; a nearly planar four-membered ring with trigonal planar carbon centers and various degrees of pyramidalization at the phosphorus atoms. The transannular C–C distance is sufficiently long for these compounds to be described as diradicals. A singlet ground state has been predicted by *ab initio* calculations for derivatives of **41** owing to the conjugative interaction between the non-bonding electron pairs on phosphorus and the unpaired electrons at the carbon atoms.[55a] Consistently, no EPR signal is observed. However, the energy difference between the singlet and triplet states is small and substituent effects may play a key role in determining relative stabilities; a silyl group at the carbon atoms stabilizes the singlet state whereas σ-attracting influences at the phosphorus favor the triplet ground state.[55c]

10.4.1.3 Charged radicals (anions and cations)

While a direct reaction between the Grignard reagent ArMgBr and phosphorus trichloride results in multiple products with P–P bonds, sterically protected triarylphosphines Ar_3P are produced by converting the Grignard reagent into $(ArCu)_x$, and reacting the resulting arylcopper reagent *in situ* with phosphorus trichloride. Subsequent one-electron oxidation, either by bulk electrolysis or with silver(I) salts AgX (X = $[ClO_4]^-$, $[PF_6]^-$, $[SbF_6]^-$), produces persistent radical cations, $[Ar_3P]^{\bullet+}$ (Ar = Dipp, Tripp; Dipp = 2,6-$i$$Pr_2C_6H_3$; Tripp = 2,4,6-$i$$Pr_3C_6H_2$) (**42**) (Scheme 10.17).[56] The solution EPR spectra show a doublet with $a(^{31}P) = 239$ (**42a**) and 237 G (**42b**), thus indicating the expected pyramidal, phosphorus-centered radicals. Frozen glass EPR experiments imply a significant flattening of the molecule (small $a_\perp$ component of the hfc tensor) that brings the *iso*-propyl substituents in *ortho* positions closer together, thus providing more shielding for the unpaired electron and, therefore, contributing to the stability of the radical cations **42**.[56b]

The formation of the persistent diphosphene radical anions, $[RPPR]^{\bullet-}$ (**43**), is observed in solution when the "dimeric" contact ion pairs, $[M^+]_2[R_4P_4]^{2-}$ (M = Na, R = phenyl; M = K, R = cyclohexyl), are treated with [2.2.2]cryptand. By contrast, in the solid state, the radical anions dimerize to re-form the ion-separated salts $[R_4P_4]$ $[(M^+)_2\{[2.2.2]cryptand\}]_2$, as established by X-ray crystallography (Scheme 10.18).[57] The EPR spectra of **43** exhibit a triplet arising from the coupling of the unpaired electron to two equivalent phosphorus atoms ($a(^{31}P) = 40.9$ (**43a**) and 45.2 (**43b**) G; $g = 2.0089$ and 2.0099, respectively) with additional splitting owing to spin density distribution onto the substituents ($a(^1H) = 3.0$ and 1.4 G (**43a**) and $a(^1H) = 3.6$ G (**43b**)).

Diphosphene radical anions that are kinetically stabilized both in solution and in the solid state, **43c** (R = Bbt = 2,4,6-$[CH(SiMe_3)_2]_3C_6H_2$) and **43d** (R = Tbt = 4-$[C(SiMe_3)_3]$-2,6-$[CH(SiMe_3)_2]_2C_6H_2$), are produced by one-electron reduction of the neutral diphosphene obtained by magnesium reduction of $RPCl_2$ (Scheme 10.18).[58] Analogously to **43a–b**, the EPR spectra of **43c** and **43d** show a triplet with $a(^{31}P) = 48.0$ and 47.6 G ($g = 2.009$ and 2.010), respectively, indicating uniform delocalization of the

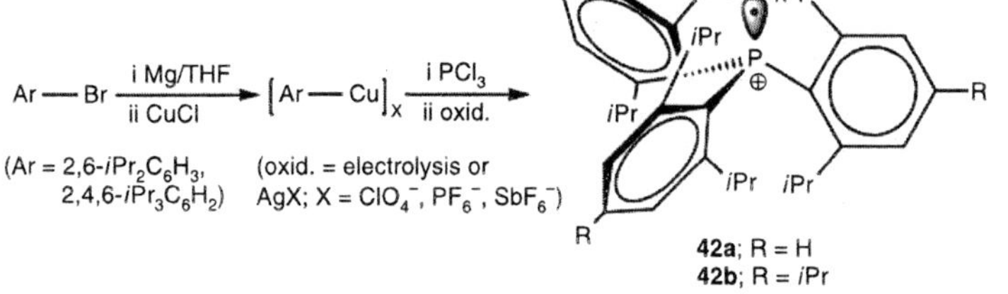

42a; R = H
42b; R = *i*Pr

Scheme 10.17

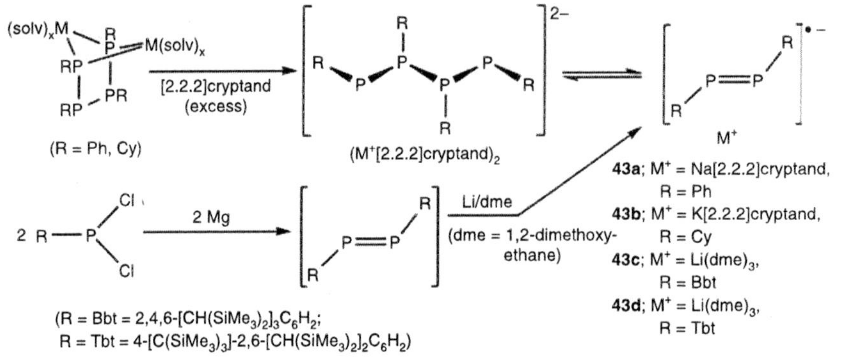

Scheme 10.18

spin density on both phosphorus centers. However, in contrast to **43a–b**, further delocalization onto the substituents is not observed. The Raman spectra of **43c–d** suggest significant elongation of the P–P bond compared to that in the neutral precursors by displaying a strong Raman line at about 540 cm^{-1} for the ν_{P-P} stretching vibration, cf. about 610 cm^{-1} for the P=P double bond in RP=PR (R = Bbt, Tbt) and 530 cm^{-1} for the P–P single bond in Ph_2P–PPh_2. Elongation and, hence, weakening of the P–P bond in the radical anions **43** is consistent with DFT calculations that indicate a SOMO composed of an antibonding P–P π^* orbital.[58b,58c]

10.4.2 Arsenic, antimony, and bismuth

The chemistry of stable, arsenic-, antimony-, and bismuth-centered radicals has remained virtually unexplored. The EPR spectra of the persistent arsenic and antimony congeners of the triarylphosphine radical cation, $[Ar_3P]^{\bullet+}$ (**42b**) (Scheme 10.17), indicate the expected pyramidal structure, but no further discussion of these species was presented.[56a]

An antimony-containing congener of the diphosphene radical anion **43c**, $[BbtSbSbBbt]^{\bullet-}$ (**44**), is obtained from a reaction analogous to that shown for the synthesis of **43c** in Scheme 10.18.[58b] The crystal structure of **44** reveals an elongation of about 0.05 Å in the Sb–Sb bond length (2.7511(4) Å) compared to the value of 2.7037(6) Å in the distibene [BbtSb=SbBbt], consistent with a SOMO comprised of an E–E π^* orbital as predicted for both **43** and **44** (E = P, Sb) by DFT calculations. Only a broad signal is observed in the EPR spectrum of **44** in solution ($g = 2.097$), while in the solid state a complicated spectrum is obtained due to coupling of the unpaired electron with two equivalent antimony centers each with two EPR-active isotopes (^{121}Sb, $I = 5/2$, 57.25 %; ^{123}Sb, $I = 7/2$, 42.75 %).[58b]

10.5 Group 16 element radicals

10.5.1 Sulfur

The majority of the advances in the chemistry of stable, heavy group 16 radicals in recent years are based primarily on dithiadiazolyls and related heterocyclic CNS/Se radicals (Chapters 5 and 14). Apart from these intriguing compounds, which show potential for the construction of novel one-dimensional metals or materials with unique magnetic properties, examples of odd-electron group 16 species within the scope of this chapter are sparse.

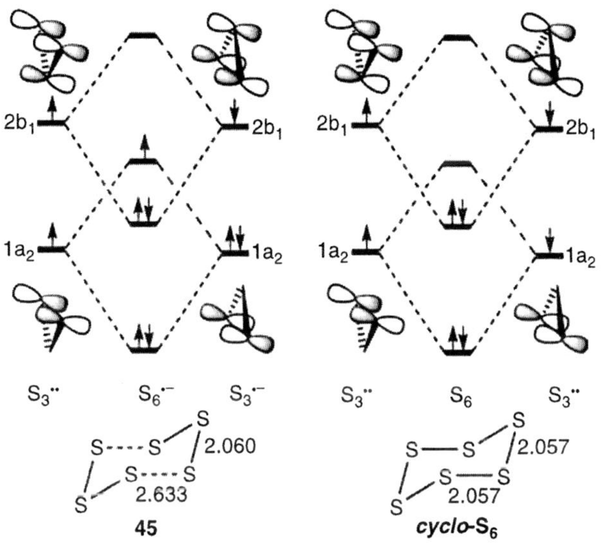

*Figure 10.6 Orbital interaction diagram for the formation of the cyclic radical anion $[S_6]^{\bullet-}$ (**45**) and* cyclo-S_6 *from two S_3 units. (Reproduced with permission from [59]. Copyright Wiley-VCH Verlag GmbH & Co. KGaA.)*

The stable cyclic radical anion $[S_6]^{\bullet-}$ (**45**) is formed unexpectedly as the $[Ph_4P]^+$ salt from the reaction between $[Ph_4P]N_3$, Me_3SiN_3 and hydrogen sulfide (H_2S).[59] The crystal structure displays the six-membered $[S_6]^{\bullet-}$ ring in a chair conformation with two long central S–S bonds (2.633 Å) (Figure 10.6), thus indicating a relatively weak union of two S_3 fragments in each of which the average S–S bond length is 2.060 Å (cf. 2.057 Å in *cyclo*-S_6). The solid state EPR spectrum of **45** suggests a formulation in which $[S_6]^{\bullet-}$ is comprised of the diradical $[S_3]^{\bullet\bullet}$ and the anion radical $[S_3]^{\bullet-}$. The eigenvalues for the g tensor, $g_1 = 2.056$, $g_2 = 2.036$ and $g_3 = 2.003$, are consistent with those previously observed for the $[S_3]^{\bullet-}$ radical anion, the source of the blue color in ultramarine (Lapis Lazuli).[60] Both of these experimental observations are in agreement with the molecular orbital analysis that reveals a bonding interaction between the two S_3 fragments in **45** involving two components (Figure 10.6): (i) an electron pair bond between the $2b_1$ SOMOs of both fragments and (ii) a three-electron bond between the $1a_2$ SOMO of the biradical $[S_3]^{\bullet\bullet}$ and the $1a_2$ HOMO of the $[S_3]^{\bullet-}$ radical anion. Neutral S_3 has a closed shell ground state, but becomes a biradical after valence excitation of an electron from $1a_2$ to $2b_1$. As a comparison, the three-electron bond in **45** is converted into a more stabilizing electron pair bond in neutral *cyclo*-S_6 by the loss of the antibonding electron, thereby accounting for the disparity in the central S–S bond lengths of these two cyclic species.

10.5.2 Selenium and tellurium

The simplest organochalcogeno radicals $[RE]^{\bullet}$ (E = sulfur, selenium, tellurium; R = alkyl, aryl) have not been isolated as stable species, presumably due to the lack of steric protection from the substituent. However, the existence of the organic tellurium-centered radicals $[RTe]^{\bullet}$ (R = *n*Butyl, phenyl) as intermediates in the hydrotelluration reactions of alkenes and alkynes, has been ascertained by spin trapping experiments.[61] The EPR spectra of the radical adducts so formed, $[(RTe)(2,6\text{-}Br_2\text{-}1\text{-}NO\text{-}3\text{-}SO_3\text{-}C_6H_2)]^{\bullet-}$ (**46**; R = *n*Butyl, phenyl) (Scheme 10.19), with natural abundance tellurium (^{123}Te, $I = 1/2$, 0.9 %; ^{125}Te, $I = 1/2$, 7.0 %), show coupling only to the nitrogen atom and two hydrogens of the trapping agent

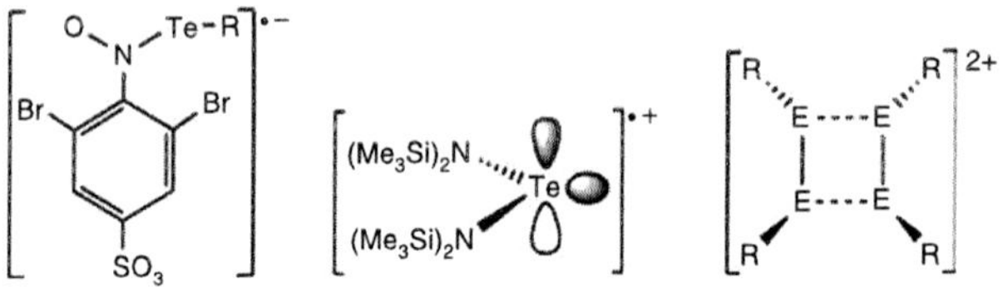

46
(R = *n*Bu, Ph)

47

48
(E = Se; R = Me, Et
E = Te; R = Et, Pr, *i*Bu)

Scheme 10.19

($a(^{14}N) = 21.6\,G$; $a(^{1}H) = 0.7\,G$). When isotope labeling is employed (92 % ^{125}Te), a complicated spectrum is observed from which hfc constants of $a(^{14}N) = 21.6\,G$, $a(^{125}Te) = 15.8\,G$, $a(^{79/81}Br) = 6.95\,G$ and $a(^{1}H) = 0.7\,G$ are obtained, thus indicating a radical with significant spin density at the tellurium center.

The first stable tellurium(III) radical cation, $[(Me_3Si)_2N]_2Te^{\bullet+}$ (**47**) (Scheme 10.19), was synthesized from the neutral precursor by one-electron oxidation with $Ag[AsF_6]$.[62] The crystal structure of this seminal paramagnetic species displayed a slight shortening of the Te–N bonds and elongation of the Si–N distances compared to those of the neutral precursor. Together with the observed changes in bond angles, these metrical parameters suggest enhanced back-bonding of the nitrogen lone pairs to the positively charged tellurium center. Only a single, broad signal ($\Delta\omega_{1/2} = 15\,G$) was observed in the EPR spectrum of **47** indicative of a small spin density on the nitrogen atoms. These data are consistent with a tellurium-centered radical with a pair of electrons in an sp^2 orbital and the odd electron located in a p orbital (Scheme 10.19). Stabilization of **47** (towards dimerization) is attributed to a combination of steric protection from the substituents and electrostatic repulsion between the positively charged tellurium centers.

One-electron oxidation of dialkyl dichalcogenides, RE–ER, with $[NO]^+[OSO_2CF_3]^-$ produces the diamagnetic dications, $[(REER)_2]^{2+}$ (E = Se, R = ethyl, Pr, *i*Bu; E = Te, R = methyl, ethyl) (**48**) (Scheme 10.19), rather than the monomeric radical cations, $[RE{=}ER]^{\bullet+}$, presumably due to lack of sufficient steric protection from the substituents that are present, for example, in **47**.[63] The chalcogen–chalcogen bond lengths in **48** are about 0.05 Å shorter than those in the neutral precursors, while the distance between the two halves of the rectangular dication is 0.6–0.7 Å longer owing to a combination of the relatively weak π^*–π^* interactions between two monomeric $[RE{=}ER]^{\bullet+}$ units and electrostatic repulsion between these two positively charged species (cf. formation of the I_4^{2+} cation[64]). DFT calculations predict that the dimerization is exothermic when solvent effects are taken into account and, consistently, no EPR signal has been observed for either the selenium or tellurium derivatives of the type **48**.[63]

10.6 Group 17 element radicals

The most common group 17 element-centered radicals are the homodiatomic dications $[E_2]^{\bullet+}$ (E = bromine, iodine) which, as has been shown for the rectangular iodine derivative, dimerize in the solid state through π^*–π^* interactions to give the diamagnetic dications $[(E_2)_2]^{2+}$.[64] For the bromine derivative only a broad, non-characteristic EPR signal has been observed, while the iodine congener shows no signal.[65]

The analogous chlorine radical cation $[Cl_2]^{\bullet+}$ has been isolated as the adduct $[Cl_2O_2]^{\bullet+}$ (**49**), which is formally a *side-on* π complex of $[Cl_2]^{\bullet+}$ with dioxygen (Figure 10.7a). The source of the intriguing paramagnetic species **49** is the reaction between chlorine (Cl_2) and an $[O_2]^+$ salt with either $[SbF_6]^-$ or $[Sb_2F_{11}]^-$ as the counter ion.[66] Similar to the dimer $[I_4]^{2+}$, the structure of **49** is a planar trapezium with

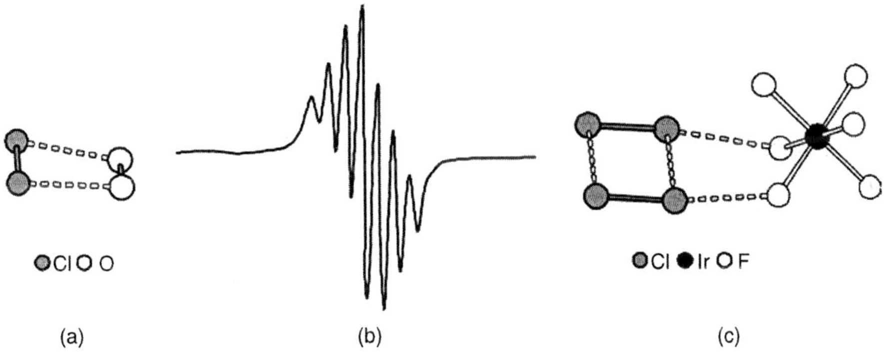

Figure 10.7 (a) $[Cl_2O_2]^{•+}$ (**49**; anion = $[Sb_2F_{11}]^-$), (b) EPR Spectrum of **49**, and (c) $[(Cl_2)_2]^+[IrF_6]^-$ (**50**·IrF_6^-). ((a) Reprinted with permission from [66]. Copyright 1999 American Chemical Society; (b) Reproduced with permission from [67]. Copyright Wiley-VCH Verlag GmbH & Co. KGaA.)

short bonds within the Cl_2 and O_2 units (1.909(1) and 1.207(5) Å, respectively), and a long Cl–O distance (about 2.41 Å) between these two units, cf. the sum of van der Waals radii for Cl and O = 3.30 Å. The Cl–Cl distance is about 0.07 Å shorter than that of Cl_2, thus indicating that most of the positive charge, and the unpaired electron, resides on the chlorine atoms. Consistently, the EPR spectrum of **49** (Figure 10.7b) shows a seven-line pattern ($g = 1.9988$) due to the coupling of the unpaired electron with two equivalent chlorine centers ($a(^{35/37}Cl) = 2.23$ G; ^{35}Cl, $I = 3/2$, 75.8 %; ^{37}Cl, $I = 3/2$, 24.2 %).[66]

Subsequent attempts to produce the monomeric radical cation $[Cl_2]^{•+}$ from the reaction between Cl_2 and IrF_6 also resulted in a dimeric product in the form of the rectangular radical cation $[(Cl_2)_2]^{•+}$ (**50**) (Figure 10.7c).[67] The structure of **50** displays two short Cl–Cl bonds and two long Cl···Cl contacts (1.941(3) and 2.936(7) Å, respectively). The short chlorine–chlorine bond lengths are intermediate between those observed for Cl_2 and $[Cl_2]^{•+}$, suggesting delocalization of the unpaired spin density over both units. The EPR spectrum of **50** as the $[IrF_6]^-$ salt shows only an unresolved broad signal.[67] However, the original EPR spectroscopic study of **50** supports the uniform delocalization of the unpaired electron over all four chlorine atoms.[68]

10.7 Summary and future prospects

In the past six to seven years a number of significant developments have occurred in the intriguing field of stable radicals, primarily through the strategy of installing extremely bulky groups on the heavy p-block element centers in order to provide kinetic stabilization of highly reactive species. For example, numerous examples of heavy group 14 analogs of classic carbon-centered radicals, as well as their isoelectronic anionic congeners from group 13, have been identified. A similar approach has also been successfully employed to generate a variety of unsaturated cyclic and acyclic (neutral and charged) radicals involving elements from groups 13–15. At the same time, several heterocyclic systems with biradicaloid character have been comprehensively characterized.

To date the primary focus has been on the refinement of synthetic approaches and the elucidation of both the molecular and electronic structures of these novel species. In the future, more attention will likely be accorded to practical applications. For example, the incorporation of these paramagnetic building blocks into oligomeric or polymeric systems with unique magnetic properties represents a major synthetic challenge. In addition, novel reactivity may be expected for both known and new acyclic and heterocyclic systems with potential biradicaloid character. For example, the unusual reactivity of the group 14 alkyne analogs $Ar'M{\equiv}MAr'$ [M = Ge, Sn; $Ar' = C_6H_3\text{-}2,6\text{-}(C_6H_3\text{-}2,6\text{-}i\text{Pr}_2)_2$] in the activation of dihydrogen (H_2) under ambient conditions[69] may be related to the singlet biradical character of the M–M bond.[70] The existence of an accessible, planar diradical of the four-membered Al_2P_2 ring in the derivative [$t\text{Bu}_2\text{PAl}(\mu\text{-P}t\text{Bu}_2)]_2$, indicated by a recent computational study,[71] awaits experimental confirmation. Finally, the suggestion that the weak transannular S–S bonds in certain sulfur–nitrogen ring systems, for example S_4N_4 and 1,5-$R_4P_2N_4S_2$,[72] is a potential source of biradicaloid character merits both experimental and computational investigation.

References

1. (a) P. P. Power, *Chem. Soc. Rev.*, **103**, 789 (2003); (b) D. Griller and K. U. Ingold, *Acc. Chem. Res.*, **9**, 13 (1976).
2. H. Grützmacher and F. Breher, *Angew. Chem. Int. Ed.*, **41**, 4006 (2002).
3. F. Breher, *Coord. Chem. Rev.*, **251**, 1007 (2007).
4. (a) E. Krause and H. Polack, *Ber. Dtsch. Chem. Ges.*, **59**, 777 (1926); (b) E. Krause and H. Polack, *Ber. Dtsch. Chem. Ges.*, **61**, 271 (1928); (c) E. Krause and P. Nobbe, *Ber. Dtsch. Chem. Ges.*, **63**, 634 (1930); (d) E. Krause and P. Nobbe, *Ber. Dtsch. Chem. Ges.*, **64**, 634 (1931).
5. J. E. Leffler, G. B. Watts, T. Tanigaki, *et al.*, *J. Am. Chem. Soc.*, **92**, 6825 (1970).
6. M. M. Olmstead and P. P. Power, *J. Am. Chem. Soc.*, **108**, 4235 (1986).
7. W. J. Grigsby and P. P. Power, *Chem. Commun.*, 2235 (1996).
8. W. J. Grigsby and P. P. Power, *Chem. Eur. J.*, **3**, 368 (1997).
9. (a) D. Scheschkewitz, H. Amii, H. Gornitzka, *et al.*, *Science*, **295**, 1880 (2002); (b) H. Amii, L. Vranicar, H. Gornitzka, *et al.*, *J. Am. Chem. Soc.*, **126**, 1344 (2004); (c) D. Scheschkewitz, H. Amii, H. Gornitzka, *et al.*, *Angew. Chem. Int. Ed.*, **43**, 585 (2004); (d) V. Gandon, J-B. Bourg, F. S. Tham, *et al.*, *Angew. Chem. Int. Ed.*, **47**, 155 (2008).
10. (a) A. Rodriguez, F. S. Tham, W. W. Schoeller and G. Bertrand, *Angew. Chem. Int. Ed.*, **43**, 4876 (2004); (b) A. Rodriguez, G. Fuks, J.-B. Bourg, *et al.*, *Dalton Trans.*, 4482 (2008).
11. N. Wiberg, T. Blank, W. Kaim, *et al.*, *Eur. J. Inorg. Chem.*, 1475 (2000).
12. (a) N. Wiberg, K. Amelunxen, H. Nöth, *et al.*, *Angew. Chem. Int. Ed. Engl.*, **36**, 1213 (1997); (b) N. Wiberg, T. Blank, K. Amelunxen, *et al.*, *Eur. J. Inorg. Chem.*, 1719 (2001).
13. P. Yang, R. Köppe, T. Duan, *et al.*, *Angew. Chem. Int. Ed.*, **46**, 3579, (2007).
14. (a) T. Chivers, D. J. Eisler, C. Fedorchuk, *et al.*, *Chem. Commun.*, 3930 (2005); (*b*) T. Chivers, D. J. Eisler, C. Fedorchuk, *et al.*, *Inorg. Chem.*, **45**, 2119 (2006).
15. (a) F. G. N. Cloke, G. R. Hanson, M. J. Henderson, *et al.*, *J. Chem. Soc., Chem. Commun.*, 1002 (1989); (b) F. G. N. Cloke, C. I. Dalby, M. J. Henderson, *et al.*, *J. Chem. Soc., Chem. Commun.*, 1394 (1990); (c) F. G. N. Cloke, C. I. Dalby, P. J. Daff and J. C. Green, *J. Chem. Soc., Dalton Trans.*, 181 (1991); (d) W. Kaim and W. Matheis, *J. Chem. Soc., Chem. Commun.*, 597 (1991).
16. W. W. Schoeller and S. Grigoleit, *J. Chem. Soc., Dalton Trans.*, 405 (2002).
17. M. Nakamoto, T. Yamasaki and A. Sekiguchi, *J. Am. Chem. Soc.*, **127**, 6954 (2005).
18. (a) A. Begum, A. R. Lyons and M. C. R. Symons, *J. Chem. Soc. A*, 2290 (1971); (b) J. R. M. Giles and B. P. Roberts, *J. Chem. Soc., Chem. Commun.*, 1167 (1981); (c) J. C. Brand and B. P. Roberts, *J. Chem. Soc., Chem. Commun.*, 109 (1984).
19. W. Uhl, L. Cuypers, W. Kaim, *et al.*, *Angew. Chem. Int. Ed.*, **42**, 2422 (2003).
20. (a) L. M. McKee, Z.-X. Wang and P. von R. Schleyer, *J. Am. Chem. Soc.*, **122**, 4781 (2000); (b) H. Binder, R. Kellner, K. Vaas, *et al.*, *Z. Anorg. Allg., Chem.*, **625**, 1059 (1999).
21. V. Ya. Lee and A. Sekiguchi, Silicon-, Germanium-, and Tin-Centered Cations, Radicals and Anions, in *Reviews of Reactive Intermediate Chemistry* (eds M. S. Platz, R. A. Moss and M. Jones, Jr.), John Wiley & Sons, Inc., New Jersey, pp. 47–120 (2007).

22. V. Ya. Lee and A. Sekiguchi, *Acc. Chem. Res.*, **40**, 410 (2007).
23. V. Ya. Lee and A. Sekiguchi, *Eur. J. Inorg. Chem.*, 1209 (2005).
24. M. M. Olmstead, L. Pu, R. S. Simons and P. P. Power, *Chem. Commun.*, 1595 (1997).
25. Y. Ishida, A. Sekiguchi, K. Kobayashi and S. Nagase, *Organomet.*, **23**, 4891 (2004)
26. A. Sekiguchi, T. Matsuno, and M. Ichinohe, *J. Am. Chem. Soc.*, **123**, 12436 (2001).
27. F. Antolini, B. Gehrhus, P. B. Hitchcock and M. F. Lappert, *Chem. Commun.*, 5112 (2005).
28. B. Gehrhus, P. B. Hitchcock, and L. Zhang, *Angew. Chem. Int. Ed.*, **43**, 1124 (2004).
29. C. Cui, M. Brynda, M. M. Olmstead, and P. P. Power, *J. Am. Chem. Soc.*, **126**, 6510 (2004).
30. H. Cox, P. B. Hitchcock, M. F. Lappert, and L. J-M. Pierssens, *Angew. Chem. Int. Ed.*, **43**, 4500 (2004).
31. A. Sekiguchi, T. Fukawa, M. Nakamoto, *et al.*, *J. Am. Chem. Soc.*, **124**, 9865 (2002).
32. A. Sekiguchi, T. Fukawa, V. Ya. Lee, and M. Nakamoto, *J. Am. Chem. Soc.*, **125**, 9250 (2003).
33. (a) C. Förster, K. W. Klinkhammer, B. Tumanskii, *et al.*, *Angew. Chem. Int. Ed.*, **46**, 1156 (2007); (b) M. Becker, C. Förster, C. Franzen, *et al.*, *Inorg. Chem.*, **47**, 9965 (2008).
34. J. Iley, in *The Chemistry of Organic Germanium, Tin and Lead Compounds* (ed. S. Patai), John Wiley & Sons, Inc., New York, 1995, Chapter 5.
35. S. Inoue, M. Ichinohe, and A. Sekiguchi, *J. Am. Chem. Soc.*, **139**, 6078 (2008).
36. A. Sekiguchi, S. Inoue, M. Ichinohe, and Y. Arai, *J. Am. Chem. Soc.*, **126**, 9626 (2004).
37. (a) T. Fukawa, V. Ya. Lee, M. Nakamoto, and A. Sekiguchi, *J. Am. Chem. Soc.*, **126**, 11758 (2004); (b) V. Ya. Lee, T. Fukawa, M. Nakamoto, *et al.*, *J. Am. Chem. Soc.*, **128**, 11643 (2006).
38. (a) S. Inoue, M. Ichinohe and A. Sekiguchi, *J. Am. Chem. Soc.*, **129**, 6096 (2007); (b) S. Inoue, M. Ichinohe and A. Sekiguchi, *Organomet.*, **27**, 1358 (2008).
39. D. Bravo-Zhivotovskii, I. Ruderfer, S. Melamed, *et al.*, *Angew. Chem., Int. Ed.*, **44**, 739 (2005).
40. R. Kinjo, M. Ichinohe, and A. Sekiguchi, *J. Am. Chem. Soc.*, **129**, 26 (2007).
41. (a) M. M. Olmstead, R. S. Simons and P. P. Power, *J. Am. Chem. Soc.*, **119**, 11705 (1997); (b) L. Pu, A. D. Phillips, A. F. Richards, *et al.*, *J. Am. Chem. Soc.*, **125**, 11626 (2003).
42. A. Armstrong, T. Chivers and R. T. Boeré, *ACS Symposium Series*, **917**, 66 (2005).
43. U. Schmidt, K. Kabitzke, K. Markau and A. Müller, *Chem. Ber.*, **99**, 1497 (1966).
44. (a) M. J. S. Gynane, A. Hudson, M. F. Lappert and P. P. Power, *Chem. Commun.*, 623 (1976); (b) M. J. S. Gynane, A. Hudson, M. F. Lappert and P. P. Power, *J. Chem. Soc., Dalton Trans.*, 2428 (1980).
45. A. Dumitrescu, V. L. Rudzevich, V. D. Romanenko, *et al.*, *Inorg. Chem.*, **43**, 6546 (2004).
46. (a) J.-P. Bezombes, P. B. Hitchcock, M. F. Lappert and J. E. Nycz, *Dalton Trans.*, 499 (2004); (b) J.-P. Bezombes, K. B. Borisenko, P. B. Hitchcock, *et al.*, *Dalton Trans.*, 1980 (2004).
47. (a) S. H. Hinchley, C. A. Morrison, D. W. H. Rankin, *et al.*, *Chem. Commun.*, 2045 (2000); (b) S. H. Hinchley, C. A. Morrison, D. W. H. Rankin, *et al.*, *J. Am. Chem. Soc.*, **123**, 9045 (2001).
48. P. Agarwal, N. A. Piro, K. Meyer, *et al.*, *Angew. Chem. Int. Ed.*, **46**, 3111 (2007).
49. (a) S. Loss, A. Magistrato, L. Cataldo, *et al.*, *Angew. Chem. Int. Ed.*, **40**, 723 (2001); (b) L. Cataldo, C. Dutan, S. K. Mistra, *et al.*, *Chem. Eur. J.*, **11**, 3463 (2005).
50. (a) M. Gouygou, C. Tachon, M. Koenig, *et al.*, *J. Org. Chem.*, **55**, 5750 (1990); (b) Y. Canac, D. Bourissou, A. Baceiredo, *et al.*, *Science*, **279**, 2080 (1998).
51. Y. Canac, A. Baceiredo, W. W. Schoeller, *et al.*, *J. Am. Chem. Soc.*, **119**, 7579 (1997).
52. (a) A. F. Armstrong, T. Chivers, M. Parvez and R. T. Boeré, *Angew. Chem. Int. Ed.*, **43**, 502 (2004); (b) A. F. Armstrong, T. Chivers, H. M. Tuononen, *et al.*, *Inorg. Chem.*, **44**, 7981 (2005).
53. S. Ito, M. Kikuchi, M. Yoshifuji, *et al.*, *Angew. Chem. Int. Ed.*, **45**, 4341 (2006).
54. H. Sugiyama, S. Ito and M. Yoshifuji, *Angew. Chem. Int. Ed.*, **42**, 3802 (2003).
55. (a) E. Niecke, A. Fuchs, F. Baumeister, *et al.*, *Angew. Chem. Int. Ed.*, **34**, 555 (1995); (b) E. Niecke, A. Fuchs and M. Nieger, *Angew. Chem. Int. Ed.*, **38**, 3028 (1999); (c) E. Niecke, A. Fuchs, M. Nieger, *et al.*, *Angew. Chem. Int. Ed.*, **38**, 3031 (1999); (d) W. W. Schoeller, C. Biegemann, E. Niecke and D. Gudat, *J. Phys. Chem. A*, **105**, 10731 (2001).
56. (a) S. Sasaki, K. Sutoh, F. Muramaki and M. Yoshifuji, *J. Am. Chem. Soc.*, **124**, 14830 (2002); (b) R. T. Boeré, A. M. Bond, S. Cronin, *et al.*, *New. J. Chem.*, **32**, 214 (2008).
57. J. Geier, J. Harmer and H. Grützmacher, Angew. *Chem. Int. Ed.*, **43**, 4093 (2004).

58. (a) T. Sasamori, E. Mieda, N. Nagahora, *et al.*, *Chem. Lett.*, **34**, 166 (2005); (b) T. Sasamori, E. Mieda, N. Nagahora, *et al.*, *J. Am. Chem. Soc.*, **128**, 12582 (2006); (c) N. Nagahora, T. Sasamori, Y. Hosoi, *et al.*, *J. Organomet. Chem.*, **693**, 625 (2008).
59. B. Neumüller, F. Schmock, R. Kirmse, *et al.*, *Angew. Chem., Int. Ed.*, **39**, 4580 (2000).
60. D. Reinen and G.-G. Lindner, *Chem. Soc. Rev.*, **28**, 75 (1999).
61. A. F. Keppler, G. Cerchiaro. O. Augusto, *et al.*, *Organometallics*, **25**, 5059 (2006).
62. M. Björgvinsson, T. Heinze, H. W. Roesky, *et al.*, *Angew. Chem., Int. Ed. Engl.*, **20**, 1677 (1991).
63. B. Mueller, H. Poleschner and K. Seppelt, *Dalton Trans.*, 4424 (2008).
64. (a) R. J. Gillespie, R. Kapoor, R. Faggiani, *et al.*, *J. Chem. Soc., Chem. Commun.*, 8 (1983); (b) C. G. Davies, R. J. Gillespie, P. R. Ireland and J. M. Sowa, *Can. J. Chem.*, **52**, 2048 (1974); (c) R. Faggiani, R. J. Gillespie, R. Kapoor, *et al.*, *Inorg. Chem.*, **27**, 4350 (1988).
65. (a) R. J. Gillespie and J. Passmore, *Adv. Inorg. Chem. Radiochem.*, **17**, 49 (1975); (b) K. O. Christie and J. S. Muirhead, *J. Am. Chem. Soc.*, **91**, 7777 (1961); (c) R. S. Eachus, T. P. Sleight and M. C. R. Symons, *Nature (London)*, **222**, 769 (1969).
66. T. Drews, W. Koch and K. Seppelt, *J. Am. Chem. Soc.*, **121**, 4379 (1999).
67. S. Seidel and K. Seppelt, *Angew. Chem., Int. Ed.*, **39**, 3923 (2000).
68. R. S. Eachus and M. C. R. Symons, *J. Chem. Soc., Dalton Trans.*, 431 (1976).
69. (a) C. Cui, M. M. Olmstead, J. C. Fettinger, *et al.*, *J. Am. Chem. Soc.*, **127**, 17530 (2005); (b) Y. Peng, M. Brynda, B. D. Ellis, *et al.*, *Chem. Commun.*, 6042 (2008).
70. Y. Hung, M. Brynda, P. P. Power and M. Head-Gordon, *J. Am. Chem. Soc.*, **128**, 7185 (2006).
71. T. Pankewitz, W. Klopper, P. Henke and H. Schnöckel, *Eur. J. Inorg. Chem.*, 4879 (2008).
72. See Ref. 130 in the review by Breher.[3]

11

Application of Stable Radicals as Mediators in Living-Radical Polymerization

Andrea R. Szkurhan, Julie Lukkarila and Michael K. Georges
Department of Chemical and Physical Sciences, University of Toronto at Mississauga, Mississauga, Canada

11.1 Introduction

The free radical polymerization process[1,2] has attracted, and will continue to attract, considerable attention from polymer chemists because it is an industrially significant process that is relatively easy to perform under a wide variety of conditions. Although the process enables the formation of a myriad of homopolymers and copolymers, it suffers from its inability to provide designer materials with complex architectures. The end products of free radical polymerizations are "dead" or non-reactive polymer chains as a result of unpredictable, irreversible, and unavoidable chain termination reactions, the Achilles heel of this polymerization process.[3]

Historically, anionic polymerization has been the method of choice for the production of polymeric materials with intricate structural design because termination and chain transfer reactions, for the most part, are nonexistent. Polymer chain ends remain active even at the end of the polymerization, permitting continued manipulation of the initially formed polymer chains. The anionic polymerization process, however, is incompatible with many functional groups (carbonyl, nitrile, hydroxyl, amine) and requires rigorous purification of monomers and solvents and the exclusion of oxygen and water. To address the limitations of both the free radical and anionic polymerization processes, living free radical polymerization processes have been developed, the three most prominent being Stable Free Radical Polymerization (SFRP), also known as Nitroxide-Mediated Polymerization (NMP),[4,5] Atom Transfer Radical Polymerization (ATRP)[6,7] and Reversible Addition Fragmentation Chain Transfer (RAFT) polymerization.[8] Reviews of these processes are available elsewhere.[9–11]

This chapter details the use of stable radicals, in particular nitroxides, as mediators in living-radical polymerization. The recent use of other stable free radicals and their application as mediators for the

SFRP process is also included. A background on free radical polymerization, along with an overview of the pioneering work that led to the discovery of stable free radical polymerization (SFRP) process, is briefly discussed. Selected literature is used to highlight the application of the SFRP process to aqueous polymerization processes and the synthesis of novel materials, including polymers with unique and complex architectures. There has been no attempt to cover all the literature in the area; the huge popularity of the subject matter and concomitant large number of publications in the area makes that task too unwieldy. Apologize are made in advance to those authors whose papers have been omitted.

11.2 Living polymerizations

Living polymerizations are chain growth processes in which termination and chain transfer reactions are absent, all chains are initiated simultaneously and each initiated chain continues to add monomer until all the monomer is consumed.[12–15] At 100 % conversion, the propagating chain centers remain active; upon addition of a second monomer, polymerization begins again to produce a diblock copolymer. This procedure can be repeated again to form a triblock copolymer and so on. Due to a high initiation efficiency and a fast exchange reaction between active and dormant species, the average molecular weight of the final polymer produced corresponds to the ratio between the initial monomer concentration, $[M]_0$, and the initiator concentration, [I]. Moreover, since chain transfer and termination reactions are minimal, molecular weight distributions (M_w/M_n) approach the ideal value of one. Living radical polymerization processes attempt to mimic the anionic process as closely as possible in the characteristics listed above.

11.2.1 Living-radical polymerization background

The beginning of living-radical polymerization began in 1956 with the seminal work of Otsu.[16–19] In an attempt to control radical polymerizations Otsu focused on the use of dithiocarbonates, **1**, for which, in 1982, he coined the term iniferters, due to their ability to act as *ini*tiator-trans*fer*-*ter*minator agents as illustrated in Scheme 11.1.[20] The S–S bonds of the iniferters homolytically cleave when heated to form a radical capable of initiating a polymer chain and reversibly terminating a propagating chain end, essentially all of the necessary components for a living-radical polymerization system.

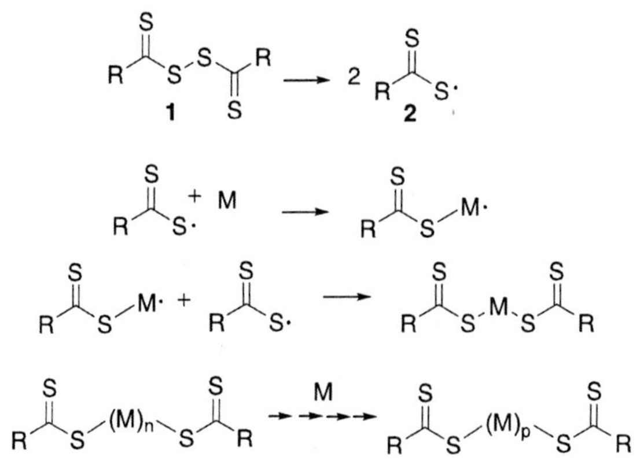

Scheme 11.1 Iniferter chemistry, where R = $N(CH_3)_2$

In the early pioneering years of living-radical polymerizations Otsu *et al.*[19] and Clouet *et al.*[21] performed an extensive amount of work with iniferters, in the process developing a myriad of unique block copolymers. However, while the concept of how a living-radical polymerization could be performed was clearly evident in the this chemistry, the work generally produced polymers with broad molecular weight distributions, since the dithiocarbamyl radical **2**, which reversibly terminates the propagating polymer chain, also initiates new chains and does so throughout the course of the polymerization. In addition, the dithiocarbamyl radical can lose carbon disulfide (CS_2) to form a nitrogen-centered radical that has been implicated in initiating new chains.[22]

In the 1970s, Braun investigated the use of pinacol molecules to initiate the polymerization of vinyl monomers, such as, styrene and methyl methacrylate.[23,24] In 1981 he reported the use of disubstituted tetraarylethanes as free radical initiators to form low molecular weight oligomers of methyl methacrylate[25–27]and styrene[28] that were terminated on both ends by the molecular fragments originating from the disubstituted tetraarylethane. These oligomers were able to function as initiators to reinitiate polymerization in the presence of excess monomer and grow to a higher molecular weight upon heating.[29,30] The growing polymer chains, however, would lose their living ends rather quickly through termination reactions preventing the formation of very high molecular weight products.[31]

11.3 Stable free radical polymerization

In 1985, the use of nitroxides as reversible radical trapping agents for carbon-centered radicals was demonstrated by Solomon and Moad.[32,33] 2,2,6,6-Tetramethylpiperidinyl-N-oxy (TEMPO) was used as a trapping agent for acrylate and methacrylate monomers initiated with azobisisobutyronitrile (AIBN) to generate alkoxyamines adducts such as **3**. The alkoxyamines were subsequently used as initiators for polymerizations of acrylates at temperatures between 80 and 100 °C,[34,35] resulting in the formation of low molecular weight oligomers with molecular weight distributions (M_w/M_n) of ~2. And there the area of living-radical polymerization remained for a number of years.[36] In 1992 Otsu *et al.* reported that up to that point the goal of achieving narrow molecular weight distributions polymers by a free radical process had not been achieved.[37] That changed shortly thereafter and a new era of radical polymerization began.

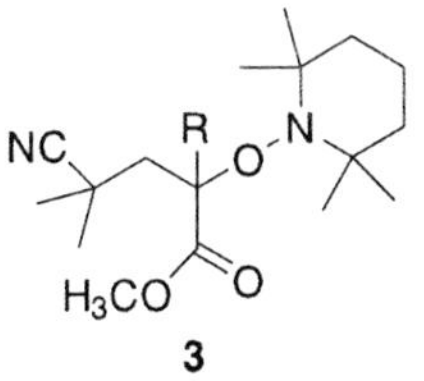

3

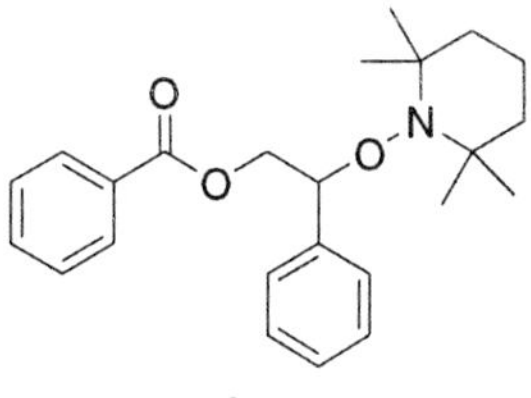

4

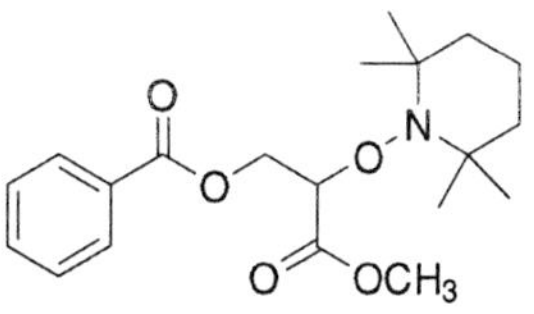

5

11.3.1 Background of the work performed at the Xerox Research Centre of Canada

In 1990, at the Xerox Research Centre of Canada (XRCC), one of the authors of this review (M. K. Georges), and a colleague were assigned a task of synthesizing polymers for a new toner program; M. K. Georges was assigned the free radical polymerization process, his colleague the anionic polymerization process. The monomer choices were the same for each approach and included styrene and its derivatives, 1,3-butadiene and various alkyl acrylates. None of the polymers synthesized in the first year by the free radical polymerization process exhibited any of the sought after program requirements, while the corresponding polymers synthesized by the anionic process performed exceptionally well. Management, however, was concerned with the high cost of the anionic process. Without any direct evidence, it was concluded that two of the main reasons for the superior properties of the materials prepared by the anionic polymerization process were the well defined molecular weights and the narrow molecular weight distributions. Attempts to obtain these two attributes with iniferter chemistry were not successful. It became increasingly obvious as the project wore on that a more stable free radical was required for the process. With that in mind, during a discussion with a colleague who worked on electron spin resonance, the idea of using nitroxides presented itself. Nine months and possibly 250 experiments later, the first example of a living-radical polymerization of styrene was realized.

In 1993, the polymerization of styrene initiated with benzoyl peroxide (BPO) in the presence of TEMPO at 130 °C was reported to yield moderately high molecular weight polystyrene with narrow molecular weight distributions (M_w/M_n ~1.3) that exhibited a linear increase in molecular weight with conversion.[38] The critical decision in the success of this work was to start with styrene as the monomer, as opposed to acrylates or methacrylates, which would probably have led to failure. The process was labeled the stable free radical polymerization (SFRP) process based on the use of the stable radical TEMPO to control the polymerization.

11.3.2 General considerations and mechanism

The general mechanism of living polymerizations mediated by stable radicals by now is well known and has been extensively studied but is presented in Scheme 11.2 for completeness, where M is a monomer unit, T is a nitroxide, and P_n represents a polymer chain with n repeat units. The process begins with an initiator, which upon heating dissociates to form radicals that add to monomer. The SFRP process was initially carried out at temperatures in the range 125–135 °C to allow all the polymer chains to initiate at more or less the same time. An excess of the stable radical caps most of the initiated chains preventing the reaction from going out of control. The C–O bond formed between the propagating radical site and the stable radical is labile at the aforementioned temperatures and, upon continued heating, undergoes reversible homolytic cleavage to reform the stable radical and an active radical site on the end of the polymer chain. The polymer active site reacts with another nitroxide molecule to reform the dormant polymer chain or with monomer to undergo chain extension.

The equilibrium reaction between the dormant and active species significantly favors the dormant species, with $K = k_d/k_c = 2.1 \times 10^{-11}$ M at 125 °C,[39] where k_d is the dissociation rate constant and k_c is the

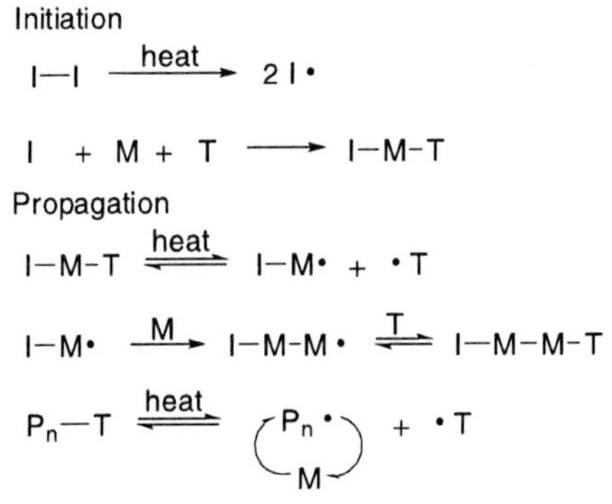

Scheme 11.2 Stable free radical polymerization process

recombination rate constant. At any given time most of the polymer chains are in the dormant form, reducing the overall concentration of propagating radicals, thereby dramatically reducing irreversible termination by radical chain combination. Although some termination continues to occur throughout the course of the polymerization,[40] the rate at which termination occurs decreases as the polymerization proceeds since termination by coupling generally involves at least one short chain.[41] In the absence of other side reactions that generate new chains, polymer growth from the initiating species occurs in a "living" fashion. Ultimately, the SFRP process enables a linear increase in the molecular weight of the polymer chain with conversion, while producing polymers with very low M_w/M_n. Since termination reactions cannot be completely avoided in any free radical polymerization process, these are not truly "living" systems. Comprehensive kinetic analyses of the SFRP process have been reported[42–48] using a plethora of methods, such as, HPLC,[49,50] ESR,[51–53] GPC/SEC,[54–56] alkoxyamine fluorescence quenching[57] and ^{31}P[58] and ^{1}H NMR.[59]

11.3.3 Unimolecular initiators

While conventional radical initiators can be used in stable free radical polymerization systems, their use often requires tweaking the system to obtain the correct ratio of initiator to mediator, since the initiation efficiency of various primary initiators differs and any large excess of free nitroxides adversely affects the rates of polymerization.[60] An alternative to primary initiators are alkoxyamines,[33–36] adducts of an initiator fragment (optional), a monomer, and a nitroxide (for example **3–5**). The C–O bond of the alkoxyamine is labile and upon heating dissociates to form an initiating species and a stable radical mediator in a 1 : 1 ratio, although this is not necessarily the right ratio of the propagating chain to nitroxide for a well controlled polymerization.[61]

There are several methods to preparing alkoxyamine unimer initiators that involve the controlled generation and trapping of carbon-centered radicals by stable radicals. A commonly used alkoxyamine, 2-phenyl-2-(2,2,6,6-tetramethylpiperidine-1-oxy)ethyl benzoate (BST) **4**, has been synthesized by the reaction of benzoyl peroxide with an excess of styrene in the presence of TEMPO at 80 °C.[47,61,76] Synthesis of the acrylate derivative of **4** has been accomplished by oxymercuration of *t*-butyl acrylate to generate a mercury(II) acetate intermediate, followed by an oxidative demercuration in the presence of TEMPO. A subsequent benzoylation gives the desired alkoxyamine product **5**.[62]

Other alkoxyamine derivatives have been synthesized that do not contain the benzoyloxy moiety. A reaction of di-*tert*-butyl peroxide with ethyl benzene generates the corresponding benzylic radical, which is trapped by TEMPO to form the alkoxyamine. This reaction can be performed either thermally[63] or photolytically,[64] with the latter reaction allowing for an increase in product yield.

Another alkoxyamine synthesis uses Jacobsen's catalyst, a reagent typically used in the epoxidation of olefins. The epoxidation mechanism proceeds via a radical intermediate that generally collapses to give the corresponding epoxide. However, the radical intermediate can be trapped by a stable nitroxide radical and then reduced to form the corresponding alkoxyamine.[65]

Alkoxyamine synthesis has also been attained through atom transfer radical addition (ATRA) in which halogen abstraction from an alkyl halide by a copper complex yields a radical that can be trapped by a stable radical, such as a nitroxide.[66] Wang and Zu reported the synthesis of the alkoxyamine composed of the adduct of the benzoyloxy radical, derived from BPO, and TEMPO, simply by heating BPO in the presence of TEMPO.[67] Braslau and Hill described the synthesis of arylethyl-functionalized N-alkoxyamines and highlighted their uses as initiators for polymerizations.[68] More recently, a method for preparing alkoxyamines starting with 4-hydroxyTEMPO dissolved in a simple ketone in the presence of hydrogen peroxide, followed by the addition of copper(I) chloride (CuCl), gave varying yields of the alkoxyamine,[69] while the synthesis of SG1-based alkoxyamines using a photodecomposition of either azo compounds or dithiocarbamates in the presence of SG1 **6** [N-tert-butyl-N-[1-diethylphosphono-(2,2-dimethylpropyl) nitroxide] (also referred to as DEPN) were reported.[70]

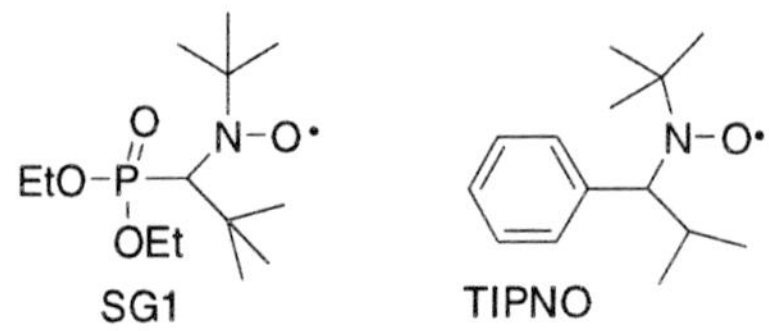

SG1 **6** TIPNO **7**

An extensive amount of work has been performed on the use of alkoxyamines prepared with sterically bulky nitroxides. The C–O bond dissociation energy of these alkoxyamines is significantly decreased when compared to that of TEMPO derived alkoxyamines. A number of papers have shown how the steric bulk of a nitroxide affects the rate of C–O bond dissociation of alkoxyamines prepared with styrene.[71–74] An increase in the ring size of cyclic nitroxides also causes a significant decrease in the C–O bond dissociation energy when compared to TEMPO.[75] Seven- and eight-membered ring nitroxides have large k_d values as compared to TEMPO and exhibit increased polymerization rates for styrene.[76,77] Interestingly, even with their larger k_d values, these nitroxides do not allow controlled polymerizations of *n*-butyl acrylate, giving instead polymerizations that stop at low polymer conversion or polymerizations that proceed to high conversion but yield polymers with broad molecular weight distributions ($M_w/M_n > 2$).

A number of recent studies on the other factors that affect C–O bond homolysis of alkoxyamines have been reported. Studies with aromatic alkoxyamines have shown that there is a competition between N–O and O–C homolysis depending on the alkyl substituent on the oxygen of the nitroxide. Primary or secondary alkyl groups result in N–O cleavage while tertiary or benzyl groups allow O–C cleavage.[78] The identity of the counterion of the carboxylate salts of SG1-derived alkoxyamines has been shown to affect the C–O homolysis rate of the RS/SR diastereoisomers but appears to have little effect on the SS/RR isomers.[79] Also, the effect of the penultimate unit in the polymer on the C–O bond homolysis rates in SG1-based alkoxyamines[80] and alkoxyamines based on imidazoline and imidazole nitroxides[81] have also been reported.

Other sterically bulky nitroxides, for example **8–11**, have been more successful in allowing the polymerization of acrylates.[82–85] The steric bulk of **11**, for example, reduces the bond dissociation energy enough that polymerization of *n*-butyl acrylate occurs at relatively low temperatures (70–105 °C).[86,87] In spite of the low reaction temperatures these reactions were quite fast, not due as suggested by the authors to the higher k_d of the N–O bond but instead due to the lower rate of recombination (k_c) of the nitroxide with the transient macroradical,[82] a conclusion supported with results obtained by Debuigne *et al*. using the diadamantyl nitroxide **12**.[88]

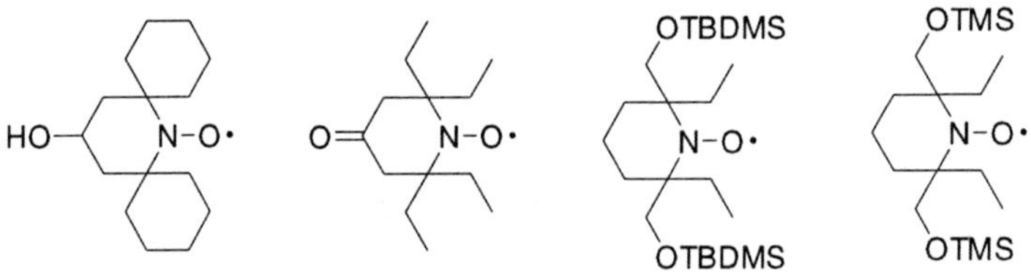

8 **9** **10** **11**

The synthesis and application of a nitroxide containing one adamantyl substituent to living radical polymerization has been reported.[89,90] Braslau concluded that steric bulk and an inherent instability of the nitroxide were both factors in whether a nitroxide could mediate the polymerization of acrylate monomers. This prompted the synthesis of 1-adamantyl-3-methyl-2-phenylazabutane-1-nitroxide, **13**, an acyclic nitroxide containing both an α hydrogen and an adamantyl group. The corresponding alkoxyamine was investigated as an initiator for the polymerization of styrene, *N,N*-dimethyl acrylamide, *tert*-butyl acrylate and *n*-butyl acrylate.

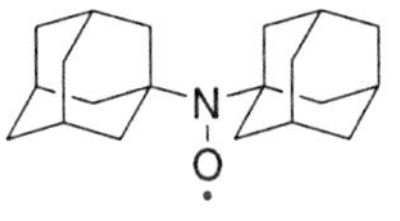

12

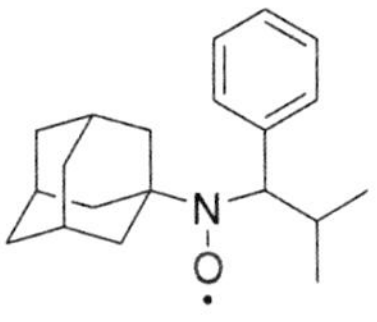

13

Debuigne *et al.*[88] synthesized diadmantyl nitroxide **12**, a sterically bulky nitroxide with no α hydrogen atoms, and found that while this nitroxide enabled low temperature (100 °C) polymerization of styrene, it failed to enable the polymerization of *n*-butyl acrylate. It was further shown that free nitroxide accumulated in the reaction solutions over time in both acrylate and styrene polymerizations, but more so in the acrylate polymerizations. These results prompted the authors to further argue that the more important requirement for whether a nitroxide will enable the polymerization of acrylates is related to its inherent stability or instability.

11.3.4 Persistent radical effect

The main reaction that takes place in the stable free radical polymerization is the dynamic equilibrium that occurs between the propagating chain end, the stable mediating radical, and the dormant polymer chain, which allows for the repeated generation of a propagating chain end that can add monomer following a reversible termination reaction with the mediating stable radical. The kinetics of this equilibrium operates on a phenomenon known as the persistent radical effect (PRE).[47,91,92] In addition to the main reversible termination reaction that takes place between the propagating chain end and the stable radical, other side reactions, such as termination by chain–chain coupling, occur. In the early stages of polymerization, a small number of the propagating chains undergo termination by coupling at diffusion-controlled rates, since the radicals are small and the reaction medium is non-viscous, resulting in dead oligomeric chains. Unlike the propagating chain radicals, the radical mediator is stable and does not terminate, resulting in an increase in its concentration, leading to a more efficient formation of the dormant polymer species, which in turn reduces the amount of termination by chain–chain coupling. Overall, the PRE gives rise to a small excess of mediating radicals and ensures good control over the polymerization process. A small amount of excess nitroxides is thus often added when alkoxyamines are used to initiate polymerization to help establish this excess concentration of nitroxide.[61]

11.3.5 Requirements of stable radicals as mediating agents

In order for a radical to be able to mediate a polymerization, it must be relatively stable under the reaction conditions and must react with the propagating chain in a reversible manner. In addition, it must not react with itself or with monomer to initiate new chains nor participate in side reactions, such as β-hydrogen atom abstraction.

11.3.6 Nitroxides as mediating agents

The stability of nitroxides is attributed to the delocalization of three electrons in the N–O bond. The high delocalization energy (~30.4 kcal/mol) provides a high thermodynamic stability to the radical center and prevents the dimerization of two nitroxide radicals.[93] The overall thermodynamic stability of different nitroxide radicals is influenced by the substituents on the carbons attached to the nitrogen. Thus, for example, if all four groups attached to the carbons attached to the nitroxide nitrogen are alkyl, the nitroxide is stable. Replacing one of the alkyl groups with a hydrogen or attaching a phenyl group directly to nitrogen renders the nitroxide relatively unstable.[93] In the former case abstraction of the alpha hydrogen of one nitroxide molecule by another results in a disproportionate reaction producing a hydroxylamine and a nitrone. In the later case, the reaction follows the path depicted in Scheme 11.3.

11.3.7 Nitroxides and their ability to moderate polymerizations

The cyclic nitroxide TEMPO was initially shown to be an efficient mediator for the homopolymerization of styrene and the copolymerization of styrene and acrylates in which the acrylate concentration was 50 % or less on a molar basis. Attempts at the homopolymerization of acrylates and methacrylates were uniformly unsuccessful proceeding to about 5 % conversions and producing low molecular weight oligomers. This inability of TEMPO to work efficiently with acrylate monomers has been well documented.[82,94–99] It has been suggested that the failure of acrylate homopolymerizations in the presence of TEMPO is due to the strong C–O bond formed between the acrylate chain end and nitroxide.[100] This causes an unfavorably low value of the equilibrium constant (K_{eq}), attributed to a very low dissociation rate constant (k_d) relative to a very high recombination rate constant (k_c). As such, the cleavage of the TEMPO–acrylate bond is not much faster than monomer conversion and the rate of recombination of TEMPO with the active acrylate polymer chain end is fast. This prevents both a linear increases in molecular weight with conversion and low M_w/M_n when TEMPO is the mediating nitroxide. With this in mind, a series of nitroxides were developed that formed alkoxyamines with styrene and acrylate monomers with relatively large k_d values. Alkoxyamines, comprised of a benzyl radical with the acyclic nitroxides 2,2,5-trimethyl-4-phenyl-3-azahexane-3-oxy (TIPNO) **7**[101] or SG1 **6**,[102,103] were shown to have k_d values of 3.6×10^{-3} s^{-1} and 3.3×10^{-4} s^{-1}, respectively, as compared to 1.1×10^{-3} s^{-1} for TEMPO.[51] Both TIPNO and SG1 were shown early on to be very effective in mediating the polymerization of acrylates and other non-styrenic monomers, specifically acrylamides, 1,3-dienes, and acrylonitriles.[104,105] A strong case has been made for the importance of k_d in successful polymerizations and the need for k_p/k_d to be $\leq 6.0 \times 10^5$ l mol^{-1}, in addition to the need to be careful with experimental conditions such as the purity of the monomers and the amount of reagent in a given-sized reaction vessel.[106]

It has also been proposed that the difficulty associated with TEMPO-mediated polymerizations of acrylates is directly related to the excess free nitroxide that is generated throughout the course of the

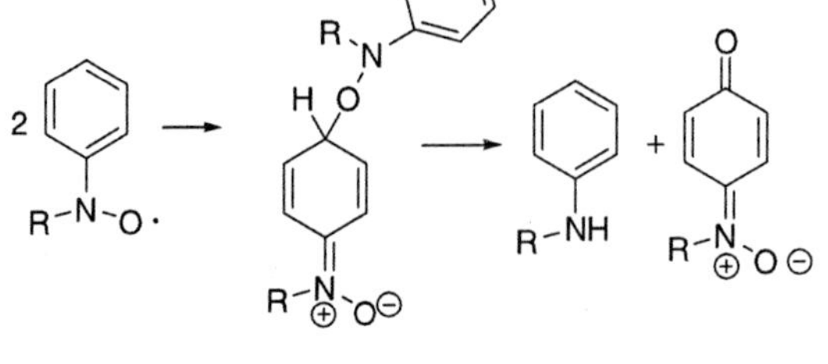

Scheme 11.3 Decomposition pathway for N–phenyl containing nitroxide

polymerization as a consequence of premature chain termination. In the case of styrene polymerizations a spontaneous autoinitiation reaction between two styrene molecules[107,108] generates enough new radicals to consume the excess TEMPO, allowing the polymerization to proceed uninhibited. A corresponding autoinitiation mechanism does not exist for acrylates, resulting in a build up of free nitroxide in the reaction solution. The excess nitroxide shifts the equilibrium between the TEMPO-capped dormant polymer chains and the uncapped propagating polymer chains to the TEMPO-capped polymer, effectively inhibiting the polymerization. This argument suggests that an additive that can react with TEMPO and remove it from the reaction solution should allow the polymerization of acrylates to proceed in the presence of TEMPO. That was indeed the case with the addition of small amounts of ascorbic acid[109] and the *in situ* formation of ene-diols.[110]

In fact, it can be argued that the effectiveness of TIPNO and SG1 in acrylate polymerizations is due to an inherent instability of these nitroxides. The nitroxides contain hydrogen on the carbon α to the nitrogen,[111] enabling them to react with each other to disproportionate into a hydroxylamine and a nitrone.[93] This self-initiated degradation of the nitroxide effectively removes excess nitroxide from the reaction solution, allowing acrylate polymerizations to proceed successfully.[112] Ideally, the rate of decomposition of the nitroxide should be about slightly less than twice the rate of polymer chain termination, since each termination reaction produces two molecules of nitroxide.

The syntheses of new nitroxides have been reported and include imidazoline-containing nitroxides,[113] piperidinyl-N-oxyl radicals with bulky substituents, in particular eight-membered rings,[114] ketone bearing nitroxides,[115] and water soluble nitroxides based on TIPNO that allow polymerization of sodium styrenesulfonate in water below 100 °C.[116]

The successful polymerization of methacrylates mediated by nitroxides has proven even more challenging than acrylate polymerizations. The preferred reaction of most nitroxide with a methacrylate monomer is a β-hydrogen abstraction leading to a terminated polymer chain and an hydroxylamine.[73,117,118] Nitroxide-mediated methyacrylate polymerizations also exhibit a large equilibrium constant (K), resulting in a high concentration of propagating radical chains. This leads to an increase in the amount of irreversible termination reactions and the accumulation of excess free nitroxide, which inhibits the polymerization. Charleux *et al*. addressed these problems by polymerizing methyl methacrylate (MMA) in the presence of SG1 and a small amount of styrene (4.4–8.8 mol %), a monomer with a low K value.[119,120] However, care needs to be taken with the amount of SG1 in that the presence of excess SG1 leads to an increase in disproportionation, with the amount of disproportionation increasing with an increasing amount of SG1.[121]

Homopolymerization of MMA without added styrene has been performed using the indole-based nitroxide, DPAIO, **14**.[122] DPAIO features include a slow homolysis rate, a fast recombination rate and a low propensity to abstract hydrogen, thus minimizing the amount of disproportionation. Polymerizations of MMA at 100 °C with DPAIO proceed to 80 % conversion yielding homopolymers with polydispersities as low as 1.4. Actual molecular weights were often lower than the theoretical molecular weights, a result attributed to a combination of a slower initiation rate and a slow decomposition of the alkoxyamine.

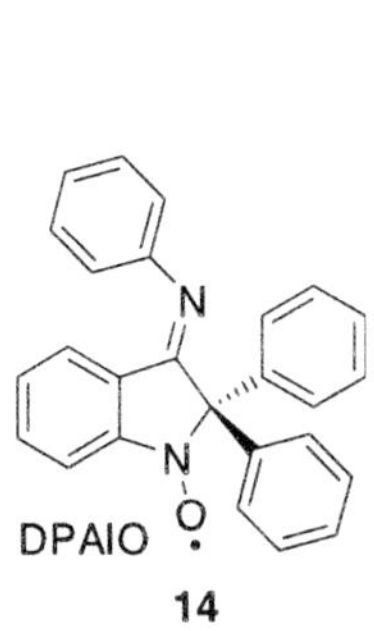

14

Polymerizations of methacrylate performed in the presence of sulfuric acid have been reported, with alkoxyamines prepared with methyl methacrylate, TEMPO, and either benzoyl peroxide or AIBN. Reaction times are relatively short (45–120 min) at reaction temperaturtes of 70 °C and 130 °C yielding polymers with $M_w/M_n = 1.09–1.28$.[123]

11.3.8 Rate enhancement of stable free radical polymerization through the use of additives

Initial polymerizations of styrene performed by SFRP using TEMPO as the moderating nitroxide were quite slow, taking over 40 hours to reach conversions of 76 %.[38] To increase the rate of polymerization, additives were added to the reaction solutions to reduce the amount of excess free nitroxide, which increases over time due to unavoidable termination by radical chain–chain coupling. Additives that work well for styrene polymerization include camphorsulfonic acid (CSA),[124,125] acetic anhydride,[126] 2-fluoro-1-methylpyridinium *p*-toluenesulfonate (FMPTS),[127] and dextrose.[128] Ascorbic acid[109] and ene-diols[110] are effective for acrylate polymerizations. More recently, the use of DMF as an additive for the polymerization of *tert*-butyl acrylate initiated by 4-oxo-TEMPO-capped polystyrene macroinitiator has been shown to be effective.[129]

To circumvent the use of additives, a desirable goal in any polymerization, Studer *et al.* employed a combination of alkoxyamines to polymerize styrene, with one alkoxyamine containing TEMPO (low k_d) and the other containing a sterically bulky nitroxide (higher k_d, lower k_c), to very good effect.[82] It was subsequently shown that polymerizations of styrene can proceed quickly and efficiently using a primary initiator, such as BPO, in the absence of an additive, if the ratio of TEMPO to BPO is varied according to the targeted molecular weight.[60]

11.4 Non-nitroxide-based radicals as mediating agents

The uses of other stable radicals, some isolatable, others generated *in situ*, have been investigated, with varying degrees of success for living radical polymerizations.

11.4.1 Triazolinyl radicals

Triazolinyl radicals with the general structure **15**, are five-membered ring structures that contain a constrained hydrazyl moiety that increases the molecules' stability by allowing π-orbital overlap between the two adjacent nitrogen atoms.[130,131] Triazolinyl radicals offer some degree of control in the free radical polymerization of styrene and a variety of methacrylates.[132–136] For example, triazolinyl-mediated polymerizations using the stable radicals **16** or **17**, or a mixture of both, as mediators have been shown to be quite effective in the synthesis of both homopolymers and block copolymers.

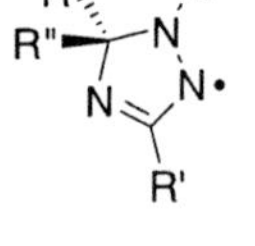

15

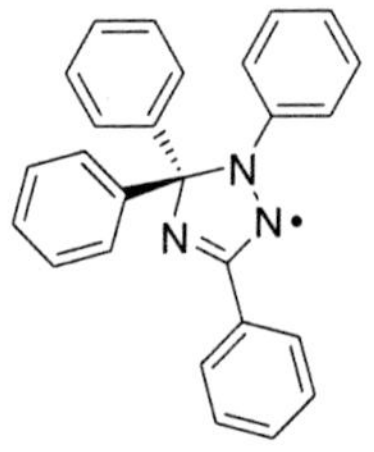

16

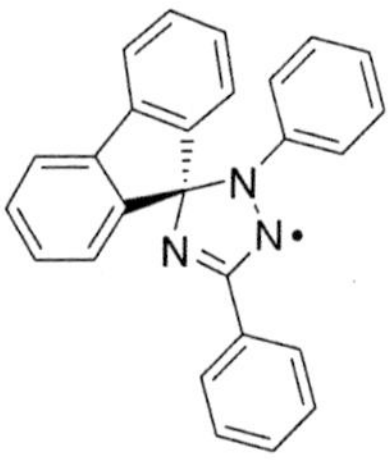

17

Triazolinyl radical **16** decomposes at elevated temperatures generating a phenyl radical that is capable of initiating polymerization.[132] More importantly, this slow, high temperature decomposition prevents the accumulation of excess triazolinyl radicals that occurs as a result of unavoidable chain termination reactions. In comparison to living-radical polymerizations performed using nitroxides, triazolinyl-mediated polymerizations typically produce polymers with broader molecular weight distributions, a result of premature chain termination that occurs in the early stages of the polymerizations.

11.4.2 Verdazyl radicals

Verdazyls are stable radicals that contain a constrained resonance-delocalized hydrazyl moiety in their ring, similar to the above mentioned triazolinyl radicals. The most common verdazyls have the general structures **18** and **19**, where X = oxygen or sulfur and R is typically methyl, ethyl or isopropyl. The R′ group can be quite varied and includes hydrogen, alkyl, phenyl, substituted benzenes, imidazoles, pyridines and the like.[137–144]

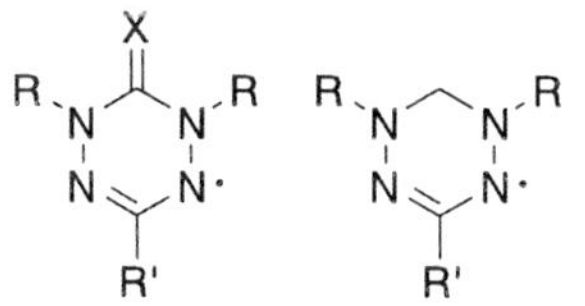

18 **19**

Although verdazyl radicals were originally used as polymerization inhibitors[145] and are important in the study of molecular magnets,[146,147] they also have also been employed as mediators in living-radical polymerizations. 1,3,5-Triphenylverdazyl **18** was originally used for the polymerization of styrene and methyl methacrylate at 60 °C, but reaction times were long and conversions were low.[148] Polymerizations at 110 °C allowed a linear increase in molecular weight with conversion, although the molecular weight distributions were typically broad ($M_w/M_n > 2$) and approximately 60 % of the chain ends were dead, a result of bimolecular termination.[149]

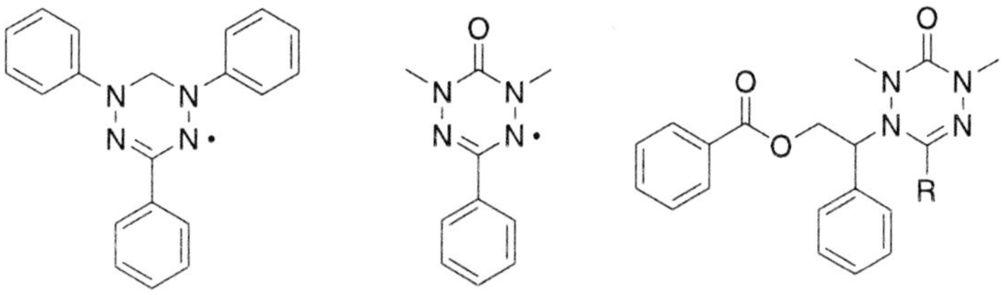

18 **19** **20**

A series of 6-oxoverdazyl radicals, such as **19**, proved more successful for the polymerization of styrene[150] and *n*-butyl acrylate.[151] Adducts of the verdazyl radical with a monomer unit (**20**, where R = hydrogen, ethyl, phenyl) were used as initiators for the polymerizations. The success of **19** to moderate the aforementioned polymerizations can be attributed in part to an inherent instability of the verdazyl radical resulting in the formation of an azomethine imine.[152] As a result an excessive amount of verdazyl does not accumulate in the polymerization solution as a consequence of unavoidable coupling reactions between polymer chains.

11.4.3 Other radicals as mediators

This section describes radical mediators that are best describes as persistent, rather than truly stable, since they are formed *in situ* and, although they are relatively long lived, they cannot be isolated.

11.4.3.1 Borinate radicals

Borinate radicals are relatively stable species formed by the decomposition of peroxyborinate precursor molecules.[153] Peroxyborinates decompose quite readily at ambient temperatures by homolytic cleavage of their peroxide bond to yield an alkoxy radical and a borinate radical, as for example **21**, the latter exhibiting stability due to the back-donation of electron density into the empty *p*-orbital of boron.

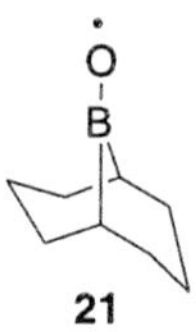

21

Peroxyborinates can be used as initiators in a similar fashion to the nitroxide-derived alkoxyamines. The alkoxy radical produced from the homolytic cleavage of the peroxyborinate initiates polymerization while the borinate radical forms a weak, reversible bond with the propagating radical chain.

Polymerizations performed using peroxyborinate initiators with ethyl methacrylate give polymers that increase in molecular weight with time; however, the molecular weight distributions are typically broad (M_w/M_n ~2.5), in part because the system does not allow for a simultaneous initiation of all polymer chains.[153] An 8-boro-indane/oxygen system containing a boron bridging atom between a six-membered ring and a more reactive five-membered ring undergoes reaction with oxygen to expand the five-membered ring to a seven-membered ring containing a peroxide moiety. A reaction of the resulting peroxyborinate with methyl methacrylate showed improvement in the molecular weight distributions of the resulting homopolymer, (M_w/M_n ~1.3–1.6).[154]

11.4.3.2 (Arylazo)oxyl radicals

(Arylazo)oxyl radicals, such as **22**, are formed *in situ* either by an electron transfer reaction between an activated carbon–halogen compound and an (arenediazo)oxyl anion or by the one-electron oxidation of an arenedihyponitrite or cyanate anion with an arenediazonium ion. Both reactions give carbon-centered radicals as byproducts which can be used to initiate the polymerization, while the (arylazo)oxyl radicals serves to reversibly terminate the growing polymer chain.[155]

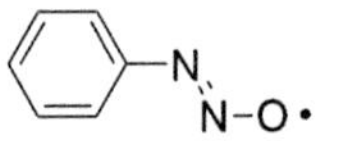

22

Polymerizations of methyl methacrylate and *n*-butyl acrylate at temperatures between 25 and 40 °C with PhN_2BF_4/NaN_2O_2 as the initiating/mediating system produce polymers that increase in molecular weight with conversion but which exhibit broad molecular weight distributions ($M_w/M_n > 3$).[156] Block copolymers prepared using this system contained significant amounts of homopolymers.

11.4.3.3 Cobalt-mediated radical polymerizations

Cobalt-mediated polymerization was first reported by Wayland in 1994 for the polymerization of acrylates using a (tetramesitylporphyrinato)cobalt (TMPCo•) complex (**23**).[157,158] The polymerizations proceed by a reversible mechanism involving a dormant cobalt(III)-capped polymer chain and the active free radical propagating polymer chain to give polymers with $M_w/M_n = 1.25$. Much higher molecular weights ($M_n = 1 - 5.5 \times 10^5$) and faster reaction rates were realized using the octabromo derivative of TMPCo•.[158] The addition of the initiator 2,2′-azo-bis(4-methoxy-2,4-dimethyl valeronitrile (V-70) aids the formation of the organo–cobalt mediating agent by serving as a hydrogen source for the TMPCo• to TMPCoH transformation. This latter compound reacts readily with acrylate monomers to form an organocobalt complex which then initiates and controls the polymerization.[159]

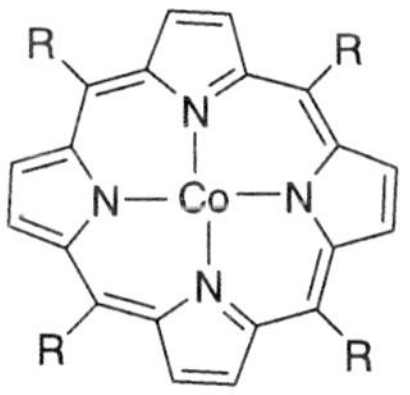

23

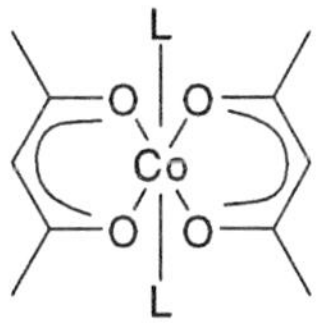

24

An advantage of the cobalt-mediated radical polymerization is its ability to control the polymerization of reactive monomers, such as vinyl acetate. Cobalt bis(acetylaceotnate) complexes of general structure **24** allow for the production of polymers with M_n up to 99 000 g mol^{-1} and M_w/M_n between 1.1 and 1.3.[160] A detailed account of the role of the vazo initator in vinyl acetate polymerization has been presented by Wayland.[161]

The ability to polymerize acrylonitrile[162] allows the synthesis of block copolymers of poly(vinyl acetate-*b*-polyacrylonitrile), which can be readily hydrolyzed to give water soluble blocks of poly(vinyl alcohol-*b*-poly(acrylic acid).[163] The cobalt(II) tetramesitylporphyrin complex also enables the polymerization of vinyl acetate but the yields are generally low, ~20 %.[164] On the other hand the Co(acac) was shown not to be as efficient for *n*-butyl acrylate polymerization as the porhorincobalt complex.[165]

Acrylate polymerizations with organo–cobalt porphyrin complexes have been shown to proceed by both a reversible transfer and a degenerative transfer mechanism.[166,167] For vinyl acetate polymerization the mechanism has been shown to vary according to the reaction conditions. In bulk, a degenerative chain transfer mechanism predominates but in the presence of water or pyridine a reversible termination mechanism prevails.[168,169]

11.4.3.4 Organotellurium-mediated living-radical polymerizations

First introduced by Yamago *et al.*,[170–172] the organotellurium-mediated LRP (TERP) was shown to work well with a variety of monomers, including styrene, acrylates vinyl acetate and methacrylates, although the molecular weight distributions for methacrylates were broad. The addition of a small amount of dimethyl ditelluride $(MeTe)_2$,[171,173] which serves to deactivate the propagating chain in the methacrylate polymerizations, was shown to enable better control and much improved molecular weight distributions.[174]

The tellurium structures, R–TeL, contain an R group, which acts as the initiator for the polymerization, and an L non-transferable substituent, which remains with the tellurium. The TeL moiety reacts reversibly with the propagating chain. Structures **25–30** are typical.

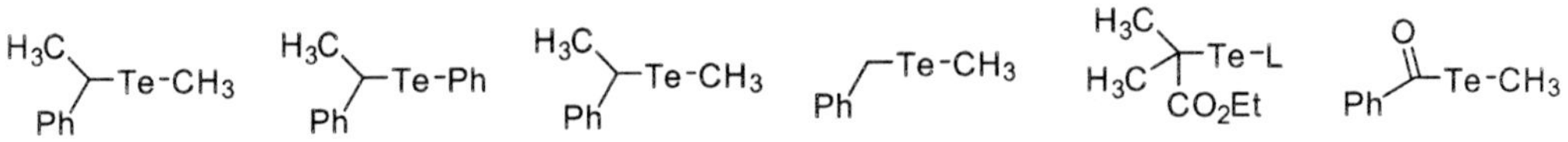

25 **26** **27** **28** **29** **30**

A recent study has shown that the L group can affect both the rates of polymerization and molecular weight distributions in styrene polymerizations. Thus, in the case of **29**, improved results were obtained when L was phenyl, *p*-methoxyphenyl, *p*-trifluoromethylphenyl in comparison to when it was 2-furyl and n-butyl or methyl.[175]

Initial work with styrene concluded both reversible termination and degenerative transfer (DT) mechanisms operate, with the latter dominating.[173] It was subsequently shown that the degenerative transfer mechanism also predominates for acrylate, methacrylate, and vinyl acetate polymerizations.[176] However, in the case of vinyl acetate, the polymers exhibit significant head-to-head addition and broad molecular weight distributions at high monomer conversion; the latter result due to a small degenerative chain transfer constant for vinyl acetate.

Organostibines[177–179] and organobismuthines[180] complexes have also shown to work well but they also tend to be more sensitive to oxygen than organtellurium complexes and are more problematic to work with.

11.4.3.5 Reverse chain transfer catalyzed polymerization (RTCP)

Iodide-mediated radical polymerization,[181–183] involving an alkyl iodide as an initiating dormant species, and a conventional radical initiator to produce $P^\bullet$, has recently been improved upon with the use of germanium in the form of germanium(IV) iodide (GeI_4) and germanium(II) iodide (GeI_2),[184] tin in the form of tin(II) iodide (SnI_2) and tin(IV) iodide (SnI_4), and phosphorous[185] in the form of phosphorus(III) iodide (PI_3). The role of the metal halides, illustrated with GeI_4, is shown in Scheme 11.4. The GeI_4, acting as a chain transfer agent, serves to deactivate the propagating radical chain, while P-I is activated with $^\bullet GeI_3$ in a reversible chain transfer process. Reaction temperatures for styrene are 80 °C or 100 °C with reaction times varying from 4 to 24 hours depending upon the system. M_w/M_n's are typically in the 1.15–1.25 range.

RTCP has recently been extended to the use of N-iodosuccinimide (NIS) as a catalyst to deactivate the propagating polymer chain in a reversible manner. In these polymerizations, dicumyl peroxide or AIBN was used to initiate $P^\bullet$, which reacts with NIS to produce the $NS^\bullet$ species that then reacts reversibly with P-I. These polymerizations have been shown to work quite well for styrene and a variety of methacrylates.[186]

11.5 Aqueous stable free radical polymerization processes

A distinct advantage that the SFRP processes have over anionic polymerization is a lack of sensitivity to water, thus allowing them to proceed under miniemulsion, emulsion, and suspension conditions. This

Scheme 11.4 Reverse chain transfer catalyzed polymerization illustrated with GeI_4

provides the opportunity to produce block copolymers and polymers with complex architectures using a process that enables the industrial production of large-scale materials and eliminates the high viscosity and heat dissipation issues that exist in bulk polymerizations. Heterogeneous processes also offer the advantage of being able to reach high molar mass polymers with high conversion at a faster rate of polymerization as compared to bulk or solution systems.

11.5.1 Living-radical miniemulsion polymerization

Mechanistically, miniemulsion systems are much simpler than emulsion systems since they begin with preformed, small, stabilized droplets. Nucleation and further polymerization of these droplets allow for the formation of polymer particles. An issue with miniemulsion systems is the need for a small molecule hydrophobe, such as hexadecane, to impart stability to the small monomer droplets against diffusional degradation (Ostwald ripening). The hydrophobe is often regarded as an impurity in the final latex particle. Furthermore, the use of a high shear device is required for the formation of the small droplets, adding difficulty to large-scale syntheses. Nevertheless, colloidally stable latexes are readily achieved using living-radical techniques under these conditions.

Initial SFR miniemulsion polymerizations[187,188] were performed using either BPO[189,190] or potassium persulfate (KPS)[191] as the initiator, TEMPO as the stable radical mediator, and hexadecane as the costabilizer. To avoid the use of aqueous phase initiation and remove the need for hexadecane, miniemulsion polymerizations were subsequently performed using long-chain alkoxyamines or TEMPO-terminated polystyrene oligomer with good success.[192,193]

While initial miniemulsion polymerizations were performed between 125–135 °C in a pressurized reaction vessel, the use of SG1 (**6**) allowed the temperature to be dropped to 90 °C.[194] Miniemulsion polymerizations using TEMPO have been accomplished at 100 °C through the use of rate enhancing additives; however, the polymers tended to exhibit a broad molecular weight distributions (M_n/M_w ~1.4–1.6) attributed to a low activation rate due to the lower temperature.[195]

Various reaction parameters have been reported to affect miniemulsion polymerizations. Cunningham *et al*. found an increase in the rate of polymerization with an increase in the concentration of the surfactant sodium dodecyl benzenesulfonate (SDBS) at a constant particle size of ~120 nm. No decrease in the degree of livingness was observed.[196,197] Nakamura *et al*.[198] reported that particle sizes of greater than 170 nm give polymerization rates similar to bulk polymerizations and good control while smaller sizes give faster reaction rates but less livingness. The results were explained by suggesting that a fraction of the free TEMPO becomes deactivated by being absorbed at the interface between the organic and aqueous phase resulting in a decrease in the deactivation rate.[199] The loss in livingness was further noted in the preparation of block copolymers [200,201] Shifts to higher molecular weights of the initial homopolymer were evident but the gel permeation chromatography plots showed a low molecular weight shoulder attributed to dead homopolymer.

The concentration of a styrene-TEMPO macroinitiator has also been shown to affect the polymerization, with concentrations below 0.02 M giving poor results due to the contribution of autoinitiation and the suppression of deactivation due to the interfacial activity of TEMPO.[202,203] The latter argument was also used to explain why bimolecular termination is higher in miniemulsion systems than in solution and increases with decreasing particle size.[204] Also, in agreement with earlier studies at lower temperatures, Alam *et al*.[205] reported that the spontaneous thermal rate of initiation is 3–15 times faster in miniemulsion than in bulk at temperatures of 110 °C and 125 °C.

There has been a lot of work dealing with compartmentalization, with a conclusion that for a styrene TEMPO system at 125 °C and [PS–TEMPO] = 0.02 M, a particle size of less than 70 nm is required to see slowing of the rate of polymerization due to compartmentalization of both the propagating chain and the free nitroxide.[206,207] The slowing of the rate of propagation is attributed to an increase in the deactivation of the propagating chain by free nitroxide, thus preventing free nitroxide from accumulating. Similar results were reported by Tobita *et al.*[208] who, along with Zutterlund, related the results to a confined space effect.

On the other hand, Charleux[209] considered that compartmentalization of free nitroxide does not occur and the lower amount of free nitroxide is due to the compartmentalization of the propagating chain, which in turn reduces the amount of irreversible termination, the process that generates the free nitroxide. In a recent paper, it was shown that, in fact, SG1 rapidly diffuses between the various phases of an emulsion system and can not be considered compartmentalized.[210] Thus, the choice of nitroxide may prove to critical to the degree of compartmentalization.

11.5.2 Emulsion polymerization

The establishment of a viable stable free radical emulsion polymerization process has been harder to realize than that of miniemulsion polymerization. Mechanistically, emulsion polymerization is more complex than miniemulsion polymerization because it involves the very intricate process of particle nucleation, which is ultimately responsible for particle formation. The final product of an emulsion is a latex, which consists of nano-sized particles suspended in water. Gilbert[211] and Charleux *et al.*[212] have written noteworthy reviews of controlled radical polymerization under emulsions conditions, with the latter review focusing on the use of SG1.

Early attempts at conducting a conventional emulsion polymerization using living-radical techniques were fraught with problems and produced emulsions with poor colloidal stability, as seen by the formation of large particles or phase separation of the latex into an upper foamy oily layer and a lower aqueous layer.[213,214] When colloidal stability was achieved, it was often at the expense of the system's livingness and resulted in polymers with poor molecular weight control or broad molecular weight distributions.

Stable emulsion latexes were first realized using a seeded emulsion polymerization system. A polystyrene seed latex, having an average particle diameter of 90 nm and prepared by a conventional radical polymerization process, was swollen with styrene and a living-radical polymerization was conducted inside the swollen particles using an alkoxyamine initiator.[215] While a high conversion latex was obtained, the reaction was slow and the molecular weight distributions were broad with tailing towards the lower molecular weight region.

Nicolas *et al.*[216] reported a two-step emulsion polymerization using a water soluble sodium carboxylate group containing alkoxyamine. In the first step, a seed latex was created by a polymerization of *n*-butyl acrylate under very dilute emulsion polymerization conditions (0.7 % monomer loading). A batch polymerization was then performed using the seed latex from the first step. In a subsequent paper the use of a SG1-terminated poly(acrylic acid) macroinitiator was reported.[217] Emulsion polymerization of styrene or *n*-butyl acrylate conducted under batch conditions (20 % monomer loading) without any added surfactant yielded colloidally stable latexes. However, molecular weight distributions increased throughout the polymerization (final M_w/M_n between 2.0 and 3.0) due to the formation and accumulation of dead chains.

A nanoprecipitation process to create emulsion particles was reported by Szkurhan *et al.*[218] A preformed TEMPO-terminated oligomer, which serves as both a macroinitiator and particle stabilizer, dissolved in acetone, is precipitated in an aqueous solution of poly(vinyl alcohol). Removal of acetone and subsequent heating of the remaining aqueous phase in the presence of styrene yields stable latexes with particles of ~500 nm. The polymer exhibits excellent chain growth over time while maintaining narrow molecular weight distributions.

Cunningham *et al.*[219] modeled the thermodynamics of monomer transfer between droplets and particles and concluded that TEMPO-mediated *ab initio* emulsion polymerizations are not feasible due to a propensity for polymerization to occur in the large monomer droplets leading to coagulum in the final product. On the other hand, Cunningham *et al.*[220] have reported a successful surfactant free emulsion polymerization of styrene conducted at 90 °C with potassium persulfate and SG1, although molecular weight distributions were somewhat broad, M_w/M_n ~1.8.

11.5.3 Other aqueous polymerization processes

Stable free radical suspension polymerizations have been performed in which suspension beads of poly(styrene)-*b*-poly(styrene-co-acrylonitrile) and poly(styrene)-*b*-poly(styrene-*co*-butyl methacrylate) were prepared.[221,222] A two-step process was used to create stable suspension beads in which the aqueous phase first had to be prepared in a separate step requiring the surfactants to be boiled. Although stable polymer beads were obtained using this process, the polymers produced showed very broad molecular weight distributions, which were attributed to the presence of dead chains that were calculated to be 36 % of the total number of chains at 72 % monomer conversion, a rather significant contribution.

Dispersion polymerization is a heterogeneous process in which the monomer and polymeric steric stabilizer are both soluble in the continuous phase. As the polymerization proceeds, the polymer produced becomes insoluble in the continuous phase and precipitates, adsorbing stabilizer onto itself, resulting in a stabilized polymer particle. The resulting product is a stable colloidal dispersion of polymer latex particles that range in size from 500 nm to 1 μm. Unlike the other aqueous processes where the continuous phase is water, the continuous phase in a dispersion polymerization is typically an aqueous solution of an alcohol, such as *n*-pentanol or *n*-butanol.

A stable free radical dispersion polymerization of styrene has been accomplished using TEMPO as the stable radical mediator. Dispersion polymerizations were performed in an aqueous ethylene glycol solution.[223] The polymers produced were living and produced polymers with narrow molecular weight distributions (M_w/M_n ~1.2); however, high polymer conversion was difficult to obtain with typical conversions being in the 20–60 % range. Also, the resulting dispersions produced very large particles (>1 μm) when compared to dispersion polymerizations performed under the same conditions using conventional radical polymerization. Living-radical dispersion polymerization has also been performed in water using a surfactant free approach for the polymerization of *N,N*-dimethylacrylamide.[224] This was accomplished using a two-step process wherein a short poly(acrylic acid) oligomer was first synthesized using SG1 as the mediating radical. This oligomer was then used as a polyelectrolyte macroinitiator in the dispersion polymerization of poly(*N,N*-dimethylacrylamide). Since the polymerization was performed at an alkaline pH, the poly(acrylic acid) segment resulting from the oligomer acts as a charged stabilizer for the resulting particles. The final polymer obtained is a amphiphilic diblock copolymer containing a short poly(acrylic acid) block and a long poly(*N,N*-dimethylacrylamide) block. The resulting dispersion gave a stable fluid latex with particles of 100 nm diameter.

11.6 The application of stable free radical polymerization to new materials

11.6.1 Statistical copolymers

Statistical copolymers are produced by the copolymerization of two or more monomers that have similar reactivity. Although statistical copolymers can be produced by conventional radical polymerization, the final product obtained is a mixture of copolymers. Since chain initiation and termination occur throughout the course of conventional polymerizations, as the process proceeds the monomer feed ratio changes. Thus, polymer chains initiated at different times during the course of the polymerization contain different monomer ratios. Conversely, statistical copolymers produced by living-radical polymerization yield copolymers that contain the same composition, since all chains are initiated simultaneously and grow at the same rate. Using nitroxide-mediated polymerization styrene has been copolymerized with acrylates,[225–229] methacrylates,[225,229,230] other styrene derivatives,[225,231–237] acrylonitrile,[226,238–240] vinyl pyridine,[241–243] isoprene,[239] malimides,[244,245] *N,N*-dimethylacrylamide,[224] acrylic acid,[246] methacrylic acid,[247] and maleic anhydride.[248] SFRP statistical copolymers have also been created using monomer combinations that do not include styrene, as, for example, *N*-isopropylacrylamide with *N*-*tert*-butylacrylamide.[249]

11.6.2 Block copolymers

Block copolymer formation beginning with nitroxide-terminated polystyrene has been extensively studied,[101,234,235,250–258] as have block copolymers prepared by nitroxide-terminated homopolymers other than polystyrene. 4-OxoTEMPO-terminated *n*-butyl acrylate homopolymers were used to prepare poly(*n*-butyl acrylate-*b*-styrene) and poly(*n*-butyl acrylate-*b*-*t*-butyl acrylate).[259] In a similar manner TIPNO-terminated poly(*n*-butyl acrylate) gave poly(*n*-butyl acrylate-*b*-styrene).[101] The synthesis of poly(2-vinylpyridine-*b*-acrylonitrile) was reported by Lokaj *et al.*[260]

Amphiphilic diblock copolymers containing segments of monomethoxypoly(ethylene glycol) and polystyrene (MPEG-b-PS) were synthesised using a MPEG-TEMPO macroinitiator.[261] Block copolymers containing methacrylates have also been synthesized by preparing a first block of a random copolymer of MMA with styrene and chain extending this SG1-terminated polymer with either styrene or *n*-butyl acrylate.[119] Another unique block copolymer was formed by the polymerization of *n*-butyl acrylate in the presence of SG1, which was then chain extended with a mixture of methyl methacrylate and *N,N*-dimethyl acrylamide to create a random copolymer as the second block. An *n*-butyl acrylate macroinitator containing an alkoxyamine SG1 linkage on both ends was also synthesized and chain extended with methyl methacrylate and *N,N*-dimethyl acrylamide to give a triblock copolymer.[262]

The RAFT and SFRP processes have been combined to make amphiphilic and thermosensitive graft copolymers, such as P(St-*co*-(p-CMS))-*g*-PNIPAAM.[263] Other amphiphilic block copolymers containing segments of monomethoxypoly(ethylene glycol) and polystyrene (MPEG-b-PS) using TEMPO have been reported by Shoaeifar *et al.*[264] The formation of block copolymers containing protic functionalities has also been accomplished. An amphiphilic block copolymer was synthesized beginning with the polymerization of acrylic acid in the presence of a small amount of styrene to form a SG1-terminated polymer that was then chain extended with styrene.[265] The synthesis of tapered blocks has been reported by Jabber *et al.*[266]

Triblock copolymers of styrene/isoprene/styrene and styrene/butadiene/styrene using TEMPO as the nitroxide mediator have been prepared with narrow molecular weight distributions ($M_w/M_n \sim 1.3$).[267] Semicrystalline poly(octadecyl acrylate-*b*-methyl acrylate) and poly(methyl acrylate-*b*-octadecyl acrylate-*b*-methyl acrylate) copolymers have recently been prepared using SG1 as the mediating nitroxide.[268] Nitroxide-mediated polymerization of styrene, *n*-butyl acrylate, *t*-butyl acrylate, isoprene, and dimethylacrylamide to form symmetrical ABA triblock copolymers was executed using a TIPNO-terminated bidirectional alkoxyamine.[269] Triblocks, comprising poly(*n*-butyl acrylate) as a first or central block and polystyrene or poly(methyl methacrylate) as a second or outer block, have also been made using SG1 under emulsion conditions to give nanostructured latex particles.[270]

As an extension to the block copolymer work the synthesis of comb/brush structures,[271–276] star polymers[277–282] and hyperbranched polymers have been reported.[283] Two recent reviews are recommended. One covers click chemistry and includes a section on the use of living-radical polymerizations processes to make graft copolymers[284]; the second summarizes the progress in the field of polymer conjugates[285] and again includes a section on the use of living-radical polymerization processes to obtain polymer conjugates.

11.7 Conclusions

Clearly there has been enormous progress in recent years in the development, understanding, and expansion of living-radical polymerization processes. The question may still remain, what is the future of living-radical polymerizations? I am not sure that the answer is evident at this point. Much progress has been made with respect to the development and understanding into the process and this bodes well for the future. The future lies with the skill and the imagination of the scientists working in the field and we leave it in their very capable hands to determine the future.

List of abbreviations

BPO	Benzoyl peroxide
GPC	Gel permeation chromatography
k_d	Dissociation rate constant
k_c	Recombination rate constant
K_{eq}	Equilibrium constant
KPS	Potassium persulfate
M_n	Number average molecular weight
M_w	Weight average molecular weight
MMA	Methyl methacrylate
PDI	Polydispersity Index (PDI = M_w/M_n)
PRE	Persistent radical effect
SFRP	Stable free radical polymerization
TEMPO	2,2,6,6-Tetramethylpiperdinyl-*N*-oxy
TEMPO-TMS	*trans*-2,6-diethyl-2,6-bis(1-trimethylsiloxanoxyethyl)-1-(1-phenylethoxy)piperidine-*N*-oxy

References

1. Bamford, H,; Barb, W. G.; Jenkins, A. D.; and Onyon, P. F. *The Kinetics of Vinyl Polymerization by Radical Mechanisms*. Butterworths, London, 1958.
2. Morawetz, H. *Polymers. The Origins and Growth of a Science*. John Wiley & Sons, Inc., New York, 1985.
3. Moad, G.; and Solomon D. H. *The Chemistry of Free Radical Polymerization*. Pergamon Press, Oxford, England, 1995.
4. M. K. Georges, R. P. N Veregin, P. M. Kazmaier and G. K. Hamer, *Trends Polym. Sci.*, **2**, 66–72 (1994).
5. V. Sciannamea, R. Jerome and C. Detrembleur, *Chem. Rev.*, **108**, 1104–1126 (2008).
6. J.-S. Wang and K. Matyjaszewski, *J. Am. Chem. Soc.*, **117**, 5614–5615 (1995).
7. K. Matyjaszewski and J. Xia, *Chem. Rev.*, **101**, 2921–2990 (2001).
8. J. Chiefari, Y. K. B. Chong, F. Ercole, *et al.*, *Macromolecules*, **31**, 5559–5562 (1998).
9. Y. Yagci and M. A. Tasdelen, *Prog. Polym. Sci.*, **31**, 1133–1170 (2006).
10. W. A. Braunecker and K. Matyjaszewski, *Prog. Polym. Sci.*, **32**, 93–146 (2007).
11. T. Vogler and A. Studer, *Synthesis*, **13**, 1979–1993 (2008).
12. M. Szwarc, *Nature*, **178**, 1168–1169 (1956).
13. M. Szwarc, M. Levy and R. Milkovich, *J. Am. Chem. Soc.*, **78**, 2656–2657 (1956).
14. O. W. Webster, *Science*, **251**, 887–893 (1991).
15. H. Fischer, *Criteria for Livingness and Control in Nitroxide-Mediated and Related Radical Polymerizations*. ACS Symp. Ser.,, **854**, 10–23 (2003).

16. T. Otsu, *J. Polym. Sci.*, **21**, 559–561 (1956).
17. Otsu, T., *J. Polym. Sci.*, **26**, 236–239 (1957).
18. T. Otsu and K. Nayatani, *Makromol. Chem.*, **27**, 149–156 (1958).
19. For a comprehensive review of iniferter chemistry see: T. Otsu and A. Matsumoto, *Advances in Polymer Science*, **136**, 75–137 (1998).
20. T. Otsu and M. Yoshida, *Makromol. Chem.Rapid Commun.*, **3**, 127–132 (1982).
21. For a review, see G. Clouet, *Rev. Macromol. Chem. Phys.*, **31**, 311–340 (1991).
22. S. R. Turner and R. W. Blevins, *Macromolecules*, **23**, 1856–1859 (1990).
23. D. Braun and K. H. Becker, *Makromol. Chem.* **147**, 91–99 (1971).
24. D. Braun and K. H. Becker, *Ind. Eng. Chem. Prod. Res. Dev.*, **10**, 386–388 (1971).
25. A. Bledzki and D. Braun, *Makromol. Chem.*, **182**, 1047–1056 (1981).
26. D. Braun and R. Rengel, *Die Angew. Makromol. Chem.*, **98**, 265–277 (1981).
27. A. Bledzki, D. Braun, W. Menzel and K. Titzschkau, *Makromol. Chem.*, **184**, 287–294 (1983).
28. A. Bledzki and D. Braun, *Makromol. Chem.*, **187**, 2599–2608 (1986).
29. A. Bledzki and D. Braun, *Makromol. Chem.*, **184**, 745–754 (1983).
30. D. Braun, H. J. Lindner and H. Tretner, *Eur. Poly. J.*, **25**, 725–730 (1989).
31. For a review see: D. Braun, *Macromol. Symp.*, **111**, 63–71 (1996).
32. Solomon, D. H.; Rizzardo, E.; and Cacioli, P. European Patent 135280, 1985.
33. G. Moad, E. Rizzardo and D. H. Solomon, *Macromolecules*, **15**, 909–914 (1982).
34. Solomon, D. H.; Rizzardo, E.; and Cacioli, P. US Patent 4 581 429, 1986.
35. E. Rizzardo, *Chem, Aust.*, **54**, 32–33 (1987).
36. For a history of the work performed at CSIRO see: G. Moad, E. Rizzardo and S. H. Thang, *Acc. Chem. Res.*, **41**, 1133–1142 (2008).
37. E. Endo, K. Murata and T. Otsu, *Macromolecules*, **23**, 5554–5556 (1992).
38. M. K. Georges, R. P. N. Veregin, P. M. Kazmaier and G. K. Hamer, *Macromolecules*, **26**, 2987 (1993).
39. T. Fukuda, T. Terauchi, A. Goto, *et al.*, *Macromolecules*, **29**, 6393–6398 (1996).
40. For a review on the inevitability of termination in these living-radical polymerization systems see: K. Matyjaszewski, S. Gaynor, D. Greszta, *et al.*, *J. Phy. Org. Chem.*, **8**, 306–315 (1995).
41. P. A. Clay and R. G. Gilbert, *Macromolecules*, **28**, 552–569 (1995).
42. H. Fischer, *Macromolecules*, **30**, 5666–5672 (1997).
43. H. Fischer, *J. Polym. Sci. Part A: Polym. Chem.*, **37**, 1885–1901 (1999).
44. M. Souaille and H. Fischer, *Macromolecules*, **33**, 7378–7394 (2000).
45. M. Souaille and H. Fischer, *Macromolecules*, **34**, 2830–2838 (2001).
46. M. Souaille and H. Fischer, *Macromolecules*, **35**, 248–261 (2002).
47. A. Goto and T. Fukuda, *Prog. Polym. Sci.*, **29**, 329–38 (2004).
48. T. Fukuda, *J. Polym. Sci. Part A: Polym. Chem.*, **42**, 4743–4755 (2004).
49. W. G. Skene, S. T. Belt, T. J. Connolly, *et al.*, *Macromolecules*, **31**, 9103–9105 (1998).
50. G. Moad and E. Rizzardo, *Macromolecules*, **28**, 8722–8728 (1995).
51. S. Marque, C. Le Mercier, P. Tordo and H. Fischer, *Macromolecules*, **33**, 4403–4410 (2000).
52. S. A. F. Bon, G. Chambard and A. L. German, *Macromolecules*, **32**, 8269–8276 (1999).
53. R. P. N. Veregin, M. K. Georges, G. K. Hamer and P. M. Kazmaier, *Macromolecules*, **28**, 4391–4398 (1995).
54. A. Goto, T. Terauchi, T. Fukuda and T. Miyamoto, *Macromol. Rapid Commun.*, **18**, 673–681 (1997).
55. A. Goto and T. Fukuda, *Macromolecules*, **30**, 5183–5186 (1997).
56. T. Fukuda and A. Goto, *Macromol. Rapid Commun.*, **18**, 683–688 (1997).
57. O. G. Ballesteros, L. Maretti, R. Sastre and J. C. Scaiano, *Macromolecules*, **34**, 6184–6187 (2001).
58. D. Bertin, D. Gigmes, S. Marque and P. Tordo, *e-Polymers*, **002**, 1–9 (2003).
59. L. C. Li., G. K. Hamer and M. K. Georges, *Macromolecules*, **39**, 9201–9207 (2006).
60. M. Dollin, A. R. Szkurhan and M. K. Georges, *J. Polym. Sci. Part A: Polym. Chem.*, **45**, 5487–5493 (2007).
61. C. J. Hawker, *J. Am. Chem. Soc.*, **116**, 11185–11186 (1994).
62. J. L. Lukkarila, G. K. Hamer and M. K. Georges, *Tetrahedron Lett.*, **45**, 5317–5319 (2004).
63. I. Q. Li, B. A. Howell, M. T. Dineen, *et al.*, *Macromolecules*, **30**, 5195–5199 (1997).

64. T. J. Connolly, M. V. Baldovf, N. Mohtat and J. C. Scaiano, *Tetrahedron Lett.*, **37**, 4919–4922 (1996).
65. J. Dao, D. Benoit and C. J. Hawker, *J. Polym. Sci. Part A: Poly. Chem.*, **36**, 2161–2167 (1998).
66. K. Matyjaszewski, B. E. Woodworth, X. Zhang, *et al.*, *Macromolecules*, **31**, 5955–5957 (1998).
67. D. Wang and Z. Wu, *Macromolecules*, **31**, 6726–3610 (1998).
68. N. L. Hill and R. Braslau, J. *Polym. Sci. Part A: Polym. Chem.*, **45**, 2341–2349 (2007).
69. A. Dichtl, M. Seyfried and K. U. Schoening, *Synlett.*, **12**, 1877–1881 (2008).
70. Y. Guillaneuf, J. L. Couturier, D. Gigmes, *et al.*, *J. Org. Chem.*, **73**, 4728–4731 (2008).
71. G. Moad and E. Rizzardo, *Macromolecules*, **28**, 8722–8728 (1995).
72. S. Marque, J. Sobek, H. Fischer, *et al.*, *Macromolecules*, **36**, 3440–3442 (2003).
73. J. Sobek, R. Martschke and H. Fischer, *J. Am. Chem. Soc.*, **123**, 2849–2857 (2001).
74. Y. Miura, N. Nakamura and I. Taniguchi, *Macromolecules*, **34**, 447–455 (2001).
75. G. Moad, A. G. Anderson, F. Ercole, *et al.*, *Controlled Radical Polymerization*, ACS Symp. Ser., **685**, 332–360 (1998).
76. T. Schulte and A. Studer, *Macromolecules*, **36**, 3078–3084 (2003).
77. See also: C. C. Chang, K. O. Siegenthaler and A. Studer, *Helv. Chim. Acta.*, **89**, 2200–2210 (2006).
78. D. Gigmes, A. Gaudel-Siri, S. R. A. Marque, *et al.*, *Helv. Chim. Acta*, **89**, 2312–2326 (2006).
79. D. Bertin, D. Gigmes, S. R. A. Marque, *et al.*, *Chem. Phys. Chem*, **9**, 272–281 (2008).
80. D. Bertin, P. E. Dufils, I. Durand, *et al.*, *Macromol. Chem. and Phys.*, **209**, 220–224 (2008).
81. E. Bagryanskaya, D. Bertin, D. Gigmes, *et al.*, *Macromol. Chem and Phys.*, **209**, 1345–1357 (2008).
82. K. O. Siegenthaler and A. Studer, *Macromolecules*, **39**, 1347–1352 (2006).
83. Y. Miura, N. Nakamura, I. Taniguchi and A. Ichikawa, *Polymer*, **44**, 3461–3467 (2003).
84. C. Wetter, J. Gierlich, C. A. Knoop, *et al.*, *Chem. Eur. J.*, **10**, 1156–1166 (2004).
85. C. A. Knoop and A. Studer, *J. Am. Chem. Soc.*, **125**, 16327–16333 (2003).
86. A. Studer, K. Harms, C. Knoop, *et al.*, *Macromolecules*, **37**, 27–34 (2004).
87. A. Studer and T. Schulte, *Chem. Rec.*, **5**, 27–35 (2005).
88. A. Debuigne, D. Chan-Seng, L. Li, *et al.*, *Macromolecules*, **40**, 6224–6232 (2007).
89. O. Lagrille, N. R. Cameron, P. A. Lovell, *et al.*, *J. Polym. Sci., Part A: Polym. Chem.*, **44**, 1926–1940 (2006).
90. R. Braslau, G. O'Bryan, A. Nilsen, *et al.*, *Synthesis*, **9**, 1496–1506 (2005).
91. C. H. J. Johnson, G. Moad, D. H. Solomon, *et al.*, *Aust. J. Chem.*, **43**, 1215–1230 (1990).
92. H. Fischer, *Chem. Rev.*, **101**, 3581–3610 (2001).
93. L. B. Volodarsky, V. A. Reznikov, and V. I. Ovcharenko, *Synthetic Chemistry of Stable Nitroxides*, CRC Press, Ann Arbour, MI, 1994, pp 5–6
94. M. K. Georges, P. G. Odell, N. A. Listigovers and M. H. Quinlan, *Solvent Free Polymerization and Processes*, ACS Symp. Ser., **713**, 80–95 (1998).
95. J. C. Hawker, A. Bosman and E. Harth, *Chem. Rev.*, **101**, 3661–3688 (2001).
96. N. R. Cameron and R. Alistair, *Macromolecules*, **35**, 9890–9895 (2002).
97. C. A. Knoop and A. Studer, *J. Am. Chem. Soc.*, **125**, 16327–16333 (2003).
98. D. H. Solomon, *J. Polym. Sci., Part A: Polym. Chem.*, **43**, 5748–5764 (2005).
99. D. Bertin, D. Gigmes, S. R. A. Marque and P. Tordo, *Macromolecules*, **38**, 2638–2650 (2005).
100. G. S. Anachenko, M. Souaille, H. Fischer, *et al.*, *J. Polym. Sci., Part A: Polym. Chem.*, **40**, 3264–3283 (2002).
101. D. Benoit, V. Chaplinski, R. Braslau and C. J. Hawker, *J. Am. Chem. Soc.*, **121**, 3904–3920 (1999).
102. D. Benoit, S. Grimaldi, S. Robin, *et al.*, *J. Am. Chem. Soc.*, **122**, 5929–5939 (2000).
103. Y. Guillaneuf, J. L. Couturier, D. Gigmes, *et al.*, *J. Org. Chem.*, **73**, 4728–4731 (2008).
104. D. Benoit, E. Harth, P. Fox, *et al.*, *Macromolecules*, **33**, 363–390 (2000).
105. D. Bertin, D. Gigmes, C. Le Mercier, *et al.*, *J. Org. Chem.*, **69**, 4925–4930 (2004).
106. F. Chauvin, P. E. Dufils, D. Gigmes, *et al.*, *Macromolecules*, **39**, 5238–5250 (2006).
107. F. R. Mayo, *J. Am. Chem. Soc.*, **90**, 1289–1295 (1968).
108. Y. K. Chong, E. Rizzardo and D. H. Solomon, *J. Am. Chem. Soc.*, **105**, 7761–7762 (1983).
109. M. K. Georges, J. L. Lukkarila and A. R. Szkurhan, *Macromolecules*, **37**, 1297–1303 (2004).
110. A. Debuigne, T. Radhdkrishnan and M. K. Georges, *Macromolecules*, **39**, 5359–5363 (2006).
111. A. Nielson and R. Brauslau, J. *Polym Sci., Part A: Polym. Chem.*, **44**, 697–717 (2006).

112. P. Lacroix-Desmazes, J. F. Lutz, F. Chauvin, *et al.*, *Macromolecules*, **34**, 8866–8871 (2001).
113. J. A. Li, X. L. Zhu, J. Zhu and Z. P. Cheng, *J. Macromol. Sci., Part A: Pure Appl. Chem.*, **44**, 41–46 (2006).
114. A. Mannan, A. Ichikawa and Y. Miura, *Polymer*, **48**, 743–749 (2007).
115. G. O'Bryan, A. Nilsen and R. Braslau, *Macromolecules*, **40**, 7848–7854 (2007).
116. R. Nicolay, L. Mar, P. Hemery and K. Matyjaszewski, *Macromolecules*, **40**, 6067–6075 (2007).
117. C. Burguiere, M. A. Dourges, B. Charleux and J. P. Vairon, *Macromolecules*, **32**, 3883–3890 (1999).
118. G. S. Ananchenko and H. Fischer, *J. Polym. Sci., Part A: Polym. Chem.*, **39**, 3604–3621 (2001).
119. B. Charleux, J. Nicolas and O. Guerret, *Macromolecules*, **38**, 5485–5492 (2005).
120. J. Nicolas, C. Dire, L. Müeller, *et al.*, *Macromolecules*, **39**, 8274–8282 (2006).
121. R. McHale, F. Aldabbagh and P. B. Zetterlund, *J. Polym. Sci., Part A: Polym. Chem.*, **45**, 2194–2203 (2007).
122. Y. Guillaneuf, D. Gigmes, S. R. A. Marque, *et al.*, *Macromolecules*, **40**, 3108–3114 (2007).
123. O. E. Ansong, S. Jans, Y. Wei, *et al. Polym. Int.*, **57**, 863–871 (2008)
124. M. K. Georges, R. P. N. Veregin, P. M. Kazmaier, *et al.*, *Macromolecules*, **27**, 7228–7229 (1994).
125. R. P. N. Veregin, P. G. Odell, L. M. Michalak and M. K. Georges, *Macromolecules*, **29**, 4161–4163 (1996).
126. E. E. Malmstrom, C. J. Hawker and R. D. Miller, *Tetrahedron*, **53**, 15225–15236 (1997).
127. P. G. Odell, R. P. N. Veregin, L. M. Michalak and M. K. Georges, *Macromolecules*, **30**, 2232–2237 (1997).
128. B. Keoshkerian, M. K. Georges, M. Quinlan, *et al.*, *Macromolecules*, **31**, 7559–7561 (1998).
129. K. H. Kuo, W. Y. Chiu and K. C. Cheng, *Polym. Int.*, **57**, 730–737 (2008).
130. F. A. Neugebauer, H. Fischer and R. Kreiger, *Angew. Chem. Intl. Ed. Engl.*, **28**, 491–492 (1989).
131. F. A. Neugebauer and H. Fischer, *Tetrahedron*, **51**, 12883–12898 (1995).
132. D. Colombani, M. Steenbock, M. Klapper and K. Müllen, *Macromol. Rapid Commun.*, **18**, 243–251 (1997).
133. M. Steenbock, M. Klapper and K. Müllen, *Macromol. Chem. Phys.*, **199**, 763–769 (1998).
134. A. Dasgupta, T. Brand, M. Klapper and K. Müllen, *Polym. Bull.*, **46**, 131–138 (2001).
135. N. S. Khelfallah, M. Peretolchin, M. Klapper and K. Müllen, *Polym. Bull.*, **53**, 295–304 (2005).
136. M. Steenbock, M. Klapper, K. Müllen, *et al.*, *Macromolecules*, **31**, 5223–5228 (1998).
137. R. Kuhn and H. Trischmann, *Angew. Chem. Intl. Ed. Engl.*, **2**, 155–157 (1963).
138. F A Neugebauer, *Angew. Chem. Intl. Ed. Engl.*, **12**, 455–464 (1973).
139. F. A. Neugebauer and H. Fischer, *Angew. Chem. Intl. Ed. Engl.*, **19**, 724–725 (1980).
140. F. A. Neugebauer, H. Fischer and R. Siegel, *Chem. Ber.*, **121**, 815–822 (1988).
141. F. A. Neugebauer, H. Fischer and R. Kreiger, *J. Chem. Soc. Perkins Trans.*, **2**, 535–544 (1993).
142. C. L. Barr, P. A. Chase, R. G. Hicks, *et al.*, *J. Org. Chem.*, **64**, 8893–8897 (1999).
143. For a review of verdazyls see: B. D Koivisto and R. G. Hicks, *Coordination Chem. Rev.*, **249**, 2612–2630 (2005).
144. For a second review of verdazyls see: R. G. Hicks, *Org. Biomol. Chem.*, **5**, 1321–1338 (2007).
145. M. Kinoshita and Y. Miura, *Makromol. Chem.*, **124**, 211–221 (1969).
146. J.-Z. Wu, E. Bouwman, J. Reedijk, *et al.*, *Inorg. Chim. Acta.*, **361**, 326–330 (2003).
147. J. B. Gilroy, B. D. Koivisto, R. McDonald, *et al.*, *J. Mater. Chem.*, **16**, 2618–2624 (2006).
148. B. Yamada, H. Tanaka, K. Konishi and T. Otsu, *J. Macromol. Sci. Pure Appl. Chem.*, **A31**, 351–366 (1994).
149. B. Yamada, Y. Nobukane and Y. Miura, *Polym. Bull.*, **41**, 539–544 (1998).
150. S. J. Teertstra, E. Chen, D. Chan-Seng, *et al.*, *Macromol. Symp.*, **248**, 117–125 (2007).
151. E. K. Y. Chen, S. J. Teertstra, D. Chan-Seng, *et al.*, *Macromolecules*, **40**, 8609–8616 (2007).
152. A. Yang, T. Kasahara, E. K. Y. Chen, *et al.*, *Eur. J. Org. Chem.*, 4571–4574 (2008).
153. T. C. Chung, W. Janvikul and H. L. Lu, *J. Am. Chem. Soc.*, **118**, 705–706 (1996).
154. T. C. Chung and H. Hong, *Advances in Controlled/Living Radical Polymerization*, ACS Symp. Ser., **854**, 481–495 (2003).
155. J. D. Druliner, *Macromolecules*, **24**, 6079–6028 (1991).
156. Druliner, J. D.; and Fryd, M. US Patent, 5,059,657, 1991
157. B. B. Wayland, G. Posznil, S. L. Mukerjee and J. Fryd, *J. Am. Chem. Soc.*, **116**, 7943–7944 (1994).
158. B. B. Wayland, L. Basickes, S. L. Mukerjee, *et al.*, *Macromolecules*, **30**, 8109–8112 (1997).
159. Z. Lu, J. Fryd and B. B. Wayland, *Macromolecules*, **37**, 2686–2687 (2004).
160. A. Debuigne, J.-P. Caille and R. Jérôme, *Angew. Chem. Int. Ed.*, **44**, 1101–1104 (2005).

161. Li, S.; B. deBruin, C.-H. Peng, *et al.*, *J. Am. Chem. Soc.*, **130**, 13373–13381 (2008).
162. A. Debuigne, C. Michaux, C. Jérôme, *et al.*, *Chem. Eur. J.*, **14**, 7623–7637 (2008).
163. A. Debuigne, J. Warant, R. Jérôme, *et al.*, *Macromolecules*, **41**, 2353–2360 (2008).
164. C. Peng, J. Scricco, S. Li, *et al.*, *Macromolecules*, **41**, 2368–2373 (2008).
165. H. Kaneyoshi and K. Matyjaszewski, *Macromolecules*, **38**, 8163–8169 (2005).
166. B. B. Wayland, C.-H. Peng, X. Fu, *et al.*, *Macromolecules*, **39**, 8219–8222 (2006).
167. C.-H. Peng, M. Fryd and B. B. Wayland, *Macromolecules*, **40**, 6814–6819 (2007).
168. S. Maria, H. Kaneyoshi, K. Matyjaszewski and R. Poli, *Chem. Eur. J.*, **13**, 2480–2492 (2007).
169. A. Debuigne, Y. Champouret, R. Jérôme, *et al.*, *Chem. Eur. J.*, **14**, 4046–4059 (2008).
170. S. Yamago, K. Iida and J. Yoshida, *J. Am. Chem. Soc.*, **124**, 2874–2875 (2002).
171. S. Yamago, K. Iida and J. Yoshida, *J. Am. Chem. Soc.*, **124**, 13666–13667 (2002).
172. S. Yamago, K. Iida, M. Nakajima and J. Yoshida, *Macromolecules*, **36**, 3793–3796 (2003).
173. A. Goto, Y. Kwak, T. Fukuda, *et al.*, *J. Am. Chem. Soc.*, **125**, 8720–8721 (2003).
174. K. Kwak, M. Tezuka, A. Goto, *et al.*, *Macromolecules*, **40**, 1881–1883 (2007).
175. E. Kayahara, S. Yamago, K. Kwak, *et al.*, *Macromolecules*, **41**, 527–529 (2008).
176. K. Kwak, A. Goto, T. Fukuda, *et al.*, *Macromolecules*, **39**, 4671–4679 (2006).
177. S. Yamago, B. Ray, K. Iida, *et al.*, *J. Am. Chem. Soc.*, **126**, 13908–13909 (2004).
178. Y. W. Kwak, A. Goto, T. Fukuda, *et al.*, *Z. Phys. Chem.*, **219**, 283–293 (2005).
179. B. Ray, M. Kotani and S. Yamago, *Macromolecules*, **39**, 5259–5265 (2006).
180. S. Yamago, E. Kayahara, M. Kotani, *et al.*, *Angew. Chem., Int. Ed.*, **46**, 1304–1306 (2007).
181. T. Yutani and M. Tatemoto, Eur. Pat. Appl. 048937OA1, 1991.
182. T. Higashimura, M. Sawamoto, M. Kamigaito and M. Kato, M. Jpn. Pat. Appl. JP07126322A, 1995.
183. K. Matyjaszewski, S. Gaynor and J. S. Wang, *Macromolecules* **28**, 2093–2095 (1995).
184. A. Goto, H. Zushi, N. Hirai, *et al.*, *J. Am. Chem. Soc.*, **129**, 13347–13354 (2007).
185. A. Goto, N. Hirai, Y. Tsujii and T. Fukuda, *Macromol. Symp.*, **261**, 18–22 (2008).
186. A. Goto, N. Hirai, T. Wakada, *et al.*, *Macromolecules*, **41**, 6261–6264 (2008).
187. For a review see: M. F. Cunningham, *Prog. Polym. Sci.*, **33**, 365–398 (2008).
188. A second review see: P. B. Zetterlund, Y. Kagawa and M. Okubo, *Chem. Rev.*, **108**, 3747–3794 (2008).
189. T. Prodpran, V. L. Dimonie, E. D. Sudol and M. S. El-Asser, *Proc. Am. Chem. Soc., Div. Polym. Mater. Sci. Eng.*, **80**, 534–535 (1999).
190. T. Prodpran, V. L. Dimonie, E. D. Sudol and M. S. El-Asser, *Macromol. Symp.*, **155**, 1–14 (2000).
191. P. J. MacLeod, B. Keoshkerian, P. G. Odell and M. K. Georges, *Proc. Am. Chem. Soc., Div. Polym. Mater. Sci. Eng.*, **80**, 539–540 (1999).
192. B. Keoshkerian, P. J. MacLeod and M. K. Georges, *Macromolecules*, **34**, 3594–3599 (2001).
193. G. Pan, E. D. Sudol, V. L. Dimonie and M. S. El-Aasser, *Macromolecules*, **35**, 6915–6919 (2002).
194. C. Farcet, M. Lansalot, B. Charleux, *et al.*, *Macromolecules*, **33**, 8559–8570 (2000).
195. M. F. Cunningham, D. C. T. Ng, S. G. Milton and B. Keoshkerian, *J. Polym. Sci., Part A: Polym. Chem.*, **44**, 232–242 (2006).
196. M. F. Cunningham, M. Lin and B. Keoshkerian, *JCT Res.*, **1**, 33–39 (2004).
197. M. Lin, J. C. C. Hsu and M. F. Cunningham, *J. Polym. Sci., Part A: Polym. Chem.*, **44**, 5974–5986 (2006).
198. T. Nakamura, P. B. Zetterlund and M. Okubo, *Macromol. Rapid. Commun.*, **27**, 2014–2018 (2006).
199. M. N. Alam, P. B. Zetterlund and M. Okubo, *Polymer*, **49**, 3428–3435 (2008).
200. J. Wakamatsu, M. Kawasaki, P. B. Zetterlund and M. Okubo, *Macromol. Rapid. Commun.*, **28**, 2346–2353 (2007).
201. P. B. Zetterlund, T. Nakamura and M. Okubo, *Macromolecules*, **40**, 8663–8672 (2007).
202. M. N. Alam, P. B. Zetterlund and M. Okubo, *Polymer*, **49**, 883–892 (2008).
203. M. N. Alam, P. B. Zetterlund and M. Okubo, *Polymer*, **49**, 3428–3435 (2008).
204. M. N. Alam, P. B. Zetterlund and M. Okubo, *J. Polym. Sci., Part A: Polym. Chem.*, **45**, 4995–5004 (2007).
205. M. N. Alam, P. B. Zetterlund and M. Okubo, *Polymer*, **49**, 883–892 (2008).
206. P. B. Zetterlund and M. Okubo, *Macromolecules*, **39**, 8959–8967 (2006).
207. P. B. Zetterlund and M. Okubo, *Macromol. Theory Simul.*, **16**, 221–226 (2007).

208. H. Tobita and F. Yanase, *Macromol. Theory Simul.*, **16**, 476–488 (2007).
209. B. Charleux, *Macromolecules*, **33**, 5358–5365 (2000).
210. G. Delaittre and B. Charleux, *Macromolecules*, **41**, 2361–2367 (2008).
211. R. G. Gilbert, *Aust. J. Chem.*, **59**, 693–711 (2006).
212. B. Charleux and J. Nicolas, *Polymer*, **48**, 5813–5833 (2007).
213. C. Marestin, C. Nöel, A. Guyot and J. Claverie, *Macromolecules*, **31**, 4041–4044 (1998).
214. J. Cao, J. He, C. Li and Y. Yang, *Polym. J.*, **33**, 75–80 (2001).
215. S. A. F. Bon, M. Bosveld, B. Klumperman and A. L. German, *Macromolecules*, **30**, 324–326 (1997).
216. J. Nicolas, B. Charleux, O. Guerret and S. P. Magnet, *Angew. Chem. Int. Ed.*, **43**, 6186–6189 (2004).
217. G. Delaittre, J. Nicolas, C. Lefay, M. Save and B. Charleux, *Chem. Comm.*, **5**, 614–616 (2005).
218. A. R. Szkurhan and M. K. Georges, *Macromolecules*, **37**, 4776–4782 (2004).
219. J. Pohn, C. Buragin, M. K. Georges, *et al.*, *Macromol. Theory Simul.*, **17**, 73–85 (2008).
220. R. W. Simms, M. D. Hoidas and M. F. Cunningham, *Macromolecules*, **41**, 1076–1079 (2008).
221. G. Schmidt-Naake, M. Drache and C. Taube, *Angew. Makromol. Chem.*, **265**, 62–68 (1999).
222. C. Taube and G. Schmidt-Naake, *Macromol. Mater. Eng.*, **279**, 26–33 (2000).
223. L. I. Gabaston, R. A. Jackson and S. P. Armes, *Macromolecules*, **31**, 2883–2888 (1998).
224. G. Delaittre, M. Save and B. Charleux, *Macromol. Rapid Commun.*, **28**, 1528–1533 (2007).
225. C. J. Hawker, E. Elce, J. Dao, *et al.*, *Macromolecules*, **29**, 2686–2688 (1996).
226. T. Fukuda, T. Terauchi, A. Goto, *et al.*, *Macromolecules*, **29**, 3050–3052 (1996).
227. B. Lessard, A. Graffe and M. Maric, *Macromolecules*, **40**, 9284–9292 (2007).
228. L. Couvreur, B. Charleux, O. Guerret and S. Magnet, *Macromol. Chem. Phys.*, **204**, 2055–2063 (2003).
229. Y. Miura, N. Nakamura, I. Taniguchi and A. Ichikawa, *Polymer*, **44**, 3461–3467 (2003).
230. J. Cai, L. Xu, Z. Mao, *et al.*, *J. Appli. Polym. Sci.*, **102**, 3118–3122 (2006).
231. G. G. Barclay, M. King, A. Orellana, *et al.*, *Micro- and Nanopatterning Polymers*, ACS Symp. Ser., **706**, 144–160 (1998).
232. E. Yoshida, *J. Poly. Sci., Part A: Poly. Chem.*, **34**, 2937–2493 (1996).
233. E. Yoshida and Y. Takiguchi, *Polym. J.*, **31**, 429–434 (1999).
234. P. M. Kazmaier, K. Daimon, M. K. Georges, *et al.*, *Macromolecules*, **30**, 2228–2231 (1997).
235. H. Okamura, Y. Takatori, M. Tsunooka and M. Shirai, *Polymer*, **43**, 3155–3162 (2004).
236. D. J. Woo, J. Kim, M. H. Suh, *et al.*, *Polymer*, **47**, 3287–3291 (2006).
237. M. C. Bignozzi, C. K. Ober, A. J. Novembre and C. Knurek, *Polym. Bull.*, **43**, 93–100 (1999).
238. M. Baumert and R. Mülhaupt, *Macromol. Rapid. Commun.*, **18**, 787–794 (1997).
239. P. J. MacLeod, G. K. Hamer, P. Lukse and M. K. Georges, *Functional Polymers*, ACS Symp. Ser., **704**, 28–37 (1998).
240. E. Megiel and A. Kaim, *J. Polym. Sci., Part A: Polym. Chem.*, **46**, 1165–1177 (2008).
241. J. L. Pozzo, H. Bouas-Laurent, A. Deffieux, *et al.*, *Mol. Cryst. Sci. Technol., Sect. A*, **298**, 437–443 (1997).
242. Z. Chen, F. Wang, J. Yan, *et al.*, *J. Appl. Polym. Sci.*, **91**, 1842–1847 (2004).
243. Y. Higaki, H. Otsuka and A. Takahara, *Trans. Mater. Res. Soc. Japan*, **29**, 141–144 (2004).
244. G. Schmidt-Naake and S. Butz, *Macromol. Rapid Commun.*, **17**, 661–665 (1996).
245. J. Lokaj, P. Vlcek and J. Kriz, *J. Appl. Polym. Sci.*, **74**, 2378–2385 (1999).
246. B. Lessard, S. C. Schmidt and M. Maric, *Macromolecules*, **41**, 3446–3454 (2008).
247. C, Dire, B Charleux, S. Magnet and L. Couvreur, *Macromolecules*, **40**, 1897–1903 (2007).
248. D. Benoit, C. J. Hawker, E. E. Huang, *et al.*, *Macromolecules*, **33**, 1505–1507 (2000).
249. O. Gibbons, W. M. Carroll, F. Aldabbagh and B. Yamada, *J. Polym. Sci., Part A: Polym. Chem.*, **44**, 6410–6418 (2006).
250. D. Bertin and B. Boutevin, *Polym. Bull.*, **37**, 337–344 (1996).
251. E. Yoshida and T. Fuji, *J. Polym. Sci., Part A: Polym. Chem.*, **35**, 2371–2378 (1997).
252. S. Jousset, S. O. Hammouch and J. M. Catala, *Macromolecules*, **30**, 6685–6687 (1997).
253. K, Ohno, M. Ejaz, T. Fukuda, *et al.*, *Macromol. Chem. Phys.*, **199**, 291–297 (1998).
254. M. Mariani, M. Lelli, K. Sparnacci and M. Laus, *J. Polym. Sci., Part A: Polym. Chem.*, **37**, 1237–1244 (1999).
255. L. I. Gabaston, S. A. Furlong, R. A. Jackson and S. P Armes, *Polymer*, **40**, 4505–5414 (1999).

256. X. Wan, Y. Tu, and Q. Zhang, *Chin. J. Polym. Sci.*, **16**, 377–380 (1998).
257. X. Wan, Y. Tu, and Q. Zhang, *Polym. Int.*, **49**, 243–247 (2000).
258. M. C. Bignozzi, C. K. Ober and M. Laus, *Macromol. Rapid Commun.*, **20**, 622–627 (1999).
259. N. A. Listigovers, M. K. Georges, P. G. Odell and B. Keoshkerian, *Macromolecules*, **29**, 8992–8993 (1996).
260. J. Lokaj, L. Polikova, P. Holler, *et al.*, *J. Appl. Polym. Sci.*, **105**, 1616–1622 (2007).
261. P. Shoaeifar, M Abbasian and A. A. Entez'Aami, *J. Polym. Res.*, **14**, 45–52 (2007).
262. T. N. T. Phan, S. Maiez-Tribut, J. P. Pascault, *et al.*, *Macromolecules*, **40**, 4516–4523 (2007).
263. S. M. Wang, Z. Cheng, J. Zhu, *et al.*, *J. Polym. Sci., Part A: Polym. Chem.*, **45**, 5318–5328 (2007).
264. P. Shoaeifar, M. Abbasian and A. A. Entez'Aami, *J. Polym. Res.*, **14**, 45–52 (2007).
265. C. Dire, B. Charleux, S. Magnet and L. Couvreur, *Macromolecules*, **40**, 1897–1903 (2007).
266. R. Jabbar, A. Graffe, B. Lessard and M. Maric, *J. Appl. Polym. Sci.*, **109**, 3185–3195 (2008).
267. M. K. Georges, G. K. Hamer and N. A. Listigovers, *Macromolecules*, **31**, 9087–9089 (1998).
268. K. Karaky, G. Clisson, G. Reiter and L. Billon, *Macromol. Chem. Phys.*, **209**, 715–722 (2008).
269. J. Ruehl, A. Nilsen, S. Born, *et al.*, *Polymer*, **48**, 2564–2571 (2007).
270. J. Nicolas, A. V. Ruzette, C. Farcet, *et al.*, *Polymer*, **48**, 7029–7040 (2007).
271. R. B. Grubbs, C. J. Hawker, J. Dao and J. M. J. Fréchet, *Angew. Chem. Int. Ed. Engl.*, **36**, 270–272 (1997).
272. J. Nicolas, P. Couvreur and B. Charleux, *Macromolecules*, **41**, 3758–3761 (2008).
273. U. M. Stehling, E. E. Malmstroem, R. M. Waymouth and C. J. Hawker, *Macromolecules*, **31**, 4396–4398 (1998).
274. M. Abbasian and A. A. Entezami, *Polym. Adv. Technol.*, **18**, 306–312 (2007).
275. D. Gromadzki, R. Makuska, M. Netopilik, *et al.*, *Eur. Polym. J.*, **44**, 59–71 (2008).
276. D. B. Hua, W. C. Deng, J. Tang, *et al.*, *Int. J. Biol. Macromol.*, **43**, 43–47 (2008).
277. P. Rempp, E. Franta and J. Herz, *Adv. Polym. Sci.*, **86**, 145–173 (1988).
278. C. J. Hawker, *Angew. Chem. Int. Ed. Engl.*, **34**, 1456–1459 (1995).
279. A. J. Pasquale and T. E. J. Long, *Polym. Sci., Part A: Polym. Chem.*, **39**, 216–223 (2001).
280. A. W. Bosman, A. Heumann, G. G. Klaerner, *et al.*, *J. Am. Chem. Soc.*, **123**, 6461–6462 (2001).
281. T. Tsoukatos, S. Pispas and N. Hadjichristidis, *J. Polym. Sci., Part A: Polym. Chem.*, **39**, 320–325 (2001).
282. S. Abraham, J. H. Choi, C. S. Ha and I. Kim, *J. Polym. Sci., Part A: Polym. Chem.*, **45**, 5559–5572 (2007).
283. C. J. Hawker, J. M. J. Fréchet, R. B. Grubbs and J. Dao, *J Am. Chem. Soc.*, **117**, 10763–10764 (1995).
284. W. H. Binder and R. Sachsenhofer, *Macromol. Rapid Commun.*, **29**, 952–981 (2008).
285. J.-F. Lutz and H. G. Börner, *Prog. Polym. Sci.*, **33**, 1–39 (2008).

12

Nitroxide-Catalyzed Alcohol Oxidations in Organic Synthesis

Christian Brückner

Department of Chemistry, University of Connecticut, Connecticut, USA

12.1 Introduction

The oxidation of secondary amines that contain no α hydrogen atoms, such as 2,2,6,6-tetramethylpiperidine (**1**), leads to the formation of nitroxides, stable free radical compounds such as the prototypical 2,2,6,6-tetramethylpiperidine-1-oxyl (TEMPO, **2**) (Scheme 12.1).[1] An one-electron oxidation converts nitroxide **2** into the oxoammonium cation **3**, a strongly oxidizing species. Reactions of a primary or secondary alcohol with **3** oxidize the alcohol to the corresponding aldehyde or ketone, respectively, and reduce **3** to the corresponding hydroxylammonium salt **4**. If this reaction takes place in the presence of an oxidant that is capable of the two-electron oxidation needed to convert the hydroxylamine back to the oxoammonium salt, TEMPO becomes an oxidation catalyst. The latter oxidation by the so-called secondary oxidant* may require a cocatalyst, commonly bromide. It is this catalytic cycle that enables nitroxides to be used as oxidation catalysts in organic synthesis.[2,3] The nitroxide-catalyzed reactions – largely independent from the secondary oxidant used – tolerate the presence of alkanes, double and triple bonds, protected alcohols, including acetals and silanes, epoxides, amides, pyridine-type nitrogens, azides, sulfur heterocycles, esters, isonitriles, and other functional groups. The oxidations are rarely accompanied with C–C bond cleaving side reactions.

Recent and comprehensive reviews of nitroxide-catalyzed and oxoammonium oxidation reactions are available.[2,3] As such, this review is focused mainly on the application of nitroxide-catalyzed reactions in organic synthesis, using recent examples from the literature. Emphasis is placed on the scope and limitations of this reagent in the preparation of complex organic molecules. Excluded here is the use of oxoammonium compounds as stoichiometric oxidants although they are the active species in the catalytic cycle (Scheme 12.1). As well, the oxidation of non-alcoholic substrates is not reviewed; the reader is referred to a number of classic or timely reviews on many aspects of the use of nitroxides and/or oxoammonium

*This oxidant is sometimes also referred to as primary, terminal, or co-oxidant. For a discussion of possible secondary oxidants, see Section 12.9.

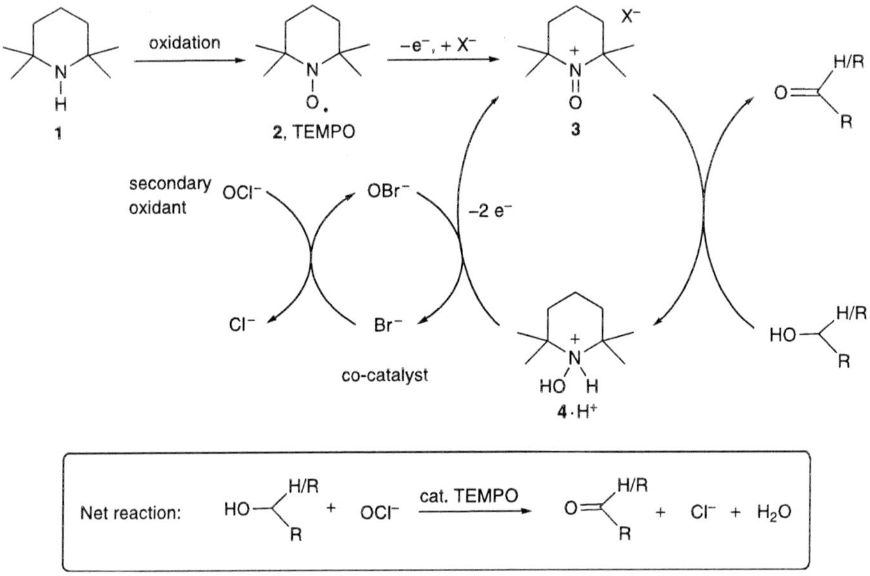

Scheme 12.1

salts as oxidants,[2–12] their stability,[13] industrial use,[14] or use in carbohydrate chemistry.[15] Apologies are made to all the contributors to the field whose work is not included.

12.2 Mechanism of TEMPO-catalyzed alcohol oxidations

The oxidation of an alcohol to an aldehyde or ketone is deceptively simple. It involves the abstraction of two hydrogens, that is, the abstraction of two electrons and two protons, from the substrate. However, the exact details of the nitroxide-catalyzed oxidation process remain unclear. Some of the uncertainties arise from the fact that the oxidizing species, the oxoammonium cation, can be formulated in two limiting resonance forms, **3A** and **3B**, and either form has been used to rationalize experimental findings.[16,17]

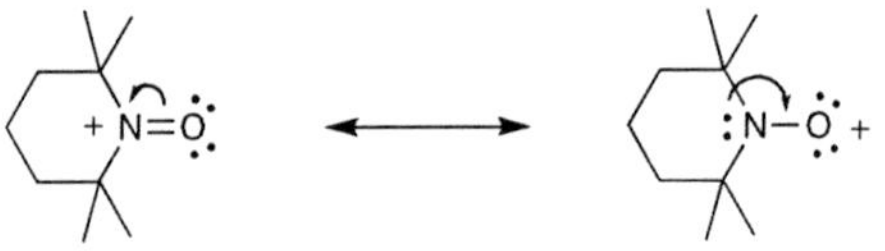

3A 3B

An investigation of the pH dependence of the rate of oxoammonium oxidations of simple alcohols in water and aqueous acetonitrile reveals the presence of several rate regimes, interpretable with the presence of different mechanisms at different pH values.[18] The mechanism dominating under acidic conditions is generally slower than the mechanism dominating under basic conditions, and secondary alcohols are oxidized faster than primary alcohols. The reactions under basic conditions are faster and show the reverse rate selectivity, that is, primary alcohols are oxidized faster than secondary alcohols. Since the majority of nitroxide-catalyzed reactions are performed under basic conditions, only the details of this reaction are considered. Oxidations of alcohols in base may also involve an alcoholate as the nucleophile, a factor that might explain the rate acceleration under basic conditions for the generally more acidic primary alcohols.[18,19]

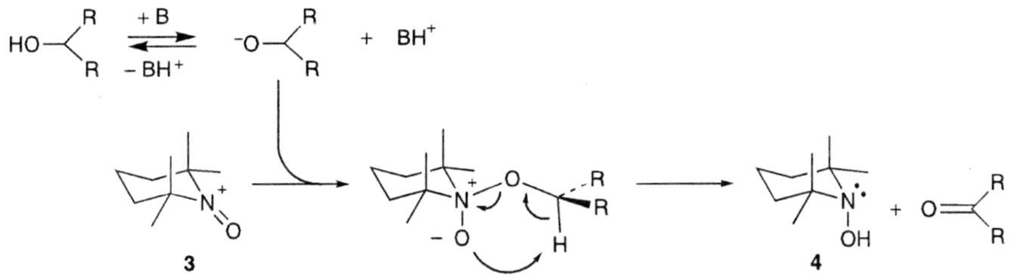

Scheme 12.2

Scheme 12.2 shows a possible mechanism for the alcohol oxidation under basic conditions. Deuterium isotope studies established that the rate determining step is the breaking of the carbon–hydrogen bond, possibly in the form of a hydride abstraction.[18,19] Under acidic or neutral conditions, the relative reactivity of the substrates is in accord with the known strengths of the carbon–hydrogen bonds on the hydroxyl-bearing carbon.[18] Recent molecular mechanics calculations shed further light on this mechanism.[20]

The details of the nitroxide-catalyzed oxidation reactions are undoubtedly more complex, however, as steric effects may inhibit or reinforce the pK_a– or C–H bond strength-controlled rates. Counterions have an effect on the rates.[21] Also, it remains to be explained why alcohols containing a β oxygen or a β nitrogen are, with exceptions,[22] inert toward oxoammonium salts in acidic media, though they react in basic media.[23]

The active catalytic species in the nitroxide-catalyzed reactions, the oxoammonium salt, can also be prepared, isolated and used as an attractive stoichiometric oxidant.[2,10] Remarkable in the context of this discussion is that reactions using stoichiometric quantities of oxoammonium salts present at times distinctly different reactivity profiles when compared to catalytic reactions. For instance, some of the reactions of stoichiometric oxoammonium salts, such as reactions with double bonds, phenols, or with electron-rich heterocyclic systems have not been reported for the catalyzed reactions, although it should be noted that in only a few cases the two methods were directly compared against each other.[24] Also, the oxo-functionalization of methylene groups α to a keto group or the conversion of 1,3-diketones to 1,2,3-triketones proceeds typically only with oxoammonium salts,[25,26] but exceptions are known.[27] On the other hand, some general limitations are shared by both reactions: unprotected amines cause problems, (activated) double bonds may react slowly, sulfides may be oxidized, and benzyl ethers may react; some of these expected side reactions are discussed in Section 12.11.

12.3 Nitroxides used as catalysts

12.3.1 Monomeric nitroxides

Though many nitroxides are known,[28,29] only a small fraction of these nitroxides have been investigated as oxidation catalysts, and the vast majority of nitroxide-catalyzed reactions were performed using TEMPO (**2**), and its 4-substituted derivatives.[2] Though some minor reactivity differences exist among them, their most significant differences lie in their physical properties, such as solubility, extraction properties, and volatility. While TEMPO is low melting (36–38 °C) and fairly volatile, its 4-acetylamino derivative, for instance, is a crystalline and non-volatile solid.[30] Thus these derivatives offer a number of practical advantages with respect to, for instance, catalyst recovery.

The search for nitroxide catalysts that are more regioselective, faster reacting or possessing higher turnover numbers is ongoing. Thus, a series of bicyclic and tricyclic piperidine-based nitroxides **5–7** was investigated.[31–34] In direct comparisons with TEMPO, the oxidation yields were appreciably better with **6**, especially with sterically hindered alcohols. On the downside, their syntheses are significantly more complex. Limited data suggest phosphate ester **8** is also a stable and effective catalyst.[35] On the other hand, neither the piperazine-based nitroxide **9**, nor the morpholine-based systems **10** or **11** proved to be competent catalysts.[36,37] Likewise, the oxazole- and pyrrolidine-based systems **12–14**,[38,39] and related systems,[40,41] failed to bring improvements over the oxidation catalysis by the parent TEMPO and closely related derivatives. Lastly, stereoselective reagents are sought. Some of the chiral nitroxides investigated are discussed in Section 12.8.

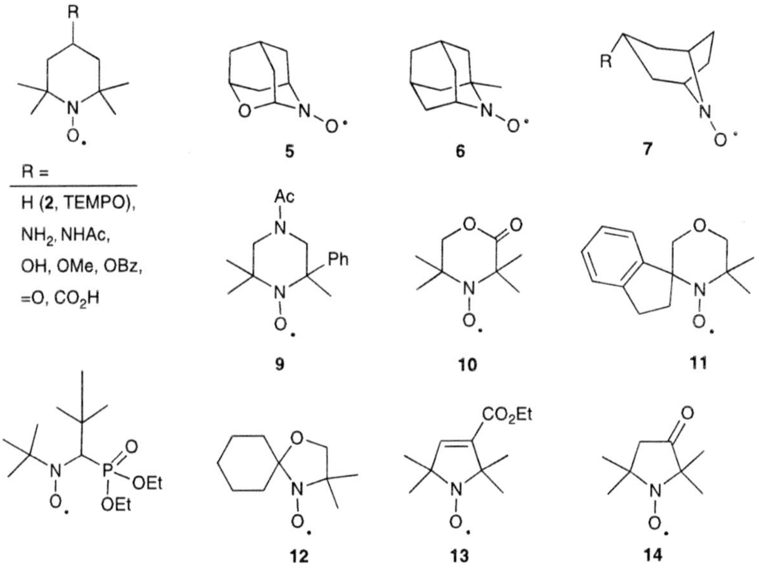

12.3.2 Ionic liquid nitroxides

A number of nitroxide-ionic liquid conjugates were prepared in recent years, representative for which are imidazolium derivatives **15**[42] and **16**.[43] These catalysts, particularly in combination with an ionic liquid secondary oxidant such as **17**,[44] promise simple product isolation by extraction with a solvent that is immiscible with the ionic liquid.

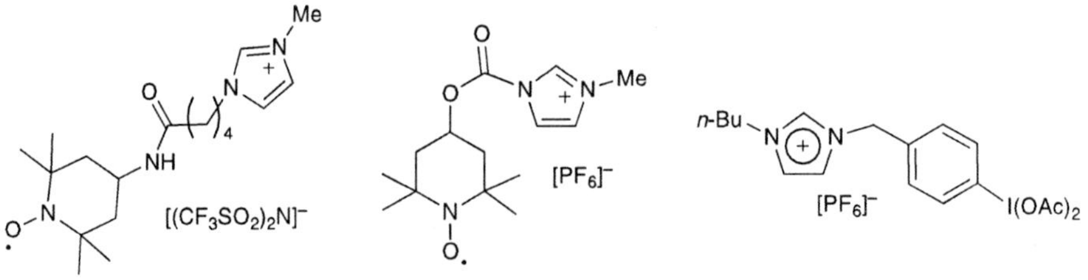

15 **16** **17**

12.3.3 Supported nitroxides

A further step toward the procedural simplification of product isolation after a nitroxide-catalyzed oxidation would be the ability to filter a supported catalyst (or secondary oxidant) from the reaction mixture. The ability to reuse solid phase nitroxides repeatedly adds to their advantages.[45–47] Multiple examples of supported nitroxides were prepared, with the solid support ranging from simple or advanced organic polymers to modified silica gels to zeolites, though not all were tested in series of successive oxidations.[48] Shown below are a number of representative systems that vary in their polymer support, linker, or nitroxide; note that most examples are based on 4-substituted TEMPO derivatives (**18**,[49,50] **19**,[51] **20**,[52] **21**,[8,9,53] and **22**[54,55]). The supported catalysts are prepared either by covalent attachment of a nitroxide to a polymer, or the corresponding nitroxide-modified monomers are polymerized. Their solubility and swellability properties vary. While most reactions are performed by suspending the polymer, polyethylene glycol derivative **22**, and related systems, are soluble but can be readily precipitated from the reaction mixture.[56]

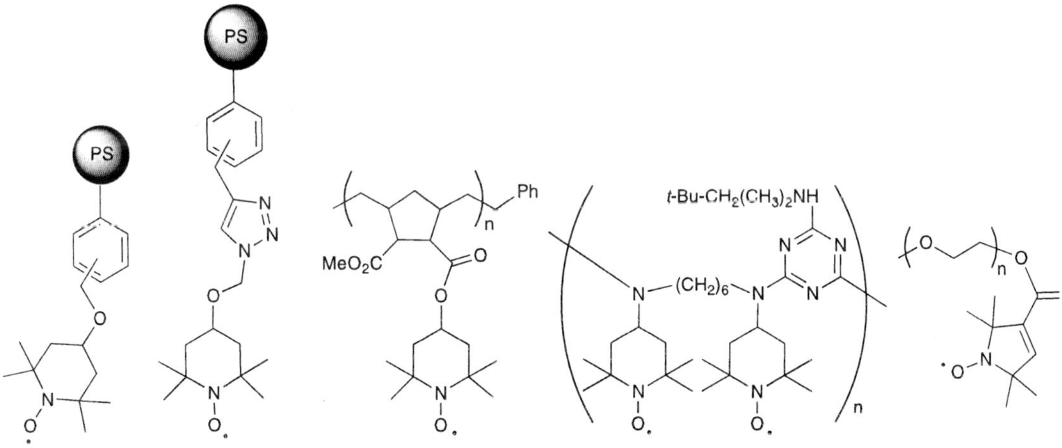

18 **19** **20** **21** **22**

TEMPO residues, or the corresponding piperidine derivatives,[46] have also been bonded covalently in a multitude of ways to silica gel surfaces (**23–25**)[57–60] or mesoporous silicas (**23**),[61,62] or nitroxide-modified silanes were polycondensed into organically modified silica gels (ormosils).[47,63,64] The catalyst derived from SBA-15 (**23**) can be used, in the presence of the cocatalyst sodium nitrite, in oxidations that utilize air as secondary oxidant.[62]

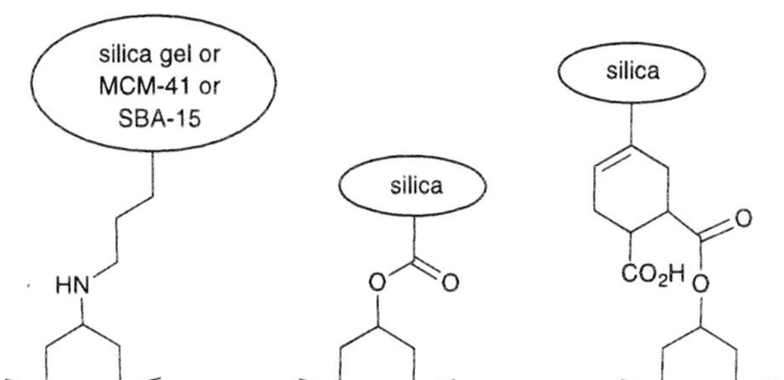

23 **24** **25**

12.4 Chemoselectivity: oxidation of primary vs secondary alcohols

Nitroxide-catalyzed reactions are extensively used to oxidize primary and secondary alcohols to aldehydes or carboxylic acids and ketones, repectively, in the presence of a large number of other functional groups.[2] The observed chemoselectivity is mainly a matter of relative oxidation rates. Under basic conditions,

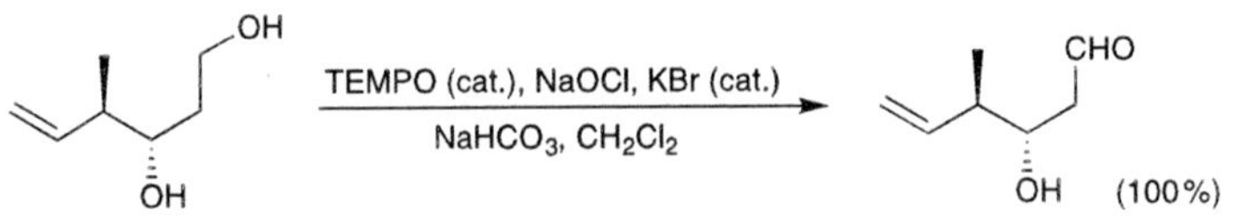

Scheme 12.3

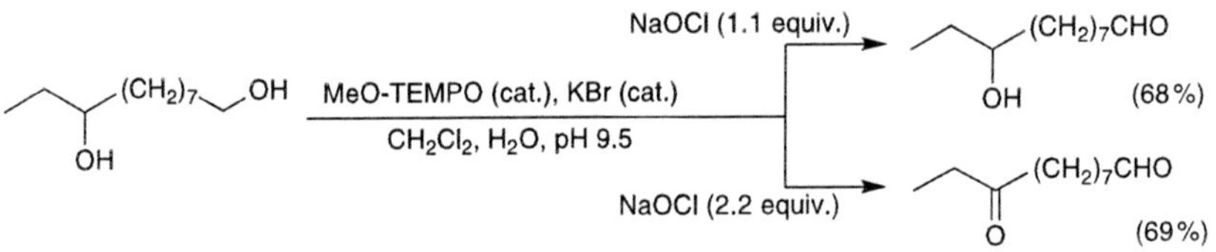

Scheme 12.4

primary alcohols react much faster than secondary alcohols,[6,45,65] and complete chemoselectivity can be observed in many reactions (see, for example, Schemes 12.4, 12.21, 12.22, 12.29, 12.30, 12.34, 12.35, 12.37–12.39, and 12.51). A prototypical example of a selective oxidation of a primary alcohol in the presence of a secondary alcohol is shown in Scheme 12.3. Optically active 4-methyl-5-hexene-1,3-diol is oxidized to the corresponding hydroxy aldehyde in quantitative yield.[66] The typical reaction conditions are also mild enough to not cause any racemization of chiral carbons in the molecule (see also, for example, Schemes 12.31, 12.34, 12.37–12.39, 12.41, and 12.51).

Stoichiometric control can be exerted to oxidize the secondary alcohol subsequently to the primary alcohol. Thus, oxidation of 1,9-undecanediol results, depending on the stoichiometric ratio of substrate to secondary oxidant, in the formation of either 9-hydroxyundecanal or 9-oxoundecanal (Scheme 12.4).[67]

12.5 Chemoselectivity: oxidation of primary vs benzylic alcohols

Benzylic alcohol oxidations are facile (see, e.g., Scheme 12.36).[68] Primary benzylic alcohols are oxidized faster than secondary benzylic alcohols, even if the secondary alcohol is located at a double-benzylic position.[69] Under certain conditions, benzylic alcohols can be oxidized with high selectivity (albeit in low yield) in the presence of primary alcohols, as the example shown in Scheme 12.5 demonstrates.[70] However, the absence of multiple examples of this reaction leaves open the question of the generality of this particular selectivity.

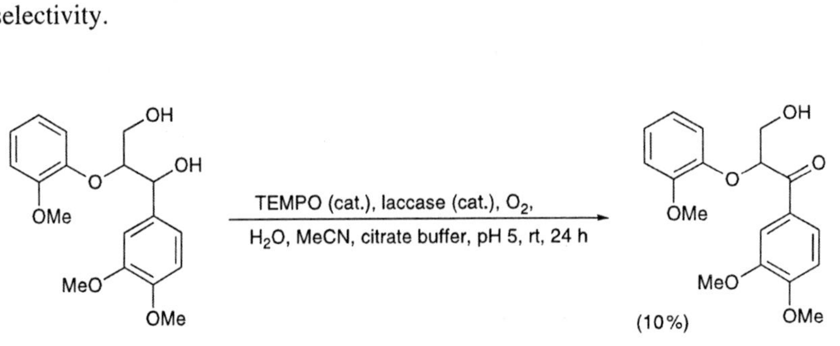

Scheme 12.5

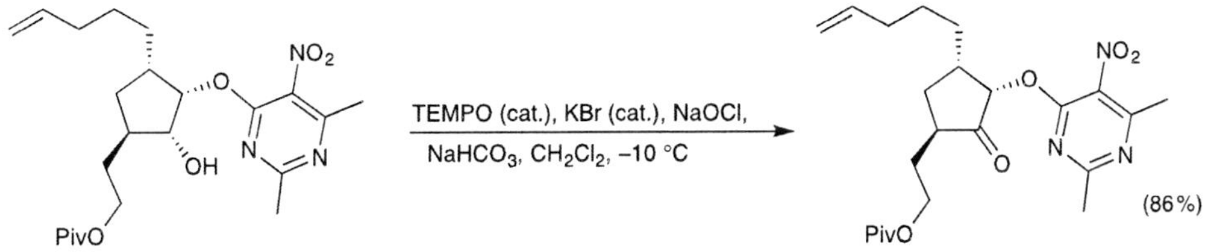

Scheme 12.6

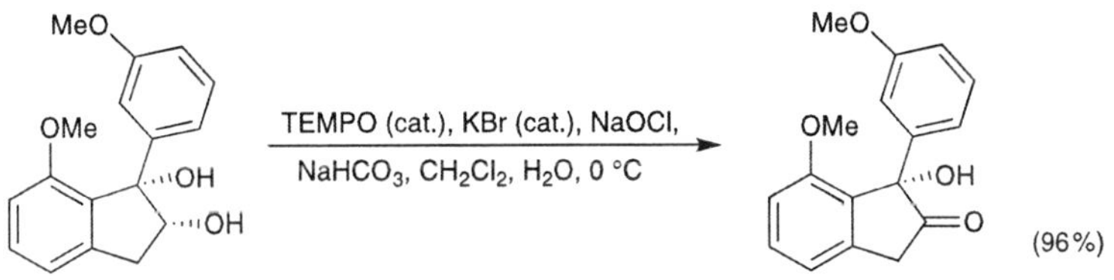

Scheme 12.7

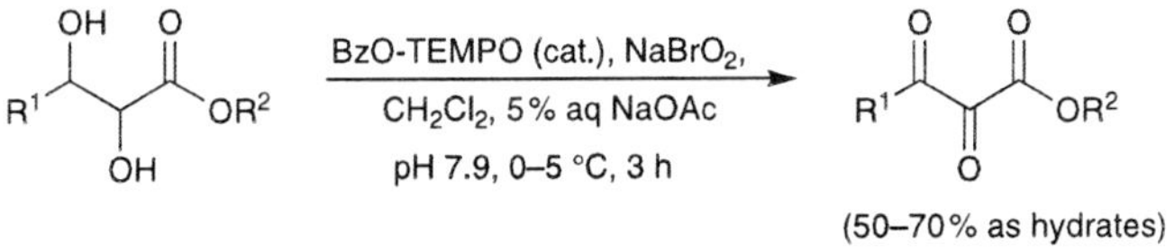

Scheme 12.8

12.6 Oxidation of secondary alcohols to ketones

Nitroxides are, in the absence of primary alcohols, potent catalysts for the conversion of secondary alcohols to ketones. Although much less frequently used than the conversion of primary alcohols to aldehydes, these reactions are also characterized by a similarly large functional group tolerance as the primary alcohol oxidations, as the reactions shown in Schemes 12.6[71] and 12.7[72] demonstrate.

The nitroxide-catalyzed reactions are powerful enough for the generation of 1,2,3-triketones (α,β-diketo esters) (Scheme 12.8). Thus, the 4-benzyloxy-derivatized TEMPO-catalyzed oxidation of an α,β-dihydroxy ester, using sodium bromite as the secondary oxidant, generates the triketone without concomitant carbon–carbon bond cleavage.[27] More commonly, however, stoichiometric quantities of oxoammonium salts have been used as reagents for the conversion of activated methylene groups to di- and triketones.[2]

12.7 Oxidations of alcohols to carboxylic acids

12.7.1 Oxidations leading to linear carboxylic acids

A twofold stoichiometric excess of some secondary oxidants, such as bleach, may lead to an intentional over-oxidation of primary alcohols to carboxylic acids. The conversion of the alcohol to the aldehyde is nitroxide catalyzed, while the second oxidation is caused directly by the secondary oxidant. Alternatively, more general and efficient nitroxide-catalyzed oxidations of primary alcohols to carboxylic acids are carried

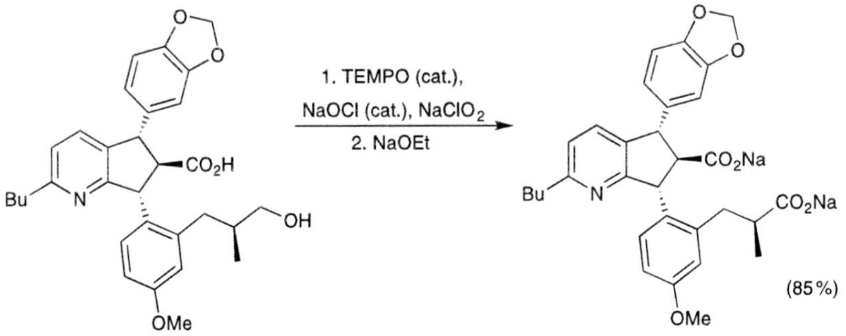

Scheme 12.9

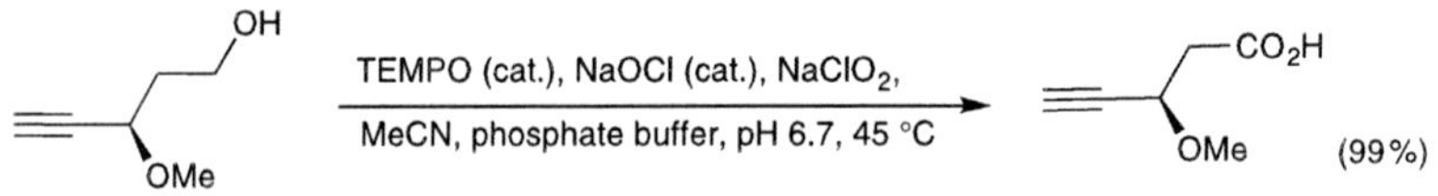

Scheme 12.10

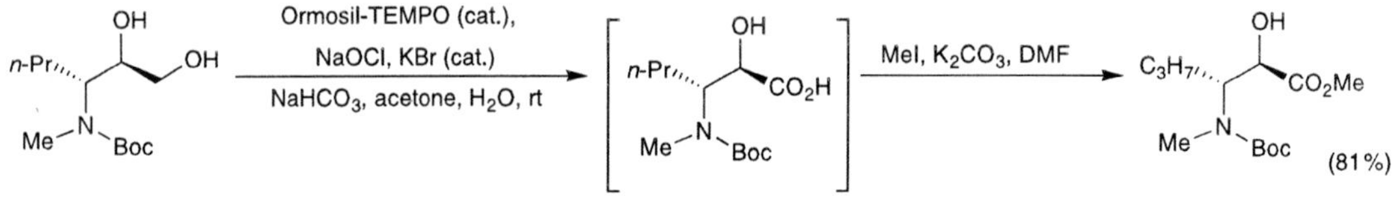

Scheme 12.11

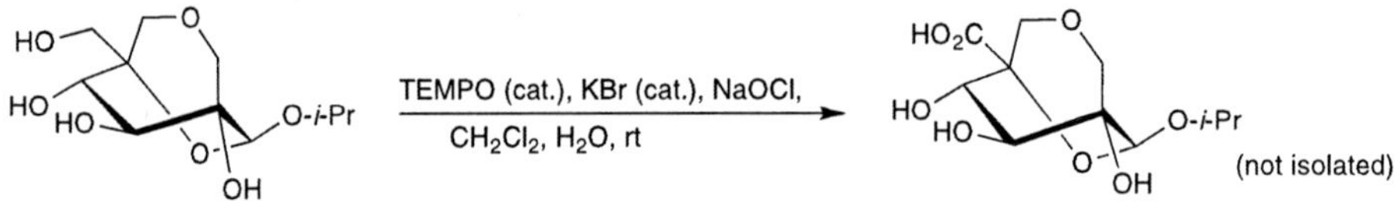

Scheme 12.12

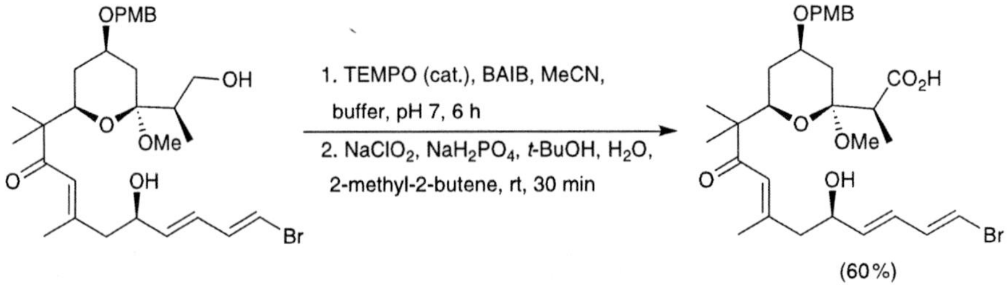

Scheme 12.13

out using sodium chlorite ($NaClO_2$) as a secondary oxidant in the presence of catalytic amounts of sodium hypochlorite (Scheme 12.9)[73,74].

The conversion of a primary alcohol to a carboxylic acid can take place in the presence of protected phenols, protected and heteroaromatic nitrogens, and alkynes[75] (Schemes 12.9 and 12.10). Notably, the conversion of a primary alcohol to a carboxylic acid in the presence of a secondary alcohol may be more rapid than the oxidation of a secondary alcohol present in the substrate. Thus, a 1,2-diol was converted to an α-hydroxy carboxylic acid (Scheme 12.11)[76] or the primary alcohol on a tricyclic polyol was converted to an acid in the presence of a number of free and protected secondary and tertiary alcohols (Scheme 12.12),[77] or even allylic alcohols (Scheme 12.13).[78]

The primary alcohol-to-carboxylic acid conversion can also be performed in a two-step, one-pot approach (Scheme 12.13).[78] The primary alcohol is oxidized to the aldehyde with sodium hypochlorite. Then sodium chlorite is added to complete the second oxidation step. It is found that a slightly lower pH, of 4–6, enhances the reactivity of the latter step.[79] For further examples of tandem reactions, see Section 12.10.

Primary alcohols in carbohydrates are particularly readily converted in aqueous solutions to carboxylic acids, as in the classic example of generating uronates (Scheme 12.14).[80] Here, the intermediate carboxylic acids were esterified and the secondary alcohols acetylated to yield the final products. Note also that next to the secondary alcohols, the acetamido and azide groups are resistant to oxidation.

In the same vein, simple alkyl glycosides, disaccharides, and cyclodextrins have been subjected to TEMPO-catalyzed electro-oxidations to provide uronic acid derivatives in modest yields (Scheme 12.15).[81] Ultrasonic irradiation appreciably increases the oxidation rates in the catalyzed oxidations of methyl α-D-glucopyanoside and sucrose.[82]

Similar to the glycosides, the C6 carbon of nucleosides is also readily susceptible to their conversion into carboxylic acids. Typical examples, the two-step oxidations of a range of nuceic base derivatives, are shown in Scheme 12.16.[83] Protected amines, amide nitrogens, and silane protecting groups are resistant toward these oxidation conditions.

A number of polysaccharides, including protected and unprotected cyclodextrins,[84–86] have been oxidized under the general sodium hypochlorite conditions.[15,87–89] In general, these oxidations closely parallel the oxidations of monomeric glycosides. Amylose, for instance, is oxidized under basic conditions on C6 to the corresponding carboxylic acid salt (Scheme 12.17); upon neutralization, the polyglucuronic acids are obtained.[89]

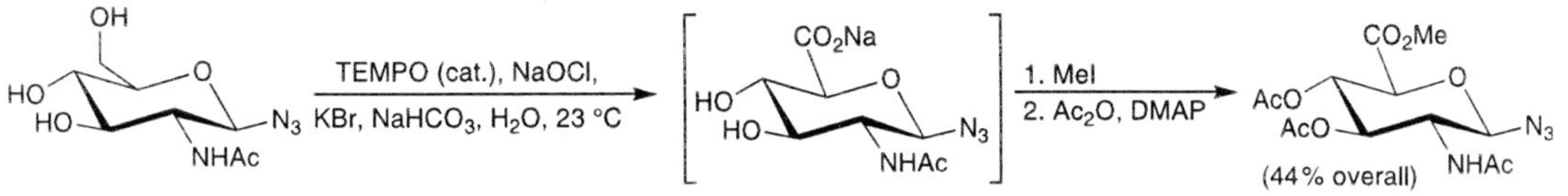

Scheme 12.14

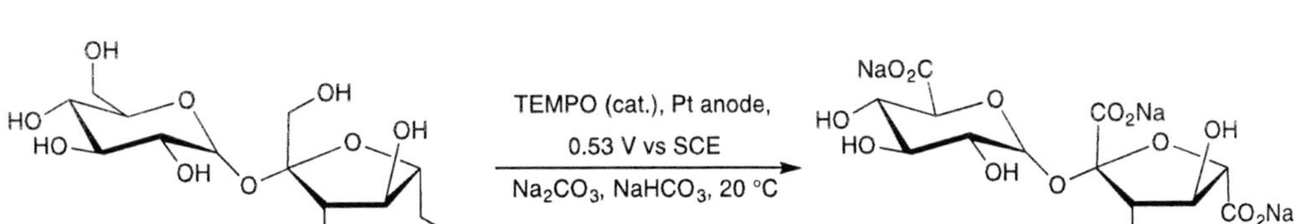

Scheme 12.15

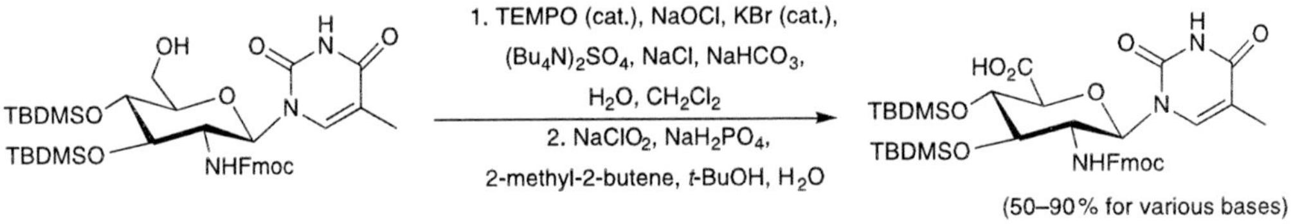

Scheme 12.16

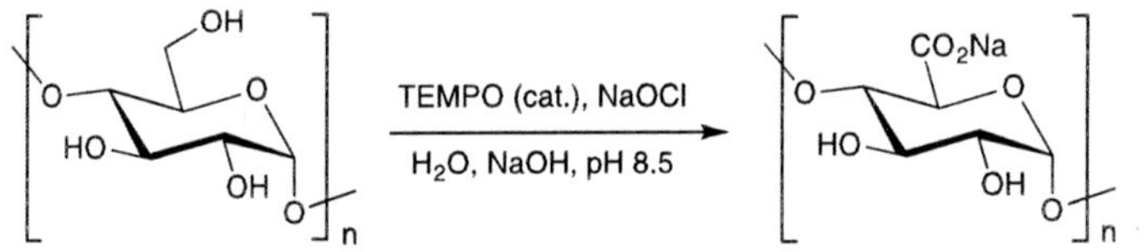

Scheme 12.17

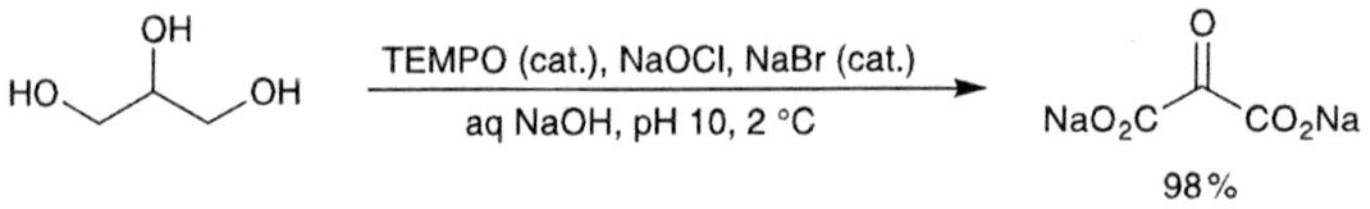

Scheme 12.18

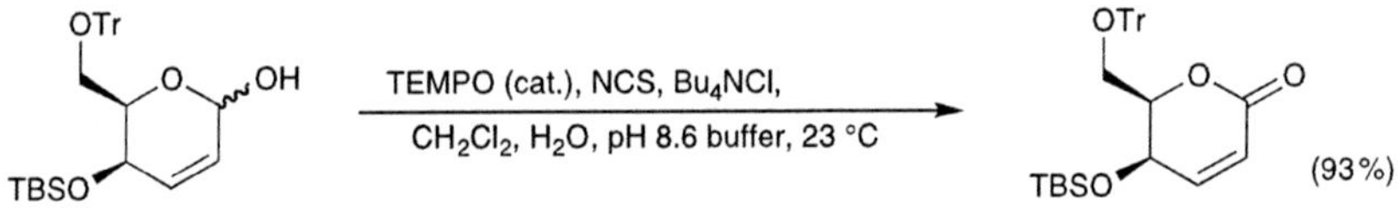

Scheme 12.19

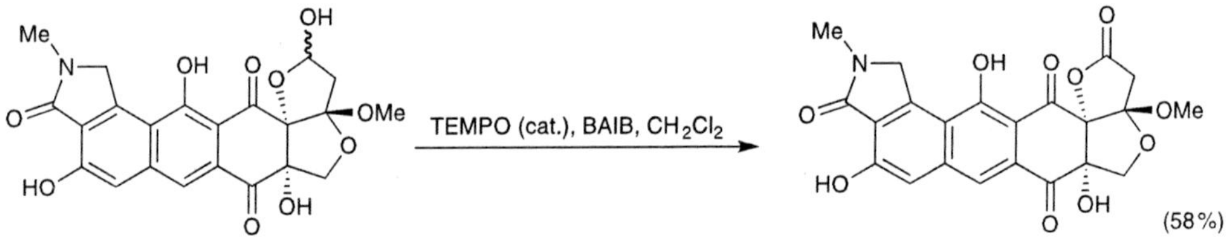

Scheme 12.20

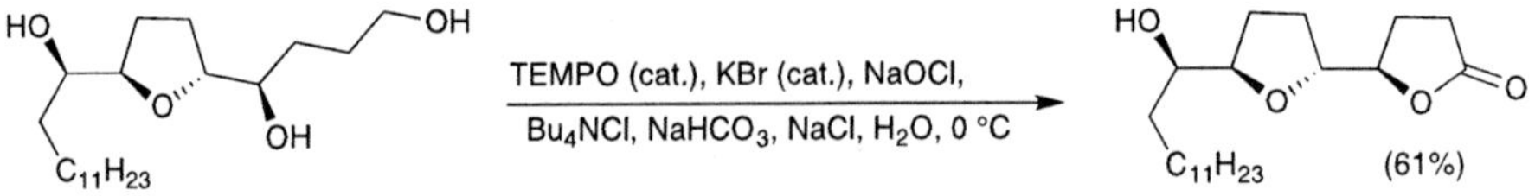

Scheme 12.21

Cellulose rendered water soluble by partial acetylation can be oxidized to partially acetylated polyglucuronic acids.[90] Water insoluble polysaccharides, such as cellulose, are only oxidized on the edges of the fibers.[91,92] Pretreated celluloses give more uniform oxidations.[93,94]

The nitroxide-catalyzed oxidation of both terminal carbons of simple aldose sugars to the corresponding glycaric acids is possible.[95] Suitable secondary oxidants are sodium hypochlorite, chlorine, or bromine (Scheme 12.29).[95,96] These glycaric acid preparations are carried out between 0 and 5 °C with the pH carefully held above 11.5. When the pH drops below 11.5, appreciable carbon–carbon cleavage is observed.[97] As the oxidation of glycerol demonstrates, even the exhaustive oxidation of primary and secondary alcohols of glycerol to the sodium salt of ketomalonic acid is possible in high yields, even though the pH in this reaction was held at 10 (Scheme 12.18).[98]

12.7.2 (Diol) oxidations leading to lactones

(Cyclic) hemiacetals are susceptible to their nitroxide-catalyzed oxidation to esters (lactones). The examples given in Schemes 12.19 and 12.20 again demonstrate the wide functional group tolerance of this method.[99,100] The organic secondary oxidants used here vary from the traditionally used bleach, and are discussed in Section 12.9.

The possibility of generating lactones from diols exists. Thus, if a nitroxide-mediated oxidation of a primary alcohol generates an aldehyde that is in a position to a second hydroxy group to form a stable intramolecular hemiacetal, as is the case for 1,4- (Scheme 12.21)[101] and 1,5-diols (Scheme 12.22),[102] this hemiacetal forms, and is subsequently (nitroxide-catalyzed) oxidized to the corresponding lactone.[103] In the strictest sense, this process represents the conversion of a primary alcohol into a carboxylic acid derivative. Even the efficient formation of a seven-membered lactone from an 1,6-diol is feasible (Scheme 12.23).[104] This reaction for the formation of medium-sized rings is synthetically quite useful.[45] Acetals, such as methylidene, benzylidene and isopropylidene acetals, are generally stable toward oxidation (Schemes 12.9, 12.22, 12.23). Also, the oxidation of a hemiacetal to a lactone can be performed with high specificity in the presence of a secondary alcohol (Scheme 12.21).

6,3-Lactones are formed from a series of partially protected thioglucosides (Scheme 12.24).[105] Remarkably, sulfur oxidation does not occur if the reaction times are carefully controlled. In a similar fashion, a 6,1-lactone was prepared from a glucose derivative, and a 6,2-lactone from a mannose derivative.[105]

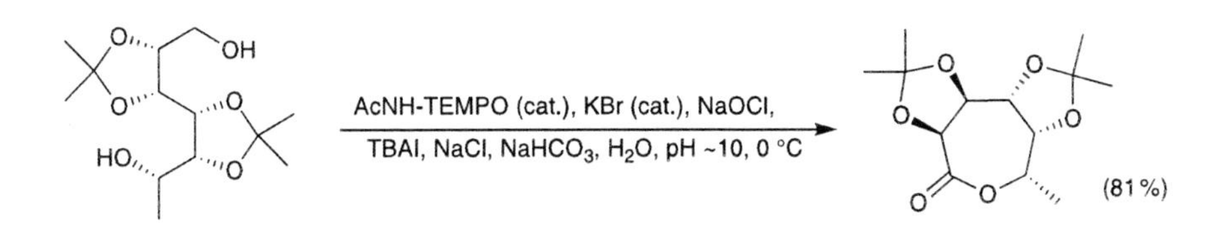

Scheme 12.22

Scheme 12.23

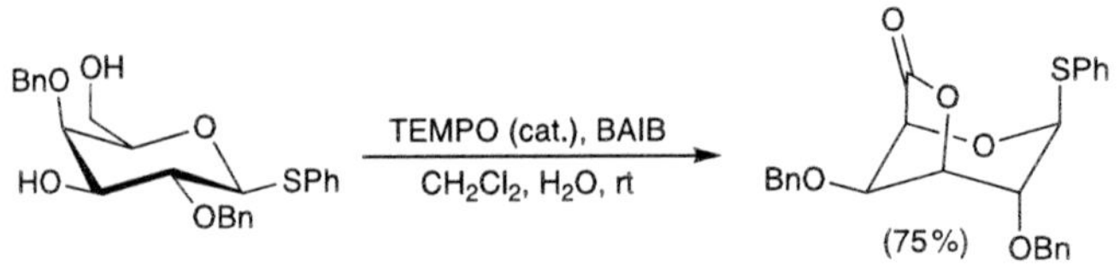

Scheme 12.24

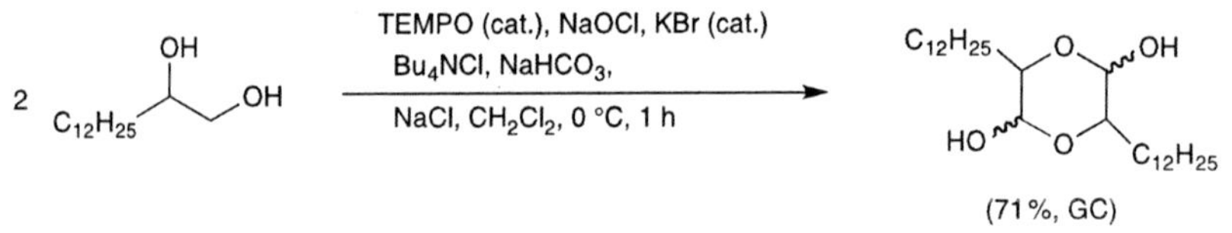

Scheme 12.25

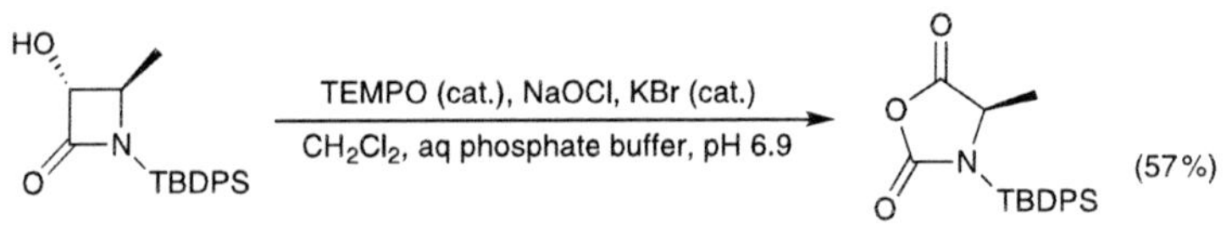

Scheme 12.26

In the TEMPO-catalyzed oxidation of 1,2-diols, cyclic hemiacetal-type dimerization products have been isolated (Scheme 12.25).[106] Presumably, stoichiometric control of the secondary oxidant prevented the oxidation of the hemiacetals to the corresponding lactone moieties.

Anhydrides/lactams can also form via an interesting ring expansion reaction that occurs when hydroxy-β-lactams are subjected to TEMPO-catalyzed reaction conditions at essentially neutrality using bleach as the secondary oxidant.[107,108] The hydroxy-β-lactam is oxidized to a N-carboxyanhydride (Scheme 12.26). Presumably, the intermediate product is an α-keto lactam that undergoes a Baeyer–Villiger-like reaction. Whether the latter step is induced by the oxoammonium oxidant or by the secondary oxidant is not known. This reaction also represents the rare case in which a carbon–carbon bond is broken during any nitroxide-catalyzed oxidation.

12.8 Stereoselective nitroxide-catalyzed oxidations

The conversion of an sp^3 alcohol carbon to an sp^2 carbonyl carbon destroys a stereocenter. Nevertheless, two possibilities exist in which a stereoselective reaction can take place. Both involve the enantioselective destruction of one stereocenter of an enantiomeric pair of stereocenters. One possibility involves the enantioselective oxidation of one isomer of racemic mixture, thus enriching the mixture with the non-oxidized chiral species (Scheme 12.27).[109] However, until most recently using catalyst **27**,[110] the ee values achieved have not been large enough for this method to be of practical value. In addition, the azaadamantane-derived catalysts of type **27** have not yet been tested in the synthesis of complex natural products.

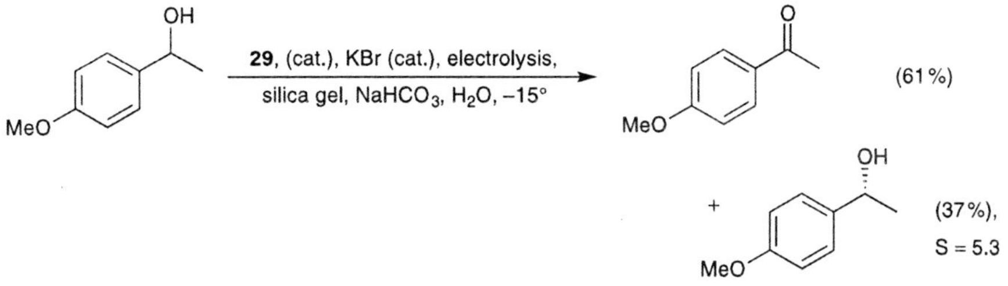

Scheme 12.27

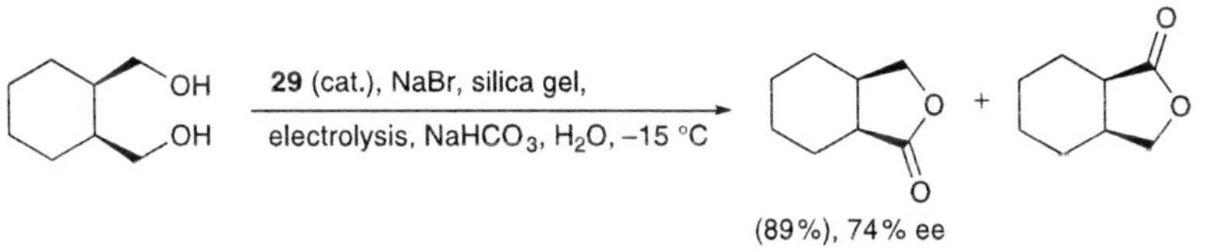

Scheme 12.28

Alternatively, one of two alcohol functionalities of a *meso*-compound may be selectively oxidized (Scheme 12.28).[111] The latter reaction theoretically allows a stereochemical reaction with 100 % conversion to a chiral product and is, therefore, of greater interest, though little explored.[41,111,112] Again, the ee's achieved to date are not practical.

Most stereoselective nitroxide-catalyzed oxidations were carried out using optically active nitroxides, examples of which are shown below,[29,37,113–120] or by using an achiral nitroxide in the presence of an optically active auxiliary (such as sparteine).[121,122] Similarly, an electrode surface can be modified with an optically active nitroxide to perform stereoselective electrochemical oxidations.[109,112,123,124] In all cases, however, the ee values achieved are modest, at best. In cases in which the chiral element is fairly remote to the oxidizing site (e.g., TEMPO derivative **26**[125]), this may not be too surprising. However, even in cases in which the chirality is much closer to the active site of the nitroxide (such as in nitroxide **28**[41,126]) or in cases in which the nitroxide is altogether C_2-symmetric (**29**[118] or the derivatives **30** to **32**[116,127]), the outcomes in terms of stereoselectivity are by and large disappointing.

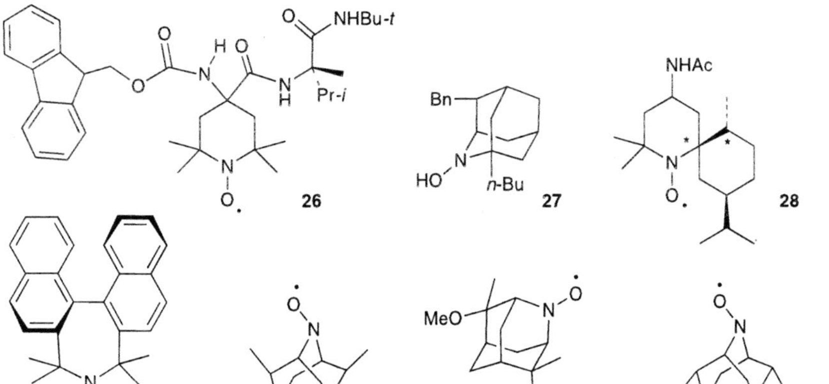

12.9 Secondary oxidants used in nitroxide-catalyzed reactions

Nitroxide-catalyzed reactions work in concert with a secondary oxidant. The secondary oxidant has two roles. In addition to converting the hydroxylamine back to the oxoammonium species, it also has to be competent in the one-electron activation of a nitroxide to the oxoammonium salt (Scheme 12.1). A range of oxidants is suitable and each oxidant is endowed with its own set of advantages and limitations. Novel systems, such as $NaNO_2/HCl$, are still being developed.[128] Here, only the main and most attractive secondary oxidants are discussed.[2]

12.9.1 Elemental halogens

Elemental chlorine, bromine, and iodine have been used as secondary oxidants in nitroxide-catalyzed oxidations. For instance, the oxidation of various sugars to their corresponding glycaric acids with chlorine or bromine has been carried out (Scheme 12.29).[96] The advantages of using halogens are the low costs of the oxidant chlorine, its atom economy, and the potential of the use of an aqueous solvent. Also, the main byproducts of these oxidants are alkaline halides (e.g., sodium chloride), and, in some cases, the products could be precipitated from water by a simple pH adjustment.[96] Obviously, these oxidants are unsuitable for any reactions involving substrates that are readily halogenated or susceptible to chlorine-induced oxidations. The main reason for using iodine rather than chlorine or bromine is that this solid is more readily delivered and tends to reduce side reactions resulting from the halogenation of electron-rich systems.[129]

12.9.2 Sodium hypochlorite (bleach)

By far the most popular, least costly, and most generally applicable secondary oxidant is hypochlorite in the form of (fresh)[60] bleach, a commercially available, slightly basic, 0.3–2 M aqueous solution of sodium hypochlorite. The oxidations are generally carried out in a vigorously stirred two-phase, water–methylene chloride system or in aqueous mixtures with acetone or acetonitrile at room temperature to 0 °C. Trifluorotoluene was suggested as organic phase to improve the yield of some oxidations.[130] As the rate and specificity of nitroxide-catalyzed oxidations are strongly dependent on the pH, the use of phosphate or carbonate buffers is employed to maintain the pH of the reaction typically between 8 and 10. At pH values above about 10, the reaction becomes prohibitively slow, whereas at lower pH values side reactions become prominent. If in large-scale reactions the buffer capacity of the chosen buffer is insufficient, base must be added to maintain the pH at the desired value.[95] The common cocatalyst bromide is also water soluble, as is sodium chlorite (for oxidations to carboxylic acids, see Section 12.7). The reaction times are short and the reaction is frequently complete within minutes to hours.

Sonocatalysis may increase the rates of oxidation of very hydrophilic substrates.[131] Likewise, the addition of a phase transfer agent (e.g., Aliquat 336, Bu_4NCl or Bu_4NBr)[45,132] may accelerate slow reactions

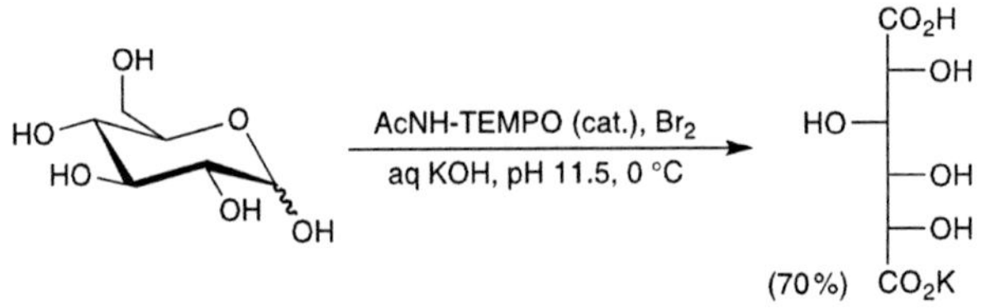

Scheme 12.29

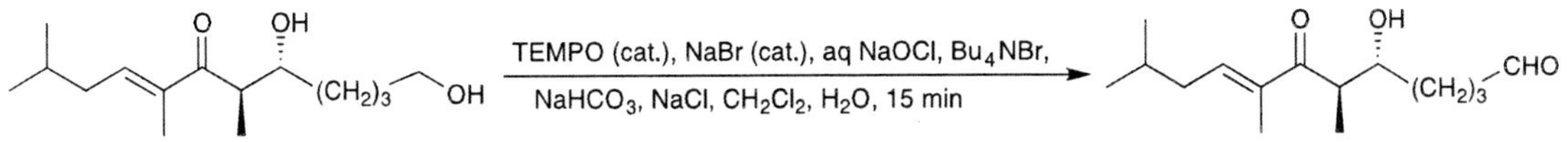

Scheme 12.30

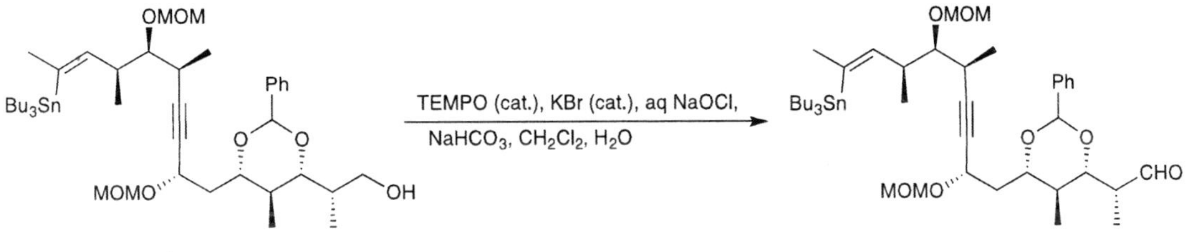

Scheme 12.31

of, for instance, very hydrophobic substrates, or may allow for the formation of carboxylic acids from alcohols or aldehydes.[45] Lithium hypochlorite,[67] calcium hypochlorite (stable solid!),[80,133,134] or *tert*-butyl hypochlorite[135] can also used as sources of hypochlorite. Reactions on a large scale have been described,[60,136] and continuous-process oxidation schemes have been devised.[137]

The power of this method can be gleaned from a number of typical examples that illustrate its speed, selectivity and functional group compatibility of the reactions (Schemes 12.30[138] and 12.31[139]). Notably, no racemization of chiral centers was observed, even when located adjacent to carbonyl groups (Scheme 12.30).

The main limitations to the use of hypochlorite are *N*-chlorination of some amides or halogenations of electron-rich aromatic rings that can take place.[136,137] Reduction of the reaction temperature (to $-5\,^{\circ}$C)[140] and the use of acetone instead of water/methylene chloride as a solvent were reported to minimize such side reactions.[141] The presence of the aqueous phase may complicate the isolation of very hydrophilic products.

12.9.3 Bis(acetoxy)iodobenzene (BAIB)

A secondary oxidant of rapidly rising popularity for nitroxide-catalyzed oxidations is the hypervalent iodine reagent bis(acetoxy)iodobenzene (BAIB).[142,143] The key advantage of using BAIB is that it does not require buffered aqueous solutions and can be used in entirely organic systems (e.g., methylene chloride, acetonitrile), thus allowing for anhydrous conditions,[142] though its use is not limited to anhydrous solvents.[144] The byproducts of the reaction (iodobenzene and acetic acid) can be removed by column chromatography or by vacuum distillation. Surprisingly, despite the fact that this reaction is carried out in an essentially acidic medium (acetic acid), the characteristic rate difference of the TEMPO-catalyzed oxidations under basic conditions (primary alcohols – fast, and secondary alcohols – slow) is maintained. Thus, primary alcohols can be selectively oxidized in the presence of secondary alcohols, even when the latter are allylic or phenolic (see also Schemes 12.13, 12.20, and 12.51). The reaction rate can be increased with a small amount of acetic acid[142] or water,[145] though the side product acetic acid may also cause undesired side reactions.[146] The reaction is tolerant to the presence of a host of functional groups, including terminal, two- and three-substituted and/or conjugated alkenes, epoxides, thioethers, selenides,

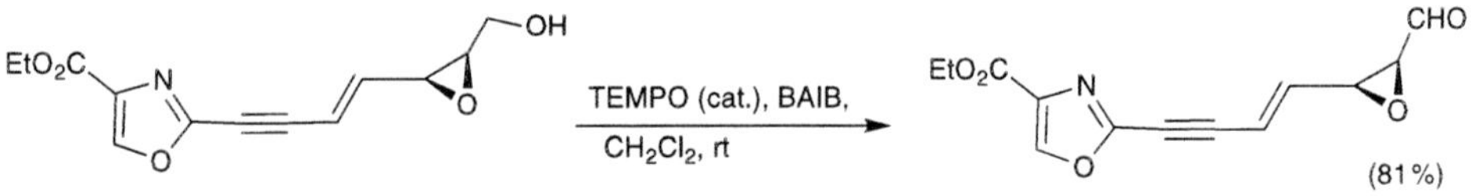

Scheme 12.32

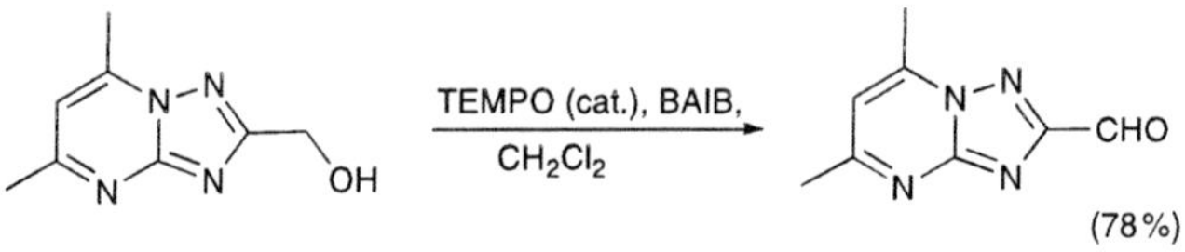

Scheme 12.33

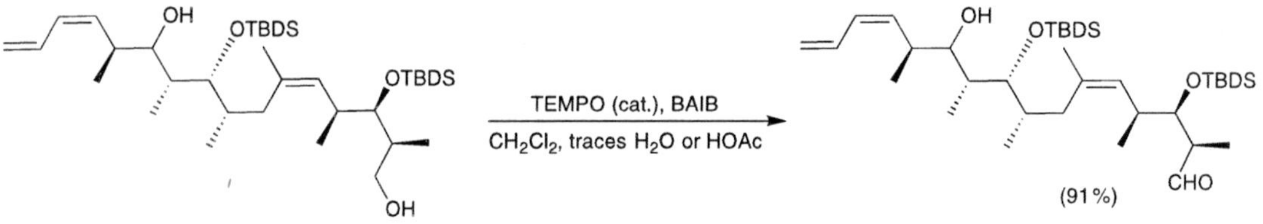

Scheme 12.34

electron-rich aromatic rings, and silane protecting groups, and the reaction conditions will not racemize chiral centers (Schemes 12.13, 12.20, 12.24, 12.32,[147] 12.33,[148] and 12.34[145]).

The costs of the secondary oxidant and the fact that its use is not atom-economic may be its greatest disadvantages. Nonetheless, the reaction has been carried out on a 1.3-molar scale (Scheme 12.34).[145,148] A number of BAIB derivatives that are also suitable as secondary oxidants have become known,[149,150] including polymeric[151] and ionic liquid derivatives.[44,152]

12.9.4 Oxygen (air)

As a particularly environmentally friendly secondary oxidant, the use of oxygen (in the form of air or in pure form), at ambient pressure or slightly above, is most desirable.[8,9] However, oxygen is unsuitable for the direct conversion of TEMPO derivatives to oxoammonium salts and, therefore, requires a cocatalyst, the most common being copper(I) or copper(II) ions in a number of "free" (chloride or perchlorate) or complexed forms in dimethylformamide, acetonitrile, or an ionic liquid.[153–157] Other metal ion-based cocatalysts for the TEMPO-catalyzed oxygen-mediated oxidation of alcohols are ceric ammonium nitrate,[158] ferric chloride/sodium nitrite,[159] cobalt and manganese nitrates,[56,57,160] or polyoxometalates.[53,161] However, the use of metal-based cocatalysts (or, not detailed here, the use of metal-based secondary oxidants)[162] erodes the environmental advantage of nitroxides, namely their ability to mediate a transition metal-free oxidation. Also, reactions with aliphatic alcohols are relatively slow. These disadvantages notwithstanding, the TEMPO/copper(II)/ligand/air oxidations have been used in the high yield conversion of complex natural products (Scheme 12.35).[157]

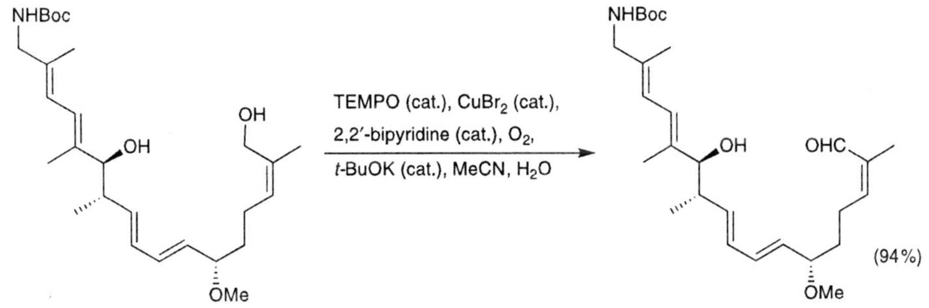

Scheme 12.35

A special case of the air oxidations are those involving laccases as cocatalysts. These lignin-degrading enzymes expressed by white rot fungi can oxidize nitroxides to oxoammonium ions, and they can be rejuvenated by air.[163] TEMPO/laccase has been used for the oxidation of benzylic alcohols (Scheme 12.5) and sugar derivatives,[164] and mechanistic studies of this reaction have become available.[70,165]

12.9.5 Peroxides

The use of cost effective and environmentally benign hydrogen peroxide as secondary oxidant would also be desirable. Aside from an example in an hydrogen bromide/ionic liquid system[166] or in the presence of methyltrioxorhenium as cocatalysts,[167] it is, however, not a generally applicable secondary oxidant in TEMPO-catalyzed reactions.

Nitroxide-catalyzed reactions using *m*-chloroperoxybenzoic acid (*m*-CPBA) as a secondary oxidant were the first nitroxide-catalyzed reactions to be discovered.[127,168] The reactions are somewhat simplified in that the reactions can be carried out using 2,2,6,6-tetramethylpiperidine hydrochloride (**1**·H^+) because *m*-CPBA also oxidizes this piperidine to TEMPO (Scheme 12.1).[168] Bromide appears to be obligatory as a cocatalyst.[169] The reaction works best only for the oxidation of secondary alcohols to ketones, since some over-oxidation of primary alcohols to carboxylic acids may be observed.[168] Also, known peracid oxidations such as the Baeyer–Villiger oxidation of ketones to esters or lactones and the epoxidation of alkenes may occur, requiring a careful selection of the alcoholic substrates to be oxidized, though this reaction can also be taken advantage of (see Section 12.10 on Tandem Reactions).

Oxone™ (DuPont; 2 $KHSO_5 \cdot KHSO_4 \cdot K_2SO_4$) is a cost effective solid peroxymonosulfate-based oxidant, which is reduced to sulfate during the reaction. It can be used in purely organic solvents when used in conjunction with a cocatalyst and phase transfer agent (Scheme 12.36).[68] Other peracid ions, such as peracetate, perborate, percarbonate, and peroxydisulfate, have been explored primarily for the oxidation of carbohydrates.[170]

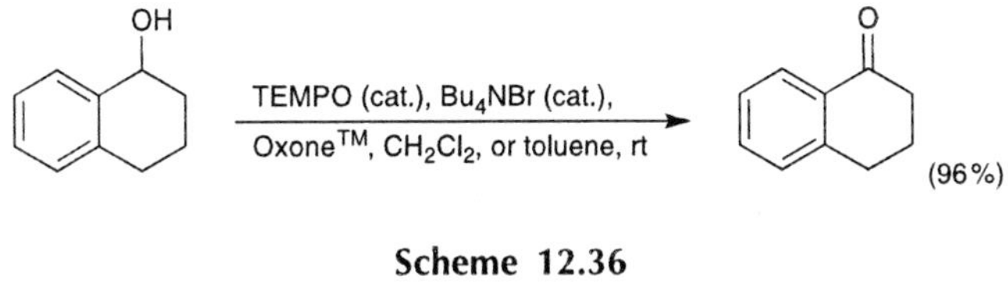

Scheme 12.36

12.9.6 Other organic secondary oxidants

Trichlorocyanuric acid (TCCA)[110,171] and *N*-chlorosuccinimide (NCS)[69] have also been used as secondary oxidants. The selectivity of the TEMPO-catalyzed reactions are generally preserved, and these reagents were successfully applied toward the synthesis of multifunctional molecules (Schemes 12.37,[172] 12.38,[173] and 12.39[174]). One advantage of these reactions is that the oxidant is a solid and a more careful stoichiometric control over the oxidation can be achieved than, for instance, when using bleach. They can be used in pure organic solvents or in buffered aqueous/organic solvent mixtures. Their disadvantages include

Scheme 12.37

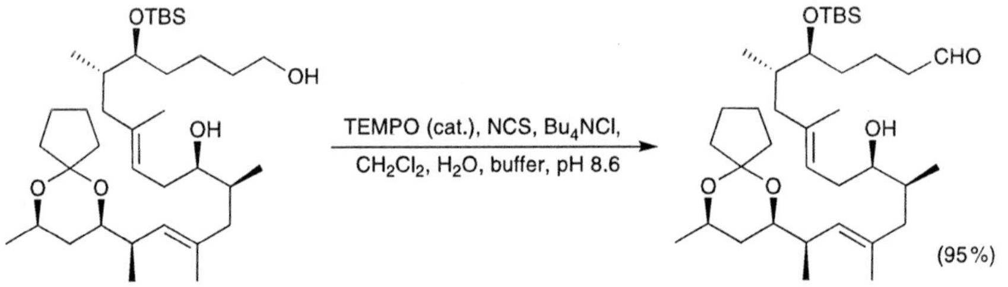

Scheme 12.38

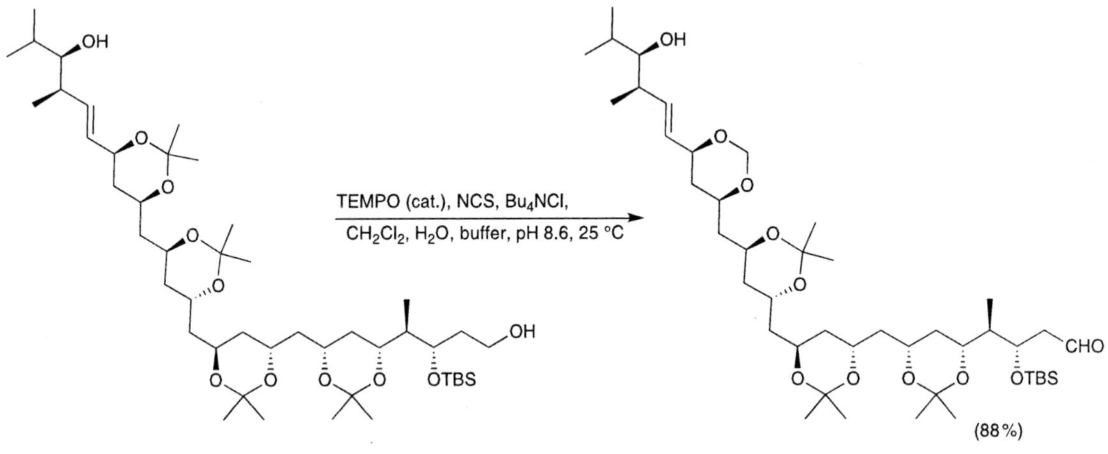

Scheme 12.39

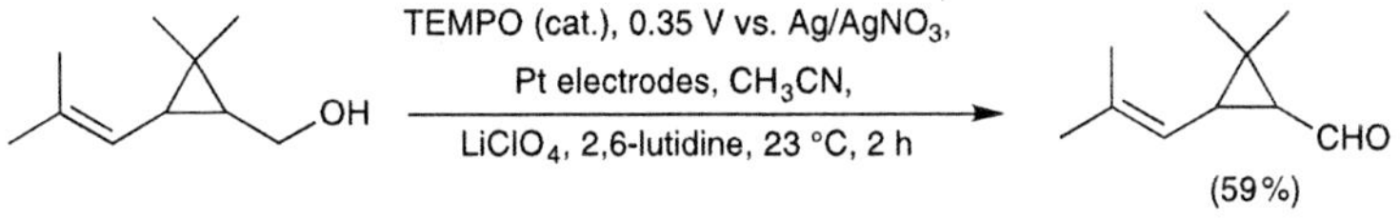

Scheme 12.40

poor atom economy and the need to remove the organic side products of the oxidants from the reaction mixtures.

12.9.7 Anodic, electrochemical oxidation

Nitroxide-catalyzed electrochemical oxidations combine the chemical specificity of oxoammonium ion oxidations with the benefits of electrochemical syntheses: the option to vary the oxidation potential (i.e., to adjust to the oxidation potential of the specific nitroxide used) and the lack of byproducts resulting from a secondary chemical oxidant (Scheme 12.40).[175] Among the disadvantages of electrochemical oxidations are that they suffer from difficulties involving the choice of the cell architecture, the electrode material and its surface modification, including the use of electrode-bound nitroxides,[176] and that they require the use of a polar solvent containing high concentration of an electrolyte to carry the current, potentially complicating the isolation of the product. Only a few reactions have been carried out using relatively simple substrates. Thus, the full potential of the electrosynthetic method for organic synthesis has yet to be realized.

12.10 Use of nitroxide-catalyzed oxidations in tandem reactions

A tandem reaction is loosely defined here as one in which the nitroxide-catalyzed reaction is coupled with a second reaction that takes place in the same flask. In the simplest case, the alcohol oxidation elicits spontaneously a sequential reaction, such as shown in the lactone cyclization reactions (Section 12.7.2). An interesting tandem reaction of an ynamide alcohol to generate a pyrrolizidinone and that was initiated by a nitroxide-catalyzed oxidation is shown in Scheme 12.41.[177] Mechanistically, this reaction proceeds through the ring opening of an amide-substituted oxetene intermediate that is formed through a [2 + 2] cycloaddition pathway between the alkyne and the aldehyde generated *in situ* during the initial oxidation. Another one-pot method can be used to convert 2,3-epoxy alcohols into isoxazole derivatives.[178]

The reactions that take place in tandem may also take place in two distinct steps. For instance, intermediate aldehydes have been trapped by a Wittig reaction (Schemes 12.42[179] and 12.43[180]).

The nitroxide/hypochlorite- and chlorite-mediated alcohol-to-carboxyl oxidations that have been carried out in a two-step, one-pot fashion may also be considered tandem reactions (Section 12.7.1).[79]

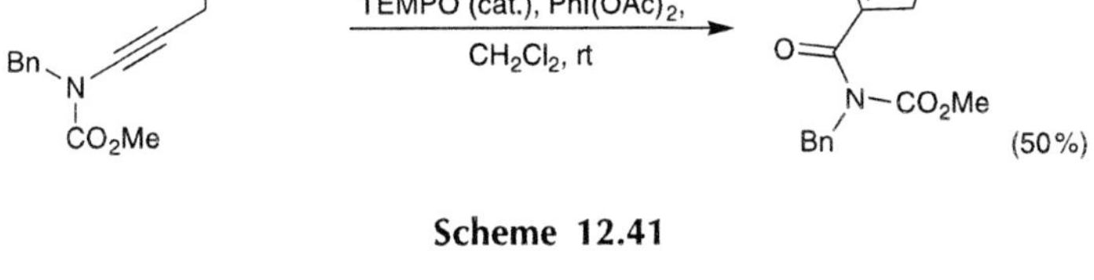

Scheme 12.41

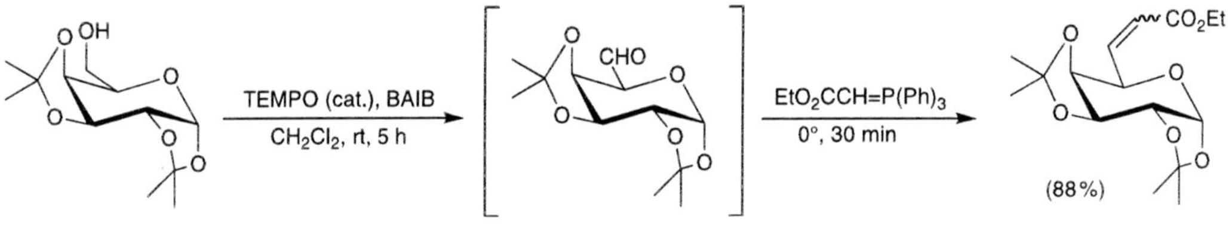

Scheme 12.42

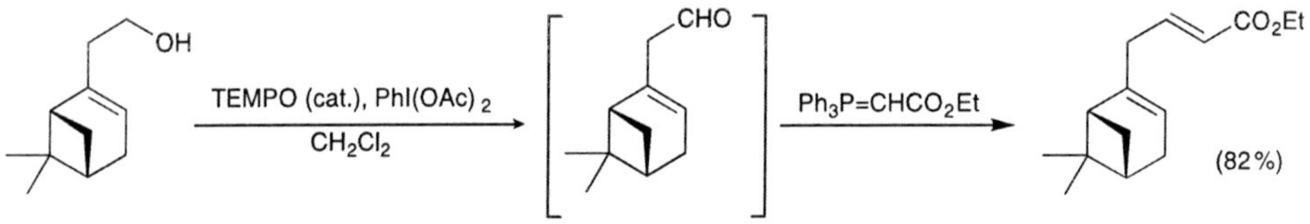

Scheme 12.43

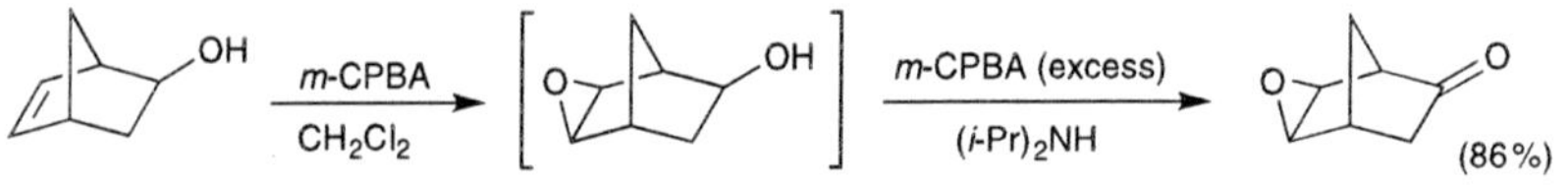

Scheme 12.44

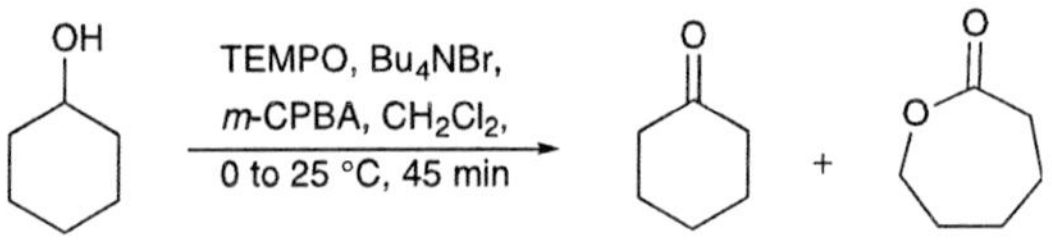

Scheme 12.45

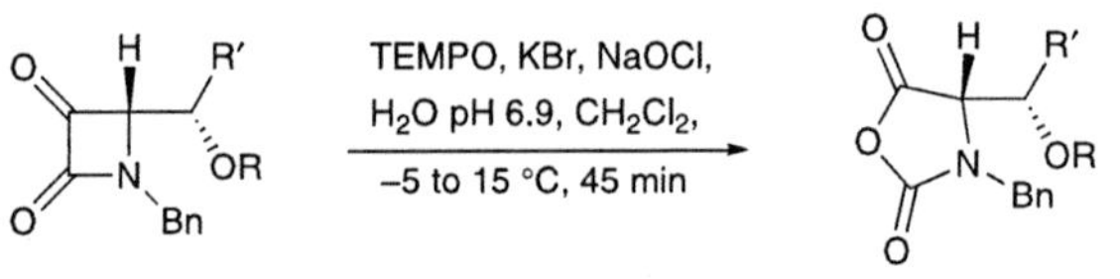

Scheme 12.46

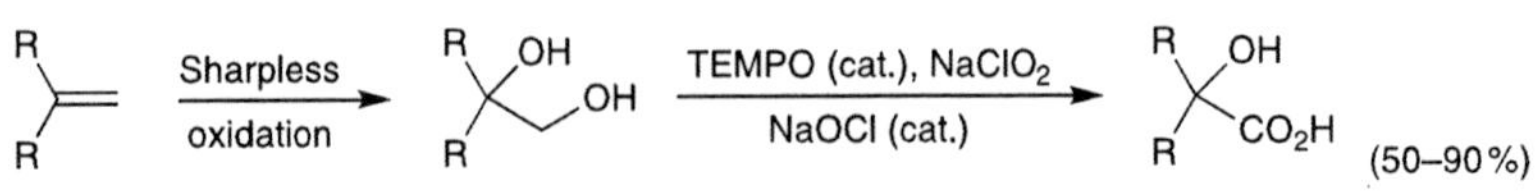

Scheme 12.47

In another variant, the second reaction is caused by the secondary oxidant, as shown for the combined oxidation–epoxidation reaction of 5-norbornene-2-ol (Scheme 12.44).[168]

If *m*-CPBA is used as the secondary oxidant in the nitroxide-catalyzed conversion of cyclohexanol to cyclohexanone, the formation of ε-caprolactone is observed (Scheme 12.45).[168] Thus, the initial catalytic oxidation of the secondary alcohol to a ketone is followed by a *m*-CPBA-induced Baeyer–Villiger-type oxygen insertion.

TEMPO, used in combination with the co-oxidant bleach, was shown to convert an α-keto-β-lactam to an anhydride in high yields (Scheme 12.46),[107] thus exhibiting an unusual Baeyer–Villiger-type oxidation behavior that is, however, most likely substrate specific (for a related example, see Scheme 12.26).

The TEMPO-catalyzed reaction can also be the second of two sequential, one-pot reaction steps. For instance, a Sharpless asymmetric dihydroxylation reaction can be combined with a TEMPO/NaOCl/$NaOCl_2$ oxidation to yield optically active α-hydroxy acids in good yields (Scheme 12.47).[181] The reaction is not completely stereospecific but, remarkably, no carbon–carbon bond cleavage occurs.

All these examples demonstrate the power of tandem reactions that are facilitated by the wide functional group tolerance of the nitroxide-catalyzed oxidations.

12.11 Predictable side reactions

By and large, the nitroxide-catalyzed alcohol oxidations are specific for alcohols, though this specificity is limited by the presence of sulfides, select activated double bonds, or free amines. The side reactions that can be expected are discussed below.

12.11.1 Oxidations of sulfur

Sulfides are rapidly converted to sulfoxides using catalytic amounts of nitroxides and a secondary oxidant (Scheme 12.48). In fact, the reaction is fast enough to allow the oxidation of sulfides to take place in the presence of primary and secondary alcohols without them being affected.[106,182] Alas, the inverse relative reactivity is also observed. Alkyl and aryl thioglycosides do not react with TEMPO-catalyzed BAIB oxidations under carefully controlled conditions,[105,183] thus allowing the conversion of a primary alcohol to a carboxylic acid in the presence of a thioether (Scheme 12.24).[103] The choice of the secondary oxidant is crucial as phenylthioglycosides[184] and phenyl sulfides[182,184] give sulfoxides using a TEMPO-catalyzed sodium hypochlorite system.[185] The sulfur in 1,3-thiazoles is not oxidized.[186] Thus, secondary oxidant, reaction conditions, and substrate potentially influence the outcome of this oxidation.

12.11.2 Oxidations of nitrogen

Oxoammonium salts, even when generated as part of a catalytic cycle, react in a less well understood manner with unprotected amine nitrogens. Pyridine and amide nitrogens are inert. At least in some cases,

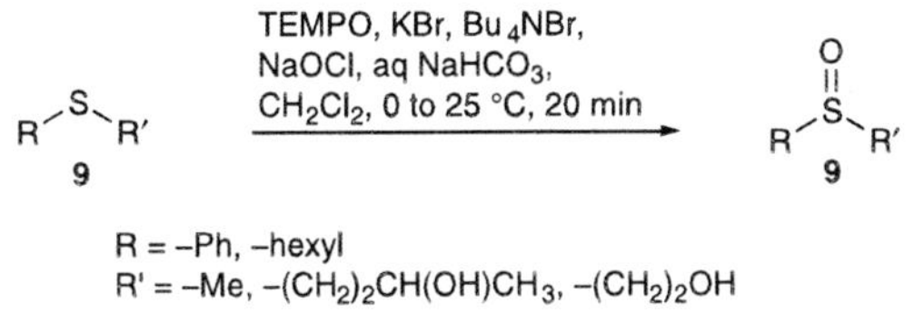

Scheme 12.48

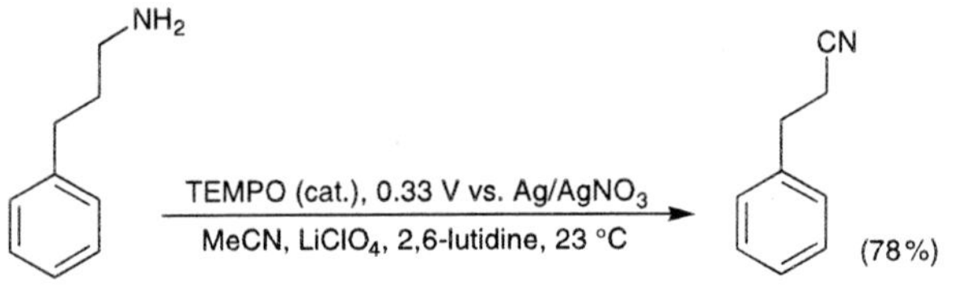

Scheme 12.49

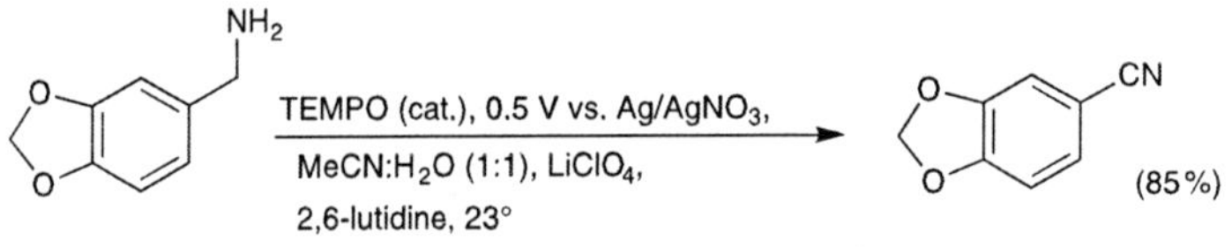

Scheme 12.50

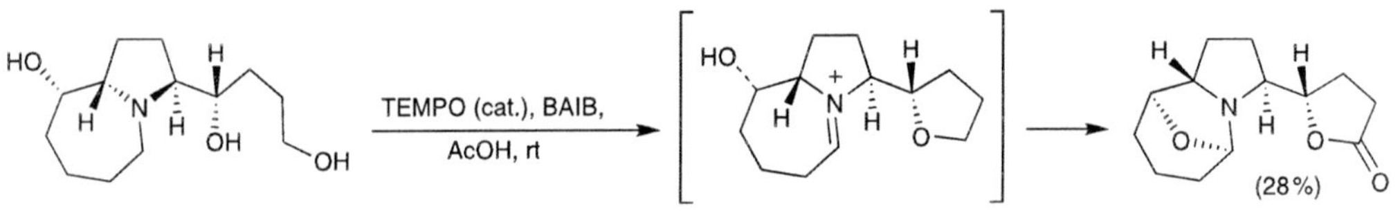

Scheme 12.51

the reactions involve an oxidation of amines to yield imines (see, e.g., Scheme 12.51).[187] *N*-Alkyl-*N*-methylanilines have been converted to *N*-alkylformanilides.[188] The scope and limits of these reactions are not yet clear. The synthetically most successful work was carried out in an electrochemical system, converting primary (Scheme 12.49) and benzylic amines (Scheme 12.50) to nitriles.[187,189]

An interesting reaction cascade elicited by a nitroxide-catalyzed BAIB oxidation is shown in Scheme 12.51.[190] The reaction illustrates several features that are typical for nitroxide-catalyzed reactions: lactone formation, resistance of secondary alcohols to oxidation, and amine-to-imine oxidation, all combined with a (nitroxide independent) tandem ring closure reaction.

12.11.3 Oxidations of carbon

Most carbons in organic molecules are inert towards nitroxide-catalyzed reactions but exceptions are known. For instance, select activated double bond systems are oxidized in nitroxide-catalyzed electro-oxidations, to provide, depending on the substrates, either enones[191] or aromatic systems (Scheme 12.52).[192] Ether carbons are also generally stable, but again, exceptions were reported for select benzyl ethers.[70,193,194]

12.12 Comparison with other oxidation methods

Nitroxide-catalyzed reactions are, next to the Dess–Martin oxidations and the Moffatt–Swern oxidations, a third family of purely organic, transition metal-free oxidants. The major advantage of the nitroxide-based reactions is that they are catalytic reactions using, in part, cost effective and benign secondary oxidants.

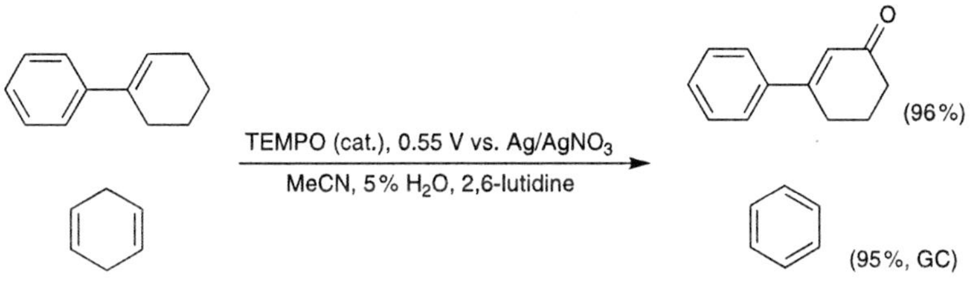

Scheme 12.52

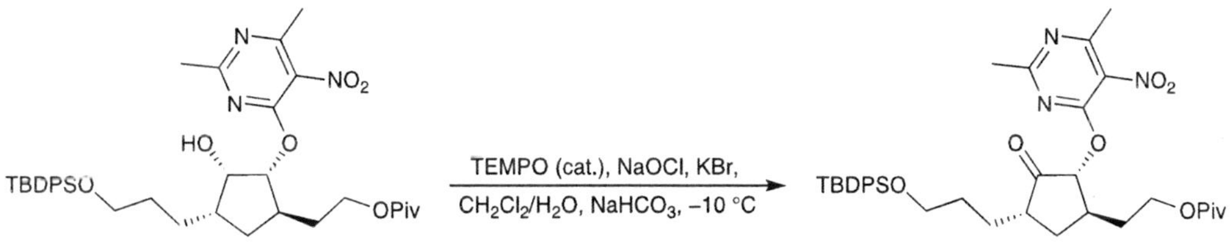

Scheme 12.53

Dess–Martin alcohol oxidations are carried out using periodinane,[195] which may not always be safe.[196] Moffatt–Swern oxidations require carefully dried reagents and solvents, generally low temperatures, and the byproduct dimethylsulfide has an obnoxious smell,[197] whereas nitroxide-catalyzed oxidations do not generate unpleasant odors and do not require anhydrous conditions. Isomerizations were observed under Swern conditions that did not occur using nitroxide-catalyzed sodium hypochlorite oxidation conditions (Scheme 12.53).[71]

Among the metal-based oxidants that are in direct competition to the nitroxide-based systems are the chromium(VI)-based reagents[198] and manganese dioxide.[199,200] While these oxidants are inexpensive, well studied, and many quite selective reactions have been developed, they all involve (toxic) metal ions that must be removed from the reaction mixtures and properly disposed of. Both methods suffer from their exclusive use in stoichiometric oxidations. Manganese dioxide oxidations also suffer from a restricted applicability to activated alcohols.[201] Using chromium(VI) reagents, side reactions such as isomerizations and oxygen migrations and the production of chromic esters have been reported. Also, TEMPO-based oxidations gave reportedly higher yields compared to pyridinium chlorochromate oxidations.[202]

12.13 Nitroxide-catalyzed oxidations and green chemistry

When considering a particular reaction for use, first and foremost its suitability for the desired transformation has to be considered. If several equivalent reactions are available, increasingly the principles of Green Chemistry are used to select one reaction over another. How, using the 12 principles of Green Chemistry,[203] do nitroxide-catalyzed oxidations stack up?

1. *Prevent waste:* Aqueous nitroxide-mediated oxidations using hypochlorite (bleach), oxygen or chlorine as secondary oxidants generate relatively benign aqueous solutions of (sodium) chloride as waste. Many nitroxide-catalyzed reactions are also tolerant to very high reactant concentrations, minimizing the quantities of solvent used.

2. *Design safer chemicals and products:* While TEMPO itself is volatile, TEMPO derivatives such as most 4-substituted TEMPO derivatives are not, minimizing possible pollution/exposure.
3. *Design less hazardous chemical syntheses:* The use of aqueous nitroxide-mediated oxidations using hypochlorite (bleach) or ambient pressure air/oxygen as secondary oxidants in aqueous solutions offer a low-hazard oxidation method.
4. *Use renewable feedstocks:* Nitroxide-mediated oxidations proved to be of particular utility in the functionalization of carbohydrates and polysaccharides, opening the way toward novel compounds and materials made from renewable feedstocks.
5. *Use catalysts, not stoichiometric reagents:* Nitroxides are good catalysts of high specificity, combined with high turnover numbers. In part, the reactions also offer excellent catalyst recovery.
6. *Avoid chemical derivatives:* Nitroxide-mediated oxidations show a number of remarkable chemical selectivities, thus avoiding protecting groups in the synthesis of complex molecules.
7. *Maximize atom economy:* Nitroxide-catalyzed oxidations using, especially, halogens, oxygen or hypochlorites as secondary oxidants are superbly atom-economic reactions.
8. *Use safer solvents and reaction conditions:* The potential for nitroxide-catalyzed reactions to be performed in aqueous solutions at ambient temperatures and pressures follows this guideline.
9. *Increase energy efficiency:* As most nitroxide-catalyzed reactions are performed at ambient pressure and temperature (or only slightly below), they are not particularly energy demanding.
10. *Design chemicals and products to degrade after use:* We are not aware of the study of the (bio)degradation of nitroxides but nitroxides may allow the preparation of degradable materials (see principle 4).
11. *Analyze in real time to prevent pollution:* In-process real-time monitoring of aqueous nitroxide-catalyzed bleach/chlorine oxidations is possible by means of monitoring the pH of the reaction solution.
12. *Minimize the potential for accidents:* The potential for using stable, crystalline nitroxides, aqueous systems at ambient temperatures and using secondary oxidants such as air, bleach or Oxone™ are steps toward accident minimization (see also principle 8).

In summary, while the use of nitroxide-catalyzed oxidations is not *per se* a green process, many aspects of these reactions allow the "greening" of oxidations.

Acknowledgements

Thanks to Professor James M. Bobbitt (University of Connecticut, USA) and Dr. Nabyl Merbouh (Simon Fraser University, Canada) for their collaboration in our joint forays into nitroxide-catalyzed and oxoammonium-based oxidations. The work was supported by the University of Connecticut.

References

1. M. B. Neiman, É. G. Rozantzev, and Y. G. Mamedova, *Nature*, **196**, 472 (1962).
2. J. M. Bobbitt, C. Brückner, and N. Merbouh, *Org. React.*, **74**, 106–424 (2009).
3. T. Vogler and A. Studer, *Synthesis*, 1979–1993 (2008).
4. V. A. Golubev, Y. N. Kozlov, A. N. Petrov, and A. P. Purmal, *Progr. Reaction Kinetics*, **16**, 35–54 (1991).
5. J. M. Bobbitt and M. C. L. Flores, *Heterocycles*, **27**, 509–533 (1988).
6. A. E. J. de Nooy, A. C. Besemer, and H. van Bekkum, *Synthesis*, 1153–1174 (1996).
7. W. Adam, C. R. Saha-Möller, and P. A. Ganeshpure, *Chem. Rev.*, **101**, 3499–3548 (2001).

8. R. A. Sheldon, I. W. C. E. Arends, G. J. ten Brink, and A. Dijksman, *Acc. Chem. Res.*, **35**, 774–781 (2002).
9. R. A. Sheldon and I. W. C. E. Arends, *Adv. Synth. Catal.*, **346**, 1051–1071 (2004).
10. N. Merbouh, J. M. Bobbitt, and C. Brückner, *Org. Prep. Proced. Int.*, **36**, 3–31 (2004).
11. M. V. N. De Souza, *Mini-Rev. Org. Chem.*, **3**, 155–165 (2006).
12. Y. Iwabuchi, *J. Synth. Org. Chem., Jpn.*, **66**, 1076–1084 (2008).
13. A. Nilsen and R. Braslau, *J. Polymer Sci., Part A: Polym. Chem.*, **44**, 697–717 (2006).
14. R. Ciriminna and M. Pagliaro, *Org. Process Res. Dev.*, **14**, 245–251 (2010).
15. P. L. Bragd, H. van Bekkum, and A. C. Besemer, *Top. Catal.*, **27**, 49–66 (2004).
16. T. Takata, Y. Tsujino, S. Nakanishi, *et al.*, *Chem. Lett.*, 937–938 (1999).
17. P. P. Pradhan, J. M. Bobbitt, and W. F. Bailey, *Org. Lett.*, **8**, 5485–5487 (2006).
18. V. A. Golubev, V. N. Borislavskii, and A. L. Aleksandrov, *Bull. Acad. Sci.USSR, Chem. Ser.*, **9**, 1874 (1977).
19. M. F. Semmelhack, C. R. Schmid, and D. A. Cortés, *Tetrahedron Lett.*, **27**, 1119–1122 (1986).
20. W. F. Bailey, J. M. Bobbitt, and K. B. Wiberg, *J. Org. Chem.*, **72**, 4504–4509 (2007).
21. Y. Liu, H. Guo, and Z. Liu, *Huaxue Xuebao*, **49**, 187–192 (1991).
22. M. G. Banwell, V. S. Bridges, J. R. Dupuche, *et al.*, *J. Org. Chem.*, **59**, 6338–6343 (1994).
23. J. M. Bobbitt, *J. Org. Chem.*, **63**, 9367–9374 (1998).
24. S. Shibuya and Y. Iwabuchi, *J. Org. Chem.*, **73**, 4750–4752 (2008).
25. V. A. Golubev and R. V. Miklyush, *Zh. Org. Khim.*, **8**, 1376–1377 (Engl. transl.) (1972).
26. T. Ren, Y. Liu, and Q. Guo, *Bull. Chem. Soc. Jpn.*, **69**, 2935–2941 (1996).
27. T. Inokuchi, P. Liu, and S. Torii, *Chem. Lett.*, 1411–1414 (1994).
28. L. B. Volodarsky, V. A. Raznikov, and V. J. Ovcharenko, *Synthetic Chemistry of Stable Nitroxides*, CRC Press, Boca Raton, FL, USA, 2000.
29. J. F. W. Keana, *Chem. Rev.*, **78**, 37–64 (1978).
30. J. M. Bobbitt and N. Merbouh, *Org. Synth.*, **82**, 80–86 (2005).
31. M. Shibuya, M. Tomizawa, I. Suzuki, and Y. Iwabuchi, *J. Am. Chem. Soc.*, **128**, 8412–8413 (2006).
32. M. Shibuya, M. Tomizawa, Y. Sasano, and Y. Iwabuchi, *J. Org. Chem.*, **74**, 4619–4622 (2009).
33. R. M. Dupeyre and A. Rassat, *Tetrahedron*, **34**, 1901–1907 (1978).
34. Y. Demizu, H. Shiigi, T. Oda, *et al.*, *Tetrahedron Lett.*, **49**, 48–52 (2008).
35. Y. Kashiwagi, T. Nishimura, and J. Anzai, *Electrochim. Acta*, **47**, 1317–1320 (2002).
36. S. D. Rychnovsky, R. Vaidyanathan, T. Beauchamp, *et al.*, *J. Org. Chem.*, **64**, 6745–6749 (1999).
37. S. D. Rychnovsky, T. Beauchamp, R. Vaidyanathan, and T. Kwan, *J. Org. Chem.*, **63**, 6363–6374 (1998).
38. J. A. Cella, J. A. Kelley, and E. F. Kenehan, *Tetrahedron Lett.*, **33**, 2869–2872 (1975).
39. T. Inokuchi, S. Matsumoto, and S. Torii, *J. Org. Chem.*, **56**, 2416–2421 (1991).
40. V. A. Golubev, G. N. Voronina, and E. G. Rozantsev, *Bull. Acad. Sci. USSR, Chem. Ser.*, 146 (1972).
41. Z. Ma, Q. Huang, and J. M. Bobbitt, *J. Org. Chem.*, **58**, 4837–4843 (1993).
42. M. Kuroboshi, J. Fujisawa, and H. Tanaka, *Electrochemistry*, **72**, 846–848 (2004).
43. X. E. Wu, L. Ma, M. X. Ding, and L. X. Gao, *Synlett*, 607–610 (2005).
44. W. Qian, E. Jin, W. Bao, and Y. Zhang, *Angew. Chem. Int. Ed.*, **44**, 952–955 (2005).
45. P. L. Anelli, C. Biffi, F. Montanari, and S. Quici, *J. Org. Chem.*, **52**, 2559–2562 (1987).
46. T. Fey, H. Fischer, S. Bachmann, *et al.*, *J. Org. Chem.*, **66**, 8154–8159 (2001).
47. R. Ciriminna, C. Bolm, T. Fey, and M. Pagliaro, *Adv. Synth. Catal.*, **344**, 159–163 (2002).
48. M. Benaglia, A. Puglisi, and F. Cozzi, *Chem. Rev.*, **103**, 3401–3429 (2003).
49. S. Weik, G. Nicholson, G. Jung, and J. Rademann, *Angew. Chem. Int. Ed.*, **40**, 1436–1439 (2001).
50. K. Yasuda and S. V. Ley, *J. Chem. Soc., Perkin Trans. 1*, 1024–1025 (2002).
51. A. Gheorghe, A. Matsumo, and O. Reiser, *Adv. Synth. Catal.*, **348**, 1016–1020 (2006).
52. C. Tanyeli and A. Gumus, *Tetrahedron Lett.*, **44**, 1639–1642 (2003).
53. A. Dijksman, I. W. C. E. Arends, and R. A. Sheldon, *Synlett*, 102–104 (2001).
54. P. Ferreira, W. Hayes, E. Phillips, *et al.*, *Green Chem.*, **6**, 310–312 (2004).
55. P. Ferreira, E. Phillips, D. Rippon, *et al.*, *J. Org. Chem.*, **69**, 6851–6859 (2004).
56. M. Benaglia, A. Puglisi, O. Holczknecht, *et al.*, *Tetrahedron*, **61**, 12058–12064 (2005).
57. M. Gilhespy, M. Lok, and X. Baucherel, *Chem. Commun.*, 1085–1086 (2005).

58. N. Tsubokawa, T. Kimoto, and T. Endo, *J. Mol. Catal. A: Chem.*, **101**, 45–50 (1995).
59. C. Bolm and T. Fey, *Chem. Commun.*, 1795–1796 (1999).
60. A. Michaud, G. Gingras, M. Morin, *et al.*, *Org. Process Res. Dev.*, **11**, 766–768 (2007).
61. D. Brunel, F. Fajula, J. B. Nagy, *et al.*, *Appl. Catal., A*, **213**, 73–82 (2001).
62. B. Karimi, A. Biglari, J. H. Clark, and V. Budarin, *Angew. Chem. Int. Ed.*, **46**, 7210–7213 (2007).
63. D. Avnir, *Acc. Chem. Res.*, **28**, 328–334 (1995).
64. M. Gilhespy, M. Lok, and X. Baucherel, *Catal. Today*, **117**, 114–119 (2006).
65. A. E. J. de Nooy, A. C. Besemer, and H. van Bekkum, *Tetrahedron*, **51**, 8023–8032 (1995).
66. U. P. Dhokte, V. V. Khau, D. R. Hutchison, and M. J. Martinelli, *Tetrahedron Lett.*, **39**, 8771–8774 (1998).
67. P. L. Anelli, S. Banfi, F. Montanari, and S. Quici, *J. Org. Chem.*, **54**, 2970–2972 (1989).
68. C. Bolm, A. S. Magnus, and J. P. Hildebrand, *Org. Lett.*, **2**, 1173–1175 (2000).
69. J. Einhorn, C. Einhorn, F. Ratajczak, and J.-L. Pierre, *J. Org. Chem.*, **61**, 7452–7454 (1996).
70. F. d'Acunzo, P. Baiocco, M. Fabbrini, *et al.*, *Eur. J. Org. Chem.*, 4195–4201 (2002).
71. F. Busqúe, S. A. Hopkins, and J. P. Konopelski, *J. Org. Chem.*, **67**, 6097–6103 (2002).
72. N. Z. Burns and P. S. Baran, *Angew. Chem. Int. Ed.*, **47**, 205–208 (2008).
73. M. Zhao, J. Li, E. Mano, *et al.*, *Org. Synth.*, **81**, 195 (2005).
74. M. Zhao, J. Li, E. Mano, *et al.*, *J. Org. Chem.*, **64**, 2564–2566 (1999).
75. P. Wipf and T. H. Graham, *J. Am. Chem. Soc.*, **126**, 15346–15347 (2004).
76. M. L. Testa, R. Ciriminna, C. Hajji, *et al.*, *Adv. Synth. Catal.*, **346**, 655–660 (2004).
77. A. J. Herrera, M. T. Beneitez, L. Amorim, *et al.*, *Carbohydr. Res.*, **342**, 1876–1887 (2007).
78. I. Paterson, G. J. Florence, A. C. Heimann, and A. C. Mackay, *Angew. Chem. Int. Ed.*, **44**, 1130–1133 (2005).
79. A. Zanka, *Chem. Pharm. Bull.*, **51**, 888–889 (2003).
80. Z. Györgydeák and J. Thiem, *Carbohydr. Res.*, **268**, 85–92 (1995).
81. M. Schämann and H. J. Schäfer, *Eur. J. Org. Chem.*, 351–358 (2003).
82. S. Brochette-Lemoine, D. Joannard, G. Descotes, *et al.*, *J. Mol. Catal. A: Chem.*, **150**, 31–36 (1999).
83. R. A. Goodnow, A. R. Richou, and S. Tam, *Tetrahedron Lett.*, **38**, 3195–3198 (1997).
84. C. Fraschini and M. R. Vignon, *Carbohydr. Res.*, **328**, 585–589 (2000).
85. K. Schnatbaum and H. J. Schäfer, *Synthesis*, 864–872 (1999).
86. T. Kraus, M. Buděšínský, and J. Závada, *Eur. J. Org. Chem.*, 3133–3137 (2000).
87. P. S. Chang and J. F. Robyt, *J. Carbohydr. Chem.*, **15**, 819–830 (1996).
88. A. E. J. de Nooy, A. C. Besemer, and H. van Bekkum, *Carbohydr. Res.*, **269**, 89–98 (1995).
89. V. Crescenzi, D. Delicato, and M. Dentini, *J. Carbohydr. Chem.*, **16**, 697–701 (1997).
90. S. Gomez-Bujedo, E. Fleury, and M. R. Vignon, *Biomacromol.*, **5**, 565–571 (2004).
91. L. E. Lillo and B. Matsuhiro, *Carbohydr. Polym.*, **51**, 317–325 (2003).
92. T. Saito, M. Yanagisawa, and A. Isogai, *Cellulose*, **12**, 305–315 (2005).
93. A. Isogai and Y. Kato, *Cellulose*, **5**, 153–164 (1998).
94. C. Tahiri and M. R. Vignon, *Cellulose*, **7**, 177–188 (2000).
95. N. Merbouh, J. F. Thaburet, M. Ibert, *et al.*, *Carbohydr. Res.*, **336**, 75–78 (2001).
96. N. Merbouh, J. M. Bobbitt, and C. Brückner, *J. Carbohydr. Chem.*, **21**, 65–77 (2002).
97. M. Ibert, F. Marsais, N. Merbouh, and C. Brückner, *Carbohydr. Res.*, **337**, 1059–1063 (2002).
98. R. Ciriminna and M. Pagliaro, *Adv. Synth. Catal.*, **345**, 383–388 (2003).
99. D. Domon, K. Fujiwara, Y. Ohtaniuchi, *et al.*, *Tetrahedron Lett.*, **46**, 8279–8283 (2005).
100. T. Siu, C. D. Cox, and S. J. Danishefsky, *Angew. Chem. Int. Ed.*, **42**, 5629–5634 (2003).
101. H. Zhang, M. Seepersaud, S. Seepersaud, and D. R. Mootoo, *J. Org. Chem.*, **63**, 2049–2052 (1998).
102. I. Kadota, H. Takamura, K. Sato, and Y. Yamamoto, *J. Org. Chem.*, **67**, 3494–3498 (2002).
103. T. Miyazawa and T. Endo, *J. Am. Chem. Soc.*, **50**, 3930–3931 (1985).
104. M. G. Banwell, A. M. Bray, A. J. Edwards, and D. J. Wong, *New J. Chem.*, **25**, 3–7 (2001).
105. L. J. van den Bos, R. E. J. N. Litjens, R. J. B. H. N. van den Berg, *et al.*, *Org. Lett.*, **7**, 2007–2010 (2005).
106. R. Siedlecka, J. Skarzewski, and J. Mlochowski, *Tetrahedron Lett.*, **31**, 2177–2180 (1990).
107. C. Palomo, J. M. Aizpurua, C. Cuevas, *et al.*, *J. Org. Chem.*, **61**, 4400–4404 (1996).
108. T. B. Durham and M. J. Miller, *J. Org. Chem.*, **68**, 27–34 (2003).

109. M. Kuroboshi, H. Yoshihisa, M. N. Cortona, *et al.*, *Tetrahedron Lett.*, **41**, 8131–8135 (2000).
110. M. Tomizawa, M. Shibuya, and Y. Iwabuchi, *Org. Lett.*, **11**, 1829–1831 (2009).
111. H. Tanaka, Y. Kawakami, K. Goto, and M. Kuroboshi, *Tetrahedron Lett.*, **42**, 445–448 (2001).
112. Y. Kashiwagi, F. Kurashima, S. Chiba, *et al.*, *Chem. Commun.*, 114–115 (2003).
113. N. Naik and R. Braslau, *Tetrahedron*, **54**, 667–696 (1998).
114. Y. Kashiwagi, F. Kurashima, C. Kikuchi, *et al.*, *Tetrahedron Lett.*, **40**, 6469–6472 (1999).
115. J. Einhorn, C. Einhorn, F. Ratajczak, and J.-L. Pierre, *Synth. Commun.*, **30**, 1837–1848 (2000).
116. B. Graetz, S. Rychnovsky, W.-H. Leu, *et al.*, *Tetrahedron: Asymmetry*, **16**, 3584–3598 (2005).
117. T. Shibata, K. Uemae, and Y. Yamamoto, *Tetrahedron: Asymmetry*, **11**, 2339–2346 (2000).
118. S. D. Rychnovsky, T. L. McLernon, and H. Rajapakse, *J. Org. Chem.*, **61**, 1194–1195 (1996).
119. K. Wright, A. de Castries, M. Sarciaux, *et al.*, *Tetrahedron Lett.*, **46**, 5573–5576 (2005).
120. R. Braslau, V. Chaplinski, and P. Goodson, *J. Org. Chem.*, **63**, 9857–9864 (1998).
121. Y. Kashiwagi, Y. Yanagisawa, F. Kurashima, *et al.*, *Chem. Commun.*, 2745–2746 (1996).
122. E. M. Belgsir and H. J. Schäfer, *Chem. Commun.*, 435–436 (1999).
123. Y. Kashiwagi, S. Chiba, and J. Anzai, *New J. Chem.*, **27**, 1545–1549 (2003).
124. Y. Kashiwagi, F. Kurashima, C. Kikuchi, *et al.*, *Heterocycles*, **53**, 1583–1587 (2000).
125. F. Formaggio, M. Bonchio, M. Crisma, *et al.*, *Chem. Eur. J.*, **8**, 84–93 (2002).
126. Z. Ma and J. M. Bobbitt, *J. Org. Chem.*, **56**, 6110–6114 (1991).
127. B. Ganem, *J. Org. Chem.*, **40**, 1998–2000 (1975).
128. X. Wang, R. Liu, Y. Jin, and X. Liang, *Chem. Eur. J.*, **14**, 2679–2685 (2009).
129. R. A. Miller and R. S. Hoerrner, *Org. Lett.*, **5**, 285–287 (2003).
130. G. Pozzi, F. Cinato, F. Montanari, and S. Quici, *Chem. Commun.*, 877–878 (1998).
131. S. Brochette-Lemoine, S. Trombotto, D. Joannard, *et al.*, *Ultrason. Sonochem.*, **7**, 157–161 (2000).
132. T. Markidis and G. Kokotos, *J. Org. Chem.*, **67**, 1685–1688 (2002).
133. F. Lin, W. Peng, W. Xu, *et al.*, *Carbohydr. Res.*, **339**, 1219–1223 (2004).
134. T. Inokuchi, S. Matsumoto, T. Nishiyama, and S. Torii, *J. Org. Chem.*, **55**, 462–466 (1990).
135. C. S. Rye and S. G. Withers, *J. Am. Chem. Soc.*, **124**, 9756–9767 (2002).
136. R. H. Yu, R. P. Polniaszek, M. W. Becker, *et al.*, *Org. Process Res. Dev.*, **11**, 972–980 (2007).
137. E. Fritz-Langhals, *Org. Process Res. Dev.*, **9**, 577–582 (2005).
138. A. N. Hulme and G. E. Howells, *Tetrahedron Lett.*, **38**, 8245–8248 (1997).
139. E. de Lemos, F. H. Poree, A. Commercon, *et al.*, *Angew. Chem. Int. Ed.*, **46**, 1917–1921 (2007).
140. Z. J. Song, M. Zhao, R. Desmond, *et al.*, *J. Org. Chem.*, **64**, 9658–9667 (1999).
141. D. W. Konas and J. K. Coward, *J. Org. Chem.*, **66**, 8831–8842 (2001).
142. A. De Mico, R. Margarita, L. Parlanti, *et al.*, *J. Org. Chem.*, **62**, 6974–6977 (1997).
143. G. Piancatelli and F. Leonelli, *Org. Synth.*, **83**, 18–23 (2005).
144. J. B. Epp and T. S. Widlanski, *J. Org. Chem.*, **64**, 293–295 (1999).
145. S. J. Mickel, G. H. Sedelmeier, D. Niederer, *et al.*, *Org. Process Res. Dev.*, **8**, 113–121 (2004).
146. D. J. Vugts, L. Veum, K. al-Mafraji, *et al.*, *Eur. J. Org. Chem.*, 1672–1677 (2006).
147. I. V. Hartung, U. Eggert, L. O. Haustedt, *et al.*, *Synthesis*, 1844–1850 (2003).
148. D. Camp, C. F. Matthews, S. T. Neville, *et al.*, *Org. Process Res. Dev.*, **10**, 814–821 (2006).
149. H. Tohma and Y. Kita, *Adv. Synth. Catal.*, **346**, 111–124 (2004).
150. A. Moroda and H. Togo, *Tetrahedron*, **62**, 12408–12414 (2006).
151. K. Sakuratani and H. Togo, *Synthesis*, 21–23 (2003).
152. W. Qian, E. Jin, W. Bao, and Y. Zhang, *Tetrahedron*, **62**, 556–562 (2006).
153. G. Ragagnin, B. Betzemeier, S. Quici, and P. Knochel, *Tetrahedron*, **58**, 3985–3991 (2002).
154. D. Geißlmeir, W. G. Jary, and H. Falk, *Monatsh. Chem.*, **136**, 1591–1599 (2005).
155. N. Jiang and A. J. Ragauskas, *J. Org. Chem.*, **71**, 7087–7090 (2006).
156. S. Mannam, S. K. Alamsetti, and G. Sekar, *Adv. Synth. Catal.*, **349**, 2253–2258 (2007).
157. A. Fürstner, C. Aissa, C. Chevrier, *et al.*, *Angew. Chem. Int. Ed.*, **45**, 5832–5837 (2006).
158. S. S. Kim and H. C. Jung, *Synthesis*, 2135–2137 (2003).
159. N. Wang, R. Liu, J. Chen, and X. Liang, *Chem. Commun.*, 5322–5324 (2005).

160. F. Minisci, F. Recupero, G. F. Pedulli, and M. Lucarini, *J. Mol. Catal. A: Chem.*, **204–205**, 63–90 (2003).
161. N. Jiang and A. J. Ragauskas, *Org. Lett.*, **7**, 3689–3692 (2005).
162. R. S. Kumar, K. Karthikeyan, and P. T. Perumal, *Can. J. Chem.*, **86**, 720–725 (2008).
163. R. ten Have and P. J. M. Teunissen, *Chem. Rev.*, **101**, 3397–3413 (2001).
164. M. Marzorati, B. Danieli, D. Haltrich, and S. Riva, *Green Chem.*, **7**, 310–315 (2005).
165. I. W. C. E. Arends, Y.-X. Li, R. Ausan, and R. A. Sheldon, *Tetrahedron*, **62**, 6659–6665 (2006).
166. N. Jiang and A. J. Ragauskas, *Tetrahedron Lett.*, **46**, 3323–3326 (2005).
167. W. A. Herrmann, J. P. Zoller, and R. W. Fischer, *J. Organomet. Chem.*, **579**, 404–407 (1999).
168. J. A. Cella, J. A. Kelley, and E. F. Kenehan, *J. Org. Chem.*, **40**, 1860–1862 (1975).
169. S. D. Rychnovsky and R. Vaidyanathan, *J. Org. Chem.*, **64**, 310–312 (1999).
170. P. L. Bragd, A. C. Besemer, and H. van Bekkum, *Carbohydr. Polym.*, **49**, 397–406 (2002).
171. L. De Luca, G. Giacomelli, S. Masala, and A. Porcheddu, *J. Org. Chem.*, **68**, 4999–5001 (2003).
172. B. G. Vong, S. H. Kim, S. Abraham, and E. A. Theodorakis, *Angew. Chem. Int. Ed.*, **43**, 3947–3951 (2004).
173. A. B. Benowitz, S. Fidanze, P. L. C. Small, and Y. Kishi, *J. Am. Chem. Soc.*, **123**, 5128–5129 (2001).
174. C. J. Sinz and S. D. Rychnovsky, *Tetrahedron*, **58**, 6561–6576 (2002).
175. M. F. Semmelhack, C. S. Chou, and D. A. Cortés, *J. Am. Chem. Soc.*, **105**, 4492–4494 (1983).
176. J. M. Bobbitt, Z. Ma, D. Bolz, *et al.*, *Supported Reagents and Catalysts in Chemistry*, Royal Society of Chemistry, London, UK, 1998.
177. K. C. M. Kurtz, R. P. Hsung, and Y. Zhang, *Org. Lett.*, **8**, 231–234 (2006).
178. E. Marotta, L. M. Micheloni, N. Scardovi, and P. Righi, *Org. Lett.*, **3**, 727–729 (2001).
179. J. M. Vatèle, *Synlett.*, 2055–2058 (2006).
180. J. M. Vatèle, *Tetrahedron Lett.*, **47**, 715–718 (2006).
181. F. J. Aladro, F. M. Guerra, F. J. Moreno-Dorado, *et al.*, *Tetrahedron Lett.*, **41**, 3209–3213 (2000).
182. R. Siedlecka and J. Skarzewski, *Synthesis*, 401–404 (1994).
183. J. D. C. Codée, B. Stubba, M. Schiattarella, *et al.*, *J. Am. Chem. Soc.*, **127**, 3767–3773 (2005).
184. J.-Y. Huang, S.-J. Li, and Y.-G. Wang, *Tetrahedron Lett.*, **47**, 5637–5640 (2006).
185. M. E. Scott and M. Lautens, *Org. Lett.*, **7**, 3045–3047 (2005).
186. M. Brünjes, G. Sourkouni-Argirusi, and A. Kirschning, *Adv. Synth. Catal.*, **345**, 635–642 (2003).
187. M. F. Semmelhack and C. R. Schmid, *J. Am. Chem. Soc.*, **105**, 6732–6734 (1983).
188. Y. Kashiwagi and J. Anzai, *Chem. Pharm. Bull.*, **49**, 324–326 (2001).
189. Y. Kashiwagi, F. Kurashima, C. Kikuchi, *et al.*, *J. Chin. Chem. Soc.*, **45**, 135–138 (1998).
190. S. G. Pyne, A. S. Davis, N. J. Gates, *et al.*, *Synlett*, 2670–2680 (2004).
191. T. Breton, D. Liaigre, and E. M. Belgsir, *Tetrahedron Lett.*, **46**, 2487–2490 (2005).
192. T. Breton, D. Liaigre, and E. M. Belgsir, *Electrochem. Commun.*, **7**, 1445–1448 (2005).
193. F. d'Acunzo, P. Baiocco, and C. Galli, *New J. Chem.*, **27**, 329–332 (2003).
194. N. S. Cho and C. H. Park, *J. Korean Chem. Soc.*, **39**, 657–665 (1995).
195. D. B. Dess and J. C. Martin, *J. Am. Chem. Soc.*, **113**, 7277–7287 (1991).
196. M. Fieser, *Reagents for Organic Synthesis*, Vol. 17, John Wiley & Sons, Inc., New York, NY, USA, 1994.
197. T. T. Tidwell, *Org. React.*, **39**, 297–572 (1990).
198. F. A. Luzzio, *Org. React.*, **53**, 1–221 (1998).
199. A. J. Fatiadi, *Synthesis*, 65–103 (1976).
200. A. J. Fatiadi, *Synthesis*, 133 (1976).
201. L. Blackburn, X. Wei, and R. J. K. Taylor, *Chem. Commun.*, **1999**, 1337–1338 (1999).
202. J. Zakrzewski, J. Grodner, J. M. Bobbitt, and M. Karpińska, *Synthesis*, 2491–2494 (2007).
203. P. Anastas and J. Warner, *Green Chemistry: Theory and Practice*, Oxford University Press, New York, USA, 1998.

13

Metal–Nitroxide Complexes: Synthesis and Magnetostructural Correlations

Victor Ovcharenko
International Tomography Center, Novosibirsk, Russia

13.1 Introduction

Interactions between the paramagnetic ions of transition metals and stable radicals are convenient and effective methods for the synthesis of multispin molecules. These metal-nitroxide systems are often called heterospin systems because they contain paramagnetic centers with both d (or f) odd electrons and centers with p electrons. Paramagnetic centers can differ in the electron spin, g factor, the character of electron spin density delocalization, and heat capacity. The presence of several paramagnetic centers in heterospin molecules has stirred growing interest in their magnetic properties because these compounds are convenient objects for studying the subtle peculiarities of exchange interactions and revealing valuable magnetostructural correlations. That is why coordination chemistry of transition metals with stable nitroxides is an actively developing direction in modern chemistry that contributes to problem solving in the field of molecular magnetism. The results of studies on the synthesis and properties of metal complexes with nitroxides have been considered from different viewpoints and at different times in many monographs and reviews.[1–19]

Importantly, this approach offers the potential for control over the structure of the designed heterospin molecule and the character of interactions between the odd electrons of paramagnetic centers, which is necessary for constructing justifiable magnetostructural correlations. This is one of the factors responsible for the continuous growth of the number of new metal–nitroxide complexes. The Cambridge Structural Database alone contains more than one thousand structurally defined metal-nitroxide systems.[20] Therefore, currently it is impossible and not necessary to collect all available literature data on the synthesis and properties of metal-nitroxide systems in one publication. In any account, a comprehensive review of this kind will already have become incomplete by the time of its publication because new heterospin compounds based on transition metal compounds with nitroxides are being added daily to the list. However, the available, continuously growing collection of data on metal-nitroxide systems allows the sorting out of both elaborate and new scientific directions based on these compounds. This present review largely reflects the scientific preferences of the author and presents an attempt to highlight those aspects of the problem that generally receive less attention from researchers.

13.2 Two types of nitroxide for direct coordination of the metal to the nitroxyl group

Since the discovery of stable nitroxides, it was quickly realized that, apart from having high kinetic stability in both the solid state and solution, they were capable of reacting in such a way that their free valence remained intact. This stimulated studies of coordination compounds where the metal was directly coordinated to the nitroxyl >N–$^•$O group. The direct coordination of the nitroxyl group, which favors direct exchange, created great interest because the energy of interactions between odd electrons in the {M–O$^•$–N<} exchange cluster (M is the paramagnetic metal ion) was maximum in this case. With the maximum exchange energy, the temperature of the potential magnetic anomaly for the heterospin system was also expected to be high.

In principle, many variants of M–O$^•$–N< coordination are possible. They can be conventionally divided into two groups if the synthetic approaches used are correlated with the molecular structure of nitroxide. Nitroxides can also be divided into two types according to complexation:

(a) nitroxides containing only >N–$^•$O, but no other functional groups; and
(b) nitroxides containing other donor functional groups along with >N–$^•$O.

13.2.1 Complexes containing only >N–$^•$O as a coordinating group

The synthesis of complexes in which only the nitroxide can coordinate requires a certain methodological approach because these compounds are extremely weak donors in accordance with thermodynamic data for complexes in solution.[21–25] Therefore, the syntheses of complexes with these nitroxides are generally performed with reagents that are rather strong Lewis acids, for example, anhydrous halides, perchlorates, hexafluoroacetyl acetonates, or halogenated metal carboxylates.

Studies on the synthesis and magnetochemistry of complexes with metal-nitroxide coordination were started in 1967, with complexes of formula $CoX_2(L^1)_2$, (X = Cl, Br, I). These heterospin complexes were synthesized using carefully dehydrated reagents and solvents. Dilution of the $CoX_2(L^1)_2$ solution led to increased dissociation of the complexes, which was indicative of their low thermodynamic stability. There was a strong antiferromagnetic interaction between the odd electrons of >N–$^•$O groups and Co(II).[26]

The $[CuCl_2(L^2)]$ and $[PdX(L)^-]_2$ dimer complexes (X = Cl, Br and L = L^1, L^2) were also synthesized in the absence of moisture (Scheme 13.1). $[CuCl_2(L^2)]$ was diamagnetic because of strong antiferromagnetic exchange[27] and $[PdX(L)^-]_2$ because of the reduction of nitroxide in the course of the reaction and further η^2-coordination of hydroxylamine anions.[28–31] Quantum chemical calculations confirmed that the organic ligand in palladium(II) complexes was in reduced form.[32] The $CuBr_2(L^2)$ diamagnet, which is rather stable in the absence of moisture and in which the nitroxide is in the very unusual η^2-coordination mode, was described in.[33] Quantum mechanical calculation for this complex showed that the binding electrons of the coordination unit were considerably delocalized, and singlet–triplet splitting was at least 5000 cm^{-1}.[32]

Considerable (in energy) antiferromagnetic exchange interaction, which generally led to complete coupling between the spins of coordinated nitroxyl groups and metal ions, was also recorded for compounds

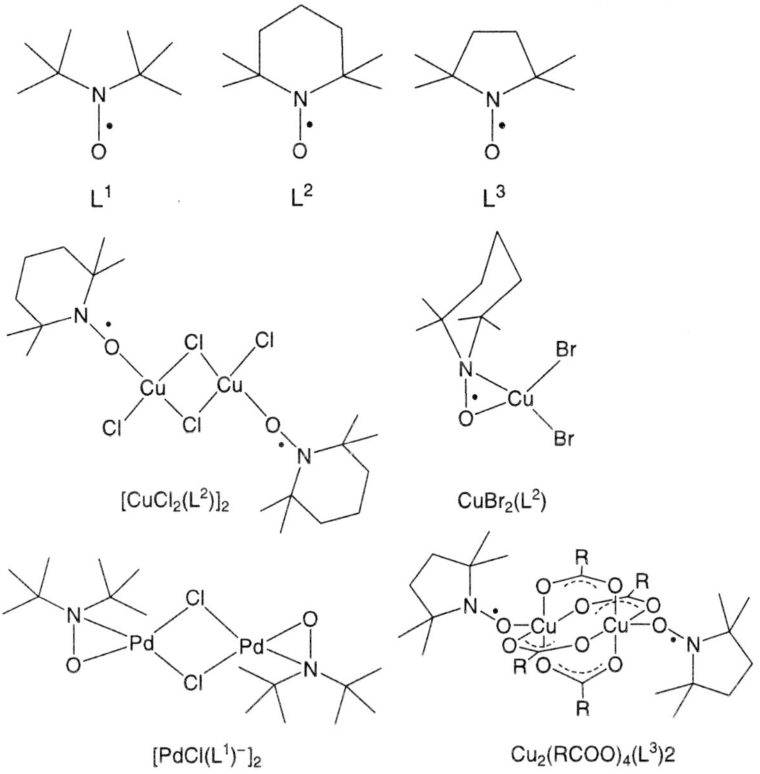

Scheme 13.1

of dehydrated metal perchlorates $M(ClO_4)_n(L^2)_2$[33–35] and binuclear metal carboxylates $M_2(O_2CR)_4L_2$ (M = Cu, Rh, or Mo; L = L^1, L^2, or L^3; and R = CCl_3, CBr_3, CF_3, C_3F_7, or C_6F_5).[36–42] Because of the weak donor properties of the >N–$^{\bullet}$O group, it was necessary to use halogenated carboxylates.[23,37] For example, $Rh_2(O_2CR)_4$ (R = CH_3, C_3H_7) did not react with L^2.

A representative group of heterospin compounds of nitroxide radicals with metal hexafluoroacetyl acetonates were obtained. Scheme 13.2 shows the structural formulas of some of these compounds. The $Cu(hfac)_2L^1$ complex, which was thermodynamically unstable in solution, completely decomposed after the addition of pyridine to the solution.[43] The enthalpy of formation measured for $Cu(hfac)_2L^2$ was of the order of −11.7 kcal/mol.[21] This is certainly a small value. Nevertheless, it had a certain insignificant contribution from the antiferromagnetic exchange interaction between the odd electrons of the metal and the coordinated >N–$^{\bullet}$O group; the magnetic interaction can be regarded as an analog of a chemical bond because the interaction gives two electron spins with opposite projections. For example, in the reaction that gave $VO(hfac)_2L^2$, it was up to 0.3 kcal/mol.[44] For $Pd(hfac)(L^2)^-$, the diamagnetism was a consequence of the reduction of the ligand in the course of the reaction.[45] The diamagnetism of $Cu(hfac)_2L^1$ and $VO(hfac)_2L^2$, however, was determined by the mutual compensation of spins. Similarly, in other complexes of paramagnetic metals (copper(II), nickel(II), cobalt(II), and manganese(II)), the strong antiferromagnetic interaction between the odd electrons of the metal ion and coordinated nitroxyl groups resulted in spin coupling in the coordination unit.[46–49]

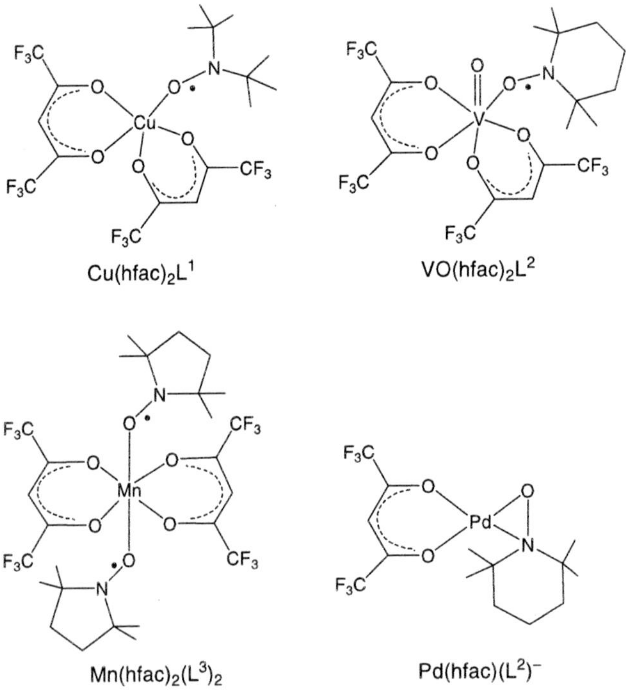

Scheme 13.2

In general, studies of metal-nitroxide systems with nitroxides from group (a) led to the following conclusions:

- As complexing agents, strong acceptors are needed because the >N–•O group, which is the product of the oxidation of the corresponding hydroxyl precursor, is a weak donor.
- Heterospin complexes should be synthesized in anhydrous media and stored in a dry atmosphere.
- The complexes have low thermodynamic stability and strongly dissociate in solution; therefore, it is recommended to have an excess of nitroxide in the reaction mixture in order to have an increased equilibrium fraction of heterospin complexes to isolate them as solids.
- The formation of metal-nitroxide systems can be hindered by redox processes.

13.2.2 Complexes containing >N–•O and other functional groups as donor fragments

Nitroxides containing other functional groups along with the >N–•O group are certainly more numerous than nitroxide ligands lacking such additional functionality; coordination compounds of the former are also much more numerous and better defined. However, not all of them can be discussed here. Possible variants of complexes with nitroxides and nitronyl (or imino) nitroxides can conventionally be divided into eight subgroups, as shown in Scheme 13.3. Note that in subgroups **A**, **B**, and **E**, the pyrrolidine heterocycle was taken as an arbitrary example. Alternative hetercycles include pyrroline, imidazoline, piperidine, or in principle any other heterocycle. Moreover, nitroxides of one of these groups can be acyclic.[50]

In complexation with nitroxides from subgroup **A** in solids and solutions, the paramagnetic ligand is primarily coordinated through the donor atoms of the Z functional group. The chemical behavior of the

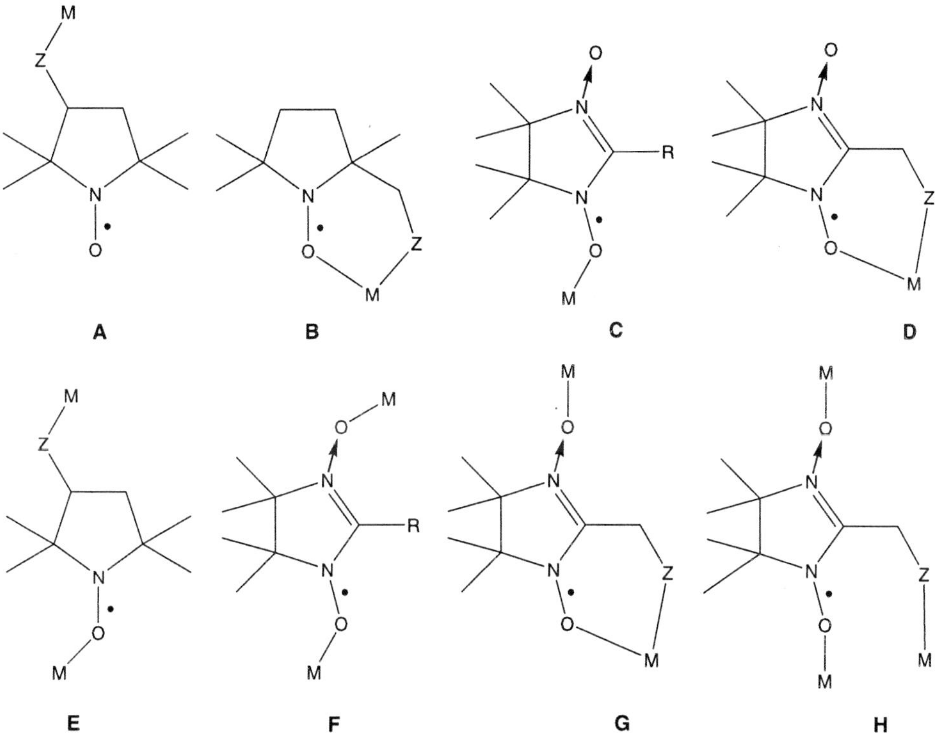

Scheme 13.3

ligands and the composition of the coordination compounds do not differ from those for diamagnetic organic ligands having the same Z functional groups. This situation is also trivial for the magnetic properties of the produced complexes because they are equivalent to the mechanical mixing of nitroxides with metal salts. Therefore, metal complexes with **A**-type nitroxides are not reviewed here. Variants **B**, **C**, and **D** are not considered either because they are related to metal complexes with nitroxides containing only >N–•O as a functional group (Section 13.2.1). Complexes of **E**, **F**, **G**, and **H** types, in which nitroxides perform the bridging function, are of interest in view of their ability to form high-dimensional structures. Molecular magnets capable of cooperative ordering at $T \geq 4.2\,K$ were found in series of these compounds. These types of radical and their metal complexes are now being extensively studied.

13.3 Ferro- and ferrimagnets based on metal–nitroxide complexes

Before discussing magnets, it is worthwhile to mention the work by Anderson and Kuechler[51], which stimulated the development of synthetic studies of ferro- and ferrimagnets based on metal–nitroxide complexes. As noted in Section 13.2.1, antiferromagnetic spin coupling occurs during the coordination of the >N–•O group in metal complexes with nitroxides from group (a), which certainly destroys prospects for real molecular magnets. Indeed, for example, copper(II) or VO(II) compounds form diamagnetic 1 : 1 heterospin complexes with nitroxides. The $Cu(hfac)_2L^4$ chain polymer complex described in[51] attracted the attention of researchers for several reasons (Scheme 13.4). Importantly, the exchange interaction in

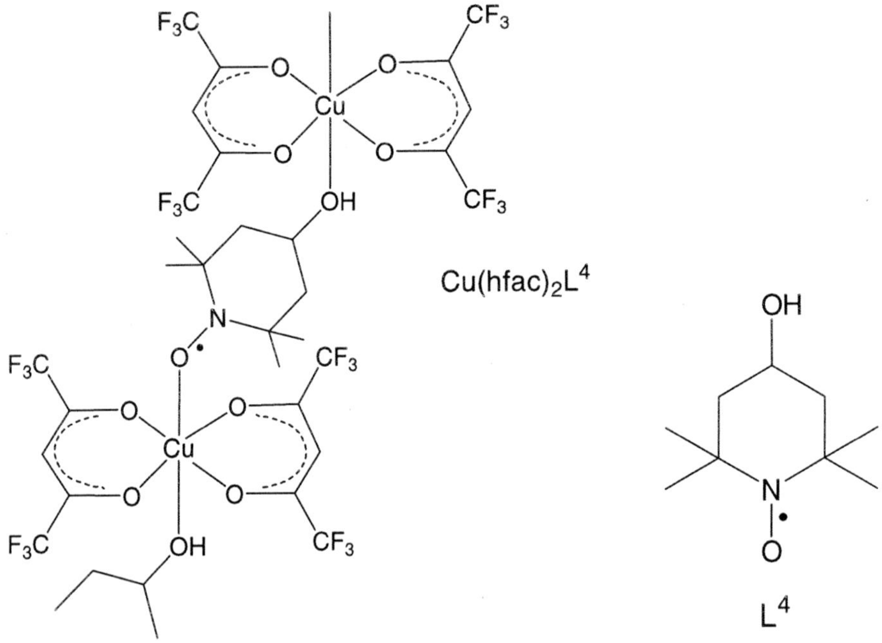

Scheme 13.4

the {Cu(II)–O$^•$–N<} exchange cluster was ferromagnetic (J = 13 ± 5 cm^{-1}); that is, the ground state was the triplet state (here and below the Hamiltonian is recorded as $\hat{H} = -2J\hat{S}_1\hat{S}_2$). This effect, that is, the possibility of ferromagnetic exchange in the {Cu(II)–O$^•$–N<} exchange cluster or mutual repulsion of odd electrons during direct coordination of the paramagnetic organic fragment to the paramagnetic metal ion, was explained later.[52–56] The discovery of this phenomenon stimulated the development of the molecular design of heterospin molecular magnets. Molecular magnets will be returned to later; now, another important result of studies reported in[51] is considered.

From a chemical viewpoint, Cu(hfac)$_2$L^4 can be synthesized with the use of water-containing reagents. This indicates that if a nitroxide contains a strong and readily coordinated donor group, direct coordination between the metal and the nitroxyl group can appear in the course of crystallization, that is, in the formation of close packing. Moreover, the product can either be an insoluble polymer phase or simply the most insoluble solid phase in the given synthetic system.

Possible difficulties in the use of this synthetic approach are evident. Firstly, along with the functional group on the nitroxide, other donor molecules present in the reaction mixture can take part in coordination to the metal ion.[57–60] For example, using water-containing solvents or reagents in the reaction of Cu(hfac)$_2$·H_2O with L^5 led to the formation of Cu(hfac)$_2$L^5(H_2O), in which the surroundings of the central atom included a water molecule (Scheme 13.5). This molecule was hydrogen bonded with the oxygen atom of the nitroxyl group L^5 of the adjacent molecule of the complex, which led to the formation of polymer chains in the solid. Nevertheless, there is no direct coordination between the metal and the nitroxyl group in the solid complex. Scheme 13.6 shows the case with donor groups blocking all coordination sites, which hindered the formation of a polymer structure.

The potential development of the polymer chain restricts the formation of molecules that are converted into a multinuclear structure. Scheme 13.7 shows different variants of heterospin complexes with

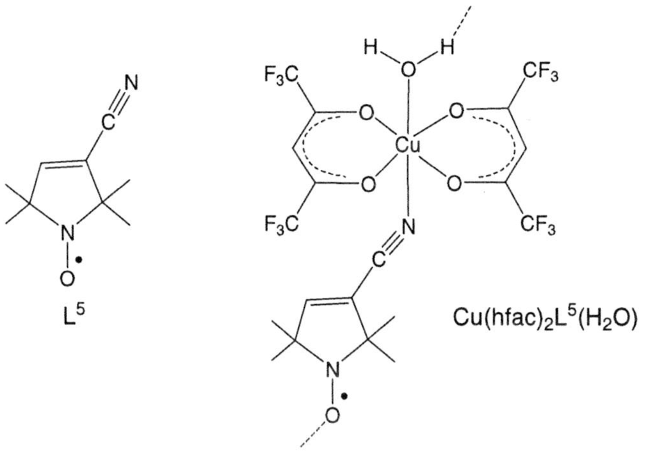

Scheme 13.5

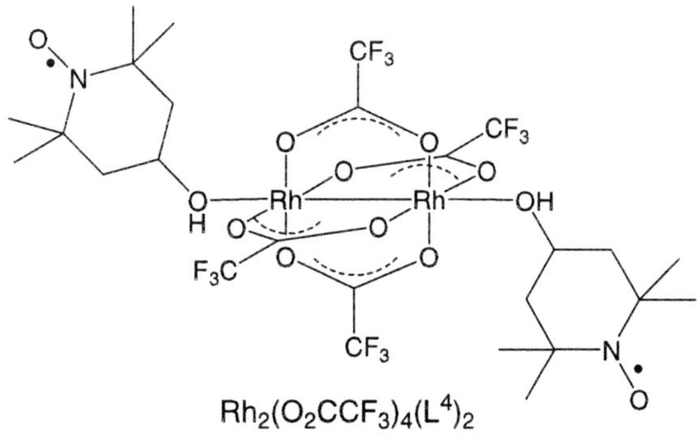

Scheme 13.6

mononitroxides, namely: the tetranuclear $[Cu(hfac)_2]_4(L^6)_2$ complex[61]; various types of the trinuclear $[Cu(hfac)_2]_3(L^n)_2$ complex, $n = 7 - 10$[62–69]; and binuclear $[Ni_2en_4L^{11}](ClO_4)_3$[70] and hexanuclear $Hg_6(C_6F_4)_6L^{12}$ complexes.[71] For copper(II) complexes, the formation of closed multinuclear molecules is promoted by the anomalously high stereochemical nonrigidity of the bishexafluoroacetyl acetonate molecule, which is capable of performing the function of "caps," preventing chain propagation. If the reaction of L^6 is performed with $Mn(hfac)_2$ instead of $Cu(hfac)_2$, the terminal metal ions have coordination number 6, which leads to the formation of heterospin compounds $\{[Mn(hfac)_2]_3(L^6)_2\}$, in which (in the solid state) the binuclear cyclic fragments are joined into chains via $Mn(hfac)_2$ terminal matrices.[72]

The same is also possible for complexes with biradical molecules. Scheme 13.8 shows examples of these heterospin complexes, namely, the binuclear $[Mn(hfac)_2]_2L^{13}$,[73] tetranuclear $[Cu(hfac)_2]_4L^{14}$,[74] heptanuclear $[Cu(hfac)_2]_7(L^{15})_2$,[75] and hexanuclear $[Cu(hfac)_2]_4\{Hg[(L^{15})^-]_2\}$[76,77] complexes.

Cis-coordination of nitroxides can favour the formation of multinuclear macrocycles, for example, the 36-membered ring in $[Mn(hfac)_2L^{12}]_6$ and the 42-membered ring in $[Cu(hfac)_2L^{16}]_6$, respectively,[78,79] or spheroid structures such as the structures in $\{[Ni(L^{17})_2]_3[Fe(CN)_6]_2\}$[80–82] (Scheme 13.9). These heterospin complexes have a large spin even with metal–nitroxide antiferromagnetic exchange but are not liable to anomalies related to the magnetic phase transformation. Thus, for polycrystalline $[Mn(hfac)_2L^{12}]_6$, the dependence of magnetization on the magnetic field is well described with the Brillouin function with $S = 12$; that is, despite the large value of S, the substance behaves as an ordinary paramagnet in the temperature range 5–300 K.[78]

The methodological approach suggested in[51] and leading to coordination of the nitroxyl group was also applicable to the synthesis of layered polymeric heterospin complexes based on metal bischelates with spin labeled deprotonated β-diketones[83–86] and enaminoketones.[87,88] The scheme that shows the formation of a layered structure for bichelates with nitroxides based on the piperidine and imidazolidine heterocycles $Cu(L^{18})_2$ and $Co(L^{19})_2$, respectively, is depicted in Scheme 13.10. These compounds are synthesized under

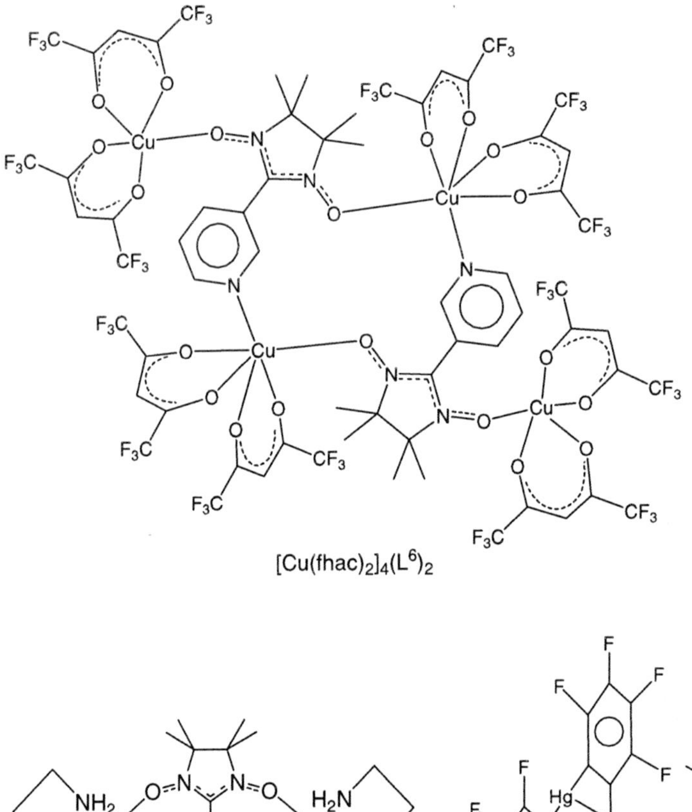

$[Cu(fhac)_2]_4(L^6)_2$

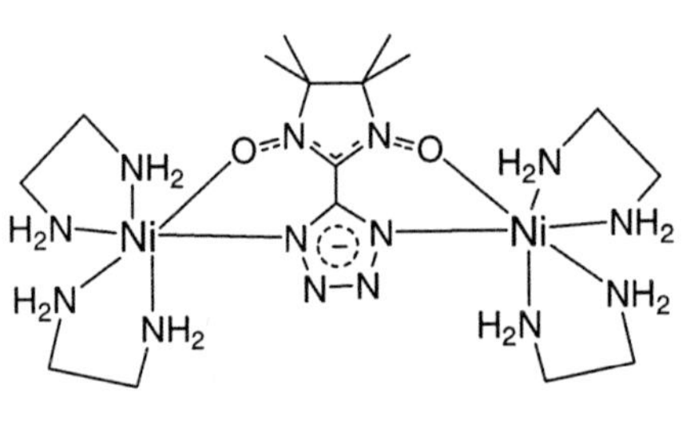

$[Ni_2en_4L^{11}]^{3+}$

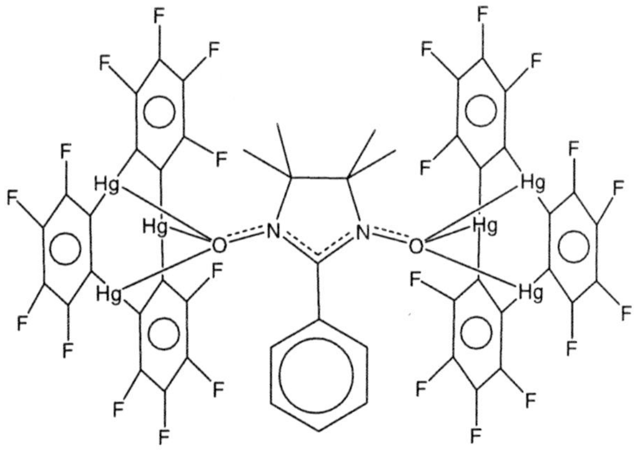

$Hg_6(C_6F_4)_6L^{12}$

Scheme 13.7

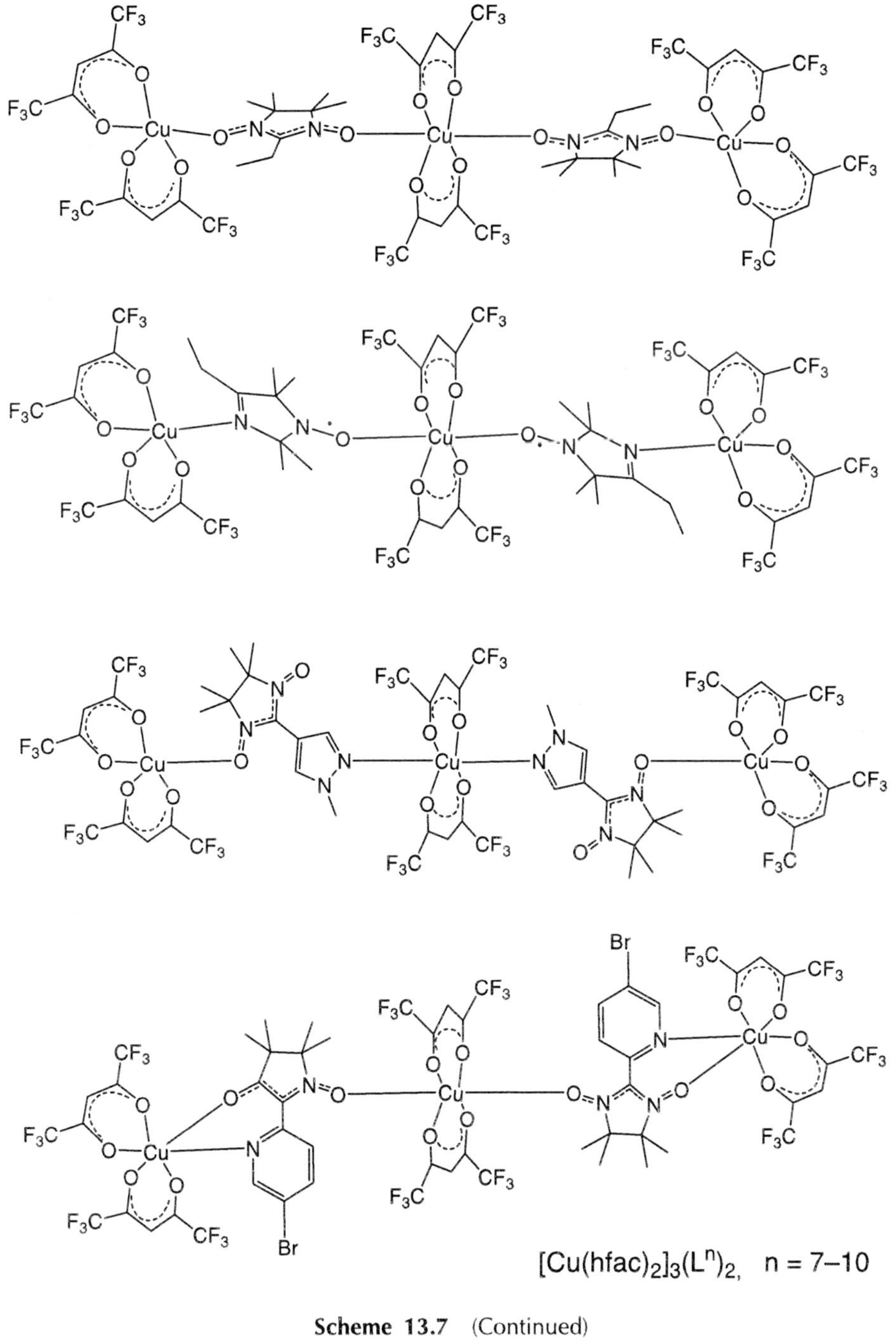

$[Cu(hfac)_2]_3(L^n)_2$, n = 7–10

Scheme 13.7 (Continued)

the normal conditions without taking any special precautions. Direct coordination between the metal and the nitroxyl group appears during the formation of the solid.

13.3.1 Molecular magnets based on 1-D systems

Studies in the field of molecular magnets involve the syntheses of open shell organic, organometalic, or coordination compounds; their subsequent crystallization from solution into solids, which ideally exhibit a magnetic phase transition to the ferro-, ferromagnetic, or weak ferromagnetic state, was recorded below a critical temperature (T_c). It is highly desirable that these molecules crystallize to form layered or

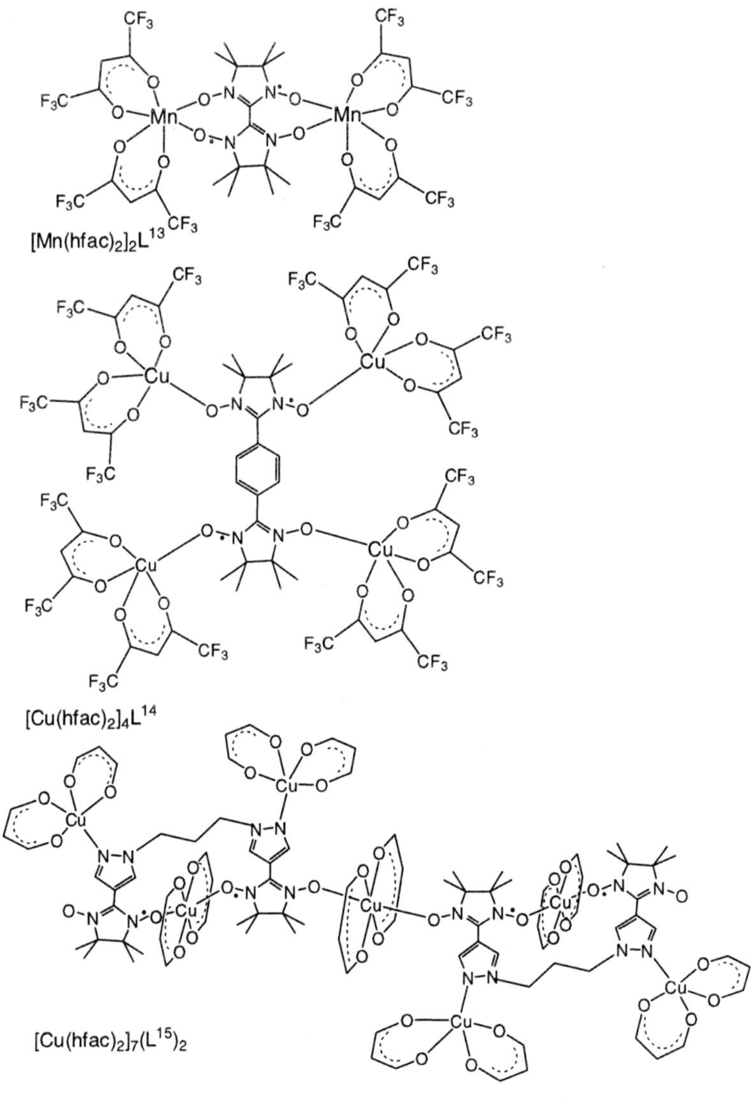

Scheme 13.8

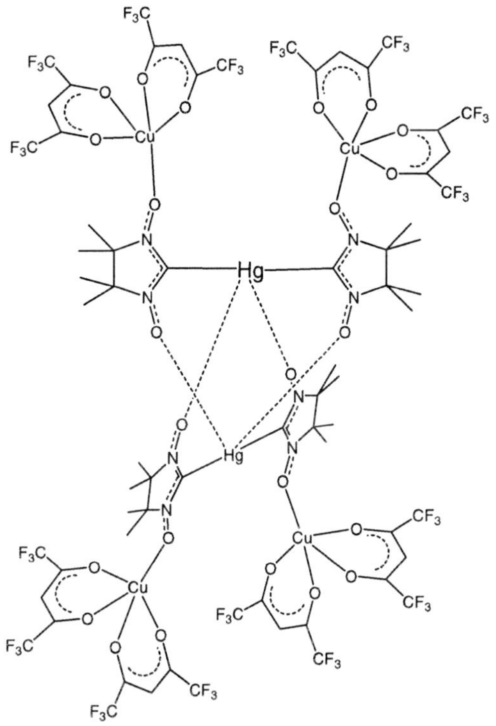

$[Cu(hfac)_2]_4\{Hg[(L^{15})^-]_2\}_2$

Scheme 13.8 (Continued)

framework polymers; this aspect of molecular magnetism research draws heavily on the concepts and principles of supramolecular chemistry. The formation of layered or framework structures in solid molecular ferromagnets is a favorable condition for a magnetic phase transition to the magnetically ordered state. However, for chemical design of molecular magnets, this is a favorable but not sufficient condition. For a phase transition, the paramagnetic centers in these polymers should be bound by an effective magnetic exchange pathway. The higher the efficiency of exchange channels in exchange interactions between the odd electrons of the paramagnetic centers, the higher the T_c that can be achieved. Success in this field can be realized through systematic studies on the effect of molecular design, as well as the conditions of synthesis and crystallization, on the parameters of the resulting molecular magnet (Curie temperature, spontaneous magnetization, etc.). The search for materials with desirable magnetic properties can be then be rationalized in terms of magnetostructural correlations. Put another way, the term "molecular magnet," can be interpreted as follows: the word "magnet" has the conventional meaning, and the attribute "molecular" reflects the methodology of the approach (the structure of the individual starting molecules and/or ions should allow the formation of a structure favorable for a magnetic phase transition in the solid state).

Consideration of polymeric systems is begun firstly by considering selectively synthesized 1-D metal-nitroxide systems, in which exchange interactions between paramagnetic centers *a priori* should be strong.

The interaction between $Mn(hfac)_2{\cdot}2H_2O$ and L^n produced chain polymers $Mn(hfac)_2L^n$, where L^n are nitronyl nitroxides, containing R = ethyl (L^7), phenyl (L^{12}), hydrogen (L^{15}), methyl (L^{20}), *n*-propyl (L^{21}),

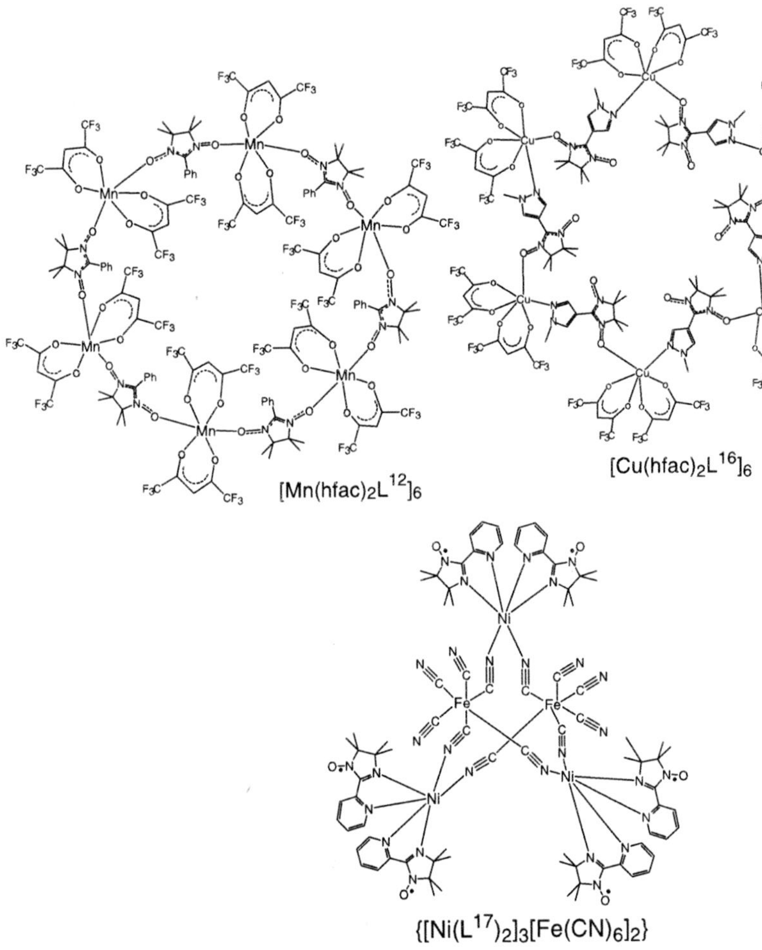

Scheme 13.9

or *i*-propyl(L^{22}) in the second position of the imidazoline ring.[89–91] All complexes in the solid state are formed from polymer chains, in which manganese(II) coordinates the paramagnetic ligands in the *trans* or *cis* positions (Scheme 13.11).

A study of the magnetic properties of an orientated $Mn(hfac)_2L^{22}$ single crystal showed that at low temperatures, the compound was magnetized a a ferromagnet, although inside the $(–Mn–L^{22}–)_n$ polymer heterospin chains the interaction between the odd electrons of manganese(II) and L^{22} was strongly antiferromagnetic, of the order of $-330\,cm^{-1}$. The easy magnetization axis of the single crystal coincided with one of the crystallographic directions of the unit cell. A magnetic phase transition with $T_c = 7.6\,K$ was recorded when magnetization occurred along this axis.[89,90] Magnetic phase transitions with $T_c = 8.1\,K$ and $T_c = 8.6\,K$, respectively, were also found in solid $Mn(hfac)_2L^7$ and $Mn(hfac)_2L^{21}$, in which the bridging bidentate L^7 (or L^{21}) in the zigzag polymer chains was *cis*-coordinated to the manganese(II) ion, and the spins of nitroxides experienced antiferromagnetic interactions with the spins of the metal.[91] A magnetic phase transition was also observed for $Ni(hfac)_2L^{20}$ as well as chain polymer complexes of rare earth element hexafluoroacetyl acetonates with nitronyl nitroxides.[92]

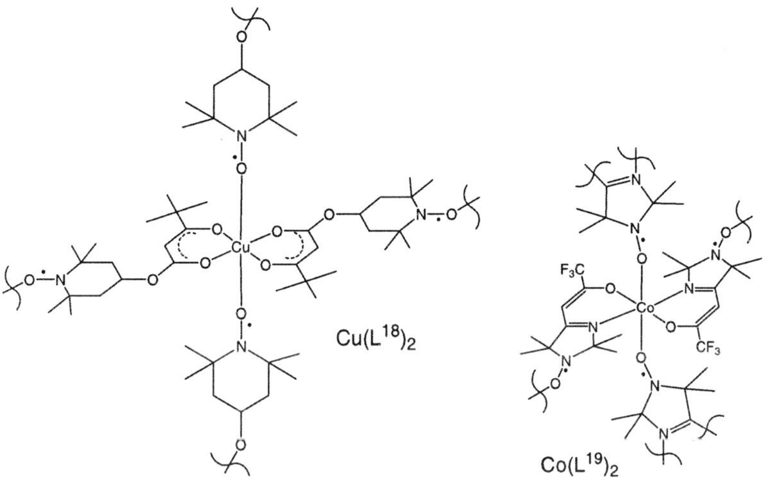

Scheme 13.10

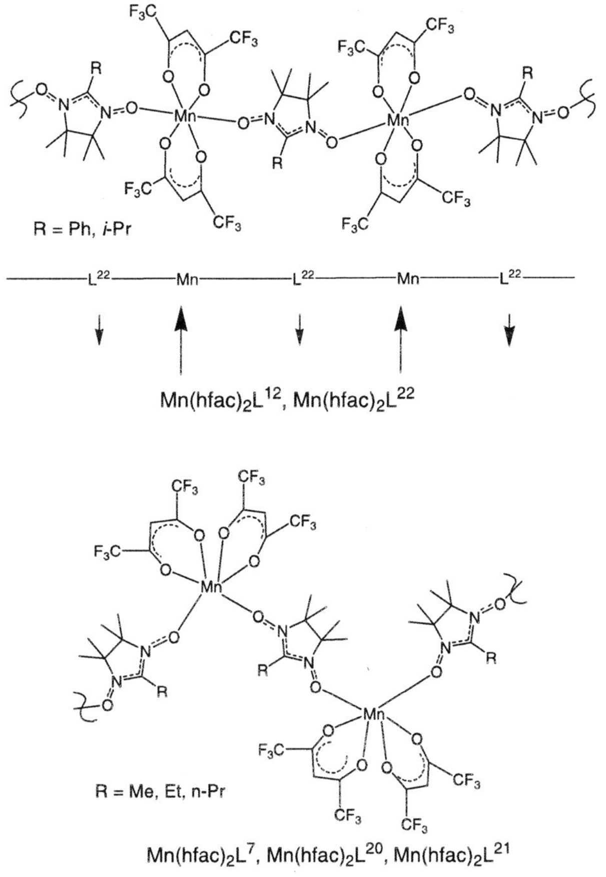

Scheme 13.11

The synthesis of complexes with nitronyl nitroxides should be conducted at moderate temperatures of up to ~ 50–60°C because these radicals are apt to lose oxygen on heating and transform into the corresponding imino nitroxides.[93–95] That is why prolonged boiling of $Mn(hfac)_2$ solutions with L^{12} was accompanied by the reduction of nitronyl nitroxide to the corresponding imino nitroxide and further formation of mixed ligand complexes of $Mn(hfac)_2$ with imino nitroxide or nitronyl nitroxide and the products of its reduction.[96,97]

As noted by Caneschi, complexes of $Mn(hfac)_2$ with nitronyl nitroxides could serve as instructive examples of one-dimensional Heisenberg magnets because magnetic anisotropy was negligibly small for both manganese ions and nitroxides.[98] All strong exchange interactions in these complexes are concentrated inside chains; the interactions between the paramagnetic centers of adjacent chains are extremely weak, since they are separated by long distances (at least 10 Å) from one another. The three-dimensional ordering in heterospin manganese complexes with nitronyl nitroxides is favored by the fact that the strong antiferromagnetic interaction of the spins of the metal and radical creates a large uncompensated moment, and at ~4–8 K, the interchain dipole–dipole interaction induces three-dimensional magnetic ordering. Since the interchain interaction can be regulated, in an infinite number of variants, through substituents in the side chain of the nitroxide, the magnetic properties of heterospin compounds can also be adjusted.

A special situation occurs when the metal-nitroxide interaction becomes essentially anisotropic in heterospin chains of this kind, as in the case of cobalt(II) complexes. The anisotropy of the exchange metal-radical interaction in the chains leads to the creation of a barrier that hinders the reorientation of magnetization, that is, retards magnetic relaxation. This is a non-trivial effect. The chain polymer heterospin complex $Co(hfac)_2L^{23}$, where L^{23} is 4′-methoxyphenyl-substituted nitronyl nitroxide, became the first example of a magnet based on truly isolated chains (single chain magnet, SCM).[99] Later publications described similar SCMs based not only on heterospin complexes of $Co(hfac)_2$ with nitronyl nitroxides[100], but also on the chain polymer complex of a rare earth element $Dy(hfac)_3L^{23}$, where L^{23} is nitronyl nitroxide with a bulky substituent (4′-phenoxyphenyl) in the side chain.[101] This bulky substituent was intentionally chosen to achieve substantial weakening of the interchain exchange interaction.

The possibility of slow relaxation of magnetization in 1-D systems was predicted long ago. For its implementation in Ising ferro- or ferromagnets, however, two conditions should be satisfied: (i) strong exchange anisotropy inside the chain and (ii) a very low ratio between the interchain and intrachain exchange ($J_{inter}/J_{intra} < 10^{-4}$). The latter circumstance dictated the choice of a bulky substituent in[101]; it was noted that the design of a heterospin compound with hysteresis anomalies above the boiling point of liquid helium but without three-dimensional magnetic ordering was a rather intricate problem. Chain polymer heterospin complexes of transition metals with organic radicals, whose non-magnetic shell can be varied in size due to organic substituents, are excellent constructional systems for reaching the desired balance between the anisotropy of exchange, the J_{inter}/J_{intra} ratio, and the structure. These specific magnets (SCMs), also called molecular magnetic wires, were a breakthrough in the field of molecular magnetism. SCMs were also obtained on heterometallic chains formed by different[102–104] and identical[105] metal ions. The possibility of creating new data storage devices[105] based on these compounds and related, single molecule magnets (SMM)[106–108] is now being actively discussed.

13.3.2 Molecular magnets based on 2-D systems

In the $Mn(hfac)_2L^n$ systems discussed above, the interchain exchange interaction was weak, and bulk magnetization appeared only due to the uncompensated magnetic moment created in the polymer chain by strong antiferromagnetic interaction between the spins of the coordinated nitronyl nitroxide and manganese(II) ion. For compounds of this kind, T_c can be raised by increasing the intrachain interaction of spins and/or decreasing the interchain distance, that is, increasing the interchain interaction of the magnetic moments.[109] It was difficult to increase the intrachain exchange interaction because it was already high enough, $J > |300 - 400|\ cm^{-1}$, as a result of direct coordination of nitroxides. Therefore, it was reasonable to replace the hfac ligands by other ligands, making interchain exchange interaction more effective. This stimulated studies of mixed ligand complexes of manganese(II) pentafluorobenzoate (pfbz) and perfluorocarboxylates with nitronyl nitroxides.[109,110] The idea proved fruitful and afforded magnetoactive heterospin complexes, among which $[Mn(pfbz)_2]_2L^7$ and $[Mn(pfbz)_2]_2L^{20}$ experienced a magnetic phase transition at $T_c \sim 25\ K$.[109] Perfect single crystals could not be grown. A hypothetical scheme of organization of their heterospin layers was suggested based on physical measurements.[110]

Many 2-D molecular magnets have been obtained based on complexes with spin labeled imidazole and benzimidazole,[13,111,112] and their structure and magnetic properties were studied systematically in comparison with those of their molecular and 1-D heterospin complexes.[113–118]

Luneau *et al.* developed an effective procedure for the synthesis of layered polymer heterospin magnets. Manganese(II) acetate, HL^{24} (or HL^{25}), and the sodium salt of a precipitator anion (e.g., $NaClO_4$) in a ratio of 2 : 3 : 1 were mixed in sequence in methanol.[13,111] The corresponding $[Mn_2L_3]ClO_4$ formed in good yields. BF_4, PF_6, BPh_4, or $ROSO_3$ (but not nitrates or chlorides) could be used as an anion. The formation of a polymer structure with a honeycomb motif was favored by the symmetric (relative to the donor groups) structure of the deprotonated L^{24} and L^{25} (Scheme 13.12). On Scheme 13.12, an intersection unit

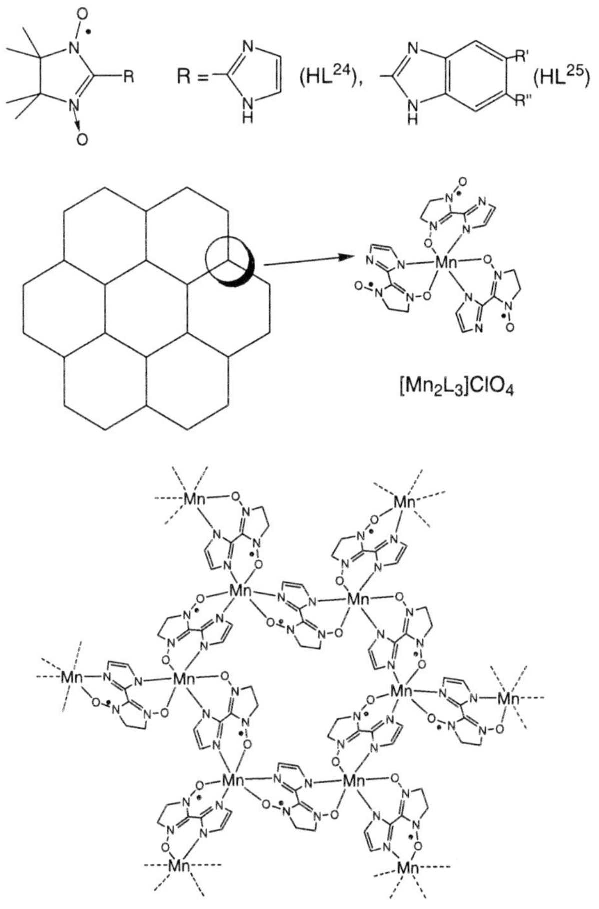

Scheme 13.12

(a {MnL_3} fragment) is isolated from the honeycomb motif, and a scheme of the formation of a hexanuclear metallocycle is shown. In these compounds, the anions lie in the interlayer space. The coordination units of the {MnL_3} fragments have a *mer* arrangement of the donor atoms. The Δ–{MnL_3} and Λ–{MnL_3} enantiomer units alternate in sequence in the polymer layers. The authors noted that the ΔΛ and ΛΔ combinations were favorable for the formation of layered structures, while ΔΔ and ΛΛ could lead to frameworks.[119] The highest temperature at which spontaneous magnetization was recorded was ~40 K.

A layered polymer structure is inherent in the $Ni(L^{26})_2$ bischelate with a deprotonated enaminoketone derivative of 3-imidazoline nitroxide.[120] This complex exists in two polymorphic modifications which differ in the position of the coordinated oxygen atoms of the >N–•O groups (*trans* positions in the α modification and *cis* in the β modification, Scheme 13.13). For β-NiL_2, whose magnetic properties are determined by the competition of the two major channels of exchange interactions (antiferromagnetic ($J \sim -115\,cm^{-1}$)

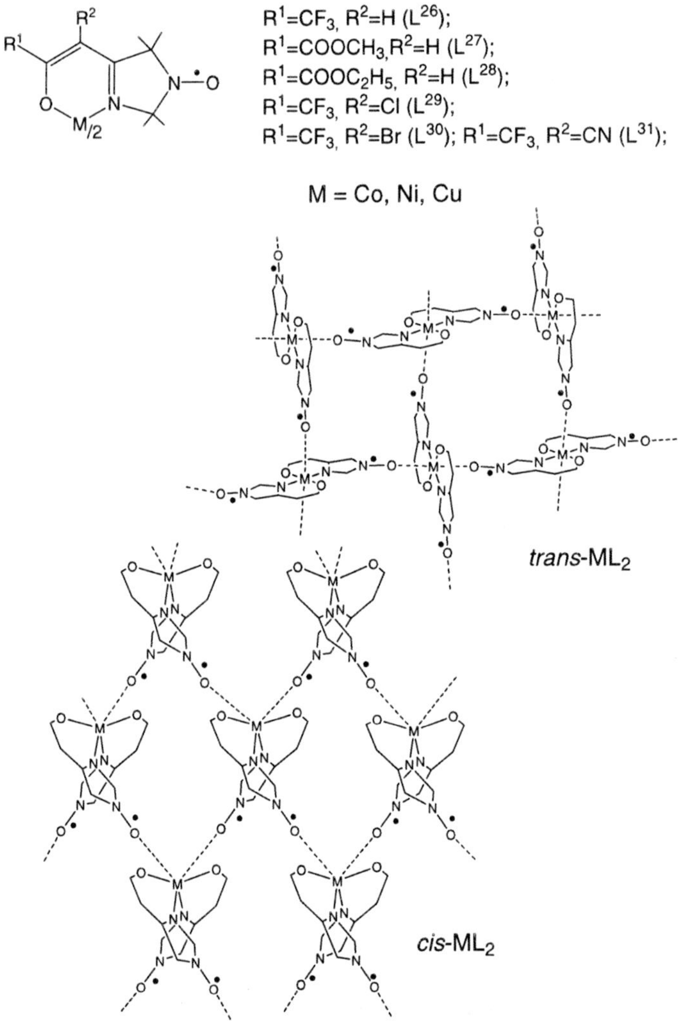

Scheme 13.13

and ferromagnetic ($J \sim 10\,\mathrm{cm}^{-1}$) channels), a magnetic phase transition to the antiferromagnetic state was recorded at $T_N = 14\,\mathrm{K}$. The desired modification could be obtained by selecting the crystallization conditions due to the stereochemical nonrigidity of $M(L^{26})_2$ bischelates, where M = nickel or cobalt.

Using spin labeled ligands capable of being deprotonated allows for the removal of unnecessary diamagnetic anion ligands from the solid heterospin compound. The "correctly" designed deprotonated nitroxide molecule should perform a number of functions: it should compensate the charge on the metal ion, perform the bridging function between adjacent metal ions, and create an effective exchange pathway. A large group of these molecular ferromagnets with a layered polymer structure was described: $Ni(L^{26})_2(ROH)_2$ and $Co(L^{26})_2(ROH)_2$, where R = CH_3, C_2H_5, n-C_3H_7, n-C_4H_9, i-C_4H_9, n-C_5H_{11}, C_3H_5; $Ni(L^{29})_2(CH_3OH)_2$, or $Co(L^{29})_2(CH_3OH)_2$.[8,121–123] For all compounds, a magnetic phase transition to a magnetically ordered state was recorded at 4–8 K. Scheme 13.14 illustrates the polymer layers in these complexes.

A specific feature of $M(L^n)_2(ROH)_2$ is the presence of alcohol molecules that are not crystal molecules in the solid compounds. In the solid, the metal ion coordinates the oxygen atoms of the hydroxly groups of two ROH (Scheme 13.14) and the {RO_H-ML_2-$_HOR$} fragments are then linked through {$_R$OH... O•—N<} hydrogen bonds with two adjacent {RO_H–ML_2–$_HOR$} fragments. As a result, the bridging molecules of alcohols link the {ML_2} fragments into polymer layers; that is, the 2-D structure is formed by hydrogen bonds,[123,124] which can be treated as a certain analogy to the formation of complex protein structures.

It is noteworthy that in $M(L^n)_2(ROH)_2$, each metal ion is surrounded by a large number of atoms inherent in organic molecules (C, H, N, O). For example, in $NiL_2(n$-$C_5H_{11}OH)_2$, each nickel(II) ion is surrounded by 90 carbon, hydrogen, nitrogen, and oxygen atoms; despite this, the compound is capable of cooperative magnetic ordering at low temperatures. Moreover, the removal of alcohol molecules and a transition to layered polymer molecules in solid $M(L^n)_2$ leads to the impossibility of ferromagnetic ordering

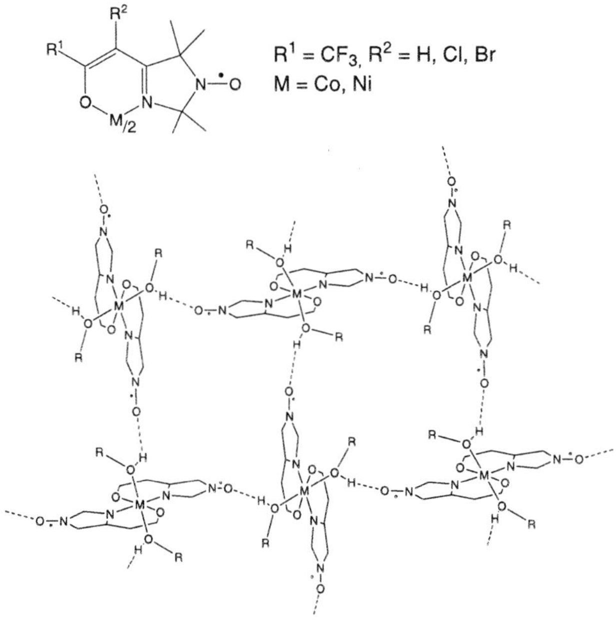

Scheme 13.14

in the latter.[120,123] This is a paradox at first sight because the transition from $M(L^n)_2$ to $M(L^n)_2(ROH)_2$ is actually accompanied by a drastic increase in the fraction of the pure organic component in the compound and a decrease in the number of paramagnetic centers per unit volume of the solid complex. Nevertheless, this increase gives rise to the ability to undergo a magnetic phase transition.

A comparison of the magnetic properties of the isostructural $Ni(L^n)_2(ROH)_2$ and $Ni(L^n)_2(H_2O)_2$ complexes showed that the replacement of alcohol by water molecules in the compounds was extremely undesirable. The replacement of alcohol by water in the $>N^{\bullet}{-}O\ldots HO_R{-}M^{II}{-}_ROH\ldots O{\bullet}{-}N<$ exchange channel led to a drastic decrease in the exchange interaction energy between the odd electrons of the paramagnetic centers and hence to a loss of the ability to undergo a magnetic phase transition, despite the structural identity of exchange channels in $Ni(L^n)_2(ROH)_2$ and $Ni(L^n)_2(H_2O)_2$.[125]

Complexes with alcohols were grown as large single crystals. This made it possible to study the anisotropy of their magnetic properties, determine easy magnetization axes, and correlate their direction with the microstructure of the crystals (Figure 13.1). Figure 13.1 shows data for $Ni(L^{26})_2(CH_3OH)_2$ and $Ni(L^{26})_2(C_2H_5OH)_2$ single crystals as an example. $Ni(L^{26})_2(CH_3OH)_2$ behaves as a typical weak ferromagnet, while $Ni(L^{26})_2(C_2H_5OH)_2$, as a metamagnet. Their easy magnetization axis coincides with the crystallographic direction of the twofold axis (**y**). In classical concepts, the coincidence of these directions is an essential requirement for single crystals belonging to the $P2_1/c$ space group.

When the alcohol solution contained both $Ni(L^{26})_2$ and $Co(L^{26})_2$, the solid products were the single crystals of solid solutions, for example, $Ni_xCo_{1-x}(L^{26})_2(C_2H_5OH)_2$. The fractions of the components of the solid could be smoothly changed in the mother solution by varying the initial $Ni(L^{26})_2/Co(L^{26})_2$ ratio in the mother solution, which made it possible to control the transition temperature and the anomalies of the critical magnetic fields.[126] Note that, for molecular magnets, the formation of solid solutions can be an extremely effective tool for control over the magnetic characteristics of the solid.

For heterospin complexes of transition metals with 3-imidazoline nitroxides, it was also shown that by varying the R^1 and R^2 substituents in the side chain of the ligand (Schemes 13.13 and 13.14), the structural dimensionality of the solid can be controlled. If R^1 is not an electron withdrawing group (e.g., alkyl, aryl), the solid complex has a molecular structure with a square or tetrahedral environment of the metal ion. If, however, R^1 is a strong electron withdrawing group (e.g., $-CF_3$ or $-COOC_2H_5$), then the acceptor properties of the metal ion also increase, the coordination number of the ion is completed to six, and a layered polymer structure is formed. When the acceptor ability of the metal ion is not high enough to reach a 2-D structure of the solid, additional acceptor groups can be introduced, for example, halogen atoms at the R^2 position.[126–135] This strategy afforded the single crystals of true ferromagnets that were polymers in solid $Cu(L^n)_2$, where $n = 28 - 31$. For this family of heterospin complexes, the saturation magnetization corresponds to the theoretical limit for compounds with $S = 3/2$ per molecular unit, $\sim 16\ 600\,G{\cdot}cm^3/\,mol$.

2-D or 3-D heterospin polymers with the analogs of 3-imidazoline enaminoketones – 2-imidazoline enaminoketones of L^{32} and L^{33} types – have not yet been isolated, although discrete complexes (Scheme 13.15) have strong intramolecular ferromagnetic exchange interactions (at least $50-100\,cm^{-1}$).[133,136,137]

When a considerable number of coordination sites at the metal ion are blocked by the donor atoms of ligands, 2-D structures are built with the use of additional functional groups of the nitroxide molecule. One example of this approach is depicted in Scheme 13.16. Layered polymer heterospin complexes $[M(hfac)_2]_3(L^{34})_2$, which underwent a magnetic phase transition to the ferromagnetic state at 9–11 K, were formed by coordination of the nitronyl nitroxide L^{34} molecule to the metal ions of three $Mn(hfac)_2$ or $Co(hfac)_2$ matrices.[138,139]

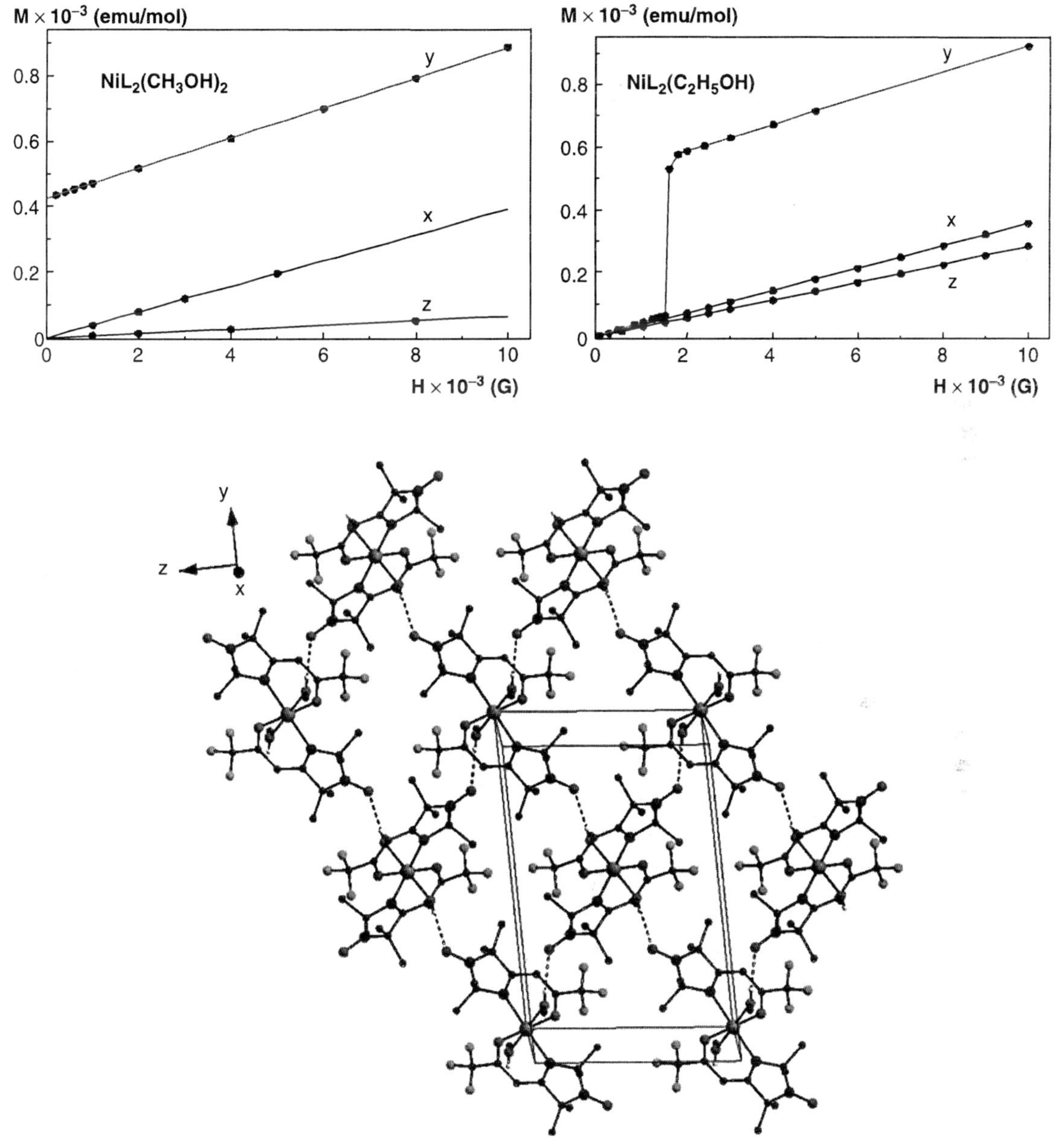

Figure 13.1 *Top: Easy magnetization axes for* $Ni(L^{26})_2(CH_3OH)_2$ *(left) and* $Ni(L^{26})_2(C_2H_5OH)_2$ *(right) single crystals. Bottom: layer fragment in* $Ni(L^{26})_2(CH_3OH)_2$.

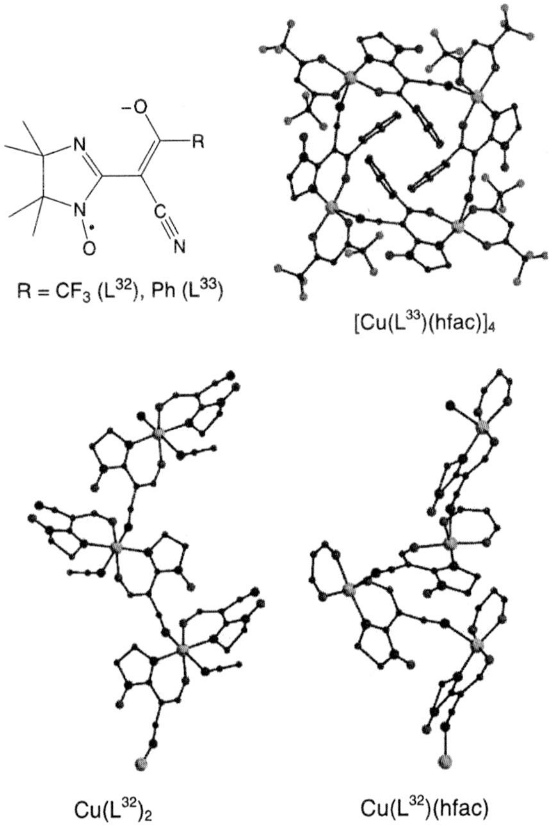

Scheme 13.15

13.3.3 Molecular magnets based on 3-D systems

The previous section discussed the family of layered polymer molecular magnets $Ni(L^{26})_2(ROH)_2$ and $Co(L^{26})_2(ROH)_2$ containing simple alcohols. Many of these compounds are kinetically unstable when stored without special precautions because of the gradual loss of alcohol. Analyses of their structures showed that the hydrocarbon tails R were directed to the interlayer space. Preliminary modeling showed that polymer layers could remain almost unperturbed and linked through polymethylene chains $(-CH_2-)_n$, lying between the coordinated hydroxyl groups of alcohols at $n = 4$ or 5. Indeed, $M(L^{26})_2(HO(CH_2)_4OH)$ and $M(L^{26})_2(HO(CH_2)_5OH)$ heterospin framework polymers, where M = nickel, cobalt, were also obtained with both 1,4-butanediol and 1,5-pentanediol and were capable of undergoing a magnetic phase transition to the magnetically ordered state at T_c ~4–8 K.[123] It is noteworthy that complexes with bifunctional alcohols possess high kinetic stability. For example, storage of $Ni(L^{26})_2(HO(CH_2)_4OH)$ and $Ni(L^{26})_2(HO(CH_2)_5OH)$ under normal conditions for 15 years did not result in any changes in their composition and properties. Large single crystals of many compounds from this series are also stable and can easily be grown to several millimeters (Figure 13.2) or even 1–2 cm along the longest direction on the external habit if grown by a skillful researcher. The possibility of synthesizing transition metal complexes with nitroxides as large single crystals also makes them convenient objects for neutron diffraction studies

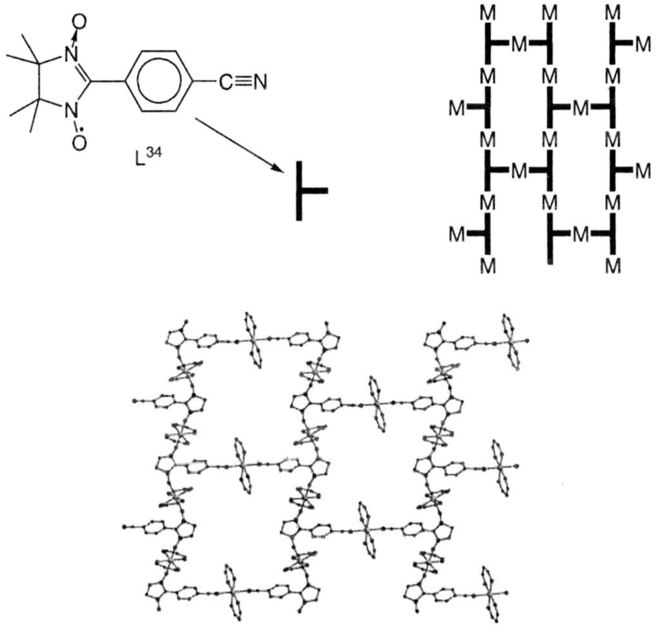

$M(hfac)_2(L^{34})_2$, M = Mn, Co

Scheme 13.16

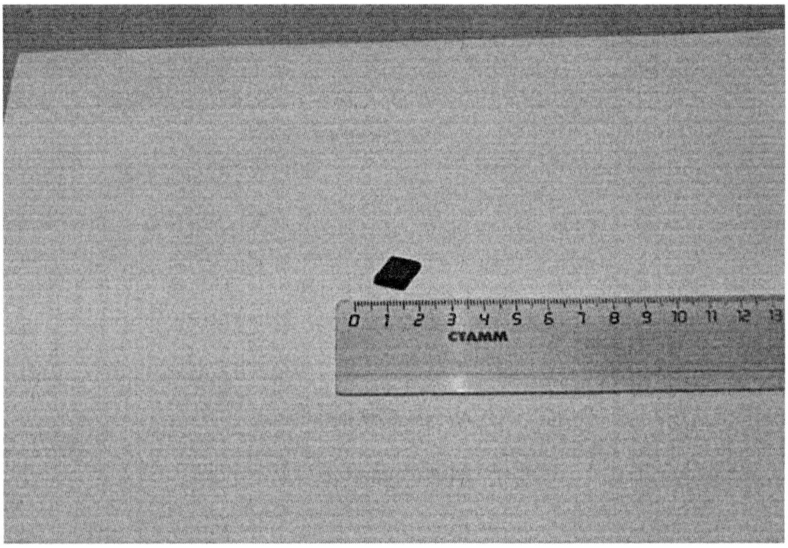

Figure 13.2 *$Ni(L^{26})_2(HO(CH_2)_4OH)$ single crystals. A full-colour version of this figure appears in the Colour Plate section of this book.*

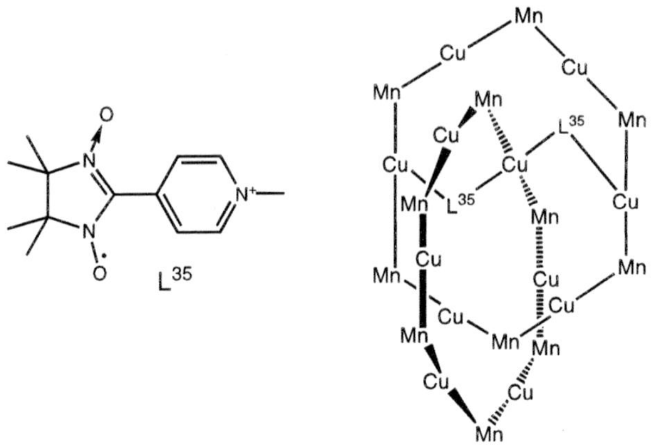

Scheme 13.17

of the anisotropy of magnetic properties[90,91,123,140,141] and spin density distribution.[127,134,142,143] These single crystals are transparent and very beautiful (Figure 13.2).

One of the most interesting findings of Kahn *et al*. was the synthesis of a multispin compound $(L^{35})_2Mn_2[CuL]_3(DMSO)_2{\cdot}2H_2O$, where R was the 2-imidazoline nitroxide cation, and L was the *o*-phenylenebis(oxamato) anion.[144,145] The three-dimensional structure of the compound is formed by a system of interpenetrating and intertwining nets, each of which represents a distorted graphite-like motif. The manganese(II) ions lie at the vertices of each hexagon of the graphite-like motif, and the copper(II) ions center the edges (Scheme 13.17). The role of a bridge between manganese(II) and copper(II) is played by the octadentate *o*-phenylenebis(oxamato) anion. In addition, each oxygen atom of the nitronyl nitroxide is coordinated to copper(II) ions from different subnets; this leads to the formation of additional polymer chains in the crystals of the compound, which is ordered as a ferrimagnet below $T_c = 22.5\,K$. Using this approach in the design of framework heterospin systems, the authors obtained a series of molecular ferromagnets $[L^+]_2M_2[CuL]_3$, where L^+ is 2-(4-N-alkylpyridinium)-4,4,5,5-tetramethylimidazoline-3-oxide-1-oxyl, and M = manganese(II), nickel(II), cobalt(II), for which $T_c = 22–37\,K$.[146]

An original strategy in the development of the syntheses of polynitroxides, preferably with >N–•O groups in the *meta* position of the benzene ring (so as to favor intraligand ferromagnetic coupling) such as L^{36}–L^{38} (Scheme 13.18),[11,147–149] and their complexations with $M(hfac)_2$[141,150–161] allowed Iwamura, Inoue *et al*. to obtain a series of molecular magnets capable of cooperative magnetic ordering at 3.4–46 K. The 3-D heterospin complex $[Mn(hfac)_2]_3(L^{37})_2$ experienced a magnetic phase transition at the highest temperature, 46 K.[141]

In summary, the data presented in this section show that, due to the high kinetic stability of nitroxides, as well as the unlimited possibilities for their chemical modification and the possibility of changing the coordination mode of radicals due to selective combination of functional groups, the design of molecular magnets based on complexes of paramagnetic metal ions with nitroxides is one of the most effective methods for the preparation of various magnetoactive compounds. The lack of coordination sites at the metal ion sometimes remains a challenge in constructing high-dimensional structures. Therefore, it would be reasonable to discuss the synthetic approach to the design of high-dimensional heterospin systems based on polynuclear compounds of metals with nitroxides, which has been developed over recent years and removed the limitations on the number of coordination sites.

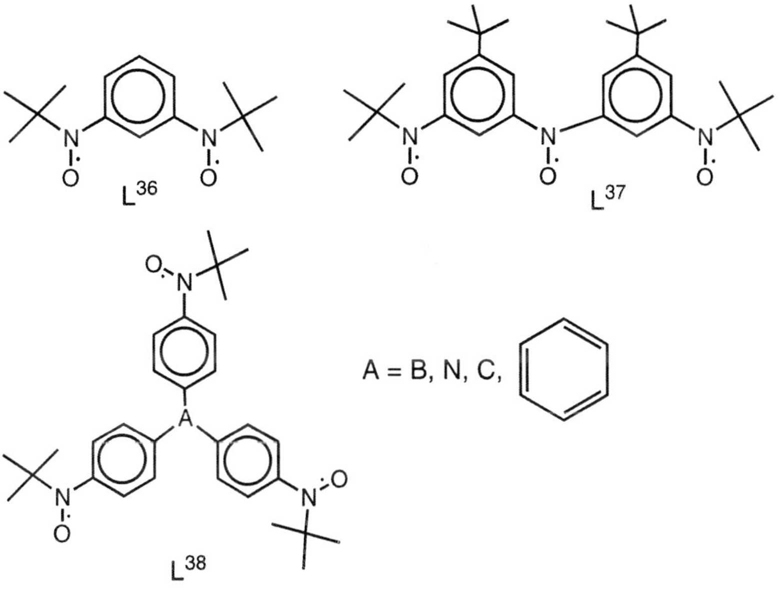

Scheme 13.18

13.4 Heterospin systems based on polynuclear compounds of metals with nitroxides

The design of molecular magnets is an actively developed field of modern chemistry because molecular (molecule-based) magnets are of interest not only for fundamental interest but also possible applications in materials science.[6–8,11–19,162–169] Various potential applications of molecular magnets have been noted; for example, original devices of a new type, transforming radiant energy into mechanical energy, and devices capable of serving as screens protecting from low frequency fields or functioning as working elements for reaching huge coercive fields at low temperatures.[7,10,16,19,162–169] Active research in the field of molecular magnetism has stimulated the generation and development of spintronics.[170,171] Problems in the development and creation of storage elements of quantum computers and logical devices, whose essential working elements are molecular nanomagnets, are being widely discussed today (e.g., 171–183 and references therein).

Among molecular nanomagnets, so-called single-molecule magnets are being studied most actively. These are generally polynuclear compounds of transition metals. Increased interest in these compounds has stimulated efforts to integrate studies in the field of the molecular design of magnets with studies in the field of the design of molecular magnets. The latter phrase is not tautology because the term "the molecular design of magnets" reflects the proper methodology of magnet assembly as an entity from the individual molecules with all paramagnetic centers of the magnet involved in cooperative interaction, forming homogeneous magnetization. In contrast, the term "the design of molecular magnets" emphasizes essentially non-homogeneous microscopic magnetization; in this case, the aim of chemical design is the synthesis of complex polynuclear compounds, in which strong magnetic interactions are concentrated. In crystals and amorphous solids, these often complex and bulky polynuclear compounds form molecular solids in which magnetic interactions between the polynuclear compounds are often one to three orders of magnitude smaller than within the polynuclear fragment.

As mentioned above, one of the most important problems in the molecular design of magnets based on coordination compounds of transition metals is to provide self-assembly of high-dimensional (layered polymer or framework) structures, which favors volume magnetization. The bridging organic or inorganic ligands should additionally perform the function of effective exchange channels between the paramagnetic centers. The most complex and occasionally insurmountable problem in the design of magnetoactive high-dimensional (2-D and 3-D) systems based on mononuclear complexes is the lack of coordination sites at the transition metal ion. Since the molecular design of magnets is mainly carried out based on metal ions from the first row of transition elements, the maxium possible number of coordination sites is six. The blocking of coordination sites by monodentate anion ligands, or ligands that are not capable of performing the bridging function, or simply donor solvent molecules, precludes the construction of extended structures in one, two, or three directions in space. This disadvantage can be avoided by using an approach based on the assembly of a high-dimensional structure whose sites contain multinuclear structures, but not the individual transition metal ions. These multinuclear structures may have the desired or preset number of terminal metal ions; by modifying the bridging ligands in the multinuclear fragment, a definite spatial orientation of these ligands can, in prinple, be set, thus directing the self-assembly of a high-dimensional structure with preset topology as desired.

Before considering the particular examples of the chemical design of multinuclear complexes with nitroxides, note that thousands of multinuclear compounds of transition metals have already been structurally defined, among which complexes with various carboxylates are most numerous.[20,172–176] The greatest progress in the development of the synthesis of molecular nanomagnets was achieved on metal carboxylates.[177–189] Special attention was paid to the fact that the development of the chemistry and technology of nanosized clusters would largely depend on progress in the understanding of the principles governing the synthesis, structure, and stability of these compounds.[176]

The results of studies of complex heterospin systems based on multinuclear pivalates of transition metals with nitroxides are summarized below.[172,190–198] Attention is concentrated on the chemical behavior of the multinuclear fragment.

13.4.1 Reactions whose products retain both the multinuclear fragment and nitroxide

The kinetically stable hexanuclear $[Mn_6(O)_2Piv_{10}(Thf)_4]$ contains four terminal manganese(II) ions, which coordinate the readily replaceable thf molecules, lying at the vertices of a flat tetrahedron.[199] The presence of coordinated thf molecules is extremely important because NIT-R, which are the products of oxidation of the corresponding dihydroxy precursors having no functional groups apart from $>N{-}^{\bullet}O$, are weak donors and can only replace weaker ligands. Indeed, the interaction of $[Mn_6(O)_2Piv_{10}(Thf)_4]$ with L^{20} in nonpolar heptane or carbon tetrachloride led to the replacement of all coordinated thf molecules and self-assembly of $[Mn_6(O)_2Piv_{10}(L^{20})_2]$ (Figure 13.3), having a framework diamond-like structure.[190–192] In the reaction of $[Mn_6(O)_2Piv_{10}(Thf)_4]$ with L^{20} in dichloromethane, only some part of coordinated thf molecules was replaced, and the $[Mn_6(O)_2Piv_{10}(Thf)_2(L^{20})Mn_6(O)_2Piv_{10}(Thf)(CH_2Cl_2)(L^{20})]$ chain polymer was always isolated as a solid irrespective of the initial $[Mn_6(O)_2Piv_{10}(Thf)_4]/L^{20}$ ratio (Figure 13.3).

The reaction of $[Mn_6(O)_2Piv_{10}(Thf)_4]$ with L^{20} in ethyl acetate led to crystallization of dumbbell-shaped $[(EtOAc)_3Mn_6(O)_2Piv_{10}(L^{20})Mn_6(O)_2Piv_{10}(EtOAc)_3]$ molecules, in which two $[Mn_6(O)_2Piv_{10}]$ hexanuclear fragments are linked by one bridging nitroxide molecule (Figure 13.3). Since all heterospin complexes formed irrespective of the initial $[Mn_6(O)_2Piv_{10}(Thf)_4]/L^{20}$ ratio, the solvent used for the synthesis obviously plays a critical role in the isolation of a particular product. This was evidenced by the fact that treatment, for example, of $[(EtOAc)_3Mn_6(O)_2Piv_{10}(L^{20})Mn_6(O)_2Piv_{10}(EtOAc)_3]$ with additional L^{20} in heptane always led to the framework compound $[Mn_6(O)_2Piv_{10}(L^{20})_2]$. Also note that both

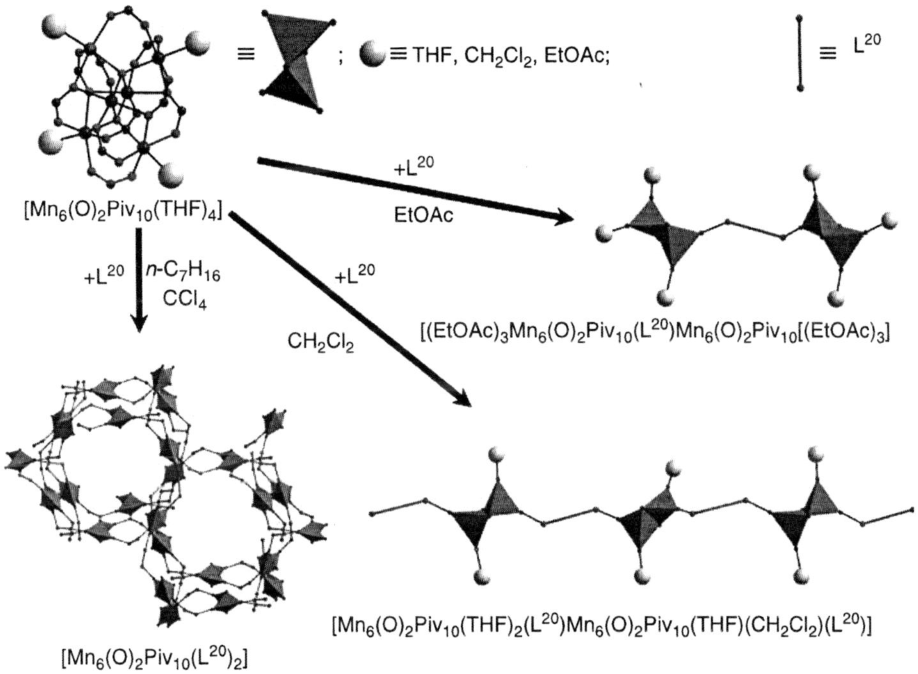

Figure 13.3 *Products of the reaction of* $[Mn_6(O)_2Piv_{10}(Thf)_4]$ *with* L^{20} *nitroxide in different solvents.*

the $[Mn_6(O)_2Piv_{10}]$ fragment and L^{20} in the compounds always remained invariable in their structural characteristics.

The introduction of an NIT-R bridge between clusters can substantially affect the magnetic properties of the compound. Suppose the starting polynuclear compound has dominant antiferromagnetic exchange interactions between the magnetic moments (large black arrows in Figure 13.4a) of the individual polynuclear fragments. This situation is most common to polynuclear compounds[7,200] and also observed for solid $[Mn_6(O)_2Piv_{10}(Thf)_4]$. As a result, at decreased temperatures, the μ_{eff} of the compound tends to zero. The annihilation of the magnetic moments of the polynuclear compounds can be avoided by linking the $[Mn^{II}_4Mn^{III}_2O_2]$ polynuclear fragments with an additional paramagnetic bridge, whose spin differs from the spin of $[Mn^{II}_4Mn^{III}_2O_2]$. In this case, even with the antiferromagnetic exchange between the adjacent spins of the polynuclear fragment and nitroxide, the difference magnetic moment can be nonzero (Figure 13.4b). This approach was effectively used by Kahn *et al.* in the design of mixed metal ferromagnetic chains.[201–209] This situation was found for the $[Mn_6(O)_2Piv_{10}(Thf)_2(NIT\text{-}Me)Mn_6(O)_2Piv_{10}(Thf)(CH_2Cl_2)(NIT\text{-}Me)]$ heterospin chain polymer complex, for which antiferromagnetic ordering was recorded below 3.5 K, and at 2 K, the spontaneous magnetic moment reached $\sim 2800\,G{\cdot}cm^3/mol$.[190,192]

However, the bridging coordination of nitroxide sets strict requirements in terms of the spatial complementarity of the polymetallic nuclei and the bridging paramagnetic ligand. Of the whole class of 2-imidazoline nitronyl nitroxides, only L^{20}, which contains a CH_3 substituent in the second position of the 2-imidazoline heterocycle, can perform the bridging function between two hexanuclear $[Mn_6(O)_2Piv_{10}]$ ligands. Further increase in the size of the substituent (R), for example, a transition to the ethyl derivative

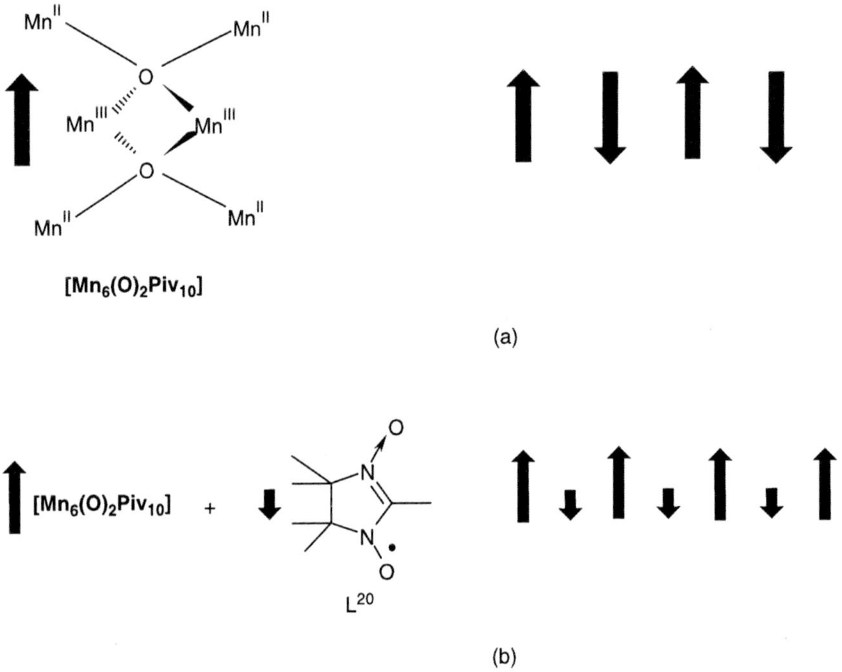

Figure 13.4 *Antiferromagnetic interaction between the magnetic moments of adjacent clusters (a). The formation of a difference magnetic moment in a nonhomogeneous antiferromagnetic chain with a nitroxide bridge (b).*

(L^7), makes the bridging coordination mode of the O•–N–C=N→O fragment impossible because of the steric requirements of the *tert*-butyl pivalate groups with the atoms of the R group.[192]

The replacement of the coordinated acetonitrile molecules by nitroxide is an effective method for the synthesis of complex multispin systems. This procedure for the synthesis of mixed metal heterospin compounds with direct coordination of the nitroxyl fragment was demonstrated for the reaction of the trinuclear $[Co_2Gd(NO_3)Piv_6(CH_3CN)_2]$ with L^{20} as an example, which gave the $[Co_2Gd(NO_3)Piv_6(L^{20})_2]$ p,d,f–heterospin complex. Each molecule of this complex contains a specific exchange cluster, which is a "chain" of five paramagnetic centers separated by a system of chemical bonds (Scheme 13.19). The

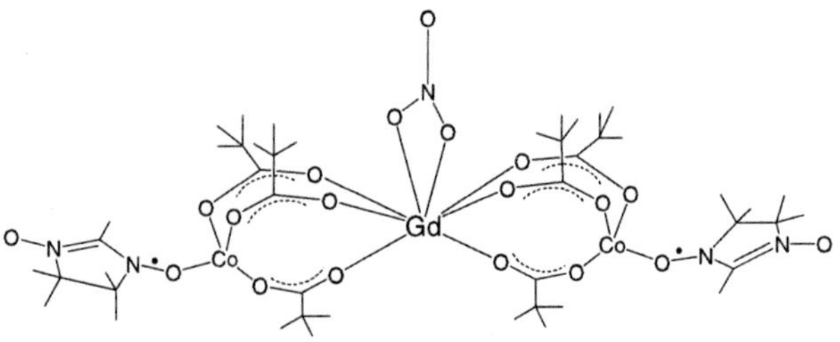

$[Co_2Gd(NO_3)Piv_6(L^{20})2]$

Scheme 13.19

five-center exchange cluster has a unique combination of odd electrons lying on the π^* orbital of nitronyl nitroxide, the *d* orbital of cobalt(II), or the *f* orbital of gadolinium(III) in one molecule.[197]

13.4.2 Transformation of polynuclear fragments in reactions with nitroxides

The synthesis of polynuclear complexes with nitroxides becomes a more complex problem when the polynculear fragment becomes kinetically unstable and can transform depending on the reaction conditions. Therefore, the kinetic mode of synthesis demands special attention, especially because the polynuclear compound should be isolated as a perfect crystal suitable for further X-ray study. This is why Winpenny has paid special attention to the fact that one of the most complex problems in the coordination chemistry of polynuclear compounds is the development of *reliable* synthetic approaches for the synthesis of a compound of definite nuclearity.[175] Publications often refer to a separate polynuclear compound having a definite structure and formed in the course of a synthesis without discussing the synthetic conditions, or the common structural bonds between frameworks of different nuclearities of the given metal, or the role of the ligand in the formation of the complex.[175] This situation (the tendency toward the transformation of polynuclear frameworks) was encountered in developing the synthesis of heterospin systems based on polynuclear nickel(II) pivalates with nitronyl nitroxides.[193]

Formally "nickel(II) pivalate" can be isolated as compounds of different nuclearities, namely, $[Ni_6(OH)_4Piv_8(HPiv)_4]$, $[Ni_7(OH)_7Piv_7(HPiv)_6(H_2O)]\cdot0.5C_6H_{14}\cdot0.5H_2O$, $[Ni_8(OH)_4(H_2O)_2Piv_{12}]$, and $[Ni_9(OH)_6Piv_{12}(HPiv)_4]\cdot HPiv\cdot3H_2O$. Due to their high solubility in apolar organic solvents, these compounds react with nitroxides and afforded heterospin compounds with L^{20}, L^{39}–L^{42} (Scheme 13.20), $[Ni_7(OH)_5Piv_9(HPiv)_2(L^{39})_2(H_2O)]\cdot0.5C_6H_{14}\cdot H_2O$, $[Ni_7(OH)_5Piv_9(HPiv)_2(L^{40})_2(H_2O)]\cdot H_2O$, $[Ni_7(OH)_5Piv_9(HPiv)_2(L^{41})_2(H_2O)]\cdot H_2O$, $[Ni_6(OH)_3Piv_9(HPiv)_4(L^{20})]\cdot1.5C_6H_{14}$, and $[Ni_4(OH)_3Piv_5(HPiv)_4(L^{42})]\cdot1.5C_7H_8$. At liquid helium temperatures, $[Ni_7(OH)_7Piv_7(HPiv)_6(H_2O)]\cdot0.5C_6H_{14}\cdot0.5H_2O$ and $[Ni_7(OH)_5Piv_9(HPiv)_2(L^{40})_2(H_2O)]\cdot H_2O$ behaved like ferrimagnets. Small deviations from collinearity during antiferromagnetic ordering of the magnetic moments in the polynuclear fragments were the source of magnetization in the complexes.

Although nickel(II) pivalates can form complexes of different nuclearities and change their nuclearity during recrystallization or reactions with nitroxides, it is very important that studies of these compounds are considerably facilitated due to their topological unity (Figure 13.5) because the polynuclear frameworks of all complexes can be produced from the hexanuclear $\{Ni_6(\mu_4\text{-}OH)_2(\mu_3\text{-}OH)_2(\mu_2\text{-}C_5H_9O_2\text{-}O,O')_6(\mu_2\text{-}C_5H_9O_2\text{-}O,O)(\mu_4\text{-}C_5H_9O_2\text{-}O,O,O',O')(C_5H_{10}O_2)_4\}$ complex shaped like an open book (Figure 13.6).

Based on this fragment, the formation of heptanuclear compounds is the result of the addition of further mononuclear fragments to the four nickel(II) ions lying at the vertices of the "open book" (Figure 13.5). After the addition of trinuclear fragments to two nickel(II) ions lying on one of the lateral edges of the "open book," a nonanuclear complex is formed. The lateral wing formed in the nonanuclear complex

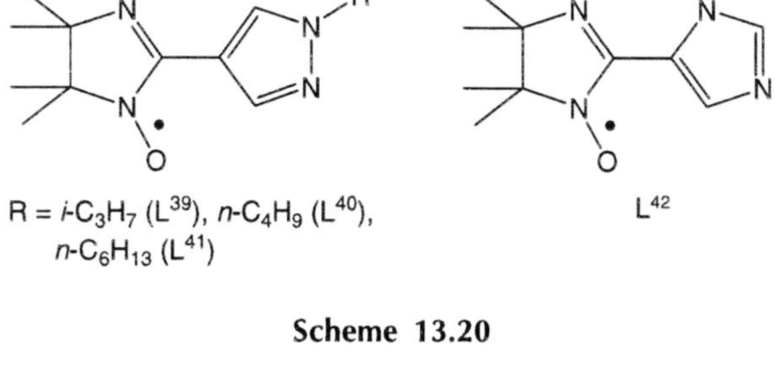

Scheme 13.20

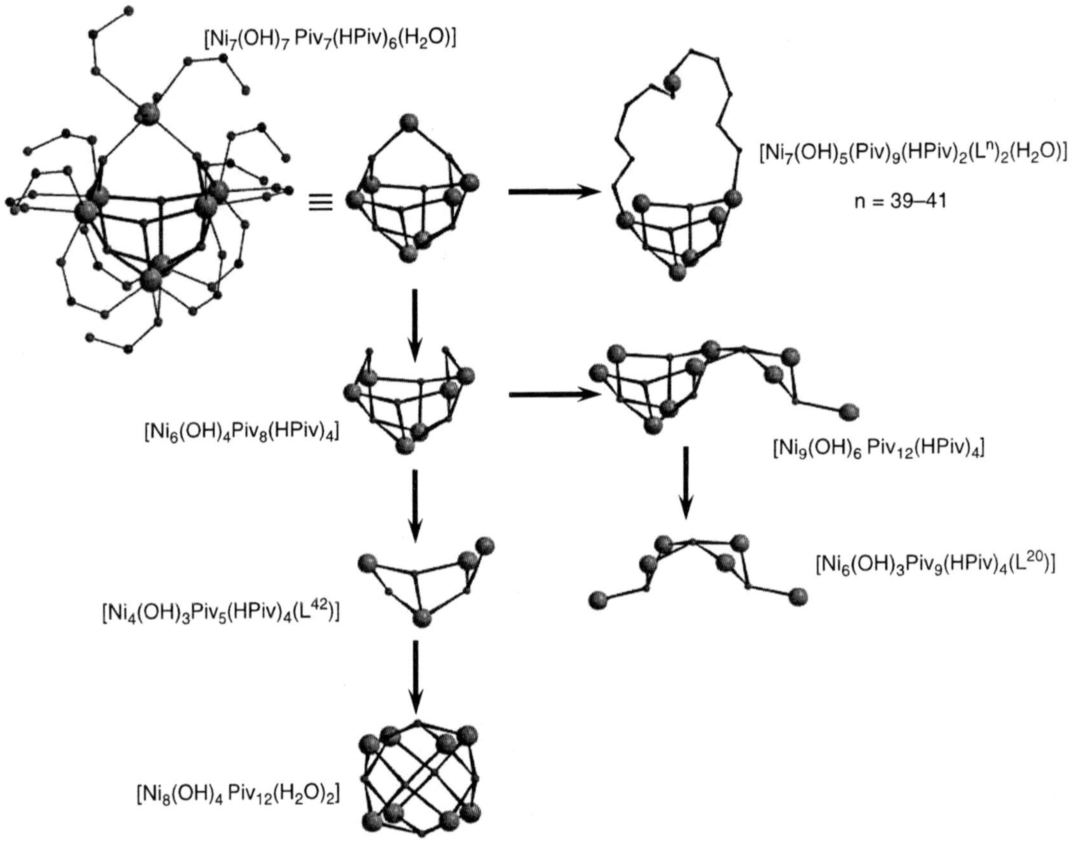

Figure 13.5 *Scheme of the transformations of the frameworks of polynuclear nickel(II) trimethylacetates and their complexes with nitronyl nitroxides.*

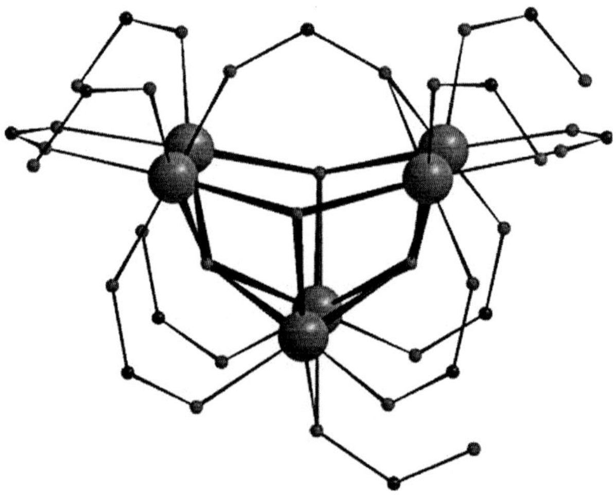

Figure 13.6 *Molecular structure of [$Ni_6(OH)_4Piv_8(HPiv)_4$]. The tert-butyl groups and hydrogen atoms are omitted.*

retains its structure in another type of hexanuclear complex, shaped like a bathtub. If, however, two edge-sharing nickel(II) ions, one of which lies at a vertex of the "open book" and the other on the binding line of the "book," are eliminated from the starting hexanuclear fragment, a tetranuclear fragment fixed in the structure of $[Ni_4(OH)_3Piv_5(HPiv)_4(L^{42})]$ is obtained. The doubling of this tetranuclear fragment leads to the formation of a highly symmetric $[Ni_8(OH)_4(H_2O)_2Piv_{12}]$ molecule, which is a new chemical variant of cubane. The understanding of the topological scheme of mutual transformation of a certain group of polynuclear fragments makes it possible to control the synthesis of these complex heterospin complexes with nitroxides, reveal the relationship between their structures, and justifiably divide the polynuclear fragment into smaller units in describing the magnetic properties of the compound.[193]

In constructing a topological scheme of this kind, the greatest difficulty is to isolate the "key" compound of the topological series, that is, the hexanuclear $[Ni_6(OH)_4Piv_8(HPiv)_4]$ complex in this case, because it is most susceptible to various transformations which depend on the conditions of synthesis. These transformations often hinder the isolation of the "key" compound as a single crystal, which can sometimes appear (and subsequently vanish with time) as an impurity in addition to the kinetically more stable solid crystals. The structure solution of the "key" polynuclear compound is vital; otherwise, it is difficult to construct even a fragment of the topological structural scheme. When this compound is detected, the preparation of compounds with the new composition and/or new polynuclear or polymer structure will possibly complement and enrich the single topological scheme, thus reflecting the electronic nature of the transition element and the types of structure formed with a particular set of ligands. An example of topological construction for a set of polynuclear carboxylates is given in.[175]

13.4.3 Transformation of both the polynuclear fragment and the starting nitroxide

Processes that involve the transformation of both the polynuclear matrix and the nitroxide molecule are most complex from a synthetic viewpoint. The amount of experimental data is yet insufficient to predict the formation of definite heterospin compounds. Studies of the products of the reaction of $[Co^{III}{}_2Co^{II}{}_4(O)_2Piv_{10}Thf_3H_2O]$, whose polymetallic fragment is almost identical to the manganese compound $[Mn^{III}_2Mn^{II}_4(O)_2Piv_{10}]$ discussed in Section 13.3.1, with L^{20} and/or its imine analog showed that redox reactions resulted in rearrangements in both the polynuclear $\{Co_6(O)_2Piv_{10}\}$ matrix and the starting nitroxide.[196] In nitroxide molecules, a methyl substituent in the second position of the 2-imidazoline ring can be oxidized, which leads to heterospin products containing L^{43}–L^{47} (Scheme 13.21).

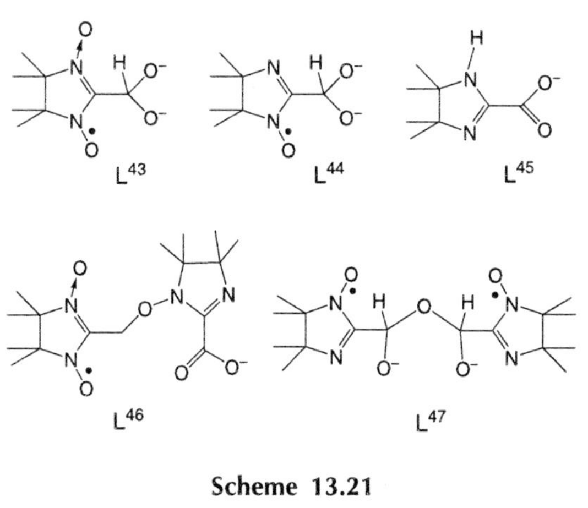

Scheme 13.21

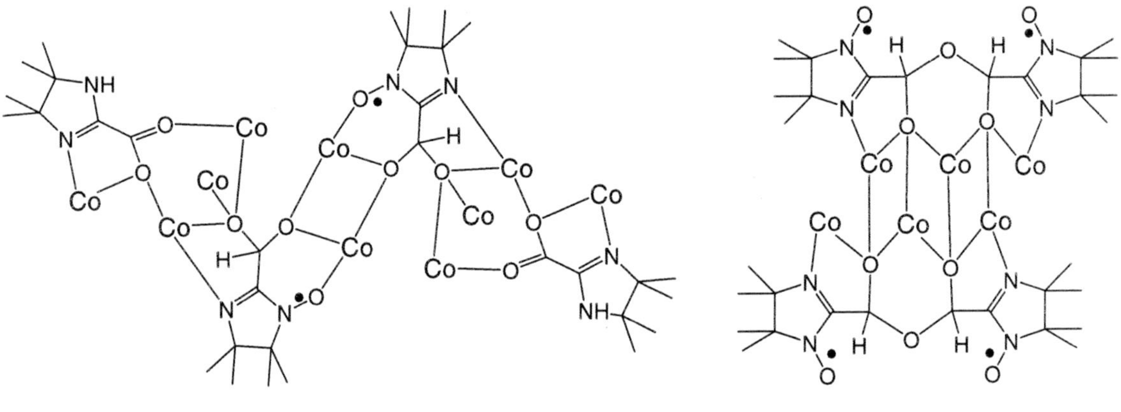

Scheme 13.22

Special attention should be paid to the spin labeled anions L^{43}, L^{44}, L^{46} and L^{47} fixed in the polynuclear matrix. These anions are not known in the form of ordinary salts or the corresponding organic derivatives, and it is difficult to suggest pure organic approaches to their synthesis. The transformation of the starting nitroxide molecule occurs on the coordinating polynuclear matrix; in the case of L^{46} and L^{47}, this is additionally a variation of template synthesis. The polynuclear matrix fixes and kinetically stabilizes the resulting organic substrate molecules. The oxidation of L^{20} and/or its imine derivative is accompanied by the reduction of cobalt(III) to cobalt(II) and the transformation of the matrix. Cobalt(III) plays an important role in these processes, creating conditions for the oxidation of the organic substrate.

Scheme 13.22 shows, in simplified form (without Piv and HPiv ligands), the polynuclear fragments of the decanuclear $[Co_{10}Piv_{14}(HPiv)_2(L^{44})_2(L^{45})_2]{\cdot}C_6H_{14}$ and hexanuclear $[Co_6Piv_8(HPiv)_2(L^{47})_2]$ molecules, which are part of a larger series of synthesized compounds including $[Co_9Piv_{12}(OH)_2(L^{43})_2]{\cdot}xC_7H_{16}$ $(x = 0.5, 1, 2)$, $[Co_8Piv_8(OH)_6(HPiv)_4(H_2O)_2(L^{46})_2]$, $[Co_6Piv_8(HPiv)_2(L^{47})_2]{\cdot}C_6H_6$, and $[Co_6Piv_8(HPiv)_2(L^{47})_2][Co_{10}Piv_{14}(HPiv)_2(L^{44})_2(L^{45})_2]{\cdot}C_7H_{16}$. These polynuclear molecules were chosen as an illustration because L^{47} in a hexanuclear complex can formally be regarded as a product of condensation of two L^{44} imino nitroxide anions, lying close to each other in the decanuclear $[Co_{10}Piv_{14}(HPiv)_2(L^{44})_2(L^{45})_2]$ matrix.

13.5 Breathing crystals

A specific class of d-block elements, having configurations from d^4 to d^8, show spin crossover effects, which are transitions of the central atom of the complex from high to low spin state at decreased temperatures or elevated pressures, or under the action of light.[7,210–213] A decrease in the temperature, as well as static loading of the crystal, lead to its compression, which, in turn, increases the energy splitting of the levels of the metal ion. This is a well known effect. In this classical variant of spin crossover, the spin multiplicity of the ground state changes within one type of paramagnetic center (metal ions in this case).

The development of syntheses of transition metal complexes with stable nitroxides led to the discovery of polynuclear complexes of copper(II), whose magnetic properties were occasionally observed to mimic the spin transitions in classical systems.[12,53,61,74,214–218] The first of these, the tetranuclear $[Cu(hfac)_2]_4(L^6)_2$ complex (Scheme 13.7), was described by Rey *et al.*[61] The classical spin crossover is impossible in copper(II) complexes with diamagnetic ligands, that is, in systems containing only d^9 paramagnetic centers. The major distinction of spin transitions in copper(II) complexes with nitroxides from classical spin

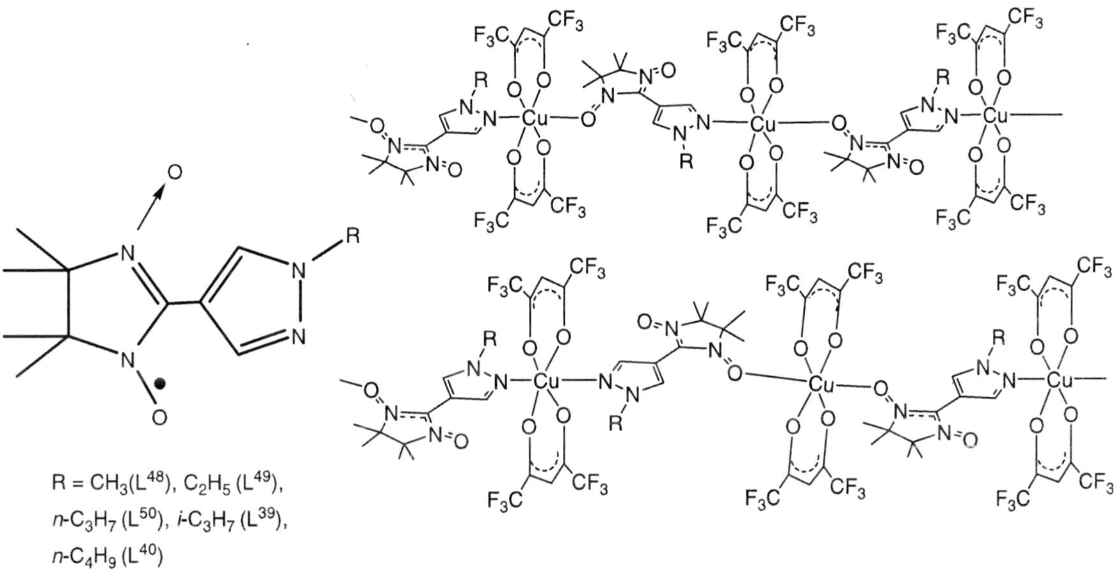

Figure 13.7 *Pyrazolyl-substituted nitronyl nitroxides and head-to-tail and head-to-head motifs in the heterospin chains of Cu(hfac)$_2$L^n breathing crystals (n = 39, 40, 48–50).*

crossover, is the possibility of changing the total spin in the heterospin exchange cluster containing two or more paramagnetic centers as a result of the structural rearrangement of the coordination environment of the metal. Studies of these compounds were hindered by the fact that, having passed the phase transition region on cooling, the crystal exploded or lost its crystal quality, which hindered its full-fledged structural analysis.[217] The situation considerably improved when copper(II) compounds with such effects and anomalously high mechanical stability were found. These were complexes of $Cu(hfac)_2$ with L^{39}, L^{40}, and L^{48}–L^{50} (nitronyl nitroxides containing a pyrazole substituent in the second position of the imidazoline ring, Figure 13.7). As a result, a family of heterospin complexes with magnetic properties similar to spin crossover was created.[215,216] The single crystals generally did not crack after the temperature range of the structural transition and the ensuing magnetic transition. The structure was determined for all heterospin complexes and their solid solutions at temperatures before and after the magnetic transition, which provided a unique opportunity for the observation of structure dynamics in the range of the phase transition for these systems. Despite the considerable (sometimes giant) changes in the unit cell volumes of the single crystals after the cooling-heating cycles, crystal compression and expansion were occasionally reversible. Due to their high mechanical stability and the ensuing reversible transitions from one polymorph to another at varying temperatures, $Cu(hfac)_2L^n$ crystals (n = 39, 40, 48–50) were called "breathing."

It is noteworthy that $Cu(hfac)_2L^n$ (n = 39, 40, 48–50), showing magnetic effects similar to spin crossover, have a polymeric chain structure in the solid phase (Figure 13.7), resulting from the bridging bidentate coordination of the paramagnetic ligands. Phase transitions rarely occur without crystal decay in these structures because the polymorphic transformation of single crystals without loss of crystal quality requires many coherent motions, not only inside polymer chains but also motions of chains as entities in crystal and especially during repeated cooling-heating cycles, which cause repeated anisotropic expansion and compression of crystals. Moreover, the synthesis and studies of these compounds require a special approach. Due to the stereochemical nonrigidity of $Cu(hfac)_2$, a significant number of compounds can be

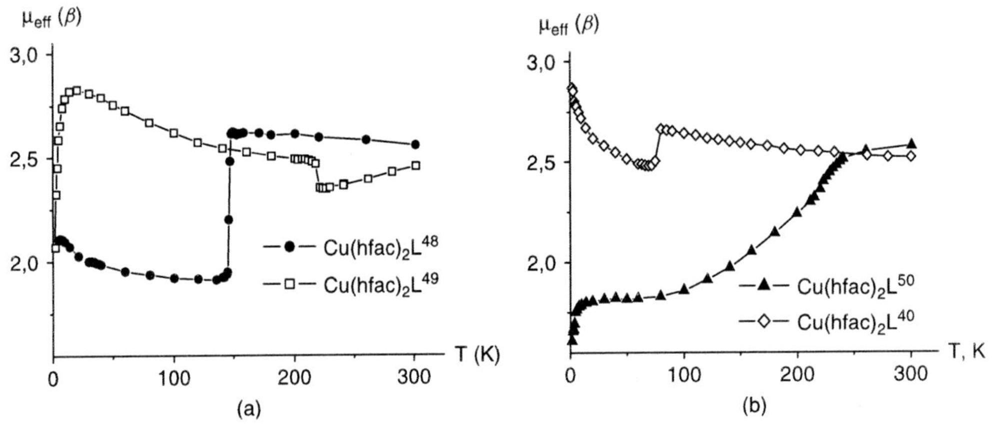

Figure 13.8 *Temperature dependences of the effective magnetic moment for Cu(hfac)$_2$L^{48} and Cu(hfac)$_2$L^{49} (a), Cu(hfac)$_2$L^{40} and Cu(hfac)$_2$L^{50} (b).*

obtained by varying the conditions of synthesis (solvent, temperature, and reagent ratio). As an extreme example, 12 different phases were obtained and structurally defined in the reaction of Cu(hfac)$_2$ with L^{48}, but only two of these exhibited magnetic anomalies on the μ_{eff}(T) curve.[79] Another difficulty was isolation of single crystals of two or more phases. Sometimes they could be separated mechanically. However, in the reactions of Cu(hfac)$_2$ with L^{48}, the single crystals of different phases were occasionally indistinguishable in both color and habit. Therefore, each single crystal had to be checked by X-ray diffraction analysis before magnetochemical experiment with crystals of the same phase.

Figure 13.8 shows the experimental temperature dependences of μ_{eff} for Cu(hfac)$_2$L^n (n = 39, 40, 48–50), which differ in the R substituent bonded to the pyrazole fragment. Although the substituents are remote (Figure 13.7) from the structurally rearranging >N–$^•$O–Cu(II)–O$^•$–N< or >N–$^•$O–Cu(II) clusters, they have a substantial effect on the character of the temperature dependence. The μ_{eff}(T) dependences can vary greatly because of different combinations of exchange interactions in the >N–$^•$O–Cu(II)–O$^•$–N< or >N–$^•$O–Cu(II) units before and after the phase transition.[12] It should be emphasized that in contrast to the classical spin crossover, spin transitions here are often accompanied by an abrupt increase in the effective magnetic moment at decreased temperatures (Figure 13.8a). This is explained by the fact that the phase transition results in a *lengthening*, not shortening, of Cu–O$_{NO}$ distances. As a consequence, the exchange interaction changes from antiferromagnetic to ferromagnetic, and μ_{eff} ceases to decrease and starts to increase at decreased temperatures.

In these systems, spin transitions are sensitive not only to variation of the R substituent in the pyrazole ring. The temperature of the transition and the form of the magnetic anomaly can be controlled by synthesizing mixed metal solid solutions M$_x$Cu$_{1-x}$(hfac)$_2$L (M = Mn, Ni, Co)[219] or, which occurs much more rarely (Figure 13.9), by forming Cu(hfac)$_2$L$_x^{48}$L$_{1-x}^{49}$ solid solutions containing different nitroxides as organic ligands.[12]

Of great interest was the family of heterospin solvates Cu(hfac)$_2$L^{40}•0.5 Solv (Solv = pentane, hexane, heptane, octane, octene, butyl chloride, butyl bromide, butyl iodide, amyl chloride, amyl bromide, amyl iodide, benzene, toluene, *o*-xylene, *m*-xylene, *p*-xylene, ethylbenzene, propylbenzene), which exhibited spin transition effects and high mechanical stability after the temperature range of the magnetic anomaly. Studies of these compounds over a wide range of temperatures (28–300 K), that is, before and after the structural transition and the ensuing magnetic phase transition, showed that the solvent molecules, incorporated in

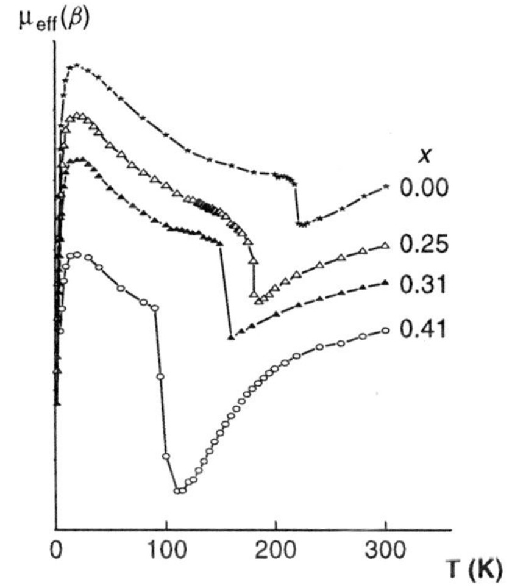

Figure 13.9 Experimental $\mu_{eff}(T)$ dependences for $Cu(hfac)_2L^{48}{}_xL^{49}{}_{1-x}$ (the curves were not fitted to the single ordinate axis).

the interchain space and involved only in van der Waals interactions with heterospin chains, significantly affected both the character of the $\mu_{eff}(T)$ dependence and the temperature range of the magnetic anomaly. The introduction of certain solvent molecules in breathing crystals actually proved a highly effective method of control over the physical parameters of the magnetic anomaly.[220] The relationship between the chemical step and the physical property was studied on specially selected series of breathing crystals. As a result, it was found that "mild" modification of T_c in $Cu(hfac)_2L^{40}{\cdot}0.5Solv$ required a substantially smaller structural step than modification by one methylene unit typically used in organic chemistry. X-ray cinema demonstrated mutual coherence of the structural dynamics of >N–•O–Cu(II)–O•–N< heterospin exchange clusters and the structural dynamics of interchain diamagnetic fragments.

Quantum chemical calculations using X-ray diffraction data allowed researchers to trace the character of changes in the parameters of exchange interactions in exchange clusters in the range of the phase transition. It was found that in the temperature range corresponding to the phase transition, the exchange parameter considerably changed in magnitude and occasionally in sign, a non-trivial finding. The smooth change in the magnetic moment on cooling, observed in the majority of experiments near T_c, was the result of the gradual increase in the fraction of the low temperature phase in the high temperature phase.

A pronounced change in the exchange integral was also recorded in a radiospectroscopic study.[221] For heterospin breathing crystals based on copper(II) complexes with nitroxides, electron paramagnetic resonance (EPR) spectroscopy can be a specific and extremely powerful experimental technique, which allows determination of the sign of the exchange integral and evaluation of the exchange parameter J by analyzing the temperature dependence of the g factor.[222–225] The unusual magnetic properties of breathing crystals also stimulated the development of the phenomenological theory of spin transitions for multispin exchange-coupled systems.[226,227] Recently, the possibility of a LIESST (Light-Induced Excited Spin State Trapping) effect was demonstrated on a breathing crystal.[228] Both initiation (which suggested preliminary preparation of a fine film or finely disperse particles of a heterospin complex on the surface of an inert and transparent support with further irradiation) and EPR recording of the effect differed substantially in the experimental technique from the classic variant.[210] To summarize the discussion of these surprising heterospin complexes, it should be emphasized that their detection is one of achievements in the field of molecular magnetism.[218]

13.6 Other studies of metal–nitroxides

The preceding sections discussed current general tendencies of research in metal-nitroxide systems. This section summarizes a variety of other specific studies which have not (yet) become objects of wide studies.

13.6.1 Analytical applications

The presence of a strong complex-forming group in the structure of a nitroxide that is soluble in organic solvents but not readily soluble in water can be used for quantitatively extracting the metal from water to organic phase. Thus, for $Pd(L^{51})_2$, the distribution coefficient is $\geq 10^3$ (Scheme 13.23). Therefore, palladium was almost entirely transferred from the aqueous solution to the organic phase after one extraction. The EPR spectrum of the organic phase contained not only the nitroxide triplet, but also quintuplet lines from the $Pd(L^{51})_2$ biradical. Since the EPR lines for mono- and polynitroxides were rather narrow and their width changed but slightly at concentrations $\leq 10^{-3}$ mol/l, the peak intensity of the biradical components of the spectrum was directly proportional to the palladium concentration. This underlies a new analytical procedure named extraction radiospectroscopy.[229] The method was subsequently considerably improved due to the addition of various chromatographic variants of organic phase separation. This allowed researchers to extract simultaneously a wide range of diamagnetic metals, with subsequent chromatographic separation of all extracted components, and ultimately determine the amount of extracted metal by EPR spectroscopy.[3]

13.6.2 NMR spectroscopy

EPR spectroscopic studies of metal-nitroxide systems are routine practice in this field. In contrast to this, NMR spectroscopy is rarely used for studying metal complexes with nitroxides[230] because of the wide NMR lines of paramagnets inherent in nitroxides. The lines can be narrowed to some extent only in very concentrated solutions[231] due to growth of the efficiency of intermolecular interactions of radicals. The lines can be narrowed much more effectively if the nitroxide is introduced in an exchange-coupled complex with a paramagnetic metal ion having small transverse relaxation time T_{2M}.[43,232–235] Researchers generally restrict themselves to the statement of the fact of coordination of nitroxide in solution.[43,232–235] In[230], the contact shifts in the NMR spectra of a series[236] of nickel(II) β-diketonate complexes were analyzed in

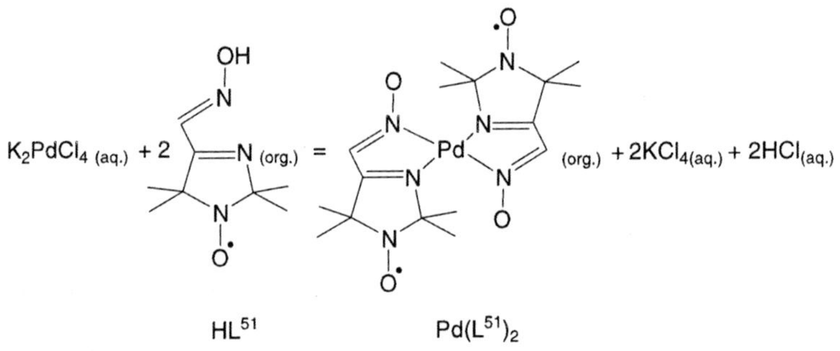

Scheme 13.23

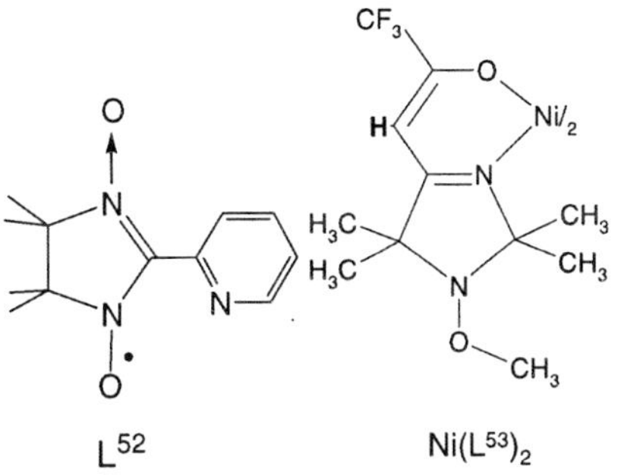

Scheme 13.24

comparison with the shifts in the spectra of similar nickel(II) complexes with L^{52} (Scheme 13.24), and a method for evaluating the exchange coupling parameter in the molecules of heterospin compounds of this kind was suggested.

Although the NMR spectra of the complexes of paramagnetic metal ions with nitroxides are complex, they convey valuable data about the magnetic properties of the molecular precursors of heterospin solids formed during crystallization from solution. The problem is not the NMR line width but rather the fact that one (especially stereochemically non-rigid) complex can exist in solution in the form of several conformations. The ratio of conformations can vary with temperature in a complex manner, and the distribution of conformations can drastically change within a small temperature range. In addition, a transition from one conformation of a heterospin complex to another can be accompanied by a change in the sign of the paramagnetic shift in the NMR spectra. Of special note is the fact that, in studies of such systems, it is possible to observe the unusual effect of multiple broadening and narrowing of NMR lines during variation of the temperature of solutions of complexes with nitroxides. An analysis of all these effects has been reported.[237] It was noted that these studies were considerably hindered by spin–spin coupling between the odd electrons of paramagnetic centers. To analyze the NMR spectra of heterospin complexes of the paramagnetic ions of transition metals with organic radicals, therefore, it was necessary to use the complete (but not reduced) Equation (13.1):

$$\Delta = \frac{P_M}{\tau_M}\left[\frac{\left(\frac{1}{T_{2M}}\right)^2 + \frac{1}{T_{2M}\cdot\tau_M} + \omega^2}{\left(\frac{1}{T_{2M}} + \frac{1}{\tau_M}\right)^2 + \omega^2}\right] \tag{13.1}$$

where τ_M is the residence time of a resonant nucleus with definite paramagnetic surroundings, P_M is the molar fraction of the paramagnetic complex in the sample, ω is the paramagnetic shift of the resonance line, and T_{2M} is the transverse relaxation time of the nucleus in the complex.

Studies of a series of heterospin complexes based on $Ni(L^{26})_2$ and its analog with the diamagnetic derivative ($Ni(L^{53})_2$, Scheme 13.24) showed that using Equation (13.1) was difficult, not only because Δ was a nonlinear function of four variables, but also because the variables themselves depended strongly on the nature of the complexes under study (metal ion, coordination polyhedron, and its structure and composition), as well as the conditions of experiment (temperature and concentrations). The non-trivial result of that work was that under the appropriate conditions, the resonance signal, whose position in the NMR spectrum was determined by the paramagnetic shift ω, could be intermittently narrowed and broadened. In other words, the $\Delta(T,\omega)$ function was modulated because of the variation of the τ_M parameter that defines exchange processes in the system whose hfi is being studied. In experiments, this effect depended on both the hyperfine splitting parameters and the possibility of varying the temperature of the complex solution within wide limits. Moreover, it appeared that in $Ni(L^{26})_2$ type heterospin complexes, different paramagnetic centers could induce paramagnetic shifts with different signs. If these peculiarities of the NMR spectra of metal-nitroxide systems are correctly taken into account, it is possible to determine the temperature range in which a certain conformation of the heterospin complex is dominant and obtain a solid molecular magnet of a definite structure in crystallization at a chosen temperature.[237]

13.6.3 Stabilization of nitroxides with β-hydrogen atoms

As is well known, nitroxides containing an hydrogen atom in the β position relative to the nitroxyl group are mostly kinetically unstable and can be transformed into various derivatives, for example, disproportionation products (Scheme 13.25).[8,238–242]

It has been shown[243,244] that nitroxides of this type can be stabilized by complexation with metals (Scheme 13.26) because the radicals acquired conformations that precluded their destruction. Although copper(II) compounds can catalyze various oxidative processes, for $Cu(hfac)_2L^{56}$ crystals grown in solution no changes in the composition and properties were detected after storage for three months under the normal conditions.

13.6.4 Increased reactivity

Nitroxides often behave as oxidants in different chemical processes.[8,245,246] For example, in the presence of copper(II) salts, di(*tert*-butyl)nitroxide accelerates the oxidation of methanol to formaldehyde with oxygen and is reduced to hydroxylamine.[247] In the reaction of copper(II) chloride ($CuCl_2$) with spin labeled thiosemicarbazone HL^{58}, the latter was quickly reduced in methanol or acetone to the corresponding hydroxylamine.[248] In the presence of copper(II) salts in alkaline media, sterically hindered hydroxylamines are quickly oxidized to the corresponding nitroxides. If oxygen is removed from the reaction mixture, these redox processes can give complexes of metals with different degrees of oxidation, for example, copper(II)

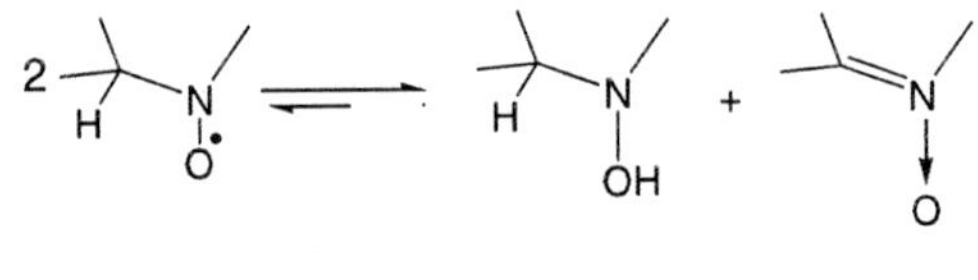

Scheme 13.25

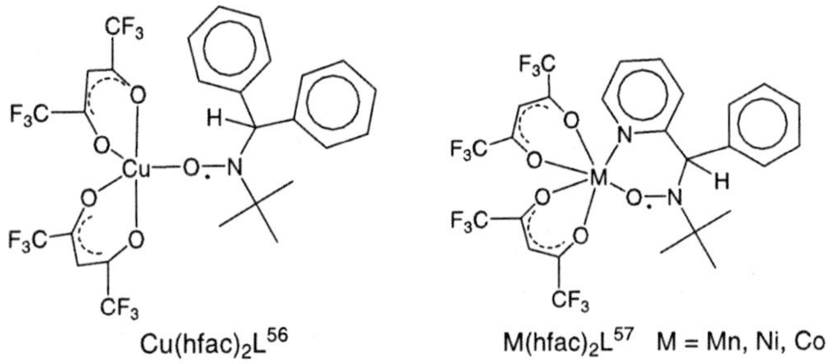

Scheme 13.26

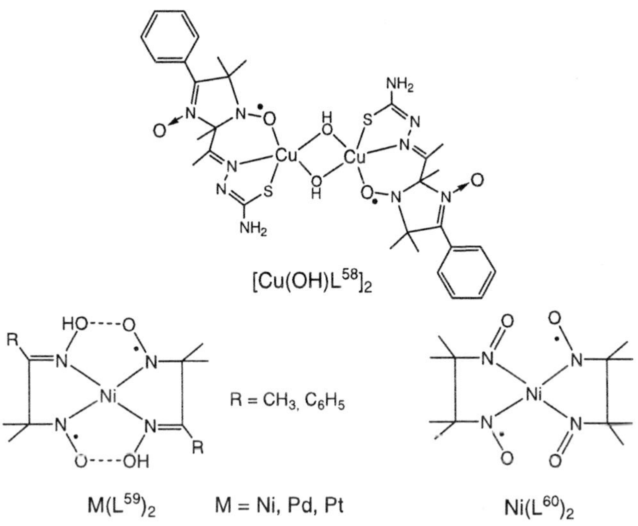

Scheme 13.27

and copper(I) compexes with nitroxides.[249] The diamagnetic $[Cu(OH)L^{58}]_2$ dimer complex obtained in an alkaline medium had an anomalously high reactivity (Scheme 13.27) and gradually decomposed the solvents such as dichloromethane and trichloromethane even in the absence of oxygen.[248]

The so-called singlet biradicals such as $M(L^{59})_2$, $Ni(L^{60})_2$ shown in Scheme 13.27 also possess high reactivity.[250–252] These are diamagnetic compounds, but the presence of paramagnetic ligands in them can be recorded by replacing one of them by a stronger ligand such as dimethylglyoxime (Hdmg).[251,252] Thus, addition of an equimolar amount of Hdmg to a solution of $Ni(L^{60})_2$ immediately caused the formation of a short-lived quintuplet with $g = 2.006$ and $a_N = 0.70$ mT, which is typical for the EPR spectra of nitronyl nitroxides. This was indicative of strong delocalization of spin density in the $^{-}$O–$^{\bullet}$N–Ni(II)–N=O fragment through the metal ion and the ensuing equivalence of N–O groups. In the series of bischelates under study, $Ni(L^{60})_2$, proved to be the most reactive compound, which reacted with ethanol, benzene, toluene, and trichlormethane.[252] In this reaction, L^{60} was converted to 3,3,4,4-tetramethyl-1,2-diazetine-1,2-dioxide. Curiously, despite complete dehydrogenation of the hydroxylamine groups of the ligand in $Ni(L^{60})_2$, the coordination unit had square geometry. The Ni–N distances in it are shorter than in classical $Ni(dmg)_2$, having intramolecular hydrogen bonds. In view of the high reactivity of $Ni(L^{60})_2$, it was a success that it was grown as single crystals. $Ni(L^{59})_2$ crystals ($R = CH_3$) were obtained unexpectedly, as a result of the destruction of the nickel(II) complex with tridentate polynitrogen aminooxime.[250] The reactivity of the compound correlated with the donor properties of solvent molecules. The higher the donor properties, the faster the destruction of $Ni(L^{59})_2$.[250]

13.6.5 Hidden exchange interactions

Progress in the chemistry of stable nitroxides has permitted the design of molecular magnets from increasingly complex organic paramagnet molecules saturated with functional groups and enabling the self-assembly of complex structures in reactions with metal ions. The exchange coupling energies (J_i) of different clusters may be close in magnitude, which hinders the unambiguous assignment of exchange parameters to the structure of certain exchange clusters. Moreover, if the values of J_i of a series of exchange clusters have opposite signs, their mutual compensation contributions smoothes the functional features on the curve of the temperature dependence χ or χT, or μ_{eff}, which hinders the extraction of exchange parameters from magnetochemical experiments. A typical example of this situation is the magnetic properties of the chain polymer heterospin complex $[Cu(hfac)_2]_7(L^{61})_2$ (Scheme 13.28). Its experimental μ_{eff}(T) curve has few distinguishing features. The μ_{eff} decreases from 6.15 to 5.68μ_B when the temperature changes from 2 to 14 K; then it slightly increases to 5.84μ_B at ~ 50 K and changes little if at all thereafter (from 5.84 to 5.88μ_B) at 300 K.[253] Based on this low-informative dependence μ_{eff}(T), it is impossible to suggest a justified scheme of exchange interactions in metal-nitroxide chains. A quantum chemical study,

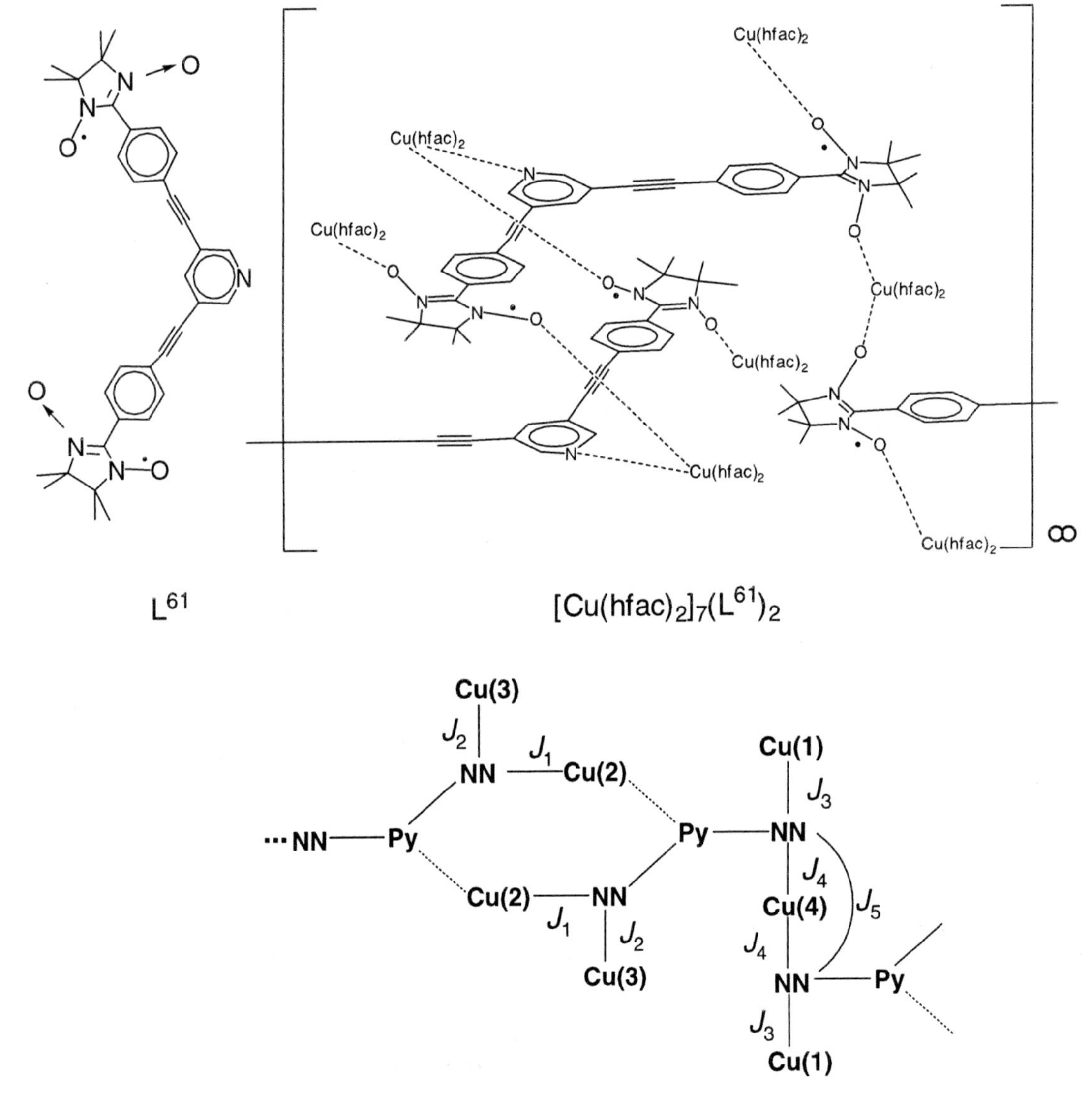

Scheme 13.28

however, revealed an extremely complex set of J_i that defined the exchange parameters of this heterospin structure.[254] It was found that the above-mentioned compensation terms formed so sophisticated combinations that they remained absolutely hidden without a quantum chemical study. In accordance with the scheme on Scheme 13.28, only significant values of J_i are given: $J_1 = 55$ and $88\,cm^{-1}$, $J_2 = -290$ and $-370\,cm^{-1}$, $J_3 = 40\,cm^{-1}$, $J_4 = 111\,cm^{-1}$, and $J_5 = -82\,cm^{-1}$. It can be seen that some of these are comparable in magnitude but opposite in sign. Therefore, quantum chemical studies of exchange interactions in complex multispin systems can be useful for understanding and reconstructing the $\mu_{eff}(T)$ or $\chi T(T)$ experimental dependence.

13.6.6 Contrast agents

In magnetic resonance tomography (MRT) – a tool allowing the rapid, non-invasive study of tissues of living organisms and effective visualization of pathological processes[255] – visualization is based on the difference between the rates of spin–lattice (T_1) and spin–spin (T_2) proton relaxations in biological tissues. The different pulse sequences used in MRT afford images, on which the contrast depends on the relaxation times T_1 or T_2.[256] However, it is not always possible to reach the required level of visualization even with modern technology. To amplify the signal from a pathological area, it is possible to use special compounds capable of penetrating in this area and enhancing contrast on T_1- or T_2-weighted images (T1-VI and T2-VI).[257] These are mainly gadolinium(III) salts with polydentate ligands, such as deprotonated diethylenetriaminepentaacetic acid (Magnevist, Omniscan)[258], or sometimes manganese or iron compounds (Teslascan, Abdoscan). However, despite the high contrast of images obtained when using these samples, their use can incur certain risks. That is why the paramagnetic metal ion that accelerates spin relaxation is generally introduced in the form of a coordination compound with a highly dentate organic ligand. These ligands should annihilate the charge on the metal ion and, additionally, hold the ion tightly in the complex, because the human organism has a very low tolerance to free transition metal ions.[259] The thermodynamic stability of the complexes can be increased only due to the chelate effect of polydentate ligands because after complexation for rare earth elements and high spin Mn^{2+} and Fe^{3+} ions, the stabilization energy with a crystal field is absent.[260–266] However, the natural limitation on the number of coordination sites restricts the theromodynamic stability of these complexes. In addition, polydentate ligands of the type of diethylenetriaminepentaacetic acid specially synthesized for these purposes are not metabolites, but alien objects for living organisms. This fully applies to various complexes of rare earth elements with nitroxides[267–269] (Scheme 13.29).

An alternative approach to the safety of contrast agents is the use of pure organic paramagnets based on non-toxic stable nitroxides. In this situation, a transition from metal-nitroxide systems to free nitroxides is

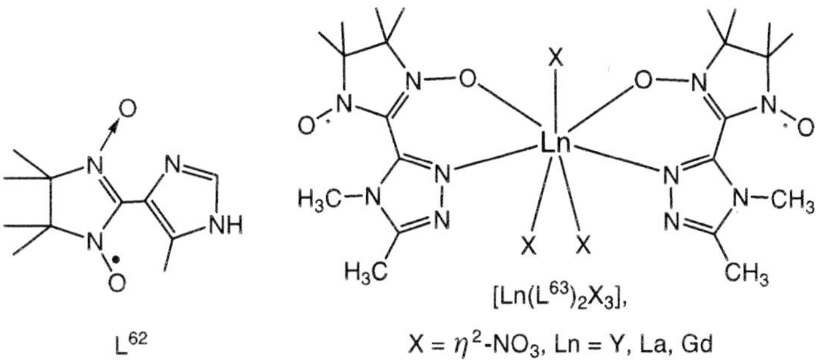

Scheme 13.29

quite justified. Note that many nitroxides were specially synthesized and checked for use as contrast agents for MRT.[270,271] However, spin labeled imidazoles bearing no charge (L^{62}, Scheme 13.29) and possessing anomalously high solubility in water and low toxicity for living organisms were found only recently.[272,273] They cannot act as effective contrast agents. It is not excluded that unification of paramagnetic metal ions with low toxic nitroxides related to L^{62} and possessing high solubility in water in a multispin complex will afford safe contrast agents capable of giving high quality magnetic resonance images in future.

13.7 Conclusions

An analysis of the accumulated material shows that synthetic strategies and major techniques were developed for the design of high-dimensional multispin systems based on coordination compounds of transition metals with nitroxides, exhibiting a tendency toward cooperative magnetic ordering. While 25 years ago approaches to the design of different classes of molecular ferromagnets were still in their infancy, today there is rich array of experimental results in this field. This fully relates to molecular magnets based on metal-nitroxide systems, many of which can now be obtained in significant amounts in the form of stable (under the normal conditions) large transparent single crystals. These compounds form a new class of light and transparent magnetoactive materials. They can be obtained under mild conditions, for example, by crystallization from solution under the normal conditions, while classic magnetic materials are generally obtained by high temperature energy consuming processes. Prospects are now being discussed for using molecular magnets based on metal-nitroxide complexes in various magnetic and photomagnetic devices, data recording and magnetic visualization systems, and spintronic devices for molecular computing. It is certainly difficult to foresee all future applications of molecular magnets. However, there is every reason to believe that the above-mentioned unique combination of physical characteristics, which is responsible for the specific properties of molecular magnets but not inherent in classic magnetic materials, will find use in the near future.

References

1. S. S. Eaton and G. R. Eaton, *Coord. Chem. Rev.*, **26**, 207–262 (1978).
2. R. S. Drago, *Coord. Chem. Rev.*, **32**, 97–110 (1980).
3. Yu. Zolotov, O. M. Petrukhin, V. Yu. Nagy, and L. B. Volodarsky, *Anal. Chim. Acta.*, **115**, 1–23 (1980).
4. S. V. Larionov, in *Imidazoline nitroxides. Synthesis, properties and applications*, (ed. L. B. Volodarsky), CRC Press, Inc., Boca Raton, Vol. 2, 81–113 (1988).
5. S. S. Eaton and G. R. Eaton, *Coord. Chem. Rev.*, **83**, 29–72 (1988).
6. A. Caneschi., D. Gatteschi, and P. Rey, *Prog. Inorg. Chem.*, **39**, 331–429 (1991).
7. O. Kahn, *Molecular magnetism*, VCH, New York, 1993.
8. L. B. Volodarsky, V. A. Reznikov, and V. I. Ovcharenko, *Synthetic chemistry of stable nitroxides*, CRC Press, Inc., Boca Raton, 1994.
9. G. A. Abakumov, *Journal of VCO Mendeleeva*, **24**, 156–160 (1979).
10. V. I. Ovcharenko and R. Z. Sagdeev, *Russian Chemical Reviews*, **68** (5), 345–363 (1999).
11. H. Iwamura, K. Inoue, and T. Hayamizu, *Pure Appl. Chem.*, **68**, 243–252 (1996).
12. V. I. Ovcharenko, K. Yu. Maryunina, S. V. Fokin, *et al.*, *Rus. Chem. Bull.*, 53, 2406–2427 (2004).
13. D. Luneau and P. Rey, *Coord. Chem. Rev.*, **249**, 2591–2611 (2005).
14. D. Luneau, A. Borta, Y. Chumakov, *et al.*, *Inorg. Chim. Acta*, **361**, 3669–3676 (2008).
15. A. Caneschi, D. Gatteschi, R. Sessoli, and P. Rey, *Acc. Chem. Res.* 22, 392–398 (1989).
16. H. Iwamura and K. Inoue, in *Magnetism: Molecules to Materials II. Molecule-Based Materials*, (eds J. S. Miller and M. Drillon), Wiley–VCH Verlag GmbH & Co. KGaA, Weinheim, 2001, pp. 61–108.

17. H. Oshio and T. Ito, *Coord. Chem. Rev.*, **198**, 329–346 (2000).
18. L. Ouahab, *Coord. Chem. Rev.*, **178-180**, 1501–1531 (1998).
19. O. Sato, J. Tao and Y.-Z. Zhang, *Angew. Chem., Int. Ed.*, **46**, 2152–2187 (2007).
20. Cambridge Structural Database, Version 5.30, November 2008 (Update February 2009).
21. Y. Y. Lim and R. S. Drago, *Inorg. Chem.*, **11**, 1334–1338 (1972).
22. P. M. Boymel, G. R. Eaton, and S. S. Eaton, *Inorg. Chem.*, **19**, 727–735 (1980).
23. C. Bilgrien, R. S. Drago, J. R. Stahlbush, and O. C. Kuechler, *Inorg. Chem.*, **24**, 4268–4272 (1985).
24. P. M. Boymel, G. A. Braden, G. R. Eaton, and S. S. Eaton, *Inorg.Chem.*, **19**, 735–739. (1980).
25. R. S. Drago, O. C. Kuechler, and M. Kroeger, *Inorg. Chem.*, **18**, 2337–2342 (1979).
26. W. Beck, K. Schmidtner, and H. J. Keller, *Chem. Ber.*, **100**, 503–511 (1967).
27. D. Jahr, K. H. Rebhan, K. E. Schwarzhans, and J. Wiedemann, *Z. Naturforsch.*, **28b**, 55–62 (1973).
28. W. Beck and K. Schmidtner, *Chem. Ber.*, **100**, 3363–3367 (1967).
29. M. Okunaka, G. Matsubayashi, and T. Tanaka, *Bull. Chem. Soc. Jpn.*, **48**, 1826–1829 (1975).
30. M. Okunaka, G. Matsubayashi, and T. Tanaka, *Bull. Soc. Chem. Jpn.*, **50**, 907–909 (1977).
31. M. H. Dickman and R. J. Doedens, *Inorg. Chem.*, **21**, 682–684 (1982).
32. M.-M. Rohmer, A. Grand, and M. Benard, *J. Am. Chem. Soc.*, **112**, 2875–2881 (1990).
33. C. M. Mikulski, J. S. Skryantz, and N. M. Karayannis, *Inorg. Nucl. Chem. Let.*, **11**, 259–263 (1975).
34. C. M. Mikulski, L. S. Gelfand, L. L. Pytlewski, *et al.*, *Trans. Met. Chem.*, **3**, 276–282 (1978).
35. C. M. Paleos, N. M. Karayannis, and M. M. Labes, *J. Chem. Soc., Chem. Commun.*, 195–196 (1970).
36. F. A. Cotton and T. R. Felthouse, *Inorg.Chem.*, **21**, 2667–2675 (1982).
37. R. M. Richman, O. C. Kuechler, S. R. Tanner, and R. S. Drago, *J. Amer. Chem. Soc.*, **99**, 1055–1058 (1977).
38. L. C. Porter, M. H. Dicbxian, and R. J. Doedens, *Inorg. Chem.*, **22**, 1962–1964 (1983).
39. L. C. Porter, M. H. Dickman, and R. J. Doedens, *Inorg.Chem.*, **25**, 678–684 (1986).
40. L. C. Porter and R. J. Doedens, *Inorg.Chem.*, **24**, 1006–1010 (1985).
41. T.-Yu. Dong, D. N. Hendrickson, O. R. Felthouse, and H.-S. Shieh, *J. Am. Chem. Soc.*, **106**, 5373–5375 (1984).
42. T. E. Felthouse, T.-Yu. Dong, D. N. Hendrickson, *et al.*, *J. Am. Chem. Soc.*, **108**, 8201–8214 (1986).
43. R. A. Zelonka and M. C. Baird, *J. Am. Chem. Soc.*, **93**, 6066–6070 (1971).
44. R. S. Drago, O. C. Kuechler, and M. Kroeger, *Inorg. Chem.*, **18**, 2337–2342 (1979).
45. L. C. Porter and R. J. Doedens, *Acta Crystallogr.*, **C41**, 838–840 (1985).
46. M. H. Dickman and R. J. Doedens, *Inorg. Chem.*, **20**, 2677–2681 (1981).
47. M. H. Dickman, L. C. Porter, and R. J. Doedens, *Inorg. Chem.*, **25**, 2595–2599 (1986).
48. L. C. Porter, M. H. Dickman, and R. J. Doedens, *Inorg.Chem.*, **27**, 1548–1552 (1988).
49. D. O. Matsunaga, D. T. McCall, M. D. Carducci, and R. J. Doedens, *Inorg. Chem.*, **29**, 1655–1659.
50. A. B. Burdukov, N. V. Pervukhina, and V. I. Ovcharenko, *Zh. Strukt. Khim. (Russ.) J. Struct. Chem.*, **36** (5), 900–907 (1995).
51. I. D. Anderson and T. C. Kuechler, *Inorg. Chem.*, **19**, 1417–1422 (1980).
52. R. N. Musin, P. V. Schastnev, and S. A. Malinovskaya, *Inorg. Chem.*, **31**, 4118–4121 (1992).
53. F. Lanfranc de Panthou., D. Luneau., R. Musin, *et al.*, *Inorg. Chem.*, **35**, 3484–3491 (1996).
54. R. N. Musin, I. V. Ovcharenko, L. Ohrström, and P. Rey, *J. Struct. Chem.*, **38**, 703–710 (1997).
55. R. N. Musin, I. V. Ovcharenko, L. Ohrström, and P. Rey, *J. Struct. Chem.*, **38**, 711–715 (1997).
56. I. V. Ovcharenko, Yu.G. Shvedenkov, R. N. Musin, and V. N. Ikorskii, *J. Struct. Chem.*, **40**, 29–34 (1999).
57. M. Dickman and R. J. Doedens, *Inorg.Chem.*, **22**, 1591–1594 (1983).
58. F. A. Cotton and T. R. Felthouse, *Inorg.Chem.*, **21**, 2667–2675 (1982).
59. R. M. Richman, O. C. Kuechler, S. R. Tanner, and R. S. Drago, *J. Amer. Chem. Soc.*, **99**, 1055–1058 (1977).
60. T. Ise, T. Ishida, and T. Nogami, *Bull. Chem. Soc. Jpn.*, **75**, 2463–2468 (2002).
61. F. Lanfranc de Panthou, E. Belorizky, R. Calemczuk, *et al.*, *J. Am. Chem. Soc.*, **117**, 11247–11253 (1995).
62. A. Caneschi, D. Gatteschi, R. Sessoli, *et al.*, *Inorg. Chem.*, **27**, 2390–2392 (1988).
63. V. I. Ovcharenko, V. N. Ikorskii, N. V. Podberezskaya, *et al.*, *Zh. Neorg. Khim.*, **32**, 1403–1406 (1987).
64. D. Luneau, P. Rey, J. Laugier, *et al.*, *J. Am. Chem. Soc.*, **113**, 1245–1251 (1991).
65. Z.-H. Jiang, Q. Yi, D.-Z. Liao, *et al.*, *Transition Met. Chem.*, **20**, 136–138 (1995).
66. C. Rajadurai, S. Ostrovsky, K. Falk, *et al.*, *Inorg. Chim. Acta*, **357**, 581–587 (2004).

67. K. Jiang, X. Wang, L. Wang, and B. Zhao, *J. Coord. Chem.*, **61**, 410–418 (2008).
68. N. V. Pervukhina and N. V. Podberezskaya, *J. Struct. Chem.*, **35**, 309–316 (1994).
69. A. B. Burdukov, V. I. Ovcharenko, N. V. Pervukhina, and V. N. Ikorskii, *Polyhedron*, **12**, 1705–1710 (1993).
70. E. V. Tretyakov, S. V. Fokin, G. V. Romanenko, and V. I. Ovcharenko, *Polyhedron*, **22**, 1965–1972 (2003).
71. M. R. Haneline and F. P. Gabbai, *Inorg. Chem.*, **44**, 6248–6255 (2005).
72. K. Okada, S. Beppu, K. Tanaka, *et al.*, *Chem. Commun.*, 2485–2487 (2007).
73. M. Tanaka, K. Matsuda, T. Itoh, and H. Iwamura, *Angew. Chem., Int. Ed.*, **37**, 810–812 (1998).
74. A. Caneschi, P. Chiesi, L. David, *et al.*, *Inorg. Chem.*, **32**, 1445–1451 (1993).
75. E. Tretyakov, S. Fokin, G. Romanenko, *et al.*, *Inorg. Chem.*, **45**, 3671–3678 (2006).
76. S. V. Fokin, G. V. Romanenko, M. Baumgarten, and V. I. Ovcharenko, *J. Struct. Chem.*, **44**, 864–869 (2003).
77. R. Weiss, N. Kraut, and F. Hampel, *J. Organomet. Chem.*, **617**, 473–482 (2001).
78. A. Caneschi, D. Gatteschi, J. Laugier, *et al.*, *J. Am. Chem. Soc.*, **110**, 2795–2799 (1988).
79. S. Fokin, V. Ovcharenko, G. Romanenko, and V. Ikorskii, *Inorg. Chem.*, **43**, 969–977 (2004).
80. K. E. Vostrikova, D. Luneau, W. Wernsdorfer, *et al.*, *J. Am. Chem. Soc.*, **122**, 718–719 (2000).
81. H. Higashikawa, K. Inoue, K. Yu. Maryunina, *et al.*, *J. Struct. Chem.*, **50**, 1155–1158 (2009).
82. K. E. Vostrikova, *Coord. Chem. Rev.*, **252**, 1409–1419 (2008).
83. C. Benelli, D. Gatteschi, C. Zanchini, *et al.*, *Inorg.Chem.*, **25**, 4242–4244 (1986).
84. J. C. Espie, J. Laugier, R. Ramasseul, *et al.*, *Nouv. J. Chim.*, **4** (4), 205–207 (1980).
85. R. Briere, A. M. Giroud, A. Rassat, and P. Rey, *Bull. Soc. Chim. Fr.*, **3-4**, 147–150 (1980).
86. A. Grand, P. Rey, and R. Subra, *Inorg.Chem.*, **22**, 391–394 (1983).
87. V. I. Ovcharenko, G. V. Romanenko, V. N. Ikorskii, *et al.*, *Inorg. Chem.*, **33**, 3370–3381 (1994).
88. G. V. Romanenko, N. V. Podberezskaya, and V. I. Ovcharenko, *Zh. Neorg. Khimii*, **37**, 1525–1527 (1992) [*Russ. J. Inorg. Chem.*, **37**, 775–778 (1992)].
89. A. Caneschi, D. Gatteschi, P. Rey, and R. Sessoli, *Inorg.Chem.*, **27**, 1756–1761 (1988).
90. A. Caneschi, D. Gatteschi, J. P. Renard, *et al.*, *Inorg.Chem.*, **28**, 1976–1980 (1989).
91. A. Caneschi, D. Gatteschi, J. P. Renard, *et al.*, *Inorg.Chem.*, **28**, 3314–3319 (1989).
92. C. Benelli, A. Caneschi, D. Gatteschi, and R. Sessoli, *Inorg.Chem.*, **32**, 4797–4801 (1993).
93. E. F. Ullman, J. H. Osiecki, D. G. B. Boocock, and R. Darcy, *J. Am. Chem. Soc.*, **94**, 7094–7059 (1972).
94. E. F. Ullman, L. Call, and J. H. Osiecki, *J. Org. Chem.*, **35**, 3623–3631 (1970).
95. P. Petrov, S. Fokin, G. Romanenko, *et al.*, *Mendeleev Communications*, 179–181 (2001).
96. M. D. Carducci and R. J. Doedens, *Inorg.Chem.*, **28**, 2492–2494 (1989).
97. A. Caneschi, D. Gatteschi, J. Laugier, *et al.*, *Inorg.Chem.*, **28**, 1969–1975 (1989).
98. A. Caneschi, D. Gatteschi, N. Lalioti, *et al.*, *J. Chem. Soc., Dalton Trans.*, 3907–3912 (2000).
99. A. Caneschi, D. Gatteschi, N. Lalioti, *et al.*, *Angew. Chem., Int. Ed.*, **40**, 1760–1763 (2001).
100. N. Ishii, T. Ishida, and T. Nogami, *Inorg. Chem.*, **45**, 3837–3839 (2006).
101. L. Bogany, C. Sangregorio, R. Sessoli, and D. Gatteschi, *Angew. Chem., Int. Ed.*, **44**, 5817–5821 (2005).
102. R. Clérac, H. Miyasaka, M. Yamashita, and C. Coulon, *J. Am. Chem. Soc.*, **124**, 12837–12844 (2002).
103. L. M. Toma, R. Lescouëzec, F. Lloret, *et al.*, *Chem. Commun.* 180–1851 (2003).
104. R. Lescouëzec, J. Vassermann, C. Ruiz-Pérez, *et al.*, *Angew. Chem., Int. Ed.*, **42**, 1483–1486 (2003).
105. T.-F. Liu, D. Fu, S. Gao, *et al.*, *J. Am. Chem. Soc.*, **125**, 13976–13977 (2003).
106. R. Sessoli, H.-L. Tsai, A. R. Schake, *et al.*, *J. Am. Chem. Soc.*, **115**, 1804–1816 (1993).
107. D. Gatteschi and R. Sessoli, *Angew. Chem., Int. Ed.*, **42**, 268–297 (2003).
108. R. Sessoli, D. Gatteschi, A. Caneschi, and M. A. Novak, *Nature*, **365**, 141–143 (1993).
109. A. Caneschi, D. Gatteschi, J.-P. Renard, *et al.*, *J. Am. Chem. Soc.*, **111**, 785–786 (1989).
110. A. Caneschi, D. Gatteschi, M. C. Melandry, *et al.*, *Inorg. Chem.*, **29**, 4228–4243 (1990).
111. K. Fegy, D. Luneau, T. Ohm, *et al.*, *Angew. Chem., Int. Ed.*, **37**, 1270–1273 (1998).
112. K. Fegy, C. Lescop, D. Luneau, and P. Rey, *Mol. Cryst Liq. Cryst. Sci. Technol.*, Sect. A: *Mol. Cryst Liq. Cryst.*, **334**, 521–532 (1999).
113. K. Fegy, N. Sanz, D. Luneau, *et al.*, *Inorg Chem.*, **37**, 4518–4523 (1998).
114. K. Fegy, D. Luneau, E. Belorizky, *et al.*, *Inorg Chem.*, **37**, 4524–4532 (1998).
115. C. Lescop, D. Luneau, E. Belorizky, *et al.*, *Inorg. Chem.*, **38**, 5472–5473 (1999).

116. C. Lescop, D. Luneau, G. Bussière, *et al.*, *Inorg. Chem.*, **39**, 3740–3741 (2000).
117. C. Lescop, E. Belorizky, D. Luneau, and P. Rey, *Inorg. Chem.*, **41**, 3375–3384 (2002).
118. C. Benelli and D. Gattesch, *Chem. Rev.*, **102**, 2369–2387 (2002).
119. W. J. Evans, R. Anwander, and J. W. Ziller, *Organometallics*, **14**, 1107 (1995).
120. V. I. Ovcharenko, G. V. Romanenko, V. N. Ikorskii, *et al.*, *Inorg. Chem.*, **33**, 3370–3381 (1994).
121. V. I. Ovcharenko, K. E. Vostrikova, G. V. Romanenko, *et al.*, *Doklady Akademii nauk SSSR*, **306**, 115–118 (1989).
122. V. I. Ovcharenko, K. E. Vostrikova, V. N. Ikorskii, *et al.*, *Doklady Akademii nauk SSSR*, **306**, 660–662 (1989).
123. V. N. Ikorskii, V. I. Ovcharenko, Y. G. Shvedenkov, *et al.*, *Inorg. Chem.*, **37**, 4360–4367 (1998).
124. V. I. Ovcharenko, V. N. Ikorskii, G. V. Romanenko, *et al.*, *Synth. Metals*, **85**, 1639–1642 (1997).
125. V. I. Ovcharenko, K. E. Vostrikova, A. V. Podoplelov, *et al.*, *Polyhedron*, **13**, 2781–2792, (1994).
126. E. Fursova, Y. Shvedenkov, G. Romanenko, *et al.*, *Polyhedron*, **20**, 1229–1234 (2001).
127. M. Bonnet, J. Laugier, V. I. Ovcharenko, *et al.*, *Mol. Cryst. Liq. Cryst.*, **305**, 401–414 (1997).
128. Y. Shvedenkov, V. Ikorskii, G. Romanenko, *et al.*, *Mol. Cryst. Liq. Cryst.*, **334**, 405–414 (1999).
129. V. N. Ikorskii, G. V. Romanenko, M. K. Sygurova, *et al.*, *J. Struct. Chem.*, **35**, 492–504 (1994).
130. A. B. Burdukov, V. I. Ovcharenko, D. A. Guschin, *et al.*, in *Coordination Chemistry at the Turn of the Century* (eds G. Ondrejovich and A. Sirota), Slovak Technical University Press, Bratislava, **4**, pp. 277–282, 1999.
131. A. B. Burdukov, D. A. Guschin, N. V. Pervukhina, *et al.*, *Crystal Engineering*, **2** (4), 265–279 (1999).
132. Y. Shvedenkov, V. Ikorskii, D. Guschini, *et al.*, *Polyhedron*, **20**, 1207–1213 (2001).
133. P. A. Petrov, S. V. Fokin, G. V. Romanenko, *et al.*, *Mendeleev Commun.*, 179–181 (2001).
134. Y. Pontillion, V. I. Ovcharenko, E. Ressouche, *et al.*, *Physica B*, **234-236**, 785–787 (1997).
135. A. B. Burdukov, V. I. Ovcharenko, D. A. Guschin, *et al.*, *Mol. Cryst. Liq. Cryst.*, **334**, 395–404 (1999).
136. P. A. Petrov, G. V. Romanenko, Y. G. Shvedenkov, *et al.*, *Russ.Chem. Bull., Int. Ed.*, **53** (1), 99–108 (2004).
137. V. A. Reznikov, G. I. Roshchupkina, D. G. Mazhukin, *et al.*, *Eur. J. Org. Chem.*, 749–765 (2004).
138. O. V. Koreneva, G. V. Romanenko, Y. G. Shvedenkov, *et al.*, *Polyhedron*, **22**, 2487–2497 (2003).
139. O. V. Koreneva, G. V. Romanenko, Y. G. Shvedenkov, *et al.*, *J. Phys. IV France*, **114**, 627–628 (2004).
140. A. Caneschi, D. Gatteschi, R. Sessoli, and P. Rey, *Mol. Cryst. Liq. Cryst.*, **176**, 329–336 (1989).
141. K. Inoue, T. Hayamizu, H. Iwamura, *et al.*, *J. Am. Chem. Soc.*, **118**, 1803–1804 (1996).
142. E. Ressouche, J. X. Boucherle, B. Gillon, *et al.*, *J. Am. Chem. Soc.*, **115**, 3610–3617 (1993).
143. E. Ressouche, A. Zheludev, J. X. Boucherle, *et al.*, *Mol. Cryst. Liq. Cryst.*, **232**, 13–26 (1993).
144. H. O. Stumpf, L. Ouahab, Y. Pei, *et al.*, *Science*, **261**, 447–449 (1993).
145. H. O. Stumpf, L. Ouahab, Y. Pei, *et al.*, *J. Am. Chem. Soc.*, **116**, 3866-3874 (1994).
146. M. G. F. Vaz, H. O. Stumpf, L. Ouahab, *et al.*, The VIth Intarnational Conference on Molecule-Based Magnets, 12–17 September 1988, Seignosse, France, p. 180, 1998.
147. K. Inoue and H. Iwamura, *Angew. Chem., Int. Ed.*, **34**, 927–928 (1995).
148. Y. Nakano, T. Yagyu, T. Hirayama, *et al.*, *Polyhedron*, **24**, 2141–2147 (2005).
149. Y. Ishimaru, M. Kitano, H. Kumada, *et al.*, *Inorg. Chem.*, **37**, 2273–2280 (1998).
150. T. Itoh, K. Matsuda, H. Iwamura, and K. Hori, *J. Am. Chem. Soc.*, **122**, 2567–2576 (2000).
151. K. Inoue, T. Hayamizu, and H. Iwamura, *Chem. Lett.*, 745–746 (1995).
152. T. Mitsumori, K. Inoue, N. Koga, and H. Iwamura, *J. Am. Chem. Soc.*, **117**, 2467–2478 (1995).
153. K. Inoue and H. Iwamura, *J. Chem. Soc., Chem. Commun.*, 2273–2274 (1994).
154. K. Inoue and H. Iwamura, *J. Am. Chem. Soc.*, **116**, 3173–3174 (1994).
155. P. Rabu, M. Drillon, H. Iwamura, *et al.*, *Eur. J. Inorg. Chem.*, 211–216 (2000).
156. K. Inoue and H. Iwamura, *Mat. Res. Soc. Symp. Proc.*, **413**, 313–320 (1996).
157. K. Inoue, F. Iwahori, A. S. Markosyan, and H. Iwamura, *Coord. Chem. Rev.*, **198**, 219–229 (2000).
158. K. Inoue and H. Iwamura, *Synthetic Met.*, **71**, 1793–1794 (1995).
159. K. Inoue, and H. Iwamura, *Adv. Mater.*, **8** (1), 73–76 (1996).
160. K. Inoue, T. Hayamizu, and H. Iwamura, *Mol. Cryst. Liq. Cryst.*, **273**, 67–80 (1995).
161. H. Iwamura, K. Inoue, and N. Koga, *J. Synth. Org. Chem., Japan*, **55** (5), 417–426 (1997).
162. J. S. Miller, *Dalton Trans.*, 2742–2749 (2006).
163. J. S. Miller, A. J. Epstein, and W. M. Reiff, *Acc. Chem. Res.*, **21**, 114–120 (1988).

164. J. S. Miller and J. L. Manson, *Acc. Chem. Res.*, **34**, 563–570 (2001).
165. L. Ouahab and T. Enoki, *Eur. J. Inorg. Chem.*, 933–941 (2004).
166. M. Ohba and H. Okawa, *Coord. Chem. Rev.*, **198**, 313–328 (2000).
167. R. Jain, K. Kabir, J. B. Gilroy, *et al.*, *Nature*, **445**, 291–294 (2007).
168. M. Verdaguer, *Science*, **272**, 698–699 (1996).
169. O. Sato, *J. Photochem. Photobiol. C: Photochem. Rev.*, **5**, 203–223 (2004).
170. V. A. Ivanov, T. G. Aminov, V. M. Novotortsev, and V. T. Kalinnikov, *Rus. Chem. Bull.*, 2357–2405 (2004).
171. C. Felser, G. H. Fecher, and B. Balke, *Angew. Chem., Int. Ed.*, **46**, 668–699 (2007).
172. E. Yu. Fursova and V. I. Ovcharenko, *Ross. Khim. Zhurn. (Zhurn. Ross. Khim. ob-va im. D. I. Mendeleeva)*, **53**(1), 23–32 (2009).
173. G. Christou, *Acc. Chem. Res.*, **22**, 328–335 (1989).
174. S. Parsons and R. E. P. Winpenny, *Acc. Chem. Res.*, **30**, 89–95 (1997).
175. G. Aromí, A. S. Batsanov, P. Christian, *et al.*, *Chem. Eur. J.*, **9**, 5142–5161 (2003).
176. E. J. L. McInnes, S. Piligkos, G. A. Timco, and R. E. P. Winpenny, *Coord. Chem. Rev.*, **249**, 2577–2590 (2005).
177. R. Sessoli, H. L. Tsai, A. R. Schake, *et al.*, *J. Am. Chem. Soc.*, **115**, 1804–1816 (1993).
178. R. Sessoli, D. Gatteschi, A. Caneschi, and M. A. Novak, *Nature*, **365**, 141–143 (1993).
179. G. Christou, *Polyhedron*, **24**, 2065–2075 (2005).
180. A. J. Tasiopoulos, W. Wernsdorfer, B. Moulton, *et al.*, *J. Am. Chem. Soc.*, **125**, 15274–15275 (2003).
181. A. J. Tasiopoulos, A. Vinslava, W. Wernsdorfer, *et al.*, *Angew. Chem., Int. Ed.*, **43**, 2117–2121 (2004).
182. M. Soler, W. Wernsdorfer, K. Folting, *et al.*, *J. Am. Chem. Soc.*, **126**, 2156–2165 (2004).
183. L. M. C. Beltran and J. R. Long, *Acc. Chem. Res.*, **38**, 325–334 (2005).
184. N. E. Chakov, W. Wernsdorfer, K. A. Abboud, *et al.*, *Dalton Trans.*, 2243–2248 (2003).
185. T. C. Stamatatos, K. A. Abboud, W. Wernsdorfer, and G. Christou, *Angew. Chem., Int. Ed.*, **46**, 884–888 (2007).
186. H. Miyasaka and M. Yamashita, *Dalton Trans.*, 399–406 (2007).
187. S. Voss, M. Burgert, M. Fonin, *et al.*, *Dalton Trans.*, **2008**, 499–505 (2008).
188. J. Kortus and A. V. Postnikov, Molecular Nanomagnets, in *Handbook of Theoretical and Computational Nanotechnology. Magnetic Nanostructures and Nanooptics* (eds M. Rieth and W. Schommers), American Scientific Publishers, pp. 503–562, 2006.
189. S. T. Bramwell, *Annu. Rep. Prog. Chem., Sect. A*, **98**, 493–504 (2002).
190. V. Ovcharenko, E. Fursova, G. Romanenko, and V. Ikorskii, *Inorg. Chem.*, **43**, 3332–3334 (2004).
191. K. Nakata, H. Miyasaka, F. Iwahori, *et al.*, *Polyhedron*, **24**, 2250–2256 (2005).
192. E. Fursova, V. Ovcharenko, K. Nosova, *et al.*, *Polyhedron*, **24**, 2084–2093 (2005).
193. V. Ovcharenko, E. Fursova, G. Romanenko, *et al.*, *Inorg. Chem.*, **45**, 5338–5350 (2006).
194. E. Y. Fursova, O. V. Kuznetsova, G. V. Romanenko, *et al.*, *J. Cluster Scien.*, **16**, 319–329 (2005).
195. E. Yu. Fursova, O. V. Kuznetsova, G. V. Romanenko, and V. I. Ovcharenko, *Russ. Chem. Bull.*, 1933–1945 (2006).
196. E. Fursova, O. Kuznetsova, V. Ovcharenko, *et al.*, *Polyhedron*, **26**, 2079–2088 (2007).
197. E. Y. Fursova, O. V. Kuznetsova, V. I. Ovcharenko, *et al.*, *Russ. Chem. Bull.*, 1805–1808 (2007).
198. E. Y. Fursova, O. V. Kuznetsova, V. I. Ovcharenko, *et al.*, *Russ. Chem. Bull.*, 1198–1204 (2008).
199. M. Murrie, S. Parsons, and R. E. P. Winpenny, *J. Chem. Soc., Dalton Trans.* **9**, 1423–1424 (1998).
200. A. L. Barra, D. Gatteschi, and R. Sessoli, *Phys. Rev. B: Condens. Matter.*, **56**, 8192–8198 (1997).
201. O. Kahn, *Acc. Chem. Res.*, **33**, 647–657 (2000).
202. Y. Pei, M. Verdaguer, O. Kahn, *et al.*, *J. Am. Chem. Soc.*, **108**, 7428–7430 (1986).
203. Y. Pei, M. Verdaguer, O. Kahn, *et al.*, *Inorg. Chem.*, **26**, 138–143 (1987).
204. O. Kahn, Y. Pei, M. Verdaguer, *et al.*, *J. Am. Chem. Soc.*, **110**, 782–789 (1988).
205. F. Lloret, K. Nakatani, Y. Journaux, *et al.*, *J. Chem. Soc., Chem. Commun.*, 642–643 (1988).
206. K. Nakatani, J. Y. Carriat, Y. Journaux, *et al.*, *J. Am. Chem. Soc.*, **111**, 5739–5748 (1989).
207. H. O. Stumpf, Y. Pei, O. Kahn, *et al.*, *J. Am. Chem. Soc.*, **115**, 6738–6745 (1993).
208. H. O. Stumpf, Y. Pei, L. Ouahab, *et al.*, *Inorg. Chem.*, **32**, 5687–5691 (1993).

209. V. Baron., B. Gillon, O. Kahn, *et al.*, *Mol. Cryst. Liq. Cryst.*, **233**, 247–256 (1993).
210. P. Gütlich, A. Hauser, and H. Spiering, *Angew. Chem., Int. Ed.*, **33**, 2024–2054 (1994).
211. P. Gütlich and H. A. Goodwin (Eds), *Spin Crossover in Transition Metal Compounds I*, Topics in Current Chemistry Series, Springer-Verlag, Berlin, Heidelberg, **233**, 2004.
212. P. Gütlich and H. A. Goodwin (Eds), *Spin Crossover in Transition Metal Compounds II*, Topics in Current Chemistry Series, Springer-Verlag, Berlin, Heidelberg, **234**, 2004.
213. P. Gütlich and H. A. Goodwin (Eds), *Spin Crossover in Transition Metal Compounds III*, Topics in Current Chemistry Series, Springer-Verlag, Berlin, Heidelberg, **235**, 2004.
214. F. Iwahory, K. Inoue, and H. Iwamura, *Mol. Cryst. Liq. Cryst.*, **334**, 533–538 (1999).
215. V. I. Ovcharenko, S. V. Fokin, G. V. Romanenko, *et al.*, *J. Struct. Chem.*, **43**, 153–167 (2002).
216. V. I. Ovcharenko, S. V. Fokin, G. V. Romanenko, *et al.*, *Mol. Phys.*, **100** (8), 1107–1115 (2002).
217. P. Rey and V. I. Ovcharenko, in *Magnetism: Molecules to Materials IV*, (eds J. S. Miller and M. Drillon, Wiley-VCH Verlag GmbH & Co. KGaA, Weinheim, p. 41–63, 2003.
218. C. Hirel, L. Li, P. Brough, *et al.*, *Inorg. Chem.*, **46**, 7545–7552 (2007).
219. K. Maryunina, S. Fokin, V. Ovcharenko, *et al.*, *Polyhedron*, **24** (16-17), 2094–2101 (2005).
220. V. I. Ovcharenko, G. V. Romanenko, K. Y. Maryunina, *et al.*, *Inorg. Chem.*, **47**, 9537–9552 (2008).
221. S. L. Veber, M. V. Fedin, A. I. Potapov, *et al.*, *J. Am. Chem. Soc.*, **130**, 2444–2445 (2008).
222. M. V. Fedin, S. L. Veber, I. A. Gromov, *et al.*, *J. Phys. Chem. A*, **110**, 2315–2317 (2006).
223. M. V. Fedin, S. L. Veber, I. A. Gromov, *et al.*, *J. Phys. Chem. A.*, **111**, 4449–4455 (2007).
224. M. V. Fedin, S. L. Veber, I. A. Gromov, *et al.*, *Inorg. Chem.* **26**, 11405–11415 (2007).
225. S. L. Veber, M. V. Fedin, K. Y. Maryunina, *et al.*, *Inorg. Chim. Acta*, **361**, 4148–4152 (2008).
226. V. A. Morozov, N. N. Lukzen, and V. I. Ovcharenko, *Russ. Chem. Bull., Russ. Ed.*, 849–852 (2008).
227. V. A. Morozov, N. N. Lukzen, and V. I. Ovcharenko, *J. Phys. Chem. B.*, **112**, 1890–1893 (2008).
228. M. Fedin, V. Ovcharenko, R. Sagdeev, *et al.*, *Angew. Chem., Int. Ed.*, **47**, 6897–6899 (2008).
229. S. V. Larionov, V. I. Ovcharenko, R. A. Sadykov, and L. B. Volodarskii, *Izv. Akad. Nauk SSSR, Ser. Khim.*, 1922–1925 (1975).
230. T. Yoshida and S. Kaizaki, *Inorg. Chem.*, **38**, 1054–1058 (1999).
231. J. Goldman, T. E. Petersen, K. Torssell, and J. Becher, *Tetrahedron*, **29**, 3833–3843 (1973).
232. D. Wilbur and R. Kreilick, *J. Chem. Phys.*, **52**, 1643–1646 (1970).
233. R. Z. Sagdeev, Y. N. Molin, R. A. Sadikov, *et al.*, *J. Magn. Reson.*, **9**, 13–26 (1973).
234. P. F. Richardson and R. W. Kreilick, *J. Am. Chem. Soc.*, **99**, 8183–8187 (1977).
235. P. F. Richardson, R. W. Kreilick, *J. Phys. Chem.*, **82**, 1149–1151 (1978).
236. T. Yoshida, T. Suzuki, K. Kanamori, and S. Kaizaki, *Inorg. Chem.*, **38**, 1059–1068 (1999).
237. R. Z. Sagdeev, V. K. Voronov, A. V. Podoplelov, *et al.*, *Russ. Chem. Bull.*, **50**, 2078–2086 (2001).
238. L. J. Berliner, *Spin labeling. Theory and Applications*, Academic Press, New York, 1976.
239. H. G. Aurich, Nitroxides, in *The Chemistry of Functional Groups* (ed. S. Patai), John Wiley & Sons Ltd, Chichester, UK, 1982.
240. J. F. W. Keana, *Chem. Rev.*, **78**, 37–64 (1978).
241. D. F. Bowman, T. Gillan, and K. U. Ingold, *J. Am. Chem. Soc.*, **93**, 6555–6561 (1971).
242. R. Briere and A. Rassat, *Tetrahedron*, **32**, 2891–2898 (1976).
243. V. I. Ovcharenko, F. Lanfranc de Pantou, V. A. Reznikov, *et al.*, *Inorg. Chem.*, **34**, 2263–2264 (1995).
244. F. A. Villamena, M. H. Dickman, and D. R. Crist, *Inorg. Chem.*, **37**, 1454–1457 (1998).
245. R. Ziessel, G. Ulrich, R. C. Lawson, and L. Echegoyen, *J. Mater. Chem.*, **9**, 1435–1448 (1999).
246. M. Kadirov, E. Tretyakov, Y. Budnikova, *et al.*, *J. Electroanal. Chem.*, **624**, 69–72 (2008).
247. W. Brackman and C. J. Graasbeek, *Recueil trav.chim.*, **85** (2), 257–267 (1966).
248. V. I. Ovcharenko, T. N. Sorokina, V. N. Ikorskii, *et al.*, *Bull. Acad. Scin. USSR, Div. Chem. Scien. (Engl. Transl.)*, **35**, 148–153 (1986).
249. A. B. Burdukov, V. I. Ovcharenko, V. N. Ikorskii, *et al.*, *Inorg. Chem.*, **30**, 972–976 (1991).
250. E. O. Schlemper and R. K. Murmann, *Inorg. Chem.*, **22**, 1077–1081 (1983).
251. V. N. Kirichenko, S. V. Larionov, I. A. Mikhilov, *et al.*, *Russ. J. Inorg. Chem.*, **29**, 1624–1626 (1984).
252. S. V. Fokin, G. V. Romanenko, and V. I. Ovcharenko, *Mendeleev Commun.*, 127–128 (2001).

253. C. Rajadurai, V. Enkelmann, V. Ikorskii, *et al.*, *Inorg. Chem.*, **45**, 9664–9669 (2006).
254. E. V. Gorelik, V. I. Ovcharenko, and M. Baumgarten, *Eur. J. Inorg. Chem.*, 2837–2846 (2008).
255. P. A. Rinck, *Magnetic Resonance in Medicine*, Blackwell Wissenschafts, 2001.
256. P. R. Moran, N. G. Kumar, N. Karstaedt, and S. C. Jackels, *Magn. Reson. Imaging*, **4**, 229–235 (1986).
257. P. C. Lauterbur, M. H. Mendonçsa-Dias, and A. M. Rudin, Augmentation of tissue proton spin-lattice relaxation rates by *in vivo* addition of paramagnetic ions, in *Frontiers of Biological Energetics* (eds P. O. Dutton, J. Leigh, and A. Scarpa), Academic Press, New York, 1978.
258. A. E. Merbach and É. Tóth, *The Chemistry of Contrast Agents in Medical Magnetic Resonance Imaging*, John Wiley & Sons Ltd, Chichester, UK, 2001.
259. P. Caravan, J. J. Ellison, T. J. McMurry, and R. B. Lauffer, *Chem. Rev.*, **99**, 2293–2352 (1999).
260. J. B. Livramento, É. Tóth, A. Sour, *et al.*, *Angew. Chem., Int. Ed.*, **44**, 1480–1484 (2005).
261. J. Xu, D. G. Churchill, M. Botta, and K. N. Raymond, *Inorg. Chem.*, **43**, 5492–5494 (2004).
262. V. C. Pierre, M. Botta, S. Aime, and K. N. Raymond, *J. Am. Chem. Soc.*, **128**, 9272–9273 (2006).
263. M. Polášek, M. Šedinová, J. Kotek, *et al.*, *Inorg. Chem.*, **48**, 455–465 (2009).
264. M. Polášek, J. Kotek, P. Hermann, *et al.*, *Inorg. Chem.*, **48**, 466–475 (2009).
265. L. Frullano, B. Tejerina, and T. J. Meade, *Inorg. Chem.*, **45**, 8489–8491 (2006).
266. E. Balogh, M. Mato-Iglesias, C. Platas-Iglesias, *et al.*, *Inorg. Chem.*, **45**, 8719–8728 (2006).
267. J.-P. Sutter, M. L. Kahn, S. Golhen, *et al.*, *Chem. Eur. J.*, **4**, 571–576 (1998).
268. M. L. Kahn, R. Ballou, P. Porcher, *et al.*, *Chem. Eur. J.*, **8**, 525-531 (2002).
269. M. L. Kahn, J.-P. Sutter, S. Golhen, *et al.*, *J. Am. Chem. Soc.*, **122**, 3413–3421 (2000).
270. G. Sosnovsky, N. U. M. Rao, S. W. Li, and H. M. Swartz, *J. Org. Chem.*, **54**, 3667–3674 (1989).
271. J. F. W. Keana, L. Lex, J. S. Mann, *et al.*, *Pure Appl. Chem.*, **62**, 201–205 (1990).
272. V. Ovcharenko, E. Fursova, T. Tolstikova, *et al.*, *Doklady Chemistry*, **404**, 171–173 (2005).
273. A. A. Savelov, D. A. Kokorin, E. Y. Fursova, and V. I. Ovcharenko, *Doklady Chemistry*, **416**, 241–243 (2007).

14

Rechargeable Batteries Using Robust but Redox Active Organic Radicals

Takeo Suga and Hiroyuki Nishide

Department of Applied Chemistry, Waseda University, Tokyo, Japan

14.1 Introduction

Robust organic radicals bearing an unpaired electron, such as 2,2,6,6-tetramethylpiperidinyl-*N*-oxy (TEMPO), galvinoxyl, and tris(pentachlorophenyl)methyl radical, have been examined and used as a *spin source*, where "robust" signifies enough stability to be isolated and to exist for appreciable lengths of time under ambient conditions.[1,2] For instance, nitroxide (N–O•) derivatives provide dynamic information of target biological molecules by spin labeling Electron Paramagnetic Resonance (EPR) spectroscopic techniques.[3] For many years, robust radicals have been incorporated into well defined π-conjugated polymers, coordination metal–organic and hydrogen bonding frameworks, with the goal of investigating intra/intermolecular high spin interactions and realizing new molecule-based magnets.[4] In these studies and applications, researchers have mainly paid attention to the chemical stability or non-reactivity of the organic radicals, achieved by steric protection and/or delocalization of the unpaired electron.[5] On the other hand, most organic radicals such as $\cdot CH_3$, and $\cdot CCl_3$ are short-lived, intrinsically "highly reactive". Curiously, some of the robust nitroxide radicals, such as TEMPO, are durable but reactive under specific reaction conditions, for example, in the presence of radical intermediates or propagating radical during the polymerization. Thus, nitroxides can be often applied as spin traps or capping agents for controlled radical polymerization (nitroxide-mediated radical polymerization).[6] TEMPO derivatives have also been studied as redox mediators/catalysts for the oxidation of alcohols to carbonyl compounds.[7] In this chapter, the focus is on "reactivity", especially the redox activity of robust organic radicals and their application as electrode active and charge storage materials in rechargeable batteries.[8]

The redox reactions of nitroxide radicals will be introduced firstly, followed by the mechanism and performance of organic radical batteries, and then molecular design and synthetic methods of a series of radical polymers. Our successful approaches toward a totally organic polymer-based battery will also be described.

14.2 Redox reaction of organic radicals

The rapid and reversible redox process associated with nitroxide radicals has been extensively applied, in conjunction with chemical oxidizing agents, to the oxidation of alcohols since the 1980s.[9] Figure 14.1 represents the mediation mechanism, in which the *electrochemical* oxidation of TEMPO affords the corresponding oxoammonium salt, which can transform alcohols into carbonyl compounds.[10] In the absence of an oxidizable substrate, the nitroxide/oxoammonium couple (Figure 14.1, dotted square) can be regarded as a possible charge storage prototype.

Nitroxide radicals display two redox couples, as illustrated in Figure 14.2. On the anodic side, nitroxides can be oxidized to form the corresponding oxoammonium cation, which is ascribed to the p-type redox reaction. On the cathodic side, the nitroxide radical is reduced to the aminoxy anion, which corresponds to the n-type doping of the material.[11]

The most important features of the redox reactions of nitroxides are its remarkably rapid and reversible one-electron transfer for the oxidation process. Electrochemical measurements (Nicholson methods) revealed that the heterogeneous electron transfer constants (k_O) of p-type redox reaction for a series of nitroxide derivatives (about 10^{-1} cm/s)[12] were much higher than that of other organic-based redox couples such as disulfide compounds (about 10^{-8} cm/s).[13] X-ray crystal structures of TEMPO and the corresponding oxoammonium cation reveal relatively small structural changes (0.1 Å decrease in both the N–O bond length and deviation of the nitrogen atom from the C_2O plane, respectively) upon oxidation,

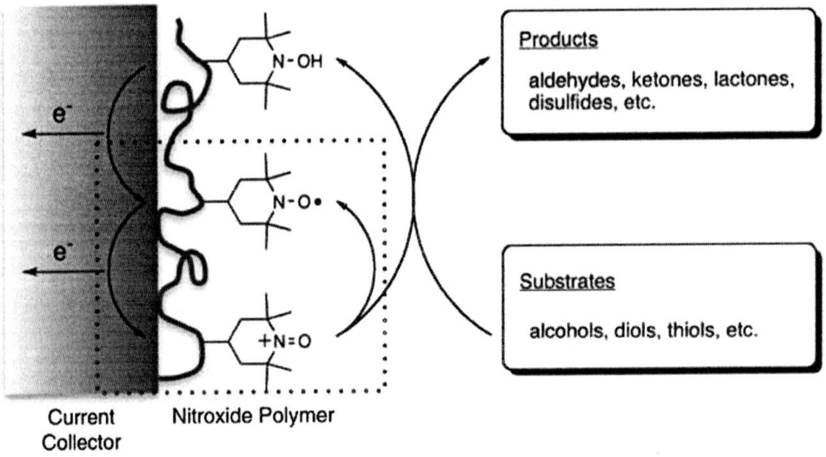

Figure 14.1 *Electrocatalytic oxidation of alcohols using poly(TEMPO-substituted acrylamide) mediator. The concept of "charge storage in the radicals" can be found in the dashed square.*

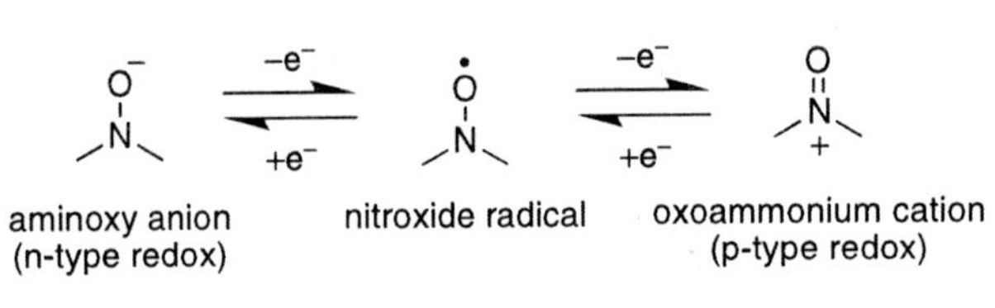

Figure 14.2 *Redox processes of the nitroxide radical.*

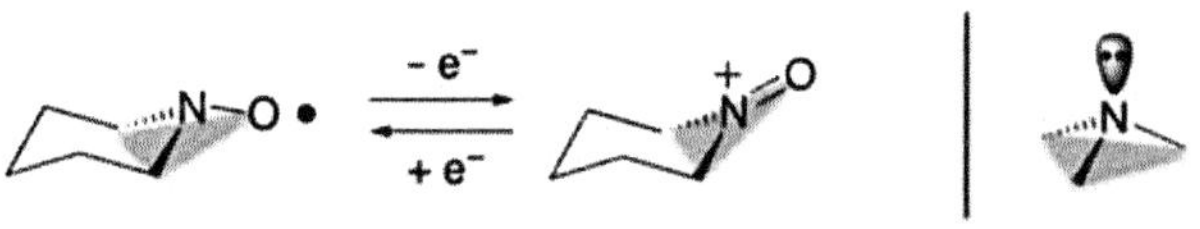

	Nitroxide	Oxoammonium cation	cf) Amine
Electronic state of N atom	sp^3/sp^2	sp^2	sp^3
N-O bond length (Å)	1.28	1.18	-
C-N-O bond angle (°)	112	122	106.7
Deviation from C_2O plane (Å)	0.177	0.007	-

Figure 14.3 *Structural implications of TEMPO/TEMPO$^+$ BF_4^-.*

which gives rise to both the high redox stability (cyclability) and rapid electron transfer process of the nitroxide/oxoammonium (NO•/N^+=O) redox couple (Figure 14.3).[14,15]

The electrode kinetics (electron transfer process) of the nitroxide radical polymer layer confined at a current collector surface was also characterized by normal pulse voltammetry, the results of which suggested a rapid charge propagation (apparent diffusion coefficient, $D_{app} \sim 10^{-10}\ cm^2/s$) throughout the polymer layer in a micrometer scale.[16] The amorphous, solvated, and slightly swollen structure of the radical polymer ensures good counterion mobility during the electrode process. Based on such a rapid and simple one-electron transfer process, nitroxides are applicable for an electrode active and charge storage material, promising high power rate performance and long cyclability of the battery.

14.3 Mechanism and performance of an organic radical battery

A rechargeable battery, such as a lithium ion battery, is one of the most important energy storage devices in a wide range of applications from mobile phones to automotives. In general, a battery consists of two electrodes, one for a cathode and one for an anode, in contact with an electrolyte solution. Reversible redox reactions at both electrodes can store charge (energy), in conjunction with counterion transport.[17,18] A rechargeable battery is characterized by its high energy density, constant output voltage, and no self-discharge compared with capacitors (electronic double layer capacitor (EDLC) or electrochemical capacitors). Conventional rechargeable batteries show lower power rate performance compared with capacitors. Common inorganic-based electrode materials are widely used in commercially available rechargeable batteries. Organic-based or polymer-based electrodes are relatively new concepts, but promise potential advantages, such as lightweight, environmentally-benign characteristics, mechanical flexibility, and processing compatibility. In this section, the synthesis and battery performance of radical polymers are described.

For battery applications, the TEMPO radical must be immobilized at the current collector or electrode to prevent it from diffusing away from the electrode into the electrolyte (self-discharge).[19] This is most easily accomplished through binding the radical to a polymer film. A typical example is poly(2,2,6,6-tetramethylpiperidinyloxy-4-yl methacrylate) **1**.[8] This polymer populates the radical moiety densely in an aliphatic and non-conjugated structure, which allows for quantitative redox reactions under a constant electronic potential, in contrast to those of the π-conjugated conducting polymers.[20] Figure 14.4 shows the charging and discharging mechanism of a prototype organic radical-based battery, where the p-type nitroxide polymer serves as a cathode operated in conjunction with a lithium or carbon anode. During the charging process, the p-type radical polymer ($NO^\bullet$) in the cathode is oxidized to the oxoammonium

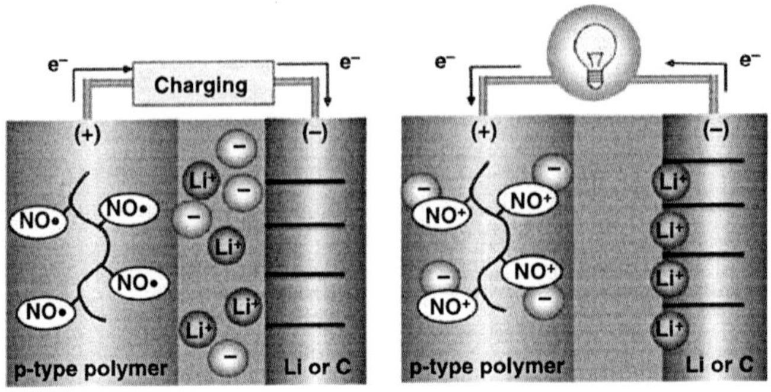

Figure 14.4 Charging and discharging mechanism of a prototype "organic radical battery" based on a radical polymer cathode.

cation (NO^+). During the discharging process, the nitroxide radical is regenerated by reduction of the oxoammonium cation. A battery composed of the radical polymer electrode as the electrode active or charge storage component is referred to as an "organic radical battery" herein.

The radical polymer **1** is obtained via conventional radical polymerization of the precursor monomer and the subsequent chemical oxidation ($M_w \sim 10^4$, Scheme 14.1). The radical precursor (a hindered amine) is a well known antioxidant and light stabilizer for plastics and commodity materials, and is considered to be a safe and non-toxic material. Polymer **1** is thermally stable ($T_{d10\%} = 263\,°C$), and the radical moieties on the polymer are also stable up to 200 °C. The radical density (~1.0 per repeating unit) remains unchanged for over one year under ambient conditions. The polymer is amorphous ($T_g = 70\,°C$) and has good processability and molding ability. This polymer displays appropriate solubility in organic solvents and is insoluble in electrolyte solutions, such as ethylene carbonate and diethyl carbonate with lithium hexafluorophosphate ($LiPF_6$).

The radical polymer itself does not possess sufficient electric conductivity, so it must be mixed with 20–50 wt- % graphite fibers to create the cathode electrode. A scanning electron microscopy (SEM) image of the composite electrode (Figure 14.5 inset) reveals that a graphite fiber with a diameter of 150 nm is thoroughly covered with a thin polymer layer (thickness 50–100 nm). The test cell was fabricated by stacking the radical polymer electrode with a separator film and lithium metal or a graphite carbon as the anode. Ethylene carbonate containing $LiPF_6$ was used as the electrolyte (Figure 14.5).

Charge–discharge curves of the battery composed of the polymer **1** cathode and lithium anode displayed a plateau voltage at about 3.5 V, in good agreement with the formal redox potential of **1** vs Li, which

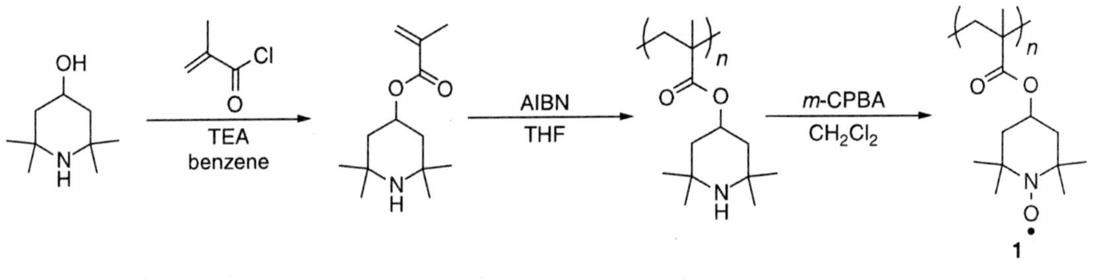

Scheme 14.1

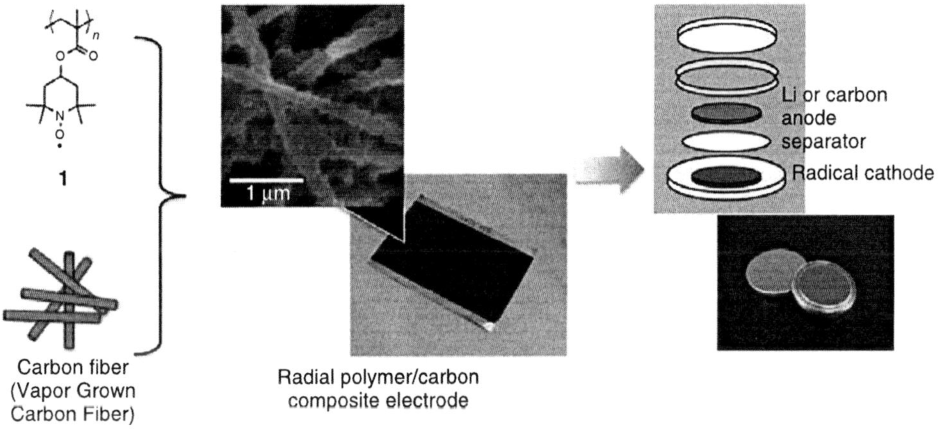

Figure 14.5 *A typical fabrication method of the prototype "organic radical battery". A full-colour version of this figure appears in the Colour Plate section of this book.*

was obtained from cyclic voltammetry. The charge/discharge capacities were about 110 mAh/g (Coulomb efficiency ~100 %), which was close to 100 % utilization of the loaded amount of radical polymer, based on theoretical capacity of **1** (111 mAh/g).[21] The battery exhibited a surprisingly high current capability, allowing rapid charging within about one minute and large discharge currents (50 C, where 1 C is defined as the current density at which the charging and discharging of the battery takes one hour) without substantial loss of output voltages (Figure 14.6), which was in contrast to conventional lithium ion batteries. The cycle performance during charging and discharging of the battery was extremely stable, and no significant deterioration in the capacity was observed for more than 1000 cycles. This surprisingly high current capability and long cyclability can be attributed to the rapid and reversible one-electron transfer reaction of the radical, no significant structural change of the radical upon electron transfer, and to the nanometer-sized, amorphous electrode structure.

The battery performance based on the polymer **1** cathode is presented on an energy/power density diagram (Ragone plot) in Figure 14.7, along with data for conventional secondary batteries (lithium ion, nickel metal hydride batteries) and electric double layer capacitor (EDLC). The radical battery is characterized

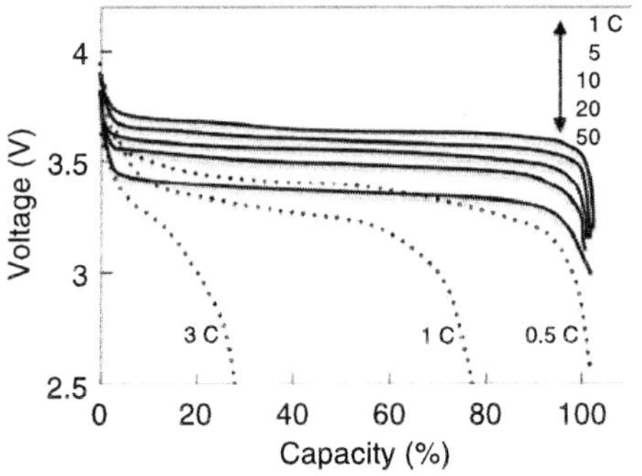

Figure 14.6 *Current rate performance of the organic radical battery (solid line, 1–50 C rate) and a conventional lithium ion battery (dashed line, 0.5–3 C rate).*

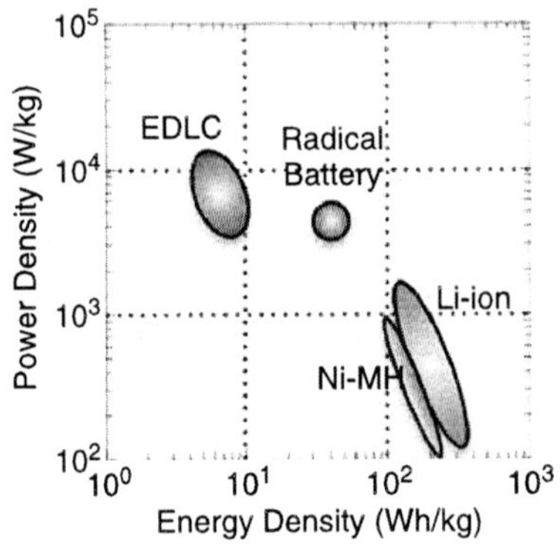

Figure 14.7 Projection of the "organic radical battery" in the energy and power density diagram (Ragone plot).

by both remarkably high power density and relatively high energy density (i.e., high capacity). By using its high power performance and relatively high capacity, the radical battery has been tested, for example, as an uninterruptible power supply system for the data backup of personal computers and computer servers during power failure. Optimization of the composition of polymer/carbon electrode and electrolyte, as well as fabrication method of the battery, has led to a 100-mAh class of aluminum laminated film packaged organic battery.[8,22,23] Applications for which high power capabilities, rather than high energy density, are important, such as the subbattery in electronic devices and motor drive assistance in electric vehicles, would be appropriate for organic radical batteries in the future.

14.4 Molecular design and synthesis of redox active radical polymers

As described in Section 14.3, redox active radical groups are immobilized in the form of electrode materials to impede self-discharge accompanied by the dissolution into the electrolyte solution. One of the simplest ways for immobilization is to incorporate the radical group into the (aliphatic) polymer structure, in which the radical groups are electrochemically isolated and non-interacting. For several years, various combinations of the durable radicals and polymer structures have been extensively investigated. In this section, we focus on the p-type redox active radical polymers, and summarize the molecular design and synthetic methodology of these polymers in view of the polymer backbone as a radical carrier (Table 14.1).

Conventional radical polymerization of radical-containing monomers does not yield radical polymers due to the high reactivity of radical pendant groups, which terminates the polymerization. Therefore, polymerization procedures are categorized by (i) radical polymerization of the precursor monomers, followed by chemical oxidation, and (ii) anionic, cationic, and transition metal-catalyzed polymerization of the radical monomers.

14.4.1 Poly(methacrylate)s and poly(acrylate)s

Poly(2,2,6,6-tetramethylpiperidinyloxy-4-yl methacrylate) **1** was obtained by the polymerization of 2,2,6,6-tetramethylpiperidine methacrylate, a material known as a light stabilizer, followed by the oxidation reaction of the precursor polymer with *m*-chloroperbenzoic acid.[8] The radical density of the obtained polymer often remained lower (80–90 %) than the theoretical density due to incomplete polymer reaction. The

Table 14.1 *P-type redox active radical polymers*

	Polymer	*Monomer* and polymn type	Oxidation	M_n (M_w/M_n)	E_p[f] (V vs. Li/Li^+)[a]	Solubility[b]	Theoretical capacity Q(mAh/g)[c]	Charge capacity (mAh/g)	Ref.
[redacted]	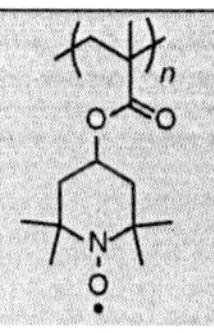	*methacrylate* 1) radical (AIBN)	*m*CPBA	27,000 (3.3)	3.58	insoluble	111	77–110	*8, 19, 20*
		2) anionic (PhMgBr)	–	40,000 (1.2)	3.58	soluble	111	–	*10, 21*
		3) group transfer	–	18,000 (3.2)	3.62	soluble	111	105[d]	*22*
2	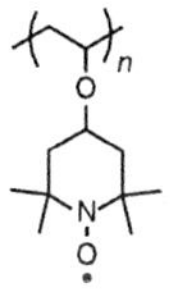	*vinyl ether* cationic (BF_3OEt_2)	–	–	3.55	insoluble	135	114	*23,24*
3	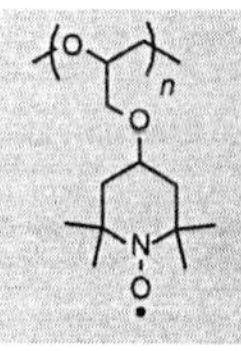	*epoxide* anionic (*t*BuOK)	–	32,000 (1.8)	3.54	insoluble	117	78	*26*
4	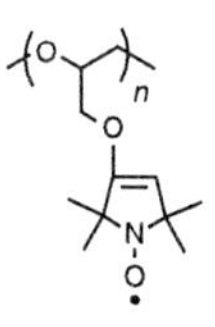	*epoxide* anionic coodinated ($ZnEt_2/H_2O$)	–	400,000 (1.7)	3.62	insoluble	131	88	*27*
5	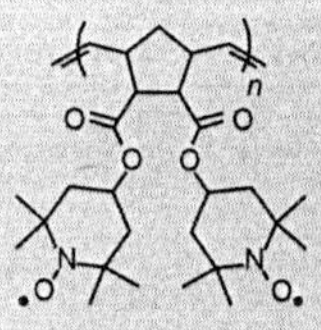	*norbornene* ROMP (Grubbs 2nd generation cat.)	–	13,000 (1.2)	3.60	soluble	109	109[d]	*29–32*
6	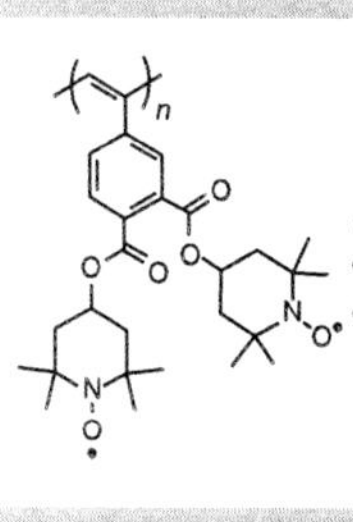	*acetylene* coodinated (Rh cat.)	–	84,000 (3.4)	3.60	insoluble	97	96	*33*
7	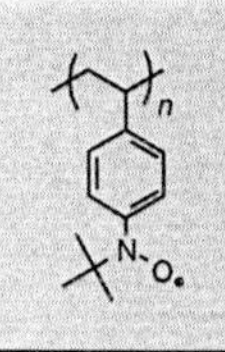	*styrene* radical (AIBN)	$(C_4H_9)_4NF$ Ag_2O	21,000 (1.6)	3.50	insoluble	141	–	*34*

[a] Formal redox potentials for the radical polymers were standardized by Li/Li^+ reference electrode.
[b] Solubility in the electrolyte solution (e.g. ethylene carbonate/diethyl carbonate with 1.0 M $LiPF_6$.)
[c] $Q = 96485 \times a/(F_w \times 3600)$ (mAh/g), where *a*: the number of radicals in the repeating unit; F_w = formula weight of the repeating unit.
[d] chemically- or photo-crosslinked.

polymer **1** obtained by this method was insoluble in the electrolyte solution (e.g., ethylene carbonate/diethyl carbonate with 1.0 M $LiPF_6$ and $LiN(CF_3SO_2)_2$), and applicable as an electrode active material. Anionic polymerization of the radical monomer, 4-methacryloyl-2,2,6,6-tetramethylpiperidinyl-*N*-oxy, using the Grignard reagent (PhMgBr) or *sec*-BuLi as an initiator produced the corresponding radical polymer **1** with quantitative radical density.[10,24] Group transfer polymerization of the radical monomer also afforded the corresponding radical polymer **1**.[25] The polymers **1** obtained from the latter two polymerization techniques, were both quite soluble in the electrolyte solution, and needed to be cross-linked.

14.4.2 Poly(vinyl ether)s and poly(allene)s

Poly(TEMPO-substituted vinyl ether) **2** was obtained via cationic polymerization of the corresponding radical monomer using boron trifluoride etherate ($BF_3{\cdot}OEt_2$) as an initiator.[26] The radical monomer was easily synthesized from 4-hydroxy-TEMPO and vinyl acetate using iridium or palladium catalysts. The polymer **2** has higher theoretical capacity (135 mAh/g) than the poly(methacrylate) derivative **1**, and showed quantitative redox behavior even in the high loaded amount of radical (70 wt-% radical) in the carbon composite electrode. The polymer **2** was also partially hydrophilic, and was also employed as a cathode active material in the aqueous-based electrolyte, which was advantageous in view of environmentally-benign characteristics and the improvement of safety issues.[27]. Allenyl ether monomer bearing 2,2,6,6-tetramethylpiperidine was polymerized via conventional radical polymerization, followed by chemical oxidation, to yield the corresponding radical polymer.[28]

14.4.3 Poly(cyclic ether)s

Anionic ring-opening polymerization of epoxides bearing TEMPO or the proxyl group under bulk condition using *t*-BuOK yielded the corresponding polyethers **3** and **4**, which were insoluble but swollen in the electrolyte solution. The ionophoric polyether backbone with a low glass transition temperature (18 °C) was designed to improve compatibility with the electrolyte solution, enabling high power capability and high utilization of radical polymer in the carbon composite electrode. The radical density of the polymer **3** was 84 % per repeating unit.[29] Anionic coordinated ring-opening polymerization of the glycidyl ether derivatives using $ZnEt_2/H_2O$ as an initiator also afforded the corresponding radical polymers. The proxyl group was persistent under the reaction condition.[30] Oxetane (four-membered cyclic ether) monomers containing nitroxide group could not be polymerized via cationic ring-opening polymerization.

14.4.4 Poly(norbornene)s

Ring-opening metathesis polymerization (ROMP) of TEMPO-substituted norbornene using the second generation Grubbs catalyst was successfully achieved to yield the corresponding polymer **5** without any side reactions.[31,32] Moreover, the polynorbornene backbone has olefin sites, which tuned the solvent solubility of the polymer in combination with a bis(azide) crosslinker via photo-crosslinking. Poly(TEMPO-substituted norbornene) provided design flexibility and good processability of the redox active films, leading to an organic-based, flexible paper battery.[33] Coumarin was also examined as photo-crosslinker in this polymer.[15] Ring-opening metathesis polymerization of a series of norbornene derivatives was also reported.[34,35]

14.4.5 Poly(acetylene)s

A series of acetylene monomers bearing TEMPO were polymerized via rhodium-catalyzed coordination polymerization, to give high molecular weight polymers (e.g. polymer **6**). However, 30–40 % of the radical sites were redox inactive, which suggested an unfavored polymer chain structure (the radical sites were less accessible and compatible with electrolyte).[36]

14.4.6 Poly(styrene)s

Poly(nitroxylstyrene)s **7** and **8** were obtained via conventional radical polymerization of the silyl-protected styrenic monomers, followed by the deprotection with tetrabutylammonium fluoride ($(C_4H_9)_4NF$) and chemical oxidation with silver oxide or fresh lead oxide.[37] Poly(nitronylnitroxyl styrene) **10** was also obtained from the silyl-protected monomer in the same procedure.[38] Poly(galvinoxylstyrene) **9** was synthesized via radical polymerization of the corresponding precursor monomer, hydrogalvinoxylstyrene, and chemical oxidation with potassium ferricyanate under basic conditions (Section 14.5).[39,40]

14.4.7 Combination of radicals with biopolymers and ionic liquids

Besides synthetic polymers bearing nitroxide groups, other approaches for immobilization of radicals have been reported. Pyridinium-based cationic lipids bearing the TEMPO group were synthesized, and incorporated into DNA via complex formation.[41] Imidazolium-based ionic liquids bearing the TEMPO group were also synthesized[42]; these were insoluble in the electrolyte solution and applicable for an electrode active material.[43]

14.5 A totally organic-based radical battery

In the previous sections, the focused has been on only p-type redox active radical polymers, especially the TEMPO-substituted polymers, as a cathode active material in the rechargeable battery. In this section, we demonstrate, for the first time, a "totally organic-based radical battery" composed of radical polymer-based cathode and anode. For this purpose, the development of n-type redox active radical polymers was inevitable, but it was even more challenging. No other organic radicals except for TEMPO have been reported as an electrode active material in extensive studies, and, even in TEMPO derivatives, an n-type redox process of the radical remained irreversible. The electrochemical reversibility of these redox couples originates both from the "stability" of the radical ($R^\bullet$) and the oxidized or reduced form (R^+ or R^-). Judicious molecular design of the radical polymer to stabilize n-type redox couples is required.

Not many, but some n-type redox polymers bearing quinone, quinoline, and oxadiazole derivatives and n-dopable polythiophenes have been reported as an electron accepting and transporting material in the organic light emitting diodes, organic thin-film transistors, and photovoltaics. However, the use of such n-type redox activity of the polymers has been limited to transient (short time-scale) charge separation and transport, and the use of these materials for long term charge storage has not been examined except for a few reports on capacitors based on n-dopable polythiophenes.[44] In this section, our recent successful approaches to explore n-type or bipolar (both p- and n-type) redox active radical polymers with emphasis on battery applications are demonstrated (Table 14.2).

We focused on a reversible and rapid one-electron redox reaction between galvinoxyl radicals and galvinolate anions, where the reduction of the neutral radicals is an n-type redox reaction. Poly (galvinoxylstyrene) **9** displayed a reversible redox couple at 3.15 V vs Li/Li^+ in the presence of organic base additive, such as tetrabutylammonium hydroxide or alkali butoxide. The polymer **9** was used as an anode active material in conjunction with p-type redox active radical cathode (e.g. poly(TEMPO-substituted norbornene) **5**), to fabricate a "totally organic radical battery" (Type I, Figure 14.8a). A plateau voltage of the battery was 0.66 V, in good agreement with the potential gap between both polymers. This battery

Table 14.2 *N-type redox active radical polymers*

	Polymer	*Monomer* and polymn type	Oxidation	M_n (M_w/M_n)	E_n[f] (V vs. Li/Li$^+$)[a]	Solubility[b]	Theoretical capacity Q(mAh/g)[c]	Charge capacity (mAh/g)	Ref.
8	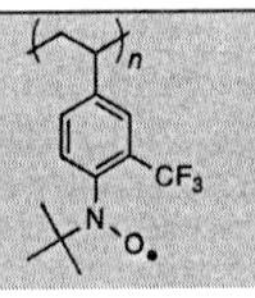	*styrene* radical (AIBN)	$(C_4H_9)_4NF$ Ag_2O	36,000 (1.4)	2.00	insoluble	104	–	*34*
9	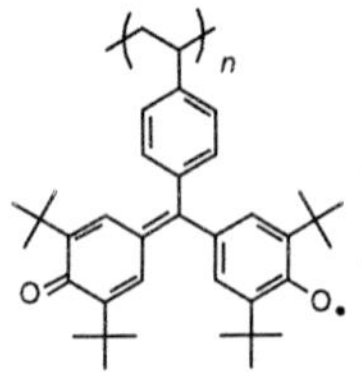	*styrene* radical (AIBN)	–	13000 (1.2)	3.15	soluble	51	42[a]	*36*
10	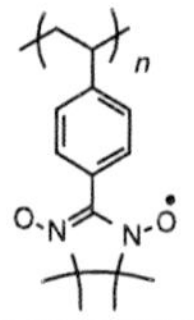	*styrene* radical (AIBN)	$(C_4H_9)_4NF$ MnO_2	52,000 (3.2)	3.50[b] (2.15)	insoluble	103	91[b] (82)	*35*

[a]crosslinked with 10 mol%tetraethyleneglycol diacrylate.
[b]Formal potential and charge capacity for p-type redox of nitronylnitroxide.

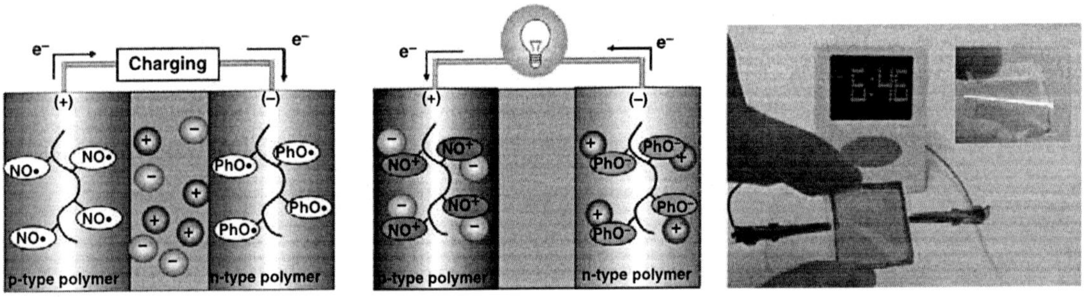

(a) (b)

Figure 14.8 *(a) Charging and discharging mechanism of a "totally organic-based radical battery" composed of polymer **5** cathode and **9** anode (Type I). (b) A see-through paper-like battery; blue in the charged state (galvinolate anion) and light-yellow in the discharged state (galvinoxyl). A full-colour version of part (b) of this figure appears in the Colour Plate section of this book.*

exhibited strikingly high current capability, allowing rapid charging within 10 seconds and large discharge currents (360 C). This totally polymer-based "radical battery" offers flexible device formability, enabling thin and paper-like batteries. More excitingly, a see-through flexible battery was also fabricated with the polymer **5** cathode and **9** anode, in which the blue color in the charged state resulting from the galvinolate anion reverted dramatically to a light-yellow color of the galvinoxyl radical in the discharged state (the color change accompanied by the redox reaction can be used as the indicator of the charging level, Figure 14.8b).

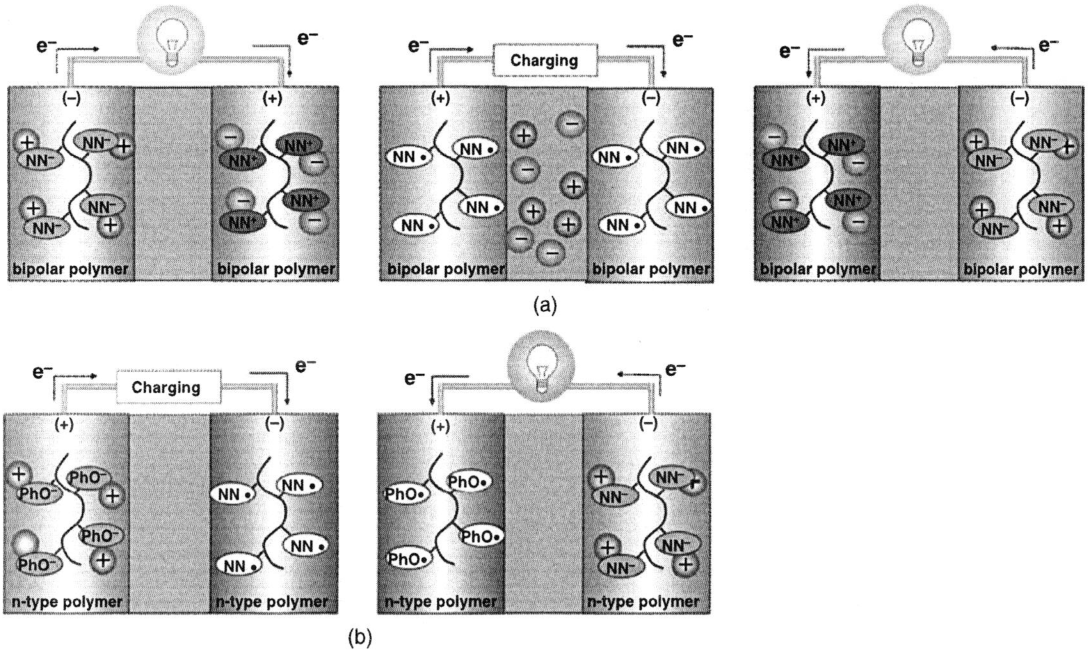

Figure 14.9 *(a) Symmetric configuration with both polymer* ***10*** *cathode and anode (a pole-less battery, Type II). (b) A "rocking-chair configuration" with the polymer* ***9*** *cathode and* ***10*** *anode (Type III).*

In the polystyrene derivatives **7** and **8**, p-type redox (or cathode active material) to n-type redox (or anode active material) switching was achieved by introducing electron withdrawing groups, such as the trifluoromethyl group, proximally to the nitroxide.

Poly(nitronylnitroxylstyrene) **10** exhibited two reversible redox couples at −0.61 and 0.72 V vs Ag/AgCl under the basic conditions, which was assigned to the n-type reduction and p-type oxidation of the nitronylnitroxide, respectively. The bipolar (p- and n-type) redox activity was used as both cathode and anode active material in a totally organic-based rechargeable battery with a symmetric configuration. Charge–discharge curves of a battery composed of the polymer **10** for both electrodes gave a plateau voltage of 1.3 V, in good agreement with the potential gap of formal potentials ($E_p{}^f - E_n{}^f$) for the p-type oxidation and n-type reduction of nitronylnitroxide. The reversed bias also enabled charging of the battery, and the battery direction was reversible (pole-less battery, Type II, shown in Figure 14.9a).

The development of n-type redox active radical polymers is also of practical importance in view of the cell configuration. The n-type polymer cathode enabled a "rocking-chair"-type battery, in which only cations were exchanged at both electrodes, giving rise to the ultimate reduction of the electrolyte down to the minimum amount necessary to allow current flow. Finally, the totally polymer-based, rocking-chair battery was fabricated with the n-type **9** cathode and the n-type **10** anode, which is an unprecedented battery configuration for organic batteries (Type III shown in Figure 14.9b).

14.6 Conclusions

Robust but redox active organic radicals can be employed as a new electrode active and charge storage material, because these radicals have fully reversible one-electron redox reactions featuring fast electrode kinetics, high cyclability, and relatively high battery electrode capacity. Our exploration of the n-type redox active radical polymers enabled a "totally organic-based radical battery" with various configurations including a pole-less or rocking-chair battery. The radical polymer-based electrodes provide good processability and shape flexibility, which could lead to paper-like and wearable energy storage devices, widely applicable in battery on the chip and RFID tags. The integrated printable electronics technology will enable facile fabrication of the battery in the future.

As seen in this chapter, inspired from the rapid and reversible redox activity, the robust organic radicals are now being explored for other electronics applications. For instance, the photo-crosslinked poly(TEMPO norbornene) was employed as the counter electrode to improve the performance of the poly(viologen)-based electrochromic display.[45,46] Recently, a battery-inspired non-volatile organic memory by sandwiching a dielectric material (poly(vinylidene fluoride) with the radical polymers **5** and **9** was reported. The "radical memory" showed unique electroconductive bistability (ON-OFF ratio: 10^4, $>$1000 cycles), and opened up the high potential of a redox active radical even in a dry electronic device.[47,48] The mediator activity of the TEMPO derivatives has also been proposed as an alternative system for iodide/triiodide redox couple in the dye-sensitized solar cells.[49] In view of "redox activity", robust organic radicals could be re-examined as new functional materials, which promotes us to challenge yet unexplored functions of the organic radical molecules.

References

1. For recent reviews see: R. G. Hicks, *Org. Biomol. Chem.*, **5**, 1321–1338 (2007).
2. (a) W. A. Walters, *The Chemistry of Free Radicals*, Oxford University Press, London (1948). (b) E. G. Rozantsev, and V. D. Sholle, *Synthesis*, 190 (1971).
3. J. F. W. Keana, *Chem. Rev.*, **78**, 37 (1978).
4. (a) A. Rajca, *Chem. Rev.*, **94**, 871–893 (1994). (b) P. M. Lahti (ed.) *Magnetic Properties of Organic Materials*, Marcel Dekker, Inc., New York, 1999. (c) H. Murata, and H. Nishide, Persistent High-Spin Organic Polyradicals, in *Carbon-based Magnetism* (eds T. Makarova and F. Palacio), Elsevier, Amsterdam (2006), pp. 53–74.
5. (a) L. B. Volodarsky, and V. A. Reznikov, *Synthetic Chemistry of Stable Nitroxides*, CRC Press, Boca Raton, Florida (1993). (b) H. G. Aurich, *The Chemistry of Functional Groups, Nitrones, Nitronates, and Nitroxides*, John Wiley & Sons Ltd, Chichester (1989).
6. C. J. Hawker, A. W. Bosman and E. Harth, *Chem. Rev.*, **101**, 3661–3688 (2001).
7. (a) H. Zimmer, D. C. Lankin, and S. W. Horgan, *Chem. Rev.*, **71**, 229–246 (1971). (b) A. E. J. deNooy, A. C. Besemer, and H. van Bekkum, *Synthesis*, 1153–1174 (1996). (c) R. A. Sheldon, I. Arends, G. J. Ten Brink, and A. Dijksman, *Acc. Chem. Res.*, **35**, 774–781 (2002).
8. (a) H. Nishide, and K. Oyaizu, *Science*, **319**, 737–738 (2008). (b) H. Nishide, and T. Suga, *Electrochem. Soc. Interface*, **14**(4), 32–36 (2005). (c) H. Nishide, S. Iwasa, Y.-J. Pu, *et al.*, *Electrochim. Acta*, **50**, 827–831 (2004). (d) K. Nakahara, S. Iwasa, M. Satoh, *et al.*, *Chem. Phys. Lett.*, **359**, 351–354 (2002).
9. (a) M. F. Semmelhack, and C. R. Schmid, *J. Am. Chem. Soc.*, **105**, 4492 (1983). (b) T. Miyazawa, T. Endo, S. Shiihashi, M. Okawara, *J. Org. Chem.*, **50**, 1332 (1985). (c) T. Osa, U. Akiba, I. Segawa, and J. M. Bobbitt, *Chem. Lett.*, 1423 (1988). (d) A. Merz, and H. Bachmann, *J. Am. Chem. Soc.*, **117**, 901 (1995). (e) A. Dijksman, A. Marino-González, A. M. I. Payeras, *et al.*, *J. Am. Chem. Soc.*, **123**, 6826–6833 (2001).
10. F. Maccorquodale, J. A. Crayston, J. C. Walton, and D. J. Worsfold, *Tetrahedron Lett.*, **31**, 771 (1990).
11. Most nitroxides such as TEMPO show an irreversible n-type redox reaction, especially in protic media, the reduction of nitroxide radical forms the corresponding hydroxylamine (N–OH). But some nitroxides with an electron withdrawing group, and nitronylnitroxide can achieve reversible n-type redox reaction as discussed in Section 14.5.
12. T. Suga, Y.-J. Pu, K. Oyaizu, and H. Nishide, *Bull. Chem. Soc. Jpn*, **77**, 2203–2204 (2004).
13. (a) S. J. Visco, C. C. Mailhe, and L. C. DeJonghe, *J. Electrochem. Soc.*, **136**, 661 (1989). (b) N. Oyama, T. Tatsuma, T. Sato, and T. Sotomura, *Nature*, **373**, 598 (1995).
14. Y. Yonekuta, K. Oyaizu, and H. Nishide, *Chem. Lett.*, **36**, 866–867 (2007).

15. For instance, a two-electron redox reaction of disulfide compounds ($2RS^- \rightarrow RS - SR + 2e^-$) is coupled with bond formation or cleavage, which restricts the rapid electron transfer and stability of the redox couples.
16. K. Oyaizu, Y. Ando, H. Konishi, and H. Nishide, *J. Am. Chem. Soc.*, **130**, 14459–14461 (2008).
17. M. Winter, and R. J. Brodd, *Chem. Rev*. **104**, 4245–4269 (2004).
18. M. S. Whittingham, *Chem. Rev*. **104**, 4271–4301, (2004).
19. A dissolved radical serves as a redox mediator or shuttle, which carries the charge between the two electrodes, leading to the self-discharge of the battery: C. Buhrmester, L. M. Moshurchak, R. L. Wang, and J. R. Dahn, *J. Electrochem. Soc*. **153**, A1800 (2006).
20. The small charge storage capacity and the fluctuating voltage of the π-conjugated conducting polymers originate from the low doping levels and the potential shift via doping process: D. MacInnes, Jr., M. A. Druy, P. J. Nigrey, *et al*., *J. Chem. Soc., Chem. Commun*., 1981, 317.; J. C. Carlberg, and O. Inganas, *J. Electrochem. Soc*. **144** (4), pp. L61–L64 (1997).
21. Theoretical redox capacity (Q) was calculated from the formular weight (F_w) of the monomer repeating unit, and the number of radicals in the repeating unit (a), using the following equation: $Q = 96485 \times a/(F_w \times 3600)$ (mAh/g).
22. (a) K. Nakahara, J. Iriyama, S. Iwasa, *et al*., *J. Power Sources*, **165**, 870–873 (2007). (b) K. Nakahara, J. Iriyama, S. Iwasa, *et al*., *J. Power Sources*, **163**, 1110–1113 (2007). (c) K. Nakahara, J. Iriyama, S. Iwasa, *et al*., *J. Power Sources*, **165**, 398–402 (2007).
23. (a) J.-K. Kim, G. Cheruvalley, J.-W. Choi, *et al*., *J. Electrochem. Soc*., **154**, A839–A843 (2007). (b) J.-K. Kim, G. Cheruvalley, J.-W. Choi, *et al*., *Electrochim. Acta*, **178**, 1546–1551 (2007). (c) J.-K. Kim, G. Cheruvalley, J.-H. Ahn, *et al*., *J. Ind. Eng. Chem*., **14**, 371–376 (2008). (d) H.-Q. Li, Y. Zou, and Y.-Y. Xia, *Electrochim. Acta*, **52**, 2153–2157 (2007).
24. J. Allgaier, and H. Finkelman, *Makromol. Chem. Rapid Commun*., **14**, 267–271 (1993).
25. L. Bugnon, C. J. H. Morton, P. Novak, *et al*., *Chem. Mater*., **19**, 2910–2914 (2007).
26. M. Suguro, S. Iwasa, Y. Kusachi, *et al*., *Macromol. Rapid Commun*., **28**, 1929–1933 (2007).
27. K. Koshika, N. Sano, K. Oyaizu, and H. Nishide, *Chem. Commun*., 836–838 (2009).
28. X. Zhang, H. Li, L. Ki, G. Lu, *et al*., *Polymer*, **49**, 3393–3398 (2008).
29. T. Suga, K. Yoshimura, and H. Nishide, *Macromol. Symp*. **245-246**, 416–422 (2006).
30. K. Oyaizu, T. Suga, K. Yoshimura, and H. Nishide, *Macromolecules*, **41**, 6646–6652 (2008).
31. Previously, Tanyeli, *et al*. reported the polymerization of norbornene monomer bearing the reduced *N*-hydroxypiperidine group. C. Tanyeli, and A. Gusmus, *Tetrahedron Lett*., **44**, 1639–1642 (2003).
32. T. Suga, S. Kasatori, K. Yoshimura, and H. Nishide, *Polym. Prep., Jpn*, **54(2)**, 4567–4568 (2005).
33. T. Suga, H. Konishi, and H. Nishide, *Chem. Commun*., 1730–1732 (2007).
34. (a) T. Katsumata, M. Satoh, J. Wada, *et al*., *Macromol. Rapid. Commun*., **27**, 1206–1211 (2006).
35. T. Katsumata, J. Qu, M. Shiotsuki, *et al*., *Macromolecules*, **41**, 1175–1183 (2008).
36. (a) J. Qu, T. Katsumata, M. Satoh, *et al*., *Chem. Eur. J*., **13**, 7965–7973 (2007). (b) J. Qu, T. Fujii, T. Katsumata, *et al*., *J. Polym. Sci. A. Polym. Chem*., *45*, 5431–5445 (2007).
37. T. Suga, Y.-J. Pu, S. Kasatori, and H. Nishide, *Macromolecules*, **40**, 3167–3173 (2007).
38. T. Suga, S. Sugita, H. Ohshiro, *et al*., *Angew. Chem. Int. Ed*., submitted.
39. T. Suga, H. Ohshiro, S. Sugita, *et al*., *Adv. Mater*., **21**, 1627–1630 (2009).
40. T. Kaneko, H. Tatsumi, T. Aoki, *et al*., *J. Polym. Sci. A. Polym. Chem*., **37**, 189 (1999).
41. J. Qu, R. Morita, M. Satoh, *et al*., *Chem. Eur. J*., **14**, 3250–3259 (2008).
42. W. Qian, E. Jin, W. Bao, and Y. Zhang, *Tetrahedron*, **62**, 556–562 (2006).
43. S. H. Lee, J.-K. Kim, G. Cheruvally, *et al*., *J. Power Sources*, **184**, 503–507 (2008).
44. Even for n-dopable polythiophene, the small charge storage capacity and the fluctuating voltage have impeded their use in a rechargeable battery, where a constant output voltage is required.
45. Y. Takahashi, N. Hayashi, K. Oyaizu, *et al*., *Polym. J*., **40**, 763–767 (2008).
46. Y. Takahashi, K. Oyaizu, K. Honda, and H. Nishide, *J. Photopolym. Sci. and Tech*., **20**, 29–34 (2007).
47. Y. Yonekuta, K. Susuki, K. Oyaizu, *et al*., *J. Am. Chem. Soc*., **129**, 14128–14129 (2007).
48. Y. Yonekuta, K. Honda, and H. Nishide, *Polym. Adv. Tech*., **19**, 281–284 (2008).
49. Z. Zhang, P. Chen, T. N. Murakami, *et al*., *Adv. Func. Mater*. **18**, 341–346 (2008).

15

Spin Labeling: A Modern Perspective

Lawrence J. Berliner
Department of Chemistry and Biochemistry, University of Denver, Denver, USA

15.1 Introduction

Spin labeling has been an accepted physical biochemical technique for over 40 years, having been introduced by Harden McConnell[1] who coined the name as it fit within the definition of other probes that were called reporter groups by Burr and Koshland.[2] The approach was to place a spectroscopic probe in a biological macromolecule or system (such as an enzyme or protein) and to monitor spectroscopic variables in the probe that were uniquely sensitive to the immediate environment in the biosystem. The reporter group scheme is depicted in Figure 15.1. While in the past these probes were fluorophores, an NMR isotope such as ^{19}F or ^{13}C, resonance Raman probe, and so on, for electron paramagnetic resonance (EPR) the godsend was the stable paramagnetic nitroxyl (nitroxide) *[NOTE: Nitroxide is the parent name used by Chemical Abstracts Service for R_2N–O•; however the correct IUPAC name is aminoxyl.]* moiety. These molecules are classified as spin *labels*, when a covalent linkage to the biological system is employed (i.e., with proteins, polymers, and nucleic acids) while a spin *probe* involves non-covalent interactions (cells, membranes, liquid crystals, and some polymer systems). EPR has the advantage of being a non-invasive, non-destructive technique. One wonders how the field of site-specific mutation would have ever been accepted if the provinciality over the possibility of structural/environmental perturbation was never resolved. The spin label technique has flourished since then with several texts devoted specifically to the subject. Spin labels have now enjoyed a several decade history in demonstrating the applicability of paramagnetic nitroxides to problems of structure and function in enzymes, membranes, cells, and animals.[3–7] While there are comparable applications to other macromolecular systems, such as polymers, dendrimers and the like, this chapter focuses mostly on biological examples. It focuses primarily on applications to protein structure, with an emphasis on distance measurements. In particular, the modern approach called site directed spin labeling (SDSL) is highlighted. In addition, how spin labeling can be used to monitor local pH at the molecular level is illustrated. One other biopolymer, polynucleotides, and DNA, are briefly addressed. Not covered are applications to lipids and membranes, *in vivo* approaches, and chemical polymers.

Figure 15.1 *Schematic representation of an enzyme–substrate complex in native protein (top left), protein containing reporter group (solid black area) adjacent to substrate binding area (top right), and reporter group distant from substrate binding area (bottom right). (Reprinted from [2], Copyright 1964, with permission from National Academy of Sciences, U.S.A.)*

It has been quoted many times, after the continuous criticism of the media, that the reporter group must "*report the news*" not "*make the news*".[3] That is, it is important to ensure that the sometimes bulky nitroxide spin label does not perturb the macromolecular system under study. On the other hand, there must be an ever so slight perturbation in order to "sense" the physical environment. Interestingly, the NMR community has adopted spin labeling and now frequently uses spin labels as useful probes in determining intramolecular distances. In fact, the perturbation artifact argument is routinely ruled out by measuring several functional and physical aspects of the biological system before and after labeling.

15.2 The early years

Although the method is technically not restricted to just nitroxide radicals (also known as nitroxyls) it is fair to say that more than 99.5 % of all papers using EPR reporter group techniques employ nitroxides as spin labels. The technique, as are many in biophysical/biomedical spectroscopy, was initially limited by the availability and synthesis of the appropriate organic radicals. It was not until Rozantsev and Neiman at the Institute of Chemical Physics in Moscow introduced a vast series of stable nitroxides with structures that were adaptable to biological systems that spin labeling took off.[8] Rozantsev later summarized this vast amount of work in a translated book.[9] A great deal of activity occurred at Stanford in the McConnell group, first with proteins, then with membrane systems, through the second half of 1960s that was quickly adopted in several other laboratories in the world. In fact, during that period the first *in vivo* toxicity study was carried out at Stanford in Harden McConnell's laboratory circa 1966 where an unspecified person poured an unspecified, but large amount of TEMPONE into a beaker housing a goldfish with no adverse affects.[6] This was truly a golden period where many of the applications of spin labeling and nitroxides that we use today were conceived.

15.3 Advantages of nitroxides

What are the specific advantages to nitroxides that made them so endearing to the EPR field? First of all, they are remarkably stable in almost any solvent and over a wide pH range in aqueous solution. The nitroxide moiety is quite tolerant to synthetic manipulations at other sites on the typically piperidine, pyrroline or pyrollidine ring. The generalized structures, including some nitroxide rings that can be an intergral part of a lipid or polymer chain, are depicted in Figure 15.2.

Freezing and thawing or distilling/boiling have no adverse affects on the stability of the paramagnetic group. Since EPR does not require optical transparency and does not suffer from the magnetic susceptibility issues frequently encountered in NMR, it is possible to work with totally opaque solutions, solids or mixtures thereof.

EPR sensitivity is, by definition, 600–700 times higher per spin compared with proton NMR, with solution detectability down to almost nanomolar levels for narrow linewidth signals. Since the EPR spectral line shape is a major component of spin label spectral analysis, particularly with respect to nitroxides tumbling motion, it is fairly easy to distinguish free "unattached" or unreacted labels in a sample where incomplete dialysis or separation was impossible. Since this one-electron radical is a paramagnet, there is a distance-dependent relaxation effect on neighboring (proton) nuclei up to distances of about 20 Å that may be quantitated in correlative NMR experiments with proteins or nucleic acids. This is based on the same dipole–dipole interaction that is measured in nucleus–nucleus interactions and is analogous to Forster energy transfer distance measurements in fluorescence experiments. Some applications of this phenomenon are discussed later in this chapter.

The only real drawback of nitroxide spin labels is their susceptibility to reduction to the diamagnetic hydroxylamine in the presence of organic or biological reducing agents. Synthetic procedures involving,

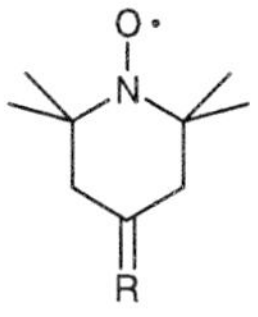

I
(a piperidine nitroxide)

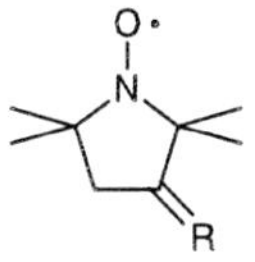

II
(a pyrrolidine nitroxide)

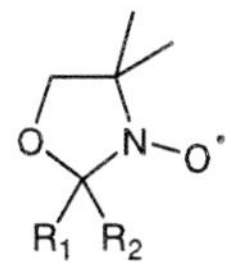

III
(a doxyl nitroxide)

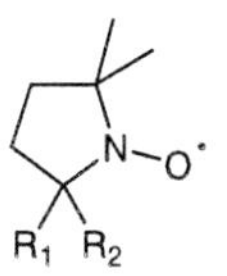

IV
(a proxyl nitroxide)

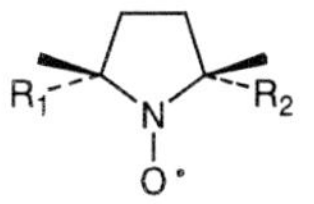

Va
(a *cis*-azethoxyl nitroxide)

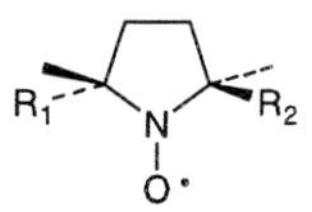

Vb
(a *trans*-azethoxyl nitroxide)

Figure 15.2 *Classes of nitroxide structures. The majority of nitroxides fall within one of the three classes: (I) 2,2,6,6-tetramethyl-piperidine-N-oxyl, (II) the 2,2,5,5-tetramethyl-pyrrolidine-N-oxyl, and (III) the 4,4-dimethyloxazolidine-N-oxyl (doxyl) nitroxides, respectively. Two other versatile classes of nitroxides are the side chain substituted 2,2,5,5-tetramethyl-pyrrolidine-N-oxyl (proxyl) nitroxides (IV) and the cis- and trans-azethoxyl nitroxides (V). (From [4] with permission.)*

for example, sodium tetrahydroborate ($NaBH_4$) reduce the nitroxides; however, the radical can be easily regenerated in mild hydrogen peroxide (H_2O_2) or exposure to oxygen (O_2). Since blood and tissue contain significant amounts of ascorbate and other reducing agents, special care must be taken to ensure that a label does not enter the cytoplasm, where bioreduction is quite rapid, particularly for the piperidine nitroxides. On the other hand, the ease by which a nitroxide moiety may be either one-electron reduced or oxidized allows it unique physiological properties, with potentially imminent clinical applications in the pathological events of trauma and oxidative stress (Chapter 17).

15.4 Applications of spin labeling to biochemical and biological systems

This can literally encompass everything large and small: proteins and enzymes, lipids and membranes, nucleic acids, pharmaceuticals, drug-receptor interactions, cells and cell membranes, polymers, animals. A number of applications are presented here, some of which are historical in context.

15.4.1 Stoichiometry and specificity: proteins and enzymes

A major obstacle with reporter group (labeling) has been the ability to obtain unique, specific covalent modification(s) yielding a 1:1 nitroxide label to macromolecule complex stoichiometry. That is, the labeling must be *specific* to one amino acid and not spread over several different residues. This would make the spectral results totally ambiguous and any eventual conformational conclusions, and especially distance measurements, will weigh the "closest" spins more highly than the distant spins, albeit with a degree of inaccuracy as well. In distance measurements, this would be particularly devastating if the site of interest were a more distant site. There have been a series of spin labels based on known protein modification reagents, but all of these required detailed structure and stoichiometric analysis to insure that the labeling was indeed complete and specific.

The most useful amino acid candidate for labeling purposes is the thiol side chain (–SH) of cysteine (Cys). However, it is more the exception than the rule to find free disulfide (cystine) bridged cysteines in proteins since they are frequently used as structural building blocks. Yet, most thiol (cysteine) reagents are highly reactive and specific versus other protein modification reagents.[10] In particular, the methanethiomethane sulfonate group is uniquely specific (and reversible!) in its chemistry with cysteine thiols. A major breakthrough occurred in 1982 with the first spin labeled analogs of methanemethylthiol thiosulfonate, $R\text{–}S\text{–}SO_2\text{–}CH_3$, where R is a pyrrolinyl, pyrrolidinyl or piperidinyl nitroxide group.[11] A schematic of the biochemistry of the label MTSSL is shown in Figure 15.3.

For spin labeling methodology to be applicable to proteins in general, the desirable situation would be where one could place the nitroxides moiety anywhere in the protein sequence. This became feasible in the late 1980s with the advent of site specification mutation methodology. The field blossomed after Hubbell and coworkers coined "site-directed spin labeling" (SDSL), which offered the possibility of probing the local environment at a specific residue site in a protein or protein complex of interest.[13] The method is especially powerful in distinguishing aspects of secondary structure (alpha helix vs beta sheet vs random structure), identifying helix–helix interfaces, discerning membrane bound vs exposed segments, and, as will be discussed later, the intramolecular distance measurements within a protein. The method involves site-directed "cysteine scanning" at all or targeted regions of interest in a protein molecule. Hence, it becomes possible to examine several different positions in a protein conformationally over a significant portion of the protein sequence.

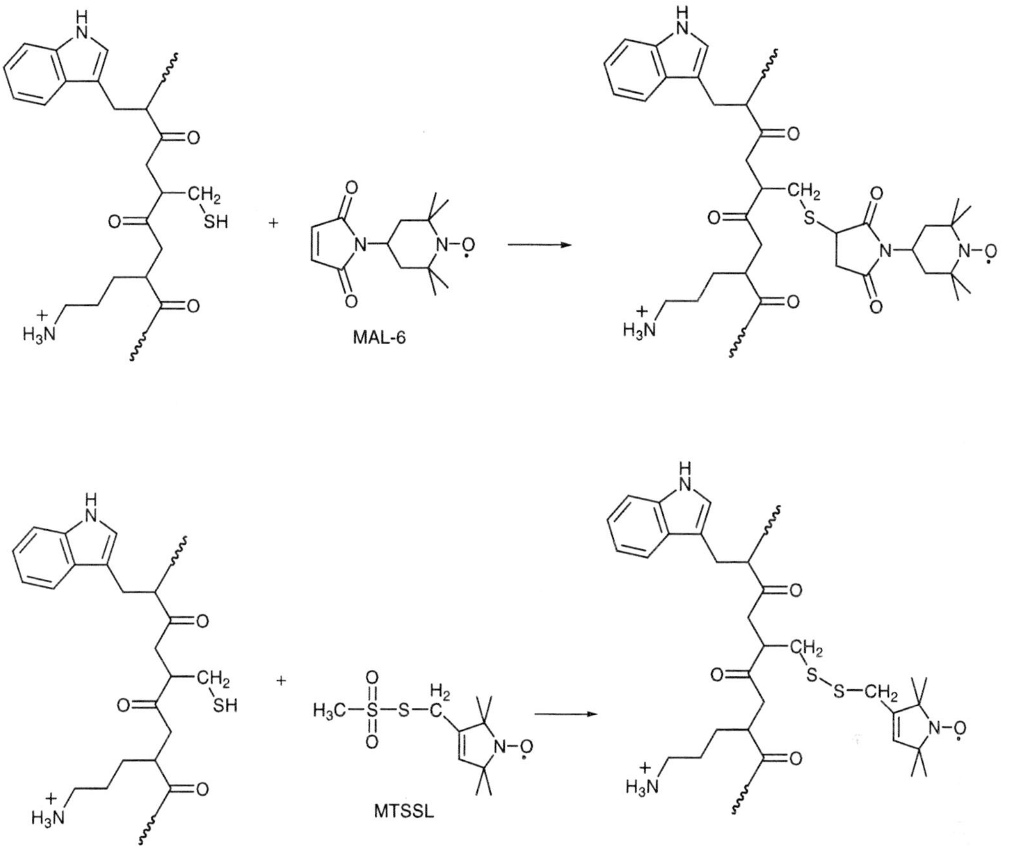

Figure 15.3 *Labeling chemistry of the piperidine maleimide spin label (MAL-6) and MTSSL with cysteine side chains. The MTSL moiety can be removed with DTT or mercaptoethanol to yield the thiol nitroxides and free cysteine. The molecular volume of the MTSL moiety is similar to that of the tryptophan side chain. (Reprinted with permission from [12]. Copyright 1998 Plenum Press.)*

15.4.2 The reporter group approach: who makes the news?

While NMR spectroscopists are "purists" in that they observe natural spins (nuclei) without altering the physical or chemical environment, spin labels and fluorescent probes are sometimes bulky structures that are introduced exogenously. As previously mentioned, spin labeling was continuously accused of perturbing the system by "making the news" instead of reporting the news.[3] For some years even the fluorescence community accused the spin labelers, while neglecting the typically very bulky bi- and tricyclic aromatic molecules needed in luminescence labeling studies. When substituting a nitroxide side chain at a specific position in a protein, a minor structural perturbation might occur, but evidence has continually shown that the majority of these labeling studies, including where they were placed at internal positions in a protein, did not result in altering structure/function changes, although the thermal stability of the protein may be altered slightly. The fact is that it is necessary to slightly perturb or interact with the macromolecular environment in order to report meaningful information.

15.5 Distance measurements

As mentioned earlier, it is possible to estimate intramolecular distances in a protein by prudent placement of two paramagnets at specific sites. If the paramagnetic centers are (preferably) undergoing little or no motion and within circa 10–15 Å of one another, their intramolecular separation can be calculated from the dipole–dipole interaction. When one of the paramagnets is a transition metal ion and the other is a spin label, the observation is an apparent decrease in the nitroxide spectrum due to very efficient relaxation (as manifested in line broadening) from the presence of the metal ion.

15.5.1 Metal–spin label distance measurements

The theory was worked out by Leigh for metal–nitroxide interactions over thirty years ago.[14] With a properly calibrated system, the distance could virtually be picked off of a calibration curve! When the nitroxide is fairly mobile there are considerations which make the calculated distance less accurate. Firstly, the "position" of the nitroxide is a weighted average of its allowed tumbling volume in space with the shorter distances emphasized due to the $1/r^6$ behavior of the dipole–dipole interaction. Secondly, the "effective" rotational correlation time (i.e., tumbling time) of the nitroxide group must be estimated, since rapid motion can average dipolar interactions to zero. The principal requirements in Leigh's theory then rely on the following assumptions[4,14]:

(1) Each radical interacts with one metal ion at a distance, r, and the radial vector between them is fixed in the free radical frame.
(2) The free radical orientation is isotropic (i.e., all orientations with respect to the laboratory frame are of equal probability).
(3) The A (or T) component of the dipolar Hamiltonian dominates in determining the spin label lineshape, that is, $\omega\tau_c > 1$.
(4) The T_1 of the metal cation is of the order of the inverse of the dipolar interaction; furthermore, it is determined by the lattice, and has no angular dependence nor dependence on the nitroxide group specifics.
(5) No motional modulation of the dipolar interaction.

Each nitroxide line $g(H)$ is broadened by the dipolar interaction with the metal ion according to Equation 15.1:

$$g(H) = \delta H/(H - H_o)^2 + \delta H^2 \tag{15.1}$$

where $\delta H = C(1 - 3\cos^2\theta_r')^2 + \delta H_o$ and $C = g\beta\mu^2 T_{1k}/\hbar r^6$. When δH_o is the natural line width in the absence of the paramagnetic metal, T_{1k}, the spin lattice relaxation time of the metal ion, and θ_r' defines the angle between the radial vector of length r and the applied field.

15.5.2 Spin label–spin label distance measurements

Since incorporating a paramagnetic metal ion is challenging unless the system under investigation possesses an intrinsic metal ion site, most of the more recent activity has been directed towards doubly spin labeled proteins. A depiction of this measurement is given in Figure 15.4. The majority of doubly labeled SDSL studies fall into one of three limiting cases, as outlined by Hustedt and Beth.[15] (1) The nitroxides are rigidly bound to the protein or protein complex, adopting a unique inter-nitroxide distance and orientation with respect to each other. The protein has a global rotational correlation time of approximately 1 μs or longer.

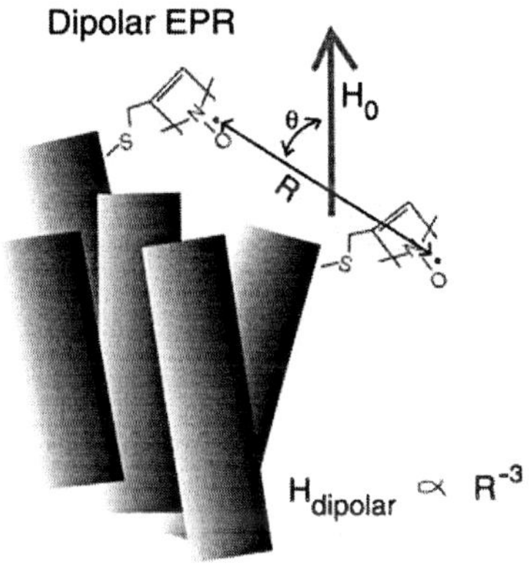

Figure 15.4 *Schematic depiction of a doubly labeled five-helix protein. The inter-nitroxide distance is determined from the spin–spin dipolar interaction. (Reprinted with permission from Academic Press.)*

The EPR spectrum will consequently be insensitive to global motions at most conventional microwave frequencies. (2) The nitroxides adopt a static distribution of orientations and distances with respect to each other. Neither the global rotation of the protein, nor significant local rotational motions of the nitroxides, occur on a timescale less than 1μs. (3) The nitroxide side chains undergo significant local motions or the protein is undergoing global rotation on a timescale at which the EPR spectrum is sensitive.

It would be ideal for every system to fall into case 1, since quantitative methods for determining accurate and precise distance and orientation information have already been developed for this case.[17,18] The other cases involve aspects of global rotational diffusion, internal protein dynamics, side chain dynamics, and inherent orientational disorder. It would be desirable to adjust the experimental conditions if possible to reduce rotational dynamics in order to improve the resolution of the dipolar splittings. The global rotational diffusion time can be increased by cooling or freezing the sample in solvent systems that form a glass,[19] or by adding reagents, such as glycerol or sucrose, that can increase significantly the viscosity.[20] All of these methods should slow internal protein and local probe motion without altering the distribution of interprobe distances and orientations.

15.5.3 Example of strong dipolar interactions

If two nitroxide groups are rigidly (and uniquely) bound to a large macromolecule and are situated within about <12–15 Å of one another, large dipolar splittings can be observed. While such splittings have been seen for frozen biradical nitroxides, it was not until 1984 that a resolvable dipolar splitting between two distinctly placed spin labels in a protein was demonstrated.[21] The specific system under investigation was the important glycolytic enzyme, glyceraldehyde-3-phosphate dehydrogenase (GAPDH), whose mechanism of action involves the coenzyme NAD^+, and substrate glyceraldehyde-3-phosphate. In the X-band EPR experiment a nitroxide NAD^+ coenzyme analog was placed in the active site of each GAPDH subunit in the tetrameric enzyme structure. Figure 15.5 depicts the spin label and the coordinate system for the distance measurements. The topography is such that the nitroxide moieties on the NAD^+ analog situate in one dimer about a twofold rotation axis, where they are about 11–13 Å apart across the R-axis in Figure 15.6; their intersubunit distances across the other two axes in the unit cell exceed 25 Å, where no dipolar interaction was detectable. The work was subsequently revisited and more thoroughly measured and analyzed at several frequencies up to 94 GHz.[17] This particular example was truly groundbreaking for a spin label distance application to a relevant biochemical system.

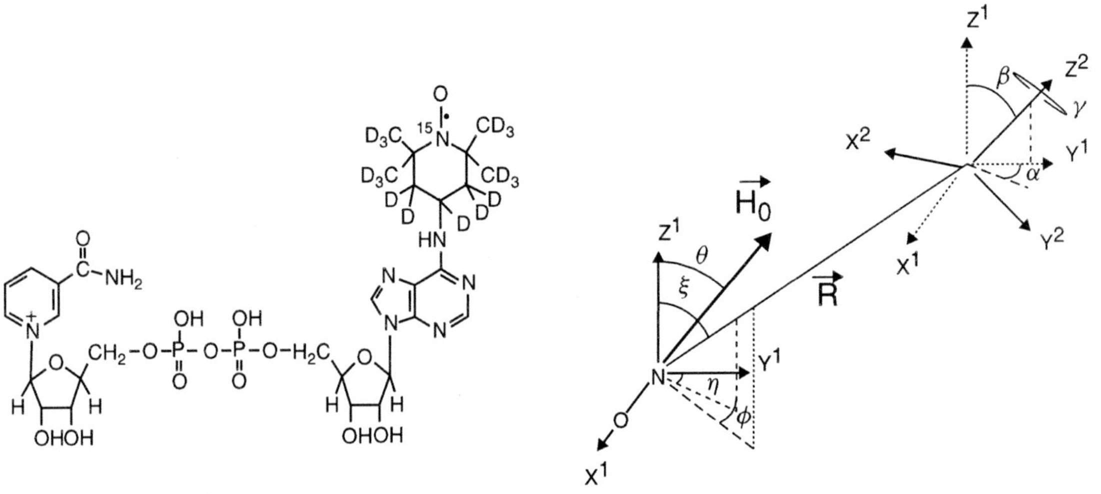

Figure 15.5 *Structure of the spin labeled NAD analog anti-[^{15}N,D_{17}]-N^6-SL-NAD^+ (left), and a diagram illustrating the two angles (ξ, η) relating the interelectron vector, R, to the axis frame defined by nitroxide 1, and the three angles (α, β, γ) relating the orientation of nitroxide 2 to nitroxide 1 (right). The angles θ and φ define the orientation of the external magnetic field, H, with respect to the axis frame of nitroxide 1. (Adapted from [17], Copyright 1997, with permission from Elsevier.)*

15.5.4 Multiple-quantum EPR and distance measurements

The determination of intra- and intermolecular distances has become an important application of contemporary EPR spectroscopy.[16] Particularly challenging problems encompass the structure of protein complexes and the functional dynamics of proteins that are neither soluble nor can be crystallized, such as the plethora of membrane proteins.

Distances can be determined from the dipole–dipole interaction between two paramagnets that are strategically placed within 8–25 Å of each other. This methodology, initially developed by Rabenstein and Shin,[21] involves incorporating a spin label at two specific cysteines followed by line shape analysis for dipolar contributions. This method, which is accurate to within 10 %, yields fairly accurate distances. However, more recent instrumental and methodological advances have extended this distance up to 50 Å and beyond. The Freed laboratory at the NIH Center at Cornell University (ACERT) has developed capabilities for higher sensitive pulsed EPR methods that preclude the need for freezing samples in order to enhance the dipolar interactions.[22,23] It uses double quantum coherence (DQC), a powerful application of pulsed EPR for distance measurements between interacting spins.[22–24] The DQC EPR experiment (Figure 15.7) permits detection of weak dipolar interactions between two nitroxides. Oscillations in echo amplitude versus echo time are due to this dipolar interaction.[25] A detailed analysis of DQC indicates that distances can be measured to ∼ 80 Å, comparable to the capabilities of fluorescence energy transfer.[23]

A recent and exciting approach by Budil and Freed has shown that, using a new method based on double electron–electron resonance (DEER) spectroscopy, it is possible to measure forces between small protein domains using a doubly labeled model peptide from the α-helical coiled-coil leucine zipper.[26] The equilibrium distribution of distances between the two rigidly attached nitroxide spin labels yielded a closing force of 100 ± 10 pN between monomers, which agreed quite well with theoretical predictions. Hence, this methodology can be extended into molecular assemblies and molecular motors where accurate distance changes are important.

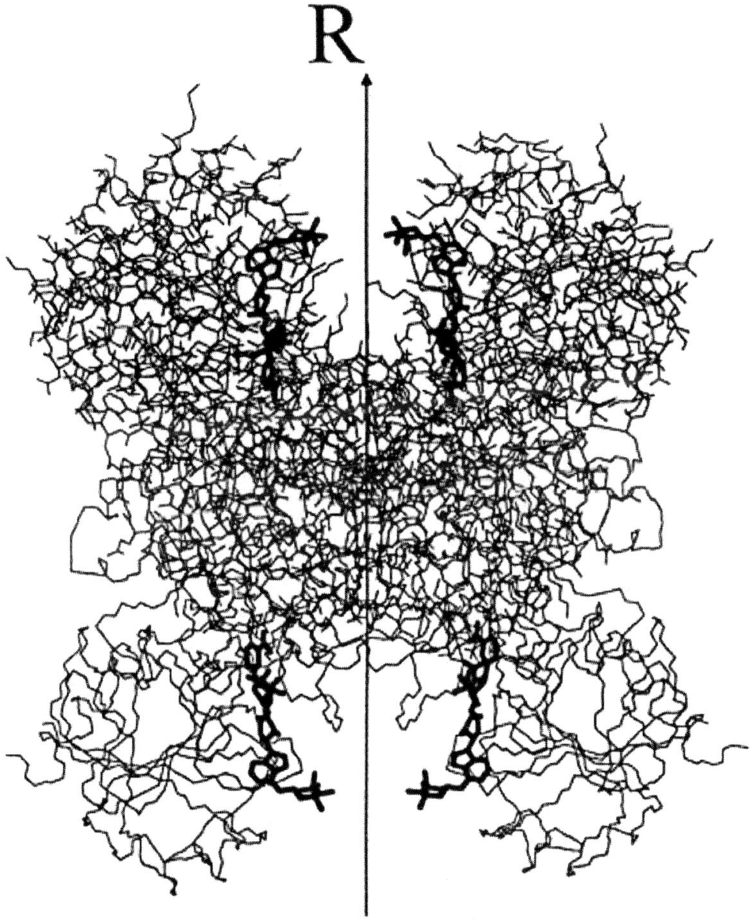

Figure 15.6 Structure of the N6-SL-NAD+-GAPDH tetramer as determined from molecular modeling. All four N6-SL-NAD′ cofactors are shown in bold. The lower pair of GAPD subunits shows the backbone only. The upper pair shows the backbone and side chains. The crystallographic R axis extends upward through the tetramer. (Adapted from [17], Copyright 1997, with permission from Elsevier.)

15.6 Site directed spin labeling (SDSL): how is it done?

For site directed spin labeling it is necessary to be able to spin label a protein with "distance markers" that can analyze aspects of conformation and polarity in the cases of both singly (and multiply) labeled proteins. Briefly, one (or occasionally two) thiol group(s) must be incorporated in a protein sequence and subsequently be covalently labeled with a uniquely specific reactive, reversible spin label. As mentioned earlier, Berliner and coworkers introduced the MTSL label and demonstrated its efficacy with the thiol protease, papain.[11] This was followed by the groundbreaking work of Hubbell and coworkers who recognized that it was possible potentially to genetically engineer cysteine residues into proteins at will.[13] The technique is now relatively standard with a series of protocols and parameters for any protein system. The following, taken in part from Feix and Klug,[12] summarizes the SDSL paradigm and the information that can be learned from it.

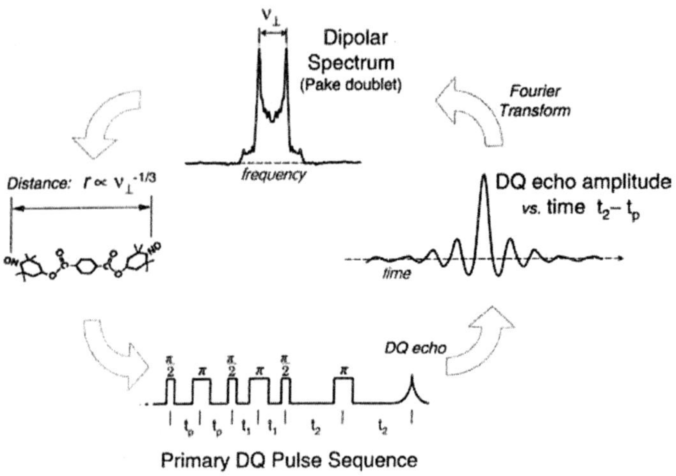

Figure 15.7 The pulsed DQC ESR experiment. (Left) A doubly labeled molecule is dissolved in a frozen amorphous matrix. (Bottom) The six-pulse DQC sequence preserves contributions to the double quantum echo signal arising solely from the dipolar interaction between the pair of nitroxide spin labels. The time intervals t_p and t_2 are varied such that the total time of the pulse sequence remains constant. (Right) The maximum of the echo signal is recorded as a function of $t_2 - t_p$. (Top) A Fourier transform of this echo signal yields the Pake-type dipolar spectrum in the frequency domain. The separation $\nu_\perp$ of the two sharp peaks is directly related to the distance, r, between the labels, $r \alpha \nu_\perp^{-1/3}$. (The echo signal and the dipolar spectrum shown here were measured for the molecule on the left, for which $r \approx 16.2 \pm 0.5$ Å.). (Reprinted from [23] with permission from AAAS.)

15.6.1 The SDSL paradigm

- Clone gene for the protein under study into an appropriate vector.
- Construct a "wild type" background containing no reactive cysteines.
- Site-specifically introduce unique cysteines as eventual labeling sites at desired positions in the protein structure.
- Verify the functional viability, expression, and global structure of the cysteine-containing mutants.
- Isolate and label with a SH-specific nitroxide.
- Characterize the local structure and conformational dynamics of the spin labeled sites by EPR spectroscopy.

15.6.2 SDSL parameters

- Motion – for example, ΔH_0^{-1} (Inverse width of the center line). Proportional to rotational motion (increases as motion gets faster).
- Accessibility – Π(x). Proportional to the bimolecular collision rate of a spin label with a given paramagnetic agent. (NiEDDA, NiAA, CrOx for polar; O_2 for non-polar environments.)
- Bilayer depth – Φ. Linearly proportional to depth of nitroxide relative to the membrane surface. Calibrated for a given membrane system with lipid spin labels of known depths.
- Dipolar coupling – Measure of distance between two nitroxides in a double-labeled system.

The real success of this method has been its ability to permit protein structure determination almost entirely from the EPR results. That is, it is possible to determine local rigidity and proximity to other structural elements, accessibility to solvent or the non-polar interior, and quantitative conformational movements form double labeling studies. In fact, the trends and cycles in increasing and decreasing exposure (accessibility) can trace an alpha-helical backbone or beta-sheet structure. By using MTSL labels of varying tether length and rigidity, these results can be confirmed unambiguously.

15.7 Other spin labeling applications

Spin-label-based distance measurements can be applied to other macromolecular systems, such as polymers, dendrimers, and the like but, as noted in the Introduction (Section 15.1), the topics in this chapter focus on biochemical/biological applications. The field of lipids and membranes, which is technically a spin probe approach, is vast and is not covered here. An excellent book has appeared recently that covers both the theory and applications.[27]

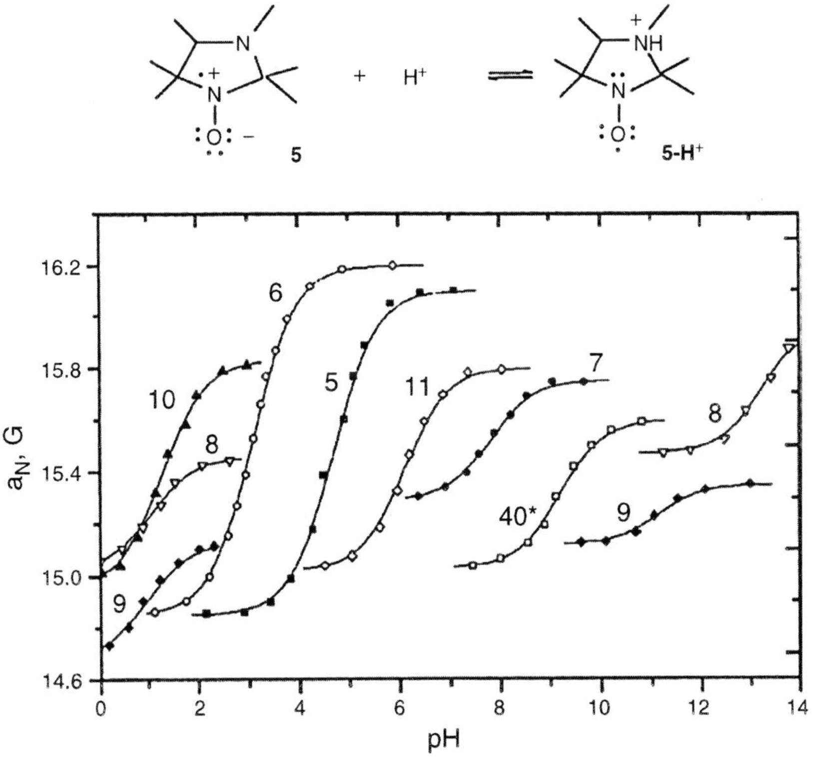

Figure 15.8 Imidazoline nitroxide protonation reaction (top) and a plot of pH induced changes in the hyperfine splitting for eight different imidazoline nitroxides (bottom). (Reprinted with permission from [28]. Copyright 1998 Plenum Press.)

15.7.1 pH sensitive spin labels

When probing the local environment of a protein, or for that matter a cellular or tissue region in an animal, the ability to assess the "polarity" on the basis of the local pH would be advantageous. Khramtsov, in collaboration with Volodarsky and others in Novosibirsk, has synthesized and developed a series of imidazoline nitroxides that can be protonated.[28,29] These labels have been tested over an array of pH ranges, each one, as shown in Figure 15.8, acting as a "pH indicator" probe which can be covalently attached to a protein, perhaps based on a dithio label or a methylthiolsulfonate analog (i.e., a pH sensitive MTSL). Consequently, the local pH near the labeling site and/or the environment of the labeled protein can be determined. In a fascinating *in vivo* study, the pH in the stomach of a laboratory animal was determined non-invasively.[30]

15.7.2 Spin labeled DNA – structure, dynamics and sequence analysis

Nucleic acid structures pose some major challenges. To study DNA structure and conformation, introducing a probe at some sensitive site, such as a purine or pyrimidine base, must be executed carefully to sense local

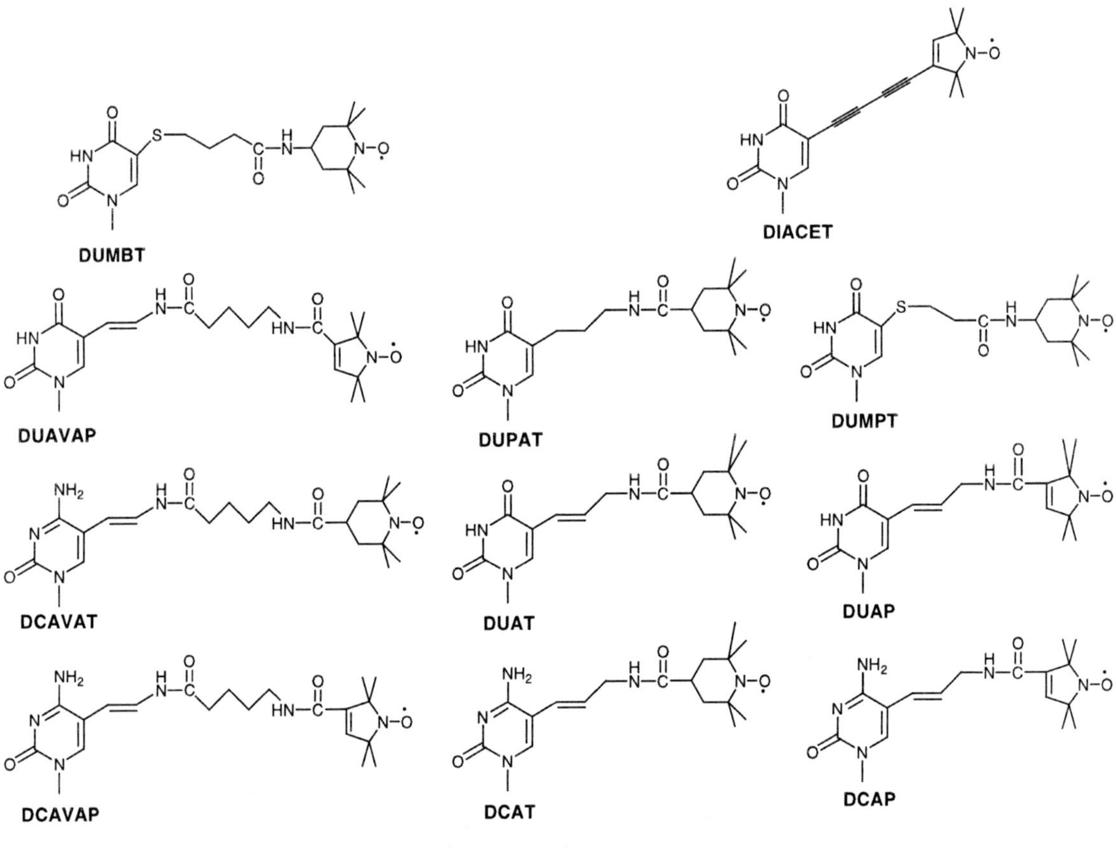

Figure 15.9 Representative spin labeled pyrimidine bases that can be incorporated into nucleic acid structures. (Adapted from [31] with permission.)

structure, global motion, DNA wobbling, and so on, without imparting significant structural perturbations. Bobst has been one of the pioneers in the field, both in spin label synthesis and applications.[31] A selected sample of a range of modified pyrimidine bases is shown in Figure 15.9. These labels offer properties ranging from very rigid, with restricted tumbling volumes, to very flexible. The analysis of both local and global motions within a polynucleotide system is complex. Figure 15.10 depicts a motional model of the dynamic components that contribute to the EPR spectrum. Needless to say, the plethora of modified bases, the methodology to incorporate them into various polynucleotides, and so on, is a great feat.

An impressive application to DNA analysis has been the ability to detect sequence "tracts" within a DNA molecule by EPR. If this were eventually commercialized, the applications of spin labeling and EPR in areas such as forensics might evolve as a sensitive alternative strategy to current methods.[32]

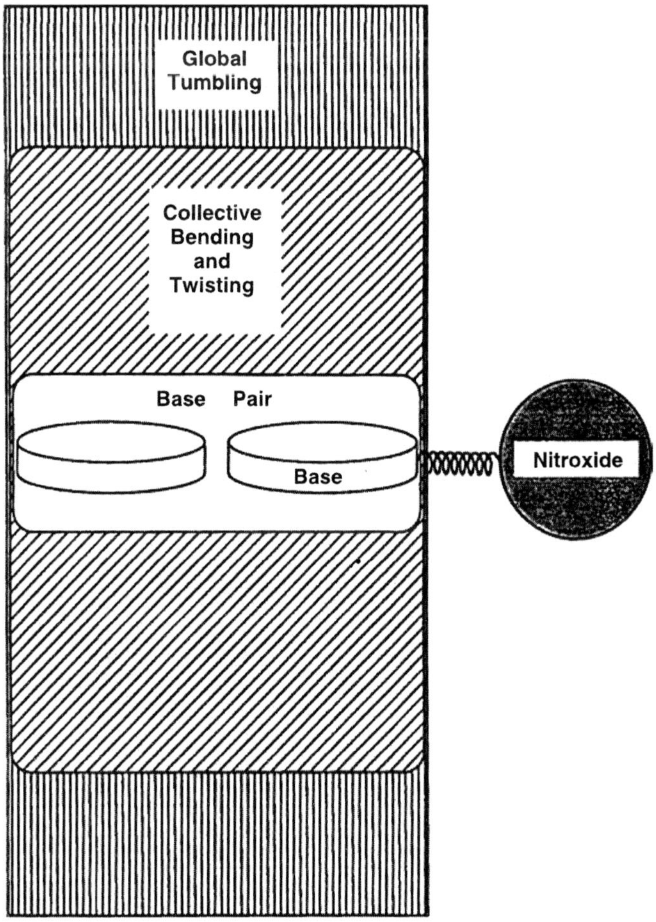

Figure 15.10 *Schematic diagram of a nitroxide (reporter group) monitoring dynamic processes in DNA according to the degree of coupling to each mode. The nitroxide ring is couple to the base motion through a tether linkage. The labeled base is coupled to another base which together experience base pair motion. This base pair is coupled to the collective bending and twisting of all base pairs in the helix. Collective motions are also coupled to the global tumbling. (Adapted with permission from [31]. Copyright 1998 Plenum Press.)*

15.8 Conclusions

The spin label technique offers some unique advantages over other biophysical methods of probing conformation in biological systems. There are no real molecular weight limits; optical transparency or physical homogeneity of the sample is not required; sensitivity is much higher than, for example, NMR; intramolecular distances can be assessed; and the tools of molecular biology now allow virtually any position in a protein or enzyme to be labeled. This technology leaves great promise for future studies of proteins and enzymes. Hence, we began with simple nitroxides based on five- and six-membered ring heterocyclic structures (Figure 15.2) that were modified to be crude chemical modification reagents. The field then matured with spin labels that were substrates, inhibitor or cofactor analogs of specific biochemical systems. Most recently, SDSL has allowed the modification of any site along a polypeptide chain in a protein of choice with reversible labels that probe conformation and can also be used in intramolecular distance measurements.

This chapter has covered both the historical development of spin labeling, from its beginnings to the recent "state of the art" and begins to address where the field is going and what are the big unresolved issues. From this author's perspective, it is imperative that other spectroscopic techniques be combined with EPR. As to addressing the great frontiers in biological systems, the aspects of free radical biology and medicine have great potential for the use of spin labels. Nitrone spin traps, which are precursors to nitroxides (i.e., spin labels) yield nitroxide adducts upon reaction with organic radicals of all sorts. The resultant EPR spectrum can frequently be characterized to identify the radical that combined with the nitrone. By strategically placing nitrone "spin labels" at particular sites in a biological system, there is the possibility of trapping radicals at some local site of interest.

The major problem remains with sensitivity. While the EPR signal of a radical is upwards of 600 to 1000 times more sensitive than, for example, a proton NMR spectrum, the minimum concentration limits are still too high for many biological problems. By reducing the line width, as was shown with ^{15}N perdeuterated nitroxides, the signal-to-noise ratio can be increased several fold, although less than an order of magnitude. Where the field will lead depends tremendously on synthetic organic chemistry and, perhaps, the development of yet other stable paramagnetic species beyond nitroxides. The development of other radical centers that afford the same sensitivity and dynamic specificity as nitroxides remains a future challenge. For example, trityl based radicals have line widths narrower than nitroxides by at least an order of magnitude. It only remains for the clever synthetic organic chemists to develop versatile analogs of these compounds for a new generation of spin labels to emerge. Lastly, combining EPR with NMR, fluorescence, and other physical techniques may open the way for other more powerful probes of biological systems.

References

1. T. J. Stone, T. Buckman, P. L. Nordio, and H. M. McConnell, *Proc. Nat. Acad. Sci. USA*, **54**, 1010–1017 (1965).
2. M. Burr and D. E. Koshland, Jr., *Proc. Nat. Acad. Sci. USA*, **52** 1017–1024 (1964).
3. L. J. Berliner, *Spin Labeling: Theory and Applications*, Academic Press, New York, (1976).
4. L. J. Berliner, *Spin Labeling II: Theory and Applications*, Academic Press, New York, (1979).
5. L. J. Berliner and J. Reuben, *Spin Labeling: Theory and Applications*, Biological Magnetic Resonance, Volume 8, Plenum Publishing Corp., New York, (1989).
6. L. J. Berliner and J. Reuben, *Spin Labeling: The Next Millenium*, Biological Magnetic Resonance, Volume 14, Plenum Publishing Corp., New York, (1998).
7. G. I. Likhtenshtein, *Spin Labeling Methods in Molecular Biology*, John Wiley & Sons, Inc., New York, (1976).
8. E. G. Rosantzev, and M. B. Neiman, *Tetrahedron*, **20**, 131–137 (1964).
9. E. G. Rozantsev, *Free Nitroxyl Radicals*, Plenum Press, New York, (1970).

10. L. J. Berliner, *Ann. NY Acad. Sci*., **414**, 153–161 (1983).
11. L. J. Berliner, J, Grunwald, H. O. Hankovszky, and K. Hideg, *Anal. Biochemistry*, **119**, 450–455 (1982).
12. J. B. Feix and C. S. Klug, Site-directed spin labeling of membrane proteins and peptide-membrane interactions, in Spin *Labeling: The Next Millennium* (eds L. J. Berliner and J. Reuben), Biological Magnetic Resonance, Volume 14, pp 251–281, Plenum Press, New York, (1998).
13. C. Altenbach, T. Marti, H. G. Khorana and W. L. Hubbell, *Science*, **248**, 1088–1092 (1990).
14. J. S. Leigh, Jr., *J. Chem. Phys*., **52**, 2608–2612 (1970).
15. E. J. Hustedt, and A. H. Beth, *Ann. Rev. Biophys. Biomol. Struct*., **28**, 129–153 (1999).
16. W. Xiao and Y. K. Shin, EPR, in *Distance Measurements in Biological Systems by EPR* (eds G. R. Eaton, S. S. Eaton and L. J. Berliner), Vol. **19**, Kluwer Academic/Plenum Publishing Corp., New York (2000).
17. E. J. Hustedt, A. I. Smirnov, C. F. Laub, and A. H. Beth, *Biophys. J*. **74**, 1861–1877 (1997).
18. E. J. Hustedt, and A. H. Beth, in *Distance Measurements in Biological Systems by EPR* (eds G. R. Eaton, S. S. Eaton and L. J. Berliner), Vol. **19**, Kluwer Academic/Plenum Publishing Corp., New York (2000).
19. P. Hanson, D. J. Anderson, G. Martinez, *et al*., *Mol. Phys*., **95**, 957–966 (1998).
20. D. Rabenstein, and Y.-K. Shin, *Proc. Natl. Acad. Sci. USA*, **92**, 8239–8243 (1995).
21. A. H. Beth, B. H. Robinson, C. E. Cobb, *et al*., *J. Biol. Chem*., **259**, 9717–9728 (1984).
22. P. P. Borbat, and J. H. Freed, in *Distance Measurements in Biological Systems by EPR* (eds G. R. Eaton, S. S. Eaton and L. J. Berliner), Vol. **19**, pp. 383–459, Kluwer Academic/Plenum Publishing Corp., New York (2000).
23. P. P. Borbat, A. J. Costa-Filho, K. A. Earle, *et al*., *Science*, **291**, 266–269 (2001).
24. P. Borbat, and J. H. Freed, *Chem. Phys. Lett*., **313**, 145–154 (1999).
25. J. H. Freed, in *Spin Labeling: Theory and Applications* (ed. L. J. Berliner), pp. 53–132, Academic Press, New York, (1976).
26. S. V. Gullà, G. Sharma, P. Borbat, *et al*., *J. Am. Chem. Soc*., **131**, 5374–5375 (2009).
27. M. A. Hemminga and L. J. Berliner, *ESR Spectroscopy in Membrane Biophysics*, Biological Magnetic Resonance, Volume 27, Springer Publishing Corp., New York (2007).
28. V. V. Khramtsov and L. B. Volodarsky, in Spin *Labeling: The Next Millennium* (eds L. J. Berliner and J. Reuben), Biological Magnetic Resonance, Volume 14, pp 109–180, Plenum Press, New York, (1998).
29. V. V. Khramtsov, I. A. Grigor'ev, M. A. Foster, *et al*., *Cell. Mol. Biol*., **46**, 1361–1374. (2000).
30. M. A. Foster, I. A. Grigor'ev, D. J. Lurie, *et al*., *Magn. Reson. Med*., **49**, 558–67. (2003).
31. R. S. Keyes and A. M. Bobst, in Spin *Labeling: The Next Millennium* (eds L. J. Berliner and J. Reuben), Biological Magnetic Resonance, Volume 14, pp 283–338, Plenum Press, New York, (1998).
32. A. M. Bobst, and J. D Hester, *Detection of nucleic acid target sequences by electron paramagnetic resonance spectroscopy*, US Patent Application number 10/182,559 (2002).

16

Functional *in vivo* EPR Spectroscopy and Imaging Using Nitroxide and Trityl Radicals

Valery V. Khramtsov and Jay L. Zweier

Dorothy M. Davis Heart & Lung Research Institute, The Ohio State University, Ohio, USA

16.1 Introduction

Magnetic resonance spectroscopies in combination with functional molecular probes are powerful tools in biochemical and biomedical research. Broad clinical applications of nuclear magnetic resonance, NMR, and magnetic resonance imaging (MRI) are possible due to the sufficient depth of magnetic field penetration in living tissues and existence of endogenous NMR-sensitive nuclei, up to 100 M concentration in the case of water protons. However, low intrinsic NMR sensitivity and overlap of the various endogenous NMR signals limit functional *in vivo* NMR and MRI applications beyond the anatomical resolution. Electron paramagnetic resonance (EPR) spectroscopy has greater sensitivity than NMR for the same probe concentration, due to the larger magnetic moment of the electron compared with that of the proton. Over the past decade low-field L-band EPR spectrometers, designed to increase the depth of microwave penetration and decrease non-resonant losses, have become commercially available allowing experiments with isolated organs and small animals such as mice. Moreover, RF-EPR instruments with lower fields (down to 300 MHz), as well as instrumentation for spatial and spectral–spatial EPR imaging (EPRI), have been constructed.[1–4] However, the potential of EPR-based techniques is still far from being maximally defined, predominantly because of the requirement of the development of stable paramagnetic probes. In return, the use of exogenous EPR probes and lack of endogenous EPR signals provide EPR techniques with additional advantage in signal specificity. Nitroxyl radicals (NRs) and triarylmethyl radicals (TAMs) represent two main classes of soluble paramagnetic materials used for EPR spectroscopy and imaging applications. The synthesis of stable organic NRs,[5,6] in which the unpaired electron is localized at the sterically protected NO group ($\rho_O^\pi \approx 0.6$; $\rho_N^\pi \approx 0.4$), revolutionized numerous areas of EPR applications.

Half a century of continuous progress in the nitroxide chemistry has resulted in design of specific NRs for spin labeling,[7] site-directed spin labeling,[8] EPR oximetry,[9,10] pH,[11] thiols[12,13] and NO measurements,[14–16] including *in vivo* functional EPR/EPRI application.[1,17–19] TAM radicals represent a fundamentally different basic structure of stable organic radicals, with approximately 60 % of the unpaired electron localized at the sterically protected central carbon atom and the remaining 40 % delocalized over the three aryl substituents.[20] NRs have the advantages of well developed chemistry, resulting in variability of structure, solubility, functionality, and ability to be targeted. On the other hand, TAMs have advantages over NRs in that they offer extreme stability toward tissue redox processes, longer relaxation times, and narrower line widths, making them particularly attractive for imaging applications.[21]

16.2 Nitroxyl radicals

The first stable di-*tert*-alkyl nitroxide of piperidine type (TEMPO)[5] (Scheme 16.1) was synthesized fifty years ago. The stability of the N–O group allowed chemical reactions of the nitroxides which do not involve the radical center allowing further NR derivitization.[6,22] In the early 1960s, EPR spectral sensitivities of the NRs to their local environment, including viscosity[23] and polarity,[24] were reported. Later on, McConnell and colleagues established the spin labeling technique, demonstrating the EPR spectral sensitivity of biologically relevant macromolecules labeled with NRs to molecular motion and microenvironment.[25,26] The combination of EPR spin labeling with site-directed mutagenesis reactions led to the site-directed spin labeling (SDSL) technique currently widely used in the study of protein structure and dynamics.[8,27,28] The sensitivity of NR EPR spectra to pair-wise Heisenberg spin exchange[29] or magnetic dipole–dipole interactions,[30] provide experimental tools for the measurement of intermolecular distances and the location of paramagnetic species, such as metal ions and oxygen. NRs were the first paramagnetic probes used for EPR oximetry based on the spin exchange phenomenon between diradical oxygen molecule and NRs.[10,31–33]

Specific chemical reactions of NRs further extend their functional applications. The most well known reaction of NRs, their one-electron reduction to EPR silent hydroxylamines, is largely responsible for the biodegradation of the nitroxides in living tissues[34] and, in general, significantly limits many biological applications of NRs. On the other hand, the rate of the NRs' reduction provides information on the redox state of living tissues.[17] The other reactions of the NRs, which do not involve radical center but can be followed by EPR,[35] include reactions of the specific NRs with protons, thiols, and nitric oxide (NO). The functionally-enhanced NRs for EPR detection of corresponding biologically relevant molecules were developed, namely pH-, SH-, and NO-sensitive spin probes.[35,36] The main types of the nitroxides discussed in this chapter are shown in Scheme 16.1.

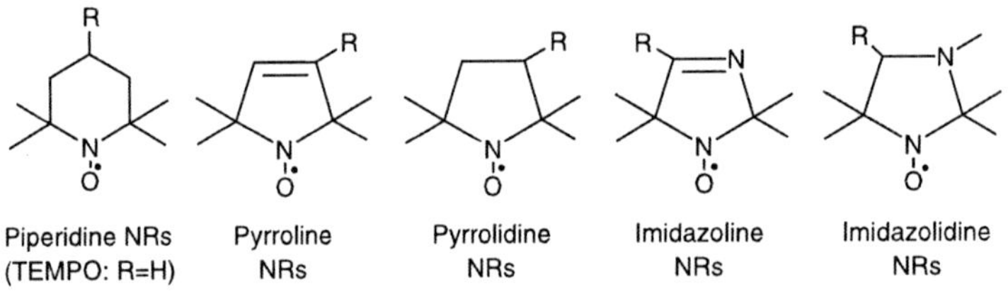

Scheme 16.1 Types of nitroxide radicals (NRs)

16.3 Triarylmethyl (trityl) radicals

Trityl radicals have an even longer history than NRs, since the initial report by Gomberg in 1900 of the synthesis of triphenylmethyl radical.[37] Nevertheless, only recently have these compounds, with sterically protected trivalent carbon, regained attention as the basic structural fragment for the synthesis of stable organic radicals. By the late 1990s, Nycomed Innovation AB further refined Gomberg's original trityl radical in order to avoid hydrogen hyperfine coupling and enhance its stability and water solubility.[21,38] The placement of heteroatoms in these systems eliminates hydrogen atoms in close proximity to the radical center, and consequently the numerous hyperfine splittings and broadening of the EPR signal caused by coupling of the radical electron with the hydrogen nuclear spins. A new family of trityl spin probes was synthesized, also known as tetrathiatriarylmethyl (TAM), bearing four sulfur atoms on the phenyl ring. The most representative members are TAM derivatives containing carboxyl groups, namely cTAM, deuterated cTAM, and the more hydrophilic Oxo63 and Oxo31 derivatives (Scheme 16.2). Recently, creative efforts have been employed for improved synthesis of these complex molecules.[39,40] cTAM can now be synthesized on a large scale in an efficient way.[40]

TAMs display extraordinary stability toward tissue redox processes with life time up to 24 hours in human blood,[21] surpassing the behavior of NRs in this regard. The EPR spectra of these TAM derivatives[21,39,40] display a very narrow single line which is generally not broadened by interaction with proteins and other biological molecules, making these molecules particularly attractive for imaging applications using EPRI and proton electron double resonance imaging (PEDRI) techniques.[41,42] In the latter case, the long relaxation time of TAMs makes them easily saturatable by radio frequency (RF) irradiation and provides an advantage over NRs for PEDRI applications allowing for enhancement of sensitivity and resolution with less RF heating of the sample.[21,33,43] Applications of TAM radicals include EPR oximetry[21,33,43] and recently a reported sensitivity to the superoxide anion[44,45] and pH.[46,47]

In this chapter, functional applications of EPR-based spectroscopy and imaging of the NR and TAM probes are reviewed, with particular emphasis on *in vivo* EPR measurement of oxygen, pH, tissue redox, and intracellular GSH content.

16.4 *In vivo* EPR oximetry using nitroxyl and trityl probes

The measurements of molecular oxygen concentration in living tissues are of crucial importance for monitoring the energetic metabolism from a physiological and pathological point of view. In certain stress conditions, for example, high exercise levels, interruption of normal blood supply or biochemical shock, the

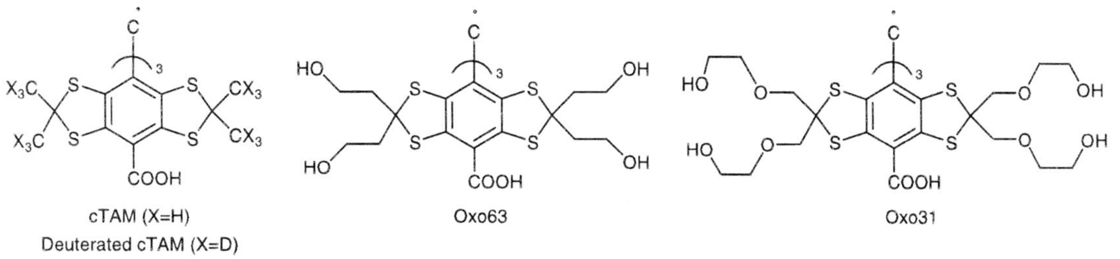

Scheme 16.2 Representative chemical structures of triarylmethyl radicals (TAMs)

oxygen homeostasis, at least locally, may be compromised. The effectiveness of cancer therapy is affected by the oxygenation status of normal and tumor tissues.[48,49] Oxygenation status of ischemic tissue in stroke and myocardial infarction is critically important but is measured only with difficulty and invasiveness that frustrates clinical application.

16.4.1 Magnetic resonance approaches for *in vivo* oximetry

A reasonable depth of penetration of magnetic field in the living tissues makes NMR and low field EPR techniques the most appropriate approaches for non-invasive *in vivo* oxygen measurements.[50] Magnetic resonance methods rely on the effect of paramagnetic oxygen on the relaxation times of the molecules excited by electromagnetic radiation. The effect of oxygen on NMR relaxation times was described when NMR was first discovered by Bloch *et al.*[51]. After inventing MRI, oxygen was the primary candidate to be considered as potential contrast agent.[52,53] However, this possibility was ruled out due to the small effect of oxygen on the T_1 and T_2 values of the protons in blood and loss of its paramagnetism upon complexation by carrier molecules.[54] When the other parameters are kept constant, MRI oximetry is possible via water T_2 measurements, due to the difference in magnetic properties of oxyhaemoglobin (diamagnetic) and deoxyhaemoglobin (paramagnetic). The blood oxygen level dependent (BOLD) effect[55] has been extensively used for functional MRI of the brain despite the relatively low sensitivity of effective T_2 to tissue oxygenation and dispersion of the data. Several NMR techniques for oxygen measurement were developed relying on exogenous fluorinated probes, including ^{19}F NMR spectroscopy/imaging using perfluorocarbon (PFC) emulsions[56,57] and fluorinated nitroimidazoles.[58] However, spin lattice relaxation rates of fluorinated probes may also depend on other physiological or histological parameters.[57]

EPR oximetry is one of the most promising and rapidly developing techniques for measurement of oxygen in living tissues.[9] While oxygen in its triplet ground state demonstrates a strong X-band EPR signal in the gas phase of 120 lines at low pressure,[59] no EPR spectra have ever been reported for dissolved oxygen due to line broadening. Hence, biological EPR oximetry relies on exogenous paramagnetic probes.

16.4.2 Nitroxide probes for EPR oximetry

Nitroxyl radicals were the first paramagnetic probes used for EPR oximetry.[10,31,60,61] The physical basis of the spin label EPR oximetry method relies on Heisenberg exchange between NR and dioxygen, a stable diradical. Both the longitudinal (T_1) and transverse (T_2) relaxation time of the NRs can be rather close to the bimolecular collision rate of dissolved oxygen in aqueous solutions and be significantly affected by the encounter rate. The encounter rate (w) is governed by the Smoluchowski equation, $w = 4\pi \cdot r \cdot D [O_2]$, where r is the interaction distance and D is the diffusion constant of oxygen, which is much greater than the diffusion constant of the NR. According to the early work of Pake and Tuttle (1959)[62], the exchange-induced EPR line broadening is proportional to the radical–radical collision rate and, therefore, to oxygen concentration. The observation of line width broadening of the TEMPO radical by dissolved oxygen in various solvents was reported by Povich in 1975.[63] Two years later, Backer *et al.*[31] were the first to apply the T_2 oximetry method based on oxygen-induced line broadening of the NR to follow mitochondrial respiration in samples containing about 100 liver cells. The T_1-sensitive EPR oximetry methods were introduced by Hyde *et al.*[64,65] T_1-sensitive experiments include pulsed saturation recovery, continuous wave saturation and rapid passage displays.[66] T_1-sensitive methods might have an advantage for highly viscous environments or spin labeled macromolecules because for the nitroxides $T_2 \ll T_1$, and for the optimal sensitivity the collision rate needs to be comparable to the relaxation time. However, for most biological applications in low viscosity solutions, T_2-sensitive methods are preferred because they are instrumentally easy. An advantage of EPR oximetry is that it is based on pure physical interaction between the molecules of paramagnetic probe and oxygen and does not interfere with the oxygen metabolism, therefore providing a basis for non-invasive oxygen measurements in biological systems, including that *in vivo*.

EPR spectra of NRs show typical triplet patterns (doublet for ^{15}N–O isotopically substituted NRs) due to hyperfine interaction with a nitrogen nucleus possessing a non-zero nuclear spin, $I = 1$ ($I = 1/2$ for the ^{15}N isotope). A typical effect of oxygen on the individual line of the triplet EPR spectrum of the NRs is shown in Figure 16.1 for the CTPO radical (Scheme 16.3). At low oxygen levels each component of the CTPO triplet spectrum shows additional partially resolved superhyperfine splittings with 12 protons of four methyl groups and proton at carbon C4 of the heterocycle. An increase in oxygen concentration results first in the masking the superhyperfine structure and then in the broadening of the enveloped EPR line, clearly seen in Figure 16.1 as an increase of the peak-to-peak line width (ΔH_{pp}). Note that NRs with larger number of protons in the radical heterocycle, for example, a TEMPO radical of the piperidine type (Scheme 16.1), do not reveal superhyperfine structure even in the absence of oxygen. The line broadening

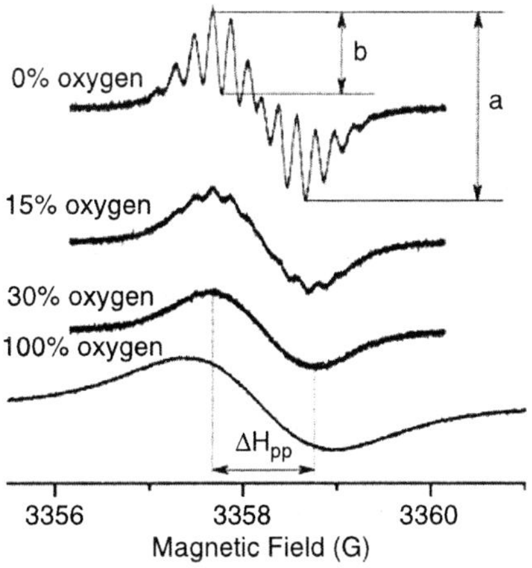

Figure 16.1 *EPR spectra of 50μM aqueous solution of the CTPO radical measured at 37 °C at various oxygen concentrations. Spectral parameters: microwave power, 2.4 mW; time constant 5.12 ms; sweep time, 81.92 s; number of points, 4096; modulation amplitude, 0.1 G; (0.8 G for spectrum at 100 % oxygen); sweep width, 4 G (6 G for spectrum at 100 % oxygen). Temperature and oxygen content of the sample during the experiment was controlled by Temperature and Gas Controller (Noxygen, Germany). Line width at 100 % oxygen is 1.60 ± 0.03 G.*

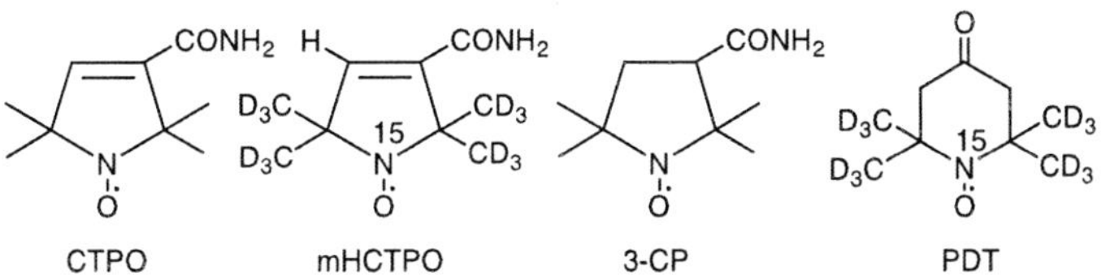

Scheme 16.3 Chemical structures of NRs: 3-carbamoyl-2,2,5,5-tetramethylpyrrolin-1-yloxy (CTPO), 4-protio-3-carbamoyl-2,2,5,5-tetraperdeuteromethyl-3-pyrrolinyl-1-^{15}N-oxy (mHCTPO), 3-carbamoyl-2,2,5,5-tetramethylpyrrolidin-1-yloxy (3-CP), and 4-oxo-2,2,6,6-tetramethylpiperidine-d_{16}-1-^{15}N-oxyl (perdeuterated tempone, PDT)

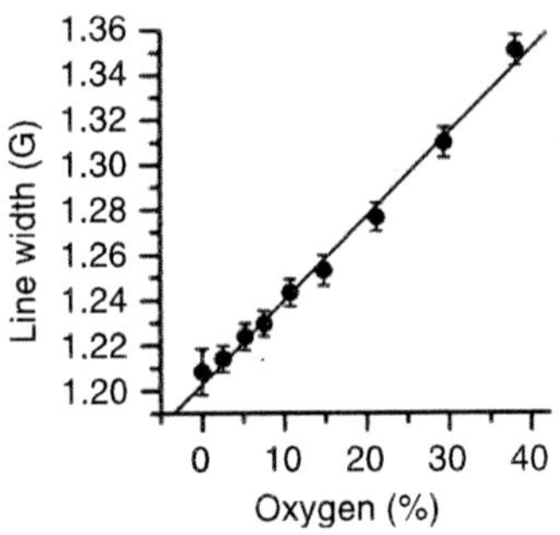

Figure 16.2 The dependence of the EPR spectrum line width of a 200μM aqueous solution of TEMPO on oxygen concentration measured at 24 °C. Spectral parameters were as follows: microwave power, 10 mW; modulation amplitude, 0.5 G. Line width at 100 % oxygen (not shown) was 1.70 $\pm$ 0.03 G.

effect of oxygen on the ΔH_{pp} of the NRs slightly varies with the radical structure and on average is about 500 mG/100 % oxygen or about 450 mG/mM of oxygen (Figures 16.1 and 16.2). Therefore the NRs with the typical line width of about 1 G in oxygen free solutions are practically insensitive to oxygen concentrations below 2 % O_2, as seen in Figure 16.2 for the TEMPO radical ($\Delta H_{pp} = 1.21$ G at $[O_2] = 0$).

An example of L-band EPR oximetry application for monitoring oxygen depletion in ischemic myocardium obtained in isolated rat heart[67,68] using TEMPO as a spin probe is demonstrated in Figure 16.3. The spectral data showed a gradual decrease in the line width over duration of ischemia by about 0.5 G, approaching the line width observed in the absence of oxygen. Significantly slower oxygen consumption was observed in cardiopleged hearts (solid line)[67] subjected to global ischemia than that observed in noncardiopleged hearts (dashed line)[68] characterized by exponential decay time constants of 14.3 min and 1.9 min, respectively. This would be expected since in the cardiopleged hearts contractile function is arrested, which results in a much lower rate of oxygen consumption.

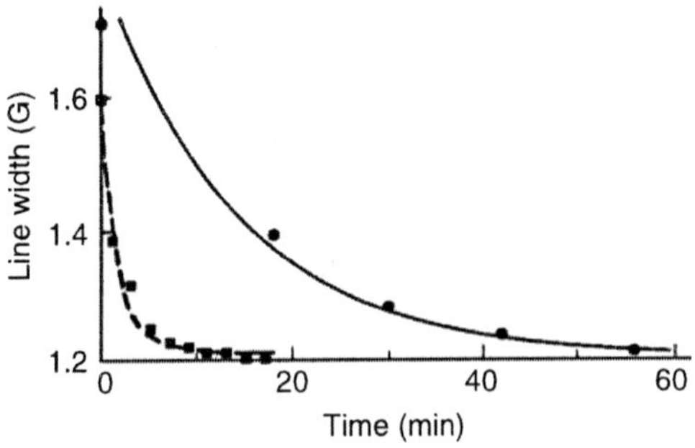

Figure 16.3 Kinetics of oxygen consumption in ischemic cardiopleged (•)[67] and non-cardiopleged (■)[68] hearts measured as changes in L-band EPR line width of TEMPO spin label. The isolated rat hearts were perfused by the method of Langendorff with Krebs bicarbonate-buffered perfusate solution equilibrated with 95 % O_2/5 % CO_2. The hearts were preloaded with infusion of 1 mM TEMPO reaching maximal EPR signal after 10 min of the NR administration.[67,68] Kinetics of oxygen consumption by myocardium was modeled with a single exponential function yielding exponential decay time constants, $\tau = 1.9$ min and $\tau = 14.3$ min, for non-cardiopleged (dashed line) and cardiopleged (solid line) hearts, respectively.

The use of perdeuterated NRs provides almost one order of magnitude enhancement in sensitivity to low oxygen concentrations due to narrow line width (ΔH_{pp}).[67,69,70] Alternatively, measurement of "the depth of resolution" of the superhyperfine structure of the NRs (e.g., parameter a/b for the CTPO shown in the Figure 16.1) provides similar or even better sensitivity to oxygen than perdeuterated NRs, and has been used for oxygen measurements in cellular and enzymatic systems.[31,60,71] However, among the drawbacks of both approaches are the need to limit NR concentration (ideally $<100\,\mu M$) and modulation amplitude, which compromise spectral intensity. Therefore, application of these approaches in *in vivo* systems using low field EPR can be complicated by overlapping of the line broadening effects induced by respective variations of oxygen and NR concentrations. Typically, the line broadening effect of the NR concentration on its line width is about 100–200 mG/mM, and may interfere with accurate oxygen measurements at high NR concentrations.[72]

Halpern *et al*.[73] proposed the application of a selectively deuterated CTPO radical, mHCTPO (Scheme 16.3), with only one hydrogen superhyperfine splitting in order to discriminate the contributions of variations in oxygen and NR concentrations on EPR line shape. Figure 16.4 shows the low field component of the EPR spectrum of the mHCTPO obtained from a large FSa tumor under two conditions: the mouse was first bathed for four minutes in a gas atmosphere of oxygen; next it was bathed in air for four minutes. Variations in both oxygen and radical concentrations affect "the depth of resolution" of the doublet spectrum characterized by the parameter R. Conversely, an increase *only* in spin label concentration but *not* oxygen results in narrowing of the hydrogen hyperfine splitting parameter (a_H). Measurement of both parameters, R and a_H, allows for discrimination between these two factors and quantitatively address the oxygen concentration in tumor tissue. As can be seen from EPR spectral analysis, despite compensatory vasoconstriction thought to prevent changes in oxygenation with the administered oxygen atmosphere, there are clear changes in the tumor oxygenation, from 42 ± 8 t with oxygen breathing to 7 ± 8 t in air (Figure 16.4).

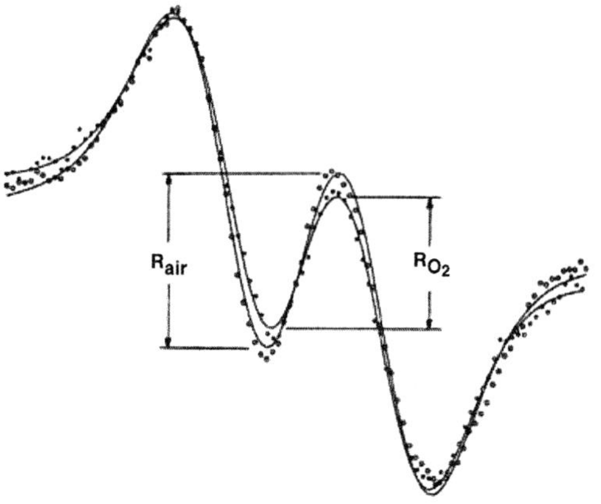

Figure 16.4 *EPR spectra of the mHCTPO radical measured in the FSa tumor in the living mouse. NR was administered i.p. at 30 mM as a 0.7 ml bolus. ○, Tumor spectrum from the anesthetized animal breathing air; •, tumor spectrum from the anesthetized animal breathing nearly 100 % O_2 with a 2-min equilibration between spectra. The R value for the 100 % O_2-inspired gas is 0.29 ± 0.02, while that for air breathing is 0.37 ± 0.02 (2-min measurements), corresponding to pO_2 of 42 ± 8 t and 7 ± 8 t. Despite the small apparent changes in the spectra, the large number of contributing spectral points gives significant pO_2 differences. (Reprinted with permission from [10]. Copyright 1994 National Academy of Sciences.)*

The well developed chemistry of the NRs[74,75] allows manipulation of their structure and properties, including the presence of charge, hydrophilic or hydrophobic groups, and the ability to be targeted. Small neutral NRs normally penetrate cellular membranes easily and are equally distributed throughout the intracellular and extracellular environments. Conversely, charged NRs will not cross the plasma membrane, and thus can be used to measure oxygen concentrations in the extracellular compartment.[76] Glockner *et al.*[32] demonstrated simultaneous measurement of intracellular and extracellular oxygen concentrations in a single sample: extracellular measurement was achieved using positively charged nitroxides encapsulated in liposomes, while intracellular oxygen was determined using a membrane-permeable nitroxide along with an extracellular broadening agent. Moreover, NRs can be incorporated in liposomes or linked to carrier molecules[77] to achieve organ or tissue selectivity. In general, NRs have low toxicity and can be administered to the animal by infusion or by intraperitoneal, intravenous or intratissue injection. To protect NRs against reduction and enhance their sensitivity to oxygen, Liu *et al.*[78] developed proteinaceous microspheres filled with the NRs dissolved in an organic liquid. Encapsulation of the NRs into the micosphere greatly increased the sensitivity of the EPR line width to [O_2] because of the higher solubility of oxygen in organic solvents. The authors used encapsulated NRs to measure the changes in oxygen concentration *in vivo* for 70 minutes after intravenous injection of the microspheres into a mouse. A potential concern of this approach is the leakage, as well as the release of the organic liquid from the microspheres when they are eventually metabolized, resulting in possible toxicity to the tissue.

The distribution of the NRs in living tissue can be measured using EPRI techniques. Moreover, spectral–spatial imaging can be performed; this contains a complete spectral profile, as a function of field, at each spatial voxel element. Application of one to three sets of gradient coils allows spectral–spatial EPRI in one, two or three spatial dimensions resulting in 2-, 3- or 4-D spectral–spatial images, respectively.[79] Because the spatial and spectral dimensions are fully separable, information about local line width, and hence local oxygen concentration, can be derived independently from local spin density. The continuous wave EPR imaging (CW EPRI) technique works better for the NRs imaging while pulsed Fourier transform imaging (FT EPRI) is limited to the probes with narrow line width of the order of 100 mG or less.[80]

Figure 16.5 demonstrates *in vivo* spectral–spatial EPRI in a rat tail using EPRI and a soluble NR, 3-CP (Scheme 16.3). Both the cross-sectional anatomy and oxygen distribution are evident. The oxygen map

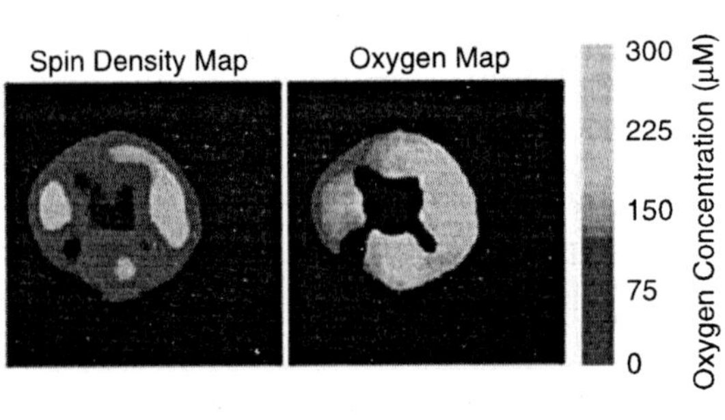

Figure 16.5 Images of the NR spin density and oxygen distribution in the cross-section of a rat tail reconstructed from 3-D spectral–spatial data obtained in vivo using L-band EPR imaging. Water soluble probe, 3-CP (Scheme 16.3) was infused intravenously, 200 mg kg^{-1}, and after a 25 min period for the label to distribute and reach maximum the images were obtained. Total image acquisition time was 20 min. Certain anatomical features (major blood vessels) and oxygen gradients are prominent in the image. (Reprinted with permission from [69]. Copyright 2000 Wiley-Liss.)

shows the differences in the oxygen perfusion. The major blood vessels and muscle bundles separated by bone were visible. Oxygen gradients from the center of the four major vessels into the distal regions can be seen in the oxygen map.

The oxygenation status of normal and tumor tissues is an important parameter, being in part responsible for the effectiveness of cancer therapy. Halpern *et al.*[10] demonstrated the efficiency of low field 250 MHz EPR oximetry in combination with a mHCTPO probe to report oxygen concentration in the body water of murine FSa and NFSa fibrosarcomas 7 cm deep in the tissues of a living animal (Figure 16.4). An oximetric 2-D (1 spectral/1 spatial) spectral image of the tumor was also obtained allowing, in principle, direct assessment of tumor hypoxia to determine the usefulness of radiation and chemotherapy adjuvants directed to hypoxic cell compartments. Hydrophobic perdeuterated NR, PDT (Scheme 16.3), has been applied to measure oxygenation level in lipophilic compartments with higher oxygen solubility in radiation induced fibrosarcoma (RIF-1) in mice.[82] The data showed a threefold lower level of oxygenation of the tumor tissue compared with that of the normal muscle.[82]

16.4.3 TAM oximetric probes

Nitroxyl radicals were also the first compounds tested in dynamic nuclear polarization (DNP) to probe the oxygen environment.[83] However, bioreduction of the nitroxides into EPR silent products and comparatively broad spectral lines complicate their application, particularly for EPR oxygen mapping. In this respect, triarylmethyl radicals developed for biomedical applications by Nycomed Innovation AB[38] (Scheme 16.2), have the advantage of extraordinary stability *in vivo* and a very narrow single EPR line of about 100 mG or less. The oxygen broadening of the TAMs in water is about 500 mG/mM of oxygen[21] similar to that for the NRs. On the other hand, the concentration broadening of the TAMs is about 10 mG/mM,[21] which is one order of magnitude less than that for the NRs.[72] These properties make TAM radicals superior oximetric probes for *in vivo* EPR/EPRI and DNP/PEDRI applications.[21,43]

Images of a tumor obtained using 250 MHz EPRI when the mouse was either breathing air or carbogen (95 % oxygen, 5 % carbon dioxide) are shown in Figure 16.6.[84] The images obtained using TAM Oxo31 probe (Scheme 16.2) show the heterogeneity in the distribution of oxygen in the tumor with higher [O_2] values after animal was carbogen-breathing for 10 minutes. Note that continuous wave spectral–spatial EPRI is based on measurement at numerous magnetic field gradients which typically requires a long acquisition time, 20 minutes in Figures 16.5 and 16.6, which is often unpractical for biological use.

Recent developments in PEDRI[85,86] demonstrated that this method allowed simultaneous co-registration of free radical distribution and anatomic information. Since PEDRI is based on proton MRI, it circumvents the resolution limitations of EPRI that occur due to very broad line widths of most paramagnetic labels, and inherently offers high spatial resolution and rapid image data collection. Therefore, PEDRI is now recognized as a powerful alternative to conventional EPRI. One of the successful applications of functional PEDRI is oxygen mapping. It is based on the paramagnetic character of dissolved oxygen which affects the EPR line width of free radical probes and, as a consequence, alters the RF power saturability of the probes. Therefore, PEDRI with variable saturation power offers a reliable method of imaging oxygen concentrations *in vivo* using TAM probes.[21,33,43,87] PEDRI oxygen images of a mouse with a tumor obtained using Oxo63 (Scheme 16.2) are shown in Figure 16.7. Spatially resolved functional information (oxygen map) was extracted from two PEDRI images collected with interleaving pulse sequences in which the EPR irradiation was applied at two different power levels, 45 and 3 W. The expanded tumor region clearly shows heterogeneity in oxygen distribution, severely hypoxic regions, and an increase in the oxygen level

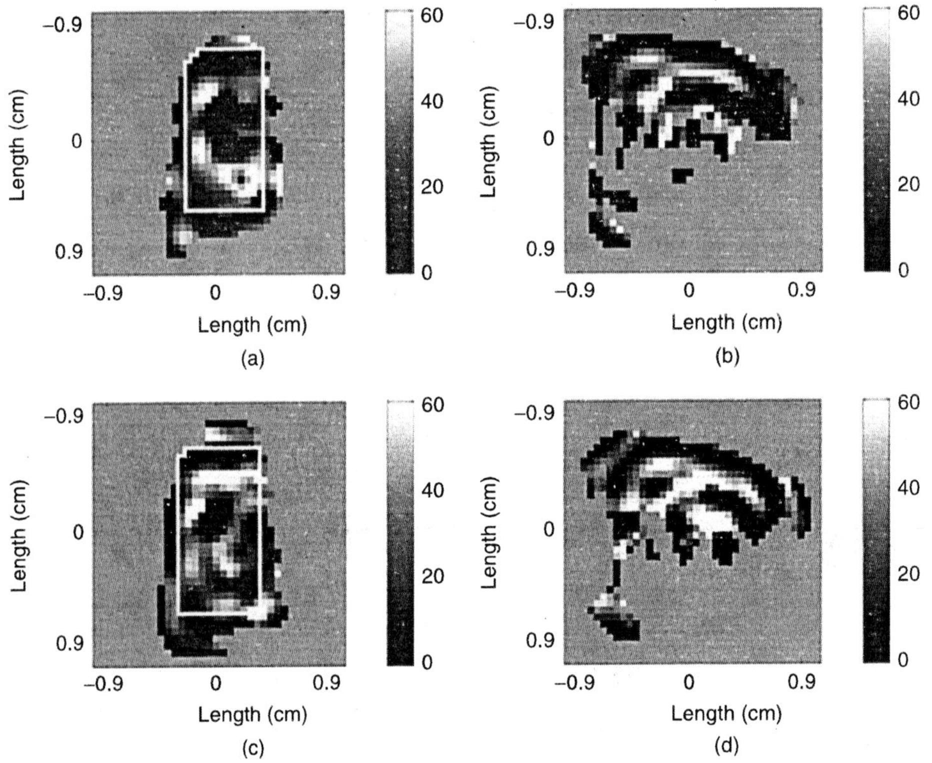

Figure 16.6 Oxygen image of an FSa tumor in the leg of a mouse obtained by 250 MHz EPRI using Oxo31 TAM probe (Scheme 16.2) with $\Delta H_{pp} = 87$ mG at $[O_2] = 0$. The distribution of oxygen is shown in a coronal (**a**) and sagittal (**b**) slice of the tumor in the leg of an air-breathing mouse. The images (**c**) and (**d**) represent the corresponding slices in the same tumor after carbogen-breathing for 10 min. The intensity bar to the right of each image quantifies oxygenation in torr. Image acquisition time was approximately 20 min. The oxygen distributions (**c**) and (**d**) are similar to that in (**a**) and (**b**). (Reprinted with permission from [84]. Copyright 2003 Wiley-Liss.)

in response to carbogen-breathing. The study shows that PEDRI can be used for dynamic studies of oxygen measurement.

The synthetic chemistry of TAM probes is still in its infancy. The first reports on the synthesis of TAMs useful for EPRI and PEDRI appeared in the patent literature.[38] Recently, creative efforts have been reported for the synthesis of these complex molecules.[39,40,88] The cTAM radical (Scheme 16.2) can now be synthesized efficiently on a large scale[40]. Fluorinated TAMs possessing a high affinity to fluorous media were specially designed[89] for assessment of tumor oxygenation using biocompatible perfluorocarbon emulsions. Note that the sensitivity of line broadening of TAMs is higher in perfluorocarbon liquids than in water in agreement with the higher solubility of oxygen in these solvents. Recently dual function pH and oxygen probes were also proposed[46,47] (Section 16.5).

The applications of other oxygen-sensitive paramagnetic materials include particulate probes such as lithium phthalocyanine particles[90,91] and carbonaceous materials (chars, coals, carbon blacks).[92] It should be noted that particulate probes such as lithium phthalocyanine and synthetic char are suitable for measurements of oxygen partial pressure, whereas soluble probes such as nitroxides and trityl compounds measure dissolved oxygen concentration.[9,91]

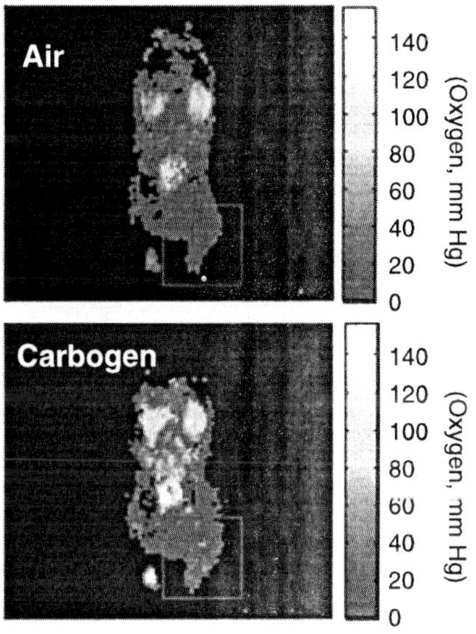

Figure 16.7 PEDRI oxygen images obtained using Oxo63 probe ($\Delta H_{pp} = 165\,mG$ at $[O_2] = 0$). The images were obtained from a mouse bearing the SCC tumor (squamous cell carcinoma) under air and carbogen-breathing conditions. The imaging times ranged between 4 and 8 min. The tumor region is shown within the square. (Reprinted with permission from [33]. Copyright 2002 National Academy of Sciences.)

16.5 EPR spectroscopy and imaging of pH using nitroxyl and trityl probes

The critical role of pH status in physiology and pathophysiology of living organisms is well recognized. At the microscopic scale, local pH drastically affects the vital activities of cell, cellular organelles, and enzymes. Therefore, spatially and temporarily addressed pH measurements *in vivo* are of considerable clinical relevance. For *in vivo* pH measurements, ^{31}P-NMR has proven to be the most suitable non-invasive approach. However, pH assessment using ^{31}P-NMR and inorganic phosphate (P_i) has its own limitations which are rarely discussed, including the lack of resolution (about 0.2–0.3 pH units and even less at lower pH), the fact that P_i concentrations vary with metabolism and ischemia, and the chemical shift dependence on ionic strength.[93,94] Because of these problems, exogenous pH probes are being designed for NMR spectroscopy to improve detection of myocardial acidosis[93] and extracellular pH in tumors.[95,96] Upon application of exogenous probes, low field EPR spectroscopy has higher sensitivity compared with NMR for the same probe concentration, and reasonable depth of penetration in living tissues.

16.5.1 pH-sensitive nitroxyl radicals

The first pH effect on the EPR spectra of stable NRs was observed in very strong acids and was related to protonation of the nitroxyl fragment itself.[97,98] Synthesized by Ullman and Osiecki in 1970,[99] 2-imidazoline NRs or imino nitroxyl radicals (INR, Scheme 16.4) were apparently the first reported stable pH-sensitive NRs with pK_a close to physiological range. A large pH effect on nitrogen hyperfine splitting ($\Delta a_{N1} \approx 2\,G$/pH unit) was observed upon protonation of the imino nitrogen of the INR1 and INR2. However, INRs did not find applications as pH probes, probably due to the relative complexity of their EPR spectra and extremely fast reduction to EPR silent products in biological fluids.[16,100,101]

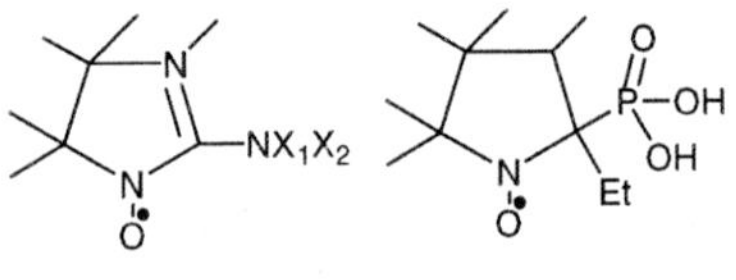

Scheme 16.4 The chemical structures of the pH-sensitive NRs of 2-imidazoline type, INR1 ($X_1 = X_2 = H$, $pK_a = 6.4$) and INR2 ($X_1 = X_2 =$ methyl, $pK_a = 6.8$),[99] and phosphorus-containing NR of pyrrolidine type

Comparatively small pH effects on the EPR spectra of the NRs of the piperidine and pyrrolidine types with ionizable functional groups were obtained,[102–107] apparently due to the long distance between radical center and ionizable group. The phosphorus-containing radical (Scheme 16.4) provided significantly larger pH effects for phosphorus hfs ($\Delta a_F \approx 2.5$ G) in the acidic region at pH < 4. This is not surprising due to the proximity of the phosphorus atom to the exchangeable proton and high phosphorus hyperfine splitting ($a_P \approx 50$ G). However, the synthesis of the phosphorus-containing NRs with higher pKa values is required for their applications as spin pH probes in biological systems.

Stable NRs of the imidazoline and imidazolidine types have been proven to be the most useful spin probes for EPR spectroscopy and imaging of pH,[2,11] due to the large effect of pH on their EPR spectra and large number of structures synthesized. Scheme 16.5 illustrates the chemical origin of the pH effect on EPR spectra of the imidazoline and imidazolidine radicals, ATI and HMI, respectively.

For the NR having an ionizable group in its structure with an equilibrium constant (K_a), the ratio of concentrations of its neutral (R) and charged (RH^+) states is described by the Henderson–Hasselbalch equation, $[H^+] = K_a[RH^+]/[R]$. This provides the basis for EPR measurements of pH using NRs with distinguishable EPR spectra of R and RH^+ forms. One of the great strengths of the technique is that it is ratiometric, or in other words, the pH measurement is independent of the concentration of the NR but rather depends on the ratio, $[RH^+]/[R]$. In general, spectral simulation is required for accurate $[RH^+]/[R]$ determination. In practice, two convenient spectral parameters can be used as markers of pH. The first parameter is the ratio of peak intensities of RH^+ and R spectral components resolved upon detection by high frequency EPR (Figure 16.8a) and partly resolved in X (9.5 GHz)-band EPR spectra[36] (Figure 16.8b). The second parameter is the nitrogen hyperfine splitting (a_N), measured as a distance between unresolved spectral components, for example, distance between low and centralfield spectral component of X-band EPR spectra (Figure 16.8b), and being used almost exclusively as a highly sensitive pH marker in numerous applications. Note that sensitivity of the latter experimental parameter to pH strongly depends both on EPR frequency and spectrometer settings (e.g., modulation amplitude) and can be optimized.[109]

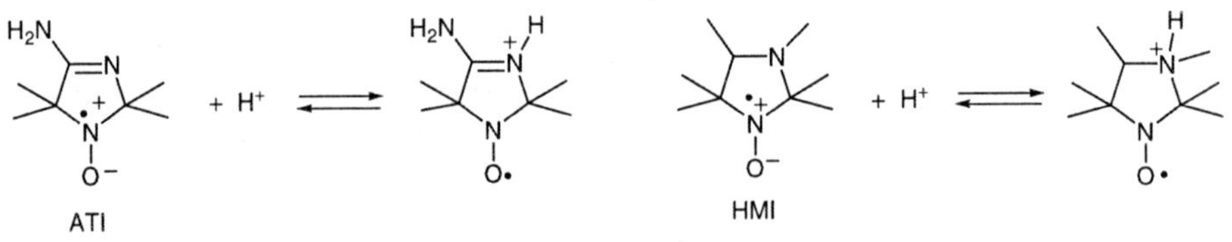

Scheme 16.5 Reversible protonation of the nitrogen atom N3 of the imidazoline NR, ATI ($pK_a = 6.1$), and imidazolidine radical, HMI ($pK_a = 4.7$). Two main resonance structures are shown illustrating higher unpaired electron density on nitrogen atom N1 in the unprotonated form. Protonation results in an EPR-detected difference in hyperfine splitting (a_N), and g-factor ($\Delta a_N = (0.8 - 1.2)$ G and $\Delta g \approx 0.0002$) between R and RH+ forms[108]

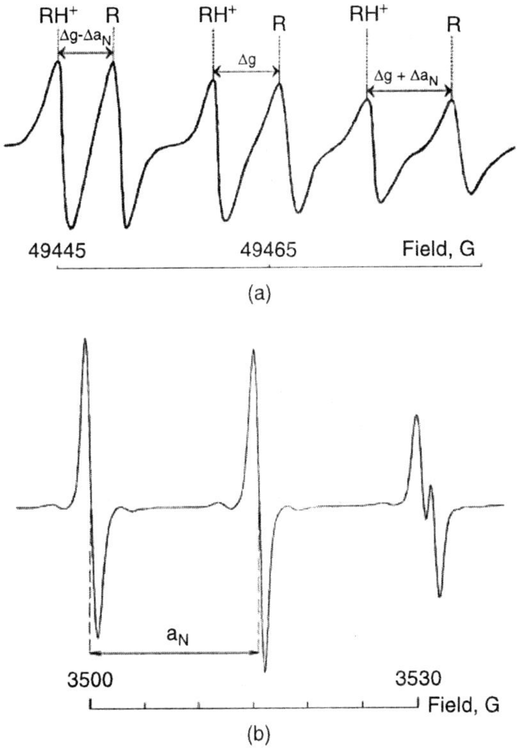

Figure 16.8 *(a) The 140 GHz EPR spectrum of a 0.5 mM aqueous solution of the imidazolidine radical HMI (Scheme 16.6) at pH 4.7. The dotted lines depict the positions of the peaks corresponding to protonated, RH^+, and neutral, R, forms of the radical. Note that the contribution of g-factor change in the spectral shift between two triplet forms is equal to 6.7 G and significantly overcomes the corresponding shift due to the change in nitrogen hfs, 1.25 G. (Reprinted from [109], Copyright 2004, with permission form IOS Press.) (b) X-band, 9.9 GHz EPR spectrum of 0.5 mM aqueous solution, pH 6.1, of imidazoline spin probe ATI (Scheme 16.5). Spectral splitting between RH^+ and R forms is observed at high field component ($\Delta g + \Delta a_N \approx 0.4\,G + 0.8\,G = 1.2\,G$) but not resolved at central ($\Delta g \approx 0.4\,G$) and low field ($\Delta g - \Delta aN \approx 0.4\,G - 0.8\,G = -0.4\,G$) components. (Reprinted from [109], Copyright 2004, with permission form IOS Press.)*

A wide variety of pH-sensitive NRs have been developed with different ranges of pH sensitivity, labeling groups, lipophilicity, and stability towards bioreduction.[35,102,110–112] These spin pH probes, together with low field EPR-based techniques, offer unique opportunities for non-invasive pH assessments in living animals in compartments with widely varying pH ranges. The potential applications are enormous, as tumors may have specific pH values compared to surrounding tissues, local skin treatments can have specific effects on local skin pH at various levels, and local areas of infection or inflammation can exhibit specific localized reductions in pH allowing infection to be imaged and localized.

The low depth of microwave penetration does not allow X-band application *in vivo* but was found to be very effective for non-invasive pH measurements *in vitro*, for example, inside biodegradable polymers used as drug delivery systems.[113] Spectral–spatial X-band EPR using ATI and HMI probes (Scheme 16.5) has been applied for non-invasive direct and depth specific measurement of pH within rat and human skin obtained from cosmetic surgery.[114]

L-band EPR spectroscopy (1.2 GHz) using spin pH probes has been shown to be a valuable tool for *in vivo* monitoring of microacidity in small animals such as rodents.[2,18,115–117] The first *in vivo* applications of spin pH probes were performed using L-band technique in mice.[115,116] Mäder *et al*. demonstrated that L-band EPR spectroscopy is a valuable tool for *in vivo* monitoring of pH-induced degradation of an implanted polymer in mice.[115] Gallez *et al*.[116] took advantage of the oral administration of the HMI ($pK_a = 4.7$, Scheme 16.5) probe to monitor the pH value inside the stomach of mice after administration of different antacids. The HMI spin pH probe has subsequently been applied to *in vivo* studies of stomach acidity in rats using L-band EPR, longitudinally detected EPR (LODEPR), and PEDRI.[117] The structure of the nitroxide API with properties optimized for monitoring stomach acidity is shown in Scheme 16.6.[18] The introduction of bulky groups in the vicinity of the NO fragment enhanced API stability towards reduction, resulting in about a fourfold increase of its lifetime *in vivo* compared with that for HMI. In addition, the presence of two ionizable groups in the API structure – the imino nitrogen N3 and the pyridine nitrogen – significantly extended its range of pH sensitivity (Figure 16.9b). Moreover, the hydrophilic character of pyridine, hydroxy, and amino groups of the API probe prevents its penetration through biomembranes and redistribution from the stomach. Figure 16.9 demonstrates the 300 MHz EPR spectrum of the API probe, a calibration curve for hyperfine splitting (a_N), and a typical example of real-time measurements of the stomach acidity of living rats.[18] The long lifetime of the API probe *in vivo* allows for monitoring of drug-induced perturbation of stomach acidity and its normalization afterwards for one hour or longer periods, therefore demonstrating the applicability of pH-sensitive NRs to the studies of drug pharmacology and disease in living animals.

L-band EPR spectroscopy using pH-sensitive NRs is also a useful tool to monitor myocardial pH in isolated rat hearts. The ATI probe ($pK_a = 6.1$, Scheme 16.5) has a convenient range of pH sensitivity (Figure 16.10a) to study ischemia-induced acidosis and has been applied in isolated rat hearts[11,118] (Figure 16.10b). Interestingly, L-band spectral measurement at modulation amplitude, 2G, significantly larger than the line width, 0.8 G, is strongly recommended. The latter results in an increased signal-to-noise ratio, a symmetric EPR spectrum, and a smooth dependence of nitrogen hyperfine splitting on pH in the range from 5.2 to 7.2 (Figure 16.10a). Figure 16.10b shows the typical decrease in myocardial pH measured by EPR and ATI probe. The hydrophilic ATI probe apparently measures extracellular pH mostly from the aorta, as confirmed by the spatial image of the radical using L-band imager (data not shown). Nevertheless, the EPR data are in reasonable agreement with the data observed by the ^{31}P-NMR technique for ischemia-induced intracellular pH changes.[119,120] It seems to be very attractive to use recently developed EPR/NMR co-imaging[121] to monitor simultaneously the values of extra- and intracellular pH. Intracellular and extracellular targeting is an important direction in the development of pH spin probes.

The penetration depth for L-band frequency, 1.2 GHz, in aqueous samples is somewhat greater than 1 cm and, in general, allows pH mapping by L-band EPR spectroscopy *in vivo* in small animals such as

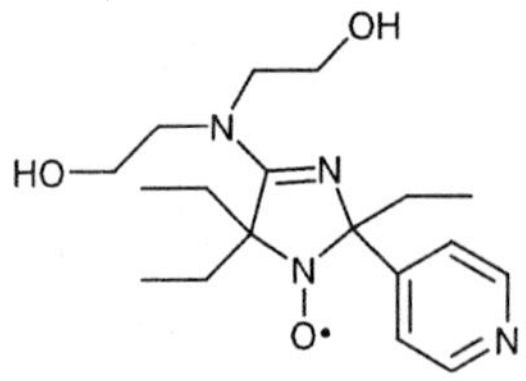

Scheme 16.6 The chemical structure of API probe, 4-[bis(2-hydroxyethyl)amino]-2-pyridine-4-yl-2,5,5-triethyl-2,5-dihydro-1H-imidazol-oxyl, synthesized to optimize NR properties for the measurement of stomach acidity *in vivo*[18]

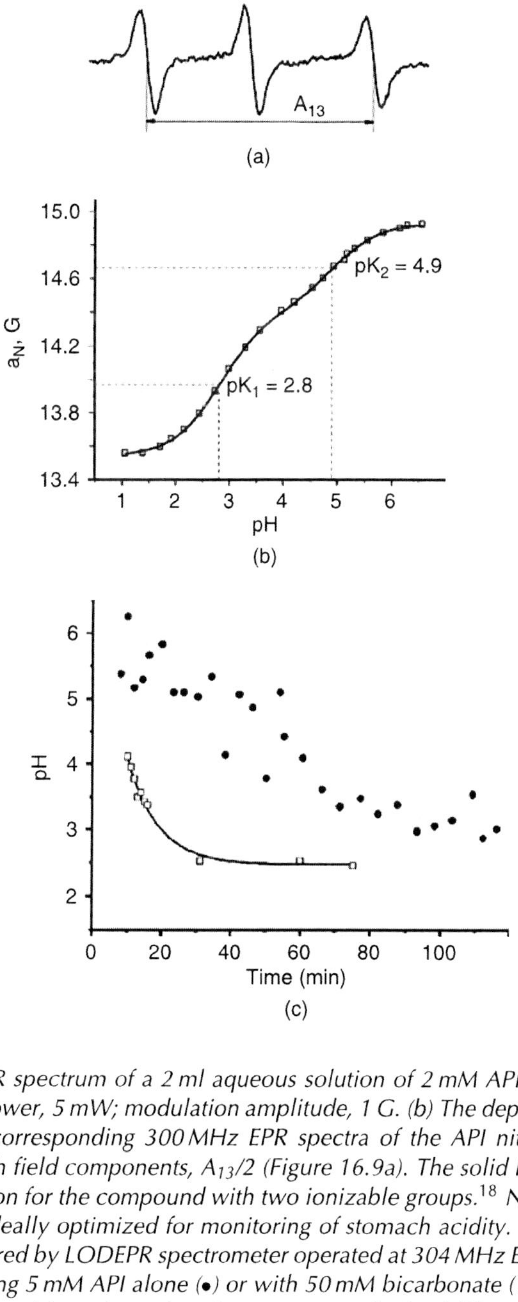

Figure 16.9 *(a) 300 MHz EPR spectrum of a 2 ml aqueous solution of 2 mM API probe. Spectrometer settings were as follows: microwave power, 5 mW; modulation amplitude, 1 G. (b) The dependence of hyperfine splitting (a_N) on pH calculated from corresponding 300 MHz EPR spectra of the API nitroxide measured as a half a distance between low and high field components, $A_{13}/2$ (Figure 16.9a). The solid line was calculated according to the standard titration equation for the compound with two ionizable groups.[18] Note convenient pH sensitivity range from pH 1.8 to pH 6, ideally optimized for monitoring of stomach acidity. (c) Time dependencies of pH changes in the stomach measured by LODEPR spectrometer operated at 304 MHz EPR excitation frequency, after giving 3 ml of gavage containing 5 mM API alone (•) or with 50 mM bicarbonate (□). After dosing the rats were placed in the LODEPR coil assembly and a series of spectra were obtained. The pH values from each spectrum were determined using corresponding calibration curve for a_N.*

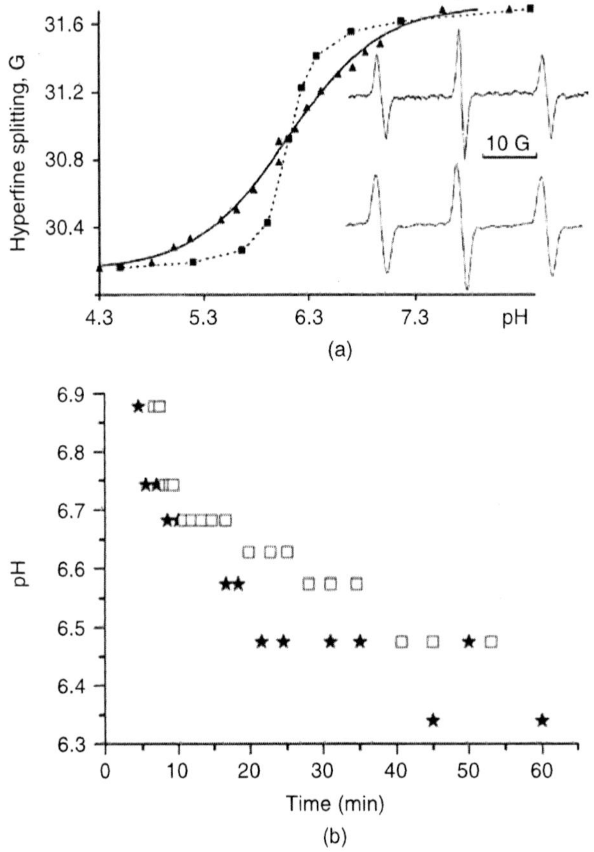

Figure 16.10 *(a) The pH dependences of hyperfine splitting, (a_N) measured as a half of the distance between the low and high field components of the L-band EPR spectra of the ATI radical measured at different modulation amplitudes, 0.5 G (■) and 2 G (▲). Insert: The EPR spectra of the 0.5 mM aqueous solutions of the radical R1 measured at pH 6.1, and modulation amplitudes 0.5 G (top) and 2 G (bottom), showing about a twofold increase in signal-to-noise ratio for the spectrum measured at higher modulation. Note that the shift of the low and high field EPR lines between RH^+ and R forms of the radical of the top EPR spectrum is comparable with the line width, resulting in significant disturbance of the EPR line shape and corresponding narrowing of the a_N titration curve (■). (b) Acidification of extracellular medium of the rat heart located directly in the resonator of the L-band EPR spectrometer during 1 h of global ischemia. The ATI probe (5 mM, 2 ml) in 0.9 % NaCl (★) or in perfused buffer (□) was administered to the heart at the onset of ischemia.*

rodents. However, loss in EPR sensitivity at low frequency, aggravated by the necessity of acquisition of numerous spectral projections with variable strength of field gradients, significantly limits spatial and temporal resolution of the approach. The application of low field EPR spectral detection at high modulation amplitude provides a significant improvement in sensitivity, allowing for easy conversion of the "position" image of the low or high field spectral component (Figure 16.10a) in the "pH map". Figure 16.11a shows a 3-D image (1-spectral/2-spatial) of a phantom sample of three capillaries with aqueous solutions of ATI probe with a good spatial (0.2 mm) and functional (0.2 pH units at pH close to the pK of the radical) resolution obtained for five minutes acquisition time.

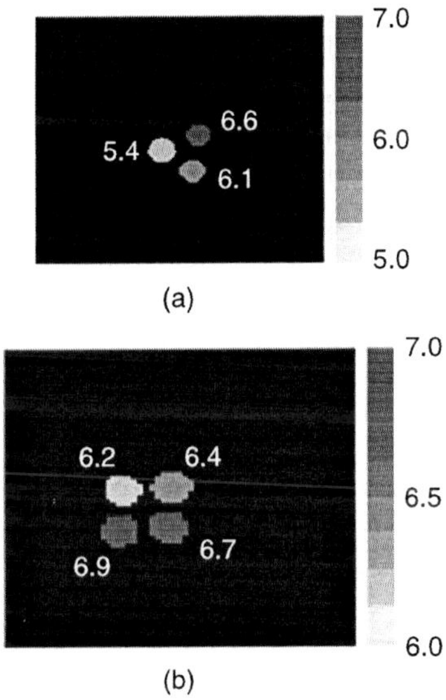

Figure 16.11 *(a) L-band pH mapping. Cross-sectional 3-D (1-spectral/2-spatial) image of a phantom prepared using capillary tubes of 2 mm diameter filled with 1 mM solutions of the radical ATI prepared at different pH (indicated in the figure). The intensity bar to the right of the image quantifies pH values. The data acquisition parameters were: acquisition time, 5 min; projections, 64; maximum gradient, 30 G/cm; field of view, 24 mm. (b) pH mapping by functional PEDRI of a phantom prepared using capillary tubes of 5 mm diameter filled with 1 mM aqueous solution of ATI prepared at different pH (indicated in the figure). The intensity bar to the right of the image quantifies pH values. The pH map was calculated from two PEDRI images acquired at the operating EPR frequency 567 MHz and different EPR excitation fields, 214.16 G and 214.88 G, which correspond to peak amplitudes of DNP spectra of the RH^+ and R forms of the ATI probe, respectively. Total acquisition time, 8.8 s. Field of view, 30 mm × 30 mm, with resolution 64 × 64.*

Figure 16.11b demonstrates the capacity of pH mapping using the recently proposed functional PEDRI[122] approach. Normally, PEDRI is acquired at fixed EPR field, resulting in loss of EPR spectral information. In the proposed functional approach, PEDRI images were acquired at two different EPR excitation fields which correspond to peak amplitudes of DNP spectra of the RH^+ and R forms of the pH probe. This permits extraction of the $[RH^+]/[R]$ ratio and, therefore, the pH values from the NMR signal enhancements observed at each pixel of only two PEDRI images. Note that pH mapping using functional PEDRI (Figure 16.11b) allows for about a 30-fold decrease in acquisition time compared with EPRI (Figure 16.11a) for the phantoms with the same NR probe and similar spatial and functional resolutions. This is particularly important for *in vivo* applications where the experimental window and stability of the NRs are limited. In addition, PEDRI allows for functional (pH map) and anatomical resolution in one experimental set-up otherwise available only in EPR/NMR co-imaging.[121]

16.5.2 Dual function pH- and oxygen-sensitive trityl radicals

TAMs containing ionizable groups may demonstrate pH-sensitive magnetic resonance parameters similar to the phenomenon described for pH-sensitive NRs. The first observations of the pH sensitivity of EPR spectra of TAMs were reported for the derivatives containing carboxyl groups.[46]

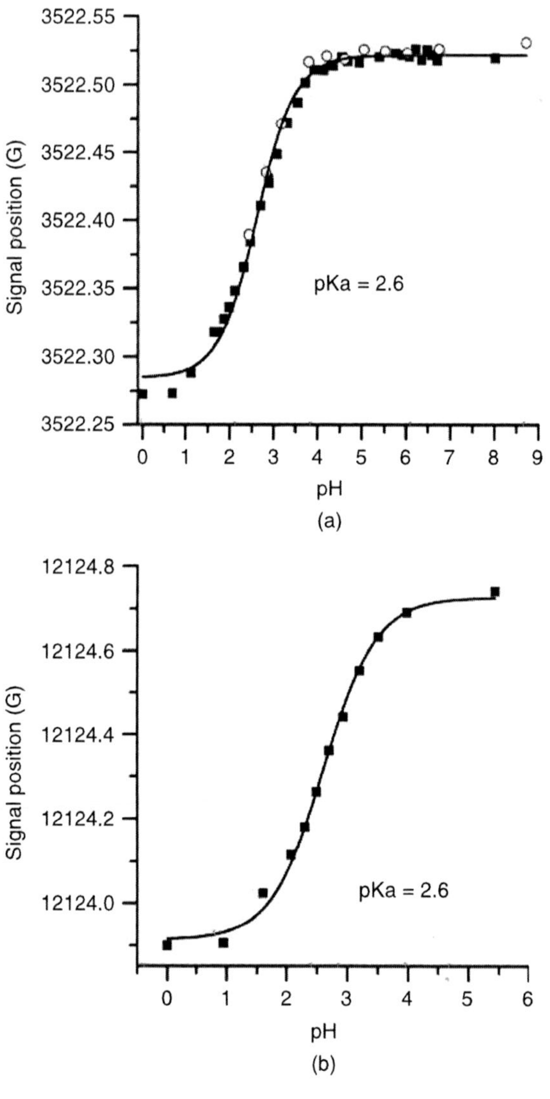

Figure 16.12 *pH dependence of the EPR spectral line position of 50μM solution of Oxo63 TAM in 1.5 mM sodium citrate buffer measured by X-band (a) and Q-band (b) EPR spectroscopy. The symbols (■) and (○) denote the data obtained upon titration from alkaline pH into direction to acidic pH and in the reverse direction, respectively. The values of g-factors for neutral RH and deprotonated (R^-) forms of the radical were found to be equal to 2.00329 and 2.00315, correspondingly. (Reprinted with permission from [46]. Copyright 2007 American Chemical Society.)*

The pH-dependent shift of the EPR line position of Oxo63 TAM (Scheme 16.2) measured at two different EPR frequencies is shown in Figure 16.12. This line shift is reversible in agreement with the reversible deprotonation of the COOH group. The value of the line shift is proportional to the EPR frequency being equal to 0.27 G and 0.81 G for 10 GHz X-band and for 35 GHz Q-band spectra, respectively, which is characteristic for the changes in g-factor. The low pK_a value of Oxo63 derivative limits its application to

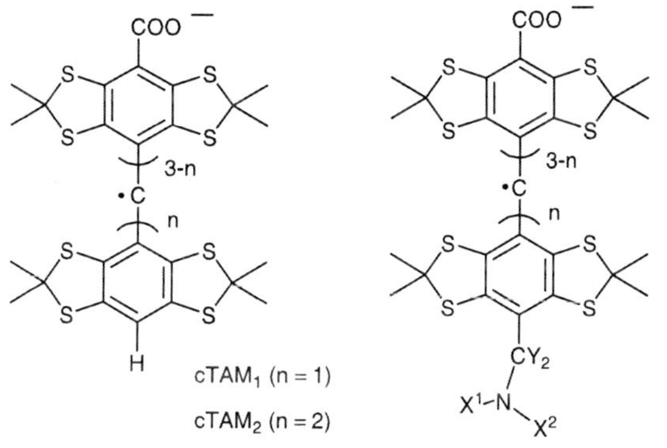

	n	X¹	X²	Y
$aTAM_1$	3	H	H	H
$aTAM_2$	2	tBu	H	H
$aTAM_3$	1	tBu	H	H
$aTAM_4$	1	Et	Et	H
$aTAM_5$	1	Et	Et	D

Scheme 16.7 Chemical structures of TAM derivatives containing carboxyl groups, $cTAM_1$ and $cTAM_2$, amino groups, $aTAM_1$, and both carboxyl and amino groups, $aTAM_2$–$aTAM_5$

pH range from 2 to 4, which still could be useful, for example, for the studies of the stomach acidity.[18] However, another limitation of this particular pH probe is its frequency-dependent pH shift, which becomes impractically small at low field (about 30 mG at L-band). Note that pH-sensitive NRs keep their sensitivity at low field EPR due to frequency-independent pH effect on the hyperfine splitting (a_N). A similar effect was observed for carboxyl group containing TAM derivatives (Scheme 16.7) with one ($cTAM_1$) or two ($cTAM_2$) protons attached to the aryl groups. $cTAM_1$ and $cTAM_2$ demonstrate doublet and triplet EPR spectral patterns, respectively, with the hydrogen hyperfine splitting (a_H) reversibly changed in acidic pH range owing to reversible deprotonation of carboxyl groups.[46] The observed a_H changes were comparatively small (about 10–20 mG) in agreement with the long distance between COOH group and hydrogen atom, being located at different phenyl rings. Nevertheless, the data supported the principal of using hyperfine splitting parameter of TAM derivatives as a pH marker.

Recently synthesized TAMs containing amino groups[47] represent the first pH-sensitive trityl probes with reasonably valuable spectral properties for application in physiological range of pH from 6.8 to 9.0. The presence of nitrogen and hydrogen atoms in direct proximity to protonatable amino groups resulted in strong pH-induced changes of the corresponding hyperfine splittings, Δhf s $\approx$ 300–1000 mG (Figure 16.13).

The superposition of two forms of the $aTAM_4$ radical at pH around pK_a is particularly obvious for the outermost spectral components, as shown in Figure 16.14a. The high field fractions of the integrated EPR spectra of $aTAM_4$ shown in Figure 16.14b demonstrate isosbestic point characteristics for a chemical equilibrium between two forms of the radical. The pH dependence of the fraction of the $aTAM_4$ protonated form shown in Figure 16.14c represents a typical titration curve with $pK_a = 8.0 \pm 0.1$ at 37 °C. The ratio of the spectral amplitudes of the high field components shown in Figure 16.14a is a convenient experimental parameter for aqueous acidity measurements using $aTAM_4$ pH probe allowing for pH measurement in physiologically relevant pH range of 6.8 to 9.0.[47] Note that changes in oxygen concentration do not influence the $[RH^+]/[R]$ ratio of $aTAM_4$ but result in line broadening in a linearly dependent manner on oxygen concentration (6 mG/per % $[O_2]$).[47] The independent character of pH and $[O_2]$ effects on the EPR spectra of $aTAM_4$ provides dual functionality to this probe, allowing an extraction of both parameters from a single spectrum.

TAM radicals, such as Oxo63, were found to be useful probes for EPR oximetry due to both their high sensitivity to oxygen-induced line broadening ($\approx$5.3 mG per % $[O_2]$[21]) and the presence of a single spectral line. The appearance of complex spectral patterns for dual function probes, for example, for $aTAM_4$, decreases their value for oxygen measurements, particularly for oxygen mapping. Further simplification

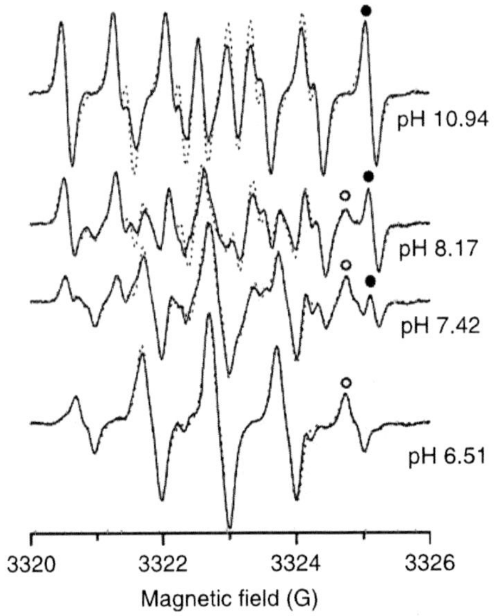

Figure 16.13 X-band EPR spectra of 50μM aTAM$_4$ at different pH in 1.5 mM Na-pyrophosphate buffer under nitrogen atmosphere, 37 °C, measured at microwave power 0.63 mW, modulation amplitude 0.1 G, and sweep width 8 G. Dotted lines represent the calculated EPR spectra with peak-to-peak line widths, $\Delta H_L = 96$ mG, $\Delta H_G = 97$ mG, and the following hfs constants: $a_N = 0.78$ G, a_{H1} (CH_2) = 0.95 G and a_{H2} (CH_2) = 2.05 G for deprotonated form and $a_N = 1.01$ G, a_{H1} (CH_2) = 0.97 G, a_{H2} (CH_2) = 1.05 G and $a_H(-NH^+Et_2) = 0.15$ G for the protonated form of the radical. At intermediate pH EPR spectra were calculated as superposition of two forms of aTAM$_4$, protonated and deprotonated ones, shifted as total one relatively to the other by 20 mG due to g-factor difference. The high field spectral component of protonated and deprotonated components are marked by the symbols (○) and (•), correspondingly. (Reprinted with permission from [47]. Copyright 2008 American Chemical Society.)

of spectral properties of dual function TAM probes is desired, ideally providing pH-sensitive doublet hyperfine splitting. The development of TAM derivatives based on more hydrophilic structures, such as Oxo63, is also important for elaboration of non-toxic dual function probes.

16.6 Redox- and thiol-sensitive nitroxide probes

Regulation of tissue redox status is important for maintenance of normal physiological conditions in the living body. Disruption of redox homoeostasis may lead to oxidative stress and can induce many pathological conditions, such as cancer, neurological disorders, and ageing. The intracellular thiols, and particularly the redox couple of glutathione (GSH), and its disulfide form (GSSG), are considered the major regulators of the intracellular redox state.[123] Therefore, non-invasive spectroscopic evaluation and imaging of tissue redox status and, in particularly, GSH redox status, could have clinical applications.[17,19,124–127]

16.6.1 Nitroxides as redox-sensitive EPR probes

NRs introduced into biologically relevant systems are predominantly observed in the radical and hydroxylamine forms and exist in the redox equilibrium, as shown in Scheme 16.8. For the most biologically

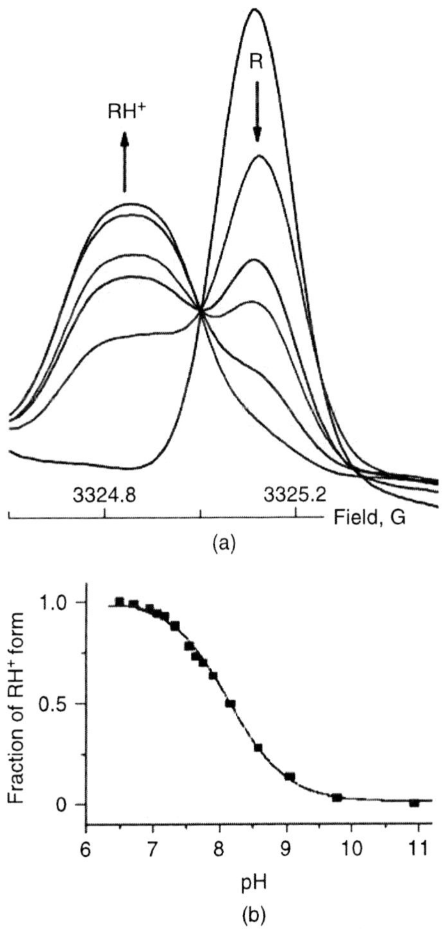

Figure 16.14 *High field components of the absorption EPR spectra of 50μM aTAM$_4$ radical in 1.5 mM pyrophosphate buffer at various pH values, 37 °C (a) which demonstrate isosbestic point characteristics for the chemical equilibrium between protonated and unprotonated aTAM$_4$ radicals (Scheme 16.7). The arrows indicate the direction of signal intensity changes upon acidification of the sample. (b) The pH dependence of the fraction of aTAM$_4$ radical protonated form obtained from the EPR spectra. Line represents the best fits of titration equation to the experimental data yielding pK$_a$ value 8.0 ± 0.1.*

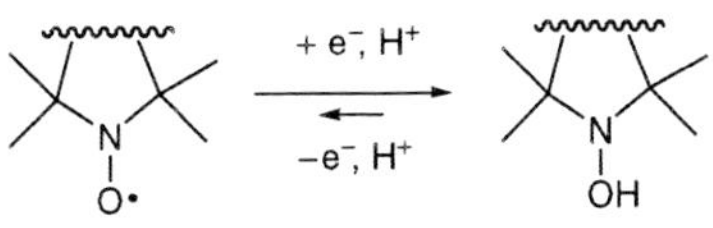

Scheme 16.8 Illustration of the nitroxide/hydroxylamine redox couple. In general, for the most biologically relevant samples one-electron reduction of the nitroxides prevails and the equilibrium is strongly shifted towards hydroxylamine form[34]

relevant experimental situations this equilibrium is strongly shifted towards hydroxylamine formation, and therefore NRs undergo reduction to the EPR silent product.[34] The reduction of NRs to EPR silent hydroxylamines in many cases significantly limits their applications in biological systems. On the other hand, EPR-measured rates of NR decay provide information on tissue redox capacity. The reduction of the NRs to hydroxylamines by cells is primarily intracellular, and therefore the rate of the NRs' reduction is determined to a great extent by intracellular redox status. Hence, the EPR-measured rate of NRs reduction depends on overall tissue redox status, allowing for the differentiation of normal and pathological states.[17,125,126]

Ischemic heart disease represents a pathologic condition with compromised redox state. Das *et al.*[125] used TEMPO to determine changes in the intracellular redox environment of the heart during ischemia and reperfusion, and the effects of the red-wine-derived polyphenolic antioxidant, resveratrol, on such changes. Figure 16.15 shows typical EPR-measured kinetics of TEMPO reduction during the global ischemic phase, supporting the conclusion that the reducing capacity of the resveratrol-treated heart is significantly higher than the vehicle-control-treated heart. Note that ischemia/reperfusion lowered the ratio of GSH/GSSG measured *in vitro* while resveratrol significantly improved the ratio.[125]

The application of L-band EPRI for redox mapping of the tumor in living mice[17,128] using 3-CP nitroxide is demonstrated in Figure 16.16. A significant decrease in the rate of loss of EPR probe signal after mice are treated with the GSH-depleting agent, BSO, clearly demonstrates a central role of glutathione in the reduction of the NR. Note that, in general, appreciable chemical reduction of the NRs by GSH is not observed over several hours.[129–131] However, GSH significantly contributes to the reduction of the NRs in biological samples indirectly by acting as a secondary source of reducing equivalents.[17,131,132]

In summary, low field EPR and EPRI techniques using NRs provide a useful tool for quantitative assessment and mapping of the redox environment in living tissues.

16.6.2 Disulfide nitroxide biradicals as GSH-sensitive EPR probes

The redox couple of glutathione (GSH) and its disulfide form (GSSG) is considered the major regulator of the intracellular redox state.[123] Therefore, glutathione redox status *in vivo* might be a useful indicator of disease risk in humans.

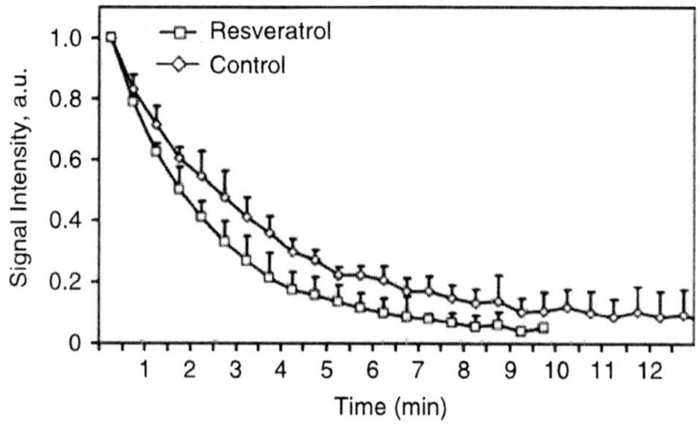

Figure 16.15 Typical kinetics of the decay of the L-band EPR spectral intensity of the TEMPO nitroxide during ischemia of the isolated heart. The heart was perfused with Krebs–Henseleit bicarbonate buffer containing 0.2 mM TEMPO nitroxides for 15 min and global ischemia was performed for 30 min. For the resveratrol groups, resveratrol (2.5 mg/kg body wt/day) was fed by gavaging for 10 days before the experiment. (Adapted from [125], Copyright 2008, with permission from Elsevier.)

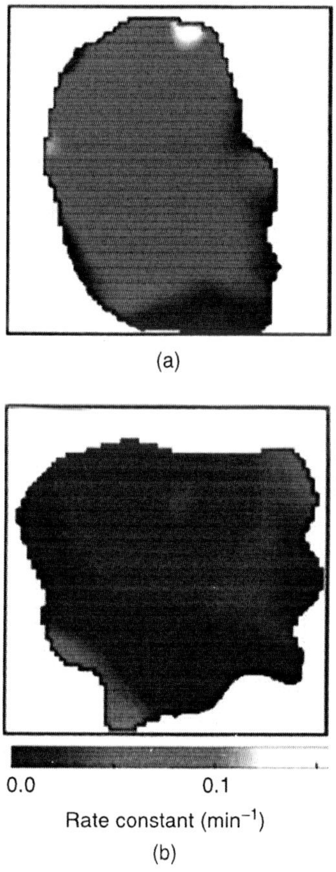

Figure 16.16 Redox mapping of tumor of untreated (a) and BSO-treated (b) mice. 2-D spatial mapping of pseudo-first order rate constants of the nitroxide 3-CP reduction obtained in RIF-1 tumors implanted in the upper leg of a mouse and measured using in vivo L-band EPR imaging.[128] The rate of reduction was significantly slower in tumors of mice treated for 6 h with 2.25 mmol/kg of L-buthionine-S,R-sulfoximine (BSO), a GSH depleting agent. (Reprinted from [17], Copyright 2002, American Association for Cancer Research.)

The use of fluorometric, photometric, and chromatographic assays for GSH measurements is mostly limited to *in vitro* or *ex vivo* systems, due to the invasiveness involved and the limited depth of penetration of light.[133–136] NMR based approaches normally detect endogenous GSH but have somewhat low sensitivity and the spectral assignment is complicated due to the overlapping of numerous resonances.[137–140]

EPR spectroscopy in combination with thiol-specific nitroxides allows determination of the accessible thiol groups in various biological macromolecules, such as human plasma low density lipoproteins[141] and erythrocyte membranes.[142] This approach normally requires purification of the sample from the unbound label and cannot be used *in vivo*. Moreover, application of thiol-specific mononitroxides for the EPR measurement of glutathione or cysteine is hardly possible due to insignificant EPR spectral changes of the label upon binding to low molecular weight compounds. The latter limitation was overcome by the development of the disulfide nitroxyl biradicals, DNB[12,143,144] (Scheme 16.9). The principal advantage of the DNB reagents is the large EPR spectral changes that accompany their reaction with low molecular weight thiols such as GSH (Figure 16.17).

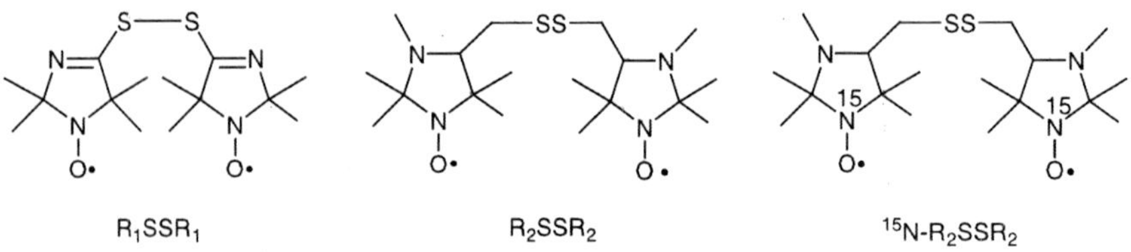

Scheme 16.9 Structures of the disulfide nitroxyl biradicals (DNB): bis(2,2,5,5-tetramethyl-3-imidazoline-1-oxyl-4-il)-disulfide (R_1SSR_1),[12] bis(2,2,3,5,5-pentamethyl-1-oxyl-imidazolidine-4-ylmethyl)-disulfide (R_2SSR_2)[144] and its ^{15}N-substituted analog, bis(2,2,3,5,5-pentamethyl-1-oxyl-imidazolidine-4-ylmethyl)-disulfide 1−^{15}N (^{15}N-R_2SSR_2)[19]

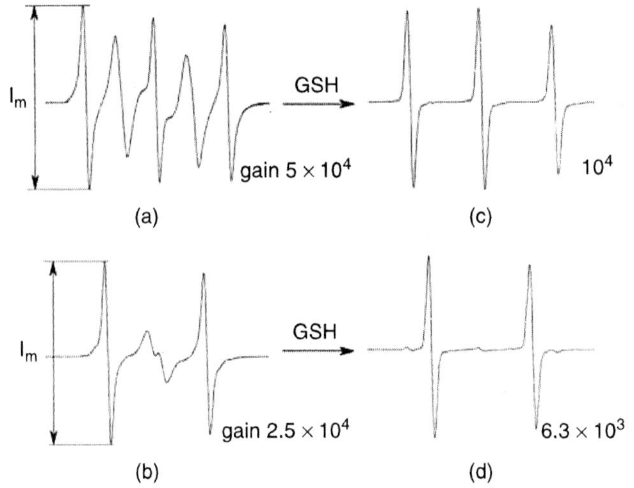

Figure 16.17 *The L-band EPR spectra of 0.1 mM aqueous solutions of R_2SSR_2 (a, c) and ^{15}N-R_2SSR_2 (b, d) before (a, b) and after (c, d) addition of 1 mM GSH. The EPR spectra of the RSSR and ^{15}N-R_2SSR_2 labels have additional "biradical" components due to intramolecular spin exchange between two radical fragments of the DNB. The splitting of the disulfide bonds in the reaction of thiol–disulfide exchange with GSH results in the disappearance of the "biradical" components and corresponding increase of the peak intensity of the "monoradical" spectral component, I_m. The maximal increase of the I_m corresponding to complete splitting of the disulfide bond in the excess of GSH is equal to 4.6- and 3.0-fold for R_2SSR_2 and ^{15}N-R_2SSR_2, respectively.*

DNB labels, being paramagnetic analogs of widely used disulfide Ellman's reagent,[133] react with thiols, BSH, via the reversible reaction of thiol–disulfide exchange (Equation 16.1):

$$\text{RS–SR} + \text{BSH} \Longleftrightarrow \text{RSSB} + \text{RSH} \qquad (16.1)$$

EPR spectra of the DNBs are significantly affected by intramolecular spin exchange between two radical fragments, as shown in Figures 16.17a and 16.17b. Figures 16.17c and 16.17d demonstrate the typical changes of the EPR spectra of the DNB labels upon addition of GSH. The decrease of biradical spectral components and simultaneous increase of the monoradical ones is consistent with the splitting of the disulfide bond of the biradical. The DNB labels, being lipophilic compounds, diffuse easily across cellular membranes. The biradical R_1SSR_1 reacts with intracellular GSH at physiological pH within a few seconds, therefore providing a fast and reliable EPR approach for determination of GSH in optically non-transparent samples. The approach is based on a dominant contribution of the GSH in the intracellular pool of the fast reacting thiols. The R_1SSR_1 DNB label has been used to measure GSH in various cells and cellular homogenates,[12,145–149] animal and human blood,[150,151] and isolated organs.[152] Note that GSH detection using fast reacting DNB label, R_1SSR_1, requires an excess of the label over the thiol. The method, being simple and reliable for *in vitro* systems, cannot be used *in vivo* due to its invasiveness because vital thiols are consumed during the measurement.

The application of slow reacting DNB labels, R_2SSR_2[144] and ^{15}N-R_2SSR_2,[19] allows for quantitative measurements of GSH content by analysis of the kinetics of their EPR spectral changes. The kinetics approach looses the attractive simplicity of the static EPR measurements using DNB label but gains a decisive advantage by using low concentrations of the label compared with the GSH content. This important advantage makes the approach less invasive and, therefore, applicable *in vivo*. Another favorable aspect is the observation of extremely low reaction rates of the DNB with the protein thiols, for example, the rate constants of the R_2SSR_2 reaction with SH groups of human serum albumin and hemoglobin were less or about 1 % of the corresponding values for GSH.[144] Taking into account that GSH is a major intracellular thiol compound present in cytosol in concentrations from 1 to 10 mM, a predominant contribution of GSH to the reaction shown in Equation 16.1 might be expected. However, for the quantitative *in vivo* application of DNB for GSH measurement the possible reduction of the radical fragment has to be taken into account.[13,19]

The first *in vivo* EPR measurements of GSH were performed in ovarian xenograft cisplatin-sensitive and cisplatin-resistant ovarian tumors grown in nude mice using L-band EPR spectroscopy and DNB label ^{15}N-R_2SSR_2. Intracellular GSH has been shown to be one of the major factors modulating tumor response to a variety of commonly used anti-neoplastic agents, such as cisplatin.[153] The use of a ^{15}N-substituted compound resulted in decreasing the number of EPR spectral lines and about a twofold increase in the SNR (cf. spectra in Figures 16.17a and 16.17b) compared with non-substituted R_2SSR_2 label which is important for *in vivo* applications. Figure 16.18a shows typical kinetics of the change of the amplitude of the L-band EPR low field spectral component, I_m (Figure 16.17b), measured after injection of the DNB into ovarian xenograft tumors grown in nude mice. In general, the kinetics should be analyzed in terms of the two main contributing reactions, namely splitting of the disulfide bond and reduction of the NO fragment. Both reactions can result in I_m increase. However, the maximal reduction-induced I_m increase (about 14 % for the ^{15}N-R_2SSR_2[19]) is much less than maximal I_m increase due to the reaction with GSH (about 200 %, cf. Figures 16.17b and 16.17d). It provides a simple way to estimate comparative contribution of the DNB reduction to the overall I_m increase. For example, in the kinetics shown in the Figure 16.18a, the observed *in vivo* maximal increase of the I_m exceeded the possible reduction-induced I_m increase by one order of magnitude or more. It allows, in first approximation, to describe the initial exponential I_m increase only by reaction of the DNB with GSH neglecting the contribution of the reduction. Comparison of the concentrations of GSH obtained *in vivo* and *in vitro* (Figure 16.18b) shows reasonably good agreement further supporting validity of this approach. The observed concentrations of GSH in cisplatin-resistant tumors (3.3 mM) are almost twice that in cisplatin-sensitive tumors (1.8 mM) and are in agreement with previously reported data.[154,155]

In summary, it seems likely that the use of *in vivo* EPR to measure redox processes, and GSH redox particularly, will become an increasingly used and valuable tool.

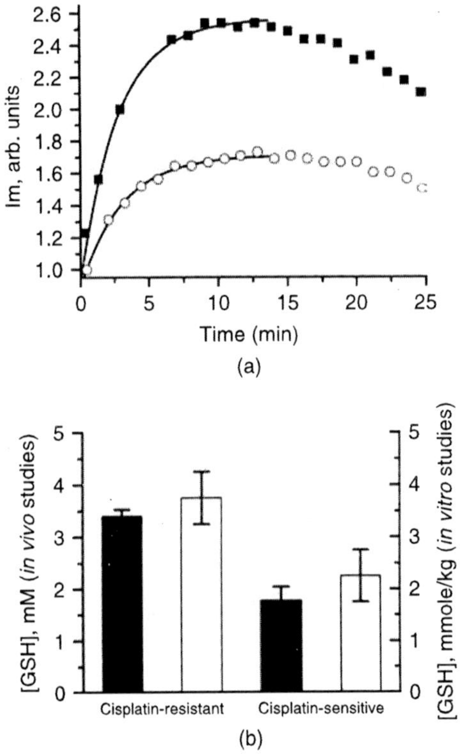

Figure 16.18 (a) The kinetics of the I_m peak intensity change measured in vivo in ovarian tumor-bearing mice located directly in the resonator of L-band EPR spectrometer after 15μl intratumoral injection of 50 mM DNB label, ^{15}N-R_2SSR_2. The representative kinetics observed in the cisplatin-resistant (■) and cisplatin-sensitive (○) ovarian tumors are shown. Lines are the fit of the initial part of the kinetic curve by the monoexponent supposing $k_{obs}(pH\ 7.1, 37\,^{\circ}C) = 2.6\,M^{-1}\,s^{-1}$ and yielding $[GSH]_r = 3.3$ mM and $[GSH]_s = 1.8$ mM for the cisplatin-resistant and cisplatin-sensitive ovarian tumors, respectively. (b) GSH concentration measured in cisplatin-sensitive and cisplatin-resistant ovarian tumors in in vivo (filled bars) and in vitro (empty bars). In the case of in vitro studies GSH concentration was determined by EPR using DNB in the homogenates from the same tumors. (Adapted from [125], Copyright 2008, with permission from Elsevier.)

16.7 Conclusions

In the past decade functional EPR-based spectroscopy and imaging applications have moved closer to the biomedical field. However, the bottleneck of *in vivo* EPR applications is the lack of appropriate exogenous paramagnetic probes. The development of stable organic free radicals, the NRs and TAMs, provides a wide range of paramagnetic probes variable in solubility and tissue redistribution, spectral and functional sensitivity, ability to be targeted, and their lifetimes in living tissues. Spectral sensitivities of the NR and TAM probes towards biologically relevant parameters of their microenvironment, such as content of oxygen, protons (pH), thiols (GSH), and other reducing species, provide unique functionality to *in vivo* EPR spectroscopy and imaging.

Acknowledgements

This work was partly supported by NIH grants KO1 EB03519, R21 CA132068, and R21 HL091423.

References

1. G. He, A. Samouilov, P. Kuppusamy and J. L. Zweier, *Mol. Cel.l Biochem.*, **234-235**, 359–367 (2002).
2. V. V. Khramtsov, I. A. Grigor'ev, M. A. Foster, *et al.*, *Cell. Mol. Biol.*, **46**, 1361–1374 (2000).
3. D. J. Lurie, *Br. J. Radiol.*, **74**, 782–784 (2001).
4. I. Nicholson, F. J. Robb, S. J. McCallum, *et al.*, *Phys. Med. Biol.*, **43**, 1851–1855 (1998).
5. O. A. Lebedev and S. N. Kayanovskii, *Trudy po. Khimii. Khim. Technologii. (Gorkii)*, **8**, 649–652 (1959).
6. M. B. Neiman, E. G. Rozatzev and Y. G. Mamedova, *Nature*, **196**, 472–474 (1962).
7. L. J. Berliner (Ed), *Spin Labeling. The Next Millennium*, pp. 423, Plenum Press, New York, 1998.
8. W. L. Hubbell, D. S. Cafiso and C. Altenbach, *Nat. Struct. Biol.*, **7**, 735–739 (2000).
9. H. M. Swartz, *Antioxid. Redox Signal.*, **6**, 677–686 (2004).
10. H. J. Halpern, C. Yu, M. Peric, *et al.*, *Proc. Natl. Acad. Sci. USA*, **91**, 13047–13051 (1994).
11. V. V. Khramtsov, *Curr. Org. Chem.*, **9**, 909–923 (2005).
12. V. V. Khramtsov, V. I. Yelinova, L. M. Weiner, *et al.*, *Anal. Biochem.*, **182**, 58–63 (1989).
13. H. M. Swartz, N. Khan and V. V. Khramtsov, *Antioxid. Redox Signal.*, **9**, 1757–1771 (2007).
14. T. Akaike, M. Yoshida, Y. Miyamoto, *et al.*, *Biochemistry*, **32**, 827–832 (1993).
15. J. Joseph, B. Kalyanaraman and J. S. Hyde, *Biochem. Biophys. Res. Commun.*, **192**, 926–934 (1993).
16. Y. Woldman, V. V. Khramtsov, I. A. Grigor'ev, *et al.*, *Biochem. Biophys. Res. Commun.*, **202**, 195–203 (1994).
17. P. Kuppusamy, H. Li, G. Ilangovan, *et al.*, *Cancer Res.*, **62**, 307–312 (2002).
18. D. I. Potapenko, M. A. Foster, D. J. Lurie, *et al.*, *J. Magn. Reson.*, **182**, 1–11 (2006).
19. G. I. Roshchupkina, A. A. Bobko, A. Bratasz, *et al.*, *Free Rad. Biol. Med.*, **45**, 312–320 (2008).
20. S. Xia, F. A. Villamena, C. M. Hadad, *et al.*, *J. Org. Chem.*, **71**, 7268–7279 (2006).
21. J. H. Ardenkjaer-Larsen, I. Laursen, I. Leunbach, *et al.*, *J. Magn. Reson.*, **133**, 1–12 (1998).
22. O. A. Lebedev, M. L. Khidekel and G. A. Razuvaev, *Dokl. Akad. Nauk SSSR*, **140**, 1327–1331 (1961).
23. N. Edelstein, A. Kwok and A. H. Maki, *J. Chem. Phys.*, **41**, 179–183 (1964).
24. A. V. Il'yasov, *J. Struct. Chem.*, **3**, 84–86 (1962).
25. T. J. Stone, T. Buckman, P. L. Nordio and H. M. McConnell, *Proc. Natl. Acad. Sci. USA*, **54**, 1010–1017 (1965).
26. O. H. Griffith and H. M. McConnell, *Proc. Natl. Acad. Sci. USA*, **55**, 8–11 (1966).
27. C. Altenbach, S. L. Flitsch, H. G. Khorana and W. L. Hubbell, *Biochemistry*, **28**, 7806–7812 (1989).
28. W. L. Hubbell, H. S. McHaourab, C. Altenbach and M. A. Lietzow, *Structure*, **4**, 779–783 (1996).
29. Y. N. Molin, K. M. Salikhov and K. I. Zamaraev, *Spin Exchange. Principles and Applications in Chemistry and Biology*, Springer-Verlag, Berlin, 1980.
30. M. D. Rabenstein and Y. K. Shin, *Proc. Natl. Acad. Sci. USA*, **92**, 8239–8243 (1995).
31. J. M. Backer, V. G. Budker, S. I. Eremenko and Y. N. Molin, *Biochim. Biophys. Acta*, **460**, 152–156 (1977).
32. J. F. Glockner, S. W. Norby and H. M. Swartz, *Magn. Reson. Med.*, **29**, 12–18 (1993).
33. M. C. Krishna, S. English, K. Yamada, *et al.*, *Proc. Natl. Acad. Sci. USA* **99**, 2216–2221 (2002).
34. N. Kocherginsky and H. M. Swartz, *Nitroxide spin labels. Reactions in biology and chemistry*, pp. 270, CRC Press, Boca Raton, 1995.
35. V. V. Khramtsov and L. B. Volodarsky, in *Spin labeling. The next Millennium* (ed. L. J. Berliner), pp. 109–180, Plenum Press, New York, 1998.
36. V. V. Khramtsov, I. A. Grigor'ev, M. A. Foster and D. J. Lurie, *Antioxid. Redox Signal.*, **6**, 667–676 (2004).
37. M. Gomberg, *J. Am. Chem. Soc.*, **22**, 757–771 (1900).
38. S. Anderson, K. Golman, F. Rise, *et al.*, *US Patent*, 5,530,140 (1996).
39. T. J. Reddy, T. Iwama, H. J. Halpern and V. H. Rawal, *J. Org. Chem.*, **67**, 4635–4639 (2002).

40. I. Dhimitruka, M. Velayutham, A. A. Bobko, *et al.*, *Bioorg. Med. Chem. Lett.*, **17**, 6801–6805 (2007).
41. D. J. Lurie, D. M. Bussell, L. H. Bell and J. R. Mallard, *J. Magn. Reson.*, **76**, 366–370 (1988).
42. D. J. Lurie, J. M. S. Hutchison, L. H. Bell, *et al.*, *J. Magn. Reson.*, **84**, 431–437 (1989).
43. K. Golman, J. S. Petersson, J. H. Ardenkjaer-Larsen, *et al.*, *J. Magn Reson. Imaging*, **12**, 929–938 (2000).
44. V. K. Kutala, N. L. Parinandi, J. L. Zweier and P. Kuppusamy, *Arch. Biochem. Biophys.*, **424**, 81–88 (2004).
45. V. K. Kutala, F. A. Villamena, G. Ilangovan, *et al.*, *J. Phys Chem. B*, **112**, 158–167 (2008).
46. A. A. Bobko, I. Dhimitruka, J. L. Zweier and V. V. Khramtsov, *J. Am. Chem. Soc.*, **129**, 7240–7241 (2007).
47. I. Dhimitruka, A. A. Bobko, C. M. Hadad, *et al.*, *J. Am. Chem. Soc.*, **130**, 10780–10787 (2008).
48. M. Hockel, C. Knoop, K. Schlenger, *et al.*, *Radiother. Oncol.*, **26**, 45–50 (1993).
49. M. Hockel, B. Vorndran, K. Schlenger, *et al.*, *Gynecol. Oncol.*, **51**, 141–149 (1993).
50. D. Grucker, *Progress in Nuclear Magnetic Resonance Spectroscopy*, **36**, 241–270 (2000).
51. F. Bloch, W. W. Hansen and M. Packard, *Phys. Rev.*, **70**, 474–485 (1946).
52. R. J. Alfidi, J. R. Haaga, S. J. El-Yousef, *et al.*, *Radiology*, **143**, 175–181 (1982).
53. I. R. Young, D. R. Bailes, M. Burl, *et al.*, *J. Comput. Assist. Tomogr.*, **6**, 1–18 (1982).
54. R. C. Brasch, D. A. London, G. E. Wesbey, *et al.*, *Radiology*, **147**, 773–779 (1983).
55. S. Ogawa, T. M. Lee, A. R. Kay and D. W. Tank, *Proc. Natl. Acad. Sci. USA*, **87**, 9868–9872 (1990).
56. L. C. Clark, Jr., J. L. Ackerman, S. R. Thomas, *et al.*, *Adv. Exp. Med. Biol.*, **180**, 835–845 (1984).
57. R. P. Mason, W. Rodbumrung and P. P. Antich, *NMR Biomed.*, **9**, 125–134 (1996).
58. E. O. Aboagye, A. B. Kelson, M. Tracy and P. Workman, *Anticancer Drug Des.*, **13**, 703–730 (1998).
59. M. Tinkham and M. W. P. Strandberg, *Phys. Rev.*, **99**, 537–539 (1955).
60. C. S. Lai, L. E. Hopwood, J. S. Hyde and S. Lukiewicz, *Proc. Natl. Acad. Sci. USA*, **79**, 1166–1170 (1982).
61. G. Bacic, M. J. Nilges, R. L. Magin, *et al.*, *Magn. Reson. Med.*, **10**, 266–272 (1989).
62. G. E. Pake and T. R. Tuttle, *Phys. Rev. Lett.*, **3**, 423–425 (1959).
63. M. J. Povich, *J. Phys. Chem.*, **79**, 1106–1109 (1975).
64. J. S. Hyde, J.-J. Jin, J. B. Felix and W. L. Hubbell, *Pure. Appl. Chem.*, **62**, 255–260 (1990).
65. W. Froncisz, C. S. Lai and J. S. Hyde, *Proc. Natl. Acad. Sci. USA*, **82**, 411–415 (1985).
66. J. S. Hyde and W. K. Subszynski, in *Spin Labeling: Theory and Application* (eds L. J. Berliner and J. Reuben), pp. 399–425, Plenum Press, New York, 1989.
67. P. Kuppusamy, M. Chzhan, K. Vij, *et al.*, *Proc. Natl. Acad. Sci. USA*, **91**, 3388–3392 (1994).
68. J. L. Zweier and P. Kuppusamy, *Proc. Natl. Acad. Sci. USA*, **85**, 5703–5707 (1988).
69. S. S. Velan, R. G. Spencer, J. L. Zweier and P. Kuppusamy, *Magn. Reson. Med.*, **43**, 804–809 (2000).
70. B. Gallez, G. Bacic, F. Goda, *et al.*, *Magn. Reson. Med.*, **35**, 97–106 (1996).
71. T. Sarna, A. Duleba, W. Korytowski and H. Swartz, *Arch. Biochem. Biophys.*, **200**, 140–148 (1980).
72. H. J. Halpern, M. Peric, C. Yu and B. L. Bales, *J. Magn. Reson., Ser. A.*, **103**, 13–22 (1993).
73. Y. J. Lin, B. A. Teicher and H. J. Halpern, *Journal of Labelled Compounds and Radiopharmaceuticals*, **XXVIII**, 621–631 (1990).
74. L. B. Volodarsky, V. A. Reznikov and V. I. Ovcharenko, *Synthetic chemistry of stable nitroxides*, CRC Press, Boca Raton, FL, 1994.
75. K. Hideg, T. Kalai and C. P. Sar, *J. Heterocycl. Chem.*, **42**, 437–450 (2005).
76. J. E. Baker, W. Froncisz, J. Joseph and B. Kalyanaraman, *Free. Rad. Bio.l Med.*, **22**, 109–115 (1997).
77. B. Gallez, R. Debuyst, R. Demeure, *et al.*, *Magn. Reson. Med.*, **30**, 592–599 (1993).
78. K. J. Liu, M. W. Grinstaff, J. Jiang, *et al.*, *Biophys. J.*, **67**, 896–901 (1994).
79. P. Kuppusamy and J. L. Zweier, *NMR Biomed.*, **17**, 226–239 (2004).
80. R. Murugesan, J. A. Cook, N. Devasahayam, *et al.*, *Magn. Reson. Med.*, **38**, 409–414 (1997).
81. P. Kuppusamy, R. A. Shankar and J. L. Zweier, *Phys. Med. Biol.*, **43**, 1837–1844 (1998).
82. P. Kuppusamy, M. Afeworki, R. A. Shankar, *et al.*, *Cancer Res.*, **58**, 1562–1568 (1998).
83. T. Guiberteau and D. Grucker, *J. Magn. Reson.*, **124**, 263–266 (1997).
84. M. Elas, B. B. Williams, A. Parasca, *et al.*, *Magn. Reson. Med.*, **49**, 682–691 (2003).
85. D. J. Lurie, G. R. Davies, M. A. Foster and J. M. Hutchison, *Magn. Reson. Imaging.*, **23**, 175–181 (2005).
86. H. Li, G. He, Y. Deng, *et al.*, *Magn. Res. Med.*, **55**, 669–675 (2006).
87. D. Grucker and J. Chambron, *Magn. Reson. Im.*, **11**, 691–696 (1993).

88. Y. Liu, F. A. Villamena, J. Sun, *et al.*, *J. Org. Chem.*, **73**, 1490–1497 (2008).
89. B. Driesschaert, N. Charlier, B. Gallez and J. Marchand-Brynaert, *Bioorg. Med. Chem. Lett.*, **18**, 4291–4293 (2008).
90. K. J. Liu, P. Gast, M. Moussavi, *et al.*, *Proc. Nat.l Acad. Sci. USA*, **90**, 5438–5442 (1993).
91. T. Presley, P. Kuppusamy, J. L. Zweier and G. Ilangovan, *Biophys. J.*, **91**, 4623–4631 (2006).
92. R. B. Clarkson, B. M. Odintsov, P. J. Ceroke, *et al.*, *Phys. Med. Biol.*, **43**, 1907–1920 (1998).
93. S. Pietri, S. Martel, M. Culcasi, *et al.*, *J. Biol. Chem.*, **276**, 1750–1758 (2001).
94. R. J. Gillies, J. R. Alger, J. A. den Hollander and R. G. Shulman, in *Intracellular pH: Its Measurement, Regulation and Utilization in Cellular Functions* (eds R. Nuccitelli and D. W. Deamer), pp. 79–104, Alan R. Liss, New York, 1982.
95. R. J. Gillies, Z. Liu and Z. Bhujwalla, *Am. J. Physiol.*, **267**, C195–203 (1994).
96. R. J. Gillies, N. Raghunand, M. L. Garcia-Martin and R. A. Gatenby, *IEEE Eng. Med. Biol. Mag.*, **23**, 57–64 (2004).
97. B. M. Hoffman and T. B. Eames, *J. Am. Chem. Soc.*, **91**, 2169–2170 (1969).
98. V. Malatesta and K. U. Ingold, *J. Am. Chem. Soc*, **95**, 6404–6407 (1973).
99. E. F. Ullman and J. H. Osiecki, *J Org Chem*, **35**, 3623–3631 (1970).
100. R. F. Haseloff, S. Zollner, I. A. Kirilyuk, *et al.*, *Free. Rad. Res.*, **26**, 7–17 (1997).
101. A. A. Bobko, E. G. Bagryanskaya, V. A. Reznikov, *et al.*, *Free Rad. Biol. Med.*, **36**, 248–258 (2004).
102. V. V. Khramtsov and L. M. Weiner, in *Imidazoline Nitroxides* (ed. L. B. Volodarsky), Vol. 2, pp. 37–80, CRC Press, Boca Raton, FL, 1988.
103. J. C. Hsia and J. M. Boggs, *Biochim. Biophys. Acta*, **266**, 18–25 (1972).
104. A. T. Quintanilha and R. J. Mehlhorn, *FEBS Lett.*, **91**, 104 (1978).
105. C. R. Nakaie, G. Goissis, S. Schreier and A. C. Paiva, *Braz. J. Med. Biol Res*, **14**, 173–180 (1981).
106. G. A. A. Saracino, A. Tedeschi, G. D'Errico, *et al.*, *J. Phys. Chem. A.*, **106**, 10700–10706 (2002).
107. A. E. Mathew and J. R. Dodd, *J. Heterocycl. Chem.*, **22**, 225–228 (1985).
108. V. V. Khramtsov, L. M. Weiner, I. A. Grigor'ev and L. B. Volodarsky, *Chem. Phys. Lett.*, **91**, 69–72 (1982).
109. V. V. Khramtsov, I. A. Grigor'ev, M. A. Foster, *et al.*, *Spectroscopy*, **18**, 213–225 (2004).
110. I. A. Kirilyuk, A. A. Bobko, I. A. Grigor'ev and V. V. Khramtsov, *Org. Biomol. Chem.*, **2**, 1025–1030 (2004).
111. I. A. Kirilyuk, A. A. Bobko, V. V. Khramtsov and I. A. Grigor'ev, *Org. Biomol. Chem.*, **3**, 1269–1274 (2005).
112. M. A. Voinov, J. F. Polienko, T. Schanding, *et al.*, *J. Org. Chem.*, **70**, 9702–9711 (2005).
113. K. Mader, S. Nitschke, R. Stosser and H. H. Borchert, *Polymer*, **38**, 4785–4794 (1997).
114. C. Kroll, W. Hermann, R. Stosser, *et al.*, *Pharm. Res*, **18**, 525–530 (2001).
115. K. Mader, B. Gallez, K. J. Liu and H. M. Swartz, *Biomaterials*, **17**, 457–461 (1996).
116. B. Gallez, K. Mader and H. M. Swartz, *Magn. Reson. Med.*, **36**, 694–697 (1996).
117. M. A. Foster, I. A. Grigor'ev, D. J. Lurie, *et al.*, *Magn. Reson Med.*, **49**, 558–567 (2003).
118. V. V. Khramtsov, I. A. Grigor'ev, I. A. Kirilyuk, *et al.*, *Free Rad. Biol. Med.*, **33** (Suppl. 2), S423–S424 (2002).
119. J. L. Zweier, P. Wang, A. Samouilov and P. Kuppusamy, *Nat. Med.*, **1**, 804–809 (1995).
120. T. F. Rehring, J. I. Shapiro, B. S. Cain, *et al.*, *Am. J. Physiol.*, **275**, H805–813 (1998).
121. A. Samouilov, G. L. Caia, E. Kesselring, *et al.*, *Magn. Reson. Med.*, **58**, 156–166 (2007).
122. V. V. Khramtsov, G. L. Caia, K. Shet, *et al.*, *J. Magn. Reson.*, **202**, 267–273 (2010).
123. F. Q. Schafer and G. R. Buettner, *Free. Rad. Biol. Med.*, **30**, 1191–1212 (2001).
124. G. He, V. K. Kutala, P. Kuppusamy and J. L. Zweier, *Free. Radic. Biol. Med.*, **36**, 665–672 (2004).
125. S. Das, N. Khan, S. Mukherjee, *et al.*, *Free. Rad. Biol. Med.*, **44**, 82–90 (2008).
126. N. Ojha, S. Roy, G. He, *et al.*, *Free. Rad. Biol. Med.* **44**, 682–691 (2008).
127. F. Hyodo, B. P. Soule, K. Matsumoto, *et al.*, *J. Pharm. Pharmacol.*, **60**, 1049–1060 (2008).
128. P. Kuppusamy and M. C. Krishna, *Curr. Top. Biophys.*, **26**, 29–34 (2002).
129. J. Glebska, J. Skolimowski, Z. Kudzin, *et al.*, *Free. Rad. Biol. Med.*, **35**, 310–316 (2003).
130. E. Finkelstein, G. M. Rosen and E. J. Rauckman, *Biochim. Biophys. Acta*, **802**, 90–98 (1984).
131. A. A. Bobko, I. A. Kirilyuk, I. A. Grigor'ev, *et al.*, *Free Rad. Biol. Med.*, **42** (2007).
132. K. Takeshita, A. Hamada and H. Utsumi, *Free. Rad. Biol. Med.*, **26**, 951–960 (1999).

133. G. L. Ellman, *Arch. Biochem. Biophys.*, **82**, 70–77 (1959).
134. N. Patsoukis and C. D. Georgiou, *Anal. Bioanal. Chem.*, **383**, 923–929 (2005).
135. F. Tietze, *Anal. Biochem.*, **27**, 502–522 (1969).
136. G. T. Yamashita and D. L. Rabenstein, *J. Chromatogr.*, **491**, 341–354 (1989).
137. J. A. Willis and T. Schleich, *Biochim. Biophys Acta.*, **1265**, 1–7 (1995).
138. M. Terpstra, P. G. Henry and R. Gruetter, *Magn. Reson. Med.*, **50**, 19–23 (2003).
139. J. C. Livesey, R. N. Golden, E. G. Shankland, *et al.*, *Int. J. Radiat. Oncol. Biol. Phys.*, **22**, 755–757 (1992).
140. D. I. Potapenko, E. G. Bagryanskaya, I. A. Grigoriev, *et al.*, *Magn. Reson. Chem.*, **43**, 902–909 (2005).
141. M. Kveder, A. Krisko, G. Pifat and H. J. Steinhoff, *Biochim. Biophys. Acta*, **1631**, 239–245 (2003).
142. M. Soszynski and G. Bartosz, *Free. Rad. Biol. Med.*, **23**, 463–469 (1997).
143. V. V. Khramtsov, V. I. Yelinova, I. Glazachev Yu, *et al.*, *J. Biochem. Biophys. Methods*, **35**, 115–128 (1997).
144. N. Khan, V. V. Khramtsov and H. M. Swartz, in *Methods in Redox Signaling*, (ed. D. K. Das), Mary Ann Liebert, Inc., New Rochelle, NY, 81–89 (2009).
145. A. Balcerczyk and G. Bartosz, *Free. Rad. Res.*, **37**, 537–541 (2003).
146. A. Balcerczyk, A. Grzelak, A. Janaszewska, *et al.*, *Biofactors*, **17**, 75–82 (2003).
147. A. Bratasz, V. V. Khramtsov and P. Kuppusamy, *Free. Rad. Biol. Med.* **35** (Suppl. 1), S147 (2003).
148. A. Bratasz, N. M. Weir, N. L. Parinandi, *et al.*, *Proc. Natl. Acad. Sci. USA*, **103**, 3914–3919 (2006).
149. L. M. Weiner, H. Hu and H. M. Swartz, *FEBS Lett.*, **290**, 243–246 (1991).
150. V. Yelinova, Y. Glazachev, V. Khramtsov, *et al.*, *Biochem. Biophys. Res. Commun.*, **221**, 300–303 (1996).
151. V. I. Yelinova, V. V. Khramtsov and A. L. Markel, *Biochem. Biophys. Res. Commun.*, **263**, 450–453 (1999).
152. H. Nohl, K. Stolze and L. M. Weiner, *Methods. Enzymol*, **251**, 191–203 (1995).
153. C. A. Rabik and M. E. Dolan, *Cancer Treat. Rev.*, **33**, 9–23 (2007).
154. A. Bratasz, K. Selvendiran, T. Wasowicz, *et al.*, *J. Translat. Med.*, **6** (2008), doi: 10.1186/1479-5876-6-9.
155. F. Y. Lee, A. Vessey, E. Rofstad, *et al.*, *Cancer Res.*, **49**, 5244–5248 (1989).

17

Biologically Relevant Chemistry of Nitroxides

Sara Goldstein[1] and Amram Samuni[2]

[1]*Institute of Chemistry and the Accelerator Laboratory, The Hebrew University of Jerusalem, Jerusalem, Israel*

[2]*Department of Molecular Biology, The Hebrew University of Jerusalem – Hadassah Medical School, Jerusalem, Israel*

17.1 Introduction

The ever-increasing knowledge of the involvement of radicals in diverse pathological processes has expanded the search for more efficient antioxidants that can diminish radical-induced damage. Accordingly, the activity of a broad array of potentially useful antioxidants has been extensively studied. These antioxidants include enzymes such as superoxide dismutase (SOD) and catalase, which remove superoxide ($HO_2^{\bullet}/O_2^{\bullet-}$) and H_2O_2, respectively; chelators that render transition metal ions redox inactive thus pre-empting Fenton-like reactions; and low molecular weight antioxidants (LMWA) that react with toxic radicals terminating radical chain reactions and restituting impaired cellular sites. A more restrictive class of antioxidants is confined to natural or synthetic, hydrophilic or lipophilic LMWA alone, for example, ascorbate, urate, trolox, vitamin E, β-carotene, butylated hydroxytoluene, carnosine, polyphenols, flavonoids, and thiols. However, the protective effect of administered SOD against oxidative stress is limited by its short half-life in the circulation and its inability to cross cell membranes.[1,2] LMWA, which are diamagnetic, operate primarily by reducing deleterious oxidants. They are progressively depleted from the tissue, particularly under oxidative stress, and yield secondary radicals that might also induce biological damage.

Cyclic nitroxides ($R_2NO^{\bullet}$) are cell-permeable stable radicals of diverse size, charge, and lipophilicity stabilized by methyl groups at the α position in five-membered pyrrolidine, pyrroline, or oxazolidine and six-membered piperidine ring structures. The structures of several nitroxides used in our studies are given in Figure 17.1.

Cyclic nitroxides have been used for years as biophysical probes to monitor membrane dynamics,[3] cellular pH,[4] redox state *in vivo*,[5] as potential contrast agents for magnetic resonance imaging (MRI),[6] and catalysts for specific oxidation of alcohols and sugars.[7–9] The observation[10,11] that cyclic nitroxides catalyze

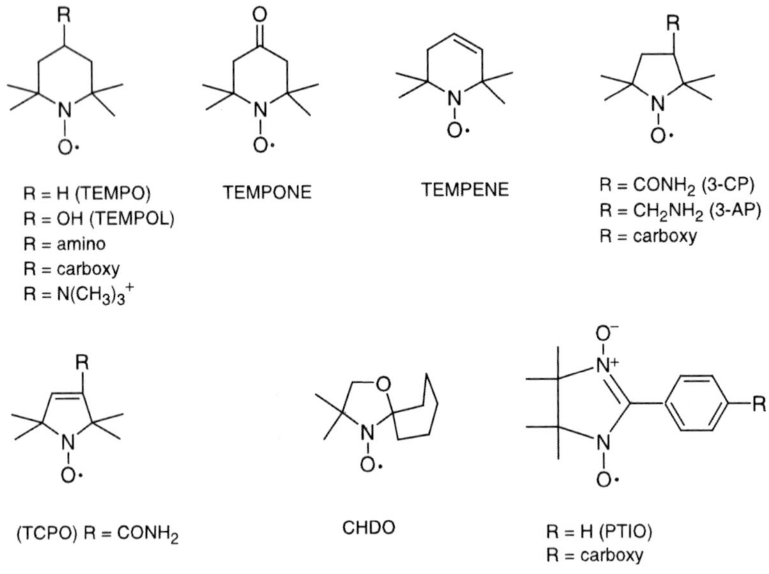

Figure 17.1 *Structures of selected nitroxide radicals.*

the removal of superoxide prompted numerous studies of their antioxidative activity. Cyclic nitroxides, which are paramagnetic and far more selective than LMWA, do not react with most diamagnetic species, including H_2O_2, and even with radicals such as O_2 and $^{\bullet}NO$ (excluding the latter in the case of aromatic[12] and nitronyl nitroxides).[13–15] These radicals prevent oxidative damage in various biological systems ranging from molecular via cellular to laboratory animal levels (recent reviews are given elsewhere).[16–18] The protective effects of nitroxides have, in part, been attributed to their ability to catalyze the dismutation of superoxide radicals[10,11,19,20] and scavenge a large variety of deleterious radicals, such as peroxyl radicals ($RO_2^{\bullet}$),[21] $^{\bullet}NO_2$,[22,23] $CO_3^{\bullet-}$,[20,23,24] and thiyl radicals.[25–28] Their antioxidative effects are often associated with a one-electron exchange among their reduced (R_2NOH) and oxidized ($R_2N^+{=}O$) states (Scheme 17.1).

Hydroxylamines have extremely weak O–H bonds, for example, 69.6 and 71.8 kcal/mol for hydroxylamines derived from TEMPO and TEMPONE, respectively,[29,30] and therefore the nitroxides never participate in hydrogen abstraction reactions. However, being radicals, they react rapidly with other radicals, and in this respect the nitroxides are antioxidants. Nitroxides can react with diverse biologically relevant radical oxidants and reductants while being recycled through the oxoammonium cation and hydroxylamine derivatives (Scheme 17.1). They can be consumed when their reaction with biological molecules or certain

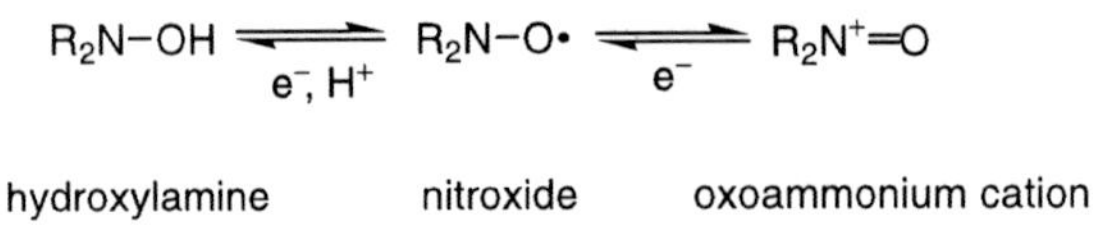

Scheme 17.1 One-electron oxidation and reduction of nitroxide to oxoammonium cation and hydroxylamine, respectively

radicals lead to the formation of the respective amine, as is the case with thiyl radicals,[26–28] or when the oxoammonium cation is highly unstable as in the case of TEMPONE[22] and 4-amino-TEMPO (unpublished results). The highly oxidizing oxoammonium cation can mediate selective oxidation of primary alcohols in mono- and polysaccharides that are catalyzed by nitroxides.[9,31] The formation of the oxoammonium cation could be also responsible for the pro-oxidative activity and potential adverse effects of nitroxides that otherwise act as antioxidants.[32]

Nitroxides also differ from common LMWA by their multifunctionality, that is, they operate through diverse modes of action.[23,32,33] While this multifunctionality contributes toward the biological activity of nitroxides, it greatly complicates the study and elucidation of the mechanisms underlying their effects. In contrast to common LMWA, the catalytic nature of the chemistry of nitroxides allows their self-replenishment. Such recycling plays an important role particularly, since the nitroxide administered into the tissues is reduced to the respective hydroxylamine by cellular reducing equivalents. There is an equilibrium between the nitroxide and the hydroxylamine, which is dependent on the oxygen status and redox status of the tissue milieu.[34] While reduction of nitroxide to hydroxylamine can be reversed, such recycling is achievable via different mechanisms. The elucidation of the mechanisms underlying nitroxides catalytic activity requires characterization of their apparent self-replenishment as well as detailed kinetic studies of the individual reaction steps. Here, the main mechanisms by which cyclic nitroxides detoxify biologically deleterious radicals, which are key intermediates in many inflammatory and degenerative diseases, are reviewed.

17.2 Mechanisms of nitroxide reactions with biologically relevant small radicals

The reactions of $^{\bullet}NO_2$, $CO_3^{\bullet-}$, $HO_2^{\bullet}$ and $RO_2^{\bullet}$ radicals (denoted as $X^{\bullet}$) with nitroxides lead to a common intermediate, which can oxidize ferrocyanide, NADH and 2,2′-azino-bis(3-ethylbenzothiazoline-6-sulfonate) ($ABTS^{2-}$).[15,19–22] In some cases the intermediate was identified spectrophotometrically as the oxoammonium cation.[15,21,22] The kinetic results demonstrate that the formation of the oxoammonium cation proceeds via an inner sphere electron transfer mechanism (Equations 17.1 and 17.2).

$$R_2NO^{\bullet} + X^{\bullet} \rightleftharpoons R_2NO{-}X \qquad (17.1)$$

$$R_2NO{-}X \rightleftharpoons R_2N^{+}{=}O + X^{-} \qquad (17.2)$$

This conclusion was derived using Marcus theory of oxidation–reduction reactions[35] for the formation of the oxoammonium cation via Equation 17.3.

$$R_2NO^{\bullet} + X^{\bullet} \rightleftharpoons R_2N^{+}{=}O + X^{-} \qquad (17.3)$$

Assuming that Equation 17.3 takes place via an outer sphere electron transfer mechanism, k_3 can be calculated using the Marcus equation, $k_3 = (k_{aa}k_{bb}K_3f_3)^{1/2}$. Here k_3 is the electron transfer rate constant for the cross reaction, k_{aa} and k_{bb} are the self-exchange rate constants for the reactants, K_3 is the cross reaction equilibrium constant, and $\ln f_3 = (\ln K_3)^2/4\ln(k_{aa}k_{bb}/10^{22})$. Using the oxidation potentials for the redox couples $R_2N^{+}{=}O/R_2NO^{\bullet}$ and $X^{\bullet}/X^{-}$ (Table 17.1), it is possible to calculate K_3 as $\Delta E_3{}^{o} = (RT/nF)\ln K_3$ and $\Delta E_3{}^{o} = E^{o}(X^{\bullet}/X^{-}) - E^{o}(R_2N^{+}{=}O/R_2NO^{\bullet})$. The self-exchange rate constant for $X^{\bullet}/X^{-}$ is significantly low,[36–39] and since that for $R_2N^{+}{=}O/R_2NO^{\bullet}$ cannot exceed $1 \times 10^{10}\ M^{-1}s^{-1}$, the calculated values of k_3 are orders of magnitude lower than the experimental ones determined using pulse radiolysis.[15,19–22]

Therefore, the reaction of $R_2NO^{\bullet}$ with $X^{\bullet}$ must take place via an inner sphere electron transfer mechanism (Equations 17.1 and 17.2). In the case of carbon-centered radicals, which are important in biological systems

Table 17.1 The oxidation potentials vs NHE for the redox couples $R_2N^+{=}O/R_2NO^\bullet$ and $X^\bullet/X^-$

	$E_{1/2}{}^{oxd}$ (mv)	E° (mv)
TEMPO	740,[64] 722,[46] 719[68]	
4-COOH-TEMPO	805[20]	
4-COO⁻-TEMPO	771[20,68]	
TEMPOL	825,[64] 810,[46] 808[68]	
4-amino-TEMPO	817,[20] 826,[46] 851[68]	
4-$N(CH_3)_3{}^+$-TEMPO	940[20]	
TEMPONE	913,[46] 918[20]	
TEMPENE	795[20]	
3-COOH-proxyl	870[20]	
3-COO⁻-proxyl	792,[46] 772[69]	
3-AP	853[46]	
3-CP	861,[46] 872[69]	
TCPO	955[20]	
CHDO	900[46]	
PTIO	915[15]	
carboxy-PTIO	936[15]	
$HOO^\bullet$		750[70]
$CH_3OO^\bullet$		770[71]
t-$BuOO^\bullet$		710[39]
$CO_3{}^{\bullet-}$		1590[70]
$^\bullet NO_2$		1040[70]

at low concentrations of oxygen, some of the reactions proceed via the formation of stable adducts and others via electron transfer mechanism.[40–44]

The reactivity of the nitroxides towards $^\bullet NO_2$ and $CO_3^{\bullet-}$ radicals is almost the same for piperidine, pyrrolidine, and oxazolidine nitroxides, and does not depend on the pH or buffer concentration, namely $k_o = k_1k_2/(k_{-1} + k_2) = (2.5 - 8.7) \times 10^8\ M^{-1}s^{-1}$ (Table 17.2).[20,22]

Table 17.2 The rate constant k_o ($M^{-1}s^{-1}$) of the reaction between nitroxide and $^\bullet NO_2$ or $CO_3{}^{\bullet-}$ as determined by pulse radiolysis[15,20,22]

Nitroxide	$^\bullet NO_2$	$CO_3{}^{\bullet-}$
TEMPO	$(7.1 \pm 0.2) \times 10^8$	$(4.0 \pm 0.1) \times 10^8$
TEMPOL	$(8.7 \pm 0.2) \times 10^8$	$(4.0 \pm 0.1) \times 10^8$
4-amino-TEMPO	$(5.5 \pm 0.2) \times 10^8$	$(3.9 \pm 0.1) \times 10^8$
TEMPONE	$(7.1 \pm 0.2) \times 10^8$	$(4.0 - 4.8) \times 10^8$
4-$N(CH_3)_3{}^+$-TEMPO	$(4.0 \pm 0.1) \times 10^8$	$(6.3 \pm 0.1) \times 10^8$
3-CP	$(7.1 \pm 0.2) \times 10^8$	$(4.0 \pm 0.1) \times 10^8$
3-carboxy-proxyl	$(4.9 \pm 0.2) \times 10^8$	$(2.4 \pm 0.1) \times 10^8$
TCPO	$(3.2 \pm 0.1) \times 10^8$	$(2.7 \pm 0.1) \times 10^8$
CHDO	$(2.5 \pm 0.1) \times 10^8$	$(2.6 \pm 0.1) \times 10^8$
PTIO	$(2.0 \pm 0.1) \times 10^7$	ND
carboxy-PTIO	$(1.5 \pm 0.1) \times 10^7$	ND

ND – not determined.

Table 17.3 *The rate constants k_o ($M^{-1}s^{-1}$) for the reaction of nitroxides with peroxyl and $HO_2^{\bullet}$ radicals*[15,19–21]

Nitroxide	t-BuOO•	$CH_3OO^{\bullet}$	$CH_2(OH)OO^{\bullet}$	HOO•
TEMPO	$(2.8 \pm 0.2) \times 10^7$	$(5.1 \pm 0.1) \times 10^7$	$(1.0 \pm 0.1) \times 10^8$	$(1.1 \pm 0.1) \times 10^8$
TEMPOL	$(3.3 \pm 0.2) \times 10^6$	$(5.4 \pm 0.2) \times 10^6$	$(4.4 \pm 0.1) \times 10^7$	$(2.7 \pm 0.3) \times 10^7$
4-amino-TEMPO	$(1.0 \pm 0.2) \times 10^6$	$(1.5 \pm 0.2) \times 10^6$	ND	$<1 \times 10^7$
TEMPONE	$(2.8 \pm 0.2) \times 10^5$	$(5.4 \pm 0.2) \times 10^5$	ND	$<6 \times 10^6$
TEMPENE	ND	ND	ND	$(7.9 \pm 0.2) \times 10^7$
3-CP	$(8.1 \pm 0.6) \times 10^5$	$(1.1 \pm 0.1) \times 10^6$	$(9.0 \pm 0.1) \times 10^6$	$(1.1 \pm 0.1) \times 10^6$
3-AP	$(9.6 \pm 0.6) \times 10^5$	$(1.5 \pm 0.6) \times 10^6$	ND	$(1.1 \pm 0.2) \times 10^6$
CHDO	$(5.0 \pm 0.5) \times 10^4$	$(6.7 \pm 0.5) \times 10^4$	ND	$(1.6 \pm 0.2) \times 10^5$
PTIO	ND	ND	ND	$<2 \times 10^4$

ND – not determined.

In contrast, the reactivity of nitroxides towards $HO_2^{\bullet}$ and $RO_2^{\bullet}$ is highly dependent on the ring size and, to a lesser extent, on the side chain. Furthermore, in the case of piperidine nitroxides, the rate of formation of $R_2N^+{=}O$ is both H^+ and general acid catalyzed (Equations 17.4 and 17.5).[20,21]

$$R_2NOX + H^+ \longrightarrow R_2N^+{=}O + HX \tag{17.4}$$

$$R_2NOX + HA \longrightarrow R_2N^+{=}O + HX + A^- \tag{17.5}$$

The uncatalyzed rate constants (k_o) derived for $HO_2^{\bullet}$ and alkyl peroxyl radicals are given in Table 17.3, demonstrating that the most efficient scavenger is TEMPO, which has the lowest oxidation potential among the studied nitroxides (Table 17.1).

In the case of thiyl radicals ($RS^{\bullet}$), the reaction also proceeds via the formation of an adduct with the general structure R_2NOSR'. However, the decomposition of the adduct does not yield the respective oxoammonium cation but instead the respective amine via a complex mechanism (Scheme 17.2).[28]

The rate constant for the reaction of nitroxides with thiyl radicals is independent of the structure of the nitroxide and thiyl radicals and is similar to that determined for $^{\bullet}NO_2$ and $CO_3^{\bullet-}$ radicals, $k = (5-7) \times 10^8\ M^{-1}s^{-1}$ (Table 17.4).[28]

At physiological pH the unstable adduct R_2NOSR' decomposes mainly via heterolysis of the N–O bond to yield the respective amine and sulfinic acid (R′S(O)OH). In acidic solutions the protonated form of the adduct decomposes via homolysis of the N–O bond to yield the reducing sulfenyl radical ($R'SO^{\bullet}$) and the aminium cation radical ($R_2NH^{\bullet+}$), which can oxidize thiols[45] forming the respective amine and thiyl radicals (Scheme 17.2).

17.3 Nitroxides as SOD mimics

The ability of nitroxide to catalyze superoxide dismutation has been previously demonstrated by the persistence of its concentration under a flux of superoxide.[10,11] It has been demonstrated that piperidine and pyrrolidine nitroxides catalyze the dismutation of superoxide by utilizing the $R_2NO^{\bullet}/R_2N^+{=}O$ redox couple (Equations 17.6 and 17.7).[19,20,46]

$$R_2NO^{\bullet} + HO_2^{\bullet} \xrightarrow{H^+} R_2N^+{=}O + H_2O_2 \tag{17.6}$$

$$R_2N^+{=}O + O_2^{\bullet-} \longrightarrow R_2NO^{\bullet} + O_2 \tag{17.7}$$

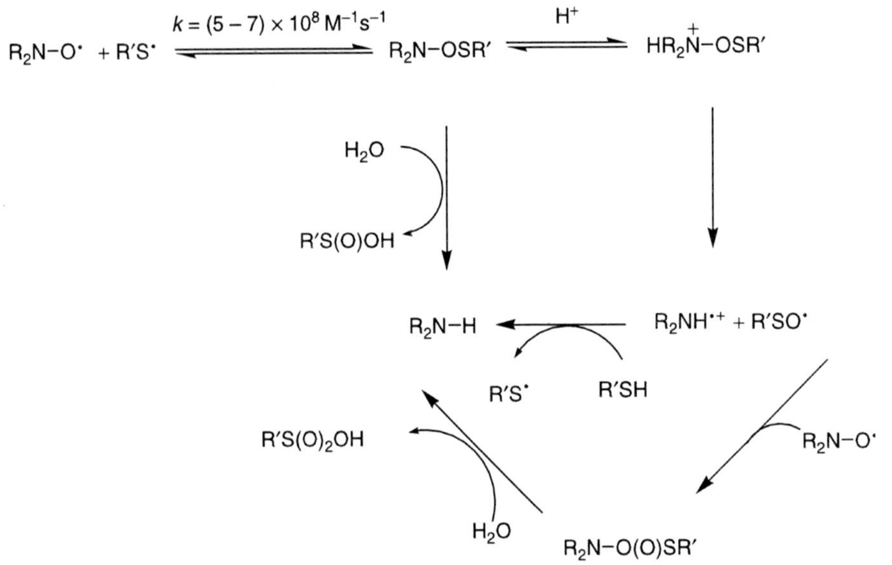

Scheme 17.2 Mechanism of nitroxides reaction with thiyl radicals

Table 17.4 *Rate constants for the reaction of R′S• with nitroxide (10^8 M^{-1} s^{-1}) as determined by pulse radiolysis*[28]

	GS• [a]	CysS• [a]	PenS• [a]
3-CP	5.2 ± 0.6^b	6.6 ± 0.2^c	5.7 ± 0.2^b
TEMPOL	5.1 ± 0.5^c	5.6 ± 0.2^c	4.9 ± 0.2^b
TEMPO	7.1 ± 0.6^c	ND	ND

[a]GSH = glutathione, CysSH = cysteine, PenSH = penicillamine.
[b]pH 5.6–6.1.
[c]pH 6.0–7.1.
ND – not determined.

The rate constant (k_o) for the uncatalyzed Equation 17.6 (Table 17.3) does not exceed 2×10^8 $M^{-1}s^{-1}$ for piperidine nitroxides at the best conditions, for example, low pH or high buffer concentration.[20]

The rate constant of the reaction of $HO_2^•$ with nitroxides correlates well with the one-electron oxidation potential $E^o(R_2N^+{=}O/R_2NO^•)$ of the nitroxide. For the five-membered ring compounds rather inefficient H^+ catalysis and no buffer-induced catalysis were observed, implying that the latter is too slow at feasible buffer concentrations.[19,20,46] This finding implies that either (i) an adduct is formed, but at feasible pH values the H^+ catalysis is too slow in comparison with the non-catalyzed decomposition; or (ii) no adduct is formed on the reaction path and H^+ catalysis occurs by way of a third order concerted reaction, where $HO_2^•$, the nitroxide and H^+ simultaneously form the transition state. If no adduct exists, the uncatalyzed reaction would have to be a one-step electron transfer. Such a reaction does not necessarily have to be an outer sphere electron transfer; since there are no steric or spin restrictions, the reactants can come quite close for substantial molecular overlap in the transition state to occur. If so, the rate

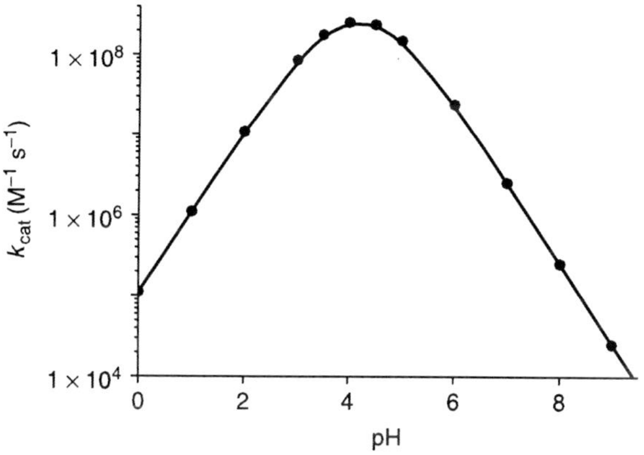

Figure 17.2 *Dependence of k_{cat} on the pH. The dependence of the catalytic rate constant of superoxide dismutation by piperidine nitroxides calculated using Equation 17.8 and $k_6 = 2 \times 10^8\,M^{-1}s^{-1}$, $k_7 = 3.5 \times 10^9\,M^{-1}s^{-1}$, $pK_a = 4.8$.*

constant may be significantly higher than that implied by the Marcus theory for an outer sphere electron transfer reaction.

The rate constant of Equation 17.7 is independent of the structure of the nitroxide and varies between 1.5×10^9 and $5 \times 10^9\ M^{-1}s^{-1}$ for piperidine and pyrrolidine derivatives.[19,20] Equations 17.6 and 17.7 are both pH dependent ($pK_a(HO_2^\bullet) = 4.8$) and with increasing the pH the rate of Equation 17.6 decreases and that of Equation 17.7 increases. Under limiting concentrations of the nitroxide and assuming the steady state approximation for $R_2N^+{=}O$, rate equation 17.8 is obtained where $[O_2^{\bullet-}]_T = ([O_2^{\bullet-}] + [HO_2^\bullet])$ and $[R_2NO^\bullet]_o = ([R_2NO^\bullet] + [R_2N^+{=}O])$.

$$-\frac{d[O_2^{\bullet-}]_T}{dt} = \frac{2\,k_6\,k_7}{(1+[H^+]/K_a)k_6 + (1+K_a/[H^+])k_7} \times [R_2NO^\bullet]_o[O_2^{\bullet-}]_T = k_{cat}[R_2NO^\bullet]_o[O_2^{\bullet-}]_T \quad (17.8)$$

The simulated curve of, k_{cat} vs pH using $k_6 = 2 \times 10^8\ M^{-1}s^{-1}$ and $k_7 = 3.5 \times 10^9\ M^{-1}s^{-1}$ is given in Figure 17.2.

The dependence of k_{cat} on the pH displayed a bell-shaped curve having a maximum around pH 4, and demonstrates that the catalytic activity of the nitroxide under physiological conditions predominantly depends on the rate of Equation 17.6. The best SOD mimic among the nitroxides are the six-membered ring derivatives, but their catalytic activity at pH 7.4 is about $k_{cat} \sim 1\times 10^6\ M^{-1}s^{-1}$, that is, three orders of magnitude lower than that of Cu,Zn-SOD. It is worthwhile to note that in the pH range 4–5, which is closer to that of biological membranes, their catalytic activity is about an order of magnitude lower than that of Cu,Zn-SOD.

17.4 Nitroxides as catalytic antioxidants in biological systems

Nitroxides can afford protection catalytically acting both as reducing or oxidizing agents, while being continuously recycled through the oxoammonium cation or the hydroxylamine, respectively. The rapid reduction of the nitroxide in biological systems into hydroxylamine apparently limits their application[34,47–49] although hydroxylamine itself can provide some protection[50–52] partially as a hydrogen atom donor as in the cases of $CO_3^{\bullet-}$,[24] $^\bullet OH$[53] and peroxyl radicals.[54,55]

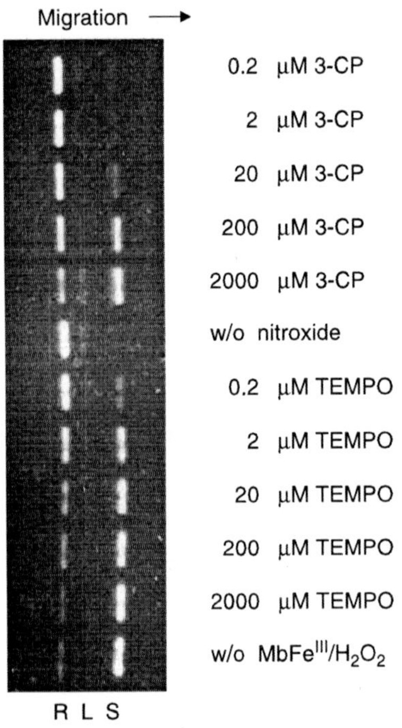

Figure 17.3 *DNA scission induced by MbFe^{IV} and MbFe^{IV} in the absence and presence of nitroxides. The system contained 1 μg/mL pUC19 plasmid DNA, 40 μM MbFe^{III}, 2 mM H_2O_2, 20 μM EDTA in 2 mM phosphate buffer pH 7.6. The reaction was started by the addition of 2 mM H_2O_2. Following 30 min incubation the samples were treated with proteinase K to degrade the protein and subjected to electrophoresis on agarose. The migrating bands of supercoiled circularly closed, relaxed nicked circular, and the linear DNA are denoted S, R, and L respectively.*

In several experimental systems nitroxides provided significant protection in a concentration-dependent manner, even at μM levels against radical-induced damage to various biological targets.[21,54,56–60] The protective effect of nitroxides against damage induced by H_2O_2 in the presence of heme-protein also demonstrates their catalytic nature at significantly low concentration (unpublished results). Incubation of metmyoglobin (MbFe^{III}) with H_2O_2 can lead to biological damage (e.g., scission of single and double DNA strand breaks) mediated by ferrylmyoglobin (MbFe^{IV}) and, particularly, by the globin radical ($^{\bullet}$MbFe^{IV}). Nitroxides were shown to prevent DNA breakage in a dose-dependent manner where even 2μM TEMPO provided significant protection (Figure 17.3). In this reaction system the nitroxide is being recycled by H_2O_2, as seen in Scheme 17.3.

In fact, several pathways are suggested for nitroxide recycling (Scheme 17.4).

(i) Reduction of the oxoammonium cation back to the parent nitroxide via one-electron transfer mechanism. This has been demonstrated in the case of H_2O_2 (Equation 17.6)[19] and in the case of 3-CP$^+$ reacting with CH_3OH.[21]

(ii) Reduction of the oxoammonium cation to the respective hydroxylamine via two-electron transfer mechanism as in the case of NAD(P)H[19,46,61] and piperidine derivatives, for example, TEMPO$^+$, reacting with alcohols.

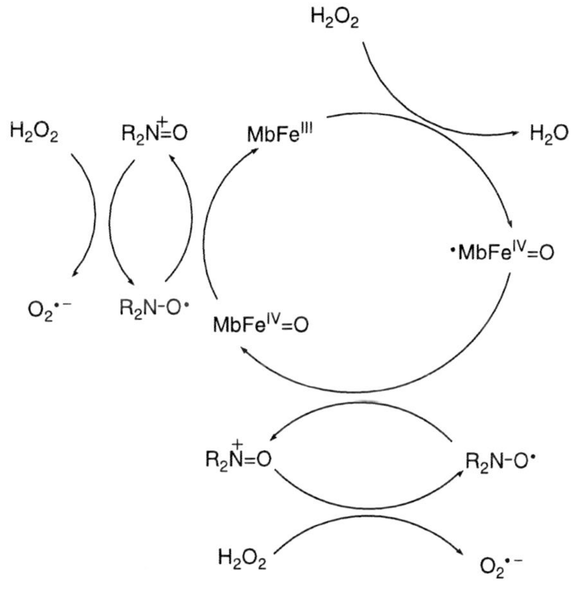

Scheme 17.3 Recycling of nitroxide in the $MbFe^{III}/H_2O_2$ system

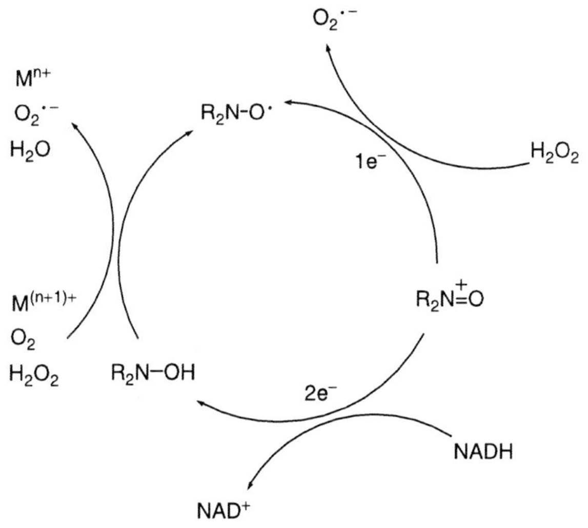

Scheme 17.4 Recycling of nitroxide in the presence of biologically relevant oxidants and reductants

(iii) Reduction of the oxoammonium cation and oxidation of the hydroxylamine to the parent nitroxide via comproportionation (Equation 17.9).[62–64]

$$R_2N^+{=}O + R_2NOH \rightleftharpoons 2\ R_2NO^{\bullet} + H^+ \qquad (17.9)$$

The rate of Equation 17.9 is pH dependent, increasing with the pH since the fastest reaction of the oxoammonium cation is with the deprotonated form of the hydroxylamine.[64] In the case of

TEMPO and TEMPOL, the rate constant of the comproportionation reaction varies between 20 and $50\,M^{-1}s^{-1}$ at physiological pH range[62,64] and, therefore, cannot greatly contribute toward recycling of the nitroxides in the tissue.

(iv) Oxidation of hydroxylamine to the respective nitroxides[34,65,66] by biologically relevant oxidants such as oxygen,[65] H_2O_2[19] and metal ions[67] (Scheme 17.4).
These mechanisms can, at least partly, explain why a residual concentration of nitroxide in the μM range persisting in the tissue demonstrates significant protection in a catalytic manner.

17.5 Conclusions

Most reactions of nitroxides with biologically relevant paramagnetic species involve the oxidation or reduction of $RNO^{\bullet}$ to $RN^{+}{=}O$ and RNOH respectively. Nitroxide reactions with $NO_2^{\bullet}$, $CO_3^{\bullet -}$, $^{\bullet}OH$, $HO_2^{\bullet}$, organic peroxyl radicals, $Fe^{IV}{=}O$ and globin radical yield the oxoammonium cation, whereas nitroxide oxidation of reduced transition metals, such as iron(II) or copper(I), leads to the formation of hydroxylamine. The catalytic activity of nitroxide requires its rapid and continuous recycling, which enables it to operate catalytically and induce biological effects even at extremely low residual levels. Accordingly, several different mechanisms can underlie the recycling of the nitroxide antioxidant, including the reduction of $R_2N^{+}{=}O$ by reductants abundant in the cell and oxidation of the cellular pool of R_2NOH by O_2, H_2O_2, and transition metal ions. Consequently, although most of the nitroxide administered is rapidly converted into R_2NOH, the low traces of $R_2NO^{\bullet}$ thus formed are still biologically effective.

Acknowledgements

This work was supported by the Israel Science Foundation of the Israel Academy of Sciences.

References

1. B. Halliwell, and J. M. C. Gutteridge, *Free Radicals in Biology and Medicine*, Oxford University Press, 1989.
2. J. M. C. Gutteridge, and B. Halliwell, *Antioxidants in Nutrition, Health and Disease*, Oxford University Press, 1994.
3. C. S. Lai, L. E. Hopwood, and H. M. Swartz, *Exp. Cell Res.*, **130**, 437–442 (1980).
4. P. D. J. Morse, and H. M. Swartz, *Magn. Res. Med.*, **2**, 114–127 (1985).
5. H. M. Swartz, N. Khan, and V. V. Khramtsov, *Antioxid. Redox Signal.*, **9**, 1757–1771 (2007).
6. F. Hyodo, K.-I. Matsumoto, A. Matsumoto, *et al.*, *Cancer Res.*, **66**, 9921–9928 (2006).
7. J. M. Bobbitt, *J. Org. Chem.*, **63**, 9367–9374 (1998).
8. W. F. Bailey, J. M. Bobbitt, and K. B. Wiberg, *J. Org. Chem.*, **72**, 4504–4509 (2007).
9. P. L. Bragd, A. C. Besemer, and H. van Bekkum, *Carbohydr. Polym.*, **49**, 397–406 (2002).
10. A. Samuni, C. M. Krishna, P. Riesz, *et al.*, *J. Biol. Chem.*, **263**, 17921–17924 (1988).
11. A. Samuni, C. M. Krishna, J. B. Mitchell, *et al.*, *Free Rad. Res.*, **9**, 241–249 (1990).
12. E. Damiani, L. Greci, and C. Rizzoli, *J. Chem. Soc. Perkin Trans. 2*, 1139–1144 (2001).
13. T. Akaike, M. Yoshida, Y. Miyamoto, *et al.*, *Biochemistry*, **32**, 827–832 (1993).
14. J. Joseph, B. Kalyanaraman, and J. S. Hyde, *Biochem. Biophys. Res. Commun.*, **192**, 926–934 (1993).
15. S. Goldstein, A. Russo, and A. Samuni, *J. Biol. Chem.*, **278**, 50949–50955 (2003).
16. B. P. Soule, F. Hyodo, K. Matsumoto, *et al.*, *Free Radic. Biol. Med.*, **42**, 1632–1650 (2007).

17. B. P. Soule, F. Hyodo, K. I. Matsumoto, *et al.*, *Antioxid. Redox Signal.*, **9**, 1731–1743 (2007).
18. V. E. Kagan, J. F. Jiang, H. Bayir, and D. A. Stoyanovsky, *Free Radic. Biol. Med.*, **43**, 348–350 (2007).
19. S. Goldstein, G. Merenyi, A. Russo, and A. Samuni, *J. Am. Chem. Soc.*, **125**, 789–795 (2003).
20. S. Goldstein, A. Samuni, K. Hideg, and G. Merenyi, *J. Phys. Chem. A*, **110**, 3679–3685 (2006).
21. S. Goldstein, and A. Samuni, *J. Phys. Chem. A*, **111**, 1066–1072 (2007).
22. S. Goldstein, A. Samuni, and A. Russo, *J. Am. Chem. Soc.*, **125**, 8364–8370 (2003).
23. O. Augusto, D. F. Trindade, E. Linares, and S. M. Vaz, *An. Acad. Bras. Cienc.*, **80**, 179–189 (2008).
24. S. Goldstein, A. Samuni, and G. Merenyi, *Chem. Res. Toxicol.*, **17**, 250–257 (2004).
25. A. P. Darmanyan, D. D. Gregory, Y. Guo, and W. S. Jenks, *J. Phys. Chem. A*, **101**, 6855–6863 (1997).
26. G. G. Borisenko, I. Martin, Q. Zhao, *et al.*, *J. Am. Chem. Soc.*, **126**, 9221–9232 (2004).
27. G. G. Borisenko, I. Martin, Q. Zhao, *et al.*, *J. Biol. Chem.*, **279**, 23453–23462 (2004).
28. S. Goldstein, A. Samuni, and G. Merenyi, *J. Phys. Chem. A*, **112**, 8600–8605 (2008).
29. L. R. M. Mahoney, G. D. Mendenhall, and K. U. Ingold, *J. Am. Chem. Soc.*, **95**, 8610–8614 (1973).
30. R. Amorati, M. Lucarini, V. Mugnaini, *et al.*, *J. Org. Chem.*, **68**, 1747–1754 (2003).
31. S. Lemoine, C. Thomazeau, D. Joannard, *et al.*, *Carbohydrate Res.*, **326**, 176–184 (2000).
32. Y. Aronovitch, D. Godinger, A. Israeli, *et al.*, *Free Radic. Biol. Med.*, **42**, 1317–1325 (2007).
33. S. M. Vaz, and O. Augusto, *Proc. Natl. Acad. Sci. USA*, **105**, 8191–8196 (2008).
34. H. M. Swartz, *Free Rad. Res.*, **9**, 399–405 (1990).
35. R. A. Marcus, *J. Phys. Chem.*, **67**, 853–857 (1963).
36. J. Lind, X. Shen, G. Merenyi, and B. O. Jonsson, *J. Am. Chem. Soc.*, **111**, 7654–7655 (1989).
37. H. H. Awad, and D. M. Stanbury, *J. Am. Chem. Soc.*, **115**, 3636–3642 (1993).
38. S. C. Schindler, Jr., E. W. Castner, Creutz, and C. Sutin, N., *Inorg. Chem.*, **32**, 4200–4208 (1993).
39. T. N. Das, T. Dhanasekaran, Z. B. Alfassi, and P. Neta, *J. Phys. Chem. A*, **102**, 280–284 (1998).
40. K. D. Asmus, S. Nigam, and R. L. Willson, *Int. J. Radiat. Biol. Relat. Stud. Phys. Chem. Med.*, **29**, 211–219 (1976).
41. D. J. Kieber, and N. V. Blough, *Anal. Chem.*, **62**, 2275–2283 (1990).
42. A. L. J. Beckwith, V. W. Bowry, and K. U. Ingold, *J. Am. Chem. Soc.*, **114**, 4983–4992 (1992).
43. V. W. Bowry, and K. U. Ingold, *J. Am. Chem. Soc.*, **114**, 4992–4996 (1992).
44. G. Ananchenko, and H. Fischer, *J. Chem. Soc., Perkin Trans. 2*, 1887–1889 (2001).
45. A. Wójcik, S. Naumov, B. Marciniak, and O. Brede, *J. Phys. Chem. B*, **110**, 12738–12748 (2006).
46. M. C. Krishna, D. A. Grahame, A. Samuni, *et al.*, *Proc. Natl. Acad. Sci. USA*, **89**, 5537–5541 (1992).
47. M. Semjurc, S. Pecar, K. Chen, *et al.*, *Biochim. Biophys. Acta*, **1073**, 329–335 (1991).
48. A. A. Bobko, I. A. Kirilyuk, I. A. Grigor'ev, *et al.*, *Free Radic. Biol. Med.*, **42**, 404–412 (2007).
49. J. Trnka, F. H. Blaikie, R. A. J. Smith, and M. P. Murphy, *Free Radic. Biol. Med.*, **44**, 1406–1419 (2008).
50. R. Zhang, A. Pinson, and A. Samuni, *Free Radic. Biol. Med.*, **24**, 66–75 (1998).
51. S. Xavier, K. Yamada, A. M. Samuni, *et al.*, *Biochim. Biophys. Acta*, **1573**, 109–120 (2002).
52. J. S. Zigler, C. Qin, T. Kamiya, *et al.*, *Free Radic. Biol. Med.*, **35**, 1194–1202 (2003).
53. A. Samuni, S. Goldstein, A. Russo, *et al.*, *J. Am. Chem. Soc.*, **124**, 8719–8724 (2002).
54. U. A. Nilsson, G. Carlin, and F. A. Bylund, *Chem. Biol. Interact.*, **74**, 325–342 (1990).
55. R. Amorati, M. Lucarini, V. Mugnaini, *et al.*, *J. Org. Chem.*, **68**, 1747–1754 (2003).
56. T. Offer, and A. Samuni, *Free Radic. Biol. Med.*, **32**, 872–881 (2002).
57. D. Fedeli, E. Damiani, L. Greci, *et al.*, *Mutat. Res.*, **535**, 117–125 (2003).
58. E. Venditti, A. Scirè, F. Tanfani, *et al.*, *Biochim. Biophys. Acta*, **1780**, 58–68 (2008).
59. G. G. Falcioni, R. Damiani, E. Santroni, *et al.*, *Free Rad. Res.*, **28**, 507–516 (1998).
60. A. Lewinska, M. Wnuk, E. Slota, and G. Bartosz, *Mutat. Res.*, **649**, 7–14 (2008).
61. M. C. Krishna, A. Samuni, J. Taira, *et al.*, *J. Biol. Chem.*, **271**, 26018–26025 (1996).
62. V. A. Golubev, V. D. Sen, I. V. Kulyk, and A. L. Aleksandrov, *Izv. Akad. Nauk SSSR Ser Khim*, **10**, 2235–2239 (1975).
63. T. H. Zauche, and J. H. Espenson, *Int. J. Chem. Kinet.*, **31**, 381–385 (1999).
64. A. Israeli, M. Patt, M. Oron, *et al.*, *Free Radic. Biol. Med.*, **38**, 317–324 (2005).

65. K. Chen, J. F. Glockner, P. D. Morse, and H. M. Swartz, *Biochem. Biophys. Acta*, **970**, 270–277 (1988).
66. S. M. Hahn, M. C. Krishna, A. M. DeLuca, *et al*., *Free Radic. Biol. Med*., **28**, 953–958 (2000).
67. P. Bar-On, M. Mohsen, R. L. Zhang, *et al*., *J. Am. Chem. Soc*., **121**, 8070–8073 (1999).
68. J. E. Baur, S. Wang, and M. C. Brandt, *Anal. Chem*., **68**, 3815–3821 (1996).
69. S. Morris, B. Sosnovsky, Hui. B., *et al*., *J. Pharm. Sci*., **80**, 149–152 (1991).
70. D. M. Stanbury, *Adv. Inorg. Chem*., **33**, 69–138 (1989).
71. G. Merenyi, J. Lind, and L. Engman, *J. Chem. Soc. Perkin Trans. 2*, 2551–2553 (1994).

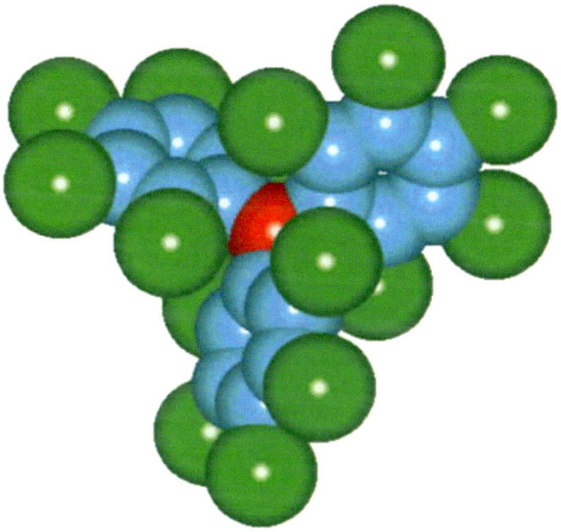

Figure 2.1 *CPK representation of perchlorotriphenylmethyl radical (**1•**) showing the high steric shielding of the central carbon atom (in red) surrounded by the six bulky chlorine atoms at the ortho positions.*

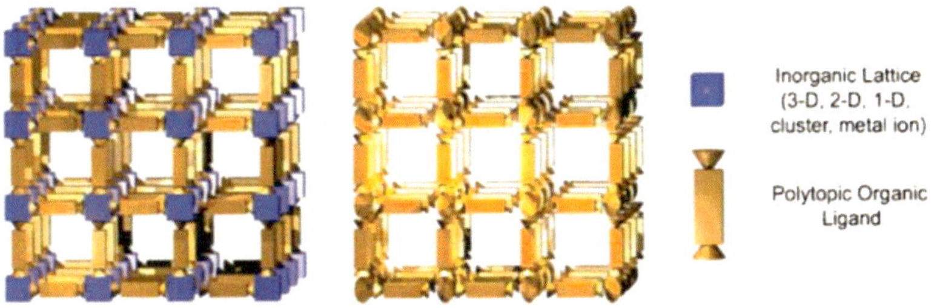

Figure 2.11 *Left: Coordination polymer created by an inorganic subunit (0-D clusters or isolated metal ions) connected through polytopic organic ligands. Right: Purely organic open framework formed by organic molecules interlinked through non-covalent bonds such as hydrogen bonds or $\pi-\pi$ interactions. (Reprinted with permission from [53]. Copyright 2007 Royal Society of Chemistry.)*

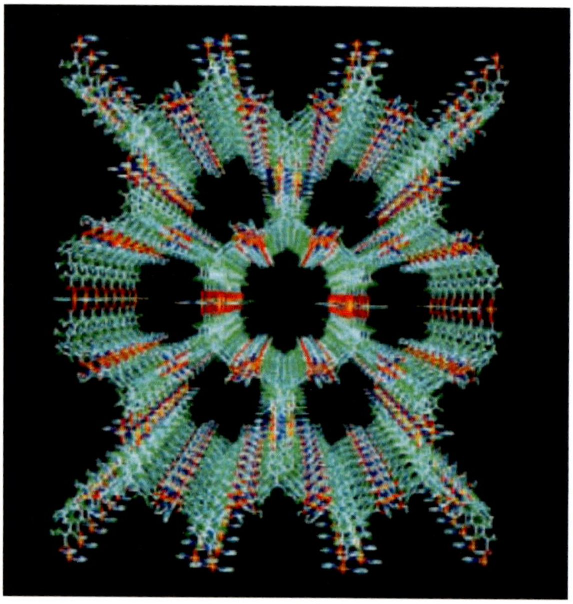

Figure 2.12 *Illustration of the nanochannel-like structure of* ***MOROF-1****. The superposition of the honeycomb (6,3) layers creates large pores of dimensions of 3.1 and 2.8 nm between opposite vertices. (Reprinted with permission from [56]. Copyright 2004 Royal Society of Chemistry.)*

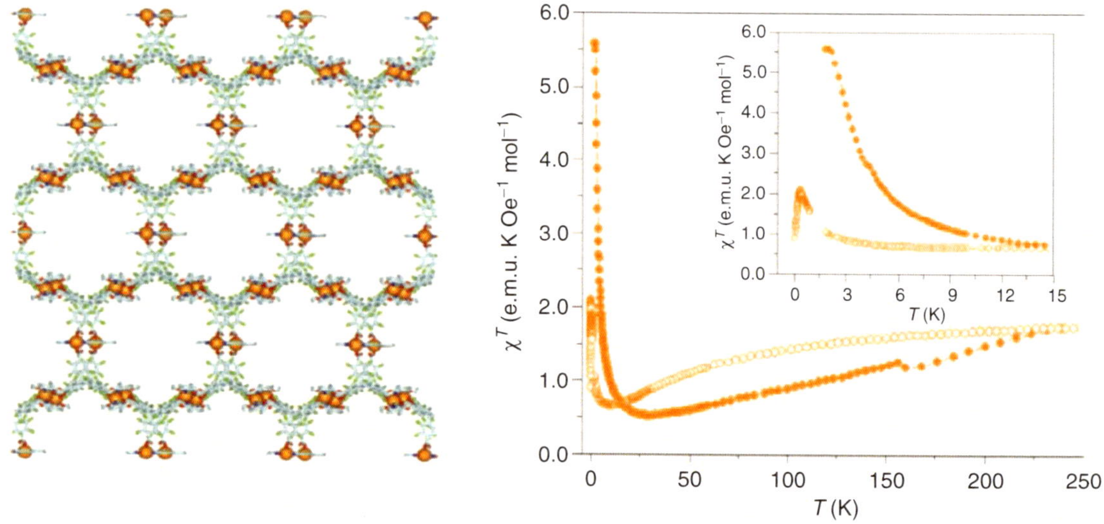

Figure 2.13 *Left: A honeycomb (6,3) layer showing the organization of the copper ions (orange spheres) and the central methyl carbon (magenta spheres) of the* **13** *radicals. Right: χT as a function of the temperature for as-synthesized (•) and evacuated (○)* ***MOROF-1****. Inset: Magnetic field dependence of the magnetization at 2 K. (Reprinted with permission from [56]. Copyright 2004 Royal Society of Chemistry.)*

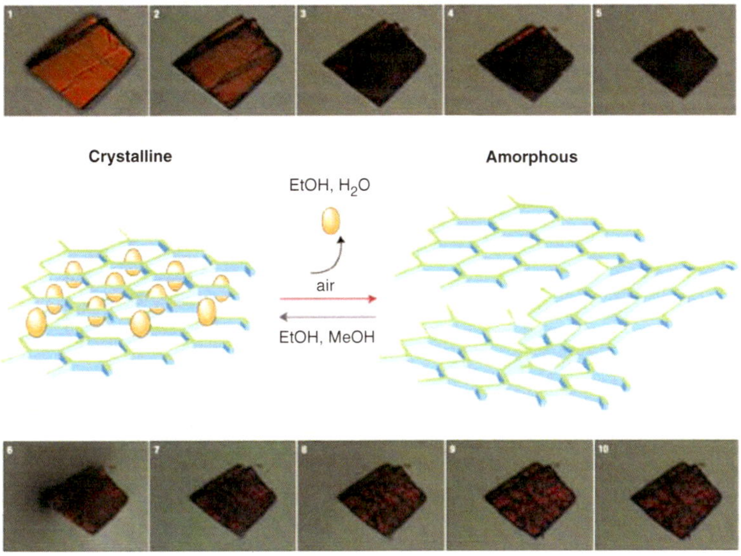

Figure 2.14 Images of a single crystal of **MOROF-1** followed with an optical microscope. The top series show the 'shrinking' process, in which a crystal of **MOROF-1** exposed to the air experiences a volume decrease of around 30%. In the lower series, the same crystal exposed against to ethanol liquid begins to swell. The scheme represents the structural changes of **MOROF-1** in contact or not with ethanol or methanol solvent. (Reprinted with permission from [56]. Copyright 2004 Royal Society of Chemistry.)

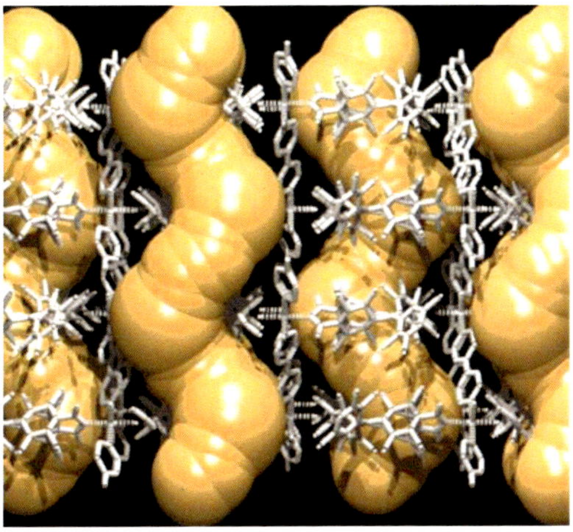

Figure 2.15 Illustration of the helical nanochannel-like structure of **MOROF-2**. (Reprinted with permission from [56]. Copyright 2004 Royal Society of Chemistry.)

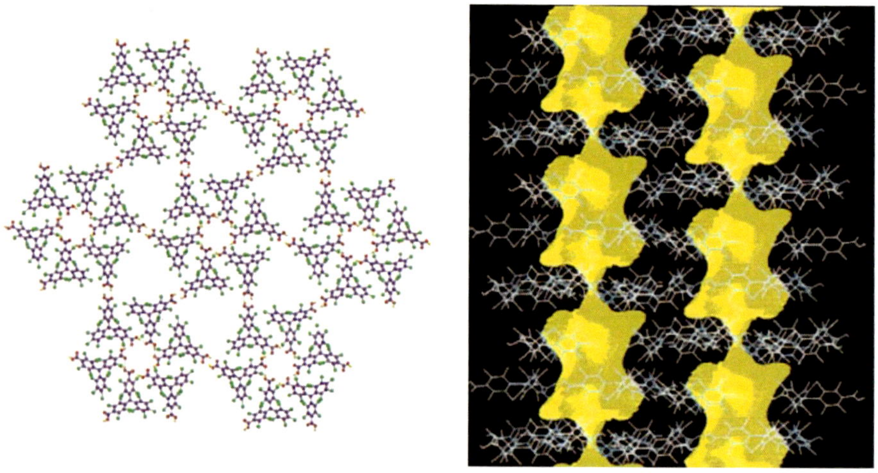

Figure 2.16 *Left: Crystal structure of* **POROF-1** *obtained from the crystallization of radical* **12** *showing the 2-D hydrogen bonded layer. The repetitive* $R_6^6(24)$ *hydrogen bonded hexamers generate polar windows due to the presence of six carboxylic groups, whereas the linking of each hexamer with six more identical units in a hexagonal topology originates six trigonal-shaped hydrophobic voids. Right: Space-filling view along the b axis of the large nanocontainers formed along the one-dimensional channel. (Reprinted with permission from [63a]. Copyright 2004 American Chemical Society.)*

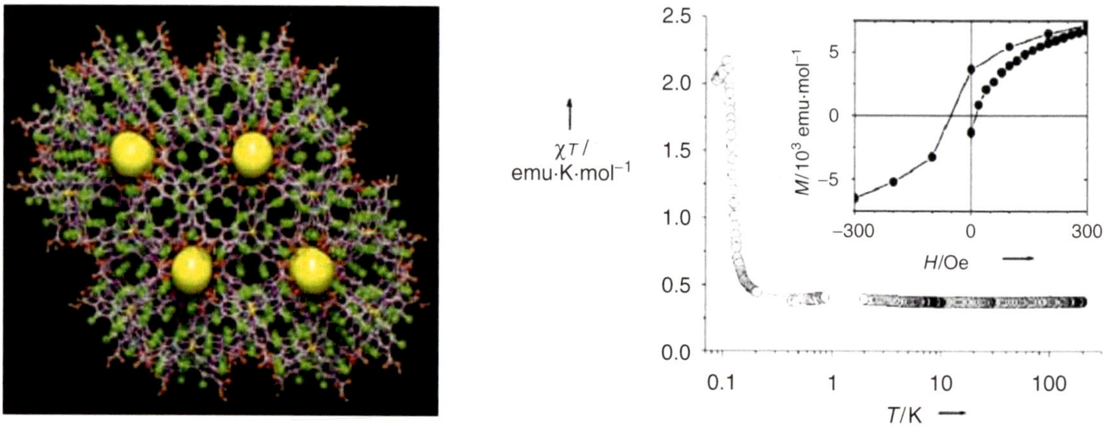

Figure 2.17 *Left: Crystal packing of* **POROF-2**, *showing the tubular hydrogen bonded channels. The empty space is represented as yellow spheres. Right: Temperature dependence of* χT *up to 200 K, measured with an applied magnetic field of 200 Oe. Inset: magnetic hysteresis obtained at 0.08 K. (Reprinted with permission from [53]. Copyright 2007 Royal Society of Chemistry.)*

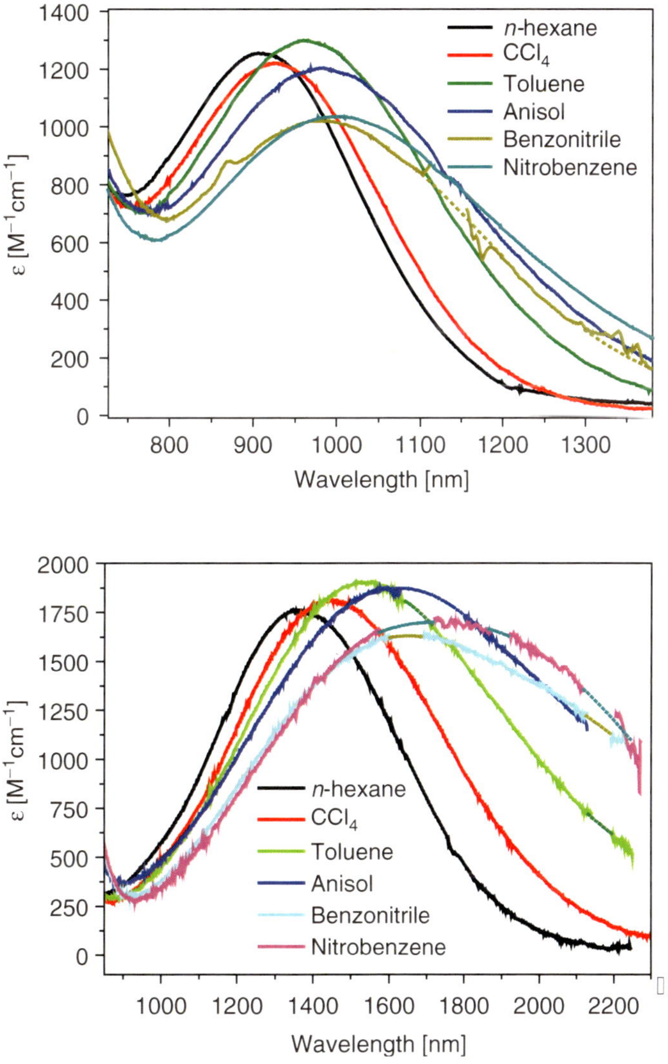

Figure 2.22 *IVT bands of D-A dyads **17** (top) and **18** (bottom) in some selected solvents. (Reprinted with permission from [82]. Copyright 2007 American Chemical Society.)*

Figure 2.32 *Colors shown by dichloromethane solutions of the different oxidation states of the radical* ***18****. From left to right: H-****18****,* ***[K(18crown-6)]$^+$ 18$^-$****,* ***18*** *and* ***18$^+$ BF_4^-****. (Adapted with permission from [15]. Copyright Wiley-VCH Verlag GmbH & Co. KGaA.)*

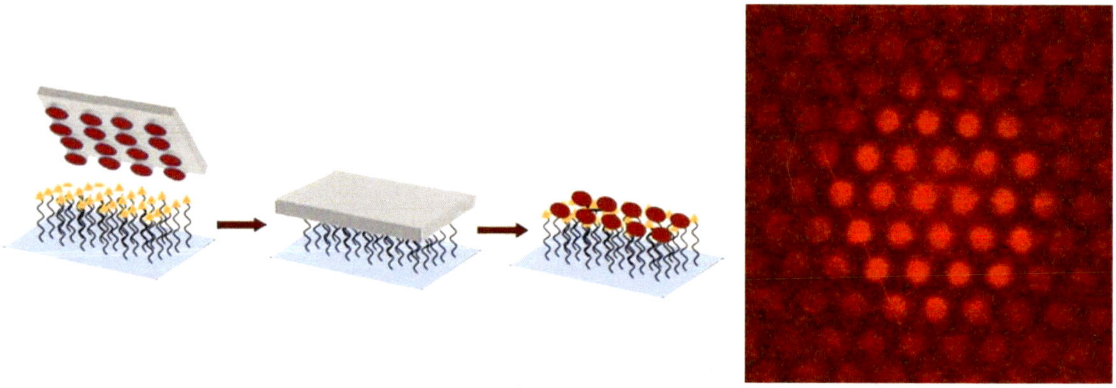

Figure 2.35 *Left: Schematic representation of the microcontact printing of a functionalized PTM radical on an adhesive pre-functionalized SAM. Right: Confocal microscopy image (λ_{exc} = 488 nm) of a glass surface patterned with a PTM radical derivative. The diameter of the fluorescent dots is 5 μm. (Reprinted with permission from [101]. Copyright 2007 Wiley-VCH Verlag GmbH & Co. KGaA.)*

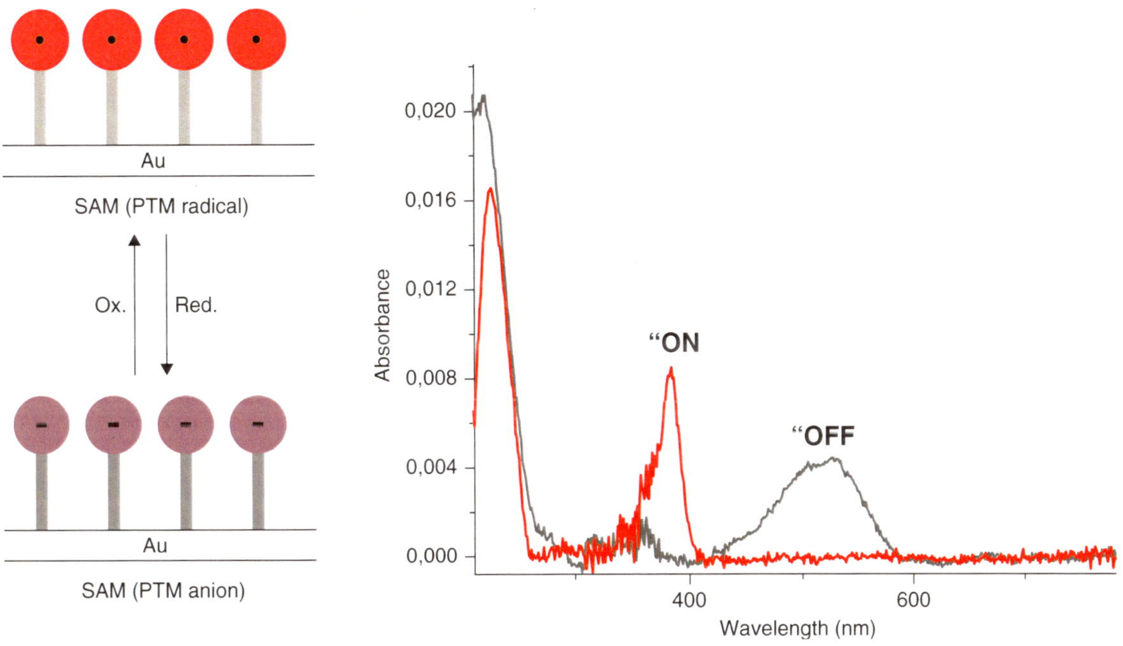

Figure 2.36 *Left: Representation of the electroactive PTM SAM. Right: Absorbance spectra of the PTM anion SAM on silicon (OFF state, black line) before and after oxidation to the PTM radical SAM (ON state, red line). (Reprinted with permission from [101]. Copyright 2007 Wiley-VCH Verlag GmbH & Co. KGaA.)*

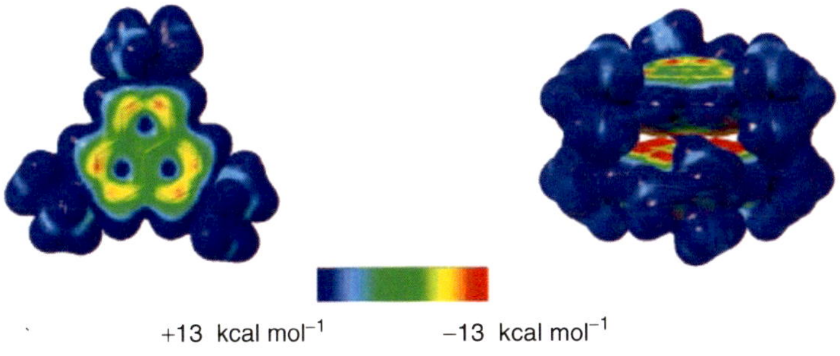

Figure 3.10(b) *Electrostatic potential surfaces of the monomer and the* π *dimer of* ***TBPLY*** *calculated by quantum chemical calculations, which indicate "aromaticity generation" by the* π *dimerization (Reprinted with permission from [15]. Copyright 2006 American Chemical Society.)*

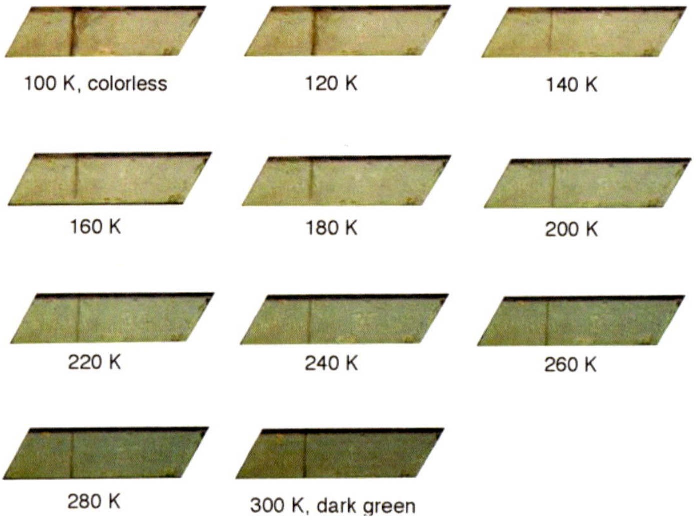

Figure 3.19(a) *Pictures of **TBDAP** single crystal showing a continuous color change from colorless to dark green depending on the temperature. (Reprinted by permission from Macmillan Publishers Ltd,[29] copyright 2008.)*

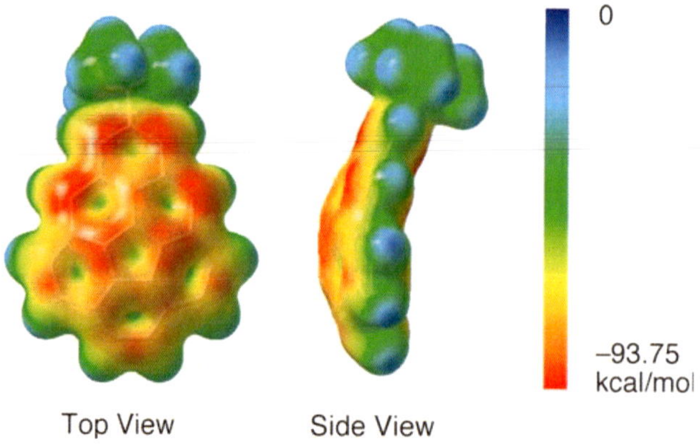

Figure 3.56(c) *Electrostatic potential surfaces of phenalenyl-fused corannulene anion. Calculations carried out at the RB3LYP/6-311G**//RB3LYP/6-311G** level of theory. (Reprinted with permission from [80]. Copyright 2008 American Chemical Society.)*

Figure 6.2 *The Di(1-Adamantyl)iminoxyl Radical.*

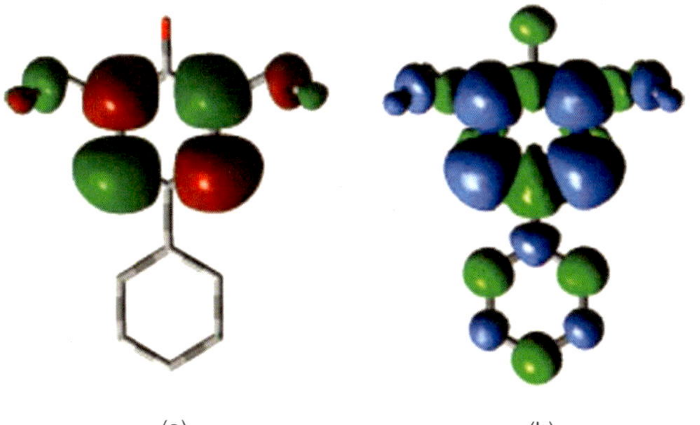

Figure 7.4 *(a) SOMO and (b) spin density plot for 1,5-diisopropyl-3-phenyl-6-oxoverdazyl (blue = positive spin density, green = negative spin density).*[51] *(Reprinted with permission from [51]. Copyright 2007 American Chemical Society.)*

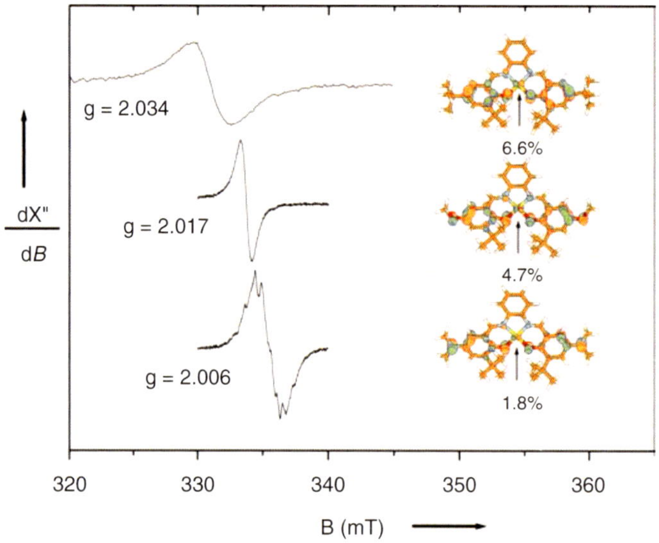

Figure 8.14 *Isotropic EPR spectra of 1 mM dichloromethane solutions of: (a)* $[Ni^{II}(L_9)^{\bullet}]^+$; *(b)* $[Ni^{II}(L_9{}^{OMe})^{\bullet}]^+$; *and (c)* $[Ni^{II}(L_9{}^{NMe})^{\bullet}]^+$. *([90] Reproduced by permission of the Royal Society of Chemistry.)*

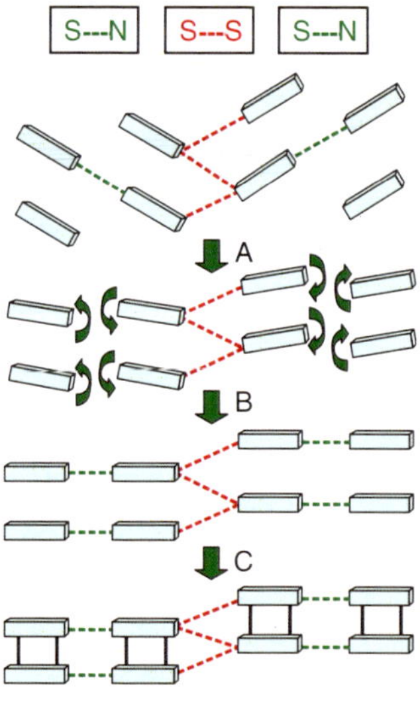

Figure 9.29 *Schematic illustration of the conversion of the high temperature (monomeric) to low temperature (π dimeric) forms of 1,3,2-dithiazolyl radical* ***71***.[243] *(Reprinted with permission from [243]. Copyright 2004 American Chemical Society.)*

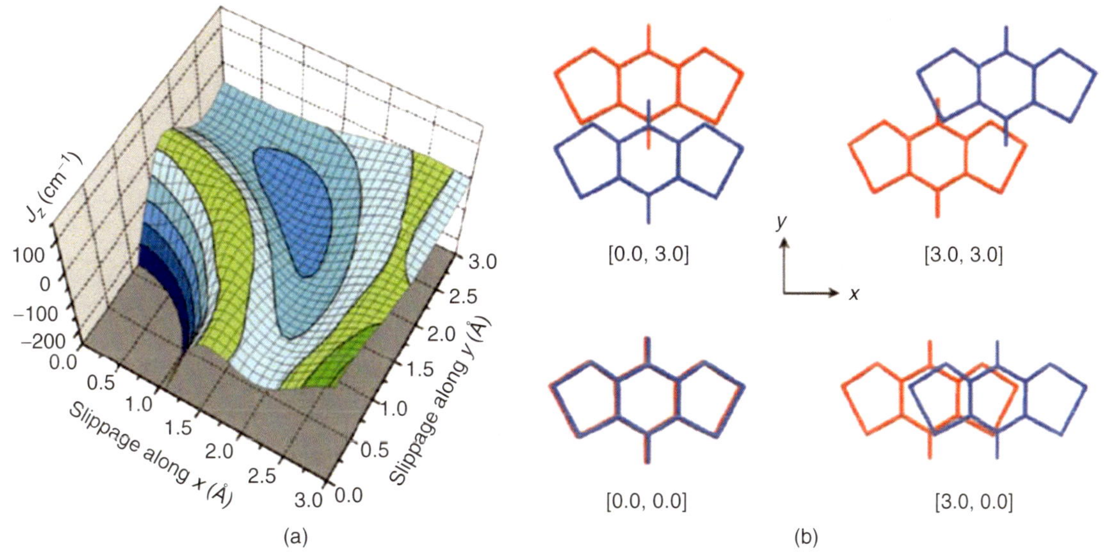

Figure 9.31 *(a) Surface plot of calculated intermolecular magnetic exchange (J_π) between two π stacked bis(dithiazolyl) radicals* ***97a*** *($R_1 = R_2 = H$) as a function of displacement of the two radicals along x and y vectors. (b) Definition of x and y displacement vectors.[223] (Reprinted with permission from [223]. Copyright 2009 American Chemical Society.)*

Figure 13.2 *$Ni(L^{26})_2(HO(CH_2)_4OH)$ single crystals.*

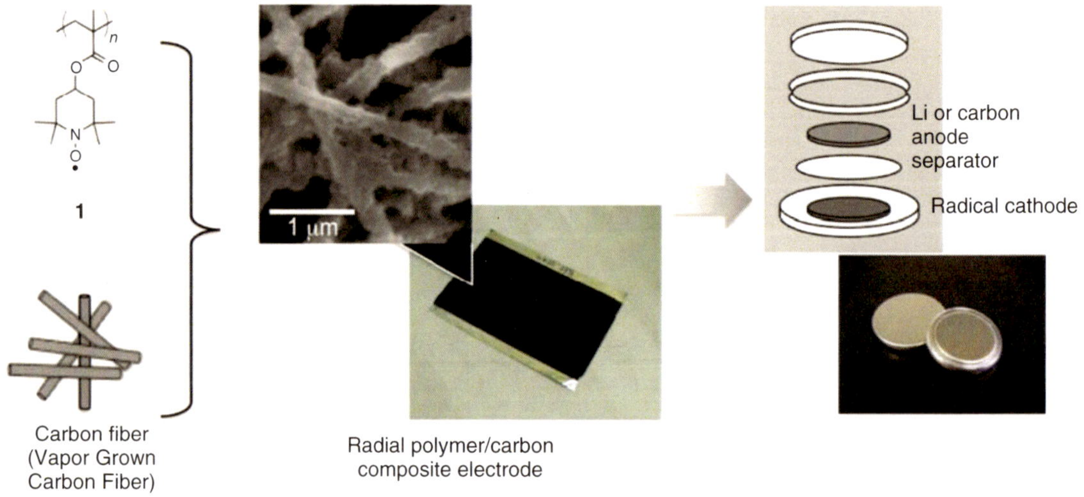

Figure 14.5 *A typical fabrication method of the prototype "organic radical battery".*

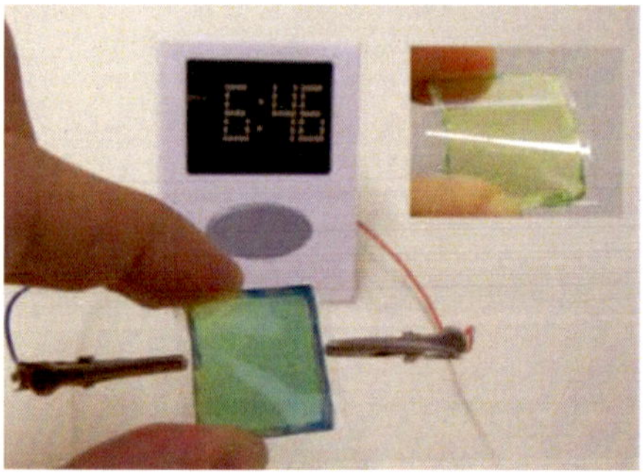

Figure 14.8(b) *A see-through paper-like battery; blue in the charged state (galvinolate anion) and light-yellow in the discharged state (galvinoxyl).*

Index